Index of Applications

Agriculture & Animals

Animal pens, 2, 3, 4, 6, 15, 17–18, 20, 28, 35, 49–50
Animal shelter funding, 572
Bats in a cave, 279
Bird population, 279
Counting farm animals, 342
Dinosaurs, 443
Dog/cat food ingredients, 351
Dolphin speed, 387
Filling an aquarium, 317
Fish management, 276, 293
Horse coat conditioner, 312
Mass of a chicken, 443
Mass of a spider, 441
Puppies
  prices of, 302
  sizes of, 264
Snail's pace, 448
Tomato plant growth rate, 444
Wheat harvest, 546, 563–564, 566, 569

Arts & Leisure

Balloons, 101–102
Book discounts, 304
Camera lens f-stops, 523
CD prices, 365
Concert seating, 154
Cribbage, 251
Dice rolls, 204
DVDs, 133, 267, 365
Gift certificates, 46, 132
Kite making, 462
Movie times, 16, 17
Party costs, 125, 126–127, 130–131, 133, 138, 142, 151, 154, 156, 171, 175, 177, 182, 203
Photograph sizes, 281, 283, 289
Pulitzer Prize winners, 251
Restaurants
  prices at, 225
  seating at, 35
  tipping at, 30, 132
Ticket sales, 195, 227, 248, 337, 341, 382, 573
TV ratings, 275
Weddings, 175, 180
World's largest mural, 121

Astronomy

Black holes, 444
Distances within the solar system, 440, 444, 528, 577
Message to extraterrestrials, 389, 401
Moon's orbit, 448
Spacecraft path, 144
Sun's size, 443

Business & Economics

Annual meeting budget, 553
Cash in store, 379, 380
Conference registration fees, 331–332
Job sharing, 333
Sales commissions, 238, 324, 329, 331
Stock prices, 46, 63, 65, 308
Total quantity sold, 552
Wage packages, 328, 335
Web sales, 66–67, 71

Engineering

Access ramps, 237, 267, 275, 278, 293, 315
Egyptian surveying tool, 458
Highway grade, 263, 279, 318
Highway paving, 365
Roof building, 210–211
Staircase slope, 215, 227, 259, 276, 315
Traffic flow through doors, 552, 557–558, 561, 567, 569–570, 574

Environmental Science

Air pollution, 514
Crude oil reserves, 539–540
Garbage production, 540
Landfill size, 539
Natural gas reserves, 571
Projected human population, 302, 304, 318
Silver reserves, 537–538
Storm surge, 278–279
Tree growth, 444
Valdez oil spill, 293

Food & Drink

Bulk food purchases, 46, 130, 132
Candy costs, 132, 365
Carrot bags, 373
Coffee
  cost of, 51, 387
  weight of, 312
Cookie baking, 253
Corn flake servings, 572
Milk spills, 96
Molasses spills, 106
Nut mixture, 451
Onion bags, 336
Orange juice mixture, 316
Party punch, 105
Peanuts, 351, 450
Pie portions, 571
Pizzas, 14
Potato bags, 351, 354
Sodium content, 118
Soft drinks
  sales of, 315
  spills of, 106
Subway sandwiches, 529
Sugar cube drop, 479, 492, 500
Surplus food giveaways, 253

General & Physical Science

Chemical formulas, 259, 261, 268, 311–313
Color chart, 321
Distance to horizon, 478, 484
Earth's formation, 439
Electrical charges, 49, 56, 63
Electrical circuit, 193, 483, 552, 560
Elevations, 62, 63, 64
Falling objects, 483, 492, 493, 500

(continued)

(continued)

Focal distance of lens, 572
Half-life of uranium, 439
Heat transfer, 561
Hours of sunlight, 201
Jet stream, 338
Object trajectory, 492, 500, 511
Refrigeration cycle, 561
Scientific notation, 438–442, 444
Speed of light, 444, 577
Temperature change, 560
Temperature conversion, 193, 216, 237
Units of measurement, 264–267, 278, 285–286, 311–312, 315, 341, 450
Water spray, 500
Water temperatures, 237, 309, 312–313, 318
Wind speed, 338, 365, 385

General Interest

Academic grades, 173, 176, 181, 255, 294, 298, 303, 310–313, 319
Ages
  and birth years, 463
  drivers', 295, 303
Box dimensions, 89, 106
Burning time of candles, 203
Center of population, 304
City layout, 36–37
Coins, 143, 248, 305–306, 340–341, 351
Connecting dots, 525
National debt, 448
Native American leaders, 255
Pen costs, 319
Scheduling, 253, 367, 370, 382
Sewing a hem, 571
Shadow lengths, 285–286, 290–291
Shoe sizes, 520
Stamp costs, 312
State capitals, 201
U.S. currency, 373

Geometry

Angles, 269, 280, 284, 349, 352, 373, 385, 494, 528
Area, 79, 97, 98, 99, 100, 536
Circle, 98–99, 105, 121, 527
Cone, 267
Cube, 106, 428–429
Cylinder, 101, 105–106
Parallelogram, 292, 298
Perimeter, 15, 97–98, 104–106, 118, 351, 385, 495, 528, 536
Rectangle, 98, 100, 106, 188, 192, 233, 281–282, 421, 450
Rectangular prism, 101
Similar figures, 281–283, 289–290, 314
Slope and square, 254
Sphere, 101–102, 106
Square, 98, 406–407, 421, 455–456
Surface area, 100–101, 103, 105–106, 119, 122
Trapezoid, 89, 99–100, 106, 463
Triangles, 98, 239, 289–291, 453–455, 457, 461, 504, 523
Volume, 96, 100–103, 105–106, 119, 122–123

Health & Medicine

Bacteria, 93
Calories, 118, 244, 248, 252, 254, 293, 315, 380, 385, 529
Exercise heart rate, 193
Exercise pulse rate, 132
Growth rate of child, 388
Intravenous feeding, 266
Length of HIV virus, 439
Medicine dosage, 260, 267, 272, 277, 279
Record number of births, 385
Scheduling surgical procedures, 527
Wheelchair ramp, 267, 275

Household Maintenance

Air compressor rental, 133
Dry ice for freezer, 230
Electricity costs, 115, 180
Fertilizer mixture, 259, 266
Filling a swimming pool, 569
Heat pump output, 309
House height, 462
Ladder safety, 270, 280, 458–459, 462, 523, 526
Leaking faucet, 262, 318
Natural gas costs, 115, 183–184
Plumber fees, 182
Pressure washer, 131, 154
Roof, cost of, 201
  measuring, 214, 462–463
Sidewalk repair, 106, 123, 182
Utility payments, 204
Ventilation fan, 569
Waste water disposal costs, 197
Water costs, 132, 197, 238
Water usage, 573

Personal Finance

ATM card, 51, 238
Car loan, 318–319
Car registration costs, 201
Copy machine card, 130, 211, 226, 233
Credit card fees, 31–32, 43, 46, 107–110, 138, 215, 450, 577
Discounted prices, 155
Financial aid for college, 540
Home sales, 519
Household income, 293, 295
Investment earnings, 305, 307, 312, 318, 388
Jewelry costs, 373
Loan cost, 238, 318
Medicare payment, 132
Monthly expenditures, 268
Pick-up truck value, 221, 238
Property tax, 180
Rental costs, 268, 302
Sales tax, 130, 132
Shirt costs, 92
Social Security payment, 132
Telephone bills, 131, 237
Telephone card, 46, 52, 155, 226
Transcript costs, 133, 226
Tuition costs, 8, 15, 17, 130, 132, 155, 180
Wages, 313, 315, 317, 324, 327, 328

Sports

Baseball, 303
Basketball, 237, 386

(continued on back endpaper)

INTRODUCTORY ALGEBRA

Everyday Explorations

Fourth Edition

ALICE KASEBERG
formerly of Lane Community College

HOUGHTON MIFFLIN COMPANY BOSTON NEW YORK

About the Cover

The skier on the cover reminds us that cross country skiing is not always a conservative sport. Many volcanic mountains have slopes angled between 30 and 40 degrees from horizontal. Guides to recreational trail construction suggest that novice skiers have trouble on slopes exceeding 10 percent while experienced skiers can handle short slopes of 40%. Being comfortable with changing units of measure, say from degrees to percent, as well as understanding slope in a variety of applications is an important part of the everyday mathematics in this textbook. (Divide the slope in degrees by 90 degrees to obtain percent.)

Cover photograph ©Alec Pytlowany/Masterfile

p. 125 Figure 1 Adapted with permission from Michelle Hymen, "Partying for Profit," The Register Guard, May, 10, 1994. **p. 257** Figure 1 Kevork Djansezian, AP/Wide World Photos. **p. 389** Figure 1 Reprinted with permission of Cornell University.

Copyright ©2008 by Houghton Mifflin Company. All rights reserved.

No part of this work may be reproduced or transmitted in any form or by any means, electronic or mechanical, including photocopying and recording, or by any information storage or retrieval system without the prior written permission of Houghton Mifflin Company unless such copying is expressly permitted by federal copyright law. Address inquiries to College Permissions, Houghton Mifflin Company, 222 Berkeley Street, Boston, MA 02116-3764.

Printed in the U.S.A.

Library of Congress Catalog Card Number: 2006939891

ISBNs:

Instructor's Annotated Edition:
ISBN-13: 978-0-618-92005-1
ISBN-10: 0-618-92005-6

For orders, use student text ISBNs:
ISBN-13: 978-0-618-91878-2
ISBN-10: 0-618-91878-7

2 3 4 5 6 7 8 9—CRK—11 10 09 08 07

To my grandparents, for their belief in education

My paternal grandparents chose tuition over farm payments, sending their son to a university in the depth of the 1930s depression. My maternal grandfather, a widower and civil servant, sent all four daughters through that same university, where the youngest one met and married the farmers' son.

Contents

Preface ix

To the Student xxiii

1 Algebraic Representations 1

A systematic plan is presented for thinking about many mathematical problems. Algebra is approached in four ways—numerically, symbolically, verbally, and visually.

1.1 Problem-Solving Steps and Strategies 2

1.2 Numeric Representations 7

1.3 Verbal Representations 17

Mid-Chapter Test 26

1.4 Symbolic Representations 27

1.5 Visual Representations: Rectangular Coordinate Graphs 36

Chapter Summary 48

Review Exercises 49

Chapter Test 52

2 Operations with Real Numbers and Expressions 54

Success in mathematics requires a solid foundation. This chapter summarizes number operations, properties, notation, unit analysis, and formulas needed for algebra.

2.1 Addition and Subtraction with Integers 55

2.2 Multiplication and Division with Positive and Negative Numbers 65

2.3 Properties of Real Numbers Applied to Simplifying Algebraic Fractions and Adding Like Terms 73

Mid-Chapter Test 83

2.4 Exponents and Order of Operations 84

2.5 Unit Analysis and Formulas 94

2.6 Inequalities, Intervals, and Line Graphs 107

Chapter Summary *115*
Review Exercises *116*
Chapter Test *121*
Cumulative Review of Chapters 1 and 2 *122*

▶ 3 Solving Equations and Inequalities in One Variable 125

Types of equations and methods of writing equations from word problems open this chapter. Linear equations and inequalities are then solved with symbols, tables, and graphs.

3.1 Linear Equations in One and Two Variables 126
3.2 Solving Equations with Algebraic Notation 134
3.3 Solving Equations with Tables, Graphs, and Algebraic Notation 144
Mid-Chapter Test *155*
3.4 Solving Linear Equations with Variables on Both Sides of the Equation 156
3.5 Solving Linear Inequalities in One Variable 166
Chapter Summary *177*
Review Exercises *178*
Chapter Test *182*
Cumulative Review of Chapters 1 to 3 *183*

▶ 4 Formulas, Functions, Linear Equations, and Inequalities in Two Variables 185

Previous work with linear equations is formalized by studying formulas, functions, slope, intercepts, and equations of graphs. Inequalities are solved with graphs and symbols.

4.1 Solving Formulas 186
4.2 Functions and Graphs 194
4.3 Linear Functions: Slope and Rate of Change 204
Mid-Chapter Test *216*
4.4 Linear Equations: Intercepts and Slope 216
4.5 Linear Equations 227
4.6 Inequalities in Two Variables 239
Chapter Summary *249*
Review Exercises *250*
Chapter Test *255*
Cumulative Review of Chapters 1 to 4 *256*

▶ 5 Ratios, Rates, and Proportional Reasoning 257

Proportional reasoning is one of the most common applications of linear equations. Ratios, proportions, and unit analysis are examined in a variety of settings including geometry. The chapter introduces work with averages and with the rational expressions that will be studied in Chapter 9.

5.1 Ratios, Rates, and Percents 258
5.2 Proportions and Proportional Reasoning 269
5.3 Proportions in Similar Figures and Similar Triangles 280
Mid-Chapter Test *292*
5.4 Averages 293
5.5 Writing Equations from Word Problems with Quantity-Rate Tables 305
Chapter Summary *314*
Review Exercises *315*
Chapter Test *319*
Cumulative Review of Chapters 1 to 5 *320*

▶ 6 Systems of Equations and Inequalities 321

This chapter presents four different ways to solve systems of two linear equations, as well as three different solution outcomes. Work is then extended to three equations and to systems of inequalities.

6.1 Solving Systems of Equations with Graphs 322
6.2 Setting Up Systems of Equations 332
6.3 Solving Systems of Equations by Substitution 343
Mid-Chapter Test *353*
6.4 Solving Systems of Equations by Elimination 354
6.5 Solving Systems of Equations in Three Variables 367
6.6 Solving Systems of Linear Inequalities by Graphing 374
Chapter Summary *383*
Review Exercises *384*
Chapter Test *387*
Cumulative Review of Chapters 1 to 6 *388*

▶ 7 Polynomial Expressions and Integer Exponents 389

Multiplication and factoring are used to change the form of polynomial expressions in order to make it easier to do operations.

7.1 Operations on Polynomials 390
7.2 Multiplication of Binomials and Special Products 402

7.3 Factoring Trinomials 411
Mid-Chapter Test 421
7.4 Factoring Special Products and Greatest Common Factors 422
7.5 Exponents 428
7.6 Scientific Notation 437
Chapter Summary 445
Review Exercises 446
Chapter Test 449
Cumulative Review of Chapters 1 to 7 450

8 Squares and Square Roots: Expressions and Equations 452

Expressions containing squares and square roots are examined, along with their graphs and their use in solving related equations.

8.1 Pythagorean Theorem 453
8.2 Square Root Expressions and Properties and the Distance Formula 464
8.3 Solving Square Root Equations and Simplifying Expressions 474
Mid-Chapter Test 484
8.4 Graphing and Solving Quadratic Equations 485
8.5 Solving Quadratic Equations by Taking the Square Root or by Factoring 496
8.6 Solving Quadratic Equations with the Quadratic Formula 504
8.7 Range, Box and Whisker Plots, and Standard Deviation 513
Chapter Summary 521
Review Exercises 523
Chapter Test 526
Cumulative Review of Chapters 1 to 8 527

9 Rational Expressions and Equations 530

Rational expressions are the algebraic equivalent of fractions. Because fraction skills are essential for work with rational expressions, this chapter provides additional review of the basic operations with fractions.

9.1 Rational Functions: Graphs and Applications 531
9.2 Simplifying Rational Expressions 540
9.3 Multiplication and Division of Rational Expressions 546
Mid-Chapter Test 553

9.4 Finding the Common Denominator and Addition and Subtraction of Rational Expressions 554

9.5 Solving Rational Equations 562

Chapter Summary 570

Review Exercises 571

Chapter Test 573

Final Exam Review 575

Appendix: Graphing Calculator Basics for the TI-83/84 579

Answers to Selected Odd-Numbered Exercises and Tests 585

Glossary/Index 620

Preface

Why *Introductory Algebra: Everyday Explorations*?

My purpose in writing Introductory Algebra has been to present algebra with multiple representations within everyday applications using appropriate calculator operations. *Introductory Algebra: Everyday Explorations* is based on

- the premise that concept development and understanding of mathematical thinking are facilitated by number patterns, problem solving, exploration, and discovery,
- agreement with the standards advocated by organizations such as the American Mathematical Association of Two Year Colleges (AMATYC), the Mathematical Association of America (MAA), and the National Council of Teachers of Mathematics (NCTM),
- the motivation of everyday applications,
- the availability of technology and its considered use, and
- the mastery of certain basic skills.

What is the Everyday Exploration Approach?

Students make tuition payments, pay utility bills, use credit cards, order merchandise on-line or from a catalog, rent moving trucks, use a photocopy machine, and so forth. This is everyday mathematics. By exploring these settings, Introductory Algebra encourages additional observations outside the classroom while identifying the rules and equations that form the mathematical basis inside the classroom.

From one edition to the next, what never changes is the four-fold representation (numeric, visual, verbal, and symbolic) with problem solving. In the first chapter, the input-output tables for the everyday settings lay the ground work for multiple representations and, at the same time, lay a solid foundation on which to build the function concept.

Why Start with Problem Solving?

We need to do something new, but not intimidating, for all students the first day. The problem solving setting in Section 1.1 is new to everyone; hence, providing a level playing field. The problem is open-ended and has no single correct answer. Though the opening section may seem like a "side track" or perhaps redundant of later material, it is important to catch the interest of the reviewing student and yet not discourage the less well-prepared student.

How Does the Book Accommodate the Varied Mathematical Preparation of Students?

For the student new to algebra, it is important to remove the mystery. Included within this text are a number of important and often overlooked ideas, such as:

- What are the ways to write multiplication?
- What are the different uses for parentheses?
- What numbers are hidden in algebraic expressions?

For the student reviewing algebra, Introductory Algebra includes problem solving and projects. Some are gathered over 35 years experience in mathematics education and some are freshly inspired by the Internet.

For both sets of students, the Glossary/Index provides instant definitions for essential vocabulary.

Both sets of students should take advantage of the marked "core" exercises in planning their homework. Core exercises are clearly identified by blue numbers throughout the text, making them easy for instructors to assign to students (see Exercises 9-10 on page 6 for examples of core exercises). Core plus extra skill and review are suggested for the student new to algebra. A sampling of core exercises plus exercises after the core and problem solving are recommended for the reviewing student. All students will benefit from working in groups on writing and project exercises.

What Has Changed with the New Edition?

With each edition comes the challenge of what can be done to improve the book. As in the past, reviewers and current users have contributed ideas for improvements for the fourth edition. These include:

- Introduce area and perimeter concept in the first chapter to ease the transition to applying formulas in the second chapter.
- Introduce equations in the first chapter. Use equations and formulas to describe relationships in the first two chapters. Solve equations in later chapters.
- Add subscripts to locate left and right sides of an equation in a graphical solution and to aid in solving systems of equations by graphing.
- Add a cumulative review to each chapter. Look at some of these as trial questions for inclusion as future examples or in exercise sets.

In addition, the fourth edition

- Includes 240 Instructor Extras in the margin to provide additional in-class group practice or a source for quizzes, review, or tests.
- Blends short historical comments about people, mathematical history, or mathematical notation within the reading or as margin boxes. Some reference a website.
- Increases the number of writing exercises.
- Identifies and adds new connections to other chapters.
- Features a graphing calculator appendix personally crafted to fit my philosophy for introducing this incredible tool.

What Makes *Introductory Algebra: Everyday Explorations* Special?

Inclusion or mention of a variety of hands-on materials and projects give instructors permission to personalize their instruction with their favorite activities, materials, and projects.

Frequent mention of the importance of group work and working together supports a variety of instructional techniques.

In the section on measures of dispersion (a square root application) I have included the box and whisker plot. Since their introduction in 1979, box plots have become a preferred method for comparing sets of data outside the mathematical and statistical fields. As an Internet challenge, pick a topic and search on it with "box and whisker plot" or just "box plot". You will quickly observe why box plot deserves to be an everyday application.

What If I Am a New Instructor?

Because many new instructors have little teaching experience and adjunct teachers have minimal time to prepare, the explorations in many sections suggest how to introduce a topic.

Whatever you do, share why you are excited about mathematics. You may be one of the many instructors inspired by fractals or finding the largest prime number of the Mersenne type. You may be inspired by mathematics and music or mathematics and art.

Here are the topics in the first 15 pages of the Instructor's Resource Manual:

Preface

Thriving with Change
Get Your Colleagues on Board
Expectations: You and Your Students

Developmental Algebra

Course Planning
Week-by-Week Course Plans

Strategies for Teaching and Learning

- Problem Solving
- Discovery Learning
- Four-fold Approach
- Vocabulary
- Planning Assignments
- Guided Reading
- Applications
- Projects
- Tests
- Testing and Assessment Strategies
- Learning from Tests
- Quizzes
- Portfolios
- Graphing Calculators

Tips on Classroom Management

- Be Prepared!
- Start Each Class with a Warm-up
- The First Class
- Taking Attendance and Returning Homework
- In Class
- Paying Attention to Your Timing
- Maintain a Positive Atmosphere
- Organizing Groups
- Getting Students Acquainted
- Monitoring Groups: Listen; Don't Interrupt
- Calculator Teaching Aids
- Questions on Homework
- Coaching on Homework

The remaining IRM is Section-by-Section Lessons, Extra Activities, and Tips by Topic. For a copy of the Instructor's Resource Manual, contact your local Houghton Mifflin sales representative.

With the remainder of this Preface and in To the Student are additional elements for effective teaching and learning. Encourage your students to read their section at the times suggested within the material.

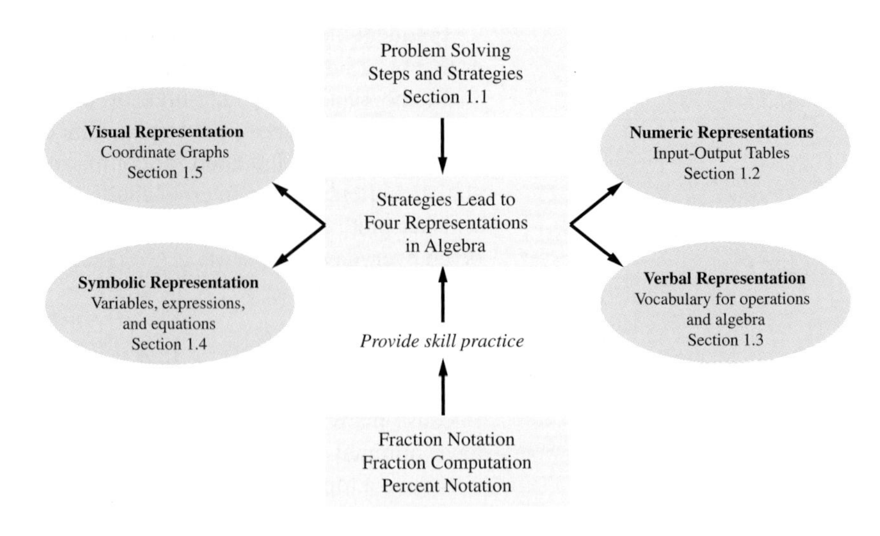

Pedagogy

Objectives

The learning outcomes for each section are listed at the beginning of the section. They serve as a summary for both students and instructors and coordinate with the titles on the examples.

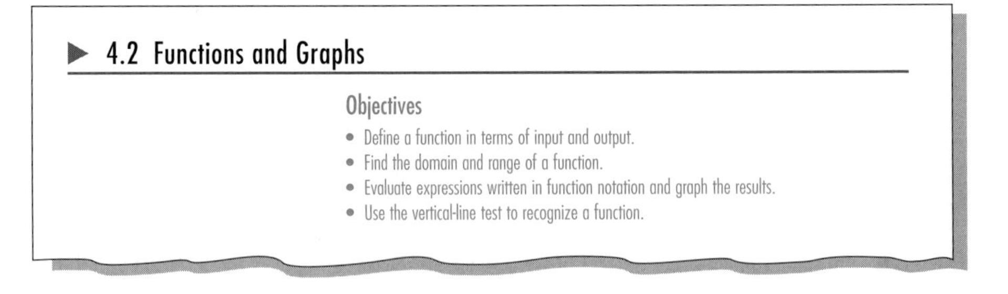

Warm-Ups

The Warm-up at the beginning of each section is designed to serve as a class opener, reviewing important concepts, beginning exploration, and linking prior and upcoming topics. Warm-ups tend to be skill-oriented; they generally connect to the algebra needed to solve text examples. The answers to the Warm-up appear in the Answer Box at the end of the section.

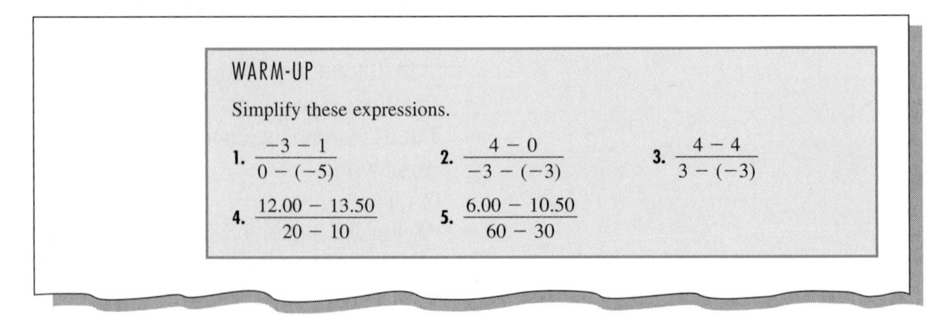

Small-Group Work

Some sections contain introductory questions or activities. These are intended to be done in class in small groups. In Section 1.1, for example, Exercise 17 (Numbers in Words) calls attention to the fact that students have different backgrounds and experiences. The exercise demonstrates how each student may contribute to the class and, in turn, learn from others. It is important to emphasize that students improve their own understanding by helping others.

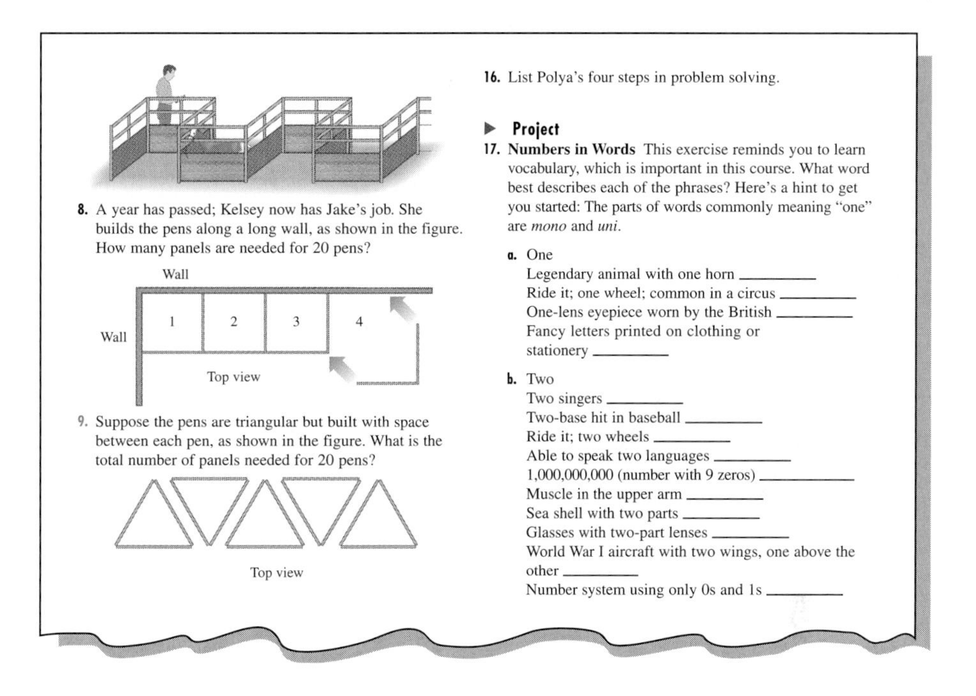

8. A year has passed; Kelsey now has Jake's job. She builds the pens along a long wall, as shown in the figure. How many panels are needed for 20 pens?

9. Suppose the pens are triangular but built with space between each pen, as shown in the figure. What is the total number of panels needed for 20 pens?

16. List Polya's four steps in problem solving.

▶ **Project**

17. Numbers in Words This exercise reminds you to learn vocabulary, which is important in this course. What word best describes each of the phrases? Here's a hint to get you started: The parts of words commonly meaning "one" are *mono* and *uni*.

a. One
Legendary animal with one horn ________
Ride it; one wheel; common in a circus ________
One-lens eyepiece worn by the British ________
Fancy letters printed on clothing or stationery ________

b. Two
Two singers ________
Two-base hit in baseball ________
Ride it; two wheels ________
Able to speak two languages ________
1,000,000,000 (number with 9 zeros) ________
Muscle in the upper arm ________
Sea shell with two parts ________
Glasses with two-part lenses ________
World War I aircraft with two wings, one above the other ________
Number system using only 0s and 1s ________

▶ Problem Solving

George Polya's four-step approach to problem solving—understanding the problem, making a plan, carrying out the plan, and checking the solution—is introduced in Section 1.1 and revisited where appropriate. The text then focuses on planning strategies. The strategies of *trying a simpler problem*, *using manipulatives*, and *drawing a picture* lay the foundation for Section 1.1. *Making a table of inputs and outputs* is introduced formally in Section 1.2. *Looking for a number pattern* starts in Section 1.3. *Making a graph* first appears in Section 1.4. *Working backwards* is the fundamental idea in solving equations and formulas in Sections 3.2, 4.1, and 8.3. *Choosing a test number or ordered pair and checking it* is used in drawing a line graph in Sections 2.6 and 3.5, in identifying half-planes for two-variable inequalities in Section 4.6, and in solving systems of inequalities in Section 6.6. *Guessing and checking*, which is a natural extension of choosing a test number for inequalities, is essential in building and solving systems of equations in Sections 6.2 and 6.5. *Making a systematic list* is an essential component of factoring in Sections 7.3 and 7.4.

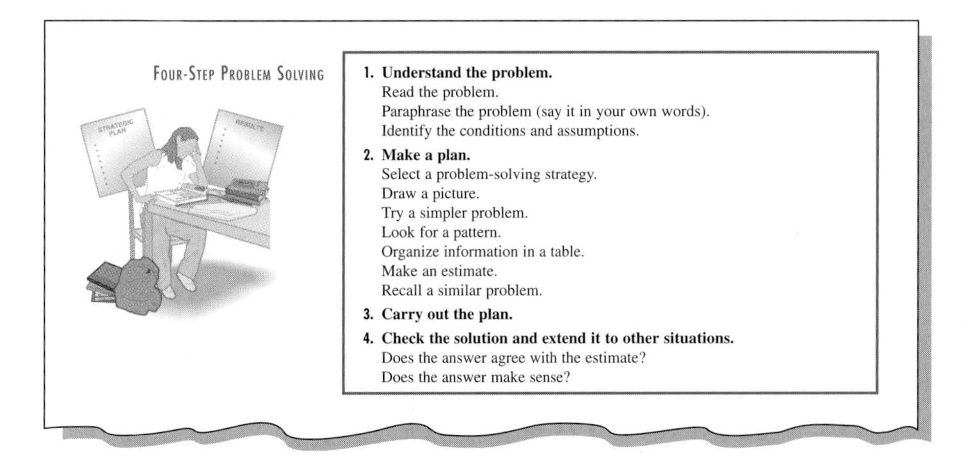

FOUR-STEP PROBLEM SOLVING

1. **Understand the problem.**
   Read the problem.
   Paraphrase the problem (say it in your own words).
   Identify the conditions and assumptions.
2. **Make a plan.**
   Select a problem-solving strategy.
   Draw a picture.
   Try a simpler problem.
   Look for a pattern.
   Organize information in a table.
   Make an estimate.
   Recall a similar problem.
3. **Carry out the plan.**
4. **Check the solution and extend it to other situations.**
   Does the answer agree with the estimate?
   Does the answer make sense?

Explorations

Some examples are intended to be used in class for individual or group exploration. The solutions to these exploratory examples are included in the Answer Box at the end of the section.

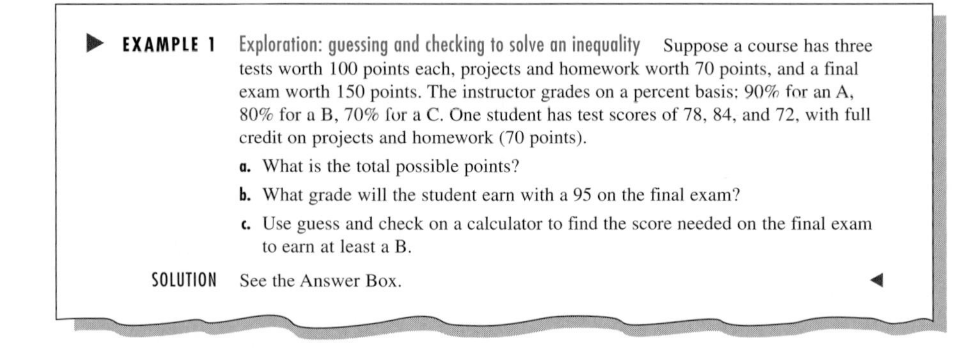

▶ **EXAMPLE 1** Exploration: guessing and checking to solve an inequality Suppose a course has three tests worth 100 points each, projects and homework worth 70 points, and a final exam worth 150 points. The instructor grades on a percent basis: 90% for an A, 80% for a B, 70% for a C. One student has test scores of 78, 84, and 72, with full credit on projects and homework (70 points).

a. What is the total possible points?

b. What grade will the student earn with a 95 on the final exam?

c. Use guess and check on a calculator to find the score needed on the final exam to earn at least a B.

SOLUTION See the Answer Box. ◀

Hands-On Materials

The text supports use of an assortment of hands-on materials. Section 1.1 opens with examples in which toothpicks model petting-zoo panels. Chapter 2 recommends use of a variety of hands-on materials, including colored plastic chips for integers and algebra tiles for adding like terms.

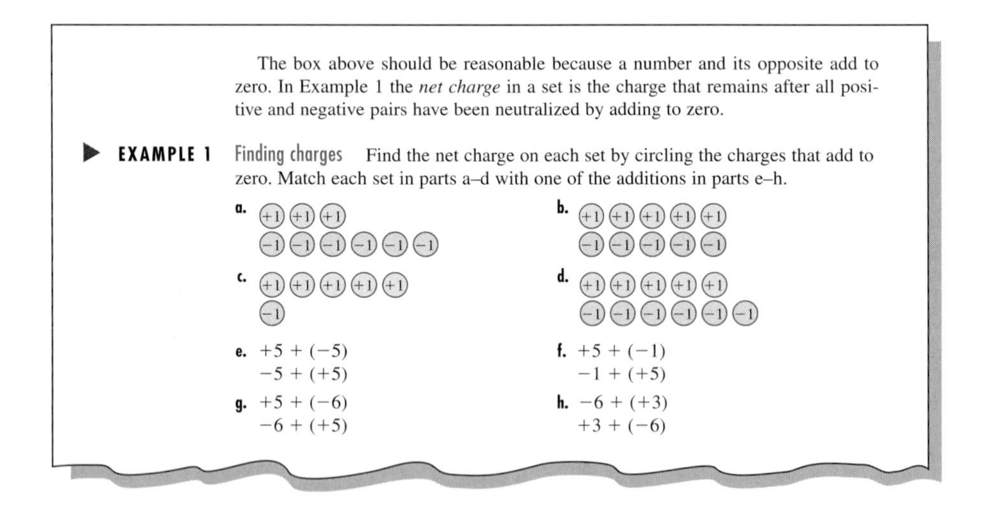

The box above should be reasonable because a number and its opposite add to zero. In Example 1 the *net charge* in a set is the charge that remains after all positive and negative pairs have been neutralized by adding to zero.

▶ **EXAMPLE 1** Finding charges Find the net charge on each set by circling the charges that add to zero. Match each set in parts a–d with one of the additions in parts e–h.

a. (+1) (+1) (+1)
(−1) (−1) (−1) (−1) (−1) (−1)

b. (+1) (+1) (+1) (+1) (+1)
(−1) (−1) (−1) (−1) (−1)

c. (+1) (+1) (+1) (+1) (+1)
(−1)

d. (+1) (+1) (+1) (+1) (+1)
(−1) (−1) (−1) (−1) (−1) (−1)

e. $+5 + (-5)$
$-5 + (+5)$

f. $+5 + (-1)$
$-1 + (+5)$

g. $+5 + (-6)$
$-6 + (+5)$

h. $-6 + (+3)$
$+3 + (-6)$

Examples

Each example begins with a title, which states the purpose of the example. Usually these titles relate back to the objectives for the section. Several examples link with others elsewhere in the textbook.

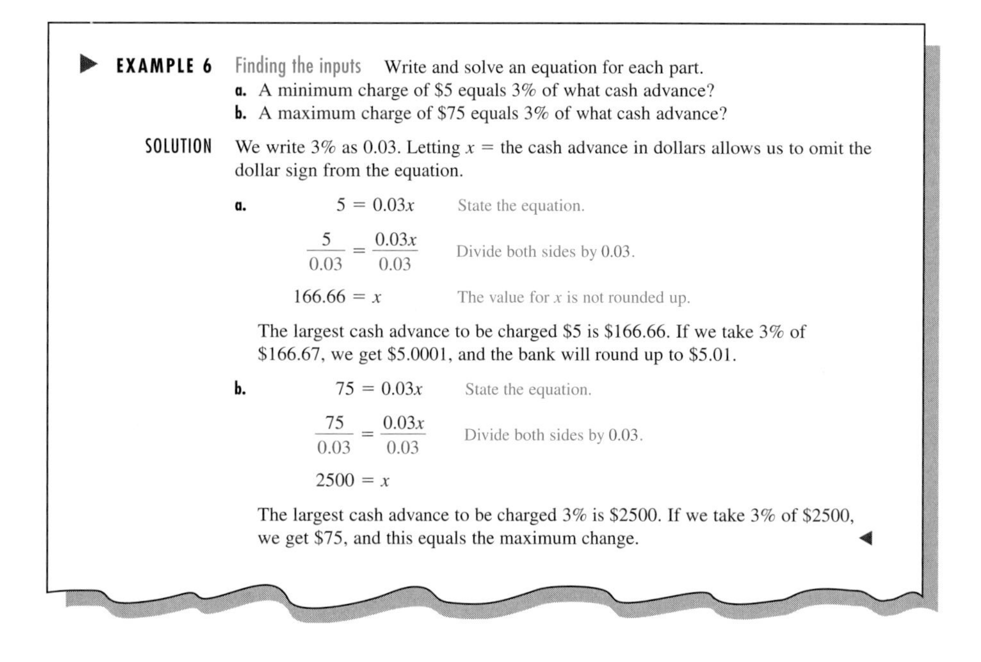

EXAMPLE 6 Finding the inputs Write and solve an equation for each part.

a. A minimum charge of \$5 equals 3% of what cash advance?
b. A maximum charge of \$75 equals 3% of what cash advance?

SOLUTION We write 3% as 0.03. Letting x = the cash advance in dollars allows us to omit the dollar sign from the equation.

a. $5 = 0.03x$ State the equation.

$\frac{5}{0.03} = \frac{0.03x}{0.03}$ Divide both sides by 0.03.

$166.66 = x$ The value for x is not rounded up.

The largest cash advance to be charged \$5 is \$166.66. If we take 3% of \$166.67, we get \$5.0001, and the bank will round up to \$5.01.

b. $75 = 0.03x$ State the equation.

$\frac{75}{0.03} = \frac{0.03x}{0.03}$ Divide both sides by 0.03.

$2500 = x$

The largest cash advance to be charged 3% is \$2500. If we take 3% of \$2500, we get \$75, and this equals the maximum change. ◀

Think about it

"Think about it" questions are included within the reading material to encourage students to relate examples to prior material, to extend examples, and to practice verbalization skills. Answers to the questions are provided in the Answer Box.

THINK ABOUT IT 1: Adding parentheses may be essential in doing Example 1 on a calculator. Try these different sequences of keystrokes and compare answers. Which is correct?

a. 10 000 000 [÷] 60 [×] 24 [×] 365 [ENTER]
b. 10 000 000 [÷] 60 [÷] 24 [÷] 365 [ENTER]
c. 10 000 000 [÷] [(] 60 [×] 24 [×] 365 [)] [ENTER]

Applications

To encourage creative thinking and depth in understanding, the text often poses a variety of questions about a single application setting. In addition, several applications, such as the credit card fee schedule, are repeated throughout the text so that students may observe the continuity and connections among topics.

EXAMPLE 4 Writing ordered pairs, then an equation On August 24, 2005, the Associated Press (AP) reported that flying the president's airplane, Air Force One, cost \$6029 per hour in fuel, compared with \$3974 in the prior year. Assume that the cost of fuel is a function of time in years. To simplify the numbers, assume that 1990 is year 0, the year Air Force One was placed into service.

a. Record the data in ordered pairs.
b. Find a linear equation for the function.
c. What is the meaning of the slope and y-intercept?
d. Is a linear function for fuel cost a reasonable assumption?

Tables and Graphs

Extensive use is made of data in tabular form. Tables encourage organization of information and promote observation of patterns. They also prepare students for functions and spreadsheet technology. Where appropriate, a graph is related to the table, to underscore the connections among algebra, geometry, statistics, and the real world. Number patterns and their corresponding equations and graphs are employed to emphasize the fact that algebra is the transition language between arithmetic and analysis.

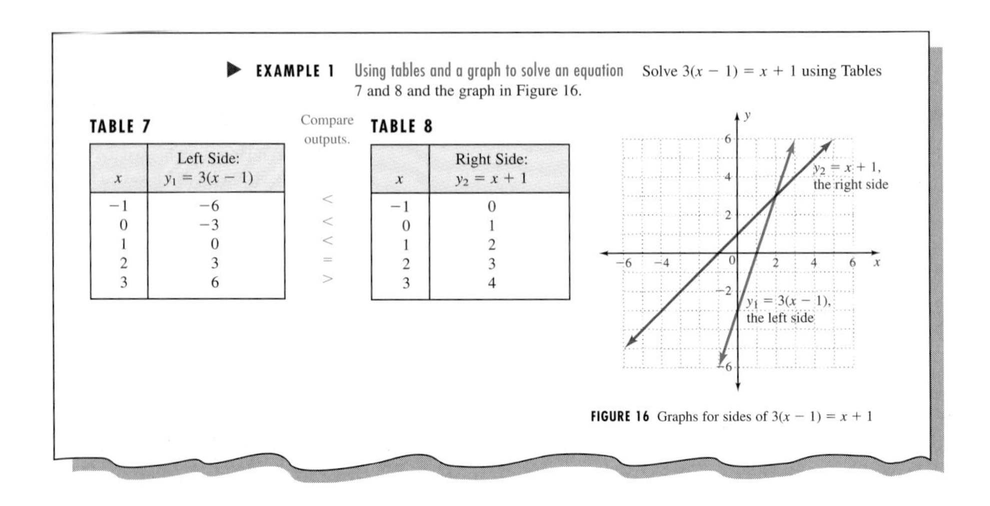

▶ **EXAMPLE 1** Using tables and a graph to solve an equation Solve $3(x - 1) = x + 1$ using Tables 7 and 8 and the graph in Figure 16.

TABLE 7

x	Left Side: $y_1 = 3(x - 1)$
−1	−6
0	−3
1	0
2	3
3	6

Compare outputs.

<
<
<
=
>

TABLE 8

x	Right Side: $y_2 = x + 1$
−1	0
0	1
1	2
2	3
3	4

FIGURE 16 Graphs for sides of $3(x - 1) = x + 1$

Calculator Techniques

At the very least, a scientific calculator is required for this text. General keystrokes for scientific calculators are supplied where appropriate.

The use of graphing calculator technology—even if it is only by the instructor—enhances learning, as it gives students an understanding of basic concepts. Calculator suggestions are provided throughout the text in Graphing Calculator Technique boxes in addition to the Graphing Calculator Basics Appendix.

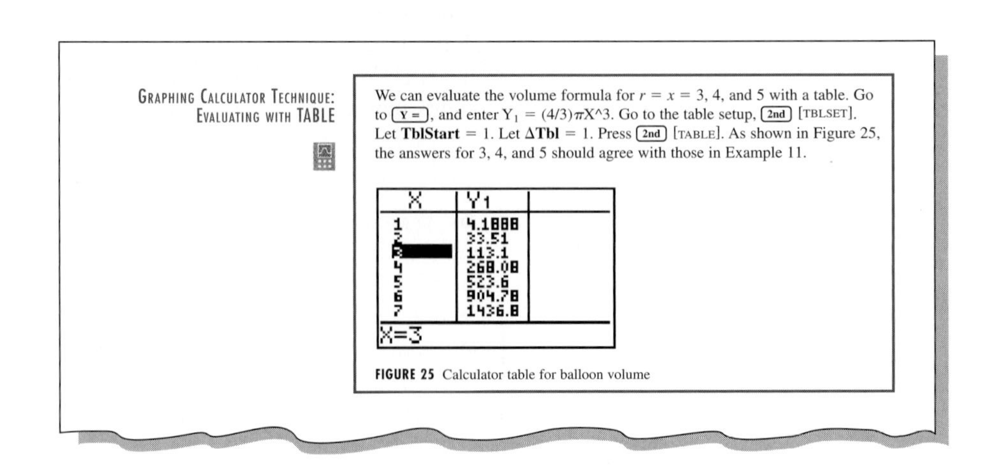

GRAPHING CALCULATOR TECHNIQUE: EVALUATING WITH TABLE

We can evaluate the volume formula for $r = x = 3$, 4, and 5 with a table. Go to [Y=], and enter $Y_1 = (4/3)\pi X$^3. Go to the table setup, [2nd] [TBLSET]. Let **TblStart** = 1. Let **ΔTbl** = 1. Press [2nd] [TABLE]. As shown in Figure 25, the answers for 3, 4, and 5 should agree with those in Example 11.

X	Y1
1	4.1888
2	33.51
3	113.1
4	268.08
5	523.6
6	904.78
7	1436.8

X=3

FIGURE 25 Calculator table for balloon volume

Answer Boxes

Answers to the Warm-up and Explorations, as well as the "Think about it" questions, are placed in the Answer Box at the end of the section (just before the exercises). By providing answers as feedback, the Answer Box permits the text to be used in class or as a laboratory manual for group work or independent study.

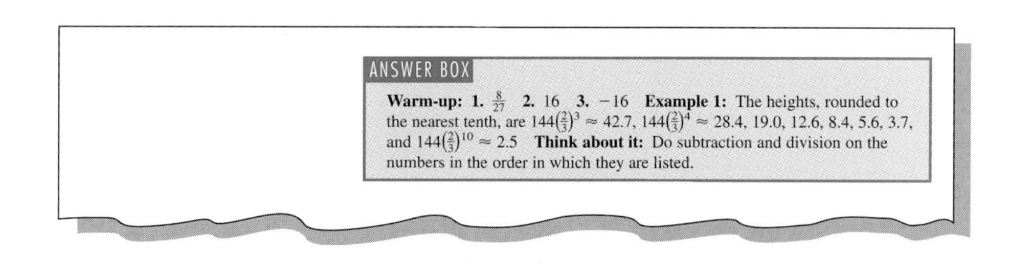

ANSWER BOX

Warm-up: 1. $\frac{8}{27}$ **2.** 16 **3.** -16 **Example 1:** The heights, rounded to the nearest tenth, are $144(\frac{2}{3})^3 \approx 42.7$, $144(\frac{2}{3})^4 \approx 28.4$, 19.0, 12.6, 8.4, 5.6, 3.7, and $144(\frac{2}{3})^{10} \approx 2.5$ **Think about it:** Do subtraction and division on the numbers in the order in which they are listed.

Exercises

Tables and graphs in the exercises give students practice in skills such as solving equations. They also help students to learn graphing technology.

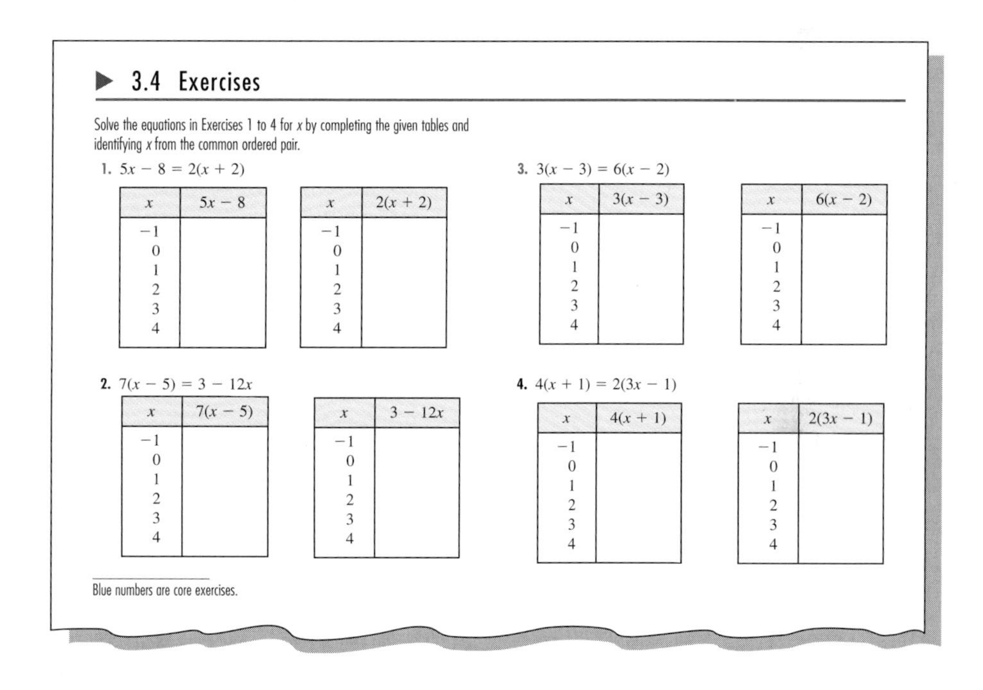

▶ 3.4 Exercises

Solve the equations in Exercises 1 to 4 for x by completing the given tables and identifying x from the common ordered pair.

1. $5x - 8 = 2(x + 2)$

x	$5x - 8$
-1	
0	
1	
2	
3	
4	

x	$2(x + 2)$
-1	
0	
1	
2	
3	
4	

2. $7(x - 5) = 3 - 12x$

x	$7(x - 5)$
-1	
0	
1	
2	
3	
4	

x	$3 - 12x$
-1	
0	
1	
2	
3	
4	

3. $3(x - 3) = 6(x - 2)$

x	$3(x - 3)$
-1	
0	
1	
2	
3	
4	

x	$6(x - 2)$
-1	
0	
1	
2	
3	
4	

4. $4(x + 1) = 2(3x - 1)$

x	$4(x + 1)$
-1	
0	
1	
2	
3	
4	

x	$2(3x - 1)$
-1	
0	
1	
2	
3	
4	

Blue numbers are core exercises.

Projects

Projects are intended for group work or for individual effort. They may be more complicated problems related to the topic at hand, activity-based problems using manipulatives, or real-world applications that require research outside class. Selected projects might be due a week or so after homework is completed for any given section. Projects suited to in-class group work are marked with an asterisk in the Index of Projects, included in the Annotated Instructor's Edition and repeated in the *Instructor's Resource Manual*. Chapter Projects are included in the Review Exercises.

99. Stacking Up Coins

a. Arrange 25 coins into four stacks that fit the following conditions: The second stack is 3 times the first stack. The third stack is 1 less than the second. The fourth stack is 2 more than the first. How many coins are in each stack? Write an equation that would solve the same problem.

b. Arrange 28 coins into four stacks that fit the following conditions: The third stack is 3 more than the second stack. The first stack is twice the second stack. The fourth stack is 1 more than the first stack. How many coins are in each stack? Write an equation that would solve the same problem.

c. Describe a strategy to arrange the coins into the requested stacks.

▶ **Project**

100. Equivalent Equations Which equation, if any, in each set is not equivalent to the other three? Show clearly how you made your choice.

a. $2x = 6$, $6 \div 2 = x$, $2 \div x = 6$ (for x not zero), $6 = 2x$

b. $x + 3 = -2$, $x = -6$, $x = -2 \cdot 3$, $x \div (-2) = 3$

c. $5x + 4 = 24$, $5x = 20$, $24 = 5x + 4$, $4 - 24 = 5x$

d. $4 + x = 9$, $9 = x + 4$, $9 - 4 = x$, $9 - x = 4$

e. $x - 6 = -3$, $x + 3 = 6$, $x - 3 = 0$, $x = -9$

f. $2x + 3 = 15$, $2x = 12$, $3 - 15 = 2x$, $15 = 3 + 2x$

▶ Mid-Chapter Test

To keep students engaged and build their confidence, a Mid-Chapter Test is included in each chapter. This test gives students the opportunity to check their progress. All answers appear in the back of the book, in the answer section.

▶ 3 Mid-Chapter Test

1. State whether each equation is an identity or a conditional equation.
   a. $2x + 4 = -1$ b. $2x + 3x = 5x$
   c. $4(x - 2) = 4x - 8$ d. $2x = x$

2. Which pairs of equations are not equivalent?
   a. $4x + 5 = 29$, $4x = 24$
   b. $\frac{1}{2}x = 16$, $x = 8$
   c. $2x + 3x = 10$, $5x^2 = 10$
   d. $3x - 2 = 8$, $3x = 6$

In Exercises 3 to 5, is the given input x a solution to the equation?

3. $3x - 4 = -16$, $x = -4$
4. $4x - 3 = -9$, $x = -3$
5. $-8x + 5 = 1$, $x = \frac{1}{2}$

In Exercises 6 to 8, write equations. Let x be the unknown number.

6. Five more than twice a number is 10.
7. How many credit hours can be taken for $545 if tuition is $85 per credit hour and fees are $35?
8. How many minutes were used if there is $7.20 left on a $20 telephone card and each minute costs $0.04?

In Exercises 9 and 10, write each equation in words.

9. $5 = 3x - 4$
10. $2(2 + x) = -3$

In Exercises 11 to 18, solve for x.

11. $x - 4 = 3$ 12. $\frac{2}{3}x = 24$
13. $2x + 3 = -7$ 14. $\frac{1}{2}x - 8 = -1$
15. $3x = \frac{1}{2}$ 16. $3 - 2x = 8$
17. $3(2 + x) = 18$ 18. $4 - 2(x - 1) = -4$

▶ Chapter Summary, Review Exercises, and Chapter Test

Every chapter ends with a Chapter Summary, Review Exercises, and a Chapter Test. The student is provided with answers to the odd-numbered Review Exercises and answers to all of the Chapter Test questions.

▶ 2 Chapter Summary

Vocabulary

For definitions and page references, see the Glossary/Index.

absolute value
adding like terms
additive inverse
algebraic fractions
associative properties
base
capacity
circumference
commutative properties
composite numbers
compound inequality
continued equality
cubed
distributive property of multiplication over addition
equivalent expressions
evaluate
exponent
factors
formula
greatest common factor
grouping symbol
inequality
infinite
infinity sign
interval
like terms
multiplicative inverse
non-negative
order of operations
pi
power
prime number
radical
reciprocal
simplification property of fractions
simplify
squared
surface area
term
unit analysis
volume

Cumulative Review

To help students maintain skills, a set of Cumulative Review Exercises is placed at the end of Chapters 2 to 8.

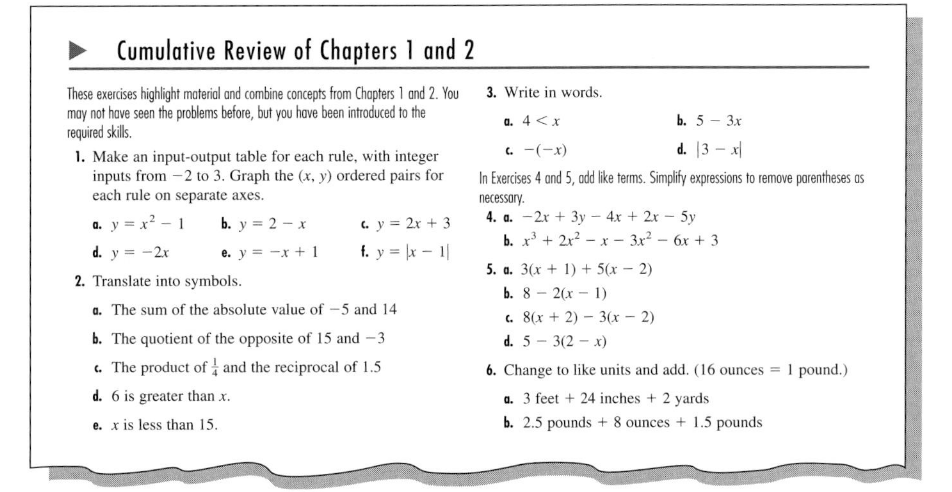

Cumulative Review of Chapters 1 and 2

These exercises highlight material and combine concepts from Chapters 1 and 2. You may not have seen the problems before, but you have been introduced to the required skills.

1. Make an input-output table for each rule, with integer inputs from -2 to 3. Graph the (x, y) ordered pairs for each rule on separate axes.

a. $y = x^2 - 1$ **b.** $y = 2 - x$ **c.** $y = 2x + 3$
d. $y = -2x$ **e.** $y = -x + 1$ **f.** $y = |x - 1|$

2. Translate into symbols.

a. The sum of the absolute value of -5 and 14
b. The quotient of the opposite of 15 and -3
c. The product of $\frac{1}{4}$ and the reciprocal of 1.5
d. 6 is greater than x.
e. x is less than 15.

3. Write in words.

a. $4 < x$ **b.** $5 - 3x$
c. $-(-x)$ **d.** $|3 - x|$

In Exercises 4 and 5, add like terms. Simplify expressions to remove parentheses as necessary.

4. a. $-2x + 3y - 4x + 2x - 5y$
b. $x^3 + 2x^2 - x - 3x^2 - 6x + 3$

5. a. $3(x + 1) + 5(x - 2)$
b. $8 - 2(x - 1)$
c. $8(x + 2) - 3(x - 2)$
d. $5 - 3(2 - x)$

6. Change to like units and add. (16 ounces = 1 pound.)

a. 3 feet + 24 inches + 2 yards
b. 2.5 pounds + 8 ounces + 1.5 pounds

Final Exam Review

A Final Exam Review, containing exercises requiring both short and long answers, follows the last chapter. The review is divided by chapter and may be used as a source of additional exercises or cumulative review material.

Glossary/Index

For the convenience of students and instructors, essential vocabulary is defined and referenced by page in the combined Glossary/Index. Vocabulary is considered essential if it is listed in the Chapter Summary. Nonessential terms are referenced only by page in the Glossary/Index.

Instructor Resources

Annotated Instructor's Edition

The Instructor's Edition provides the complete student text with solutions next to each respective exercise.

Complete Solutions Manual

The Complete Solutions Manual provides worked-out solutions to all of the problems in the text.

Instructor's Resource Manual

This manual provides structured lesson and group-activity suggestions for each section in the textbook, incorporates materials from the textbook with supplemental projects and activities, and suggests core homework assignments and guided-discovery questions.

Test Bank

The Test Bank includes multiple tests per chapter as well as final exams. The tests are made up of a combination of multiple-choice, free-response, true/false, and fill-in-the-blank questions.

Computerized Testing

Contains all Test Bank questions electronically so you can quickly create customized tests in print or online.

Course Management System

This Course Management System allows you to assign, collect, grade, and record homework assignments via the web. This proven homework system has been enhanced to include links to other electronic content. The Course Management System is more than a homework system, it is a complete learning system for math students.

Online Teaching Center

Along with direct access to the student website, this. website offers numerous resources for instructors. Available at **college.hmco.com/pic/kasebergintro4e.**

Student Resources

Student Solutions Manual

The Student Solutions Manual provides worked-out solutions to the odd-numbered problems in the text.

Online Study Center

Additional student resources can be accessed via this website. Available at **college.hmco.com/pic/kasebergintro4e.**

Additional Resources

NEW! Math Study Skills Workbook, 3/E by Paul D. Nolting
This workbook is designed to reinforce skills and minimize frustration for students in any math class, lab. or study skills course. It offers a wealth of study tips and sound advice on note taking, time management, and reducing math anxiety. In addition, numerals opportunities for self-assessment enable students to track their own progress.

Acknowledgments

Reviewers and class-testers are the appreciated but unrecognized co-authors of a single-author textbook. Their comments, questions, and suggestions have contributed significantly to this and earlier editions. First let me thank the reviewers leading to this edition:

Jodi Edington, *Kalamazoo Valley Community College*
Ion Georgiou, *Foothill College*
Laura Moore-Mueller, *Green River Community College*
Ellen Musen, *Brookdale Community College*
Patricia Rhodes, *Treasure Valley Community College*
Patrick Woomer, *Adirondack Community College*
Monica Zore, *Marian College*

Then let me thank those who provided valuable input through the focus group and Web survey:

Barbara S. Allen, *Delta College*
Chris Allgyer, *Mountain Empire Community College*
Wayne Barber, *Chemeketa Community College*
Judy Chilcott, *Truckee Meadows Community College*
Terry Darling, *Lewis and Clark Community College*
Hortencia I. Garcia, *Eastern Connecticut State University*
Tom Grogan, *Cincinnati State University*
Judith Hector, *Walters State Community College*
Stephen Paul Hess, *Grand Rapids Community College*
Tracey Hoy, *College of Lake County*
Doug Mace, *Kirtland Community College*
Annette Magyar, *Southwestern Michigan University*
Tammi Marshall, *Cuyamaca College*
William S. Newhall, *Truckee Meadows Community College*
Michael Norris, *Los Medanos College*
William A. Prescott, Jr., *University of Maine at Machias*
Carol Rardin, *Central Wyoming College*
Marcell Romancky, *Kirtland Community College*
Michael Satnik, *Central Washington University*
Barbara D. Sehr, *Indiana University–Kokomo*
Sister Monica Zore, *Marian College*

Finally, I would like to thank the following reviewers and class-testers for their helpful comments and significant contributions to the first, second, and third editions:

Carol Achs, *Mesa Community College*
Rick Armstrong, *Florissant Valley Community College*
Mayme Kay Banasiak, *University of Tennessee–Chattanooga*
Linda Bastian, *Portland Community College–Sylvania Campus*
Paula Castagna, *Fresno City College*
Deann Christianson, *University of the Pacific*
Jennifer Dollar, *Grand Rapids Community College*
Dennis C. Ebersole, *Northampton Community College*
Grace P. Foster, *Beaufort County Community College*
Dave Gillette, *Chemeketa Community College*
Judith H. Hector, *Walters State Community College*
Nancy Henry, *Indiana University–Kokomo*
Tracey Hoy, *College of Lake County*
Donna L. Huck, *North Central High School, Spokane*
Mark Hugen, *Cerritos College*
Charlotte Hutt, *Southwest Oregon Community College*
Virginia Lee, *Brookdale Community College*
Marveen McCready, *Chemeketa Community College*
Elizabeth Mefford, *Walter State Community College*
Charles Miller, *Albuquerque Technical Vocational Institute*
Alice A. Mullaly, *Southern Oregon State College*
Susan D. Poston, *Chemeketa Community College*
Douglas Robertson, *University of Minnesota*
Gil Rodriguez, *Los Medanos College*
Larry Smyrski, *Henry Ford Community College*
John Thickett, *Southern Oregon State College*
Susan M. White, *DeKalb College*
Tom Williams, *Rowan-Cabarrus Community College*
Robert Wynegar, *University of Tennessee, Chattanooga*

Jennifer Dollar, Grand Rapids Community College, Sally Jackman, Richland College, and Kathleen Offenholley, Brookdale Community College have been invaluable eyes in producing this edition. Nevertheless, I am responsible for any errors that remain.

This project started in 1992, and it still has the full support of my husband, Rob Bowie. To say that I could not have done it without him is a total understatement.

A special thank you to Richard Stratton, Lynn Cox, and Katherine Greig at the Houghton Mifflin Company. Their support to me and for this fourth edition has been greatly appreciated. Merrill Peterson at Matrix Productions was most helpful as I negotiated the changes in the production industry. My gratitude to Merrill and everyone for making this the best edition yet.

To the Student

Getting an education is one of the most valuable activities you will carry out in your life. Although poor economic circumstances may require you to relocate or retrain, you will not lose your education in a stock market crash. Your education will give you the skills to learn, again and again.

Whether your current educational goal is obtaining a degree or advancing in your chosen career, you should know exactly why you are taking this course and why it is important for you to learn and succeed. If your goal and purpose are not clear, you may find it difficult to make the daily commitment needed to study, learn, and succeed.

What to Expect from This Text

Introductory Algebra: Everyday Explorations presents the skills, concepts, problem-solving opportunities, and applications now considered fundamental to an Introductory Algebra course. The subtitle, Everyday Explorations, stresses how tuition payments, utility bills, credit cards, merchandise orders, truck rentals, and photocopy machines (to name a few) provide settings for your algebraic experience. You are encouraged to look for these and other settings outside the classroom while identifying their rules and equations inside the classroom.

Alternative Approaches

Consider how you best learn directions to a friend's house—in words over the phone (verbally), from a map (visually), or from having been there (kinesthetically). As a student of algebra, you may prefer words (a verbal approach); drawings, pictures, and graphs (a visual approach); working with objects in your hands (a kinesthetic approach); or a combination of these approaches. The best way to achieve success in mathematics is to learn new concepts using the approach that you find easiest and then reinforce your learning with other approaches.

To help you achieve success with algebra, *Introductory Algebra: Everyday Explorations* presents concepts in as many ways as space permits. Because of the variety of approaches, those of you who have had algebra before will find that some concepts are presented quite differently than the way you learned them the first time. Acknowledging alternative approaches is a way of validating your own discoveries.

Independent Thinking

Although you will find the examples helpful, this text is designed to encourage your independent thinking. Look for patterns and relationships; discover concepts for yourself; seek out applications that are meaningful to you. Try to work through examples and explorations on your own first, before you look at the solution.

The more involvement you have with the material, the more useful it will be and the longer you will remember it.

Groups

You are encouraged to work with others throughout this course. One of the most important benefits of working on mathematics with other people in and out of class is that you clarify your own understanding when you explain an idea to someone else. This is especially true for the kinesthetic learner.

Beginning the New Term: First Day

Problem Solving

Section 1.1 introduces problem-solving steps in a mathematical context. Right now, think about these four basic problem-solving steps in the context of your planning for the next few months:

1. *Understand the problem.* Your problem is that you don't have enough time to do everything you would like to do.
2. *Make a plan.* One time management strategy is to make a list of everything you have to do and when you have to do it. Make a chart showing each waking hour for the next seven days. For each hour, write in what you plan to do with that time. (There are many problem-solving strategies for planning. For a more complete listing, see Problem Solving in the Preface.)
3. *Carry out the plan.* Follow your plan (your schedule) for a week. Write notes on it to indicate when you varied from the plan and why.
4. *Check and refine.* After one week, review the plan. Did you get everything done that you needed to do? What do you need to change in your plan? Are your school load and work load reasonable? Redo the schedule to accommodate needed changes.

Time Management

Make sure your course load is sensible.

Make sure your course load is sensible. Exceptionally few people can productively manage 60 hours per week of classes, study, and work. Check how sensible your plan for this term is:

> Multiply your number of credits hours by 3, and then add your number of hours of employment.

For most people, 40 to 45 hours is a reasonable commitment to school and employment.

Make or buy a calendar for the term, with spaces large enough for noting assignments, tests, appointments, and errands. Keep the calendar with you at all times.

Are You in the Right Course?

Have you studied the prerequisite material recently?

Each mathematics course has one or more prerequisite courses. Having passed a placement test does not ensure that you are prepared to succeed. If you have studied the background material recently, then usually with time, effort, confidence, and patience you will be able to learn the new material. If you have had a semester or a quarter or a summer break since your last math course, it is necessary to review. If the review provided in the text is not sufficient for you to recall, say, operations with fractions, you should immediately seek advice from your instructor or outside help. Use your prior book as a reference. If you took the prerequisite course more than a year ago, you may want to retake it before going on.

After the First Class

As you review your course syllabus, write test dates and other deadlines on your calendar. If you are working or taking classes at two schools, make sure there are no schedule conflicts with final exams. Talk with your instructor this week to resolve any scheduling problems.

Beginning the New Term: First Week

Here are a number of ways to be successful as a student.

Get a Good Start

Have homework ready to turn in as you walk into class.

To succeed, you need to attend class, read the book before class, and do the homework in a timely manner. Plan your study time. Some students set up their schedules to have the hour after math class free, to review notes and start the assignment.

Success also depends on being prepared with the proper equipment: an appropriate calculator, a six-inch ruler also marked in centimeters, and graph paper. Do all your graphs on graph paper.

Keep in mind that your first homework paper is a "grade application," just as a cover letter and resume are part of a job application. First impressions count. Neatness and completeness make a lasting impression on the instructor. So does having homework ready to turn in as you walk into class.

Stick to Your Plan

Successful students plan their time carefully. Because studying is most productively done in the daytime and in hour-long segments, plan to study between classes. Unless you have a pressing need to leave campus, stay an extra hour and study again after your last class. Not only will a schedule help you be more efficient; it will also remind you of your priorities and prevent you from avoiding tasks that need to be done.

Prepare for the Next Day's Class

Each of the following steps will get you progressively more prepared for your next class.

- Skim the textbook section first. Read objectives. List unfamiliar words and identify new skills or concepts.
- Scan the section and look for definitions of vocabulary.
- Write vocabulary words and definitions on note cards.
- Outline the section, including summaries of skills and applications. (For your convenience in outlining, the objectives, headings, and example titles are in color in the text.)
- Read through the steps in several examples.
- Try the homework ahead of time.

Take Notes

Observe the five R's of note-taking (the Cornell system):

Don't recopy notes; summarize them.

1. *Record.* Write down the ideas and concepts in a lecture. (Reading the text ahead of time will help identify these items.) Don't recopy notes.

2. *Reduce*. Summarize notes immediately after class (or as soon as possible); highlight important items.
3. *Recite*. Say out loud in your own words the main ideas and concepts.
4. *Reflect*. Think about how the material fits in with what you already know.
5. *Review*. Once a week, go over the ideas from each class so far in the term.

Beginning the New Term: First Month

Homework

Highlight exercises that were difficult and that you want to review again later.

Do the homework as one of the steps in your learning—not just as a requirement of the course. Work on assignments as soon as possible, right after class or early in the day or weekend. This gives you the option of going back later and spending more time on a difficult exercise. During long study periods, build in breaks to keep yourself fresh: Work for an hour, do another subject for an hour, and then come back.

Make the homework meaningful. Write notes to yourself on homework papers. Highlight exercises that were difficult and that you want to review again later. Highlight formulas or key steps. Re-read the objectives. Summarize the definitions and solution methods in your own words to be sure you understand. Describe how the current section fits in with prior sections.

Using the Answer Box and Answer Section Effectively

Practice working quickly. Do not work with the answers in front of you. Wait to check your answers until you have finished several exercises or half or more of the homework assignment. Let your own reasoning tell you whether something is correct.

Stuck on the Homework?

Suppose you took notes in class, read the section, and still are stumped by an exercise. If you understand the directions but can't get the problem to work, try it again on a clean sheet of paper. If you don't know how to do an exercise, summarize the relevant information and drawings and go on to another exercise. Be sure to read the exercise aloud before you give up. Sometimes we hear things that we miss when reading silently.

Sometimes we get too close to a problem and overlook the obvious. A fresh point of view may help. Come back later. If necessary, call another student. If you are off-campus, call your instructor during office hours to get a hint or suggested strategy. Many instructors also welcome e-mail questions.

Obtain help from your teacher or from the resource center as you need it. Don't wait until just before a test.

Falling Behind?

Do current assignments first.

If you find yourself falling behind, let your instructor know that you are trying to catch up. Set up a plan that allows two to three days for each missed assignment. Most important, do current assignments first, even if you have to skip a few problems because you missed material. Work immediately after the class session. By doing the current assignment first, you will stay with the class. If you gradually complete missed work, within a reasonable amount of time you will be completely caught up. Do not skip class because you are behind or confused.

Forgetting Material?

Many students select one or two exercises from each section and write them on 3″ × 5″ cards, with complete solutions on the back. These "flash cards" may then be shuffled and practiced at any time for review. Cards provide an excellent way to study for tests and the final exam. Include vocabulary words in your card set.

Strategies for Taking and Learning from Tests

Prepare Yourself Academically

1. Attend class, and do the homework completely and regularly. If there are exercises on the homework that you do not know how to do, get help—from a classmate, the teacher, or another appropriate source.

Work under time pressure on a regular basis.

2. Work under time pressure on a regular basis. Set yourself a limited amount of time to do portions of the homework. Use a time limit when doing review exercises or the practice tests at the middle and end of each chapter. Working in one- or two-hour blocks of time is usually more productive than spending all afternoon and evening on math one day a week.

Prepare Yourself Physically

3. Get a good night's sleep. Being rested helps you think clearly, even if you know less material.

Prepare Yourself Mentally

Psych yourself up! This is especially important if you have your test later in the day.

Get everything ready for the next day.

4. If you have a test at 8:00 a.m., use the last few minutes before bed to get everything ready for the next day. Make your lunch, set your books or pack on a chair by the door, set out your umbrella or appropriate weather gear, and make sure you have change for the bus or train or that your car's tires, battery, and gasoline level are okay.
5. Plan 10 or 15 minutes of quiet time before the test. Try to arrive early, if possible.
6. Mentally picture yourself taking the test.
   a. Imagine writing your name on the test.
   b. Imagine reading through the test completely to see where the instructor put various types of questions.
   c. Imagine writing notes, formulas, or reminders to yourself on the test.
   d. Imagine working your favorite type of problem first.

Take the Test Right

7. Arrive early. Be ready—pencil sharpened and homework papers ready to turn in.
8. Concentrate on doing the steps that you imagined in item 6 above.

Work problems you know how to solve first.

9. Work quickly and carefully through those problems you know how to solve. Don't spend over two minutes on one problem until you have tried every problem.
10. Be confident that, having prepared for the test, you can succeed.

Learn from the Test

11. After you turned in the test, did you remember information that would have helped you on the test? Would reading through the test more thoroughly at the start have given you time to recall information you needed?
12. Before you forget, look up and write down anything that you needed to know for the test but did not know.
13. When you get the test back, look at each item you missed. Which ones did you know how to do, and which ones did you not know how to do? Figure out what caused you to miss the ones you knew how to do. Carefully re-work on paper the ones you did not know how to do, getting help as needed.
14. Write down what you will do differently in preparing for the next test.

CHAPTER 1

Algebraic Representations

Have you ever been unable to follow someone's directions? What works best for you? A map? A list of street names with right or left turns? A number of blocks and turning at familiar buildings? Or do you take a taxi (and once you have been there, you can always find it again)? We find our way around mathematics in different ways. One or two ways may work best for you; other ways work best for other students or even your own child. As an experiment, which way (visual or verbal) makes more sense to you as a preview of the first chapter of this book?

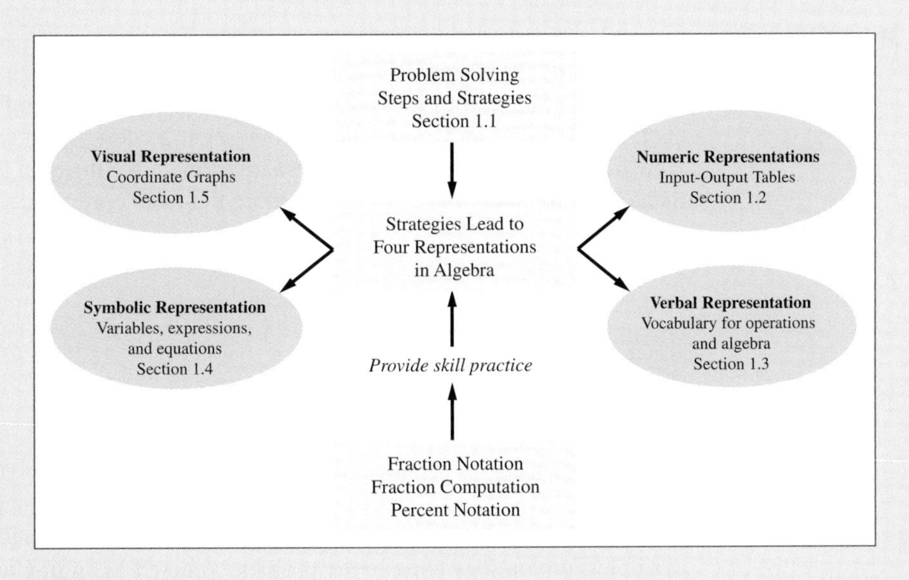

FIGURE 1 Chapter 1 overview

In Chapter 1, we use problem solving and strategies to help find our way in algebra, and we record our results in four representations: *numerically, verbally, symbolically,* and *visually.* Section 1.1 introduces us to problem solving and strategies. Sections 1.2 to 1.5 present numeric, verbal, symbolic, and visual representations. Along the way, we practice essential skills in working with fractions and percents.

▶ 1.1 Problem-Solving Steps and Strategies

Student Note:
Use the Objectives to guide your study. Ask yourself: What do I need to learn from this section? How do these ideas fit with what I already know? Study the section until you can answer yes to "Do I know the objectives?"

Student Note:
Warm-ups review prior material or preview the current lesson. Answers to Warm-ups and selected examples are given in the Answer Box just before the Exercises.

Objectives

- Identify and apply Polya's four steps for problem solving.
- Identify conditions and assumptions in problem solving.

WARM-UP

Give the number most likely to be next in each pattern.

1. 2, 4, 6, 8, __

2. 3, 5, 7, 9, __

3. 4, 7, 10, 13, __

MATHEMATICS, ESPECIALLY ALGEBRA, was developed by people trying to solve real problems and to describe the world around them. New mathematics is being developed even now, and algebra is the language used to express these new ideas.

LINKED EXAMPLE: PETTING ZOO PENS Linked examples introduce settings used again in future sections.

▶ **EXAMPLE 1** Petting Zoo Pens Exploration Jake is setting up the petting zoo at Central Park. He needs to build 20 pens for the animals. For the sides of the pens he has a large number of straight, movable panels of one length, as shown in Figure 2. The panels fasten together only at the ends. How might he design and arrange the pens? How many panels will he need? These questions, like many in real life, have more than one correct answer, and there is more than one correct way of getting the answer.

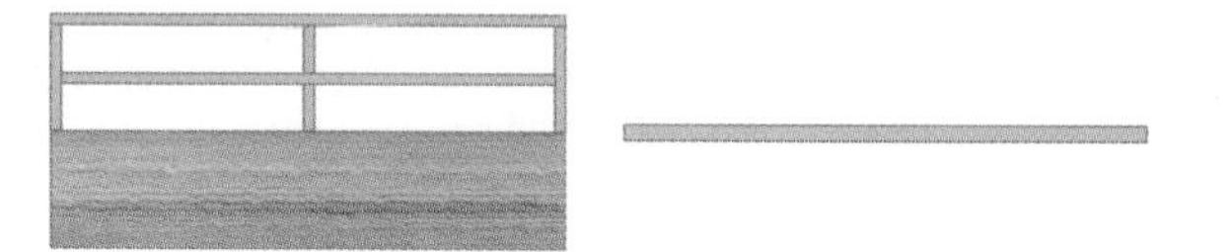

FIGURE 2 Panels for the petting zoo pens

SOLUTION You will find one solution to this exploration spread throughout the next several examples, or you may explore the problem yourself as an in-class activity. ◀

Student Note:
The end of an Example is marked by a ◀.

We will examine problem-solving techniques throughout the course. We start in this section by dividing problem solving into four steps. This four-step approach was first published by George Polya in *How To Solve It* in 1945.

▶ Problem-Solving Step 1: Understand the Problem

The first step in solving a problem is to understand the problem. In the exploration, Jake is given the task of building pens from panels. If he is new to the job and is given no instructions, he must first decide how to arrange the panels.

Jake must understand the conditions and make some assumptions. A **condition** is a *requirement or restriction stated within a problem setting*. An **assumption** is *something not stated but taken as a fact*.

We now look at the conditions and assumptions in Jake's problem.

EXAMPLE 2 Understanding the problem of building pens What conditions and assumptions must Jake consider?

SOLUTION Several conditions are stated in the problem: The panels are straight and movable. The panels are of one length and fasten only at the ends. It is not clear how the pens are to be arranged or even how many sides a pen should have. Jake must make assumptions in order to solve this problem. Jake might assume that each pen will hold only one animal, that any panel can also be a gate, and that the pens should be arranged for public viewing. What other assumptions might he make? ◀

THINK ABOUT IT: What assumptions and conditions might you encounter on your first day of class?

Student Note:
"Think about it" questions check your understanding, stress key ideas, or make connections between ideas. Possible answers are listed in the Answer Box.

Understanding the problem means understanding the questions, the given information (conditions), and any assumptions you have to make. Often you will have to read a problem several times to understand it clearly; then you will have to read it again to gather details. It is a good idea to say the problem in your own words.

Problem-Solving Step 2: Make a Plan

Once you understand the problem, you need to plan how you will solve the problem. To make a plan, you think about your assumptions, develop a set of strategies for solving the problem, and search your memory for similar problems and the strategies you used to solve those problems.

EXAMPLE 3 Making a plan for building pens Suppose Jake assumes that the pens are to be connected and set out in one long line. Come up with a plan for predicting the number of panels needed, without actually building the pens.

SOLUTION Part of making a plan is deciding which strategy to use. One strategy would be *drawing a picture* of a set of 20 pens and counting the panels. Another strategy would be *trying a simpler problem*, perhaps drawing a picture of 1, 2, and 3 pens. We could then *look for a pattern* that would allow us to find the number of panels for any number of pens, without either drawing or building them all. Finally, we might *organize information by using a table*. ◀

▶ Polya's research revealed that good problem solvers form at least one plan and think about a number of strategies before they start solving a problem. If the problem involves computation, *making an estimate* is also important. After a few lessons we might ask, *Have we seen a similar problem*?

Problem-Solving Step 3: Carry Out the Plan

Now, using what we understand about the problem, we can apply our strategies.

EXAMPLE 4 Drawing a picture of building pens Assuming that the pens will be square and in one long line, draw a picture and count the number of panels.

SOLUTION The picture in Figure 3 shows the pens. If you count 10 panels for the first three pens, you are counting correctly. How many panels are there for 20 pens? See the Answer Box.

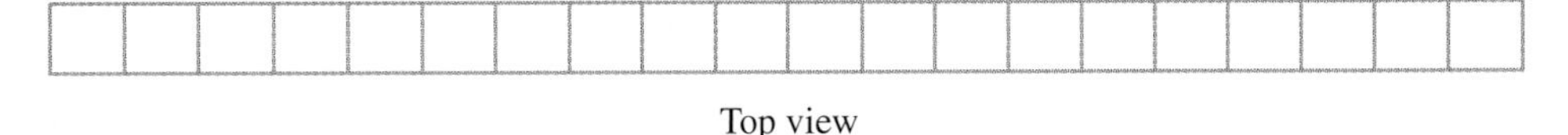

Top view

FIGURE 3 Twenty pens in a line ◀

▶ Drawing and counting panels for 20 pens may lead to error. By drawing a few of the pens, we can *do a simpler problem*. We then can organize our information in an **input-output table,** which is *a tabular form for describing or summarizing numerical relationships*. Think of the left side, or input, as the source of numbers and the right side, or output, as the number matched with the input to its left by some rule. In short, the output *depends* on the input.

▶ EXAMPLE 5 Making a table about building pens If Jake first builds 1 pen, then attaches a second pen, and then adds a third, as shown in Figure 4, what is the pattern in the total number of panels used?

SOLUTION Table 1 organizes the information. The headings in the table tell us to write the number of pens and the total number of panels used in building that number of pens.

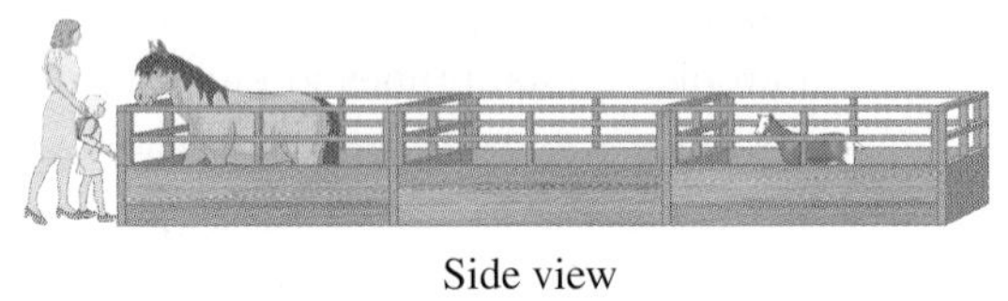

FIGURE 4 Two views of three pens

TABLE 1 Pens and Panels

Input: Number of Pens	Output: Total Number of Panels
1	4
2	7
3	10

From the figure and table we see that each new pen adds 3 more panels. How can we predict the number of panels for 20 pens? See the Answer Box for one way. ◀

We will practice making tables in Section 1.2 and study ways to find rules beginning in Section 1.3.

▶ Problem-Solving Step 4: Check the Solution and Extend It to Other Situations

Go back to the exploration in Example 1. Does the number of panels needed for 20 pens make sense in the problem? Example 4 used a picture, and Example 5 used a table of numbers. Because the results are the same with two different solution methods for our assumptions, we may be reasonably sure the answer is correct.

Finally, we may extend our solution method to other situations. What other assumptions and solutions are possible? Must we use square pens? Which pen designs use more panels? Which pen designs use fewer panels? Can we tie each animal to a panel and use fewer panels? Is there a space limitation? Will the design fit in the space? When reporting an answer, always list the assumptions you made in order to solve the problem.

▶ The Four-Step Cycle

Think of the problem-solving approach as a cycle with four steps: understand, make a plan, carry out the plan, and check. If the answer does not check, start the process again. If you find that you did not fully understand the problem, cycle back to the understanding step. If your strategies do not work, cycle back and make a new plan.

It is important to go through each step of the process. You may have to go through all or part of the process more than once.

FOUR-STEP PROBLEM SOLVING

1. **Understand the problem.**
   Read the problem.
   Paraphrase the problem (say it in your own words).
   Identify the conditions and assumptions.
2. **Make a plan.**
   Select a problem-solving strategy.
   Draw a picture.
   Try a simpler problem.
   Look for a pattern.
   Organize information in a table.
   Make an estimate.
   Recall a similar problem.
3. **Carry out the plan.**
4. **Check the solution and extend it to other situations.**
   Does the answer agree with the estimate?
   Does the answer make sense?

If you are working in groups, a good strategy is to work through the entire problem individually, check your work, see if you can answer the problem differently using different assumptions, and finally explain your thinking to others in the group and share information. Together you may think of solutions that no one thought of working alone. *Even if you are not working in a group in class,* working together outside class is highly recommended for checking homework exercises and for understanding concepts.

ANSWER BOX

Warm-up: 1. 10 **2.** 11 **3.** 16 **Think about it:** You might assume that your mathematics instructor will state the conditions for succeeding in the course. And when he or she does so, you learn what these conditions include, such as requirements for attendance, assignments, tests, and class participation. In turn, the instructor will assume that you have enrolled or intend to enroll in the course. Two conditions for enrollment may be to complete a registration process and to pay tuition. **Example 4:** 61 panels
Example 5: We add 3 panels for each new pen. The number of panels required for 20 pens in a straight line is 4 (for the first pen) plus 3 per pen times 19 additional pens; $4 + 3 \times 19$ is 61 panels. We assume that $4 + 3 \times 19$ means to multiply 3 and 19 and then add 4.

▶ 1.1 Exercises

In Exercises 1 to 4, suggest an assumption that could be made. Answers may vary.

1. You take out a student loan to attend school.
2. You write your first name but not your last name on your test.
3. You leave this message on your instructor's voice mail: "This is Anna; I won't be in class today."
4. You allow your daughter to go to a movie.

Blue numbers are core exercises.

Exercises 5 to 10 offer other solutions to the petting zoo problem. Think about the conditions and assumptions in each exercise.

5. Jake's assistant Kelsey suggests using triangular pens, as shown in the figure. How many panels would be needed to set up 20 triangular pens?

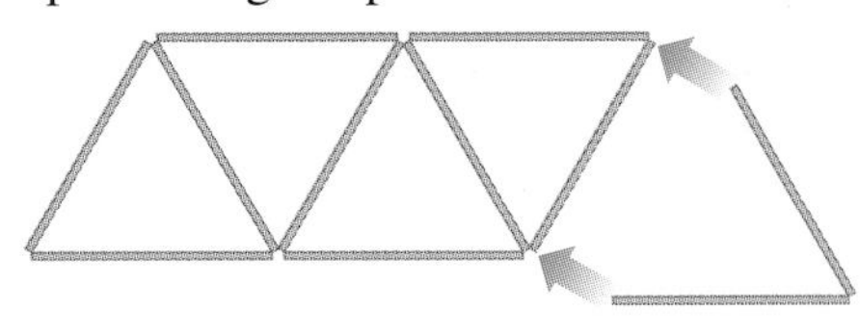

Top view

6. Suppose the pens are square but built with space between each pen, as shown in the figure. What is the total number of panels needed for 20 pens?

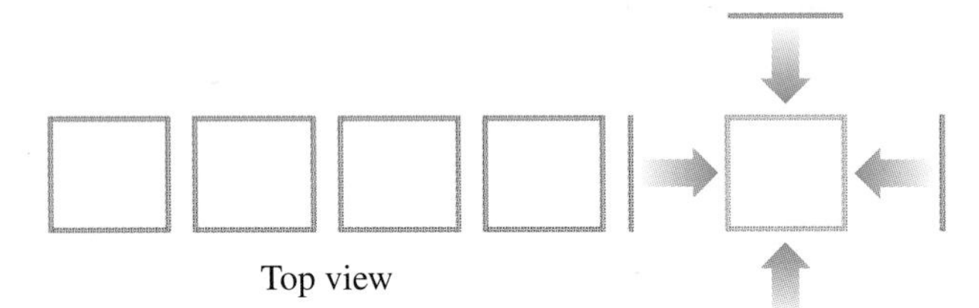

Top view

7. Jake's other assistant, Dean, assumes that the panels can be fixed firmly to the floor and designs a zig-zag pattern, as shown in the figure. What other important assumption has Dean made? How many panels are needed for 20 "open" pens?

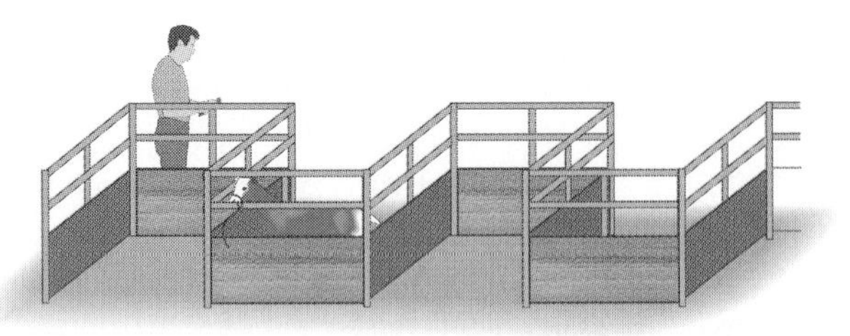

8. A year has passed; Kelsey now has Jake's job. She builds the pens along a long wall, as shown in the figure. How many panels are needed for 20 pens?

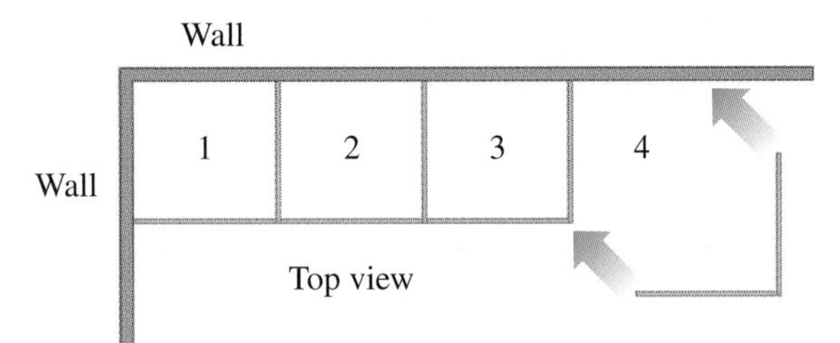

Top view

9. Suppose the pens are triangular but built with space between each pen, as shown in the figure. What is the total number of panels needed for 20 pens?

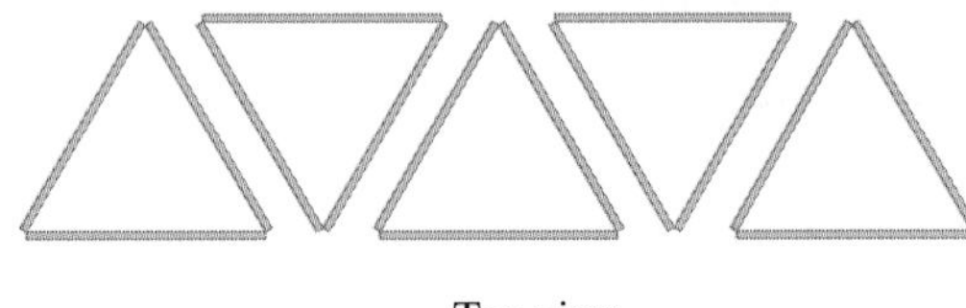

Top view

Blue numbers are core exercises.

10. Cassandra is facilities manager for the park featured in Example 1. When Jake comes to ask for more money to buy additional panels, she suggests that he build the pens in a double row, as shown in the figure. How many panels will he now need for the 20 pens?

Top view

Answers to Exercises 11 to 14 may vary.

11. Write two conditions you face in your life that may affect your performance in this course.

12. Write two conditions for the course, as stated in your syllabus.

13. Write one assumption you have made about your course work.

14. Write one assumption your instructor might have made about you or your background as related to this course.

15. List five strategies that may form steps in a problem-solving plan.

16. List Polya's four steps in problem solving.

▶ Project

17. **Numbers in Words** This exercise reminds you to learn vocabulary, which is important in this course. What word best describes each of the phrases? Here's a hint to get you started: The parts of words commonly meaning "one" are *mono* and *uni*.

a. One
Legendary animal with one horn ________
Ride it; one wheel; common in a circus ________
One-lens eyepiece worn by the British ________
Fancy letters printed on clothing or stationery ________

b. Two
Two singers ________
Two-base hit in baseball ________
Ride it; two wheels ________
Able to speak two languages ________
1,000,000,000 (number with 9 zeros) ________
Muscle in the upper arm ________
Sea shell with two parts ________
Glasses with two-part lenses ________
World War I aircraft with two wings, one above the other ________
Number system using only 0s and 1s ________

c. Three
Three babies at one birth __________
Three-sided geometric shape __________
Stable camera stand __________
Three-wheeled toy to ride __________

d. Four
Group of four singers __________
Four of these make a gallon in liquid measure __________
Four babies at one birth __________

e. Five
Five-sided geometric shape __________
Olympic event (fence, swim, run, shoot, ride horse) __________

f. Seven
Month when school typically begins __________
Track event with seven activities __________

h. Eight
Ocean animal with eight legs __________
Full set of eight notes in music __________
Halloween month __________
Eight-sided geometric shape __________

h. Ten
Ten years __________
Last month of calendar year __________
Dot in our number system; the __________ point
Track event with ten activities __________

i. One hundred
A hundred years __________
One-hundredth of a dollar __________
One-hundredth of a meter __________

h. One thousand
A thousand thousands __________
A thousand years __________

k. Find words with parts meaning six or nine.

▶ 1.2 Numeric Representations

Student Note:
To prepare for each class, read the Objectives, take notes on the vocabulary, look at subject headings, and read quickly through application examples.

Objectives

- Build an input-output table from a problem setting.
- Use the basic language for sets of numbers.
- Change notation involving fractions and decimals.
- Order numbers on a number line.
- Add, subtract, multiply, and divide numbers in fraction and mixed-number notation.

Student note:
Now is the time to check fraction skills. If your skills are good, improve your skills by helping others. If you need help, get it now and plan extra time for practice.

WARM-UP

We add, subtract, multiply, and divide $\frac{2}{3}$ and $\frac{1}{4}$ as follows. (The = symbol means "is equal to," and the dot commonly replaces a multiplication sign.) To add or subtract, we must first change to a common denominator.

$\frac{2}{3} + \frac{1}{4} = \frac{8}{12} + \frac{3}{12}$, then add numerators, $\frac{11}{12}$

$\frac{2}{3} - \frac{1}{4} = \frac{8}{12} - \frac{3}{12}$, then subtract numerators, $\frac{5}{12}$

$\frac{2}{3} \cdot \frac{1}{4} = \frac{2}{12}$, then change to lowest terms, $\frac{1}{6}$

$\frac{2}{3} \div \frac{1}{4} = \frac{2}{3} \cdot \frac{4}{1}$, multiply, $\frac{8}{3}$ or $2\frac{2}{3}$

Now, add, subtract, multiply, and divide $\frac{3}{4}$ and $\frac{1}{5}$.
What assumptions did you make?

AS A CHILD, was counting your first mathematics experience? It should not be surprising that number patterns and input-output tables (such as Table 1) are the first numeric representations in algebra. This section continues input-output tables and defines sets of numbers. Because of the vital role of fractions in algebra, you are also invited to use Examples 6 to 9 to review notation, order, and operations with fractions.

Input-Output Tables, Continued

Throughout this course, input-output tables give a numeric representation of a relationship. In Example 1 the input is the number of pens, and the output is the number of panels needed to build pens in the shape described.

EXAMPLE 1 Linked Example: Petting Zoo Pens; (Building an input-output table from a figure) As you may recall from Section 1.1, Kelsey, Jake's assistant, suggested using triangular pens, as shown in Figure 5. Make an input-output table in which the number of pens is the input and the total number of panels is the output.

SOLUTION Kelsey will need three panels for the first pen and two for each additional pen. The input-output table is shown in Table 2.

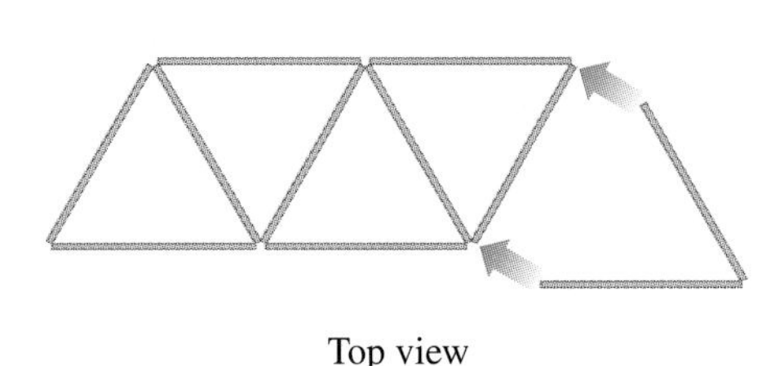

Top view

FIGURE 5 Triangular pens

TABLE 2 Triangular Pens and Panels

Input: Number of Pens	Output: Total Number of Panels
1	3
2	5
3	7
4	9

Because the input-output table represents a relationship between the input and the output, we need to build the habit of reading from the input number to the output number.

EXAMPLE 2 Reading an input-output table on tuition Table 3 shows the credit hours taken (input) and the resulting tuition paid (output).

a. What is the cost of a 3-hour writing course?
b. What is the cost of a 4-hour mathematics course?
c. What is the cost of a 5-hour chemistry course?

TABLE 3 Tuition at College

Input: Credit Hours	Output: Tuition Paid
1	\$168
2	\$336
3	\$504
4	\$672

SOLUTION **a.** From Table 3, a 3-hour writing course costs \$504.

b. From Table 3, a 4-hour mathematics course costs \$672.

c. We have at least two methods for answering the third question.

Method 1: The first and each additional credit hour costs \$168. If the fifth credit hour costs another \$168, the cost for 5 credit hours would be \$672 for the first 4 credit hours and \$168 for the fifth credit hour, for a total of \$840.

Method 2: If we divide each tuition amount by the number of credit hours, we find each credit hour to be worth $168. The product of 5 credit hours at $168 per credit hour is $840.

The second method illustrates that the multiplication of the input (credit hours) by a cost per credit hour gives the output (tuition). Both methods assume that tuition rates continue in the same pattern for any possible number of credit hours. ◀

In Example 3 the inputs are grouped. With grouped inputs, we need to find the group that includes an input number and read across to the output.

▶ EXAMPLE 3 Reading an input-output table with grouped inputs From Table 4, find the shipping and handling charges for the following merchandise costs.

a. $69.95 **b.** $39.95 **c.** $379.95 **d.** $200.00

TABLE 4 J.C. Penney Shipping and Handling (S & H) Charges

Merchandise Cost	S & H Charge	Merchandise Cost	S & H Charge
Up to $40	$7.95	$100.01–$150	$17.95
$40.01–$60	$9.95	$150.01–$200	$20.95
$60.01–$80	$12.95	$200.01–$300	$22.95
$80.01–$100	$14.95	Over $300	$24.95

SOLUTION

a. $69.95 is between $60.01 and $80, so the charge is $12.95.

b. $39.95 is smaller than $40, so the charge is $7.95.

c. $379.95 is over $300, so the charge is $24.95.

d. $200 is the end of the group $150.01–$200, so the charge is $20.95. ◀

▶ Sets of Numbers

All possible numbers in each column of the input-output table can be described with sets of numbers. A **set** is *a collection of objects or numbers*. In order to define the common sets of numbers used in algebra, we need to define opposite numbers and the square root symbol.

OPPOSITE NUMBERS Two numbers are **opposites** if they are *on different sides of zero and the same distance from zero on a number line*, as shown in Figure 6. The opposite of 5 is -5. The opposite of -2 is 2. We use opposites to define the set of integers and will learn more about opposites in Section 2.1.

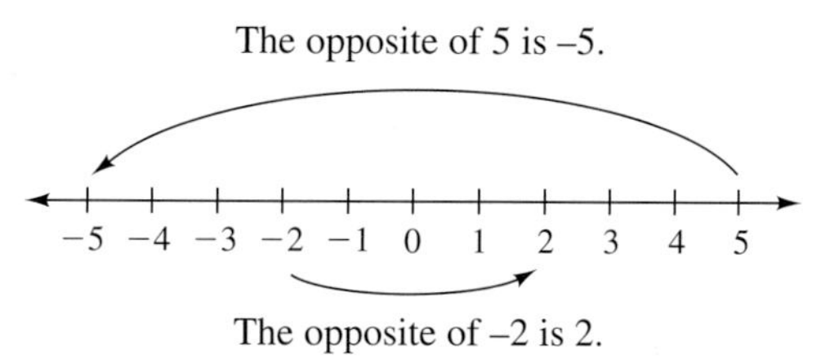

FIGURE 6

SQUARE ROOT SYMBOL The **square root symbol,** $\sqrt{\ }$, *asks for the positive number that, when multiplied by itself, gives the number inside*; $\sqrt{16}$ is 4. We will use this definition for now. You will learn more about the square root symbol and square roots in Section 8.2.

THINK ABOUT IT 1: This is boldface type. *This is italic type.* Write a sentence about how definitions are presented within a paragraph in this textbook. Use the words *boldface* and *italic*. ◀

SETS OF NUMBERS A listing of the contents of a set is usually placed in braces, { }. For example, the **natural numbers** (or counting numbers) are *the numbers in the set* {1, 2, 3, 4, . . . }. The three dots indicate that the number list continues without end.

The **whole numbers** are *the natural numbers and the number zero*—that is, the numbers in the set {0, 1, 2, 3, 4, . . . }. The **integers** consist of *the natural numbers, the opposites of the natural numbers, and zero*—that is, the numbers in the set { . . . , −3, −2, −1, 0, 1, 2, 3, . . . }. The integers are the numbers we write on the number line; see Figure 7.

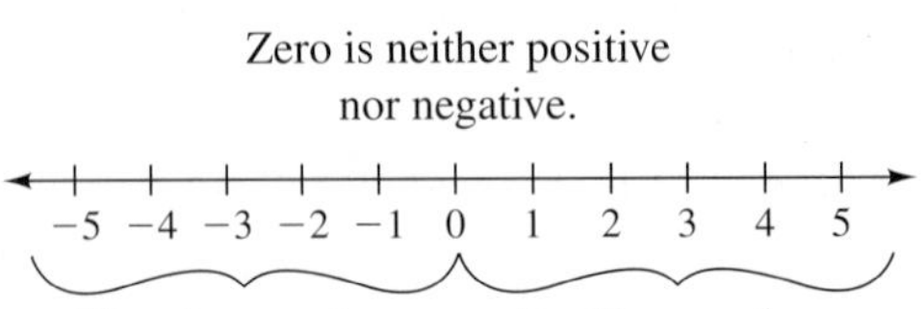

FIGURE 7

The numbers greater than zero (those that lie to the right of zero on the number line) are called **positive numbers**. *The numbers less than zero* (to the left of zero on the number line) are called **negative numbers**. Mathematicians agree that *a number without a sign,* say 5, *is considered positive,* or +5. The positive sign looks just like an addition sign, although it is sometimes printed smaller. Zero is neither positive nor negative.

THINK ABOUT IT 2: How do we read the numbers −3, −2, and −1? ◀

We use the integers to build a rational number. **Rational numbers** are *the set of numbers that can be written by dividing one integer by another integer, so long as we do not divide by zero*. We say that division by zero is **undefined** (*has no mathematical meaning*). Fractions are a common notation for a rational number. The fraction $\frac{3}{4}$ shows the division of 3 by 4.

▶ **EXAMPLE 4** Showing that a number is a rational number If needed, rewrite each number in a notation that shows it is a rational number and tell what two integers are divided.

a. $\frac{7}{8}$ **b.** $\frac{9}{2}$ **c.** $1\frac{1}{2}$ **d.** 5

e. 0.35 **f.** $-\frac{4}{5}$ **g.** \$1.25

SOLUTION **a.** 7 divided by 8 **b.** 9 divided by 2

c. $1\frac{1}{2}$ is $\frac{3}{2}$ as an improper fraction, 3 divided by 2

d. 5 is an integer, but when divided by the integer 1, $\frac{5}{1}$ is in rational-number notation.

e. 0.35 is in decimal notation and means 35 hundredths, or $\frac{35}{100}$. As a rational number, it is 35 divided by 100.

f. $-\frac{4}{5}$ is negative 4 divided by 5. (See p. 70 for more on signs and fractions.)

g. \$125 divided by 100 ◀

▶ **EXAMPLE 5** Matching numbers with names Write each number with every set to which it belongs: 5, 1.5, −3, $-\frac{7}{5}$, $\sqrt{9}$, −1.3, 0.2, $1\frac{1}{3}$, 0.

a. natural numbers

b. whole numbers

c. integers

d. rational numbers

Student Note:
Knowing the names of sets of numbers is important for understanding exercises and directions. To help you learn and remember new material, make 3- by 5-inch flash cards for notation, symbols, and vocabulary.

SOLUTION **a.** The natural numbers include 5 and $\sqrt{9}$ (or 3).

b. The whole numbers are zero and the natural numbers, so they include 0, 5, and $\sqrt{9}$.

c. The integers are the whole numbers and their opposites, so they include -3, 0, 5, and $\sqrt{9}$.

d. All of the numbers are rational numbers. ◀

THINK ABOUT IT 3: What sets of numbers are inputs in Example 1? Outputs? Inputs in Example 2? Outputs? Inputs in Example 3? Outputs? ◀

The most general set of all numbers we study at this time is the real numbers. Each real number can be located by a point on the number line. **Real numbers** are *the set containing both rational and irrational numbers.* **Irrational numbers** are *those numbers whose decimal notation cannot be written by dividing two integers.* The **infinite** (*without bound*) set of irrational numbers includes $\sqrt{2}$, $\sqrt{3}$, $\sqrt{5}$, and π (the Greek letter pi, as used in the formulas for the circumference and area of a circle; see Section 2.5). We study irrational numbers in Chapter 8.

THE SET OF REAL NUMBERS

The set of real numbers includes rational numbers and irrational numbers.

Rational numbers can be written by dividing one integer by another integer. Sets of rational numbers include

- natural numbers $\{1, 2, 3, \ldots\}$
- whole numbers $\{0, 1, 2, 3, \ldots\}$
- integers $\{\ldots, -3, -2, -1, 0, 1, 2, 3, \ldots\}$

as well as numbers in special notation, such as

- fractions $\left\{\frac{1}{2}, \frac{1}{4}, \ldots\right\}$
- square roots with exact decimal values $\{\sqrt{1}, \sqrt{4}, \sqrt{9}, \sqrt{16}, \ldots\}$

Irrational numbers cannot be written by dividing two integers. They include

- square roots without exact decimal values $\{\sqrt{2}, \sqrt{3}, \sqrt{5}, \ldots\}$
- pi, written with the Greek letter π, which is equal to 3.14159265 . . .

▶ Changing Notation

THE EQUALS SIGN The **equals sign,** =, placed between two numbers *says that the numbers have the same value* (*are equal*). Thus, $\frac{1}{2} = 0.5$ indicates that the fraction notation $\frac{1}{2}$ is equal in value to the decimal notation 0.5. We use the equals sign between steps in changing notation.

NOTATION The **numerator** is the *top number in fraction notation,* and the **denominator** is *the bottom number in fraction notation.* For fractions of physical objects, the numerator tells how many parts and the denominator tells what kind of part.

Think about the many notations we have to write numbers: fractions, decimals, improper fractions, mixed numbers, and so forth. The number 1.5, which is in decimal notation, can also be written as a mixed number, $1\frac{1}{2}$, or as an improper fraction, $\frac{3}{2}$ (or as a percent, 150%, in Section 1.4). It is easy to forget how to change from one number notation to another. The next two examples review fractions and decimals.

HISTORICAL NOTE:

Robert Recorde, in 1557, was the first to use the equals sign. Sir Isaac Newton (1642–1727) and Gottfried Liebniz (1646–1716) used the equals sign in their important mathematical works and are credited with its continuing to be the symbol as we define it today. See **http://members.aol.com/jeff570/mathsym.html,** maintained by Jeff Miller.

▶ **EXAMPLE 6** Changing between mixed numbers and improper fractions Write parts a and b as improper fractions. Write parts c and d as mixed numbers.

a. $2\frac{3}{4}$ **b.** $1\frac{2}{5}$ **c.** $\frac{77}{20}$ **d.** $\frac{55}{28}$

SOLUTION **a.** The mixed number $2\frac{3}{4}$, means 2 plus $\frac{3}{4}$.

$2\frac{3}{4} = \frac{8}{4} + \frac{3}{4}$ The whole number 2 is $\frac{8}{4}$.

$2\frac{3}{4} = \frac{11}{4}$

The shortcut of multiplying 2 times 4 and adding 3 is changing the whole number into fourths and then adding the numerator. In algebra, it is convenient to know that a mixed number means the sum of the whole number and the fraction part.

b. $1\frac{2}{5} = \frac{5}{5} + \frac{2}{5}$ The whole number 1 is $\frac{5}{5}$.

$1\frac{2}{5} = \frac{7}{5}$

c. $\frac{77}{20} = 3 + \frac{17}{20}$ Dividing 77 by 20 gives the answer 3 plus a remainder of 17.

$\frac{77}{20} = 3\frac{17}{20}$

d. $\frac{55}{28} = 1 + \frac{27}{28}$ Dividing 55 by 28 gives the answer 1 plus a remainder of 27.

$\frac{55}{28} = 1\frac{27}{28}$ ◀

▶ **EXAMPLE 7** Changing between decimals and fractions Write parts a and b as fractions. Write parts c, d, and e as decimals.

a. 0.19 **b.** 0.875 **c.** $\frac{3}{8}$ **d.** $\frac{6}{5}$ **e.** $\frac{2}{3}$

SOLUTION **a.** There are two decimal places in 0.19, so the decimal is read "nineteen hundredths."

$0.19 = \frac{19}{100}$ Write 19 over 100.

b. There are three decimal places in 0.875, so the decimal is read "eight hundred seventy-five thousandths."

$0.875 = \frac{875}{1000}$

To change fractions to **lowest terms** (*the condition of a fraction when the numerator and denominator have no common factors*), first factor the numerator and denominator. Because a number divided by the same number equals 1, we can eliminate factors common to both parts of the fraction.

$\frac{875}{1000} = \frac{\boxed{25} \cdot \boxed{5} \cdot 7}{\boxed{25} \cdot \boxed{5} \cdot 8}$ *Note:* $\frac{25}{25} = 1$ and $\frac{5}{5} = 1$

$\frac{875}{1000} = \frac{7}{8}$

Factors are *two or more numbers being multiplied*. The example shows that factors are not necessarily prime numbers (a **prime number** is *a number greater than one with no integer factors except 1 and itself*).

c. To change $\frac{3}{8}$ to a decimal, divide the numerator by the denominator.

$\frac{3}{8} = 0.375$

d. To change $\frac{6}{5}$ to a decimal, divide the numerator by the denominator.

$\frac{6}{5} = 1.2$

e. To change $\frac{2}{3}$ to a decimal, divide the numerator by the denominator.

$\frac{2}{3} = 0.66666666\ldots$

For a repeating decimal, draw a line over the repeating part and drop the rest of the repeat.

$\frac{2}{3} = 0.\overline{6}$ ◀

For now, we will delay changing a repeating decimal back to a fraction. Algebra provides a convenient way to do this for unfamiliar decimal notations. See page 366.

Order of Numbers

We change notation so that we can compare numbers, as in placing them in order on a number line.

EXAMPLE 8 **Placing numbers in order on a number line** Draw a number line with -5 on the left and 5 on the right. Show the locations for $3\frac{1}{4}$, $\sqrt{25}$, $-\frac{18}{5}$, $-\frac{1}{2}$, $\frac{11}{4}$, -3.5, 1.5, and $-\sqrt{4}$.

SOLUTION The negative number -3.5 is between -4 and -3. The negative number $-\frac{18}{5}$ is $-3\frac{3}{5}$, left of -3.5. The negative number $-\frac{1}{2}$ is between -1 and 0. The improper fraction $\frac{11}{4}$ is $2\frac{3}{4}$. Finding square roots, we have $-3.5\sqrt{4} = -2$ and $\sqrt{25} = 5$. The positions of the numbers are shown in Figure 8.

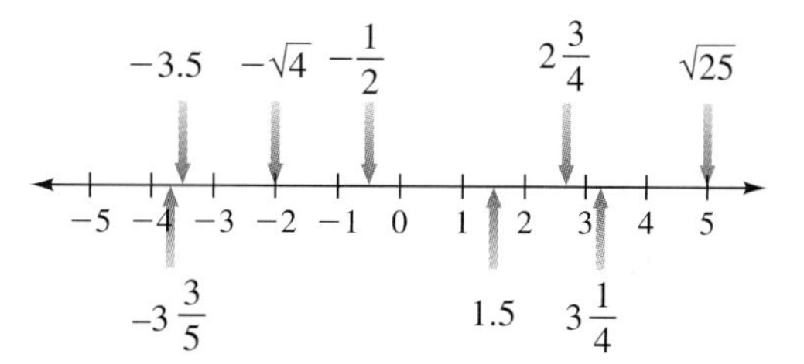

FIGURE 8 Numbers in order on a number line ◀

Add, Subtract, Multiply, and Divide: Operations with Fractions

As shown in Example 7, fractions and division are closely related. Think of $\frac{8}{5}$ as 8 divided by 5 so that you will know that the notation $\frac{a}{b}$ means a divided b. In the next few weeks, as you solve equations, you will use all four operations with fractions. Now is the time to refresh your skills with the fraction operations—without a calculator.

EXAMPLE 9 **Calculating with fractions** Do the calculation. Then, describe each operation in a word setting.

a. $2\frac{3}{4} + 1\frac{2}{5}$

b. $2\frac{3}{4} - 1\frac{2}{5}$

c. $2\frac{3}{4} \cdot 1\frac{2}{5}$

d. $2\frac{3}{4} \div 1\frac{2}{5}$

SOLUTION (The mixed numbers were changed to improper fractions before computation to show similarities among the operations.)

Student Note:
For a more detailed example of least common denominator, see Section 9.4, Example 1, page 554.

a. $2\frac{3}{4} + 1\frac{2}{5} = \frac{11}{4} + \frac{7}{5}$ Change to common denominators.

$\frac{55}{20} + \frac{28}{20} = \frac{83}{20}$ or $4\frac{3}{20}$

If the Winters family eats $2\frac{3}{4}$ pizzas and the Summers family eats $1\frac{2}{5}$ pizzas (see Figure 9), adding gives what they eat together.

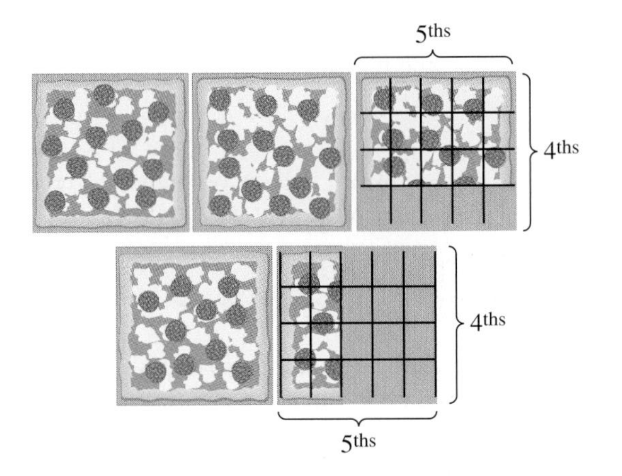

FIGURE 9 Giant pan pizza with 20ths shown

b. $2\frac{3}{4} - 1\frac{2}{5} = \frac{11}{4} - \frac{7}{5}$ Change to common denominators.

$\frac{55}{20} - \frac{28}{20} = \frac{27}{20}$ or $1\frac{7}{20}$

Subtracting gives how much more pizza is eaten by the Winters family than by the Summers family.

c. $2\frac{3}{4} \cdot 1\frac{2}{5} = \frac{11}{4} \cdot \frac{7}{5}$ Multiply across.

$\frac{77}{20} = 3\frac{17}{20}$

Multiplying gives the miles walked at $2\frac{3}{4}$ miles per hour in $1\frac{2}{5}$ hours after eating the pizza.

d. $2\frac{3}{4} \div 1\frac{2}{5} = \frac{11}{4} \div \frac{7}{5}$ Change to multiplication.

$\frac{11}{4} \cdot \frac{5}{7} = \frac{55}{28}$ or $1\frac{27}{28}$

Dividing gives how long it takes to walk $2\frac{3}{4}$ miles at $1\frac{2}{5}$ miles per hour. ◀

THINK ABOUT IT 4: Add, subtract, multiply, and divide $3\frac{1}{5}$ and $1\frac{2}{3}$.

In Example 9 part d, the number $\frac{5}{7}$ is said to be the reciprocal of $\frac{7}{5}$. When we multiply $\frac{5}{7}$ and $\frac{7}{5}$, we get 1. **Reciprocals** are two numbers that *multiply to give 1*. You will learn more about reciprocals on page 19 (Section 1.3) and on page 69 (Section 2.2).

THINK ABOUT IT 5: What is the reciprocal of $1\frac{1}{2}$? Check your answer by multiplying $1\frac{1}{2}$ and its reciprocal.

ANSWER BOX

Warm-up: $\frac{19}{20}, \frac{11}{20}, \frac{3}{20}, \frac{15}{4}$ **Think about it 1: Boldface** is used to highlight a word, and *italic* is used for its definition. **Think about it 2:** negative three, negative two, and negative one, respectively. **Think about it 3:** Integers for inputs and outputs in Examples 1 and 2. Rational numbers in decimal notation for inputs and outputs in Example 3. **Think about it 4:** $4\frac{13}{15}, 1\frac{8}{15}, 5\frac{1}{3}, 1\frac{23}{25}$ **Think about it 5:** Because $1\frac{1}{2} = \frac{3}{2}$, its reciprocal is $\frac{2}{3}$. $\frac{3}{2}$ times $\frac{2}{3} = 1$.

▶ 1.2 Exercises

▶ Input-Output Tables

In Exercises 1 to 4, draw the next figure in each pattern, and explain your reasoning. Make an input-output table for the figure. Predict the output when the input is 10.

1. The input is the number of dots; the output is the number of non-overlapping segments (pieces of lines) separated by the dots.

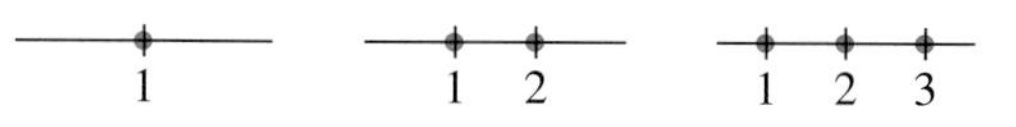

2. The input is the number of rays (arrows); the output is the number of non-overlapping angles formed. There is one angle formed by the first ray—a full rotation of 360 degrees.

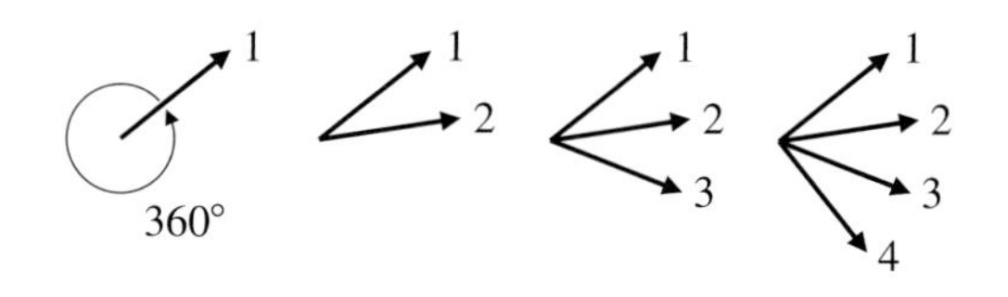

3. The input is the number of lines; the output is the number of regions separated (like pieces of pie) by the lines.

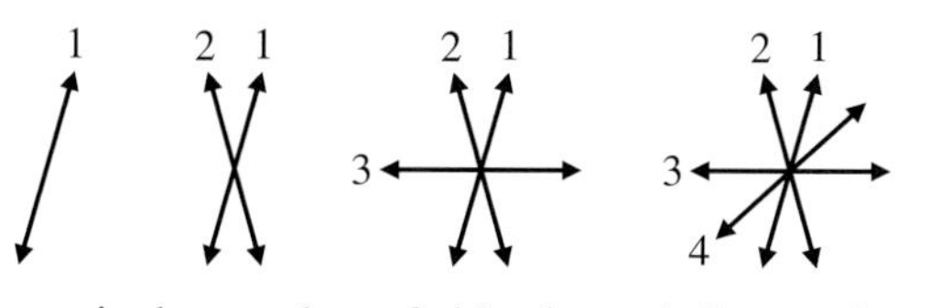

4. The input is the number of sides in each figure; the output is the number of vertices (corners) in each figure.

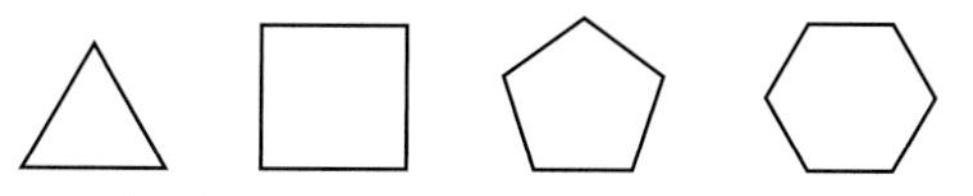

5. **Perimeter** is *distance around the outside of a flat object.* The five triangular pens shown have perimeter equal to the length of seven panels. Make an input-output table for inputs 1, 2, 3, 4, 5, and 10 pens and the perimeter of that number of pens with the triangular design. Is the perimeter the same as the total number of panels, given in Table 2? Why?

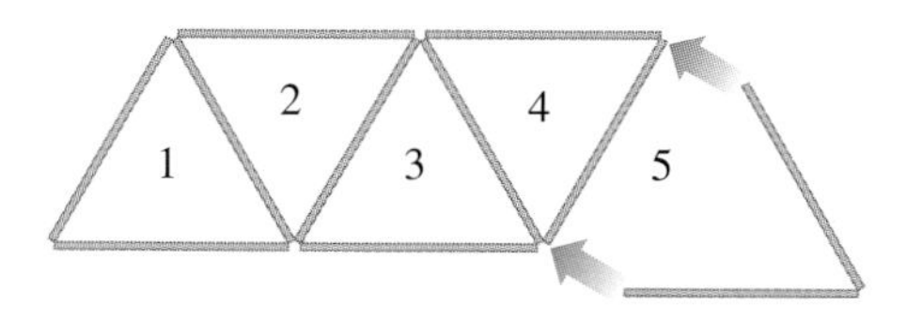

Top view

6. Refer to Example 2.
   a. If the tuition cost remains \$168 per credit hour, what is the cost of 12 credit hours?
   b. If the tuition cost was \$1680, how many credit hours were taken?
   c. Suppose one \$20 fee is added to the tuition cost (at \$168 per credit hour). How many credit hours were taken if the total cost was \$1364?
   d. Suppose a \$30 fee is added to the tuition cost of \$168 per credit hour. If the total of tuition and fees was \$1878, find the number of credit hours.

7. Make an input-output table showing the cost for 14 to 20 credit hours if the tuition is \$67 per credit hour for 1 to 18 credit hours and \$1500 for more than 18 credit hours.

8. Make an input-output table for the following tuition model: Tuition is \$75 per credit hour for 1 to 11 hours; \$900 total for 12 to 18 hours; and an additional \$40 per credit hour for each hour above 18 hours. Let inputs be the even numbers from 2 to 20 hours and outputs be the total tuition cost.

9. L.L. Bean Shipping and Handling (S & H)

Order Subtotal	S & H Charge
Up to \$25	\$3.95
\$25.01 to \$50	\$5.95
\$50.01 to \$100	\$7.95
\$100.01 to \$150	\$9.95
\$150.01 and up	\$11.95

What is the S & H charge for these order subtotals?

a. \$35 b. \$100
c. \$200 d. \$20.95

10. Hearth Song Shipping and Handling (S & H)

Merchandise Total	S & H Charge
\$14.99 and under	\$4.99
\$15.00–\$24.99	\$6.99
\$25.00–\$49.99	\$8.99
\$50.00–\$74.99	\$10.99
\$75.00–\$99.99	\$12.99
\$100.00–\$149.99	\$14.99
\$150.00–\$199.99	\$16.99
\$200.00 and over	\$19.99

What is the S & H charge for these merchandise totals?

a. \$35 b. \$100
c. \$200 d. \$20.95

▶ Sets of Numbers

In Exercises 11 to 16, write a sentence that uses the word in its mathematical meaning. Then write a second sentence in which the word's meaning is not mathematical.

11. set 12. real
13. rational 14. input
15. whole 16. negative

Blue numbers are core exercises.

Exercises 17 and 18 could be group activities.

17. **a.** Which set of real numbers includes positive integers but not negative integers or zero?

b. What kind of numbers might be used to write elevations below sea level?

c. Which set of real numbers includes $\frac{2}{3}$, $\frac{5}{3}$, and 3.03?

d. When we combine the even numbers and the odd numbers, we get which set of real numbers?

18. From the set $\{\frac{2}{3}, -2, 0, 2, \frac{6}{2}\}$, select a number that fits each statement.

a. Division by this number is undefined.

b. This rational number is not defined in the set of integers.

c. This integer is not defined in the set of natural numbers.

19. Are the natural numbers positive or negative?

20. Give an example of a negative rational number that is not an integer.

▶ Changing Notation

It is best to do Exercises 21 to 30 by hand to be sure that you know how to do them.

21. Change these to fractions:

a. 0.30 **b.** 0.05
c. 0.125 **d.** 0.02

22. Change these to fractions:

a. 0.40 **b.** 0.08
c. 0.625 **d.** 0.56

23. Change these to decimals:

a. $\frac{7}{16}$ **b.** $\frac{4}{5}$
c. $\frac{18}{25}$ **d.** $\frac{9}{20}$

24. Change these to decimals:

a. $\frac{3}{4}$ **b.** $\frac{1}{5}$
c. $\frac{5}{8}$ **d.** $\frac{7}{25}$

25. Change these to mixed numbers:

a. $\frac{16}{9}$ **b.** $\frac{27}{8}$
c. $\frac{20}{7}$ **d.** $\frac{16}{6}$

26. Change these to mixed numbers:

a. $\frac{25}{8}$ **b.** $\frac{31}{7}$
c. $\frac{9}{6}$ **d.** $\frac{22}{10}$

27. Change these to improper fractions:

a. $1\frac{4}{5}$ **b.** $2\frac{1}{3}$
c. $4\frac{3}{4}$ **d.** $3\frac{1}{2}$

28. Change these to improper fractions:

a. $3\frac{5}{6}$ **b.** $2\frac{3}{8}$
c. $2\frac{5}{9}$ **d.** $3\frac{2}{3}$

29. Change these fractions to lowest terms:

a. $\frac{25}{35}$ **b.** $\frac{25}{32}$
c. $\frac{42}{63}$ **d.** $\frac{27}{36}$

30. Change these fractions to lowest terms:

a. $\frac{16}{28}$ **b.** $\frac{36}{45}$
c. $\frac{30}{54}$ **d.** $\frac{9}{25}$

▶ Order of Numbers

For Exercises 31 to 38, use a ruler to draw a number line. Choose which integers to show. Mark and label the location of the following sets of numbers. A hint for Exercises 37 and 38 appears in Exercise 48.

31. $2.5, -1, 0, -0.5, -4, 2, -1.5$

32. $1.25, -3.25, 4, -3, -4.25, 3$

33. $-3, 2\frac{1}{4}, -\frac{1}{2}, \frac{3}{4}, 1.5, -2\frac{1}{2}$

34. $2, -2\frac{1}{4}, \frac{1}{2}, -\frac{1}{4}, 2.5, -1.5$

35. $-1\frac{1}{2}, -3\frac{3}{4}, \sqrt{16}$, opposite of $\sqrt{9}$, opposite of -2

36. $-1\frac{1}{4}, -2$, opposite of $-\frac{3}{4}$, $\sqrt{36}$, opposite of $\sqrt{4}$, opposite of -1.5

37. $1.25, \frac{8}{7}, 1\frac{1}{6}$

38. $-2.5, -2\frac{4}{7}, \frac{-16}{7}$

▶ Computation with Fractions

In Exercises 39 to 46, add, subtract, multiply, and divide these pairs of fractions:

39. $\frac{1}{2}$ and $\frac{1}{3}$ (That is, find $\frac{1}{2} + \frac{1}{3}$, $\frac{1}{2} - \frac{1}{3}$, $\frac{1}{2}$ times $\frac{1}{3}$, and $\frac{1}{2}$ divided by $\frac{1}{3}$.)

40. $\frac{3}{4}$ and $\frac{1}{3}$

41. $\frac{3}{5}$ and $\frac{1}{3}$

42. $\frac{1}{2}$ and $\frac{2}{5}$

43. $1\frac{1}{3}$ and $\frac{5}{6}$

44. $1\frac{1}{4}$ and $\frac{3}{8}$

45. $3\frac{1}{5}$ and $2\frac{1}{2}$

46. $2\frac{2}{3}$ and $1\frac{5}{8}$

47. Application: Movie Times and Fractions

a. The Gateway Theater is open 10 hours on weekdays. How many times can a new movie be shown if the movie time plus ads, trailers, and clean-up is $2\frac{1}{4}$ hours? How many more minutes of advertising can be sold and enable the theater to fit in the same number of shows?

Blue numbers are core exercises.

b. The Shopping Plaza Theater is open 12 hours on weekends. How many times can a new children's movie be shown if the movie time plus advertising, trailers, and clean-up is $1\frac{3}{4}$ hours? To add another show for how many additional minutes will the theater workers need to be hired?

▶ Writing

48. What is it about the fraction notation for $\frac{5}{8}$ that tells how to find $\frac{5}{8}$ in decimal notation? Why are number values easier to compare in decimal notation than in fraction notation? *Hint:* Look at Exercises 35 and 36.

49. A friend has become too dependent on a calculator for doing fractions and is stumped on how to change $1\frac{3}{25}$ to a decimal. Explain two ways in which this might be done without a calculator.

50. Both $\frac{21}{12}$ and $\frac{28}{16}$ give $1\frac{3}{4}$ when changed to mixed numbers. Explain why. Change $1\frac{3}{4}$ to an improper fraction. Explain why you did not get one of the first two fractions. What is a fourth improper fraction with $1\frac{3}{4}$ as its mixed number? Explain how to find a fraction with a denominator 200 that equals $1\frac{3}{4}$.

▶ Problem Solving: Guess and Check Skills

Use the following hint to do Exercises 51 and 52. Guess a pair of numbers that fits the first condition; see how closely your guess fits the second condition.

51. a. Find two numbers that add to 12 and multiply to 27.

b. Find two numbers that add to 10 and multiply to 21.

52. a. Find two numbers that multiply to 36 and add to 13.

b. Find two numbers that multiply to 24 and add to 10.

▶ Project

53. Tuition Tables Make a tuition table for your school. Make a second table for nonresidents (out-of-district, out-of-state and/or international students) who attend your school. Use the Internet to research the tuition at a university you would like to attend. Make tables for that university.

▶ 1.3 Verbal Representations

Objectives

- Identify and apply vocabulary for basic operations and algebraic notation.
- Find the rules for input-output tables and express them in words and in equations.
- Change word phrases to expressions.
- Change sentences to equations.

WARM-UP

Make an input-output table for each figure.

1. Let the inputs be 1 to 4 pens and the outputs be the total number of panels.

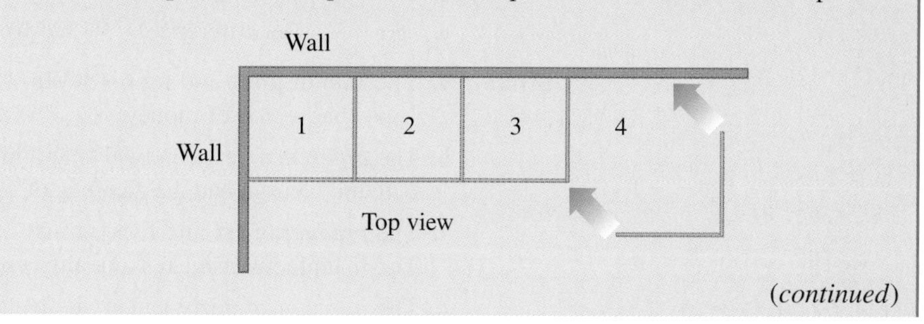

(continued)

(concluded)

2. Let the input be the position of the design and the output be the **perimeter** (*distance around the outside of a flat object*). Assume the sides are of equal length. State your assumptions.

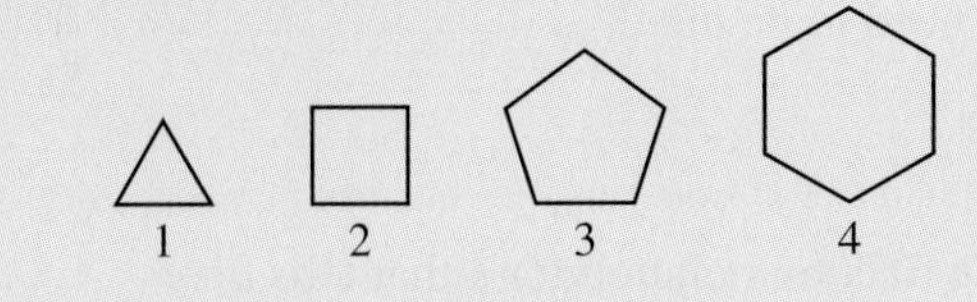

3. Let the input be the position of the design from left to right and the output be the **area** (*measure of the surface enclosed within a flat object*) as measured in squares for each design. How are the outputs the same as, and how are they different from, the numbers in Exercise 1?

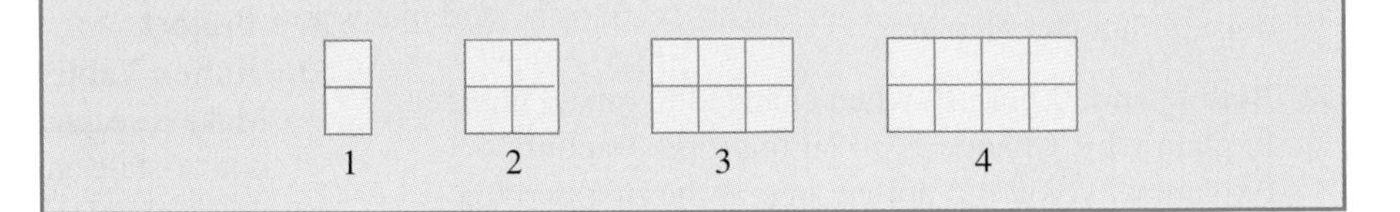

"Words, words, words. I'm so sick of words."

—Eliza Doolittle in *My Fair Lady,* lyrics by Alan Jay Lerner

Student Note:
To learn vocabulary words, write words on one side of small cards and definitions on the other.

AS AN ALGEBRA STUDENT, you may become frustrated with the quantity of special terms and rules for using the algebraic language. This section introduces basic algebraic vocabulary and words associated with number operations in the context of input-output rules. Be patient. Learning the words takes time but leads to clear thinking, effective communication of ideas, and long-term success.

Answers to Basic Operations

In algebra, we use common words such as *sum*, *difference*, and *product* in special ways. We have words such as *quotient* that appear only in mathematics.

DEFINITIONS: ANSWERS TO OPERATIONS

A **sum** is the answer to an addition problem.
A **difference** is the answer to a subtraction problem.
A **product** is the answer to a multiplication problem.
A **quotient** is the answer to a division problem.

EXAMPLE 1 Practicing vocabulary Use the first three words in two sentences. In the first sentence, the word should have its mathematical meaning. In the second sentence, the meaning of the word should not be mathematical. Use the fourth word in a mathematical sentence only.

a. sum **b.** difference **c.** product **d.** quotient

SOLUTION

a. The *sum* of three and four is seven.
Luisa has a *sum* of money.

b. The *difference* between twelve and nine is three.
Bill and George had a *difference* of opinion.

c. The *product* of six and five is thirty.
The manufacturer makes a quality *product*.

d. The *quotient* of thirty and six is five. ◀

Example 1 contains two agreements. An **agreement** is *a common way to do something*—say within this book, within a course, or within the mathematics community. The first agreement is regarding subtraction of two numbers: "The difference between twelve and nine" is written 12 subtract 9, not 9 subtract 12. The second is regarding the division of two numbers: "The quotient of thirty and six" is written 30 divided by 6, not 6 divided by 30.

Variables and Equations

Variable and *equation* are special in the study of algebra.

DEFINITION OF VARIABLE AND EQUATION

> A **variable** is a letter or symbol that can represent any number from some set. An **equation** is a statement of equality between two quantities. An equals sign, =, separates the quantities.

It might be said that *variable* is the most important word in algebra. The use of variables in algebra makes it possible to:

- Describe rules or relationships in equations instead of words.
- Make general statements about sets of numbers. Where possible, we *name the set of numbers associated with the variable.*
- Talk about a quantity or value without knowing the number. Where possible, we *say (or define) what the variable represents.*

When we write equations from sentences, words such as *equal, is,* and *to get* show where to place the equals sign. A false statement, $4 = 5$, is not an equation.

EXAMPLE 2 Writing with variables Rewrite these statements using the suggested variables and, as needed, equations.

a. In this text, we agree to write the difference between two numbers as the first number subtract the second number. Let a and b be the two numbers.

b. In textbooks, letters used as variables are printed in italic type. Let a and b be two variables.

c. Rational numbers are defined as the set of numbers that can be written by dividing one integer by another, so long as we do not divide by zero. Let a and b be the integers being divided.

d. A **property** is *a statement that is always true for a given set of numbers*. It is a property of real numbers that a number (except zero) divided by itself equals one. Let a be the number.

e. A property of rational numbers is that division by a number (except zero) is the same as multiplication by the reciprocal of that number. Let a and b be the two parts of the rational number.

SOLUTION

a. The difference between a and b is $a - b$.

b. Letters, such as a and b, used as variables are printed in italic type: a and b.

c. Rational numbers are the set of numbers that can be written $\frac{a}{b}$, where a and b are integers and b is not equal to zero.

d. $\frac{a}{a} = 1$ for all real numbers except $a = 0$.

e. Division by $\frac{a}{b}$ is equal to multiplication by the reciprocal, $\frac{b}{a}$, for all rational numbers except $a = 0$ and $b = 0$.

THINK ABOUT IT: In Example 2, parts c, d, and e, the sets of numbers for the variables are named or known by definition. What were they?

Input-Output Tables, Continued

DESCRIBING A RULE IN WORDS The **input-output rule** tells us *what to do to the input to get the output*. To write a rule, look for one or two operations (addition, subtraction, multiplication, or division) that, when applied to each input, give the matching output.

> To find a rule, look for a pattern. Follow these steps:
>
> **1.** Match each input with an output.
>
> **2.** Describe what we do to the input number to obtain the output.

In Exercise 1 of the Warm-up, we return to the petting zoo pens of Section 1.1. Kelsey assumes she can build all the pens along a wall. The input-output table from her sketch, shown in Figure 10, is given in Table 5. The inputs are 1 to 4 pens, and the outputs are the total number of panels.

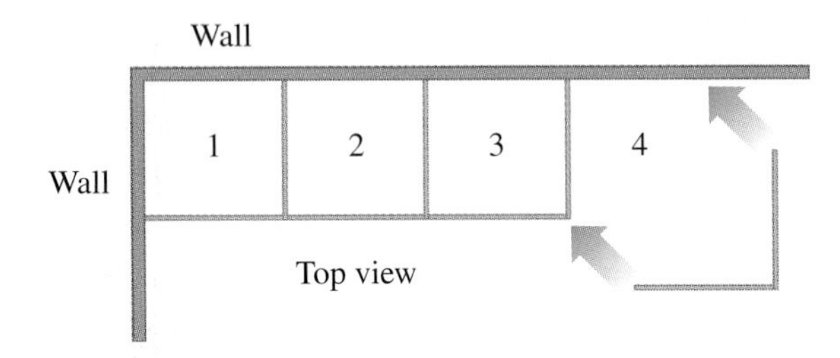

FIGURE 10 Pens by wall

TABLE 5 Pens by Wall and Panels

Input: Number of Pens	Output: Total Number of Panels
1	2
2	4
3	6
4	8

▶ EXAMPLE 3 Finding a rule

a. Show the input-output matching from Table 5.

b. Suggest a rule to predict the number of panels needed for 10, 50, and 100 pens.

c. What rule describes the matching in part a? Describe your rule in a sentence using *input* and *output*.

SOLUTION **a.** In Table 5, one pen takes 2 panels, so the input 1 matches with the output 2. Two pens take 4 panels, so the input 2 matches with the output 4. Similarly, the input 3 matches with 6, and the input 4 with 8.

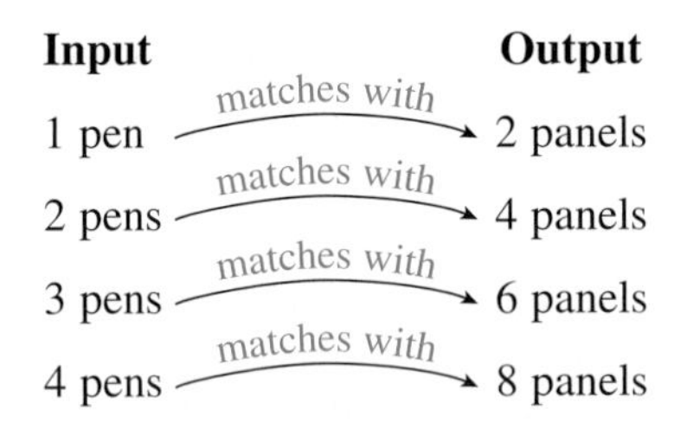

b. The numbers suggest that the output is two times the input: 10 pens take 20 panels, 50 pens take 100 panels, and 100 pens take 200 panels.

c. All except one of the following statements describe the rule for the input table in part a. Find which statement does *not* state the rule.

1. Add the input to itself to get the output.

2. Multiply the input by itself to get the output.

3. The output is twice the input.

4. Double the input to get the output.

5. The output is two times the input.

See the Answer Box for the answer. ◀

WRITING A RULE IN ALGEBRAIC NOTATION To change a rule into algebraic notation, we start by replacing the words *input* and *output* by variables. In Example 4, we write our rule using variables.

▶ EXAMPLE 4 Writing rules in algebraic notation Describe each choice in the solution to part c of Example 3 as an equation in algebraic notation. Let n be the input variable. Let y be the output variable.

SOLUTION Statement 1 is described by $y = n + n$. Statements 3, 4, and 5 are described by $y = 2 \times n$, or $y = 2n$. Statement 2 is described by $y = n \times n$ (which is not the same as $n + n$ or $2 \times n$ except when n equals 2). ◀

Because the letter x is commonly used as a variable, mathematicians agree to avoid the use of $\times$ as a multiplication sign. Multiplication can be written many ways. Note the use of parentheses, ( ).

WAYS TO WRITE MULTIPLICATION

> When two numbers are multiplied, we use a dot or parentheses: $3 \cdot 4$ or $\frac{3}{4}\left(\frac{5}{8}\right)$
>
> When a number is multiplied by a letter, as in 2 times n, we may write $2n$, $2 \cdot n$, or $2(n)$; $2n$ is the most common form.
>
> When two different letters are multiplied together, as in a times b, we write $a \cdot b$ or, more commonly, ab.
>
> When the two letters are the same, as in n times n, we write n^2.
>
> When we press the multiplication key on a calculator, we get the asterisk on the display. Many computer programs also use [*] for multiplication.

▶ In Example 1, we associated the number of pens (input) with the total number of panels (output) needed to build the pens. In Example 5, we associate the position (input) of a design with the perimeter (output) of the design.

▶ EXAMPLE 5 Finding a rule Look at the designs in Figure 11. The numbers below the designs show which is the first, second, third, and fourth position in the pattern.

a. Make an input-output table in which the position of the design is the input and the perimeter of the design is the output.
b. Describe the input and output matching needed to find a rule.
c. Write a sentence describing the rule for obtaining the output from the input.
d. Use the rule to predict outputs for inputs of 50 and 100.
e. Define variables. Write the rule as an equation.

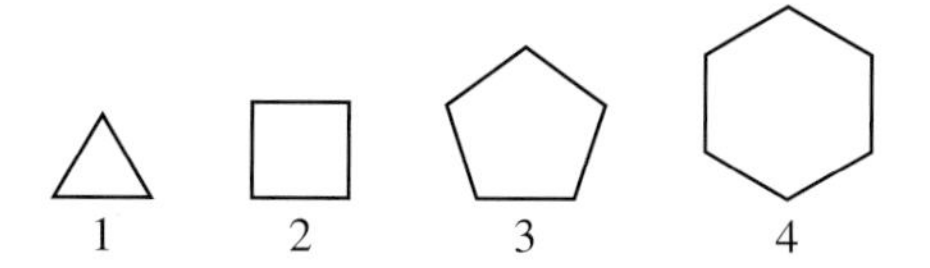

FIGURE 11 Perimeter of figures

SOLUTION **a.** **TABLE 6** Perimeter

Input	Output
1	3
2	4
3	5
4	6
5	7

b. The first design has three sides, so in Table 6, 1 matches with 3. The second design has four sides, so 2 matches with 4. The third design has 5 sides, so 3 matches with 5. Similarly, 4 matches with 6, and 5 with 7.

c. Which statement does not state the rule?

1. Add 2 to the input to get the output.
2. The output is 2 greater than the input.
3. The output is 2 more than the input.
4. The input less 2 gives the output.
5. The input plus 2 gives the output.

The answers to parts c, d, and e are in the Answer Box. ◀

▶ Writing Expressions and Equations

VOCABULARY We used variables to describe our inputs above. In $6 + x$, $5 - y$, and πr, the x, y, and r are variables, like n. The numbers 6, 5, and π are constants.

DEFINITION OF CONSTANT

> A **constant** is a number, letter, or symbol whose value is fixed.

When written together, as in $6 + x$, $5 - y$, and πr, the variables, constants, numbers, and operations form expressions.

DEFINITION OF EXPRESSION

> An **expression** is any combination of signs, numbers, constants, and variables with operations such as addition, subtraction, multiplication, or division.

There is one special type of constant: a numerical coefficient. Examples of numerical coefficients are the 2 in the expression $2n$ and π in πr.

DEFINITION OF NUMERICAL COEFFICIENT

> A **numerical coefficient** is the sign and number multiplying the variable or variables.

A number or numerical coefficient without a positive or negative sign is assumed to be positive. Thus, 2 means positive 2. *A variable without a numerical coefficient is assumed to have a positive 1 coefficient*. Thus, n means positive $1n$.

▶ **EXAMPLE 6** Practicing vocabulary Identify the numerical coefficients, variables, and constants in these expressions.

a. $2x + 1$ **b.** πr^2 **c.** x **d.** $-3x + 25$

SOLUTION **a.** The 2 is the numerical coefficient of the variable x. Both 2 and 1 are constants.

b. The π (or pi) is both a numerical coefficient and a constant. The variable is r.

c. The variable x is assumed to have as a numerical coefficient the constant 1.

d. The -3 is the numerical coefficient of the variable x. Note that the negative sign on the 3 is part of the coefficient. Both -3 and 25 are constants. ◀

MORE WORDS FOR OPERATIONS The words sum, difference, product, and quotient were defined as the answers in addition, subtraction, multiplication, and division, respectively. Table 7 summarizes other words and phrases for these operations.

Student Note:
The parentheses on Table 7 all enclose additional information. None are showing a multiplication.

TABLE 7 Words for Operations

Addition (sum)	Subtraction (difference)	Multiplication (product)	Division (quotient)
plus added to increased by greater than* more than*	$a - b$ a less b a minus b a decreased by b $b - a$ a less than b** a fewer than b**	times double a number ($2n$) twice a number ($2n$) triple a number ($3n$) half of a number ($\frac{1}{2}n$) $\frac{1}{2}$ of 10, 5% of 20 5 items at $4 each	divide in half ($\div 2$) divide by a half ($\div \frac{1}{2}$)

*In some settings, these words may be interpreted as subtraction or inequalities (see Chapter 2).
**In a subtraction setting, *than* indicates that the numbers are to be written in the reverse of the word order.

▶ **EXAMPLE 7** Writing expressions from words Write these subtractions as expressions.

a. The difference in age between 6 and a number, n
b. Fifteen dollars less ten dollars
c. Six less than n
d. x fewer than 4
e. x decreased by 5

SOLUTION In parts a, b, and e, write the numbers or variables in the order in which they are stated. Parts c and d contain *than*, so the order in which the numbers or variables should be written is the reverse of the word order. See the Answer Box for the answers. ◀

▶ **EXAMPLE 8** Writing expressions from words Write these phrases as expressions. Let n be the input number.

a. One less than twice the input
b. The difference between half the input and five
c. The quotient of three and double the input
d. 4 credit hours at n dollars per credit hour
e. The product of four and the input is then decreased by one

SOLUTION Parts a, b, and e contain a subtraction and a multiplication. Part c contains a division and a multiplication, and part d contains a multiplication. See the Answer Box for the answers. ◀

▶ **EXAMPLE 9** Writing equations from word problems Define variables.

a. The input is the number of payments. Each payment is $35. Write an equation for the total value of x payments.
b. The input is the number of rides on the subway. Each ride costs $1.25. Write an equation for the total cost of x rides.
c. The situation is the same as in part b, but you start with a prepaid ticket worth $20. Write an equation for the value on the ticket after x rides on the subway.

SOLUTION **a.** Let x be the number of payments and v be the total value of the payments in dollars: $v = 35x$. If v is in dollars, then no $ sign is needed in the equation.

b. Let x be the number of rides and c be the total cost in dollars: $c = 1.25x$.

c. Let x be the number of rides and v be the remaining value in dollars. The total cost of the x rides in part b is subtracted from the value of the prepaid ticket: $v = 20 - 1.25x$. ◀

ANSWER BOX

Warm-up: 1. See Table 5. **2.** See Table 6. Assume the sides are of equal length, 1. **3.** Output is 2 squares, 4 squares, 6 squares, 8 squares. The numbers are the same as in Exercise 1, but the first describes lengths (panels) and the other describes area (in squares). It is not unusual to find repeated number patterns. **Example 3: c.** Statement 2 does not state the rule. **Example 5: c.** Statement 4 does not state the rule. **d.** The output for 50 is 52 and for 100 is 102. **e.** $y = 2 + n$ or $y = n + 2$ **Example 7: a.** $6 - n$ **b.** \$15 − \$10 **c.** $n - 6$ **d.** $4 - x$ **e.** $x - 5$ **Example 8: a.** $2n - 1$ **b.** $\frac{1}{2}n - 5$ or $\frac{n}{2} - 5$ **c.** $3 \div 2n$ is correct, but fraction notation is preferred; $\frac{3}{2n}$ **d.** $4n$ **e.** $4n - 1$ **Think about it: c.** a and b are integers. **d.** a is any real number except 0 **e.** $\frac{a}{b}$ is rational, so a and b are any integer except 0.

▶ 1.3 Exercises

▶ Operations

Write the phrases in Exercises 1 to 20 using the correct operation symbol ($+, -, \cdot, \div$), and then do the operation. Write answers as fractions and decimals, where appropriate. Round decimals to the nearest thousandth. Assume the second number is subtracted from or divided into the first number.

1. Find the product of $\frac{3}{4}$ and $\frac{1}{2}$.
2. Find the quotient of 5 and $\frac{1}{2}$.
3. Find the sum of $\frac{1}{3}$ and $\frac{1}{2}$.
4. Find the difference between $\frac{3}{4}$ and $\frac{1}{3}$.
5. Find the quotient of 5 and $\frac{2}{15}$.
6. Find the sum of 0.25 and $\frac{3}{4}$.
7. Find the product of $\frac{3}{5}$ and $\frac{4}{9}$.
8. Find the difference between $\frac{7}{5}$ and $\frac{3}{4}$.
9. Find the difference between $2\frac{2}{3}$ and $1\frac{1}{4}$.
10. Find the product of $2\frac{2}{3}$ and $1\frac{1}{4}$.
11. Find the quotient of $2\frac{2}{3}$ and $1\frac{1}{4}$.
12. Find the sum of $2\frac{2}{3}$ and $1\frac{1}{4}$.
13. Find the difference between $\frac{4}{5}$ and $\frac{1}{3}$.
14. Find the product of $\frac{5}{6}$ and $\frac{3}{4}$.
15. Find the sum of $1\frac{1}{2}$ and $2\frac{1}{4}$.
16. Find the quotient of $1\frac{1}{2}$ and $2\frac{1}{4}$.
17. Find the sum of $1\frac{1}{2}$ and $2\frac{1}{2}$.
18. Find the product of $1\frac{1}{4}$ and 8.
19. Find the difference between $1\frac{1}{4}$ and $\frac{3}{8}$.
20. Find the quotient of $1\frac{1}{2}$ and $2\frac{1}{2}$.

▶ Variables and Equations

In Exercises 21 to 23, use one or more variables to translate the statements into algebraic notation.

21. We agree to write the quotient between two numbers as the first number divided by the second number.
22. "The product of a number and its reciprocal is 1" is a definition.
23. The letters at the beginning of the alphabet are often used as constants for agreements, definitions, and properties. The letters at the end of the alphabet and the first letter or sound of key words are often used for variables in word problems.
24. Rewrite the definition of a rational number using the word *quotient*.

Identify the constants, numerical coefficients, and variables in the expressions in Exercises 25 and 26.

25. **a.** $2\pi r$
    b. $1.5x$
    c. $-4n + 3$
    d. $x^2 - 9$
26. **a.** πd
    b. $\frac{x}{2}$
    c. $x^2 - 4$
    d. $-2n - 1$

In Exercises 27 to 30, write a sentence that uses the word in its mathematical meaning. Write a second sentence in which the word's meaning is not mathematical.

27. expression
28. constant
29. variable
30. pattern

Blue numbers are core exercises.

31. Which of the following are equations?

a. $y = x - 1$ **b.** $3 + x$ **c.** $3 + 1 = 4$

d. $5 - 1 = 6$ **e.** $x - 4$ **f.** $3 + x = 4$

32. Which of the following are equations?

a. $x = 0$ **b.** $4 + x = 5$ **c.** $\frac{4}{5} = 0.8$

d. $3x$ **e.** $9 - 3 = 12$ **f.** $x + 4 = 7$

▶ Input-Output Tables

In Exercises 33 to 36, build an input-output table for the given equation. The variable x represents inputs, and the variable y represents the outputs. Use the first three natural numbers as inputs.

33. $y = 3x$

34. $y = 4x$

35. $y = 2.5x$

36. $y = 3.5x$

For Exercises 37 and 38, build an input-output table. Write an equation for its rule.

37. Input: number of hours, t
Output: distance traveled, D, in miles
Rule: Distance traveled is 55 miles per hour times the time t.
Use inputs $\{1, 2, 3, t\}$

38. Input: number of cans, n
Output: total cost, C, in dollars
Rule: Cost is \$0.86 per can.
Use inputs $\{1, 2, 3, n\}$

▶ Writing Expressions and Equations

In Exercises 39 to 42, write an expression in symbols for each word phrase. Let n be the input: To find a difference or quotient, write the numbers in the order in which they are stated.

39. a. The product of three and the input

b. The quotient of eight and the input

c. The difference between the input and four

d. The quotient of the input and five

e. 15 pounds at n dollars per pound

40. a. The difference between four and the input

b. The quotient of the input and eight

c. The product of the input and three

d. The sum of three and the input

e. n credit hours at \$89 per credit

41. a. Three plus twice the input

b. The difference between four and triple the input

c. The product of the input and seven, increased by four

d. Multiply the input by itself

e. n ounces at \$0.79 per ounce

42. a. The quotient of the input and three, decreased by two

b. Three times the input, less two

c. Add two to twice the input

d. The product of seven and half the input

e. 2.5 yards at n dollars per yard

In Exercises 43 to 46, what is the rule in words? What is the output when the input is 100? Write the rule as an equation. (To save space, the input-output tables are shown horizontally instead of vertically).

43.

Input, x	2	3	4	20	100
Output, y	10	15	20	100	?

44.

Input, x	2	3	4	20	100
Output, y	14	21	28	140	?

45.

Input, x	3	4	5	100
Output, y	1	2	3	?

46.

Input, x	7	8	9	100
Output, y	4	5	6	?

▶ Writing

47. Explain why x does not necessarily represent a positive number.

48. Explain why $-x$ does not necessarily represent a negative number.

49. Write down the definition of the term *equation*, and use it to explain why $9 - 3 = 12$ is not an equation.

50. Instant Message. Can the letters in an Instant Message be variables? Explain in terms of U, R, GTG, POS, w/o, or other letters.

51. Compare the rule statements in Exercises 37 and 38 with the statements in Exercises 39 to 42. Describe the difference between the wording for equations and that for expressions.

Blue numbers are core exercises.

52. Problem Solving: Squares and Perimeter The small squares in the designs in Warm-up 3, page 18, make rectangles and are named 1 by 2, 2 by 2, 3 by 2, and 4 by 2. Draw a 5 by 2 rectangle. Draw a rectangle with the number of little squares (area) equal to the number for its perimeter.
When the rectangles have the same length sides, as in 2 by 2, the rectangles are called squares. Draw little squares forming a big square such that the number of little squares equals the number for the big square's perimeter.

▶ Project

53. Numbers as Words II The project continues Exercise 17, page 6. Complete that exercise, and then find the words for the following definitions.

a. Two thin metal strips fastened together and used to detect small changes in temperature

b. Set of three musical notes

c. Olympic event involving skiing and shooting

d. Flag with three colors in stripes, such as the French flag

e. Eyeglass lens with two parts (top and bottom)

f. Square dance for four couples

▶ 1 Mid-Chapter Test

1. Input-Output Rule

a. In the table, what is the output for $x = 3$?

b. Complete the table for a rule with one operation.

Input, x	2	3	4	??	100
Output, y	3	4.5	6	45	??

c. In words, what is the rule that changes the input to the output?

d. What is the rule as an equation?

2. Input-Output Table For the figure shown.

a. Make a table in which the number of sides is the input and the number of triangles inside is the output.

b. Draw the next figure in the pattern. (*Hint:* It should contain 5 triangles.)

c. Predict how many triangles would be inside a 20-sided figure.

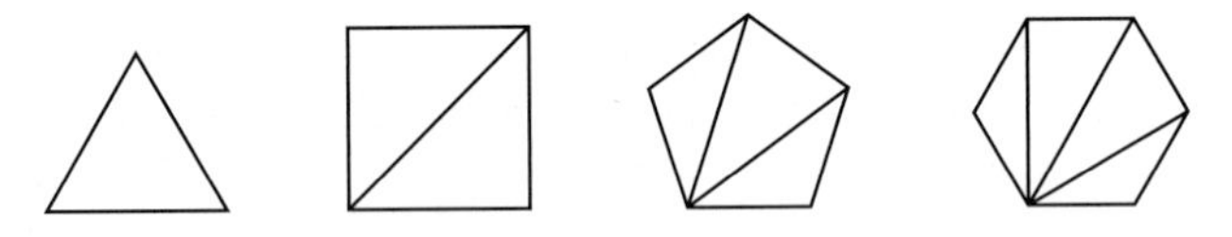

d. Write an equation relating the input and output.

e. Explain the meaning of each variable in your equation.

f. What set of numbers is appropriate for the inputs?

3. Tell whether each statement is true or false. If the statement is false, explain why.

a. A whole number is a rational number.

b. A number may be both an integer and a rational number.

c. Zero is a natural number.

d. Dividing any two integers gives another integer.

4. Write each of these sentences as an equation or an expression.

a. The difference between 5 and a number

b. The product of a number and 3.5 is 105.

c. A number increased by 7

d. The quotient of 6 and a number is 30.

e. Explain why you chose to write an equation or an expression for each.

5. Match each word with the phrase that suggests its definition. Choose from: constant, quotient, variable, expression, numerical coefficient.

a. represents a number or set of numbers

b. the sign and number multiplying a variable

c. any combination of signs, numbers, constants, and variables

d. a fixed value

6. The reciprocal of $\frac{7}{5}$ is $\frac{5}{7}$ because their product is 1. Write this statement using two variables, a and b, where each variable represents an integer and neither variable represents zero.

7. Change $2\frac{2}{5}$ to an improper fraction and to a decimal.

8. Change 0.65 to a fraction.

9. Add, subtract, multiply, and divide the fractions $\frac{6}{7}$ and $\frac{2}{3}$.

10. a. Find the sum of $2\frac{1}{3}$ and $3\frac{3}{4}$.

b. Find the product of $1\frac{1}{2}$ and $1\frac{1}{4}$.

c. Find the difference between $3\frac{1}{4}$ and $1\frac{3}{4}$.

d. Find the quotient of $3\frac{1}{2}$ and $1\frac{3}{4}$.

▶ 1.4 Symbolic Representations

Objectives

- Write rules for input-output tables in words and in equations.
- Change notation among percents, fractions, and decimals.
- Write equations from rules containing percents.
- Read and interpret data in multiple-rule tables.
- Change sentences with two or more operations to equations.

> **WARM-UP**
>
> **1.** Make an input-output table. Let the natural numbers 1 to 4 be the inputs. Let the area (number of squares) in each figure below be the output.
>
> 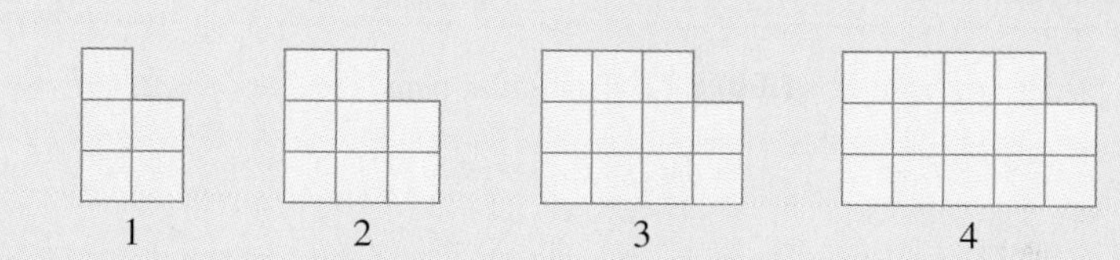
>
>
> **2.** Make an input-output table. Let the natural numbers 1 to 4 be the inputs. Let the perimeter (distance around the outside) in each figure in Exercise 1 be the output.
>
> **3.** Divide these numbers by 100.
>
> **a.** 87.5 **b.** 79 **c.** 120 **d.** 0.5

IN SYMBOLIC REPRESENTATION, we write variables, expressions, and equations in algebraic notation or symbols. We introduced the notation with the vocabulary in the last section: x and y are variables, $-3x^3$ is an expression, and $x = 4$ is an equation. In this section we work with more complicated expressions and equations and, for skill review, include examples on percent notation and operations.

▶ Algebraic Notation

Historical records show that some 3600 years ago, both the ancient Egyptians and Babylonians did problems that clearly reflect algebraic thinking. The representations for the problems and solutions were in words and sentences. It was not until less than 500 years ago, while Europeans explored and began their settlement of North and South America, that mathematicians such as René Descartes (1596–1650, France) began to use letters as variables as commonly seen today. Modern-day algebra courses ask you to translate between word representation and symbolic representation just a few days after you begin your study. Don't be too hard on yourself if it takes a while to grasp some of the ideas involved!

▶ Expressing Rules for Input-Output Tables

We now return to the triangular petting zoo pens for an example of finding a rule and writing it in symbols by *comparing the outputs of one table with known outputs of another table*.

▶ **EXAMPLE 1** Linked Example: Petting Zoo Pens; Finding a Rule by Comparing Outputs Kelsey's triangular pens are shown in Figure 12. Table 8 shows the number of pens as input and the total number of panels as output. Suppose that, by listing more inputs, Kelsey finds that 20 pens require 41 panels.

a. For each additional pen in the input column, how much does the output change? *Hint:* Another petting zoo pen changed by the same number (see Table 5, page 20).

b. Add a column to Table 8 with outputs given in Table 5, and compare the outputs. Predict the number of panels needed for 50 and 100 pens.

c. Describe the pattern between inputs and outputs in words and with an equation.

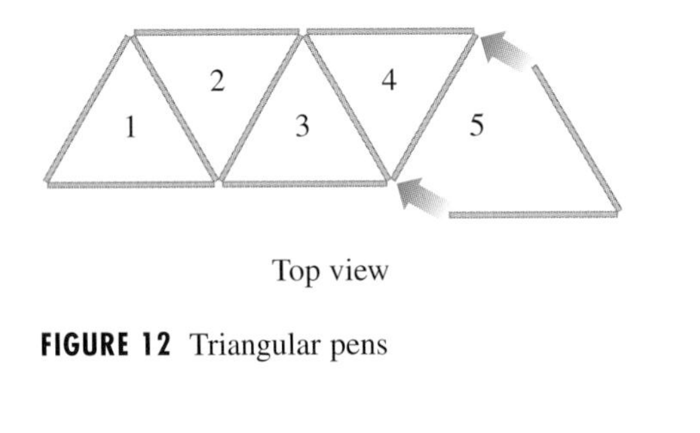

FIGURE 12 Triangular pens

TABLE 8 Triangular Pens and Panels

Input: Number of Pens	Output: Total Number of Panels
1	3
2	5
3	7
4	9
20	41
n	

SOLUTION **a.** Although we are matching inputs and outputs across the table, it is natural to look down the columns. For each input change of 1, the outputs change by 2. This makes sense because we are adding two panels with each pen. Repeated addition of 2 has multiplication by 2 as a shortcut. Thus we compare the output column with $y = 2n$, a rule that also shows a change of 2 for each input change of 20 (the rule for the pens built next to a wall, Section 1.3, page 20).

b. Table 9 is Table 8 along with outputs for $y = 2n$. Now look across the rows. Each output in the second column of Table 9 is 1 more than the output in the third column, $y = 2n$. We can predict that 50 pens will require 1 more than 2 times 50, or 101 panels. Also, 100 pens will require 1 more than 2 times 100 pens, or 201 panels.

Student Note:
The first pen takes 3 panels, and each remaining pen takes 2; thus, for n pens, $L = 3 + 2(n - 1)$ is a reasonable rule. Algebra shows that the rules are equivalent.

TABLE 9 Compare with Known Rule

Input: Number of Pens	Output: Total Number of Panels	Output: Based on Known Rule
1	3	2
2	5	4
3	7	6
4	9	8
20	41	40
n		$2n$

c. Let L be the number of panels. To find the number of panels, we add 1 to the product of 2 and the number of pens, $L = 2n + 1$ or $L = 1 + 2n$. ◀

▶ **EXAMPLE 2** Finding a rule by comparing with a known rule

a. Make an input-output table with the position of each group of squares in Figure 13 as the input and with area (number of squares in each group) as the output.

b. For a change of 1 in the input column, how much does the output change? State a known rule that shows the same change, and add a column for the known rule to the table.

c. Compare the output column and your new column. Write a sentence and an equation describing how the input number and area are related.

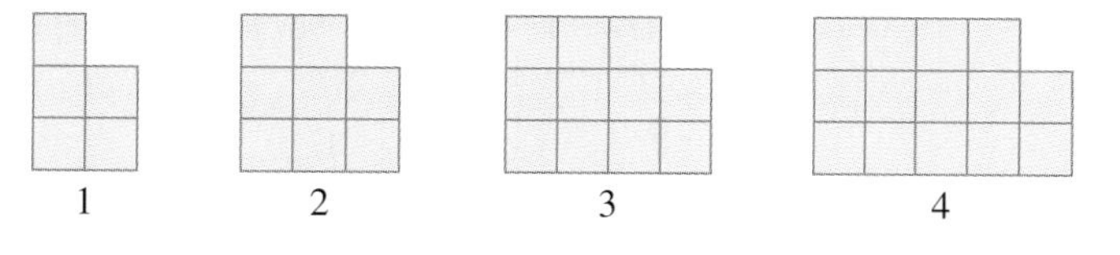

FIGURE 13

SOLUTION **a.** **TABLE 10** Area Pattern

Input: Position, n	Output: Area, A
1	5
2	8
3	11
4	14

b. For each change of 1 in the input column, the output numbers change by 3. From one position to the next, three squares have been added. Because repeated addition of 3 has multiplication by 3 as a shortcut, we write the known column with $y = 3n$, where n is the input and y is the output.

c. **TABLE 11** Compare with a Known Rule

Input: Position, n	Output: Area, A	Output with Known Rule, $y = 3n$
1	5	3
2	8	6
3	11	9
4	14	12

Look across the rows to find how the known rule compares with the area. Each area is 2 more than the output for the known rule. Thus the area is 2 more than 3 times the input (position number), $A = 2 + 3n$ or $A = 3n + 2$. ◀

THINK ABOUT IT 1: How does the rule in Example 2, $A = 2 + 3n$, fit with the shape formed by the squares in Figure 13?

▶ EXAMPLE 3 Finding a rule by comparing with a known rule

a. For Figure 13 in Example 2, make an input-output table with the position for each group of squares as input and the perimeter (distance around the outside) of the group as the output.

b. State a known rule that has the same change in output for a 1-unit change of input. Add a column for the known rule to your table.

c. Compare the known rule to the perimeter. Write a sentence and an equation describing how the position and perimeter are related.

SOLUTION **a.** Table 12 has the figure number and perimeter. Let the side of a square be of length 1.

b. Look down the columns. For a change of 1 in the input, the perimeter changes by 2. This suggests $y = 2n$ as a known rule.

c. Look across the rows to find how the known rule compares with the perimeter of the figure.

TABLE 12 Perimeter and Comparing with a Known Rule

Input: Position, n	Output: Perimeter, p	Output: Known Rule, $y = 2n$
1	10	2
2	12	4
3	14	6
4	16	8

Each perimeter is 8 more than the known rule in the same row. This suggests that the perimeter is 8 more than 2 times the input. For an input n and a perimeter, p, the equation is $p = 2n + 8$. ◀

Percents

In Sections 1.2 and 1.3 we practiced operations with fractions. Fractions are also the key to percent. **Percent*** means *per hundred or division by 100.* It may be helpful to recall the words for the decimal places; for example, 0.123 means 1 tenth, 2 hundredths, and 3 thousandths.

Percents are everywhere: interest on credit cards, automobile, and student loans; sales tax, property tax, and income tax rates; and tips at a restaurant. By agreement, we use decimals rather than percents in an algebraic representation. Examples 4 and 5 review notation with percents, and Example 6 reviews multiplication with percents.

▶ **EXAMPLE 4** Changing decimal notation to percent notation Write as a fraction then as a percent:

a. 0.79 **b.** 0.875

SOLUTION The solution steps show why familiar shortcuts work.

a. There are two decimal places in 0.79, so the decimal is read "seventy-nine hundredths."

$$0.79 = \frac{79}{100}$$ Write the 79 over 100.

$$\frac{79}{100} = 79\%$$ **Percent*** means *per hundred.* Replace the division bar and the 100 in the denominator by %.

b. There are three decimal places in 0.875, so the decimal is read "eight hundred seventy-five thousandths."

$$0.875 = \frac{875}{1000}$$ Write the 875 over 1000.

$$\frac{875}{1000} = \frac{87.5}{100}$$ Write the fraction with a denominator of 100.

$$= 87.5\%$$ Replace the /100 by %.

Change the fraction $\frac{875}{1000}$ to lowest terms. See Answer Box. ◀

Percent* may be used as an adjective, adverb, or noun. *Percentage* is only a noun. In this text, *percent* is used (correctly but colloquially) for percentage. **Percentage is most commonly used to name *the part of arithmetic dealing with percents.* (These facts are from *Webster's New World College Dictionary.*)

EXAMPLE 5 Changing percent notation to decimal notation Apply the definition of *percent* to change these percents first to fraction notation and then to decimal notation. Recall or look for a shortcut, but be able to explain why it works.

a. 37.5% **b.** 120% **c.** 0.5%

SOLUTION **a.** $37.5\% = \frac{37.5}{100}$ Replace % with division by 100. Do the division.

$37.5\% = 0.375$

b. $120\% = \frac{120}{100}$ Replace % with division by 100. Do the division.

$120\% = 1.20$

c. $0.5\% = \frac{0.5}{100}$ Replace % with division by 100. Do the division.

$0.5\% = 0.005$ ◀

A percent should be changed to a decimal before it is written with a variable or in an expression.

EXAMPLE 6 Writing percent expressions from words Write each percent phrase as an expression.

a. 20% of a number n **b.** $62\frac{1}{2}\%$ of a number n

c. $\frac{1}{2}\%$ of a number n

SOLUTION The numerator and denominator of the fraction are multiplied by 10 in parts b and c, to clear the decimal point in 62.5 and 0.5.

a. 20% of $n = \frac{20}{100}n$ or $0.20n$

b. $62\frac{1}{2}\%$ of $n = 62.5\%$ of n Change to a fraction.

$\frac{62.5}{100}n = \frac{625}{1000}n$ or $0.625n$

c. $\frac{1}{2}\%$ of $n = 0.5\%$ of n Change to a fraction.

$\frac{0.5}{100}n = \frac{5}{1000}n$ or $0.005n$ ◀

Reading Tables with Multiple Rules

LINKED EXAMPLE: CREDIT CARD FEES In Tables 8 to 12, one rule described the entire table. In many applications, rules change as inputs get larger. Consider Table 13, which describes the fees on cash advances for a Professional Association of Diving Instructors (PADI) credit card. The key to using this type of table is deciding which rule to use.

TABLE 13 Fees for Cash Advance

Input: Amount of Cash Advance	Output: Fee on Credit Card Statement
\$0 to \$166.66	\$5
\$166.67 to \$2499.99	3% of the amount
\$2500 to credit limit	\$75

EXAMPLE 7 Reading a fee schedule From Table 13, find the fee for each of these PADI card cash advances.

a. \$15 **b.** \$125 **c.** \$200 **d.** \$500 **e.** \$3000

SOLUTION Compare each dollar value with the table input and choose the matching output rule.

a. \$15 is between \$0 and \$166.66. Fee is \$5.

b. \$125 is between \$0 and \$166.66. Fee is \$5.

c. \$200 is between \$166.67 and \$2499.99, so fee is 3% of 200. As a decimal, 3% = 0.03. Fee is 0.03(200), or \$6.

d. \$500 is between \$166.67 and \$2499.99, so fee is 3% of 500. Fee is 0.03(500), or \$15.

e. \$3000 is above \$2500. Fee is \$75. ◀

Sentences and Equations with Multiple Operations

Examples 1, 2, and 3 had rules containing two operations. Examples 8 and 9 show changing between sentences and equations for rules with two or more operations.

EXAMPLE 8 *Changing sentences to equations* Change these sentences into equations with two or more operations. Define the variables, as needed.

a. The output, y, is ten less than twice the input, x.
b. The perimeter, p, of a triangle is the sum of the three sides, a, b, and c.
c. The input is divided by two and then decreased by three to give the output.
d. The perimeter of a rectangle is the sum of twice the length and twice the width.
e. The output is one hundred divided by the sum of the input and three.
f. The value in dollars, V, of a phone card is \$10 less n minutes at \$0.05 per minute.

SOLUTION
a. $y = 2x - 10$
b. $p = a + b + c$
c. Let x be the input and y be the output, $\frac{x}{2} - 3 = y$
d. Let p be the perimeter and l and w be the length and width, $p = 2l + 2w$
e. Let x be the input and y be the output, $y = \frac{100}{x + 3}$
f. $V = 10 - 0.05n$ ◀

THINK ABOUT IT 2: Why is no dollar sign needed in the equation for part f?

EXAMPLE 9 *Changing equations into sentences* Change the first two equations into words using input and output. Change the second two equations into words using appropriate words from geometry.

a. $y = 3x - 4$ **b.** $y = 5 - \frac{x}{2}$
c. $A = \pi r^2$ **d.** $A = \frac{1}{2}bh$

SOLUTION
a. The output, y, is four less than three times the input, x.
b. The output, y, is five decreased by half of the input, x.
c. The area, A, of a circle is the constant pi times the product of the radius, r, with itself.
d. The area, A, of a triangle is half the product of the base, b, and the height, h. ◀

ANSWER BOX

Warm-up: 1. See Table 10. **2.** See Table 12. **3. a.** 0.875, **b.** 0.79, **c.** 1.2 **d.** 0.005 **Think about it 1:** For position n, there are n groups of three plus two extra squares. **Example 4:** $\frac{875}{1000} = \frac{7}{8}$ **Think about it 2:** The value, V, is defined in dollars.

▶ 1.4 Exercises

▶ Input-Output Tables

For Exercises 1 to 4, make an input-output table with position as input (n): 1, 2, 3, 4, 50 and 100. The output (A) is the area. Write the rule in words and with an equation. As a check, for an input of 20, the outputs will be **1.** 61, **2.** 41, **3.** 83, **4.** 60.

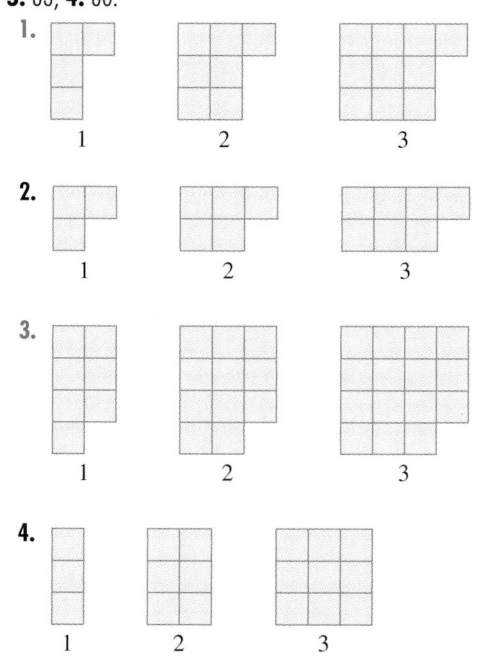

For Exercises 5 to 8, return to the figures in Exercises 1 to 4, respectively. Make input-output tables with position as input (n): 1, 2, 3, 4, 50, and 100. The output (p) is the perimeter (assume one unit is the length of one side of a square). Write the rule in words and with an equation. As a check, for an input of 20, the outputs will be **5.** 48, **6.** 46, **7.** 50, and **8.** 46.

▶ Words and Equations

Make tables for the rules in Exercises 9 to 12. Use integer inputs from 1 to 5. Write the rule as an equation.

9. Input: number of sales made
Output: total income in dollars
Rule: Income is \$250 plus \$75 per sale.

10. Input: number of days on vacation
Output: total cost of vacation in dollars
Rule: Cost is \$1590 plus \$200 per day.

11. The output is two more than four times the input.

12. The output is one more than the input times itself.

13. Change from equations into sentences.

a. $y = 10 - \frac{x}{5}$

b. (Area of a circle) $A = \pi \cdot \frac{1}{4}d^2$

c. (Circumference of circle) $C = \pi \cdot 2r$

d. (Surface area of a box with length, width, and height): $S = 2lw + 2hl + 2hw$

14. Change from equations into sentences.

a. $y = \frac{5}{x} - 10$

b. (Area of a circle) $A = \frac{\pi d^2}{4}$

c. (Circumference of a circle) $C = \pi d$

d. (Volume of a box, length, width, and height): $V = lwh$

15. Change from sentences into equations.

a. The surface area of a cube is the product of six and the area of one side $n \cdot n$.

b. The sum of a number, n, and its opposite, is zero.

c. The product of a number, n, and its reciprocal is 1.

d. The length of the diameter of a circle is the product of two and the length of the radius.

16. Change from sentences into equations.

a. The area of a parallelogram is the product of the base and the height.

b. The product of a rational number, $\frac{a}{b}$, and its reciprocal is 1.

c. The product of zero and a number is zero.

d. The product of 1 and a number, n, is n.

17. These equations are all for the area of a triangle. Write each in words.

a. $A = b\frac{h}{2}$ **b.** $A = \frac{1}{2}bh$

c. $A = \frac{b}{2}h$ **d.** $A = \frac{bh}{2}$

18. List several agreements about variables in the equations in Exercises 13 to 16.

▶ Percent

19. Change these to decimals and fractions:

a. 15% **b.** 0.5%

c. 48% **d.** 250%

Blue numbers are core exercises.

20. Change these to decimals and fractions:

a. 20% **b.** 45%

c. 0.2% **d.** 112%

21. Write 100% as a fraction and as a decimal.

22. Write 1 as a percent.

23. A study reported by *Science News* (Nov. 26, 2005, p. 347) found that 45% of all medication taken in space is sleeping pills. Write this percent as a fraction and as a decimal.

24. California community colleges are considered financially secure if college reserves are 5% of their total budget. Write this percent as a fraction and as a decimal.

In Exercises 25 to 28, change the fractions and decimals to percents

25. a. 0.9 **b.** $\frac{2}{3}$ **c.** 0.5

d. 4.9 **e.** $6\frac{1}{4}$ **f.** 9

26. a. 0.6 **b.** $\frac{1}{6}$ **c.** 0.25

d. 0.7 **e.** $2\frac{1}{2}$ **f.** 6

27. a. 1.5 **b.** $\frac{3}{4}$ **c.** 0.36

d. 5.6 **e.** 2.25 **f.** 15

28. a. 0.3 **b.** $\frac{3}{5}$ **c.** 0.06

d. 0.8 **e.** 0.15 **f.** 5

Write each percent statement in Exercises 29 and 30 as an expression.

29. a. 35% of n **b.** 10% of x

c. $87\frac{1}{2}$% of n **d.** $37\frac{1}{2}$% of x

e. $\frac{1}{2}$% of n **f.** 108% of x

30. a. 25% of x **b.** 15% of n

c. $6\frac{1}{2}$% of x **d.** 150% of n

e. $2\frac{1}{4}$% of x **f.** $12\frac{1}{2}$% of n

31. One of the dangers of a weightless environment is bone loss. An astronaut on the space station loses 1.5% of his or her hip-bone mass (m) each month. In comparison, the typical postmenopausal woman loses 1% of her hip-bone mass (m) each year. (*Science News*, Nov. 26, 2005, p. 347)

a. Write an expression for the bone mass lost by an astronaut in one month.

b. Write an expression for the bone mass remaining for the astronaut after one month. *Hint:* The astronaut starts with 100% of the bone mass.

c. Write an expression for the bone mass lost by a postmenopausal woman in a year.

d. Write an expression for the bone mass remaining for the woman after one year. *Hint:* The woman starts with 100% of the bone mass.

e. Suggest another question with this information, and answer it.

▶ Multiple-Rule Tables

32. You are one day late making your minimum monthly payment on a credit card. What is the late fee on the following balances on the day after the payment was due?

Balance on Day After Payment Due Date	Late Payment Fee
\$0–\$100	\$15
\$100.01–\$250	\$29
\$250.01 or over	\$39

a. \$800 **b.** \$90 **c.** \$245

33. Linked Example, Credit Card Fees

a. Write each of the output rules for the PADI credit card fee schedule shown in the table. Use n as the input amount, y as output.

Input: Cash Advance	Output: Fee	Output Rule
\$0 to \$166.66	\$5	
\$166.67 to \$2499.99	3% of advance	
\$2500 to credit limit	\$75	

For parts b to e, find the fee if the cash advance is:

b. \$150. **c.** \$300.

d. \$1000. **e.** \$2600.

f. Comment on the fee for a cash advance of \$5.

g. Calculate 3% of \$166.67 and 3% of \$2499.99 and suggest why these dollar amounts appear in the table.

34. A local credit union Visa card has the following payment schedule:

Credit Card Balance	Minimum Payment
\$0 to \$30	Full balance
\$30.01 to \$375	\$15
\$375.01 and over	4% of the balance

What is the minimum payment for each of the following balances?

a. \$45 **b.** \$400 **c.** \$450

d. \$700 **e.** \$15 **f.** \$385

Blue numbers are core exercises.

▶ More Input-Output Tables

35. Building Pens in Pairs Cassandra suggests that Jake build pens in *pairs*, as shown in the figure. The pairs and panels pattern is shown in the following table. Describe the rule in words and with an equation. Use it to predict the number of panels needed for 10 pairs and 25 pairs of pens.

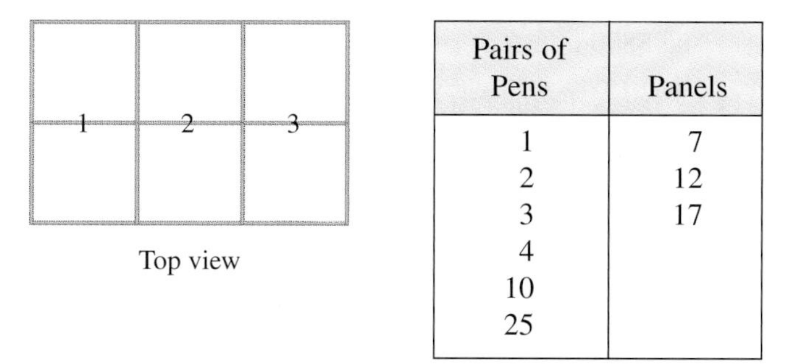

Top view

Pairs of Pens	Panels
1	7
2	12
3	17
4	
10	
25	

36. Seating at Square Tables The seating at 3 square restaurant tables placed next to each other is shown in the figure. Eight chairs may be placed at the 3 tables. Find the number of chairs for other arrangements as you complete the following table. Describe the rule.

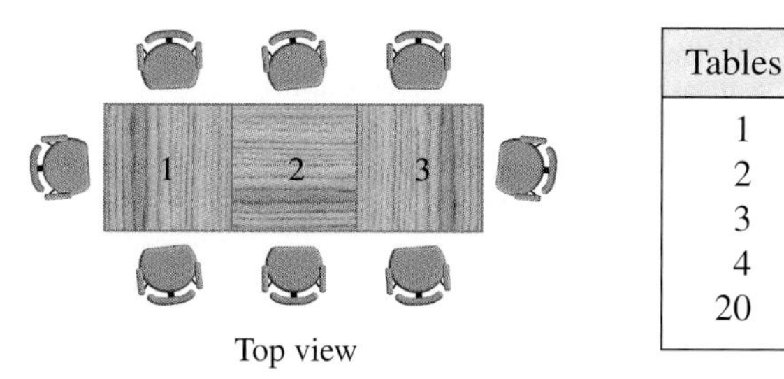

Top view

Tables	Chairs
1	4
2	6
3	8
4	
20	

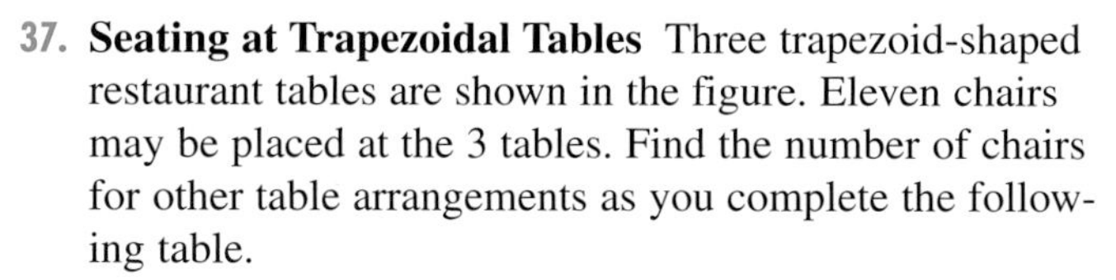

37. Seating at Trapezoidal Tables Three trapezoid-shaped restaurant tables are shown in the figure. Eleven chairs may be placed at the 3 tables. Find the number of chairs for other table arrangements as you complete the following table.

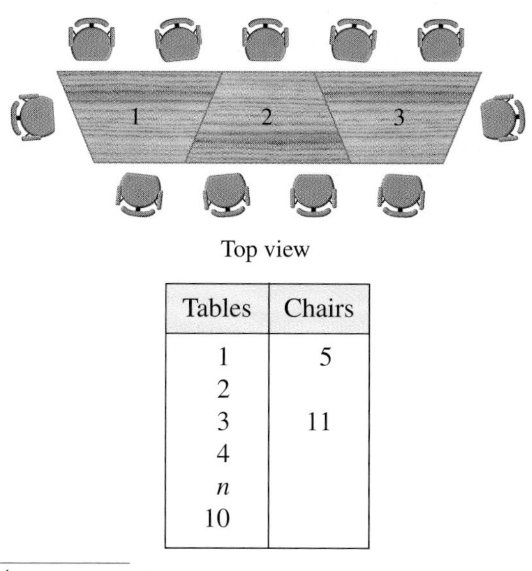

Top view

Tables	Chairs
1	5
2	
3	11
4	
n	
10	

Blue numbers are core exercises.

For Exercises 38 and 39, make an input-output table for the rule given. The **even numbers** are the integers divisible by 2. The **odd numbers** are the integers not divisible by 2.

38. The output is the sum of the input and 2 if the input is even. The output is the product of the input and 2 if the input is odd. Use integer inputs 0 through 8.

39. The output is 5 if the input is an even number. The output is twice the input if the input is an odd number. Use integer inputs $\{0, 1, 2, \ldots, 8\}$.

▶ Computation Review

In Exercises 40 and 41, use the two input values to calculate the outputs, according to the rule given at the top of each output column.

40.

Input a	Input b	Output $a + b$	Output $a - b$	Output $a \cdot b$	Output $a \div b$
9	6				
$\frac{2}{3}$	$\frac{1}{6}$				
0.5	0.25				
49	0.7				
6.25	2.5				

41.

Input a	Input b	Output $a + b$	Output $a - b$	Output $a \cdot b$	Output $a \div b$
15	5				
$\frac{3}{4}$	$\frac{2}{5}$				
0.36	0.06				
5.6	0.7				
2.25	1.5				

▶ Writing

42. On page 117 of the February 19, 2005, issue of *Science News*, we find the following phrase: "… successfully calculated answers to equations with reversed terms such as 59 − 13 and 13 − 59." Explain why this use of the word *equation* may not be appropriate. Rewrite the phrase with a better word.

43. The perimeters in Figure 13 and in Exercise 1 have the same rule, yet the shapes and areas are different. What is it about the shapes that give the same perimeter? Draw another set of shapes with the same perimeter pattern.

44. Explain how walking along square blocks in a city, say from 1st and A streets to 3rd and B Streets, illustrates how groups of squares can have different shapes but the same perimeter. *Hint:* Draw a picture.

▶ Problem Solving: Products and Sums

Exercises 45 and 46 contain modified input-output tables. Use the two input values to calculate the outputs according to the rule given at the top of each output column. You may need to find one or both inputs.

45.

Input x	Input y	Output xy	Output $x + y$
4	3	12	
		12	13
5	4		
	1	20	
		4	4
		18	9

46.

Input x	Input y	Output xy	Output $x + y$
2	6	12	
10	2		
		15	8
1	15		
	2	18	
4		4	

▶ Project

47. Credit Cards A World Points Master Credit Card promotion offered 3.9% APR (annual percentage rate for a year) on balance transfers during the first three-month period.

a. Balance transfers from another credit card are charged 3% of the transfer. Find the transfer fee on an $800 transfer.

b. What is the interest owed for one month on the total of the $800 transfer and transfer fee? *Hint:* A month is $\frac{1}{12}$ of a year.

c. A minimum payment is $30 or 4% of the balance, whichever is greater. What is the minimum payment?

d. Suppose you make only the minimum payment. Now how much is your balance?

e. Suppose you pay the minimum payment (part c), the transfer fee (part a), and interest (part b). What is the total payment? If you make no new charges, what is your balance?

f. You make no new charges and you pay a $30 minimum payment on time and all the interest due each month. In how many months will $750 be paid off?

g. A late payment is a "promotion turn-off event." The 3.9% APR rises to 15.9% APR on the transfers and new charges, and a $39 late fee is added. Suppose the balance is $824. What interest is due? What is the late fee plus interest plus minimum payment?

h. Discuss how it is possible to make a minimum payment and owe more than originally due.

▶ 1.5 Visual Representations: Rectangular Coordinate Graphs

Objectives

- Identify quadrants, axes, the origin, ordered pairs, and coordinates.
- Build a graph from an input-output table.
- Build a table and graph to fit a single rule or multiple rules.
- Read and interpret a graph.

WARM-UP

Make an input-output table for each rule using inputs 0.0, 0.5, 1.0, 1.5, 2.0, 2.5, and 4.0. Let x be input and y be output. What equation is described?

1. The output is 3 plus 2 times the input. (*Hint:* Multiplication is done before addition, as you will see in Section 2.4.)

2. The output is 3 plus the input times itself.

IN THIS SECTION, we introduce graphs as an essential visual representation in algebra. Graphs describe position and relate algebraic equations to points and lines.

Maps

Have you ever used a map to find your way? A map shows a position by two or more numbers or letters. In 1791, the Frenchman Pierre Charles L'Enfant designed the national capital, Washington, D.C., so that a number and letter could identify each street intersection. His plan, shown in Figure 14, required that the city be divided into four quadrants: N.E., N.W., S.W., and S.E.

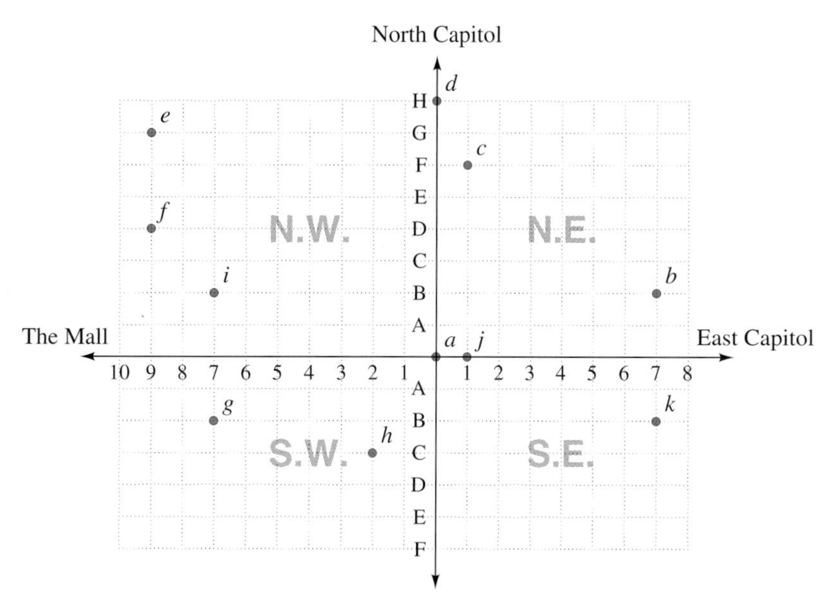

FIGURE 14 Simplified map of Washington, D.C.

In his book *What Do You Care What Other People Think?* (New York: Bantam Books, 1988), Richard Feynman* relates his problem with the quadrant system. When Feynman hired a taxi to take him from his Washington, D.C. hotel to 7th and B, the address of the National Aeronautics and Space Administration (NASA), the taxi took him to an empty lot.

EXAMPLE 1 Exploration On the map in Figure 14, find the four points that mark possible positions of the NASA building, whose address is 7th and B.

SOLUTION The possible positions are b, i, g, and k. The actual location of NASA, g, is in the S.W. quadrant. Feynman discovered the disadvantage of the street plan: Unless the quadrant is named, there are four possible positions for most addresses. ◀

▶ In Example 2, we find buildings, given the names of the two streets that intersect nearby. Notice how important the quadrants are in locating the buildings.

EXAMPLE 2 Finding positions On the map in Figure 14, find the letter that labels the position of the following buildings.

1. U.S. Capitol building, Figure 15, intersection of East Capitol and North Capitol
2. Union Station, 1st and F, N.E.

*Feynman (1918–1988) is the scientist who brought national attention to the o-ring problem leading to the 1986 Challenger space shuttle disaster.

3. Martin Luther King Memorial Library, 9th and G, N.W.

4. Food and Drug Administration, 2nd and C, S.W.

5. Federal Bureau of Investigation, 9th and D, N.W.

6. Veterans Administration, North Capitol and H

7. Supreme Court, 1st and East Capitol

FIGURE 15 U.S. Capitol Building

SOLUTION **1.** *a* **2.** *c* **3.** *e* **4.** *h* **5.** *f* **6.** *d* **7.** *j* ◀

▶ Graphs and Ordered Pairs

To identify position in mathematics, we use a plan similar to the Washington, D.C., map. Note how this plan avoids using N.E., N.W., S.W., and S.E.

RECTANGULAR COORDINATE GRAPH René Descartes (1596–1650) designed the **Cartesian coordinates** to be *a pair of numbers that indicate the position of a point on a flat surface by the point's distance from two lines*. The *flat surface* is the **coordinate plane**. The lines are two number lines placed at right angles so that they cross at zero. Figures 16 and 17 show rectangular **coordinate graphs**, also known as *Cartesian coordinate graphs*.

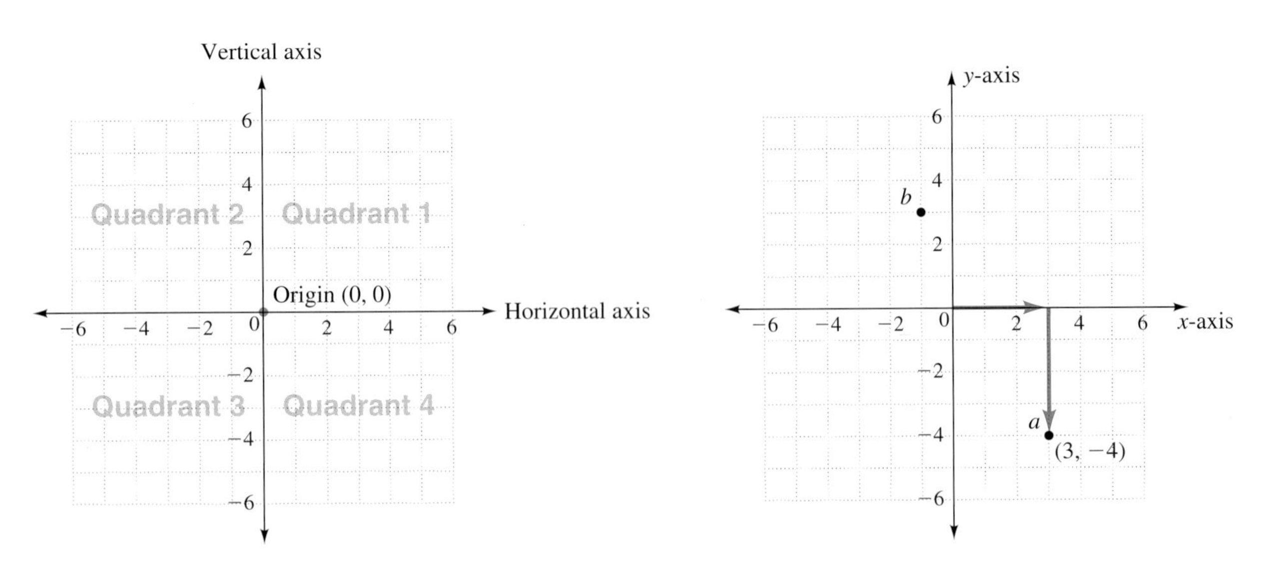

FIGURE 16 The coordinate plane

FIGURE 17 Locating $(3, -4)$

The *number line that goes left to right* is called the **horizontal axis**, as in Figure 16, or the **x-axis**, as in Figure 17. The *number line that runs up and down* is called the **vertical axis** or the **y-axis**. The *point where the axes cross* is called the **origin**. The number lines divide the plane into four *sections* called **quadrants**, as in the map of Washington, D.C., in Figure 14.

ORDERED PAIRS The Cartesian coordinates are placed in parentheses. For example, (0, 0) describes the origin. Because the order of the numbers is important, the *Cartesian coordinates* are also called **ordered pairs**.

ORDERED PAIRS

An ordered pair contains two real numbers (x, y) that describe a position on a coordinate plane. For each ordered pair:

- The first number, x, describes the horizontal distance from the origin, along the x-axis. Positive numbers are to the right; negative numbers are to the left.
- The second number, y, describes the vertical distance from the x-axis. Positive numbers are up; negative numbers are down.

Student Note:
The parentheses in the ordered pair, (x, y), group the numbers. The comma tells us that we don't multiply anything.

The numbers in an ordered pair may be zero, positive, or negative. The ordered pair $(3, -4)$ describes the position found by counting 3 to the right of the origin, then down 4, as shown in Figure 17.

THINK ABOUT IT 1: What is the ordered pair describing the point labeled b in Figure 17?

▶ **EXAMPLE 3** Matching ordered pairs to points on a graph In Figure 18, find the position for each ordered pair.

a. (4, 2) **b.** $(-2, 3)$ **c.** $(3, -2)$ **d.** $(-5, -2)$ **e.** (0, 5) **f.** (2, 0)

Name the letter shown at each position and the quadrant or axis where each point is located.

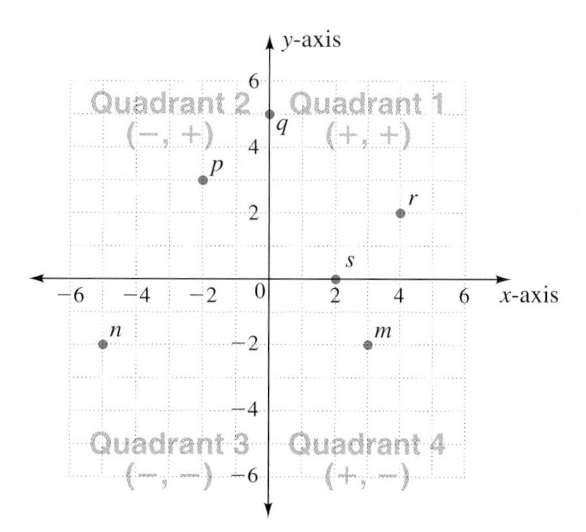

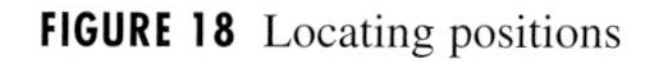
FIGURE 18 Locating positions

SOLUTION **a.** To find (4, 2), go to 4 on the horizontal axis and count up 2 in the vertical direction. The point (4, 2), labeled r, is in the first quadrant.

b. To find $(-2, 3)$, go to -2 on the horizontal axis and count up 3 in the vertical direction, to the point labeled p in the second quadrant.

c. To find (3, −2), go to 3 on the horizontal axis and count down 2 in the vertical direction, to the point labeled m in the fourth quadrant. Observe that (−2, 3) and (3, −2) do not describe the same point.

d. To find (−5, −2), go to −5 on the horizontal axis and count down 2, to the point labeled n in the third quadrant.

e. A zero in an ordered pair means no movement in one direction, so the point is on one of the axes. The point (0, 5) is on the vertical axis at q, 5 up from the origin.

f. The point (2, 0) is on the horizontal axis at s, 2 to the right of the origin. ◀

THINK ABOUT IT 2: Each of the quadrants in Figure 18 is labeled with a pair of signs, positive or negative. Compare the pairs of signs in the labels with the signs of the ordered pairs graphed in each quadrant.

▶ **EXAMPLE 4** Naming ordered pairs Name the ordered pairs shown on the graph in Figure 19. Name the quadrant in which each point is located.

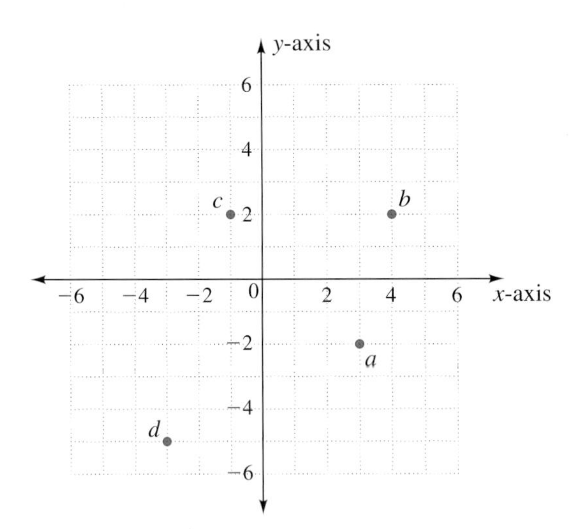

FIGURE 19 Naming ordered pairs

SOLUTION **a.** (3, −2); Quadrant 4 **b.** (4, 2); Quadrant 1

c. (−1, 2); Quadrant 2 **d.** (−3, −5); Quadrant 3 ◀

USING THE WORD *GRAPH*

The word *graph* is used as both a noun and a verb. Used as a noun, **graph** refers to *the points that have been plotted* or *the line or curve drawn through those points*. As a verb, **graph** means *to locate points on the coordinate plane as described by ordered pairs*.

Creating Graphs: The Four-Step Process

We now return to the four problem-solving steps, which we apply to building a graph. We graph data from the tables in the Warm-up.

▶ **EXAMPLE 5** Graphing in the first quadrant from an input-output table based on a rule Use the four-step process to make a first-quadrant graph with the ordered pairs in Table 14. Should the points be connected?

SOLUTION ***Understand:*** We begin building the graph in Figure 20 by labeling the inputs on the horizontal axis (or x-axis) and the outputs on the vertical axis (or y-axis). Each row of Table 14 may then be written as an ordered pair (x, y), with x used for the input and y for the output.

Plan: The graph is to be in the first quadrant. The numbers are reasonably small, so we number to 8 on the horizontal axis and to 12 on the vertical axis.

Sketch the graph: We plot the ordered pairs, as shown in Figure 20. The points appear to lie in a straight line. Because other positive inputs make sense, we draw a line (left to right) through the points and label the line with the rule, $y = 2x + 3$.

Check and extend: On the line formed by all the ordered pairs, we select another point, such as $(x, y) = (3, 9)$. We find that $x = 3$, $y = 9$ makes a true statement when we place the numbers in the rule $y = 2x + 3$; that is, $9 = 2(3) + 3$. So the numbers $x = 3$ and $y = 9$ for the ordered pair (3, 9) could be listed in the table.

TABLE 14 Ordered Pairs for $y = 2x + 3$

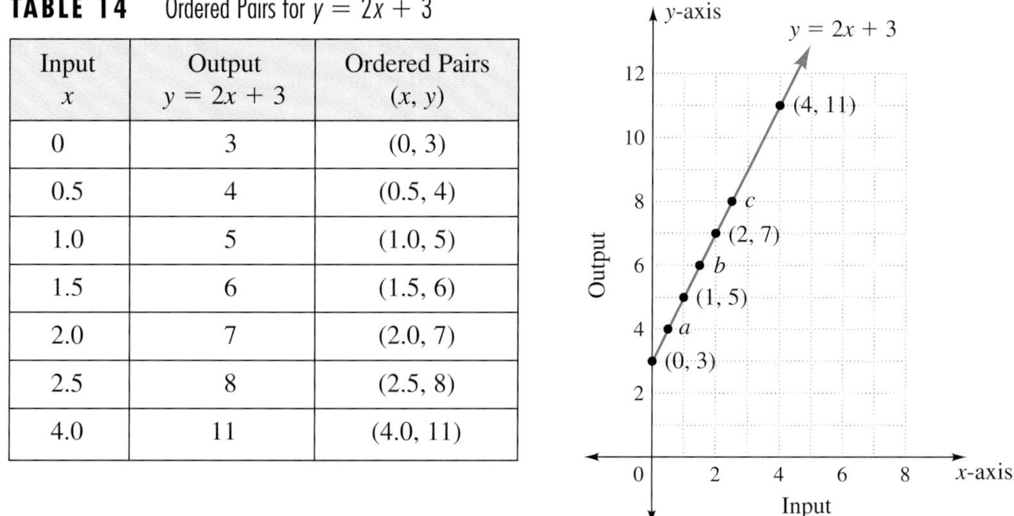

Input x	Output $y = 2x + 3$	Ordered Pairs (x, y)
0	3	(0, 3)
0.5	4	(0.5, 4)
1.0	5	(1.0, 5)
1.5	6	(1.5, 6)
2.0	7	(2.0, 7)
2.5	8	(2.5, 8)
4.0	11	(4.0, 11)

FIGURE 20 Graph of $y = 2x + 3$ ◀

THINK ABOUT IT 3: Name the ordered pairs for points a, b, and c in Figure 20.

▶ **EXAMPLE 6** Graphing in the first quadrant from an input-output table based on a rule Use the four-step process to write each row of Table 15 as an ordered pair and to make a first-quadrant graph. Should the points be connected?

SOLUTION ***Understand:*** The inputs and outputs are given. The ordered pairs (x, y) for the table values are (0, 3), (0.5, 3.25), (1.0, 4), (1.5, 5.25), (2.0, 7), and (4.0, 19).

Plan: The graph is to be in the first quadrant. The input numbers are small, so we can use zero and the positive integers from 1 to 7 on the horizontal axis. The output numbers go from 0 to 19, so we count by 2 on the vertical axis to fit the numbers 0 to 20.

Sketch the graph: We plot the ordered pairs as shown in Figure 21. Other positive inputs make sense in the rule, so we connect the points with a curve (draw left to right). We label the curve with the rule $y = x^2 + 3$.

TABLE 15 $y = x^2 + 3$

Input x	Output $y = x^2 + 3$
0	3
0.5	3.25
1.0	4
1.5	5.25
2.0	7
4.0	19

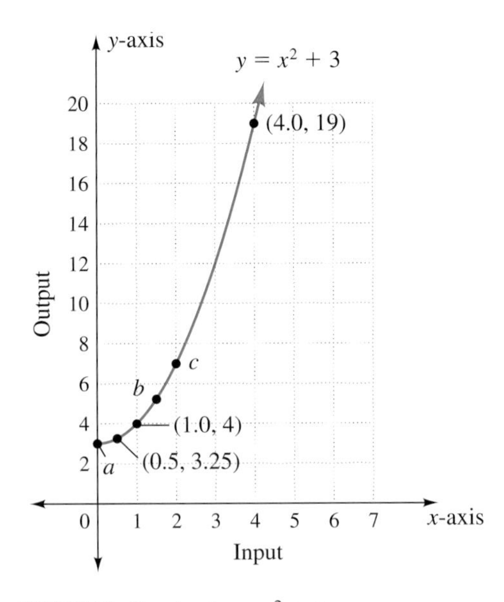

FIGURE 21 Graph of $y = x^2 + 3$

Check and extend: On the curve formed by all the points, we select another ordered pair, such as $(x, y) = (3, 12)$. We find that placing $x = 3$ and $y = 12$ in $y = x^2 + 3$ gives $12 = 3^2 + 3$, a true statement. The numbers $x = 3$ and $y = 12$ could be listed in the table. ◀

THINK ABOUT IT 4: Name the ordered pairs for points a, b, and c in Figure 21.

▶ Finding the Scale

When we decide *what numbers to use for the marks or labels on axis*, we are finding the **scale** for the axis. When numbers are very large or very small, we need to count by numbers other than 1 or 2.

One way to estimate the scale for an axis is to subtract the lowest input (or output) value in a table from the highest and divide by 10. Then round the quotient to an easy counting number, such as 5, 10, 20, 50, or 100.

SCALE

$$\text{Estimated scale} = \frac{\text{highest number} - \text{lowest number}}{10}$$

Then round to an easy counting number.

▶ EXAMPLE 7 Finding the scale for the output axis The data in Table 16 show median incomes for households in the United States by age in 2003. The input data naturally fit counting by ten. What would be an appropriate scale for the output axis, for graphing ordered pairs based on Table 16? Graph the data from the table.

SOLUTION To estimate the output scale, we subtract the lowest output from the highest and divide by 10. We then round to the nearest easy counting number.

$$\begin{aligned}\text{Estimated scale} &= \frac{\text{highest output} - \text{lowest output}}{10}\\ &= \frac{60{,}200 - 23{,}800}{10}\\ &= 3640\\ &\approx 4000\end{aligned}$$

The symbol $\approx$ means "is approximately equal to." We count by 4000 on the vertical axis. The graph is shown in Figure 22.

TABLE 16 Age and Income

Age* x (years)	Median Income y (dollars)
20	27,000
30	44,800
40	55,000
50	60,200
60	49,200
70	23,800

*The ages are midpoints of intervals: 20 stands for ages 15 to 24, 30 stands for ages 25 to 34, . . . , 70 stands for ages 65 and older.
Source: Table 675, Statistical Abstract of the United States, 2006.

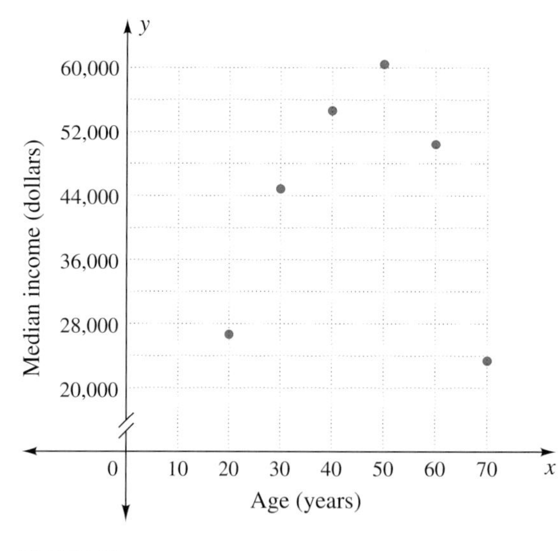

FIGURE 22

We do not connect the points in Figure 22, because the inputs represent a range of numbers and there is no rule for obtaining outputs for other inputs. ◀

Note that a *double slash*, //, is used on the vertical axis in Figure 22 to indicate that the space between zero and 20,000 is not the same as the space between each of the other numbers on the axis.

Reading Graphs

Tables and graphs help answer these questions: What is the output when the input is given? What is the input when the output is given? Graphs help with estimating numbers between the entries on a table and observing patterns in tabular data.

READING A GRAPH

- To find an output, locate the given input on the input axis, trace vertically to the graph, and then trace horizontally to the output axis.
- To find an input, locate the given output on the output axis, trace horizontally to the graph, and then trace vertically to the input axis.

LINKED EXAMPLE: CREDIT CARD FEES In Table 17, we repeat the PADI credit card fee schedule from Section 1.4. The information is graphed in Figure 23.

TABLE 17 Cash Advance and Fees

Input: Amount of Cash Advance	Output: Fee on Credit Card Statement
\$0 to \$166.66	\$5
\$166.67 to \$2499.99	3% of amount
\$2500 to credit limit	\$75

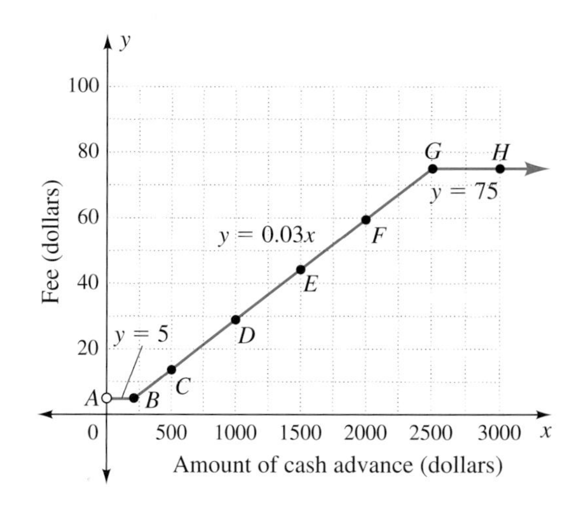

FIGURE 23 Fee based on amount of advance

▶ **EXAMPLE 8** Reading a graph Table 17 and its graph (Figure 23) show the fee charged (output) in terms of the amount of cash advance taken with a credit card.

a. Name the point (A, B, C, D, etc.) in Figure 23 that shows a \$15 fee for a \$500 cash advance. What is the ordered pair for the point?
b. Point B shows what fact from the table? Name the ordered pair for point B.
c. Find the cash advance available for a \$30 fee. What letter marks this point?
d. Why is the graph level between points A and B?
e. Using the table, find the fee for a \$1500 cash advance. Write the result as an ordered pair. At which letter would we plot the point?
f. Name the ordered pair for point A. Explain what the point means.

SOLUTION

a. Point C; the ordered pair is (500, 15).
b. A \$166.66 cash advance has a \$5 fee. The ordered pair is (166.66, 5).
c. \$1000; point D
d. The output is \$5 for all inputs between \$0 and \$166.66.
e. 0.03(\$1500) = \$45; the ordered pair is (1500, 45); point E.
f. (0, 5) suggests that for a zero cash advance there is a \$5 fee. This would not happen; thus, the graph is a small circle instead of a dot. *The small circle marks position but is not part of the graph of the rule.* ◀

Student Note:
One reason to write units (dollars) on the axes is so that dollar signs may be omitted from inside ordered pairs.

Creating a Graph

The steps in creating a graph are listed below, along with some questions and instructions that might help you complete each problem-solving step.

1. **Understand the problem.**
   Which information is the input, to be labeled on the horizontal axis?
   Which information is the output, to be labeled on the vertical axis?
   Keep in mind that *the output depends on the input*; that is, the variable placed on the vertical axis depends on the variable placed on the horizontal axis.
2. **Plan the graph.**
   Which quadrants are implied or make sense in the problem situation?
   What numbers are needed on the axes?
   Use the double slash, if needed, only between zero and the next number on the axes.
   What units are needed on the axes?
   If the data have units of measure (inches, dollars, liters, years, etc.), label the axes with the units.
3. **Sketch the graph.**
   Make a table, if needed.
   Graph the ordered pairs.
   Look for a pattern, from left to right. What shape do the points make?
   Should the points be connected?
   Label ordered pairs or rules for outputs.
4. **Check the graph, and extend the results as needed.**
   Do all the data points lie on the graph?
   Do other points on the graph make sense in the problem situation?
   How might new data or information change the graph?

Student Note.
The summary for creating a graph is placed here because the summary also helps in reading a graph.

ANSWER BOX

Warm-up: 1. See Table 14. **2.** See Table 15. **Think about it 1:** (−1, 3) **Think about it 2:** All the ordered pairs in a quadrant have the same combination of positive and negative signs. **Think about it 3:** (0.5, 4), (1.5, 6), (2.5, 8) **Think about it 4:** (0, 3), (1.5, 5.25), (2.0, 7)

▶ 1.5 Exercises

Write ordered pairs for points *A* to *I* in Exercises 1 and 2.

1\.

2\.

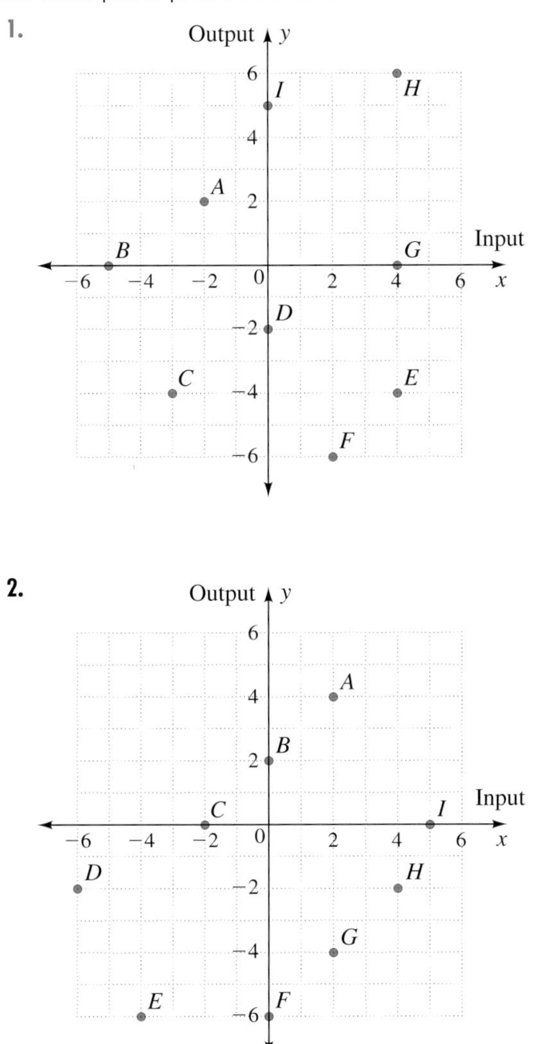

For Exercises 3 and 4, name the quadrant in which each point is located.

3\. **a.** $(-4, 3)$ **b.** $(3, -4)$
c. $(-2, -4)$ **d.** $(-3, -2)$

4\. **a.** $(4, -3)$ **b.** $(-3, 2)$
c. $(-2, -3)$ **d.** $(2, -3)$

For Exercises 5 and 6, name the axis on which each point is located.

5\. **a.** $(0, -4)$ **b.** $(0, -2)$ **c.** $(3, 0)$

6\. **a.** $(0, 4)$ **b.** $(-2, 0)$ **c.** $(-3, 0)$

In Exercises 7 to 10, draw a graph from the input-output table.

7\.

Input x	Output y
-3	5
-2	3
-1	1
0	-1
1	-3

8\.

Input x	Output y
-2	5
-1	2
0	-1
1	-4
2	-7

9\.

Input x	Output y
-3	9
-2	7
-1	5
0	3
1	1

10\.

Input x	Output y
-2	-11
-1	-8
0	-5
1	-2
2	1

In Exercises 11 to 18, graph the two ordered pairs. Then, for each exercise, draw a straight line through the two points and write ordered pairs, (x, y), for two other points on the line. Extend by guessing the rule.

11\. $(1, 2), (3, 6)$

12\. $(3, 3), (5, 5)$

13\. $(2, -1), (8, -4)$

14\. $(3, 1), (9, 3)$

15\. $(8, 4), (2, 1)$

16\. $(4, -4), (2, -2)$

17\. $(1, 3), (2, 6)$

18\. $(3, 5), (5, 7)$

In Exercises 19 to 26, make a table and a graph for each rule. Use integer inputs of 0 to 5. Write the rule on the graph.

19\. The output is 5 more than twice the input.

20\. The output is 2 more than four times the input.

21\. The output is half the input.

22\. The output is 1 more than twice the input.

23\. The output is 3 times the input minus 2.

24\. The output is 1 less than half the input.

25\. The output is the difference between 8 and the input.

26\. The output is the difference between 10 and twice the input.

Blue numbers are core exercises.

27. Bulk Food Purchases Honey-roasted peanuts in the bulk foods department cost \$6.50 per pound.

a. Use the four-step approach to build a table and graph for inputs of 0 to 4 pounds. Let outputs be total cost. Label your steps.

b. Use your graph to estimate the cost of these purchases: $2\frac{1}{2}$ pounds, $1\frac{3}{4}$ pounds, $3\frac{1}{4}$ pounds.

c. One and a half pounds of the same nut mixture is available in packages for \$9.49. Plot this information as a data point. Which is a better buy, bulk or packaged nuts?

28. Bulk Food Purchases, Continued In the bulk foods department, candy costs \$2.29 per pound.

a. Use the four-step approach to build a table and graph for inputs of 0 to 4 pounds. Let outputs be total cost. Label your steps.

b. Use your graph to estimate the total cost of these candy purchases: $2\frac{1}{2}$ pounds, $1\frac{3}{4}$ pounds, $3\frac{1}{4}$ pounds.

c. Two and a half pounds of the same candy is available in packages for \$3.98. Plot this information as a data point. Which is a better buy, bulk candy or packaged candy?

29. Telephone Card You have a prepaid \$6 Mega Mexicana phone card. Calls cost \$0.03 per minute.

a. Use the four-step approach to build a table and graph for inputs of 0 to 200 minutes. Let outputs be the value remaining on the telephone card. Label your steps.

b. Circle the point on your graph that matches the value remaining after 100 minutes have been used.

c. What is the meaning of the ordered pair (0, 6) on your graph?

d. What is the meaning of the ordered pair (200, 0) on your graph?

30. Gift Certificate You have been given a \$36 gift certificate to Koffee Klatch. Each cup of latte costs \$3.

a. Use the four-step approach to build a table and graph for inputs of 0 to 12 cups of latte, counting by 2. Let outputs be the value remaining on the gift certificate. Label your steps.

b. Circle the point on your graph that matches the value remaining after buying 4 cups of latte.

c. What is the meaning of the ordered pair (0, 36) on your graph?

d. What is the meaning of the ordered pair (12, 0) on your graph?

31. The graph shows closing prices for shares of stock for 15 days.

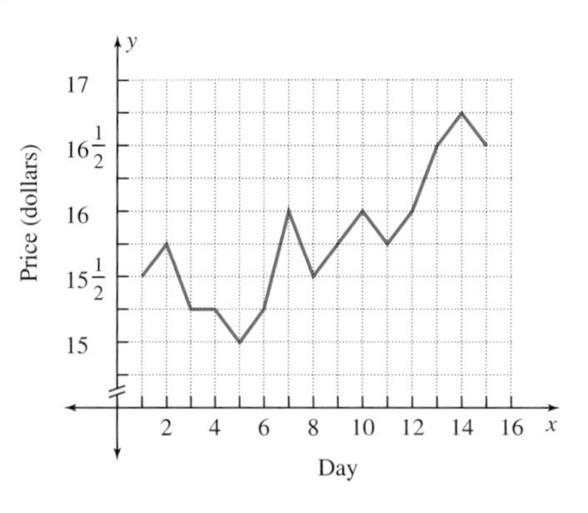

a. List the ordered pairs (prices in decimals) for the 15 days.

b. What is the highest price in the 15 days?

c. What is the lowest price?

d. With a complete sentence, describe the general trend of the prices.

32. The graph shows closing prices for shares of stock for 15 days.

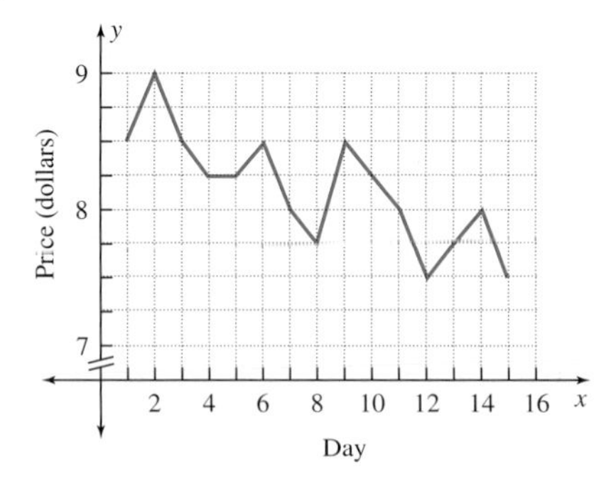

a. List the ordered pairs (prices in decimals) for the 15 days.

b. What is the highest price in the 15 days?

c. What is the lowest price?

d. With a complete sentence, describe the general trend of the prices.

33. Linked Example: Credit Card Fees Refer to Table 17 and Figure 23 from Example 8.

a. Why is the graph in Figure 23 level for inputs above \$2500?

b. Why is there a change in the graph at point B?

c. What is the fee (output) for these cash advances: \$100; \$400; \$700; \$3000?

d. What is the cash advance (input) for a \$60 fee?

Blue numbers are core exercises.

e. If the fee (output) was \$75, can we find the amount of the cash advance?

f. Why is the graph in three pieces?

g. Between which two points is (2, 5) located? Explain what the point means. Is a cash advance a good way to obtain small amounts of money?

h. Why is it possible to give the ordered pairs for points B and E?

▶ Writing

In Exercises 34 to 44, write answers in complete sentences.

34. How do you find the point described by an ordered pair?

35. How do you find the ordered pair describing a point?

36. How do you find in which quadrant an ordered pair is located?

37. How do you find on which axis an ordered pair $(a, 0)$ or $(0, b)$ is located?

38. How do you use a graph to build a table?

39. Why is (x, y) called an ordered pair?

40. Name the set of numbers reasonably expected for inputs and outputs in Exercises 31 and 32.

41. In Figure 22, the scale on the vertical axis jumped from zero to \$20,000 (as indicated by the double slash, //) and then went in steps of \$4000. What are the advantages and disadvantages of this approach?

42. Describe which variables are traditionally placed on which axes. Include the words *horizontal axis*, *vertical axis*, *output variable*, and *input variable*.

43. Make a copy of Figure 22. On the copy, make a second vertical axis on the right side and count in steps of \$20,000 (instead of \$4000). Graph the data from Table 16. Discuss what happens when too large a scale is chosen.

44. Copy the graph in Figure 23. On the copy, make a second vertical axis on the right side and count in steps of \$10 (instead of \$20). Graph the data from Table 17. Discuss what happens when too small a scale is chosen.

▶ Problem Solving: Hidden Axes

45. Find the ordered pairs for points A and B from these portions of the coordinate plane. Each space equals one. The axes are hidden, so you must reason relative positions from the given point(s).

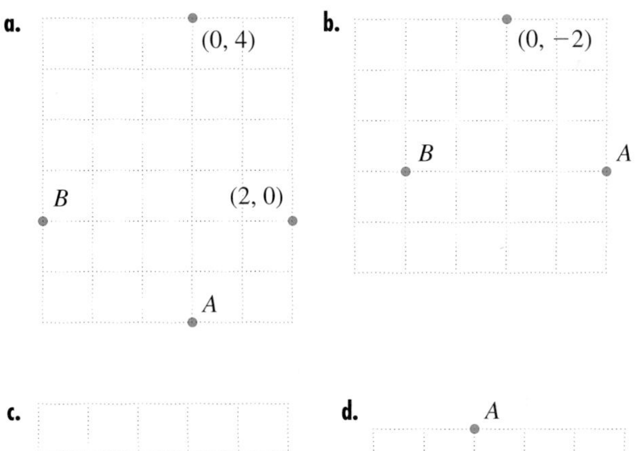

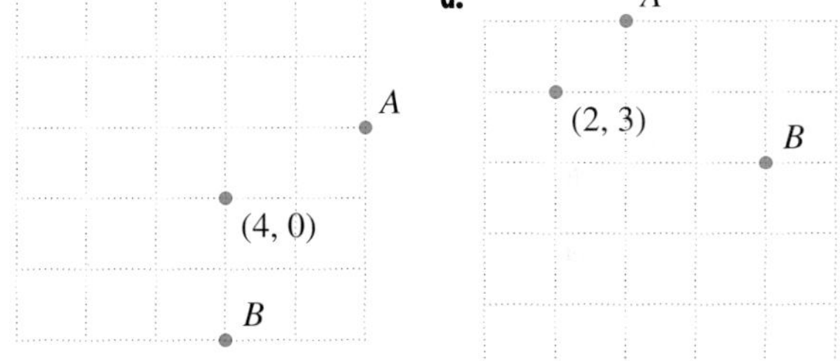

▶ Project

46. Shipping Charges The 2005 shipping and handling (S & H) charges for the National Geographic Society catalogue are based on the cost of merchandise. Use the four steps to graph the data from the table up to a merchandise total of \$250.

Merchandise Total	S & H Charges
\$0.01 to \$25.00	\$5.95
\$25.01 to \$50.00	\$9.95
\$50.01 to \$75.00	\$12.95
\$75.01 to \$100.00	\$15.95
\$100.01 to \$150.00	\$18.95
\$150.01 to \$200.00	\$20.95
Over \$200.00	10% of merchandise total to a maximum of \$75

Blue numbers are core exercises.

▶ 1 Chapter Summary

Vocabulary

For definitions and page references, see the Glossary/Index.

agreement
area
assumption
Cartesian coordinates
condition
constant
coordinate graph
coordinate plane
denominator
difference
equals sign
equation
even numbers
expression
factors
graph
horizontal axis
infinite
input-output rule
input-output table
integers
irrational numbers
lowest terms
natural numbers
negative numbers
numerator
numerical coefficient
odd numbers
opposites
ordered pair
origin
percent
percentage
perimeter
positive numbers
prime number
product
property
quadrant
quotient
rational numbers
real numbers
reciprocal
scale
set
square root symbol
sum
undefined
variable
vertical axis
whole numbers
x-axis
y-axis

Concepts

1.1 Problem-Solving Steps and Strategies

The four problem-solving steps are to understand the problem, make a plan, carry out the plan, and check the solution. We find conditions and make assumptions when we solve problems.

Problem-solving strategies include drawing a picture, trying a simpler problem, looking for a pattern, organizing information by using a table, making an estimate, and recalling a similar problem or known rule.

In a table, the output depends on the input.

1.2 Numeric Representations

Algebra is represented numerically with number patterns and tables. Input-output tables display number relationships and patterns. The language of algebra permits us to describe and use sets of numbers and to describe and solve problems.

To add or subtract fractions, first find a common denominator and then add or subtract. To multiply fractions, multiply numerators and multiply denominators:

$$\frac{a}{b} \cdot \frac{c}{d} = \frac{a \cdot c}{b \cdot d}, b \neq 0, d \neq 0$$

In algebra, division is most often written as a fraction. To divide by $\frac{c}{d}$, multiply by the reciprocal, $\frac{d}{c}$:

$$\frac{a}{b} \div \frac{c}{d} = \frac{a}{b} \cdot \frac{d}{c}, b \neq 0, c \neq 0, d \neq 0$$

Always write fractions in lowest terms.

1.3 Verbal Representations

Algebra is represented verbally with expressions and equations containing signs, numbers, constants, numerical coefficients, and variables. The multiplication of 2 times n may be written as $2n$, $2 \cdot n$, $2(n)$, or $2 * n$. The division of a by b is usually written in fraction notation as $\frac{a}{b}$.

We find a rule from an input-output table by matching each input with its output. The rule describes what operations we do to the input number to obtain the output.

1.4 Symbolic Representations

Comparing the outputs of one table with those of another table is another strategy for finding a rule.

The percent sign means per 100 or division by 100: $n\% = \frac{n}{100}$. Change percent notation to fraction notation by changing the percent sign to division by 100 and writing the resulting fraction in lowest terms. Change $n\%$ to decimal notation by writing n hundredths in decimal form.

1.5 Visual Representations: Rectangular Coordinate Graphs

An ordered pair of real numbers describes a position relative to a set of axes. We apply the problem-solving steps to graphing by considering the questions and following the instructions on page 44.

The scale markings on the axes should be equally spaced and clearly numbered. When the scale does not start at zero, we make a double slash between the origin and the first mark on the axis to indicate a break in the numbering. The axes should be labeled with units of measure. A small circle marks position but is not part of the graph of the rule.

To estimate scale, find the difference between the highest and lowest input (or output) value, divide by 10, and round to a convenient number.

▶ 1 Review Exercises

1. State an assumption you might make about each of these sentences:

a. You want to catch the 3:15 p.m. bus home from campus.

b. You want to subtract 8 and 5.

c. You want to divide 12 and 8.

d. You want to ride your bicycle to school.

2. State a condition given in each of these settings:

a. Calculating the area of a rectangle where the length is twice the width.

b. Measuring the angles in a triangle where the sum of the angles is 180°.

c. You want to go to a movie that starts at 4:45 p.m.

d. You are building a fence from 8-foot sections of prefabricated material.

3. Tell whether each statement describes a numerator (N) or denominator (D).

a. The 3 in $\frac{2}{3}$ shows the kind of fraction.

b. The 2 in $\frac{2}{3}$ shows two parts, each of whose size is one third.

c. The word is like *number*, saying "how much."

d. The word is like *denomination*, saying what kind or what size.

e. The 2 in $\frac{2}{3}$ shows two units divided into three shares.

f. The 3 in $\frac{2}{3}$ shows the size of each share.

g. When we find a common ____, we are changing fractions into the same kind.

4. Each word describes the answer to an arithmetic operation. Name the operation.

a. product

b. quotient

c. difference

d. sum

5. For each of the following numbers, name all of the number sets to which it belongs: real numbers, rational numbers, integers, whole numbers.

a. 1.5 **b.** 6

c. 0.75 **d.** $\frac{1}{2}$

e. $\frac{1}{3}$ **f.** -5

6. In the expression $3x - 4$, which number is the numerical coefficient?

7. Identify the constant in the expression $x + 3$.

8. What is the variable in the expression $x \div 3$?

9. Write each expression in words.

a. $3x - 4$

b. $x^2 + 3$

c. $x \div 3$

10. Write each sentence with an equation. Let x be the input. Let y be the output. The output is

a. Four more than the product of two and the input.

b. The difference between five and the square of the input.

c. Fifteen percent of the input.

d. 5% of the input.

e. $8\frac{1}{2}\%$ of the input.

In Exercises 11 to 14, make a table for whole-number inputs on the interval 0 to 6. Write the rule.

11. The output is half of the input.

12. The output is 4 more than the input.

13. The output is double the input if the input is even. The output is 4 if the input is odd.

14. The output is 5 if the input is even. The output is 2 less than the input if the input is odd.

Make an input-output table for Exercises 15 and 16.

15. The input, h, is the number of kilowatt hours of electricity used. The output is \$5.50 plus \$0.20 per kilowatt hour used.

16. The input, n, is the number of pounds of rice purchased. The rice costs \$0.89 per pound. The output is the total cost, including a \$0.10 "store coupon" refund.

17. Building Pens Jake's pens require the panels pattern shown in the following table. (Three pens are shown in the figure.) Complete the table. Describe the rule in words and with an equation. Use the rule to predict the number of panels needed for 20 and 50 pens.

1	2	3

Top view

Pens	Panels
1	4
2	7
3	10
4	

Student Note: Don't give up on finding rules. *Hint:* Know multiplication facts up to 12 times 12 so you can recognize multiples of numbers. If you still are unable to compare with known multiples, take heart! You will learn another way in Chapter 3.

18. a. Make a new table for Jake's pens in Exercise 17. The input is the number of pens (1 to 6 pens), and the output is the outside perimeter (in panels) of the pens. Why does the table in Exercise 17 not describe perimeter? Describe a rule for your table with a sentence and with an equation.

b. Make a new table for Jake's pens in Exercise 17. The input is the number of pens (1 to 6 pens), and the output is the area of the pens (one pen = 1 square unit).

19. Using the designs, make a table with inputs 1 to 3 with position of the design as input and with area (in number of squares) as output. Write the output for inputs of 5, 50, and 100 as ordered pairs. Write a rule for the table with a sentence and with an equation. *Hint:* For the 18th input, the area is 73.

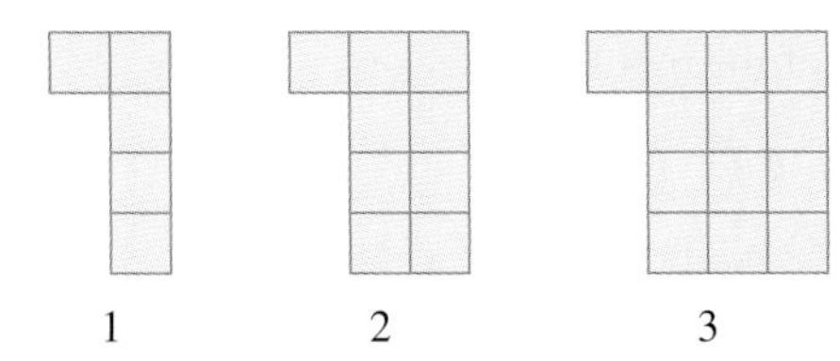

20. For the designs in Exercise 19, make a table with position of each design as input and perimeter of the design as output. Write the output for inputs of 5, 50, and 100 as ordered pairs. Write a rule for the table with a sentence and with an equation. *Hint:* For the 18th input, the perimeter is 46 units.

21. Do these operations:

a. $\frac{5}{8} - \frac{1}{6}$ **b.** $\frac{4}{10} - \frac{3}{15}$

c. $1\frac{3}{5} + 4\frac{2}{3}$ **d.** $3\frac{1}{2} - 2\frac{1}{8}$

e. $2\frac{1}{2} \cdot 3\frac{1}{4}$ **f.** $\frac{3}{10} \div \frac{4}{5}$

22. Add, subtract, multiply, and divide the two numbers. In doing subtraction or division, subtract or divide in the order listed.

a. $\frac{5}{6}$ and $\frac{3}{8}$ **b.** $2\frac{1}{6}$ and $1\frac{3}{4}$

23. Multiplication and division facts are related: If $\frac{12}{2} = 6$, then $6 \cdot 2 = 12$. If $\frac{12}{3} = 4$, then $4 \cdot 3 = 12$. If $a \neq 0$, write the multiplication fact for $\frac{a}{0} = n$. Explain whether the multiplication fact is true.

24. Change these fractions to decimals and percents:

a. $\frac{3}{4}$ **b.** $\frac{7}{8}$ **c.** $\frac{3}{20}$

d. $\frac{4}{25}$ **e.** $\frac{8}{5}$ **f.** $\frac{23}{10}$

25. Change these decimals to fractions and percents:

a. 0.25 **b.** 0.35 **c.** 0.28

d. 0.375 **e.** 0.004 **f.** 1.2

26. Change these percents to fractions and decimals:

a. 48% **b.** 125% **c.** 12.5%

d. $33\frac{1}{3}$% **e.** $66\frac{2}{3}$% **f.** 4.8%

27. Find the following percents of 360:

a. 10% **b.** 20% **c.** 50%

d. 25% **e.** 3% **f.** 0.5%

28. Use the following table to find what discount is received on each of these purchases.

Total Purchase	Discount
\$0 to \$149.99	5% of the purchase
\$150 to \$499.99	6% of the purchase
\$500 and over	\$50 or 8% of the purchase, whichever is greater

a. \$65 **b.** \$145 **c.** \$250

d. \$500 **e.** \$550 **f.** \$700

29. With x as input and y as output, write each rule in the table for Exercise 28 with an equation.

30. A natural gas bill states: "A \$3.00 charge will be assessed on past due balances between \$50 and \$176. A late charge of 1.7% will be assessed on past due balances over \$176." Find the late change on each of the following past due balances.

a. \$80 **b.** \$176

c. \$177 **d.** \$200

31. What words describe the parts of the following graph labeled a, b, and c? What ordered pairs describe the locations labeled d, e, f, and g?

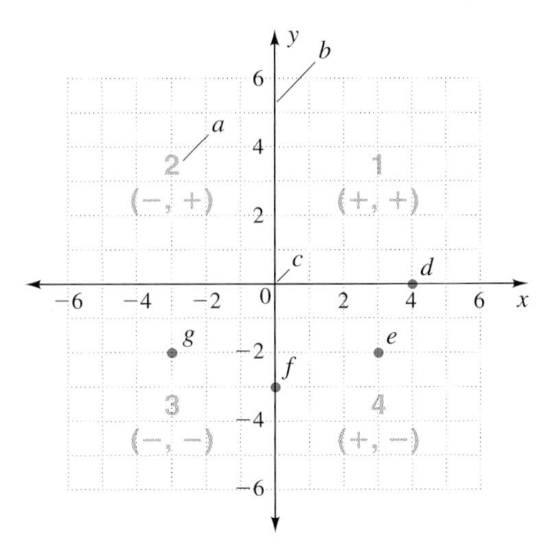

32. Graph these ordered pairs on one set of axes:

a. $(-3, -2)$	**b.** $(-2, 1)$	**c.** $(2, 3)$
d. $(3, -1)$	**e.** $(1, -2)$	**f.** $(0, -4)$
g. $(-1, 0)$	**h.** $(0, 3)$	**i.** $(1, 0)$
j. $(0, 0)$		

33. Plot the sets of ordered pairs in parts a, b, and c on one graph and the sets of ordered pairs in parts d, e, and f on another. Connect points in each part.

a. $(1, 2), (2, 3), (3, 4), (4, 5)$

b. $(1, 1), (2, 0), (3, -1), (4, -2)$

c. $(-4, 5), (-3, 4), (-2, 3), (-1, 2)$

d. $(1, 5), (2, 10), (3, 15), (4, 20)$

e. $(2, 1), (4, 2), (6, 3), (8, 4)$

f. $(1, 3), (2, 5), (3, 7), (4, 9)$

34. What ordered pair $(0, y)$ would fit each pattern in Exercise 33? Describe how you found the ordered pair.

35. For the graphs in Exercise 33, write a description of the output in words or symbols, using x as the input.

36. Give an example of an ordered pair located in:

a. Quadrant 1 and containing two natural numbers.

b. Quadrant 3 and containing two integers that are not natural numbers.

c. Quadrant 2 and containing two rational numbers that are not integers.

d. Quadrant 4 and containing two mixed numbers.

The graphs in Exercises 37 to 39 represent three different dieting experiences. From the graph in Exercises 37 to 39:

a. Estimate the total weight loss by days 10, 20, 30, and 40.

b. How much weight was lost between days 0 and 10? 10 and 20? 20 and 30? 30 and 40?

c. Suggest a description of weight loss and number of days after diet starts.

37.

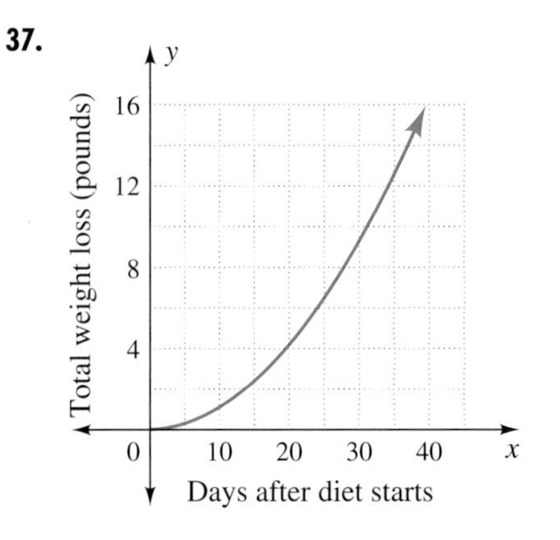

38.

39.

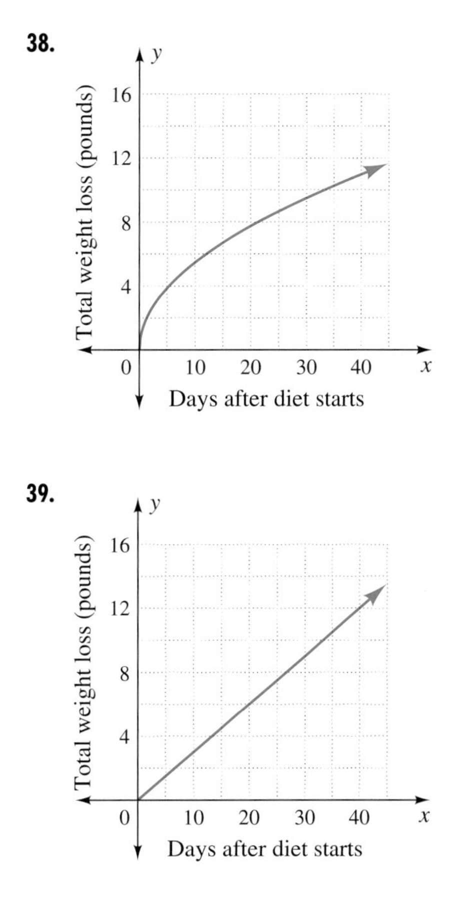

40. a. Pinto beans cost \$1.19 per pound in the bulk foods department. Use the four-step process (Section 1.5) to make a table and graph for 0 to 8 pounds. Let weight in pounds be the input and total cost in dollars be the output.

b. What is the rule describing the bulk purchase graph? Let y be the total cost of x pounds.

c. Several sizes of packaged pinto beans are also available: 1 pound at \$1.39, 2 pounds at \$2.69, and 4 pounds at \$3.99. Graph the weight and cost for the packaged choices as individual ordered pairs.

d. Discuss which might be the best buy and under what circumstances.

In Exercises 41 to 43, make an input-output table for the situation described, and graph the ordered pairs from the table.

41. Decaffeinated coffee beans cost \$24 per pound. Let the input be the number of pounds used and the output be the total cost of coffee. Show the cost for 0 to 8 pounds.

42. The first eight uses of an automatic teller machine (ATM) card are free; each additional use costs \$1.50. Let the input be the number of uses and the output be the total cost for a month. Show the cost for 0 to 12 uses of the card.

43. A 300-minute long-distance telephone card costs \$9.99.

a. What is the cost per minute?

b. Let the input be the number of minutes talked and the output be the value remaining on the card. Show the value of the card for inputs of 0 to 300 minutes.

44. Why would we not connect points in Exercise 42?

45. Why would we connect points in Exercises 41 and 43?

46. Assume that the coordinate axes are temporarily invisible and the scale on each graph is one space equals one unit. Give the coordinates of A and B in each graph.

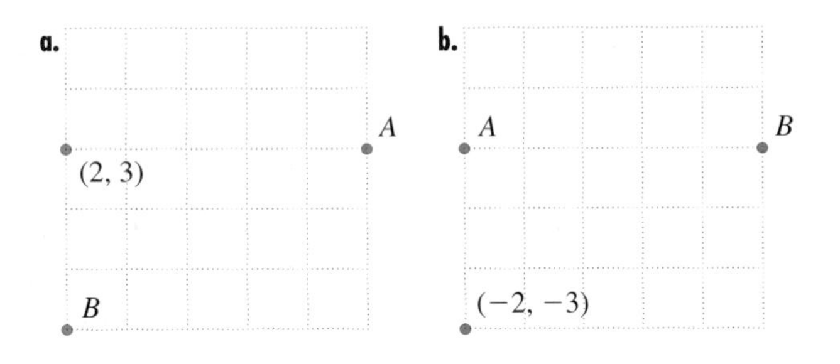

47. For Fun Plot these ordered pairs and connect them in the order listed. Do not connect ordered pairs separated by the word "lift"; lift your pencil.
(9, 5), (3, 2), (6, 8), (7, 2), (2, 5), (9, 5), lift
(−7, 6), (−3, 2), (−5, 8), (−6, 2), (−1, 6), (−7, 6), lift
(−3, −2), (−8, −6), (−3, −6), (−8, −2), (−6, −10), (−3, −2), lift
(3, −8), (8, −4), (2, −4), (6, −8), (6, −2), (3, −8)

48. For Fun Plot these ordered pairs and connect them in the order listed. Do not connect ordered pairs separated by the word "lift"; lift your pencil.
(7, −1), (6, 1), (7, 1), (7, 3), (6, 3), lift
(−5, 3), (−4, 2), (−2, 2), (−1, 3), lift
(−6, 3), (−7, 3), (−7, 1), (−6, 1), (−7, −1), lift
(1, 3), (2, 2), (4, 2), (5, 3), lift
(−3, −2), (−2, −3), (2, −3), (3, −2)

▶ Chapter Project

49. Orion Shipping and Handling Chart The table shows a partial shipping and handling (S & H) chart for Orion Telescopes and Binoculars. The table has some surprising features. Find and explain these surprises. Explore numerically and graphically so that each method supports the results shown by the other. Explain why you chose your "Total Merchandise" inputs in the numeric exploration.

Total Merchandise	S & H Charge
...	
\$300 to \$500	\$22.50
\$500.01 to \$1200	5% of merchandise total
\$1200.01 to \$4000	4% of merchandise total
Over \$4000	3% of merchandise total

Explain how you chose your scale for the horizontal axis and then for the vertical axis of your graph. What strategies for making purchases might you suggest to take advantage of your observations?

▶ 1 Chapter Test

1. List several assumptions that a student might make about a chapter test.

2. List several conditions that a teacher might place on a chapter test.

Fill in the missing word in Exercises 3 to 5.

3. The answer to a subtraction problem is the ________.

4. A listing of numbers in braces, { }, or a collection of objects is a ________.

5. The numbers described by {. . . , −3, −2, −1, 0, 1, 2, 3, . . .} are the ________.

For Exercises 6 and 7, refer to the graph below.

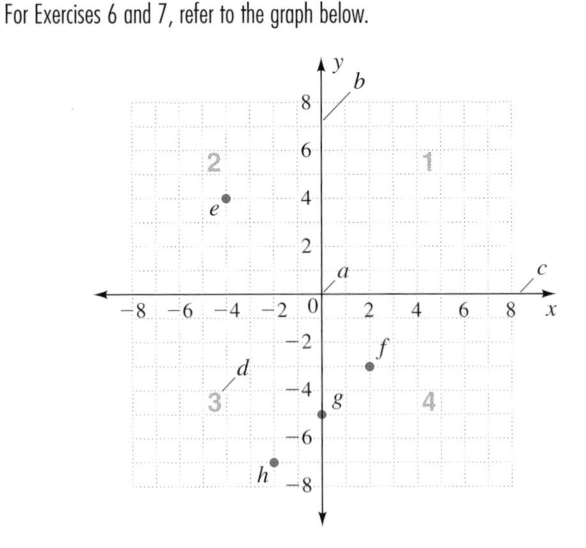

6. What words describe the parts of the graph labeled *a*, *b*, *c*, and *d*?

7. What ordered pairs describe the locations labeled *e*, *f*, *g*, and *h*?

8. Make an input-output table with integer inputs of 0 to 5 for the following rule: Output is twice the input plus 3. Write the rule in symbols. Define your variables. Graph the data.

9. EMBARQ offers "anytime" long distance at 39 cents per call plus 4 cents per minute. Use the four-step process to make a table and graph for the cost of a call from 0 to 60 minutes.

- **a.** Understand: What units go on the axes? What is an equation for the cost of a call?
- **b.** Plan: Explain how you chose quadrants and scale on axes.
- **c.** Sketch the graph: Make a table and draw a graph from the table.
- **d.** Check and extend: AT&T offers long-distance service at 5 cents per minute. For what length of call is EMBARQ the best deal?

10. Make an input-output table with integer inputs of 0 to 8 for this rule: If the input is an even number, the output is half the input; if the input is an odd number, the output is double the input.

11. Find the sum and difference of $\frac{5}{8}$ and $\frac{1}{6}$.

12. Find the product and quotient of $1\frac{1}{9}$ and $1\frac{5}{12}$.

13. Write $\frac{16}{25}$ as a decimal and as a percent.

14. Write 4.2 as a fraction and as a percent.

15. Find 9% of 35.

16. Complete the table and write the rule in words and as an equation.

x	y
1	4
2	8
3	12
4	16
5	
100	

17. Which of rules a, b, c, and d describe the table in Exercise 16?

- **a.** The output is the area of a square with side x.
- **b.** The output is the perimeter of a square with side x.
- **c.** The output is the number of dollars in x quarters.
- **d.** The output is the number of quarters in x dollars.

18. The vertical and horizontal axes have been omitted from the figure below. Use the given point to name the ordered pairs for points *A* and *B*.

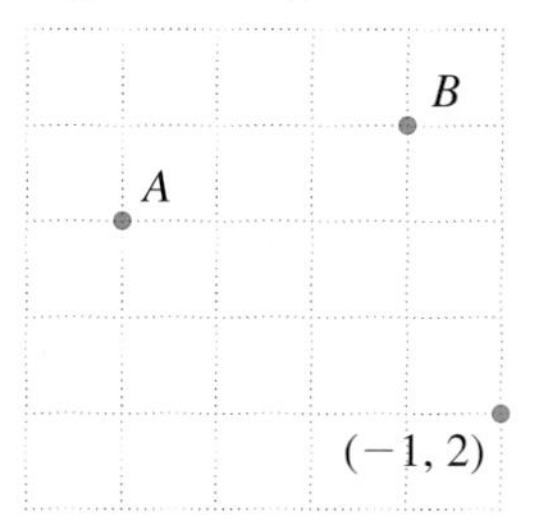

19. Name three ways in which parentheses, ( ), have been used in this chapter.

20. Describe the coincidence of alphabetical order in placing traditional variables on the axes. Include *input*, *output*, *x*, *y*, *horizontal*, and *vertical*.

CHAPTER 2

Operations with Real Numbers and Expressions

A ball is dropped from a height of 144 centimeters. After each time it hits the floor, the ball bounces two-thirds as high as it did on the last bounce (see Figure 1). What is the height of the ball after it hits the floor ten times? We solve this problem in Section 2.4.

In this chapter, we do operations with positive and negative numbers, exponents, and units of measure. We look at how numbers behave (their *properties*). We use rules that tell us which operations to do first in a problem (the *order of operations*). At the end of the chapter, we name sets of numbers with *inequalities* and *intervals*.

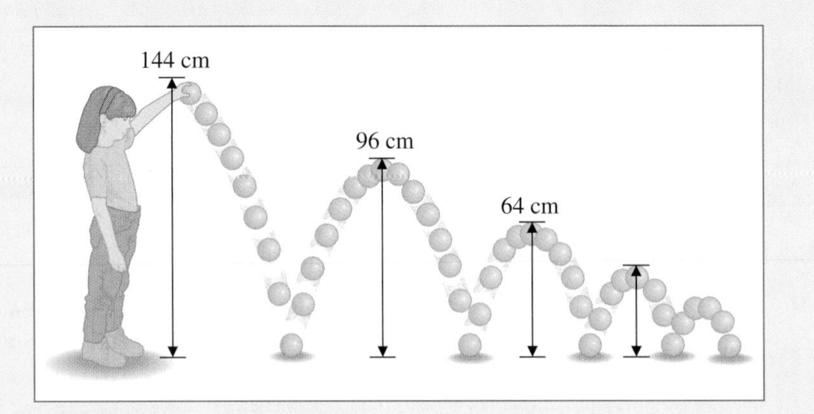

FIGURE 1 Heights of bouncing balls

▶ 2.1 Addition and Subtraction with Integers

Objectives

- Use the electrical charge model to add and subtract integers.
- Find the absolute value of a number.
- Use the absolute value of a number to add integers.
- Find the opposite (called the additive inverse) of a number.
- Change the subtraction $a - b$ to the addition $a + (-b)$.

Student Note:
If you are familiar with integer operations, focus on the deeper concepts: how all four representations (numeric, verbal, symbolic, and visual) are used in this section, how we define addition formally with absolute value, and how there are two ways of interpreting subtraction.

WARM-UP

Give examples of how integers might be used in these applications:

1. Elevations (geography) **2.** Temperature **3.** Electricity
4. Golf **5.** Stock market **6.** Football

IN THIS SECTION, we will add and subtract with integers. We will find rules for these operations that apply to all real numbers, but in this first section will use them only for integers.

▶ Integers and Opposites

INTEGERS In Section 1.2, we defined the *integers* as the set of natural numbers, their opposites, and zero. Figure 2 shows a number line with the integers from -5 to 5. The opposites of the natural numbers are shown as the negative integers.

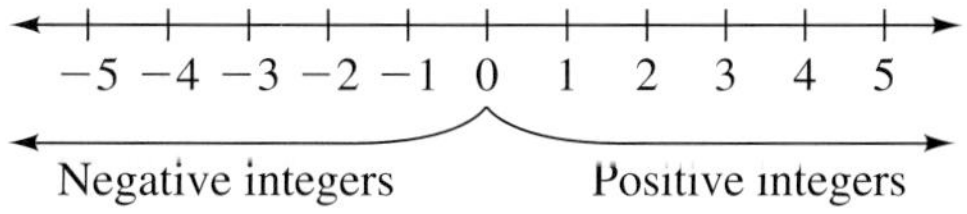

−5 −4 −3 −2 −1 0 1 2 3 4 5
Negative integers Positive integers

FIGURE 2 Integers

Until the 1700s, European mathematicians denied or questioned the existence of negative numbers. The main concern was equating zero with nothing and asking, "How could there be a quantity less than nothing?" Acceptance of negative numbers appeared in India and China in the 7th century. Today, as suggested by the Warm-up settings, as well as loans, credit card debt, bounced checks, and national debt, negative numbers are everywhere.

OPPOSITES Our first use of negative numbers is with opposite. The formal name for opposite is *additive inverse*, where *inverse* means reversed or opposite in position or direction. The additive inverse is the number that, when *added* to its opposite, gives zero. From Section 1.2:

DEFINITION OF OPPOSITES

The **opposite**, or **additive inverse**, of a number is the number that is the same distance from zero on the number line but on the other side of zero.

The opposite of a is written $-a$.
The opposite of -3 is 3.
The opposite of 5 is -5.
A number and its opposite add to zero:

$$a + -a = 0$$
$$-3 + 3 = 0$$
$$5 + (-5) = 0$$

In the box for definition of opposites, parentheses are used in a new way: to group a sign with a number or letter. It is awkward to write $+\ -a$ and $+\ -5$, so instead we write $+\ (-a)$ and $+\ (-5)$.

THINK ABOUT IT 1: Name three ways in which parentheses were used in Chapter 1.

▶ Addition of Integers: Electrical Charge Model

To see how positive and negative integers can be added, we begin with the electrical charge model. The model is a setting where we can visualize the mathematics. We break positive and negative numbers into sets of ones (see Figure 3). A circle with $+1$ means a charge of positive one. A circle with -1 means the opposite, or a charge of negative one.

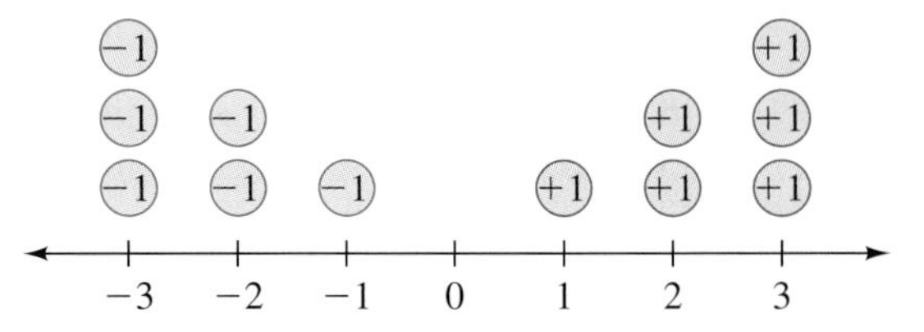

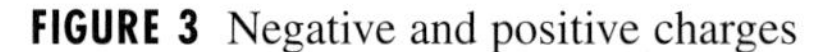

FIGURE 3 Negative and positive charges

ELECTRICAL CHARGES

> One positive charge is opposite to one negative charge. Zero is described as *neutral*, having neither a positive nor a negative charge.
>
> (+1) + (−1) = 0 or $+1 + (-1) = 0$

The box above should be reasonable because a number and its opposite add to zero. In Example 1 the *net charge* in a set is the charge that remains after all positive and negative pairs have been neutralized by adding to zero.

▶ **EXAMPLE 1** Finding charges Find the net charge on each set by circling the charges that add to zero. Match each set in parts a–d with one of the additions in parts e–h.

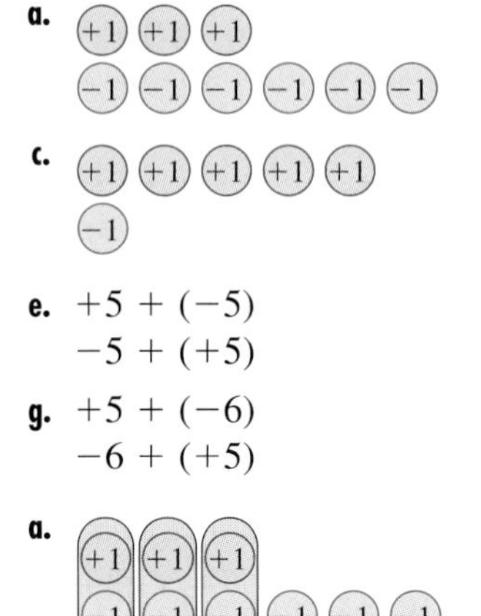

a. (+1) (+1) (+1)
(−1) (−1) (−1) (−1) (−1) (−1)

b. (+1) (+1) (+1) (+1) (+1)
(−1) (−1) (−1) (−1) (−1)

c. (+1) (+1) (+1) (+1) (+1)
(−1)

d. (+1) (+1) (+1) (+1) (+1)
(−1) (−1) (−1) (−1) (−1) (−1)

Student Note:
We place parentheses around integers to separate the addition or subtraction sign from the sign on the number.

e. $+5 + (-5)$
$-5 + (+5)$

f. $+5 + (-1)$
$-1 + (+5)$

g. $+5 + (-6)$
$-6 + (+5)$

h. $-6 + (+3)$
$+3 + (-6)$

SOLUTION **a.** (+1) (+1) (+1)
(−1) (−1) (−1) (−1) (−1) (−1)

Three pairs add to zero.
The net charge is -3.
The additions are in part h.
$-6 + (+3) = -3$ and $+3 + (-6) = -3$

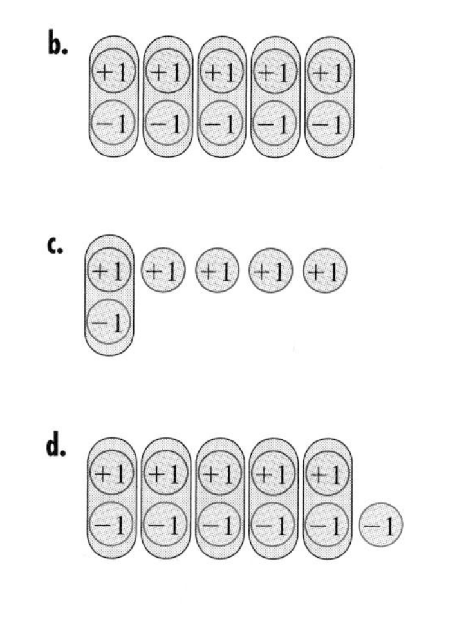

b. Five pairs add to zero.
The net charge is 0.
The additions are in part e.
$+5 + (-5) = 0$ and $-5 + (+5) = 0$

c. One pair adds to zero.
The net charge is +4.
The additions are in part f.
$+5 + (-1) = +4$ and $-1 + (+5) = +4$

d. Five pairs add to zero.
The net charge is −1.
The additions are in part g.
$+5 + (-6) = -1$ and $-6 + (+5) = -1$ ◀

THINK ABOUT IT 2: If there are more negative charges, the net charge is ____. If there are more positive charges, the net charge is ____. If the numbers of negative and positive charges are equal, the net charge is ____.

▶ **EXAMPLE 2** Drawing addition of integers Draw the charge set for each of these additions and find the net charge. Write the other addition fact described.

a. $-8 + (+3)$ **b.** $+4 + (-7)$ **c.** $-2 + (-3)$

SOLUTION

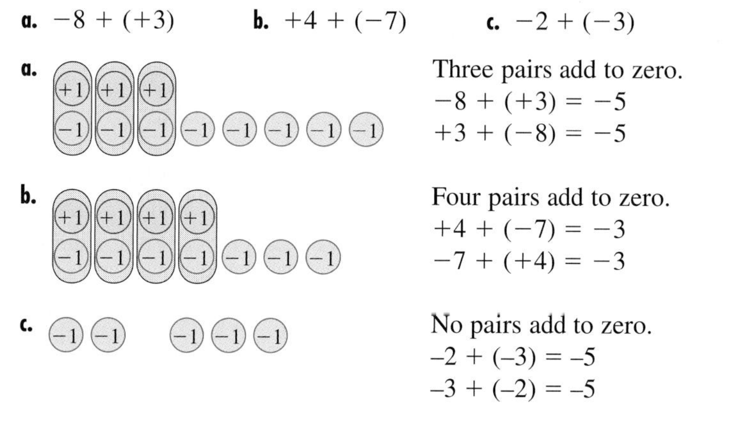

a. Three pairs add to zero.
$-8 + (+3) = -5$
$+3 + (-8) = -5$

b. Four pairs add to zero.
$+4 + (-7) = -3$
$-7 + (+4) = -3$

c. No pairs add to zero.
$-2 + (-3) = -5$
$-3 + (-2) = -5$ ◀

THINK ABOUT IT 3: Find **a.** $-6 + (+9)$, **b.** $2 + (-7)$, and **c.** $-8 + (-3)$

ADDING INTEGERS WITH THE CHARGE MODEL

To add charges:

1. Eliminate the pairs of charges that add to zero.

2. Find the net charge that remains.

This way of thinking about positive and negative numbers is called the *charge model* because it is closely related to work with charged atoms and molecules in chemistry.

▶ Addition of Integers with Absolute Value

ABSOLUTE VALUE The general rule for addition requires absolute value. When you think about which integer has more charges, −8 or +3, you are starting to think in terms of absolute value. The **absolute value** of a number is *the non-negative distance it is from zero. Non-negative means positive or zero.* The symbol for absolute value is two vertical lines placed around a number; for example, the absolute value of negative three is written $|-3|$.

▶ **EXAMPLE 3** Finding absolute value Find the absolute value and show it on a number line.

a. -3 **b.** 4

SOLUTION **a.** The absolute value of -3 is $|-3| = 3$, because -3 is a distance of three units from zero.

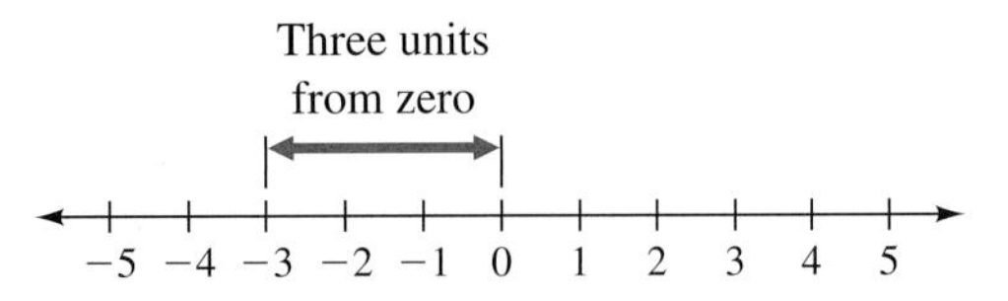

b. The absolute value of $+4$ is $|+4| = 4$, because $+4$ is a distance of four units from zero.

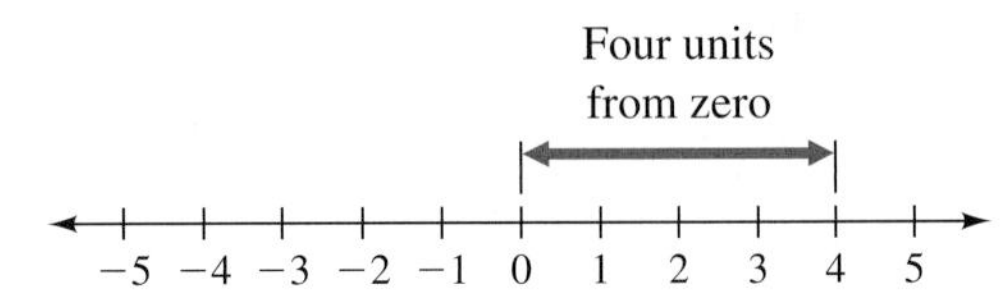

◀

THINK ABOUT IT 4: Which number is further from zero, -3 or 4? Which number has more charges, -3 or 4?

ABSOLUTE VALUE AND ADDITION In our charge model, the absolute value is the number of charges, regardless of sign.

▶ **EXAMPLE 4** Adding with absolute value Which of the numbers in each addition has the greater absolute value and by how much? When the numbers are added, which number controls the sign of the answer? Predict the sign on these sums; then do the additions.

a. $-8 + 2$ **b.** $9 + (-3)$ **c.** $-4 + (-3)$

SOLUTION **a.** In $-8 + 2$, the -8 has the greater absolute value by 6. The sign on the sum is negative.

$$-8 + 2 = -6$$

b. In $9 + (-3)$, the 9 has the greater absolute value by 6. The sign on the sum is positive.

$$9 + (-3) = +6$$

c. In $-4 + (-3)$, the -4 has the greater absolute value. Both of the numbers are negative so there are no pairs that add to zero. The sign on the sum is negative.

$$-4 + (-3) = -7$$

◀

Absolute value lets us write a formal description of adding positive and negative numbers.

ADDITION OF POSITIVE AND NEGATIVE NUMBERS USING ABSOLUTE VALUE

- To add numbers with the same sign, add their absolute values and place the common sign on the answer.
- To add numbers with opposite signs, find the positive (or zero) difference in their absolute values and place the sign from the number with the greater absolute value on the answer.

▶ **EXAMPLE 5** Adding integers Find these sums. Explain your answer in terms of absolute value.

a. $+15 + (-26)$ **b.** $-5 + (+8)$ **c.** $-8 + (-7)$

SOLUTION **a.** The signs are different. We mentally do the subtraction $|-26| - |15| = 11$. We write the sign from -26 on the difference, 11:

$$+15 + (-26) = -11$$

b. The signs are different. We mentally do the subtraction $|8| - |-5| = 3$. We write the sign from $+8$ on the difference, 3:

$$-5 + (+8) = +3$$

(In most cases, if the sign is positive, we drop the sign.)

c. The signs are alike. We mentally add $|-8| + |-7| = 15$ and write the common sign on the sum, 15:

$$-8 + (-7) = -15$$

◀

▶ Subtraction of Integers

SUBTRACTION AS REMOVING OBJECTS The charge model in Figure 4 shows subtraction as removing objects from a set.

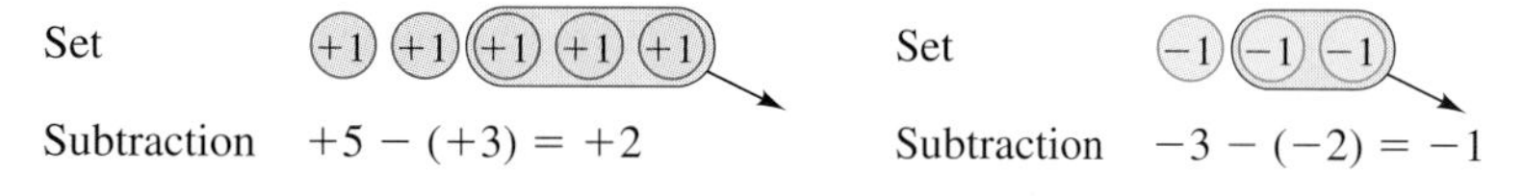

FIGURE 4 Subtraction as removing objects

These subtractions are possible because there are plenty of positive or negative charges available for the subtraction. Now we must consider how to subtract when charges are not available.

▶ **EXAMPLE 6** Exploration of subtraction What might be done when no charges are available to be removed from the set? Test your method with these problems.

a. $+5 - (-3)$ **b.** $-3 - (+2)$

SOLUTION Discuss your method and then check it:

a. $+5 - (-3) = +8$ **b.** $-3 - (+2) = -5$ ◀

Because a pair of opposite charges, one positive and one negative, adds to zero, we can add any number of pairs of opposites to the set of charges without changing its net charge. We use these steps.

SUBTRACTING INTEGERS BY REMOVING OBJECTS

To subtract $x - y$:

1. Start with x charges.
2. Add in pairs of opposite charges until there are y charges available to subtract.
3. Subtract the y charges.
4. Count the net charge.

▶ **EXAMPLE 7** Subtracting with the charge model Draw the charge sets for these subtractions. Subtract by adding in pairs of opposites.

a. $+3 - (+5)$ **b.** $-2 - (-3)$ **c.** $+3 - (-2)$

SOLUTION **a.** For $+3 - (+5)$, the charge set starts with 3 positive charges. We need a total of 5 positive charges for the subtraction, so we add 2 (or more) pairs of opposites to the three positives.

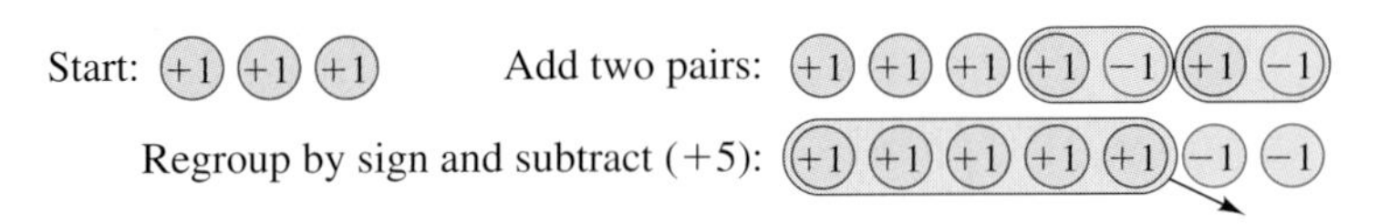

Now we have 5 positive charges and can remove them. The net charge is -2.

$$+3 - (+5) = -2$$

b. For $-2 - (-3)$, the charge set starts with 2 negative charges. We need a total of 3 negative charges for the subtraction, so we add 1 (or more) pair of opposites to the two negatives.

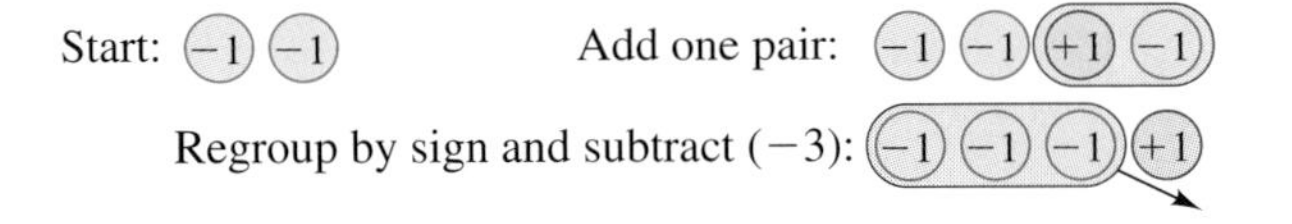

Now we have 3 negative charges and can remove them. The net charge is $+1$.

$$-2 - (-3) = +1$$

c. For $+3 - (-2)$, the charge set starts with 3 positive charges. We need 2 negative charges for the subtraction, so we add 2 (or more) pairs of opposites to the three positives.

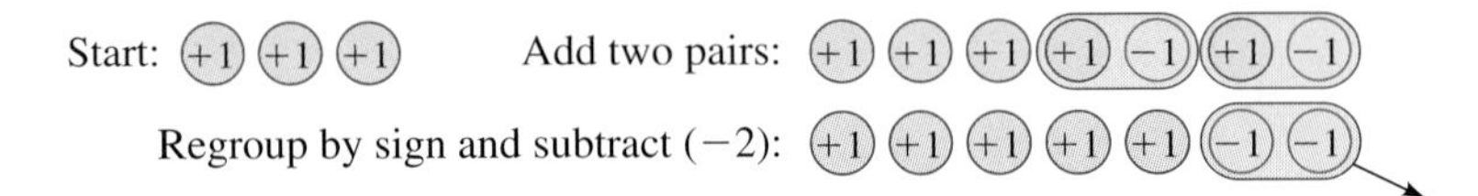

Now we have 2 negative charges and can remove them. The net charge is $+5$.

$$+3 - (-2) = +5$$

THINK ABOUT IT 5: In $-2 - (-3) = +1$ (part b of Example 7), the answer after subtraction is greater than the starting number. In $3 - (-2) = +5$ (part c of Example 7), the answer after subtraction is greater than the starting number. Why? What is $0 - (-2)$? ◀

SUBTRACTION BY ADDITION OF THE OPPOSITE NUMBER There are times when removing objects to do subtraction is not logical, and as a general method, drawing the little charges is inefficient. Return, for a moment, to opposites.

▶ **EXAMPLE 8** Finding a pattern regarding opposites and subtraction

a. What is the opposite of -2? **b.** What is the opposite of (-2)?

c. What is $0 - (-2)$? **d.** What is $3 - (-2)$?

e. What is $4 - (+1)$?

f. In which parts, among parts a to e, are you doing subtractions?

g. In which parts, among parts a to e, are you finding opposites?

SOLUTION **a.** 2 **b.** 2

c. $0 - (-2) = 0 - (-2) - 2 + 2$ Add a pair of opposites, -2 and $+2$.

$0 - (-2) = 0 + 2$ Because $-(-2)$ and -2 are opposites they add to zero.*

d. $3 - (-2) = 3 - (-2) - 2 + 2$ Add a pair of opposites, -2 and $+2$.

$3 - (-2) = 3 + 2$ Because $-(-2)$ and -2 are opposites they add to zero.*

$3 - (-2) = 5$

e. $4 - (+1)$ is just the familiar $4 - 1 = 3$, but we do it as in parts c and d to see a general rule for subtraction.

$4 - (+1) = 4 - (+1) + (+1) + (-1)$ Add a pair of opposites, $+(+1)$ and $+(-1)$.

$4 - (+1) = 4 + (-1)$ Because $-(+1)$ and $+(+1)$ are opposites they add to zero.*

f. c, d, and e **g.** a and b ◀

OPPOSITE OF AN OPPOSITE

The opposite of the opposite of a is a: $-(-a) = a$.
The opposite of the opposite of 5 is 5: $-(-5) = 5$.

We say that a and $-(-a)$ are equivalent expressions. **Equivalent expressions** are *expressions that have the same value for all replacements of the variables.* Numbers can also be equivalent.

We can generalize the results in Example 8. The expression $x - y$ is equivalent to $x + (-y)$, and $x - (-y)$ is equivalent to $x + (+y)$.

SUBTRACTION OF INTEGERS X AND Y

To subtract two numbers, change the subtraction to addition of the opposite number:

$$x - y = x + (-y) \quad \text{or} \quad x - (-y) = x + (+y)$$

Then do the addition.

▶ **EXAMPLE 9** Subtracting by adding the opposite number Change each subtraction to addition of the opposite number, and then add.

a. $6 - 5$ **b.** $2 - 4$ **c.** $-1 - (+2)$

d. $2 - (-2)$ **e.** $2 - (+3)$ **f.** $-5 - (-3)$

SOLUTION The first line in each solution is changing the subtraction to addition of the opposite number.

a. $6 - 5 = 6 + (-5)$

$6 + (-5) = +1$

b. $2 - 4 = 2 + (-4)$

$2 + (-4) = -2$

c. $-1 - (+2) = -1 + (-2)$

$-1 + (-2) = -3$

d. $2 - (-2) = 2 + (+2)$

$2 + (+2) = 4$

e. $2 - (+3) = 2 + (-3)$

$2 + (-3) = -1$

f. $-5 - (-3) = -5 + (+3)$

$-5 + (+3) = -2$ ◀

*We made two assumptions at this step: We temporarily ignored the first number and interpreted "subtract negative two" as "the opposite of negative two."

THINK ABOUT IT 6: Complete these sentences with the word *increase* or *decrease*:

a. Subtracting a positive results in a(n) ________ in the net charge.
b. Subtracting a negative results in a(n) ________ in the net charge.

▶ Application: Elevations

We use the word *elevation* when we are looking at vertical distance. We describe a *difference* in elevation with a positive number. A *change* in elevation, as when a plane is landing, can be negative. *To find the difference in elevation, we subtract the lesser height from the greater height.*

▶ **EXAMPLE 10** Finding differences in elevation Use the data in Table 1 to find the difference in elevation between the two places given.

a. Mt. McKinley in Alaska and Pike's Peak in Colorado
b. Mt. Everest in the Himalayas and the Mariana Trench in the Pacific Ocean (south of Japan and east of the Philippines)
c. Death Valley in California and the Dead Sea between Israel and Jordan

TABLE 1 Selected Elevations

Location	Elevation Relative to Sea Level (feet)	
Mt. Everest	+29,028	↑ Greater heights
Mt. McKinley	+20,320	
Pike's Peak	+14,110	
Mauna Kea	+13,710	
Sea Level	0	
Death Valley	−282	↓ Lesser heights
Dead Sea	−1312	
Ocean Floor, near Hawaii	−16,400	
Mariana Trench	−35,840	

SOLUTION **a.** The elevation of Mt. McKinley minus the elevation of Pike's Peak is

$$20{,}320 - 14{,}110 = 6210 \text{ ft}$$

b. The elevation of Mt. Everest minus the elevation of the Mariana Trench is

$$29{,}028 - (-35{,}840) = 29{,}028 + (+35{,}840) = 64{,}868 \text{ ft}$$

c. The elevation of Death Valley minus the elevation of the Dead Sea is

$$-282 - (-1312) = -282 + (+1312) = 1030 \text{ ft}$$ ◀

Calculators are convenient when we are working with large or messy numbers.

Check your solutions to Example 2 and Example 7 with a calculator. To enter a negative number into the calculator, enter (−) preceding the number. Do not use the subtraction operation for a negative sign. No sign is needed for a positive number. Compare your results with those in Figure 5 and Figure 6.

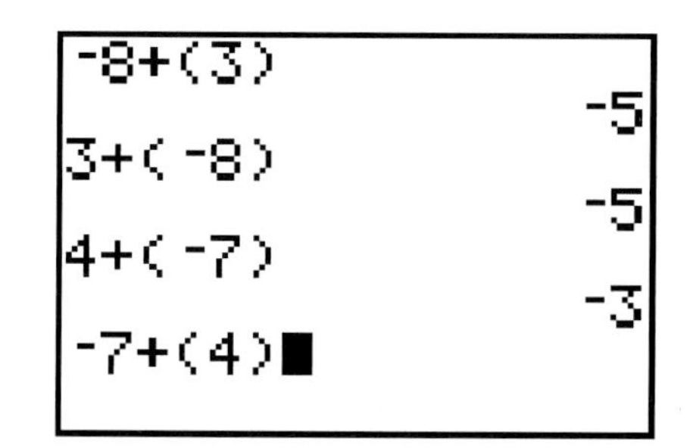

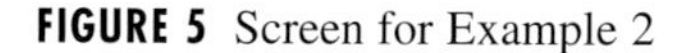

FIGURE 5 Screen for Example 2

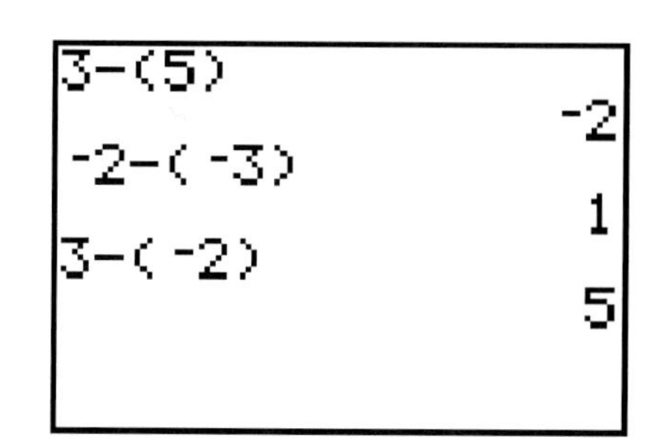

FIGURE 6 Screen for Example 7

ANSWER BOX

Warm-up: Sample answers include: 1. Elevation at sea level is zero. Mountains are above sea level; under the ocean and places such as Death Valley are below sea level. 2. The Celsius temperature scale has water freezing at 0 degrees, temperatures in cold winters may drop well below zero, water boils at 100 degrees Celsius. 3. In electricity, a neutral charge is zero, a proton has a positive 1-unit charge, an electron has a negative 1-unit charge. 4. In golf, par is zero, a birdie is -1, or one stroke below par, and a bogey is $+1$, or one stroke above par. In golf, the positive scores are less desirable. 5. In the stock market, end-of-day prices are compared with the prior day's end price (as a zero reference) with gains being positive and losses being negative. 6. In football, the gain (positive) or loss (negative) is referenced to the ball's ending position on the prior play. **Think about it 1:** Parentheses may give extra information in a sentence or mathematical statement, may show a multiplication, or may indicate ordered pairs in graphing. **Think about it 2:** negative; positive; zero **Think about it 3:** **a.** $+3$, **b.** -5, **c.** -11 **Think about it 4:** 4 is further from zero and has more charges. **Think about it 5:** Removing negative values made the quantity more positive. $0 - (-2) = 2$ **Think about it 6:** **a.** decrease **b.** increase

▶ 2.1 Exercises

What two addition facts describe each charge set in Exercises 1 to 6? What is the net charge?

1. (+1) (+1) (+1) (+1) (−1) (−1) (−1)

2. (+1) (+1) (+1) (−1) (−1) (−1)

3. (+1) (+1) (+1) (+1) (+1) (−1) (−1) (−1) (−1) (−1)

4. (+1) (+1) (+1) (+1) (+1) (+1) (+1) (+1) (−1) (−1) (−1) (−1)

5. (+1) (+1) (+1) (+1) (+1) (−1) (−1) (−1) (−1) (−1) (−1) (−1)

6. (+1) (+1) (+1) (−1) (−1) (−1) (−1) (−1) (−1) (−1) (−1)

In Exercises 7 to 14, find the sum and then check with a calculator.

7. **a.** $-8 + 3$ **b.** $4 + (-7)$ **c.** $+4 + (-4)$

8. **a.** $-5 + (-2)$ **b.** $+3 + (-3)$ **c.** $5 + (-12)$

9. **a.** $-3 + (+3)$ **b.** $-4 + (-7)$ **c.** $-12 + (+8)$

10. **a.** $5 + (+8)$ **b.** $-9 + (+9)$ **c.** $15 + (-8)$

11. **a.** $-3 + (+2)$ **b.** $+6 + (-4)$ **c.** $-14 + (-6)$

12. **a.** $+3 + (-5)$ **b.** $-4 + (-3)$ **c.** $-13 + (+6)$

13. **a.** $-8 + (-17)$ **b.** $-9 + (+16)$ **c.** $24 + (-8)$

14. **a.** $-7 + (+23)$ **b.** $12 + (-19)$ **c.** $-22 + (-9)$

Blue numbers are core exercises.

In Exercises 15 and 16, give the opposite of each number or expression.

15. **a.** 5 **b.** $-\frac{1}{2}$ **c.** 0.4 **d.** x **e.** $-2x$

16. **a.** -5 **b.** $\frac{2}{3}$ **c.** 2.5 **d.** $3x$ **e.** $-ab$

In Exercises 17 to 22, calculate the expression.

17. **a.** $|4|$ **b.** $|-6|$ **c.** $-(-5)$ **d.** $-(-2)$

18. **a.** $|-5|$ **b.** $-(-4)$ **c.** $-|-3|$ **d.** $|7|$

19. **a.** $-|7|$ **b.** $-|-8|$ **c.** $-(-3)$ **d.** $|-7|$

20. **a.** $-(-6)$ **b.** $-|4|$ **c.** $|-(-2)|$ **d.** $-|-5|$

21. **a.** $|-4|$ **b.** $|5|$ **c.** $|4 - 9|$ **d.** $-|2 + 5|$

22. **a.** $|7|$ **b.** $|-6|$ **c.** $|3 - 9|$ **d.** $-|3 + 4|$

23. What is $-(-x)$ and $-|x|$ if

a. $x = +4$ **b.** $x = -4$

24. What is $|x|$ and $-x$ if

a. $x = -3$ **b.** $x = 3$

25. Write in words the expression $-|x|$.

26. Write in words the expression $-(-x)$.

27. Explain with numeric examples why $x - (-y) = x + (+y)$ is not needed in the box for Subtraction of Integers x and y, page 61.

28. Group the expressions into two sets of equivalent expressions. $+5$, $-(+5)$, $+(-5)$, $-(-5)$, -5, 5, $+(+5)$

In Exercises 29 to 32, copy the charge figures. Circle the pairs of opposites. Finish the subtraction.

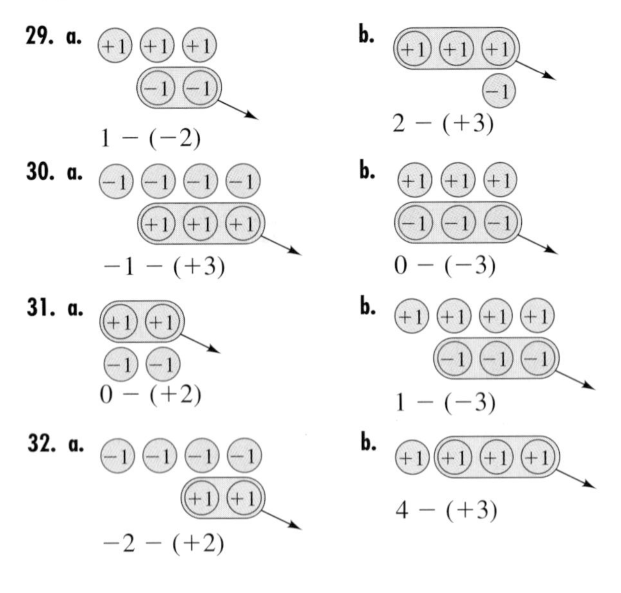

In Exercises 33 to 44, write each subtraction as addition of the opposite number. Add. Check with a calculator.

33. **a.** $8 - 11$ **b.** $-8 - (-11)$ **c.** $-16 - 3$

34. **a.** $0 - 5$ **b.** $-17 - 4$ **c.** $-4 - (-17)$

35. **a.** $12 - 7$ **b.** $-12 - 7$ **c.** $-17 - (-4)$

36. **a.** $17 - 4$ **b.** $16 - (-3)$ **c.** $-12 - 9$

37. **a.** $-4 - (-4)$ **b.** $14 - (10)$ **c.** $0 - (-5)$

38. **a.** $8 - 9$ **b.** $-4 - 5$ **c.** $26 - 9$

39. **a.** $10 - 14$ **b.** $-8 - (-9)$ **c.** $-7 - 6$

40. **a.** $-3 - (-3)$ **b.** $4 - (-5)$ **c.** $19 - (-8)$

41. **a.** $2 - 7$ **b.** $2 - 5$ **c.** $2 - (-5)$

42. **a.** $-2 - (-5)$ **b.** $-7 - (-2)$ **c.** $-5 - 17$

43. **a.** $-7 - 5$ **b.** $-5 - 7$ **c.** $-5 - (-7)$

44. **a.** $-7 - (-5)$ **b.** $5 - 7$ **c.** $-7 - 16$

45. Find the difference in elevation between the given highest and lowest points.

a. South America: Mt. Aconcagua, 6960 meters, and Valdes Peninsula, −40 meters

b. Africa: Kilimanjaro, 5896 meters, and Lake Assal, −156 meters

c. Australia: Mt. Kosciusko, 2228 meters, and Lake Eyre, −16 meters

d. The Mariana Trench (Pacific Ocean) is 10,930 meters below sea level. The Puerto Rico Trench (Atlantic Ocean) is 8600 meters below sea level. Write an appropriate subtraction, and determine how much deeper the Mariana Trench is than the Puerto Rico Trench.

46. Checkers at grocery stores must balance their cash drawers each shift. An error in either direction, cash-over (too much money) or cash-under (too little), is recorded. Write an expression using absolute value to describe the total error due to cash-over or cash-under.

a. Monday, \$15 under; Tuesday, \$10 under; Wednesday, balance; Thursday, \$5 under; Friday, \$10 over

b. Monday, \$5 under; Tuesday, \$5 over; Wednesday, \$15 over; Thursday, \$5 under; Friday, balance

c. Monday, \$5 over; Tuesday, \$12 over; Wednesday, balance; Thursday, \$7 under; Friday, \$10 under

d. Monday, balance; Tuesday, \$13 over; Wednesday, \$20 under; Thursday, \$5 under; Friday, \$8 over

Blue numbers are core exercises.

47. True or false: Opposite numbers have the same absolute value.

48. True or false: $-|x| = |-x|$ for all x.

▶ Writing

49. Explain what absolute value means.

50. Explain how to add with positive and negative numbers.

51. Explain how to subtract with positive and negative numbers.

52. Explain how the charge model shows that opposites add to zero.

53. Explain how the charge model shows that subtracting a negative increases the result.

54. Name and explain the uses of the symbols { }, ( ), and | | and the word *non-negative*.

55. Connecting with Chapter 1 In a particularly active six days of trading in 2006, Google stock closed at $440, $400, $425, $450, $435, and $435 (rounded to the nearest $5). Follow steps a to e in Exercise 56, but for the 6 days of data given here instead of 15 days.

▶ Project

56. Stock Prices Using the Internet or your library's collection of back issues of newspapers, obtain three weeks' closing prices for one share of a common stock of your choice.

a. Make a table with four headings: date, closing price, change, percent change.

b. Record the date and the closing price of the stock for 15 consecutive trading days.

c. In the third and fourth columns, write the change in closing price since the preceding day. Divide the change by the preceding day's price to get percent change.

d. Make a graph of the dates and closing prices.

e. Comment, in complete sentences, on any trend you observe.

Blue numbers are core exercises.

▶ 2.2 Multiplication and Division with Positive and Negative Numbers

Objectives

- Use the website sales model to multiply integers.
- Find the reciprocal (called the multiplicative inverse) of a real number.
- Multiply and divide rational numbers.

WARM-UP

What are the missing numbers? Guess and check. Some answers may be fractions.

1. $-3 + \square = 0$ **2.** $\square + 5 = 1$ **3.** $\square + (-2) = 0$

4. $4 + \square = 1$ **5.** $\square \cdot 3 = 1$ **6.** $6 + \square = 0$

7. $5 \cdot \square = 1$ **8.** $\frac{3}{4} \cdot \square = 0$ **9.** $\frac{1}{2} \cdot \square = 1$

10. Which of the above exercises contain opposites (additive inverses)?

11. Which of the above exercises contain reciprocals (multiplicative inverses)?

IN THIS SECTION, we will multiply and divide with integers and rational numbers. We will rewrite division by fractions as multiplication by reciprocals.

Multiplication of Integers

WEB SALES MODEL Thinking in terms of a website sales model gives us a way of figuring out what sign is needed when we multiply positive and negative integers.*

Suppose you open a sales site on the Internet. All sales and all bills are delivered by your local server. A delivery is recorded as a positive number. Sales are written as positive numbers, and bills are written as negative numbers.

The server delivers three sales for \$20 each. You write $+3(+20)$. Your net worth is increased, so $+3(+20) = +60$.

The server delivers two bills for \$10 each. You write $+2(-10)$. Your net worth is decreased, so $+2(-10) = -20$.

The name of your site is close to that of another site. You sometimes receive sales and bills for the other site. When the server takes away items delivered by mistake because of a wrong address, the removal is recorded as a negative number.

The server takes away four sales for \$5 each. You write $-4(+5)$. Your net worth is decreased, so $-4(+5) = -20$.

EXAMPLE 1 Exploration: investigating the web sales model Write each transaction in positive or negative numbers and find how the net worth changes.

a. The server brings two sales for \$35 each.
b. The server brings three bills for \$50 each.
c. The server takes away four sales for \$30 each.
d. The server takes away five bills for \$8 each.

SOLUTION See the Answer Box. ◀

When the server takes away bills, you do not need to pay them, and so your net worth is increased.

EXAMPLE 2 Finding change in net worth Use the web sales model to write each of these transactions as a multiplication of positive and negative numbers. Find the change in net worth.

a. The server brings three sales for \$50 each.
b. The server takes away four sales for \$10 each.
c. The server takes away four bills for \$25 each.
d. The server brings two bills for \$75 each.

SOLUTION The dollar signs are omitted in the computation step.

a. $+3(+50) = +150$; net worth increases \$150.

b. $-4(+10) = -40$; net worth decreases \$40.

c. $-4(-25) = +100$; net worth increases \$100.

d. $+2(-75) = -150$; net worth decreases \$150. ◀

THINK ABOUT IT: Complete these sentences based on the example.

Numbers with like signs:
a. A positive number multiplied by a positive number is ______.
b. A negative number multiplied by a negative number is ______.

Numbers with unlike signs:
c. A positive number multiplied by a negative number is ______.
d. A negative number multiplied by a positive number is ______.

---

*Thanks to Dr. Judith H. Hector, Walters State Community College, Morristown, Tennessee, for her delivery person model that led to this model.

PATTERNS IN THE PRODUCTS OF INTEGERS The web sales model suggests that we can multiply positive and negative numbers and have results that make sense. In Examples 3 and 4, we look at number patterns to confirm the rules.

▶ **EXAMPLE 3** Multiplying numbers with like and unlike signs Complete the input-output table in Table 2. What happens to the output as the input, x, gets smaller? What might we conclude about the signs from multiplying the two numbers?

TABLE 2

Input x	Output $y = +3x$
$+2$	$+3(2) = 6$
$+1$	$+3(1) = 3$
0	
-1	
-2	
-3	

First rows: We know these facts.

Remaining rows: Find a pattern to suggest these answers.

SOLUTION **TABLE 3**

Input x	Output $y = +3x$
$+2$	$+3(2) = 6$
$+1$	$+3(1) = 3$
0	$+3(0) = 0$
-1	$+3(-1) = -3$
-2	$+3(-2) = -6$
-3	$+3(-3) = -9$

In the first three rows, as the input decreases by 1, the output decreases by 3.

To continue the pattern, these three outputs must be -3, -6, and -9.

In Table 3, $+3(2)$ and $+3(1)$ have like signs, their products are positive. The numbers $+3(-1)$, $+3(-2)$, and $+3(-3)$ have unlike signs, their products are negative. ◀

▶ In Example 4, we make a table and look for a pattern, as we did in Example 3.

▶ **EXAMPLE 4** Multiplying numbers with like and unlike signs Use the first three rows of Table 4 to find a pattern in the outputs. Then use the pattern to complete the table. What does the pattern say about the signs from multiplying the two numbers?

TABLE 4

Input x	Output $y = -3x$
$+2$	$-3(2) =$ ____
$+1$	$-3(1) =$ ____
0	$-3(0) =$ ____
-1	$-3(-1) =$ ____
-2	$-3(-2) =$ ____
-3	$-3(-3) =$ ____

First three rows: Refer to Example 3 here.

Last three rows: Find a pattern to suggest these answers.

SOLUTION **TABLE 5**

Input x	Output $y = -3x$
$+2$	$-3(2) = -6$
$+1$	$-3(1) = -3$
0	$-3(0) = 0$
-1	$-3(-1) = 3$
-2	$-3(-2) = 6$
-3	$-3(-3) = 9$

In the first three rows, as the input decreases by 1, the output increases by 3.

To continue the pattern, these three outputs must be 3, 6, and 9.

As in Example 3, if two numbers have like signs, their product is positive; if two numbers have unlike signs, their product is negative. ◀

The multiplication rules are summarized in these four statements.

MULTIPLICATION

If two numbers have like signs, their product is positive:

$$8(9) = 72, \qquad -6(-7) = 42$$

If two numbers have unlike signs, their product is negative:

$$8(-7) = -56, \qquad -9(6) = -54$$

The product of zero and any real number n is zero:

$$0 \cdot n = 0 \qquad 0 \cdot 4 = 0$$

The product of 1 and any real number n is n:

$$1 \cdot n = n \qquad 1 \cdot 6 = 6$$

▶ **EXAMPLE 5** Multiplying integers Multiply the integers.

a. $3(-4)$ **b.** $-3(-4)$ **c.** $3(4)$ **d.** $-3(4)$

e. $3(0)$ **f.** $0(4)$ **g.** $3(1)$ **h.** $1(4)$

SOLUTION **a.** -12 **b.** 12 **c.** 12 **d.** -12

e. 0 **f.** 0 **g.** 3 **h.** 4 ◀

▶ Multiplication with Rational Numbers

The rules for signs on rational numbers are the same as those for signs on integers.

▶ **EXAMPLE 6** Multiplying rational numbers Find the products.

a. $-7\left(\frac{3}{8}\right)$ **b.** $\frac{1}{4}(-8)$ **c.** $-\frac{1}{7}(28)$ **d.** $-9\left(-\frac{1}{3}\right)$

e. $-\frac{3}{4}\left(-\frac{4}{5}\right)$ **f.** $-\frac{8}{3}\left(-\frac{9}{4}\right)$ **g.** $\frac{8}{3}(0)$ **h.** $-4\left(\frac{1}{8}\right)$

SOLUTION Place integers over 1 for ease of multiplication with fractions.

a. $-\dfrac{7}{1}\left(\dfrac{3}{8}\right) = -\dfrac{21}{8}$ **b.** $\dfrac{1}{4}\left(-\dfrac{8}{1}\right) = -\dfrac{8}{4} = -2$*

*This and other solutions contain **continued equalities**, *three or more equivalent expressions separated by two or more equals signs*. By now you should understand equality. Without continued equalities books are longer and heavier!

c. $-\frac{1}{7}\left(\frac{28}{1}\right) = -\frac{28}{7} = -4$

d. $-\frac{9}{1}\left(-\frac{1}{3}\right) = +\frac{9}{3} = 3$

e. $-\frac{3}{4}\left(-\frac{4}{5}\right) = +\frac{3\cdot 4}{4\cdot 5} = \frac{3}{5}$

f. $-\frac{8}{3}\left(-\frac{9}{4}\right) = +\frac{2\cdot 4\cdot 3\cdot 3}{3\cdot 4} = 6$

g. $\frac{8}{3}(0) = 0$

h. $-\frac{4}{1}\left(\frac{1}{8}\right) = -\frac{4}{8} = -\frac{1}{2}$ ◀

▶ Division with Rational Numbers

RECIPROCAL OR MULTIPLICATIVE INVERSE We used the reciprocal to divide fractions in Section 1.2.

DEFINITION OF RECIPROCALS

> The **reciprocal**, or **multiplicative inverse**, of a number n is *the number that, when multiplied by n, gives* 1. Thus, for all real numbers except zero,
>
> $$n \cdot \frac{1}{n} = 1 \qquad \frac{3}{4} \cdot \frac{4}{3} = 1$$

Because the product of a number and its reciprocal is positive one, the number and its reciprocal have the same sign; either both are positive or both are negative.

▶ **EXAMPLE 7** Finding a reciprocal Find the reciprocal of each number. Check by multiplying the number and its reciprocal.

a. 8 **b.** -6 **c.** $-\frac{2}{3}$ **d.** $1\frac{1}{4}$ **e.** 0.75

SOLUTION

a. The reciprocal of 8 is $\frac{1}{8}$.
Check: $\frac{1}{8} \cdot 8 = \frac{8}{8} = 1$ ✓

b. The reciprocal of -6 is $-\frac{1}{6}$.
Check: $\left(-\frac{1}{6}\right) \cdot -6 = +\frac{6}{6} = 1$ ✓

c. The reciprocal of $-\frac{2}{3}$ is $-\frac{3}{2}$, or $-1\frac{1}{2}$.
Check: $\left(-\frac{3}{2}\right) \cdot -\frac{2}{3} = +\frac{6}{6} = 1$ ✓

d. Because $1\frac{1}{4} = \frac{5}{4}$, its reciprocal is $\frac{4}{5}$.
Check: $\frac{4}{5} \cdot \frac{5}{4} = \frac{20}{20} = 1$ ✓

e. Because $0.75 = \frac{3}{4}$, its reciprocal is $\frac{4}{3}$, or $1\frac{1}{3}$.
Check: $\frac{4}{3} \cdot \frac{3}{4} = \frac{12}{12} = 1$ ✓ ◀

SIGNS FOR DIVISION Because the sign does not change when we find the reciprocal of a number, the rules for multiplication of positive and negative numbers apply to division of integers and of all real numbers.

SIGNS FOR MULTIPLICATION AND DIVISION

> In multiplication and division of two real numbers:
>
> - If the signs are alike, the answer is positive:
>
> $$-5(-6) = +30, \qquad \frac{21}{3} = 7, \qquad \frac{-14}{-2} = +7$$
>
> - If the signs are different, the answer is negative:
>
> $$5(-6) = -30, \qquad \frac{28}{-4} = -7, \qquad \frac{-35}{7} = -5$$

CHANGING DIVISION TO MULTIPLICATION In Section 2.1, Example 9, page 61, we subtracted integers by adding the opposite: $a - b = a + (-b)$. That is, we used the idea of inverses (opposites) to change subtraction to addition. Similarly, we use the idea of inverses (reciprocals) to change division to multiplication. In fraction notation, we divide by multiplying by the reciprocal:

$$\frac{a}{b} \div \frac{c}{d} = \frac{a}{b} \cdot \frac{d}{c}$$

▶ EXAMPLE 8 Dividing with integers and rational numbers Divide by multiplying by the reciprocal.

a. $+24 \div \left(+\frac{1}{3}\right)$ **b.** $+63 \div (-9)$ **c.** $-\frac{1}{4} \div (-8)$

d. $-36 \div 1\frac{1}{2}$ **e.** $-45 \div \left(-\frac{9}{5}\right)$ **f.** $+\frac{3}{4} \div \left(-\frac{3}{4}\right)$

Student Note:
Writing extra signs helps you pay attention to like and unlike signs.

SOLUTION Place integers over 1 for ease of multiplication.

a. $+24 \div \left(+\frac{1}{3}\right) = +\frac{24}{1} \cdot \left(+\frac{3}{1}\right)$ Reciprocal

$= \frac{72}{1} = 72$ Like signs; quotient is positive.

b. $+63 \div (-9) = +\frac{63}{1} \cdot \left(-\frac{1}{9}\right)$ Reciprocal

$= -\frac{63}{9} = -7$ Unlike signs; quotient is negative.

c. $-\frac{1}{4} \div (-8) = -\frac{1}{4} \cdot \left(-\frac{1}{8}\right)$ Reciprocal

$= \frac{1}{32}$ Like signs; quotient is positive.

d. $1\frac{1}{2} = \frac{3}{2}$ Change to improper fractions.

$-36 \div \left(+\frac{3}{2}\right) = -\frac{36}{1} \cdot \left(+\frac{2}{3}\right)$ Reciprocal

$= -\frac{72}{3} = -24$ Unlike signs; quotient is negative.

e. $-45 \div \left(-\frac{9}{5}\right) = -\frac{45}{1} \cdot \left(-\frac{5}{9}\right)$ Reciprocal

$= 25$ Like signs; quotient is positive.

f. $+\frac{3}{4} \div \left(-\frac{3}{4}\right) = +\frac{3}{4} \cdot \left(-\frac{4}{3}\right)$ Reciprocal

$= -1$ Unlike signs; quotient is negative. ◀

SIGNS ON FRACTIONS The placement of signs on fractions can be confusing. In part b of Example 8, the reciprocal of -9 was written $-\frac{1}{9}$. It might have been more natural to write $\frac{1}{-9}$ or $\frac{-1}{9}$. The three fractions are equivalent expressions; they have the same value.

SIGNS ON FRACTIONS

For all real numbers, a and b, b not zero,

$$-\frac{a}{b} = \frac{-a}{b} = \frac{a}{-b}$$

In general, we place the negative sign in front of the fraction in writing an answer. However, it is also correct—and more convenient when doing operations such as multiplication—to place the negative sign in the numerator. We rarely, if ever, leave the one negative sign in the denominator.

▶ EXAMPLE 9 Placing signs on fractions Write each as a fraction in lowest terms.

a. $9 \div (-45)$ **b.** $-9 \div 45$

SOLUTION **a.** $9 \div (-45) = \frac{9}{-45} = -\frac{1}{5}$ **b.** $-9 \div 45 = \frac{-9}{45} = -\frac{1}{5}$ ◀

Practice doing the multiplications in Example 6 with a calculator. Parts e and f are shown in Figure 7.

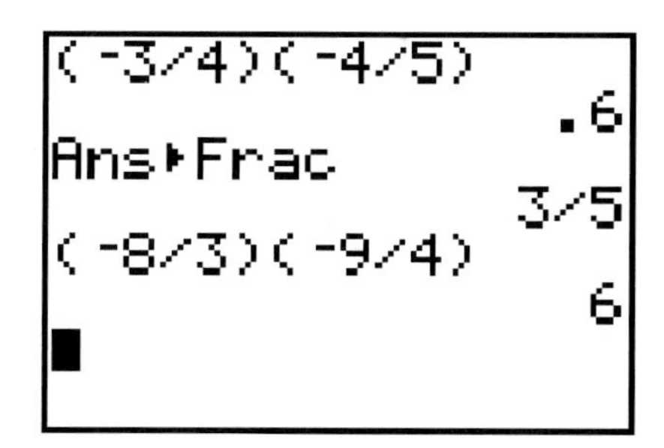

FIGURE 7 From Example 6

Obtain prior answer with 2nd ANS. Obtain fraction notation with MATH 1:Frac ENTER.

8-1▸Frac
1/8
Ans*8
1
(-6)-1▸Frac
-1/6
Ans*-6

FIGURE 8 From Example 7

The calculator reciprocal key is x^{-1}. Practice with this key by entering each number in Example 7, finding its reciprocal, and checking with a multiplication. The results for parts a and b of Example 7 are shown in Figure 8.

ANSWER BOX

Warm-up: 1. 3 **2.** −4 **3.** 2 **4.** −3 **5.** $\frac{1}{3}$ **6.** −6 **7.** $\frac{1}{5}$ **8.** 0 **9.** 2 **10.** 1, 3, 6 **11.** 5, 7, 9 **Example 1: a.** $+2(+35) = +70$ **b.** $+3(-50) = -150$ **c.** $-4(+30) = -120$ **d.** $-5(-8) = +40$ **Think about it: a.** positive **b.** positive **c.** negative **d.** negative

▶ 2.2 Exercises

In Exercises 1 to 8, write an expression and find the change in net worth for your website business.

1. The server takes away two sales for \$150 each.

2. The server brings four bills for \$20 each.

3. The server brings two sales for \$400 each.

4. The server takes away three bills for \$70 each.

5. The server takes away eight bills for \$40 each.

6. The server takes away two sales for \$300 each.

7. The server brings three bills for \$90 each.

8. The server brings four sales for \$125 each.

Complete the tables in Exercises 9 to 12. Then graph the ordered pairs.

9.

Input x	Output $5x$
2	
1	
0	
−1	
−2	

10.

Input x	Output $4x$
2	
1	
0	
−1	
−2	

11.

Input x	Output $-2x$
2	
1	
0	
−1	
−2	

12.

Input x	Output $-5x$
2	
1	
0	
−1	
−2	

Find the products in Exercises 13 to 22. Do not use a calculator. Find the sign first; then find the number.

13. a. $-7(-6)$ **b.** $6(-8)$ **c.** $3(-15)$ **d.** $-5(-7)$

14. a. $4(-7)$ **b.** $-8(-4)$ **c.** $-6(9)$ **d.** $-7(9)$

15. a. $12(-4)$ **b.** $8(-8)$ **c.** $15(4)$ **d.** $-9(-9)$

16. a. $5(-12)$ **b.** $-7(7)$ **c.** $-5(-15)$ **d.** $-11(11)$

17. a. $-3\left(\frac{1}{3}\right)$ **b.** $-3\left(-\frac{1}{3}\right)$ **c.** $2\left(-\frac{1}{2}\right)$ **d.** $-\frac{1}{4}(-4)$

18. a. $-2\left(-\frac{1}{2}\right)$ **b.** $3\left(-\frac{1}{3}\right)$ **c.** $-4\left(\frac{1}{4}\right)$ **d.** $\frac{1}{2}(-2)$

19. a. $\frac{3}{4}(-8)$ **b.** $-\frac{1}{5}(5)$ **c.** $-4\left(-\frac{3}{4}\right)$

Blue numbers are core exercises.

20. a. $-\frac{4}{5}(4)$ **b.** $-\frac{9}{2}\cdot\left(-\frac{8}{5}\right)$ **c.** $-4\left(\frac{2}{3}\right)$

21. a. $-\frac{5}{3}\left(-\frac{6}{7}\right)$ **b.** $-\frac{1}{6}(9)$ **c.** $-\frac{2}{3}(0)$

22. a. $-\frac{5}{8}(-8)$ **b.** $-2(0)$ **c.** $5\left(-\frac{2}{9}\right)$

In Exercises 23 to 26, write the reciprocal of each number or expression. Assume the variables do not equal zero.

23. a. 4 **b.** -2 **c.** $\frac{1}{2}$ **d.** $-\frac{3}{4}$ **e.** 0.5

24. a. -3 **b.** 6 **c.** $-\frac{1}{3}$ **d.** $\frac{2}{3}$ **e.** 0.25

25. a. $3\frac{1}{3}$ **b.** 6.5 **c.** x **d.** $\frac{a}{b}$ **e.** $-x$

26. a. $2\frac{3}{4}$ **b.** 8.2 **c.** n **d.** $\frac{x}{y}$ **e.** $-y$

In Exercises 27 and 28, divide using multiplication by the reciprocal.

27. a. $15 \div (-5)$ **b.** $-45 \div 9$ **c.** $42 \div (-6)$

28. a. $-18 \div 6$ **b.** $-18 \div (-9)$ **c.** $-36 \div 2$

In Exercises 29 to 36, divide and then check with a calculator. Find the sign first; then find the number.

29. a. $\frac{56}{-8}$ **b.** $\frac{-56}{-8}$ **c.** $-\frac{56}{8}$

30. a. $\frac{49}{-7}$ **b.** $\frac{-72}{-8}$ **c.** $\frac{-72}{-6}$

31. a. $\frac{-55}{-11}$ **b.** $\frac{-28}{4}$ **c.** $\frac{-27}{9}$

32. a. $\frac{36}{-4}$ **b.** $\frac{-36}{12}$ **c.** $\frac{-48}{-24}$

33. a. $24 \div \frac{2}{3}$ **b.** $24 \div \left(-\frac{3}{2}\right)$ **c.** $-16 \div \frac{3}{8}$

34. a. $15 \div \left(-\frac{1}{3}\right)$ **b.** $-\frac{2}{3} \div \frac{4}{5}$ **c.** $-\frac{5}{8} \div \left(-\frac{2}{3}\right)$

35. a. $-16 \div \frac{8}{5}$ **b.** $24 \div \left(-\frac{3}{4}\right)$ **c.** $-\frac{3}{4} \div \left(-\frac{6}{7}\right)$

36. a. $-\frac{3}{4} \div \left(-\frac{1}{4}\right)$ **b.** $\frac{2}{3} \div \left(-\frac{5}{6}\right)$ **c.** $-18 \div \frac{2}{3}$

Complete the tables in Exercises 37 and 38. Write a sentence about patterns you observe.

37.

a	b	$-\left(\frac{b}{a}\right)$	$\frac{-b}{a}$	$\frac{b}{-a}$
5	35			
-27	3			

38.

a	b	$-\left(\frac{b}{a}\right)$	$\frac{-b}{a}$	$\frac{b}{-a}$
-2	8			
-20	-4			

Blue numbers are core exercises.

39. What two expressions are correctly described by "the opposite of b divided by a"?

40. Match the symbols $-\frac{x}{y}$, $\frac{-x}{y}$, and $\frac{x}{-y}$ with the word descriptions.

a. The opposite of the quotient of x and y

b. The quotient of x and the opposite of y

c. The quotient of the opposite of x and y

Connecting with Chapter 1 In Exercises 41 and 42, make an input-output table and graph for these rules, with the integers -3 to 3 as inputs. Can the points be connected? Why or why not?

41. If the input is zero or positive, the output equals the input. If the input is negative, the output is negative one times the input (a rule used in mathematics).

42. The output is negative one times the input if the input is even. The output equals the input if the input is odd.

▶ Writing

43. Explain how to multiply two negative numbers.

44. Explain how to find the reciprocal of a number in decimal notation, such as 2.5.

45. One student says that the reciprocal of $\frac{2}{3}$ is $\frac{3}{2}$. Another student says that the reciprocal is $\frac{1}{\frac{2}{3}}$. Show that the answers are equivalent.

46. Explain why $\frac{3}{4}$ and $-\frac{4}{3}$ are not reciprocals.

▶ Problem Solving: Product and Sums

In Exercises 47 and 48, complete the tables.

47.

x	y	$x \cdot y$	$x + y$
2	-2		
3		-6	
-4		-12	
	3		0
2			-1

48.

x	y	$x \cdot y$	$x + y$
-4		12	
		-16	0
		-15	2
-5			-8
-1		6	

▶ Project

49. Greatest Product

a. List ten pairs of integers that add to 13, such as -3, 16.

b. Find the product of each pair.

c. What is the greatest product of two whole numbers that add to 13?

d. What fractional or decimal values give a larger product than the whole numbers do?

e. Repeat parts a to d for ten pairs of integers that add to 15.

f. Describe how the largest product is related to the sum of the numbers.

▶ 2.3 Properties of Real Numbers Applied to Simplifying Algebraic Fractions and Adding Like Terms

Objectives

- Use the associative and commutative properties of real numbers.
- Use the distributive property of multiplication over addition.
- Identify the greatest common factor and divide expressions.
- Simplify algebraic fractions containing variables.
- Add like terms.

WARM-UP

Describe ways to make these problems easy to work without a calculator. Then do the problems.

1. $60 + (-20) + 30 + 40 + 70 + 20$

2. $\$4.75 + \$8.98 + \$6.25$

3. $3\frac{1}{4} + 2\frac{1}{2} + 1\frac{3}{4}$

4. $5 \cdot 27 \cdot 2$

5. $7(\$19.98)$

THIS SECTION EMPHASIZES properties of real numbers and algebraic notation. We use the properties to simplify *expressions in fractional notation* (**algebraic fractions**) and to add like terms. Whenever possible, we practice using positive and negative numbers in fraction and decimal notation.

▶ Four Properties of Real Numbers

SHORTCUTS USING PROPERTIES OF REAL NUMBERS In the Warm-up, you were asked to describe ways to make the problems easy to work. Example 1 shows ways the Warm-ups might have been done.

▶ EXAMPLE 1 Finding shortcuts in operations Add or multiply the following with shortcuts.

a. $60 + (-20) + 30 + 40 + 70 + 20$

b. $\$4.75 + \$8.98 + \$6.25$

c. $3\frac{1}{4} + 2\frac{1}{2} + 1\frac{3}{4}$

d. $5 \cdot 27 \cdot 2$

SOLUTION Here are some shortcuts you may have used.

a. Change the order of the numbers. Place pairs of numbers that add to 100 or add to zero next to each other.

$$60 + (-20) + 30 + 40 + 70 + 20$$

$$(60 + 40) + ((-20) + 20) + (30 + 70) \quad \text{Add pairs first.}$$

$$100 + 0 + 100 = 200$$

b. Change the order of the numbers; look for values that add to even dollars.

$$\$4.75 + \$8.98 + \$6.25$$

$$(\$4.75 + \$6.25) + \$8.98 \quad \text{Add pairs first.}$$

$$\$11.00 + \$8.98 = \$19.98$$

c. Change the order of the numbers; look for numbers that add to 1.

$$3\tfrac{1}{4} + 2\tfrac{1}{2} + 1\tfrac{3}{4}$$

$$(3\tfrac{1}{4} + 1\tfrac{3}{4}) + 2\tfrac{1}{2} \quad \text{Add pairs first.}$$

$$5 + 2\tfrac{1}{2} = 7\tfrac{1}{2}$$

d. Change the order of the numbers. Place pairs of numbers that multiply to 10 next to each other. Multiply these pairs first.

$$5 \cdot 27 \cdot 2 = (5 \cdot 2) \cdot 27$$

$$= 10 \cdot 27 = 270$$

◀

In the following, look for formal names to your everyday shortcuts. These real number properties allow us to use shortcuts.

ASSOCIATIVE PROPERTIES OF ADDITION AND MULTIPLICATION When we added $60 + 40 + (-20) + 20 + 30 + 70$, the **associative property of addition** allowed us to add the -20 and 20 separately from the other numbers. The term *associative property* is based on the word *associate*, meaning "to select groups" or "to choose (business or social) connections" (see Figure 9).

FIGURE 9 Business associates associate

THE ASSOCIATIVE PROPERTIES OF ADDITION AND MULTIPLICATION "REGROUP"

$a + (b + c) = (a + b) + c$ for all real numbers

$a \cdot (b \cdot c) = (a \cdot b) \cdot c$ for all real numbers

The parentheses in the box show pairs of numbers to be added or multiplied first. The associative property says we can group addition or multiplication pairs in any way that is convenient.

▶ **EXAMPLE 2** Grouping numbers conveniently Using the associative property, group the numbers to take advantage of opposites, sums of 10, or products of 100. Insert parentheses or brackets, [ ], to show the groups.

a. $8 + 2 + 4 + 6 + 7 + 9 + 1$

b. $4 + (-3) + 3 + 2$

c. $7 \cdot (-2) \cdot (-50)$

SOLUTION **a.** $(8 + 2) + (4 + 6) + 7 + (9 + 1) = 10 + 10 + 7 + 10$
$= 37$

b. $4 + [(-3) + 3] + 2 = 4 + 0 + 2$
$= 6$

c. $7 \cdot [(-2) \cdot (-50)] = 7 \cdot 100$
$= 700$ ◀

It is clearer to use brackets around parentheses than to use two sets of parentheses: $[(-3) + 3]$ instead of $((-3) + 3)$.

COMMUTATIVE PROPERTIES OF ADDITION AND MULTIPLICATION When we multiplied $5 \cdot 27 \cdot 2$, we changed the order of 2 and 27, to get $5 \cdot 2 \cdot 27$. This change, which made the problem easier, was permitted by the **commutative property of multiplication**. The word *commutative* comes from the word *commute*, which means "to change position or order" (see Figure 10).

FIGURE 10 Workers commute

THE COMMUTATIVE PROPERTIES OF ADDITION AND MULTIPLICATION "CHANGE ORDER"

$a + b = b + a$ for all real numbers
$a \cdot b = b \cdot a$ for all real numbers

The commutative properties let us change the order of the numbers in addition and multiplication problems.

▶ **EXAMPLE 3** Changing number positions Show how a change of order with the commutative property makes it possible to add or multiply three numbers mentally.

a. $5 \cdot 13 \cdot 4$

b. $-4 \cdot 3 \cdot (-25)$

c. $-6 + 5 + (-4)$

SOLUTION In each problem, we change the order (commutative property) and then group two numbers (associative property) to multiply or add.

a. $5 \cdot 13 \cdot 4 = (5 \cdot 4) \cdot 13$
$= 20 \cdot 13 = 260$

b. $-4 \cdot 3 \cdot (-25) = (-4 \cdot (-25)) \cdot 3$
$= 100 \cdot 3 = 300$

c. $-6 + 5 + (-4) = (-6 + (-4)) + 5$ Group the negative numbers.
$= -10 + 5 = -5$ ◀

Not all real-world and mathematical operations are associative and commutative. If you put on your socks and then your shoes, the outcome is entirely different than if you put on your shoes followed by your socks. Finding whether subtraction and division are commutative is left to the exercises.

▶ A Fifth Property of Real Numbers

Now we return to Exercise 5 of the Warm-up.

▶ **EXAMPLE 4** Finding shortcuts in operations Estimate 7($19.98). Multiply mentally, and describe any shortcuts you use.

SOLUTION Because $19.98 is almost $20, first estimate the total as 7 times $20 = $140. Then find the exact answer by subtracting 7 times $0.02 from $140 (see Figure 11).

$$7(\$19.98) = 7(\$20.00 - \$0.02) = \$140.00 - \$0.14 = \$139.86$$

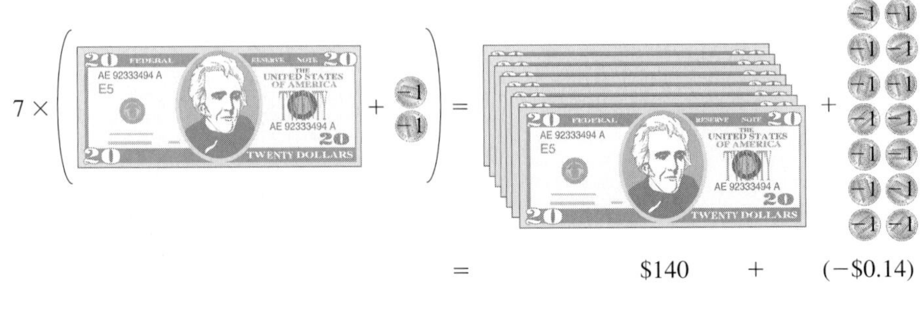

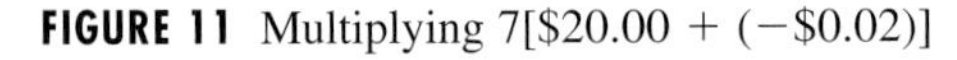

FIGURE 11 Multiplying 7[$20.00 + (−$0.02)] ◀

DISTRIBUTIVE PROPERTY OF MULTIPLICATION OVER ADDITION Note that when we multiplied 7[$20 + (−$0.02)] we made use of the **distributive property of multiplication over addition**.

The distributive property is important because it changes two factors, a and $(b + c)$, into two terms, ab and ac. Recall from Section 1.2 that **factors** are *two or more signed numbers, variables, or expressions being multiplied.* A **term** is *a signed number, variable, or expression being added or subtracted.*

DISTRIBUTIVE PROPERTY OF MULTIPLICATION OVER ADDITION

> For all real numbers a, b, and c,
>
> $$a(b + c) = ab + ac$$

The name *distributive* may have been chosen because multiplying each term in the parentheses by a is like dealing cards to each person in a game or serving cake to each guest at a party. Both dealing and serving are *distributive* actions (see Figure 12).

FIGURE 12 Card players distribute cards

In Section 2.1 we wrote subtraction as addition of the opposite number. As a result, we can distribute multiplication over subtraction:

$$a(b - c) = ab - ac$$

To read $a(b - c)$ we can say a times the quantity $b - c$.

▶ EXAMPLE 5 Using the distributive property Multiply these expressions.

a. $5(x + 3)$ **b.** $0.27(x - 175)$

c. $-(x - 4)$ **d.** $6(a + b - c)$

SOLUTION **a.** $5(x + 3) = 5 \cdot x + 5 \cdot 3 = 5x + 15$

b. $0.27(x - 175) = 0.27(x) - 0.27(175)$
$= 0.27x - 47.25$

Parentheses are used here; the multiplication dot could be confused with the decimal point.

c. $-(x - 4)$ may also be written $-1(x - 4)$.
$-1(x - 4) = -1(x) - (-1)(4)$
$= -1x - (-4) = -x + 4$

Parentheses are used here to help group the signs with the numbers.

Note that both signs on $x - 4$ changed when we multiplied by a negative.

d. $6(a + b - c) = 6 \cdot a + 6 \cdot b - 6 \cdot c = 6a + 6b - 6c$ ◀

There is no limit to the number of terms inside the parentheses.

THINK ABOUT IT 1: How many multiplications are in $a(b + c) = ab + ac$?

▶ Application: Simplification Property of Fractions

In Section 1.2, we noted that a fraction is in **lowest terms** when *the numerator and denominator have no common factors*. The **simplification property of fractions** says that *if the numerator and denominator of a fraction contain the same factor (a common factor), those factors can be eliminated.*

SIMPLIFICATION PROPERTY OF FRACTIONS

For all real numbers, a not zero and c not zero,

$$\frac{ab}{ac} = \frac{a \cdot b}{a \cdot c} = \frac{b}{c}$$

The associative property allows us to regroup $\frac{a}{a}$ from $\frac{ab}{ac}$ to give $\frac{a}{a} = 1$. The factor a is the *common factor* because it is a factor that appears in both the numerator and the denominator. The **greatest common factor** is *the largest possible factor*.

In the following examples, the word *simplify* will be used as a short way to write instructions. **Simplify** means *to use the simplification property of fractions to eliminate common factors, as well as to do the given operations.*

EXAMPLE 6 Simplifying fractions to lowest terms Name the common factors in the numerator and denominator, and simplify to lowest terms.

a. $\frac{16}{56}$ **b.** $\frac{6x}{4x}$ **c.** $\frac{12ac}{15bc}$

SOLUTION **a.** The common factor is 8.*

$$\frac{16}{56} = \frac{8 \cdot 2}{8 \cdot 7} = \frac{2}{7}$$

b. The common factors are 2 and x.

$$\frac{6x}{4x} = \frac{2 \cdot 3 \cdot x}{2 \cdot 2 \cdot x} = \frac{3}{2}$$

c. The common factors are 3 and c.

$$\frac{12ac}{15bc} = \frac{2 \cdot 2 \cdot 3 \cdot a \cdot c}{3 \cdot 5 \cdot b \cdot c} = \frac{4a}{5b}$$

◀

DIVISION WITH THE DISTRIBUTIVE PROPERTY Because we can write division by a as multiplication by the reciprocal, $\frac{1}{a}$, we can distribute division over addition or subtraction:

$$\frac{b+c}{a} = \frac{b}{a} + \frac{c}{a} \quad \text{or} \quad \frac{b-c}{a} = \frac{b}{a} - \frac{c}{a}$$

EXAMPLE 7 Using the distributive property Divide and simplify.

$$\frac{3x - xy}{x}$$

SOLUTION

$$\frac{3x - xy}{x} = \frac{3x}{x} - \frac{xy}{x} = 3 - y$$ Simplify each fraction.

An alternative method is to use the distributive property in reverse: $ab + ac = a(b + c)$. The variable x is the greatest common factor in the numerator:

$$\frac{3x - xy}{x} = \frac{x(3 - y)}{x} = 3 - y$$

◀

*As we noted in Section 1.2, common factors do not need to be prime.

Application: Adding Like Terms

THE TILE MODEL Algebra tiles are shapes that let us visualize certain algebraic ideas. Their area is the connection with algebra. In Section 1.3 we counted squares to find area. To find the area of a square or rectangle, we multiply the lengths of the sides that meet at one corner.

In Example 8, let a be the side of the large square, let b be the side of the small square, and let a and b be the sides of the rectangle.

EXAMPLE 8 Finding area Explain how to find the area of each tile in Figure 13.

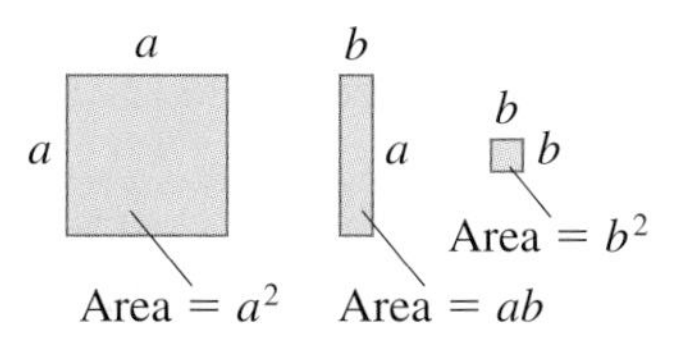

FIGURE 13 Area of algebra tiles

SOLUTION The large square has area $a \cdot a = a^2$. The small square has area $b \cdot b = b^2$. The rectangle has area ab. ◀

EXAMPLE 9 Adding area The areas of each shape are shown. Write an expression for the total area in each set.

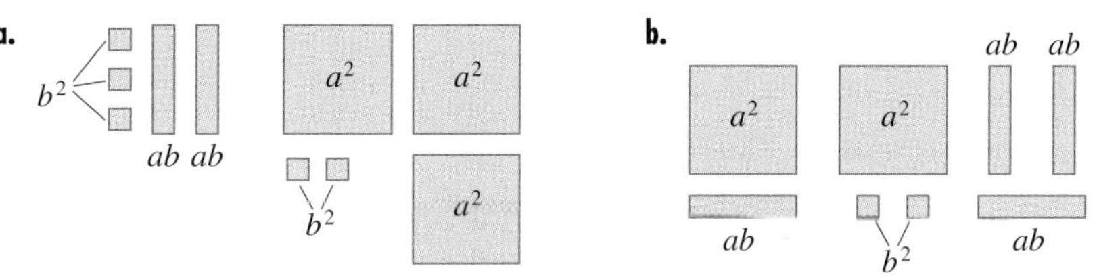

SOLUTION **a.** $a^2 + a^2 + a^2 + ab + ab + b^2 + b^2 + b^2 + b^2 + b^2 = 3a^2 + 2ab + 5b^2$

b. $a^2 + a^2 + ab + ab + ab + ab + b^2 + b^2 = 2a^2 + 4ab + 2b^2$ ◀

Because no numbers are given in Example 9, it is sensible to count the a^2 terms together because they describe the same tile shape. The ab tiles have a different shape so they cannot be counted with either the a^2 or the b^2 tiles.

Student Note:
For practice, use the glossary/index for defining numerical coefficient.

LIKE TERMS In Example 9 we say that the a^2 terms are *like terms*. The b^2 terms are *like terms*, and the ab terms are *like terms*. **Like terms** *have identical variable factors and exponents*. When we *add the numerical coefficients of terms with identical variable factors and exponents*, we are **adding like terms**.

EXAMPLE 10 Adding like terms What is the sum of all 18 tiles in parts a and b of Example 9?

SOLUTION $3a^2 + 2ab + 5b^2 + 2a^2 + 4ab + 2b^2$ Arrange by like terms.

$= 3a^2 + 2a^2 + 2ab + 4ab + 5b^2 + 2b^2$ Add the like terms.

$= (3 + 2)a^2 + (2 + 4)ab + (5 + 2)b^2$

$= 5a^2 + 6ab + 7b^2$ ◀

THINK ABOUT IT 2: How did the number properties help us to add the terms in Example 10?

ANSWER BOX

Warm-up: 1. 200 **2.** \$19.98 **3.** $7\frac{1}{2}$ **4.** 270 **5.** 4 **6.** \$139.86
Think about it 1: Three; $a \cdot (b + c)$, $a \cdot b$, $a \cdot c$ **Think about it 2:** The commutative property of addition let us change the order so that like terms were next to each other. The associative property of addition let us pair up the like terms. The distributive property did the addition: $3a^2 + 2a^2 = (3 + 2)a^2 = 5a^2$; $2ab + 4ab = (2 + 4)ab = 6ab$; $5b^2 + 2b^2 = (5 + 2)b^2 = 7b^2$

▶ 2.3 Exercises

1. Which number property describes each shortcut?

a. Group factors.

b. Group terms.

c. Group negative numbers and positive numbers and then add.

d. Change the order of the numbers to be added or multiplied.

e. Group pairs of numbers whose sums are 0, 10, or 100.

f. Group pairs of numbers whose products are 1, 10, or 100.

g. Rewrite one number of a product as the sum or difference of two numbers and then multiply.

2. Write each number as the sum or difference of two numbers that might be useful in a shortcut with the distributive property.

a. $3\frac{3}{4}$ **b.** $4\frac{1}{3}$ **c.** \$1.98 **d.** \$6.05

In Exercises 3 to 24, use shortcuts, if possible, to find the sums and products.

3. a. $2\frac{1}{2} + 1\frac{1}{3} + 3\frac{2}{3} + 5\frac{1}{2}$
b. $1\frac{2}{5} + 3\frac{1}{3} + 4\frac{3}{5}$

4. a. $6\frac{1}{4} + 3\frac{1}{2} + 5\frac{3}{4} + 2\frac{1}{2}$
b. $2\frac{1}{3} + 3\frac{7}{8} + 5\frac{2}{3}$

5. a. $4.25 + 2.98 + 1.75$
b. $2.60 + 1.30 + 3.40$

6. a. $2.75 + 6.15 + 1.85$
b. $3.15 + 2.75 + 1.25$

7. a. $-3 + (-4) + (+5)$
b. $+7 + (-8) + (-3)$

8. a. $4 + (-6) + (-1)$
b. $-5 + (-4) + (+4)$

9. a. $2 + (-5) + (-4)$
b. $-7 + (-3) + (+7)$

10. a. $-4 + (-6) + (+1)$
b. $4 + (-7) + (-9)$

11. $-4\frac{3}{4} - 10\frac{1}{2} + 3\frac{3}{4} - 2\frac{1}{2}$

12. $-1\frac{2}{3} + 5\frac{1}{3} + 2\frac{2}{3} + 1$

13. $-6 + (-7) + (-8) + 20$

14. $-18 + 7 + (-9) + (-12)$

15. a. $-4 - (-3)$ **b.** $-1 - (-3)$

16. a. $-3 - (+2)$ **b.** $-8 - 4$

17. a. $8 - (+3) - (-4)$ **b.** $-5 - (-7) + (-1)$

18. a. $6 - (-7) - (+2)$ **b.** $-4 - (+8) - (-5)$

19. a. $\frac{1}{4} \cdot 25 \cdot 8 \cdot 4$ **b.** $8(\frac{2}{3})(1.5)$

20. a. $\frac{1}{2} \cdot 15 \cdot 4 \cdot 3$ **b.** $7(\frac{4}{5})(1.25)$

21. a. $6(-8)(-1)$ **b.** $-5(5)(-1)$ **c.** $4(3)(-2)$

22. a. $5(4)(-3)$ **b.** $-9(-8)(-1)$ **c.** $-4(2)(-8)$

23. a. $-5(-6)(-7)$ **b.** $-6(9)(2)$ **c.** $-3(0)(-7)$

24. a. $-5(-8)(-8)$ **b.** $-2(-3)(0)$ **c.** $5(-2)(-9)$

In Exercises 25 to 32, explain how you would find the products mentally, using the distributive property.

25. 4 times \$4.97 **26.** 7 times \$5.99

27. 3 times \$10.98 **28.** 6 times \$7.96

29. $4(2\frac{3}{4})$ **30.** $3(1\frac{2}{3})$

31. $6(3\frac{5}{6})$ **32.** $5(1\frac{2}{5})$

Blue numbers are core exercises.

For Exercises 33 to 38, find the value of each side. Do the operations in parentheses first. Which property is used?

33. $-3(4)(-5) = -3(-5)(4)$

34. $-3(4 + 5) = -3(4) - 3(5)$

35. $-3 + 4 + 5 = 4 + (-3) + 5$

36. $-3(4 \cdot 5) = (-3 \cdot 4) \cdot 5$

37. $3 + (4 + 5) = (3 + 4) + 5$

38. $3(4 - 5) = 3(4) - 3(5)$

In Exercises 39 to 44, simplify with the distributive property.

39. a. $6(x + 2)$ **b.** $-3(x - 3)$ **c.** $-6(x + 4)$

40. a. $-(2 - x)$ **b.** $x(x - 3)$ **c.** $-(3 + x)$

41. a. $-3(x + y - 5)$ **b.** $-(x - y - z)$

42. a. $4(x - 4)$ **b.** $-4(x + 2)$ **c.** $-5(x - 3)$

43. a. $-(x - 3)$ **b.** $y(4 + y)$ **c.** $-(2 - y)$

44. a. $-5(2x - 4 + y)$ **b.** $-(x + y - z)$

Prime numbers have no positive integer factors except 1 and the number itself. The first prime number is 2. **Composite numbers** have factors other than themselves and 1. The number 1 is neither prime nor composite. In Exercises 45 and 46, find the prime factors for the composite numbers.

45. a. 45 **b.** 59 **c.** 72 **d.** 111

46. a. 39 **b.** 28 **c.** 51 **d.** 61

Simplify the expressions in Exercises 47 to 52. Assume that no variable is zero.

47. a. $\frac{-2x}{-6x}$ **b.** $\frac{-14a}{21a}$ **c.** $\frac{6x}{15xyz}$

48. a. $\frac{2xy}{-10xyz}$ **b.** $\frac{-15ab}{9ac}$ **c.** $\frac{-6ac}{-27cx}$

49. a. $\frac{15xy}{21y}$ **b.** $\frac{39abc}{13acd}$ **c.** $\frac{-12xy}{48xz}$

50. a. $\frac{9ab}{45bc}$ **b.** $\frac{72xz}{18xy}$ **c.** $\frac{16ab}{48ac}$

51. a. $\frac{ab}{bc}$ **b.** $\frac{ab}{ac}$ **c.** $\frac{ay}{by}$

52. a. $\frac{ac}{bc}$ **b.** $\frac{xy}{xyz}$ **c.** $\frac{bx}{xy}$

In Exercises 53 and 54, divide the expressions. Simplify to lowest terms. Do two parts of each exercise in two ways.

53. a. $\frac{3x + 4}{4}$ **b.** $\frac{4x + 8}{4}$

c. $\frac{x^2 + xy}{x}$ **d.** $\frac{ab - bc}{b}$

54. a. $\frac{2x + 6}{2}$ **b.** $\frac{5x + 2}{5}$

c. $\frac{x^2 + xy}{y}$ **d.** $\frac{ab - ac}{a}$

In Exercises 55 and 56, write an expression for the total area.

55.

56.

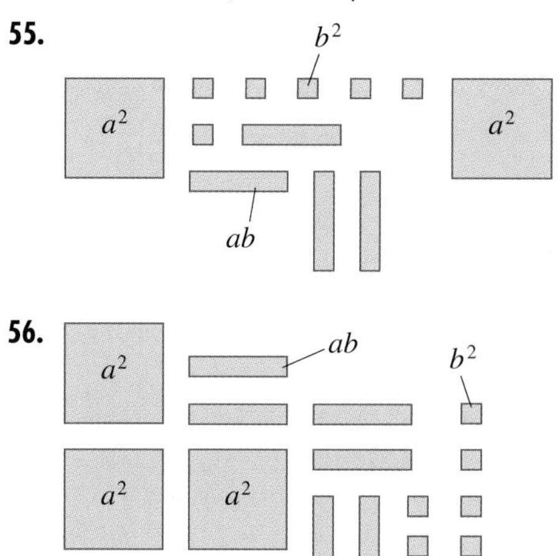

In Exercises 57 to 60, the large square has area $x \cdot x = x^2$ square units. The area of the small square is $1 \cdot 1 = 1$ square unit. The area of the rectangle is $x \cdot 1 = x$ square units. Find the total area of each set of tiles.

57.

58.

59.

60.

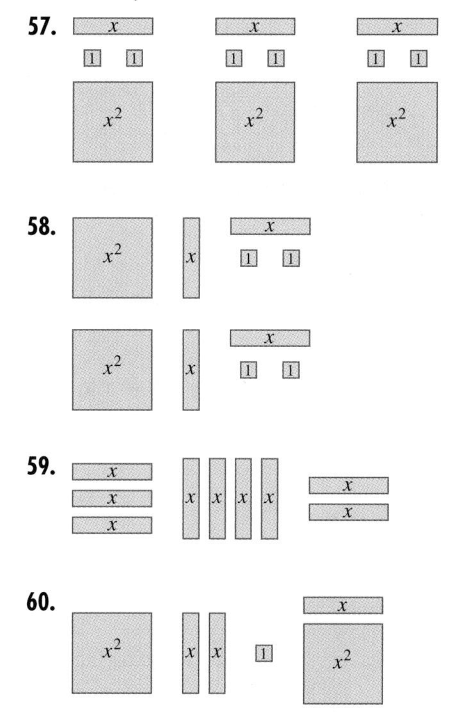

Identify the numerical coefficient in each of the terms in Exercises 61 and 62.

61. a. $-4x$ **b.** x **c.** $-x$

62. a. $-3x^2$ **b.** $5x$ **c.** y

Blue numbers are core exercises.

In Exercises 63 and 64, add the like terms, if any.

63. **a.** $4a^2 + 5a^2$ **b.** $6a^2 - 5a$ **c.** $4a^2 - 5a^2$

64. **a.** $4x + 4y$ **b.** $3x^2 - 4x^2$ **c.** $7x^2 + 7x^2$

In Exercises 65 to 68, add like terms. Multiply out parentheses first.

65. **a.** $-3x + 4x - 6x + 8x$

b. $-6y^2 + 8y^2 - 10y^2 - 3y^2$

c. $2a + 6 + 3a + 9$

d. $3(x + 1) - 2(x - 1)$

e. $3(x - 4) + 4(4 - x)$

f. $2x - 3y + 2x - 3y$

66. **a.** $-4y - 6y + 8y - 10y$

b. $+7x^2 - 12x^2 - 5x^2 + 8x^2$

c. $5x - 15 + 4x - 12$

d. $4(x + 1) - 3(x - 1)$

e. $7(x - 3) + 5(3 - x)$

f. $2x + 3y - 2x - 3y$

67. **a.** $\frac{1}{2}x + \frac{1}{4}y + \frac{1}{2}y - \frac{1}{4}x$

b. $0.5a + 0.75b - 0.5b + 1.5a$

c. $-2x + 3y + (-4x) + (-6x)$

d. $2(b + c) - 2(b - c)$

68. **a.** $\frac{1}{2}a - \frac{1}{4}b + \frac{1}{2}a - \frac{1}{2}b$

b. $0.25x + 0.5y + 0.5y - 0.75x$

c. $-4a + 8b - 6b + 9a$

d. $3(x - y) - 3(x + y)$

69. In parts a–d, what is the value of each expression if you do the operation in parentheses first? Are any of the expressions equal?

a. $8 - (5 - 3)$ **b.** $(8 - 5) - 3$

c. $16 \div (4 \div 2)$ **d.** $(16 \div 4) \div 2$

e. Is subtraction associative?

f. Is division associative?

70. In parts a–d, tell whether the statement is true or false.

a. $2 - 3 = 3 - 2$ **b.** $4 \div 5 = 5 \div 4$

c. $8 \div 2 = 2 \div 8$ **d.** $6 - 5 = 5 - 6$

e. Is subtraction commutative?

f. Is division commutative?

Blue numbers are core exercises.

▶ Error Analysis

71. Explain why the distributive property cannot apply to $a(bc)$.

72. In dividing $2x + 2y + 2$ by 2, a student writes $x + y$. Explain what is wrong, and provide the correct answer.

Explain what is wrong with the statements in Exercises 73 to 75.

73. $3 + (4 \cdot 5) = (3 + 4) \cdot 5$

74. $3(4 \cdot 5) = (3 \cdot 4) + (3 \cdot 5)$

75. $3(4 \cdot 5) = 3 \cdot 4 \cdot 3 \cdot 5$

▶ Terms and Factors

76. How many terms are in each expression?

a. xxx **b.** $w + x + y + 2z$

c. $x^4 + 4x^3 + 6x^2 + 4x + 1$ **d.** $5y^5$

e. $ax^2 + bx + c$ **f.** $2L + 2W$

77. How many factors are in each expression?

a. $4(a + b)(c + d)$ **b.** $x(x + y)$

c. $-0.5xyz$ **d.** $ab(c + d)$ **e.** $\frac{1}{2}bh$

▶ Writing

78. Explain how the distributive property changes a product into a sum.

79. Which of these activities are commutative? Give situations that might change your answers.

a. get dressed, eat breakfast

b. start car, fasten seatbelt

c. put key in ignition, start car

d. turn right, walk five steps forward

80. Explain how to change a fraction to lowest terms.

81. How do you recognize like terms?

82. How do you find the greatest common factor?

83. How do you distinguish factors from terms?

84. Explain how algebra tiles show that x^2 and x are not like terms.

85. Explain how algebra tiles show that $x^2 + x^2 \neq 2x^4$.

▶ Connecting with Chapter 1

86. Write each percent in decimal form and show how the distributive property applies to finding the sum or difference. *Hint:* The numerical coefficient of x is 1.

a. Cost of meal, x, plus 8% tax on meal.

b. Cost of shirt, x, less 5% discount.

c. House sale price, x, less the 6% realtor fee.

d. Cost of meal, x, plus 15% tip.

87. Bone Mass Returning to Exercise 31, Section 1.4, recall that bone loss is a danger in a weightless environment. An astronaut on the space station loses 1.5% of his or her hip bone mass (m) each month. In comparison, the typical postmenopausal woman loses 1% of her hip bone mass (m) each year. (*Science News*, 11/26/05, p. 347)

a. Write an expression for the hip bone mass lost by an astronaut in 1 month and for the bone mass remaining.

b. How does the distributive property help us in writing or simplifying these expressions?

c. Write an expression for the hip bone mass lost by a postmenopausal woman in a year and for the hip bone mass remaining.

88. Water Cost Using the charges shown in the table, make a table for inputs from 0 to 50 kgal (counting by 10). Use the four steps to graph the data from the New Orleans table.

New Orleans (2005, prior to Hurricane Katrina)
Residential Charge (kgal = 1000 gal)

Water Used* x (kgal)	Cost, y (dollars)
First 20 kgal	$2.31x + 5.50$
Next 980 kgal	$2.07\ (x - 20) + 2.31\ (20) + 5.50$

*Residential usage is about 3 kgal per person per month.

▶ Project

89. Distinguishing Terms from Factors

a. Make a two-column list. Name one column *Expressions with Two or More Terms* and the other *Expressions with One Term and Multiple Factors.*

b. Record each of these expressions under the appropriate heading:

$12xy$ $\quad 3x + 2y - 7z$ $\quad 2x - y$ $\quad 2 + \sqrt{3}$

abc $\quad a + b + c$ $\quad \dfrac{a + b}{c}$ $\quad \dfrac{a}{c} + \dfrac{b}{c}$

$\dfrac{12xy}{4x^2}$ $\quad \dfrac{15ab}{abc}$ $\quad \dfrac{-b}{2a}$ $\quad \frac{1}{2}h(a + b)$

$\dfrac{a}{b}$ $\quad \frac{1}{2}ah + \frac{1}{2}bh$ $\quad \dfrac{-b}{2a} + \dfrac{\sqrt{b^2 - 4ac}}{2a}$

c. Circle the expressions that are equivalent, and connect them with a line.

d. Write each of the following after its description below: $a(b + c)$, $a + b$, $xy + wz$, $(x + y)(x - y)$

2 terms, each with 2 factors
2 factors, each with 2 terms
1 term containing 2 factors
2 terms, each with 1 factor

▶ 2 Mid-Chapter Test

In Exercises 1 to 10, add, subtract, multiply, or divide, as indicated.

1. a. $2 - 5$ **b.** $-3 + 5$ **c.** $-3 - (-5)$

2. a. $3 - (-4)$ **b.** $-2 + (-5)$ **c.** $6 + (-2.5)$

3. a. $-3(2)$ **b.** $-5(-3)$ **c.** $-(-4)$

4. a. $\dfrac{27}{-3}$ **b.** $\dfrac{-28}{-2}$ **c.** $\dfrac{-32}{4}$

5. a. $4x + 5y - 2x + y$ **b.** $2x^2 - 3x + 2x(1 - x)$

6. a. $\dfrac{4xyz}{xy}$ **b.** $\dfrac{3x}{xyz}$ **c.** $\dfrac{-2y}{4xy}$

7. $-3.89c - 42.39d + 50.00c$

8. $-3x + 6 + (-24) + 27x$

9. $-\dfrac{3}{4} + \dfrac{2}{3} + \dfrac{1}{2} - \left(-\dfrac{1}{3}\right)$

10. $13(-4)5(-3)$

11. Divide the expressions in two different ways.

a. $\dfrac{4x + 2y}{2}$ **b.** $\dfrac{ab + bc}{b}$

For Exercises 12 to 15, make input-output tables, using integers from -2 to 3 for x. Graph the (x, y) pairs on coordinate axes.

12. $y = x - 1$

13. $y = -3x$

14. $y = 2x - 3$

15. $y = 3 - x$

In Exercises 16 and 17, find the difference in elevation between the highest and lowest point on each continent.

16. Asia: Mt. Everest, 8850 meters, and the Dead Sea, -400 meters

17. North America: Mt. McKinley, 6194 meters, and Death Valley, -86 meters

18. Mt. Everest is recognized as the highest mountain in the world. Mauna Kea is the inactive volcano on the island of Hawaii. Mauna Kea's volcanic base actually rises from the ocean floor. Using the data in the following table, find which mountain is actually "taller" and by how much.

Location	Elevation Relative to Sea Level (feet)
Mt. Everest	+29,028
Mauna Kea	+13,710
Sea level	0
Ocean floor, near Hawaii	−16,400

19. Explain why we can write every addition, $a + b$ or $b + a$, but only write a subtraction, $a - b$.

20. Write in words what we have done to $a - b$ to obtain $a + (-b)$.

▶ 2.4 Exponents and Order of Operations

Objectives

- Identify the base and positive integer exponent of an expression.
- Write expressions involving positive integer exponents.
- Simplify expressions containing squares and cubes.
- Simplify expressions with the order of operations.
- Identify and use grouping symbols.

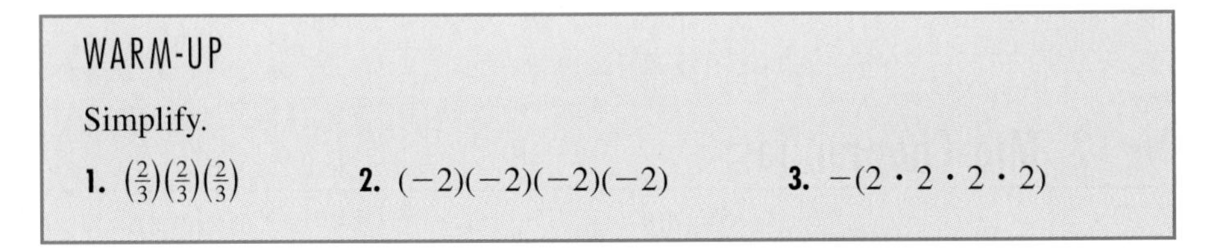
WARM-UP

Simplify.

1. $\left(\frac{2}{3}\right)\left(\frac{2}{3}\right)\left(\frac{2}{3}\right)$ **2.** $(-2)(-2)(-2)(-2)$ **3.** $-(2 \cdot 2 \cdot 2 \cdot 2)$

IN THIS SECTION, we examine properties of expressions with exponents and agreements about simplifying expressions containing two or more operations.

▶ Bases and Exponents

Let's return to the problem posed at the beginning of the chapter, on page 54.

▶ **EXAMPLE 1** Exploration of a bouncing ball pattern A ball is dropped from a height of 144 centimeters. After each time it hits the floor, the ball bounces $\frac{2}{3}$ as high as it did on the last bounce (see Figure 14). What is the height of the ball after it hits the floor ten times?

SOLUTION Start with an input-output table. Let the input be the number of times the ball has hit the floor, with zero as the starting number. Let the output be the height of the ball, with 144 centimeters as the starting height.

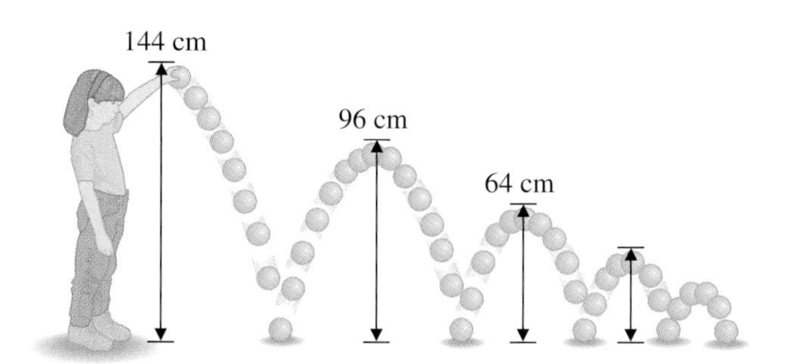

FIGURE 14 Height of bouncing ball

TABLE 6 Ball Height

Number of times ball has hit floor	Height (cm)
0	144
1	$144\left(\frac{2}{3}\right) = 96$
2	$144\left(\frac{2}{3}\right)\left(\frac{2}{3}\right) = 64$

Table 6 shows the first three entries; use your calculator to complete the table for inputs 3 to 10. See the Answer Box for outputs. *Hint:* After the sixth time the ball hits the floor, the ball's height is

$$144\left(\tfrac{2}{3}\right)\left(\tfrac{2}{3}\right)\left(\tfrac{2}{3}\right)\left(\tfrac{2}{3}\right)\left(\tfrac{2}{3}\right)\left(\tfrac{2}{3}\right) \approx 12.6 \text{ cm}$$ ◀

NAMING BASES AND EXPONENTS An exponent provides a shorter way to write an expression for the height of the ball.

$$144\left(\tfrac{2}{3}\right)\left(\tfrac{2}{3}\right)\left(\tfrac{2}{3}\right)\left(\tfrac{2}{3}\right)\left(\tfrac{2}{3}\right)\left(\tfrac{2}{3}\right) = 144\left(\tfrac{2}{3}\right)^6$$

The $\frac{2}{3}$ is the base. The **base** is *the number repeated in the multiplication. The small raised number to the right of the base* is an **exponent**. The exponent, 6, tells us the number of times the base is repeated. *The base and exponent together* are called a **power**.

$$\text{power} = \text{base}^{\text{exponent}}$$

POSITIVE INTEGER EXPONENT

In the power expression x^n, the *positive integer exponent n* indicates the number of factors of the base x. That is,

$$x^n = x \cdot x \cdot x \cdot x \cdot \cdots \cdot x$$

with n factors of x.

▶ **EXAMPLE 2** Naming bases and applying exponents Name each base; then write the expression as factors and multiply.

a. $144\left(\frac{2}{3}\right)^{10}$ **b.** $(-2)^4$ **c.** -2^4

SOLUTION **a.** The base is $\frac{2}{3}$ (144 is the numerical coefficient, which is not repeated and is not part of the base); $144\left(\frac{2}{3}\right)^{10} = 144\left(\frac{2}{3}\right)\left(\frac{2}{3}\right)\left(\frac{2}{3}\right)\left(\frac{2}{3}\right)\left(\frac{2}{3}\right)\left(\frac{2}{3}\right)\left(\frac{2}{3}\right)\left(\frac{2}{3}\right)\left(\frac{2}{3}\right)\left(\frac{2}{3}\right) \approx 2.5$.

b. The base is -2; $(-2)^4 = (-2)(-2)(-2)(-2) = 16$.

c. The base is 2, not -2; $-(2^4) = -(2 \cdot 2 \cdot 2 \cdot 2) = -16$. ◀

Mathematicians have agreed to place bases with negative signs within parentheses, as in part b of Example 2. Thus, in part c, the negative sign on -2^4 is not part of the base. The expression -2^4 means *the opposite of* 2^4.

NEGATIVE SIGNS AND BASES

Place a negative base in parentheses when applying an exponent to it:

$$(-4)^2 = (-4)(-4) = 16$$

Place a negative sign in front of the base to mean "opposite of":

$$-4^2 = \text{opposite of } 4^2 = -(4 \cdot 4) = -16$$

▶ **EXAMPLE 3** Simplifying expressions Simplify these using the definition of positive integer exponents.

a. $\left(\frac{2}{3}\right)^3$ **b.** $(-3x)^2$ **c.** $\left(\frac{x}{4}\right)^2$ **d.** $2(3a)^3$ **e.** -3^4

SOLUTION **a.** $\left(\frac{2}{3}\right)^3 = \frac{2}{3} \cdot \frac{2}{3} \cdot \frac{2}{3} = \frac{8}{27}$

b. $(-3x)^2 = (-3x)(-3x) = 9x^2$

c. $\left(\frac{x}{4}\right)^2 = \frac{x}{4} \cdot \frac{x}{4} = \frac{x^2}{16}$

d. $2(3a)^3 = 2(3a)(3a)(3a) = 54a^3$

e. $-3^4 = -3 \cdot 3 \cdot 3 \cdot 3 = -81$ ◀

▶ **EXAMPLE 4** Simplifying expressions Use the definition of positive integer exponents and $\frac{a}{a} = 1$ to simplify these expressions.

a. $\frac{x}{x^3}$ **b.** $(x^3)^2$ **c.** $\frac{a^3b}{ab^2}$ **d.** a^4a^2

SOLUTION **a.** $\frac{x}{x^3} = \frac{x \cdot 1}{x \cdot x \cdot x} = \frac{1}{x^2}$

b. $(x^3)^2 = (x^3)(x^3) = x \cdot x \cdot x \cdot x \cdot x \cdot x = x^6$

c. $\frac{a^3b}{ab^2} = \frac{a \cdot a \cdot a \cdot b}{a \cdot b \cdot b} = \frac{a^2}{b}$

d. $a^4a^2 = a \cdot a \cdot a \cdot a \cdot a \cdot a = a^6$ ◀

We **evaluate** expressions and equations when we *substitute numbers in place of the variables and simplify*. To evaluate $x + 3$ *for* $x = -2$, we replace x with -2 and obtain $-2 + 3$ or 1. In Example 5, we evaluate expressions containing exponents.

▶ **EXAMPLE 5** Evaluating expressions containing exponents Return to Example 4, let $x = -2$, $a = -1$ and $b = 3$. Evaluate each expression.

SOLUTION The $\frac{a}{a} = 1$ property permits simplification without multiplying factors.

a. $\frac{x}{x^3} = \frac{-2}{(-2)(-2)(-2)} = \frac{1}{4}$ -2 equals $-2 \cdot 1$

b. $(x^3)^2 = ((-2)^3)^2 = (-8)^2 = 64$

Because the numbers are small, multiply $-2\ (-2)(-2) = -8$ and $(-8)(-8) = 64$ rather than write out all the factors.

c. $\frac{a^3b}{ab^2} = \frac{(-1)^3(3)}{(-1)(3)^2} = \frac{(-1)(-1)(-1)(3)}{(-1)(3)(3)} = \frac{1}{3}$

Here the factors let us see the $\frac{a}{a} = 1$ property and avoid errors with signs.

d. $a^4a^2 = (-1)(-1)(-1)(-1) \cdot (-1)(-1) = 1$

List the factors and circle pairs that multiply to 1 and avoid errors with signs. ◀

SQUARES, CUBES, AND UNITS OF MEASURE When *2 is used as an exponent*, we say the base is **squared**. When *3 is an exponent*, the base is **cubed**. These words come from the measurement of area with square units (Figure 15) and of volume with cubic units (Figure 16). Area and volume will be discussed further in Section 2.5.

FIGURE 15 Square with side x **FIGURE 16** Cube with side x

Simplification of exponents follows the same pattern on units of measure as it does on variables.

EXAMPLE 6 Working with units of measure Write these expressions without parentheses.

a. A cubic foot, Figure 16 with $x = 12$ in.: $(12 \text{ in.})^3$

b. $(8 \text{ sec})^2$

c. $\left(\dfrac{5 \text{ m}}{1 \text{ sec}}\right)\left(\dfrac{1}{1 \text{ sec}}\right)$ where m stands for meters.

d. $\dfrac{3 \text{ ft}}{12 \text{ ft}^3}$ **e.** $\dfrac{4 \text{ ft}}{(12 \text{ ft})^3}$ **f.** $\dfrac{18 \text{ yd}^3}{3 \text{ yd}}$

SOLUTION

Student Note:
If you are uncomfortable with units in the denominators in parts c, d, and e, write the answers using *per*.
c. 5 m per sec^2
d. $\frac{1}{4}$ per ft^2
e. $\frac{1}{432}$ per ft^2

a. 1728 in.^3

b. 64 sec^2 (Seconds squared is used in acceleration of a car and in physics applications.)

c. $\left(\dfrac{5 \text{ m}}{1 \text{ sec}}\right)\left(\dfrac{1}{1 \text{ sec}}\right) = \dfrac{5 \text{ m}}{1 \text{ sec}^2}$

d. $\dfrac{3 \text{ ft}}{12 \text{ ft}^3} = \dfrac{3 \text{ ft}}{12 \text{ ft} \cdot \text{ft} \cdot \text{ft}} = \dfrac{1}{4 \text{ ft}^2}$

e. $\dfrac{4 \text{ ft}}{(12 \text{ ft})^3} = \dfrac{4 \text{ ft}}{(12 \text{ ft})(12 \text{ ft})(12 \text{ ft})} = \dfrac{1}{432 \text{ ft}^2}$

f. $\dfrac{18 \text{ yd}^3}{3 \text{ yd}} = \dfrac{18 \text{ yd} \cdot \text{yd} \cdot \text{yd}}{3 \text{ yd}} = 6 \text{ yd}^2$ ◀

SPOTTING THE DIFFERENCE

It is agreed that abbreviations for units—such as m for meters—are generally written without periods. (The one exception is inches, written in. to avoid confusion with the word *in.*) As a result, 5 m may be confused with $5m$, the product of 5 and the variable m. Here is how you spot the difference: If the m or other letter has a space in front of it, it refers to a unit of measurement. If there is no space in front of the letter and the letter is in italic type, it is the variable in an expression.

THINK ABOUT IT: What is the agreement about difference or quotient of two numbers?

Order of Operations

Simplifying expressions with more than one operation leads us to ask the question "How do we know which operation to do first, second, and so on?" The answer is "Use the order of operations."

The **order of operations** is *an agreement about the order in which we do mathematical operations.* This is an order mathematicians have adopted, not a property. As Example 7 shows, this order is not even programmed into all calculators.

▶ **EXAMPLE 7** Trying different orders of operations Find the value of $1 - 9 + 9 \cdot 9$ in the following ways:

a. By calculating from left to right

b. With a four-function credit-card-size or business calculator

c. With a scientific or graphing calculator

d. By doing the multiplication first and then the subtraction and addition, in order from left to right.

SOLUTION **a.** $1 - 9 + 9 \cdot 9 = 9$. When we calculate from left to right, $1 - 9 = -8$. The -8 and $+9$ add to 1, leaving $1 \cdot 9 = 9$.

b. Generally, the order of operations for a four-function calculator is the order in which the numbers and operations are entered. If we enter $1 - 9 + 9 \cdot 9$, we obtain 9, as in part a.

c. With a scientific or graphing calculator, we use $1 - 9 + 9 \cdot 9$ ENTER to obtain 73.

d. $1 - 9 + 9 \cdot 9$ Do the multiplication first.

$1 - 9 + 81$ Do the subtraction next.

$-8 + 81$ Do the addition last.

73 ◀

The correct answer in algebra is the answer found with the scientific calculator, which follows the order described in part d. This order is the one that has been agreed upon in the order of operations (see Figure 17).

The Order of Operations

1. Calculate expressions within parentheses and other grouping symbols.
2. Calculate exponents and square roots.
3. Do the remaining multiplication and division in the order of appearance, left to right.
4. Do the remaining addition and subtraction in the order of appearance, left to right.

Student Note:
Following the order of operations is essential in simplifying expressions, applying formulas, and planning solutions to equations.

FIGURE 17 Order of operations

EXAMPLE 8 Using the order of operations with exponents Which step in the order of operations explains why these expressions are equal?

a. $144\left(\frac{2}{3}\right)^3 = 144 \cdot \left(\frac{2}{3}\right)\left(\frac{2}{3}\right)\left(\frac{2}{3}\right)$

b. $\left(\frac{a^2b^3}{ab}\right)^3 = (ab^2)^3$

SOLUTION **a.** We apply the exponent before doing the multiplication between the 144 and the expression in parentheses.

b. We simplify the expression in the parentheses before applying the outside exponent. ◀

▶ In Example 9, we use the order of operations to find the value of expressions that are used in formulas elsewhere in this book. Again we expand the meaning of the word *simplify*. Here, **simplify** means *to use the order of operations to calculate the value of an expression, to apply the definition of positive integer exponents and the properties of exponents, to add like terms, to change fractions to lowest terms, or to use the real number properties* (from Section 2.3).

EXAMPLE 9 Using the order of operations to simplify expressions Simplify these expressions, commonly used for the indicated applications.

a. Surface area of a box: $2 \cdot 5 \cdot 6 + 2 \cdot 5 \cdot 8 + 2 \cdot 6 \cdot 8$

b. Area of a trapezoid: $\frac{1}{2} \cdot 7(8 + 10)$

c. Height of an object: $-\frac{1}{2}(32)(3)^2 + 64(3) + 50$

d. Rule for a sequence: $9 - 4(n - 1)$

SOLUTION **a.** $2 \cdot 5 \cdot 6 + 2 \cdot 5 \cdot 8 + 2 \cdot 6 \cdot 8$ Do multiplication left to right.

$= 60 + 80 + 96$ Do addition left to right.

$= 236$

b. $\frac{1}{2} \cdot 7(8 + 10)$ Do operations in parentheses.

$= \frac{1}{2} \cdot 7 \cdot 18$ Do multiplication left to right.

$= 63$

The commutative property of multiplication (and other properties from Section 2.3) may be applied before any step in the order of operations. Reversing the 7 and 18 allows us to do the problem mentally: $\frac{1}{2} \cdot 18 \cdot 7 = 9 \cdot 7 = 63$.

c. $-\frac{1}{2}(32)(3)^2 + 64(3) + 50$ Apply the exponent first.

$= -\frac{1}{2}(32) \cdot 9 + 64(3) + 50$ Do multiplication left to right.

$= -144 + 192 + 50$ Do addition left to right.

$= 98$

d. $9 - 4(n - 1)$ The expression in the parentheses cannot be simplified.

$= 9 + (-4)(n - 1)$ The multiplication with the -4 (distributive property) must be done before an addition or subtraction.

$= 9 + (-4n) + 4$ Add like terms. Change $+(-4n)$ back to subtraction.

$= 13 - 4n$ ◀

Caution: In part d of Example 9, a common error is to subtract $9 - 4$ first. Look for problems like this one in the exercises for this section and again in Chapter 3.

Grouping Symbols

When parentheses are needed within parentheses, we sometimes use different symbols, such as brackets or braces. *The preferred way to write multiple grouping symbols is* {[( )]}, with *parentheses* on the inside, the square-shaped *brackets* used next, and *braces* (like little wires) used outermost, although double parentheses, (( )), are acceptable.

EXAMPLE 10 Using grouping symbols Simplify the following expressions. Work the innermost parentheses, ( ), first; then the brackets, [ ]; and finally the braces, { }.

a. $\{-1 + [(9 - 9) + 8]\}^2$

b. $-1 + [9 - (9 + 8)]$

SOLUTION **a.** $\{-1 + [(9 - 9) + 8]\}^2 = \{-1 + [0 + 8]\}^2 = \{7\}^2 = 49$

b. $-1 + [9 - (9 + 8)] = -1 + [9 - 17] = -1 + [-8] = -9$ ◀

Parentheses, brackets, and braces are just one type of **grouping symbol**, *used to place terms together*. Other grouping symbols include the absolute value symbol, the square root symbol, and the horizontal fraction bar.

The *absolute value* symbol may act as a grouping symbol if it contains an expression, such as $|1 - 9|$. First calculate the expression inside the absolute value, and then find the absolute value.

The *square root* symbol, or **radical** sign, is a grouping symbol if it contains an expression rather than a single number, as in $\sqrt{8 + 1}$. The expression inside should be calculated before taking the square root. In writing an expression under a radical sign, it is important to draw the overbar over the whole expression.

The *horizontal fraction bar* acts as a grouping symbol. The numerator and denominator of the fraction are calculated separately except when eliminating common factors. (See part a in Example 11.)

EXAMPLE 11 Using special grouping symbols and the order of operations Simplify these expressions.

a. Slope of a line: $\dfrac{-3 - 6}{10 - 7}$

b. Distance formula: $\sqrt{(10 - (-2))^2 + (-3 - 6)^2}$

c. Mean absolute deviation: $\dfrac{|1.3 - 1.5| + |1.4 - 1.5| + |1.8 - 1.5|}{3}$

SOLUTION **a.** $\dfrac{-3 - 6}{10 - 7}$ Do numerators and denominators first.

$= \dfrac{-9}{3}$ Divide.

$= -3$

b. $\sqrt{(10 - (-2))^2 + (-3 - 6)^2}$ Do parenthetical expressions.

$= \sqrt{12^2 + (-9)^2}$ Apply exponents.

$= \sqrt{144 + 81}$ Add the expression under the radical sign.

$= \sqrt{225}$ Take the square root.

$= 15$

c. $\dfrac{|1.3 - 1.5| + |1.4 - 1.5| + |1.8 - 1.5|}{3}$ Do subtractions inside absolute value.

$= \dfrac{|-0.2| + |-0.1| + |0.3|}{3}$ Find the absolute value.

$= \dfrac{0.2 + 0.1 + 0.3}{3}$ Add the numerator.

$= \dfrac{0.6}{3}$ Divide.

$= 0.2$ ◀

The order of operations is often memorized by means of an acronym such as PEMDAS (parentheses, exponents, multiplication, division, addition, subtraction) or a saying such as "Please Excuse My Dear Aunt Sally." Neither of these stresses the left-to-right order for division and multiplication or subtraction and addition, though. Ask your instructor for other favorite aids to memorizing the order of operations.

Repeat Example 11 with a calculator. When using a calculator, place parentheses around the expression inside a square root sign, around numerator and denominator expressions in a fraction, and around expressions inside an absolute value sign, as shown in Figure 18. Use only parentheses on a calculator. Brackets [ ] and braces { } have special meanings. The square root expression $\sqrt{}(5^2 + 12^2)$ is written as it should be entered into the calculator, as is the expression $(-4 + 2) \div (3 - 5)$ for the fraction $\dfrac{-4 + 2}{3 - 5}$.

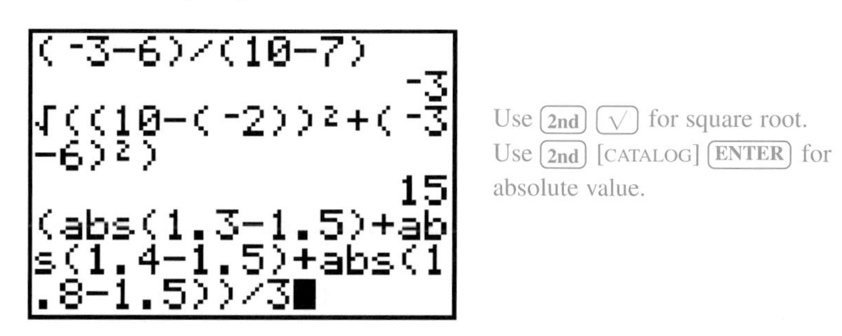

FIGURE 18 Screen for Example 11

ANSWER BOX

Warm-up: 1. $\frac{8}{27}$ **2.** 16 **3.** -16 **Example 1:** The heights, rounded to the nearest tenth, are $144(\frac{2}{3})^3 \approx 42.7$, $144(\frac{2}{3})^4 \approx 28.4$, 19.0, 12.6, 8.4, 5.6, 3.7, and $144(\frac{2}{3})^{10} \approx 2.5$ **Think about it:** Do subtraction and division on the numbers in the order in which they are listed.

▶ 2.4 Exercises

In Exercises 1 and 2, identify the base for the exponent 2, write the expression using factors, and write the expression in words.

1. **a.** $3x^2$ **b.** $-3x^2$
   c. $(-3x)^2$ **d.** ax^2
   e. $-x^2$ **f.** $(-x)^2$

2. **a.** $-b^2$ **b.** $(-b)^2$
   c. ab^2 **d.** mn^2
   e. $(ab)^2$ **f.** $-2x^2$

Blue numbers are core exercises.

Simplify the expressions in Exercises 3 to 8.

3. **a.** 3^5 **b.** 2^6 **c.** $(-2)^2$ **d.** $(-3)^3$

4. **a.** $(\frac{2}{3})^2$ **b.** $(-\frac{1}{2})^2$ **c.** $(-\frac{1}{3})^3$ **d.** $-(\frac{1}{2})^2$

5. **a.** $(\frac{1}{3})^3$ **b.** $(\frac{4}{5})^3$ **c.** $(-\frac{2}{3})^3$ **d.** $-(\frac{1}{3})^2$

6. **a.** 4^4 **b.** 5^3 **c.** $(-2)^3$ **d.** $(-3)^2$

7. **a.** $2 \cdot 4^2$ **b.** -2^2 **c.** $3(-2)^2$ **d.** $-4 \cdot 3^2$

8. **a.** $3 \cdot 5^2$ **b.** -4^2 **c.** $-3 \cdot 3^3$ **d.** $4(-3)^2$

Simplify the expressions in Exercises 9 to 26, using the definition of positive integer exponents and $\frac{a}{a} = 1$.

9. **a.** $x \cdot x \cdot x$ **b.** x^3x^2 **c.** $(a^3)^3$ **d.** $b(b^2)$

10. **a.** $x \cdot x \cdot x \cdot x$ **b.** x^2x^5 **c.** $(a^4)^3$ **d.** $a(a^3)$

11. **a.** $\frac{x^3}{x^4}$ **b.** $\frac{a^4}{a^3}$ **c.** $\frac{x^2y}{(xy)^2}$ **d.** $\frac{a^3b}{ab^2}$

12. **a.** $\frac{x^5}{x^2}$ **b.** $\frac{a}{a^2}$ **c.** $\frac{ab^2}{a^3b}$ **d.** $\frac{x^3y}{(xy)^3}$

13. **a.** $(-4x)^2$ **b.** $(3x)^3$ **c.** $(0.2x)^2$ **d.** $-(2x)^3$

14. **a.** $(-2x)^3$ **b.** $(0.4x)^2$ **c.** $-(3x)^2$ **d.** $(-2x)^2$

15. **a.** $(ab)^2$ **b.** $(xy)^3$ **c.** $(2ac)^2$

16. **a.** $(xy)^2$ **b.** $(ac)^3$ **c.** $(4ab)^3$

17. **a.** $-2(4a)^2$ **b.** $-3(-3x)^2$ **c.** $-2(-2x)^3$

18. **a.** $-3(2x)^2$ **b.** $-2(-3x)^3$ **c.** $-3(-4x)^2$

19. **a.** $\left(\frac{x}{4}\right)^3$ **b.** $\left(\frac{2a}{5}\right)^2$ **c.** $\left(-\frac{3x}{5}\right)^2$

20. **a.** $\left(\frac{n}{3}\right)^2$ **b.** $\left(\frac{3n}{4}\right)^3$ **c.** $\left(-\frac{4x}{5}\right)^3$

21. **a.** $\left(\frac{4n}{3}\right)^2$ **b.** $-\left(\frac{3x}{4}\right)^3$ **c.** $\left(-\frac{5n}{3}\right)^3$

22. **a.** $-\left(\frac{2x}{5}\right)^3$ **b.** $\left(\frac{5x}{6}\right)^2$ **c.** $\left(-\frac{4n}{3}\right)^2$

23. **a.** $(8 \text{ in.})^2$ **b.** $(0.3 \text{ sec})^2$ **c.** $(6 \text{ m})^3$

24. **a.** $(4 \text{ ft})^3$ **b.** $(5 \text{ m})^3$ **c.** $(0.3 \text{ sec})^3$

25. **a.** $\frac{3 \text{ yd}}{6 \text{ yd}^3}$ **b.** $\frac{7 \text{ m}^3}{42 \text{ m}}$ **c.** $\frac{10 \text{ in.}^2}{(4 \text{ in.})^3}$

26. **a.** $\frac{5 \text{ m}^2}{15 \text{ m}^3}$ **b.** $\frac{6 \text{ ft}}{(3 \text{ ft})^3}$ **c.** $\frac{20 \text{ km}^3}{(10 \text{ km})^2}$

Blue numbers are core exercises.

In Exercises 27 to 32, return to the stated exercise and evaluate the expressions for the given values.

27. Exercise 17 for $a = -2$ and $x = 3$

28. Exercise 18 for $x = -2$

29. Exercise 19 for $a = -1, x = -2$

30. Exercise 20 for $n = -2, x = -1$

31. Exercise 21 for $n = -1, x = -2$

32. Exercise 22 for $n = -1, x = 2$

33. $\frac{1}{2} \cdot 5 \cdot (6 + 8)$

34. $\frac{1}{2} \cdot 15 \cdot (4 + 6)$

35. $6^2 + 8^2$

36. $9^2 + 12^2$

37. $5^2 + 12^2$

38. $8^2 + 15^2$

39. $7 - 2(4 - 1)$

40. $5 - 2(6 + 2)$

41. $(7 - 2)(4 - 1)$

42. $(5 - 2)(6 + 2)$

43. $6 - 3(x - 4)$

44. $7 - 5(4 - x)$

45. $(6 - 3)(x - 4)$

46. $(7 - 5)(4 - x)$

47. $3[8 - 2(3 - 5)]$

48. $4[9 - 6(5 - 3)]$

49. $|4 - 6| + |6 - 4|$

50. $|3 - 7| - |7 - 3|$

51. $|5 - 2| + |2 - 5|$

52. $|3 - 8| - |8 - 5|$

53. **a.** $\frac{4 - 7}{2 - (-3)}$ **b.** $\frac{-2 - 3}{-2 - (-3)}$ **c.** $\frac{5 - 2}{3 - (-4)}$

54. **a.** $\frac{2 - 5}{-3 - 4}$ **b.** $\frac{7 - 2}{6 - (-3)}$ **c.** $\frac{-7 - 2}{-6 - 3}$

55. **a.** $\frac{3 + \sqrt{25}}{4}$ **b.** $\frac{9 - \sqrt{36}}{2}$ **c.** $\frac{12 + \sqrt{64}}{5}$

56. **a.** $\frac{5 + \sqrt{81}}{2}$ **b.** $\frac{14 - \sqrt{16}}{5}$ **c.** $\frac{15 + \sqrt{121}}{4}$

57. $\sqrt{(-6 - 6)^2 + (12 - 3)^2}$

58. $\sqrt{(1 - 4)^2 + (3 - 7)^2}$

59. $\sqrt{(4 - (-2))^2 + (4 - (-4))^2}$

60. $\sqrt{(3 - (-9))^2 + (-3 - 2)^2}$

Simplify the expressions in Exercises 61 to 68. Use a calculator as needed. Round to the nearest hundredth.

61. $\frac{|3 - 7.5| + |9 - 7.5| + |10 - 7.5|}{3}$

62. $\dfrac{|2 - 2.75| + |2.25 - 2.75| + |4 - 2.75|}{3}$

63. $3.14(2)^2 \cdot 6$

64. $3.14(4)^2 \cdot 6$

65. $\frac{4}{3}(3.14) \cdot 2^3$

66. $\frac{4}{3}(3.14) \cdot 4^3$

67. $-\frac{1}{2}(32)(-1)^2 + 16(-1) + 50$

68. $-\frac{1}{2}(32)(-2)^2 + 16(-2) + 50$

In Exercises 69 and 70, begin by building a table.

69. A bouncing ball bounces half the previous height with each bounce. What is the height of the fifth bounce if the ball starts from 96 centimeters?

70. A bacteria population doubles every 2 hours. If there are 1,000,000 bacteria to start with, how many will there be in 24 hours?

▶ Writing

71. What could it mean to *simplify* an expression?

72. Write the rule for the order of operations.

▶ Error Analysis

Explain the errors made in Exercises 73 to 79.

73. Simplifying $7 - 4(x - 3)$ as $3(x - 3)$

74. Simplifying $5 - 3(x - 1)$ as $5 - 3x - 3$

75. Simplifying $(3x^3)^2$ as $9x^9$

76. Simplifying $(3x^3)^2$ as $6x^6$

77. Simplifying $(-0.2x^3)^2$ as $0.4x^9$

78. Simplifying $(-0.2x^3)^2$ as $-0.4x^6$

79. Simplifying $(-0.2x^3)^2$ as $0.04x^9$

80. Problem Solving: Order of Operations

a. Check these problems. One is not correct. Change one operation sign to make it true.

$1 \cdot 9 + 9 - 8 = 10$
$1 - 9 + 9 + 8 = 9$
$1 \cdot [9 - (9 - 8)] = 8$
$-1 + 9 - 9 - 8 = 7$
$-1 - 9 \div 9 + 8 = 6$

b. Check these problems. One is not correct. Change one operation sign to make it true.

$-1 + (9 + 9) \div 9 = 1$
$1 \cdot (9 + 9) \div 9 = 2$
$1 + (9 + 9) \div 9 = 3$
$1 \cdot (9 \div 9) + \sqrt{9} = 4$
$1 + (9 \div 9) - \sqrt{9} = 5$

c. Use the four digits in the year of your birth, together with the four basic operations and the square root, as was done in parts a and b, to find 15 whole-number values between 1 and 25. Show all parentheses needed to create the values.

81. Bone Mass In the setting for Exercise 87 of Section 2.3, the astronaut loses 1.5% of her or his hip bone mass each month. A postmenopausal woman loses 1% of her hip bone mass each year.

a. Write an expression for the hip bone mass remaining for the astronaut after 2 months in space. After 4 months in space.

b. Using m for original hip bone mass, write a general formula for the hip bone mass remaining after z months for the astronaut and for the hip bone mass remaining after w years for the postmenopausal woman.

c. In how many months will the astronaut's hip bone mass drop below 90%? In how many years will the postmenopausal woman's hip bone mass drop below 90%? Explain your reasoning.

▶ Projects

82. Exploration of Number Squares and Cubes Below is a partial list of square and cubic numbers.

Square Numbers

1, 4, 9, 16, 25, 36, 49, 64, 81, 100, 121, 144, 169,
1^2 2^2 3^2 4^2 5^2 6^2 7^2 8^2 9^2 10^2 11^2 12^2 13^2

196, 225, 256, 289, 324, . . .
14^2 15^2 16^2 17^2 18^2

Cubic Numbers

1, 8, 27, 64, 125, 216, 343, 512, 729, 1000, 1331,
1^3 2^3 3^3 4^3 5^3 6^3 7^3 8^3 9^3 10^3 11^3

1728, 2197, 2744, 3375, 4096, . . .
12^3 13^3 14^3 15^3 16^3

Use these numbers to do the following.

a. Find two square numbers whose product is another square number.

b. Use the definition of a positive integer exponent to explain why the product of any two square numbers is a square number.

c. Find two cubic numbers whose product is a cubic number.

Blue numbers are core exercises.

d. Use the definition of a positive integer exponent to explain why the product of any two cubic numbers is a cubic number.

e. Suggest an equivalent expression for a^2b^2 and a^3b^3.

83. Bouncing Balls The rules of tennis require that a tennis ball bounce between 4 feet 5 inches and 4 feet 10 inches when dropped 8 feet 4 inches onto a concrete base.

a. What decimal compares a bounce of 4 feet 5 inches to the original height of 8 feet 4 inches? What percent of the original height is the bounce?

b. What decimal compares a bounce of 4 feet 10 inches to the original height of 8 feet 4 inches? What percent of the original height is the bounce?

c. Why might the height 8 feet 4 inches have been chosen as the drop height?

d. Suppose that on the first bounce the ball reaches the larger height (4 feet 10 inches). If the ball is permitted to continue bouncing, what height, in inches, will it reach on the second bounce? the third bounce? the sixth bounce? Round answers to the nearest tenth of an inch.

Using any ball and a yardstick or meterstick, plan an experiment to find the height of the ball after one bounce, the height of the ball after three bounces, and the percent rebound. A 36-inch or 1-meter starting height is recommended. State any assumptions.

e. Gather data for both the first and the third bounce. Repeat five times. Average the heights for the first bounce. Average the heights for the third bounce.

f. Calculate what percent of the original height the first bounce is. Use this percent to predict the height of the third bounce. Compare your data with your prediction.

g. Use your results to predict the height for the sixth bounce.

h. Summarize your findings.

▶ 2.5 Unit Analysis and Formulas

Objectives

- Change from one unit of measure to another using unit analysis.
- Use formulas to calculate perimeter, area, surface area, and volume.

WARM-UP

Use the simplification property of fractions on both the numbers and the variables; then multiply these fractions.

1. $\dfrac{10{,}000{,}000m}{1} \cdot \dfrac{h}{60m} \cdot \dfrac{d}{24h} \cdot \dfrac{y}{365d}$

2. $\dfrac{1000L}{1} \cdot \dfrac{1.0567q}{L} \cdot \dfrac{g}{4q}$

3. $\dfrac{f^3}{1} \cdot \dfrac{12i}{f} \cdot \dfrac{12i}{f} \cdot \dfrac{12i}{f}$

4. $\dfrac{c}{1} \cdot \dfrac{q}{4c} \cdot \dfrac{g}{4q} \cdot \dfrac{231i^3}{g} \cdot \dfrac{1}{0.0625i}$

IN THIS SECTION, we examine unit analysis, a process that helps you win arguments and read critically. The second half of the section provides practice using a variety of geometric formulas.

Using Unit Analysis

Student Note:
Unit analysis is a powerful tool for critical thinking. News broadcasts, magazines, and newspapers are filled with "facts" containing numbers and units of measure. Many of these "facts" are misleading. Unit analysis lets us examine the facts and decide whether they are reasonable.

Unit analysis is *a method for changing from one unit of measure to another or from one rate to another*. Unit analysis is a clear step-by-step process that helps us break complicated problems into manageable pieces.

Unit analysis is presented in two parts. In this section, we change from one unit of measure to another. Later, in Chapter 5, Section 5.1, we will change from one rate to another.

In unit analysis, we use $\frac{a}{a} = 1$ to eliminate unwanted units, writing facts such as 1 minute = 60 seconds as $\frac{1 \text{ minute}}{60 \text{ seconds}} = 1$. We can also use the fact that $a = \frac{a}{1}$ to write facts as fractions.

A friend claims that he has lived 10 million minutes! Is this possible? In Example 1, we will investigate this claim. We will organize our solution with the four problem-solving steps: understand, plan, carry out the plan, and check.

▶ **EXAMPLE 1** Using equivalent units in unit analysis Change 10,000,000 minutes into years.

SOLUTION *Understand:* We start with minutes and want to end with years, so we list the units-of-measure facts that move unit by unit from minutes to years:

60 minutes = 1 hour
24 hours = 1 day
365 days ≈ 1 year (We assume a year is approximately 365 days.)

Plan: Starting with minutes, we arrange the facts as fractions so that units (minutes, hours, and days) are eliminated and only years remain.

Student Note:
You may find it helpful to write facts on cards:

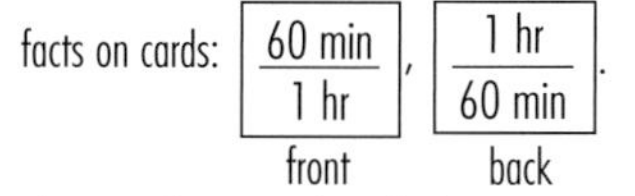

Place one fact on each card. Then arrange cards to solve problems.

Carry out the plan:

$$\frac{10{,}000{,}000 \text{ minutes}}{1} \cdot \frac{1 \text{ hour}}{60 \text{ minutes}} \cdot \frac{1 \text{ day}}{24 \text{ hours}} \cdot \frac{1 \text{ year}}{365 \text{ days}}$$

$$= \frac{10{,}000{,}000}{60 \cdot 24 \cdot 365} \text{ years}$$

$$\approx 19 \text{ years}$$

Check: We make sure that the facts are correctly written, all units except years are eliminated, and the answer is reasonable. Yes, if he is 19 years old or older, he has lived 10 million minutes! ◀

THINK ABOUT IT 1: Adding parentheses may be essential in doing Example 1 on a calculator. Try these different sequences of keystrokes and compare answers. Which is correct?

a. 10 000 000 [÷] 60 [×] 24 [×] 365 [ENTER]
b. 10 000 000 [÷] 60 [÷] 24 [÷] 365 [ENTER]
c. 10 000 000 [÷] [(] 60 [×] 24 [×] 365 [)] [ENTER]

▶ The following examples are intended to show the variety of ways we can apply unit analysis, changing systems of measure (metric to American standard), using facts more than once, and deriving surprising information from the facts.

▶ **EXAMPLE 2** Changing systems of measure How many gallons are in 1000 liters? Is this enough to fill 4 lemonade glasses, 4 children's plastic wading pools, or an Olympic swimming pool?

SOLUTION ***Understand:*** We list facts, starting with liters and ending with gallons:

1 liter = 1.0567 quarts

4 quarts = 1 gallon

Plan: Starting with liters, we arrange the facts so that liters and quarts are eliminated.

Carry out the plan:

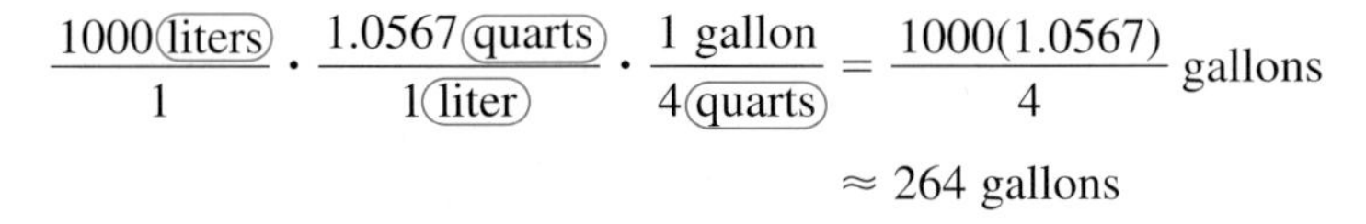

$$\frac{1000\text{(liters)}}{1} \cdot \frac{1.0567\text{(quarts)}}{1\text{(liter)}} \cdot \frac{1 \text{ gallon}}{4\text{(quarts)}} = \frac{1000(1.0567)}{4} \text{ gallons}$$

$$\approx 264 \text{ gallons}$$

Check: Many sodas are sold in 2-liter plastic bottles. Thus, 1000 liters would fill 500 plastic bottles. The best choice is 4 children's wading pools. ◀

▶ The next example uses foot and cubic foot, or $(\text{foot})^3$. If there is more than one foot in a problem, we write *feet*. You may use $\frac{\text{foot}}{\text{feet}} = 1$, but remember that $\frac{\text{foot}}{(\text{foot})^3} = \frac{1}{(\text{foot})^2}$. See also Example 6 in Section 2.4.

▶ **EXAMPLE 3** Using facts more than once How many cubic inches are there in 1 cubic foot? (See Figure 19.)

SOLUTION ***Understand:*** The term *cubic foot* means $(\text{foot})^3$. We start with cubic feet and end with cubic inches.

Facts: 1 foot = 12 inches

Plan: Our fact contains only the foot, not the cubic foot. To obtain the cubic foot, or $(\text{foot})^3$, we need to use foot as a factor three times.

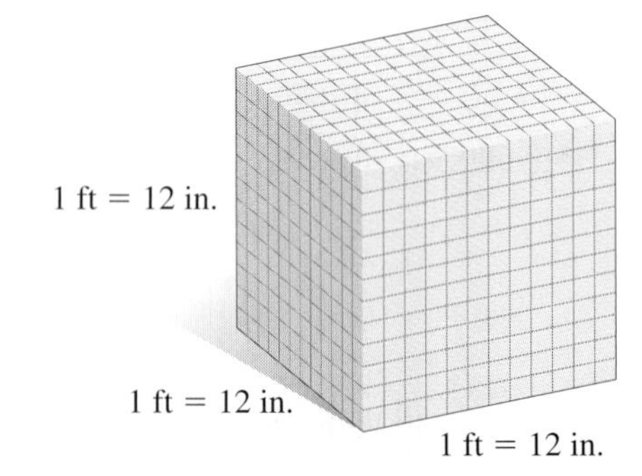

FIGURE 19

Carry out the plan:

$$\frac{1\text{(cubic foot)}}{1} \cdot \frac{12 \text{ inches}}{1\text{(foot)}} \cdot \frac{12 \text{ inches}}{1\text{(foot)}} \cdot \frac{12 \text{ inches}}{1\text{(foot)}} = \frac{12^3 \text{ inches}^3}{1}$$

$$= 1728 \text{ cubic inches}$$

Check: Our work contains foot · foot · foot and inches · inches · inches. The number part of the answer, 12^3, has three factors, like the units. ◀

▶ Example 4 is a setting familiar to parents.

▶ **EXAMPLE 4** Confirming expectations How many square inches will 1 cup of milk cover when spilled and spread to a depth of $\frac{1}{16}$ inch?

Facts: 1 gallon = 231 cubic inches

4 quarts = 1 gallon

1 quart = 4 cups

SOLUTION ***Understand:*** We want to start with the 1 cup of milk and change it into cubic inches. We will then need to get square inches, as requested.

Plan: First, we look for a unit-by-unit "path" through the facts, from cups to cubic inches. The next step is to decide what to do with $\frac{1}{16}$ inch. To obtain square inches, or $(\text{inches})^2$, we need to place $\frac{1}{16}$ inch in the denominator and let the inch eliminate one of the inches from cubic inches.

Carry out the plan:

$$\frac{1\text{ cup}}{1\text{ spill}} \cdot \frac{1\text{ quart}}{4\text{ cups}} \cdot \frac{1\text{ gallon}}{4\text{ quarts}} \cdot \frac{231\text{ cubic inches}}{1\text{ gallon}} \cdot \frac{1}{\frac{1}{16}\text{ inch}}$$

$$= \frac{231}{4 \cdot 4 \cdot \frac{1}{16}} \cdot \frac{\text{square inches}}{1\text{ spill}}$$

$$= 231\text{ square inches per spill}$$

Note that $4 \cdot 4 \cdot \frac{1}{16} = 1$.

Check: All units except square inches per spill are eliminated. ◀

THINK ABOUT IT 2: If we had placed $\frac{1}{16}$ inch in the numerator, what units would we have had?

UNIT ANALYSIS

1. ***Understand:*** Name the unit to be changed, and name the unit for the answer.
2. ***Plan:*** List facts containing the unit to be changed and all units needed to get to the unit for the answer.
3. ***Carry out the plan:*** Write the unit to be changed (or a fact containing the unit) as the first fraction. Set up a product of fractions, using the list of facts. Arrange fractions so that the units to be eliminated appear in both a numerator and a denominator. Carry out the multiplication.
4. ***Check:*** Recheck the units noted in step 1.

The milk spread in Example 4 is an area. We counted area and perimeter in Chapter 1. Now we use common formulas.

▶ Using Geometric Formulas

A **formula** is *a rule or principle written in mathematical language*. Examples 6 to 9 involve the formulas for perimeter and area. Later, we will consider the formulas for volume and surface area.

Table 7 on page 98 summarizes the formulas for two-dimensional, or flat, objects. In all the formulas, the base and height (or length and width) refer to lines that are perpendicular. *Perpendicular lines* form square corners where they cross. The square corner is a 90 degree angle and is labeled with a small square, ⊾.

PERIMETER AND AREA DEFINED The definitions in the accompanying box suggest the differences between perimeter and area.

DEFINITIONS OF PERIMETER AND AREA

Perimeter is *the distance around the outside of a flat object. The perimeter of a circle* is called the **circumference**. Perimeter is measured by units of length such as centimeters, meters, and kilometers (in the metric system) or inches, yards, and miles (in the American standard system).

Area is *the measure of the surface enclosed within a flat object*. Area is measured in square units, such as square meters or square feet.

Student Note:
Learn these formulas.

TABLE 7 Selected Geometric Formulas for Two-Dimensional Figures

Figure	Formulas
Triangle	Perimeter, $P = a + b + c$ Area $= \frac{1}{2}$ base $\cdot$ height $A = \frac{1}{2}bh$
Square	Perimeter $= 4 \cdot$ side, $P = 4s$ Area $=$ side $\cdot$ side, $A = s^2$
Rectangle	Perimeter, $P = 2l + 2w$ Area $=$ length $\cdot$ width, $A = lw$
Parallelogram	Area $=$ base $\cdot$ height $A = bh$
Trapezoid	Perimeter, $P = a + b + c + d$ Area $= \frac{1}{2}$ height $\cdot$ (sum of parallel sides) $A = \frac{1}{2}h(a + b)$
Circle	Circumference, $C = 2\pi r = \pi d$ Diameter, $d = 2r$ Area, $A = \pi r^2$, where $r =$ radius

PERIMETER In Example 5, we evaluate a formula for perimeter.

▶ **EXAMPLE 5** Finding perimeter Walking at least 10,000 steps each day is suggested for good health. A rectangular park nearby has a sidewalk around it. The park is approximately 1200 feet long and 800 feet wide (see Figure 20).

a. Which formula in Table 7 can we use to find the distance once around? Evaluate the formula.

b. How many laps around the park will give the required number of steps? State any assumptions.

c. Suggest an exercise plan.

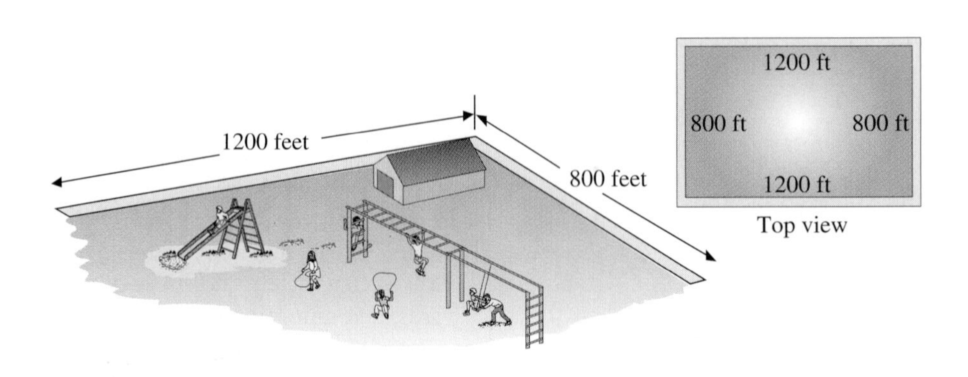

FIGURE 20 Park

SOLUTION

a. The distance around is 2 lengths and 2 widths, which is the same as the perimeter formula for a rectangle:

$$P = 2l + 2w \qquad l = 1200 \text{ ft}, w = 800 \text{ ft}$$

$$P = 2(1200 \text{ ft}) + 2(800 \text{ ft})$$

$$= 2400 \text{ ft} + 1600 \text{ ft}$$

$$= 4000 \text{ ft}$$

b. Use unit analysis on the facts: 12 inches in 1 foot, 1 lap is 4000 feet, 10,000 steps per day. Assume 32 inches per step. (For comparison, find your own step length by measuring in inches the distance you walk in 10 steps and dividing by 10.)

$$\frac{10{,}000 \text{ steps}}{1 \text{ day}} \cdot \frac{32 \text{ in.}}{1 \text{ step}} \cdot \frac{1 \text{ ft}}{12 \text{ in.}} \cdot \frac{1 \text{ lap}}{4000 \text{ ft}} = 6.67 \text{ laps per day}$$

c. Do 6 laps. Walk to the park or look at the rest of the day for other walking opportunities. ◀

USING A FORMULA

Student Note:
Following the steps as outlined will help you learn the formulas, as well as do the exercises.

To use a formula:

1. Write the formula.
2. List the facts, with units.
3. Substitute the facts into the formula.
4. Simplify the units.
5. Calculate, either by hand or by entering the entire expression into the calculator, using parentheses as needed.

Student Note:
In Examples 6, 7, and 10, the formulas were evaluated for $\pi \approx 3.14$. If a calculator π is used, answers may differ.

Several formulas, including the one for the circumference of a circle, contain the constant pi. **Pi** is *the number found by dividing the circumference of any circle by its diameter*. The symbol for pi is π. Use either the 2nd [π] key on your calculator or 3.14 to approximate pi. The calculator key will give slightly more accurate results.

Example 6 is typical of many problems in that it requires calculating parts of the perimeter separately and then adding the results.

▶ **EXAMPLE 6** Summing sections to find the perimeter. Suppose you walk on the track around a soccer field. What is the perimeter of the track shown in Figure 21? Assume that the curved parts are half-circles.

SOLUTION

This perimeter is the sum of the circumference of the curved parts (totaling a complete circle), πd, and the two straight sides, $2w$. The perimeter is

$$\text{Perimeter} = \pi d + 2w \qquad d = 73 \text{ m}, w = 100 \text{ m}, \pi \approx 3.14$$

$$P \approx 3.14(73 \text{ m}) + 2(100 \text{ m})$$

$$\approx (229.22 \text{ m}) + (200 \text{ m})$$

$$\approx 429.22 \text{ m}$$ ◀

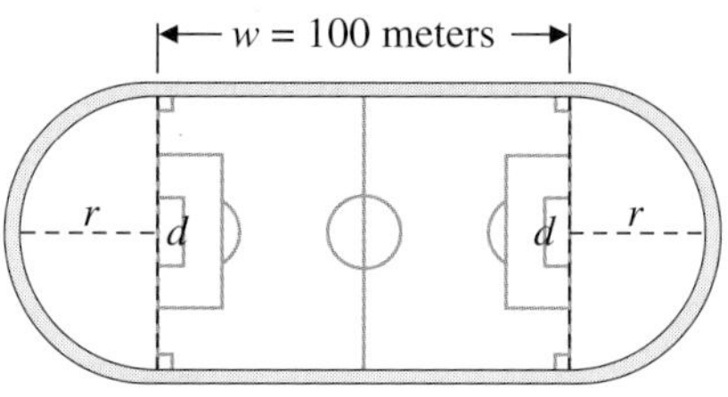

d = 73 meters

FIGURE 21 Soccer field

AREA In choosing measurements for calculating area, make sure the base and height (or length and width) are perpendicular. Look for a small square where perpendicular lines intersect.

▶ EXAMPLE 7 Summing sections to find the area The groundskeeper needs to order new turf for the area inside the track in Figure 21. Find the area.

SOLUTION The area is the sum of the area of the rectangular field and the area of the two ends, which are half circles. The perpendicular sides of the rectangle are d and w.

$$\text{Area} = \pi r^2 + dw$$
$$A \approx 3.14\left(\frac{73}{2}\text{ m}\right)^2 + (73\text{ m})(100\text{ m})$$
$$A \approx 4183\text{ m}^2 + 7300\text{ m}^2$$
$$A \approx 11{,}483\text{ m}^2$$
◀

THINK ABOUT IT 3: In Example 7, suppose that your area was 16,733 m^2 + 7300 m^2. How do you know $3.14(73\text{ m})^2 + (73\text{ m})(100\text{ m})$ is wrong? How is it fixed?

▶ EXAMPLE 8 Finding the area of a trapezoid Ray has a trapezoidal area, shown in Figure 22, of a banner to repair. What is the area in square inches?

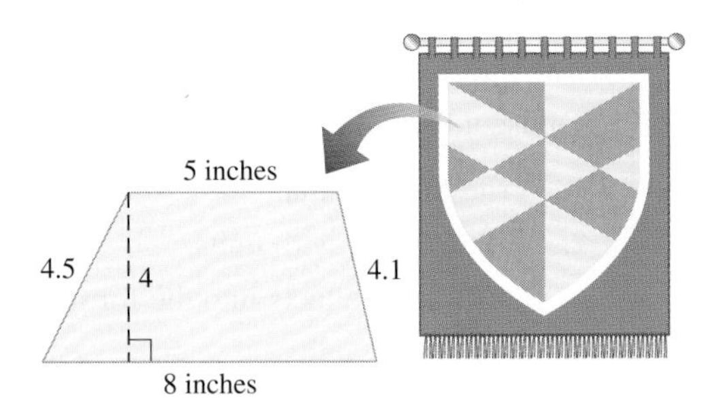

FIGURE 22 Trapezoid in banner

SOLUTION The parallel sides are $a = 5$ in. and $b = 8$ in. The height is 4 inches, perpendicular to b.

Student Note:
The lengths of the slanted sides are not needed to find the area.

$$\text{Area} = \tfrac{1}{2}h(a + b) \qquad h = 4\text{ in.}$$
$$A = \left(\tfrac{1}{2} \cdot 4\text{ in.}\right)(5\text{ in.} + 8\text{ in.})$$
$$= (2\text{ in.})(13\text{ in.})$$
$$= 26\text{ in.}^2$$
◀

SURFACE AREA AND VOLUME DEFINED Table 8 summarizes the formulas for surface area and volume of common three-dimensional, or solid, objects.

DEFINITIONS OF SURFACE AREA AND VOLUME

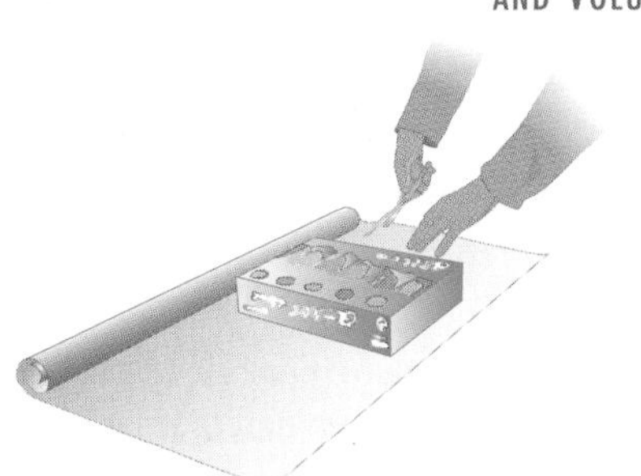

Surface area is *the area needed to cover a three-dimensional object.* Surface area is measured in square units, just like area.

Volume is *the space taken up by a three-dimensional object.* Volume is measured in cubic units, such as cubic meters or cubic feet. **Capacity** is *the amount (especially of liquids) a container holds.* We change units of capacity, such as gallons or quarts, to units of volume by describing them as cubic units. For example, 1 gallon = 231 cubic inches.

Student Note:
Three-dimensional geometric figures are shaded red, whereas two-dimensional geometric figures are shaded blue; see page 98.

TABLE 8 Selected Geometric Formulas for Three-Dimensional Figures

Figure	Formulas
Rectangular prism (box)	Surface area, $S = 2lw + 2hl + 2hw$ Volume, $V = lwh$
Cylinder	Surface area, $S = 2\pi r^2 + 2\pi rh$ Volume, $V = \pi r^2 h$, where r = radius, h = height
Sphere	Surface area, $S = 4\pi r^2$ Volume, $V = \left(\frac{4}{3}\right)\pi r^3$

▶ **EXAMPLE 9** Naming surface area and volume settings Make two lists, one headed "Surface Area" and the other "Volume" (or "Capacity"). List each setting under the heading that best describes it. The settings are gasoline in a gas tank, gift wrap on a package, soda in an aluminum can, the aluminum soda can, the space cooled by an air conditioner, the metal outside of a refrigerator, the storage part of a refrigerator, a balloon, an empty cardboard shoe box, toy wood blocks.

SOLUTION Surface area is the area needed to cover a three-dimensional object. Volume is the space taken up by a three-dimensional object (or the capacity of the object as a container). See the Answer Box. ◀

SURFACE AREA Both area and surface area are measured in square units. The difference is that surface area is for three-dimensional shapes: boxes, cylinders, and spheres.

▶ **EXAMPLE 10** Finding surface area of a storage tank If we wanted to paint a cylindrical storage tank, we would need to know the surface area to find the amount of paint needed. Find the surface area of the tank shown in Figure 23. Describe the order of operations needed to evaluate the formula.

SOLUTION Note that the height need not be shown in the vertical position. The "height" of this cylinder, 10 feet, is the distance between the circular ends. The radius is 2 feet.

$$\text{Surface area} = 2\pi r^2 + 2\pi rh \qquad r = 2\text{ ft}, h = 10\text{ ft}, \pi \approx 3.14$$

$$S \approx 2(3.14)(2\text{ ft})^2 + 2(3.14)(2\text{ ft})(10\text{ ft}) \qquad \text{Simplify powers and units.}$$

$\approx 2(3.14)(4 \text{ ft}^2) + 2(3.14)2(10) \text{ ft}^2$ Do multiplications.

$\approx 25.12 \text{ ft}^2 + 125.6 \text{ ft}^2$ Do addition.

$\approx 150.72 \text{ ft}^2$

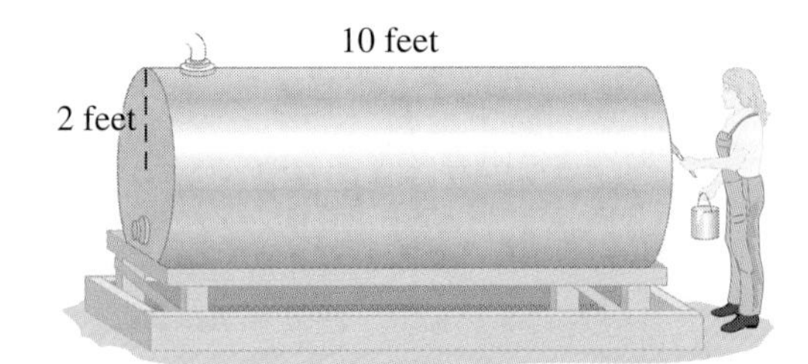

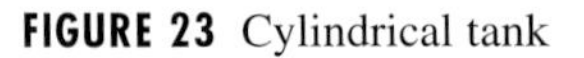

FIGURE 23 Cylindrical tank ◀

VOLUME Look for either cubic units or capacity (gallons or liters) in finding volume.

▶ **EXAMPLE 11** Finding the volume of a sphere: inflating balloons A nonsmoker in excellent physical condition, using deep inhalation and forced exhalation, might exhale 250 to 275 cubic inches of air. (Some air must remain in the lungs or they will collapse.) Suppose a balloon approximates a sphere (see Figure 24). Could someone inflate

a. an empty balloon to a radius of 3 inches in one breath?

b. an empty balloon to a radius of 4 inches in one breath?

c. a balloon from a radius of 4 inches to a radius of 5 inches in one breath?

FIGURE 24 Balloon

SOLUTION We need to compare volumes by evaluating the formula with several radii (plural of *radius*). We will *round* our answers to the nearest whole number, and we will use π from the calculator. The volume of a sphere is given by $V = \frac{4}{3}\pi r^3$.

For radius $r = 3$ in., $V = \frac{4}{3}\pi(3 \text{ in.})^3 \approx 113 \text{ in.}^3$

For radius $r = 4$ in., $V = \frac{4}{3}\pi(4 \text{ in.})^3 \approx 268 \text{ in.}^3$

For radius $r = 5$ in., $V = \frac{4}{3}\pi(5 \text{ in.})^3 \approx 524 \text{ in.}^3$

a. Reaching a radius of 3 inches is likely.

b. Only a fit person could reach a radius of 4 inches.

c. The *change in volume* is $524 - 268 = 256$ cubic inches. A fit person could inflate from a radius of 4 inches to a radius of 5 inches. ◀

Caution: The numbers in Example 11 do not take into account the resistance of the balloon against air going into it. You may want to seek medical advice before trying Example 11 as an experiment.

GRAPHING CALCULATOR TECHNIQUE: EVALUATING AN EXPRESSION ON THE COMPUTATION SCREEN

Write the expression on the screen, substituting the numbers for the variables. In Example 11, to evaluate $A = \frac{4}{3}\pi r^3$ for $r = 3$, write (4/3) π (3^3) and press [ENTER]. To evaluate the same expression for $r = 4$, use the replay option [2nd] [ENTER] to obtain prior entry. Replace the base 3 with 4, and press [ENTER]. Repeat as needed for other numbers.

This method works for an expression with any number of variables.

GRAPHING CALCULATOR TECHNIQUE: EVALUATING WITH TABLE

We can evaluate the volume formula for $r = x = 3$, 4, and 5 with a table. Go to [Y =], and enter $Y_1 = (4/3)\pi X^3$. Go to the table setup, [2nd] [TBLSET]. Let **TblStart** = 1. Let **ΔTbl** = 1. Press [2nd] [TABLE]. As shown in Figure 25, the answers for 3, 4, and 5 should agree with those in Example 11.

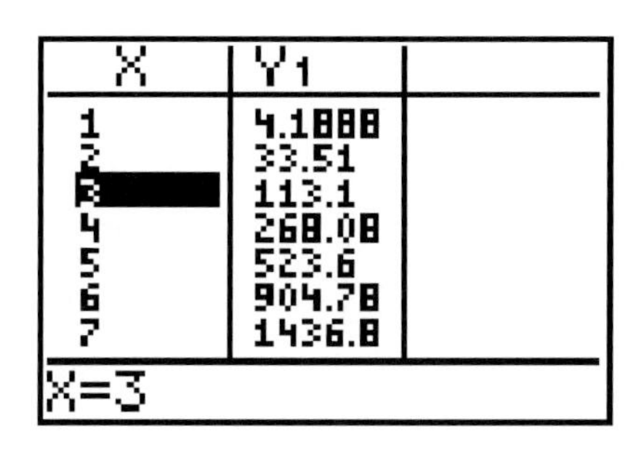

FIGURE 25 Calculator table for balloon volume

ANSWER BOX

Warm-up: 1. $1.9y$ **2.** $264g$ **3.** $1728i^3$ **4.** $231i^2$ **Think about it 1:** Both b and c are correct. **Think about it 2:** $(\text{inches})^4$ per spill **Think about it 3:** The circular part should not be larger than the rectangular part. Use $r = \frac{73}{2}$, not diameter. **Example 9:** *Surface Area*—gift wrap on a package, aluminum soda can, the metal outside of a refrigerator, a balloon, an empty cardboard shoe box. *Volume*—gasoline in a gas tank, soda in an aluminum can, the space cooled by an air conditioner, the storage part of a refrigerator, the air inside a balloon, toy wood blocks. Explain, if your answers differ from those given.

▶ 2.5 Exercises

Simplify the expressions in Exercises 1 to 8. Assume $\frac{\text{feet}}{\text{foot}} = 1$.

1. $\frac{8 \text{ ounces}}{1} \cdot \frac{1 \text{ pound}}{16 \text{ ounces}}$

2. $\frac{8 \text{ fluid ounces}}{1 \text{ cup}} \cdot \frac{4 \text{ cups}}{1 \text{ quart}}$

3. $\frac{100 \text{ yards}}{1} \cdot \frac{3 \text{ feet}}{1 \text{ yard}} \cdot \frac{12 \text{ inches}}{1 \text{ foot}}$

4. $\frac{55 \text{ mm}}{1} \cdot \frac{1 \text{ cm}}{10 \text{ mm}} \cdot \frac{1 \text{ in.}}{2.54 \text{ cm}}$

5. $\frac{\text{feet}^3}{\text{foot}}$

6. $\frac{\text{meters}^3}{\text{meters}^2}$

7. $\frac{\text{inches}^2}{\text{inches}^3}$

8. $\frac{\text{cm}}{\text{cm}^3}$

Use unit analysis to change the units in Exercises 9 to 12. Round to the nearest whole number. Use these facts again in Exercises 13 to 26.

9. How many feet in 1 kilometer?
   12 inches = 1 foot
   39.37 inches = 1 meter
   1000 meters = 1 kilometer

10. How many ounces are in 1 ton?
    1 ton = 2000 pounds
    16 ounces = 1 pound

11. How many grams are in 120 pounds?
    2.2 pounds = 1 kilogram
    1000 grams = 1 kilogram

Blue numbers are core exercises.

12. How many cups are in a 5-gallon water can?
1 gallon = 4 quarts
4 cups = 1 quart

In Exercises 13 to 26, list the facts needed and write a unit analysis to solve the problem. Unless otherwise noted, the facts you need are in the examples or earlier exercises. Round to the nearest tenth.

13. How many seconds are in 1 day?
60 seconds = 1 minute

14. Change 1,000,000 ounces into tons.

15. Change 1,000,000 seconds into days.

16. How many minutes are in 1 year?

17. Change 72 years into seconds.

18. Change 1 square meter into square centimeters.
100 centimeters = 1 meter

19. Change 1 cubic foot into cubic inches.

20. Change 1 square mile into square feet.
5280 feet = 1 mile

21. Change 1 square yard into square inches.

22. Change 1 cubic yard into cubic inches.

23. How many square feet are in 1200 square inches? Round the answer to the nearest tenth of a square foot.

24. How many square inches are in 12 square feet?

25. The difference in elevation between Mt. Everest and the Mariana Trench is 19.77 kilometers. Write this elevation difference in feet. Round to the nearest 1000.

26. How many laps to make 10,000 steps in Example 6? Let 1 step = 32 in. and 1 m = 39.37 in.

In Exercises 27 to 30, round to the nearest tenth. Use the 2nd [π].

27. Find the perimeter and area of each shape.

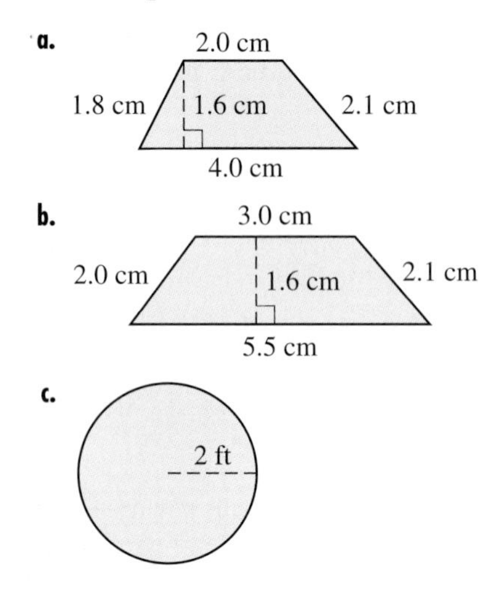

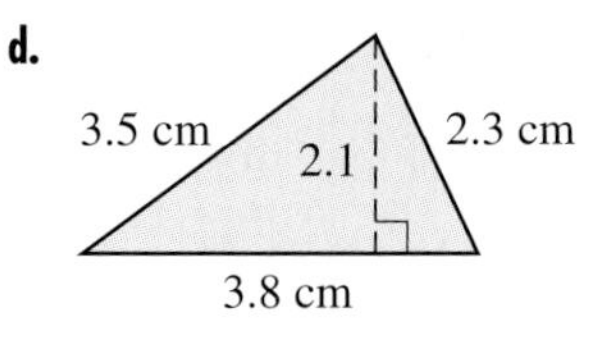

28. Find the perimeter and area of each shape.

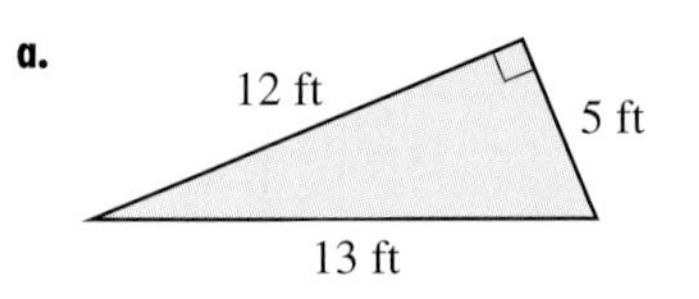

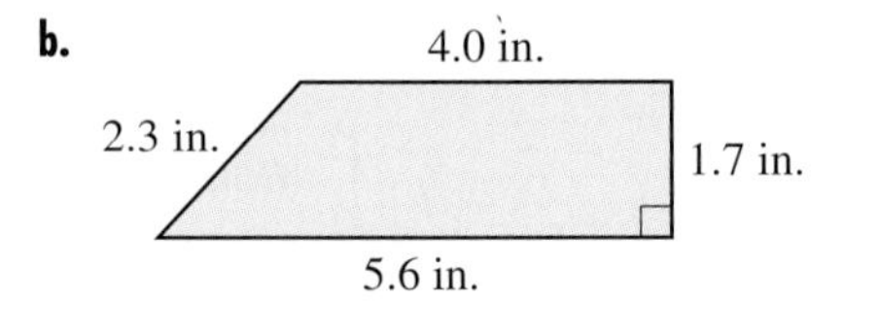

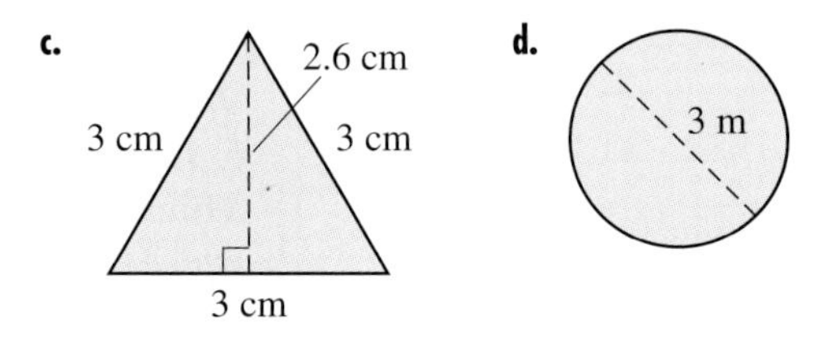

29. Find the perimeter and area of each shape.

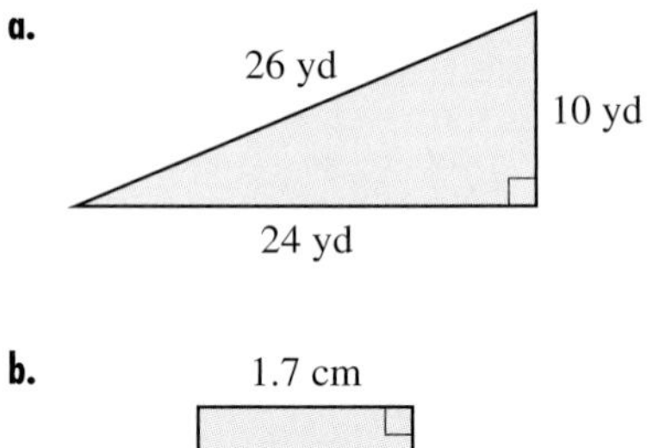

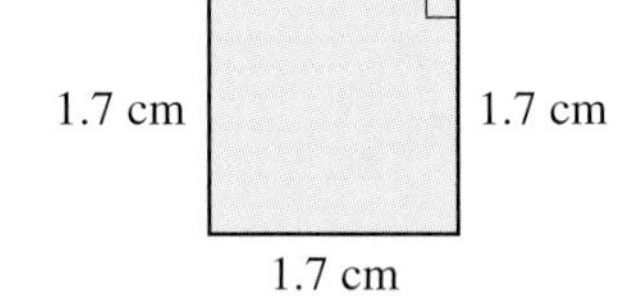

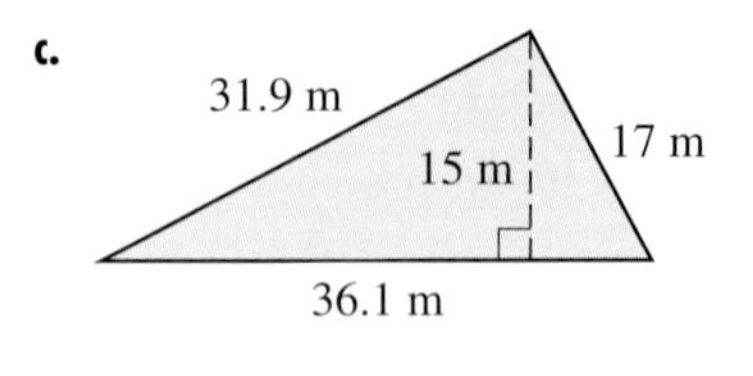

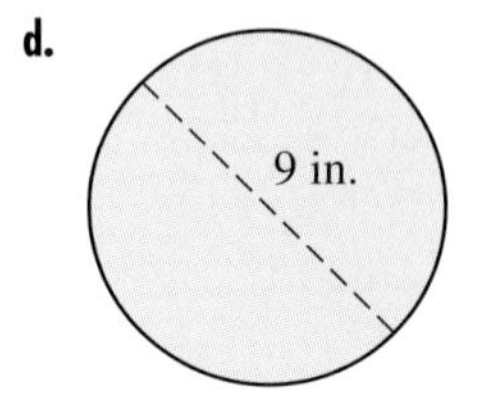

Blue numbers are core exercises.

30. Find the perimeter and area of each shape.

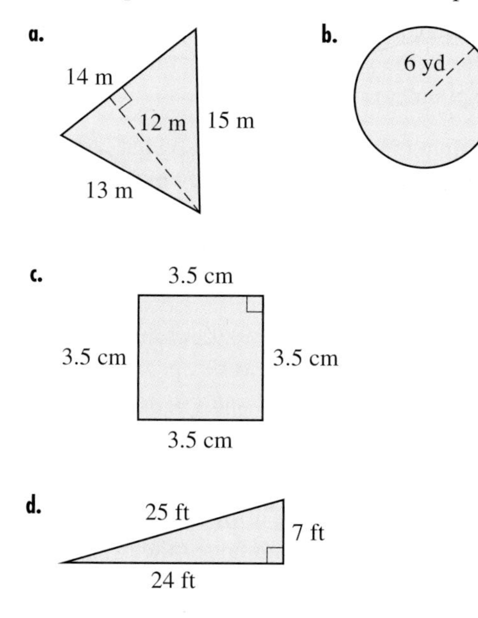

What are the surface area and volume of each shape in Exercises 31 to 36? Use $\pi \approx 3.14$. Round to the nearest tenth.

31. a.

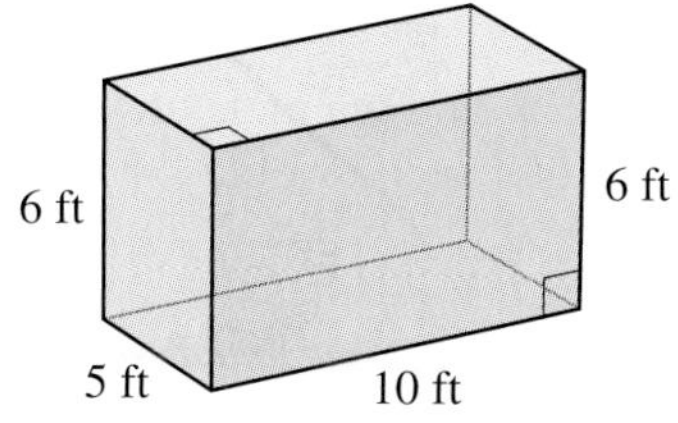

b.

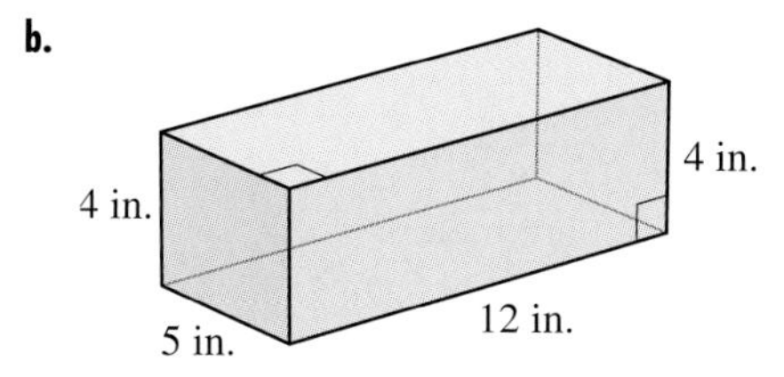

32. a.

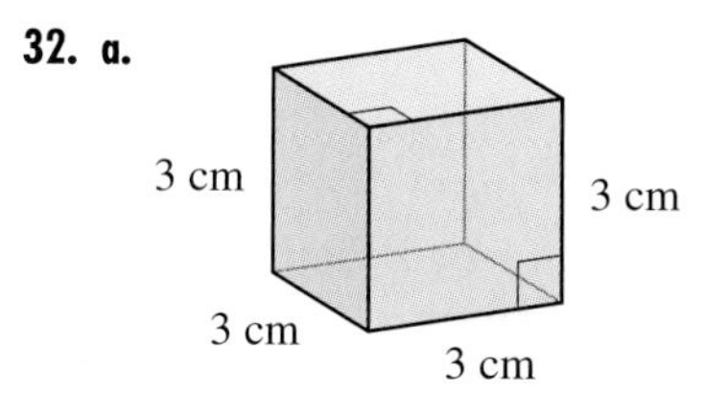

b.

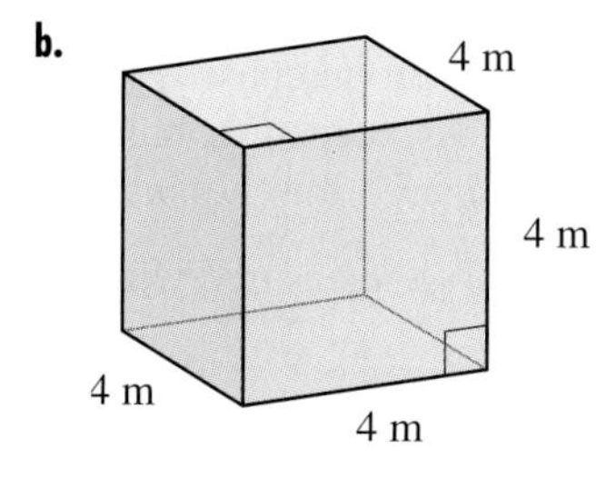

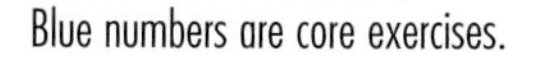

Blue numbers are core exercises.

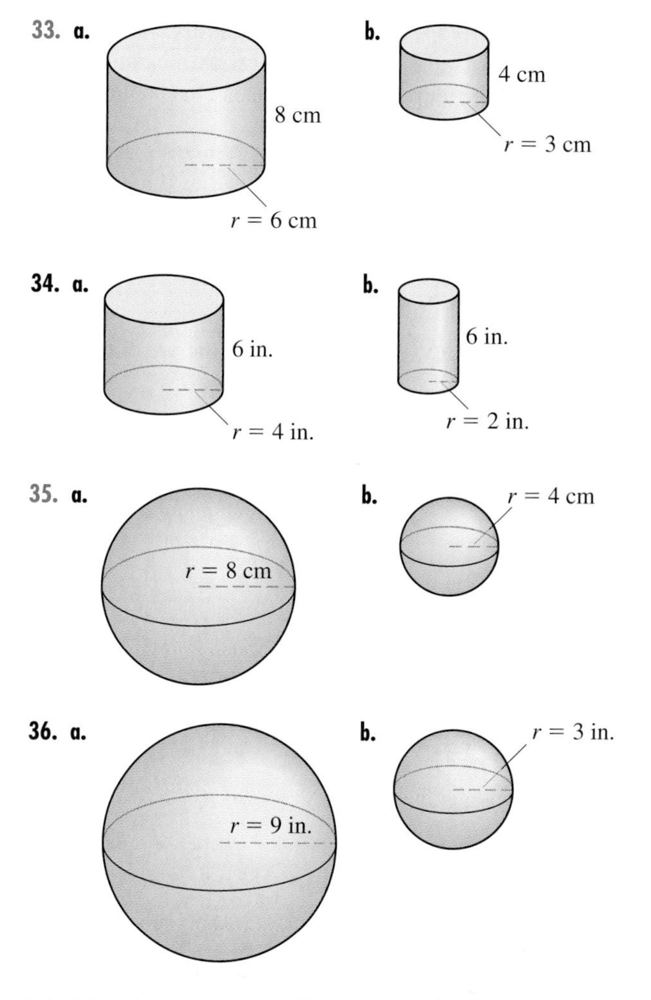

37. What is the volume of the cylinder in Example 10 in cubic feet and in gallons?

38. What is the volume of a child's wading pool 4 feet in diameter and 8 inches tall in cubic feet and in gallons? Assume a cylindrical shape.

Use unit analysis in Exercises 39 to 44.

39. For a reception, Robin needs 200 servings of punch at 12 fluid ounces per serving. How many gallons does she need to prepare?

40. Every December birdwatchers count all birds within 15-mile-diameter circles scattered across the nation. What is the area of each circle in square miles and in acres? 640 acres = 1 square mile

41. What is the volume of an athletic shoe box $4\frac{1}{2}$ inches in height, 7 inches wide, and $13\frac{1}{2}$ inches long? How many cubic inches would it take to store 1000 such boxes? How many cubic feet?

42. What is the volume of a cereal box with depth 7 centimeters, width 21 centimeters, and height 30.5 centimeters? How many cubic centimeters are needed to store 1000 such boxes? How many cubic meters?

43. Suppose a 12-fluid-ounce can of soda spills and spreads to a thickness of $\frac{1}{16}$ inch. How many square inches will the spill cover? 1 cup = 8 fluid ounces.

44. How long a sidewalk can be built with 4 cubic yards of concrete? Suppose the sidewalk is 4 inches thick and 4 feet wide.

In Exercises 45 to 48, use reasoning to find missing information. Use $\pi \approx 3.14$, and round to the nearest tenth.

45. Find the area of the blue part.

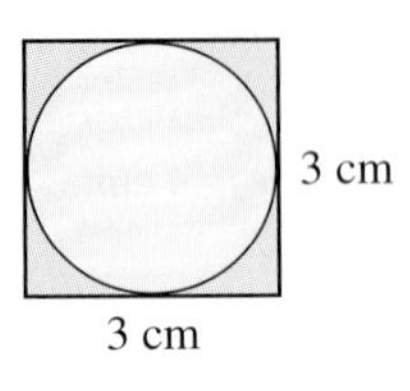

46. Find the area of the blue part.

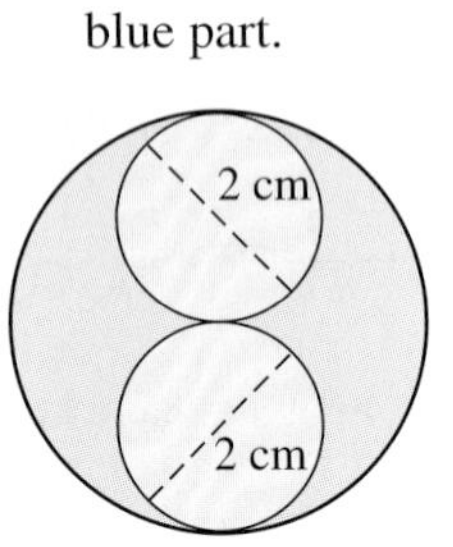

47. Find the area of the shaded part.

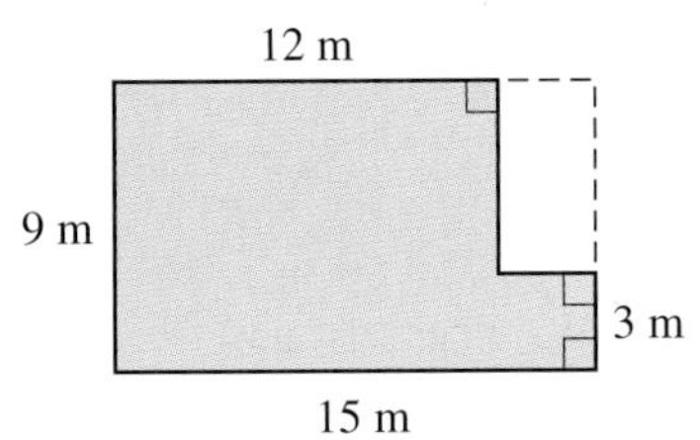

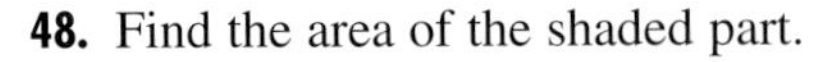

48. Find the area of the shaded part.

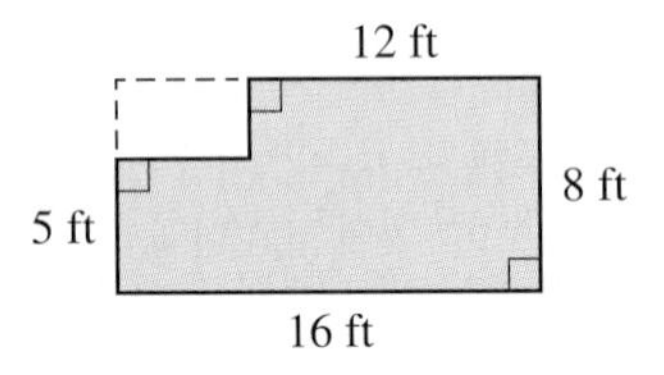

▶ Writing

49. Explain how to find the area of a trapezoid given lengths of parallel sides and height.

50. Explain how to find the volume of a sphere given radius.

51. Explain how to find the volume of a rectangular box given length, width, and height.

52. How are area and surface area the same? different?

53. Explain how finding the perimeter of a rectangle is an example of adding like terms.

54. Explain how finding the surface area of a rectangular prism (box) is an example of adding like terms.

Blue numbers are core exercises.

▶ Problem Solving: Doubling Dimensions

55. When the radius of a sphere or side of a cube doubles, what happens to the surface area? What happens to the volume? Explain your reasoning.

56. When the height of a cylinder doubles, what happens to the surface area? What happens to the volume?

▶ Projects

57. Molasses in January This exercise concerns an accident in Boston on January 15, 1919. A tank holding 2,500,000 gallons of molasses broke open, and a 6-foot-high flood poured out. Use unit analysis to answer the questions. *Facts:* 231 cubic inches = 1 gallon, 640 acres = 1 square mile.

a. If the molasses ran down a 40-foot-wide street, how long would it reach if it maintained the 6-foot thickness?

b. Suppose the molasses was stored in a cylindrical tank 50 feet high and 100 feet in diameter. What is the volume of such a tank in gallons? Round to the nearest ten thousand.

c. How many square feet would the molasses cover if it spread out to a depth of 1 inch? Would it cover a square mile?

d. How many acres would the molasses cover if it spread out to a depth of 1 inch?

58. Measuring Items Copy the following list of units of measure on your paper.

cups
gallons or liters
square yards or square meters
cubic feet
feet or meters
pounds
square feet
yards or meters
cubic yards or cubic meters

a. Identify each as a measure of length, area, volume, weight, or capacity.

b. Write each of the following items beside the appropriate unit(s) of measure: fabric area, fabric when purchased, concrete, paint when purchased, paint when applied, rope, peat moss, carpeting, butter when purchased, butter when measured for cooking.

c. Write three other items that are measured with each unit of measure.

▶ 2.6 Inequalities, Intervals, and Line Graphs

Objectives

- Write sets of numbers using inequalities.
- Write sets of numbers using intervals.
- Graph inequalities and intervals on a number line.

> **WARM-UP**
>
> The positions for 0 and 1 are marked on the number line.
>
> *a b c d e f g h i j k l*
>
> 0 1
>
> **1.** Match each number with a letter on the number line.
>
> $-0.25, 0.75, -2, 1\frac{3}{4}, 2.25, -1\frac{1}{4}, -1.5$
>
> **2.** What is the number marked by each of these letters on the number line: *d, e, g, i, k*?

IN TABLES DESCRIBING TUITION, shipping and handling charges, and credit card fees, we used sets of numbers. This section presents inequalities, intervals, and line graphs as ways to write and display these sets.

▶ Inequalities

COMPARING NUMBERS An **inequality** is *a statement that one quantity is greater than or less than another quantity*. Inequality symbols are used to compare the relative positions of two numbers on a number line or the relative sizes of two or more expressions. The symbol $>$ in $4 > -5$ indicates that 4 is greater than -5. The number 4 is to the right of -5 on the number line in Figure 26. An equivalent statement is $-5 < 4$, read "-5 is less than 4."

$-5 \quad -4 \quad -3 \quad -2 \quad -1 \quad 0 \quad 1 \quad 2 \quad 3 \quad 4 \quad 5$

FIGURE 26 Comparing -5 and 4

An easy way to read inequalities is to remember that the point of the symbol always faces the lesser number and the wide part always faces the greater number.

There are four inequality symbols:

Inequality Symbol	Meaning
$<$	is less than (to the left of another number on the number line)
$>$	is greater than (to the right of another number on the number line)
$\leq$	is less than or equal to
$\geq$	is greater than or equal to

We present all the symbols at once. Historically, however, the symbols $>$ and $<$ first appeared in print in 1631 in England, and the symbols $\geq$ and $\leq$ appeared in 1734 in France. See also **http://members.aol.com/jeff 570/relation.html**

WRITING SETS We use variables with inequalities to describe sets of numbers.

▶ **EXAMPLE 1** Writing inequalities Match each sentence in parts a to e with the inequalities shown in one of the parts f to k.

a. The set of numbers less than 1

b. The set of numbers greater than or equal to -1

c. The set of numbers greater than 1

d. The set of numbers less than or equal to 1

e. The set of numbers greater than -1

f. $x \leq 1$ or $1 \geq x$

g. $x \geq -1$ or $-1 \leq x$

h. $x > -1$ or $-1 < x$

i. $x \leq -1$ or $-1 \geq x$

j. $x > 1$ or $1 < x$

k. $x < 1$ or $1 > x$

SOLUTION **a.** k **b.** g **c.** j **d.** f **e.** h ◀

COMPOUND INEQUALITY A range of numbers, such as the set 0 to 20, may be described with a variable and two inequalities. For the set of numbers greater than or equal to zero, we write $x \geq 0$. For the set of numbers less than or equal to 20, we write $x \leq 20$.

A **compound inequality** is *two inequalities combined in one statement with two inequality signs in the same direction.* We write $x \geq 0$ and $x \leq 20$ as $0 \leq x \leq 20$. We read this as "the set of numbers between 0 and 20, including 0 and 20."

▶ **EXAMPLE 2** Writing compound inequalities Write each pair of inequalities as a compound inequality. Use only $<$ or $\leq$.

a. $x < 0$ and $x > -3$

b. $x \leq 5$ and $x > -2$

c. $x > 0$ and $x \leq 3$

d. $x < -\frac{1}{2}$ and $x \geq -\frac{3}{4}$

SOLUTION To place the x and the numbers in order, remember that the point on the inequality faces the smaller number and the x goes between the two numbers.

a. $-3 < x < 0$

b. $-2 < x \leq 5$

c. $0 < x \leq 3$

d. $-\frac{3}{4} \leq x < -\frac{1}{2}$

It is a good habit to place the smaller number to the left. ◀

LINKED EXAMPLE: CREDIT CARD FEES We return to the Professional Association of Diving Instructors (PADI) credit card fee schedule.

TABLE 9 PADI Credit Card Fees

Input: Amount of Cash Advance	Output: Fee on Credit Card Statement
\$0 to \$166.66	\$5
\$166.67 to \$2499.99	3% of the amount
\$2500 to credit limit	\$75

▶ **EXAMPLE 3** Writing inequalities: credit card fees Use inequalities or compound inequalities to describe the inputs of the credit card fee schedule in Table 9, omitting dollar signs. Then write each of the inequalities in words.

SOLUTION Table 10 shows inequalities and word descriptions for the inputs in Table 9.

TABLE 10 Input Sets as Inequalities and Words

Inequalities	Words
$0 < x \leq 166.66$	The set of numbers between 0 and 166.66, including 166.66
$166.67 \leq x \leq 2499.99$	The set of numbers between 166.67 and 2499.99, including 166.67 and 2499.99
$x \geq 2500$	The set of numbers greater than or equal to 2500

◀

The inequalities in Table 10 are satisfactory in business and finance, but not in mathematics. There are number gaps between the given input sets. Between \$166.66 and \$166.67 are all the fraction and decimal portions of one cent: $\$166.66\frac{1}{2}$, \$166.665, and \$166.669. There is another one-cent gap between \$2499.99 and \$2500. In mathematics, we prefer to include all numbers in our inequalities, leaving no gaps.

▶ **EXAMPLE 4** Removing gaps: credit card fees, continued Use inequalities to describe the inputs of the credit card fee schedule in Table 9 without any gaps between the input sets, omitting dollar signs. Describe each set in words.

SOLUTION The first inequality, $0 < x \leq 166.66$, remains the same.

In the second inequality, we start at 166.66, end at 2500, but do not want to include either. The compound inequality becomes $166.66 < x < 2500$, which excludes both 166.66 and 2500.

The third inequality, $x \geq 2500$, remains the same. The results are summarized in Table 11.

TABLE 11 Input Sets Without Gaps

Inequalities Without Gaps	Words
$0 < x \leq 166.66$	The set of numbers greater than 0 and less than or equal to 166.66
$166.66 < x < 2500$	The set of numbers greater than 166.66 and less than 2500
$x \geq 2500$	The set of numbers greater than or equal to 2500

◀

Student Note:
This is another use of parentheses and brackets.

▶ Intervals

We may also state inequalities as intervals. An **interval** is *a set containing all the numbers between its endpoints as well as one endpoint, both endpoints, or neither endpoint*. Intervals indicate the inclusion or exclusion of endpoints through the use of brackets or parentheses. Brackets, [ ], indicate that the endpoints are included in the set. Parentheses, ( ), are used when the endpoints are excluded from the set. We may mix brackets and parentheses in one interval, as shown in Table 12.

TABLE 12 Inequalities and Intervals

Inequality	Interval	Words
$3 \leq x \leq 7$	[3, 7]	The set of numbers between 3 and 7, including 3 and 7
$3 \leq x < 7$	[3, 7)	The set of numbers between 3 and 7, including 3
$3 < x \leq 7$	(3, 7]	The set of numbers between 3 and 7, including 7
$3 < x < 7$	(3, 7)	The set of numbers between 3 and 7

Caution: Interval notation may look like an ordered pair. Read carefully when you come across (a, b) to see whether the reference is to a coordinate point (a, b) or an interval (a, b) describing the set $a < x < b$. Whenever possible, we will use the words *on the interval* (a, b) to refer to interval notation in this text.

▶ **EXAMPLE 5** Writing intervals: credit card fees revisited Use intervals to describe the inputs for the credit card fees.

a. $0 < x \leq 166.66$ **b.** $166.66 < x < 2500$ **c.** $x \geq 2500$

SOLUTION **a.** $(0, 166.66]$ The left parenthesis indicates that zero is excluded.

b. $(166.66, 2500)$ The parentheses indicate that the endpoints are excluded.

c. $[2500, +\infty)$ The interval describes all numbers greater than or equal to 2500.

There is no greatest number, so we need a symbol to say that *the numbers get large without bound.* We use an **infinity sign**, ∞. Realistically, the credit card company would object to this much cash advance and put a limit, say \$2000, on the card. ◀

Infinite means *without bound.* Infinity describes a concept, not a number. Number lines and axes on coordinate graphs all have arrows on their ends because the numbers on the number line go to the right and left forever. Placing a positive sign before the infinity sign means infinite to the right on the number line; a negative sign indicates infinite to the left (see Figure 27).

$-\infty$ 0 $+\infty$

FIGURE 27 The infinite number line

We *always* write intervals with the smaller number on the left. For either $5 \geq x$ or $x \leq 5$, we write $(-\infty, 5]$. We always use a parenthesis beside the infinity sign.

▶ **EXAMPLE 6** Writing intervals Write each inequality in interval notation.

a. $x \leq 1$ **b.** $x \geq -1$ **c.** $x > -1$

d. $x \leq -1$ **e.** $x > 1$ **f.** $x < 1$

SOLUTION **a.** $(-\infty, 1]$ **b.** $[-1, +\infty)$ **c.** $(-1, +\infty)$

d. $(-\infty, -1]$ **e.** $(1, +\infty)$ **f.** $(-\infty, 1)$ ◀

▶ Line Graphs

It is sometimes useful to draw a picture of the set of numbers described by an inequality or interval. We use a line graph to do so. The graph of the inequality $1 \leq x \leq 4$ or interval $[1, 4]$ is a number line with dots (solid circles) at 1 and 4 and a line segment connecting them. (Using brackets on the number line instead of dots is also acceptable.)

Student Note:
Ask your instructor if he or she has a preference for line graph notation.

Inequality: $1 \le x \le 4$

Interval: $[1, 4]$

Words: The set of numbers between 1 and 4, including 1 and 4

Line Graph:

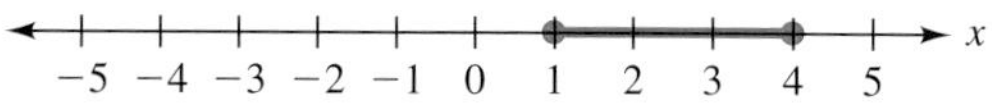

The graph of $1 < x < 4$ or its interval $(1, 4)$ has *small (open) circles* as endpoints at 1 and 4 and a line segment connecting them. (Using parentheses instead of the small circles is also acceptable.)

Inequality: $1 < x < 4$

Interval: $(1, 4)$

Words: The set of numbers between 1 and 4

Line Graph:

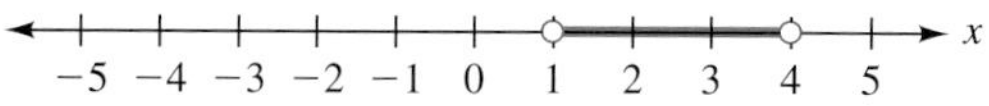

▶ **EXAMPLE 7** Drawing line graphs of inequalities Write each inequality as an interval, express it in words, and make a line graph.

a. $-2 < x \le 4$ **b.** $-3 \le x < 4$

SOLUTION Both endpoint forms are shown in the graphs.

a. Inequality: $-2 < x \le 4$

Interval: $(-2, 4]$

Words: The set of numbers between -2 and 4, including 4

Line Graph:

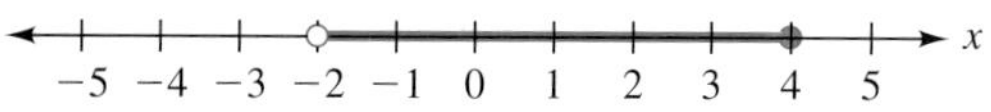

b. Inequality: $-3 \le x < 4$

Interval: $[-3, 4)$

Words: The set of numbers between -3 and 4, including -3

Line Graph: ◀

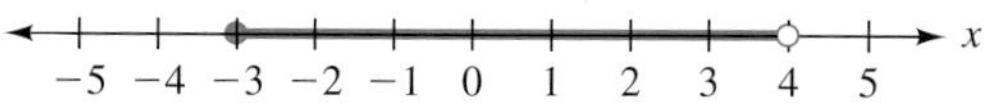

THINK ABOUT IT 1: Write the word descriptions in Example 7 in another way.

TEST POINTS To be sure that we have the line graph drawn correctly, we select a *test point* on the number line. If the test point makes the inequality true, then the graph should pass through the point. If the test point makes the inequality false, then the graph goes in the opposite direction. *Zero is usually a convenient test point for a line graph.*

▶ **EXAMPLE 8** Drawing line graphs of inequalities Write each inequality as an interval, express it in words, and make a line graph. Use a test point to check the graph.

a. $x < 4$ **b.** $x \ge -2$

SOLUTION **a.** The inequality $x < 4$ is the set of all numbers less than 4 and describes the interval $(-\infty, 4)$. The test point 0 gives $0 < 4$, which is true. Thus, the graph goes through 0. Either the parenthesis or the open circle on the 4 excludes 4 from the graph.

Inequality: $x < 4$

Interval: $(-\infty, 4)$

Words: The set of numbers less than 4

Line Graph:

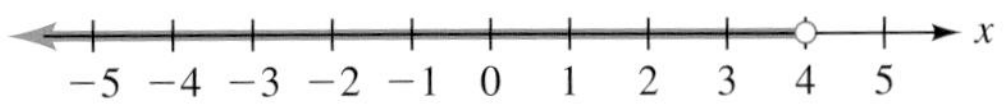

b. The inequality $x \geq -2$ is the set of all numbers greater than or equal to -2, and the interval is $[-2, +\infty)$. The test point 0 gives $0 \geq -2$, which is true. The graph goes through 0. The bracket or dot at -2 shows the inclusion of -2 in the set.

Inequality: $x \geq -2$

Interval: $[-2, +\infty)$

Words: The set of numbers greater than or equal to -2

Line Graph:

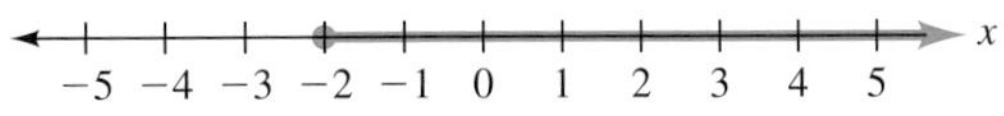

◀

THINK ABOUT IT 2: Draw each line graph in Examples 7 and 8 using brackets or parentheses, as appropriate.

SUMMARY In reading the summary chart in Table 13, observe the use of parentheses with $<$, $>$, or the infinity sign (∞) and the use of brackets with $\leq$ or $\geq$.

TABLE 13 Symbols Used in Inequalities and Intervals

Inequality Symbol	Interval Notation	Word Meaning	Line Graph Notation
$<$	(,)	is less than	open circle or (,)
$>$	(,)	is greater than	open circle or (,)
$=$		is equal to	dot
$\leq$	[,]	is less than or equal to	dot or [,]
$\geq$	[,]	is greater than or equal to	dot or [,]
$+\infty$	$, +\infty)$	positive infinity	$\rightarrow$
$-\infty$	$(-\infty,$	negative infinity	$\leftarrow$

ANSWER BOX

Warm-up: 1. *f, h, a, j, l, c, b* **2.** $-1, -\frac{1}{2}, \frac{1}{4}, 1\frac{1}{4}, 2$ **Think about it 1: a.** x is greater than -2 and less than or equal to 4. **b.** x is greater than or equal to -3 and less than 4. **Think about it 2:**

7.a.

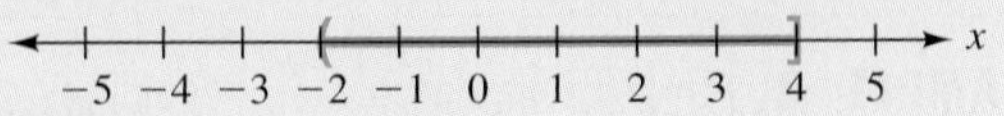

7.b.

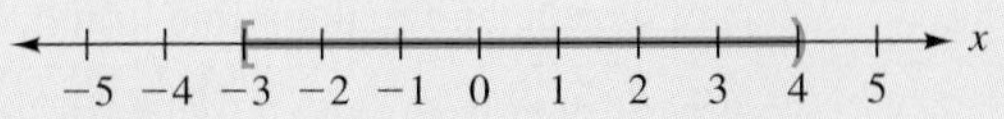

8.a.

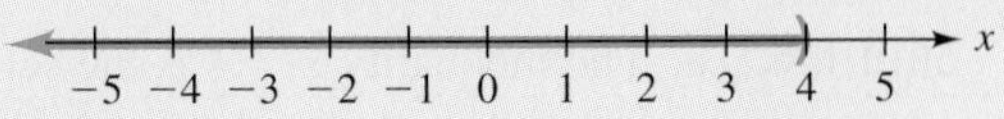

8.b.

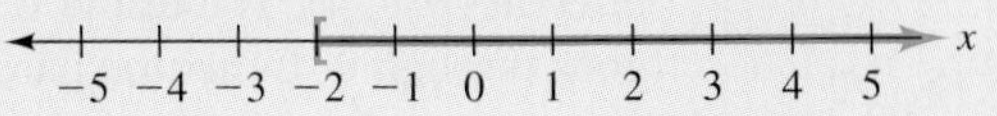

▶ 2.6 Exercises

In Exercises 1 and 2, draw a number line from -2 to 2, marked in eighths, and write the numbers where they belong on the line.

1. $-\frac{7}{8}, 0.5, -\frac{5}{4}, -1\frac{3}{4}, \frac{1}{4}, \frac{3}{2}, -\frac{1}{8}, 0.75$

2. $-\frac{6}{8}, -0.25, -1.5, 1\frac{5}{8}, \frac{1}{2}, \frac{7}{4}, -1\frac{3}{8}, \frac{3}{4}$

In Exercises 3 to 6, copy the statement and fill in the correct sign: $<$, $=$, or $>$.

3. a. $-8 \square -3$ **b.** $+4 \square -9$
c. $(-3)^2 \square 3^2$ **d.** $0.5 \square 0.5^2$
e. $6 \square -5$ **f.** $-2(6) \square -2(-5)$
g. $-6 \square -5$ **h.** $-2(-6) \square -2(-5)$

4. a. $-7 \square 5$ **b.** $2(-7) \square 2(5)$
c. $-2(-7) \square -2(5)$ **d.** $0.2^2 \square 0.2$
e. $1.5 \square 1.5^2$ **f.** $\frac{3}{4} \square (\frac{3}{4})^2$
g. $3(-7) \square 3(-6)$ **h.** $-3(-7) \square -3(-6)$

5. a. $-3.75 \square -3.25$ **b.** $3(-2) \square -3(2)$
c. $\frac{1}{2} \square -\frac{1}{2}$ **d.** $|-4| \square |2|$
e. $-2(-3) \square 2(-4)$ **f.** $(\frac{1}{2})^2 \square (-\frac{1}{2})^2$
g. $-2.5 \square -3$ **h.** $\frac{22}{7} \square \pi$

6. a. $0.5 \square 0.25$ **b.** $-0.5 \square -0.75$
c. $3(6) \square 3(-2)$ **d.** $|-3| \square -3$
e. $3(-5) \square 4^2$ **f.** $(-2)^2 \square 2^3$
g. $-2(-3)(-4) \square -4(-3)(-2)$
h. $\pi \square 3.14$

In Exercises 7 and 8, write each pair of inequalities as a compound inequality. Use only $<$ or $\leq$.

7. a. $x < 4$ and $x > 0$ **b.** $x \leq -2$ and $-5 < x$

8. a. $x > -2$ and $0 \geq x$ **b.** $x \geq -\frac{1}{2}$ and $\frac{1}{4} > x$

In Exercises 9 and 10, write each compound inequality as two separate inequalities. More than one answer is possible.

9. a. $3 < x < 8$ **b.** $-3 < x \leq -1$
c. $-2 < x < 1$

10. a. $-5 \leq x < 0$ **b.** $-\frac{1}{4} \leq x < \frac{1}{2}$
c. $-3 < x < 0$

Choose one listed inequality and one listed interval that describe each line graph in Exercises 11 to 16.

Inequality	**Interval**
(a) $-3 < x < 3$	(p) $(-\infty, -3)$
(b) $-3 \leq x \leq 3$	(q) $(-\infty, 3]$
(c) $x \geq -3$	(r) $[-3, 3]$
(d) $x > -3$	(s) $(-\infty, 3)$
(e) $x \geq 3$	(t) $[-3, +\infty)$
(f) $x < 3$	(u) $(-3, 3)$
(g) $x > 3$	(v) $(-3, +\infty)$
(h) $x \leq 3$	(w) $[3, +\infty)$

11. x
−5 −4 −3 −2 −1 0 1 2 3 4 5

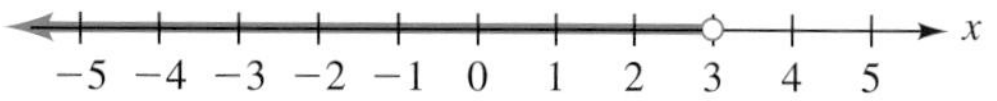

12. x
−5 −4 −3 −2 −1 0 1 2 3 4

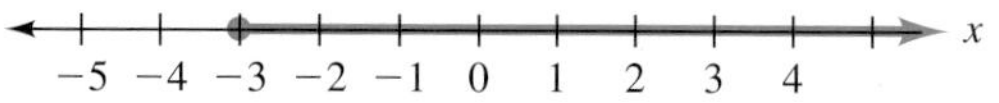

13. x
−5 −4 −3 −2 −1 0 1 2 3 4 5

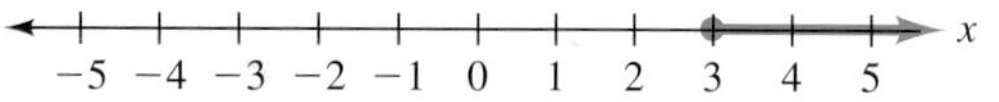

14. x
−5 −4 −3 −2 −1 0 1 2 3 4 5

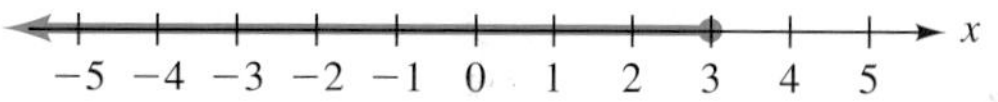

15. x
−5 −4 −3 −2 −1 0 1 2 3 4 5

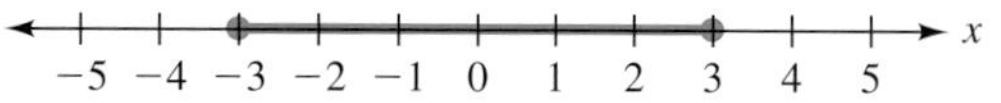

16. x
−5 −4 −3 −2 −1 0 1 2 3 4

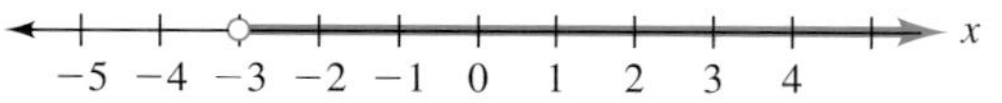

In Exercises 17 and 18, complete the tables.

17.

	Inequality	Interval	Words	Line Graph
a.	$-1 \leq x < 3$			
b.		$(-4, 1]$		
c.			x is between -3 and 5, including -3.	
d.	$x < -4$			
e.				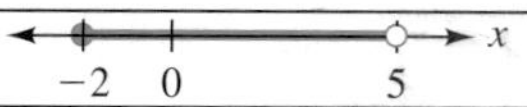 x; −2 0 5
f.	$x > -2$			
g.			x is greater than -4 and less than or equal to 2.	
h.		$[-3, +\infty)$		

Blue numbers are core exercises.

18.

	Inequality	Interval	Words	Line Graph
a.	$-4 < x < -1$			
b.		$[3, 8)$		
c.			x is greater than or equal to -2 and less than 3.	
d.	$x > -3$			
e.				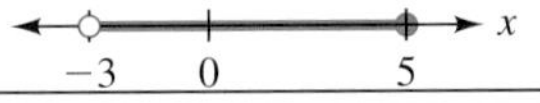
f.	$x < 2$			
g.			x is between -1 and 3, including -1.	
h.		$(-\infty, 4]$		

In Exercises 19 to 22, write inequalities to describe the situations. Do not leave any gaps between your inequalities.

19. A tax schedule. The tax, x, is

a. not over 2000

b. over 2000 but not over 5000

c. over 5000

20. A purchase discount schedule. The number of items, x, is

a. 10 or less

b. greater than 10 and less than 50

c. 50 or more

21. Customer ages, x, for restaurant meals are:

a. Small meal: less than 5 years

b. Regular meal: 5 years or more to under 65 years

c. Small meal: 65 or older

22. Age categories. The age, x, is

a. less than eighteen

b. eighteen and above but less than twenty-one

c. twenty-one and above

23. Write these birth years as intervals:

a. Seniors: from 1890 to 1941

b. War babies: from 1942 to 1945

c. Baby boomers: from 1946 to 1964

d. Echo boomers or Generation X: from 1965 to 1978

e. Millennials or Generation Y: from 1979 to 2000

▶ Writing

24. Explain the difference between $<$ and $\leq$. Write a number statement containing each.

25. What is an interval?

26. Explain the meaning of the ∞ in $(4, +\infty)$.

27. Why is $4 < x < 2$ not a true statement?

28. Why is $3 > x > 5$ not a true statement?

29. Why is $2 \leq 2$ a true statement?

30. Why is $3 \geq 3$ a true statement?

▶ Computation Review

31. Complete the table by finding n percent of each number or expression in the top row.

n percent	\$1.00	\$5.00	\$10.00	x
6%				
10%				
25%				
100%				
150%				

▶ Projects

32. Size of Numbers Make a table with the following three headings: x^2 is less than x, x^2 is equal to x, x^2 is greater than x.

a. Square each number x listed below. Compare x^2 with x and place the original number, x, under the appropriate heading in your table.

$\frac{2}{3}$, 1.5, $\left(-\frac{1}{2}\right)$, $\frac{1}{3}$, 1, 2.5, (-1), 0.5

0.1, 0, (-0.1), 2, (-2), (-2.5)

Blue numbers are core exercises.

(*Hint:* The negative numbers in the listing above are placed in parentheses to remind you to use parentheses when squaring on the calculator.)

b. In words, summarize the relationship between the size of the square of a number and the position of the original number on a number line.

For parts c and d, find a number that makes the statement true and another that makes the statement false.

c. $-x > x$

d. $\frac{1}{x} > x$

33. Utility Payments Read your utility bill (electricity, natural gas, heating oil, or water) or call the billing department of your local utility to get a current cost schedule. Write inputs in both interval and inequality form.

2 Chapter Summary

Vocabulary

For definitions and page references, see the Glossary/Index.

absolute value
adding like terms
additive inverse
algebraic fractions
associative properties
base
capacity
circumference
commutative properties
composite numbers
compound inequality
continued equality
cubed
distributive property of multiplication over addition
equivalent expressions
evaluate
exponent
factors
formula
greatest common factor
grouping symbol
inequality
infinite
infinity sign
interval
like terms
multiplicative inverse
non-negative
order of operations
pi
power
prime number
radical
reciprocal
simplification property of fractions
simplify
squared
surface area
term
unit analysis
volume

Concepts

2.1 Addition and Subtraction with Integers

(The property marked * is new and is added at this time to provide a complete list.)

Addition and subtraction are inverse operations.

Subtraction is equivalent to adding the opposite (additive inverse):

$$a - b = a + (-b)$$
$$a - (-b) = a + b$$

*The sum of a number n and 0 is n.

The sum of a number n and its opposite, $-n$, is 0.

For $a + b$, if the signs are alike, add their absolute values and place the common sign on the answer.

For $a + b$, if the signs are different, find the positive difference in their absolute values and place the sign from the number with the greatest absolute value on the answer.

2.2 Multiplication and Division with Positive and Negative Numbers

Multiplication and division are inverse operations.

Division is equivalent to multiplication by the reciprocal:

$$x \div n = x \cdot \frac{1}{n} \quad \text{or} \quad x \div \frac{a}{b} = x \cdot \frac{b}{a}$$

The product of a number n and 1 is n.

The product of a number n and 0 is 0.

In multiplication and division of two real numbers, if the signs are alike, the answer is positive. If the signs are different, the answer is negative.

2.3 Properties of Real Numbers Applied to Simplifying Fractions and Adding Like Terms

Associative property of addition:

$$(a + b) + c = a + (b + c)$$

Associative property of multiplication:

$$(a \cdot b) \cdot c = a \cdot (b \cdot c)$$

Commutative property of addition:

$$a + b = b + a$$

Commutative property of multiplication:

$$a \cdot b = b \cdot a$$

Distributive property of multiplication over addition:

$$a(b + c) = ab + ac$$

Simplification property of fractions: For all real numbers, a not zero and c not zero,

$$\frac{ab}{ac} = \boxed{\frac{a}{a}} \cdot \frac{b}{c} = 1 \cdot \frac{b}{c} = \frac{b}{c}$$

2.4 Exponents and Order of Operations

The expression x^n is the nth power of x. When the exponent n is a positive integer, x^n has n factors of the base x.

Order of operations: The order of operations for algebra and scientific calculators:

1. Calculate expressions within parentheses and other grouping symbols.
2. Calculate exponents and square roots.
3. Do remaining multiplication and division in the order of appearance, left to right.
4. Do remaining addition and subtraction in the order of appearance, left to right.

2.5 Unit Analysis and Formulas

To do a unit analysis:

1. Name the unit of measure to be changed, and name the unit of measure for the answer.
2. List facts containing the unit to be changed and all units needed to get to the unit for the answer.
3. Write the unit to be changed (or a fact containing the unit) as the first fraction. Set up a product of fractions using the list of facts. Arrange fractions so that the units to be eliminated appear in both a numerator and a denominator. Carry out the multiplication.
4. Recheck the units of measure noted in step 1.

See Table 7 on page 98 and Table 8 on page 101 for selected geometric formulas.

To use a formula:

1. Write the formula.
2. List the facts, with units.
3. Substitute the facts into the formula.
4. Simplify the units.
5. Calculate the numbers.

2.6 Inequalities, Intervals, and Line Graphs

See Table 13 on page 112.

▶ 2 Review Exercises

1. Add or subtract, as indicated.

a. $-3 + -5$ **b.** $3 + (-8)$ **c.** $4 - 17$
d. $-5 - (-18)$ **e.** $-21 + 7$ **f.** $-26 + 19$
g. $14 - (-28)$ **h.** $12 - 36$ **i.** $-32 - (-16)$
j. $-4 - (-16)$ **k.** $-11 + 22$ **l.** $8 - (-5)$

2. Add or subtract, as indicated.

a. $-4 + 9.5$ **b.** $-5 + 1.1$ **c.** $2.5 - (-6)$
d. $-3 - (-0.7)$ **e.** $-1.0 + 0.6$ **f.** $-1.3 + 0.8$
g. $2.6 - (-1.3)$ **h.** $-1.9 - 4.7$ **i.** $-0.3 + (-0.4)$
j. $\frac{5}{6} - \frac{7}{9}$ **k.** $-\frac{1}{4} + \left(-\frac{3}{4}\right)$ **l.** $\frac{5}{3} + \left(-\frac{2}{3}\right)$
m. $-\frac{2}{3} - \left(-\frac{3}{4}\right)$ **n.** $-\frac{4}{5} + \frac{5}{3}$ **o.** $-\frac{5}{6} + \left(-\frac{3}{4}\right)$

3. Multiply or divide, as indicated.

a. $-9(6)$ **b.** $-9(-6)$ **c.** $-18(-3)$
d. $-18(3)$ **e.** $-8(-7)$ **f.** $8(-7)$
g. $4 \cdot (-14)$ **h.** $-4(-14)$ **i.** $-48 \div (-24)$
j. $-48 \div 12$ **k.** $48 \div (-6)$ **l.** $-48 \div (-6)$
m. $-48 \div (-3)$ **n.** $-48 \div 3$ **o.** $48 \div (-8)$

4. Multiply or divide, as indicated.

a. $-1.0(0.6)$ **b.** $-1.3(-0.8)$
c. $2.6 \div (-1.3)$ **d.** $-1.7 \div 5.1$
e. $-0.3(-0.4)$ **f.** $0.7 \div (-1.4)$
g. $-\frac{1}{4}\left(-\frac{3}{4}\right)$ **h.** $\frac{5}{3}\left(-\frac{2}{3}\right)$
i. $-\frac{2}{3} \div \left(-\frac{4}{3}\right)$ **j.** $-\frac{4}{5} \div \frac{5}{3}$

Use real-number properties to do Exercises 5 to 10 mentally.

5. **a.** $8 + 7 + 2 + 3$ **b.** $92 + 55 + 8 + 45$

c. $1.25 + 2.69 + 3.75$ **d.** $\$5.29 + \$4.98 + \$5.02$

e. $2\frac{1}{2} + 1\frac{1}{3} + 4\frac{1}{4} + 1\frac{1}{4}$ **f.** $3\frac{7}{8} + 1\frac{1}{2} + 2\frac{1}{8} + 3\frac{1}{2}$

6. **a.** $4 + (-3) + (-7)$

b. $8 + (-3) + (-8)$

c. $-5 + 8 + 5 + (-4)$

d. $-6 + 9 + (-4) + 8$

e. $-7 + 20 + (-8) + (-5)$

f. $9 + (-3) + (-6) + (-12)$

7. **a.** $3 \cdot 5 \cdot 4$ **b.** $8 \cdot 2 \cdot 5$ **c.** $9 \cdot 4 \cdot 25$

d. $\frac{1}{2} \cdot 17 \cdot 20$ **e.** $9 \cdot \frac{1}{2} \cdot 6$ **f.** $5 \cdot \frac{1}{4} \cdot 8$

8. **a.** $5(\$8.99)$ **b.** $3(\$4.98)$ **c.** $4(\$2.03)$

d. $7(\$1.98)$ **e.** $6(\$1.03)$ **f.** $8(\$3.97)$

9. **a.** $4(3\frac{1}{4})$ **b.** $6(4\frac{1}{3})$ **c.** $8(1\frac{3}{4})$

d. $3(2\frac{2}{3})$ **e.** $9(1\frac{2}{3})$ **f.** $8(2\frac{3}{8})$

10. **a.** $(-4)(-6)(2)$ **b.** $(-3)(4)(-5)$

c. $(-2)(-4)(-6)$ **d.** $(5)(-3)(-6)(-1)$

e. $(4)(-10)(0)$ **f.** $(-3)(-7)(-3)$

11. Complete these statements.

a. $2(\_\_\_\_\_) = 2x + 2y$

b. $ac + ab = \_\_\_\_\_ (c + b)$

c. $4(\_\_\_\_\_\_\_\_\_\_) = 4x^2 - 8x + 12$

d. $3xy + 4x^2y = xy(\_\_\_\_\_)$

e. $\_\_\_\_\_ (2x + 4y - 5) = 6x + 12y - 15$

f. $15a^2bc + 5ab^2 + 10abc = 5ab(\_\_\_\_\_\_\_\_\_\_\_\_\_\_)$

g. The ________ property explains parts a to f.

12. Add like terms.

a. $12 - 3x + 5 + 6x$ **b.** $-4 - (2 - 3x)$

c. $4(2a^2 + 4) - (6 - 3a)$

d. $3a + 3(a - 2) - (2a + 5)$

e. $2a^2 + 3a + 2 - 4a^2 - a - 5$

f. $x^2 - (-3x) + 2 + 6x^2 - 2x - 1$

g. $4x^2 - 3xy - 2y^2 - 6x^2 + 6xy - y^2$

h. $a^2b + a^2b^2 + a^2 - 2a^2b^2$

i. **j.**

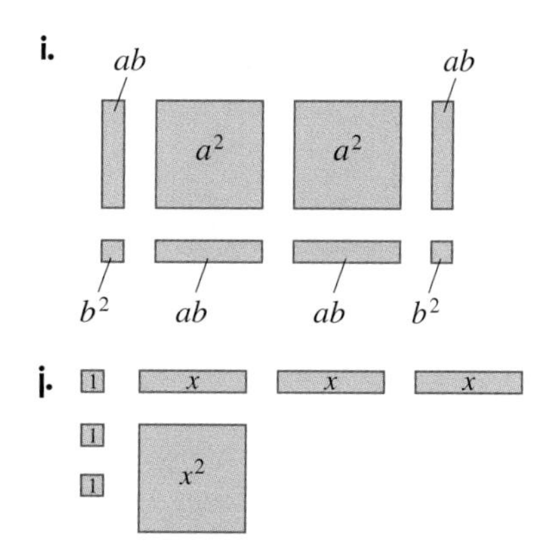

13. Simplify these expressions. Use the definition of positive integer exponents as needed. Answers should contain no parentheses.

a. $\dfrac{abc}{bcd}$ **b.** $\dfrac{4xy}{6xz}$ **c.** $\dfrac{-21cd}{14ad}$

d. $(-2x)^2$ **e.** $(-3y)^3$ **f.** $(-2y)^4$

g. $(-ab)^2$ **h.** $(ab)^2$ **i.** m^4m^5

j. m^2m^7 **k.** $m^5 \div m^2$ **l.** $m^7 \div m^4$

m. $\dfrac{3x + 6y}{3}$ **n.** $\dfrac{mn + n^2}{n}$ **o.** $\dfrac{2a + 4b}{4}$

14. Simplify. Leave expressions without parentheses.

a. $4(x - 3)$ **b.** $3(x + y - 5)$

c. $-2(x - y + 3)$ **d.** $5 - 3(x - 4)$

e. $2 - (x - 3)$ **f.** $-2(x - 4)$

g. $4 - 2(x - 4)$ **h.** $3 - (x - 4)$

15. Simplify these expressions. Answers should contain no parentheses.

a. $\left(\dfrac{4x}{y}\right)^2$ **b.** $\left(\dfrac{x}{3y}\right)^3$ **c.** $(-3ab)^2$

d. $\dfrac{3x^3}{9x}$ **e.** $\dfrac{4ab^2}{a^3b}$ **f.** $\dfrac{xy^3z}{x^2yz^2}$

g. $\dfrac{(-x)^3}{-x^2}$ **h.** $\dfrac{-3x^3}{(9x)^2}$ **i.** $\dfrac{-4x^3}{(-4x)^2}$

j. $\dfrac{2 \text{ qt}}{2 \text{ qt}}$ **k.** $\dfrac{3 \text{ ft}}{1 \text{ yd}}$ **l.** $\dfrac{125 \text{ km}^3}{(5 \text{ km})^2}$

16. Simplify these expressions.

a. $4 - 3(3 - 5)$ **b.** $(4 - 3)(3 - 5)$

c. $\sqrt{5^2 - 4(2)(-12)}$ **d.** $\dfrac{4 - \sqrt{49}}{4}$

e. $|6 - 1| - |3 - 9|$ **f.** $(7 - 2)^2 + (4 - 1)^2$

g. $\dfrac{-3 - (-5)}{-6 - 4}$ **h.** $\frac{1}{2} \cdot 11(5 + 7)$

17. Simplify.

a. $-(-2)^2$ b. $4 - (-2) + (-2)^2$

c. $5 - (-3) + (-2)^2$ d. $-(-3)^2$

e. $\sqrt{3^2 + 4^2}$ f. $\sqrt{8^2 + 6^2}$

g. $\sqrt{25^2 - 20^2}$ h. $\sqrt{15^2 - 12^2}$

i. $\sqrt{1.5^2 + 2^2}$ j. $\sqrt{10^2 - 6^2}$

k. $-|-4|$ l. $|-6 - (-5)|$

18. Simplify with a calculator, as needed. Round to a whole number.

a. $-\frac{1}{2}(32)(4)^2 + 64(4) + 60$

b. $1000\left[\left(1 + \frac{0.08}{2}\right)^2\right]^5$

19. Change the word phrases to symbols and the symbols to word phrases. Tell which expressions have the same value. What is the value?

a. The sum of 3 and 4 is divided by the difference between 5 and 8.

b. Three plus the quotient of 4 and 5 is decreased by 8.

c. The sum of 3 and 4 is divided by 5 and decreased by 8.

d. Three is added to the quotient of 4 and the difference between 5 and 8.

e. $3 + 4 \div 5 - 8$ f. $(3 + 4) \div 5 - 8$

g. $3 + 4 \div (5 - 8)$ h. $(3 + 4) \div (5 - 8)$

i. $3 + \dfrac{4}{5 - 8}$ j. $\dfrac{3 + 4}{5 - 8}$

k. $3 + \dfrac{4}{5} - 8$ l. $\dfrac{3 + 4}{5} - 8$

20. Arrange the facts listed after each problem into a unit analysis that solves the problem. Round to the nearest tenth.

a. 140 pounds is how many grams (g)?
2.2 pounds = 1 kilogram
1000 grams = 1 kilogram

b. How many yards is 100 meters?
3 feet = 1 yard
12 inches = 1 foot
39.37 inches = 1 meter

c. A polo ground is 300 yards by 200 yards. How many square feet are in this rectangular field?
1 yard = 3 feet

d. Change 1 cubic foot into cubic inches.
1 foot = 12 inches

21. The Cheez-It® snack crackers box weighs 16 ounces. The serving size is 12 crackers. There are 140 calories in each ounce. A box contains 32 servings. Find the number of calories per cracker.

22. A 15-ounce bag of Diane's Tortilla Chips® contains 15 servings. There are 80 milligrams (mg) of sodium in each serving. How many milligrams of sodium are consumed in eating a bag of chips?

23. Evaluate these formulas. Round to the nearest tenth.

a. $A = \pi r^2$, $r = 2.5$ ft

b. $A = \frac{1}{2}bh$, $b = 5$ yd, $h = 4$ yd

c. $V = \frac{4}{3}\pi r^3$, $r = 3$ m d. $V = s^3$, $s = 1.5$ cm

24. Find the perimeters and areas of these figures. Round to the nearest tenth. Use $\pi \approx 3.14$.

a.

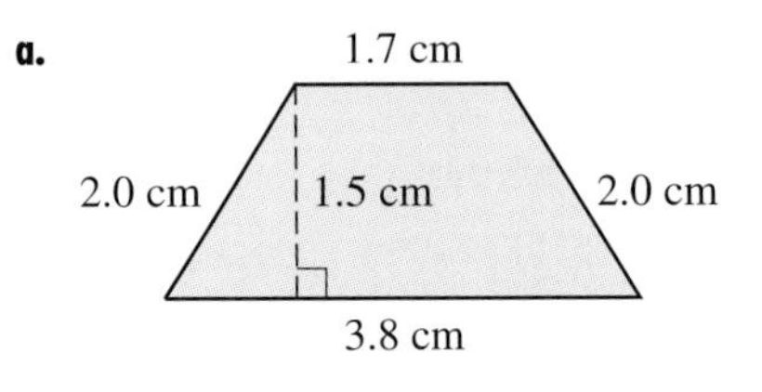

b.

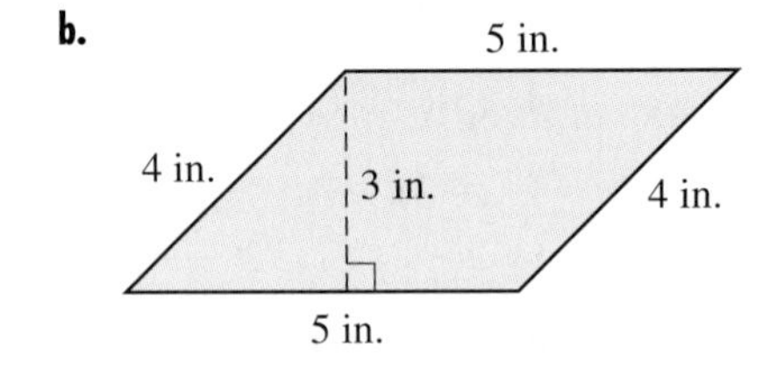

c.

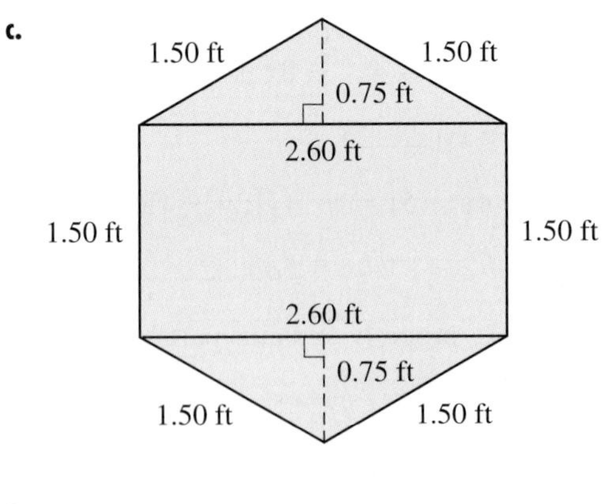

d.

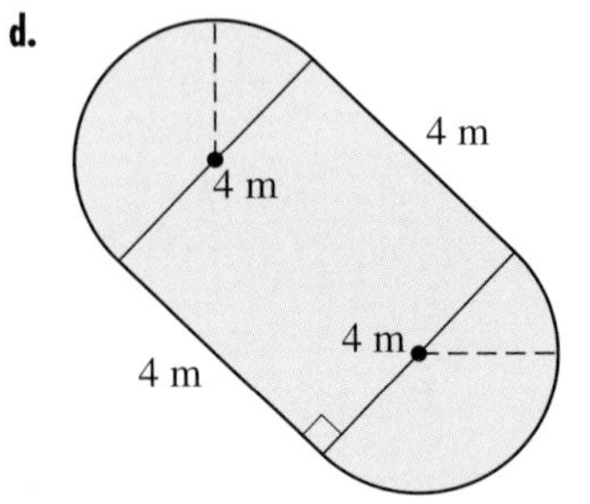

25. Find the surface areas and volumes of these figures. Round to a whole number. Use $\pi \approx 3.14$.

a.

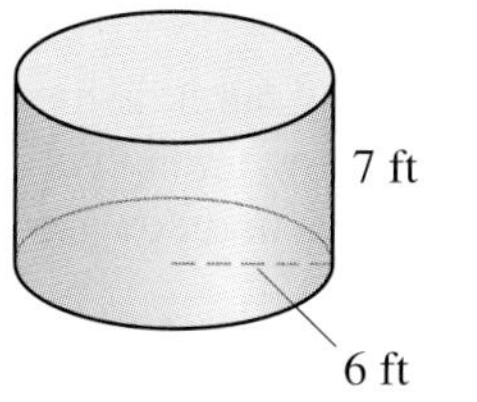

b.

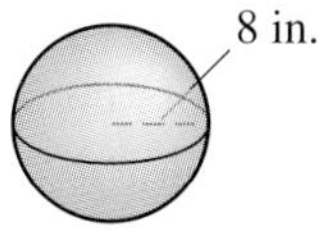

c.

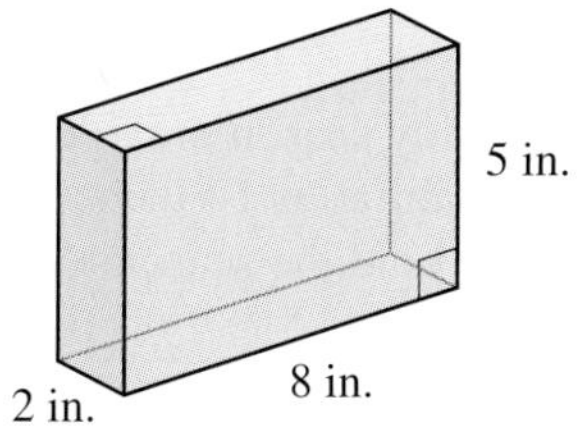

26. Find the missing numbers.

a. $-8 + \square = 0$ **b.** $5 \cdot \square = 1$

c. $\frac{3}{4} \cdot \square = 1$ **d.** $\square + (-3) = 0$

e. $\square \cdot \left(-\frac{1}{2}\right) = 1$ **f.** $5 + \square = 0$

27. Place the correct sign, $<$, $=$, or $>$, between these expressions.

a. $4 \square -3$ **b.** $2(-3) \square -2(-3)$

c. $(-2)^2 \square -2^2$ **d.** $|-4| \square |4|$

e. $-2^3 \square (-2)^3$ **f.** $|-5| \square -|5|$

g. $-\frac{1}{4} \square -\frac{1}{2}$ **h.** $-1.3 \square -1.5$

28. Choose from the inequality signs $\leq$, $<$, $>$, and $\geq$ to complete these statements. Explain why there are two correct signs for each answer.

a. $3 \square -5$ **b.** $-5 \square 2$

c. $-2 \square -2$ **d.** $\frac{1}{2} \square \frac{1}{4}$

e. $-\frac{1}{2} \square -\frac{1}{4}$

29. Give an example to show that three numbers can be regrouped to add.

30. Give an example to show that three numbers can be rearranged to multiply.

▶ Writing

31. Explain when to use the small circle and when to use the dot in a line graph.

32. Explain why $3x^2$ does not equal $(3x)^2$.

33. Explain why $-x^2$ is not the same as $(-x)^2$.

34. Explain how to find the reciprocal of 3.5.

35. Explain how to find the reciprocal of $2\frac{1}{4}$.

36. Explain why $|5|$ and $|-5|$ are equal.

37. Explain the role of inverse operations in subtraction of negatives and division of fractions.

38. Write three sentences that illustrate the difference in meaning among *base of triangle*, *base of exponent*, and any other use of *base*.

39. Write five sentences that illustrate the difference in meaning among *set of points*, *set of numbers*, *set of rules*, *set of line segments*, and *set of axes*.

40. Each row in the table below contains equivalent statements. Fill in the blanks for each row.

	Inequality	Interval	Words	Line Graph
a.	$x \leq 5$		x is less than or equal to 5	
b.	$-3 \leq x < 4$			
c.		$[-3, 5]$		
d.		$(2, 4)$		
e.		$(-5, 0]$		
f.			x is greater than -5	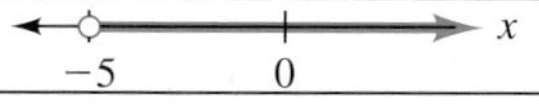
g.				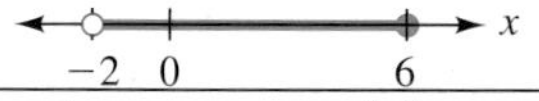
h.				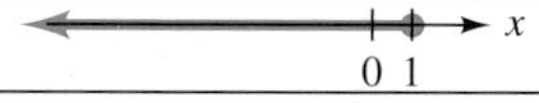
i.			x is zero or positive	
j.			x is negative	

41. Complete this table. Let x be the cost in dollars of the item.

Input: Cost of Item	Inequality	Interval
Less than \$50		
\$50 to \$500		
Over \$500		

42. Complete each table, graph the points (x, y), and then connect them.

a.

Input x	Output $y = -x - 2$
-2	
-1	
0	
1	
2	

b.

Input x	Output $y = -2x$
-2	
-1	
0	
1	
2	

c. Explain how part b suggests rules for multiplying positive and negative numbers.

43. A theater manual indicates the beam area for a spotlight is approximately $\frac{2}{3}$ that of the field area. A (6×22 ERS) spotlight set up 30 feet from the stage floor has a 5-foot field diameter. What is the diameter of the beam area?

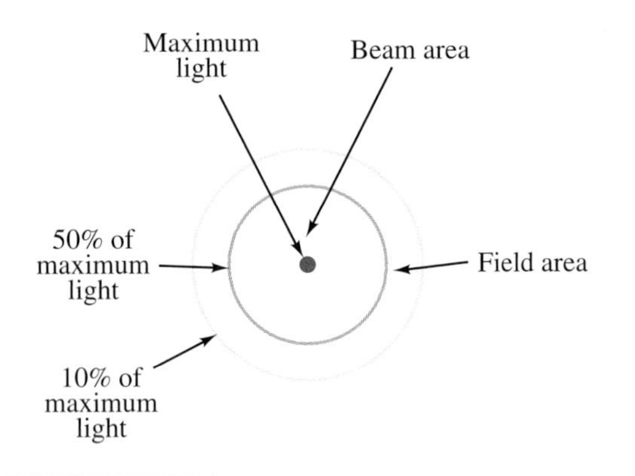

*Assume the slope of the roof does not affect the amount of rainwater.

▶ Chapter Project

44. Rainwater In order to catch rainwater with raingutters, a thrifty homeowner places 55-gallon barrels at two corners of his house, which measures 28 feet by 40 feet.*

a. If an inch of rain falls, how many cubic inches of water will go into the barrels? How many gallons? What advice do you have for this homeowner?

b. An Australian friend living in Wee Jasper has the same house, but it has a covered porch 8 feet wide all the way around it. If an inch of rain falls, how many cubic inches of water can she catch? How many gallons?

c. In part b, the Australian rainy season provides all the household water. Annual rainfall is 14 inches. Design a cylindrical water tank to hold the season's water. Assume the floor of the house is 5 feet off the ground, the slope of the porch is minimal, and the tank is filled by gravity. State any other assumptions.

d. The average American household uses 8000 gallons of water a month. Could they survive in Australia?

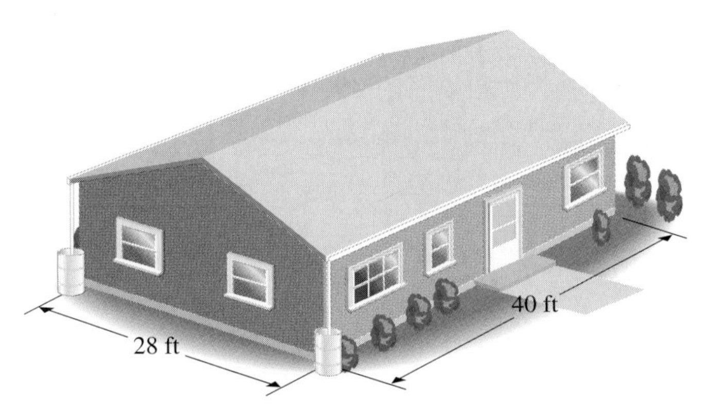

▶ 2 Chapter Test

1. Complete the table. Graph the points (x, y), and connect them.

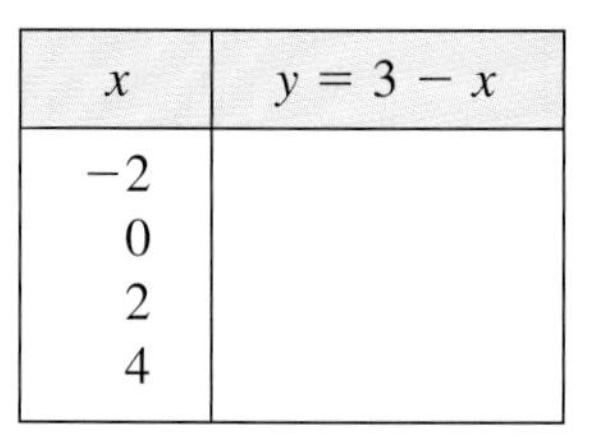

x	$y = 3 - x$
-2	
0	
2	
4	

2. Simplify.

a. $\frac{3}{4} + \frac{5}{6}$ **b.** $\frac{3}{4} - \frac{5}{6}$

c. $\frac{3}{4} \cdot \frac{-5}{6}$ **d.** $\frac{-3}{4} \div \frac{5}{6}$

3. What property allows us to change $1 + 4 + 9 + 16 + 25$ to $1 + 9 + 4 + 16 + 25$?

4. What property allows us to add the problem in Exercise 3 as $(1 + 9) + (4 + 16) + 25$ instead of following the usual left-to-right order in addition?

5. Simplify. Use the definition of positive integer exponents for Exercises g, h, and k to n.

a. $-5 + 9$ **b.** $-1.4 + 2.5 - 3.6$

c. $-4 - (-3)$ **d.** $(-3)(4)(-5)$

e. $8 - (-3)^2$ **f.** $\sqrt{26^2 - 24^2}$

g. m^2m^5 **h.** $m^5 \div m^3$

i. $36 \div 2 \cdot 2 - 3 + (3^2 - 5)$

j. $\frac{ace}{aft}$ **k.** $\frac{6x^2}{9x}$

l. $\frac{-x^3}{(-x)^2}$ **m.** $(ab)^3$

n. $\left(\frac{a}{2b}\right)^3$ **o.** $\frac{3x - 9}{3}$

p. $\frac{x^2 + 2x}{x}$

6. Add like terms. Remove parentheses as necessary.

a. $3x + 2y - 2x - 3y + 4x - 4y$

b. $x^3 - 3x^2 + x - 2x^2 + 6x - 2$

c. $2(x - 2) + 3(x - 1)$

d. $12(x - 1) - 5(x - 1)$

e.

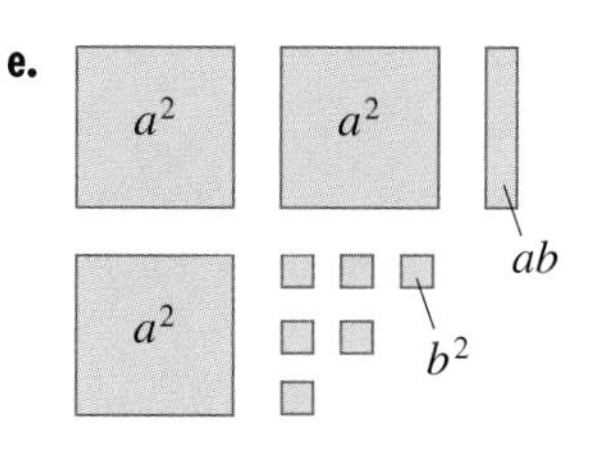

7. Complete each statement with the distributive property.

a. $3(2x + 9y) =$ ________

b. $3(2a + 9b) =$ ________

c. $2(3x^2 + 4x - 2) =$ ________

d. $ab(b - ab + a) =$ ________________

8. Are $(-x)^2$ and $-x^2$ the same? Use the word *base* to explain why or why not.

9. The formula for the area of a circle is $A = \pi r^2$. Let $\pi \approx 3.14$. Round to the nearest hundredth.

a. Find A if $r = 2$ ft. **b.** Find A if $r = 20$ ft.

c. Divide the areas to find how many times as large as the area in part a the area in part b is. Complete this statement: When the radius is multiplied by 10, the area of a circle is multiplied by ______.

10. The record for airplane flight duration is almost 5,616,000 seconds. Change the seconds into days. (The record is 2 hours less, but the above number comes out to an even number of days. The flight was refueled in the air.)

11. The area of the world's largest mural (in Long Beach Arena, CA) is 18,446,400 square inches. (12 inches = 1 foot, 1 yard = 3 feet.)

a. Change the area into square feet.

b. If the height of the rectangular mural is 35 yards, what is the length?

c. What is the perimeter of the mural in yards?

12. What is the perimeter of each figure? Let $\pi \approx 3.14$. Round to the nearest hundredth.

a.

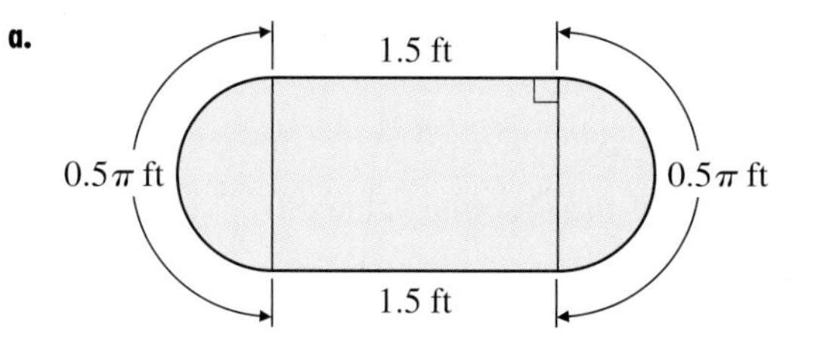

b.

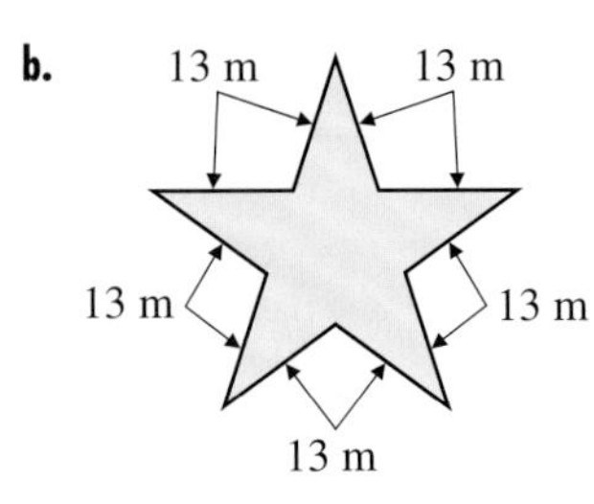

13. What are the volume and surface area of each figure? Round to the nearest whole number. Let $\pi \approx 3.14$.

a.

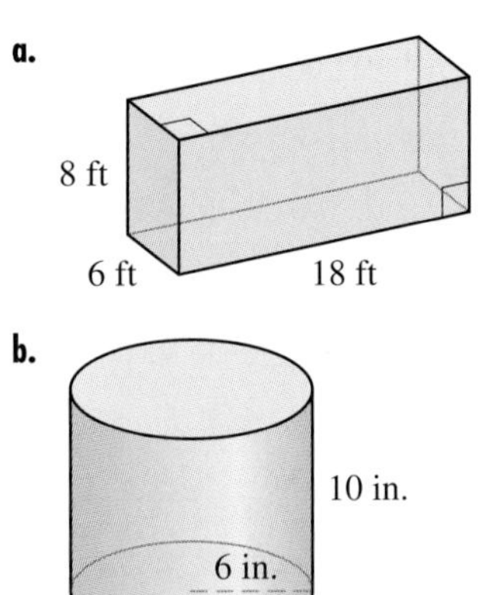

b.

In Exercises 14 and 15, complete the table.

14.

	Inequality	Interval	Line Graph
0 to 20			
More than 20 and less than 50			
50 or greater			

15.

x	y	$x + y$	$x - y$	$x \cdot y$	$x \div y$
-8	4				
-6		-8			
			9	-18	

▶ Cumulative Review of Chapters 1 and 2

These exercises highlight material and combine concepts from Chapters 1 and 2. You may not have seen the problems before, but you have been introduced to the required skills.

1. Make an input-output table for each rule, with integer inputs from -2 to 3. Graph the (x, y) ordered pairs for each rule on separate axes.

a. $y = x^2 - 1$ **b.** $y = 2 - x$ **c.** $y = 2x + 3$

d. $y = -2x$ **e.** $y = -x + 1$ **f.** $y = |x - 1|$

2. Translate into symbols.

a. The sum of the absolute value of -5 and 14

b. The quotient of the opposite of 15 and -3

c. The product of $\frac{1}{4}$ and the reciprocal of 1.5

d. 6 is greater than x.

e. x is less than 15.

3. Write in words.

a. $4 < x$ **b.** $5 - 3x$

c. $-(-x)$ **d.** $|3 - x|$

In Exercises 4 and 5, add like terms. Simplify expressions to remove parentheses as necessary.

4. a. $-2x + 3y - 4x + 2x - 5y$

b. $x^3 + 2x^2 - x - 3x^2 - 6x + 3$

5. a. $3(x + 1) + 5(x - 2)$

b. $8 - 2(x - 1)$

c. $8(x + 2) - 3(x - 2)$

d. $5 - 3(2 - x)$

6. Change to like units and add. (16 ounces = 1 pound.)

a. 3 feet + 24 inches + 2 yards

b. 2.5 pounds + 8 ounces + 1.5 pounds

7. Complete the table.

x	y	$x + y$	$x - y$	$x \cdot y$	$x \div y$
-4		2			
5				-15	
$-\frac{2}{3}$	$\frac{3}{4}$				
$\frac{7}{8}$	$-\frac{7}{10}$				
1.44	1.8				
0.25	-0.5				

8. Write the correct symbol between these expressions. Choose from <, =, and >.

a. $|-4| \square |3|$

b. $-|-6| \square -(-3)$

c. $-(-5) \square |4 - 7|$

d. $|13| \square -(-11)$

e. $-(-9) \square |-12|$

f. $-|15| \square |2 - 5|$

9. Simplify.

a. $+68 - 74 - 26 + 32 + 14$

b. $-16 + 18 - 35 + 12 - 15 - 24$

c. $-8 - (-2)$ **d.** $7 - (-3)$

e. $-4 + (-5)$ **f.** $-24 \div (-3)$

g. $(-2)(-3)(-5)$ **h.** $5 \div \dfrac{2}{3} \cdot 4$

i. $7 \div \dfrac{3}{2} \cdot 6$ **j.** $\dfrac{a}{b} \div \dfrac{a}{b}$

k. $\dfrac{x}{y} \div \dfrac{-x}{y}$ **l.** $\dfrac{6a + 2}{3}$

m. $\dfrac{xy - x^2}{x}$

10. Simplify. Use the definition of positive integer exponents as needed. Evaluate parts g and h if $x = -1, y = 2$.

a. 3^4 **b.** $\left(-\frac{1}{2}\right)^4$ **c.** $(3n)^3$

d. $5(2x)^2$ **e.** $(3x)^3$ **f.** $\left(-\frac{3}{4}b\right)^2$

g. $\left(\dfrac{2x}{3x^2}\right)^2$ **h.** $\left(\dfrac{-0.2x}{y}\right)^3$

11. Concrete is sold by the cubic yard. We wish to make a sidewalk 40 inches wide, 8 inches thick, and 20 feet long. How many cubic yards of concrete will be needed?

12. The oval shape shown is called an *ellipse*. The formula for the area of an ellipse is $A = \pi rR$. Use $\pi \approx 3.14$.

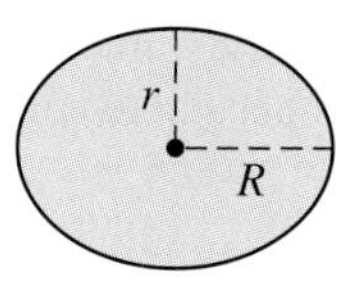

a. Find the area of an ellipse with $R = 4$ and $r = 3$.

b. If $r = R$, what shape is created?

13. A portion of a mileage chart for Pennsylvania is shown below.

	Erie	Pittsburgh	Reading	Scranton
Erie	—	135	325	300
Pittsburgh		—	255	272
Reading			—	99
Scranton				—

a. What assumption about the distances between two cities makes it possible to complete the table?

b. Complete the table.

c. Which number property is similar to our assumption about the table?

14. Adam Moving Company decides to double the dimensions (length, width, and height) of its book boxes so that movers can put more books in each box. The original book boxes were cubes, 1.5 feet on each side.

a. What is the volume of the old box?

b. Find the new length, width, and height.

c. What is the volume of the new "double" box?

d. How many times as large in volume is the new box?

e. If the original box held 30 pounds of books, what would the new box, full of books, weigh?

f. What happened to the employee who thought of this idea?

15. National Testing Service has a 100-question test. The grading is +1 point for each correct answer, 0 points for a blank, −1 for each incorrect answer.

a. Write an expression for the score if a student has 4 blanks and 90 correct responses.

b. Write an expression for the score if a student has 4 blanks and 80 correct responses.

c. Write an expression for the score if a student has 4 blanks and 70 correct responses.

d. What is the highest score possible with 1 incorrect answer?

e. Is it possible to earn a 99?

16. Write the Shipping and Handling Charge table with inequalities and intervals. Allow no gaps between intervals. State your assumptions.

Shipping and Handling Charge
Less than or equal to $24.99
$25 to $50
$50.01 to $99.99
$100 or more

17. Draw a line graph for $-3 < x \leq 2$

18. List seven ways in which parentheses have been used.

CHAPTER 3

Solving Equations and Inequalities in One Variable

Whether you are planning a birthday party (with options as shown in Figure 1), a wedding reception, or a grand opening of a new business, you are faced with a number of questions. How much will the event cost? What choices do you have within your given budget? For how many people will two different choices have the same cost? Graphs, tables, equations, and inequalities can help you answer these questions.

In this chapter, we solve equations and inequalities in one variable. We solve equations in three ways: with symbols, tables, and graphs. We focus on graphic and symbolic solutions of inequalities.

PARTY PACKAGES

- **Rock Climbing Gym:** $15 per climber for 2 hours.
- **Grand Slam USA:** $10 per person with an $80 minimum. Includes games, ice cream cake, and coupons. Optional pizza $20.
- **Fir Bowl:** $15 per hour per lane (6 children per lane maximum).
- **Lane County Ice:** $125 for up to 10 skaters. $12.50 each additional. Includes 2 pizzas, sodas, goodie bags, and a 20-minute lesson.
- **National Academy of Artistic Gymnastics:** $85 for 11 children, plus $5 each additional child. Includes invitations, instructors, and gym activities.
- **Round Table Pizza:** Using menu choices, 10 to 12 for $65.75 (drinks and pizza); 20 to 24 persons for $107.35.
- **Papa's Pizza:** $5.95 per child (includes pizza, drinks, and balloons).
- **Paintball Palace:** 2 hours of play plus shooter, mask, and 50 paintballs for the following rates: 10 to 15, $15 per person; 16 to 24, $12.50 per person; 25 to 40, $10 per person. Special offer: For groups of 15, a 16th person is free.
- **Skate World:** $82 for up to 10 children and $7 per additional skater. Includes party room, soda, and glow sticks. Deluxe package is $200 for 20 children for 2 hours of private skate or skateboard party.
- **Strike City:** $20 plus $10.25 per child. Includes 2 hours in party room, 3 games of bowling, and shoe rental.
- **Wave Pool:** $4.50 per person or $3.50 per person in reserved groups of six or more. Room is $20 per hour.

FIGURE 1

▶ 3.1 Linear Equations in One and Two Variables

Objectives

- Identify linear and nonlinear equations from their graphs.
- Identify identities and conditional equations.
- Write word sentences as linear equations in one and two variables.
- Write word sentences as equations containing parentheses.
- Write equations for applications requiring inequalities to describe inputs.

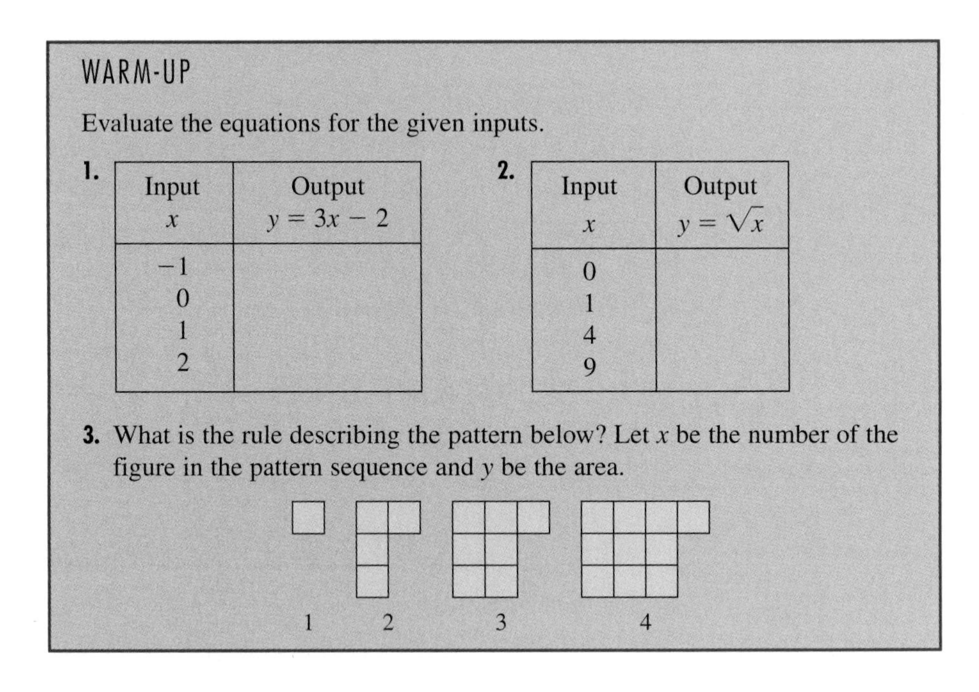

WARM-UP

Evaluate the equations for the given inputs.

1.

Input x	Output $y = 3x - 2$
−1	
0	
1	
2	

2.

Input x	Output $y = \sqrt{x}$
0	
1	
4	
9	

3. What is the rule describing the pattern below? Let x be the number of the figure in the pattern sequence and y be the area.

IN THIS SECTION, we explore reading a graph to evaluate and solve equations. We then draw graphs to illustrate linear and nonlinear equations. We define identities and conditional equations, as well as equations with multiple variables, and extend our skill in writing linear equations from word settings.

Equations

▶ EXAMPLE 1 **Exploration** A birthday party at the Strike City bowling alley costs \$20 plus \$10.25 per child. The charge includes the party room, three games of bowling, and shoe rental. The graph in Figure 2 shows the total cost for 1 to 28 children. Use the graph to estimate answers.

a. Estimate the total cost for 8 children.
b. Estimate the total cost for 20 children.
c. Explain how you found the costs in parts a and b.
d. Write an equation relating the number of children and the total cost. Define your variables.
e. What variable in your equation is known in parts a and b? Write the equation showing the total cost for 28 children.
f. Estimate how many children may be invited for \$250.
g. Estimate how many children may be invited for \$300.
h. Explain how to find the number of children in parts f and g.

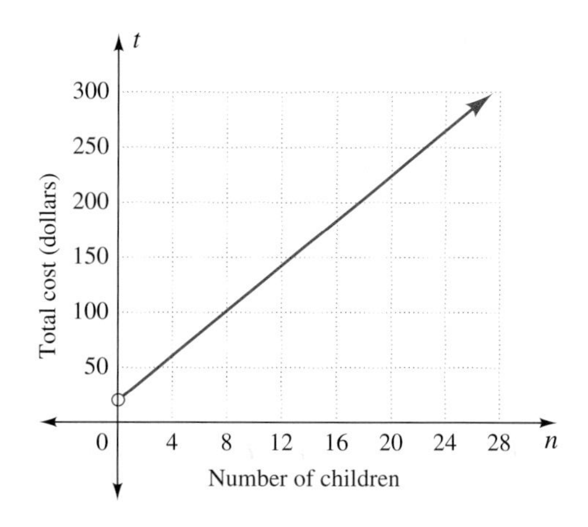

FIGURE 2 Strike City costs

i. What variable in your equation is known in parts f and g? Write the equation showing the number of children who can be invited for \$150.

SOLUTION

a. The cost is approximately \$100 for 8 children.

b. The cost is approximately \$225 for 20 children.

c. Sample: Find the number of children on the horizontal axis. Find the point on the graph above the number of children, and read the matching cost from the vertical axis.

d. Let t = total cost of party in dollars and let n = number of children. The total cost is $t = 10.25n + 20$.

e. The known variable is n, the number of children. When we know the input, n, we are *evaluating* the equation to find the output, t: $t = 10.25(28) + 20$.

f. For \$250, approximately 22 children may be invited.

g. For \$300, approximately 27 children may be invited.

h. Sample: Find the dollars on the vertical axis. Find the point on the graph to the right of the number, and read the matching number of children from the horizontal axis.

i. The known variable is t, the total cost. When we know the output, t, we are *solving the equation* to find the input, n: $150 = 10.25n + 20$ ◀

Solving an equation is *the process of finding the value of an input variable for a given output. The process includes reading from a table or graph and using algebraic operations to obtain the input variable by itself on one side of the equation.* The graph in Figure 2 gave estimated solutions to equations of interest to someone with a budget. You may know the steps to solve equations such as $150 = 10.25n + 20$. However, there are no algebraic steps to solve $3x = 3^x$. After additional practice reading graphs in the exercises, we will begin algebraic operations in Section 3.2. In Section 3.3, we will return to graphs and tables for solving equations.

LINEAR EQUATIONS Recall that an **equation** is a *statement of equality between two quantities.*

▶ **EXAMPLE 2** Reviewing graphing Graph the ordered pairs from the two Warm-up tables. Let y be the output.

SOLUTION The graph of $y = 3x - 2$, from Warm-up Exercise 1, is shown in Figure 3. The graph of $y = \sqrt{x}$, from Warm-up Exercise 2, is shown in Figure 4.

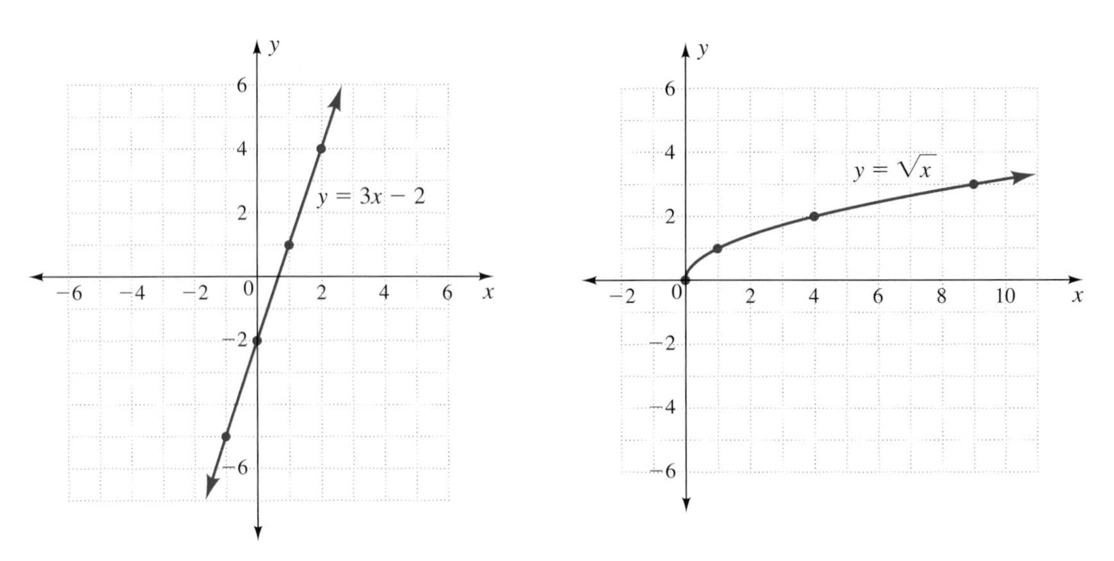

FIGURE 3 Graph of $y = 3x - 2$

FIGURE 4 Graph of $y = \sqrt{x}$ ◀

A **linear equation in two variables** describes *the set of input-output pairs, or ordered pairs* (x, y), *whose graph makes a straight line*. For example, the rule "the output is two less than three times the input" has the linear equation $y = 3x - 2$. This is the rule for Exercises 1 and 3 of the Warm-up and Figure 3.

A **nonlinear equation** is the name given to an *equation whose graph does not form a straight line*. The square root equation $y = \sqrt{x}$, graphed in Figure 4, is an example of a nonlinear equation.

One way to identify linear equations is to arrange the equations into a certain form, such as that of $y = 3x - 2$. (Vertical lines are an exception and will be presented in Chapter 4.)

LINEAR EQUATIONS

> A **linear equation in one variable** may be written in the form
>
> $$ax + b = 0$$
>
> where x is the variable and a and b are real-number constants.
>
> A **linear equation in two variables** may be written in the form
>
> $$y = ax + b$$
>
> where x and y are variables and a and b are real-number constants.

The constant multiplied may be any real number, including zero. The constant added may also be zero. It is an agreement that we write $y = ax + b$ rather than $y = b + ax$ in mathematics.*

▶ **EXAMPLE 3** **Identifying linear equations** Identify each part below as a linear equation in one variable, a linear equation in two variables, an equation but not linear, or not an equation.

a. $t = 10.25n + 20$ **b.** $-2 + 3x = 0$ **c.** $y = 5 - 0.05x$

d. $y = x^2 + 2x - 5$ **e.** $x^2 - x$ **f.** $25 = x$ **g.** $C = \pi d$, where π is a constant

*In statistics, the agreement is to write $y = a + bx$. Neither group seems willing to change.

SOLUTION

a. $t = 10.25n + 20$ is a linear equation in two variables, of the form $y = ax + b$.

b. $-2 + 3x = 0$ is a linear equation in one variable. From the commutative property of addition, it is the same as $3x + (-2) = 0$ and is of the form $ax + b = 0$.

c. $y = 5 - 0.05x$ is a linear equation in two variables. Because subtraction is the same as addition of the opposite number, the equation is also $y = 5 + (-0.05x)$. With the commutative property of addition, the equation is $y = -0.05x + 5$, the same form as $y = ax + b$.

d. $y = x^2 + 2x - 5$ has an x^2 term and thus is not a linear equation.

e. $x^2 - x$ has an x^2 term, so it is not linear. It has no equals sign and so is not an equation.

f. $25 = x$ is a one-variable linear equation. Do you know how to get $x - 25 = 0$? If not, we will learn that in the next section.

g. $C = 2\pi r$ is a linear equation in two variables, C and r; it is of form $y = ax + 0$. ◀

IDENTITIES An *identity* is a special type of equation that is true for all numbers. When we added like terms in Section 2.3, in problems like $5x + 6x = 11x$, we used the equal sign. Here the equal sign says that $5x + 6x$ has the same value as $11x$ for all values of x. Because the expressions on the left and right sides are identical in value, $5x + 6x = 11x$ is called an identity.

DEFINITION OF IDENTITY

An **identity** is formed when the expression on the left side of the equal sign is equal to the expression on the right side for all values of the variable(s).

The statement of the associative property of multiplication, $a(b \cdot c) = (a \cdot b)c$, is an identity because it is true for all real numbers a, b, and c.

CONDITIONAL EQUATIONS Most of the equations we write in algebra are conditional equations.

DEFINITION OF CONDITIONAL EQUATION

A **conditional equation** is an equation that is true for only certain values of the variable(s) in the equation.

The equations $n = 4$, $2n = 8$, and $n + 1 = 5$ are examples of (conditional) **equations in one variable** as *these statements contain only a single variable* (in this case, n). Conditional equations can have more than one variable. The equations $y = \sqrt{x}$ and $y = 3x - 2$ are *two-variable conditional equations*. The formula $V = l \cdot w \cdot h$ is an equation in *four variables*, V, l, w, and h.

▶ **EXAMPLE 4** **Naming types of equations** What kind of equation is each of the following—an identity or a conditional equation? State how many variables each conditional equation contains.

a. $x - 3 = 15$

b. $y = x - 3$

c. $a(b + c) = ab + ac$

d. $C = 2\pi r$

e. $A = \dfrac{h}{2}(a + b)$

f. $b + c = c + b$

SOLUTION

a. Conditional equation, 1 variable

b. Conditional equation, 2 variables

c. Identity. This equation states the distributive property of multiplication over addition, which is true for all real numbers.

d. Conditional equation, 2 variables. The letter π is a constant.

e. Conditional equation, 4 variables

f. Identity. This equation states the commutative property of addition, which is true for all real numbers. ◀

Writing Linear Equations in Two Variables

In two-variable equations, the input is called the **independent variable**. The output is called the **dependent variable**. For x as the input and y as the output, we say *y depends on x*.

CHOOSING VARIABLES

> In applications, look for the output, y, depending on the input, x. Write equations so that *y depends on x*.

▶ EXAMPLE 5 Writing equations in applications Write an equation that describes each of these problem situations. Define the input and output variables.

a. What is the total cost of bulk rice at \$1.29 per pound with a \$0.50 deposit on a reusable container?

b. What is the total cost of tuition at \$64 per credit hour plus \$10 in fees?

c. What is the sales tax on a purchase in a city where taxes are 6% of the price?

d. What is the value remaining on a prepaid copy machine card that costs \$5.00 and is charged \$0.05 per copy as the copy machine is used?

e. What is the total cost of a 2-hour party at the Rock Climbing Gym at \$15 per person.

SOLUTION In each equation, x is the input and y is the output.

Student Note:
By saying "y . . . in dollars" we can omit the dollar sign from the equation.

a. Total cost depends on the number of pounds purchased. Let x be the number of pounds and y be the total cost in dollars. The total cost is $y = 1.29x + 0.50$.

b. Total cost depends on the number of credits taken. Let x be the number of credits and y be the total cost in dollars. The total cost is $y = 64x + 10$.

c. Tax depends on price. Let x be the price and y be the tax in dollars. Tax is $y = 0.06x$. The phrase *of the price* reminds us to multiply the 6% by the input x.

d. The value of the card depends on the number of copies made. Let x be the number of copies and y be the value remaining on the card in dollars. The value remaining is $y = -0.05x + 5.00$. Note the $y = ax + b$ form.

e. The cost depends on the number of children attending. Let x be the number of children and y be the total cost in dollars. The total cost is $y = 15x$. ◀

COMMON PHRASES REQUIRING PARENTHESES Grouping symbols such as parentheses are needed in writing equations from sentences when an operation (multiplication or division) is done

to the sum of a and b or

to the difference of a and b.

Grouping symbols are needed when costs change for different quantities such as

> a cost for the first one (or more) minute and a different cost for each additional minute.
>
> a cost for the first one (or more) day and a different cost for each additional day.
>
> a cost for 1 to 8 children is a flat fee and different from the cost for each additional child.

Look for parentheses as grouping symbols in Example 6. Many applications require two equations and an inequality with each. Note the inequalities that define sets of inputs in parts c to f.

▶ **EXAMPLE 6** **Writing equations** Write each sentence as a one- or two-variable equation containing parentheses. Define your variables.

a. Two times the difference between a number and eight is -24.
b. The sum of eight and a number is divided by five. The quotient is 5.
c. The total cost of a telephone call is \$0.75 for the first minute and \$0.10 for each additional minute.
d. The total cost of a pressure-washer rental is \$40 for the first day plus \$25 for each additional day.
e. The National Academy of Artistic Gymnastics charges \$85 for 11 children plus \$5 for each additional child. The charge includes invitations, instructors, and gym activities.
f. Grand Slam USA charges \$10 per person with an \$80 minimum. The charge includes games, ice cream cake, and coupons.

SOLUTION Parts a and b correspond to one-variable equations. The others correspond to two-variable equations.

a. Let x = the number: $2(x - 8) = -24$.

b. Let x = the number: $\frac{8 + x}{5} = 5$.

c. Let y = the total cost in dollars: $y = 0.75$ for $x \leq 1$ minute and $y = 0.75 + 0.10(x - 1)$ for $x > 1$ minute.

d. Let y = the total cost in dollars: $y = 40$ for $x \leq 1$ day and $y = 40 + 25(x - 1)$ for $x > 1$ day.

e. For total cost y in dollars, $y = 85$ for $1 \leq x \leq 11$ children and $y = 85 + 5(x - 11)$ for $x > 11$ children.

f. For total cost y in dollars, $y = 80$ for $1 \leq x \leq 8$ children and $y = 80 + 10(x - 8)$ for $x > 8$ children ◀

THINK ABOUT IT: Compare the inputs and outputs for the party packages (parts e and f of Example 6) with the credit card fees. Predict the general appearance of a graph for parts e and f.

ANSWER BOX

Warm-up: 1. $-5, -2, 1, 4$ **2.** 0, 1, 2, 3 **3.** $y = 3x - 2$
Think about it: The inputs are grouped, and at least one group of outputs in each setting is y = a constant. The graph for parts e and f will start out flat and then rise after we get to the fee for the additional children.

▶ 3.1 Exercises

In Exercises 1 and 2, tell whether or not the equations are linear and why.

1. **a.** $x = y$ **b.** $6 = x$ **c.** $xy = 5$

d. $y = 2 + x$ **e.** $y = x^2$ **f.** $x + 3 = 5$

2. **a.** $x = 4$ **b.** $y = 3$ **c.** $\frac{1}{x} = y$

d. $3 + x^2 = y$ **e.** $3 + x = y$ **f.** $y = x$

In Exercises 3 and 4, describe the equation as an identity or a conditional equation. For conditional equations, state the number of variables in the equation.

3. **a.** $x + 4 = y$ **b.** $x + 4 = -7$

c. $A = \frac{1}{2}bh$ **d.** $a(b \cdot c) = (ab) \cdot c$

e. $3x + 4x = 7x$ **f.** $n + 0.03n = 1.03n$

4. **a.** $y = 2x + 3$ **b.** $\frac{ab}{c} = a \cdot \frac{b}{c}$

c. $\frac{-a}{b} = -\frac{a}{b}$ **d.** $5x - 2x = 3x$

e. $2x = 7$ **f.** $x + 0.05x = 1.05x$

In Exercises 5 to 8, write equations. Let x be the unknown number.

5. Four less than 3 times a number is 17.

6. Eight more than 5 times a number is 43.

7. The difference between 26 and 4 times a number is 2.

8. Subtracting three times a number from 7 is -2.

Write each sentence in Exercises 9 to 12 as an equation with x as the input, or independent variable, and y as the output, or dependent variable.

9. The output is five more than twice the input.

10. The output is three less than four times the input.

11. Five less than half the input is the output.

12. The product of negative three and the input is added to 4 to give the output.

In Exercises 13 to 22, write each equation as a sentence.

13. $3x - 5 = 16$ **14.** $x + 3 = -10$

15. $\frac{2}{3}x = 24$ **16.** $\frac{3}{4}x = 27$

17. $6 - 2x = 10$ **18.** $5 - 3x = -4$

19. $2(x - 3) = 10$ **20.** $3(2 - x) = 9$

21. $\frac{1}{2}(x + 2) = 6$ **22.** $\frac{1}{2}(x + 3) = 10$

In Exercises 23 to 34, define variables and write equations so that y depends on x. (*Hint:* Distance traveled is the product of the time and rate or speed.)

23. Find the tip if a tip is usually 15% of the cost of a meal.

24. Find the amount of tax for an 8.5% sales tax on a purchase.

25. Find the Medicare payment at 1.45% of wages.

26. Find the Social Security payment at 6.2% of wages.

27. Find the total cost for any quantity of bulk candy at $2.49 per pound.

28. Find the distance traveled at 35 miles per hour for a given number of hours.

29. Find the distance traveled in 3 hours at various speeds.

30. Find the cost of x thousand gallons of water at $2.15 per thousand gallons.

31. Find the total cost of tuition and fees for x credit hours taken. Tuition is $75 per credit hour, and fees are $32.

32. Find the value remaining on a $20 prepaid mass transit ticket when each ride costs $2.85.

33. Find the value remaining on a prepaid $26 coffee card when your usual beverages costs $4.25.

34. Find the value remaining on a $25 theater gift certificate when each ticket costs $5.

Write each sentence in Exercises 35 to 40 as a one- or two-variable equation containing parentheses. Define your variables so that y depends on x.

35. The sum of a number and 5 is multiplied by 3 to obtain 21.

36. The difference between a number and 5 is multiplied by 4 and results in 24.

37. Five times the difference between 9 and a number is -35.

38. Take a number, add 7, and then multiply the sum by 6. The result is -72.

39. Half the sum of 8 and a number gives the output.

40. The output is eight times the difference between the input and 9.

In Exercises 41 to 48, write two equations with inputs for each described with an inequality. Define your variables so y depends on x.

41. A bowling handicap is 80% of the difference between 200 and the bowling average. Above 200 there is no handicap.

42. One recommended exercise pulse rate is 60% of the difference between 220 and your age. Above age 220 there is no pulse.

Blue numbers are core exercises.

43. The total cost of a credit card call is \$0.15 per minute for the first two minutes and \$0.05 for each additional minute.

44. The total cost of an air-compressor rental is \$45 for the first day plus \$30 for each additional day.

45. The total cost of an official transcript is \$5 for the first copy and \$2 for each additional copy ordered at the same time.

46. A DVD rental is \$5 for two days and a penalty of \$6 for each additional day.

47. The total party cost at Lane County Ice is \$125 for up to 10 skaters and \$12.50 for each additional.

48. The total party cost at Skate World is \$82 for up to 10 children and \$7 for each additional.

▶ Writing

49. Write another explanation for parts c and h in Example 1, using the words *ordered pairs*.

50. What restrictions are there on the inputs, x, in Exercises 41 to 46?

51. Explain the difference between an identity and a conditional equation.

52. Is $C = 2\pi r$ a linear equation? Explain.

53. How do we identify a linear equation (either one-variable or two-variable)?

54. How can we identify the independent variable (or input) in an application?

55. What phrases in a word problem tell us that its equation may contain parentheses?

56. Computation Review Copy the numbers in the problem and do the indicated operation. Use equals signs in your work.

a. Simplify to lowest terms: $\frac{28}{35}$.

b. Change 0.8 to a percent.

c. Change $\frac{45}{6}$ to a mixed number.

d. Simplify to lowest terms: $\frac{32}{40}$.

e. Change 0.3 to a percent.

f. Simplify $4 - 2(3 - 6)$.

g. Simplify $8 - 3(2 - 8)$.

h. Change $\frac{24}{9}$ to a mixed number.

57. Connecting with Chapter 1: Predicting Graphs The three sketches of graphs illustrate most 2-hour party packages for 0 to 15 children in Figure 1, page 125. Match each graph with two to four party packages. Explain why the graphs match and which packages do not match.

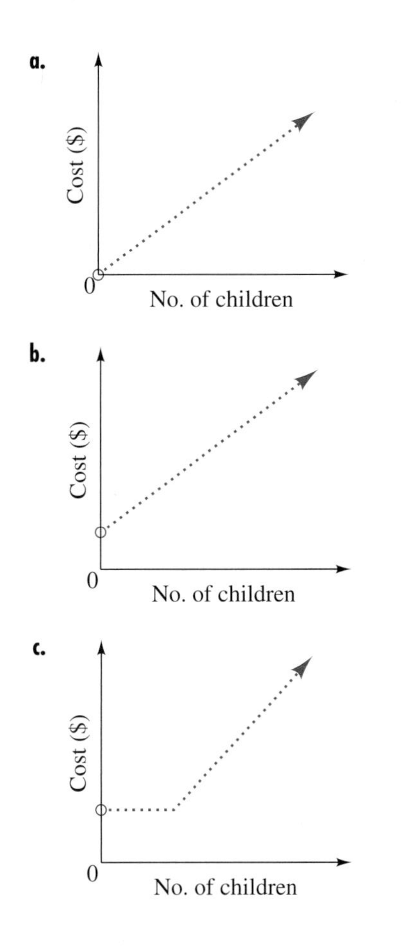

▶ Projects

58. Arriving on Time The graph shows the time taken to drive 10 miles at various rates (speeds). The equation for distance, rate, and time of travel is $d = r \cdot t$.

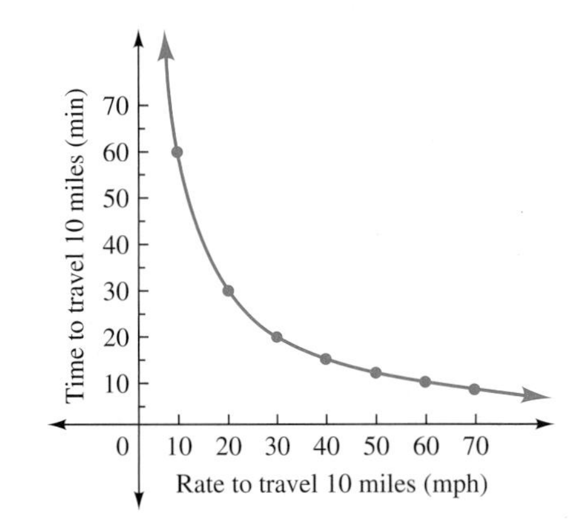

a. What is the input for the graph?

b. What is the output for the graph?

c. Write an equation for how long it takes to travel 10 miles at 10 miles per hour (mph). Answer from the graph.

d. Write an equation for how long it takes to travel 10 miles at 20 mph. Solve from the graph.

Blue numbers are core exercises.

e. Write an equation for how long it takes to travel 10 miles at 30 mph. Solve from the graph.

f. Is the time saved by increasing the speed from 10 to 20 mph the same as the time saved by increasing the speed from 20 and 30 mph? From 30 to 40 mph?

g. You need to travel 10 miles in 15 minutes. Driving 50 mph will break the speed limit. Write an equation for the speed you should travel, and solve it from the graph. Explain how to change from minutes into hours and how you might solve your equation without a graph.

59. Consecutive integers 1 Consecutive integers are *integers that follow one after another without interruption,* such as 4, 5, and 6.

a. The integers 4, 5, and 6 can be written symbolically as $x - 2$, $x - 1$, and x, where $x = 6$. Write expressions for 4, 5, and 6 with $x = 4$. Repeat with $x = 5$.

b. Consecutive *even* integers are 4, 6, and 8. The integers 4, 6, and 8 can be written symbolically as x, $x + 2$, $x + 4$, where $x = 4$. Write expressions for 4, 6, 8 with $x = 6$. Repeat with $x = 8$.

c. Evaluate x, $x + 1$, and $x + 3$ for $x = 5$. Repeat for $x = 6$. Are either answers consecutive even or consecutive odd integers?

d. Write three consecutive odd integers. Now write expressions for three consecutive odd numbers with $x =$ the first number.

e. Compare the expressions for consecutive even integers and consecutive odd integers. Show why this might be true with a number line drawing.

f. For the integers -5, -4, and -3, write as expressions where $x = -4$. Then write as expressions where $x = -1$.

▶ 3.2 Solving Equations with Algebraic Notation

Objectives

- Solve equations in one variable.
- Use the addition property of equations.
- Use the multiplication property of equations.
- Apply equation-solving properties to one- and two-operation equations.

WARM-UP

Rewrite each exercise using an inverse operation and then simplify.
Example: $3 - (-4) = 3 + (+4) = 7$

1. $3 - 4$

2. $-3 - (-4)$

3. $-3 - 4$

4. $6 \div \frac{1}{2}$

5. $-4 \div \frac{2}{3}$

6. $9 \div \frac{3}{4}$

Evaluate each expression. Describe in a complete sentence the order of operations you use to evaluate each. (*Hint:* Start with "Take the number replacing x, . . .")

7. $3x - 2$ for $x = 1$

8. $2 - 6x$ for $x = -\frac{1}{2}$

9. $4x + 1$ for $x = 2.25$

10. $\frac{2}{3}x + 4$ for $x = 27$

IN THE PADI CREDIT CARD fee schedule, the input intervals included \$166.66 and \$2500. These are not numbers that someone chose but rather the result of equation solving—the subject of this section.

The Solution to an Equation

When we find the numbers that make a conditional equation true, we solve the equation.

USEFUL TERMS

A **solution** to an equation is a value of the variable that makes the equation true.

The **solution set** is the set of all solutions to an equation.

Solving an equation is the process of finding the values of the variable for the solution set.

Whenever possible, we check our solutions to an equation by substituting the answer back into the original equation to verify that the equation is true. **Substitution** is *the replacement of variables with equivalent expressions or numbers.*

EXAMPLE 1 **Checking solutions** Substitute the number from the second equation into the first equation to see that a true equation results.

a. $4n = 28, n = 7$

b. $\frac{1}{2}x = 16, x = 8$

c. $x - 7 = 12, x = 19$

d. $2x + 3 = -7, x = -5$

e. $3 - n = 12, n = -9$

SOLUTION The symbol $\stackrel{?}{=}$ means "Does it equal?" The symbol ✓ means "Check; it does."

a. $4(7) \stackrel{?}{=} 28, 28 = 28$ ✓

b. $\frac{1}{2}(8) \stackrel{?}{=} 16, 4 = 16$, false statement

c. $19 - 7 \stackrel{?}{=} 12, 12 = 12$ ✓

d. $2(-5) + 3 \stackrel{?}{=} -7, -7 = -7$ ✓

e. $3 - (-9) \stackrel{?}{=} 12, 12 = 12$ ✓ ◀

Solving Equations by Working Backwards

One of the most effective ways to solve an equation is to work backwards. Working backwards from the answer is a common problem-solving strategy. (How often have you not known what to do with a problem and been helped by looking at the answer in the back of the book?)

EXAMPLE 2 **Exploration: Guess my number** I am thinking of a number. When I multiply it by 3 and subtract 2, I get 10. What is my number? (See Figure 5.)

FIGURE 5 "Guess my number"

SOLUTION The strategy is to work backwards. If the result was 10 and it was found by subtracting 2, then the prior number must have been $10 + 2$, or 12. If the answer after multiplying the original number by 3 is 12, then the original number must be $12 \div 3$, or 4.

Check: Multiplying 4 by 3 and subtracting 2 gives 10. We can write the check step as an equation, $3x - 2 = 10$, and substitute 4 for x: $3(4) - 2 \stackrel{?}{=} 10$. ✓ ◀

▶ Example 3 contains four one-operation exercises to work backwards.

▶ **EXAMPLE 3** Working backwards What is my number?

a. When I add 5 to my number I get 11. What is my number?
b. When I subtract 7 from my number I get 8. What is my number?
c. When I divide my number by 3, I get 6. What is my number?
d. When I multiply my number by 2, I get 15. What is my number?

SOLUTION

a. If the result is 11 after addition of 5, then the number must be 11 take away 5, or $11 - 5 = 6$.

Check: $5 + 6 \stackrel{?}{=} 11$ ✓

b. If the result is 8 after subtraction of 7, then the number was 8 plus 7, or $8 + 7 = 15$.

Check: $15 - 7 \stackrel{?}{=} 8$ ✓

c. If the result is 6 after division by 3, then the number was 6 times 3, or $6 \cdot 3 = 18$.

Check: $18 \div 3 \stackrel{?}{=} 6$ ✓

d. If the result is 15 after multiplication by 2, then the number was 15 divided by 2, or $15 \div 2 = 7.5$.

Check: $7.5 \cdot 2 \stackrel{?}{=} 15$ ✓ ◀

▶ Inverse Operations with One Operation

In algebra, we use the words *inverse operations* to describe working backwards to solve an equation. Solving equations using inverse operations usually requires writing one or more equivalent equations.

DEFINITION OF EQUIVALENT EQUATIONS

Equivalent equations are two or more equations that have the same solution set.

We say that $x + 6 = 9$ and $x = 9 - 6$ are equivalent equations because, for each equation, the solution 3 makes the equation true.

Two properties of equations—the addition property of equations and the multiplication property of equations—allow us to use inverse operations to write equivalent equations.

PROPERTIES OF EQUATIONS

The **addition property of equations** states that adding the same number to both sides of an equation produces an equivalent equation:

$$\text{If} \quad a = b, \quad \text{then} \quad a + c = b + c$$

The **multiplication property of equations** states that multiplying both sides of an equation by the same nonzero number produces an equivalent equation:

$$\text{If} \quad a = b \quad \text{and} \quad c \neq 0, \quad \text{then} \quad ac = bc$$

We state $c \neq 0$ in the multiplication property because we will be including division by c in some equations.

In Example 3, we used working backwards, or inverse operations, to solve the equations. We now use the inverse operations along with the addition and multiplication properties of equations to find equivalent equations and solve the equation.

EXAMPLE 4 **Solving equations with inverse operations and properties of equations** Solve each equation and check your solution. State the inverse operation and the property of equations used.

a. $x - 2 = -5$ **b.** $\frac{x}{5} = 2.5$

SOLUTION **a.** In $x - 2 = -5$, we subtract 2 from x. The inverse operation is to add 2.

Student Note:
The comments to the right tell what was done at each step.

$x - 2 = -5$ State the equation.

$+2 \quad +2$ Add 2 on both sides (addition property).

$x = -3$

Check: $-3 - 2 \stackrel{?}{=} -5$ ✓

b. In $\frac{x}{5} = 2.5$, we divide x by 5. The inverse operation is to multiply by 5.

$\frac{x}{5} = 2.5$ State the equation.

$\frac{5}{1} \cdot \frac{x}{5} = 5(2.5)$ Multiply both sides by 5 (multiplication property). We write $\frac{5}{1}$ instead of 5 on the left to remember to multiply the numerators and the denominators of the fractions.

$x = 12.5$

Check: $\frac{12.5}{5} \stackrel{?}{=} 2.5$ ✓ ◀

▶ The addition and multiplication properties also apply to subtraction and division. Recall that in Sections 2.1 and 2.2 we changed subtraction to *addition of the opposite* and changed division to *multiplication by the reciprocal*. Exercises 1 to 6 in the Warm-up for this section reviewed these operations.

EXAMPLE 5 **Solving equations with inverses** State the inverse operation and solve the equation.

a. $4 + x = 11$ **b.** $-2x = 12$

SOLUTION **a.** In $4 + x = 11$, we add 4 to x. The inverse operation is to subtract 4.

$4 + x = 11$ State the equation.

$-4 \quad\quad -4$ Subtract 4 from both sides (addition property).

$x = 7$

Check: $4 + 7 \stackrel{?}{=} 11$ ✓

b. In $-2x = 12$, we multiply x by -2. The inverse operation is to divide by -2.

$-2x = 12$ State the equation.

$\frac{-2x}{-2} = \frac{12}{-2}$ Divide both sides by -2 (multiplication property).

$x = -6$

Check: $(-2)(-6) \stackrel{?}{=} 12$ ✓ ◀

LINKED EXAMPLE The PADI credit card fee schedule has a $5 minimum charge, a $75 maximum charge, and 3% of the cash advance between the minimum and the maximum.

EXAMPLE 6 Finding the inputs Write and solve an equation for each part.

a. A minimum charge of $5 equals 3% of what cash advance?

b. A maximum charge of $75 equals 3% of what cash advance?

SOLUTION We write 3% as 0.03. Letting x = the cash advance in dollars allows us to omit the dollar sign from the equation.

a.

$$5 = 0.03x \quad \text{State the equation.}$$

$$\frac{5}{0.03} = \frac{0.03x}{0.03} \quad \text{Divide both sides by 0.03.}$$

$$166.66 = x \quad \text{The value for } x \text{ is not rounded up.}$$

The largest cash advance to be charged $5 is $166.66. If we take 3% of $166.67, we get $5.0001, and the bank will round up to $5.01.

b.

$$75 = 0.03x \quad \text{State the equation.}$$

$$\frac{75}{0.03} = \frac{0.03x}{0.03} \quad \text{Divide both sides by 0.03.}$$

$$2500 = x$$

The largest cash advance to be charged 3% is $2500. If we take 3% of $2500, we get $75, and this equals the maximum change. ◀

Inverse Operations with Two Operations

Many of the party package equations have two operations.

EXAMPLE 7 Exploration The birthday party at Strike City costs $20 plus $10.25 per child.

a. Describe how you might find the number of children who could be invited for a total cost (budget) of $184.

b. Write an equation for the total cost, t, for any number of children, n. Write another equation showing the $184 budget.

SOLUTION

a. If we have a $t = 184$ dollar budget we might first subtract the $20 cost from $184. It must be paid regardless of how many children attend. Then we divide the remaining money by the cost per child. Our common-sense process shows what is needed to solve a two-operation linear equation. See Exercise 75 for another right method—and a wrong method—for solving this problem.

b. $t = 10.25n + 20$; $184 = 10.25n + 20$ ◀

When an equation has more than one operation, such as $184 = 10.25x + 20$, we need to know the answer to the question "Which operation is done first in solving the equation for x?" To find the answer, we look at the order of operations on x and then reverse that order and do inverse operations. In Warm-up Exercises 7 to 10, we wrote a sentence to describe the order of operations on x. Writing such a sentence is the first step in our plan in Example 8.

EXAMPLE 8 Solving an equation Solve the equation $184 = 10.25n + 20$, using the following four-step process.

Understand: Make an estimate.

Plan: Write the order of operations on x, and then list the order backwards with the inverse operations.

Carry out the plan.
Check.

SOLUTION **_Understand:_** Because 10(20) is 200, n is below 20.

Student Note:
The plan shows the mathematical reasoning behind our thinking in Example 7.

Plan: In the equation, we multiply n by 10.25 and then we add 20. The reverse order of operations with inverses is to subtract 20 and then divide by 10.25.

Carry out the plan:

$$184 = 10.25n + 20 \quad \text{Write the equation.}$$
$$-20 \qquad\qquad -20 \quad \text{Subtract 20 from both sides.}$$
$$164 = 10.25n$$
$$\frac{164}{10.25} = \frac{10.25n}{10.25} \quad \text{Divide by 10.25 on both sides.}$$
$$16 = n$$

Check: $184 \stackrel{?}{=} 10.25(16) + 20$ ✓ ◀

SUMMARY OF SOLVING EQUATIONS

> Until you master the steps, list the order of equations on the input variable. The solution to an equation requires the opposite, or inverse, of each operation in the reverse order of operations on the input variable.

▶ In the next few problems, we will focus on the plan. It is a good idea to estimate the solution; if you do not estimate, be sure to check your answer.

▶ **EXAMPLE 9** Stating a plan Solve $3x - 2 = 1$.

SOLUTION **_Plan:_** In the equation, we multiply x by 3 and subtract 2. To solve, we will add 2 and then divide by 3.

$$3x - 2 = 1 \quad \text{Write the equation.}$$
$$3x - 2 + 2 = 1 + 2 \quad \text{Add 2 to both sides.}$$
$$3x = 3$$
$$\frac{3x}{3} = \frac{3}{3} \quad \text{Divide both sides by 3.}$$
$$x = 1$$

Check: $3(1) - 2 \stackrel{?}{=} 1$ ✓ ◀

In Example 9, the addition step is on the same line as the equation. In Example 8, the subtraction step is below the equation. Use whichever method you or your instructor prefers.

▶ In Example 10, we change a subtraction to addition of the opposite in order to write the order of operations.

▶ **EXAMPLE 10** Solving equations with subtraction of the variable term Write a plan and then solve the equation $5 = 2 - 6x$.

SOLUTION **_Plan:_** The equation may be written as $5 = 2 + (-6)x$. We multiply x by -6 and then add 2. To solve, we will subtract 2 and then divide by -6.

Student Note:
The comments to the right tell what to do in the next step. Try writing the next step before reading it.

$5 = 2 - 6x$	Change the subtraction to addition of the opposite.
$5 = 2 + (-6)x$	Subtract 2 from both sides.
$5 - 2 = 2 + (-6)x - 2$	Add like terms.
$3 = (-6)x$	Divide both sides by −6.
$\frac{3}{-6} = \frac{(-6)x}{-6}$	Simplify.
$-\frac{1}{2} = x$	

Check: $5 \stackrel{?}{=} 2 - 6\left(-\frac{1}{2}\right)$ ✓ ◀

▶ In Examples 11 and 12, we have fractions as coefficients of the variable. The examples show two different ways to solve an equation containing fractions. In Example 11, we use the fact that division by $\frac{2}{3}$ is the same as multiplication by $\frac{3}{2}$.

▶ EXAMPLE 11 Solving fractional equations Solve $\frac{2}{3}x + 4 = 22$.

SOLUTION ***Plan:*** Because $\frac{2}{3}$ is multiplied by x before 4 is added, we will subtract 4 and then divide by $\frac{2}{3}$.

$\frac{2}{3}x + 4 = 22$	Subtract 4 from both sides.
$\frac{2}{3}x + 4 - 4 = 22 - 4$	Add like terms.
$\frac{2}{3}x = 18$	Divide both sides by $\frac{2}{3}$.
$\frac{2}{3}x \div \frac{2}{3} = 18 \div \frac{2}{3}$	Change $\div \frac{2}{3}$ to $\cdot \frac{3}{2}$.
$\frac{2}{3}x \cdot \frac{3}{2} = 18 \cdot \frac{3}{2}$	Simplify.
$x = 27$	

Check: $\frac{2}{3}(27) + 4 \stackrel{?}{=} 22$ ✓ ◀

The inverse nature of multiplication and division means we can multiply by a reciprocal rather than divide by a fraction.

▶ EXAMPLE 12 Using an alternative solution method for fractional equations Solve $\frac{2}{3}x + 4 = 22$.

SOLUTION ***Plan:*** Because $\frac{2}{3}$ is multiplied by x before 4 is added, we will subtract 4 and then multiply by the reciprocal of $\frac{2}{3}$.

$\frac{2}{3}x + 4 = 22$	Subtract 4 from both sides.
$\frac{2}{3}x + 4 - 4 = 22 - 4$	Simplify.
$\frac{2}{3}x = 18$	Multiply by the reciprocal, $\frac{3}{2}$, on both sides.
$\frac{3}{2} \cdot \frac{2}{3}x = \frac{18}{1} \cdot \frac{3}{2}$	Simplify.
$1x = \frac{54}{2}$	Simplify.
$x = 27$	

Check: $\frac{2}{3}(27) + 4 \stackrel{?}{=} 22$ ✓ ◀

Because every linear equation can be written $ax + b = 0$, every linear equation will solve or simplify to just two operations using a and b.

THINK ABOUT IT: What operations on a are used to solve $ax + b = 0$? What operations on b?

GRAPHING CALCULATOR TECHNIQUE: CHECKING A SOLUTION

The $=, >, \geq, <, \leq$ options under 2nd [TEST] may be used to check equations and inequalities on the calculation screen. If you apply one of these options to any statement, the calculator will give 1 if the statement is true or 0 if the statement is false.

To check Example 10, enter $5 = 2 - 6(-1/2)$. Be sure to use a negative sign, not a subtraction sign, in front of the 1/2. The calculator screen in Figure 6 shows 1 for true.

To check Examples 11 and 12, enter $(2/3) \times 27 + 4 = 22$. The calculator screen shows 1 for true.

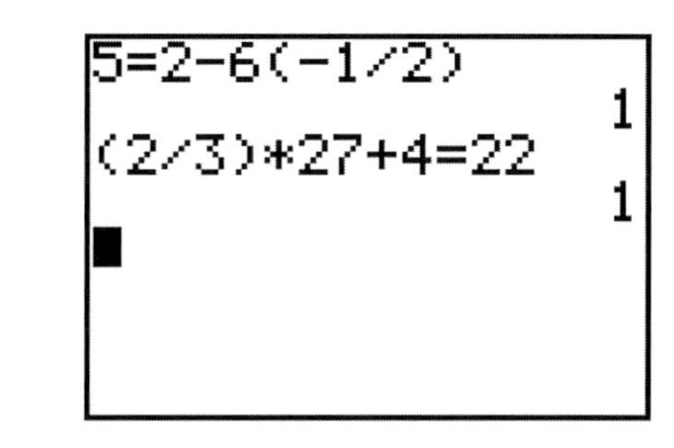

FIGURE 6

ANSWER BOX

Warm-up: 1. $3 + (-4) = -1$ **2.** $-3 + (+4) = 1$ **3.** $-3 + (-4) = -7$ **4.** $6 \times \frac{2}{1} = 12$ **5.** $-4 \times \frac{3}{2} = -6$ **6.** $9 \times \frac{4}{3} = 12$ **7.** 1. Take the number replacing x, multiply by 3, and subtract 2. **8.** 5. Take the number replacing x, multiply by a negative 6, and add 2. **9.** 10. Take the number replacing x, multiply by 4, and add 1. **10.** 22. Take the number replacing x, multiply by $\frac{2}{3}$, and add 4. **Think about it:** Multiply or divide; add or subtract.

▶ 3.2 Exercises

In Exercises 1 to 4, check these "solutions" to equations by substituting the numbers into the equation. Note any wrong "solutions." Explain how to solve the equation correctly.

1. a. $2n = 8, n = 4$

b. $x + 1 = -5, x = -6$

c. $n - 3 = 15, n = 12$

2. a. $3n = 6, n = 3$

b. $1 - x = 4, x = 3$

c. $\frac{1}{2}n = 15, n = 30$

3. a. $55t = 440, t = 20$

b. $\frac{1}{2}x = 10, x = 5$

c. $1.05x = 42, x = 40$

4. a. $5r = 320, r = 64$

b. $2x - 3 = 13, x = 8$

c. $0.9x = 7.2, x = 8$

In Exercises 5 to 12, tell what was done to the first equation to get to the second equation.

5. $2x = 10; x = 5$

6. $x + 4 = 6; x = 2$

7. $2x + 5 = 11; 2x = 6$

8. $2x - 5 = 11; 2x = 16$

9. $2x - 4 = 10; 2x = 14$

10. $x - 7 = -3; x = 4$

11. $\frac{1}{2}x = 15; x = 30$

12. $\frac{1}{2}x = 10; x = 20$

Copy each equation in Exercises 13 to 36. State the inverse operation needed to solve the equation. Solve the equation. Show the check.

13. $x - 5 = 8$

14. $x + 4 = 7$

Blue numbers are core exercises.

15. $9 = x + 12$

16. $x - 6 = -3$

17. $x - 6 = -10$

18. $17 = x - 5$

19. $2x = 26$

20. $3x = 36$

21. $3 = 8x$

22. $2 = 7x$

23. $\frac{x}{4} = 16$

24. $\frac{x}{3} = 12$

25. $\frac{-x}{12} = 4$

26. $\frac{-x}{10} = 5$

27. $-4 = \frac{1}{2}x$

28. $-8 = \frac{1}{4}x$

29. $-\frac{3}{4}x = 12$

30. $-\frac{2}{3}x = 6$

31. $\frac{3}{4}x = 18$

32. $\frac{2}{3}x = 16$

33. $\frac{2}{3} + x = 16$

34. $\frac{3}{4} + x = 18$

35. $\frac{3}{4} - x = 18$

36. $x - \frac{3}{4} = 18$

Match each equation in Exercises 37 to 42 with one of sentences a to g. Copy the equation and its sentence, and then solve the equation. Show the check.

(a) x times -4 plus 2 gives 3.
(b) x times $\frac{1}{4}$ subtract 2 gives 3.
(c) x times $\frac{1}{2}$ subtract 4 gives 3.
(d) x times 4 plus 2 gives 3.
(e) x times -2 subtract 4 gives 3.
(f) x times -2 plus 4 gives 3.
(g) x times 2 plus 4 gives 3.

37. $4 - 2x = 3$

38. $2 - 4x = 3$

39. $\frac{1}{2}x - 4 = 3$

40. $2x + 4 = 3$

41. $\frac{1}{4}x - 2 = 3$

42. $4x + 2 = 3$

Solve for x in each of the equations in Exercises 43 to 64. Show the check.

43. $4x + 3 = 23$

44. $5x - 2 = 23$

45. $3x - 2 = 43$

46. $6x + 4 = 52$

47. $10 - x = -2$

48. $4 - 2x = 10$

49. $3 = 10 - x$

50. $-3 = 4 - 2x$

51. $0 = 3 - 3x$

52. $6 = 2 - 4x$

53. $3 - 3x = 9$

54. $2 - 4x = -1$

55. $\frac{1}{2}x + 3 = -2$

56. $\frac{1}{2}x - 2 = -4$

57. $0 = \frac{3}{2}x + 3$

58. $0 = \frac{3}{2}x - 2$

59. $-4 = \frac{2}{3}x - 2$

60. $12 = \frac{2}{3}x - 2$

61. $4.2x - 3 = -9.3$

62. $2.5x + 2 = -3.5$

63. $7.5 = 4.2x - 3$

64. $5 = 2.5x + 2$

▶ Linked Example

Solve these equations in Exercises 65 to 68 to find the cash advance for a 3% credit card fee.

65. $0.03x = 9$

66. $0.03x = 15$

67. $0.03x = 31.5$

68. $0.03x = 36$

▶ Linked Example

Solve the party package equations in Exercises 69 to 74 to find the number of children, x, who can attend for the given budget. If it is necessary to round down to a whole child, show how much money will remain from the amount budgeted.

69. Rock Climbing Gym: $t = 15x$, budget = \$180

70. Rock Climbing Gym: $t = 15x$, budget = \$135

71. Papa's Pizza: $t = 5.95x$, budget = \$50

72. Papa's Pizza: $t = 5.95x$, budget = \$75

73. Strike City: $t = 10.25x + 20$, budget = \$135

74. Strike City: $t = 10.25x + 20$, budget = \$200

▶ Writing

75. Consider the following solutions to the Strike City cost budget problem (Example 7).

a. Suppose we first divide the budget, \$184, by the cost per child, \$10.25, and then subtract \$20. Discuss the reasonableness of the result.

b. Suppose we write an equation, $184 = 10.25x + 20$ for the budget problem and first divide all terms by the cost per child, 10.25. What is the meaning of each term? What is our next step?

76. When something is filled halfway, some will say that it is half full and others will say that it is half empty. Equating these two ideas, we have

$$\tfrac{1}{2}\text{ full} = \tfrac{1}{2}\text{ empty}$$

What is the result of multiplying each side by 2?

77. Explain why we can use the addition property of equations to subtract the same number on both sides of an equation.

78. Explain how to solve $ax = b$ for x.

79. Explain how to solve $x - a = b$ for x.

80. Explain how to solve $ax + b = 0$ for x.

Blue numbers are core exercises.

81. Explain why this equation-solving step is helpful. What is the next step?

$$\begin{array}{l} 5 - x = 8 \\ \quad + x \qquad + x \\ 5 \qquad = 8 + x \end{array}$$

82. Connecting to Chapter 2 In parts a to f, complete each sentence with one word.

a. Opposites add to _____.

b. Reciprocals multiply to _____.

c. Subtracting 5 from 5 gives _____.

d. Dividing 5 by 5 gives _____.

e. Subtraction is the inverse operation to _______.

f. Division is the inverse operation to __________.

g. Subtraction of x is equivalent to adding the ________ of x.

h. Division by x is equivalent to multiplication by the ________ of x.

▶ Error Analysis

83. Explain what is wrong with this equation-solving step:

$$\begin{array}{l} 5 + 2x - 7 = 15 \\ +7 \qquad\quad +7 \\ \hline 12 + 2x \qquad = 15 \end{array}$$

84. Explain what is wrong with this equation-solving step:

$$\tfrac{4}{5}x - \tfrac{4}{5} = 32 - \tfrac{4}{5}$$
$$x = 31\tfrac{1}{5}$$

▶ Other Applications

The equations in Exercises 85 to 88 have letters other than x as variables. Solve for the variable.

85. $110 = 55t$

86. $150 = 3r$

87. $212 = \frac{9}{5}C + 32$

88. $32 = \frac{9}{5}C + 32$

In Exercises 89 to 98, place the given number(s) in the formula and solve for the remaining letter.

89. $D = 55t$; if $D = 200$ solve for t

90. $D = 55t$; if $D = 450$ solve for t

91. $D = 3r$; if $D = 200$ solve for r

92. $D = 3r$; if $D = 450$ solve for r

93. The 6% tax on a purchase, p, is $T = 0.06p$. If $T = \$0.10$, solve for p.

94. The 6% tax on a purchase, p, is $T = 0.06p$. If $T = \$0.25$, solve for p.

95. Electricity cost is $C = \$5.50 + \$0.0821x$, where x is the number of kilowatt hours used. If $C = \$75$, solve for x.

96. Electricity cost is $C = \$5.50 + \$0.0821x$, where x is the number of kilowatt hours used. If $C = \$50$, solve for x.

97. The area of a triangle is $A = \frac{1}{2}bh$. If $A = 15$ and $b = 3$, solve for h.

98. The area of a triangle is $A = \frac{1}{2}bh$. If $A = 28$ and $h = 7$, solve for b.

▶ Problem Solving

99. Stacking Up Coins

a. Arrange 25 coins into four stacks that fit the following conditions: The second stack is 3 times the first stack. The third stack is 1 less than the second. The fourth stack is 2 more than the first. How many coins are in each stack? Write an equation that would solve the same problem.

b. Arrange 28 coins into four stacks that fit the following conditions: The third stack is 3 more than the second stack. The first stack is twice the second stack. The fourth stack is 1 more than the first stack. How many coins are in each stack? Write an equation that would solve the same problem.

c. Describe a strategy to arrange the coins into the requested stacks.

▶ Project

100. Equivalent Equations Which equation, if any, in each set is not equivalent to the other three? Show clearly how you made your choice.

a. $2x = 6$, $6 \div 2 = x$, $2 \div x = 6$ (for x not zero), $6 = 2x$

b. $x + 3 = -2$, $x = -6$, $x = -2 \cdot 3$, $x \div (-2) = 3$

c. $5x + 4 = 24$, $5x = 20$, $24 = 5x + 4$, $4 - 24 = 5x$

d. $4 + x = 9$, $9 = x + 4$, $9 - 4 = x$, $9 - x = 4$

e. $x - 6 = -3$, $x + 3 = 6$, $x - 3 = 0$, $x = -9$

f. $2x + 3 = 15$, $2x = 12$, $3 - 15 = 2x$, $15 = 3 + 2x$

101. Consecutive Integers 2 Parts a to f present consecutive integer number puzzle problems. In parts a to d write an equation and solve it to find the three integers.

a. The sum of three consecutive integers is 42.

b. The sum of three consecutive integers is 288.

c. The sum of three consecutive odd integers is 177.

d. The sum of three consecutive odd integers is 429.

e. A series of books is published at 7-year intervals. When the seventh book is issued, the sum of the publication years is 13,741. When was the first book published?

f. A series of books is published at 4-year intervals. When the fifth book is issued, the sum of the publication years will be 10,020. When was the first book published?

▶ 3.3 Solving Equations with Tables, Graphs, and Algebraic Notation

Objectives

- Find the solution to an equation from a table and a graph.
- Solve equations containing parentheses.

WARM-UP

Complete the tables for the given inputs.

1.

Input x	Output $y = 3x - 2$
-1	
0	
1	
2	

2.

Input x	Output $y = 2(1 - 3x)$
-1	
0	
1	
2	

IN THIS SECTION, we use tables and graphs to solve equations. We extend our algebraic equation solving to equations containing parentheses.

Why Tables and Graphs to Solve Equations?

Tables and graphs

- give a numerical and visual picture of what solving an equation means
- permit us to solve equations for which we have not yet learned the algebraic methods
- permit us to solve equations that cannnot be solved by algebraic methods
- give a hint of the numerical solution methods used in computer programs in business and science

To appreciate the need for computer methods, consider designing the path (see Figure 7) for the New Horizons* spacecraft: Leave Earth in January 2006 at the correct path to fly past Jupiter in February 2007 for a gravitational assist in velocity and a change in direction so as to travel across the solar system, reaching Pluto and its moon Charon in July 2015. There are equations for planetary motion that must be solved so that New Horizons passes just as Jupiter and Pluto are in the right position. There are equations for segments of the flight path. There are equations for how the gravity of the Sun and that of each planet act on the spacecraft. Trust me, you would not want to solve these equations by hand! Space science may not be your goal, but right now algebra in its many forms is.

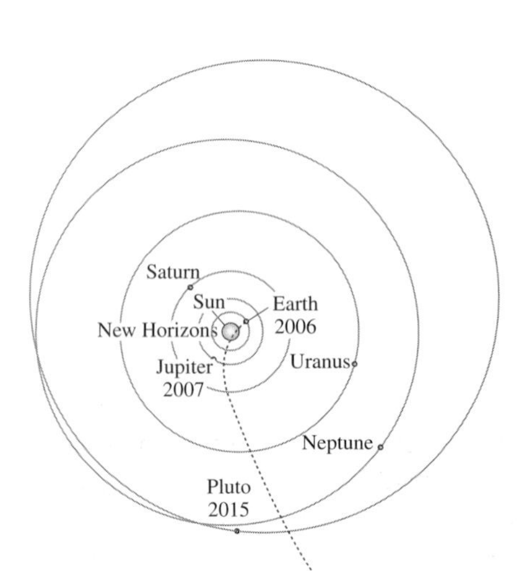

FIGURE 7 Trajectory for New Horizons

Solving Equations from Tables

When either an input or an output number is given, a two-variable equation becomes a one-variable equation. We can use a table to solve for the other variable.

*Consult the internet for information about New Horizons. (www.nasa.gov/mission_pages/newhorizons/main/index.html) and its current location (www.pluto.jhuapl.edu/mission/whereis_nh.php).

▶ EXAMPLE 1 Solving an equation from a table When $y = 4$, the equation $y = 3x - 2$ becomes $4 = 3x - 2$. Solve for x from Table 1.

TABLE 1

Input x	Output $y = 3x - 2$
-1	-5
0	-2
1	1
2	4

SOLUTION To solve $4 = 3x - 2$, we look down the output column, find $y = 4$, and then look in the input column for the solution: $x = 2$. Thus, $4 = 3x - 2$ when $x = 2$. To check, we substitute $x = 2$ in $4 = 3x - 2$.

Check: $4 \stackrel{?}{=} 3(2) - 2$ ✓ ◀

Because a table shows many input-output pairs, we can solve many different equations with the same table. We can use the table to estimate solutions when the input-output pairs are between entries in the table. To find still other solutions, we can extend the table using patterns.

▶ EXAMPLE 2 Finding and estimating solutions from tables Solve these equations from Table 2. Extend the table and estimate solutions as needed.

a. $3x - 2 = -5$

b. $3x - 2 = 2$

c. $3x \quad 2 = 9$

SOLUTION **a.** To solve $3x - 2 = -5$, we look down the output column for -5. Looking in the input column, we find $x = -1$.

Check: $3(-1) - 2 \stackrel{?}{=} -5$ ✓

TABLE 2

Input x	Output $y = 3x - 2$
-1	-5
0	-2
1	1
2	4

TABLE 3 List $y = 1$ to $y = 4$

Input x	Output $y = 3x - 2$
1	1
$1\frac{1}{3}$	2
$1\frac{2}{3}$	3
2	4

b. To solve $3x - 2 = 2$, we look down the output column for 2. The number 2 does not appear, but it would be between the outputs 1 and 4. In fact, 2 is $\frac{1}{3}$ of the distance between 1 and 4. For this linear equation, the input that goes with 2 will be $\frac{1}{3}$ of the distance from 1 to 2 or $1\frac{1}{3}$ (see Table 3).

Check: $3\left(1\frac{1}{3}\right) - 2 \stackrel{?}{=} 2$ ✓

c. To solve $3x - 2 = 9$, we extend the table. To extend, we place other inputs into the equation $y = 3x - 2$, or we look at number patterns for the input change, Δx, and the output change, Δy. As Table 4 shows, the output increases by 3 for every increase of 1 in the input, so we can continue the pattern. An output of 9 will have an input x between 3 and 4. Because 9 is $\frac{2}{3}$ of the distance between 7 and 10, the solution will be $\frac{2}{3}$ of the distance between 3 and 4, or $3\frac{2}{3}$ (see Table 5).

TABLE 4 Change in x and in y

Δx	Input x	Output $y = 3x - 2$	Δy
	-1	-5	
$+1$	0	-2	$+3$
$+1$	1	1	$+3$
$+1$	2	4	$+3$
$+1$	3	7	$+3$
$+1$	4	10	$+3$

TABLE 5 List $y = 7$ to $y = 10$

Input x	Output $y = 3x - 2$
3	7
$3\frac{1}{3}$	8
$3\frac{2}{3}$	9
4	10

Check: $3(3\frac{2}{3}) - 2 \stackrel{?}{=} 9$ ✓ ◀

The small triangle in part c of Example 2 is the Greek letter delta. The letter **delta** (Δ) before a variable means *the change in that variable.*

If it is easy to estimate solutions from a table, we do so; otherwise, we find solutions with another method.

SUMMARY: SOLVING AN EQUATION FROM A TABLE

To solve the equation $n = ax + b$ from a table for $y = ax + b$, look for n in the output column. The solution to the equation $n = ax + b$ is the input x for output n.

If n is between numbers in the output column, estimate x. If n is above or below the numbers in the output column, extend the table.

This solution method also works for other equations written in $y =$ form.

GRAPHING CALCULATOR TECHNIQUE: SOLVING AN EQUATION FROM A TABLE

Repeat Example 2 using the table feature of your graphing calculator.

Enter the equation $Y_1 = 3X - 2$ into [Y=].

For part a, press [2nd] [TBLSET]. Let **TblStart** $= -1$ and **ΔTbl** $= 1$. Press [2nd] [TABLE], find $Y_1 = 4$, and note the solution $X = 2$.

For part b, enter **TblStart** $= 1$ and **ΔTbl** $= 1/3$, as shown in Figure 8. Press [2nd] [TABLE], find $Y_1 = 2$, and note that $X = 1\frac{1}{3}$ written in decimal notation, as shown in Figure 9.

For part c, enter **TblStart** $= 3$ and let **ΔTbl** $= 1/3$ as in part b. Press [2nd] [TABLE], find $Y_1 = 9$, and note that $X = 3\frac{2}{3}$ written in decimal notation.

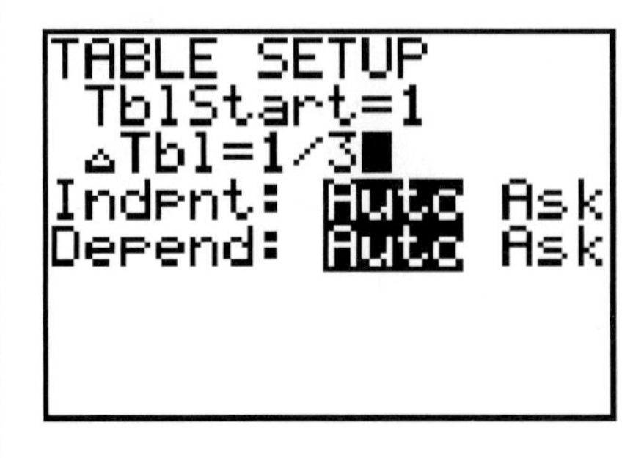

FIGURE 8 [TBLSET] screen

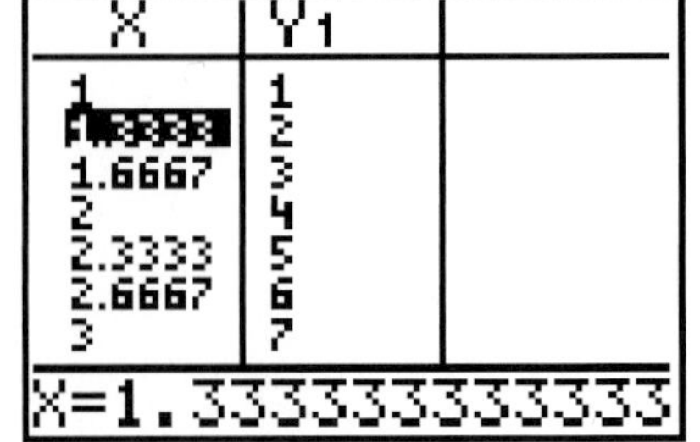

FIGURE 9 [TABLE] screen

Solving Equations from Graphs

When we solve the equation $3x - 2 = -5$ with a graph, we make two graphs. We graph the left side of the equation as $y_1 = 3x - 2$ and the right side of the equation as $y_2 = -5$ (see Figure 10). The solution is the value of x in the ordered pair $(x, -5)$ at the point of intersection. In Figure 10, the graphs cross at $(-1, -5)$, so the solution to $3x - 2 = -5$ is $x = -1$. To check, verify that $3(-1) - 2 = -5$.

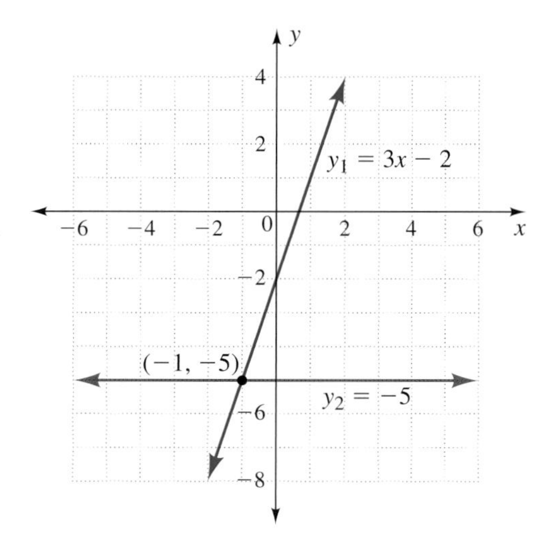

FIGURE 10 Graphs based on $y = 3x - 2$

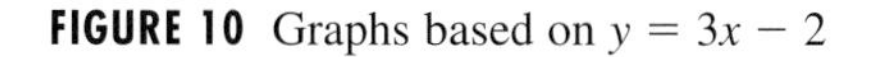

We used subscripts on y in the equations. **Subscripts** are *small numbers or letters placed to the right of variables to distinguish a particular item from a group of similar items*. In this case the items are equations, $y_1 = 3x - 2$ and $y_2 = -5$, that have different outputs for a given input, x.

We solve two equations based on $y_1 = 3x - 2$ in Example 3. Thus there are two equations and graphs labeled y_2: $y_2 = 4$ and $y_2 = 9$.

▶ **EXAMPLE 3** **Solving equations from a graph** Use the graph in Figure 11 to solve these equations.

a. $3x - 2 = 4$ **b.** $3x - 2 = 9$

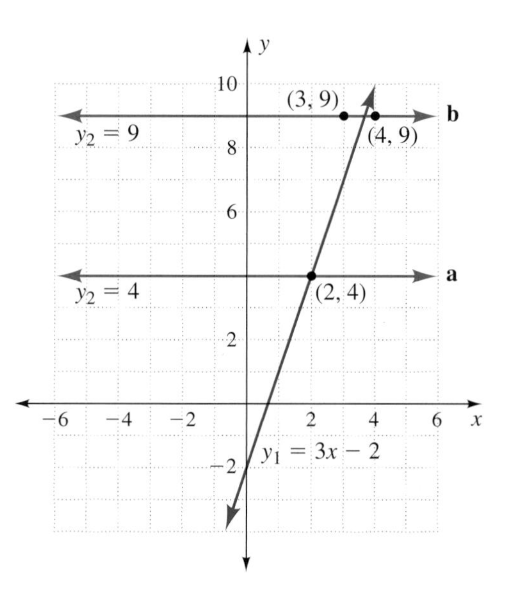

FIGURE 11 Graphs based on $y = 3x - 2$

Student Note:
Is (3, 4) an ordered pair in part b?

SOLUTION **a.** To solve $3x - 2 = 4$, we find the ordered pair for the intersection of the graphs $y_1 = 3x - 2$ and $y_2 = 4$. The graphs cross at (2, 4), so the solution to $3x - 2 = 4$ is $x = 2$.

Check: $3(2) - 2 \stackrel{?}{=} 4$ ✓

b. To solve $3x - 2 = 9$, we find the ordered pair for the intersection of the graphs $y_1 = 3x - 2$ and $y_2 = 9$. The graphs cross between (3, 9) and (4, 9), so the solution to $3x - 2 = 9$ is between 3 and 4. Using an inequality, we have $3 < x < 4$; in words, x is in the interval (3, 4).

Check: See part c of Example 2. ◀

GRAPHING CALCULATOR TECHNIQUE: SOLVING AN EQUATION FROM A GRAPH

To solve $3x - 2 = 9$, enter $Y_1 = 3X - 2$ and $Y_2 = 9$ in [Y=]. Set (WINDOW) using the interval [−10, 10] for x and [−10, 10] for y: **Xmin** = −10, **Xmax** = 10, **Xscl** = 1, **Ymin** = −10, **Ymax** = 10, **Yscl** = 1. Press (GRAPH) and (TRACE). Use the cursor (▶) to trace (see Figure 12) on 3X − 2 to near Y = 9. Under (ZOOM), choose **2 : Zoom In** (see Figure 13), and press (ENTER). Trace to near Y = 9 again. Read the corresponding X value, X ≈ 3.67 (which is close to $3\frac{2}{3}$). Press (2nd) [QUIT]. Check by evaluating 3X − 2 for X = 3.67. The result is approximately 9.

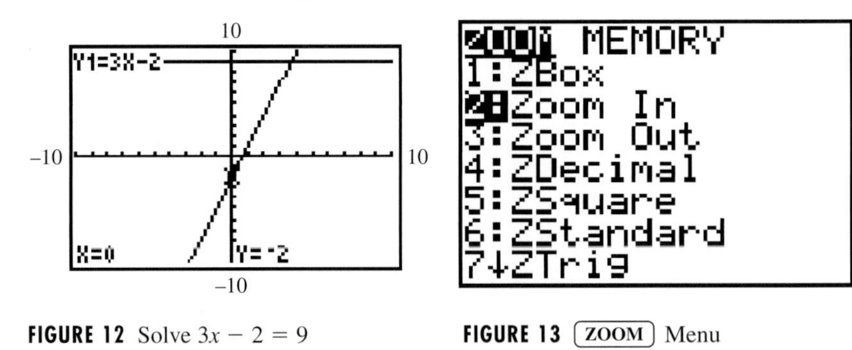

FIGURE 12 Solve $3x - 2 = 9$ **FIGURE 13** (ZOOM) Menu

Student Note:
Use (▲) or (▼) to shift from Y_1 to Y_2 in either (Y=) or (TRACE).

▶ The equations in Example 3 had either one solution or one solution on an interval. In Example 4, we have a curved graph, illustrating that some equations may have more than one solution or no solution at all.

▶ **EXAMPLE 4** Solving equations from a graph Use the graph of $y = x^2$ in Figure 14 to find the solutions to the equations. Use sets to describe the solutions.

a. $x^2 = 9$ **b.** $x^2 = 0$ **c.** $x^2 = -2$

(TRACE) 3 + 2/3 (ENTER).
The cursor goes to $(3\frac{2}{3}, 9)$.

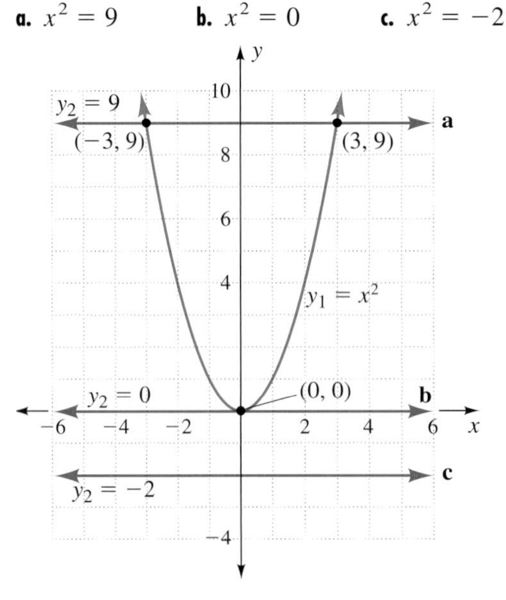

FIGURE 14 Graphs based on $y = x^2$

SOLUTION **a.** To solve $x^2 = 9$, we find the ordered pairs for the intersection of the graphs $y_1 = x^2$ and $y_2 = 9$. The graphs cross twice, at $(-3, 9)$ and $(3, 9)$. There are two solutions to $x^2 = 9$: $x = -3$ or $x = 3$. The solution set is $\{-3, 3\}$.

Check: $3^2 \stackrel{?}{=} 9$ ✓ and $(-3)^2 \stackrel{?}{=} 9$ ✓

b. To solve $x^2 = 0$, we find the ordered pair for the intersection of the graphs $y_1 = x^2$ and $y_2 = 0$. The graphs cross at $(0, 0)$. There is one solution to $x^2 = 0$: $x = 0$. The solution set is $\{0\}$.

Check: $0^2 \stackrel{?}{=} 0$ ✓

c. To solve $x^2 = -2$, we look for the intersection of the graphs $y_1 = x^2$ and $y_2 = -2$. There are no intersections. Note that the origin, $(0, 0)$, is the lowest point on the graph of $y_1 = x^2$. We say that $x^2 = -2$ has no real-number solution. ◀

When an equation such as that in Example 4 has no real-number solution, we say the solution set is empty and write the symbol { } or ∅. The symbol { } shows *a set with nothing in it*, the **empty set**. The other symbol, ∅, is also common. Choose either symbol.

THINK ABOUT IT 1: Compare parts b and c of Example 4. Are {0} and { } the same result?

SUMMARY: SOLVING AN EQUATION FROM A GRAPH

> To solve an equation from a graph, make two graphs. One graph is $y =$ the left side of the equation. The other graph is $y =$ the right side of the equation. The solution to the equation is the value of x at the point(s) of intersection (x, y) of the two graphs.

▶ Solving Equations Using the Distributive Property

In solving an equation such as $2(1 - 3x) = -10$, we use the distributive property of multiplication over addition (or subtraction) to remove the parentheses:

$$a(b + c) = ab + ac \quad \text{or} \quad a(b - c) = ab - ac$$

▶ **EXAMPLE 5** Applying the distributive property Solve the two equations. The first step is to use the distributive property. Check both answers with a table and a graph.

a. $2(1 - 3x) = 8$ **b.** $2(1 - 3x) = -10$

SOLUTION Because the left sides are the same, parts a and b could have identical solution steps. (Check this yourself.) The given solution steps are different—which do you prefer? Why?

a.

$2(1 - 3x) = 8$	Use the distributive property.
$2 - 6x = 8$	Subtract 2 from both sides.
$2 - 6x - 2 = 8 - 2$	Simplify.
$-6x = 6$	Divide both sides by -6.
$\dfrac{-6x}{-6} = \dfrac{6}{-6}$	Simplify.
$x = -1$	

b.

$2(1 - 3x) = -10$	Use the distributive property.
$2 - 6x = -10$	Add $6x$ to both sides.
$2 - 6x + 6x = -10 + 6x$	Simplify.
$2 = -10 + 6x$	Add 10 to both sides.
$2 + 10 = -10 + 6x + 10$	Simplify.
$12 = 6x$	Divide both sides by 6.
$2 = x$	

Check: Table 6 shows $y = 2(1 - 3x)$. In the first row of the table, the ordered pair $(-1, 8)$ confirms that $2(1 - 3x) = 8$ for $x = -1$ in part a. The ordered pair $(2, -10)$ confirms that $2(1 - 3x) = -10$ for $x = 2$ in part b. Figure 15 shows the graph of $y_1 = 2(1 - 3x)$ intersecting the graph of $y_2 = 8$ at $(-1, 8)$, confirming part a. Figure 15 also shows the graph of $y = 2(1 - 3x)$ intersecting the graph of $y = -10$ at $(2, -10)$, confirming part b.

TABLE 6

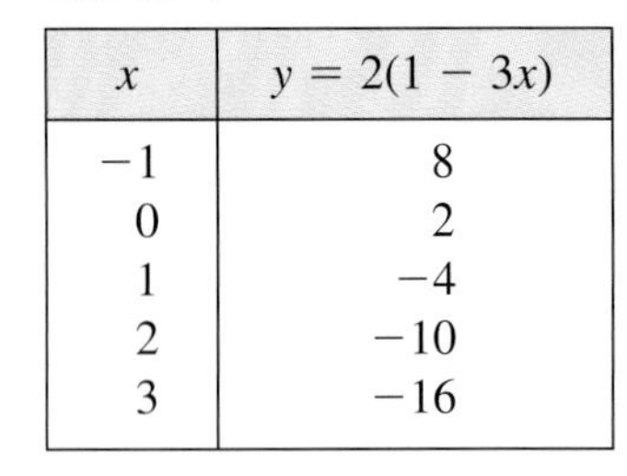

x	$y = 2(1 - 3x)$
−1	8
0	2
1	−4
2	−10
3	−16

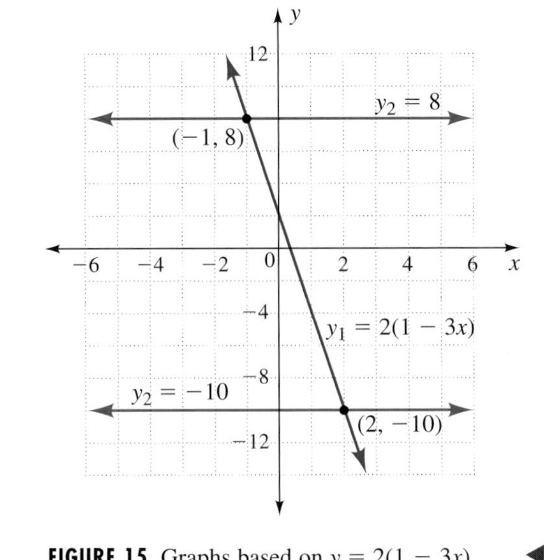

FIGURE 15 Graphs based on $y = 2(1 - 3x)$ ◀

THINK ABOUT IT 2: Try an alternative solution to each part of Example 5: First divide each side by 2. (This omits applying the distributive property.)

▶ In Example 6, we use the distributive property and addition of like terms to simplify the left side before solving the equation.

▶ **EXAMPLE 6** **Solving equations using algebraic notation** Solve for x: $13 - 4(x - 2) = 1$.

SOLUTION

$13 - 4(x - 2) = 1$	Use the distributive property.
$13 - 4x + 8 = 1$	Add like terms.
$21 - 4x = 1$	Subtract 21 from both sides (addition property).
$21 - 4x - 21 = 1 - 21$	
$-4x = -20$	Divide both sides by -4 (multiplication property).
$\dfrac{-4x}{-4} = \dfrac{-20}{-4}$	Simplify.
$x = 5$	

Check: $13 - 4(5 - 2) \stackrel{?}{=} 1$ ✓ ◀

Applications: Equations with Parentheses

EXAMPLE 7 Writing and solving equations with parentheses Describe each setting with an equation containing parentheses. Make an estimate of the answer. Solve the equation. Note the step where $ax + b = c$ is found.

a. Two-thirds of the difference between 10 and a number is 40.

b. A party at the National Academy of Artistic Gymnastics is \$85 for 11 children and \$5 for each additional child. How many children can attend on a budget of \$120?

c. Rental of a historic house for a reception is \$150 for the first three hours and \$75 per hour thereafter. The fee for a supervisor is \$50. For how many hours can the house be rented on a budget of \$500?

SOLUTION **a.** Let x be the number. In $\frac{2}{3}(10 - x) = 40$, x must be negative because $\frac{2}{3}(10) \approx 7$.

Equation	Step
$\frac{2}{3}(10 - x) = 40$	The equation contains a fraction; multiply both sides by the denominator, 3.
$3 \cdot \frac{2}{3}(10 - x) = 3 \cdot 40$	Simplify.
$2(10 - x) = 120$	Use the distributive property to get $ax + b = c$.
$20 - 2x = 120$	Subtract 20 from both sides.
$20 - 2x - 20 = 120 - 20$	Simplify.
$-2x = 100$	Divide both sides by -2.
$\frac{-2x}{-2} = \frac{100}{-2}$	Simplify.
$x = -50$	

Check: $\frac{2}{3}[10 - (-50)] \stackrel{?}{=} 40$ ✓

Student Note:
We write $5(x - 11)$, not $5x$, because the first 11 children cost \$85.

b. Let x be the total number of children; $x > 11$. In $85 + 5(x - 11) = 120$, x will be under 21 because ten additional children would cost \$50 which, added to \$85, would be over the budget.

Equation	Step
$85 + 5(x - 11) = 120$	Use the distributive property.
$85 + 5x - 55 = 120$	Add like terms to get $ax + b = c$.
$5x + 30 = 120$	Subtract 30 from both sides.
$30 + 5x - 30 = 120 - 30$	Simplify.
$5x = 90$	Divide both sides by 5.
$x = 18$	

The budget permits 18 children to attend.

Check: $85 + 5(18 - 11) < 120$ ✓

Student Note:
We write $75(x - 3)$, not $75x$, because the first 3 hours cost \$150.

c. Let x be the total number of hours. In $150 + 75(x - 3) + 50 = 500$, x is about 6.

Equation	Step
$150 + 75(x - 3) + 50 = 500$	Use the distributive property.
$150 + 75x - 225 + 50 = 500$	Add like terms to get $ax + b = c$.
$75x - 25 = 500$	Add 25 to both sides.
$75x = 525$	Divide both sides by 75.
$x = 7$	

The budget permits 7 hours of rental.

Check: $150 + 75(7 - 3) + 50 \stackrel{?}{=} 500$ ✓ ◀

THINK ABOUT IT 3: Try an alternative solution to part a of Example 7: First multiply both sides by the reciprocal of the fraction $\frac{2}{3}$.

ANSWER BOX

Warm-up: **1.** $-5, -2, 1, 4$ **2.** $8, 2, -4, -10$. **Think about it 1:** $\{0\}$ is the solution "zero"; $\{\ \}$ is the empty set or no solution. **Think about it 2:** In part a of Example 5, division by 2 gives $1 - 3x = 4$. The remainder of the solution plan is to subtract 1 and divide by -3. The solution to part b follows the same plan—divide by 2, subtract 1, and divide by -3. **Think about it 3:** Multiplication by $\frac{3}{2}$ gives $10 - x = 60$. The remainder of the solution plan is to subtract 10 and divide by -1.

▶ 3.3 Exercises

1. Solve these equations from the table.

x	$y = 10 - x$
1	9
2	8
3	7
4	6

a. $10 - x = 8$ **b.** $10 - x = 6$

c. $10 - x = 3$ **d.** $10 - x = -2$

2. Solve these equations from the table.

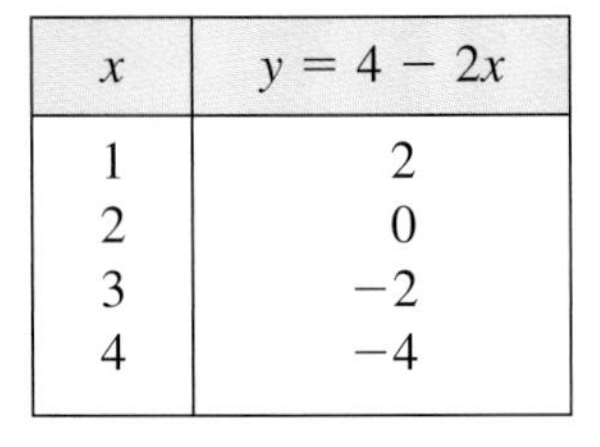

x	$y = 4 - 2x$
1	2
2	0
3	-2
4	-4

a. $4 - 2x = -4$ **b.** $4 - 2x = 0$

c. $4 - 2x = -10$ **d.** $4 - 2x = 6$

3. Solve these equations from the table.

x	$y = 2 - 4x$
1	-2
2	-6
3	-10
4	-14

a. $2 - 4x = -2$ **b.** $2 - 4x = -6$

c. $2 - 4x = 6$ **d.** $2 - 4x = -8$

4. Solve these equations from the table.

x	$y = 3 - 3x$
1	0
2	-3
3	-6
4	-9

a. $3 - 3x = -6$

b. $3 - 3x = 3$

c. $3 - 3x = -15$

d. $3 - 3x = -5$

In Exercises 5 to 8, complete the tables. Your inputs will contain fractions or decimals.

5.

x	$y = 2x + 1$
3	7
	8
4	9
	10
5	11
	12

6.

x	$y = 1 - 4x$
-3	13
	12
	11
	10
-2	9
	8

7.

x	$y = 5x - 4$
3	11
	12
	13
	14
	15
4	16

8.

x	$y = 5 - 3x$
0	5
	4
	3
1	2
	1
	0

Blue numbers are core exercises.

9. Solve these equations from the graph.

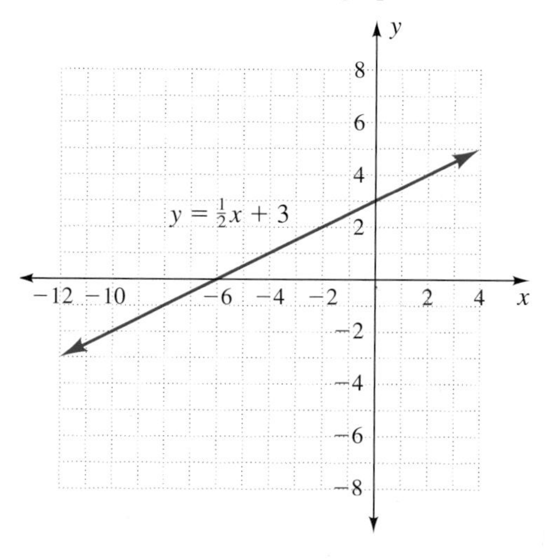

a. $\frac{1}{2}x + 3 = 1$ **b.** $\frac{1}{2}x + 3 = 0$ **c.** $\frac{1}{2}x + 3 = -2$

10. Solve these equations from the graph.

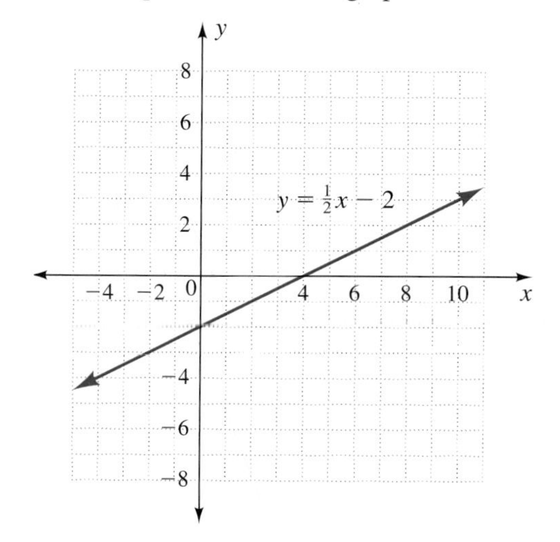

a. $\frac{1}{2}x - 2 = 3$ **b.** $\frac{1}{2}x - 2 = 0$ **c.** $\frac{1}{2}x - 2 = -4$

11. Solve these equations from the graph.

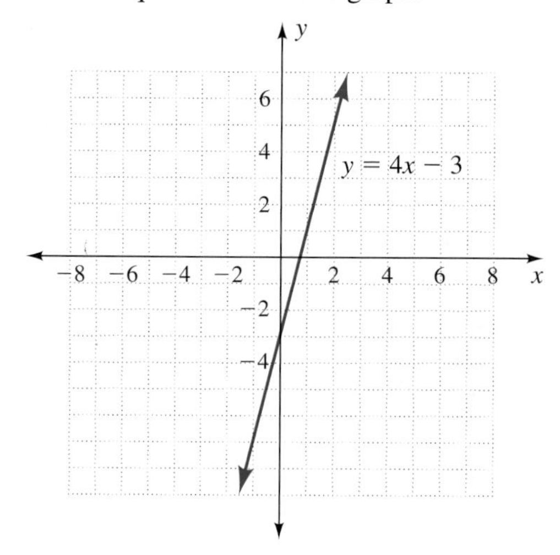

a. $4x - 3 = 1$

b. $4x - 3 = -7$

c. $4x - 3 = 5$

12. Solve these equations from the graph.

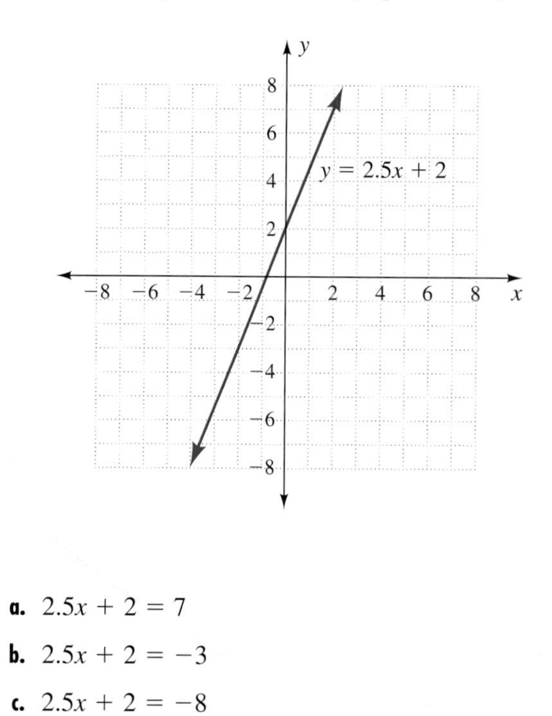

a. $2.5x + 2 = 7$

b. $2.5x + 2 = -3$

c. $2.5x + 2 = -8$

Use the graphs to find the solutions to the nonlinear equations in Exercises 13 and 14. Give answers in sets.

13.

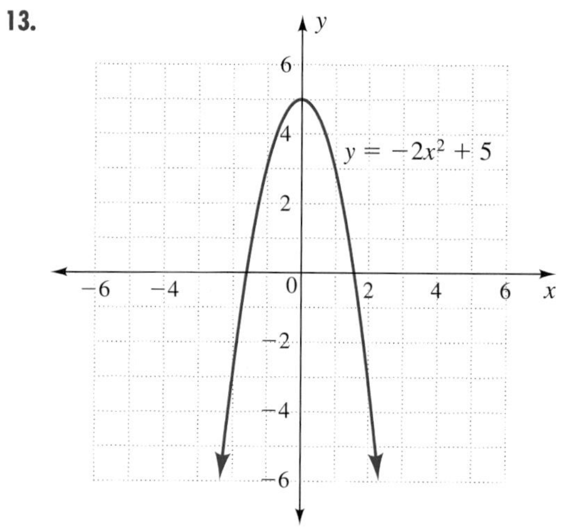

a. $-2x^2 + 5 = -3$

b. $-2x^2 + 5 = 3$

c. $-2x^2 + 5 = 5$

d. $-2x^2 + 5 = 6$

Blue numbers are core exercises.

14.

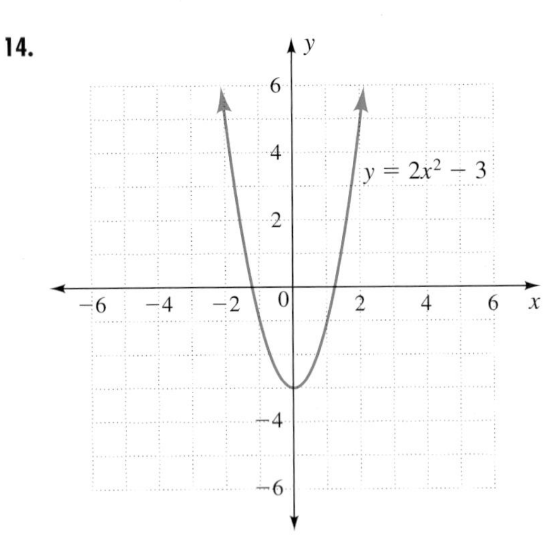

a. $2x^2 - 3 = -3$ **b.** $2x^2 - 3 = -1$

c. $2x^2 - 3 = 5$ **d.** $2x^2 - 3 = -5$

15. **a.** Make a table and graph for $y = 8 - 3(x + 2)$.

b. Solve $8 - 3(x + 2) = 11$ from the table and with algebraic symbols.

c. Solve $8 - 3(x + 2) = 0$ from the graph and with algebraic symbols.

d. Solve $8 - 3(x + 2) = -4$ with any of the three methods.

16. **a.** Make a table and graph for $y = 8 - 3(x + 4)$.

b. Solve $8 - 3(x + 4) = 2$ from the table and with algebraic symbols.

c. Solve $8 - 3(x + 4) = 0$ from the graph and with algebraic symbols.

d. Solve $8 - 3(x + 4) = -4$ with any of the three methods.

The equations in Exercises 1 to 12 were linear in the form $ax + b = c$. Solve the equations in Exercises 17 to 34. Circle the step where linear form first appears.

17. $2(x - 2) = 4$ **18.** $3(x - 1) = -3$

19. $5(x + 4) = 12$ **20.** $4(x + 5) = -10$

21. $2(2 - x) = -5$ **22.** $3(1 - x) = -5$

23. $3(x + 2) = -7$ **24.** $5(x + 2) = 8$

25. $\frac{1}{2}(6 - 3x) = 9$ **26.** $\frac{1}{3}(x + 4) = 3$

27. $\frac{2}{3}(x - 5) = -2$ **28.** $\frac{3}{8}(3 - 2x) = 3$

29. $4 - 3(x + 1) = 16$ **30.** $6 - 5(x - 1) = 31$

31. $7 - 3(4 - x) = -26$ **32.** $5 - 4(3 - x) = 17$

33. $8 - 6(x - 7) = 74$ **34.** $3 - 6(x - 4) = -3$

Blue numbers are core exercises.

Write the sentences in Exercises 35 to 40 as an equation containing parentheses. Solve the equation.

35. Take a number, add 5, and then multiply the sum by 2. The result is 14.

36. Take a number, subtract 6, and then multiply the difference by 3. The result is 15.

37. Take a number, subtract 4, and then multiply the difference by negative 2. The result is 6.

38. Take a number, add 3, and then multiply the sum by 7. The result is −21.

39. Half the difference between a number and five gives 24.

40. One-third the sum of a number and six gives −2.

In Exercises 41 to 48, define a variable, and write an equation containing parentheses to answer the question. Solve the equation.

41. A local truck rental costs $29.95 plus $0.89 cents per mile for up to 50 miles. More than 50 miles are charged an additional $0.40 per mile. How many miles are traveled if the cost is $150? Ignore other charges and fees.

42. A pressure washer costs $30 for the first 4 hours and $10 per additional hour. What is the total hours rented, x, if the total cost is $70?

43. An official transcript at LBCC costs $5 for the first copy and $1 for each additional copy. How many transcripts are ordered if the total cost is $11? Also explain how to solve without an equation.

44. At an outdoor concert, the staff adds 3 seats to each row of x seats. There are 25 rows altogether, and there are 1200 seats altogether. How many seats, x, were in each original row?

45. A party at Grand Slam USA costs $80 for the first 8 children and $10 for each additional child. How many children attend if the total cost is $150? Also explain how to solve without an equation.

46. Skate World charges $82 for 10 children and $7 for each additional child. How many children, x, can be invited on a $145 budget?

47. A party at Lane County Ice costs $125 for the first 10 children and $12.50 for each additional child. How many children attend if the total cost is $175? Also explain how to solve without an equation.

48. A party at the National Academy of Artistic Gymnastics costs $85 for the first 11 children and $5 for each additional child. How many children can attend if the total cost is $150?

▶ Writing

49. Explain how to solve $mx + b = n$ for x from a table for $y = mx + b$. Assume that the number n appears in the table.

50. Explain how to solve $mx + b = n$ for x from a graph of $y = mx + b$. Assume that the number n appears on the vertical axis.

51. Explain how to solve the linear equation $ax + b = 0$ for x. Show the solution.

52. Explain how to solve the linear equation $ax + b = c$ for x. Show the solution.

53. Explain the difference in meaning between the symbols $\{\ \}$ and $\{0\}$.

54. True or false: If the numbers in (x, y) make a true statement when substituted into an equation, then (x, y) lies on the graph of the equation.

55. True or false: All ordered pairs for points on a graph make a true statement when substituted into the equation for the graph.

56. True or false: If an ordered pair (a, b) is on a graph, then the ordered pair (b, a) is also on the graph.

▶ Error Analysis

57. Explain the error in going from the first equation to the second.

$$2(1 - 3x) = 4$$
$$2 - 6x = 8$$

58. Explain the error in checking $x = -6\frac{2}{3}$ in the equation $3(x + 12) = 16$ by entering $3(-6 + 2/3 + 12)$ on a calculator.

▶ Project

59. Percent and the Distributive Property Writing expressions involving percent often requires adding like terms and applying the distributive property. In each set of expressions, find the one expression that does not reflect the situation described. Explain why it is different.

a. An 8% sales tax is added to the price n: $n + 0.08n$, $1n + 0.08n$, $n(1 + 0.08)$, a number plus 8% of the number, $n + 8\%$

b. A 7% sales tax is added to the price, p: $1.00p + 0.07p$, $p(1 + 0.07)$, $1.07p$, $1 + 0.07p$

c. The original price, x, is discounted by 5%: $x - 0.05x$, $x - 5\%$, $x(1 - 0.05)$, $0.95x$, $1x - 0.05x$

d. The original price, n, is discounted by 20%: the number minus 20% of the number, $n - \dfrac{n}{5}$, $n - \frac{1}{5}$, $n\left(1 - \frac{1}{5}\right)$

▶ 3 Mid-Chapter Test

1. State whether each equation is an identity or a conditional equation.

a. $2x + 4 = -1$ **b.** $2x + 3x = 5x$

c. $4(x - 2) = 4x - 8$ **d.** $2x = x$

2. Which pairs of equations are not equivalent?

a. $4x + 5 = 29$, $4x = 24$

b. $\frac{1}{2}x = 16$, $x = 8$

c. $2x + 3x = 10$, $5x^2 = 10$

d. $3x - 2 = 8$, $3x = 6$

In Exercises 3 to 5, is the given input x a solution to the equation?

3. $3x - 4 = -16$, $x = -4$

4. $4x - 3 = -9$, $x = -3$

5. $-8x + 5 = 1$, $x = \frac{1}{2}$

In Exercises 6 to 8, write equations. Let x be the unknown number.

6. Five more than twice a number is 10.

7. How many credit hours can be taken for \$545 if tuition is \$85 per credit hour and fees are \$35?

8. How many minutes were used if there is \$7.20 left on a \$20 telephone card and each minute costs \$0.04?

In Exercises 9 and 10, write each equation in words.

9. $5 = 3x - 4$

10. $2(2 + x) = -3$

In Exercises 11 to 18, solve for x.

11. $x - 4 = 3$

12. $\frac{2}{3}x = 24$

13. $2x + 3 = -7$

14. $\frac{1}{2}x - 8 = -1$

15. $3x = \frac{1}{2}$

16. $3 - 2x = 8$

17. $3(2 + x) = 18$

18. $4 - 2(x - 1) = -4$

19. Use the table to solve the following equations.

a. $\frac{1}{2}x + 4 = 5$

b. $\frac{1}{2}x + 4 = 8$

c. $\frac{1}{2}x + 4 = 6.5$

d. $\frac{1}{2}x + 4 = 4$

x	$y = \frac{1}{2}x + 4$
2	5
4	6
6	7
8	8

Solve the equations in Exercises 20 and 21 from the graphs.

20.

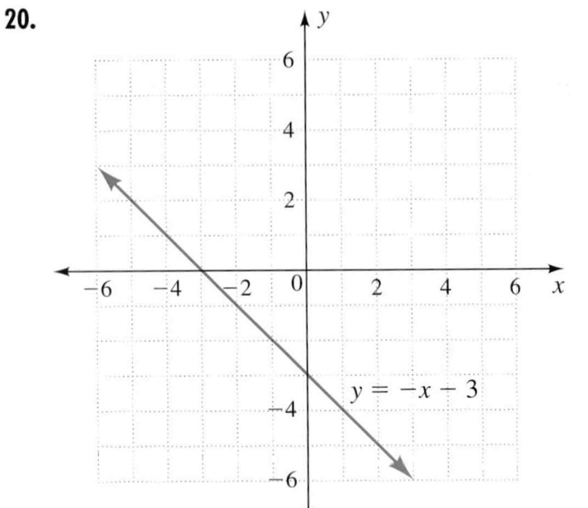

a. $-x - 3 = -5$

b. $-x - 3 = 1$

c. $-x - 3 = -2$

21.

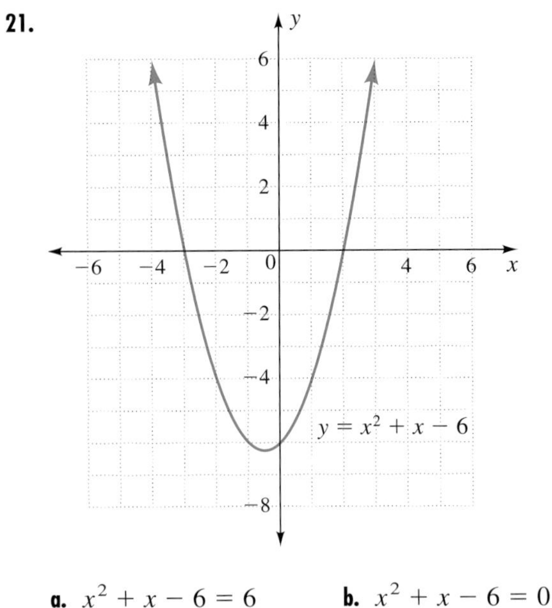

a. $x^2 + x - 6 = 6$ **b.** $x^2 + x - 6 = 0$

c. $x^2 + x - 6 = -7$

22. A student solves an equation by writing $2x = 10 = x = 5$. Why is this incorrect?

23. One of several ways to express the area, A, of a trapezoid is half the height, h, multiplied by the sum of the two parallel sides, a and b. Write this as an equation.

24. A bowling handicap is 80% of the difference between 200 and the bowling average. For averages above 200, there is no handicap. Write an equation to find the bowling average if the handicap is 32. Solve the equation.

25. With a coupon, the first four half-gallon cartons of ice cream cost \$5. Additional cartons cost \$2.19 each. For a party budget of \$24 write an equation containing parentheses to show how many cartons can be purchased. Solve the equation. What is the change?

▶ 3.4 Solving Linear Equations with Variables on Both Sides of the Equation

Objectives

- Solve equations with expressions on the left and right sides.
- Set up and solve equations from applications.

WARM-UP

Simplify:

1. $4(x + 3)$ **2.** $-2(x + 1)$ **3.** $-4(x - 2)$

4. $-(x + 1)$ **5.** $13 - 4(x - 2)$ **6.** $8 - 3(x - 1)$

7. $2x + 3x - 4x$ **8.** $5x - 2x + x$

HAVE YOU NEEDED to compare costs from two moving truck companies? Writing linear equations for each and setting them equal gives an equation with variables on both sides. The solution tells us when costs are equal. In this section, we use tables, graphs, and algebraic notation to solve equations with variables on both sides.

Student Note:
Find rental rates for your city or region online at Budget, Penske, U-Haul or other moving truck rentals. See also Example 7 and the exercises.

Solving Equations: Tables and Graphs

In the following examples, we use tables and graphs to solve equations with variables on both sides.

SOLVING A LINEAR EQUATION USING TABLES AND A GRAPH

1. Let y_1 = the left side of the equation. Build a table and plot a graph.
2. Let y_2 = the right side of the equation. Build a table and plot a graph on the same axes used for the left side.
3. Look for the ordered pair that appears in both tables and at the point of intersection. The x in the ordered pair is the solution to the equation. The y in the ordered pair is the value obtained when x is substituted into each side of the original equation.

EXAMPLE 1 *Using tables and a graph to solve an equation* Solve $3(x - 1) = x + 1$ using Tables 7 and 8 and the graph in Figure 16.

TABLE 7

x	Left Side: $y_1 = 3(x - 1)$
−1	−6
0	−3
1	0
2	3
3	6

Compare outputs.

<
<
<
=
>

TABLE 8

x	Right Side: $y_2 = x + 1$
−1	0
0	1
1	2
2	3
3	4

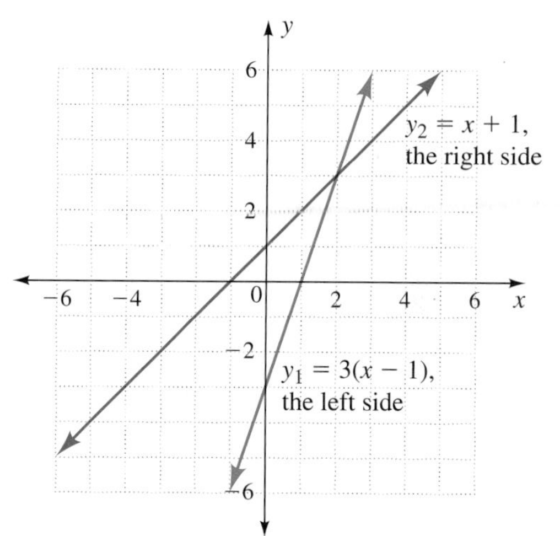

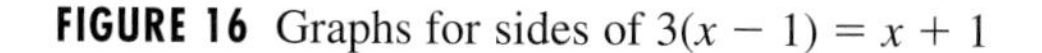

FIGURE 16 Graphs for sides of $3(x - 1) = x + 1$

SOLUTION To solve from the tables, we find the ordered pair (2, 3) appearing in both tables. Comparing the outputs by putting an inequality (or equal) sign between the tables is helpful in finding a solution. The input $x = 2$ makes the two expressions $3(x - 1)$ and $x + 1$ equal to 3. Thus, $x = 2$ solves the equation $3(x - 1) = x + 1$.

Check: $3(2 - 1) \stackrel{?}{=} 2 + 1$ ✓

To solve $3(x - 1) = x + 1$ from the graph, we look for the point of intersection of the two graphs $y = 3(x - 1)$ and $y = x + 1$. The graphs intersect at (2, 3). This means that the input $x = 2$ makes both $3(x - 1)$ and $x + 1$ equal to 3. Thus, $x = 2$ solves the equation.

Check: $3(2 - 1) \stackrel{?}{=} 2 + 1$ ✓ ◀

THINK ABOUT IT 1: Look at the inequality signs between Tables 7 and 8, and compare the signs above the equals sign with the sign below the equals sign.

▶ Example 2 shows an equation for which longer tables are needed to find a common ordered pair.

▶ EXAMPLE 2 **Using tables and a graph to solve an equation** Solve $4(x - 2) = 2(x + 1)$ with tables and a graph.

SOLUTION First we build Tables 9 and 10, compare outputs, and extend the tables until a common ordered pair is found. Both tables contain the ordered pair (5, 12). At $x = 5$, $4(x - 2) = 12$ and $2(x + 1) = 12$. Thus, $x = 5$ solves the equation $4(x - 2) = 2(x + 1)$.

TABLE 9

x	$y_1 = 4(x - 2)$
0	−8
1	−4
2	0
3	4
4	8
5	12

Compare outputs.

$<$
$<$
$<$
$<$
$<$
$=$

TABLE 10

x	$y_2 = 2(x + 1)$
0	2
1	4
2	6
3	8
4	10
5	12

Check: $4(5 - 2) \stackrel{?}{=} 2(5 + 1)$ ✓

Next we plot the graph of each side of the equation, as shown in Figure 17. The ordered pair (5, 12) locates the point of intersection of the two graphs $y_1 = 4(x - 2)$ and $y_2 = 2(x + 1)$. Thus, $x = 5$ as an input to both $4(x - 2)$ and $2(x + 1)$ gives 12 as an output, so $x = 5$ solves the equation $4(x - 2) = 2(x + 1)$.

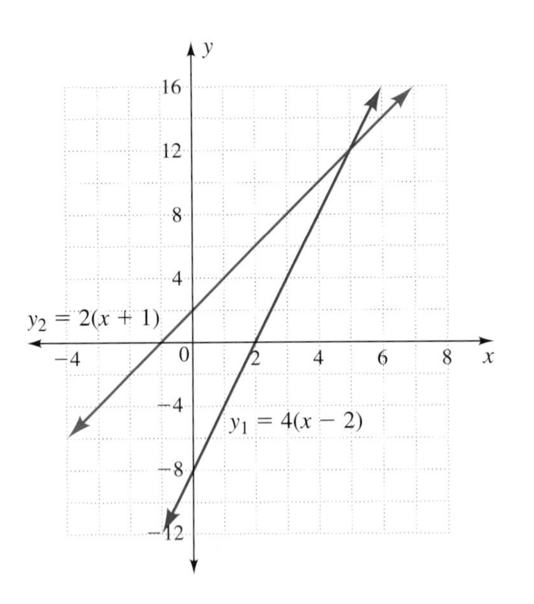

FIGURE 17 Graphs for $4(x - 2) = 2(x + 1)$

Check: $4(5 - 2) \stackrel{?}{=} 2(5 + 1)$ ✓ ◀

THINK ABOUT IT 2: Will the output for $x = 6$ in $y_1 = 4(x - 2)$ be greater than the output for $x = 6$ in $y_2 = 2(x + 1)$? How do we know without substituting $x = 6$?

GRAPHING CALCULATOR TECHNIQUE: SOLVING AN EQUATION BY GRAPH AND TABLE

Enter the left side of the equation as Y_1 and the right side of the equation as Y_2.

To solve from a graph: Evaluate each side of the equation for one or two inputs, and use the resulting ordered pairs to estimate the window settings. Graph the equations. If the lines do not intersect, reset the window to enclose the area where the lines appear to get closer together. Graph and reset the window as needed. Trace to the point of intersection. Zoom in and trace to find the ordered pair for the point of intersection.

To solve from a table: Estimate a solution. Go to [TBLSET]. For **TblStart**, choose a number near your estimate. For **ΔTbl**, choose the number you used for the scale on the x-axis. Go to [TABLE]. You want $Y_1 = Y_2$ for the same X. Continue to choose a new **TblStart** number, if needed, and a new **ΔTbl** number until the outputs from Y_1 and Y_2 are equal. The X corresponding to the equal outputs is the solution to the original equation.

▶ The graphing calculator is useful when the common ordered pair, or intersection of two graphs, has inputs between two integers. With a graphing calculator, we can trace and zoom in to a point of intersection or we can adjust the table to decimal inputs.

▶ Solving Equations with Algebraic Notation

For many equations, symbolic solutions are easier than tables or graphs. If the symbolic method becomes difficult or even impossible, we can turn to graphing calculators or computers for a solution.

SOLVING LINEAR EQUATIONS WITH ALGEBRAIC NOTATION

1. Multiply on both sides by the least common denominator to remove fractions or decimals. Add like terms.
2. Apply the distributive property to expressions containing parentheses. Add like terms.
3. Using the addition property of equations, get all the terms with the variable to be solved for onto one side of the equation. Using the addition property again, get the constant terms to the other side of the equation. Add like terms again, if necessary.
4. Solve for the variable, using the multiplication property of equations.

Caution: When an equation has a variable on both sides, you will *make fewer errors if you keep the variable's coefficient positive when you add variable terms to each side* (step 3).

▶ **EXAMPLE 3** Solving equations using algebraic notation Solve for x: $6 - 5x = 3(1 - x)$.

SOLUTION

$6 - 5x = 3(1 - x)$	Distributive property
$6 - 5x = 3 - 3x$	Add $5x$ (addition property) to obtain a positive coefficient on x.
$6 - 5x + 5x = 3 - 3x + 5x$	Add like terms to get $c = ax + b$.
$6 = 2x + 3$	Subtract 3 (addition property).
$6 - 3 = 2x + 3 - 3$	Add like terms.

Student Note:
The comments tell what to do next.

$3 = 2x$ Divide both sides by 2 (multiplication property).

$\frac{3}{2} = \frac{2x}{2}$ Simplify.

$1\frac{1}{2} = x$

Check: $6 - 5\left(1\frac{1}{2}\right) \stackrel{?}{=} 3\left(1 - 1\frac{1}{2}\right)$ ✓ ◀

In Example 4, we repeat the equation from Example 3.

EXAMPLE 4 Solving with tables and a graph Solve $6 - 5x = 3(1 - x)$ with tables and a graph.

SOLUTION Tables 11 and 12 have no ordered pair in common. The inequalities written between the tables suggest that the ordered pairs are closest between $x = 1$ and $x = 2$. On a calculator, we let $Y_1 = 6 - 5X$ and $Y_2 = 3(1 - X)$. Under [TBLSET], let **TblStart** $= -1$ and **ΔTbl** $= 0.5$. Figure 18 shows the resulting table, with ordered pair $(1.5, -1.5)$.

TABLE 11

x	$y_1 = 6 - 5x$
−1	11
0	6
1	1
2	−4
3	−9

Compare outputs.

>
>
>
<
<

TABLE 12

x	$y_2 = 3(1 - x)$
−1	6
0	3
1	0
2	−3
3	−6

The graph in Figure 19 shows the point of intersection at $\left(1\frac{1}{2}, -1\frac{1}{2}\right)$.

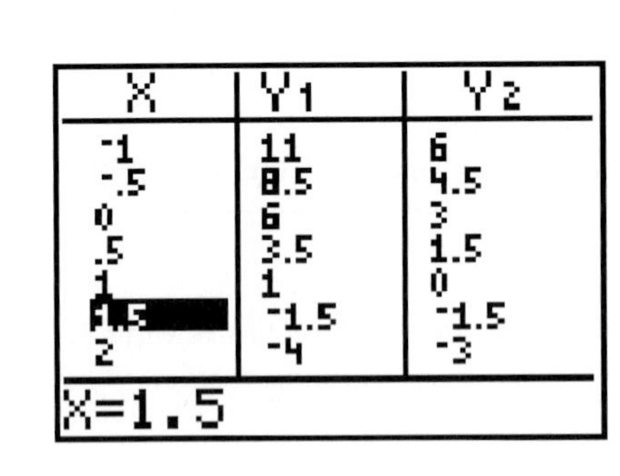

FIGURE 18 [TABLE] Screen

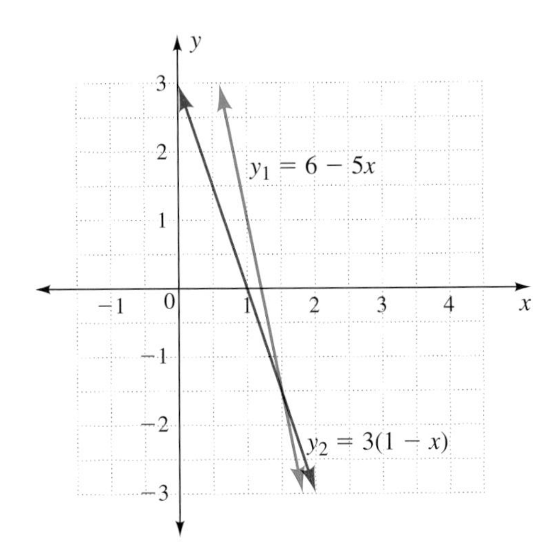

FIGURE 19 Graph of $6 - 5x = 3(1 - x)$

Check: $6 - 5\left(1\frac{1}{2}\right) \stackrel{?}{=} 3\left(1 - 1\frac{1}{2}\right)$ ✓ ◀

THINK ABOUT IT 3: What number is obtained on both sides of the Check in Example 3? What is the meaning of this number in Example 4?

▶ In Examples 5 and 6, we continue work with variables on both sides of the equation. In Example 5, we multiply both sides of the equation by the denominator to eliminate the fraction notation.

▶ EXAMPLE 5 **Solving equations containing fractions** Solve $\frac{5}{6}x = x + 4$.

SOLUTION

$\frac{5}{6}x = x + 4$	Multiply both sides by 6.
$\frac{6}{1} \cdot \frac{5}{6}x = 6(x + 4)$	Apply the distributive property.
$5x = 6x + 24$	Subtract $5x$ from both sides to get $c = ax + b$.
$-5x \quad -5x$	
$0 = x + 24$	Subtract 24 from both sides.
$-24 \quad -24$	
$-24 = x$	

Check: $\frac{5}{6}(-24) \stackrel{?}{=} (-24) + 4$ ✓ ◀

▶ In Example 6, we eliminate the fraction and use the distributive property in solving the equation.

▶ EXAMPLE 6 **Solving equations containing fractions** Solve $2(x - 7) = \frac{1}{2}x - 2$.

SOLUTION

$2(x - 7) = \frac{1}{2}x - 2$	Multiply both sides by $\frac{2}{1}$.
$\frac{2}{1} \cdot 2(x - 7) = \frac{2}{1}\left(\frac{1}{2}x - 2\right)$	Apply the distributive property.
$4x - 28 = x - 4$	Subtract x from both sides to get $ax + b = c$.
$-x \quad -x$	
$3x - 28 = -4$	Add 28 to both sides.
$+28 \quad +28$	
$3x = 24$	Divide by 3 on both sides.
$\frac{3x}{3} = \frac{24}{3}$	Simplify.
$x = 8$	

Check: $2(8 - 7) \stackrel{?}{=} \frac{1}{2}(8) - 2$ ✓ ◀

▶ Application: Moving Truck Rentals

In Example 7, we solve an equation and use intervals and inequalities to describe the results.

▶ EXAMPLE 7 **Comparing moving truck rentals** Two competing truck rental agencies offer 24-foot trucks in Miami. Budget charges \$49.95 plus \$0.89 per mile. U-Haul charges \$39.95 plus \$0.99 per mile. Other charges and options apply.

a. Write an equation to describe the total rental cost for each agency, where x is the number of miles driven. The total cost, y, is in dollars.

b. Match each of your equations to its graph in Figure 20.

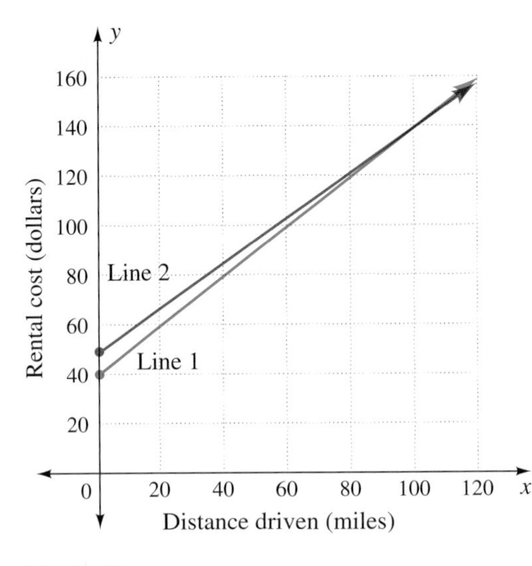

FIGURE 20

c. Solve the equation from the graph and with algebraic notation:

$$0.89x + 49.95 = 0.99x + 39.95$$

d. What is the meaning of the solution in part c?

e. For what mileage will Budget's truck be cheaper? Describe the x inputs with an interval and with an inequality.

f. For what mileage will U-Haul's truck be cheaper? Describe the x inputs with an interval and with an inequality.

SOLUTION **a.** The two rental costs are

Budget: $y = 0.89x + 49.95$

U-Haul: $y = 0.99x + 39.95$

b. Line 1 is U-Haul's graph. Line 2 is Budget's graph. The graph has miles on its horizontal axis and total dollar cost on its vertical axis.

c. The point of intersection of the graphs, approximately (100, 140), shows that $x =$ 100 miles:

Student Note:
The subtraction and division steps are not shown in part c. They are done mentally, and only the results are shown.

$$0.89x + 49.95 = 0.99x + 39.95 \quad \text{Subtract } 0.89x \text{ from both sides.}$$
$$49.95 = 0.10x + 39.95 \quad \text{Subtract 39.95 from both sides.}$$
$$10 = 0.10x \quad \text{Divide both sides by 0.10.}$$
$$x = 100$$

d. The point of intersection, (100, 140), gives the number of miles, $x = 100$, and the approximate total dollar cost, $y \approx 140$, at which the two agencies' charges are the same. The exact dollar cost is $0.89(100) + 49.95 = 138.95$.

e. Budget's will be better for longer trips, with mileage in the interval $(100, +\infty)$, or $x > 100$. Of course, infinity $(+\infty)$ is not a reasonable mileage because there is a limit on the number of miles that can be driven.

f. U-Haul's will be better for short trips, with mileage in the interval $[0, 100)$, or $0 \le x < 100$.

GRAPHING CALCULATOR SOLUTION Let $x =$ miles traveled and $y =$ total dollar cost. Enter $Y_1 = 0.89X + 49.95$ and $Y_2 = 0.99X + 39.95$. Set the window with X in the interval [20, 114], scale 20, and Y in the interval [0, 150], scale 20. Graph. Trace to the point of intersection (see Figure 21), which shows $x = 100$ and $y = 138.95$.

Student Note:
The X-interval is 94 units long, which will give a trace along whole-number inputs on TI 83 and 84 calculators.

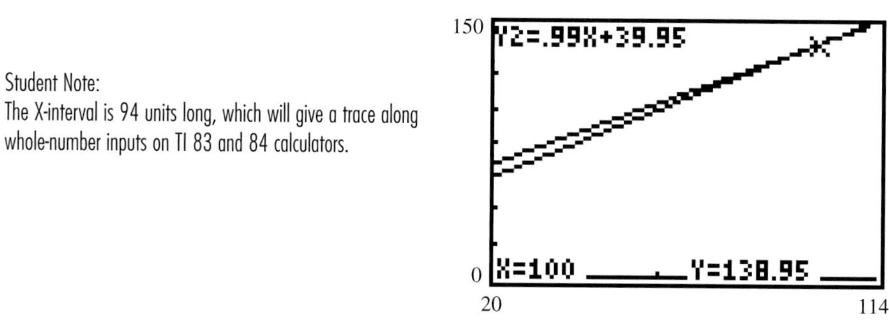

FIGURE 21 20 to 114 on X, 0 to 150 on Y ◀

THINK ABOUT IT 4: What is the meaning of the dots where the graphs intersect the y-axis in Figure 20?

ANSWER BOX

Warm-up: **1.** $4x + 12$ **2.** $-2x - 2$ **3.** $-4x + 8$ **4.** $-x - 1$ **5.** $21 - 4x$ **6.** $11 - 3x$ **7.** $1x$ **8.** $4x$ **Think about it 1:** The inequality signs are in opposite directions. **Think about it 2:** The inequality sign between the tables changes after the equal sign. **Think about it 3:** In Example 3, $6 - 5(1\frac{1}{2}) = 6 - 7\frac{1}{2} = -1\frac{1}{2}$ and $3(1 - 1\frac{1}{2}) = 3(-\frac{1}{2}) = -1\frac{1}{2}$. In Example 4, the point of intersection was $(1\frac{1}{2}, -1\frac{1}{2})$. The $-1\frac{1}{2}$ is the y coordinate of the point of intersection. **Think about it 4:** The dots mean that there is a charge if no miles are driven.

▶ 3.4 Exercises

Solve the equations in Exercises 1 to 4 for x by completing the given tables and identifying x from the common ordered pair.

1. $5x - 8 = 2(x + 2)$

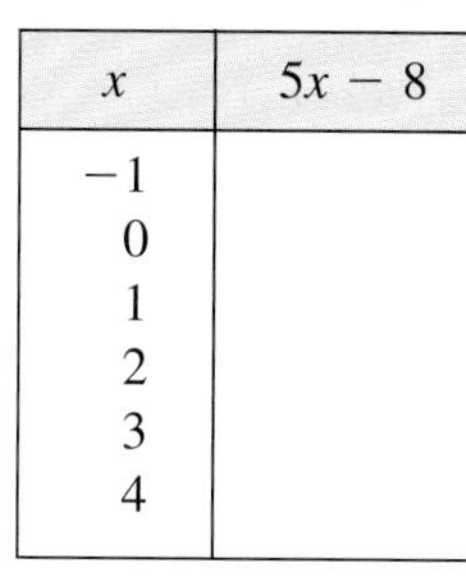

x	$5x - 8$
−1	
0	
1	
2	
3	
4	

x	$2(x + 2)$
−1	
0	
1	
2	
3	
4	

2. $7(x - 5) = 3 - 12x$

x	$7(x - 5)$
−1	
0	
1	
2	
3	
4	

x	$3 - 12x$
−1	
0	
1	
2	
3	
4	

3. $3(x - 3) = 6(x - 2)$

x	$3(x - 3)$
−1	
0	
1	
2	
3	
4	

x	$6(x - 2)$
−1	
0	
1	
2	
3	
4	

4. $4(x + 1) = 2(3x - 1)$

x	$4(x + 1)$
−1	
0	
1	
2	
3	
4	

x	$2(3x - 1)$
−1	
0	
1	
2	
3	
4	

Blue numbers are core exercises.

5. Solve with the graph.

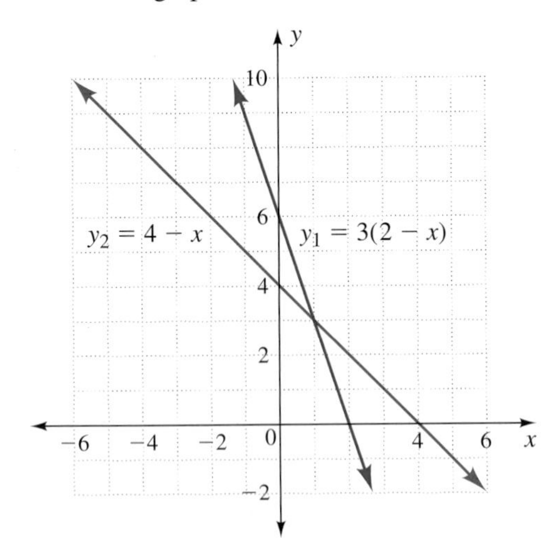

a. $3(2 - x) = 4 - x$

b. $3(2 - x) = 6$ (*Hint:* Draw $y = 6$.)

c. $4 - x = 6$

d. $4 - x = 2$ (*Hint:* Draw $y = 2$.)

e. $3(2 - x) = 0$

6. Solve with the graph:

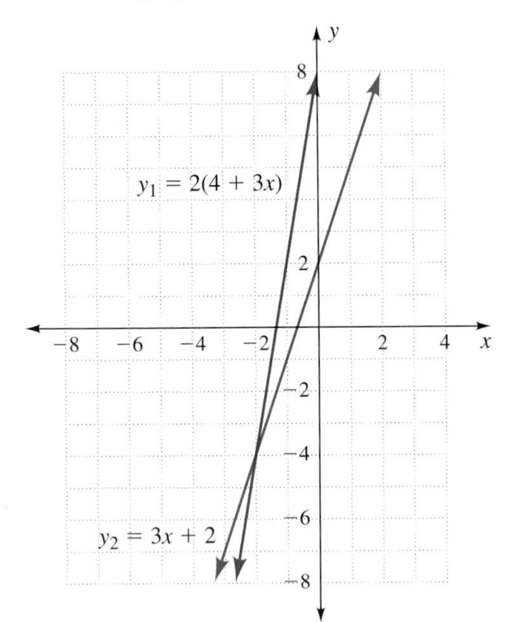

a. $2(4 + 3x) = 3x + 2$

b. $2(4 + 3x) = 8$ (*Hint:* Draw $y = 8$.)

c. $3x + 2 = 8$

d. $3x + 2 = -1$ (*Hint:* Draw $y = -1$.)

e. $2(4 + 3x) = 2$

7. Solve with the graph:

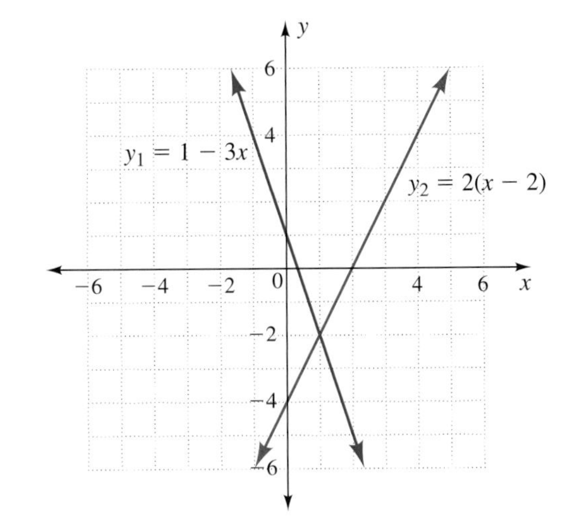

a. $1 - 3x = 2(x - 2)$

b. $2(x - 2) = 2$ (*Hint:* Draw $y = 2$.)

c. $1 - 3x = 4$ (*Hint:* Draw $y = 4$.)

d. $1 - 3x = -5$

e. $2(x - 2) = 0$

8. Solve with the graph:

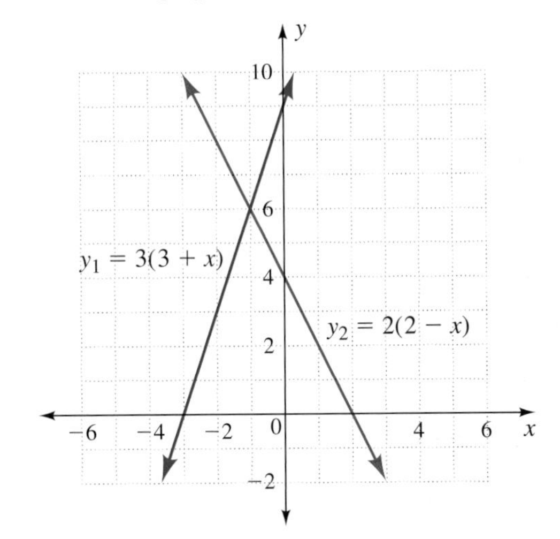

a. $3(3 + x) = 2(2 - x)$

b. $3(3 + x) = 0$

c. $2(2 - x) = 0$

d. $2(2 - x) = 4$ (*Hint:* Draw $y = 4$.)

e. $3(3 + x) = 3$

Solve the equations in Exercises 9 to 38 using any of the three methods.

9. $2(x - 3) = 0$

10. $3(x - 1) = 2$

11. $-2(x + 1) = 5$

12. $-3(x - 2) = 0$

Blue numbers are core exercises.

13. $2(3x + 1) = 5 - 3x$

14. $3(3x + 1) = 13 - 6x$

15. $4x + 6 = 2(1 + 3x) + 1$

16. $7x + 2 = 3(1 + x) + 2$

17. $7(x + 1) = 11 + x$

18. $3(x + 2) = 4 - x$

19. $2(x + 3) = 3x - 2$

20. $2(2 + x) = 13 - 4x$

21. $\frac{3}{5}x = 15$

22. $\frac{5}{8}x = 20$

23. $\frac{1}{3}x = x - 12$

24. $\frac{2}{3}x + 4 = x + 7$

25. $\frac{3}{4}x = x + 3$

26. $\frac{4}{7}x - 12 = x$

27. $\frac{1}{2}(x + 5) = x - 4$

28. $\frac{3}{5}(2 - x) = x + 6$

29. $6 - 4(x - 2) = 22$

30. $5 - 3(x - 2) = 9$

31. $2(8 - x) = 1 + 4x$

32. $3(x + 1) - 2x = 3 - 4x$

33. $2 - 5(x + 1) = 4 - 3x$

34. $5x - 2(x - 1) = 4x - 3$

35. $7 - 2(x - 1) = 5 - 3x$

36. $4 - x + 3 = -2 + 4x$

37. $5 - 2(x - 3) = 3(x - 3)$

38. $7 - 2(x - 1) = 5(x - 1)$

In Exercises 39 to 44, tell what was done to the first equation to obtain the second equation. Which property of equations or real numbers lets us do each step?

39. $2(2x + 3) = 2 + 6x + 1$
$4x + 6 = 2 + 6x + 1$

40. $7x + 2 = 3 + 3x + 2$
$4x + 2 = 3 + 2$

41. $7x + 7 = 11 + x$
$7x = 4 + x$

42. $3(x + 2) = 4 - x$
$3x + 6 = 4 - x$

43. $3x + 6 = 4 - x$
$4x + 6 = 4$

44. $4x + 6 = 6x + 2 + 1$
$4x + 6 = 6x + 3$

45. In South Chicago, a Penske 26-foot moving truck rental costs \$49.95 plus \$0.59 per mile for 6 hours. The same truck from U-Haul costs \$39.95 + \$0.99 per mile.

   a. For each agency, write an equation describing the total cost.

   b. State the equation found by setting the two cost descriptions equal. Solve and explain the meaning of the solution.

   c. Without graphing, give the ordered pair for the point of intersection of the graphs of the equations in part a. Explain what the output in the ordered pair means.

   d. Which company has the steeper graph? Why?

   e. For what distances is U-Haul the cheaper to use?

46. In New York City, a U-Haul 24-foot moving truck costs \$39.95 plus \$1.59 per mile for 6 hours. The same truck from Budget costs \$69.99 plus \$0.89 per mile. Repeat parts a to e from Exercise 45.

47. In Phoenix, a U-Haul 24-foot moving truck costs \$39.95 plus \$0.59 per mile for 6 hours. The same truck from Budget costs \$59.95 plus \$0.57 per mile.

   a. State the equation found by setting the costs equal, and solve for the number of miles.

   b. Is the solution to part a meaningful?

   c. Which is the better rental?

48. In Phoenix, a Penske 26-foot moving truck rents for \$49.95 plus \$0.59 per mile. The same truck from U-Haul costs \$39.95 plus \$0.59 per mile. Repeat parts a to c from Exercise 47.

▶ Writing

49. Describe how to find the solution to an equation of the form $ax + b = cx + d$ from tables.

50. Describe how to find the solution to an equation of the form $ax + b = cx + d$ from a graph.

51. What algebraic steps do we use in solving an equation with variables on both sides (for example, $ax + b = cx + d$)?

52. What algebraic steps do we use in solving the equation $a(x + c) = d$? Describe two ways.

53. What algebraic steps do we use in solving an equation containing a fraction (for example, $\frac{ax}{b} + c = d$)?

▶ Error Analysis

54. Explain the error in writing the second line to solve the first equation:
$$9 - 4(x + 3) = 7$$
$$5(x + 3) = 7$$

55. Explain the error in writing the second line to solve the first equation:
$$4(x - 3) = 10$$
$$x - 3 = 10 - 4$$

56. **Connecting with Chapter 1: Equations and percents** A Discover card gives cash rewards for purchases during a given year.

   a. How much is the reward on the first \$1500 at 0.25%?

   b. How much is the reward on the next \$1500 at 0.50%?

Blue numbers are core exercises.

c. Rewards of 1% start after you make purchases totaling \$3000 (parts a and b). "Rewards are redeemable in increments of \$20." What is the total spending needed to earn the first \$20 in rewards?

d. What equation describes part c?

e. What is the spending needed to earn the second \$20 in rewards during a year?

f. What equation describes part e?

g. "Earn up to 5% cash-back bonus" on select purchases requires a special sign-up fee (fee not specified). What amount of spending is needed to earn \$20 at 5%?

▶ Project

57. Equations and Identities Use guess and check or algebraic notation to solve each equation. Write one of the following statements for each equation.

Has only $x = 0$ as a solution.
Has no real-number solutions, { }.
Is an identity (true for all real numbers).
Has both $x = 0$ and $x = 1$ as a solution, $\{0, 1\}$.
Has one solution, $x = 4$.

a. $-(x - 5) = 5 - x$ **b.** $3 + 3x = 3 - 3x$

c. $3x = 5x$ **d.** $-3(x - 2) = -3x - 2$

e. $3x + 6 = 3(x + 2)$ **f.** $-2(x - 4) = -2x - 8$

g. $2x + 3x = 5x^2$ **h.** $2x + 3x = 5x$

i. $5 - 2(x - 3) = 3(x - 3)$

j. $5 - 2(x - 3) = 11 - 2x$

k. $3x + 4 = x + 2(x + 2)$

l. $4(2 - x) = 2(1 - 2x)$

m. $x(x - 1) = 0$ **n.** $3 - 2(x + 1) = 1$

o. $\frac{2}{3}(16 - x) = 8$

▶ 3.5 Solving Linear Inequalities in One Variable

Objectives

- Solve one-variable inequalities with a graph.
- Graph solutions to inequalities on a number line.
- Solve one-variable inequalities with algebraic notation.
- Write and solve inequalities for application settings.

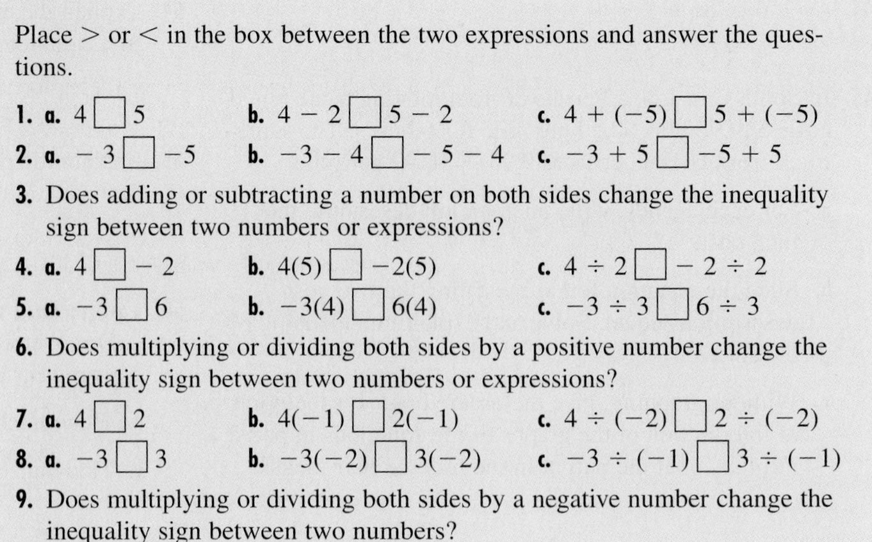

WARM-UP

Place > or < in the box between the two expressions and answer the questions.

1. a. $4 \square 5$ **b.** $4 - 2 \square 5 - 2$ **c.** $4 + (-5) \square 5 + (-5)$

2. a. $-3 \square -5$ **b.** $-3 - 4 \square -5 - 4$ **c.** $-3 + 5 \square -5 + 5$

3. Does adding or subtracting a number on both sides change the inequality sign between two numbers or expressions?

4. a. $4 \square -2$ **b.** $4(5) \square -2(5)$ **c.** $4 \div 2 \square -2 \div 2$

5. a. $-3 \square 6$ **b.** $-3(4) \square 6(4)$ **c.** $-3 \div 3 \square 6 \div 3$

6. Does multiplying or dividing both sides by a positive number change the inequality sign between two numbers or expressions?

7. a. $4 \square 2$ **b.** $4(-1) \square 2(-1)$ **c.** $4 \div (-2) \square 2 \div (-2)$

8. a. $-3 \square 3$ **b.** $-3(-2) \square 3(-2)$ **c.** $-3 \div (-1) \square 3 \div (-1)$

9. Does multiplying or dividing both sides by a negative number change the inequality sign between two numbers?

HAVE YOU EVER TRIED to figure out how many points you needed in a final exam to get a certain grade for a course? Is it necessary to get a certain number? Or is there a range of numbers that will work? Inequalities may help. In this section, we solve one-variable linear inequalities both with a graph and with algebraic notation. We apply the inequalities to calculating grades.

▶ **EXAMPLE 1** Exploration: guessing and checking to solve an inequality Suppose a course has three tests worth 100 points each, projects and homework worth 70 points, and a final exam worth 150 points. The instructor grades on a percent basis: 90% for an A, 80% for a B, 70% for a C. One student has test scores of 78, 84, and 72, with full credit on projects and homework (70 points).

a. What is the total possible points?

b. What grade will the student earn with a 95 on the final exam?

c. Use guess and check on a calculator to find the score needed on the final exam to earn at least a B.

SOLUTION See the Answer Box. ◀

The solution to the exploration is an inequality. Although only one final exam score will result in a grade of 80%, a higher final exam score will still result in at least a B. We will return to the exploration problem in Example 8, where we will solve it with symbols.

One-Variable Linear Inequalities

DEFINITION OF ONE-VARIABLE LINEAR INEQUALITY

A **linear inequality in one variable** can be written $ax + b < c$, where a, b, and c are real numbers and a is not 0.

The above definition and the properties and definitions that follow are true also for the other inequality symbols, $>$, $\geq$, and $\leq$. The $ax + b$ is the same form as $ax + b$ on linear equations.

The **solution set of a one-variable inequality** is *the set of values of the input variable that make the inequality a true statement.*

Solving One-Variable Inequalities from a Graph

In each of the next two examples, we graph the left side of the inequality and the right side on a coordinate graph. We then find the point of intersection and find the solution to the inequality on a number line below the graph.

▶ **EXAMPLE 2** Solving an inequality with a graph Solve $6 - 2x \leq 12$ with a graph. Show the solution set as a line graph, an inequality, and an interval.

SOLUTION The graphs of $y_1 = 6 - 2x$ and $y_2 = 12$ are shown in Figure 22. The point of intersection is $(-3, 12)$.

We mark $x = -3$ on a number line, using a dot because of the $\leq$ inequality sign. We then select $x = 0$ as a test point:

$$6 - 2x \leq 12 \qquad \text{Substitute } x = 0.$$

$$6 - 2(0) \leq 12$$

$$6 \leq 12$$

The inequality is true, so $x = 0$ is in the solution set. We mark 0 on the number line and draw an arrow from -3 through 0. The solution set is $x \geq -3$, or $[-3, +\infty)$.

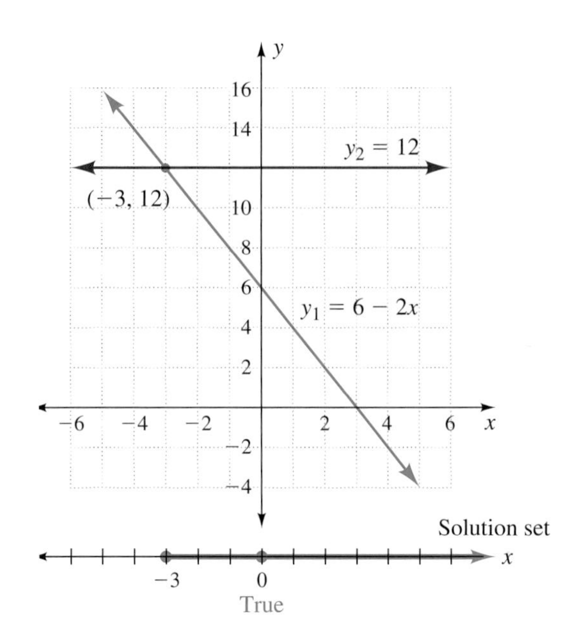

FIGURE 22 Solution to $6 - 2x \leq 12$ 

SOLVING ONE-VARIABLE INEQUALITIES WITH A GRAPH

1. Graph $y_1 =$ the left side of the inequality and $y_2 =$ the right side.
2. Find the point of intersection (x, y). Mark the x on a number line. Use a dot if the inequality is $\geq$ or $\leq$; use a small circle if the inequality is $>$ or $<$.
3. Choose a test number on either side of the x on the number line. Substitute the test number into the inequality. If the number makes the statement true, draw an arrow on the number line from x through the test number. If the number makes the statement false, draw the arrow from x in the opposite direction from the test number.
4. Write an inequality for the solution set shown on the number line.

Student Note:
$x = 0$ may be a convenient test number.

Example 3 has variables on both sides of the inequality.

▶ **EXAMPLE 3** *Solving an inequality with a graph* Solve $2x + 3 < -3x - 2$ by graphing.

SOLUTION The graphs of $y_1 = 2x + 3$ and $y_2 = -3x - 2$ are in Figure 23. The lines intersect at $(-1, 1)$. We mark $x = -1$ on a number line, using a small circle. We then choose a test number—say, $x = 0$. Next we substitute $x = 0$ into the inequality:

$$2x + 3 < -3x - 2$$

$$2(0) + 3 < -3(0) - 2$$

$$3 < -2$$

The test number gives a false statement, so we draw an arrow from $x = -1$ in the opposite direction from zero. The solution set is $x < -1$.

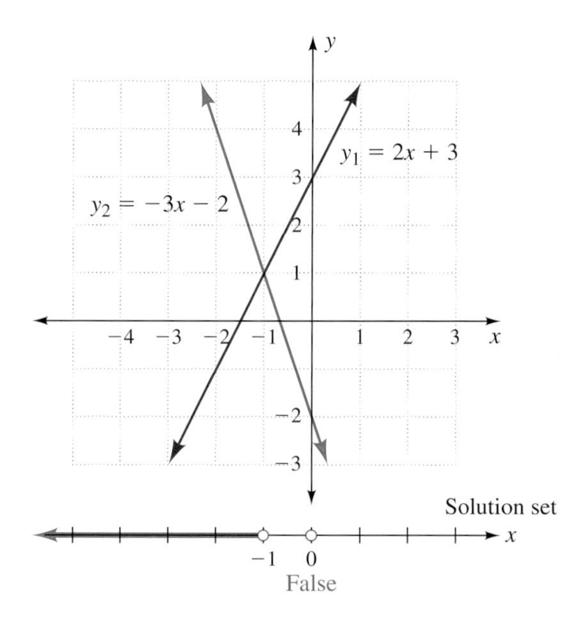

FIGURE 23 Solution to $2x + 3 < -3x - 2$ ◀

THINK ABOUT IT 1: For what values of x is the graph of $y = 2x + 3$ below the graph of $y = -3x - 2$? Could the positions of the graphs be used to solve the inequality $2x + 3 < -3x - 2$?

▶ Solving One-Variable Inequalities Using Algebraic Notation

When we solved equations with algebraic notation, we created our equivalent equations at each step in the solution. Similarly, the inequalities in each step of a solution are said to be equivalent. **Equivalent inequalities** *have the same solution set.*

In the Warm-up, we explored various operations on inequalities. It may be surprising that although $-2 < 1$, $(-3)(-2) > (-3)(1)$ because $6 > -3$. Thus, to get an equivalent inequality when we multiply or divide by a negative number, we reverse the direction of the inequality sign.

MULTIPLICATION (BY A *NEGATIVE* NUMBER) PROPERTY OF INEQUALITIES

> If the same *negative* number is used to multiply (or divide) each side of an inequality, the direction of the inequality sign is changed:
>
> If $a < b$, then $ac > bc$
>
> where a and b are real numbers and $c < 0$.

It is always a good idea to predict one number that will make an inequality true before solving it. Predicting is like choosing a test number, as we did for the line graphs in earlier examples. The prediction helps us make sure that our solution is correct.

▶ **EXAMPLE 4** Solving an inequality with algebraic notation

a. Predict a number that will satisfy $-3x < 15$.

b. Solve the inequality and make sure that the predicted number satisfies the resulting inequality.

c. Graph the solution set.

SOLUTION **a.** The product of x and -3 must be smaller than 15. Because $(-3)(-2) = 6$, we predict that $x = -2$ will satisfy the inequality and will be in the solution set.

b. $-3x < 15$ Divide both sides by -3 and reverse the inequality sign.

$\frac{-3x}{-3} > \frac{15}{-3}$ Simplify.

$x > -5$

Check: $x = -2$ satisfies $x > -5$.

c. Solution set:

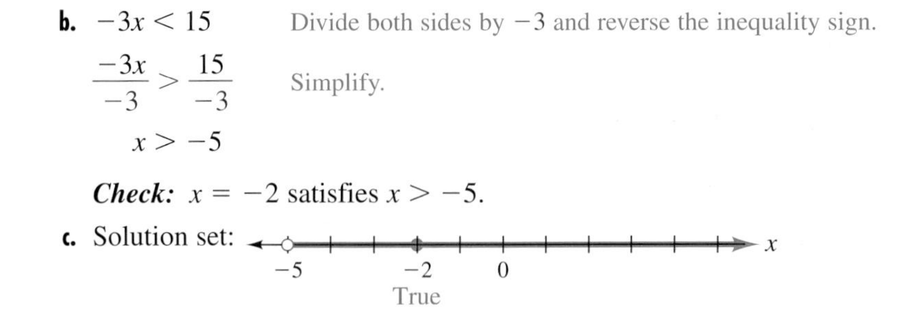

◀

The next two inequality properties state that multiplication and division by a positive number leave the inequality sign unchanged, as do all additions and subtractions.

MULTIPLICATION (BY A *POSITIVE* NUMBER) PROPERTY OF INEQUALITIES

If the same *positive* number is used to multiply (or divide) on both sides of an inequality, the direction of the inequality sign is not changed:

If $a < b$, then $ac < bc$

where a and b are real numbers and $c > 0$.

ADDITION PROPERTY OF INEQUALITIES

If the same number is added (or subtracted) on both sides of an inequality, the direction of the inequality sign is not changed:

If $a < b$, then $a + c < b + c$

for all real numbers a, b, and c.

▶ In Example 5, we return to Example 2 and solve the inequality with symbols.

▶ EXAMPLE 5 Solving an inequality Solve $6 - 2x \leq 12$. Write the solution set as an inequality and an interval, and graph the solution set.

SOLUTION We predict $x = 0$ because $6 - 2(0) \leq 12$ is true.

$6 - 2x \leq 12$ Subtract 6 from both sides.

$-2x \leq 6$ Divide by -2 and reverse the inequality sign.

$\frac{-2x}{-2} \geq \frac{6}{-2}$ Simplify.

$x \geq -3$

Solution set: $x \geq -3$, $[-3, +\infty)$

−3 0 True

Check: Our predicted number, $x = 0$, is in the solution set $x \geq -3$. ◀

In many cases, we can avoid multiplication and division by a negative number if we keep the coefficient on the variable positive. We will solve the inequality in Example 6 two ways.

EXAMPLE 6 **Solving an inequality with division by a negative** Solve for x in two different ways: $2x + 3 \le -3x - 2$. Compare the solutions with those found in Example 3.

SOLUTION First we will use division by a negative.

$$2x + 3 \le -3x - 2 \qquad \text{Subtract } 2x.$$
$$3 \le -5x - 2 \qquad \text{Add 2.}$$
$$5 \le -5x \qquad \text{Divide by } -5 \text{ and reverse the inequality sign.}$$
$$\frac{5}{-5} \ge \frac{-5x}{-5}$$
$$-1 \ge x$$

Now we will solve the equation again, keeping the coefficient on the variable positive.

$$2x + 3 \le -3x - 2 \qquad \text{Add } 3x.$$
$$5x + 3 \le -2 \qquad \text{Subtract 3.}$$
$$5x \le -5 \qquad \text{Divide by 5.}$$
$$\frac{5x}{5} \le -\frac{5}{5} \qquad \text{The inequality sign is not changed.}$$
$$x \le -1$$

The pointed end of the inequality faces x in both solutions. Thus, $-1 \ge x$ and $x \le -1$ represent the same set of numbers. In Example 3, the inequality excludes $x = -1$. Otherwise, the solutions agree. ◀

KEEPING COEFFICIENTS POSITIVE

> Keeping the variable's coefficient positive has the distinct advantage of eliminating the need to change the inequality sign. It is possible to keep the variable positive in solving most inequalities.

Applications

EXAMPLE 7 **Writing and solving inequalities: party plans** Suppose Julian wishes to find the number of children for which a party at the Pearl Street Ice Cream Parlour (at \$6.50 per child) will cost less than a party at the National Academy of Artistic Gymnastics (at \$85 plus \$5 for each child above the minimum of 11).

a. State and graph a cost equation for each location.
b. Write an inequality describing the cost comparison.
c. Solve the inequality from the graph.
d. Solve the inequality with algebraic notation.

SOLUTION **a.** The cost equations are

$$y = 6.50x$$
$$y = 85 \text{ for } x \le 11 \quad \text{or}$$
$$y = 85 + 5(x - 11) \text{ for } x > 11$$

They are graphed in Figure 24.

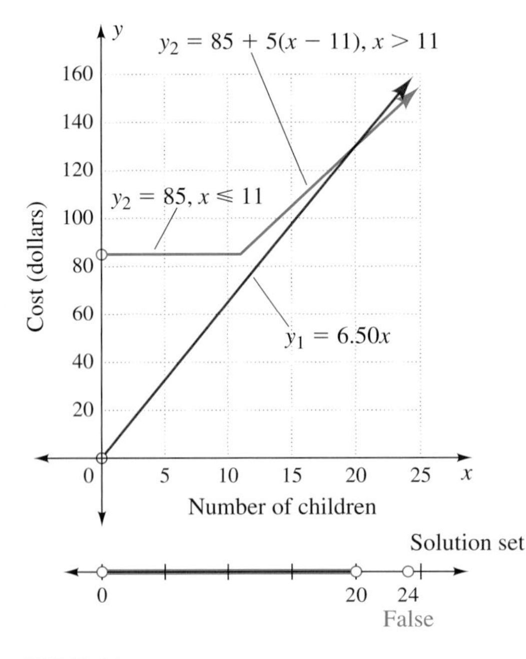

FIGURE 24

b. The Pearl Street Ice Cream Parlour will be cheaper when $6.50x < 85 + 5(x - 11)$.

c. The point of intersection of the two graphs is approximately (20, 130). We mark $x = 20$ on a number line, using a small circle. We then choose a test number—say, $x = 24$.

$6.50x < 85 + 5(x - 11)$	Substitute $x = 24$.
$6.50(24) < 85 + 5(24 - 11)$	Simplify.
$156 < 150$	False

The test number gives a false statement, so we draw a line from $x = 20$ toward $x = 0$. The solution set is $x < 20$ children. We should assume that $x > 0$.

d.

$6.50x < 85 + 5(x - 11)$	Apply the distributive property.
$6.50x < 85 + 5x - 55$	Add like terms.
$6.50x < 30 + 5x$	Subtract $5x$ on both sides.
$6.50x - 5x < 30 + 5x - 5x$	Add like terms.
$\frac{1.50x}{1.50} < \frac{30}{1.50}$	Divide both sides by 1.50.
$x < 20$	

The Pearl Street Ice Cream Parlour is cheaper for under 20 children. ◀

THINK ABOUT IT 2: For what values of x is the graph of $y = 6.50x$ below the graph of $y = 85 + 5(x - 11)$? Could the positions of the graphs be used to solve the inequality $6.50x < 85 + 5(x - 11)$?

▶ In Example 8, we return to the exploration in Example 1 to write an inequality and solve it.

EXAMPLE 8 Writing and solving inequalities: grades Suppose a course has three tests worth 100 points each, projects and homework worth 70 points, and a final exam worth 150 points. The instructor grades on a percent basis: 90% for an A, 80% for a B, 70% for a C. One student has test scores of 78, 84, and 72, with full credit on projects and homework (70 points). Write an inequality showing the points earned, the points possible, and the final exam score needed to earn at least a B. Solve the inequality.

SOLUTION The grade is based on points earned relative to total points. We add the points earned using a variable to represent the last test, and place this sum over the total possible points to obtain a percent. Because any percent larger than 80% will give a B (or A), we write an inequality using ≥ 0.80.

$$\frac{78 + 84 + 72 + 70 + x}{100 + 100 + 100 + 70 + 150} \geq 0.80 \qquad \text{Simplify.}$$

$$\frac{304 + x}{520} \geq 0.80 \qquad \text{Multiply by 520.}$$

$$\frac{(520)304 + x}{520} \geq 520(0.80)$$

$$304 + x \geq 416 \qquad \text{Subtract 304.}$$

$$304 + x - 304 \geq 416 - 304$$

$$x \geq 112$$

The student had a C+ on tests: $(78 + 84 + 72) \div 3 = 78$ average. The student needs $\frac{112}{150} = 75\%$ on the final for a B in the course. The homework helped! ◀

SOLVING ONE-VARIABLE LINEAR INEQUALITIES WITH ALGEBRAIC NOTATION

The solution set for an inequality is an inequality.

1. As with linear equations, use addition, subtraction, multiplication, and division to isolate the variable.
2. When you multiply or divide both sides by a negative number, reverse the direction of the inequality sign.
3. Try to keep the variable's coefficient positive to avoid the mistakes that tend to arise when you multiply or divide by a negative number and reverse the direction of the inequality sign.

Solving with a Graphing Calculator

The graphing calculator will provide a line graph solution to an inequality. If an input x makes the inequality true, the calculator will output $y = 1$. If an input x makes the inequality false, the calculator will output $y = 0$. The calculator will make a graph of these ones and zeros.

GRAPHING CALCULATOR TECHNIQUE: SOLVING AN INEQUALITY

Student Note:
To find inequality signs, press 2nd [TEST].

Solve $6 - 2x \leq 12$ (from Example 2 and Example 5).

Enter $Y_1 = 6 - 2X$ and $Y_2 = 12$ to view the original graphs.

Enter $Y_3 = 6 - 2X \leq 12$ to view the solution set graphed on $y = 1$.

Set a window similar to the one shown in Figure 22. Graph. The graph is shown in Figure 25.

In addition to the graphs for Y_1 and Y_2, a horizontal line will appear at $y = 1$ for $x \geq -3$.

(continued)

(*concluded*)

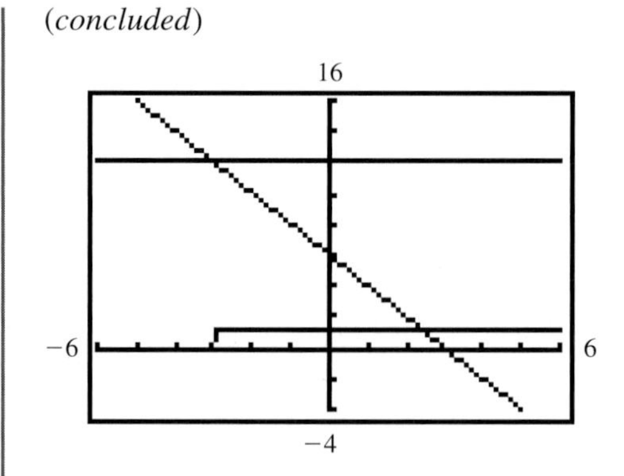

FIGURE 25 Solution to $6 - 2x \leq 12$

Student Note:
Use (▲) or (▼) to move among Y_1, Y_2, and Y_3 on the graph and (◀) or (▶) on the table.

A horizontal line will also appear along the x-axis ($y = 0$) for $x < -3$. Because $y = 0$ lies on the x-axis, points on $y = 0$ will not show unless you shut off the axes:

(2nd) [FORMAT] **Axes Off** (ENTER) (GRAPH)

Trace along Y_3 to see the ordered pairs with $y = 0$ for $x < -3$ and $y = 1$ for $x \geq -3$. If the graphical results are unclear, look at the zeros and ones with (2nd) [TABLE] under Y_3. Restore the axes with (2nd) [FORMAT] **Axes On** (ENTER) (2nd) [QUIT].

ANSWER BOX

Warm-up: 1. all < **2.** all > **3.** no **4.** all > **5.** all < **6.** no **7.** >, <, < **8.** <, >, > **9.** yes **Example 1: a.** 520 points **b.** The percent is 399/520 ≈ 77%. The student will earn a C. **c.** A score of 112 is needed on the final exam for a B. **Think about it 1:** The graph of $y = 2x + 3$ is below that of $y = -3x - 2$ for $x < -1$. This is the same solution we arrived at with a test point. **Think about it 2:** The graph of $y = 6.50x$ is below that of $y = 85 + 5(x - 11)$ for $x < 20$. This is the same solution we arrived at with the test point.

▶ 3.5 Exercises

In Exercises 1 to 8, graph each solution set on a number line and write each as an interval.

1. $x > 5$ **2.** $x < 3$

3. $x \leq -2$ **4.** $x \geq -3$

5. $x \geq 0$ **6.** $x < 0$

7. $-1 > x$ **8.** $-4 < x$

Use the graph to solve the inequalities in Exercises 9 to 12. Graph the solution set on a number line.

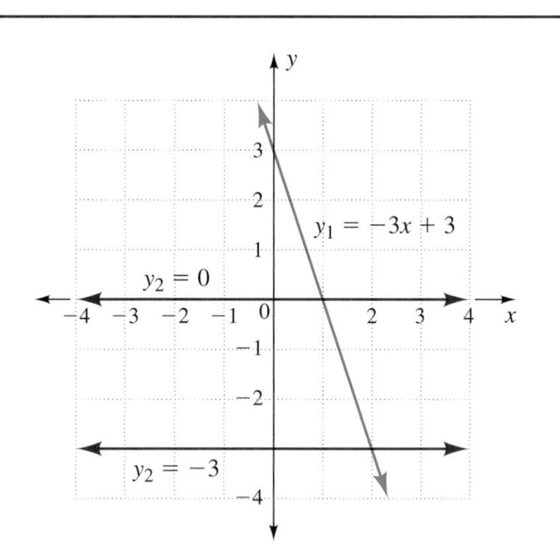

9. $-3 > -3x + 3$ **10.** $-3 < -3x + 3$

11. $0 < -3x + 3$ **12.** $0 > -3x + 3$

Use the graph to solve the inequalities in Exercises 13 to 16. Graph the solution set on a number line.

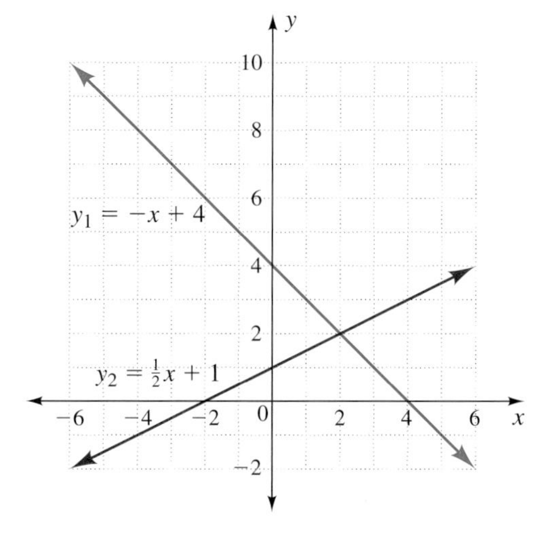

13. $-x + 4 < \frac{1}{2}x + 1$ **14.** $-x + 4 > \frac{1}{2}x + 1$

15. $-x + 4 > 0$ **16.** $\frac{1}{2}x + 1 > 0$

In Exercises 17 to 24, solve each inequality with a coordinate graph. Graph the solution set on a number line.

17. $1 < x + 3$ **18.** $2 > x - 1$

19. $3x - 5 \geq 7$ **20.** $2 - 3x \leq 5$

21. $-x < -3x + 2$ **22.** $x > 2x - 1$

23. $2 - 2x < 3 - 3x$ **24.** $2 - 2x > 3 - 3x$

In Exercises 25 to 56, solve each inequality using algebraic notation. Write the solution set as an inequality and an interval.

25. $-3 > -3x + 3$ **26.** $-3 < -3x + 3$

27. $0 < -3x + 3$ **28.** $0 > -3x + 3$

29. $-x + 4 < \frac{1}{2}x + 1$ **30.** $-x + 4 > \frac{1}{2}x + 1$

31. $-x + 4 > 0$ **32.** $\frac{1}{2}x + 1 > 0$

33. $-1 < -3x + 2$ **34.** $3 < 2x - 1$

35. $-2 \leq 2 - 2x$ **36.** $0 \leq 3 - 3x$

37. $3 - 2x \leq -5$ **38.** $7 + 2x > -3$

39. $-4 > x + 5$ **40.** $-2 < x - 4$

41. $1 - 4x < -4$ **42.** $2x - 7 > -2$

43. $-2 > 2x + 1$ **44.** $3 \leq 4 - 2x$

45. $2x < x + 5$ **46.** $x > 3x - 4$

47. $4 - 2x \geq x - 2$ **48.** $4 - 3x \leq 8 + x$

49. $3x - 4 < -2x + 1$ **50.** $3x - 3 > -x + 1$

51. $2(x + 3) < 5x$ **52.** $3(x - 1) \leq 4x$

53. $\frac{1}{2}x > 4 + x$ **54.** $\frac{1}{2}x < x + 1$

55. $-x > -\frac{1}{2}x + 1$ **56.** $-\frac{1}{2}x > 1 - x$

57. Suppose Audrey decides to have a two-hour party at Strike City. The total cost of \$10.25 per person plus \$20 party room rental is given by $y = 10.25x + 20$, where x is the number of people. Her budget is \$184. The figure shows a graph of her cost equation and budget equation.

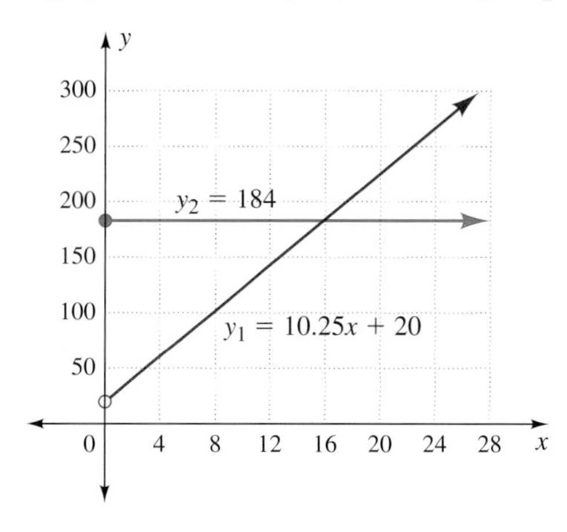

a. Name the point of intersection.

b. Write an inequality that describes the number of people Audrey can have at Strike City.

c. Solve the inequality. Show the solution set as a line graph, as an inequality, and as an interval.

58. Repeat Exercise 57 for a \$266 budget.

59. Rock Climbing Gym charges \$15 per child. Seth has to buy a cake for \$25 and has a \$250 budget. Write and solve an inequality that shows the number of children who can attend on Seth's budget.

60. Papa's Pizza offers a party at \$5.95 per child with a \$40 fee for the cake and playground supervisor. Write and solve an inequality that shows the number of children who can attend on a \$115 budget.

61. For a wedding reception, the Country Inn charges a \$350 service fee plus \$38.50 per person for food. Write and solve an inequality that shows the number of people who can attend on a \$3430 budget.

62. The Valley Inn charges \$37.50 per person and a 20% service fee. Write an inequality that shows the number of persons who can attend a wedding reception on a \$4050 budget.

Blue numbers are core exercises.

63. Using the information in Exercises 61 and 62, write and solve an inequality that shows for how many people it will cost less to use the Country Inn than the Valley Inn.

64. In Seattle, Penske charges \$49.95 plus \$0.59 per mile for a 26-foot moving truck. U-Haul charges \$39.95 plus \$0.79 per mile for the same truck. Write and solve an inequality to show for what mileage Penske is cheaper.

Exercises 65 to 68 relate to Mrs. Kay's math class, where there are 150 points on the final exam and 520 total points for the course. Just prior to the final exam, several students are thinking about their grades. For each exercise, write an inequality and solve it.

65. A student has earned 78, 84, and 72 points on tests and only 5 points on projects and homework. Is it possible for the student to earn a B or better (80%)?

66. Is it possible for the student in Exercise 65 to earn a C (70%)?

67. A student has earned the full 70 points on homework and projects and expects 135/150 on the final. She wants to know at least how many total points she must have had on the three tests to get an A (90%) in the course.

68. A student missed one test, has no homework, and has test scores of 74 and 84 on the other two tests. Is it possible for the student to get a D (60%)?

69. Test scores are 88 out of 100, 84 out of 100, and 89 out of 100. Homework is 70 out of 70. What final exam score (200 points possible) is needed to get 90% or better?

70. Test scores are 92 out of 100, 88 out of 100, and 91 out of 100. Homework is 25 out of 70. What final exam score (200 points possible) is needed to get 90% or better?

▶ Writing

71. Rewrite the following statement so that it correctly describes division of an inequality by a negative number: If $a < b$, then $ac < bc$, where a and b are real numbers and $c > 0$.

72. Describe the effect on $-3 < 4$ of multiplying by -2.

73. Describe the effect on $-8 < -6$ of dividing by -2.

74. What is the advantage of keeping the variable's coefficient positive when solving an inequality?

▶ Projects

75. Compound Inequalities Inequalities containing two inequality symbols, such as $-3 < x < 4$, are called *compound inequalities*. To solve a compound inequality, do the same operation or operations to all three parts of the inequality, to isolate the variable between the two inequality signs.

$-1 < 2x + 3 < 5$ Subtract 3.
$-4 < 2x < 2$ Divide by 2.
$-2 < x < 1$

Number-line graph of solution set:

−5 −4 −3 −2 −1 0 1 2 3 4 5

Solve each compound inequality, and sketch the solution set on a number-line graph.

a. $-4 < 3x - 1 < 5$ **b.** $-1 \le 2x + 5 \le 7$

c. $-8 \le \frac{1}{2}x - 3 \le -4$ **d.** $-3 \le \frac{1}{4}x + 2 \le -1$

e. $-3 < 4 - x < 6$ **f.** $-5 \le 1 - 3x \le 7$

(*Hint:* If you multiply or divide by a negative number, reverse the direction of all inequality signs.)

76. Inequality or Equation When we place variables in inequalities, we obtain a *conditional inequality* if the inequality is not true for some value of the variable(s). In the following review of the names of algebraic statements, x is the unknown number.

"Four is less than five" is an inequality: $4 < 5$.

"Four is less than five times a number" is a conditional inequality: $4 < 5x$.

"Four less than a number is 5" is a conditional equation: $x - 4 = 5$.

"Four less than nine is five" is an identity: $9 - 4 = 5$.

Write each statement below in algebraic notation. Explain why each is an inequality, a conditional inequality, a conditional equation, or an identity. Let x be the unknown number.

a. Six is less than eight.

b. Six is two less than eight.

c. Four greater than negative three is one.

d. Four is greater than negative three.

e. Four greater than a number is negative three.

f. Four times negative three is greater than negative thirty.

g. Negative three is greater than a number.

h. Negative three is less than three.

i. Negative three less three is negative six.

j. Negative three is greater than negative six.

k. Negative six divided by a number is two.

l. Negative six divided by negative three is greater than zero.

77. Party Packages Over Time In 1992, Skate World cost \$45 for up to 10 skaters and \$5 for each additional. Wave Pool cost \$30 plus \$3 per child.

In 2006, Skate World cost \$82 for 10 skaters and \$7 for each additional. Wave Pool cost \$20 plus \$3.50 per child.

a. Make two graphs, one for 1992 and one for 2006. Graph each location for 0 to 20 children.

b. Write inequalities to describe for what number of children each location is cheaper.

c. How are the two graphs alike? What causes the changes over time?

▶ 3 Chapter Summary

Vocabulary

For definitions and page references, see the Glossary/Index.

addition property of equations
addition property of inequalities
conditional equation
conditional inequality
consecutive integers
delta (Δ)
dependent variable
empty set
equation
equation in one variable
equivalent equations
equivalent inequalities
identity
independent variable
linear equation in one variable
linear equation in two variables
linear inequality in one variable
multiplication property of equations
multiplication (by a negative number) property of inequalities
multiplication (by a positive number) property of inequalities
nonlinear equation
solution
solution set
solution set of a one-variable inequality
solving an equation
subscript
substitution

Concepts

3.1 Linear Equations of One and Two Variables

In writing equations, always define the variables. In applications, the output, y (the dependent variable), depends on the input, x (the independent variable).

Equations contain parentheses (or other grouping symbols):

- when rules (such as those about cost) change for additional inputs.
- when an operation is applied to a sum or difference of two numbers

3.2 Solving Equations with Algebraic Notation

Solutions to equations may be found by guessing or observation, through step-by-step algebraic procedures, from an input-output table, or from a graph.

Solutions to equations may make use of any or all of these steps:

- Reverse the order of operations. Changing equations into sentences helps us recognize the order of operations.
- Apply inverse operations.
- Add (or subtract) the same number to both sides of an equation to produce an equivalent equation.
- Multiply (or divide) both sides of an equation by the same nonzero number to produce an equivalent equation.
- Multiply both sides of an equation by the reciprocal of a fraction instead of dividing by that fraction.

3.3 Solving Equations with Tables, Graphs, and Algebraic Notation

All ordered pairs (x, y) that make a two-variable equation true belong on the input-output table or graph for that equation. Likewise, all ordered pairs on the table or graph make the equation true.

The table solution to an equation is the input entry in the table that makes the left and right sides of the equation equal.

The graphical solution to an equation is the first number, x, in the ordered pair (x, y) at the point of intersection of the graphs of the left and right sides of the equation. The second number, y, in the ordered pair is the value obtained on each side when the x is substituted into the original equation.

Use the distributive property, $a(b + c) = ab + ac$, to simplify the expressions in the equation before solving.

3.4 Solving Linear Equations with Variables on Both Sides of the Equation

To solve a linear equation with variables on both sides using a table and a graph, we modifiy the process from Section 3.3:

1. Let y_1 = the left side of the equation. Build a table and plot a graph.
2. Let y_2 = the right side of the equation. Build a table and plot a graph on the same axes used for the left side.
3. Look for the ordered pair (x, y) that appears in both tables and at the point of intersection. The x is the solution to the equation. The y is the value obtained when x is substituted into each side of the original equation.

Solving a linear equation with algebraic notation may involve simplifying steps, as well as solving steps:

1. Multiply on both sides by the least common denominator to remove fractions or decimals.
2. Apply the distributive property to expressions containing parentheses. Add like terms on each side.
3. Using the addition property of equations, get all the terms with the variable to be solved for onto one side of the equation. Using the addition property again, get the constant terms to the other side of the equation. Add like terms again, if necessary. Look for linear form, $ax + b = 0$ or $ax + b = c$.
4. Solve for the variable, using the multiplication property of equations.

3.5 Solving Linear Inequalities in One Variable

The solution set for an inequality is an inequality. To solve a one-variable inequality with a graph:

1. Graph y_1 = the left side of the inequality and y_2 = the right side.
2. Find the point of intersection (x, y). Mark the x on a number line. Use a dot if the inequality is $\geq$ or $\leq$; use a small circle if the inequality is $>$ or $<$.
3. Choose a test number on either side of the x on the number line. Substitute the test number into the inequality. If the number makes the statement true, draw an arrow on the number line from x through the test number. If the number makes the statement false, draw the arrow from x in the opposite direction from the test number.
4. Write an inequality for the solution set shown on the number line.

The solution set for an inequality is an inequality. To solve one-variable linear inequalities with algebraic notation:

1. As with linear equations, use addition, subtraction, multiplication, and division to isolate the variable.
2. When you multiply or divide both sides by a negative number, reverse the direction of the inequality sign.
3. Try to keep the variable's coefficient positive to avoid the mistakes that tend to arise when you multiply or divide by a negative number and reverse the direction of the inequality sign.

▶ 3 Review Exercises

In Exercises 1 to 5, fill in the blank with one or more of the following: identity, conditional equation, equivalent equations, two-variable equation, nonlinear equation

1. $3x + 4 = 7$ and $3x = 3$ are ________________.
2. $x + 3 = 3 + x$ is a(n) ________________.
3. $y = 3x + 4$ is a(n) ________________.
4. $y = x^2 + x$ is a(n) ________________.
5. $3x - 5 = -2$ is a(n) ________________.

In Exercises 6 to 13, write equations. Let x be the unknown number.

6. Six less than three times a number is -15.
7. Four is five times a number less eleven.
8. The product of a number and negative seven is 21.
9. The quotient of a number and six is 12.
10. Three times a number is four more than twice the same number.
11. Five less than twice a number is the same number.
12. Six times the sum of two and a number is negative six.
13. Twice the difference between seven and a number is -4.

In Exercises 14 to 17, write each equation in words.

14. $\frac{x}{5} = 15$
15. $4x - 3 = 29$
16. $3(x - 4) = -18$
17. $5 - 2(x - 1) = 7 + x$

In Exercises 18 to 22, is the given input x a solution to the equation?

18. $-4x + 5 = -3, x = -2$
19. $-5x + 4 = 9, x = -1$
20. $2.5x - 3 = -1, x = 0.8$

21. $8 - 7.5x = 5, x = 0.4$

22. $2x + 1 = 4x - 3, x = 2$

In Exercises 23 to 40, solve for the variable.

23. $x + 3 = -4$

24. $x - 4 = 8$

25. $3x = 27$

26. $4x = -12$

27. $\frac{1}{2}x = 12$

28. $\frac{1}{4}x = 20$

29. $4x - 2 = 22$

30. $3x - 5 = 34$

31. $-2x + 3 = 9$

32. $-4x + 5 = 21$

33. $-2(x - 4) = 18$

34. $-6(x - 5) = 42$

35. $5 - 2(x + 1) = -11$

36. $7 - 3(x + 2) = -8$

37. $3x - 1 = x + 1$

38. $5 - 3(x - 4) = x + 9$

39. $-2(x - 3) = \frac{1}{2}x + 3$

40. $3 - (x - 4) = x + 3$

41. Use the table to solve the following equations.

a. $3x - 2 = 4$

b. $3x - 2 = 13$

c. $3x - 2 = 8$

d. $3x - 2 = 16$

x	$y = 3x - 2$
2	4
3	7
4	10
5	13

42. Solve these equations using the graph.

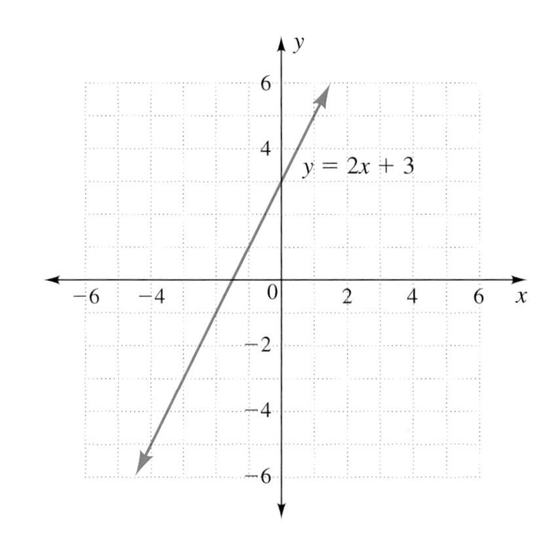

a. $2x + 3 = 5$

b. $2x + 3 = 1$

c. $2x + 3 = -2$

43. Solve these equations using the graph.

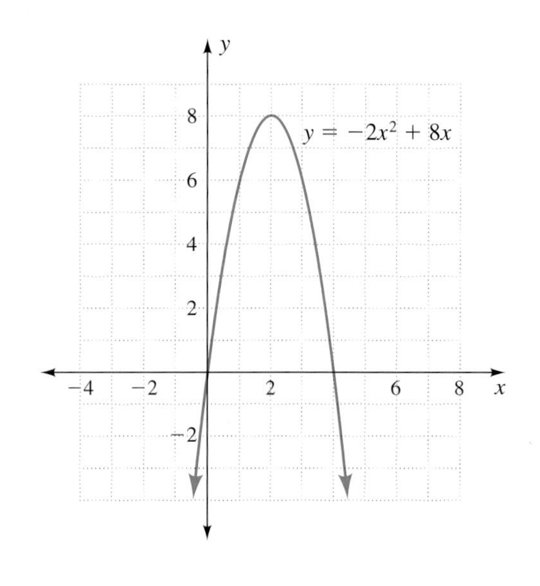

a. $0 = -2x^2 + 8x$

b. $6 = -2x^2 + 8x$

c. $8 = -2x^2 + 8x$

d. $9 = -2x^2 + 8x$

44. Use the graph as needed to do the following.

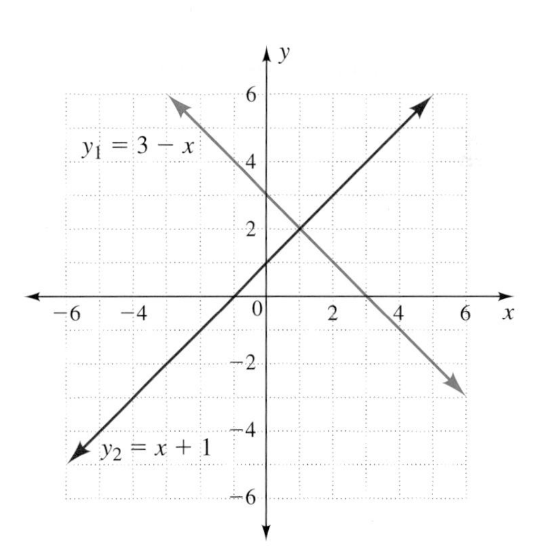

a. Find the point of intersection of the two lines.

b. Substitute the intersection point into each equation shown on the graph.

c. Solve the equation $x + 1 = 3 - x$.

d. Describe how the equation $x + 1 = 3 - x$ relates to the graph.

45. Use the graph to solve the following equations.

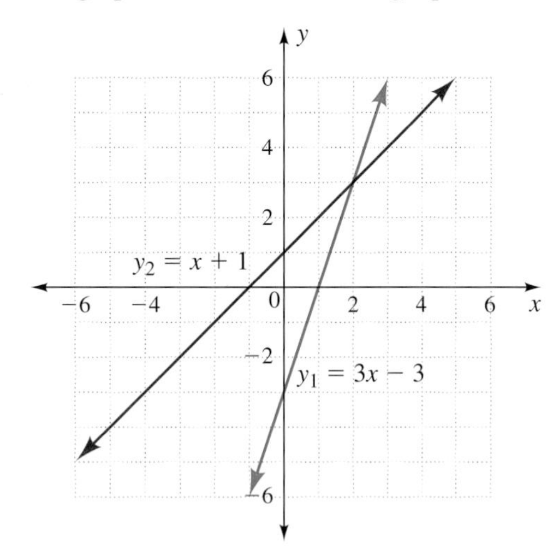

a. $3x - 3 = x + 1$

b. $x + 1 = -2$

c. $3x - 3 = -6$

46. Use the graph to solve the following equations.

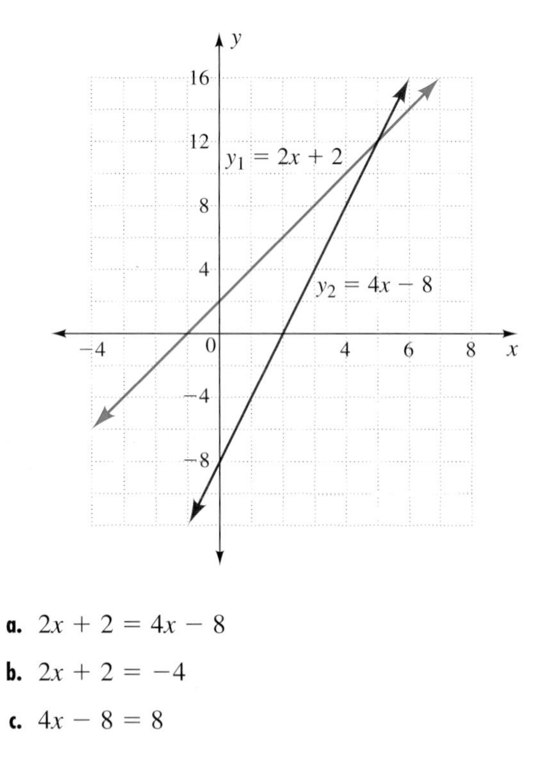

a. $2x + 2 = 4x - 8$

b. $2x + 2 = -4$

c. $4x - 8 = 8$

In Exercises 47 to 54, define variables and write equations so that y depends on x. Formulas are listed on the page facing the inside back cover.

47. Find the distance traveled in 4 hours at various speeds.

48. Find the distance traveled at 65 miles per hour for a given number of hours.

49. Find the $7\frac{1}{2}\%$ tax on a meal.

50. Find the 0.2% property tax on the value of a house.

51. Find the total cost for a given number of credits at a college that charges $300 per credit tuition and $150 in fees.

52. At Peoples Energy* in Chicago, natural gas costs $0.6578 per therm plus a $0.0031 per therm environmental charge. Find the cost of the gas. Let x = number of therms.

53. Find the value remaining after a given number of visits on a $520 health club deposit when $10 is deducted for each visit.

54. Find the value remaining after a given number of years on a $20,000 car that drops in value $2000 each year.

To answer the questions in Exercises 55 to 59, write and solve an equation containing parentheses.

55. To make room for a larger-than-expected crowd at an outdoor wedding, the chair rental company places 3 additional seats at the end of each row of x seats. There are 20 rows of seats and 440 seats altogether. How many seats were in each original row?

56. By making the seats and aisle narrower, Squeez-um Airlines was able to add 2 more seats in each of the 47 rows on its economy flights. The economy flight now holds 423 passengers. How many seats were in each of the original rows? (Assume that all rows have the same number of seats.)

57. At Con Edison, New York City in September, the basic charge for electricity is $10.39. The first 241.7 kWh costs $0.176707 per kWh. Additional kilowatt-hours cost $0.183146 per kWh. If the total for these charges was $64.51, how many kilowatt-hours were used?

58. At Peoples Energy,* natural gas delivery costs $8.95 plus 0.23151 per therm for the first 50 therms. For more than 50 therms, additional therms cost $0.12200 per therm. For $30 in late spring, how many therms are delivered?

59. Peoples Energy* adds to cost a state tax (0.10% of cost), a municipal utility tax (5.15% of cost) and a gas revenue tax ($0.024 per therm). For 55.88 therms in May, a customer pays $62.56. What was the cost, n, before these charges were added? Round down to nearest cent.

*Simplified from sample bills on Peoples Energy and Con Edison websites. (I don't make this stuff up!) Why do sample bills show May or September?

60. After each step below, write what was done to obtain the next step:

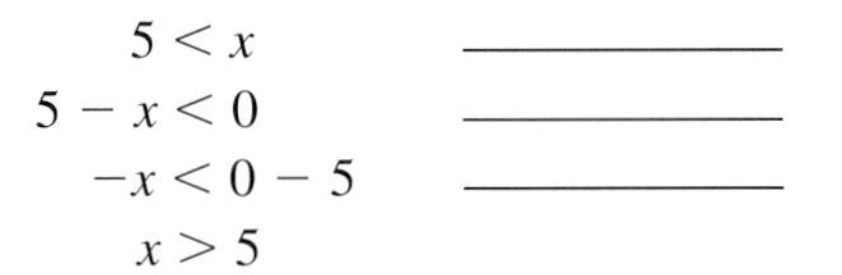

$5 < x$ __________

$5 - x < 0$ __________

$-x < 0 - 5$ __________

$x > 5$

What do these steps tell you about all statements of the form $a < b$ and $b > a$?

61. Which is a solution set for $2 < x - 3$? Draw the solution set on a number line.

a. $x > 5$ **b.** $x < 5$ **c.** $x > 1$ **d.** $x < 1$

62. Which is a solution set for $5 - x \geq 2$? Draw the solution set on a number line.

a. $x \geq 3$ **b.** $-3 \leq x$ **c.** $3 \geq x$ **d.** $x \leq 7$

In Exercises 63 and 64, sketch a copy of the graph and label each line with its equation. Solve the indicated inequality first with the graph and then using algebraic notation.

63. $y = 15 - 2x$ and $y = x - 6$; $15 - 2x < x - 6$

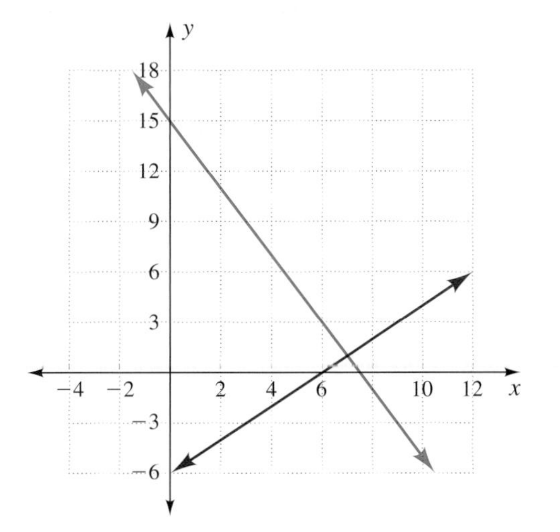

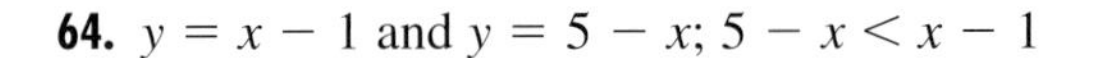

64. $y = x - 1$ and $y = 5 - x$; $5 - x < x - 1$

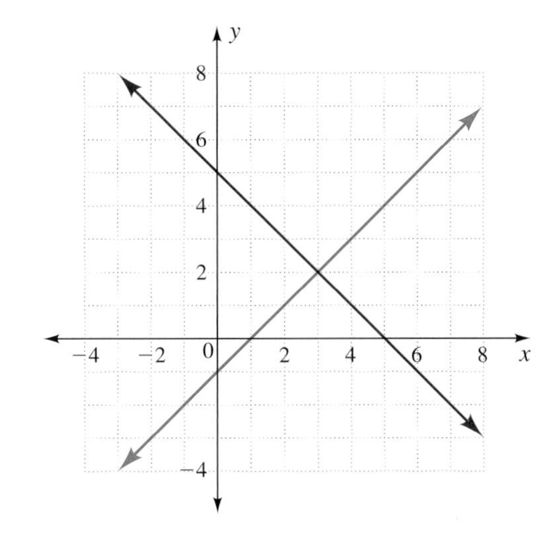

In Exercises 65 and 66, solve with a graph and with symbols.

65. $3x - 5 \leq 3 - x$ **66.** $x - 4 > 2 - x$

In Exercises 67 to 84, solve for x. Write the solution set as an inequality and an interval.

67. $1 < x - 4$

68. $5 > 2 - x$

69. $-\frac{1}{2}x > 8$

70. $4 \geq -\frac{1}{3}x$

71. $5 - 3x > 13$

72. $6 - 5x \leq 31$

73. $13 \leq 7 - \dfrac{x}{3}$

74. $5 < -10 - \dfrac{x}{3}$

75. $-2x < 1 - x$

76. $-x > x + 2$

77. $x > 2x + 3$

78. $2x - 2 < 1 - x$

79. $x + 1 \leq 3 - x$

80. $x + 5 \geq 1 - 3x$

81. $-2x + 2 > -2 - x$

82. $2x + 4 < 2 - x$

83. $3x + 2 \geq 3 - 2x$

84. $-x - 1 \leq 2x - 3$

The two students in Exercises 85 and 86 both want to earn at least a B (80%). If the final exam is worth 150 points, is it possible for each student to earn at least a B? Write an inequality for each student and solve the inequality.

85. Student 1 has earned 82 and 72 on tests (100 points each); 20, 0, 20, 20, and 18 on quizzes (20 points each); and 12 of 70 points on homework.

86. Student 2 has earned 82 and 72 on tests (100 points each); 15, 15, 15, 16, and 16 on quizzes (20 points each); and 70 of 70 points on homework.

In Exercises 87 to 90, define variables, write an equation, and solve the equation.

87. A wedding cake costs \$550, and the catering price per person is \$25. Up to how many persons may attend on a \$3500 budget?

88. An anniversary reception is planned for a location that charges a \$300 setup fee and \$24 per person. How many people can attend on a budget of \$2100?

89. One sidewalk repair bid is \$500 plus \$24 per linear foot of sidewalk. A second bid is \$640 plus \$16 per linear foot. For what lengths of sidewalk will the first bid be cheaper?

90. One plumber charges a \$50 travel fee and \$96 per hour. A second plumber charges a \$30 travel fee and \$106 per hour. For what numbers of hours will the first plumber be cheaper?

91. Consecutive Integers 3 (This exercise is related to projects in Sections 3.1 and 3.2.) Write equations and solve for parts a, b, and c.

a. An integer and twice the next consecutive integer add to 17. What are the integers?

b. The sum of three consecutive odd integers is 567. What is the first integer?

c. The sum of three consecutive even integers is –12. What is the greatest integer of the three?

d. Why would the notation x, $x + 1$, and $x + 2$ be preferable to a, b, and c for describing three consecutive integers?

▶ Chapter 3 Project

92. Paintball The Paintball Palace offers 2 hours of play plus shooter, mask, and 50 paintballs for the following rates: 10 to 15 people, \$15 per person; 16 to 24, \$12.50 per person; 25 to 40, \$10 per person. Special offer: For groups of 15, a 16th person is free.

a. Make a table showing the total cost of 12 to 30 persons attending a party at the Paintball Palace.

b. Draw a graph showing the total cost for 12 to 30 persons.

c. Write equations for each of the three rate groups. Include inequalities to describe the groups.

d. Explain why the business makes the offer of a free 16th person.

e. Explain why your 16-member group declines the special and asks to pay for the 16th person.

f. Describe possible ways to divide up the cost in case the 16th person wanted to get in free.

g. Show on your graph how many persons can play for \$300. Write the equation(s) that give this result.

h. Discuss strategies that groups should use in going to the Paintball Palace. How does the graph show that your strategies work?

▶ 3 Chapter Test

In Exercises 1 to 5, define variables and write the sentence in symbols. Solve the one-variable equations.

1. Six more than half a number is 15.

2. Three times a number less 7 is -31.

3. The output is a third of the input.

4. The output is two less than twice the input.

5. The sum of a number and -3 multiplied by 3 gives -12.

In Exercises 6 to 15 solve for x using algebraic notation.

6. $x + 8 = -3$

7. $4 - x = 5$

8. $\frac{2}{5}x = 30$

9. $-6x = 3$

10. $5 - 2x = 3$

11. $\frac{1}{2}x + 5 = -3$

12. $2(x - 2) = -x - 1$

13. $4(x - 3) = 6$

14. $4 - 2(x - 4) = 2x$

15. $-2(x - 3) = -0.5(x - 6)$

16. Circle the solution to the equation $5 - 2x = 3$ on the table and on the graph.

Input: x	Output: $y = 5 - 2x$
−1	7
0	5
1	3
2	1
3	−1

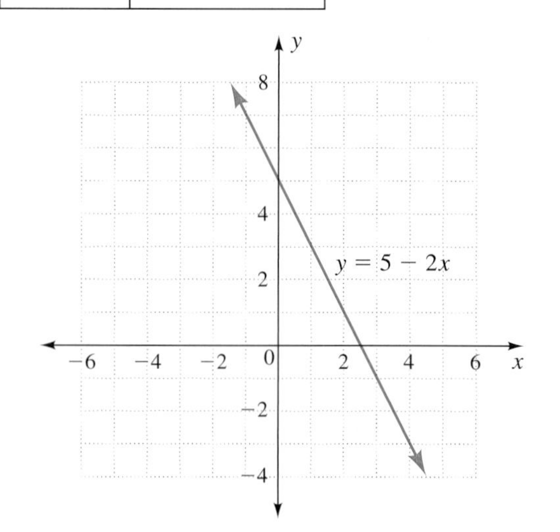

17. Explain how to estimate the solution to $5 - 2x = 0$ from the table in Exercise 16.

18. Explain how to estimate the solution to $5 - 2x = 0$ using the graph in Exercise 16.

19. The graph in the figure may be used to answer the following.

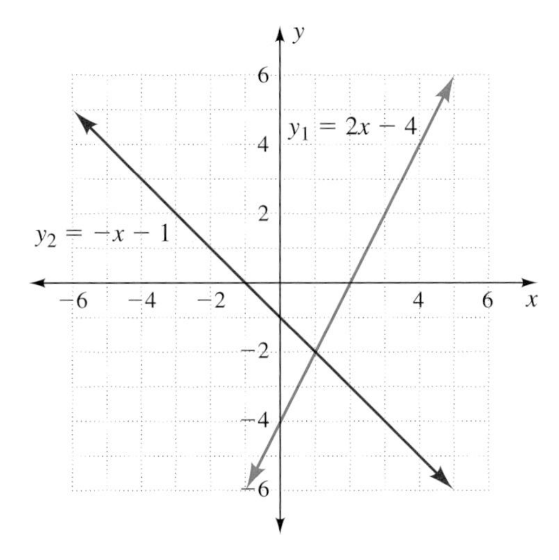

a. Find the point of intersection of the two lines.

b. Substitute the intersection point into each equation shown on the graph.

c. Solve the equation $2x - 4 = -x - 1$.

d. What does the graph indicate about the equation $2x - 4 = -x - 1$?

e. Solve the inequality $2x - 4 > -x - 1$. Write the solution as an inequality.

In Exercises 20 to 22, solve the inequality. Show the solution set as a line graph, as an inequality, and as an interval.

20. $2x + 5 < -3$

21. $3 - 2x > 11$

22. $2x + 8 \geq \frac{1}{2}(x + 1)$

23. NW Natural Gas Company charges home users a \$6 monthly fee plus \$1.36828 per therm (1 therm ≈ 100 cubic feet).

a. Make a table for the total monthly cost of 0 to 200 therms. Count by 40.

b. Graph the data from part a.

c. Write an equation to find the total cost for a month's use of gas with the input in therms.

d. Use your equation to find the amount of gas (in therms) used by a home with a bill of \$101.78.

24. Penske rentals in South Chicago cost \$29.95 plus \$0.59 per mile for a 10-foot truck. Budget offers the same truck for \$21.95 plus \$0.99 per mile. Write an inequality and find the mileage for which Budget is the lower cost. What is the cost when the rentals are equal?

25. The equation $x + 3 = 3 + x$ is an identity. What happens when you solve the identity for x? When you graph y_1 = left side of an identity and y_2 = right side, what is the result?

26. The administrators of California community colleges try to keep 3% of their annual budgets as reserves (not spent); 5% set aside is called prudent reserves. In November 2004, according to its school newspaper, one Northern California college went on the watch list because reserves dropped below 1.1 million dollars. What was the college's budget? (As a check, its prudent reserves would have been 1.8 million dollars).

Exercises 27 to 30 check your understanding of the words *simplify, solve, equation, and expression.*

27. Write two examples of steps in simplifying an expression.

28. Write two examples of steps in solving an equation.

29. Write two examples of simplifying within an equation.

30. It is possible to solve an expression?

▶ Cumulative Review of Chapters 1 to 3

1. Make an input-output table for each rule, with integer inputs from −3 to 3. Graph the (x, y) ordered pairs for each rule on separate axes.

a. $y = x + 1$ **b.** $y = |x + 2|$ **c.** $y = x^2 - 2$

2. Translate into symbols; if the statement is an equation, solve for the variable.

a. The sum of a number and six is multiplied by three.

b. The sum of a number and six, multiplied by three, is twelve.

c. The difference between fifteen and the quotient of a number and two is three.

3. Write in words:

a. $x > 5$ **b.** $-|x|$

c. $3 - 2(a + 8)$

4. Change $\frac{1}{2}$ hour + 15 minutes + $\frac{1}{3}$ hour to like units and add.

5. Add, subtract, multiply, and divide -5 and -4.

6. Add, subtract, multiply, and divide $-\frac{3}{5}$ and $\frac{5}{8}$.

7. Add, subtract, multiply, and divide 1.25 and -0.5

8. Add, and explain the properties permitting mental arithmetic: $2\frac{3}{4} + 1\frac{1}{2} + 3\frac{1}{4}$

9. a. Write an expression for four times the third power of a variable.

b. Write an expression for the third power of the product of four and a variable.

c. Identify the base in each.

10. Explain how the definition of positive integer exponents and the properties of real numbers show how to write $(3x^2)^3$ without parentheses.

11. Solve each equation.

a. $5 - 2x = 19$ **b.** $2 - 5(x + 7) = -23$

12. Solve each inequality. Write the solution as an interval and graph it on a number line.

a. $3x + 4 < -8$

b. $3(x + 5) \geq 5 - 3x + 1$

13. Budget rents a hand truck (for appliances) for \$15 and moving pads for \$12 per dozen in Miami. The same company rents a hand truck for \$12 and moving pads at \$10 per dozen in South Chicago. State equations and find how many dozen pads can be rented with a hand truck at each location for \$100.

14. U-Haul rents a 10-foot truck in Miami for \$19.95 plus \$0.99 per mile. Budget rents the same size truck for \$24.95 plus \$0.79 per mile. Write an inequality and solve it to show for what number of miles U-Haul is less expensive.

15. A natural gas company charges a \$3 late charge for past-due balances between \$50 and \$176 and a late charge of 1.7% for balances over \$176. Why does the \$3.00 late charge end at \$176? What is the balance if the late charge is \$5.10? Approximately what percent per year is 1.7% per month?

CHAPTER 4

Formulas, Functions, Linear Equations, and Inequalities in Two Variables

In amateur bowling, the members of each team are given a bonus score called a handicap. The handicap permits teams of differing ability to be compared. In league play, the average and handicap are recalculated each week.

The bowler's average, A, is used in the formula $H = 0.8(200 - A)$ to find the handicap, H. No handicap is given for averages of 200 or over. (The maximum score is 300.) The handicap is added to the bowler's points after each game. Suppose Bowler A, with an average of 150, bowls a 165 game, and Bowler B, with an average of 200, bowls a 206 game. Which bowler has the higher score under the handicap system?

This opening example makes use of a formula. Section 4.1 on formulas focuses on extending equation-solving skills. Section 4.2 introduces functions. The bowling handicap is a function of the bowling average and is described by a linear equation, the topic we discuss in Sections 4.3 to 4.5. The chapter closes, in Section 4.6, with the graphing of inequalities in two variables.

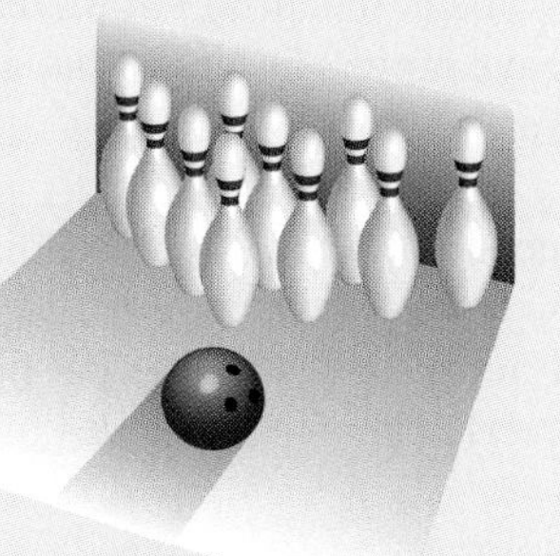

FIGURE 1 Ten pin bowling

▶ 4.1 Solving Formulas

Objectives

- Describe relationships using the phrase *in terms of.*
- Solve a formula for one variable.
- Evaluate formulas.

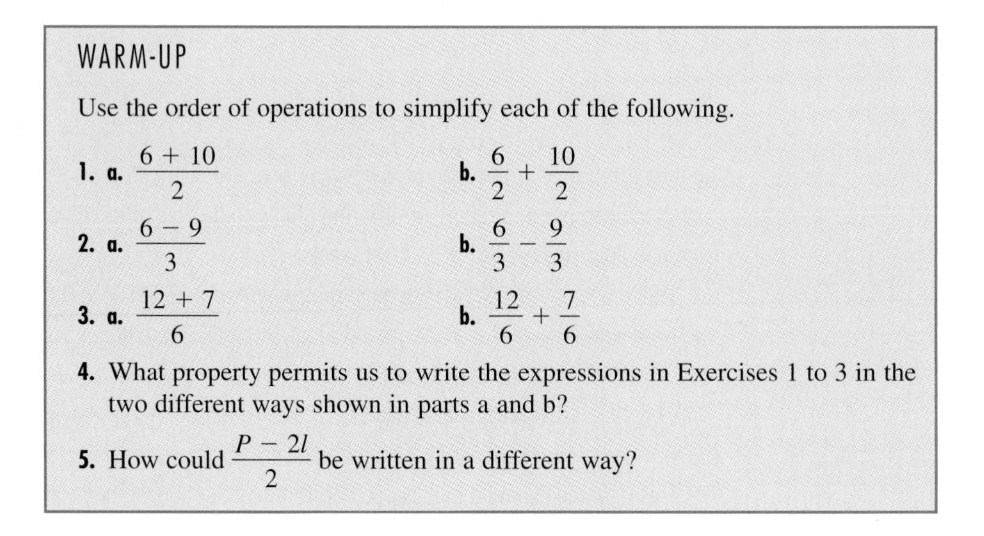

WARM-UP

Use the order of operations to simplify each of the following.

1. a. $\dfrac{6 + 10}{2}$ **b.** $\dfrac{6}{2} + \dfrac{10}{2}$

2. a. $\dfrac{6 - 9}{3}$ **b.** $\dfrac{6}{3} - \dfrac{9}{3}$

3. a. $\dfrac{12 + 7}{6}$ **b.** $\dfrac{12}{6} + \dfrac{7}{6}$

4. What property permits us to write the expressions in Exercises 1 to 3 in the two different ways shown in parts a and b?

5. How could $\dfrac{P - 2l}{2}$ be written in a different way?

IN THIS SECTION, we practice solving and evaluating formulas and dealing with subscripted variables. We evaluated formulas for geometric figures in Section 2.5. A list of other common formulas is on the page facing the back cover.

▶ **EXAMPLE 1** Exploration The international airport is 55 miles from your home, 30 miles of which is freeway. You need to be at the airport at 8:30 a.m. for a 10:30 a.m. flight. What time do you need to leave home? Discuss what assumptions you make about type of streets traveled and rate of travel.

SOLUTION To solve this problem, you need to know $d = rt$, the distance, rate, and time formula. You need to guess how fast you can drive the 55 miles. You might assume that you can drive the speed limit—say, 60 miles per hour on the freeway and 25 miles per hour on city streets. There are 30 miles of freeway and 25 miles of city streets.

Writing the miles per hour as a fraction, $\dfrac{\text{miles}}{\text{hours}}$, we find the time for the freeway travel using $d = rt$:

$$30 \text{ miles} = \frac{60 \text{ miles}}{1 \text{ hour}} \cdot t \qquad \text{Multiply by 1 hour.}$$

$$30 \text{ miles} \cdot 1 \text{ hour} = 60 \text{ miles} \cdot t \qquad \text{Divide by 60 miles.}$$

$$\frac{30 \text{ miles} \cdot 1 \text{ hour}}{60 \text{ miles}} = t \qquad \text{Simplify the left side.}$$

$$\frac{1}{2} \text{ hour} = t$$

We then find the time for the city street travel using $d = rt$:

$$25 \text{ miles} = \frac{25 \text{ miles}}{1 \text{ hour}} \cdot t \qquad \text{Multiply by 1 hour.}$$

$$25 \text{ miles} \cdot 1 \text{ hour} = 25 \text{ miles} \cdot t \qquad \text{Divide by 25 miles.}$$

$$\frac{25 \text{ miles} \cdot 1 \text{ hour}}{25 \text{ miles}} = t \qquad \text{Simplify the left side.}$$

$$1 \text{ hour} = t$$

We travel $\frac{1}{2}$ hour on the freeway and 1 hour on the city streets, for a total of $1\frac{1}{2}$ hours. We need to leave $1\frac{1}{2}$ hours before 8:30 a.m., or at 7:00 a.m. ◀

THINK ABOUT IT 1: How long will the trip take if the freeway is congested and traffic slows to an average of 30 miles per hour?

Solving Formulas for Repeated Use

In Example 1, we solved the formula $d = rt$ for the time, t, at each different rate traveled. If you plan to make repeated use of formulas, it is easier to first solve the formula for the variable you want.

EXAMPLE 2 Solving formulas Solve $d = rt$ for t.

SOLUTION

$$d = rt \qquad \text{Mark the variable } t \text{ with an arrow. Divide both sides by } r.$$

$$\frac{d}{r} = \frac{rt}{r} \qquad \text{The value of } \frac{r}{r} \text{ is 1.}$$

$$\frac{d}{r} = t$$

◀

Using the Phrase *in terms of*

When we solve a formula, we apply addition and multiplication properties to get one variable by itself on one side of the equation. We say that the variable is **in terms of** the other variable(s) when *the variable is by itself on one side of the equal sign and terms containing other variables, numbers, and operations are on the other side.*

EXAMPLE 3 Using *in terms of* Describe each equation using the phrase *in terms of*. If you recognize the formula, use words instead of variables.

a. $d = rt$ **b.** $t = \frac{d}{r}$

c. $H = 0.8(200 - A)$ **d.** $v = -gt + v_0$

SOLUTION

a. Distance is in terms of rate and time.

b. Time is in terms of distance and rate.

c. A bowling handicap is in terms of a bowling average.

d. The variable v is in terms of g, t, and v_0. This is the formula for the velocity (v) of an object thrown straight up into the air in terms of gravity (g), time (t), and initial velocity (v_0). ◀

A formula is *solved for a variable* when the variable appears by itself on one side and the other side no longer contains the variable. A variable cannot be *in terms of* itself. In Example 4, we find out whether formulas are solved for a variable.

EXAMPLE 4 Solving formulas Which of these formulas are solved for the variable on the left?

a. $A = \frac{1}{2}h(a + b)$

b. $n = \dfrac{n + 1}{2}$

c. $r = \dfrac{A - 2\pi r^2}{2\pi h}$

SOLUTION a. The formula is correctly solved for A, the area of the trapezoid. The variable a on the right is the length of one of the parallel sides of a trapezoid. *In copying a formula, do not change the form of any letter, A to a or a to A.*

b. The formula is not solved for n. The variable n appears on both sides.

c. The formula is not solved for r. The variable r appears on both sides. ◀

Steps in Solving Formulas

One difference between the equation solving we did in other sections and the formula solving here is that here we have variables remaining on both sides of the answer. Because the solutions to formulas contain variables and operations, it is important to apply the equation-solving steps carefully.

SOLVING FORMULAS

- *Keep track of the selected variable.* Mark the variable for which you are solving with an arrow, $\downarrow$.
- *Make a plan.* Write the order of operations on the selected variable. Write a plan that uses the inverse operations in the reverse order.

EXAMPLE 5 Solving and evaluating formulas

a. Solve $A = \frac{1}{2}bh$ for b in terms of A and h.

b. If the height of the triangle is 6 feet and the area is 30 square feet, what is the base?

SOLUTION a. ***Plan:*** Note that multiplication by $\frac{1}{2}$ is the same as division by 2. Thus, b is multiplied by h and the result is divided by 2. We will use the inverse operations in the reverse order: multiply by 2 and divide by h.

$$A = \frac{1}{2}\overset{\downarrow}{b}h$$ Mark b with an arrow. Multiply both sides by 2 to eliminate the $\frac{1}{2}$ on the right.

$$2 \cdot A = 2 \cdot \frac{1}{2}bh$$ Divide both sides by h to eliminate the h on the right.

$$\frac{2 \cdot A}{h} = \frac{bh}{h}$$ Simplify.

$$\frac{2A}{h} = b$$

b. In $b = \dfrac{2A}{h}$, let $A = 30$ square feet and $h = 6$ feet:

$$\frac{2(30 \text{ square feet})}{6 \text{ feet}} = 10 \text{ feet}$$ ◀

EXAMPLE 6 Solving and evaluating formulas

a. Solve $P = 2l + 2w$ for w in terms of P and l.

b. Find the width of a rectangle whose perimeter is 90 inches and length is 29 inches.

SOLUTION **a.** ***Plan:*** w is multiplied by 2 and then is added to $2l$. We must subtract $2l$ and then divide by 2.

$$\begin{aligned} P &= 2l + 2w && \text{Subtract } 2l \text{ from both sides.} \\ P - 2l &= 2w && \text{Divide both sides by 2.} \\ \frac{P - 2l}{2} &= w \end{aligned}$$

b.

$$\begin{aligned} w &= \frac{P - 2l}{2} && \text{Let } P = 90 \text{ inches and } l = 29 \text{ inches} \\ w &= \frac{90 \text{ inches} - 2(29 \text{ inches})}{2} \\ w &= 16 \text{ inches} \end{aligned}$$

◀

THINK ABOUT IT 2: What do we obtain when we divide $P - 2l$ by 2 in $\frac{P - 2l}{2}$?

▶ Multiple Methods

Because formulas can often be solved in several ways, answers may not always look the same. Students in higher mathematics classes often use their algebra skills to make their answers look like those in the answer section of the textbook. Examples 7 and 8 show two different ways to solve the area of a trapezoid formula for one of its variables. The answers will not look alike.

▶ **EXAMPLE 7** Solving a formula: method 1 With a reverse order of operations, solve $A = \frac{1}{2}h(a + b)$ for a.

SOLUTION ***Plan:*** Observe that b is added to a, then the sum is multiplied by h and divided by 2. The reverse order, with opposite operations, is to multiply by 2, divide by h, and subtract b.

$$\begin{aligned} A &= \tfrac{1}{2}h(a + b) && \text{Mark } a \text{ with an arrow. Multiply by 2.} \\ 2 \cdot A &= 2 \cdot \tfrac{1}{2}h(a + b) && \text{Simplify.} \\ 2A &= h(a + b) && \text{Divide by } h. \\ \frac{2A}{h} &= \frac{h(a + b)}{h} && \text{Simplify.} \\ \frac{2A}{h} &= a + b && \text{Subtract } b. \\ \frac{2A}{h} - b &= a + b - b && \text{Simplify.} \\ \frac{2A}{h} - b &= a \end{aligned}$$

◀

▶ **EXAMPLE 8** Solving a formula: method 2 Clearing the formula of fractions and using the distributive property to remove the parentheses, solve for a in $A = \frac{1}{2}h(a + b)$.

SOLUTION

$$A = \tfrac{1}{2}\, h(a + b) \qquad \text{Multiply by 2 to clear the fraction.}$$

$$2 \cdot A = 2 \cdot \tfrac{1}{2}\, h(a + b)$$

$$2A = h(a + b) \qquad \text{Multiply } h \text{ times } a \text{ and } b.$$

$$2A = ha + hb \qquad \text{Subtract } hb \text{ from both sides.}$$

$$2A - hb = ha \qquad \text{Divide by } h \text{ on both sides.}$$

$$\frac{2A - hb}{h} = \frac{ha}{h} \qquad \text{Simplify.}$$

$$\frac{2A - hb}{h} = a$$

◀

THINK ABOUT IT 3: Write $\dfrac{2A - hb}{h}$ in another way.

SUBSCRIPTS In Chapter 3 we wrote subscripts on the y in equations on our graphs. With the graphing calculator [Y =] option, we enter multiple equations under Y_1, Y_2, Y_3, etc., as shown in Figure 2.

```
Plot1 Plot2 Plot3
\Y1=█
\Y2=
\Y3=
\Y4=
\Y5=
\Y6=
\Y7=
```

FIGURE 2 Subscripts in [Y =] screen

The small zero to the right of the variable v in part d of Example 3 is another use of a subscript.

▶ **EXAMPLE 9** Solving formulas containing subscripts Solve $v = -gt + v_0$ for t in terms of v, g, and v_0.

SOLUTION The subscript on v_0 indicates that it stands for velocity at time $t = 0$. The two variables v_0 and v are not the same and cannot be combined; v and v_0 must be treated as different variables.

$$v = -gt + v_0 \qquad \text{Subtract } v_0 \text{ from both sides.}$$

$$v - v_0 = -gt \qquad \text{Divide both sides by } -g.$$

$$\frac{v - v_0}{-g} = t \qquad \text{Multiply numerator and denominator by } -1.$$

$$\frac{v_0 - v}{g} = t$$

◀

▶ Application

In Example 10, we return to the setting in the chapter opening.

▶ **EXAMPLE 10** Evaluating a formula: bowling handicap A ten-pin bowler's handicap, H, is 80% of the difference between 200 and the bowler's average score, A. No handicap is given for averages of 200 or over. (The maximum score is 300.) The handicap is added to the bowler's points after each game. Suppose Aleta, with an average of 150, bowls a 165 game, and Betty, with an average of 200, bowls a 206 game. Use the formula $H = 0.8(200 - A)$ to find out which bowler has the higher score under the handicap system.

Student Note:
We will find averages in Section 5.4.

SOLUTION Aleta's handicap is found by substituting her average, 150, into the formula:

$H = 0.8(200 - A)$

$H = 0.8(200 - 150)$ Find the difference in parentheses.

$H = 0.8(50)$ Simplify.

$H = 40$

Aleta's total score is $165 + 40 = 205$.

Betty has an average of 200 and so receives no handicap. Her score, 206, is still higher than Aleta's total of 205. ◀

In Example 10, we used the bowling formula in its original form. In Example 11, we solve the formula for A.

▶ **EXAMPLE 11** Solving a formula: bowling averages His computer's hard drive crashed, and Scott lost the team's bowling averages. He has last week's handicaps for the following bowlers and wants to find their averages:

Fran, 60 Miguel, 20 Scott, 0

a. Solve the bowling handicap formula, $H = 0.8(200 - A)$, for the average, A.
b. Use the formula to find the average for each bowler.

SOLUTION **a.** One solution is

$H = 0.8(200 - A)$ Multiply using the distributive property.

$H = 160 - 0.8A$ Add $0.8A$ to both sides.

$H + 0.8A = 160$ Subtract H from both sides.

$0.8A = 160 - H$ Divide by 0.8 on both sides.

$$A = \frac{160 - H}{0.8}$$

b. For Fran, with $H = 60$,

$$A = \frac{160 - H}{0.8}$$

$$A = \frac{160 - 60}{0.8} = 125$$

For Miguel, with $H = 20$,

$$A = \frac{160 - H}{0.8}$$

$$A = \frac{160 - 20}{0.8} = 175$$

Because Scott's handicap is 0, his average score could be any number 200 or larger. ◀

ANSWER BOX

Warm-up: 1. both 8 **2.** both −1 **3.** both $3\frac{1}{6}$ **4.** the distributive property of multiplication over addition, used here as division over addition or subtraction **5.** $\frac{P}{2} - \frac{2l}{2}$ or $\frac{P}{2} - l$. **Think about it 1:** If freeway traffic is moving at 30 miles per hour, it will take 1 hour to travel the 30 miles, so the trip will take a total of 2 hours. **Think about it 2:** When we divide each term by 2, we get $\frac{P}{2} - l$. **Think about it 3:** $\frac{2A}{h} - \frac{hb}{h}$ or $\frac{2A}{h} - b$.

▶ 4.1 Exercises

In Exercises 1 to 6, use the phrase *in terms of* to describe each formula or equation. If you recognize the formula, use words; otherwise, use the variables themselves.

1. $I = prt$

2. $P = 2l + 2w$

3. $r = \frac{d}{t}$

4. $C = 2\pi r$

5. $G = \frac{T_1 + T_2 + T_3 + H + E}{P}$, where G = percent earned, T = test, H = homework, E = final exam, and P = total points possible

6. The volume of a sphere: $V = \frac{4}{3}\pi r^3$

What formula is described by each statement in Exercises 7 to 10? You may need to look up the correct formula.

7. The area of a rectangle in terms of length and width

8. The time needed for a trip in terms of distance and rate

9. The area of a circle in terms of radius

10. The circumference of a circle in terms of diameter

In Exercises 11 to 16, write the statements as formulas. Some formulas may contain parentheses.

11. The area, A, is half the product of b and h.

12. The perimeter, P, is twice the sum of l and w.

13. The perimeter, P, is the sum of twice l and twice w.

14. The area, A, is half of h multiplied by the sum of a and b.

15. The number of calories used, C, is the product of the number of calories per minute, f, and the number of minutes of exercise, m.

16. The last term, L, is the sum of f and the product of d with the difference between n and 1.

Solve each formula in Exercises 17 to 48 for the indicated letter.

17. $p = 5n$ for n

18. $q = 4d$ for d

19. $A - P = H$ for A

20. $A - P = H$ for P

21. $C = 2\pi r$ for r

22. $A = lw$ for w

23. $A = bh$ for h

24. $d = rt$ for r

25. $I = prt$ for t

26. $I = prt$ for r

27. $C = \pi d$ for d

28. $A = \pi r^2$ for r^2

29. $P = R - C$ for C

30. $P = R - C$ for R

31. $PV = nRT$ for n

32. $PV = nRT$ for R

33. $C_1V_1 = C_2V_2$ for V_1

34. $C_1V_1 = C_2V_2$ for C_2

35. $P = a + b + c$ for c

36. $P = 2l + 2w$ for l

37. $A = \frac{1}{2}h(a + b)$ for h

38. $A = \frac{1}{2}h(a + b)$ for b

39. $V = \frac{1}{3}\pi r^2 h$ for r^2

40. $V = \frac{1}{3}\pi r^2 h$ for h

41. $x = \frac{-b}{2a}$ for b

42. $x = \frac{-b}{2a}$ for a

43. $y = mx + b$ for b

44. $y = mx + b$ for m

45. $d^2 = \frac{3h}{2}$ for h

46. $S = \frac{a}{1 - r}$ for a

Blue numbers are core exercises.

47. $t^2 = \dfrac{2d}{g}$ for g

48. $t^2 = \dfrac{2d}{g}$ for d

In Exercises 49 and 50, A = amount, P = principal, t = time in years, and r = percent interest.

49. a. Solve $A = P + Prt$ for r.

b. An amount of \$11,050 is received on a two-year time certificate with a \$10,000 principal. What is the rate of interest, r?

50. a. Solve $A = P + Prt$ for t.

b. An amount of \$60,125 is received on a time certificate at 6.75% interest on a \$50,000 principal. What is the number of years, t, on the certificate?

51. a. Solve $C = \frac{5}{9}(F - 32)$ for F.

b. What is the Fahrenheit temperature, F, corresponding to a C of 37° Celsius?

52. a. Solve $K = C + 273$ for C.

b. What is the Celsius temperature corresponding to absolute zero, 0 K?

c. Use the answer to part b and the answer to part a of Exercise 51 to obtain the Fahrenheit temperature corresponding to absolute zero.

53. The bowling handicap, H, in terms of bowling average, A, is $H = 0.8(200 - A)$.

a. What is the bowling handicap for a bowler with a 140 average?

b. Using an order of operations plan, solve the formula for A.

c. What is the bowling average if the handicap is 24?

54. An aerobic heart rate for exercise is $R = 0.7(220 - A)$, where A is age in years.

a. What is the heart rate for a 20-year-old?

b. Solve the formula for A.

c. What is the age for someone with a predicted aerobic heart rate of 119?

▶ Equations

55. One equation for a straight line is $y = mx + b$.

a. Solve the equation for b.

b. Find b if $x = 3$, $y = 4$, and $m = 2$.

c. Find b if $x = 3$, $y = 4$, and $m = -2$.

d. Find b if $x = 3$, $y = 4$, and $m = \frac{1}{2}$.

e. Find b if $x = 3$, $y = 4$, and $m = -\frac{1}{2}$.

Exercises 56 to 63 provide practice in solving for the variable y, a skill needed to make a table or graph. Solve for y in terms of x.

56. $xy = -4$

57. $xy = -6$

58. $3x - y = 10$

59. $2x - y = 3$

60. $x - 2y = -5$

61. $2x - 3y - 4 = 0$

62. $3x - 2y = 6$

63. $2x + 3y = 9$

▶ Projects

64. Electric Formulas The electric formula wheel, common in handbooks for electricians, shows up in surprising places. This one came from a theater reference, the *Backstage Handbook,* by Paul Carter (Shelter Island, NY: Broadway Press, c. 1994).

a. For the formula in the first quadrant, $R = \dfrac{E}{I}$, show what other formulas are equivalent by solving for the other letters.

b. For the formula $R = \dfrac{E^2}{W}$, show what formulas are equivalent (you may need to ask for suggestions on solving $R = \dfrac{E^2}{W}$ for E).

c. How many unique formulas are represented?

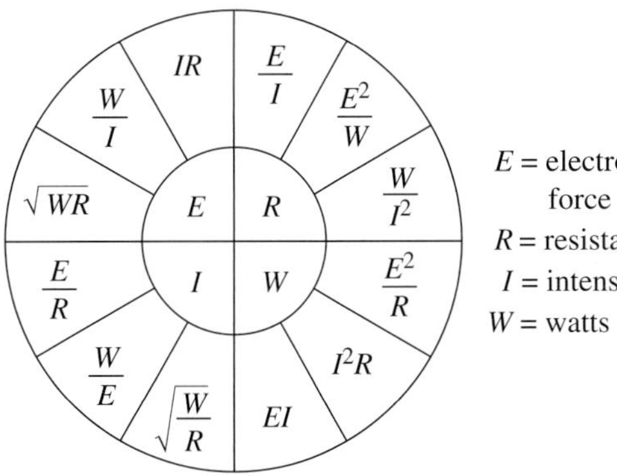

65. Subscript Research In the nutrition field vitamins are identified by subscripts: vitamin B_1, vitamin B_2, vitamin B_6, etc. In music, middle C is C_4, and a chord might be described as G_4–C_5–E_5. Research an application of subscripts in a subject of interest to you. Give several examples and some detail about the application. Explain why subscripts are necessary in the application.

Blue numbers are core exercises.

▶ 4.2 Functions and Graphs

Objectives

- Define a function in terms of input and output.
- Find the domain and range of a function.
- Evaluate expressions written in function notation and graph the results.
- Use the vertical-line test to recognize a function.

WARM-UP

Complete the table for each equation.

1. *An Absolute Value Equation,* $x = |y|$

x	y
	-2
	-1
	0
	1
	2

2. *A Squaring Equation,* $x = y^2$

x	y
	-2
	-1
	0
	1
	2

THIS SECTION introduces functions and function notation. We evaluate functions and draw graphs of functions. We look at graphs to find out whether a relationship is a function.

▶ Defining Functions

When you spend money, do you expect to spend the same amount as others for the same item? Is buying a car difficult because you are never sure what you are going to pay? Do you expect to be graded fairly or to be paid the same when doing the same work as others? In everyday life we expect that a certain action will receive a certain response. It seems only fair!

In mathematics we have similar expectations. We like to have one answer. In fact, a mathematical relationship where a given input always results in the same, single outcome is given a special name, *function*.

DEFINITION OF FUNCTION

A function is a relationship or association where for each input x, there is exactly one output y.

▶ EXPLORATION EXAMPLE 1 Identifying the input, the function, and the output For each function setting in the table, name the input, the function, and the output. Explain why the corresponding entry is not a function.

Everyday Functions	Not Functions
a. On what day of the month were you born? b. Cost of tickets (no matter who buys the ticket)	a. Name someone born on the 6th of the month. b. Cost of tickets vary (are you friends of the management?)
Mathematical Functions	**Not Functions**
c. What is the opposite of -4? d. Find $\sqrt{9}$.	c. Name an even number. d. Find a square root of 9.

SOLUTION The definition for the symbol for square root is in the Glossary/Index.
Answers in Answer Box. ◀

Calculator language may make *function* seem more familiar. You may have heard the name *four-function calculator* given to those that only add, subtract, multiply, and divide. When we use the negative *function* key (−), on a calculator, we expect the opposite of the number we enter following that entry. If, instead, the absolute value *function* of the number appears, we would assume the calculator is mal*functioning* and take it in for an exchange.

▶ Sets of Inputs (Domain) and Sets of Outputs (Range)

We find the set of inputs and the set of outputs to decide what numbers to use on the axes when we draw a graph. Similarly, we find the set of inputs and the set of outputs to work with functions. In Example 2, the function itself is not described but is merely replaced by the phrase *is a function of.*

▶ **EXAMPLE 2** Finding the input and output sets Explain why each sentence describes a function. What are the input and output phrases in each sentence? What sets describe the inputs and outputs?

a. The remaining value of a prepaid transit ticket is a function of the number of identical trips taken.

b. The total cost of a city water bill is a function of the number of thousands of gallons of water used each month.

SOLUTION

a. For each *number of trips taken* (input), there is exactly one *remaining value of the transit ticket* (output). Each identical trip should cause the same cost to be subtracted from the ticket. The inputs are the set of positive integers; the outputs are positive real numbers in dollars and cents.

b. For every *number of thousands of gallons used* (input), there is exactly one *cost* (output). Each thousand gallons of water usage should have the same cost. The inputs are positive real numbers in thousands of gallons, and the outputs are positive real numbers in dollars and cents. ◀

If we want to be more formal in describing the sets of inputs and outputs, we use the words *domain*, as in "domestic" or "home," and *range*, as in "to go out."

DEFINITION OF DOMAIN AND RANGE

The **domain** is the set of inputs to a function.
The **range** is the set of outputs from a function.

Functions Described in Ordered Pairs

To describe a function, we can use the same ordered pair (x, y) as in coordinate graphing, except we say that *y is a function of x*. Example 3 shows that not all sets of ordered pairs are functions.

EXAMPLE 3 Finding whether ordered pairs describe a function Which sets of ordered pairs describe a function? Write the input set (domain) and output set (range) for the function.

a. (5, 25), (6, 36), (7, 49)
b. (5, 6), (5, 7), (5, 8)
c. (25, 5), (25, −5), (9, 3), (9, −3)
d. (5, 6), (6, 6), (7, 6)
e. (Cuba, Maria Conchita Alonso), (Cuba, Orlando Hernandez), (Cuba, Desi Arnaz)
f. (Roberto Clemente, Puerto Rico), (Jose Feliciano, Puerto Rico), (Rita Moreno, Puerto Rico)

SOLUTION

a. Function. Each input is associated with exactly one output.
{5, 6, 7}, {25, 36, 49}

b. Not a function. For the input 5, there are three different outputs: 6, 7, and 8.

c. Not a function. For the input 25, there are two outputs: 5 and −5. For the input 9, there are two outputs: 3 and −3.

d. Function. Each input is associated with exactly one output. This example shows that the outputs do not need to be different numbers.
{5, 6, 7}, {6}

e. Not a function. For the input Cuba, there are different outputs.

f. Function. For each input name, there is one output.
{Clemente, Feliciano, Moreno}, {Puerto Rico} ◀

The ordered pairs in part a of Example 3 are from the equation $y = x^2$. Because each input x has only one square ($x \cdot x$, or x^2), we say that y is the *squaring function*.

THINK ABOUT IT 1: What rule describes each of the other parts of Example 3?

Function Notation

We can use **function notation** *to write functions in symbols.*

FUNCTION NOTATION

> The notation $f(x)$ is read "function of x" or "f of x." The f and x in $f(x)$ are not being multiplied.

Student Note:
Here is another use of parentheses.

The initial letter f (or any other letter) represents the function, and the letter x (or any other letter) in parentheses represents the input variable. The ordered pair (x, y) can be written $(x, f(x))$. The squaring function can be written $f(x) = x^2$. The square root function can be written $g(x) = \sqrt{x}$.

EXAMPLE 4 Writing function notation For each setting, choose a letter for the function and another letter for the input variable, and then write in function notation.

a. The area of a circle is a function of the radius. The area of a circle is pi times the square of the radius.

b. The total cost of a purchase is a function of the purchase price and the sales tax. The sales tax is 8.25% of the purchase price.

c. The household cost for waste water disposal is a function of the thousands of gallons of water recorded by the water meter. The total cost is \$6.00 plus \$2.384 times the number of thousands of gallons of water.

SOLUTION In each part, the first sentence states the function and the second describes the input variable.

a. Let A be the area function. Let r be the radius.

$$A(r) = \pi r^2$$

b. Let C be the total cost function in dollars. Let p be the purchase price.

$$C(p) = p + 0.0825p$$

c. Let C be the cost function of waste water disposal in dollars. Let x be the number of thousands of gallons of water.

$$C(x) = 2.384x + 6.00$$

◀

Note: As mentioned earlier, in function notation such as $f(x)$, $g(x)$, $A(r)$, $C(p)$, and $C(x)$, the parentheses do *not* imply multiplication.

THINK ABOUT IT 2: How is the product of two variables such as f and g usually written?

▶ Evaluating Functions

One of the most important reasons for studying functions is that the notation $f(x)$ is a convenient way to indicate the evaluation of a function for a certain number or expression. *Function notation has the advantage of allowing us to simultaneously name the association (rule) and the input.*

FUNCTION EVALUATION

> The notation for evaluating a function for $x = a$ is $f(a)$. To evaluate $f(a)$, we substitute a for every x in the expression named by $f(x)$ and then simplify.

▶ EXAMPLE 5 Evaluating a function: city water bill The total cost of city water is \$6.00 plus \$0.869 per thousand gallons (kgal) used. If we let x be the number of thousand gallons used in the cost function, C, we can write $C(x) = 0.869x + 6.00$. Find $C(0)$, $C(1)$, $C(2)$, $C(3)$, and $C(4)$.

SOLUTION To find $C(0)$, we substitute $x = 0$ for x in $C(x) = 0.869x + 6.00$:

$$C(0) = 0.869(0) + 6.00 = 6.00$$

Similarly, by rounding to the nearest cent, we find that

$$\text{for } x = 1, \quad C(1) = 0.869(1) + 6.00 = 6.87$$
$$\text{for } x = 2, \quad C(2) = 0.869(2) + 6.00 = 7.74$$
$$\text{for } x = 3, \quad C(3) = 0.869(3) + 6.00 = 8.61$$
$$\text{for } x = 4, \quad C(4) = 0.869(4) + 6.00 = 9.48$$

◀

In Example 6, the only new idea is the notation for evaluating the function defined by the expression $-2.5x + 20$. Everything else is from Chapter 1.

▶ **EXAMPLE 6** Evaluating a function: prepaid transit ticket After x trips at \$2.50 each, a \$20 prepaid transit ticket is worth $-\$2.50x + \$20$. If we let x be the number of trips in the value remaining function, f, we can write $f(x) = -2.50x + 20$.

a. Find $f(0)$, $f(2)$, $f(4)$, $f(6)$, and $f(8)$.
b. Summarize the results from part a in a table.
c. Make a graph from the table.

SOLUTION **a.** To find $f(0)$, we substitute $x = 0$ for x in $f(x) = -2.50x + 20$:

$$f(0) = -2.50(0) + 20 = 20$$

Similarly,

$$\text{for } x = 2, \quad f(2) = -2.50(2) + 20 = 15$$

$$\text{for } x = 4, \quad f(4) = -2.50(4) + 20 = 10$$

$$\text{for } x = 6, \quad f(6) = -2.50(6) + 20 = 5$$

$$\text{for } x = 8, \quad f(8) = -2.50(8) + 20 = 0$$

b. The evaluations are summarized in Table 1.

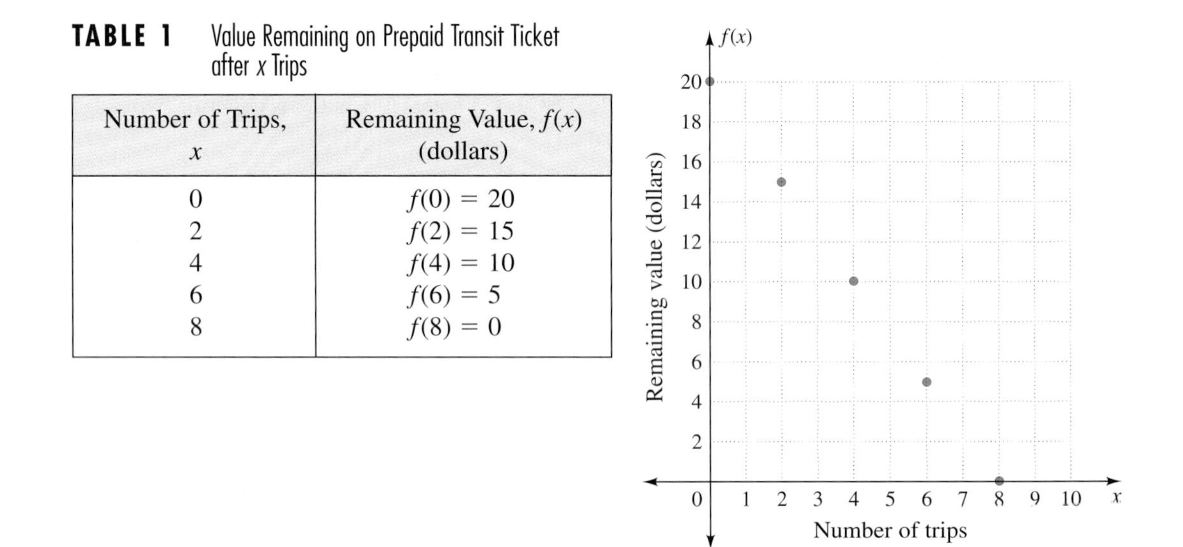

TABLE 1 Value Remaining on Prepaid Transit Ticket after x Trips

Number of Trips, x	Remaining Value, $f(x)$ (dollars)
0	$f(0) = 20$
2	$f(2) = 15$
4	$f(4) = 10$
6	$f(6) = 5$
8	$f(8) = 0$

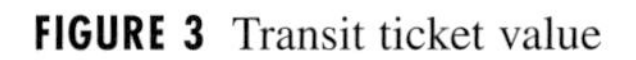

FIGURE 3 Transit ticket value

c. The ordered pairs are $(x, f(x))$. The points, shown in Figure 3, lie in a line that goes down from left to right. ◀

▶ Functions and Graphs: Vertical-Line Test

Not all mathematical associations or relations are functions. Relationships such as finding the square root of x, in which there are two possible outcomes for each input are not functions. In Example 7, we look at an equation that is not a function.

▶ **EXAMPLE 7** Identifying when an equation is not a function

a. Complete Table 2 for $x = |y|$.
b. Graph the ordered pairs (x, y).

c. If we consider x to be the input, why is the equation $x = |y|$ not a function?

TABLE 2

$x = \|y\|$	y
	-2
	-1
	0
	1
	2

SOLUTION **a.** The ordered pairs for the table are $(2, -2)$, $(1, -1)$, $(0, 0)$, $(1, 1)$, and $(2, 2)$.

b. These five points are plotted in Figure 4.

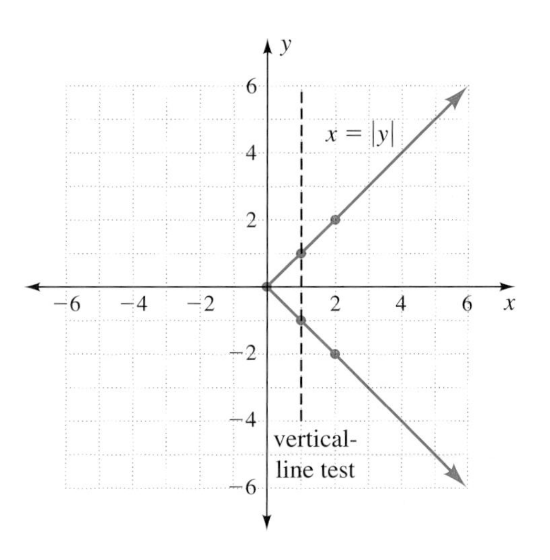

FIGURE 4 Graph of $x = |y|$

c. The equation $x = |y|$ is not a function because for $x = 1$, y is either 1 or -1. Similarly, for $x = 2$, $y = 2$ or -2. Thus, y is not a function of x. ◀

Looking at a graph is an easy way to find out if an equation is a function. In Figure 4, the points $(1, 1)$ and $(1, -1)$ are on the same vertical line. A graph having two points on the same vertical line is not a function.

VERTICAL-LINE TEST

> A graph shows a function if every vertical line intersects the graph no more than once.

▶ **EXAMPLE 8** Applying the vertical-line test Which of these graphs represent functions? If the graph does not represent a function, name two ordered pairs that show that it does not.

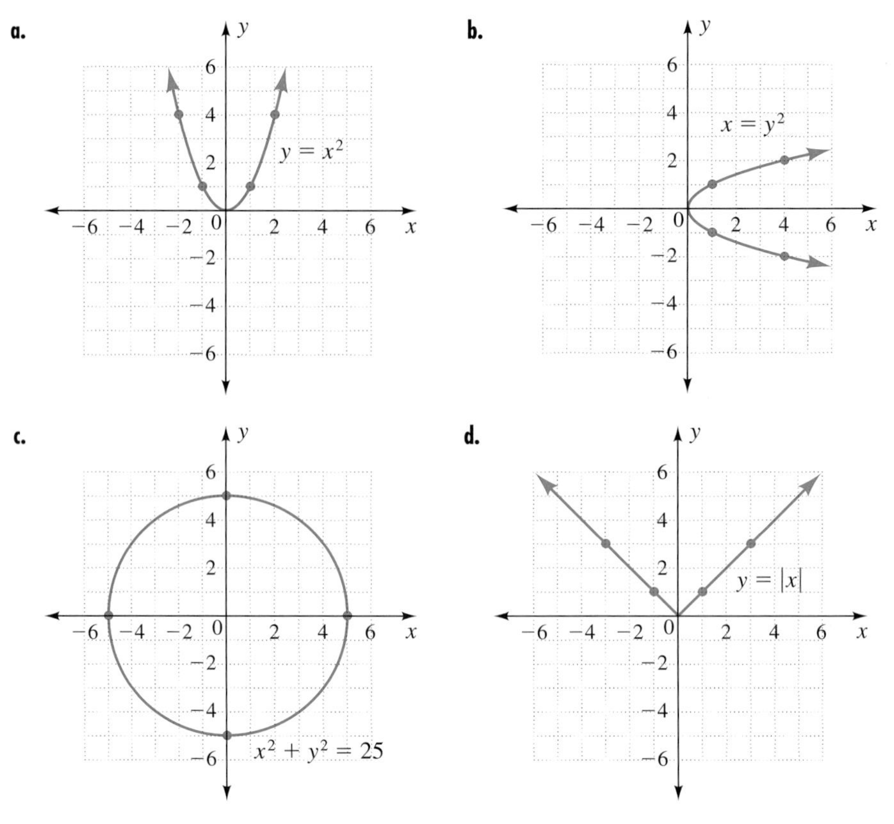

SOLUTION **a.** Function

b. Not a function. For example, (4, 2) and (4, −2) are on the same vertical line.

c. Not a function. For example, (0, 5) and (0, −5) are on the same vertical line.

d. Function ◀

THINK ABOUT IT 3: What two ordered pairs in part b of Example 8 have $x = 1$? What two ordered pairs in part c have $x = 3$? What two ordered pairs in part c have $x = -3$?

Caution: In Section 4.1, we used the phrase *in terms of*. The *in terms of* phrase describes a relationship between variables, as does a function. However, *in terms of* is more general than *is a function of* and may be used to describe all types of equations, either functions or not functions. In Example 8, equations b and c are in terms of x and y but neither one is a function.

Explore another type of function in Exercise 63, Probability Functions.

ANSWER BOX

Warm-up: 1. 2, 1, 0, 1, 2 **2.** 4, 1, 0, 1, 4 **Example 1: a.** Input is name, function is to give day of birth, output is one day (from a possible set, 1 to 31). More than one person has been born on the 6th. **b.** Input is number of tickets, function is the pricing schedule, output is the cost of the tickets. Price will not be the same to each customer. **c.** Input is −4, function is "the opposite of," output is 4. Each (even no.) is not unique. **d.** Input is 9, function is "find the positive square root," output is 3. The number 9 has two square roots, 3 and −3. **Think about it 1: b.** $x = 5$ **c.** $y^2 = x$ or y is the square root of x. **d.** $y = 6$ **e.** The output is a person born in the input country. **f.** The output is the place of birth of the person named as input. **Think about it 2:** The product of f and g is usually written $f \cdot g$ or just fg. This prevents confusion with the function notation $f(g)$. **Think about it 3:** (1, −1) and (1, 1); (3, 4) and (3, −4); (−3, 4) and (−3, −4)

▶ 4.2 Exercises

Exercises 1 to 8 give a relationship. Is the relationship a function?

1. The input is a woman with first name Barbara; the output is her profession.

2. The input is the last name Jordan; the output is a first name.

3. The input is a state of birth; the output is a person born there.

4. The input is an award winner; the output is the award won.

5. a. The output is the absolute value of the input.

b. The output is a number whose absolute value is the input.

6. a. For the input 5, the output is an integer greater than 4.

b. For any input, the output is one larger.

7. a. The output is the reciprocal of the input.

b. For the input 2, the output is a fraction less than 1.

8. a. For the input 1, the output is an integer made up of ones.

b. For any input, the output is 2.

In Exercises 9 to 16, which sets of ordered pairs are functions? For each set that is not a function, explain why. For each function, name the set of inputs (domain) and the set of outputs (range).

9. (Eden, Barbara), (Tuckman, Barbara), (McClintock, Barbara)

10. (Jordan, Barbara), (Jordan, Michael), (Jordan, Vernon)

11. (CA, Amy Tan), (CA, Maxine Hong Kingston), (CA, Ursula Le Guin)

12. (Haing Ngor, Academy Award), (Aung San Suu Kyi, Nobel Peace Prize), (Ieoh Ming Pei, Gold Medal of the American Institute of Architects)

13. a. $(5, 5), (-5, 5), (6, 6), (-6, 6)$

b. $(5, -5), (5, 5), (6, 6), (6, -6)$

14. a. $(5, 5), (5, 6), (5, 7)$

b. $(5, 6), (6, 7), (7, 8)$

15. a. $(2, \frac{1}{2}), (3, \frac{1}{3}), (4, \frac{1}{4})$

b. $(2, \frac{1}{2}), (2, \frac{1}{3}), (2, \frac{1}{4})$

16. a. $(1, 11), (1, 111), (1, 1111)$

b. $(-2, 2), (-3, 2), (-4, 2)$

17. Which is the description of a function?

a. For each input, there is exactly one output.

b. For each output, there is exactly one input.

18. Which is the description of a domain?

a. The set of inputs

b. The set of outputs

In Exercises 19 to 22, give a reason why each relationship is true. Name the set of inputs (domain) and the set of outputs (range).

19. The cost of a long distance telephone call is a function of how long the parties talk.

20. The cost of the roof of a house is a function of the area of the roof.

21. The hours of sunlight on a clear day in December is a function of the distance from the equator.

22. The capital of a state is a function of the name of the state.

In Exercises 23 to 26, define variables and write each sentence in function notation.

23. The amount earned is a function of the number of hours worked. An hourly worker earns \$8 per hour.

24. Auto registration cost is a function of the value of the car. Auto registration costs 1% of the current value of the car.

25. The circumference of a circle is a function of the diameter. Circumference is pi times diameter.

26. The volume of a sphere is a function of the radius. Volume is $\frac{4}{3}\pi$ times the cube of the radius.

Find $f(-2)$, $f(-1)$, $f(0)$, $f(1)$, and $f(2)$ for the functions in Exercises 27 to 32. Sketch a graph of each function using the ordered pairs, $(x, f(x))$.

27. $f(x) = 2x - 1$

28. $f(x) = 1 - 2x$

29. $f(x) = 2 - 3x$

30. $f(x) = \frac{1}{2}x - 1$

31. $f(x) = \frac{1}{4}x + 1$

32. $f(x) = 3 - 2x$

In Exercises 33 to 40, the functions are labeled with the letters h, H, g, and G instead of f. Find $h(4)$ and $h(-4)$ in Exercises 33 and 34.

33. $h(x) = \left|\frac{1}{2}x\right|$

34. $h(x) = |2x|$

Blue numbers are core exercises.

Find $H(4)$ and $H(-4)$ in Exercises 35 and 36.

35. $H(x) = x - x^2$

36. $H(x) = x^2 - 2x$

Find $g(-2)$ and $g(1)$ in Exercises 37 and 38.

37. $g(x) = x^2 + 1$

38. $g(x) = 1 - x^2$

Find $G(-2)$ and $G(1)$ in Exercises 39 and 40.

39. $G(x) = 2 - x^2$

40. $G(x) = x^2 + x - 1$

Explain how to tell which of the graphs in Exercises 41 to 44 represent functions.

41.

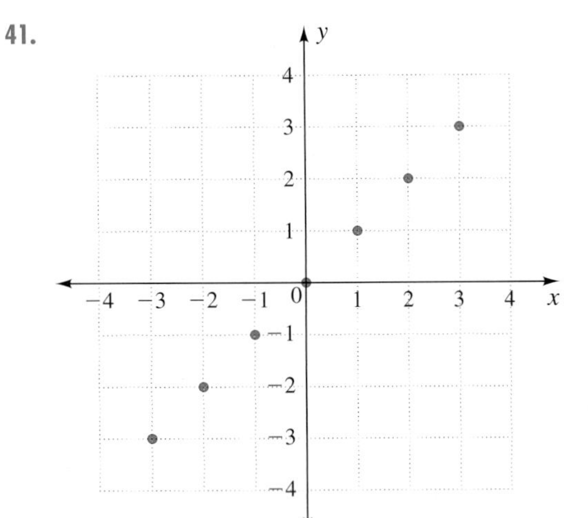

42.

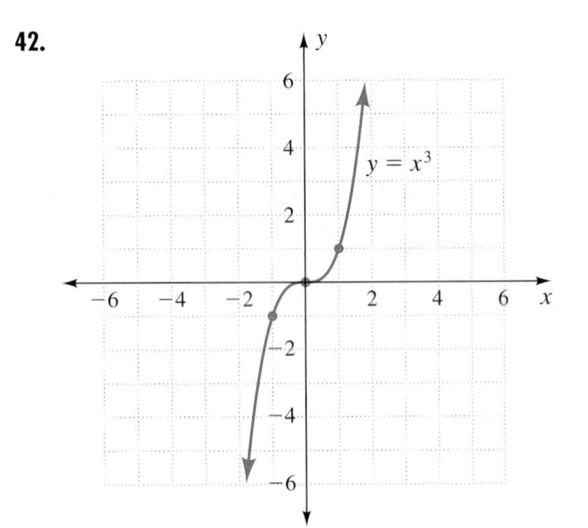

43.

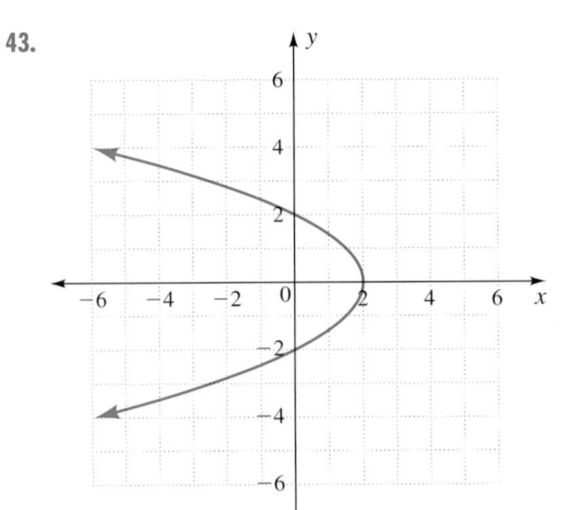

44.

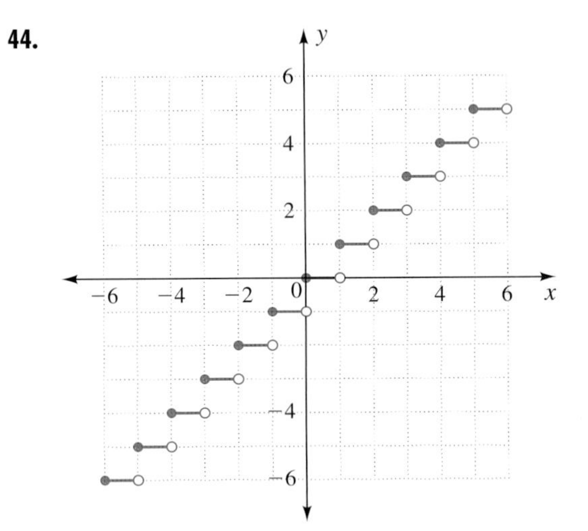

In Exercises 45 to 54, in which equations or inequalities is y a function of x? Explain your reasoning.

45. $y = x + 2$

46. $y = 1 - x$

47. $x + y = 3$

48. $x - y = 2$

49. $y = x^2$

50. $x = y^2$

51. $y^2 + 1 = x$

52. $y = x^2 + 1$

53. $y < x$

54. $y > x + 2$

▶ Writing

55. Explain how to find $f(a)$ for a function $y = f(x)$.

56. Explain how the vertical-line test identifies a graph of a function.

57. How would you explain to someone that $f(x)$ does not mean multiplication?

58. Given a value for $f(x)$, how would you find x?

Blue numbers are core exercises.

59. A slight change of wording can prevent a relationship from being a function. Paintball Palace charges group rates: 10 to 15, \$15 per person, 16 to 24, \$12.50 per person, 25 to 40, \$10 per person. Three graphs for Paintball Palace are shown. If the graph shows a function, state a sentence using *is a function of*. If the graph does not show a function, explain, in the context of the labels on the axes, why it does not.

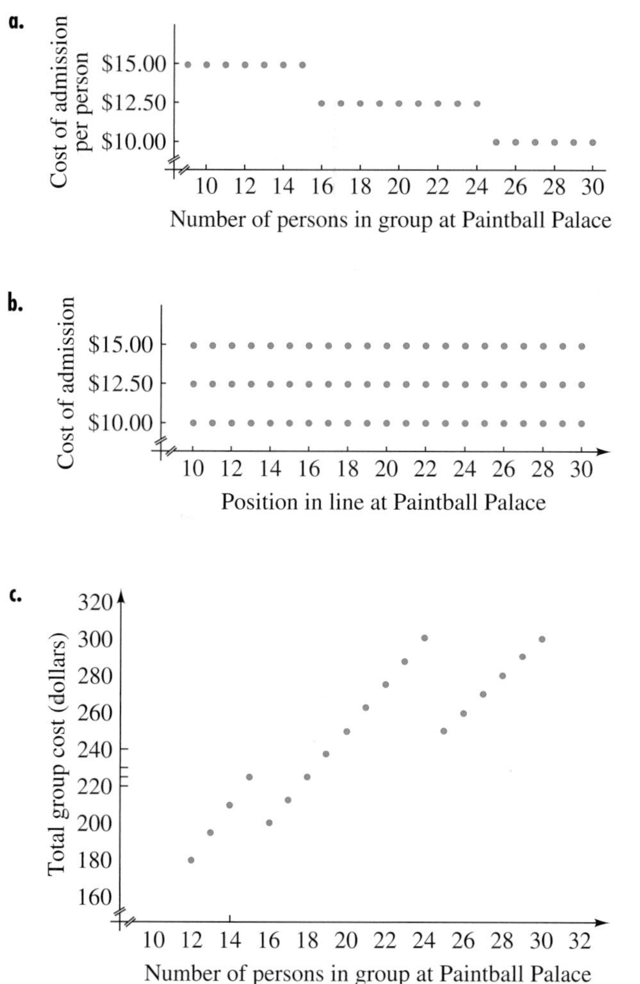

60. Refer to Example 6.

a. Explain why the remaining value is on the vertical axis.

b. Explain why a first-quadrant graph is used.

c. Explain why we do not connect the points in the graph.

d. Give ordered pairs for four other points that could be marked on the graph.

61. Linked Example

Input: Amount of Cash Advance, x	Output: Fee, $f(x)$
\$0 to \$166.66	\$5
\$166.67 to \$2499.99	3% of the amount
\$2500 to credit limit	\$75

For the fee schedule functions, find

a. $f(0)$ **b.** $f(20)$

c. $f(100)$ **d.** $f(200)$

e. $f(350)$ **f.** $f(1000)$

g. $f(1100)$ **h.** $f(3500)$

i. Explain why the rule for the interval (166.66, 2500) is $f(x) = 0.03x$.

Given the output, $f(x)$, find the input, x

j. $f(x) = 0$ **k.** $f(x) = \$5$

l. $f(x) = \$12$ **m.** $f(x) = \$21$

n. $f(x) = \$27$ **o.** $f(x) = \$36$

p. $f(x) = \$75$

▶ Projects

62. Birthday Candle Use a ballpoint pen to mark centimeters from the top of a birthday candle. Place the candle in a holder, and then mount the candle holder on a piece of styrofoam covered with aluminum foil. Make sure that the styrofoam is large enough that it will not tip over.

Predict the length of time required for 1 centimeter of the candle to burn. Repeat for 2 centimeters. Light the candle and record the time it takes for the candle to burn 1, 2, 3, and 4 centimeters. Is the distance burned a function of time? Use the data to predict the length of time required to burn 6 centimeters on a candle.

63. Probability Functions Another way to express our definition of function is to say "There is a 100% chance of one output for each input." For a probability function $P(x)$, a number (the probability) is assigned to each possible outcome of the activity. The probabilities add to 1 (that is 100%). For the activity *toss a fair coin*, the probability function is $P(\text{head}) = 50\%$ and $P(\text{tail}) = 50\%$.

a. If the activity is *toss four coins and read the top faces*, we have 16 possible outcomes. If H = head on top and T = tail (or back of the coin) on top, one of the outcomes is HHTH and another outcome is HTHH. Both HHTH and HTHH have 3 heads and 1 tail. List the 16 different outcomes, and group them according to numbers of heads and tails. What is the probability function for each group of outcomes if each outcome has a probability of $\frac{1}{16}$? *Hint:* There are five groups.

b. If the activity is *roll two six-sided dice and read the numbers of dots* on the top faces, there are 36 possible outcomes (two dots and one dot is different from one dot and two dots). List the 36 different outcomes and their sums. If the probability of each outcome is $\frac{1}{36}$, what is the probability function for the sums?

c. Make up your own problem. The problem might be related to the birth order of children, the outcomes for a well-known basketball player shooting free throws, or any other event or activity with multiple possible outcomes.

64. Utility Bills From the many utility examples in the book thus far, you should be ready to look for examples of functions on your monthly utility bills. Record the function or the data that you think give a function. Be sure to include units.

▶ 4.3 Linear Functions: Slope and Rate of Change

Objectives

- Find the slope of a line from its graph.
- Find the slope of a line from ordered pairs.
- Find the slope of a line from a table.
- Find the slopes of horizontal and vertical lines.
- Write the meaning of slope in terms of units on the axes.
- Recognize the slope concept in a variety of situations.

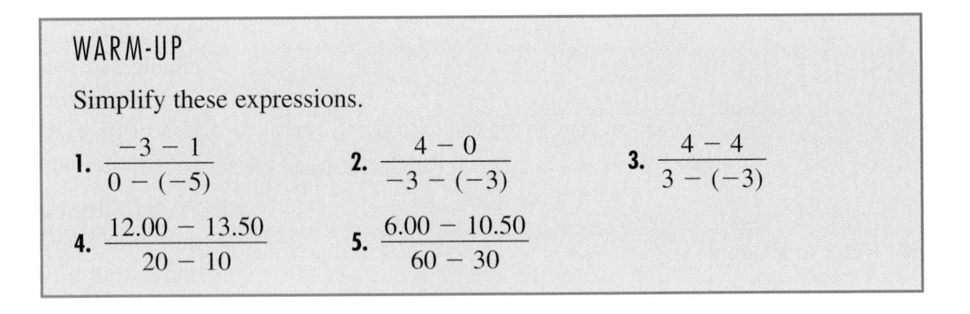

WARM-UP

Simplify these expressions.

1. $\dfrac{-3-1}{0-(-5)}$ 2. $\dfrac{4-0}{-3-(-3)}$ 3. $\dfrac{4-4}{3-(-3)}$

4. $\dfrac{12.00-13.50}{20-10}$ 5. $\dfrac{6.00-10.50}{60-30}$

IN THIS SECTION and the next two, we return to linear equations, $y = ax + b$. We define the slope, or rate of change. We look at slope visually with graphs, symbolically with a formula involving ordered pairs, and numerically with tables.

▶ EXAMPLE 1 Exploration Angie spends most of her free time outdoors and she has planned a cross country ski trip for her winter vacation. Angie plans to make a map that shows the elevation of the trail. She wants to describe the steepness, or slope, of the trail as she skis from left to right. Figure 5 shows some line segments that could appear on her map, as well as two segments that could not appear on her map. The four types of slope are shown in Figure 6.

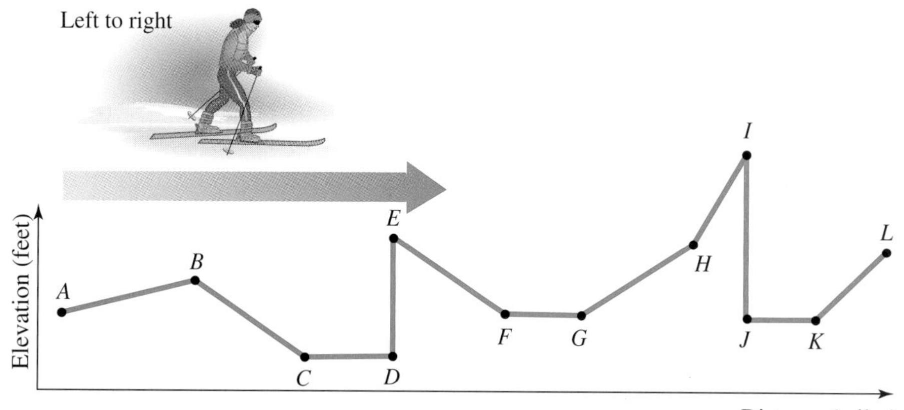

FIGURE 5 Elevation of ski trip

FIGURE 6 Type of slope

a. Use the term *positive*, *negative*, *zero*, or *undefined* to label the type of slope represented by each of the following parts of the "trail" in Figure 5: *AB*, *BC*, *CD*, *DE*, *EF*, *FG*, *GH*, *HI*, *IJ*, *JK*, *KL*.

b. Which two segments in Figure 5 are not likely to appear on a cross country ski trail? Why?

c. Label the slope of each of the following pieces of the trail as *negative*, *zero*, *positive*, or *undefined*:

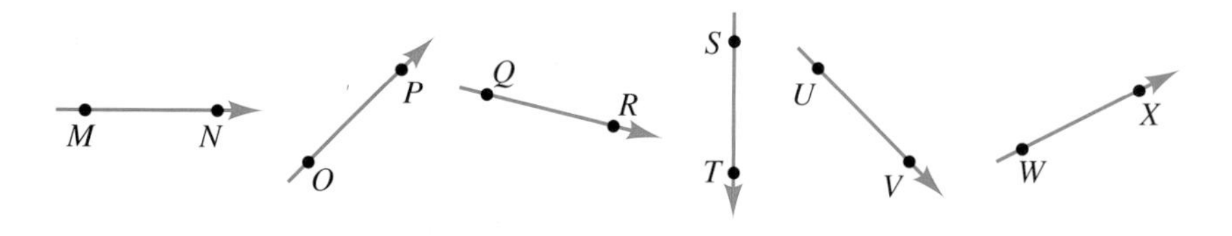

SOLUTION See the Answer Box. ◀

Finding Slope Visually from a Graph

As we saw in Example 1, the slope, or rate of change, describes the steepness and direction of a line. As we trace from left to right along a graph, the *vertical change relative to the horizontal change* defines the **slope** of the line.

FINDING SLOPE FROM A GRAPH

To find slope from a graph, divide the vertical change between two points by the horizontal change between the same points.

- A line with positive slope rises from left to right.
- A line with negative slope falls from left to right.

The definition lets us find a number to describe slope.

In Example 1, we used two letters to name a line segment. Look for this notation in other examples and exercises.

▶ **EXAMPLE 2** Finding the slope from a graph Use the graph in Figure 7 to find these slopes.

a. The slope of BC, the line segment from B to C

b. The slope of BD, the line segment from B to D

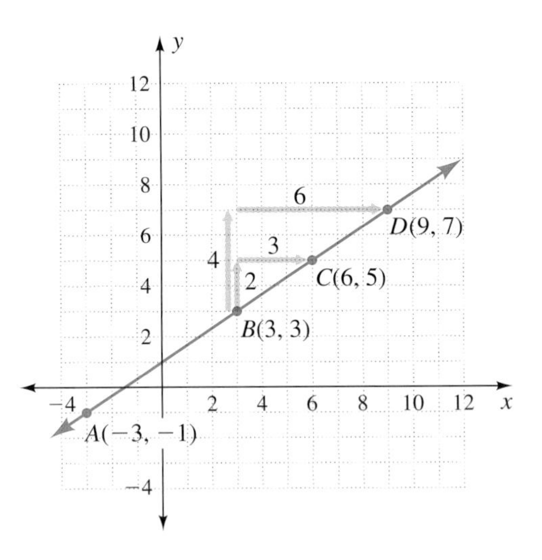

FIGURE 7 Slope on a graph

SOLUTION Both BC and BD have positive slope. To find the vertical change, we count the vertical distance between the points. To find the horizontal change, we count the horizontal distance between the points. The triangles drawn on the line with arrows show the vertical and horizontal change for each pair of points.

a. Segment BC rises 2 units for 3 units from left to right. The slope is

$$\frac{\text{Vertical change}}{\text{Horizontal change}} = \frac{2}{3}$$

b. Segment BD rises 4 units for 6 units from left to right. The slope is

$$\frac{\text{Vertical change}}{\text{Horizontal change}} = \frac{4}{6} = \frac{2}{3}$$

The slopes for the two segments are equal. ◀

▶ Example 2 suggests that we can describe slope as *rise over run*. Think about rise over run in Example 3.

▶ **EXAMPLE 3** Finding the slope from a graph Find the slope of EF from the graph in Figure 8.

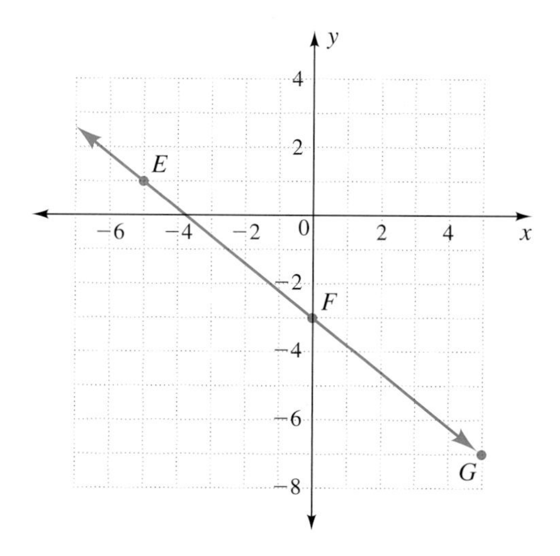

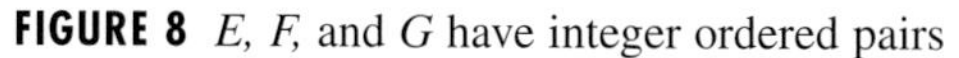

FIGURE 8 E, F, and G have integer ordered pairs

SOLUTION Line EF has a negative slope. The vertical change is 4 units, from 1 to -3, as shown in Figure 9.

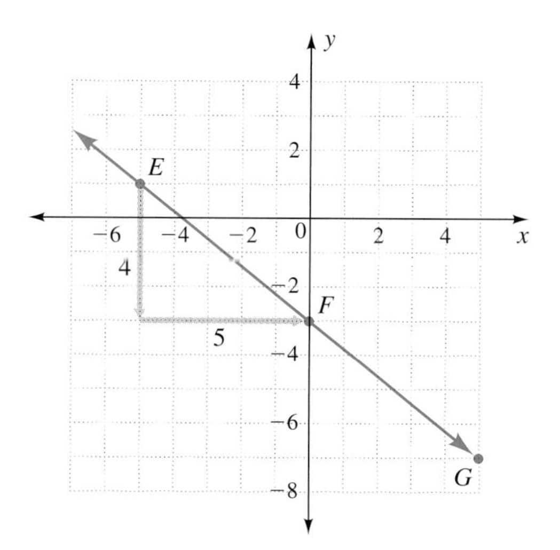

FIGURE 9 Vertical and horizontal change between E and F

Note that here the "rise" is down. The *horizontal change*, or **run**, is 5 units, from -5 to 0. The slope is

$$\frac{\text{Vertical change}}{\text{Horizontal change}} = \frac{\text{rise}}{\text{run}} = -\frac{4}{5}$$ ◀

For a negatively sloped line, the *vertical change*, or **rise**, is down. We associate negative slopes with decreasing values as we move from left to right.

NEGATIVE SLOPE

> An equation with a negative slope decreases as we trace its graph from left to right.

Finding Slope Symbolically from Ordered Pairs

In Example 4, we check our results from Example 3 with a second method, using a formula to find the slope from the ordered pairs for E and F.

SLOPE FORMULA

Student Note:
Subscripts mean that two different ordered pairs are described.

For ordered pairs (x_1, y_1) and (x_2, y_2),

$$\text{Slope} = \frac{y_2 - y_1}{x_2 - x_1}$$

Generally, we let the ordered pair (x_1, y_1) be the point on the left and the ordered pair (x_2, y_2) be the point on the right.

▶ **EXAMPLE 4** Finding slope from ordered pairs Find the slope of EF, using the ordered pairs $E(-5, 1)$ and $F(0, -3)$.

SOLUTION We let $(x_1, y_1) = (-5, 1)$ and $(x_2, y_2) = (0, -3)$.

$$\text{Slope } AB = \frac{y_2 - y_1}{x_2 - x_1} = \frac{-3 - 1}{0 - (-5)} = \frac{-4}{5}$$ ◀

Student Note:
Remember that the negative sign in a fraction may be placed anywhere: in front, in the numerator, or in the denominator.

$$-\frac{a}{b} = \frac{-a}{b} = \frac{a}{-b}$$

THINK ABOUT IT 1: Does the slope depend on the order of the points? Repeat Example 4, but let $(x_1, y_1) = (0, -3)$ and let $(x_2, y_2) = (-5, 1)$.

Finding Slope Numerically from a Table

A third method of finding slope is from a table. In Example 5, we find slope by first calculating the change, or difference, between the numbers in each column of the table and then dividing to obtain the slope. The results of the subtractions are listed in new columns headed Δx for "change in x" and Δy for "change in y."

In a table for a linear function, the slope, $\frac{\Delta y}{\Delta x}$, is constant.

▶ **EXAMPLE 5** Finding slope from a table Table 3 contains several ordered pairs from an equation. Complete the Δx and Δy columns in the table and find the slope for each pair of changes in x and y. State whether the table represents a linear equation.

TABLE 3

Δx	Input, x	Output, y	Δy	Slope
	-3	-1		
	3	3		
	6	5		
	9	7		

SOLUTION We complete the Δx and Δy columns by subtracting to find the change between entries in the table, as shown in Table 4.

TABLE 4

Δx	Input, x	Output, y	Δy	Slope
	-3	-1		
6	3	3	4	$\frac{4}{6} = \frac{2}{3}$
3	6	5	2	$\frac{2}{3}$
3	9	7	2	$\frac{2}{3}$

The slope of the line containing all the ordered pairs in the table is $\frac{2}{3}$. The table describes a linear equation. ◀

As mentioned earlier, the small triangle used in the table is the capital Greek letter *delta*. **Delta**, Δ, indicates *a change in the value of the variable that follows it.*

In Example 2 and Example 5, different ordered pairs on the same line gave the same slope. This suggests two conclusions:

- The slope of a linear function is constant.
- If the slope between points A and B is the same as the slope between points A and C, then points A, B, and C all lie on the same line.

▶ **EXAMPLE 6** Showing that three points lie on a line Do $E(-5, 1)$, $F(0, -3)$, and $G(5, -7)$ lie on the same line?

SOLUTION We must find the slope of any two segments—say, EF and EG. Let $(x_1, y_1) = (-5, 1)$ and $(x_2, y_2) = (0, -3)$.

$$\text{Slope } EF = \frac{y_2 - y_1}{x_2 - x_1} = \frac{-3 - 1}{0 - (-5)} = -\frac{4}{5}$$

Let $(x_1, y_1) = (-5, 1)$ and $(x_2, y_2) = (5, -7)$.

$$\text{Slope } EG = \frac{y_2 - y_1}{x_2 - x_1} = \frac{-7 - 1}{5 - (-5)} = -\frac{8}{10} = -\frac{4}{5}$$

The segments have point E in common and have slope $-\frac{4}{5}$. The points lie on the same line. (See Figure 9 to check the results.) ◀

You may find the arithmetic easier if you place the three ordered pairs in a table and use the method shown in Example 5.

▶ Special Slopes: Horizontal and Vertical Lines

In Example 7, we use ordered pairs to find the slopes of horizontal and vertical lines (checking our observations about ski trail elevations in Example 1).

▶ **EXAMPLE 7** Finding slopes of horizontal and vertical lines Use the graph in Figure 10 to find these slopes.

a. The slope of the horizontal line containing A and B

b. The slope of the vertical line containing B and C

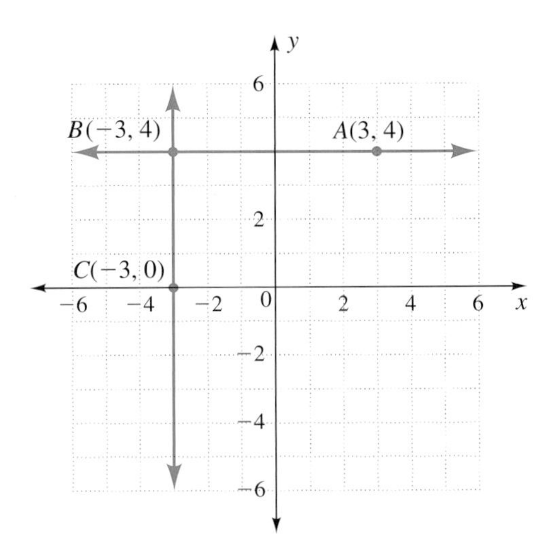

FIGURE 10 Horizontal and vertical Lines

SOLUTION **a.** Segment BA connects $(-3, 4)$ and $(3, 4)$.

$$\text{Slope } BA = \frac{y_2 - y_1}{x_2 - x_1} = \frac{4 - 4}{3 - (-3)} = \frac{0}{6} = 0$$

A horizontal line has **zero slope**.

b. Segment CB connects $(-3, 0)$ and $(-3, 4)$.

$$\text{Slope } CB = \frac{y_2 - y_1}{x_2 - x_1} = \frac{4 - 0}{-3 - (-3)} = \frac{4}{0}$$

The zero in the denominator means division by zero. Since division by zero is not defined, we say that a vertical line has **undefined slope**. ◀

THINK ABOUT IT 2: Why are vertical lines not functions?

Applications

In Example 8, we use slope to describe the steepness of a roof. The A-frame house in Figure 11 is popular in climates with lots of snow. The slope of the roof is quite large so that snow does not pile up on the roof. In hot climates, where there is no snow, the roof is relatively flat (Figure 12). Because we are given the measurements of the roof instead of ordered pairs, *rise over run* is the appropriate way to find the slope of a roof. The *rise* is the vertical distance; the *run* is half the horizontal distance across the bottom of the roof.

FIGURE 11 A-frame roof

FIGURE 12 Tropical roof

▶ **EXAMPLE 8** Finding the meaning of slope: building roofs

a. What is the slope of the roof in Figure 11?
b. What is the slope of the roof in Figure 12?
c. What units describe the slope?
d. Is a positive or negative slope appropriate for a roof?

SOLUTION **a.** Slope $= \dfrac{\text{rise}}{\text{run}} = \dfrac{24 \text{ ft}}{12 \text{ ft}} = \dfrac{2}{1}$

b. Slope $= \dfrac{\text{rise}}{\text{run}} = \dfrac{2 \text{ ft}}{12 \text{ ft}} = \dfrac{1}{6}$

c. The units for the slope of a roof are feet over feet.

d. Because one side of the roof has a positive slope and the other a negative slope, we usually omit the positive and negative signs. ◀

Because slope is the change in y divided by the change in x, *the meaning of slope in applications is the units for y divided by the units for x.*

▶ **EXAMPLE 9** Finding the meaning of slope: photocopy card A prepaid photocopy machine card costs \$15.00. Each photocopy is \$0.15. Table 5 lists the value remaining on the card after x copies are made. The letters in the table match positions on the graph in Figure 13.

a. Find the slope between points A and B.
b. Find the slope between points C and D.
c. What is the meaning of the slope? Why is it negative?

TABLE 5 Photocopy Card Value

Copies, x	Card Value, y	Points on Graph
0	\$15.00	
10	13.50	*A*
20	12.00	*B*
30	10.50	*C*
40	9.00	
50	7.50	
60	6.00	*D*

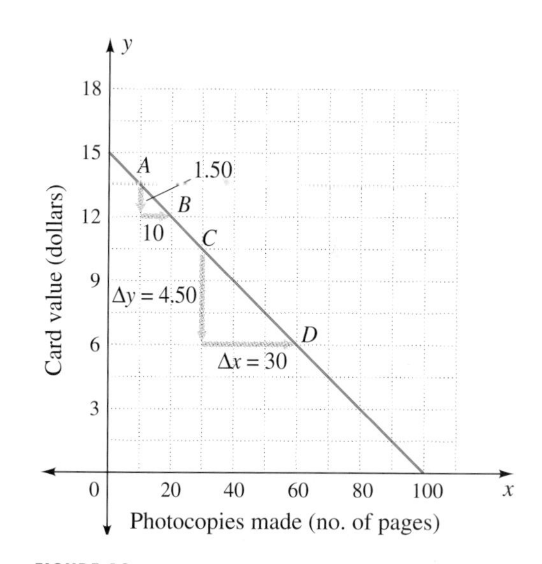

FIGURE 13 Photocopy card value

SOLUTION **a.** Slope $AB = \dfrac{y_2 - y_1}{x_2 - x_1} = \dfrac{\$12.00 - \$13.50}{(20 - 10) \text{ copies}} = \dfrac{-\$1.50}{10 \text{ copies}} = -\$0.15$ per copy

b. Slope $CD = \dfrac{y_2 - y_1}{x_2 - x_1} = \dfrac{\$6.00 - \$10.50}{(60 - 30) \text{ copies}} = \dfrac{-\$4.50}{30 \text{ copies}} = -\$0.15$ per copy

c. The vertical axis is labeled with card value in dollars. The horizontal axis is labeled with the number of photocopies. The units for slope are $\dfrac{\text{dollars}}{\text{number of copies}}$, or $-\$0.15$ per copy. The slope is negative because the vertical axis is labeled with the value of the card and the card value decreases with each additional photocopy made. ◀

Here is what we found so far about slope:

SUMMARY

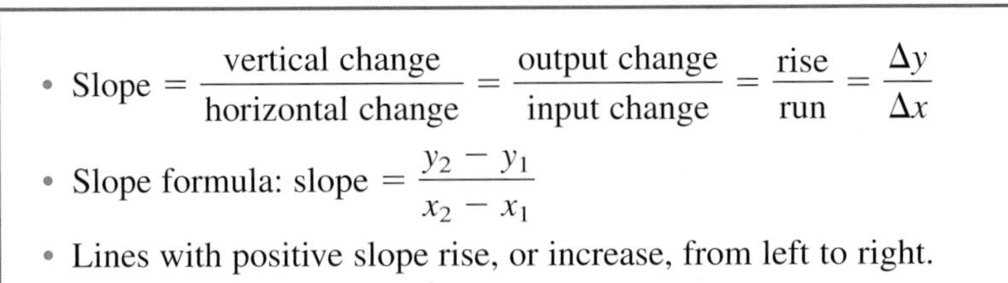

- Slope $= \dfrac{\text{vertical change}}{\text{horizontal change}} = \dfrac{\text{output change}}{\text{input change}} = \dfrac{\text{rise}}{\text{run}} = \dfrac{\Delta y}{\Delta x}$
- Slope formula: slope $= \dfrac{y_2 - y_1}{x_2 - x_1}$
- Lines with positive slope rise, or increase, from left to right.
- Lines with negative slope fall, or decrease, from left to right.
- Horizontal lines have zero slope.
- Vertical lines have undefined slope.
- The units for slope are the units on the vertical axis divided by the units on the horizontal axis.

ANSWER BOX

Warm-up: 1. $-\frac{4}{5}$ **2.** undefined **3.** 0 **4.** -0.15 or $-\frac{3}{20}$ **5.** -0.15 or $-\frac{3}{20}$ **Example 1: a.** positive, negative, zero, undefined, negative, zero, positive, positive, undefined, zero, positive **b.** *DE* and *IJ*; they are cliffs **c.** zero, positive, negative, undefined, negative, positive **Think about it 1:** No; $\dfrac{y_2 - y_1}{x_2 - x_1} = \dfrac{1 - (-3)}{-5 - 0} = \dfrac{4}{-5} = -\dfrac{4}{5}$. **Think about it 2:** A vertical line does not pass the vertical-line test; for each x, there are many y values.

▶ 4.3 Exercises

In Exercises 1 to 6, tell whether the slope of each line will be positive, negative, zero, or undefined. Use the rise and run in the graph to find the slope of each line.

1.

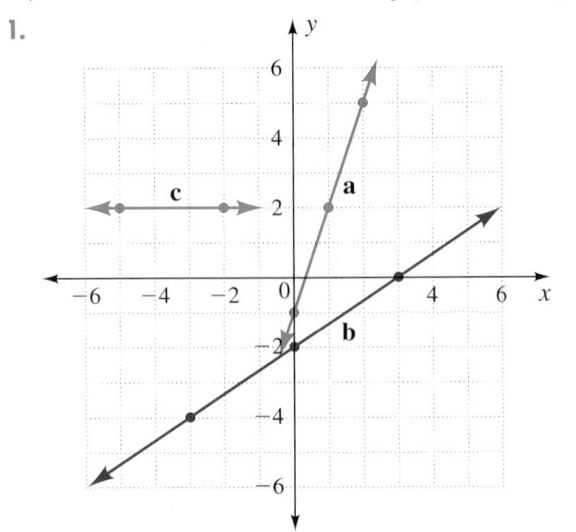

2.

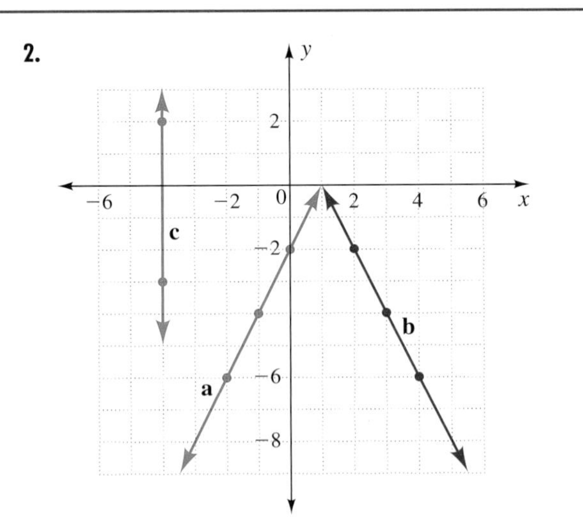

Blue numbers are core exercises.

3.

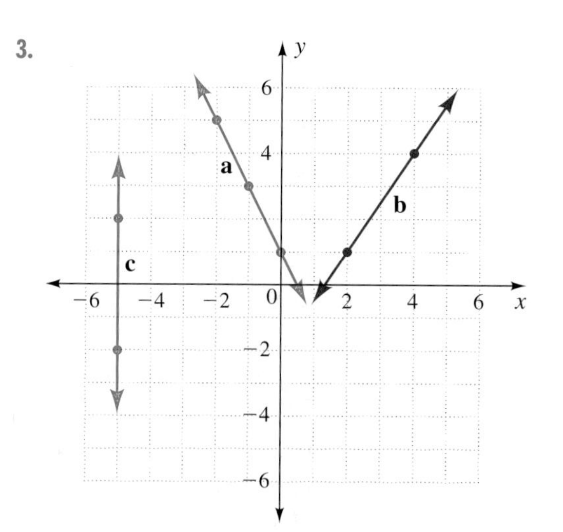

4.

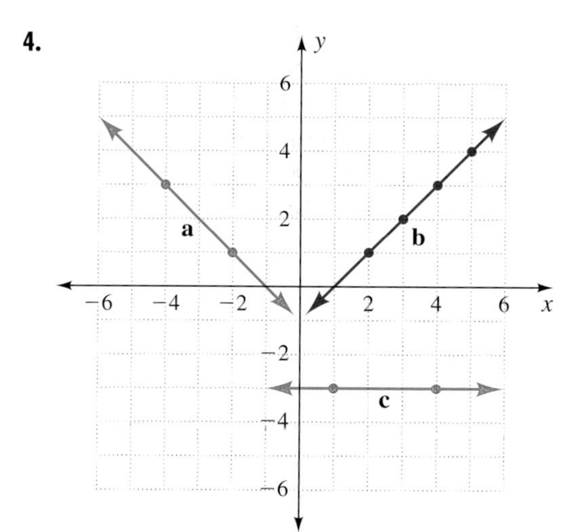

5.

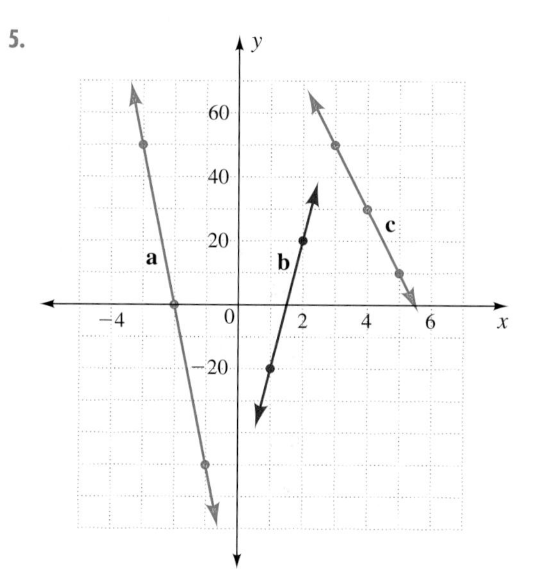

6.

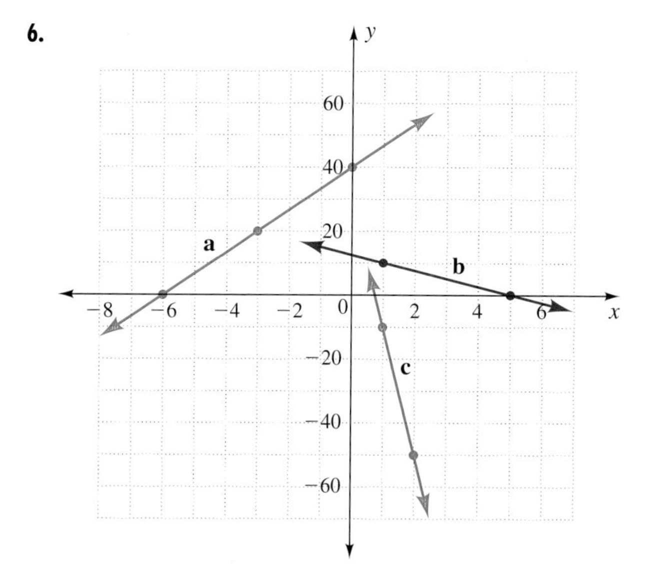

In Exercises 7 to 22, use the slope formula to find the slope. (*Hint:* To learn the formula, write the complete formula each time you use it.)

7. (0, 2) and (4, 3)

8. (2, 0) and (4, 3)

9. (−2, 3) and (0, −4)

10. (−3, 2) and (4, 0)

11. (4, 3) and (4, 4)

12. (4, 3) and (3, 3)

13. (0, 2) and (−2, 2)

14. (4, 4) and (−4, 4)

15. (−2, 3) and (4, −1)

16. (3, −4) and (−5, 2)

17. (2, −3) and (4, −3)

18. (−3, −2) and (−3, 5)

19. (0, 4) and (5, 0)

20. (−3, 0) and (0, 2)

21. (3, −4) and (3, 4)

22. (3, −2) and (−3, −2)

In Exercises 23 to 38, tell whether the slope of the graph formed from the table will be positive, negative, zero, or undefined. State whether the table represents a linear equation. For linear equations, find the slope using Δx and Δy. State the units on the slope, if any.

23.

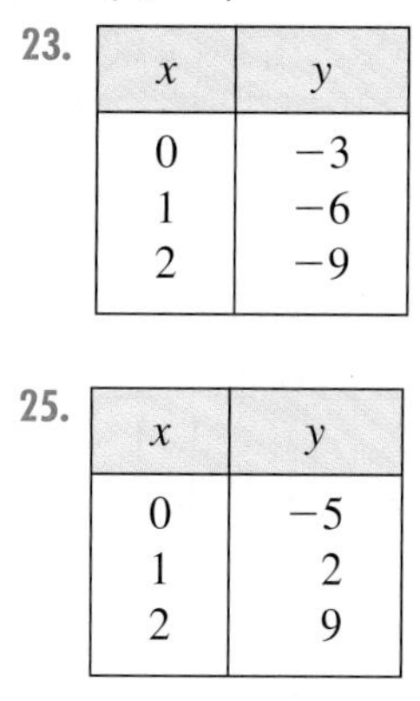

x	y
0	−3
1	−6
2	−9

24.

x	y
0	−9
1	−4
2	1

25.

x	y
0	−5
1	2
2	9

26.

x	y
0	2
1	8
2	14

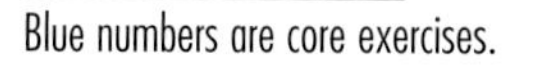

Blue numbers are core exercises.

27.

Hours	Earnings
2	$18
4	36
6	54
8	72

28.

Cookies	Calories
12	900
18	1350
24	1800
30	2250

29.

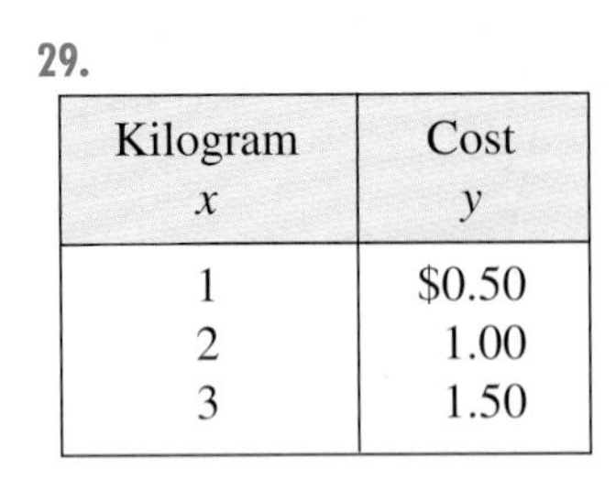

Kilogram x	Cost y
1	$0.50
2	1.00
3	1.50

30.

Gallons	Cost
2	$3.00
3	4.50
4	6.00
5	7.50

31.

Pounds x	Cost y
1	$0.32
3	0.96
5	1.60

32.

Credit	Cost
1	$ 24
5	120
8	192
10	240

33.

Time (sec)	Distance (ft)
0	0
1	16
2	64
3	144

34.

Length (ft)	Width (ft)
5	10
8	7
9	6
12	3

35.

Copies	Value
0	$15.00
10	12.50
20	10.00
25	8.75

36.

Rides	Value
0	$20.00
2	16.50
4	13.00
10	2.50

37.

Time (hr)	Distance (mi)
1	40
2	70
3	90
4	100

38.

Radius (ft)	Area (sq ft)
1	3.14
2	12.57
3	28.27
4	50.27

Use the slope formula to find the slope in Exercises 39 to 44.

39. (a, b) and (c, d)

40. $(0, 0)$ and (m, n)

41. $(a, 0)$ and $(0, b)$

42. (m, n) and (p, q)

43. (a, b) and (a, c)

44. (a, b) and (c, b)

For Exercises 45 to 48, find the slope of the roof or roof support.

45.

46.

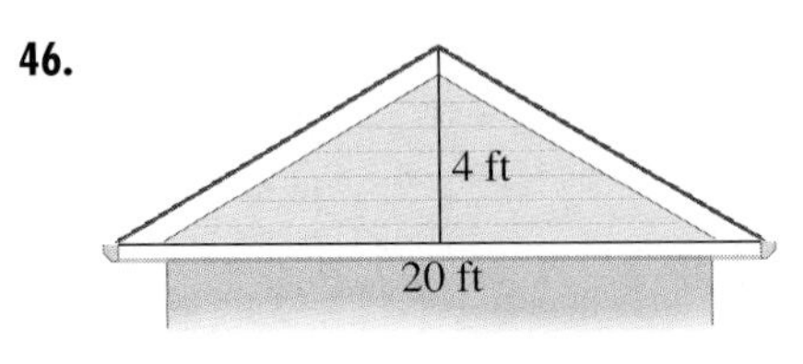

47.

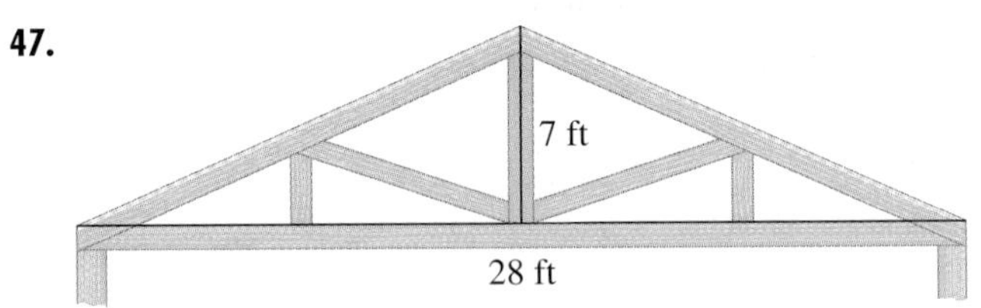

48.

For the applications in Exercises 49 to 54, do the following:

(a) Make a table for inputs 0, 1, and 2.
(b) Find the slope and write the meaning of the slope.
(c) Write an equation describing the output as a function of the input. Define input and output variables as needed.

49. The output is the total cost of g gallons of water at $1.55 per gallon.

50. The output is the total calories in n cookies at 65 calories per cookie.

51. The output is the total earnings for h hours at $6.25 per hour.

Blue numbers are core exercises.

52. The output is the total miles traveled in h hours at 55 miles per hour.

53. The output is the total kilometers traveled in h hours at 80 kilometers per hour.

54. The output is the total cost for x credit hours at \$100 per credit hour.

55. Jane purchased a pick-up for \$30,000. The value of the pick-up over several years is shown in the table.

Year, x	Value, y
0	\$30,000
1	24,000
2	21,600
3	1000

a. Draw a graph of the data.

b. What is the slope from year 0 to year 1?

c. What is the slope from year 1 to year 2?

d. What would be the meaning of a line drawn between (2, 21,600) and (2, 1000)?

e. What is the slope of the line from (2, 1000) to (3, 1000)?

f. Describe a likely story about Jane's pick-up.

In Exercises 56 to 61, arrange the slopes from flattest to steepest.

56. $\frac{4}{3}, \frac{3}{4}, 1$

57. $-1, -\frac{3}{1}, -\frac{1}{3}$

58. $-\frac{1}{2}, 0, -\frac{2}{1}$

59. $\frac{2}{5}, \frac{6}{5}, \frac{1}{2}$

60. $1, \frac{3}{4}, \frac{3}{2}, \frac{3}{5}$

61. $-2, -\frac{1}{2}, -\frac{2}{3}, -\frac{3}{2}$

62. Error Analysis Explain what is wrong with using this expression to find the slope between (a, b) and (c, d):
$$\frac{b-a}{d-c}.$$

63. Explain what is wrong with using this expression to find the slope between (a, b) and (c, d):
$$\frac{c-a}{d-b}.$$

64. In calculating the slope between (8, 2) and (4, 5), one student started with (5 − 2) divided by (4 − 8). Another student started with (2 − 5) divided by (8 − 4). Will they both obtain the correct slope? Explain why or why not.

▶ Writing

65. What does the slope tell you about the graph of a line?

66. Explain how to tell if a line has a positive slope.

67. Explain how to tell if a line drawn from a table will have a negative slope.

68. Explain how to find slope given two ordered pairs.

69. Explain how to find slope from a graph.

70. Explain how the units on the axes give meaning to the slope of a line.

71. Linked Example Find the slope between these sets of points on the three parts of the graph of the credit card cash advance fee schedule.

a. (10, 5), (50, 5)

b. (600, 18), (1000, 30)

c. (2600, 75), (3000, 75)

Explain why the slopes make sense in the problem setting. (See the schedule and graph on page 43 if needed.)

We return to slope and staircases in Section 5.1. This project makes a good preview to that section.

▶ Project

72. Staircase Slope On a staircase, the riser is the vertical distance (rise) on each step and the tread is the horizontal distance (run) between two risers. Some steps have a slight overhang, which must be ignored in measuring the tread.

a. From memory, estimate the rise to run ratio for steps in a nearby staircase. Look at a ruler in making your estimates. Record your estimate first as a fraction and then as a decimal.

b. Measure and record the riser and the tread.

c. Write the slope for the stairs first as a fraction and then as a decimal.

d. Use your results in part b to estimate the total rise and run of the staircase.

e. Comment on how your estimate compares with the actual measurements.

Extension: Find five other staircases, in locations such as an office building, a tourist attraction, and a concert hall. Measure the riser and the tread, and calculate the slope of the stairs. Discuss the relationship between the steepness of the stairs and the purpose of the stairs.

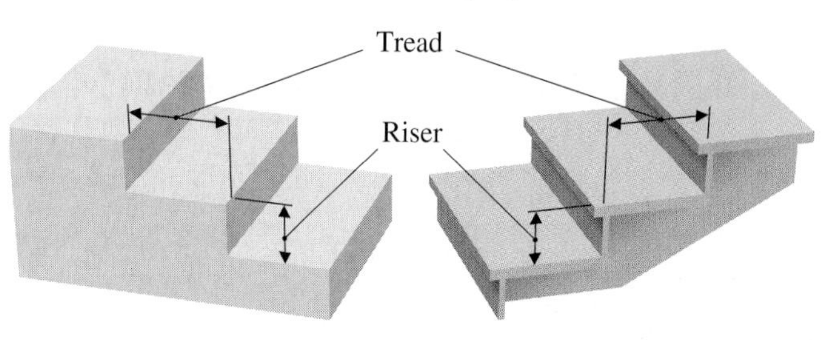

Blue numbers are core exercises.

▶ 4 Mid-Chapter Test

In Exercises 1 and 2, write each formula in words, using the phrase *in terms of*. The meanings of the variables are given following the formula.

1. $C = \frac{5}{9}(F - 32)$, Celsius and Fahrenheit temperatures

2. $V = \frac{4}{3}\pi r^3$, volume and radius of a sphere

In Exercises 3 and 4, solve the equations for *b*.

3. $3 = 4(-2) + b$ **4.** $-4 = \frac{2}{3}(-6) + b$

In Exercises 5 to 7, solve the formulas for the indicated variable.

5. Temperature scales: $C = K - 273$ for K

6. Distance seen on the moon from a height of h feet: $d^2 = \dfrac{3h}{8}$ for h

7. Last term in an arithmetic sequence: $l = a + (n - 1)d$ for d

8. Answer the following questions for the set of ordered pairs (2, 3), (3, 3), (4, 3), (5, 4), (6, 4), (7, 4).

a. Does the set describe a function?

b. What is the set of inputs (domain)?

c. What is the set of outputs (range)?

9. a. Sketch a graph of the set of ordered pairs (2, 3), (2, 4), (2, 5).

b. Complete this statement: A relationship is a function if for each input there is _____ output.

c. Explain, in terms of inputs and outputs, why the set of ordered pairs in part a is not a function.

d. Show whether or not the set of ordered pairs in part a passes the vertical-line test.

10. Find $f(6)$ for

a. $f(x) = 3x$ **b.** $f(x) = \frac{1}{2}x + 2$

c. $f(x) = \frac{1}{2}(x + 2)$ **d.** $f(x) = x^2 - 2$

11. Graph $f(x) = 3 - 4x$.

12. Find the slope between the points $(-3, 8)$ and $(4, 6)$.

13. Find the slope between the points $(-3, 8)$ and $(4, 8)$.

14. Find the slope and give its meaning.

Episode Number	Number of Survivors
1	25
2	23
3	21

15. Find the slope for each side of the absolute value graph.

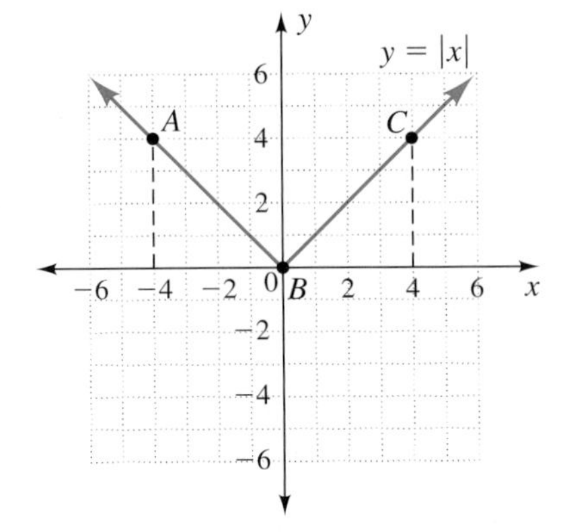

16. The slopes of the four lines in the figure are -2, $\frac{-1}{2}$, $\frac{3}{4}$, and $\frac{5}{4}$. Match each line with its slope.

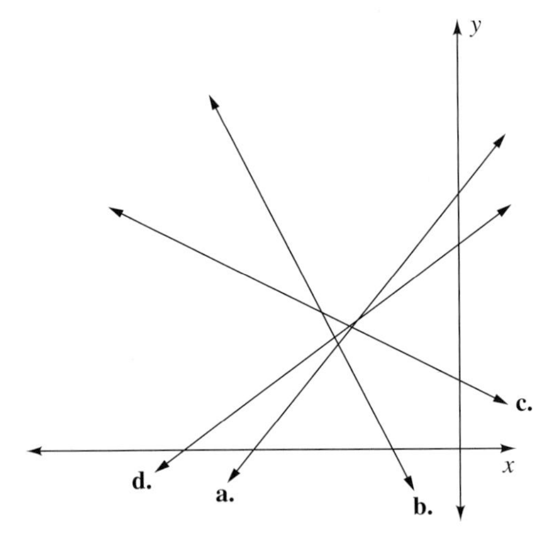

▶ 4.4 Linear Equations: Intercepts and Slope

Objectives

- Find horizontal-axis and vertical-axis intercepts from an equation.
- Find the slope and vertical-axis intercept from a linear equation.
- Interpret the meaning of intercepts in a problem setting.

- Write the equations for horizontal and vertical lines.
- Use slope and a point to graph a line.

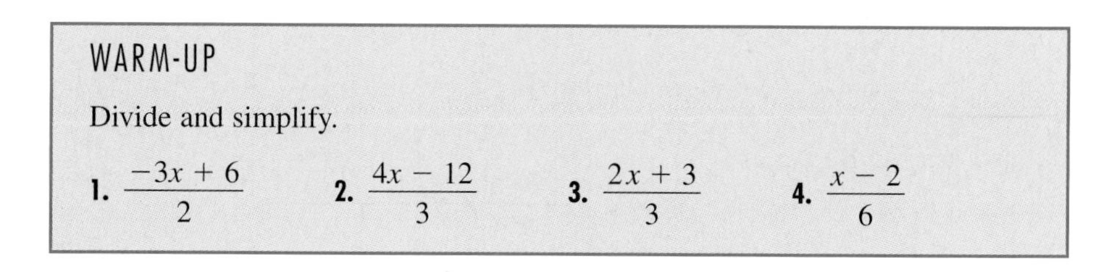

WARM-UP

Divide and simplify.

1. $\dfrac{-3x + 6}{2}$ **2.** $\dfrac{4x - 12}{3}$ **3.** $\dfrac{2x + 3}{3}$ **4.** $\dfrac{x - 2}{6}$

IN THIS SECTION, we name the intersections of a line and the axes. We explore what the equation of a line tells us about the intersections and the slope. We find the equations of horizontal and vertical lines and graph equations, given various types of information.

Special Intersections: Graphs with the Axes

Student Note:
We use the letter a here, so in the linear equation $y = ax + b$ we will need to replace a with another letter, traditionally m, giving $y = mx + b$. See page 219.

When a graph intersects one or both of the axes, the ordered pairs $(a, 0)$ and $(0, b)$ have special significance. We consider these pairs now.

The **horizontal-axis intercept point**, defined by the pair $(a, 0)$, is *the point where a graph crosses the horizontal axis* (see Figure 14). Because we commonly label the horizontal, or input, axis with x, $(a, 0)$ is also called the ***x*-intercept point**. The ***x*-intercept** is *the number a*. Remember, the x-intercept point is where the output is zero: $y = 0$.

The **vertical-axis intercept point**, defined by the pair $(0, b)$, is *the point where a graph crosses the vertical axis* (see Figure 14). Because we commonly label the vertical, or output, axis with y, $(0, b)$ is also called the ***y*-intercept point**. The ***y*-intercept** is *the number b*. Remember, the y-intercept is where $x = 0$.

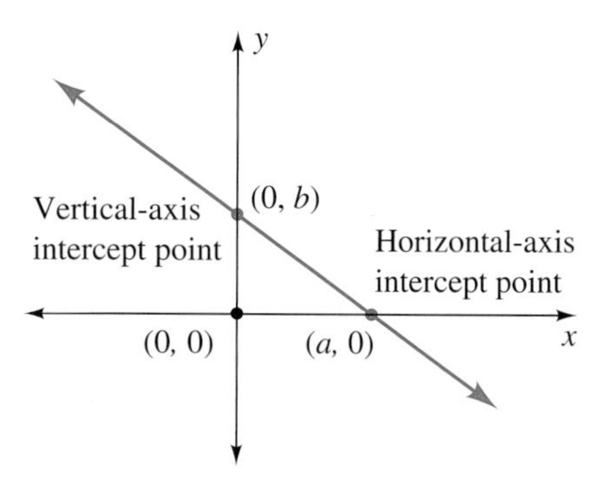

FIGURE 14 Intercepts on axes

▶ **EXAMPLE 1** Solving equations for horizontal and vertical intercepts

a. Make a table and graph for $3x + 2y = 12$.

b. Solve $3x + 2y = 12$ for its horizontal-axis intercept, using the table and graph. Check by solving with algebraic notation.

c. Solve $3x + 2y = 12$ for its vertical-axis intercept, using the table and graph. Check by solving with algebraic notation.

SOLUTION **a.** The table appears in Table 6 and the graph in Figure 15.

TABLE 6 $3x + 2y = 12$

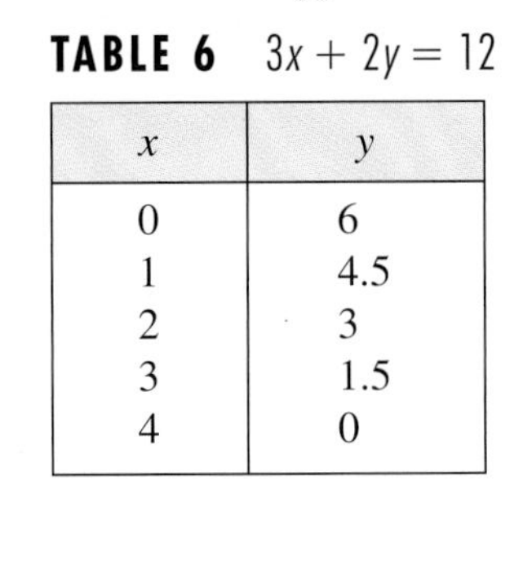

x	y
0	6
1	4.5
2	3
3	1.5
4	0

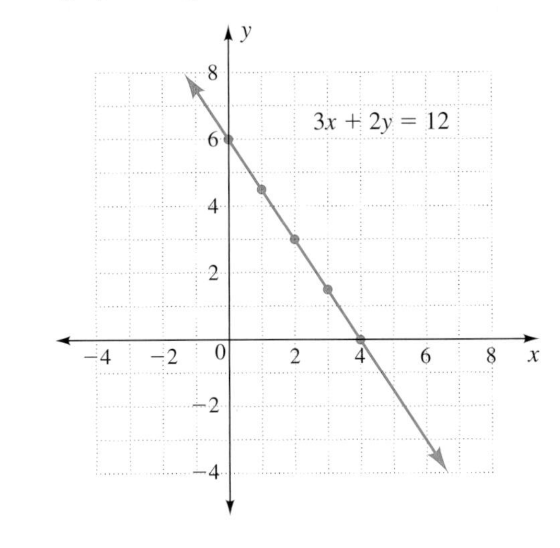

FIGURE 15 Intercepts for $3x + 2y = 12$

Student Note:
The comments tell what to do next.

b. *To find the horizontal-axis intercept*, we need the ordered pair $(a, 0)$. Looking in the table for $y = 0$, we find $x = 4$. On the graph, we look for the point of intersection of the graph of $3x + 2y = 12$ with the horizontal axis, or x-axis. The intersection is $(4, 0)$. Again, if $y = 0$, then $x = 4$.

With algebraic notation, we have

$$3x + 2y = 12 \quad \text{Let } y = 0.$$

$$3x + 2 \cdot 0 = 12 \quad \text{Divide both sides by 3.}$$

$$\frac{3x}{3} = \frac{12}{3} \quad \text{Simplify.}$$

$$x = 4$$

Check: $3(4) + 2(0) \stackrel{?}{=} 12$ ✓

Student Note:
The comments tell what to do next.

c. *To find the vertical-axis intercept*, we need the ordered pair $(0, b)$. Looking in the table for $x = 0$, we find $y = 6$. On the graph, we look for the point of intersection of the graph of $3x + 2y = 12$ with the vertical axis. If $x = 0$, then $y = 6$.

With algebraic notation, we have

$$3x + 2y = 12 \quad \text{Let } x = 0.$$

$$3(0) + 2y = 12 \quad \text{Divide both sides by 2.}$$

$$\frac{2y}{2} = \frac{12}{2} \quad \text{Simplify.}$$

$$y = 6$$

Check: $3(0) + 2(6) \stackrel{?}{=} 12$ ✓ ◀

Linear Equations: Slope and *y*-intercept

In the linear equation $y = mx + b$, you may already understand what m and b tell us about the graph of the equation. Use Examples 2 and 3 to explore and confirm your understanding. In Example 2, we multiply x by three different numbers and look at the effects on the graph.

▶ **EXAMPLE 2** Exploring equations and graphs

a. Which line in Figure 16 is the graph of each of the following: $y = x$, $y = 2x$, and $y = 5x$?

b. What is the slope of each graph?

c. Compare the steepness of the graphs.
d. What is the y-intercept point for each graph?

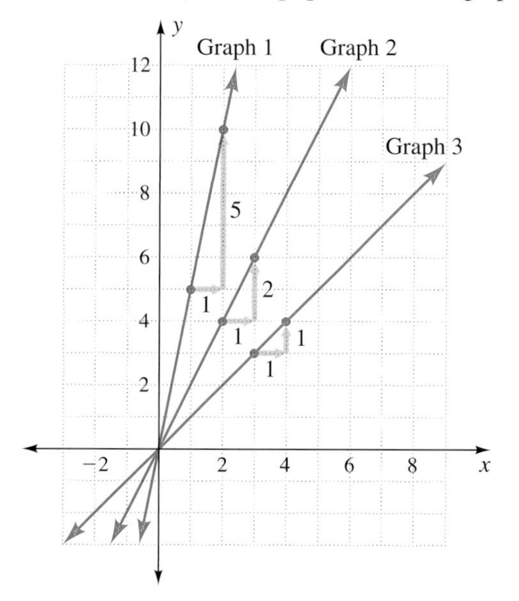

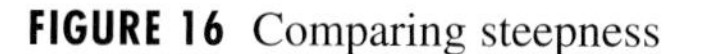

FIGURE 16 Comparing steepness

SOLUTION **a.** $y = x$ is graph 3, $y = 2x$ is graph 2, and $y = 5x$ is graph 1. The answers to b, c, and d are in the Answer Box. ◀

Slope and the Linear Equation

> Except for a vertical line, the linear equation always contains a multiplication of the input variable by the slope. The slope is represented by the letter m in $y = mx + b$. The slope constant is $m = \dfrac{\Delta y}{\Delta x}$.

The source of the number b is most easily seen by comparing graphs of several linear equations with the same slope.

▶ EXAMPLE 3 Exploring equations and graphs

a. Which line in Figure 17 is the graph of each of the following: $y = 2x$, $y = 2x + 1$, and $y = 2x - 3$?
b. What is the y-intercept of each equation?
c. What is the source of the number b in $y = mx + b$?

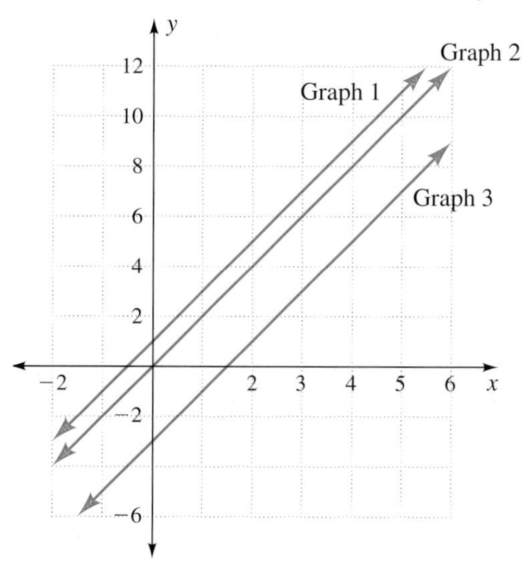

FIGURE 17 Comparing y-intercepts

SOLUTION **a.** $y = 2x$ is graph 2, $y = 2x + 1$ is graph 1, and $y = 2x - 3$ is graph 3. See the Answer Box for the other answers. ◀

We can summarize our findings about a linear equation as follows:

SLOPE AND y-INTERCEPT IN A LINEAR EQUATION

> The nonvertical, linear equation is $y = mx + b$, where the number m is the slope and the number b is the vertical-axis intercept.

Because the vertical axis is frequently labeled y, we usually call b the y-intercept rather than the vertical-axis intercept.

▶ Find the Slope and y-intercept from Linear Equations

Some equations must first be solved for y in order to find the slope and y-intercept. In Example 4, we read the slope and y-intercept from equations in $y = mx + b$ form.

▶ **EXAMPLE 4** **Finding the slope and y-intercept from an equation** Find the slope and y-intercept for each equation, labeling your answers with m and b. First write the equation in $y = mx + b$ form, if it is not already in this form.

a. $y = -2x + 4$
b. $y = 2 - 4x$
c. $2y + 3x = 6$
d. $4x - 3y = 12$
e. $y = \$15.00 - \$0.15x$
f. $y = 40 + 25(x - 1)$

SOLUTION **a.** $m = -2, b = 4$

b. $y = -4x + 2$; $m = -4, b = 2$

c. $2y + 3x = 6$ Subtract $3x$ from both sides.

$2y = -3x + 6$ Divide both sides by 2.

$y = \dfrac{-3x + 6}{2}$ Change the right side to $mx + b$.

$y = -\frac{3}{2}x + 3$

$m = -\frac{3}{2}, b = 3$

d. $4x - 3y = 12$ Subtract 12 from both sides.

$4x - 12 - 3y = 0$ Add $3y$ to both sides.

$4x - 12 = 3y$ Divide both sides by 3.

$\dfrac{4x - 12}{3} = y$ Change the left side to $mx + b$.

$\frac{4}{3}x - 4 = y$

$m = \frac{4}{3}, b = -4$

e. $y = -\$0.15x + \$15.00$; $m = -\$0.15$, $b = \$15.00$

f. $y = 40 + 25(x - 1)$ Apply the distributive property.

$y = 40 + 25x - 25$ Combine like terms.

$y = 25x + 15$

$m = 25, b = 15$ ◀

▶ Even if linear equations do not contain x and y, we can still find the slope and vertical-axis intercept. If the output is not identified, look for a variable that is a function of or depends on another variable. The output will equal the slope times the input variable plus the vertical-axis intercept.

▶ EXAMPLE 5 Finding slope and y-intercept from a formula Find the slope and vertical-axis intercept for each formula.

a. $C = 49.95 + 0.89n$ (The truck rental cost depends on the \$49.95 charge plus \$0.89 per mile.)

b. $c = 60 + 96h$ (The plumber's bill includes a \$60 fee to show up and a \$96 charge per hour.)

c. $C = \pi d$ (The circumference of a circle depends on the diameter.)

SOLUTION **a.** The slope is 0.89 or $\frac{89}{100}$ the vertical-axis intercept is 49.95.

b. The slope is 96; the vertical-axis intercept is 60.

c. The slope is π; the intercept is 0. ◀

▶ Using Slope in Graphing Lines

In many applications, we are given the slope and a point and need to draw a line.

> To draw a line with a given slope, graph the point and use the slope to find a second point.

▶ EXAMPLE 6 Drawing lines, given a point and the slope

a. Draw a line through $(-1, 3)$ with a -2 slope.

b. Draw a line through $(2, 1)$ with a $\frac{1}{2}$ slope.

SOLUTION **a.** A slope of -2 is the same as $-\frac{2}{1}$. To draw a line through $(-1, 3)$ with slope $-\frac{2}{1}$, place point A at $(-1, 3)$. Count 2 units down from A and then count 1 unit to the right to mark point B. Draw a line through A and B, as shown in Figure 18.

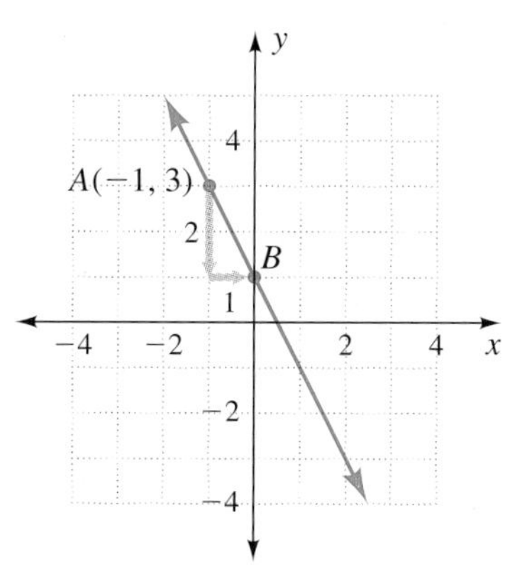

FIGURE 18 Use slope to find second point

b. To draw a line through (2, 1) with slope $\frac{1}{2}$, place point A at (2, 1). Count 1 unit up and 2 units to the right to mark point B. Draw a line through A and B, as shown in Figure 19.

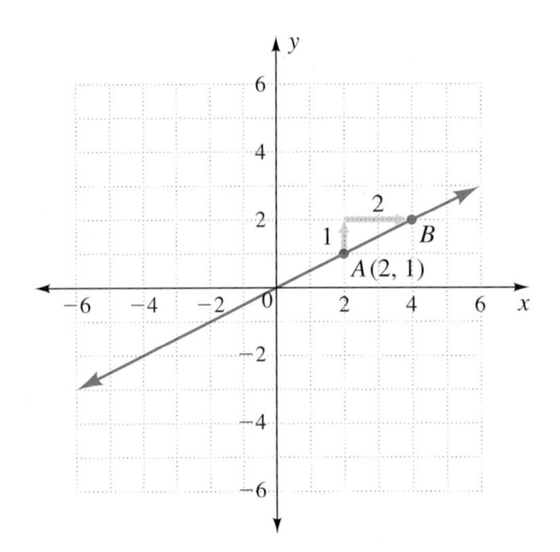

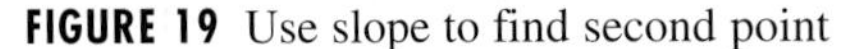

FIGURE 19 Use slope to find second point ◀

USING SLOPE TO GRAPH LINES

> To draw a line given a point (a, b) and a slope, start by graphing (a, b). Place the sign of the slope in the numerator (vertical change). Find a second point by first counting the vertical change (up or down) and then counting the horizontal change (denominator) to the right. Draw the line through the two points.

▶ Special Lines: Horizontal and Vertical

Figure 20 reminds us:

- All horizontal lines have constant outputs.
- All vertical lines have constant inputs.

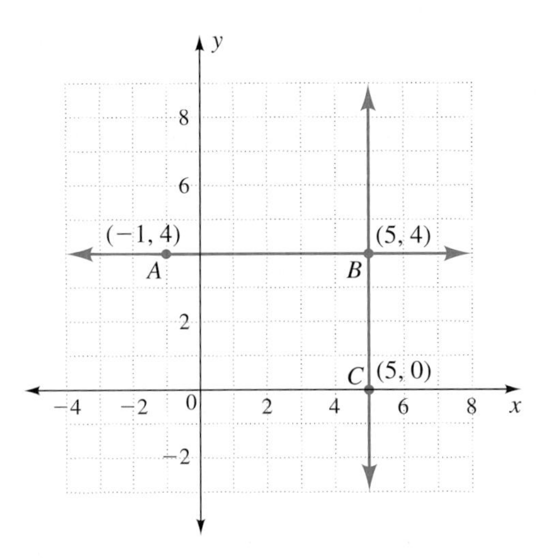

FIGURE 20 Horizontal and vertical lines

In Examples 7 and 8, we find the equations of a horizontal line and a vertical line.

▶ EXAMPLE 7 Finding the equation of a horizontal line Points $A(-1, 4)$ and $B(5, 4)$ are shown in Figure 20.

a. Find the slope of AB.

b. Find the equation of the line through AB.

SOLUTION **a.** The slope is

$$m = \frac{y_2 - y_1}{x_2 - x_1} = \frac{4 - 4}{5 - (-1)} = \frac{0}{6} = 0$$

b. From the graph in Figure 20, the y-intercept is $b = 4$. Using m and b, we find the equation:

$y = mx + b$ Substitute $m = 0$ and $b = 4$.

$y = 0x + 4$ Simplify.

$y = 4$ ◀

▶ EXAMPLE 8 Finding the equation of a vertical line Points $B(5, 4)$ and $C(5, 0)$ are shown in Figure 20.

a. Find the slope of BC.

b. List three other ordered pairs on the line containing BC.

c. Explain what is true about all ordered pairs on the line containing BC.

d. Suggest an equation for the vertical line passing through BC.

SOLUTION **a.** Slope $BC = \frac{y_2 - y_1}{x_2 - x_1} = \frac{0 - 4}{5 - 5} = \frac{-4}{0}$

Because of the division by zero, the slope is undefined.

b. Other points include (5, 1), (5, 2), and (5, 3).

c. All ordered pairs on the line containing BC have $x = 5$.

d. The equation $x = 5$ is a rule that is satisfied by all points on the vertical line through (5, 0). ◀

Note that a vertical line is not a function and cannot be written in the form $f(x) = mx + b$ or $y = mx + b$. The equation $x = 5$ cannot be entered into [Y =] *on the calculator.*

Equations of Horizontal and Vertical Lines

> The equation of a horizontal line is $y = b$, where $(0, b)$ is its y-intercept.
>
> The equation of a vertical line is $x = a$, where $(a, 0)$ is its x-intercept.

▶ Linear Functions and Intercepts

We close the section by returning to functions. Nonvertical straight lines may be described as linear functions.

Linear Functions

> A linear function is any association between x and y that can be written
>
> $y = mx + b$
>
> We may write a linear function as $f(x) = mx + b$, where $y = f(x)$.

Function notation gives us a way to describe how to find the horizontal- and vertical-axis intercepts.

> The horizontal-axis, or x-axis, intercept a is where $y = 0$. To find a, solve $f(x) = 0$; see Figure 21.
>
> The vertical-axis, or y-axis, intercept b is where $x = 0$; see Figure 21. To find b, let $b = f(0)$.

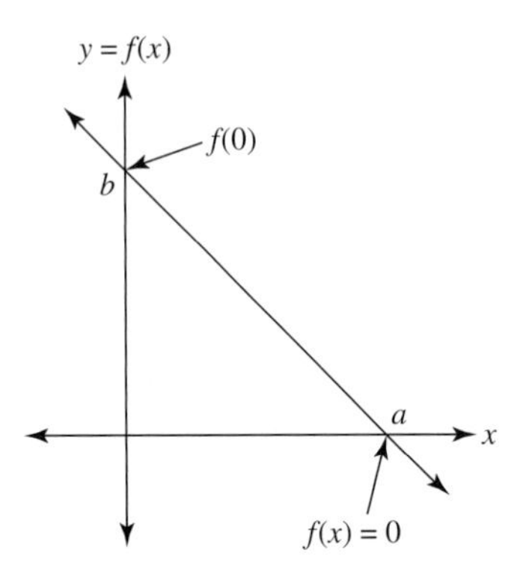

FIGURE 21 Intercepts in function notation

▶ **EXAMPLE 9** Finding intercepts with function notation For the transit ticket function, $f(x) = -2.50x + 20$, use algebra to find

a. the horizontal-axis intercept

b. the vertical-axis intercept

SOLUTION **a.** The horizontal-axis intercept is where $f(x) = 0$:

$$0 = -2.50x + 20 \qquad \text{Add } 2.50x \text{ to both sides.}$$

$$2.50x = 20 \qquad \text{Divide both sides by 2.50.}$$

$$\frac{2.50x}{2.50} = \frac{20}{2.50} \qquad \text{Simplify.}$$

$$x = 8$$

It takes 8 trips to use up the initial \$20 value of the transit ticket.

b. The vertical-axis intercept is at $f(0)$: $f(0) = -2.50(0) + 20 = 20$, the initial value of the ticket. ◀

ANSWER BOX

Warm-up: 1. $-\frac{3}{2}x + 3$ **2.** $\frac{4}{3}x - 4$ **3.** $\frac{2}{3}x + 1$ **4.** $\frac{1}{6}x - \frac{1}{3}$
Example 2: b. The slopes are 1 for $y = x$, 2 for $y = 2x$, and 5 for $y = 5x$. **c.** The graph of $y = x$ is the flattest, and the graph of $y = 5x$ is the steepest. **d.** All three graphs have the origin (0, 0) as the y-intercept point.
Example 3: b. For $y = 2x$, the y-intercept is 0; for $y = 2x + 1$, the y-intercept is 1; and for $y = 2x - 3$, the y-intercept is -3. **c.** In $y = mx + b$, the number b is the y-intercept and $(0, b)$ is the y-intercept point.

▶ 4.4 Exercises

In Exercises 1 to 8, find the x-intercept and the y-intercept point for each equation.

1. $3x + 5y = 15$
2. $5x - 2y = 20$
3. $y = 3x - 24$
4. $y = 5x + 35$
5. $x = 6 - 2y$
6. $x = 9 - 3y$
7. $4x - 3y = 12$
8. $2x + 3y = 18$

9. For each part, list the equations in order from flattest graph to steepest graph.

   a. $y = 2x, y = 4x, y = \frac{1}{2}x$

   b. $y = -3x, y = -x, y = -\frac{1}{3}x$

   c. $y = 3x, y = 1.5x, y = x, y = 2.5x$

   d. $y = -0.25x, y = -4x, y = -0.5x, y = -x$

10. For each part, tell which two equations have the same slope and which two equations have the same y-intercept.

   a. $y = 2x + 1, y = 2x - 1, y = -2x + 1$

   b. $y = 3x + 3, y = -3x - 3, y = -3x + 3$

   c. $y = -\frac{1}{2}x - 1, y = \frac{1}{2}x + 1, y = -\frac{1}{2}x + 1$

In Exercises 11 to 26, find the slope and y-intercept for each equation.

11. $y = 2x - \frac{1}{2}$
12. $y = 3x - \frac{4}{3}$
13. $y = 15 - 4x$
14. $y = 4 - 3x$
15. $y = -\frac{3}{4}x$
16. $y = \frac{1}{2}x$
17. $2x = y + 4$
18. $3x - y - 6$
19. $2x + 3y = 12$
20. $3y + 5x = 15$
21. $5y - 2x = 10$
22. $x + 3y = 6$
23. $x - 4y = 4$
24. $4x - 3y = 24$
25. $y = 12 - 0.30x$
26. $y = (x - 500) + 50$

In Exercises 27 to 38, find the slope and vertical-axis intercept for each equation. Assume that the output variable is on the left.

27. $D = 55t$
28. $C = 2\pi r$
29. $C = 8 + 2\pi r$
30. $D = 225 + 45t$
31. $p = 2.98n + 0.50$
32. $c = 0.08p + 25$
33. $V = 50 - 0.29n$
34. $V = 20 - 0.19n$
35. $H = 0.8(200 - A)$
36. $R = 0.6(220 - A)$
37. $C = 65 + 0.15(d - 100)$
38. $c = 30 + 10(h - 4)$

For the applications in Exercises 39 to 42, do the following:
(a) Make a table for inputs 0, 1, and 2.
(b) Find the slope and its meaning.
(c) Find the vertical-axis intercept and its meaning.
(d) Write an equation describing the output as depending on the input. Define input and output variables as needed.

39. The output is the total cost per semester of x hours on the computer. Each semester, students are charged a \$3 fee plus \$1 for each hour.

40. The output is the total cost of a taxi ride of x miles. The taxi driver charges a \$2 fee plus \$3 for each mile traveled.

41. The output is the total cost of a meal, where x is the price of the meal and a tip of 15% of the price of the meal is left for the serving person.

42. The output is the total cost of a shirt, where x is the price of the shirt and a sales tax of 7% of the price is added.

A slope and an ordered pair are given in Exercises 43 to 52. Plot the point, and draw a line with the given slope through the point.

43. $\frac{1}{2}$ and $(3, -4)$
44. $-\frac{4}{3}$ and $(1, 2)$
45. -4 and $(2, 3)$
46. $\frac{2}{3}$ and $(4, -1)$
47. $-\frac{3}{2}$ and $(2, 0)$
48. 2 and $(-1, 3)$
49. 0 and $(0, 4)$
50. undefined slope and $(3, 0)$
51. undefined slope and $(-2, 0)$
52. 0 and $(-2, -2)$

In Exercises 53 to 60, sketch a graph and find an equation of each line, as described.

53. horizontal line through $(-2, 4)$
54. vertical line through $(3, 5)$
55. vertical line through $(4, 3)$
56. horizontal line through $(-5, 2)$
57. vertical line through $(3, 0)$
58. horizontal line through $(0, -3)$
59. horizontal line through $(4, 0)$
60. vertical line through $(0, -4)$

For the functions in Exercises 61 to 66, find x if $f(x) = 0$, and $f(0)$. Also write your results as ordered pairs.

61. $f(x) = 3(x - 4) + x$

Blue numbers are core exercises.

62. $f(x) = 5(x - 8) - x$

63. $f(x) = 9 - 3(x - 1)$

64. $f(x) = 8 - 2(x + 4)$

65. $f(x) = 2x + 3(x - 5)$

66. $f(x) = 5x - 2(x - 5)$

For the functions in Exercises 67 to 70, find the horizontal-axis and vertical-axis intercepts, and explain their meaning, if any.

67. Total tuition in dollars, where x = number of credit hours taken: $f(x) = 115x + 48$

68. Value remaining (in dollars) on a Coffee Corner gift certificate, where x = number of lattes: $f(x) = 15 - 3x$

69. Value remaining (in dollars) on a phone card, where x = number of minutes used: $f(x) = 19.50 - 0.05x$

70. Total cost of transcripts, where x = number of transcripts ordered: $f(x) = 5 + 2(x - 1)$

▶ Writing

71. Which expression or equation gives the x-intercept: $f(0)$ or $f(x) = 0$? Why?

72. Which expression or equation gives the y-intercept: $f(0)$ or $f(x) = 0$? Why?

73. Explain how to find a horizontal-axis intercept.

74. Explain how to find a vertical-axis intercept.

75. Explain how to find the equation of a vertical line, given

a. a point on the line

b. its x-intercept

76. Explain how to find the equation of a horizontal line, given

a. a point on the line

b. its y-intercept

77. Explain the two uses of parentheses in $(x, f(x))$.

78. Which of the following are conditions for $f(x)$ to pass through the origin? For $f(x)$ to pass through the y-axis? For $f(x)$ to pass through the x-axis?

a. $f(0) = 0$ **b.** $f(x) = 0$

c. $f(0)$ **d.** $f(x) = 0$ at $x = 0$

79. Connecting with Chapters 1 and 2: Parentheses Explain the use of parentheses in each part:

a. $4 - (5 + 3)$ equals $4 - 5 - 3$

b. Slope (in a linear equation) describes a rate of change of the input variable or the steepness of a line.

c. Finding 20% of 15 is written 0.20(15).

d. We plot (3, 4) and then count spaces to draw slope.

e. The inequality $x \leq 10$ can be written $(-\infty, 10]$

f. $a + (-a) = 0$

g. The x-intercept can be found by solving $f(x) = 0$.

h. Find $f(2)$ for $f(x) = x - 3$.

80. The value (in dollars) remaining on a photocopy card after x copies is $y = -0.15x + 15.00$. Solve the equations in parts a, b, and c.

a. $10.50 = -0.15x + 15.00$

b. $y = -0.15(0) + 15.00$

c. $0 = -0.15x + 15.00$

d. How many copies have been made if the card value is \$6.75?

e. The equation in which part—a, b, or c—finds the x-intercept?

f. The equation in which part—a, b, or c—finds the y-intercept?

81. A prepaid telephone card costs \$12. Each minute costs \$0.03.

a. Find the value remaining on the card after 0, 100, 200, 300, 400, and 500 minutes.

b. How many minutes were used if \$2 remains on the card?

c. What is the x-intercept of a graph of the setting?

d. What is the meaning, if any, of the x-intercept?

e. What is the y-intercept of a graph of the setting?

f. What is the meaning, if any, of the y-intercept?

g. What equation describes the value remaining on the card after x minutes?

▶ Project

82. Slope and Scale on Axes Although the following graphs below appear to have the same slope, the three lines actually have different slopes because of the different scales on the axes. List the slope for each graph.

Blue numbers are core exercises.

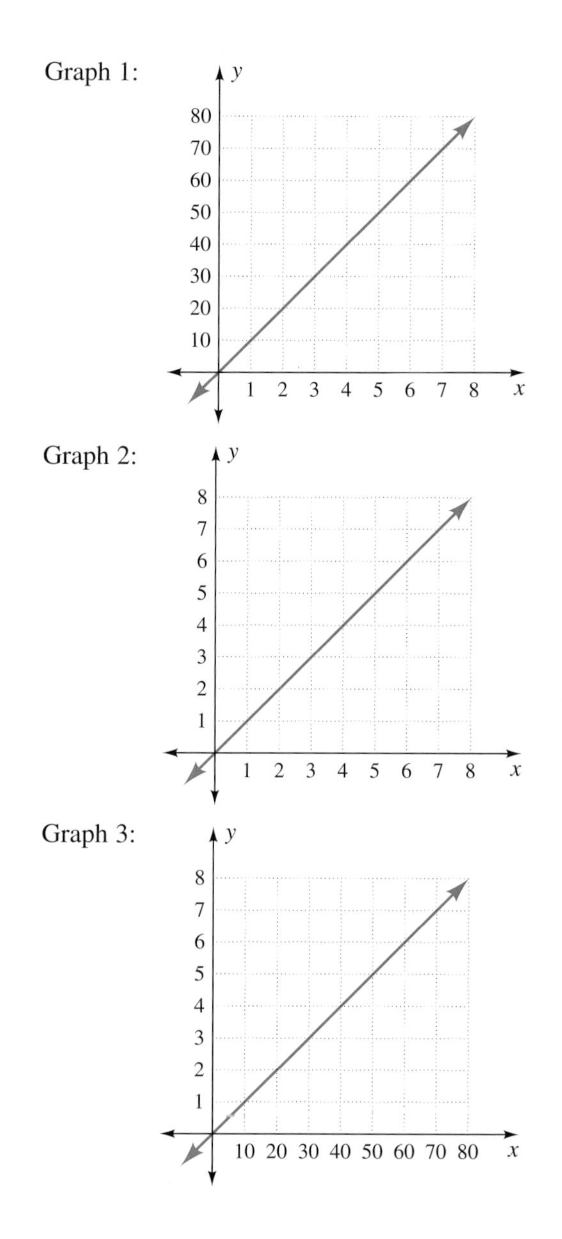

Tell which graph—graph 1, 2, or 3—is most appropriate for each of the following input-output situations. For parts a to c, assume the scale on the y-axis is in feet.

a. The height, y, of a wheelchair ramp of length x feet

b. The height, y, climbed up a rock cliff for x horizontal feet traveled

c. The height, y, climbed up a steep staircase for x horizontal feet traveled

For parts d to f, assume the scale on the vertical axis is in dollars.

d. The total cost, y, of x items at the dollar store

e. The total cost, y, of x pieces of candy at 10 cents each

f. The total cost, y, of x movie tickets at \$10 each

For parts g to i, assume the scale on the vertical axis is in miles.

g. The distance, y, traveled in x hours by the NASA space shuttle between the assembly building and the launch pad

h. The distance, y, traveled by bicycle in x hours

i. The distance, y, paved by a highway construction crew in x hours

Extension: Make up one description for each graph, stating your assumptions about the units on the axes.

▶ 4.5 Linear Equations

Objectives

- Write a linear equation from a slope and vertical-axis intercept and draw its graph.
- Find a linear equation from ordered pairs, data, a table, or a graph.
- Write equations for parallel and perpendicular lines.

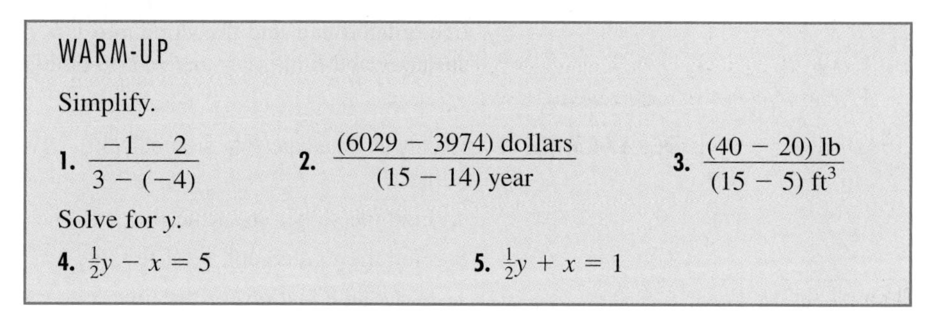

WARM-UP

Simplify.

1. $\dfrac{-1 - 2}{3 - (-4)}$ **2.** $\dfrac{(6029 - 3974)\text{ dollars}}{(15 - 14)\text{ year}}$ **3.** $\dfrac{(40 - 20)\text{ lb}}{(15 - 5)\text{ ft}^3}$

Solve for y.

4. $\frac{1}{2}y - x = 5$ **5.** $\frac{1}{2}y + x = 1$

IN SECTIONS 4.3 AND 4.4, we examined slope and intercepts of a linear equation. In this section we write linear equations when we know or believe the relationship between the variables to be linear.

Finding Linear Equations from Slope and y-intercept

In Example 1 we substitute the slope and y-intercept into $y = mx + b$.

EXAMPLE 1 Finding an equation and graphing, given the slope and y-intercept For the slope and y-intercept listed, write a linear equation and then graph the line.

a. $m = -2, b = 3$

b. $m = \frac{1}{4}, b = -3$

SOLUTION **a.** $y = -2x + 3$. To graph, start with the intercept, $b = 3$, and plot the point $(0, 3)$. Using the slope to find a second point on the graph, move 2 units down and 1 unit to the right from the intercept. Draw a line through the two points.

b. $y = \frac{1}{4}x - 3$. To graph, start with the intercept, $b = -3$, and plot the point $(0, -3)$. Using the slope to find a second point on the graph, move 1 unit up and 4 units to the right from the intercept. Draw a line through the two points.

The graphs of the lines are shown in Figure 22.

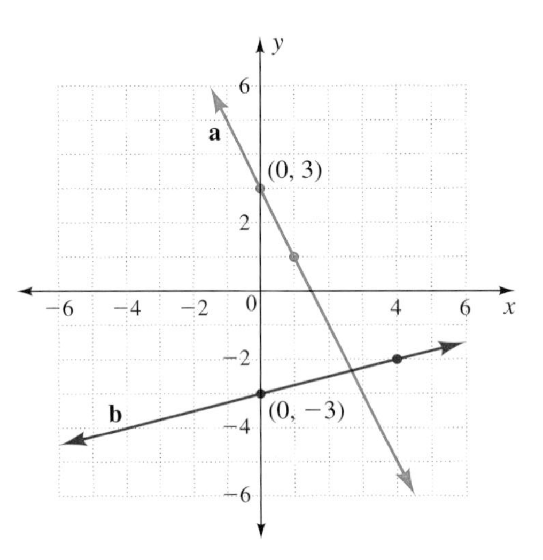

FIGURE 22 Graph from y-intercept and slope ◀

Finding Linear Equations from Ordered Pairs

In Example 2, we find the equation of a line, $y = mx + b$, given two ordered pairs. We use the slope formula to find m from two ordered pairs. To find b, we substitute one ordered pair and the slope into $y = mx + b$ and then solve for b. We then substitute m and b into $y = mx + b$ to build an equation.

EXAMPLE 2 Finding an equation from ordered pairs The ordered pairs $(-4, 2)$ and $(3, -1)$ lie on a line.

a. Find the slope, m, of the line.

b. Find the y-intercept, b, of the line.

c. Write an equation for the line.

SOLUTION **a.** Slope $= \dfrac{y_2 - y_1}{x_2 - x_1} = \dfrac{-1 - 2}{3 - (-4)} = \dfrac{-3}{7}$

b. To find the y-intercept, we substitute the slope and one of the ordered pairs into $y = mx + b$.

$$
\begin{aligned}
y &= mx + b && \text{Substitute } m = -\tfrac{3}{7} \text{ and } (x, y) = (-4, 2).\\
2 &= -\tfrac{3}{7}(-4) + b && \text{Simplify.}\\
2 &= \tfrac{12}{7} + b && \text{Subtract } \tfrac{12}{7} \text{ from both sides.}\\
2 - \tfrac{12}{7} &= b && \text{Simplify.}\\
\tfrac{14}{7} - \tfrac{12}{7} &= b\\
b &= \tfrac{2}{7}
\end{aligned}
$$

c. We substitute the slope and y-intercept into $y = mx + b$:

$$
\begin{aligned}
y &= mx + b && \text{Substitute } m = -\tfrac{3}{7} \text{ and } b = \tfrac{2}{7}.\\
y &= -\tfrac{3}{7}x + \tfrac{2}{7}
\end{aligned}
$$

◀

▶ In Example 2, we substituted one ordered pair and the slope into $y = mx + b$ and solved for b. The solving step may be avoided if we first solve $y = mx + b$ for b to create a formula for the y-intercept, as shown in Example 3.

▶ **EXAMPLE 3** **Finding a shortcut for calculating b** Make a formula for the y-intercept.

SOLUTION We can solve $y = mx + b$ directly for b:

$$
\begin{aligned}
y &= mx + b && \text{Subtract } mx \text{ from both sides.}\\
y - mx &= b
\end{aligned}
$$

◀

THE y-INTERCEPT FORMULA

> The vertical-axis intercept, b, for a linear equation is
>
> $$b = y - mx$$
>
> where m is the slope and (x, y) is any point on the line.

▶ Finding Linear Equations from Data or a Graph

The source of the ordered pairs needed to build the linear equation may be data, a graph, or (as we saw in finding slope, Section 4.3) a table.

WRITING ORDERED PAIRS FROM DATA In Examples 4 and 5, we find a linear equation from application data.

▶ **EXAMPLE 4** **Writing ordered pairs, then an equation** On August 24, 2005, the Associated Press (AP) reported that flying the president's airplane, Air Force One, cost \$6029 per hour in fuel, compared with \$3974 in the prior year. Assume that the cost of fuel is a function of time in years. To simplify the numbers, assume that 1990 is year 0, the year Air Force One was placed into service.

a. Record the data in ordered pairs.
b. Find a linear equation for the function.
c. What is the meaning of the slope and y-intercept?
d. Is a linear function for fuel cost a reasonable assumption?

SOLUTION

a. With 1990 as year 0, 2004 is year 14 and 2005 is year 15. The ordered pairs are (14, 3974) and (15, 6029).

b. Slope $m = \dfrac{6029 - 3974}{15 - 14} = \dfrac{2055}{1}$

We substitute one ordered pair, (14, 3974), and the slope, 2055, into the equation for the vertical-axis intercept:

$$b = y - mx$$
$$b = 3974 - (2055)(14)$$
$$b = -24{,}796$$

The linear equation is $y = 2055x - 24796$

c. The slope is 2055 dollars per year. The intercept is negative \$24,796.

d. It is unlikely that the cost of fuel in 1990 was negative (someone pays you to use up fuel). Applying a linear function, especially back to 1990, was a poor choice. ◀

▶ **EXAMPLE 5** Writing ordered pairs from a graph, then finding an equation The owner's manual for a Kenmore freezer recommends placing dry ice in the freezer if the power is off for longer than 24 hours. Assume that the amount of dry ice is a linear function of the size of the freezer. The data are shown as a graph in Figure 23.

a. Select two points from the graph and write them as ordered pairs.

b. Find a linear equation to calculate the amount of dry ice needed.

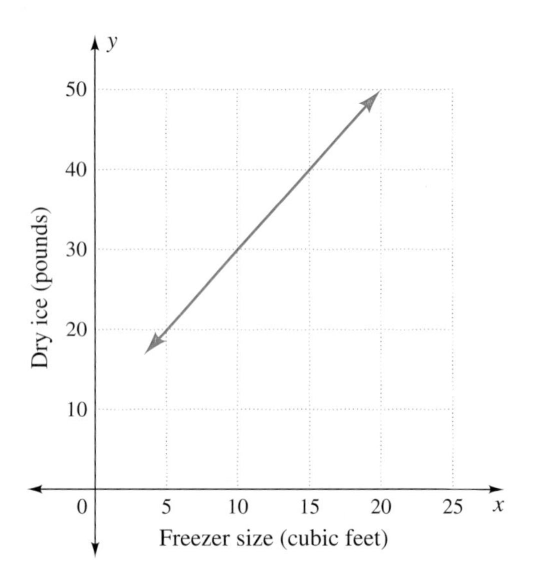

FIGURE 23 Emergency dry ice for freezer

SOLUTION

a. Two ordered pairs are (5, 20) and (15, 40).

First we find the slope:

$$\text{Slope} = m = \frac{(40 - 20)\text{ lb}}{(15 - 5)\text{ ft}^3} = \frac{20\text{ lb}}{10\text{ ft}^3} = 2\text{ lb per ft}^3$$

Then we substitute $m = 2$ and (5, 20) into the equation:

$$b = y - mx$$
$$b = 20 - 2(5)$$
$$b = 10$$

Next we substitute $m = 2$ and $b = 10$ into $y = mx + b$ to get the equation for pounds of dry ice as a function of cubic feet: $y = 2x + 10$ ◀

THINK ABOUT IT: Extend the line in Figure 23. What is the x-intercept? Does it have any meaning? What is the y-intercept? Does it have any meaning?

Building a Linear Equation

To find the equation of a line from two ordered pairs (x_1, y_1) and (x_2, y_2):

1. Find the slope from a table, from a graph, or by using

$$m = \frac{y_2 - y_1}{x_2 - x_1}$$

2. Find the y-intercept, from a table, from a graph, or by using $b = y - mx$, the slope, and either ordered pair.
3. Substitute the numbers for m and b into $y = mx + b$.
4. Use another ordered pair, if available, to check.

We can find a linear equation from data using *a statistical function*, **linear regression**, on the calculator.

Graphing Calculator Technique: Linear Regression

Student Note:
Use (−) to enter negative numbers, not the subtraction sign.

Place the ordered pairs from Example 2 in the lists provided under the statistics function and then find the equation for the straight line with the linear regression option. Here's how it's done:

To start, clear old data as needed: STAT **4 : ClrList** 2nd [L1], 2nd [L2] ENTER.

Let x be in list 1 and y in list 2. Using STAT **1 : Edit** to obtain the list screen, enter the data points $(-4, 2)$ and $(3, -1)$ (Figure 24).

Choose the linear regression option under the statistical calculations (Figure 25): STAT **CALC 4 : LinReg** 2nd [L1], 2nd [L2] ENTER.

Naming the lists is optional if you always use L1 and L2.

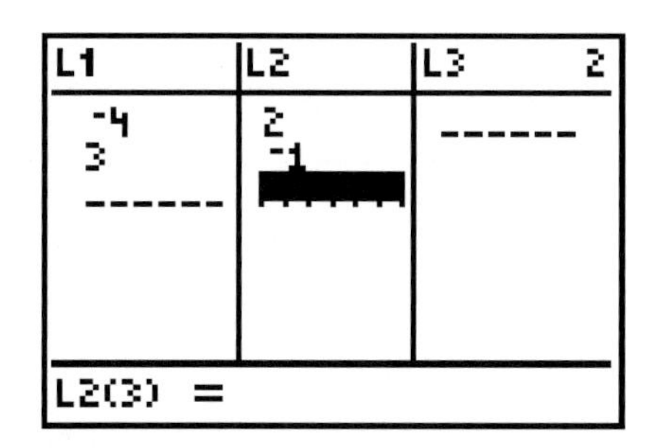

FIGURE 24 Ordered pairs in lists

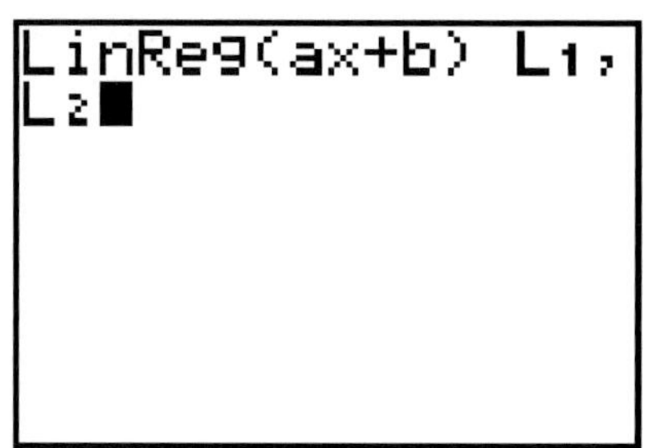

FIGURE 25 Linear regression on lists

The regression (Figure 26) gives the slope and the y-intercept:

$$a \approx -0.43 \qquad \text{and} \qquad b \approx 0.29$$

Write the regression equation in the form $y = ax + b$:

$$y = -0.43x + 0.29$$

(continued)

(concluded)
As shown in Figure 27, you can recall the slope, a, and y-intercept, b, listed under statistical variables, VARS **5 : Statistics** with the **EQ** (equation) option. Change first a and then b into fractions, to obtain

$$y = -\frac{3}{7}x + \frac{2}{7}$$

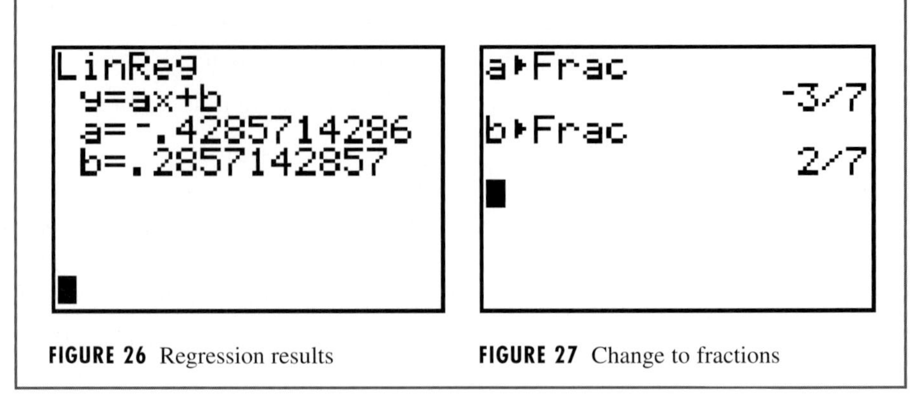

FIGURE 26 Regression results **FIGURE 27** Change to fractions

We now consider two important sets of linear equations, parallel and perpendicular lines.

Finding Linear Equations: Parallel Lines

Knowing the position of the slope number in an equation helps us to identify parallel lines. **Parallel lines** *are two or more lines in the coordinate plane with the same slope but different y-intercepts.* In geometry, we use parallel lines to determine the nature or properties of shapes. In other applications, parallel lines indicate the same rate of change but different starting or initial values (vertical-axis intercepts).

▶ EXAMPLE 6 Finding parallel lines Which lines are parallel?

a. $2x + y = 4$ **b.** $y - 2x = 3$ **c.** $\frac{1}{2}y - x = 5$ **d.** $\frac{1}{2}y + x = 1$

SOLUTION The parallel lines will have the same slope. To compare the slopes of the lines, we must change each equation into an equation of the form $y = mx + b$.

a. $2x + y = 4$ Subtract $2x$ from both sides.

$y = -2x + 4$ The slope is -2.

b. $y - 2x = 3$ Add $2x$ to both sides.

$y = 2x + 3$ The slope is 2.

c. $\frac{1}{2}y - x = 5$ Add x on both sides.

$\frac{1}{2}y = x + 5$ Multiply both sides by 2.

$y = 2x + 10$ The slope is 2.

d. $\frac{1}{2}y + x = 1$ Subtract x from both sides.

$\frac{1}{2}y = -x + 1$ Multiply both sides by 2.

$y = -2x + 2$ The slope is -2.

The lines in parts a and d have slope -2 and are parallel. The lines in parts b and c have slope 2 and are parallel. ◀

EXAMPLE 7 Applying parallel lines: photocopy costs revisited In Example 9 of Section 4.3, the prepaid photocopy machine card cost \$15.00 and each photocopy cost \$0.15. Figure 28 shows a graph for the card's value after photocopies have been made. Suppose the card value changes from \$15.00 to \$30.00 but the copies remain the same price.

a. What is the new equation for the card's value?

b. Why does the graph (see Figure 29) change?

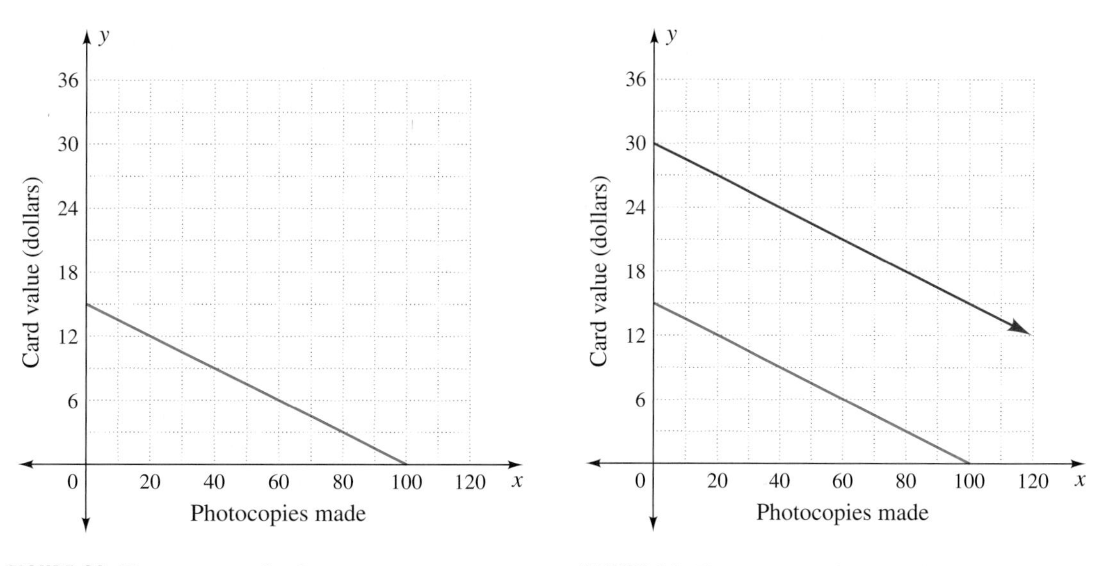

FIGURE 28 Photocopy card value

FIGURE 29 Photocopy card, new value

SOLUTION **a.** The slope, or rate, remains \$0.15 per copy. The initial cost is \$30.00 instead of \$15.00. From $y = mx + b$, the equation is therefore $y = -0.15x + 30.00$.

b. The graph, in Figure 29, shifts up because the y-intercept is now \$30.00. The new graph is parallel to the old graph. ◀

Finding Linear Equations: Perpendicular Lines

*Two lines that cross at a right angle** are **perpendicular lines**. In designing computer graphics and in finding areas of rectangles and triangles, we use perpendicular lines to show that right angles are formed. Every pair of horizontal and vertical lines is perpendicular. Example 8 suggests the numerical relationship between the slopes of perpendicular lines.

EXAMPLE 8 Exploring perpendicular lines Lines AO and BO in Figure 30 intersect at the origin.

a. What is the slope of AO?

b. What is the slope of BO?

c. On a separate piece of paper, trace the axes and lines AO and BO. Place the traced figure over the book so that it matches the original figure. Press with a pencil point at O while turning the x-axis 90° (a right angle) counterclockwise. Describe the new positions of the positive x-axis and AO.

---

*The definition of perpendicular lines assumes that the axes are numbered with the same scale.

d. What can you conclude about the lines AO and BO?
e. What can you conclude about the slopes of perpendicular lines?

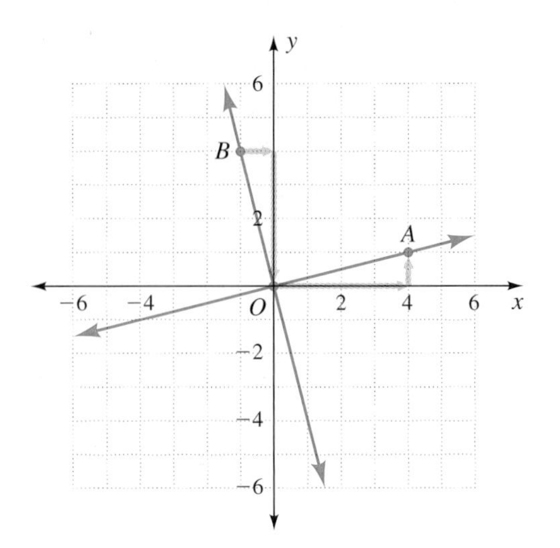

FIGURE 30 Rotate axes

SOLUTION **a.** $\frac{1}{4}$

b. $-\frac{4}{1}$

c. The positive x-axis should lie over the original positive y-axis. The line AO now matches BO.

d. Because AO and BO matched when we turned our traced figure 90°, the exploration suggests that the lines cross at a right angle and so are perpendicular.

e. The slopes, $\frac{1}{4}$ and $-\frac{4}{1}$, are opposite in sign and reciprocals of each other. ◀

Our exploration suggests that the lines in Figure 30 are perpendicular. The slopes of the lines were opposite reciprocals. Recall that reciprocals multiply to 1. These ideas suggest the following conclusion.

PERPENDICULAR LINES

> Two lines are perpendicular if
>
> **1.** their slopes multiply to -1 (that is, their slopes are opposite reciprocals) or
>
> **2.** their slopes are the same as those of the horizontal and vertical axes.

▶ **EXAMPLE 9** Finding slopes of lines perpendicular to a given line What would be the slope of a line perpendicular to each of the following?

a. $y = 3x + 2$ **b.** $y = \frac{1}{2}x - 3$ **c.** $y = -1.5x$ **d.** $y = 3$

SOLUTION **a.** The slope of $y = 3x + 2$ is 3. The slope of a perpendicular line would be the opposite reciprocal of 3: $-\frac{1}{3}$.

Student Note:
To make perpendicular lines appear at right angles on a graphing calculator, use **5 : ZSquare** under ZOOM.

b. The slope of $y = \frac{1}{2}x - 3$ is $\frac{1}{2}$. The slope of a perpendicular line would be the opposite reciprocal of $\frac{1}{2}$: $-\frac{2}{1}$, or -2.

c. The slope of $y = -1.5x$ is -1.5, or $-\frac{3}{2}$. The opposite reciprocal is $\frac{2}{3}$.

d. The line $y = 3$ is a horizontal line with slope 0. Any vertical line $x = a$ would be perpendicular. The slopes of all vertical lines are undefined. ◀

Summary Example

We now combine several ideas by finding equations of lines, given information about slopes and/or intercepts. In some problems, we use the given information to find the slope or y-intercept. In one problem, the required equation is that for a vertical line in the form $x = a$.

EXAMPLE 10 Writing linear equations Find the equation of a line with the given characteristics.

a. slope 0 and y-intercept at -1
b. parallel to $y = 3x - 4$ and y-intercept at -2
c. parallel to $y = -\frac{1}{2}x + 5$ and passing through the origin
d. slope undefined and x-intercept at -1
e. perpendicular to $y = 3x - 4$ and containing $(0, 5)$
f. perpendicular to $y = -\frac{2}{5}x + 2$ and with y-intercept point $(0, -3)$

SOLUTION

a. Let $m = 0$ and $b = -1$ in $y = mx + b$; $y = 0x - 1$; $y = -1$.

b. Because the line is parallel, the slope is the same: $m = 3$. The y-intercept is $b = -2$. The line is $y = 3x - 2$.

c. Because the line is parallel, the slope is the same: $m = -\frac{1}{2}$. The origin lies on the y-axis, and so $b = 0$. The line is $y = -\frac{1}{2}x + 0$, or $y = -\frac{1}{2}x$.

d. Because the slope is undefined, the line is vertical. If the line has an x-intercept at -1, then $x = -1$ at that point and the equation is $x = -1$.

e. A line perpendicular to $y = 3x - 4$ has slope equal to the opposite reciprocal of 3, which is $-\frac{1}{3}$. The point $(0, 5)$ is on the y-axis, so $b = 5$. The equation is $y = -\frac{1}{3}x + 5$.

f. A line perpendicular to $y = -\frac{2}{5}x + 2$ has a slope of $\frac{5}{2}$. The y-intercept point is $(0, -3)$, so $b = -3$. The line is $y = \frac{5}{2}x - 3$. ◀

ANSWER BOX

Warm-up: 1. $-\frac{3}{7}$ **2.** 2055 dollars per year **3.** 2 lb per ft^3 **4.** $y = 2x + 10$ **5.** $y = -2x + 2$ **Think about it:** Neither the x-intercept (-5 cubic feet) nor the y-intercept (10 pounds) has any meaning; clearly a freezer of zero cubic feet would require no dry ice.

4.5 Exercises

In Exercises 1 to 8, write an equation for the given data.

1. $m = \frac{1}{2}, b = 3$

2. $m = \frac{3}{2}, b = 4$

3. slope $= \frac{2}{3}$, y-intercept $= -2$

4. slope $= -4$, y-intercept $= \frac{1}{2}$

5. slope $= 5$, y-intercept $= \frac{1}{4}$

6. slope $= \frac{3}{4}$, y-intercept $= -3$

7. $m = -\frac{3}{2}$, $(0, 1)$

8. $m = \frac{4}{5}$, $(0, -2)$

Blue numbers are core exercises.

In Exercises 9 to 20, write an equation for the line passing through the two given ordered pairs.

9. $(1, 1), (3, 9)$
10. $(2, 3), (4, 7)$
11. $(2, -2), (5, -8)$
12. $(-2, -4), (1, 2)$
13. $(-3, 1), (0, -1)$
14. $(-4, -2), (0, 5)$
15. $(13, 6), (10, 0)$
16. $(3, 6), (0, 0)$
17. $(-5, 6), (-4, -2)$
18. $(3, -4), (-1, 4)$
19. $(5, 2), (3, 3)$
20. $(2, 3), (7, 1)$

In Exercises 21 to 32, a slope and ordered pair are given. Write a linear equation having the slope and containing the given point.

21. $m = 4, (3, -1)$
22. $m = -2, (2, -3)$
23. $m = -1, (4, -2)$
24. $m = 5, (1, 3)$
25. $m = \frac{1}{2}, (2, 4)$
26. $m = \frac{2}{3}, (6, -2)$
27. $m = \frac{4}{5}, (-10, 3)$
28. $m = \frac{1}{3}, (-9, 1)$
29. $m = \frac{5}{3}, (-3, 1)$
30. $m = \frac{5}{2}, (-4, 2)$
31. $m = -2, (1.5, 3)$
32. $m = 2, (-2.5, 4)$

In Exercises 33 to 42, find the slope and y-intercept for each graph and write an equation. Note the labeling on the axes in selecting your variables.

33.

34.

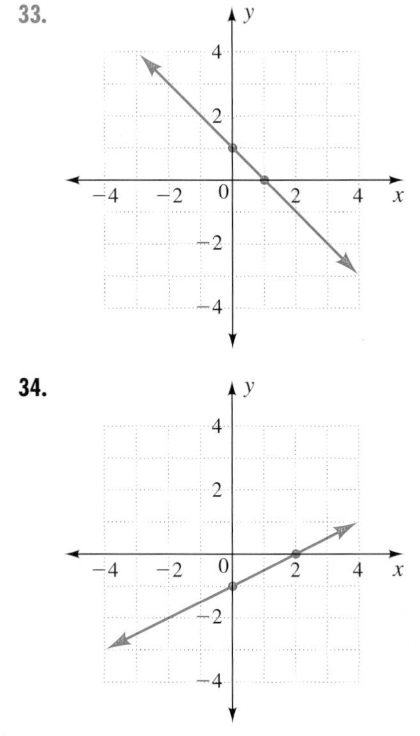

Blue numbers are core exercises.

35.

36.

37.

38.

39.

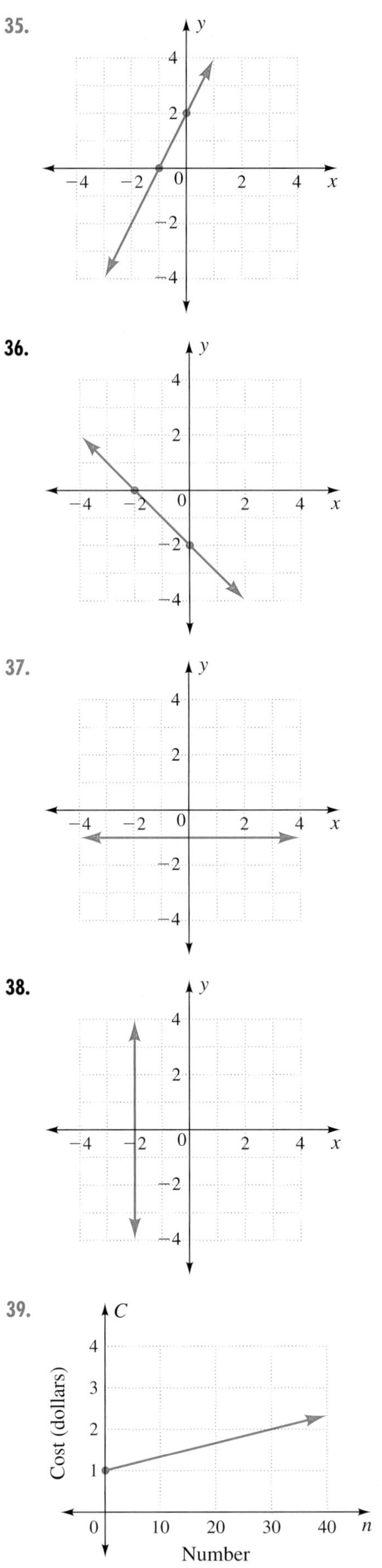

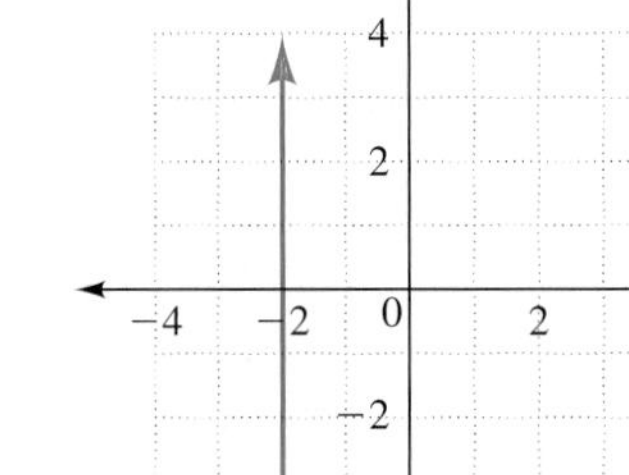

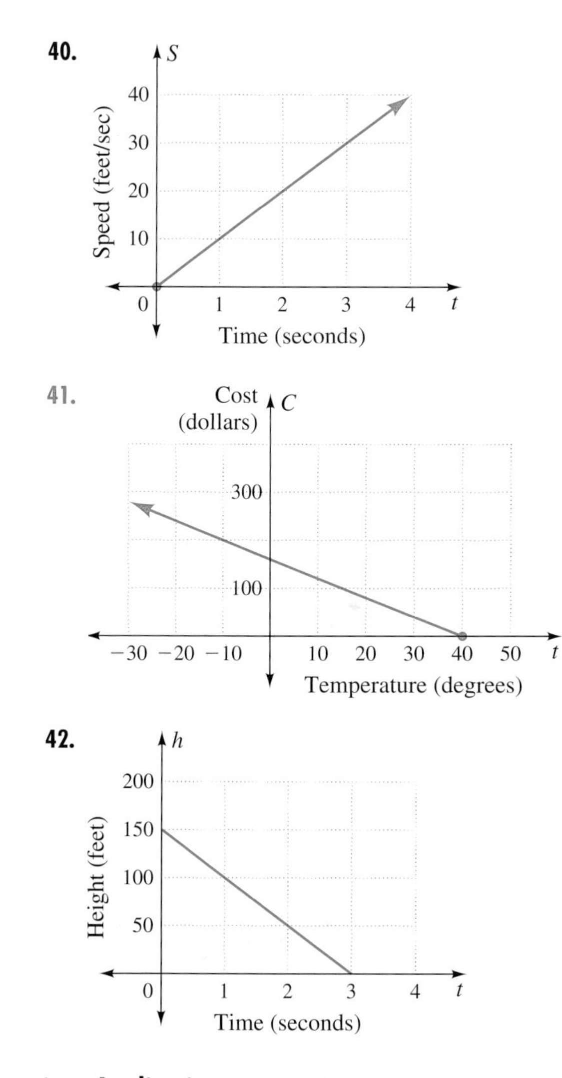

▶ Applications

43. Late one afternoon, a 30-minute call cost \$1.50. On another afternoon, an 8-minute call cost \$0.40. Assume that the cost (in dollars) of a call is a linear function of the time in minutes. Find a linear equation that gives the cost of an afternoon call.

44. One Sunday, a 44-minute call cost \$2.15. On another Sunday, a 10-minute call cost \$0.79. Assume that the cost in dollars of a call is a linear function of the time in minutes. Find a linear equation that gives the cost of a Sunday call.

45. In each region of the 64-team NCAA basketball tournament, the number-1 team is matched with the number-16 team, the number-2 team is matched with the number-15 team, and so forth. Let x be the ranking of the first team in the match-up, and let y be the ranking of the opponent. What linear equation describes the match-up?

46. An access ramp needs to rise 2.5 feet and so requires a 30-foot length. Another ramp built on the same design needs to rise 1.5 feet and requires an 18-foot length. What linear equation describes the length as a function of rise?

Blue numbers are core exercises.

47. Water freezes at 0° Celsius and 32° Fahrenheit. Water boils at 100°C and 212°F. Let the Celsius temperature C be the input and the Fahrenheit temperature F be the output.

a. Write an ordered pair (C, F) for the temperature at which water freezes and another ordered pair for the temperature at which water boils. Assume the temperature scales are linear functions of each other. Using the ordered pairs, find the slope and units for the slope.

b. What is the y-intercept?

c. Write the linear equation.

48. Experiments show that a heat pump has an output of 36,000 Btu/hr at 48°F outside temperature. The same heat pump has an output of 15,000 Btu/hr at 18°F outside temperature. Assume that the output in Btu/hr is linear for input temperatures between 18°F and 48°F.

a. Find an equation that gives the Btu/hr output at any temperature output.

b. Why might there be limitations on the inputs for this equation?

c. What is the y-intercept, and does it have any meaning?

d. What is the x-intercept, and does it have any meaning?

Use the following figure to do Exercises 49 to 54. For Exercises 49 to 52, match a line graph in the figure with each slope and point given.

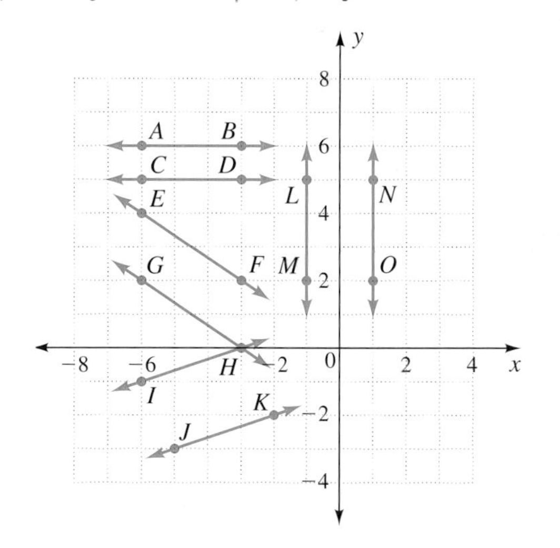

49. slope $= -\frac{2}{3}$, point $= (-6, 2)$

50. slope $= 0$, point $= (-6, 6)$

51. slope is undefined, point $= (1, 5)$

52. slope $= \frac{1}{3}$, point $= (-5, -3)$

53. Write the equation of each line: *HI*, *JK*, *LM*, *NO*.

54. Write the equation of each line: *AB*, *CD*, *EF*, *GH*.

In Exercises 55 to 58, four equations are shown. Write each equation in $y = mx + b$ form if it is not already in this form. Which lines are parallel? Which are perpendicular?

55. **a.** $y = 2x + 3$ **b.** $y = \frac{1}{2}x + 3$

c. $y = -\frac{1}{2}x + 3$ **d.** $2y = x + 4$

56. **a.** $y = -\frac{1}{3}x + 2$ **b.** $y = 3x + 4$

c. $y = -3x + 2$ **d.** $3x + y = 4$

57. **a.** $y = \frac{1}{3}x - 4$ **b.** $x - \frac{1}{3}y = 6$

c. $3y - x = 2$ **d.** $y + \frac{1}{3}x = 4$

58. **a.** $2y + x = 4$ **b.** $1 + y = \frac{1}{2}x$

c. $3 = \frac{1}{2}x - y$ **d.** $-2x = 1 - y$

In Exercises 59 to 66, sketch a graph and find an equation of each line, as described.

59. parallel to $y = 4x + 1$ through the origin

60. perpendicular to $y = 4x + 1$ through $(0, 5)$

61. perpendicular to $y = 2x - 3$ through the origin

62. parallel to $y = 2x - 3$ through $(0, -1)$

63. parallel to $y = \frac{1}{3}x - 5$ through $(0, 4)$

64. perpendicular to $y = \frac{1}{3}x - 5$ through the origin

65. perpendicular to $y = -\frac{3}{4}x + 2$ through $(0, -2)$

66. parallel to $y = -\frac{3}{4}x + 2$ through the origin

In Exercises 67 to 72, answer the following questions.
(a) Which fact gives the slope?
(b) Which fact gives the y-intercept?
(c) Write the equation using $y = mx + b$.

67. Hwang prepays $50 on racquetball court rental of $2 per hour. The equation describes the prepaid amount in dollars that remains after x hours of rental time.

68. Yolanda's $500 monthly expense account is set up through an automatic teller machine (ATM). She withdraws funds, using the $40 Fast Cash Option. The equation describes the amount in dollars that remains in her account after x withdrawals during the month.

69. Carmen rents a Cessna 152 for $126 per hour plus a $28 insurance fee. The equation describes the total cost in dollars of x hours flying time.

70. Alberto earns a weekly salary of $250 plus 10% of his sales volume. The equation describes the total weekly earnings for x dollars in sales.

Blue numbers are core exercises.

71. What words identify the slope in Exercises 67 to 70?

72. What words identify the y-intercept in Exercises 67 to 70?

In Exercises 73 to 78, will the described change give a parallel line or a steeper line? Write the new equation.

73. The total rental cost in dollars on a truck is $C = 0.79n + 35$, where n is in miles. The fixed cost increases from $35 to $45.

74. The monthly cost in dollars of water is $C = 0.65g + 5.15$, where g is in thousands of gallons. The basic charge for water service rises $2, but the cost per thousand gallons remains $0.65.

75. The total cost in dollars of gasoline is $C = 3.10g$, where g is in gallons. The gasoline rises in price by $0.10 per gallon.

76. The value in dollars remaining on a transit ticket is $V = 20 - 0.80n$, where n is the number of rides. Rosa buys a $30 ticket instead of a $20 ticket.

77. The total monthly cost in dollars of a loan is $C = 0.01x + 3$, where x is the amount borrowed. The service fee rises from $3 to $5.

78. The distance traveled at 40 miles per hour is $d = 40t$, where t is in hours. Duane increases his speed by 10 miles per hour.

▶ Writing

For Exercises 79 to 84, explain how to find the requested information.

79. a linear equation from two ordered pairs

80. a linear equation from its graph

81. the slope and y-intercept from $y = cx + d$

82. the slope and y-intercept from $ax + cy = d$

83. the slope of a line perpendicular to $y = \dfrac{ax}{b} + d$

84. the slope of a line parallel to $ax + by = c$

85. Slope can be found directly from intercepts.

a. Use the slope formula to find the slope of the line through $(a, 0)$ and $(0, b)$.

b. Explain how to find the slope of a line, given its x- and y-intercept points.

c. Try your rule on $(-5, 0)$ and $(0, -4)$. Check the slope another way.

86. Find the equation of a line passing through $(0, 8)$ and $(3, 5)$. Explain why $b = y - mx$ was not needed to find the y-intercept.

▶ **Project**

87. Slope and Right Triangles The sides forming the right angle in a right triangle are perpendicular. Use the slopes of AB, AC, and BC to show that each set of ordered pairs forms a right triangle. Name the perpendicular sides.

a. $A(0, 2)$, $B(3, 6)$, $C(7, 3)$

b. $A(-2, -1)$, $B(3, -3)$, $C(5, 2)$

c. $A(-2, 3)$, $B(0, -1)$, $C(2, 5)$

d. $A(-3, -4)$, $B(-1, -7)$, $C(0, -2)$

▶ 4.6 Inequalities in Two Variables

Objectives

- Find out if an ordered pair is a solution of a two-variable inequality.
- Find solutions to a two-variable inequality from a graph.
- Graph a two-variable linear inequality.

WARM-UP

These exercises provide practice with the properties of inequalities (Section 3.5).

1. Solve each of these for y in terms of x.

a. $2x - y < 4$ **b.** $3x + 2y \leq 6$

c. $2x - 3y > 6$ **d.** $2y - x \geq 4$

2. Change each of these into a form matching $ax + by < c$. For this exercise, the coefficient of x may be a fraction or negative number.

a. $y < 3x - 1$ **b.** $y \geq -4x + 2$

c. $y > \frac{1}{2}x + 3$ **d.** $y \leq \frac{1}{3}x - 3$

HAVE YOU EVER needed to make a choice between two items because of a budget or caloric limit? Inequalities in two variables permit us to show the relationship between our choices. In this section, we practice operations with inequalities, find out which ordered pairs are solutions to inequalities, and graph two-variable linear inequalities.

▶ Inequalities in Two Variables

DEFINITION OF TWO-VARIABLE LINEAR INEQUALITY

A **linear inequality in two variables** can be written $ax + by < c$, where a, b, and c are real numbers and a and b are not both 0.

The above definition and the properties and definitions that follow are true for the other inequality symbols, $>$, $\geq$, and $\leq$.

In the Warm-up exercises, you practiced changing two-variable inequalities into their two most common forms. To use a graphing calculator, we must solve an inequality for y. The inequality form $ax + by < c$ is generally for noncalculator use and most commonly has a positive integer for a, the coefficient on x. Example 1 shows how to change from one inequality form to another.

EXAMPLE 1 Changing the form of an inequality

a. Solve for y: $2x - y < 4$.

b. Change to $ax + by$ form: $y \le \dfrac{3x}{2} + 3$.

SOLUTION

a.

$2x - y < 4$ — Subtract $2x$ on both sides.

$-y < -2x + 4$ — Multiply by -1 on both sides and reverse the direction of the inequality sign.

$-1(-y) > -1(-2x + 4)$ — Simplify.

$y > 2x - 4$

b.

$y \le \dfrac{3x}{2} + 3$ — Multiply both sides by 2.

$2y \le 3x + 6$ — Subtract $3x$ on both sides.

$-3x + 2y \le 6$ — Multiply both sides by -1 and reverse the direction of the inequality sign.

$-1(-3x + 2y) \ge -1(6)$ — Simplify.

$3x - 2y \ge -6$ ◀

Because we can solve $ax + by = c$ for the intercepts by letting $x = 0$ or $y = 0$, the form $ax + by < c$ is useful for graphing by hand. In Example 2, we review finding the intercepts of equations.

EXAMPLE 2 Finding intercept points What are the x-intercept point and y-intercept point for each of these equations?

a. $y - x = 2$ **b.** $y - x = 0$ **c.** $40x + 20y = 160$

SOLUTION The x-intercept point has $y = 0$. The y-intercept point has $x = 0$.

a.

$y - x = 2$ — Substitute $y = 0$.

$0 - x = 2$

$x = -2$ — The x-intercept point is $(-2, 0)$.

$y - x = 2$ — Substitute $x = 0$.

$y - 0 = 2$

$y = 2$ — The y-intercept point is $(0, 2)$.

b.

$y - x = 0$ — Substitute $y = 0$.

$0 - x = 0$

$x = 0$ — This line passes through the origin. The ordered pair $(0, 0)$ is both the x-intercept point and the y-intercept point.

c.

$40x + 20y = 160$ — Substitute $y = 0$.

$40x + 20(0) = 160$

$x = 4$ — The x-intercept point is $(4, 0)$.

$40x + 20y = 160$ — Substitute $x = 0$.

$40(0) + 20y = 160$

$y = 8$ — The y-intercept point is $(0, 8)$. ◀

▶ Graphing a Two-Variable Inequality

TESTING ORDERED PAIRS An ordered pair is a solution to a two-variable inequality if it makes the inequality a true statement. In Example 3, we test ordered pairs (x, y) to see if they make an inequality true.

▶ **EXAMPLE 3** Finding out if ordered pairs are in a solution set Substitute each ordered pair into the inequality $y \geq x + 2$ to see if it makes the inequality true. Plot the ordered pairs, and compare the positions of the points to that of the line $y = x + 2$.

a. $(0, 0)$ **b.** $(2, 1)$ **c.** $(0, 2)$

d. $(-3, 2)$ **e.** $(0, 4)$ **f.** $(-3, -1)$

SOLUTION **a.** The point $(0, 0)$ makes a false inequality, $0 \geq 0 + 2$.

b. The point $(2, 1)$ makes a false inequality, $1 \geq 2 + 2$.

c. The point $(0, 2)$ makes a true inequality, $2 \geq 0 + 2$.

d. The point $(-3, 2)$ makes a true inequality, $2 \geq -3 + 2$.

e. The point $(0, 4)$ makes a true inequality, $4 \geq 0 + 2$.

f. The point $(-3, -1)$ makes a true inequality, $-1 \geq -3 + 2$.

The "false" points $(0, 0)$ and $(2, 1)$ are both to the right of the line. The "true" points $(0, 2)$ and $(-3, -1)$ lie on the line $y = x + 2$ (see Figure 31). The points $(-3, 2)$ and $(0, 4)$ are also true and lie to the left of the line.

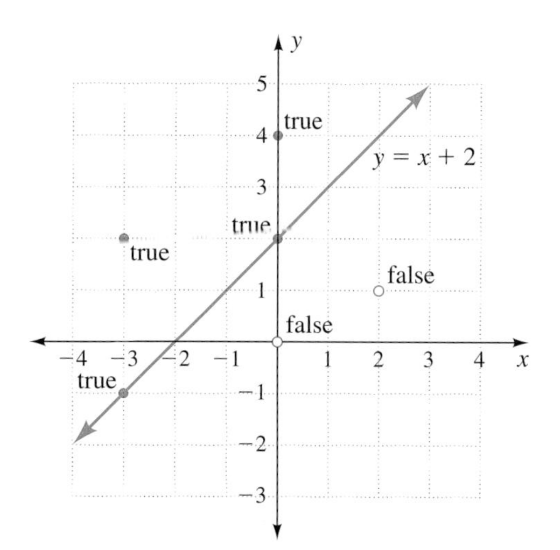

FIGURE 31 Test points for $y \geq x + 2$ ◀

GRAPHING SOLUTIONS Example 3 suggests that the solution set to two-variable inequalities will be a region on the coordinate plane. A **half-plane** is *the region on one side of a line*. The graph of every line determines two half-planes. Only one of these half-planes will make an inequality true. To find which half-plane is in the solution set, we use an ordered pair or a test point.

The line between the half-planes is called the **boundary line**. Both the half-plane and the boundary line must be considered in graphing the solution set. The inequality sign tells us whether the boundary line is part of the solution set.

Boundary Lines

> If the inequality contains $>$ or $<$, the boundary line on the half-plane is not in the solution set and we draw a dashed line.
>
> If the inequality contains $\geq$ or $\leq$, the boundary line on the half-plane is in the solution set and we draw a solid line.

▶ **EXAMPLE 4** Finding a solution set Graph $y \geq x + 2$ on coordinate axes. Use a test point to find the region described by the inequality.

SOLUTION The graph of $y \geq x + 2$ combines the graph of a boundary line, $y = x + 2$, with the half-plane given by $y > x + 2$.

We graph the boundary line $y = x + 2$ in Figure 32. Because the inequality contains $\geq$, we use a solid line for the boundary. Ordered pairs on the boundary line make the inequality true.

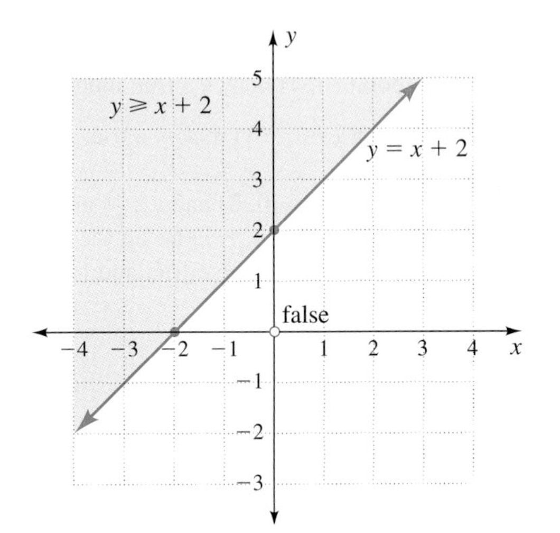

FIGURE 32 Graph for $y \geq x + 2$

We next select a test point—say, (0, 0)—and substitute it into the inequality.

$y \geq x + 2$ Let $x = 0$, $y = 0$.

$0 \geq 0 + 2$ Simplify.

$0 \geq 2$

The inequality is false. We shade the region on the opposite side from (0, 0). The solution set is the boundary line and the half-plane to its left. ◀

There are two special ideas to remember about test points:

First, always substitute the test point into the original inequality.

Second, the easiest test point to check in any inequality is the origin, (0, 0). The origin is a good test point unless it is on the boundary line.

▶ **EXAMPLE 5** Graphing a solution set where the boundary line passes through the origin Graph the inequality $y - x > 0$.

SOLUTION To find the boundary line, we replace the inequality with an equal sign and solve for y:

$$y - x = 0$$

$$y = x$$

We graph the boundary line $y = x$ in Figure 33 with a dashed line.

The boundary line passes through the origin, so we choose another test point—say, (0, 5).

$$y - x > 0$$

$$5 - 0 > 0$$

$$5 > 0$$

The inequality is true, so we shade the (0, 5) side of the boundary line.

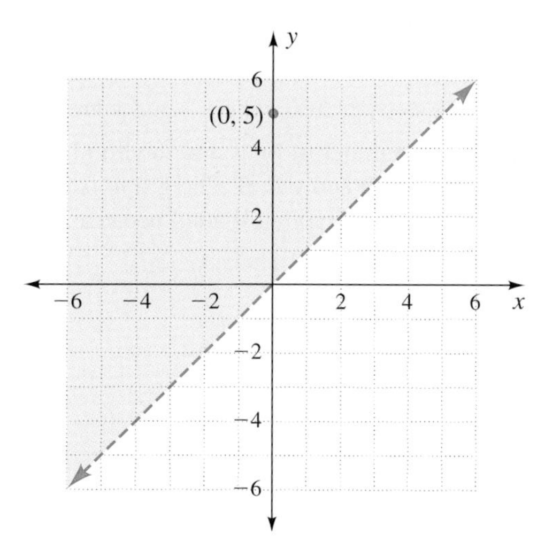

FIGURE 33 Graph for $y - x > 0$ ◀

In summary:

Graphing a Two-Variable Linear Inequality

1. Graph the boundary line formed by replacing the inequality sign with an equal sign. Use a dashed line if the inequality is $<$ or $>$. Use a solid line if the inequality is $\leq$ or $\geq$.
2. Select a test point not on the boundary line. Substitute the ordered pair for the test point into the inequality.
3. If the test point in step 2 creates a true statement, shade the half-plane that contains the test point. If the test point creates a false statement, shade the half-plane that does not contain the test point.

▶ Applications

Examples 6 and 7 illustrate applications where we must write an inequality and where we must think about the domain and range of the problem situation in drawing our graph.

▶ **EXAMPLE 6** Writing and graphing an inequality Drive-Thru-Mart keeps the currency in its overnight cash drawer limited to \$100. Suppose the cash drawer has x one-dollar bills, y five-dollar bills, and no larger bills.

Student Note:
Questions a to d are designed to help you understand the problem setting.

a. What ordered pair describes the number of one-dollar bills that can be held with 10 fives? no fives?
b. What ordered pair describes the number of fives that can be held with 10 ones? no ones?
c. Write an expression for how much money there is in x ones.
d. Write an expression for how much money there is in y fives.
e. What inequality describes the possible numbers of the two types of bills that can be held with the \$100 limit?
f. What is one limit on the domain for this application? on the range? Graph the inequality from part e.

SOLUTION **a.** With 10 fives, there can be up to 50 ones: (50, 10). With no fives, there can be up to 100 ones: (100, 0).
b. With 10 ones, there can be up to 18 fives: (10, 18). With no ones, there can be up to 20 fives: (0, 20).
c. $1x$
d. $5y$
e. $1x + 5y \leq 100$, where x and y are integers
f. The number of ones and the number of fives must be zero or a positive integer, so $x \geq 0$ and $y \geq 0$. The graph is in Figure 34. The origin makes the inequality true. The graph is shaded because the dots will be close together.

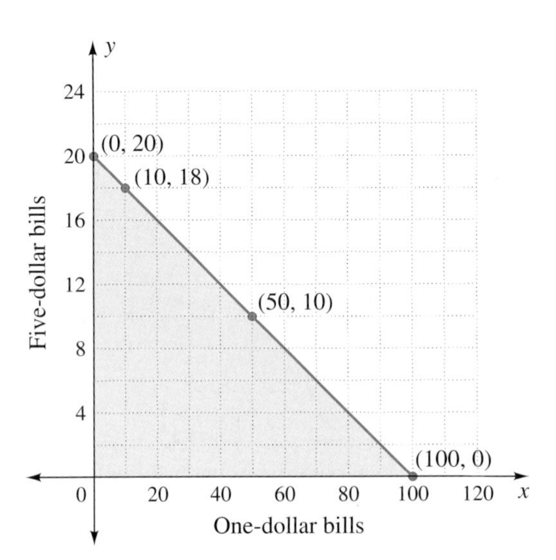

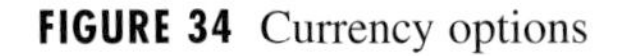
FIGURE 34 Currency options ◀

▶ **EXAMPLE 7** Writing and graphing an inequality Sens-a-diet allows a daily maximum of 160 calories in snacks. Caramel candies have 40 calories each, and ginger snaps have 20 calories each. Let x be the number of caramels and y be the number of ginger snaps.

Student Note:
Questions a to d are designed to help you read the problem and understand the setting.

a. What ordered pair describes the number of ginger snaps that can be eaten with 3 caramels? 0 caramels?
b. What ordered pair describes the number of caramels that can be eaten with 4 ginger snaps? 0 ginger snaps?
c. Write an expression for the number of calories in x caramels.
d. Write an expression for the number of calories in y ginger snaps.
e. What inequality describes the numbers of the two types of snacks that can be eaten with the 160-calorie maximum?
f. What is one limit on the domain for this application? on the range? Graph the inequality from part e.

SOLUTION **a.** (3, 2), (0, 8) **b.** (2, 4), (4, 0)

c. $40x$ **d.** $20y$

e. $40x + 20y \leq 160$, where x and y are integers

f. *To graph the inequality by hand*, we first determine the domain and range. The number of candies and the number of ginger snaps must be zero or positive, so the domain is $x \geq 0$ and the range is $y \geq 0$. The solution set is in the first quadrant and includes the axes. We then locate the boundary line $40x + 20y = 160$ by plotting the four ordered pairs from parts a and b. The line is shown in Figure 35. Then we choose a test point—say, (3, 1).

$$40x + 20y \leq 160 \qquad \text{Substitute (3, 1).}$$
$$40(3) + 20(1) \leq 160 \qquad \text{Simplify.}$$
$$140 \leq 160$$

The inequality is true, so we shade on the (3, 1) side of the boundary line and show dots for the integer solutions.

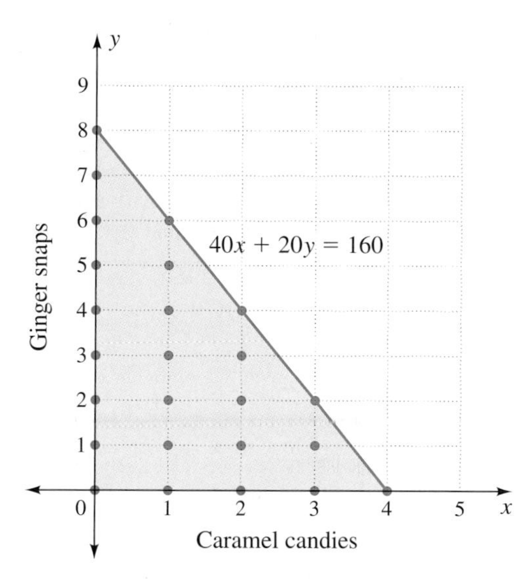

FIGURE 35 Snack options

To graph the inequality on a calculator, we first solve the inequality for y:

$$40x + 20y \leq 160 \qquad \text{Subtract } 40x \text{ on both sides.}$$
$$20y \leq -40x + 160 \qquad \text{Divide by 20 on both sides.}$$
$$y \leq -2x + 8$$

Then we enter $Y_1 = -2X + 8$, using a window that shows the first quadrant and the intercepts in parts a and b. Next we select an ordered pair as a test point—say, (3, 1).

$$y \leq -2x + 8 \qquad \text{Substitute } x = 3, y = 1.$$
$$1 \leq -2(3) + 8$$
$$1 \leq 2$$

The point is true, so we shade the half-plane containing (3, 1), which is below the line. ◀

GRAPHING CALCULATOR TECHNIQUE: SHADING OPTION

Choose the shading option with the Y = key. Enter $Y_1 = -2X + 8$. Place the cursor to the left of Y_1 and press ENTER until you see the shading below the line option, ◣. Set $0 \le x < 5$ and $0 \le y < 9$ in the viewing window. The equation and shading option shown in Figure 36 will produce the graph of $y \le -2x + 8$.

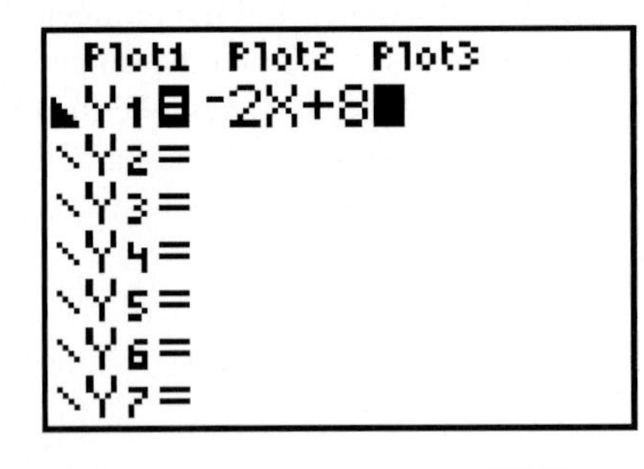

FIGURE 36 Shading option on Y =

ANSWER BOX

Warm-up: 1. a. $y > 2x - 4$ **b.** $y \le -\frac{3x}{2} + 3$ **c.** $y < \frac{2x}{3} - 2$ **d.** $y \ge \frac{1}{2}x + 2$ **2. a.** $-3x + y < -1$ **b.** $4x + y \ge 2$ **c.** $-\frac{1}{2}x + y > 3$ **d.** $-\frac{1}{3}x + y \le -3$ (The answers to Exercise 2 are commonly written without fractions or negative integers on the x term: **2. a.** $3x - y > 1$ **b.** $4x + y \ge 2$ **c.** $x - 2y < -6$ **d.** $x - 3y \ge 9$)

▶ 4.6 Exercises

In Exercises 1 to 6, identify and carry out the steps needed to change the first inequality into the second.

1. $-3x + y < -1$ to $3x - y > 1$

2. $-2x - y < 2$ to $2x + y > -2$

3. $-\frac{1}{2}x + y > 3$ to $x - 2y < -6$

4. $-\frac{1}{3}x + y \le -3$ to $x - 3y \ge 9$

5. $2x - 3y > 6$ to $y < \frac{2x}{3} - 2$

6. $2y - x \ge 4$ to $y \ge \frac{1}{2}x + 2$

In Exercises 7 to 14, which ordered pairs are solutions to the inequality?

7. $x + y > 3$; $(-2, 3)$, $(4, 0)$, $(1, 4)$

8. $x - y < 2$; $(4, 1)$, $(1, 4)$, $(-2, 1)$

9. $\frac{1}{2}x + y \ge 2$; $(-2, 3)$, $(4, -2)$, $(6, -1)$

10. $y - \frac{1}{2}x \le 3$; $(3, 2)$, $(0, 4)$, $(5, 2)$

11. $2x - 3y > 4$; $(0, 0)$, $(1, -1)$, $(3, 0)$

12. $2x + y \le 1$; $(0, 1)$, $(1, 0)$, $(-1, 1)$

13. $y \le -2$; $(-2, 3)$, $(-2, 0)$, $(0, -2)$

14. $x < 4$; $(3, 5)$, $(5, 4)$, $(-5, 0)$

In Exercises 15 to 28, write the intercepts for the boundary line and then graph the inequality on the coordinate plane.

15. $y < -3x + 2$

16. $y < 2x - 11$

17. $y \ge 4x - 1$

18. $y \ge x + 3$

19. $y \le 2 - 2x$

20. $y \le 3 - 3x$

21. $2x + y > 5$

22. $2x - y < 4$

23. $2 - 2y < 4x$

24. $6 + 3y \ge 2x$

25. $x > 4$

26. $y \ge -3$

27. $y \le 4$

28. $x < -2$

Blue numbers are core exercises.

In Exercises 29 to 36, write the equation of the boundary line and then write the inequality represented by the shaded coordinate plane.

29.

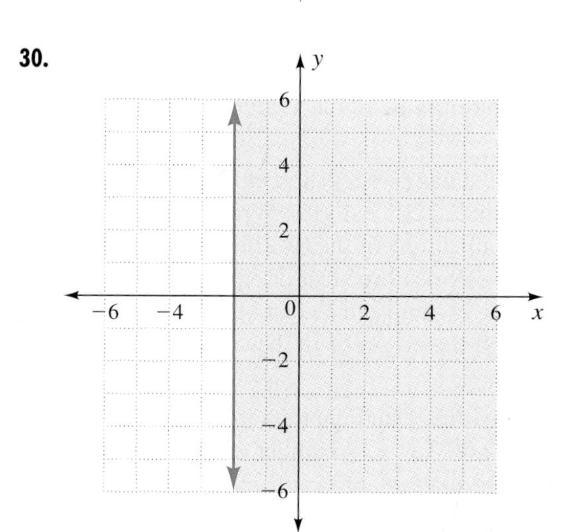

30.

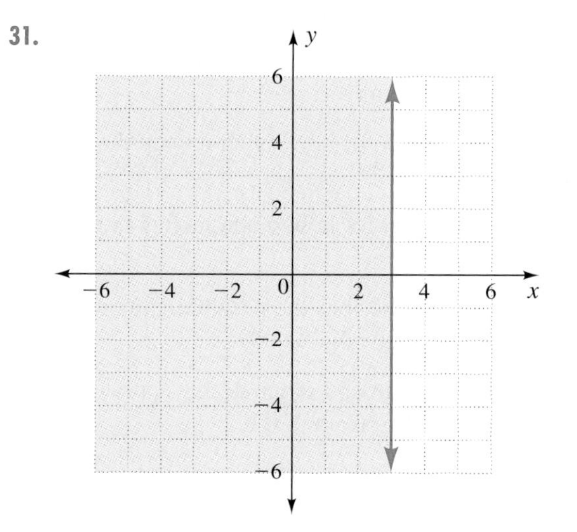

31.

32.

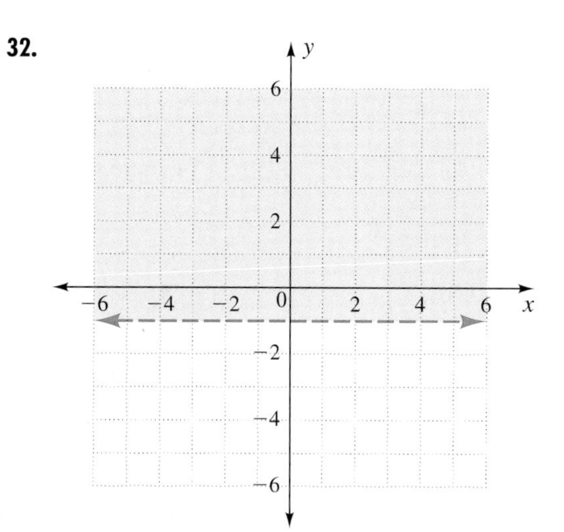

33.

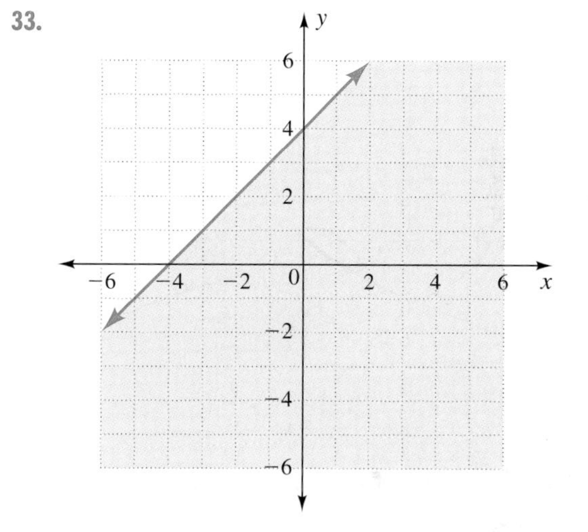

34.

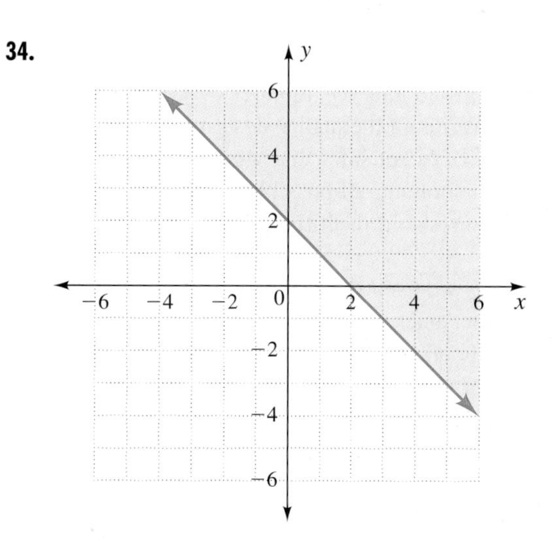

Blue numbers are core exercises.

35.

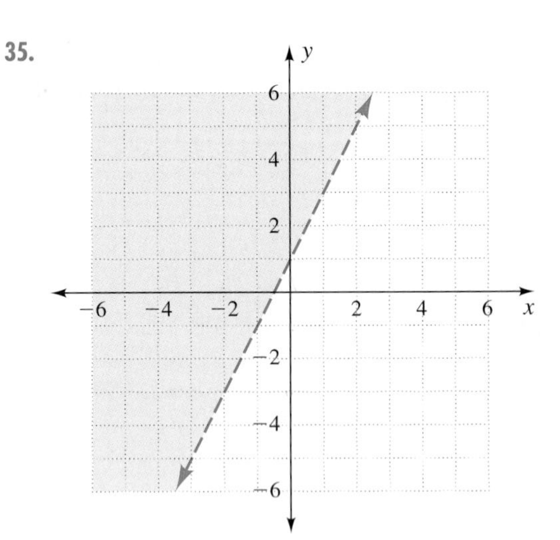

36.

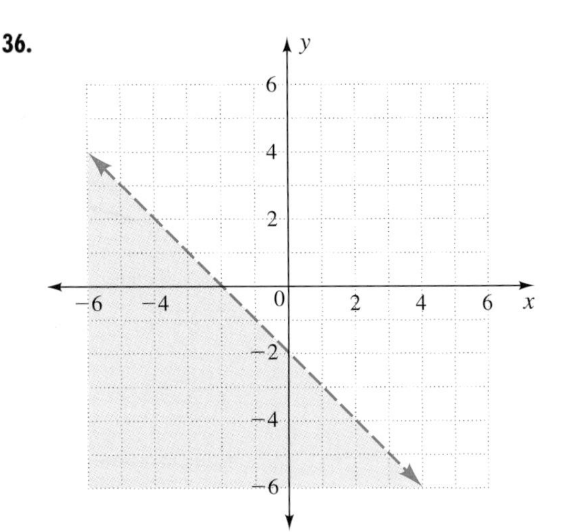

37. To meet expenses, a local theater group has a ticket sales goal of \$2400. Regular tickets sell for \$16, and student/senior tickets sell for \$12. Let x = number of regular tickets sold and y = number of student/senior tickets sold. Write an inequality describing possible combinations of ticket sales that would meet the goal. Draw a graph showing all possible combinations of ticket sales that would meet the goal.

38. In Example 7, we used the inequality $40x + 20y \leq 160$ to describe the snacks allowed with a 160-calorie maximum. If we increase the calories to 240, how will the graph change? (*Hint:* Find the new intercepts.)

39. A dieter is allowed 140 calories for a snack.

a. What are three possible combinations of apricots at 20 calories each and tangerines at 35 calories each?

b. Plot the boundary line, and describe the region that shows sensible solutions.

c. What inequality describes the snacks? What is one limit on the domain? on the range?

40. A dieter limits a snack to 60 calories.

a. What are three possible combinations of small carrots at 20 calories each and medium celery stalks at 3 calories each?

b. Plot the boundary line, and describe the region that shows sensible solutions.

c. What inequality describes the snacks? What is one limit on the domain? on the range?

41. A cup of raw raspberries contains 30 mg of ascorbic acid (vitamin C). A cup of raw sliced peaches contains 10 mg of ascorbic acid. Write an inequality to show the combinations of the two fruits that will provide at least 60 mg of ascorbic acid. What are the intercepts with the axes? Will the origin be in the solution set?

42. A teaspoon of sugar contains 15 calories. A grape contains approximately 3.5 calories. Write an inequality to show the combinations of the two items that will provide an amount of calories less than or equal to the 160 calories found in a 12-fluid-ounce can of cola. Estimate the intercepts. Will the origin be in the solution set?

43. Sesha has only dimes and quarters in her pocket. She has at most \$2.00. Let x = number of dimes and y = number of quarters.

a. What ordered pair describes the number of dimes with 4 quarters? 0 quarters?

b. What ordered pair describes the number of quarters with 15 dimes? 0 dimes?

c. Write an expression for how much money there is in x dimes.

d. Write an expression for how much money there is in y quarters.

e. What inequality describes the possible number of coins that could make the \$2.00?

f. What are the domain and range in this application? Graph the inequality from part e.

44. Describe the region shown in the figure. What party scenario might the figure describe?

Blue numbers are core exercises.

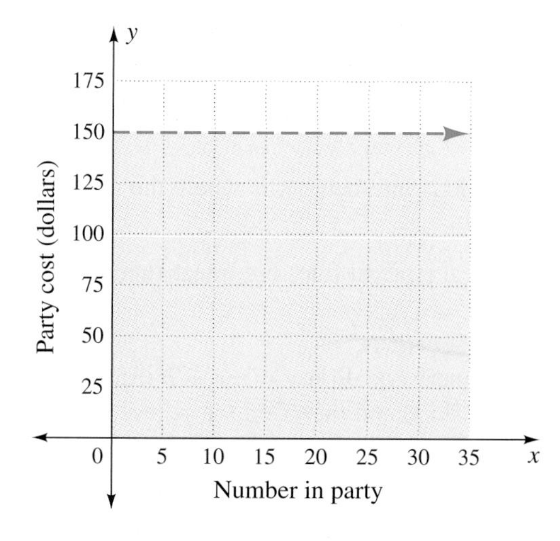

▶ Writing

45. Explain how to find which side of a boundary line to shade when graphing the solution set to an inequality.

46. Explain why some boundary lines are dashed and others are solid.

47. Explain the difference between the graphs of $x \geq 3$ as a one-variable inequality and as a two-variable inequality.

48. Explain how to find out if an ordered pair is a solution of a two-variable inequality.

▶ Project

49. Quadrants For parts a to c, identify the regions described by the inequalities.
(*Hint:* $x \geq 0$ and $y \geq 0$ describe points in the first quadrant, or on the positive portions of the axes.)

a. $x > 0$ and $y \leq 0$

b. $x < 0$ and $y \geq 0$

c. $x < 0$ and $y < 0$

What inequalities describe the regions in parts d to f?

d. The second quadrant without axes

e. The fourth quadrant without axes

f. The third quadrant together with the negative x-axis and the negative y-axis

▶ 4 Chapter Summary

Vocabulary

For definitions and page references, see the Glossary/Index.

boundary line
delta, Δ
domain
function
function notation
half-plane
horizontal-axis intercept point
in terms of
linear inequality in two variables
linear regression
parallel lines
perpendicular lines
range
rise
run
slope
slope formula
undefined slope
vertical-axis intercept point
vertical-line test
x-intercept
x-intercept point
y-intercept
y-intercept formula
y-intercept point
zero slope

Concepts

4.1 Formulas

When solving formulas, observe the order of operations on the selected variable and write a plan that uses the inverse operations in the reverse order.

Solving formulas results in answers in expression form. Solving one-variable equations results in numerical answers.

4.2 Functions and Graphs

The function notation $f(x)$ is read "function of x" or "f of x." The f and x in $f(x)$ are not being multiplied. The notation for evaluating a function for an input a is $f(a)$. To evaluate $f(a)$, substitute a for every x in the expression and then simplify.

4.3 Linear Equations: Slope and Rate of Change

Because the scales on the axes affect the appearance of steepness of a line, we use a number, *slope*, to describe the steepness. The slope describes a rate of change.

$$\text{Slope} = \frac{\text{vertical change}}{\text{horizontal change}} = \frac{\text{rise}}{\text{run}}$$

$$= \frac{\text{change in output}}{\text{change in input}} = \frac{\Delta y}{\Delta x}$$

$$m = \frac{y_2 - y_1}{x_2 - x_1} \quad \text{slope formula}$$

Lines that rise from left to right have *positive slope.*

Lines that drop from left to right have *negative slope.*

A horizontal line has *zero slope*, because the change in y is zero.

A vertical line has *undefined slope*, because the change in x is zero.

The slope of a line is constant.

If the slope between points A and B is the same as the slope between points A and C, then points A, B, and C all lie on the same line.

The units on slope are the units on the vertical axis divided by the units on the horizontal axis.

4.4 Linear Equations: Intercepts and Slope

The equation for a linear function is $y = mx + b$, where the number m is the slope and the number b is the vertical-axis intercept, or y-intercept.

The equation of a vertical line is $x = a$, where (a, y) is any point on the line.

The equation of a horizontal line is $y = b$, where (x, b) is any point on the line.

To draw a line with a given slope and ordered pair, graph the point and then use the slope to find a second point.

The horizontal-axis intercept, a, is where the output is zero, $y = 0$, or $f(x) = 0$.

The vertical-axis intercept, b, is where the input is zero, $x = 0$, or $f(0)$.

All nonvertical straight lines are linear functions.

4.5 Linear Equations

To find the linear equation $y = mx + b$ from ordered pairs, first find the slope and then find the y-intercept. Find m using the definition or slope formula. Find b using $b = y - mx$, where (x, y) is any point on the line.

Use data, a table, or a graph to write ordered pairs and then, as needed, find the slope and the y-intercept.

Parallel lines have the same slope but different y-intercepts.

Two lines are perpendicular

- if their slopes multiply to -1 (that is, their slopes are opposite reciprocals) or
- if their slopes are the same as those of the horizontal and vertical axes.

4.6 Inequalities in Two Variables

To solve $y > ax + b$, graph $y = ax + b$ as a boundary line and then substitute in a test point to find out which side of the boundary line to shade. Use a solid boundary line for inequalities containing $\geq$ or $\leq$. Use a dashed boundary line for inequalities containing $>$ or $<$.

4 Review Exercises

In Exercises 1 to 4, solve the equations for the indicated variable.

1. $6 = 4(-1) + b$ for b

2. $-5 = \frac{1}{2}(-6) + b$ for b

3. $37 = \frac{5}{9}(F - 32)$ for F

4. $100 = \frac{5}{9}(F - 32)$ for F

5. Describe the formula in Exercise 8 below, using the phrase *in terms of.*

6. Describe the formula in Exercise 9 below, using the phrase *in terms of.*

In Exercises 7 to 22, solve the formulas for the indicated variable.

7. $W = hp$ for h

8. $A = \dfrac{bh}{2}$ for b

9. $I = \dfrac{AH}{T}$ for T

10. $R = \dfrac{E^2}{W}$ for W

11. $PV = nRT$ for T

12. $W = I^2R$ for R

13. $ax + by = c$ for x

14. $P = 2l + 2w$ for l

15. $C = 35 + 5(k - 100)$ for k

16. $P_1V_1 = P_2V_2$ for P_2

17. $y_2 - y_1 = m(x_2 - x_1)$ for m

18. $A = \frac{1}{2}h(b_1 + b_2)$ for b_1

19. $\dfrac{P_1V_1}{T_1} = \dfrac{P_2V_2}{T_2}$ for V_2

In Exercises 20 to 22 which formula, if any, in each set is not equivalent to the first one? Explain how each equivalent formula was obtained from the first formula.

20. a. $P = R - C$
$P - C = R$
$P + C = R$
$P - R = -C$

b. $C = p + 0.08p$
$C = 1.08p$
$C = p(1 + 0.08)$
$\frac{C}{1.08} = p$

21. a. $P = 2l + 2w$
$P - 2l = 2w$
$2P = l + w$
$\frac{P}{2} = l + w$

b. $d = rt$
$\frac{d}{r} = t$
$\frac{d}{t} = r$
$dr = t$

22. a. $I = prt$
$\frac{I}{pr} = t$
$\frac{I}{pt} = r$
$\frac{I}{rt} = p$

b. $A = \frac{a + b + c}{3}$
$\frac{A}{3} = a + b + c$
$3A = a + b + c$
$A = \frac{a}{3} + \frac{b}{3} + \frac{c}{3}$

In Exercises 23 and 24, a formula and values are given. Solve for the remaining variable.

23. If $A = \frac{1}{2}h(a + b)$ with $A = 27$, $h = 3$, and $a = 9$, what is b?

24. If $A = \frac{a + b + c}{3}$ with $A = 80$, $a = 75$, and $b = 78$, what is c?

In Exercises 25 to 28, find $f(0)$, $f(3)$, $f(-5)$, and $f(a)$ for each function.

25. $f(x) = 7 - 2x$ **26.** $f(x) = 3x^2$

27. $f(x) = x^2 + x$ **28.** $f(x) = -3x - 4$

29. This function refers to slang names for money, with the inputs in "bits":

(2, 25), (4, 50), (6, 75), (8, 100)

a. What is the set of inputs, or domain?

b. What is the set of outputs, or range?

c. Extra: What are the outputs?

30. This function refers to the card game cribbage:

(5, 10), (6, 9), (7, 8), (8, 7), (9, 6), (10, 5)

a. What is the domain?

b. What is the range?

c. Extra: What is the rule?

31. This function is common in describing nails:

(1, 2), (2, 6), (3, 10), (4, 20), (5, 40), (6, 60)

The input number is the length of the nail.

a. What is the domain?

b. What is the range?

c. Extra: What unit describes the nail?

32. This function is common in metric measurement:

(deka, 10), (hecto, 10^2), (kilo, 10^3), (mega, 10^6), (giga, 10^9), (tera, 10^{12})

The input is the prefix name.

a. What is the domain?

b. What is the range?

c. Extra: What is the output?

In Exercises 33 to 38, which rules are functions? The parentheses show possible input-output ordered pairs.

33. What writing prize did the person win? (Gwendolyn Brooks in 1950, Pulitzer Prize), (Alice Walker in 1983, Pulitzer Prize), (Charles Gordone in 1970, Pulitzer Prize)

34. Who was born in the given year? (1918, Paul Harvey), (1918, Ann Landers), (1918, Abigail Van Buren)

35. What was the person's year of birth? (Delores Huerta, 1930), (Cesar Chavez, 1927), (Philip Vera Cruz, 1905)

36. In what country was this person a political leader? (Emiliano Zapata, Mexico), (Simon Bolivar, Venezuela), (Salvador Allende Gossens, Chile)

37. What is the last name of a composer whose first initial is G? (G, Bizet), (G, Donizetti), (G, Gershwin), (G, Handel)

38. What is each person's given name? (Adams, John), (Adams, John Quincy), (Adams, Abigail)

In Exercises 39 and 40, make a chart with these four headings: Function, Sketch of graph with inputs -3 to $+3$, Domain, Range. List these functions in the lefthand column, and then complete the chart.

39. a. The squaring function, $f(x) = x^2$

b. $f(x) = x$

c. $f(x) = 2$

40. a. The absolute value function, $f(x) = |x|$

b. $f(x) = -x$

c. $f(x) = -2$

41. Simplify these expressions.

a. $\frac{4 - (-2)}{6 - 4}$ **b.** $\frac{3 - (-4)}{-6 - 4}$

c. $\frac{0 - (-2)}{6 - (-4)}$ **d.** $\frac{1 - 3}{-6 - 9}$

42. Find the slope of the line passing through the two points.

a. $(2, -3)$ and $(5, -1)$

b. $(0, -4)$ and $(-5, -4)$

c. $(2, 3)$ and $(2, -4)$

d. $(-3, -2)$ and $(4, 1)$

43. Draw a line with a $-\frac{5}{3}$ slope passing through $(5, 2)$.

44. Draw a line with a $\frac{-2}{5}$ slope passing through $(-2, 5)$.

45. Draw a line with a $\frac{2}{3}$ slope passing through $(-1, 4)$.

46. Draw a line with a $\frac{4}{3}$ slope passing through $(2, -3)$.

47. Explain how to tell if a line has negative slope.

48. Explain how to tell from a table if the slope will be positive.

49. Explain how to draw a line with slope $\frac{a}{b}$ through a point (x, y).

50. Choose two ordered pairs, graph them, predict the slope of the line passing through them, and then calculate the slope.

Complete each sentence in Exercises 51 to 54 by choosing one of the following:
(a) let $x = a$.
(b) substitute $y = 0$ into the equation and solve for x.
(c) let $y = b$.
(d) let $x = b$.
(e) substitute $x = 0$ into the equation and simplify.

51. To find the horizontal intercept for $y = mx + b$,

52. To find the vertical-axis intercept for $y = mx + b$,

53. To find the equation of a vertical line through $(a, 0)$,

54. To find the equation of a horizontal line through $(0, b)$,

55. What is $f(0)$ on the graph of a function?

56. What is $f(x) = 0$ on the graph of a function?

In Exercises 57 to 62, solve for x where $f(x) = 0$.

57. $f(x) = 2x - 5$

58. $f(x) = -2x + 7$

59. $f(x) = -3x - 4$

60. $f(x) = 3x + 9$

61. $f(x) = \frac{1}{2}x - 6$

62. $f(x) = \frac{1}{4}x + 8$

In Exercises 63 to 72, change the equation into $y = mx + b$ form, if possible, and find the slope, m, and the constant term (y-intercept), b.

63. $3x + 5y = 15$

64. $3y - 9x = -6$

65. $5x - 2y = 10$

66. $4x - y = -10$

67. $4y - 3x = 8$

68. $3y + 2x = 9$

69. $y + 3 = 0$

70. $x + 3 = 0$

71. $x = 3$

72. $2 - y = 0$

Find the slope and vertical-axis intercept for each formula in Exercises 73 and 74. Assume the output variable is on the left.

73. $c = 2 + 1.5(n - 1)$

74. $S = 25(c + 3)$

In Exercises 75 to 82, find the equation and sketch the graph. You may put multiple graphs on the same axes.

75. Vertical line through $(4, -2)$

76. Horizontal line through $(3, -1)$

77. Vertical line through $(-2, 1)$

78. Horizontal line through $(-1, 3)$

79. Horizontal line through $(0, -2)$

80. Vertical line through $(3, 0)$

81. Horizontal line through $(2, 0)$

82. Vertical line though $(0, -1)$

In Exercises 83 to 90, use the slope and y-intercept to build an equation. Find an equivalent equation containing no fractions. Sketch a graph.

83. $m = 2, b = 4$

84. $m = 0, b = 3$

85. $m = 3, b = 0$

86. $m = -2, b = -3$

87. $m = \frac{1}{2}, b = 2$

88. $m = -3, b = \frac{1}{2}$

89. $m = -3, b = \frac{1}{4}$

90. $m = \frac{1}{4}, b = -1$

In Exercises 91 and 92, find the slope, y-intercept, and equation for each line.

91.

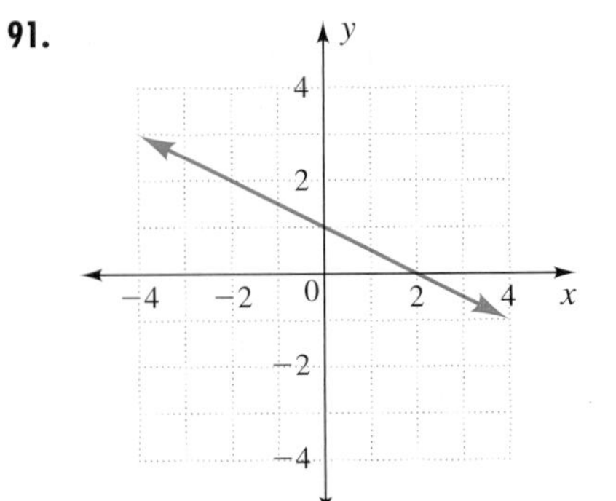

92.

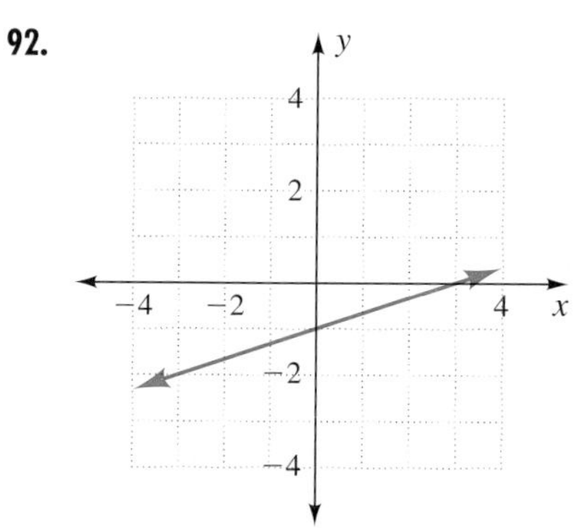

In Exercises 93 to 96, determine whether the data in the table are linear. If they are linear, give the slope and equation.

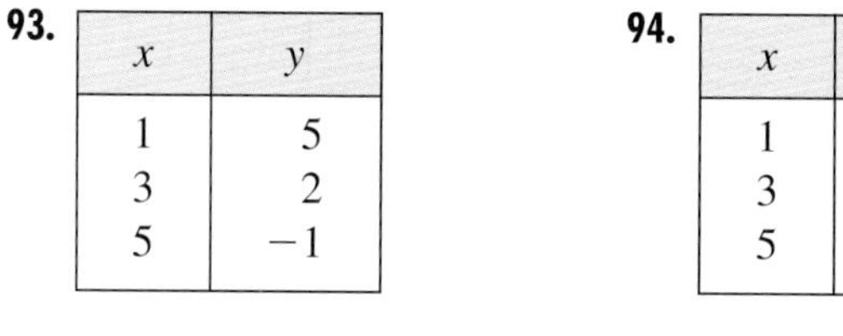

93.

x	y
1	5
3	2
5	−1

94.

x	y
1	4
3	7
5	11

95.

x	y
1	5
3	8
5	13

96.

x	y
1	6
3	1
5	−4

For Exercises 97 to 100, do the following.

(a) Write each problem situation using data points. Let time be the input. Assume a linear relationship.

(b) What is the slope of the line through these points?

(c) What is the meaning of the slope?

(d) What is the equation?

97. 14 minutes at \$0.95 and 5 minutes at \$0.50 (long-distance phone calls)

98. 2 minutes at \$1.45 and 7 minutes at \$1.45 (hotel phone calls)

99. 12 feet takes 12 hours and 32 feet takes 17 hours (sidewalk repair)

100. 6 dozen in 2 hours and 12 dozen in 3 hours (cookie baking)

101. Which one of these equations is not equivalent to the others?

a. $y = mx + b$ **b.** $y = b + mx$

c. $mx + b = y$ **d.** $y + b = mx$

e. $b + mx = y$

102. Do the ordered pairs $(-2, 4)$, $(0, 3)$, and $(4, 1)$ lie on a straight line? If so, what is the equation of the line?

In Exercises 103 to 106, find the equation and sketch the graph.

103. Line parallel to $y = 3x + 2$ through the origin

104. Line parallel to $y = \frac{1}{2}x - 4$ through $(0, 5)$

105. Line perpendicular to $y = -\frac{2}{3}x + 3$ through $(0, -2)$

106. Line perpendicular to $y = -5x - 2$ through the origin

For Exercises 107 to 109, write an equation based on the following information: A chemistry book suggests using this procedure to estimate the number of calories you need each day. Multiply your weight in pounds by an activity factor and then subtract 10 calories for each year over the age of 35 (to a maximum of 400 calories). The activity factors are 10 calories per pound for physically inactive; 15 calories per pound for moderately active; and 20 calories per pound for very active.

107. The calories needed, in terms of weight, by a moderately active 40-year-old person

108. The calories needed, in terms of weight, by an inactive 25-year-old

109. The calories needed, in terms of weight, by a very active 50-year-old

110. In the early 1980s, the federal government gave away surplus cheese, butter, and powdered milk. Eligibility guidelines for monthly family income were as follows: one person, \$507; two, \$682; three, \$857; four, \$1032; five, \$1207; six, \$1382; seven, \$1557; eight, \$1732. Are the data linear? Explain why or why not. If they are linear, fit a linear equation for maximum income in terms of number of people in the household.

111. The Budget Rental cost schedule for a 10-foot truck in New York City is shown in the table.

Miles, x	Cost, y (dollars)
5	\$24.44
10	28.89
15	33.34
20	37.79
25	42.24

a. What is the input, or independent variable?

b. What is the output, or dependent variable?

c. Are the data linear?

d. What is the cost of driving 12.5 miles?

e. How many miles may be driven for \$57?

f. What is the y-intercept? What does it mean?

g. What is the slope of the line? What does it mean?

h. What is the equation of the line?

i. What is the x-intercept? What does it mean?

112. Which ordered pair or pairs make $y \leq x - 3$ true?

a. (4, 2) **b.** (−1, −6)

c. (3, 8) **d.** (0, 0)

113. Which ordered pair or pairs make $y > 3x - 6$ true?

a. (0, −6) **b.** (−2, −12)

c. (2, 1) **d.** (3, 4)

In Exercises 114 to 119, graph the inequalities. Note that the boundary equations appeared in Exercises 63 to 68.

114. $3x + 5y > 15$ **115.** $3y - 9x \geq -6$

116. $5x - 2y < 10$ **117.** $4x - y \leq -10$

118. $4y - 3x \leq 8$ **119.** $3y + 2x > 9$

120. A diet has a 400-calorie snack limit. Vincente chooses to eat olives for his snack. Green olives contain 20 calories each, and ripe (black) olives contain 25 calories each. Define the variables. What are four different combinations of olives with fewer than 400 calories? What are four different combinations with exactly 400 calories? Plot your solutions on a graph. Shade the region of the graph that shows all combinations that satisfy the diet.

121. Matching Which inequality, $x \geq 0$, $x \leq 0$, $x > 0$, $x < 0$, $y \geq 0$, $y \leq 0$, $y > 0$, or $y < 0$, describes each of the following half-planes?

a. All points on the x-axis or in Quadrant 1 or 2

b. All points on the y-axis or in Quadrant 1 or 4

c. All points to the left of the y-axis in Quadrants 2 and 3

d. All points below the x-axis in Quadrants 3 and 4

e. All points to the right of the y-axis in Quadrants 1 and 4

f. All points on the x-axis or in Quadrant 3 or 4

122. Write a description similar to those in parts a to f of Exercise 121.

a. $x \leq 0$

b. $y > 0$

▶ Chapter 4 Project

123. Slope and Geometric Properties A square is placed on the coordinate axes (see the figure), with one corner at the origin and two sides along the horizontal and vertical axes. The length of each side is n, as indicated by the coordinates at points A, B, C, and D. Two *diagonal* lines connect the corners: AC and BD.

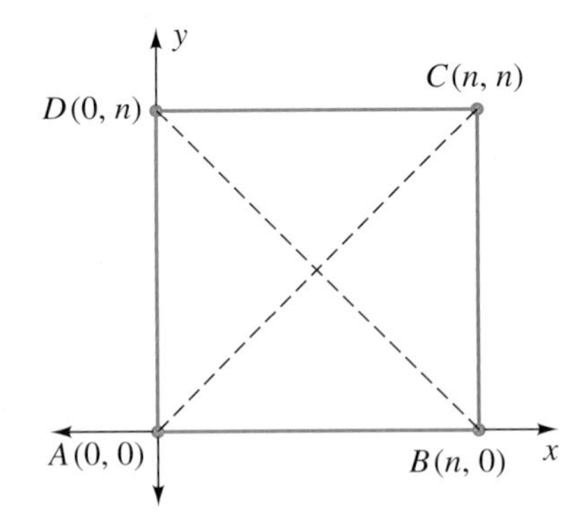

a. Find the slopes of the diagonals.

b. What can you conclude about the diagonals of a square?

c. Use the slope formula to find the slope of each side of the square: AB, BC, CD, and AD. Make at least three observations about pairs of lines and their slopes.

d. For another figure, $E(4, 1)$, $F(-1, 4)$, $G(-4, -1)$, and $H(1, -4)$, show whether or not $EFGH$ has perpendicular diagonals.

e. What additional fact is needed for $EFGH$ to be a square?

f. Use the slope formula to find the slope of each side: EF, FG, GH, and EH. Make at least three observations about pairs of lines and their slopes.

g. Write an equation for the line passing through each of EF, FG, GH, and EH.

h. Write each equation in part g without fractions.

▶ 4 Chapter Test

1. Solve $5 = \frac{1}{2}(-8) + b$ for b.

2. Write a formula describing your course grade in terms of tests, homework, and final exam.

3. Solve each formula for the indicated variable.

a. $C = \pi d$ for d **b.** $A = \frac{1}{2}bh$ for h

c. $y = mx + b$ for b **d.** $P_1V_1 = P_2V_2$ for V_2

4. Which are functions?

a. To what Native American nation did the person belong? (Sitting Bull, Sioux), (Crazy Horse, Sioux), (Cochise, Apache), (Geronimo, Apache), (Captain Jack, Modoc)

b. What person played the indicated instrument? (trumpet, Davis), (trumpet, Armstrong), (trumpet, Gillespie)

c. (2, 3), (3, 4), (4, 5)

d. (2, 3), (2, 4), (2, 6)

5. a. Find the slope for each line in the figure.

b. Which line—graph 1, 2, 3, or 4—shows the distance traveled by a bicyclist averaging 10 miles per hour?

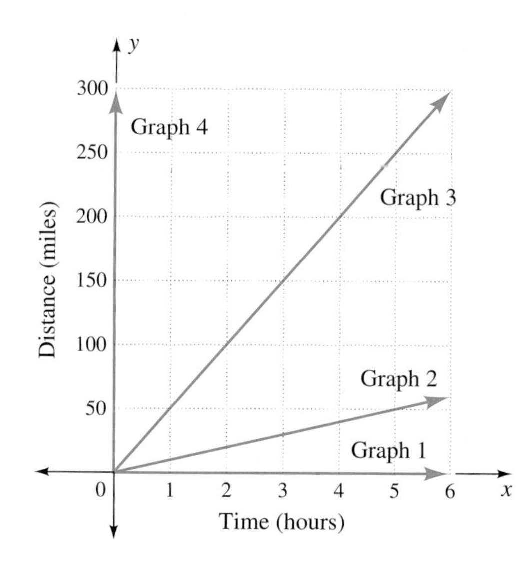

6. Find the slope of the line passing through the two points.

a. $(5, -3)$ and $(2, -1)$ **b.** $(1, -4)$ and $(0, -4)$

c. $(1, 3)$ and $(2, -4)$ **d.** $(4, -3)$ and $(4, 1)$

7. What are the slope and y-intercept for each equation?

a. $y = 5 - 2x$ **b.** $6y + 3 = 2x$

8. Draw a line with a $-\frac{3}{7}$ slope.

9. Draw a line with a $\frac{5}{3}$ slope passing through $(2, -5)$.

10. Match each function expression or equation with one of the following descriptions: $f(x) = 0$, $f(a)$, or $f(0)$.

a. the x-intercept of the graph of a function

b. the y-intercept (or vertical-axis intercept) of the graph of a function

c. the function $f(x)$ evaluated at $x = a$

In Exercises 11 and 12, write a linear equation for the given data. Write an equivalent equation containing no fractions.

11. $m = 5, b = -1$ **12.** $m = \frac{1}{3}, b = 2$

13. State whether the pair of equations describes parallel lines, perpendicular lines, or neither.

a. $y = 3x + 2$ and $y = -3x + 2$

b. $y = 2x + 3$ and $y = 2x - \frac{1}{3}$

c. $y = -\frac{1}{2}x + 3$ and $y = 2x + 3$

d. $y = 2x - 3$ and $y = \frac{1}{2}x + 3$

14. Explain how to find a linear equation from a slope and one ordered pair.

15. Explain how to find the equation of a line passing through $(0, k)$ that is parallel to $y = ax + d$.

16. Explain how to find the equation of a horizontal line through (j, k).

17. Graph $y \geq 2x + 6$.

18. Graph $2x + 3y < 12$.

19. Penske Rentals offers a 10-foot truck at all locations for the costs shown in the table.

Miles, x	Cost, y
5	\$32.90
10	35.85
15	38.80
20	41.75
25	44.70

a. What is the input, or independent variable?

b. What is the output, or dependent variable?

c. Are the data linear?

d. What is the cost for driving 22 miles?

e. How many miles can be driven for \$57?

f. What is the y-intercept? What does it mean?

g. What is the slope of the line? What does it mean?

h. What is the equation of the line?

i. What is the x-intercept? What does it mean?

20. Explain the use of parentheses in $f(x) = 2(x - 3) + 4$ on $(-2, 5]$

▶ 4 Cumulative Review of Chapters 1 to 4

For Exercises 1 to 5, follow the form of the example, as needed.

Example: Add, subtract, multiply, and divide $2\frac{2}{3}$ and $1\frac{1}{4}$.

$$2\tfrac{2}{3} + 1\tfrac{1}{4} = \tfrac{8}{3} + \tfrac{5}{4} = \tfrac{32}{12} + \tfrac{15}{12} = \tfrac{47}{12} = 3\tfrac{11}{12}$$

$$2\tfrac{2}{3} - 1\tfrac{1}{4} = \tfrac{8}{3} - \tfrac{5}{4} = \tfrac{32}{12} - \tfrac{15}{12} = \tfrac{17}{12} = 1\tfrac{5}{12}$$

$$2\tfrac{2}{3} \cdot 1\tfrac{1}{4} = \tfrac{8}{3} \cdot \tfrac{5}{4} = \tfrac{2}{3} \cdot \tfrac{5}{1} = \tfrac{10}{3} = 3\tfrac{1}{3}$$

$$2\tfrac{2}{3} \div 1\tfrac{1}{4} = \tfrac{8}{3} \div \tfrac{5}{4} = \tfrac{8}{3} \cdot \tfrac{4}{5} = \tfrac{32}{15} = 2\tfrac{2}{15}$$

1. Add, subtract, multiply, and divide $\frac{1}{3}$ and $\frac{3}{7}$.
2. Add, subtract, multiply, and divide $3\frac{1}{5}$ and $1\frac{3}{4}$.
3. Add, subtract, multiply, and divide $+4.8$ and -6.4.
4. Add, subtract, multiply, and divide $8x$ and $-2x$.
5. Add, subtract, multiply, and divide $4x^2$ and $6x$.
6. **a.** Make a table and graph showing the total cost, in dollars, of a meal with a 22% total tax and tip. Use the input set $\{0, 10, 20, 30\}$.
   b. What is the vertical-axis intercept point? What does this point mean?
   c. What equation describes the relationship in part a?
   d. What is the meaning of the slope in this setting?
7. Unit analysis: How many days are there in 1,000,000 minutes?

For Exercises 8 and 9, use the formula for the area of a trapezoid, $A = \frac{1}{2}h(a + b)$, where a and b are the parallel sides.

8. Find the area if $h = 7$ ft, $a = 12$ ft, and $b = 21$ ft.
9. The area is 91 square inches, and the sum of the parallel sides is 13 inches. Find the height, h.

For Exercises 10 to 13, solve for x in each equation.

10. $\frac{2}{3}x = 64$
11. $2(x + 5) = x + 3$
12. $10 - x = 3x - 6$
13. $9 - 2(x - 3) = 21$
14. Solve for y: $3x + 2y = 18$.
15. **a.** What is the slope of the graph of the equation in Exercise 14?
    b. What is the vertical-axis intercept for this graph?
16. Solve for x: $5 - 3x \geq -4$. Show the solution set on a line graph.
17. Draw a coordinate graph of the solution set of $y \leq -x + 5$.
18. Find $f(0)$ for $f(x) = 3x - 5$.
19. What is $f(0)$ on the graph of a function?
20. Suppose $f(x) = \frac{1}{2}x + 2$. Solve for x in $f(x) = 0$.
21. Find $f(-2)$ if $f(x) = x + 6$.
22. Solve $s = \dfrac{a}{1 - r}$ for a.
23. Find the slope, y-intercept, and equation of the line containing the points $(-4, 0)$ and $(2, -6)$.
24. Find the **a.** slope, **b.** y-intercept, and **c.** equation of the line graphed in the figure.

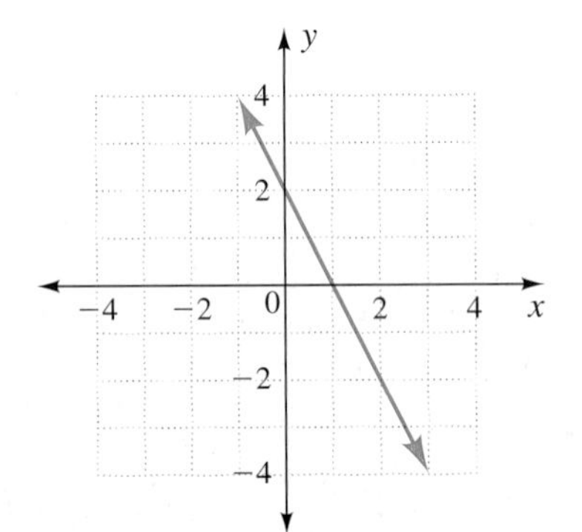

    d. State the inequality defined by the boundary and the half-plane containing the origin.
25. Arrange the terms *perpendicular lines*, *parallel lines*, *slope*, *vertical lines*, and *horizontal lines* in a concept map (see Figure 1, Chapter 1). Use within your map connecting statements that include the words *zero*, *undefined*, *opposite*, *reciprocal*, and *same*.

CHAPTER 5

Ratios, Rates, and Proportional Reasoning

The nose wheel of the plane in Figure 1 is raised off the ground. Loading or unloading the plane caused the balance point to shift behind the main landing gears, making the plane tilt. The balance point in an airplane is called the center of gravity. Center of gravity is just one of many applications of averages, a concept we will explore in Section 5.4.

In this chapter, we study ratios, percents, and proportions and relate them to various applications, including rates, similar triangles, and word problems. We extend proportional reasoning to averages in statistics and geometry.

FIGURE 1 Federal Express DC-10 at Los Angeles International Airport, August 21, 1991

▶ 5.1 Ratios, Rates, and Percents

Objectives

- Write ratios in any one of three forms.
- Simplify ratios.
- Change from one rate to another with unit analysis.
- Write percents as ratios and find percent change.
- Apply continued ratios.

WARM-UP

1. Identify the greatest common factors and simplify to lowest terms. Leave as improper fractions.

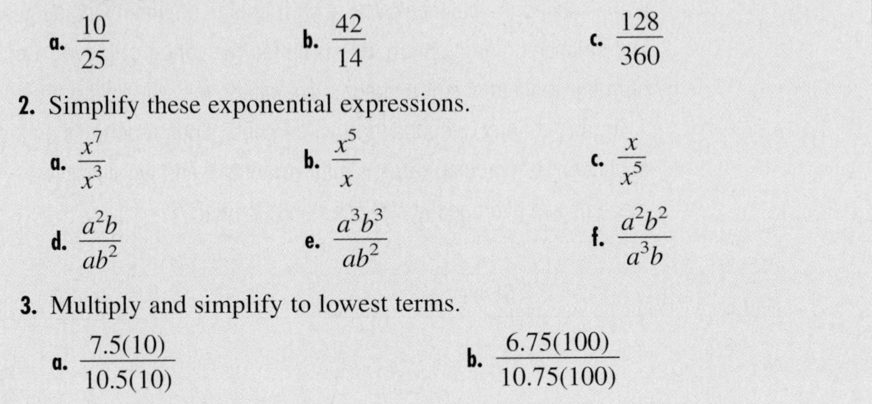

a. $\frac{10}{25}$ **b.** $\frac{42}{14}$ **c.** $\frac{128}{360}$

2. Simplify these exponential expressions.

a. $\frac{x^7}{x^3}$ **b.** $\frac{x^5}{x}$ **c.** $\frac{x}{x^5}$

d. $\frac{a^2b}{ab^2}$ **e.** $\frac{a^3b^3}{ab^2}$ **f.** $\frac{a^2b^2}{a^3b}$

3. Multiply and simplify to lowest terms.

a. $\frac{7.5(10)}{10.5(10)}$ **b.** $\frac{6.75(100)}{10.75(100)}$

IN THIS SECTION, we examine ratios, percents, and rates. We return to unit analysis, using it to simplify ratios and to change from one rate to another.

▶ Ratios

Our lives are filled with numbers, and we are continually comparing them, using words like *fast*, *slow*, *large*, *small*, *high*, *low*, *increase*, *decrease*, *near*, and *far*. A ratio is one way we have to compare two (or more) numbers.

DEFINITION OF RATIO

> A **ratio** is a comparison of two (or more) like or unlike quantities.

You will see in Examples 13 how percents and continued ratios permit us to compare three or more numbers.

There are three common ways to write the ratio of two numbers: as a fraction, with the word *to*, or with a colon.

▶ **EXAMPLE 1** Writing ratios Write each ratio in two other ways.

a. $\frac{3}{4}$ **b.** 5 to 2 **c.** 4 : 5

SOLUTION **a.** 3 to 4, 3 : 4 **b.** 5 : 2, $\frac{5}{2}$ **c.** $\frac{4}{5}$, 4 to 5 ◀

In writing continued ratios (three or more numbers), use the colon, as in 3 : 4 : 5. Engineers and contractors use the same form for describing concrete; thus 1 : 2 : 3 means 1 part cement, 2 parts sand, 3 parts round rock. Gardeners recognize the three numbers 16-16-16 or 4-6-4 as the ratio of nitrogen, phosphate, and soluble potash in fertilizer and plant food. Chemical ratios look messy but are easier once the names of the letters are learned. H_2O is water, with the ratio of 2 atoms of hydrogen to 1 atom of oxygen; C_2H_6O denotes the ratio of 2 atoms of carbon, 6 of hydrogen, and 1 of oxygen in dimethyl ether (an ingredient in hair spray).

▶ EXAMPLE 2 Understanding continued ratios In concrete, the 1 : 2 : 3 ratio represents 1 part cement to 2 parts sand to 3 parts round rock.

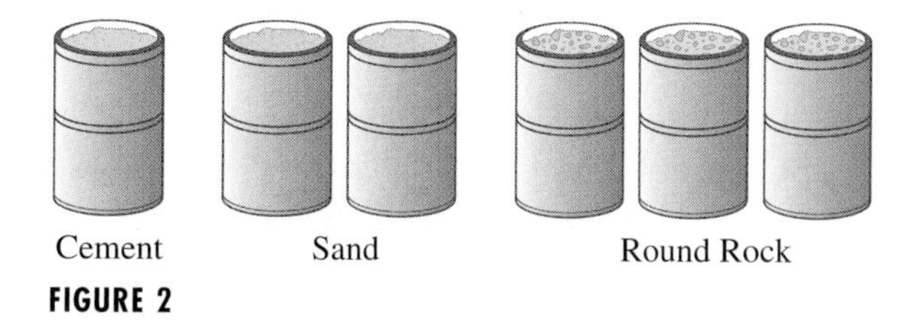

FIGURE 2

a. How many parts in total are represented?
b. What is the ratio of round rock to cement?
c. What is the ratio of sand to cement?
d. In a wheel barrow, how many shovels of sand and round rock will have to be added to 1 shovel of cement?

SOLUTION
a. $1 + 2 + 3 = 6$ parts
b. 3 to 1
c. 2 to 1
d. 2 shovels of sand and 3 of round rock ◀

If ratios of two numbers have the same decimal notation or simplify to the same fraction, they are equivalent.

EQUIVALENT RATIOS

Two ratios are **equivalent ratios** if they simplify to the same number.

Any ratio that is not in lowest terms must be simplified. The same steps used to simplify a fraction are used to simplify a ratio.

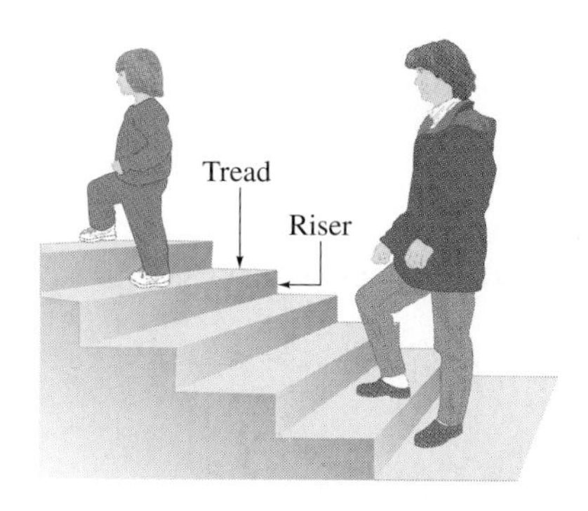

FIGURE 3 Stair slope

▶ Simplifying Ratios to Lowest Terms

Ratios are common in construction. We found the slopes of roofs in Section 4.3 and ratios of ingredients in concrete in Example 2. Here we simplify the slopes of stairs, the ratio formed by dividing the height of the *riser* by the width of the *tread* (see Figure 3). In Example 3, look for how the decimals are removed before the slope ratios are simplified.

▶ EXAMPLE 3 Simplifying ratios containing decimals: stair ratios Write each ratio of riser to tread in fraction notation. Simplify to lowest terms, if possible.

a. 5 to 12
b. 7.5 to 10.5
c. 6.75 to 10.75

SOLUTION **a.** $\frac{5}{12}$

b. First multiply the numerator and denominator by 10 to clear the decimals. Then factor the numerator and denominator and eliminate common factors.

$$\frac{7.5}{10.5} = \frac{7.5(10)}{10.5(10)} = \frac{75}{105} = \frac{3 \cdot 5 \cdot 5}{3 \cdot 5 \cdot 7} = \frac{5}{7}$$

c. First multiply the numerator and denominator by 100 to clear the decimals. Then factor the numerator and denominator and eliminate common factors.

$$\frac{6.75}{10.75} = \frac{6.75(100)}{10.75(100)} = \frac{675}{1075} = \frac{25 \cdot 27}{25 \cdot 43} = \frac{27}{43}$$ ◀

Student Note:
Stair calculator is at
http://www.blocklayer.com/Stairs/StairsEng.aspx?gad=st

SIMPLIFYING RATIOS WITH UNITS In Example 4, the units are alike and drop out when we build ratios. The ratios are from dosage computation—an important topic for nurses, medical record clerks, and pharmacists.

▶ **EXAMPLE 4** Simplifying ratios with like units Adult dosage applies to a person who weighs at least 150 pounds, is at least 150 months of age, or has a body surface area of at least 1.7 square meters. (Body surface area estimates are important in burn treatment.) Suppose a 60-pound 84-month-old has a body surface area of 1 square meter.

a. What is the ratio of the child's weight to adult weight?
b. What is the ratio of the child's age to adult age?
c. What is the ratio of the child's body surface area to adult body surface area?
d. With decimal notation, compare the ratios from parts a, b, and c.

HISTORICAL NOTE
The abbreviation for pound is lb, from the Roman unit of mass, *libra*.

SOLUTION **a.** $\frac{60 \text{ lb}}{150 \text{ lb}} = \frac{2 \cdot 30}{5 \cdot 30} = \frac{2}{5}$

b. $\frac{84 \text{ mo}}{150 \text{ mo}} = \frac{6 \cdot 14}{6 \cdot 25} = \frac{14}{25}$

c. $\frac{1 \text{ m}^2 \text{ body surface area}}{1.7 \text{ m}^2 \text{ body surface area}} = \frac{1(10)}{1.7(10)} = \frac{10}{17}$

d. We use division to change the ratios to decimals. The weight ratio is 0.4 to 1. The age ratio is 0.56 to 1. The body surface area ratio is 0.59 to 1. The weight ratio would give the lowest dosage, and the body surface area ratio would give the highest dosage. ◀

If the units in the parts of a ratio are different, they must be made the same in order to compare the parts. In Example 5, we change to like units with unit analysis in order to simplify ratios and compare their parts.

▶ **EXAMPLE 5** Simplifying ratios with unlike units Use unit analysis to change the units on each ratio to like units. Simplify the ratio. Use the simplified ratio to compare the measures.

a. 100 centimeters : 2 meters
b. 1500 grams to 1 kilogram
c. Ostrich height to smallest hummingbird length, 9 feet to $2\frac{1}{4}$ inches

SOLUTION In each part below, the useful fact contains both units.

a. A useful fact is 100 cm = 1 m.

$$\frac{100 \text{ cm}}{2 \text{ m}} \cdot \frac{1 \text{ m}}{100 \text{ cm}} = \frac{100}{200} = \frac{1}{2}$$

The units can all be eliminated. We see that 100 centimeters is half of 2 meters.

b. A useful fact is 1000 g = 1 kg.

$$\frac{1500\text{ g}}{1\text{ kg}} \cdot \frac{1\text{ kg}}{1000\text{ g}} = \frac{1500}{1000} = \frac{3}{2}$$

The units can all be eliminated. We see that 1500 grams is $\frac{3}{2}$, or $1\frac{1}{2}$, times 1 kilogram.

c. A useful fact is 12 in. = 1 ft.

$$\frac{9\text{ ft}}{2\frac{1}{4}\text{ in.}} \cdot \frac{12\text{ in.}}{1\text{ ft}} = \frac{108}{2.25} = \frac{48}{1}$$

The units can all be eliminated. We divide 108 by 2.25 with a calculator, to get 48. The ostrich's height is 48 times the length of the hummingbird. ◀

Special applications in fertilizer (5-5-5) and chemical formulas ($C_{10}H_{14}C_2$) are not simplified (to 1-1-1 or $C_5H_7C_1$). More on this in the exercises.

SIMPLIFYING RATIOS WITH ALGEBRAIC EXPRESSIONS In Example 6, the ratios contain expressions.

▶ EXAMPLE 6 Simplifying ratios with algebraic expressions Write each ratio as a fraction. Then find and cross out the greatest common factor.

a. $3x^2 : 15x$

b. $24x^2 : 16x^3$

c. $35(x + 2) : 7$

SOLUTION **a.** $3x$ is the greatest common factor.

$$\frac{3x^2}{15x} = \frac{3x \cdot x}{3x \cdot 5} = \frac{x}{5}$$

b. $8x^2$ is the greatest common factor.

$$\frac{24x^2}{16x^3} = \frac{8x^2 \cdot 3}{8x^2 \cdot 2x} = \frac{3}{2x}$$

c. 7 is the greatest common factor.

$$\frac{35(x+2)}{7} = \frac{5 \cdot 7(x+2)}{7} = \frac{5x+10}{1}$$ ◀

▶ Rates and Unit Analysis

Rates are *special ratios for comparing quantities with different units.* A rate often contains the word *per*, which means "for each" or "for every."

Some rates are abbreviated with just a single letter for each word, such as mpg and rpm in Example 7. When we change a rate into a ratio, we place the word that follows *per* in the denominator.

▶ EXAMPLE 7 Writing rates as ratios Write these rates in fraction notation.

a. mph (miles per hour)
b. liters per minute
c. mpg (miles per gallon)
d. rpm (revolutions per minute)
e. dollars per hour

SOLUTION Observe the abbreviations used within the fraction form.

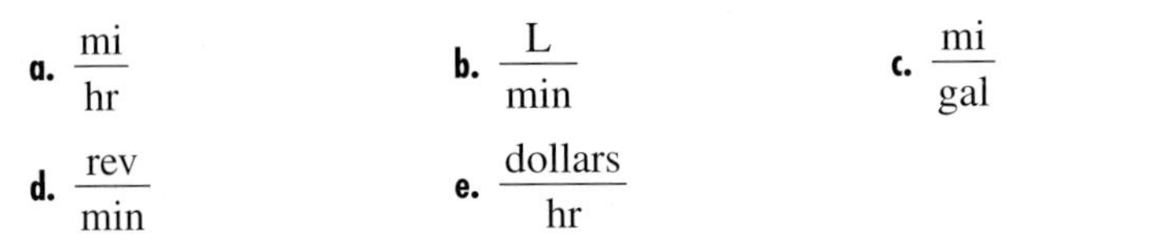

a. $\frac{\text{mi}}{\text{hr}}$ **b.** $\frac{\text{L}}{\text{min}}$ **c.** $\frac{\text{mi}}{\text{gal}}$

d. $\frac{\text{rev}}{\text{min}}$ **e.** $\frac{\text{dollars}}{\text{hr}}$ ◀

▶ To change from one rate to another, we use unit analysis. We usually have to change two different kinds of units. Here are the steps in unit analysis, from Section 2.5.

UNIT ANALYSIS: STEP BY STEP

1. Name the unit to be changed, and name the unit for the answer.
2. List facts containing the unit to be changed and all units needed to get to the unit for the answer.
3. Write the unit to be changed (or a fact containing the unit) as the first fraction. Set up a product of fractions, using the list of facts. Arrange fractions so that the units to be eliminated appear in both a numerator and a denominator. Carry out the multiplication.
4. Recheck the units noted in step 1.

In Example 8, we change miles per hour into feet per minute.

▶ **EXAMPLE 8** Changing a rate with unit analysis Change 100 miles per hour to feet per minute.

SOLUTION We use the four steps in unit analysis.

Step 1: We need to change miles to feet and hours to minutes.

Step 2: Useful facts:

5280 ft = 1 mi

60 min = 1 hr

Step 3: We start with 100 miles per hour as a fraction.

$$\frac{100 \text{ mi}}{1 \text{ hr}} \cdot \frac{5280 \text{ ft}}{1 \text{ mi}} \cdot \frac{1 \text{ hr}}{60 \text{ min}} = \frac{100 \cdot 5280 \cdot 1 \text{ ft}}{1 \cdot 1 \cdot 60 \text{ min}}$$

$$= \frac{8800 \text{ ft}}{1 \text{ min}}$$

Step 4: We see that 100 miles per hour is 8800 feet per minute. ◀

THINK ABOUT IT 1: Is 8800 feet per minute faster or slower than one mile per minute?

▶ In Step 3 of Example 8, the units that we eliminated were not next to each other. If you want the units to be next to each other in a rate problem, place the rate to be changed in the center and write the facts as fractions on each side.

$$\frac{5280 \text{ ft}}{1 \text{ mi}} \cdot \frac{100 \text{ mi}}{1 \text{ hr}} \cdot \frac{1 \text{ hr}}{60 \text{ min}}$$

We will use this suggestion in Example 9, where we change drops per second into cups per hour.

▶ **EXAMPLE 9** Changing a rate with unit analysis Suppose a faucet leaks 1 drop per second. Find out how many cups per hour are wasted.

SOLUTION *Step 1:* We need to change drops into cups and seconds into hours.

Step 2: Useful facts:

An experiment indicates that there are 16 drops per teaspoon.

3 tsp = 1 tablespoon (tbsp)

16 tbsp = 1 cup

60 sec = 1 min

60 min = 1 hr

Step 3: We place the 1 drop per second fraction in the center of the paper and write the other facts as fractions on each side of it.

Change time ← Start here → Change quantity

$$\frac{60 \text{ min}}{1 \text{ hr}} \cdot \frac{60 \text{ sec}}{1 \text{ min}} \cdot \frac{1 \text{ drop}}{1 \text{ sec}} \cdot \frac{1 \text{ tsp}}{16 \text{ drops}} \cdot \frac{1 \text{ tbsp}}{3 \text{ tsp}} \cdot \frac{1 \text{ cup}}{16 \text{ tbsp}}$$

$$= \frac{60 \cdot 60 \text{ cup}}{16 \cdot 3 \cdot 16 \text{ hr}}$$

$$\approx 4.7 \text{ cups per hour}$$

◀

▶ Ratios and Percents

FINDING RATIOS FROM PERCENTS Recall from Section 1.4 that *percent* means "per hundred," so n percent is the *ratio of n to 100.* The quantity 15 percent means 15 per hundred and is written $\frac{15}{100}$ as a fraction or ratio.

In highway construction, the *slope* of a road is its **grade**. The grade is written as a percent. Highway signs, like the one in Figure 4, are placed near the tops of mountain passes to warn truckers and motorists about the steepness of the downhill grade. A 6% grade indicates that there is a drop of 6 vertical feet for every 100 feet traveled horizontally. To change a grade to a slope ratio, rewrite the percent as a fraction and simplify to lowest terms.

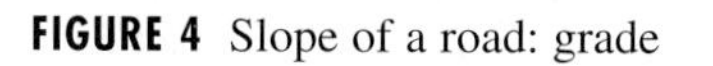

FIGURE 4 Slope of a road: grade

▶ **EXAMPLE 10** Finding ratios from percents: slope of a road grade Write each road grade as a ratio and simplify.

a. 6% **b.** 7% **c.** 5%

SOLUTION **a.** The grade is 6%:

$$\frac{6}{100} = \frac{3 \cdot \boxed{2}}{50 \cdot \boxed{2}} = \frac{3}{50}$$

b. The grade is 7%, or $\frac{7}{100}$.

c. The grade is 5%:

$$\frac{5}{100} = \frac{\boxed{5} \cdot 1}{\boxed{5} \cdot 20} = \frac{1}{20}$$

◀

FINDING PERCENT CHANGE To find the **percent change**, we *subtract the original number from the new number and divide the difference by the original number.* If the answer is positive, the percent change is an increase. If the answer is negative, the percent change is a decrease.

$$\text{Percent change} = \frac{\text{new number} - \text{original number}}{\text{original number}}$$

In Example 11, we calculate percent change.

▶ **EXAMPLE 11** Finding percent change Estimate the percent change in prices. Then find the percent change to the nearest tenth of a percent.

a. Dockers® slacks, originally \$56, on sale for \$48
b. Three-piece bathroom-rug set: 1986, \$8; 1995, \$25.97
c. Three-piece bathroom-rug set: 2003, \$49.97; 2006, \$14.99

SOLUTION **a.** The difference in prices is \$8, so the change is $\frac{1}{7}$ of the original price.

$$\frac{\$48 - \$56}{\$56} = \frac{-8}{56} \approx -0.143$$

The sale results in a 14.3% reduction in price.

b. The difference in prices is about \$18, or a little more than twice \$8. The percent change will be about 200%.

$$\frac{\$25.97 - \$8.00}{\$8.00} = \frac{17.97}{8.00} \approx 2.246$$

The 1995 price is 224.6% higher than the 1986 price.

c. The difference in prices is \$35; almost three-fourths the original price. The percent change will be almost 75%.

$$\frac{\$14.99 - \$49.97}{\$49.97} = \frac{-34.98}{49.97} \approx -0.70$$

The 2006 price is 70% lower than the 2003 price. ◀

THINK ABOUT IT 2: If you estimated 300% in part b of Example 11, you compared the 1995 price to the 1986 price, instead of comparing the change in price to the 1986 price. Calculate the percent formed by dividing the new price by the old price in each part of Example 11.

▶ **EXAMPLE 12** Using percent change For the given percent changes, find the ending size or value.

a. A 10-inch (at the shoulder) puppy grew another 50% of its height.
b. A 10-inch puppy grew another 150% of its height.
c. A 10-inch puppy grew another 200% of its height.
d. A \$100 stock lost 10% of its value.
e. A \$100 stock lost 80% of its value.
f. A \$90 stock grew by 10% of its value.

SOLUTION We change the percents to decimals and adjust the original by the change.

a. $10 + 10(0.50) = 10 + 5 = 15$ inches

b. $10 + 10(1.50) = 10 + 15 = 25$ inches

c. $10 + 10(2.00) = 10 + 20 = 30$ inches

d. $100 - 100(0.10) = 100 - 10 = \$90$

e. $100 - 100(0.80) = 100 - 80 = \$20$

f. $90 + 90(0.10) = 90 + 9 = \$99$ ◀

▶ Example 13 illustrates how we compare many parts to a total by working with percents. The percent approach may be easier to use than a continued ratio in such cases.

▶ EXAMPLE 13 Finding percents Here is how Monique spent her time during one week. What percent of her week was spent on each activity?

Classes:	12 hours
Study:	30 hours
Work:	20 hours
Sleep:	56 hours (8 hours each night)
Transportation:	7 hours
Other:	remaining time (walking between classes, with family, eating, etc.)

SOLUTION 24 hours per day $\times$ 7 days per week = 168 hours per week

Classes:	$12 \div 168 = 0.071 = 7.1\%$
Study:	$30 \div 168 = 0.179 = 17.9\%$
Work:	$20 \div 168 = 0.119 = 11.9\%$
Sleep:	$56 \div 168 = 0.333 = 33.3\%$
Transportation:	$7 \div 168 = 0.042 = 4.2\%$
Other:	$168 - (12 + 30 + 20 + 56 + 7) = 43$ hr remaining
	$43 \div 168 = 0.256 = 25.6\%$

Check: The percents add to 100%. ✓ ◀

FINDING PERCENT

> To find the percent, we divide the part by the total.

THINK ABOUT IT 3: For Example 13, write the hours for the six activities as a continued ratio.

ANSWER BOX

Warm-up: 1. a. $5, \frac{2}{5}$ **b.** $14, \frac{3}{1}$ **c.** $8, \frac{16}{45}$ **2. a.** x^4 **b.** x^4 **c.** $\frac{1}{x^4}$ **d.** $\frac{a}{b}$ **e.** a^2b **f.** $\frac{b}{a}$ **3. a.** $\frac{5}{7}$ **b.** $\frac{27}{43}$ **Think about it 1:** Faster, $\frac{8800\text{ ft}}{1\text{ min}} \cdot \frac{1\text{ mi}}{5280\text{ ft}} \approx 1.67$ mi per min **Think about it 2: a.** $\frac{48}{56} \approx 85.7\%$ **b.** $\frac{25.97}{8} \approx 324.6\%$ **c.** $\frac{14.99}{49.97} \approx 30\%$ **Think about it 3:** 12 : 30 : 20 : 56 : 7 : 43

▶ 5.1 Exercises

In Exercises 1 and 2, write each ratio in two other ways.

1. a. 5 to 3 **b.** $\frac{3}{2}$ **c.** 4 : 9

2. a. $\frac{4}{5}$ **b.** 8 : 5 **c.** 7 to 11

In Exercises 3 and 4, identify common factors and simplify to lowest terms.

3. a. $\frac{15}{35}$ **b.** $\frac{48}{16}$ **c.** $\frac{5280}{3600}$

4. a. $\frac{84}{32}$ **b.** $\frac{52}{26}$ **c.** $\frac{1024}{288}$

For Exercises 5 and 6, write the stair slopes in fraction notation and simplify to lowest terms. All measures are in inches, so units may be disregarded.

5. a. 3 to 15 **b.** 7.5 to 10 **c.** 6.75 to 10.5 **d.** 7.25 to 9.75

6. a. 6 to 12 **b.** 7 to 10.5 **c.** 7.75 to 10.25 **d.** 6.75 to 10.25

Blue numbers are core exercises.

For Exercises 7 to 10, find the slope ratio for each equation.

7. **a.** $y = 3x$ **b.** $y = 4 + 2x$ **c.** $y = 2 - 4x$

8. **a.** $y = -5x$ **b.** $y = \frac{1}{2}x$

c. $y = -\frac{1}{4}x$

9. **a.** $y = 2.98x + 5.00$ **b.** $y = 20 - 1.50x$

10. **a.** $y = 5 - 0.05x$ **b.** $y = 0.65x + 5.15$

In Exercises 11 and 12, change the units to like units with a unit analysis multiplication, and simplify the ratio.

Useful facts: 1000 milliliters = 1 liter
1000 grams = 1 kilogram
1000 meters = 1 kilometer
100 centimeters = 1 meter
16 ounces = 1 pound
1 yard = 36 inches
1 yard = 3 feet
1 mile = 5280 feet
4 quarts = 1 gallon
2 pints = 1 quart

11. **a.** $\frac{1}{2}$ foot to 1 inch

b. 3000 grams to 1 kilogram

c. 32 ounces to 5 pounds

d. $2\frac{2}{3}$ yards to 2 feet

e. 200 milliliters to 20 liters

f. 2 years to 150 months

g. 20 minutes to $\frac{1}{4}$ hour

12. **a.** 1 foot to 3 inches

b. 2 meters to 1200 centimeters

c. 1500 meters to 1 kilometer

d. 2 quarts to 4 gallons

e. 2 liters to 2000 milliliters

f. 4 years to 150 months

g. 120 minutes to $\frac{1}{2}$ hour

For Exercises 13 to 18, write a ratio from the sentence and simplify it, if possible. More facts are on the inside of the back cover.

13. Sergei adds 3 cans of water to each can of frozen orange juice.

14. Frantisek adds 2 cans of milk to a can of soup.

15. Mikael adds a half pint of oil to 1 gallon of gasoline.

16. Sylvia adds 1 tablespoon of plant fertilizer to a gallon of water.

17. Betty orders an intravenous feeding of 240 cc in 16 hours.

18. Johann fills an intravenous feeding order for 240 mL in 12 hours.

In Exercises 19 to 22, simplify to lowest terms.

19. **a.** $12x : 4x^2$ **b.** $2xy$ to $6x^2y$ **c.** $x : 3x^4$

20. **a.** $18x : 6x^3$ **b.** $6x^2y$ to $9xy^2$ **c.** $x : 4x^3$

21. **a.** $24(x + 2) : 6$ **b.** $15(x - 1) : 3$ **c.** $12(x + 3) : 4$

22. **a.** $28(x - 3) : 7$ **b.** $21(x + 2) : 3$ **c.** $18(x - 2) : 3$

In Exercises 23 and 24, write a sentence using each of the rates in an appropriate setting.

23. **a.** stitches per inch **b.** feet per second

c. miles per gallon **d.** revolutions per hour

24. **a.** revolutions per minute **b.** gallons per hour

c. kilometers per hour **d.** calories per day

25. Which of the following are correct calculator entries for evaluating $\dfrac{60 \cdot 60}{16 \cdot 3 \cdot 16}$?

a. 60 [×] 60 [÷] 16 [×] 3 [×] 16 [ENTER]

b. 60 [×] 60 [÷] 16 [÷] 3 [÷] 16 [ENTER]

c. 60 [×] 60 [÷] [(] 16 [×] 3 [×] 16 [)] [ENTER]

d. 60 [×] 60 [×] [(] 16 [×] 3 [×] 16 [)] [x^{-1}] [ENTER]

26. Which of the following are correct calculator entries for evaluating $\dfrac{100 \cdot 5280}{60 \cdot 60}$?

a. 100 [×] 5280 [×] 60 [x^{-1}] [×] 60 [x^{-1}] [ENTER]

b. 100 [×] 5280 [÷] 60 [×] 60 [ENTER]

c. 100 [×] 5280 [÷] [(] 60 [×] 60 [)] [ENTER]

d. 100 [×] 5280 [÷] 60 [÷] 60 [ENTER]

In Exercises 27 and 28, change miles per hour to feet per second. Start with a fraction showing miles per hour, and then set up the unit analysis product of fractions.

Useful facts: 5280 feet = 1 mile
60 seconds = 1 minute
60 minutes = 1 hour

27. **a.** 30 miles per hour **b.** 55 miles per hour

28. **a.** 80 miles per hour **b.** 500 miles per hour

Blue numbers are core exercises.

In Exercises 29 and 30, change feet per second to miles per hour (mph).

29. a. 88 feet per second **b.** 66 feet per second

30. a. 220 feet per second **b.** 4.4 feet per second

31. Use unit analysis to find the number of feet you have driven in a 4-second reaction time at 60 miles per hour. Guess before you start: Will it be more than, equal to, or less than a residential city block of length 334 feet?

32. A lawn mower cuts 100 feet in 30 seconds. How many miles per hour is this?

33. The oil in a lawn mower needs to be changed every 25 hours. If the mower travels 100 feet in 30 seconds and shuts off when stopped, how many miles will it travel between oil changes?

34. How far is it possible to bicycle in 10 minutes at 20 miles per hour?

35. How far is it possible to walk in 30 minutes at 6 miles per hour?

36. In 2005 the Helheim glacier in Greenland flowed at a rate of 14 kilometers per year. The rate was *three times the rate recorded in 2001, 1996, and 1988*. Change this rate to inches per hour.

37. Change 250 mL in 12 hours to microdrops per minute. (60 microdrops is 1 mL and 1 hr is 60 min.)

38. An infant of 1 year should receive how many milligrams (mg)? (Adult dosage is 500 mg, 150 mo = 1 adult dosage, and 1 yr = 12 mo.)

39. Change 300 mg to mL when the contents label indicates 5 mL = 0.1 g. (1 g = 1000 mg).

40. Normally units follow the number but the fact 1 gram = 15 grains is written 1 g = gr 15. Change gr 3 to milligrams. (1 g = 1000 mg).

41. Wizard Island in Crater Lake, Oregon, is a small cinder cone. The sides of the cone form a 62.5% grade. What is the slope in fraction notation?

42. Lookout Mountain in Chattanooga, Tennessee, has an inclined railway with a grade near the top of 72.7%. What is the slope in fraction notation? Is the grade closest to $\frac{3}{4}$, $\frac{5}{6}$, or $\frac{6}{7}$?

In Exercises 43 to 48, tell what percent grade is described.

43. A wheelchair access ramp with a slope of 1 to 12

44. A loading ramp with a slope of $\frac{1}{4}$

45. An access ramp with a slope of $\frac{1}{8}$

46. A freeway ramp with a slope of 1 to 20

47. 30 feet per mile

48. 250 feet per mile

In Exercises 49 to 52, estimate the percent change between 1986 and 2003. Then find the percent change to the nearest tenth of a percent. Finally, find the percent the 2003 price is of the 1986 price. Repeat for the 2003 and 2006 prices.

49. Conair folding hair dryer: 1986, $12; 2003, $19.99; 2006, $17.99

50. S&W tomatoes, 28-oz can: 1986, $1; 2003, $1.79; 2006, $1.48

51. True Temper plastic rake: 1986, $5; 2003, $10.99; 2006, $11.89

52. Comet powdered cleaner: 1986, $0.34; 2003, $0.89; 2006, $0.99

In Exercises 53 to 56, find the ending size or value for the given percent change.

53. a. The cost of a $15 meal is increased by 25% tax and tip.

b. An $18.75 belt is on sale for 25% off.

54. a. A $100 stock loses 20% of its value.

b. An $80 stock gains 20% of its value.

55. a. A 12-inch puppy grows another 50% of its height.

b. An $18 DVD is on sale for 50% off.

56. a. An 8-inch puppy grows another 40% of its height.

b. An $11.20 box of chocolates is on sale for 40% off.

57. In Exercises 53 to 56, ignore the settings and look at the numbers. The answer to part a was the original number in part b. Explain why the answer to part b was not the original number in part a.

58. Copy the formula on page 264 for percent change. Solve the formula for *new number*.

59. Tom was just informed of a 20% increase in his rent.

a. If his old rent was $450, what is his new rent?

b. If his new rent is $450, what was his old rent? Guess and check. Write an equation using the percent change formula. Solve your equation.

60. Alex just received a 25% salary increase.

a. If her old monthly salary was $3200, what is her new salary?

b. If her new monthly salary is $3200, what was her old salary? Guess and check. Write an equation using the percent change formula. Solve your equation.

Blue numbers are core exercises.

61. Marguerite spends the following each month: rent, \$360; utilities, \$132; food, \$240; clothes, \$84; travel (including bus pass), \$96; entertainment, \$48; renters' insurance, \$60; books and magazines, \$24; savings, \$156. What percent of her total monthly income does she spend on each item?

62. Jasmin spends the following each month: rent, \$400; utilities, \$140; food, \$320; clothes, \$130; car payment, \$300; car insurance, \$150; gasoline, \$60; savings, \$20; other, \$80. What percent of her total monthly income does she spend on each item?

63. The numbers on plant food are not a chemical formula but percents of the volume of the three main ingredients: nitrogen, phosphate, and potash. Hence 5-5-5 is not the same as 16-16-16 because the members of first set add to 15% of the total volume and those of the second set add to 48% of the total volume. The percent of other ingredients are 85% and 52%, respectively. What percent other ingredients are found in these bags or containers?

a. 10-10-10 (evergreen shrub)

b. 30-3-3 (turf builder)

c. 22-4-14 (lawn winterizer)

d. 10-15-10 (plant food, added to water)

e. 4-6-4 (plant food, placed on dirt)

64. Another formulation for concrete is 1 : 2 : 4 (cement, sand, round rock).

a. What is the ratio of sand to total?

b. What is the ratio of cement to round rock?

c. You get interrupted while mixing concrete. You put in 2 buckets of cement, 3 buckets of sand, and 4 buckets of round rock. How many more buckets of which ingredients need to be added?

d. What is the total cost if you order enough of each to mix with 20 cubic ft of cement? Cement costs \$6.87 per cubic foot, sand is \$1.68 per cubic ft, and round rock is \$4.43 per cubic ft.

65. During one day, an accounting consultant spent 4 hours on project A, 3 hours on project B, and 1 hour on project C. Write a continued ratio and then percents to describe the time spent on the projects. Find the amount charged to each project if the billing for the day was \$1000.

66. A \$1 million estate is to be divided among Goodwill, Habitat for Humanity, and United Way in the ratio 2 : 2 : 1. What percent goes to each? How much will each receive?

67. Chemistry Notation can make problems look harder than they are. A molecule of nicotine, $C_{10}H_{14}N_2$, has 10 atoms of carbon, 14 atoms of hydrogen, and 2 atoms of nitrogen. What is the continued ratio of atoms in the following chemical formulas? (Ca is calcium, Na is sodium, and O is oxygen)

a. $C_9H_8O_4$ (the active ingredient in aspirin)

b. $NaHCO_3$ (baking soda)

c. $CaCO_3$ (chalk)

d. $C_{12}H_{22}O_{11}$ (sugar)

e. $C_8H_{10}N_4O_2$ (caffeine)

68. Chemical Unit Analysis The *mole* referred to in this exercise is a chemical unit of measure, not a small furry animal. Do nothing with the chemical formulas.

a. Apply unit analysis to 25.0 grams CH_4 to find grams HCN. (Chemists assume that the environment has an ample source of nitrogen, N.) *Hint:* Start with 25.0 grams CH_4 over 1. Round answer to the nearest tenth.

1 mole CH_4 is 16.0 grams CH_4
1 mole HCN is 27.0 grams HCN
2 moles CH_4 is equated with 2 moles HCN

b. Apply unit analysis to 7.0 grams Na to find grams NaCl. (Chemists assume that the environment has an ample source of chlorine, Cl.) *Hint:* Start with 7.0 grams Na over 1. Round answer to the nearest tenth.

1 mole NaCl is 58.5 grams NaCl
2 moles Na is equated with 2 moles NaCl
1 mole Na is 23.0 grams Na

69. In Example 3, the stair ratios were 5 to 12, 7.5 to 10.5, and 6.75 to 10.75. Arrange the slopes from flattest to steepest. Explain how you compared them.

70. Ratio Error Analysis Class A has a ratio of 5 to 6 of girls to boys. Class B has a ratio of 4 to 5 of girls to boys. A student suggests that the ratio of girls to boys in the combined classes is 9 to 11.

a. Explain when the ratio 9 to 11 would be correct.

b. List possible numbers of girls and boys in Class A.

c. List possible numbers of girls and boys in Class B.

d. Find the ratio of girls to boys with 22 students in Class A and 36 in Class B.

Blue numbers are core exercises.

e. Find the ratio of girls to boys with 44 students in Class A and 27 in Class B.

f. Explain why a simplified ratio does not indicate the original number of students.

▶ Projects

71. Circumference and Diameter Use a ruler and string to measure the diameter and circumference of ten circular objects, and record your results in a table. Measure in centimeters.

a. Graph your results, with diameter as input and circumference as output.

b. Draw a straight line that seems to pass through most of the points.

c. Where should your line cross the y-axis? the x-axis?

d. Pick two points and calculate the slope of the line.

e. What should the slope be? Why?

72. Price Changes Example 11 and Exercises 49 to 52 used 1986 and 2003 prices for selected items. Find prices of comparable products today, and calculate the percent change since 2003. Discuss why items got cheaper or more expensive.

73. Circle Graphs A protractor can be used to divide a half circle into 180 degrees (or 180°) and a full circle into 360°. Find the degrees needed to show each of the percents in Example 13 on a circle graph. (*Hint:* Multiply each percent by 360°.) Draw the circle graph.

74. Slopes and Angles The figure shows a protractor drawn on grid paper, with 0 to 10 marked on horizontal and vertical lines. The angle between line AB and the horizontal is about 26.5° because line AB passes between 26° and 27°. From the horizontal and vertical lines we know that line AB has a slope (rise over run) of $\frac{5}{10} = \frac{1}{2} = 0.5$. Thus, we can say that a line with a slope of $\frac{1}{2}$ makes a 26.5° angle with the horizontal. By using a ruler to connect A and points on the vertical scale, we can read other angle measures and slopes.

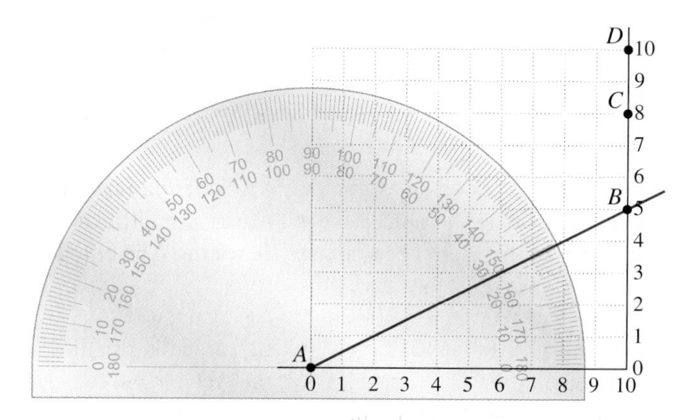

a. What angle does line AC make with the horizontal? What is the slope of AC?

b. What angle does line AD make with the horizontal? What is the slope of AD?

Estimate the slopes for the angles in parts c to e:

c. 10° **d.** 20° **e.** 30°

Estimate the angles for the slopes in parts f to h:

f. $\frac{1}{10}$ **g.** $\frac{3}{10}$ **h.** $\frac{7}{10}$

i. Is the relationship between angle and slope a linear function?

▶ 5.2 Proportions and Proportional Reasoning

Objectives

- Write equivalent ratios as proportions.
- Check proportions using cross multiplication.
- Solve percent problems using proportions.
- Set up proportions from applications.
- Solve proportions and equations using cross multiplication, as appropriate.

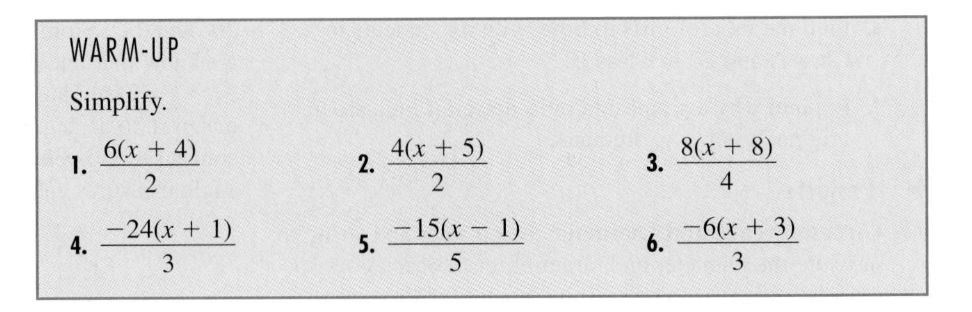

WARM-UP

Simplify.

1. $\dfrac{6(x + 4)}{2}$
2. $\dfrac{4(x + 5)}{2}$
3. $\dfrac{8(x + 8)}{4}$
4. $\dfrac{-24(x + 1)}{3}$
5. $\dfrac{-15(x - 1)}{5}$
6. $\dfrac{-6(x + 3)}{3}$

HAVE YOU EVER FALLEN off a ladder or known someone who did? In this section, we look at proportions and apply them to solving ladder safety and other application problems. The emphasis is on numeric and algebraic work.

Proportions and Proportional Statements

From Nuisance Wildlife Control Operators in New York State website:

> I was [inside and] using a 20-foot straight ladder but the roof was 15 feet in that spot. Instead of getting a shorter ladder, I set the long one against the rafter and started up. Just as I reached the rafter the ladder slipped, and down I went. End result: cracked rotator cup in my elbow, dislocated toe, multiple fractures in my feet, large gash across my knee.

Both the California State Compensation Insurance Fund and the Farm Employer's Labor Service websites note that 609 workers were killed and 272,000 injured in ladder-related accidents in 2002. It is unclear whether these data include the annual 300 deaths and 130,000 injuries of people doing projects at home, as noted by the U.S. Department of Agriculture website (**www.aphis.usda.gov/mrpbs/safety-security/library/ladder_safety.dpf**).

Safety guidelines for home and professional ladders warn people to set the top of a straight ladder four times as high as the distance the foot of the ladder is from the wall (see Figure 5). The height-to-base ratio is 4 to 1. This 4 : 1 ratio is also the slope of the ladder.

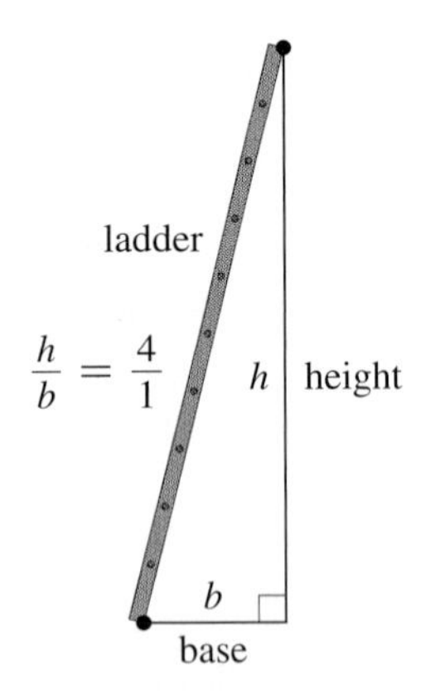

FIGURE 5

EXAMPLE 1 Exploration: safe-ladder ratio Each of the following ladders reaches 12 feet up a wall. Write a simplified height-to-base ratio for each ladder position. Compare the simplified ratio with 4 : 1. How is an accident likely to happen if the ratio is not safe?

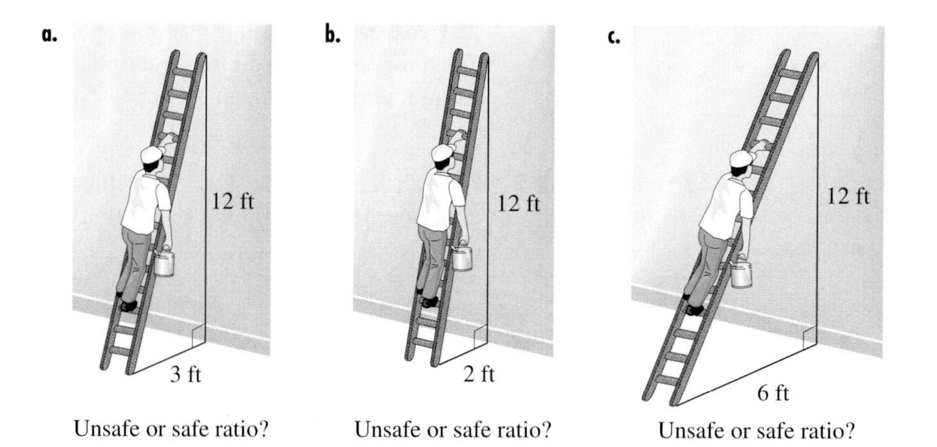

Unsafe or safe ratio? Unsafe or safe ratio? Unsafe or safe ratio?

SOLUTION **a.** 12 : 3 = 4 : 1, a safe ratio

b. 12 : 2 = 6 : 1, not a safe ratio; the ladder may tip over backwards because the ratio is larger than 4 : 1.

c. 12 : 6 = 2 : 1, not a safe ratio; the bottom of the ladder may slide away from the wall because the ratio is smaller than 4 : 1. ◀

▶ Proportions and Cross Multiplication

In part a of Example 1, the safe-ladder ratios form a proportion:

$$12 : 3 = 4 : 1$$

or, in fraction notation,

$$\frac{12}{3} = \frac{4}{1}$$

This leads to the definition of a proportion.

DEFINITION OF PROPORTION

> Two equal ratios form a **proportion**:
>
> $$\frac{a}{b} = \frac{c}{d}$$
>
> where $b \neq 0$ and $d \neq 0$.

▶ **EXAMPLE 2** Finding a property of proportions What equation results from multiplying both sides of the proportion $\frac{a}{b} = \frac{c}{d}$ by bd, the product of the denominators?

SOLUTION

$$\frac{a}{b} = \frac{c}{d} \qquad \text{Multiply both sides by } bd.$$

$$bd\left(\frac{a}{b}\right) = bd\left(\frac{c}{d}\right) \qquad \text{Simplify, and arrange the variables alphabetically.}$$

$$ad = bc$$ ◀

CROSS MULTIPLICATION PROPERTY OF PROPORTIONS

> For a proportion $\frac{a}{b} = \frac{c}{d}$,
>
> $$ad = bc$$
>
> where $b \neq 0$ and $d \neq 0$.

Cross multiplication *sets diagonal products equal*, with the equation $ad = bc$. The process is called cross multiplication because lines drawn between the parts being multiplied form an X-shaped cross (see Example 3).

▶ **EXAMPLE 3** Finding out if two ratios form a proportion Use cross multiplication to find out if either of the following is a proportion.

a. $\frac{9}{12} \stackrel{?}{=} \frac{6}{8}$ **b.** $\frac{16}{15} \stackrel{?}{=} \frac{15}{14}$

SOLUTION **a.** Yes, this is a proportion. Both diagonals multiply to 72.

$$\frac{9}{12} \times \frac{6}{8} \qquad 72 \quad 72$$

b. No, this is not a proportion. The diagonals multiply to different numbers.

$$\frac{16}{15} \times \frac{15}{14} \qquad 224 \quad 225$$

◀

THINK ABOUT IT 1: How would simplifying each ratio show whether a proportion was formed?

▶ Example 4 shows that a set of four numbers may be arranged into a proportion in several ways.

▶ **EXAMPLE 4** Building proportions: medication dosages Make proportions showing that a child of age 96 months should receive a dosage of 160 milligrams if 250 milligrams is the dosage for persons 150 months or older.

SOLUTION Some proportions are

$$\frac{96 \text{ mo}}{150 \text{ mo}} = \frac{160 \text{ mg}}{250 \text{ mg}}$$

$$\frac{96 \text{ mo}}{160 \text{ mg}} = \frac{150 \text{ mo}}{250 \text{ mg}}$$

$$\frac{250 \text{ mg}}{150 \text{ mo}} = \frac{160 \text{ mg}}{96 \text{ mo}}$$

◀

The last two proportions are the most common forms.

THINK ABOUT IT 2: Write a sentence to describe each proportion in Example 4.

The units in Example 4 draw attention to the importance of matching the units in a proportion.

MATCHING UNITS IN PROPORTIONS

When proportions contain units, the units must match—either side to side or within numerator and denominator pairs.

In either case, the product of the units will be identical for the two multiplications when we cross multiply.

JUST FOR FUN

What *four consecutive letters of the alphabet form a word?* You may scramble the order of the letters. (*Note:* "Figh" is not a fruit, and "d'cab" is not what you call to get a ride to d'airport!) Find a proportion that gives the word upon cross multiplication.

Solving Percent Problems Using Proportions

We now turn to using proportions to solve percent problems. Many percent problems are phrased in sentences, as shown on the left below. The equation for each sentence is shown on the right.

Sentence	**Equation**
15% of what number is 45?	$0.15x = 45$
72 is what percent of 80?	$72 = \frac{x}{100} \cdot 80$
15 is what percent of 5?	$15 = \frac{x}{100} \cdot 5$
What number is 150% of 18?	$x = 1.50(18)$

Writing these sentences as equations requires that we remember to change percents into decimals. Some students like to use equations. Other students prefer to write the sentences as proportions. The above problems all fit this form:

$$\text{If } a \text{ is } n\% \text{ of } b, \text{ then } \frac{n}{100} = \frac{a}{b}.$$

The symbol $n\%$ is equal to the ratio $\frac{n}{100}$, which forms the left side of the proportion. The right side is the percent quantity over the base quantity. Note that *n is out of 100* and *the percent quantity is out of the base, b.*

▶ EXAMPLE 5 Writing percent problems as proportions Change each percent problem to a proportion and solve.

a. 15% of what number is 45?
b. 72 is what percent of 80?
c. 15 is what percent of 5?
d. What number is 150% of 18?

SOLUTION **a.** Write 15% as $\frac{15}{100}$. The percent quantity is 45. The base is x.

$$\frac{15}{100} = \frac{45}{x} \quad \text{Cross multiply.}$$

$$15x = (100)(45) \quad \text{Divide both sides by 15.}$$

$$x = \frac{4500}{15} \quad \text{Simplify.}$$

$$x = 300$$

Check: 15% of 300 is 45. ✓

b. Write "what percent" as $\frac{x}{100}$. The percent quantity is 72. The base is 80.

$$\frac{x}{100} = \frac{72}{80} \quad \text{Cross multiply.}$$

$$80x = (100)(72) \quad \text{Divide both sides by 80.}$$

$$x = \frac{7200}{80} \quad \text{Simplify.}$$

$$x = 90$$

Check: 72 is 90% of 80. ✓

c. Write "what percent" as $\frac{x}{100}$. The percent quantity is 15. The base is 5.

$$\frac{x}{100} = \frac{15}{5} \quad \text{Cross multiply.}$$

$$5x = (100)(15) \quad \text{Divide both sides by 5.}$$

$$x = \frac{1500}{5} \quad \text{Simplify.}$$

$$x = 300$$

Check: 15 is 300% of 5. ✓

d. Write 150% as $\frac{150}{100}$. The percent quantity is x. The base is 18.

$$\frac{150}{100} = \frac{x}{18} \quad \text{Cross multiply.}$$

$$100x = (150)(18) \quad \text{Divide both sides by 100.}$$

$$x = \frac{2700}{100} \quad \text{Simplify.}$$

$$x = 27$$

Check: 27 is 150% of 18. ✓ ◀

Applications

In Example 6, we set up a proportion with a variable and cross multiply to solve for the variable.

EXAMPLE 6 Setting up and solving a proportion: medication dosage How many milligrams of medication should a child of 80 lb receive if an adult of 150 lb receives 200 mg.

SOLUTION

$$\frac{80 \text{ lb}}{x \text{ mg}} = \frac{150 \text{ lb}}{200 \text{ mg}} \quad \text{Set up the proportion and cross multiply.}$$

$$80(200) = 150x \quad \text{Divide by 150.}$$

$$x = 106.7$$

Check: 80 lb(200 mg) $\stackrel{?}{=}$ 150 lb(106.7 mg) ✓

The child's dosage is 106.7 mg. ◀

Note that in Example 6, the units in each ratio were the same: lb over mg. In the safe-ladder ratio problems, the 4-to-1 ratio was height over base.

EXAMPLE 7 Setting up and solving a proportion: safe-ladder ratio A ladder needs to reach 22 feet up a wall. How far from the wall must the ladder be in order to meet the 4-to-1 safe-ladder ratio?

SOLUTION We set up the proportion as two ratios with height over base.

$$\frac{\text{height}}{\text{base}} = \frac{\text{height}}{\text{base}}$$

$$\frac{22 \text{ ft}}{x \text{ ft}} = \frac{4}{1} \quad \text{Cross multiply.}$$

$$4x = 22 \quad \text{Divide by 4.}$$

$$x = 5.5$$

The ladder must be 5.5 feet from the wall.

Check: 22 ft(1) = 4(5.5 ft) ✓ ◀

▶ In Example 8, we return to slope. Recall that slope may be defined as rise over run or the vertical change over the horizontal change. Rise over run is like height over base in Example 7.

▶ **EXAMPLE 8** Setting up and solving a proportion: slopes and ramps The slope of a wheelchair access ramp should be 1 to 10. The ramp needs to rise 20 inches. What horizontal length is needed to build the ramp?

SOLUTION A sketch like Figure 6 may be helpful. We start with a line of slope 1 to 10 and extend it an unknown distance to a vertical line of 20 inches.

$$\frac{\text{rise}}{\text{run}} = \frac{\text{rise}}{\text{run}}$$

$$\frac{1}{10} = \frac{20 \text{ in.}}{x} \quad \text{Cross multiply.}$$

$$1x = 200$$

Check: $1(200 \text{ in.}) = 10(20 \text{ in.})$ ✓

The horizontal length is 200 inches. Using unit analysis to change inches to feet, we have

$$200 \text{ in.} \cdot \frac{1 \text{ ft}}{12 \text{ in.}} = 16\tfrac{2}{3} \text{ ft}$$

The horizontal length of the ramp is $16\frac{2}{3}$ feet.

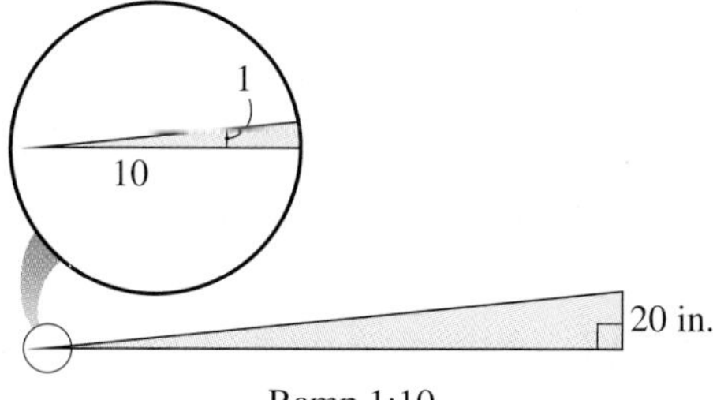

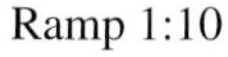

FIGURE 6 Slope of a ramp ◀

THINK ABOUT IT 3: Use a proportion to change 200 inches into feet.

▶ Proportions and Predictions

Statistics is a field of mathematics where we gather information and then draw conclusions or make predictions based on that information. How the information is gathered (*random surveys* or *random samples*) is an entire course by itself, but as the next two examples show, proportional reasoning plays an important role in the field. (See also Project 91 in the Exercises.)

Example 9 applies proportions to TV ratings.

▶ **EXAMPLE 9** Setting up and solving a proportion: TV ratings Suppose 500 U.S. households are surveyed randomly and 6 are found to be watching NBC's 8 p.m. program. Estimate the total number of households watching NBC nationwide if there are 93 million households.

SOLUTION We have two kinds of data: data on those who are watching NBC and data on totals. We set up the proportion with this in mind.

$$\frac{6 \text{ NBC households}}{500 \text{ households surveyed}} = \frac{x \text{ NBC households}}{93 \text{ million households}} \qquad \text{Cross multiply.}$$

$$6(93 \text{ million}) = 500x \qquad \text{Divide by 500.}$$

$$x = 1.116 \text{ million}$$

About 1 million households appear to be watching the NBC program. ◀

▶ In Example 9, we assumed that the households surveyed were typical of the nation as a whole. In wildlife population sampling, it is assumed that when randomly tagged animals are released, they will mix back into the general population. Any subsequent random sample will have tagged animals in the same proportion as the total population.

▶ **EXAMPLE 10** **Setting up and solving a proportion: fish population** At random locations in a certain lake, fish management personnel catch and tag 50 fish. They return the fish unharmed to the lake. They come back to the same locations two weeks later and catch 100 fish. If 4 of the fish carry a tag and 96 do not, estimate the total population of fish.

SOLUTION

$$\frac{50 \text{ initial tagged}}{x \text{ total population}} = \frac{4 \text{ caught with tag}}{100 \text{ total caught}} \qquad \text{Cross multiply.}$$

$$50 \cdot 100 = 4 \cdot x \qquad \text{Divide by 4.}$$

$$x = 1250$$

Check: $500(100) \stackrel{?}{=} 4(1250)$ ✓

This solution uses a ratio of tagged fish to total population. Other solutions are possible, using tagged to untagged or untagged to total population. ◀

▶ A Caution on Solving Equations

Example 11 shows that cross multiplication is not a good general equation-solving tool because it only applies to proportions.

▶ **EXAMPLE 11** **Solving equations** Solve each equation for x without using cross multiplication. Then identify the equation that is a proportion and solve it again with cross multiplication.

a. $x + \frac{1}{2} = 5x$

b. $\dfrac{x-1}{4} = \dfrac{x+1}{5}$

c. $\dfrac{3x}{4} = 2 + \dfrac{x+1}{5}$

SOLUTION **a.** $x + \frac{1}{2} = 5x$ Subtract x from both sides.

$\frac{1}{2} = 4x$ Multiply both sides by $\frac{1}{4}$.

$\frac{1}{8} = x$

Check: $\frac{1}{8} + \frac{1}{2} \stackrel{?}{=} 5\left(\frac{1}{8}\right)$ ✓

b. $\frac{x-1}{4} = \frac{x+1}{5}$ Multiply both sides by 20.

$\frac{20(x-1)}{4} = \frac{20(x+1)}{5}$ Simplify.

$5(x-1) = 4(x+1)$ Apply the distributive property.

$5x - 5 = 4x + 4$ Subtract $4x$ from both sides and add 5 to both sides.

$x = 9$

Check: $\frac{9-1}{4} \stackrel{?}{=} \frac{9+1}{5}$ ✓

c. $\frac{3x}{4} = 2 + \frac{x+1}{5}$ Multiply both sides by 20.

$\frac{60x}{4} = 40 + \frac{20(x+1)}{5}$ Simplify.

$15x = 40 + 4(x+1)$ Apply the distributive property.

$15x = 40 + 4x + 4$ Subtract $4x$ from both sides and add like terms.

$11x = 44$ Divide both sides by 11.

$x = 4$

Check: $\frac{3(4)}{4} \stackrel{?}{=} 2 + \frac{(4)+1}{5}$ ✓

Only the equation in part b is a proportion, which can be solved with cross multiplication.

$\frac{x-1}{4} = \frac{x+1}{5}$ Cross multiply.

$5(x-1) = 4(x+1)$ Apply the distributive property.

$5x - 5 = 4x + 4$ Subtract $4x$ on both sides.

$x - 5 = 4$ Add 5 to both sides.

$x = 9$

Check: $\frac{9-1}{4} \stackrel{?}{=} \frac{9+1}{5}$ ✓ ◀

ANSWER BOX

Warm-up: 1. $3x + 12$ **2.** $2x + 10$ **3.** $2x + 16$ **4.** $-8x - 8$ **5.** $-3x + 3$ **6.** $-2x - 6$ **Think about it 1:** Equivalent ratios would simplify to the same number or divide to the same decimal. **Think about it 2:** The ratio of the child's age, 96 months, to the adult age, 150 months, is the same as the ratio of the child's dosage, 160 mg, to the adult dosage, 250 mg. The ratio of the child's age, 96 months, to the child's dosage, 160 mg, is the same as the ratio of the adult age, 150 months, to the adult dosage, 250 mg. The ratio of the adult dosage, 250 mg, to the adult age, 150 months, is the same as the ratio of the child's dosage, 160 mg, to the child's age, 96 months. **Think about it 3:** Our fact is 12 in. = 1 ft, so 200 in. = x ft. One proportion is $\frac{12 \text{ in.}}{200 \text{ in.}} = \frac{1 \text{ ft}}{x}$. Other proportions are possible. Cross multiplication gives $12x = 200$ or $x = 16\frac{2}{3}$; thus, 200 in. = $16\frac{2}{3}$ ft.

▶ 5.2 Exercises

In Exercises 1 to 6, simplify each ratio, compare it to 4 : 1, and state whether it is a safe-ladder ratio. If the ratio represents an unsafe ladder position, guess whether the ladder will slip or tip.

1. 15 feet to 5 feet

2. 18 feet to 5 feet

3. 18 feet to 4 feet

4. 20 feet to 4 feet

5. 20 feet to 5 feet

6. 18 feet to 4.5 feet

In Exercises 7 to 12, which are proportions and which are false statements?

7. $\frac{6}{8} = \frac{15}{20}$

8. $\frac{8}{10} = \frac{12}{15}$

9. $\frac{4}{6} = \frac{6}{9}$

10. $\frac{9}{12} = \frac{15}{18}$

11. $\frac{9}{21} = \frac{21}{35}$

12. $\frac{9}{6} = \frac{15}{10}$

13. Make four different proportions with the numbers 2, 4, 5, and 10. Use cross multiplication to check.

14. Make four different proportions with the numbers 2, 3, 4, and 6. Use cross multiplication to check.

Solve the proportions in Exercises 15 to 22.

15. $\frac{3}{4} = \frac{x}{15}$

16. $\frac{5}{3} = \frac{27}{x}$

17. $\frac{x}{5} = \frac{2}{3}$

18. $\frac{5}{x} = \frac{3}{8}$

19. $\frac{7}{3} = \frac{5}{x}$

20. $\frac{2}{5} = \frac{x}{12}$

21. $\frac{4}{x} = \frac{3}{7}$

22. $\frac{x}{6} = \frac{7}{5}$

In Exercises 23 to 38, write the percent problems as equations or proportions and solve. Round to the nearest tenth or tenth of a percent.

23. 45% of what number is 36?

24. 65% of what number is 39?

25. 56 is what percent of 84?

26. 125 is what percent of 225?

27. What number is 60% of 42?

28. What number is 35% of 70?

29. 56 is what percent of 40?

30. 72 is what percent of 60?

31. What number is 18% of 25?

32. What number is 90% of 45?

33. 64% of what number is 16?

34. 32% of what number is 24?

35. 104 is what percent of 80?

36. 45 is what percent of 75?

37. 28 is what percent of 40?

38. 112 is what percent of 64?

Solve the unit problems in Exercises 39 to 44 using the following conversion information. Use either a proportion or unit analysis. Round to the nearest hundredth.

1 acre = 43,560 square feet
1 liter = 1.0567 quarts
1 kilogram = 2.2 pounds
1 kilometer = 1000 meters
1 square mile = 640 acres
1 meter = 39.37 inches
1 kilometer = 0.621 mile

39. How many meters are in 65 inches?

40. How many kilograms are in 160 pounds?

41. How many kilometers are in 15 miles?

42. How many liters are in 16 quarts?

43. How many square feet are in 640 acres?

44. How many square miles are in 10,000 acres?

In Exercises 45 to 58, set up a proportion and solve. Round to the nearest hundredth unless otherwise noted.

45. The base of a ladder can sit securely $6\frac{1}{2}$ feet from a wall. How far up the wall can the ladder reach and still satisfy the 4 : 1 safe-ladder ratio?

46. How far from a wall must the base of a ladder be set in order to safely reach 21 feet up a wall?

47. An access ramp needs to rise 54 inches. If the ramp is to have a slope ratio of 1 : 8, what horizontal distance is needed to build the ramp? Give the answer in feet.

48. An access ramp needs to rise 18 inches. If the ramp is to have a slope ratio of 1 : 8, what horizontal distance is needed to build the ramp? Give the answer in feet.

49. A staircase has a slope of 7 to 11. What horizontal distance is needed for an 8-foot vertical distance?

50. A staircase has a slope of 6.5 to 11.5. What horizontal distance is needed for an 8.5-foot vertical distance?

51. If a 10-foot storm surge (high wave of water) from Hurricane Andrew hits the Louisiana coast and comes inland 13 miles, what is the average slope of the coastal region at that location? (*Hint:* The units must be the same for a sensible answer.) Leave the answer as a fraction.

Blue numbers are core exercises.

52. If a 10-foot storm surge hits the Oregon coast and comes inland $\frac{1}{8}$ mile, what is the average slope of the coastline at that location? (*Hint:* The units must be the same.) Leave the answer as a fraction.

53. A 40-pound child should receive how many units of penicillin if the dosage is 500,000 units for a 150-pound adult? Round to the nearest ten thousand.

54. How many milligrams (mg) of atropine sulfate should be given to a 6-month infant if the dosage is 0.4 mg for a patient of 150 months?

55. An adult pain reliever is 325 mg. A low dose of the same product is 81 mg. If an adult is assumed to weigh 150 pounds, for what weight is the low dose designed?

56. If a 70-kg adult receives 10 grains (or gr 10), how many grains would be administered to a child who weighs 21 kg?

57. A ski lift goes up a 62.5% grade. If the ski lift covers a 3000-foot horizontal distance, what is the change in elevation?

58. If a ski lift rises 9000 vertical feet, find the horizontal distance covered if the average slope is 62.5%.

59. A narrow mountain road has a 9% grade. If this grade covers a 2-mile horizontal distance, what is the rise in elevation to the nearest foot?

60. A highway sign indicates a 6% downhill grade. If this 6% grade is 7 miles of horizontal distance, what is the change in elevation, to the nearest foot?

61. Bird Population A wildlife management team traps pheasants in nets and tags them at randomly located areas in a fire-damaged setting. They tag 75 birds altogether. Two weeks later they trap again, and they capture 10 tagged birds and 55 untagged birds. Use a proportion to estimate the population. Assume that the birds didn't learn to avoid the nets after being caught the first time.

62. Fish Population Suppose 15,000 hatchery fish are released in a river. At maturity, these fish return to the river along with the native fish. Assume that the hatchery fish have a 5% survival and return rate. Of 85 mature fish caught by people fishing along the river, 82 have the clipped fin of the hatchery fish. Use a proportion to estimate the number of native fish in the river.

63. Bat Population Researchers sample several locations on the ceiling of a cave and find an average of 18 adult bats per square foot. They estimate the dimensions of the cave to be 30 feet by 75 feet.

a. Use a proportion to estimate the number of adult bats in the cave.

b. Find the number of mosquitoes eaten in 30 nights if each bat consumes 500 mosquitoes per night.

▶ Writing

64. Explain how the units help us place numbers into a proportion.

65. Explain how to show that two fractions form a proportion.

66. Explain how to change "Find the percent (n) that y is of x" into a proportion.

67. Explain how to change "20% of what number (n) is x?" into a proportion.

68. Explain how to change "k is 20% of what number (n)?" into a proportion.

69. Explain how to determine whether an equation is a proportion.

70. Alternative Notation Instead of writing the data as $\frac{a}{b} = \frac{c}{d}$, another way to determine whether the data form a proportion is to use the **inner-outer product method.**

Inner product = 144

$8 : 12 = 12 : 18$

Outer product = 144

Find the products. Which of these are proportions?

a. $15 : 21 = 30 : 35$ **b.** $8 : 20 = 10 : 25$

c. $12 : 14 = 18 : 21$ **d.** $9 : 6 = 12 : 8$

e. $20 : 12 = 25 : 15$ **f.** $10 : 14 = 24 : 28$

In Exercises 71 to 76, solve each equation by first multiplying by the indicated number. Solve the proportions again using cross multiplication.

71. $\frac{2x}{9} = \frac{x+2}{6}$, 18 **72.** $\frac{3}{4}x = \frac{x}{5} + 3$, 20

73. $\frac{x}{2} = 5x + 4$, 2 **74.** $\frac{x}{3} = \frac{3x}{2} + 1$, 6

75. $\frac{1}{2} + \frac{x}{3} = \frac{5x}{2} + \frac{4}{3}$, 6 **76.** $\frac{x+4}{5} = \frac{x-3}{2}$, 10

Solve for x in the proportions in Exercises 77 to 90.

77. $\frac{x-1}{2} = \frac{x+3}{3}$ **78.** $\frac{x-4}{3} = \frac{x-1}{4}$

79. $\frac{x+4}{15} = \frac{x+2}{10}$ **80.** $\frac{x-6}{10} = \frac{x-4}{6}$

Blue numbers are core exercises.

81. $\dfrac{x+7}{9} = \dfrac{x+3}{6}$

82. $\dfrac{x-1}{8} = \dfrac{x+2}{12}$

83. $\dfrac{3x}{2} = \dfrac{6x+3}{5}$

84. $\dfrac{4x}{3} = \dfrac{3x-1}{2}$

85. $\dfrac{2x-3}{7} = \dfrac{x+2}{4}$

86. $\dfrac{5x-1}{9} = \dfrac{x+7}{3}$

87. $\dfrac{4x+2}{5} = \dfrac{x-1}{2}$

88. $\dfrac{x-3}{4} = \dfrac{3x+1}{7}$

89. $\dfrac{x+2}{2} = \dfrac{3x+1}{5}$

90. $\dfrac{x}{3} = \dfrac{3x-5}{8}$

▶ Projects

91. Population Estimates Place a large number (at least 100, but uncounted) of like objects (dry beans, marbles, coins) in a paper bag.

a. Remove a handful, count the objects, and "tag" them with some sort of label. Return the tagged objects to the bag and mix thoroughly.

b. Remove another handful, and count the tagged objects.

c. Set up a proportion to estimate the total number of objects in the bag. State your assumptions.

d. How could you improve your estimate? What other ways are there to count the objects? Count the objects or use another method to estimate the total objects.

92. Clock Angles Examine the accompanying clock faces. In answering the questions, explain your reasoning carefully and completely.

a. What is the measure of the angle formed by the hands of a clock at 7:30? *Hint:* The minute hand sweeps through 360° in completing a full circle.

b. What is the measure of the angle at 4:35?

c. What is the rate in degrees per hour for the minute hand? the hour hand? the second hand?

93. Safe-Ladder Experiment Your task is to find out for what decimal ratios of height to base a ladder—modeled with meter sticks—is safe. You will need three stiff, smoothly finished meter sticks and tape to secure one stick to a wall.

a. Fasten the first meter stick vertically to a wall. The lower end of the meter stick should be a measured distance—say, 20 centimeters—above the floor. Secure the meter stick with tape at the upper end and lower end only. Lay the second meter stick on the floor so that it is perpendicular to the wall and directly below the first stick. Lean the third meter stick (the ladder) so that it forms a steep triangle with the other two meter sticks. Press gently inward and downward on the midpoint of the meter stick. If the meter stick (ladder) does not move, the ladder is in a relatively "safe" position. If the base of the ladder moves toward the wall or away from the wall, the ladder is unsafe.

Set up a table with four headings: Height of triangle, Base of triangle, Safe or unsafe, Ratio of height to base. For the position in which you have put the ladder, record on the table the height of the triangle, the base of the triangle, and whether the ratio is safe or unsafe. Complete the last column with a calculator. Repeat for 10 or more other positions of the ladder.

b. Describe the type of accident modeled when the base of the ladder slides toward the wall.

c. Describe the type of accident modeled when the base slides away from the wall.

Blue numbers are core exercises.

▶ 5.3 Proportions in Similar Figures and Similar Triangles

Objectives

- Identify similar figures.
- Identify corresponding parts of similar figures.
- Write and solve proportions to find unknown lengths in similar triangles.

WARM-UP

Solve Exercises 1 to 6 for x.

1. $\dfrac{20}{12} = \dfrac{x}{7.5}$ **2.** $\dfrac{6}{8} = \dfrac{x}{20}$ **3.** $\dfrac{5.5}{3} = \dfrac{x}{34}$

4. $\dfrac{10 - x}{4} = \dfrac{10}{9}$ **5.** $\dfrac{4}{3} = \dfrac{x + 4}{5}$ **6.** $\dfrac{n}{3} = \dfrac{n + 3}{8}$

In Exercises 7 to 10, use this line segment:

A ———— B ———— C

Hint: If $AB = 20$ and AC is 25, then $BC = 25 - 20$.

7. If $AB = x$ and $AC = 8$, what expression describes BC?

8. If $AC = x$ and $AB = 8$, what expression describes BC?

9. If $AB = x$ and $BC = 8$, what expression describes AC?

10. If $AC = x$ and $BC = 8$, what expression describes AB?

IN THIS SECTION, we relate proportions to geometric figures. We use slope and similar triangles in applications in photography and forestry.

Similar Figures

EXAMPLE 1 **Exploring the height of a rectangle** June wishes to enlarge a photograph to fit the bottom of a scrapbook page. The photograph's size is 12 centimeters (base) by 7.5 centimeters (height). The enlargement is to have a base of 20 centimeters. To finish the rest of the page layout, she needs to know the height of the photograph after the enlargement.

a. Estimate the height of the enlarged photo.

b. Find the height, using Figure 7 and the concept of similarity. In Figure 7, a diagonal is drawn from the lower left to the upper right corner of a rectangle with the same shape as the photograph. The diagonal is then extended until it crosses a vertical line at width 20 centimeters. The height of the vertical line at width 20 is the height of the enlarged photograph.

c. What is the slope of the diagonal line in Figure 7?

d. Let x be the missing height. Use the slope to write a proportion to find the missing height. Solve the proportion.

Student Note:
Observe that the diagonal lines in Figures 7 and 8 pass through the origin.

SOLUTION See the Answer Box. ◀

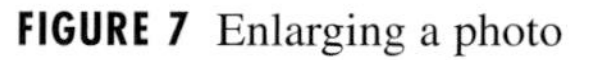

FIGURE 7 Enlarging a photo

▶ In Section 4.3, we found that a straight line has a constant slope. The fact that there is a constant slope ratio all along the diagonal line helps explain why we can apply a proportion in part d of Example 1. In Example 2, we check that the constant slope controls the shape of the rectangle as it is enlarged (or reduced). In part c, two letters are used to identify line segments.

▶ EXAMPLE 2 Exploring ratios of height to base

a. What is the slope of the diagonal line on the grid in Figure 8?

Student Note:
When we choose a rectangle in a computer illustrator program and enlarge with the cursor, the sides of the rectangle have a constant ratio. When we select a side and change the shape of the rectangle, the ratio of the sides changes.

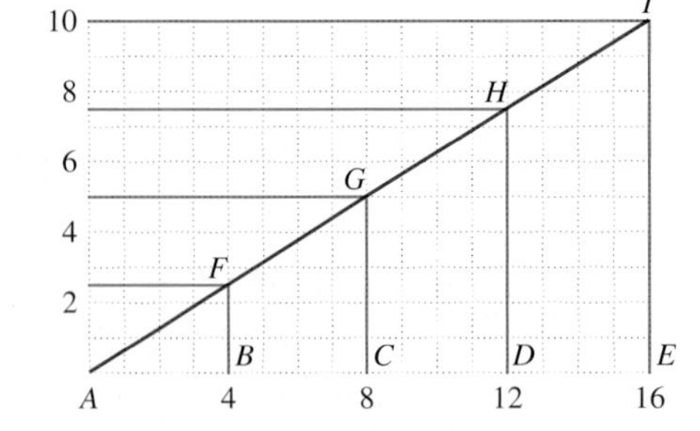

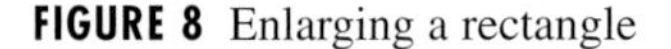

FIGURE 8 Enlarging a rectangle

b. Which horizontal and vertical lines on the grid match the original photo in Figure 7?

c. Find these ratios: BF to AB, CG to AC, DH to AD, EI to AE.

d. What do you observe about the ratios? (*Hint:* If necessary, change the ratios to decimal notation.)

e. Complete this sentence about the rectangles drawn on the grid: For each 4-unit increase in the (horizontal) base, the rectangle increases ___ units in (vertical) height.

SOLUTION See the Answer Box. ◀

Example 2 suggests that enlargement (or reduction) of a photograph will make a rectangle that is the same shape as the original but of a different size. There are two important ideas here.

First, we say that the heights of the two rectangles are corresponding sides. **Corresponding** means *in the same relative position.* The bases are also corresponding sides. Second, the right angles forming the corners of the rectangle remain square corners during the enlargement. These two ideas give us the basis for a definition of similar figures.

DEFINITION OF SIMILAR FIGURES

> **Similar figures** have corresponding sides that are proportional and corresponding angles that are equal.

In Example 3, we use ratios to test for similarity between figures.

▶ EXAMPLE 3 Identifying similar figures

Write ratios of corresponding lengths, and find out if the two figures in the pair are similar.

a.

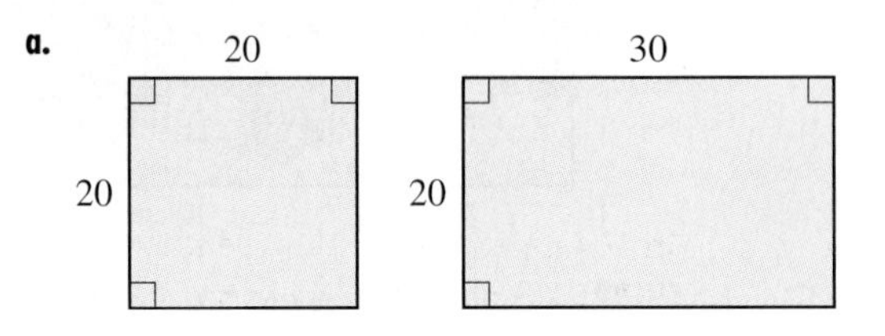

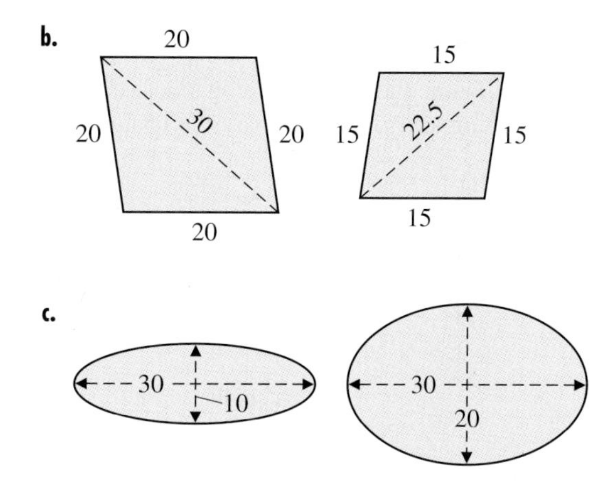

SOLUTION **a.** The figures are both rectangles and all angles are equal. However, the ratios of their sides, $\frac{20}{20}$ and $\frac{20}{30}$, are not equal. The rectangles are not similar.

b. The figures contain triangles, so if the corresponding sides of the triangles are proportional, the larger figures are similar. The outer sides are in a ratio of 20 to 15. The longest diagonals are in a ratio of 30 to 22.5. To find out whether the ratios are equal, we cross multiply

$$\frac{20}{15} \stackrel{?}{=} \frac{30}{22.5}$$

and get 450 in both directions. The figures are similar.

c. These two figures are ellipses. Within the figures, the horizontal distances are equal and the vertical distances are different. They are not similar. ◀

▶ If we are told that two figures are similar, we can assume that the corresponding sides are proportional. Enlarging or reducing a rectangle gives a similar rectangle.

▶ **EXAMPLE 4** **Using proportions to find unknown lengths** A photograph is $3\frac{1}{2}$ inches by 5 inches. Find the length of the shorter side if the longer side is

a. enlarged to 8 inches **b.** reduced to 3 inches

SOLUTION Because enlargement or reduction creates a similar figure, we can use a proportion. We set up a proportion, with short side to long side in each ratio.

a. $\frac{3.5}{5} = \frac{x}{8}$, $x = 5.6$ in. **b.** $\frac{3.5}{5} = \frac{x}{3}$, $x = 2.1$ in. ◀

Similar Triangles

Enlargements or reductions are similar to the original rectangles because they satisfy two conditions: the ratio of any height-to-base pair is constant and all angles are 90°. Only one of the two conditions has to be satisfied for triangles to be similar.

DEFINITION OF SIMILAR TRIANGLES

> Similar triangles have equal corresponding angles or proportional corresponding sides.

In Example 5, the sides are identified by two letters. The angles are identified by the letter at the **vertex**, or *corner of the triangle.*

▶ EXAMPLE 5 Identifying similar triangles

a. Name the corresponding sides of the triangles. Give the ratios of the corresponding sides. Indicate why the triangles are similar.

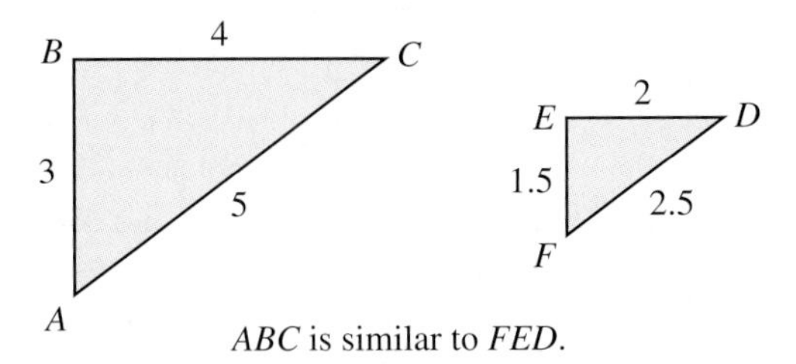

ABC is similar to *FED*.

b. Name the corresponding angles of the triangles, and indicate why the triangles are similar.

Student Note:
The marks in the angles indicate corresponding angles. They may be used without naming the angle measures.

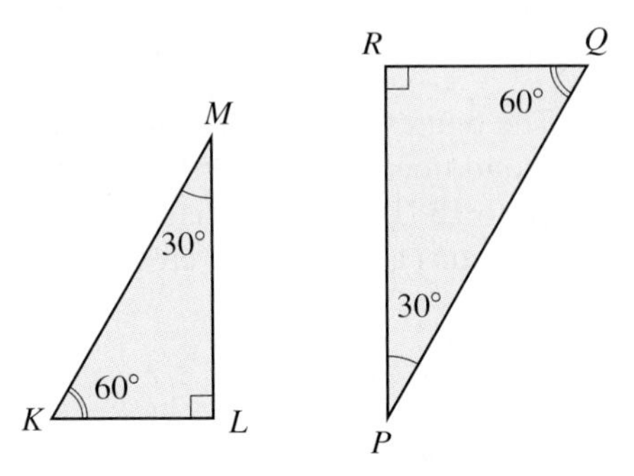

KLM is similar to *QRP*.

SOLUTION **a.** The longest sides of both triangles are corresponding, so *AC* corresponds with *FD*. The middle-length sides give the next correspondence: *BC* and *ED*. Finally, the shortest sides give the third correspondence: *AB* and *FE*.

The ratio *AC* to *FD* is 5 to 2.5; the ratio simplifies to 2 to 1.

The ratio *BC* to *ED* is 4 to 2; the ratio simplifies to 2 to 1.

The ratio *AB* to *FE* is 3 to 1.5; the ratio simplifies to 2 to 1.

The triangles are similar because the ratios of corresponding sides are the same: 2 to 1.

b. Angle *L* corresponds with angle *R* because they are both right angles.

Angle *M* corresponds with angle *P* because they are both 30°.

Angle *K* corresponds with angle *Q* because they are both 60°.

The two triangles are similar because their corresponding angles are equal. ◀

All three corresponding angles in the figure in part b of Example 5 were labeled. This was not necessary. The fact that *the angle measures in a triangle sum to 180°* permits us to identify similar triangles with just two pairs of equal corresponding angles.

▶ EXAMPLE 6 Identifying similar triangles

Explain why the pairs of triangles are not similar.

a.

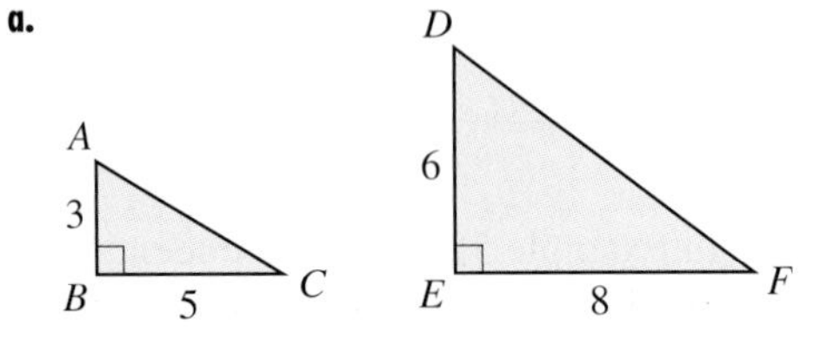

b.

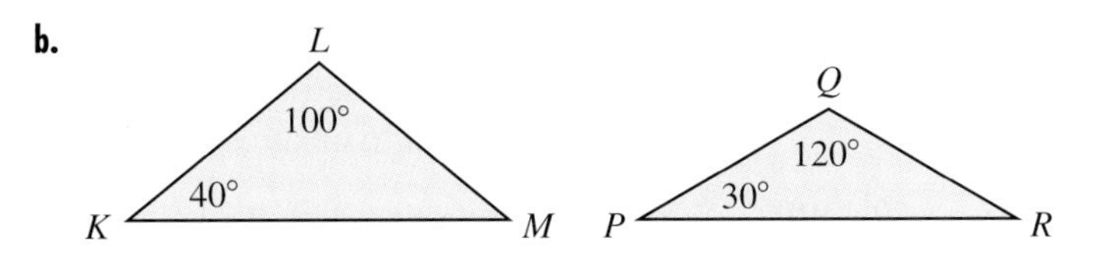

SOLUTION **a.** The ratio of the shortest sides is $AB : DE = 3 : 6$, which simplifies to $1 : 2$. The ratio of the middle-length sides is $BC : EF = 5 : 8$. The ratios of corresponding sides are different, so the triangles are not similar.

b. Because the sum of the measures of the angles of a triangle is 180°, angle M has measure 40° and angle R measure 30°. The triangles do not have equal corresponding angles and are not similar. ◀

In Example 7, we use proportions and the relationships of similar triangles to find missing sides.

EXAMPLE 7 Using proportions to find unknown lengths In part a of Example 6, if EF remains 8, what length would DE have to be for triangle DEF to be similar to triangle ABC?

SOLUTION The ratio of base to height for triangle ABC is 3 to 5. Let x be the unknown side, DE. The proportion is

$$\frac{3}{5} = \frac{x}{8} \quad \text{Cross multiply.}$$

$$5x = 24 \quad \text{Divide both sides by 5.}$$

$$x = 4.8$$ ◀

Similar Triangles and Indirect Measurement

The proportionality of similar triangles gives us a powerful tool for finding unknown lengths. The following examples involve *indirect measurement*—finding the measure of objects we are able to see but not actually measure. The proportional relationship between heights and shadows in Example 8 was known to the ancient Greeks.

EXAMPLE 8 Using proportions to find unknown lengths: trees and shadows If a tree casts a 34-foot shadow along the ground while a $5\frac{1}{2}$-foot-tall person casts a shadow 3 feet long, how tall is the tree? (Because of space limitations, the triangles shown in Figure 9 are similar but are not to scale with each other.)

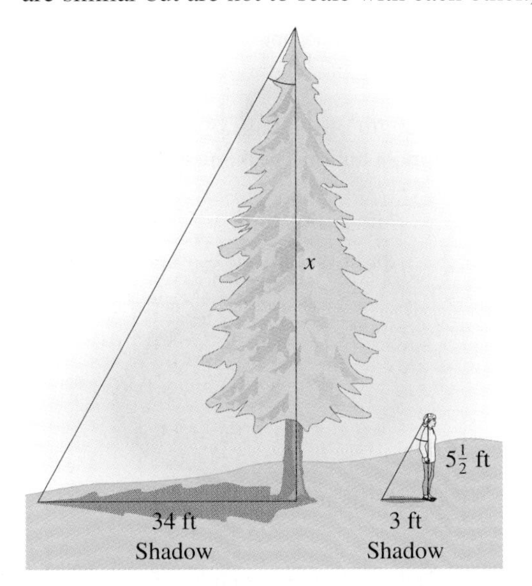

FIGURE 9 Similar triangles in tree and shadow

Geometry fact: The sun's rays form equal angles at the top of each triangle. It is assumed that the tree and the person are at right angles to the ground. This information is sufficient to create similar triangles.

SOLUTION We form a proportion with the heights and shadows:

$$\frac{\text{tree height}}{\text{tree shadow}} = \frac{\text{person height}}{\text{person shadow}}$$

$$\frac{x}{34 \text{ ft}} = \frac{5.5 \text{ ft}}{3 \text{ ft}}$$

$$x(3 \text{ ft}) = (34 \text{ ft})(5.5 \text{ ft})$$

$$x \approx 62.3 \text{ ft}$$

The tree is approximately 62.3 feet tall. ◀

▶ In Example 9, the triangles overlap. The large triangle has the streetlight as its height, and its base extends to the tip of the shadow on the ground. The base is the sum 12 ft + 8 ft. The small triangle has the person as the height and the shadow as its base.

▶ EXAMPLE 9 Using proportions to find unknown lengths: streetlight and shadows A 6-foot-tall person has an 8-foot shadow formed by a streetlight. The person is standing 12 feet from the streetlight. Figure 10 shows a side view of the situation. How high is the streetlight?

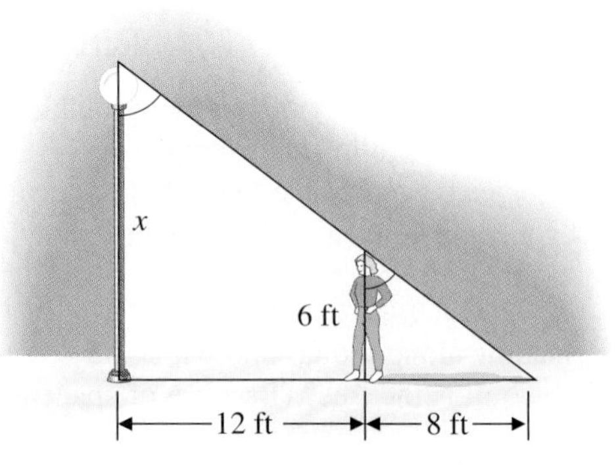

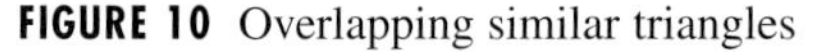

FIGURE 10 Overlapping similar triangles

SOLUTION The length of the base of the larger triangle is the sum of the 12 feet between the lamppost and the person and the 8-foot shadow.

$$\frac{\text{height of light}}{\text{base of triangle}} = \frac{\text{person height}}{\text{person shadow}}$$

$$\frac{x}{(12 + 8) \text{ ft}} = \frac{6 \text{ ft}}{8 \text{ ft}}$$

$$x(8 \text{ ft}) = (6 \text{ ft})(20 \text{ ft})$$

$$x = 15 \text{ ft}$$

The streetlight is 15 feet high. ◀

▶ Not all similar triangles are right triangles. In Example 10, we are given overlapping similar triangles and must find a missing height.

▶ **EXAMPLE 10** Using proportions to find unknown lengths Set up a proportion and solve for the height indicated by x in Figure 11.

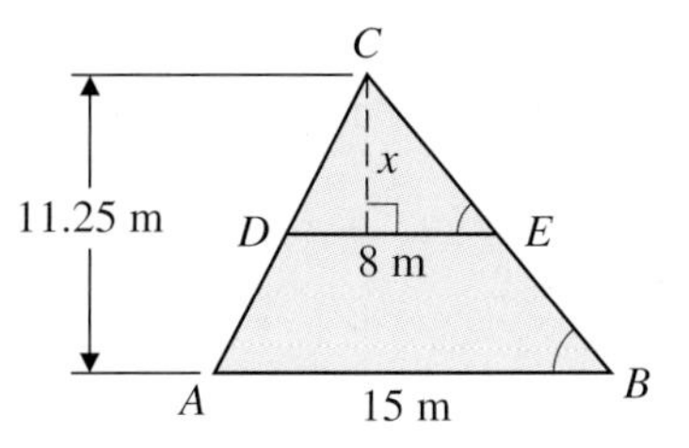

FIGURE 11 *ABC* is similar to *DEC*.

SOLUTION The height-to-base ratio of the large triangle is 11.25 meters to 15 meters. The base of the small triangle is 8 meters.

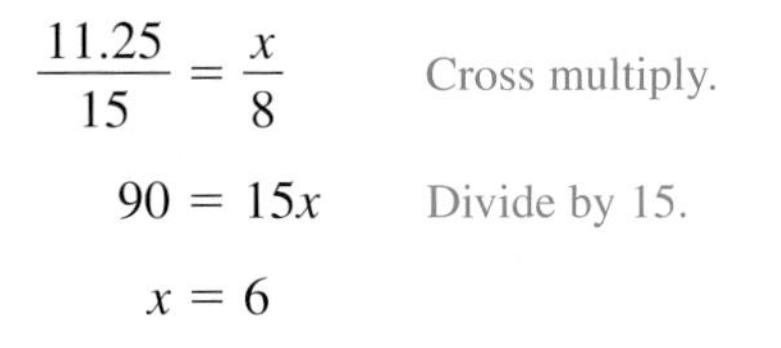

$$\frac{11.25}{15} = \frac{x}{8} \qquad \text{Cross multiply.}$$

$$90 = 15x \qquad \text{Divide by 15.}$$

$$x = 6$$

The height of the small triangle is 6 meters. ◀

In Example 11a, the unknown length is part of one side of overlapping similar triangles. If the whole side is S and one part is x, the other part is $S - x$.

▶ **EXAMPLE 11** Finding unknown lengths Find length x in each figure. Round answers to the nearest tenth.

a.

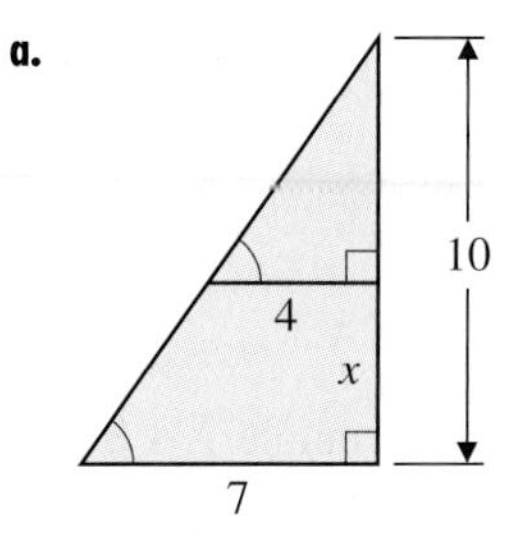

b.

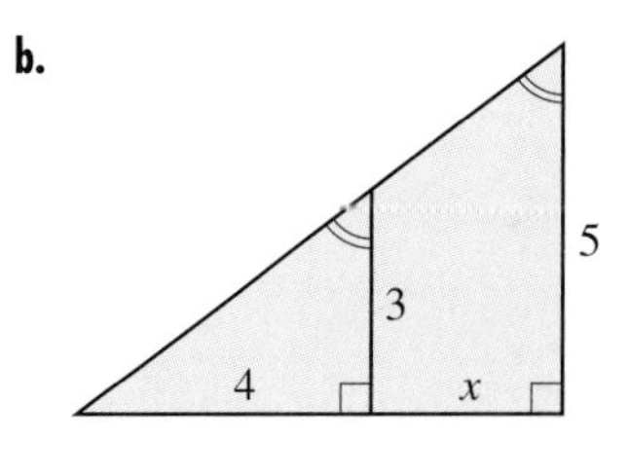

SOLUTION **a.** The height of the small triangle is $10 - x$.

$$\frac{10}{7} = \frac{10 - x}{4} \qquad \text{Cross multiply.}$$

$$7(10 - x) = 40 \qquad \text{Apply the distributive property.}$$

$$70 - 7x = 40 \qquad \text{Subtract 70 on both sides.}$$

$$-7x = -30 \qquad \text{Divide by } -7.$$

$$x \approx 4.3$$

b. The base of the large triangle is $x + 4$.

$$\frac{5}{x + 4} = \frac{3}{4} \qquad \text{Cross multiply.}$$

$$3(x + 4) = 20 \qquad \text{Apply the distributive property.}$$

$$3x + 12 = 20 \qquad \text{Subtract 12 on both sides.}$$

$$3x = 8 \qquad \text{Divide by 3.}$$

$$x \approx 2.7$$

◀

▶ In Example 12, information about the lengths of the sides is contained in the ordered pairs that label the corners of the triangles. We solve for the missing information in two ways: first using similar triangles and second using slope.

▶ EXAMPLE 12 Finding unknown lengths: ordered pairs Find the ordered pairs that describe points A and B in Figure 12.

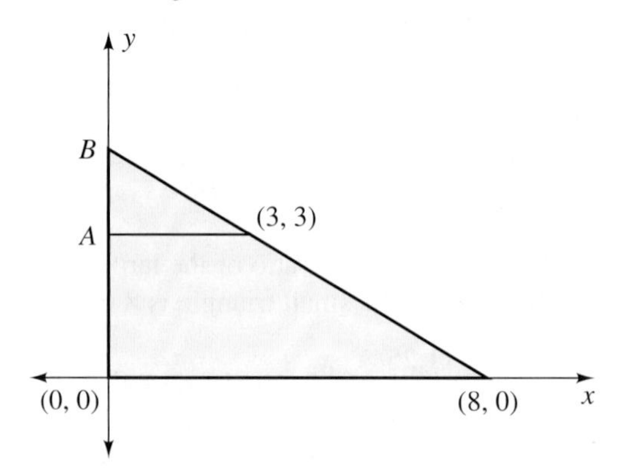

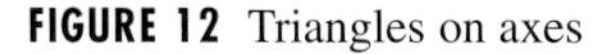

FIGURE 12 Triangles on axes

SOLUTION USING SIMILAR TRIANGLES Point A is (0, 3) because it is on the vertical axis passing through (0, 0) and it is on the same horizontal line as (3, 3). Point B is also on the vertical axis.

Let n be the distance from A to B. From A to the origin is 3. The height of the large triangle is $n + 3$. The base of the small triangle on top is 3. The base of the large triangle (resting on the x-axis) is 8.

$$\frac{n}{n+3} = \frac{3}{8} \qquad \text{Cross multiply.}$$

$$8n = 3(n+3) \qquad \text{Apply the distributive property.}$$

$$8n = 3n + 9 \qquad \text{Subtract } 3n \text{ on both sides.}$$

$$5n = 9 \qquad \text{Divide by 5.}$$

$$n = 1.8$$

The height of the large triangle is $1.8 + 3 = 4.8$. B is at (0, 4.8).

SOLUTION USING SLOPE As mentioned in the prior solution, point A is (0, 3). Because a straight line has constant slope, we can use the slope formula twice to obtain two different ratio expressions and then set them equal to make a proportion to find point B.

Between (3, 3) and (8, 0), the slope is

$$\frac{y_2 - y_1}{x_2 - x_1} = \frac{0-3}{8-3} = \frac{-3}{5}$$

Between point B, $(0, y)$, and (3, 3), the slope is

$$\frac{y_2 - y_1}{x_2 - x_1} = \frac{3-y}{3-0} = \frac{3-y}{3}$$

We set the two slope ratios equal and solve:

$$\frac{-3}{5} = \frac{3-y}{3} \qquad \text{Cross multiply.}$$

$$5(3-y) = -9 \qquad \text{Apply the distributive property.}$$

$$15 - 5y = -9 \qquad \text{Subtract 15 on both sides.}$$

$$-5y = -24 \qquad \text{Divide by } -5 \text{ on both sides.}$$

$$y = 4.8$$

B is at (0, 4.8). ◀

ANSWER BOX

Warm-up: **1.** $x = 12.5$ **2.** $x = 15$ **3.** $x \approx 62.3$ **4.** $x \approx 5.6$ **5.** $x \approx 2.7$ **6.** $n = 1.8$ **7.** $8 - x$ **8.** $x - 8$ **9.** $x + 8$ **10.** $x - 8$
Example 1: **a.** Under 15 cm; the length does not quite double between 12 cm and 20 cm, so the height can be no more than twice 7.5 cm. **b.** Slightly more than 12 cm; measuring the picture shows that the height is only slightly more than the original width. **c.** $\frac{7.5}{12}$ **d.** $\frac{7.5}{12} = \frac{x}{20}$, $x = 12.5$ cm **Example 2:** **a.** $\frac{5}{8}$ **b.** AD and DH **c.** $\frac{2.5}{4}, \frac{5}{8}, \frac{7.5}{12}, \frac{10}{16}$ **d.** All are 0.625. **e.** 2.5

▶ 5.3 Exercises

1. If the designer in Example 1 reduced the 7.5-centimeter by 12-centimeter photograph to an 8-centimeter base, estimate what the new height would be. Find the height with a proportion.

2. If a designer enlarged a $3\frac{1}{2}$-inch by 5-inch photograph so that its shortest side was 14 inches, estimate what the new base would be. Find the base with a proportion.

For Exercises 3 to 6, trace the rectangles, and use the designer's diagonal method to enlarge or reduce them to the given base. Measure and label the heights.

3. Enlarge to a 2.5-inch base.

4. Reduce the rectangle in Exercise 3 to a 1-inch base.

5. Reduce the rectangle to a 1-inch base.

6. Enlarge the rectangle in Exercise 5 to a 3-inch base.

In Exercises 7 to 10, use ratios to find out if the triangles are similar. If they are similar, name the corresponding sides of the triangles.

7.

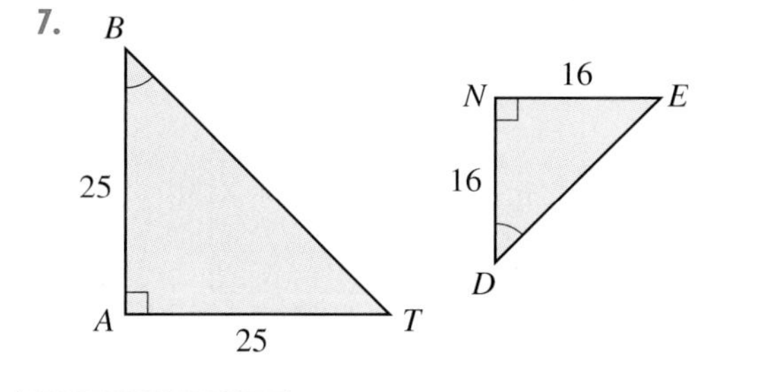

8. **9.** **10.**

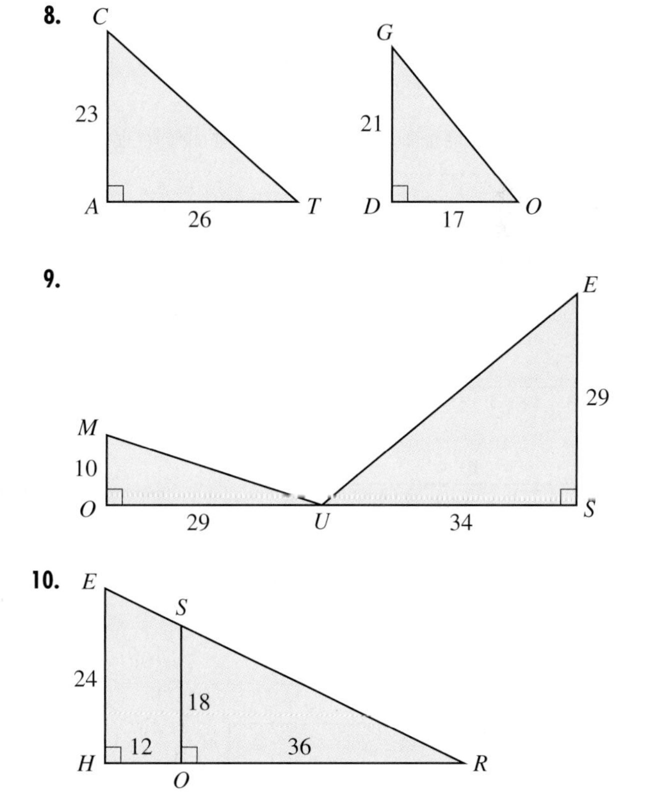

In Exercises 11 to 14, use ratios to show whether the figures are similar. Name two pairs of corresponding line segments for each pair of similar figures.

11. **12.**

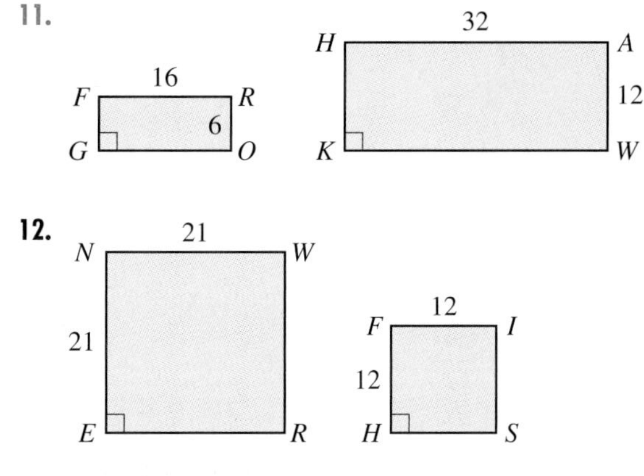

Blue numbers are core exercises.

13.

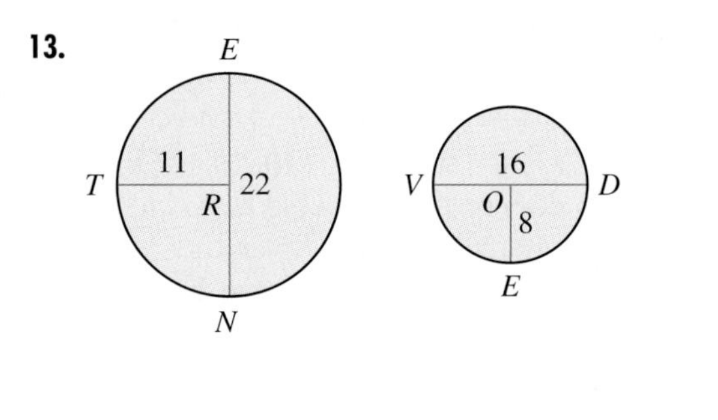

14.

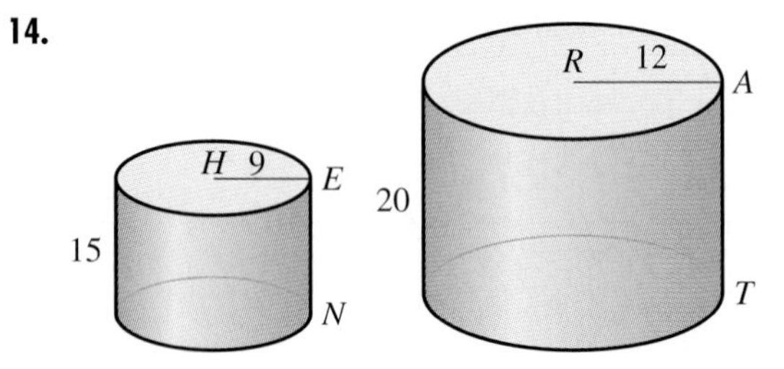

In Exercises 15 to 18, find the unknown side, n, in each pair of similar figures.

15. **16.** **17.** **18.**

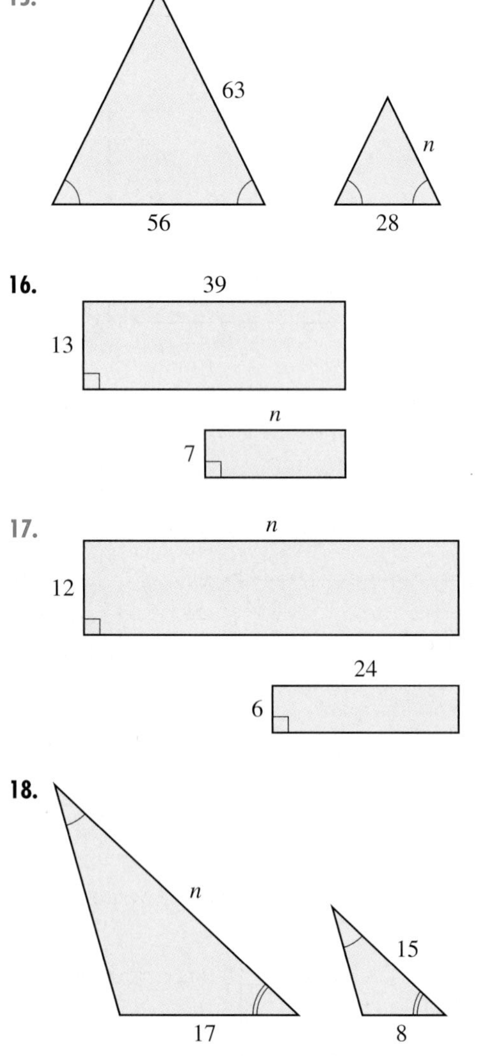

In Exercises 19 to 24, round to the nearest tenth.

19. At 3:00 p.m., a 30-foot tree casts a 35-foot shadow. A person 4 feet tall will cast how long a shadow?

20. At 5:00 p.m., a 30-foot tree casts a 125-foot shadow. How long is the shadow of a person 5.5 feet tall?

21. The shadow of a flagpole at 10 a.m. is 4 feet. The shadow of a 7-foot person is 1.4 feet. How tall is the flagpole?

22. The shadow of a tree at 2 p.m. is 8 feet. The shadow of a nearby 10-inch-tall squirrel is 2 inches. How tall is the tree?

23. The person who estimates the amount of wood in a tree is called a timber cruiser. A timber cruiser holds her arm parallel to the ground. In her hand she holds a stick, vertically, 27 inches from her eye (see the figure). A 14-inch length on the stick lines up with the top and bottom of a tree. The distance from the cruiser to the tree is 78 feet. How tall is the tree?

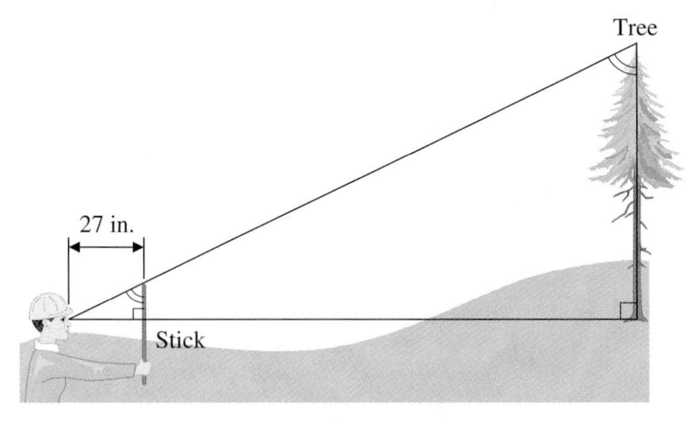

24. The timber cruiser in Exercise 23 lines up another tree in the same way. The second tree matches up with 30 inches on the stick when she is 60 feet from the base of the tree. How tall is the second tree?

Use proportions to find x in Exercises 25 to 30.

25. **26.**

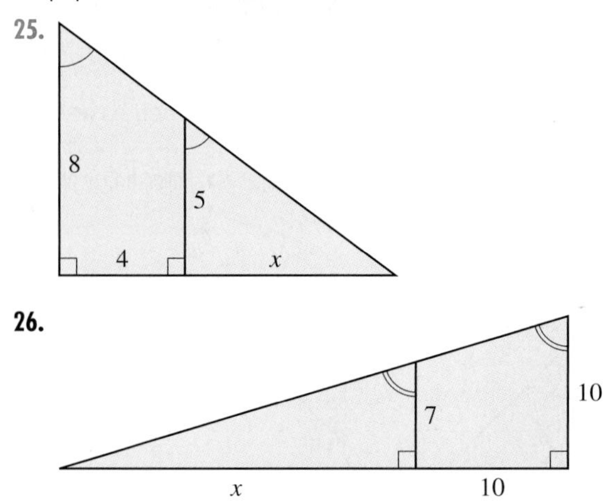

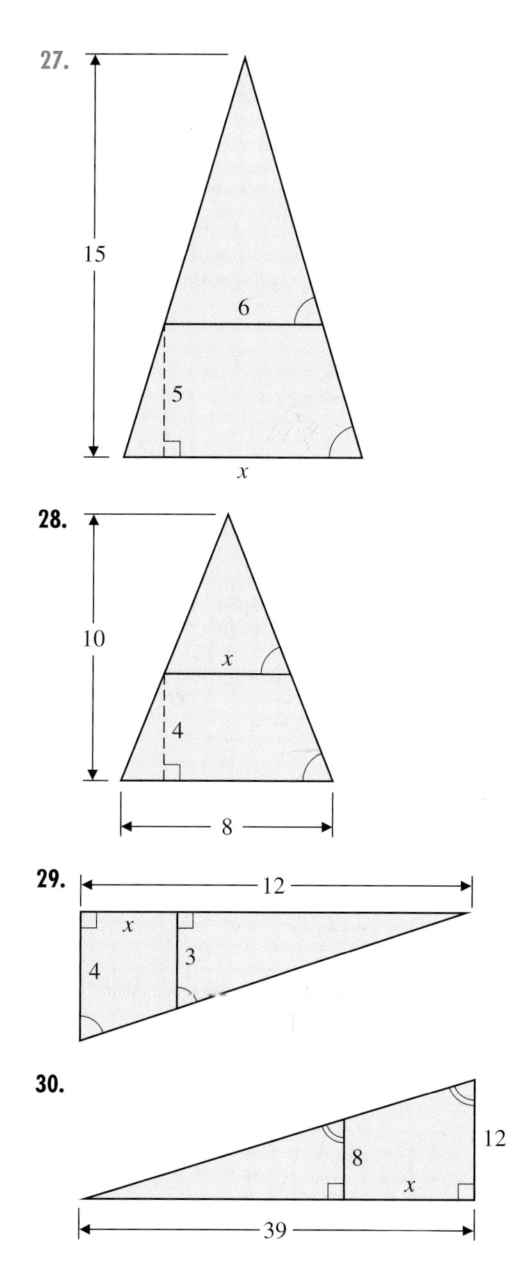

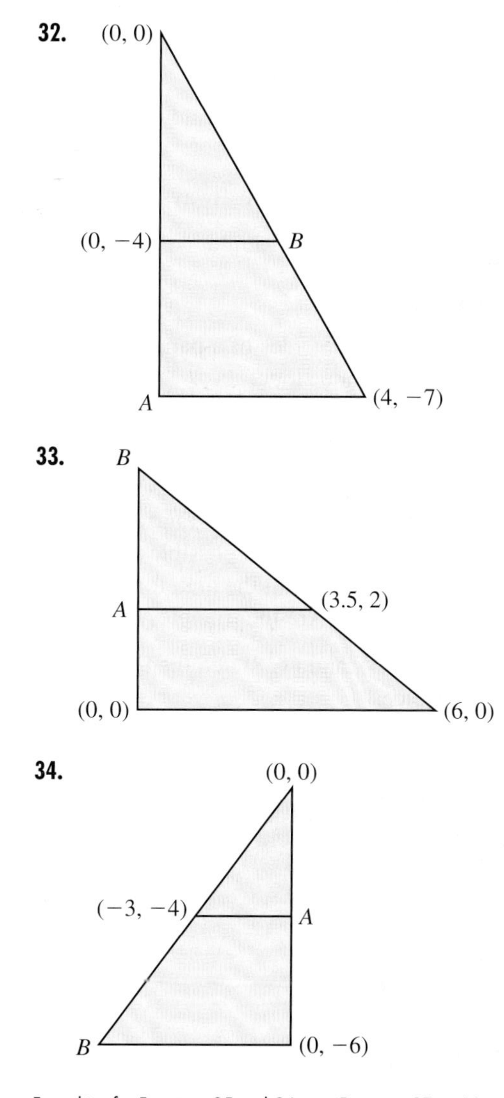

Identify the coordinates labeled *A* and *B* in Exercises 31 to 34. Use properties of similar triangles as needed.

31.

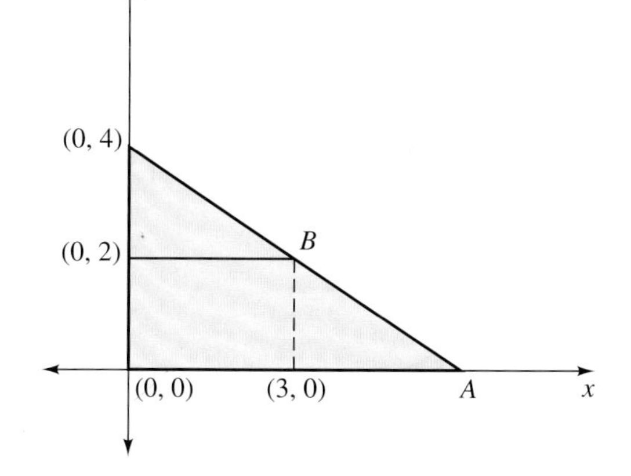

For a hint for Exercises 35 and 36, see Exercises 37 to 40.

35. A 5-foot parking meter has a 4-foot shadow from a nearby streetlight. The streetlight is 15 feet tall. How far is the streetlight from the parking meter?

36. Late one evening, a 6-foot person is standing 4 feet from a streetlight. The streetlight is 18 feet tall. How long is the person's shadow?

For Exercises 37 to 40, use this line segment:

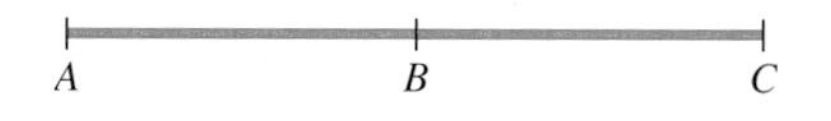

37. If segment AC is 10 units and BC is x units, write an expression for AB.

38. If segment AB is 10 units and BC is x units, write an expression for AC.

39. If segment AC is x units and AB is 10 units, write an expression for BC.

40. If segment AB is x units and AC is 10 units, write an expression for BC.

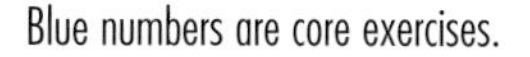

Blue numbers are core exercises.

▶ Writing

41. State two ways to identify similar triangles.

42. Describe how to use similar triangles to find ordered pairs.

43. Name three geometric figures that are always similar.

44. Explain why rectangles are not all similar.

▶ Problem Solving

45. Parallelograms The opposite sides of a parallelogram are parallel and have the same slope.

a. Three of the four vertices (corners) of a parallelogram are given: (0, 0), (3, 4), and (5, 0). Plot them. Find a fourth vertex. There are three different possible fourth vertices. Find all three, and plot them. Find the area of the triangle formed by the three possible fourth vertices. Compare that area with the area formed by the original three vertices. Are the triangles similar?

b. Repeat for (1, 1), (5, 3), and (1, 8) and then for three points of your choice.

▶ Project

46. Similar Rectangles and the Graphing Calculator Window The standard window for the TI-83 calculator is a horizontal setting of $-10 \le x \le 10$ and a vertical setting of $-10 \le y \le 10$. Thus, the height of the standard window is 20 units, and the width is also 20 units.

a. Why are the graphs on a standard window distorted?

A window setting with a horizontal to vertical ratio of 3 to 2 will produce a graph with little distortion. An example is a horizontal setting of $-15 \le x \le 15$ and a vertical setting of $-10 \le y \le 10$, which has a width of 30 units and a height of 20 units.

$$\frac{\text{Horizontal width}}{\text{Vertical height}} = \frac{30 \text{ units}}{20 \text{ units}} = \frac{3}{2}$$

In parts b to i, use a horizontal to vertical ratio of 3 to 2 to find the missing window size or setting.

b. Horizontal is 9 units wide. Find the height.

c. Vertical is 50 units tall. Find the width.

d. Vertical is 40 units tall. Find the width.

e. Horizontal is 18 units wide. Find the height.

f. $-8 \le x \le 16$, $-5 \le y \le$ ____

g. $-10 \le x \le 11$, $-6 \le y \le$ ____

h. $-20 \le x \le 20$, $-10 \le y \le$ ____

i. $0 \le x \le 100$, $0 \le y \le$ ____

▶ 5 Mid-Chapter Test

For Exercises 1 and 2, write the ratio in simplified form.

1. $12xy$ to $15y^2z$

2. $\dfrac{24(x - 2)}{8}$

Change the ratios in Exercises 3 and 4 to like units, and simplify.

3. 18 inches : 6 feet

4. 3 meters to 75 centimeters

In Exercises 5 and 6, use either proportions or equations to find the missing number.

5. 16% of what number is 11.2?

6. 360 is 250% of what number?

7. What percent increase is a salary change from \$24,000 to \$30,000?

8. What percent decrease is a CD price change from \$15.99 to \$13.99?

In Exercises 9 and 10, describe in a complete sentence a setting for each rate.

9. gallons per mile

10. pounds per week

Solve the equations in Exercises 11 to 15.

11. $\dfrac{5}{x} = 500{,}000$

12. $\dfrac{3}{5} = \dfrac{16}{x}$

13. $\dfrac{x}{12} = \dfrac{10}{8}$

14. $\dfrac{x + 5}{8} = \dfrac{2x - 1}{12}$

15. $\dfrac{5x - 2}{3x} = \dfrac{4}{3}$

16. Explain why $x - 3 = \frac{13}{3} - \frac{1}{3}x$ is not a proportion. Solve the equation.

17. Solve for b: $\dfrac{a}{b} = \dfrac{c}{d}$.

18. Use similar triangles and proportions to find x in the figure.

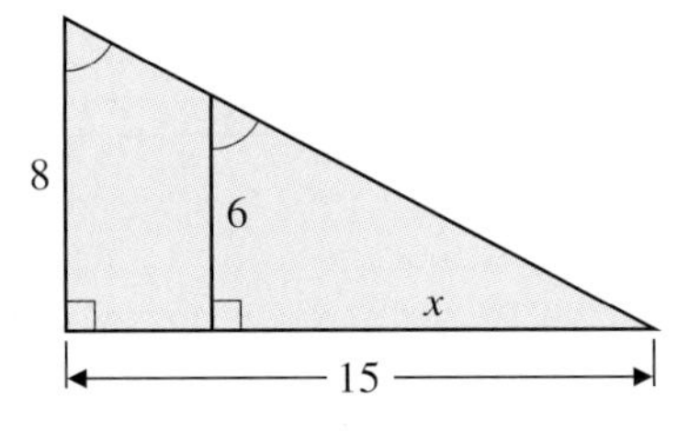

19. Use similar triangles and proportions as needed to find coordinates A and B in the figure.

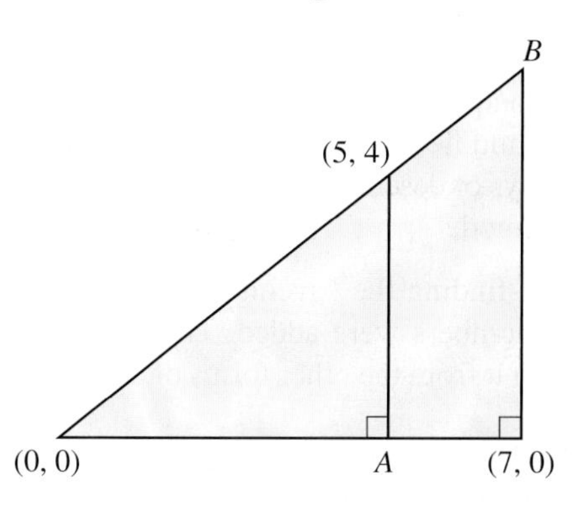

20. Calories in a diet are distributed in a ratio of 2 : 3 : 5 for fat, carbohydrate, and protein, respectively. Write the continued ratio as percents. If 1500 calories are to be consumed, how many may be allocated to each source?

21. Art glass needs to "cool" in a kiln at a temperature just below its melting point for 20 minutes per millimeter at its thickest part. How long a cooling period is needed for a bowl that measures 13 mm at its thickest part?

22. An access ramp needs to rise 48 inches. If the ramp is to have a slope ratio of 1 : 12, what horizontal distance is needed to build the ramp?

23. Ten thousand hatchery trout, each with a clipped fin, are released in a lake. Two weeks later, fishing season starts. Within two days, 260 trout are caught, and 250 have a clipped fin. Estimate the number of native trout (not clipped) originally in the lake.

24. An airplane is traveling 300 miles per hour. The pilot announces that the plane is 40 miles from the airport. In how many minutes will the flight be at the airport?

25. The lawsuit against Exxon, after the 1989 Valdez oil spill in Alaska, resulted in a \$5 billion judgment. A newspaper report indicated that the 5.9% annual interest on the judgment during the appeal process was \$9.40 per second.

a. What is 5.9% of \$5 billion?

b. Use unit analysis to show whether the newspaper statement was correct.

▶ 5.4 Averages

Objectives

- Find the mean, median, and mode of a set of data.
- Find the weighted average of a set of data.
- Find the midpoint of a line segment, given its endpoints as ordered pairs.
- Find the centroid of a figure, given its corners as ordered pairs.

WARM-UP

1. Solve for x:

$$\frac{0.78 + 0.70 + 0.90 + x}{4} = 0.80$$

2. Solve for x:

$$\frac{1.93 + 1.5x}{4} = 0.80$$

IN THIS SECTION, we look at the concept of average. We work with numbers and practice writing and solving equations. We relate averages both to geometry and to the proportions we studied in Sections 5.1 to 5.3.

Averages

An average describes the "center," or middle, of a set of numbers. In some settings, we think of average as being "normal." In other settings, an average helps us compare an individual with a group or compare one group with another group. In geometry, the average is a description of middle.

We now define three important ways of describing the middle of a set of numbers—with the mean, the median, and the mode.

MEAN Usually, people say they are finding the "average" when they add numbers together and divide by how many numbers were added. This average is called the mean or mean average, to distinguish it from the other forms of averages (see median and mode, below).

FINDING THE MEAN

> The **mean** of a set of numbers is the sum of the numbers divided by the number of numbers in the set.

In Example 1, we find a mean average of incomes.

EXAMPLE 1 Finding the mean: mean income Suppose the annual incomes of five families are \$10,000, \$10,000, \$12,000, \$13,000, and \$100,000, respectively. Find the mean, and explain why the mean does not give a good description of this set of families.

SOLUTION The mean is

$$(\$10{,}000 + \$10{,}000 + \$12{,}000 + \$13{,}000 + \$100{,}000) \div 5 = \$29{,}000$$

The mean appears to indicate that all the families have an income well above the 2001 poverty level for families of three persons, \$14,128. Yet in reality, four of the five families are below the poverty level. ◀

▶ In Section 3.5, we calculated the score needed on a final exam to earn a specific grade. In Example 2, each test and the total homework are worth the same number of points, so we can find the mean average of the percent scores.

EXAMPLE 2 Finding the mean: class average Suppose an instructor gives equal weight to homework, each of two midterms, and the final exam. A student has 78% and 70% on the midterms and 90% on homework. What percent does this student need on the final to have an 80% mean average for the class?

SOLUTION

$$\frac{0.78 + 0.70 + 0.90 + x}{4} = 0.80 \quad \text{Add the numbers in the numerator.}$$

$$\frac{2.38 + x}{4} = 0.80 \quad \text{Multiply both sides by 4.}$$

$$2.38 + x = 3.20 \quad \text{Subtract 2.38 from both sides.}$$

$$x = 0.82$$

The student needs 82% on the final exam. ◀

Note that we could add the percents in Example 2 because the instructor gives equal importance, or weight, to each. In general, *do not average percents unless they are percents of the same number or are assigned weights* (see Examples 6 and 7).

MEDIAN A very high or low number, as in Example 1, can cause the mean not to be close to most of the numbers. Because the mean does not always give a good description, statisticians invented other averages, such as the median and the mode.

FINDING THE MEDIAN

> The **median** is found by selecting the middle number when the numbers are arranged in numerical order (from smallest to largest or largest to smallest). If there is no single middle number, the median is the mean of the two middle numbers.

▶ **EXAMPLE 3** Finding the median: income Find the median of each set of incomes.

a. \$10,000, \$10,000, \$12,000, \$13,000, \$100,000
b. \$50,000, \$20,000, \$8000, \$16,000

SOLUTION **a.** The median of \$10,000, \$10,000, \$12,000, \$13,000, and \$100,000 is \$12,000, because the numbers are already arranged in order and \$12,000 is the middle number.

b. The second set of numbers must be rearranged into order: \$8000, \$16,000, \$20,000, and \$50,000. The set has no middle number. In this case, we find the mean of the two middle numbers and use it as the median. The median is (\$16,000 + \$20,000) ÷ 2 = \$18,000. ◀

THINK ABOUT IT 1: The graph in Figure 22 of Section 1.5 gives median incomes by age. Does a 50% increase in income for Microsoft founder Bill Gates impact this graph?

As suggested by Example 3, the median is more descriptive than the mean when one number in the set of data is some distance from the rest of the numbers.

MODE The mode is a third form of average. The mode is useful when the average needs to describe a most popular or most common item. In the set of numbers {2, 2, 2, 2, 3, 4}, the mode is 2 because 2 appears most often.

FINDING THE MODE

> The **mode** of a set is the number that occurs most often.

▶ Data on a Table

Table 1 shows a tally by states of the age for unrestricted operation of a private passenger car. Many states with an age of 18 allow the applicant to drive at a lower age if he or she has taken a driver's education course.

TABLE 1 Age for Unrestricted Operation of Private Passenger Cars, 1991

Driver's Age (years)	States (including the District of Columbia)
15	||
16	𝍸 𝍸 𝍸 𝍸
16.5	|
17	|||
18	𝍸 𝍸 𝍸 𝍸 ||||
19	|

Data from *The World Almanac and Book of Facts 1992* (New York: Pharos Books, 1991), p. 678.

▶ EXAMPLE 4 Finding median and mode from a table What are the median and mode for the data on driving age in Table 1?

SOLUTION The tally places the data in order by age, so we count to the middle tally mark. The middle of the 51 tally marks is the 26th tally mark, because it has 25 marks before it and 25 after it. The 26th tally mark is the last mark for age 17, so age 17 is the median. The mode is the number that is tallied most often: age 18. ◀

In Example 4, the minimum age for driving is almost evenly split between age 16 and age 18. If the tally showed the same number of states for ages 16 and 18, the data would be **bimodal**—that is, *having two modes.*

THINK ABOUT IT 2: How will the table change if lawmakers decide to reduce the number of teen deaths from auto accidents?

Finding the mean for a large quantity of data is time-consuming, so when possible we group like numbers, as in Table 1. In Example 5, we multiply the number of times each age occurs by the age, instead of adding the ages separately.

▶ EXAMPLE 5 Finding the mean from a table Find the mean age for Table 1.

SOLUTION To find the mean age for Table 1, we add the 51 ages and divide the total by 51:

$$\frac{15(2) + 16(20) + 16.5(1) + 17(3) + 18(24) + 19(1)}{51}$$

Entering the entire expression in a calculator and pressing ENTER will yield a result of approximately 17.03 years. ◀

Compare the results in Examples 4 and 5 with those in the Graphing Calculator Technique box below.

GRAPHING CALCULATOR TECHNIQUE: CALCULATOR MEAN AND MEDIAN FOR TABLE 1

Press STAT **1 : Edit**. Under L1, place the various ages. Under L2, place the number of states with each age. L2 now gives the frequency of each item in L1 (see Figure 13). Omitting L2 indicates that the set contains only one of each item in L1.

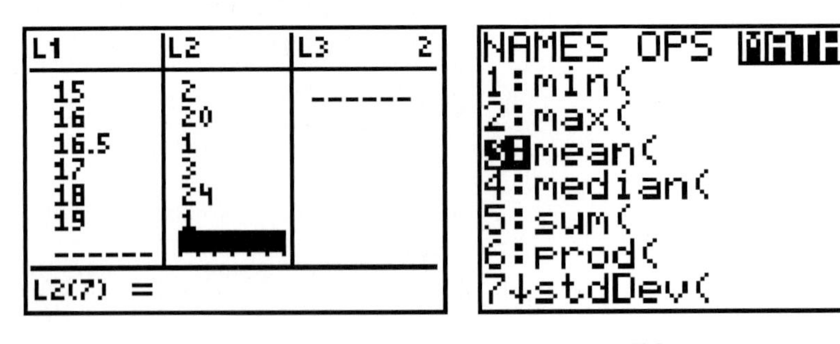

FIGURE 13 Age and tally in list **FIGURE 14** 2nd [LIST] MATH menu

Press 2nd [QUIT].

To find the mean, use 2nd [LIST] **MATH** and choose **3 : mean**. Then press ENTER (see Figure 14). When **mean(** appears, press 2nd [L1] , 2nd [L2]) ENTER.

(continued)

(concluded)

To find the median, use [2nd] [LIST] **MATH** and choose **4 : median**. Then press [ENTER]. When **median(** appears, press [2nd] [L1] [,] [2nd] [L2]) [ENTER].

Figure 15 shows the results for both the mean and the median.

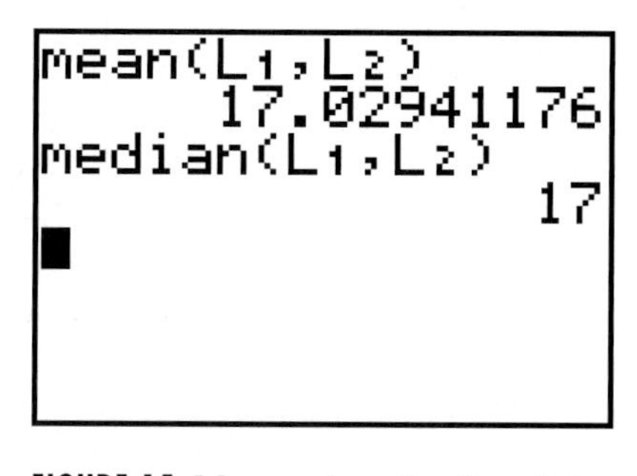

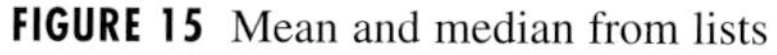

FIGURE 15 Mean and median from lists

▶ Weighted Averages

There are times when some numbers in a set are more important than others. For example, a final exam may have the same number of points as a midterm but be worth twice as much in calculating a grade for the course.

To *give importance to numbers,* we assign a **weight**. A **weighted average** is *an average found by multiplying each number by its weight, adding the products, and dividing by the total weight* (rather than the total number of numbers). Here is a formula:

$$\text{Weighted average} = \frac{\text{sum of the product of each number and its weight}}{\text{total weight}}$$

A table can help organize the weights and the products. In the table in Example 6, we multiply the numbers in the rows (across the table) and then add the numbers in the last column to get the total.

▶ **EXAMPLE 6** Finding weighted grade averages Suppose the instructor in Example 2 gives homework half the weight of the midterm and gives the final one and a half times the weight of the midterm. Find the weighted average by completing Table 2. The weighted average, which is placed at the bottom of the Score column, is the sum of the $w \cdot s$ column divided by the sum of the weights. Use the weighted average formula to set up and solve an equation to find what score the student needs on the final exam to average 80%.

TABLE 2

	Weight, w	Score, s	$w \cdot s$
Midterm	1	0.78	
Midterm	1	0.70	
Homework	0.5	0.90	
Final	1.5	x	
Total		☐	

SOLUTION The completed table is shown in Table 3.

TABLE 3

	Weight, w	Score, s	$w \cdot s$
Midterm	1	0.78	0.78
Midterm	1	0.70	0.70
Homework	0.5	0.90	0.45
Final	1.5	x	$1.5x$
Total	4.0	(0.80)	$1.93 + 1.5x$

↑ Weighted average (the circled 0.80)

The weighted average formula gives the same information:

$$\frac{1(0.78) + 1(0.70) + 0.5(0.90) + 1.5x}{1 + 1 + 0.5 + 1.5} = 0.80$$ Add the numbers in the numerator and denominator.

$$\frac{1.93 + 1.5x}{4} = 0.80$$ Multiply both sides by 4.

$$1.93 + 1.5x = 4(0.80)$$ Subtract 1.93 from both sides.

$$1.5x = 1.27$$ Divide by 1.5 on both sides.

$$x \approx 0.85$$

The student needs an 85% on the final for an 80% average. ◀

▶ **EXAMPLE 7** **Finding a weighted average** Suppose the instructor in Example 2 makes the homework worth 10% of the grade, each midterm 20%, and the final 50%. Complete Table 4 and set up a weighted average equation to find what score is needed on the final exam to average 80%.

TABLE 4

	Weight, w	Score, s	$w \cdot s$
Midterm	0.20	0.78	
Midterm	0.20	0.70	
Homework	0.10	0.90	
Final	0.50	x	
Total		( )	

SOLUTION The completed table is shown in Table 5.

TABLE 5

	Weight, w	Score, s	$w \cdot s$
Midterm	0.20	0.78	0.156
Midterm	0.20	0.70	0.140
Homework	0.10	0.90	0.090
Final	0.50	x	$0.5x$
Total	1.00	(0.80)	$0.386 + 0.5x$

↑ Weighted average (the circled 0.80)

The weighted average formula gives the same information:

$$\frac{0.20(0.78) + 0.20(0.70) + 0.10(0.90) + 0.50x}{0.20 + 0.20 + 0.10 + 0.50} = 0.80$$ Add the numbers in the numerator and denominator.

Student Note:
Division by 1 is shown to indicate the general process.

$$\frac{0.386 + 0.5x}{1} = 0.80$$ Multiply both sides by 1.

$$0.386 + 0.5x = 1(0.80)$$ Subtract 0.386 from both sides.

$$0.5x = 0.414$$ Divide by 0.5 on both sides.

$$x \approx 0.83$$

The student needs an 83% on the final for an 80% average. ◀

Examples 6 and 7 show that weights can be either numbers or percents.

THINK ABOUT IT 3: Why was the score needed by the student in Example 7 lower than that needed by the student in Example 6?

In a weighted average, each item's influence on the average depends on the quantity of that item present. We work with other tables in Section 5.5.

Averages in Geometry

The mean average appears in two important applications in geometry: midpoint and centroid. These topics are presented here, to give more practice with ordered pairs and finding the mean.

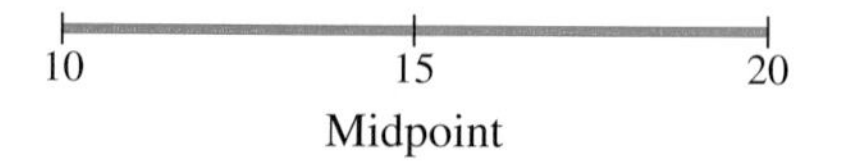

FIGURE 16 Midpoint of line segment

MIDPOINT The **midpoint of a line segment** is *the center, or the point halfway between its endpoints.* Figure 16 shows a line of length 10 units, with a midpoint 5 units from both ends. The point labeled 15 is halfway between a point labeled 10 and another labeled 20. The midpoint is the average of the coordinates of the endpoints; we can find the midpoint of any line segment by calculating this average.

The **midpoint on a coordinate graph** is *the mean in both the x and the y direction.* Thus, the midpoint of (x_1, y_1) and (x_2, y_2) is the average of x_1 and x_2 and the average of y_1 and y_2. Figure 17 shows the midpoint of the line connecting (x_1, y_1) and (x_2, y_2). The formula for finding midpoints is

$$\text{Midpoint} = \left(\frac{x_1 + x_2}{2}, \frac{y_1 + y_2}{2}\right)$$

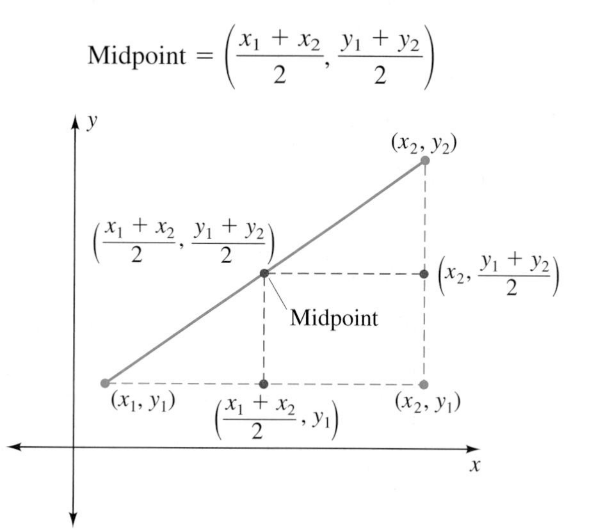

FIGURE 17 Midpoint of line segment on a coordinate graph

▶ **EXAMPLE 8** Finding midpoints Find the midpoints for the segments formed by the following coordinates. First use the formula. Then check the answer by sketching the line segment on a graph and plotting the midpoint.

a. (0, 2) and (5, 0)
b. (0, 0) and (5, 2)
c. (7, 3) and (9, −1)

SOLUTION **a.** For (0, 2) and (5, 0), the midpoint is

$$\left(\frac{x_1 + x_2}{2}, \frac{y_1 + y_2}{2}\right) = \left(\frac{0 + 5}{2}, \frac{2 + 0}{2}\right) = \left(\frac{5}{2}, 1\right) = (2.5, 1)$$

b. For (0, 0) and (5, 2), the midpoint is

$$\left(\frac{x_1 + x_2}{2}, \frac{y_1 + y_2}{2}\right) = \left(\frac{0 + 5}{2}, \frac{0 + 2}{2}\right) = \left(\frac{5}{2}, 1\right) = (2.5, 1)$$

c. For (7, 3) and (9, −1), the midpoint is

$$\left(\frac{x_1 + x_2}{2}, \frac{y_1 + y_2}{2}\right) = \left(\frac{7 + 9}{2}, \frac{3 + (-1)}{2}\right) = (8, 1)$$

The midpoint for the first two segments is the same, (2.5, 1). It is labeled M_a and M_b in Figure 18. The midpoint for the third segment is (8, 1) and is labeled M_c.

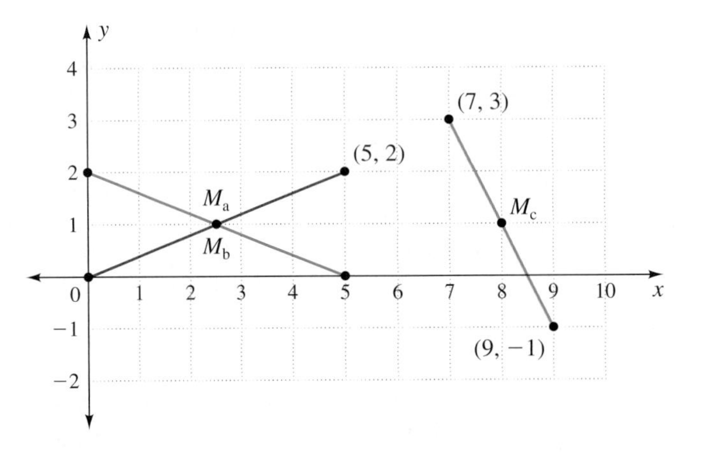

FIGURE 18 Midpoint on line segments ◀

CENTROID The midpoint locates the middle of a line, whereas the **centroid** is *the center of a flat or solid geometric shape.* In more complicated structures such as the human body, bicycles, and airplanes, the center is called the *center of mass* or *center of gravity.*

The airplane in the photograph on the chapter opener (page 257), became so out of balance that it tipped. Not only was the situation embarrassing; it was also a graphic illustration of how important the location of the center of gravity is. The *center of gravity is the average position of weight.* The pilot determines the center of gravity of the load from the weight of the luggage and cargo, the weight of the fuel, and an average weight assigned to each passenger.

FINDING THE CENTROID

> The centroid of a rectangle, square, or triangle is the average of the coordinates of its corners (vertices).

▶ **EXAMPLE 9** Finding the centroid Find the position of the centroid for the triangle in Figure 19.

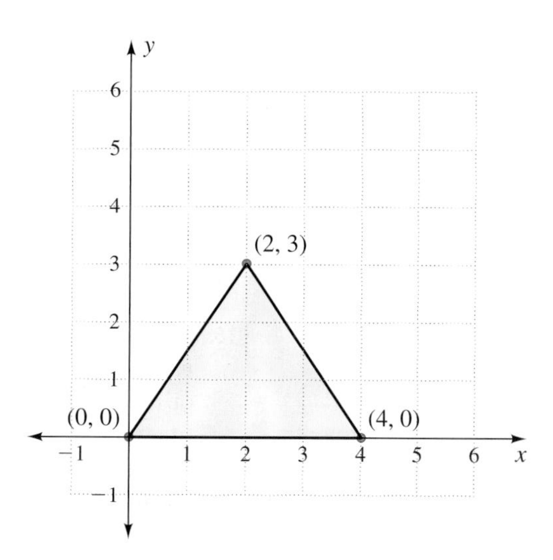

FIGURE 19

SOLUTION The average of the x-coordinates of the vertices is

$$x_c = \frac{0 + 4 + 2}{3} = \frac{6}{3} = 2$$

The average of the y-coordinates of the vertices is

$$y_c = \frac{0 + 0 + 3}{3} = 1$$

The coordinates of the centroid are $(x_c, y_c) = (2, 1)$. ◀

Student Note:
Artist Alexander Calder was a master of balance. See his hanging mobiles (**www.calder.org/SETS_SUB/site_index.html**).

Centroids (and centers of gravity) are important because the behavior of an object in motion is dependent on where the centroid is located. Thus, the flight of an airplane, the rotation of a tire, the spinning of a washing machine, and the wild motion of a carnival ride depend on a correct centering of weight. Many sports (figure skating, platform diving, pole vaulting, gymnastics, and ski jumping) require careful control of the body's center to achieve top performance.

ANSWER BOX

Warm-up: 1. $x = 0.82$ **2.** $x \approx 0.847$ **Think about it 1:** No **Think about it 2:** Ages will rise **Think about it 3:** The final exam has a larger weight in Example 7 than in Example 6, so the 83%, although lower than the 85%, is worth more in the grade calculation.

▶ 5.4 Exercises

In Exercises 1 to 4, calculate the mean, median, and mode. Make an observation about the data.

1. Quiz scores
   a. 80, 80, 80 85, 90, 95
   b. 96, 90, 85, 80, 80, 79
   c. 85, 82, 80, 65, 0, 0
   d. 0, 80, 85, 85, 90, 95
   e. What would you say to each student?
2. Advertised puppy prices, in dollars
   a. Jack Russell female: 450, 450, 350, 350, 400, 225
   b. Jack Russell male: 350, 350, 450, 450, 250, 250, 250
   c. Golden retriever male: 750, 450, 450, 700, 350, 400, 400, 700
   d. Golden retriever female: 750, 525, 700, 400, 450, 450, 700
3. Advertised rent, in dollars, for one-bedroom apartments
   a. In a city with population 150,000: 375, 600, 475, 475, 475, 435, 435, 595, 545, 375, 375, 575, 475, 415, 485, 525
   b. In a city with population 530,000: 535, 525, 525, 500, 550, 525, 525, 535, 625, 525
4. Asking prices of used sport utility vehicles
   a. Three-year-old Chevy Suburban: \$28,988, \$20,988, \$23,988, \$23,750, \$28,750
   b. Three-year-old Ford Expedition: \$19,975, \$25,998, \$22,988, \$20,995, \$23,000

▶ Mean

5. Explain how to find the mean from a set of data.
6. Choose one: The mean is influenced by (all, most, few) measurements in the set. Explain.
7. If the mean is the same for two sets of data, does this imply that the numbers in the sets are the same? Explain.
8. Comment, using complete sentences, on the effect of one low grade or one high grade on the mean of the test scores.
9. How can the formula for the area of a trapezoid be rewritten to show a mean average? Explain your formula in words.

▶ Median

10. Explain how to find the median from a set of data.
11. Choose one: The median (is, is not) influenced by one large or small measurement. Explain.
12. Is it possible to have $\frac{3}{4}$ of the students above the median test score? Explain.
13. Why might the *median of a set of numbers* and the *median of a triangle* be given such similar names?
14. The guard rail or strip of ground between the traffic lanes of a freeway is called the *median*. Explain why this is an appropriate word.

▶ Mode

15. Explain how to find the mode from a set of data.
16. *Mode* has the same root as *modern*, *model*, and *a la mode*, which are associated with current, fashionable, or most popular styles. Explain how *mode* as an average fits with these other words.

▶ Mean, Median, and Mode

17. How does one large piece of data affect the mean? the mode? the median?
18. If a fly ball or strikeout is worth 0 and a hit is worth 1, is a baseball batting average (hits divided by times at bat) a median, mode, or mean?
19. When might the median provide a better description of the average than the mean?
20. When might the mean provide a better description of the average than the median?
21. What can you conclude about a set of data if the mean is larger than the median?
22. What can you conclude about a set of data if the median is larger than the mean?
23. List a set of numbers for which the mean is larger than the median. List a set of numbers for which the mean is smaller than the median. Explain how you found your list.
24. *Sampling* involves selecting a number of objects from a set. Explain why a test must sample what a student knows.

Blue numbers are core exercises.

25. The accompanying table includes the data from Table 1, as well as data for 2002 and 2006.

Driver's Age (years)	States in 1991	States in 2002	States in 2006
15	2	1	0
16	20	11	8
16.5*	1	10	10
17	3	17	19
17.5*	0	4	1
18	24	8	13
19	1	0	0

*Also includes fractional parts of a year other than $\frac{1}{2}$.

a. What is the mode for the data on driving age in 1991? in 2002? in 2006?

b. Describe what happened to driver's age between 1991 and 2002.

c. Find the median and mean for each of the years.

d. Although the median and mean did not change significantly between 1991 and 2006, why might the 2006 ages be better?

26. A baseball batting average is the ratio of hits to times at bat:

$$\text{Batting average} = \frac{\text{number of hits}}{\text{number of times at bat}}$$

Find the batting average, in decimal notation, for each of the following players. Round to the nearest thousandth.

a. Roberto Clemente: 209 hits in 585 times at bat

b. Juan Gonzalez: 152 hits in 584 times at bat (The hits included 43 home runs.)

c. Ken Griffey, Jr.: 174 hits in 565 times at bat

d. Ivan Calderon: 141 hits in 470 times at bat

e. Hugh Duffy (1894): 236 hits in 539 times at bat

f. Rogers Hornsby (1924): 227 hits in 536 times at bat

Exercises 27 and 28 provide practice in calculating with weighted numbers.

27. Three colleges are competing in a track and field meet. A first-place finish is worth 5 points; second place, 3 points; third place, 1 point. GRCC has 4 first places, 3 seconds, and 2 thirds. LCC has 5 first places and 4 thirds. BCC has 1 first, 7 seconds, and 4 thirds. What is each team's total score?

28. A basketball team has 10 free throws at 1 point each, 25 field goals at 2 points each, and 6 three-point shots. The opposing team has 16 free throws, 20 field goals, and 4 three-point shots. What is each team's total score?

The student scores in Exercises 29 to 32 are from a course with three tests and a final exam. The final is worth twice as much as each test. Estimate an average for each student. Find a weighted average for each student.

29. Tests: 0.85, 0.70, 0.80; final: 0.95

30. Tests: 0.75, 0.70, 0.75; final: 0.68

31. Tests: 0.80, 0.80, 0.70; final: 0.60

32. Tests: 0.90, 0.80, 0.60; final: 0.85

The students in Exercises 33 to 36 are in the same course as those in Exercises 29 to 32. What final exam score does each student need to earn a 0.80 weighted average? A perfect final exam score is 1.00.

33. Tests: 0.76, 0.81, 0.72

34. Tests: 0.60, 0.65, 0.70

35. Tests: 0.50, 0.70, 0.70

36. Tests: 0.60, 0.70, 0.75

Find the midpoints of the line segments connecting the sets of points given in Exercises 37 and 38. (*Hint:* A sketch on coordinate axes may confirm that the midpoints are reasonable.)

37. **a.** $(0, 4), (5, 2)$ **b.** $(-1, 3), (3, -3)$

38. **a.** $(2, 3), (8, 12)$ **b.** $(2, -5), (-4, 3)$

Find the midpoints of the line segments connecting the sets of points given in Exercises 39 and 40. Assume that the variables a and b are positive numbers.

39. **a.** $(a, 0), (0, b)$ **b.** $(a, a), (0, 0)$ **c.** $(0, b), (0, 0)$

40. **a.** $(a, 0), (0, a)$ **b.** $(0, 0), (a, 0)$ **c.** $(a, b), (0, 0)$

Find the midpoint of each side of the triangles in Exercises 41 to 44. Find the centroid of each triangle.

41.

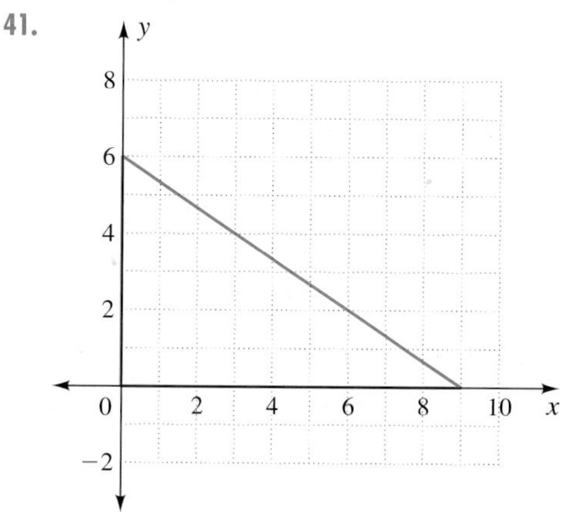

Blue numbers are core exercises.

42.

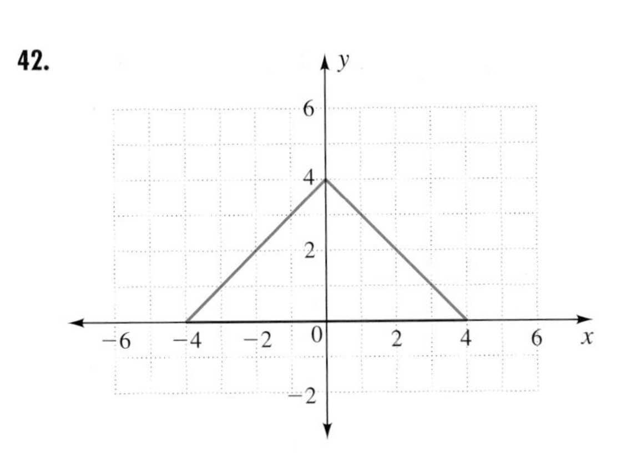

43.

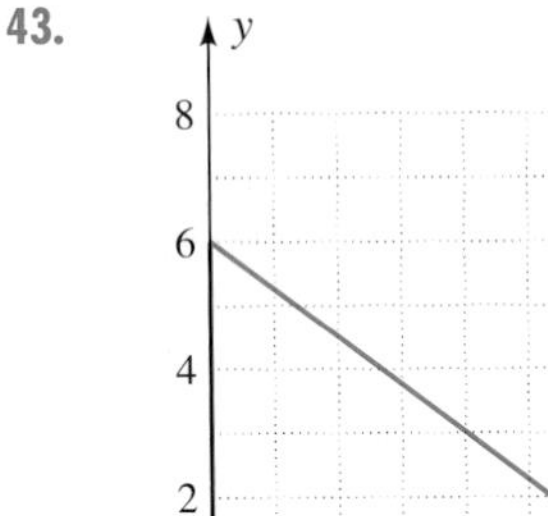

44.

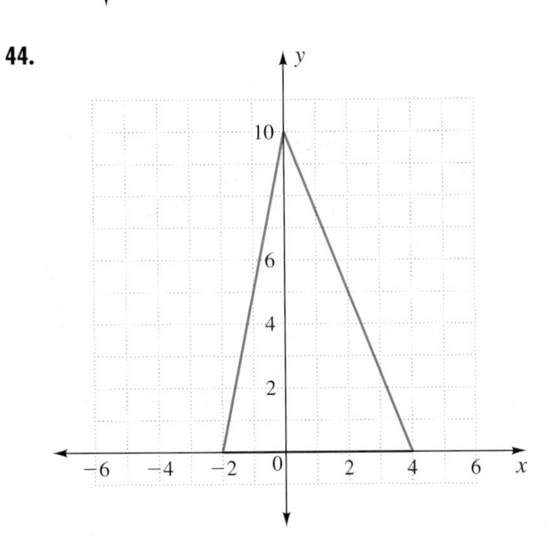

▶ Projects

45. Bookstore Discount A local bookstore offers a plan to "Buy 12 books, get one free." The store records each purchase on a card. After you buy 12 books, the store averages the purchases and you get to spend the average on your free book.

a. What is the value of the free book for a card with 55, 6, 15, 40, 8, 35, 42, 6, 20, 15, 10, and 15? a card with 10, 18, 40, 34, 12, 8, 50, 30, 45, 20, 10, and 5?

b. What is the total value of the two free books from part a?

c. Suppose you were able to sort out the purchases and place the 12 largest on one card and the others on the second card. Could you increase the total dollar value of the two free books?

d. Describe a way, if any, to sort the purchases to your advantage.

46. Center of Population The U.S. Center of Population is like a centroid of population. It would be the balance point for the United States if each person (assume equal weight) were standing at her or his home location on a rigid plate the size of the country.

a. Trace a small map of the United States, and guess where the population center started and how it moved during the period from 1790 to 2000. Explain your guesses.

b. The center is recalculated after each ten-year census. Go to **www.census.gov/geo/www/cenpop/meanctr.pdf** and print the map. Explain changes in direction.

47. Centroid Exploration

a. Cut a large triangle from a stiff piece of notebook-size cardboard. Hold a pencil in a vertical position, and move the triangle around on the eraser until the triangle balances. Mark this point on the triangle. The balance point is the centroid of the triangle.

b. Measure the sides of the triangle, and mark the mid-point of each side. Connect each midpoint with the vertex opposite that side. The intersection of these three lines is the centroid. Does this centroid match that found in part a?

c. Use a paper punch to put a hole near each corner of the cardboard triangle. Place a pencil point through one of the holes, and hang a string with a weight tied at the bottom from the pencil point. The string and the triangle must be able to turn freely. Mark the point where the string crosses the opposite side of the triangle. Connect the hole and the mark with a line. Repeat for each of the other two holes. The intersection of the lines is the centroid. How close is this experimental centroid to the balance point? Where does the string cross the opposite side?

d. Place the triangle on a piece of graph paper. Find an ordered pair for each corner and use the averaging method to find the centroid.

e. How do your results in parts a to d agree?

Blue numbers are core exercises.

5.5 Writing Equations from Word Problems with Quantity-Rate Tables

Objectives

- Distinguish between quantity and rate in problem situations.
- Explain when quantity sums and average rate are meaningful.
- Estimate the average rate, given the quantities and rates.
- Use quantity-rate tables to summarize word problems and set up equations.
- Estimate a quantity (or rate) needed to find a given average rate.

WARM-UP

Solve for x.

1. $200(55) + x(60) = 29{,}000$
2. $9(35) + 140x = (9 + x)(103)$
3. $24(1.875) + 3x = (24 + x)(2.50)$
4. $24(1.875) + 3x = (24 + x)(3.00)$
5. $0.07x + 0.12(8000 - x) = 800$

THIS SECTION BUILDS ON equation-writing skills. We use our experience with ratios and rates to tell what kind of information (a quantity or a rate) is given by a number within a word-problem setting. We use proportional reasoning to estimate outcomes.

Distinguishing Between Quantity and Rate

In the poem "Smart" by Shel Silverstein,* a father gives his child a dollar bill. The child, thinking in terms of quantity, trades the dollar bill for two quarters, then the quarters for three dimes, then the dimes for four nickels, and finally the nickels for five pennies. You can imagine the father's reaction.

In the poem, the child measured "worth" in terms of the quantity of coins. **Quantity** is a number, word, or phrase that *answers the question "How many?" or "How much?"*

His father measured "worth" as the value of the coins. Value is a type of rate. **Rate**, as in the poem, may be *monetary worth*: cost per item, 10 cents per dime. Rate may be a *percent*: an investment yielding 5.5%. Other *rates* include miles per hour and pounds per square inch. The word *per* is a clue to recognizing a rate, but it is not the only word indicating a rate.

Rate may also be *numbers given to specific results*: points assigned to letter grades to find a grade point average, different orders in which runners finish a track meet, and types of scores in a basketball game.

EXAMPLE 1 Distinguishing between quantity and rate Which of these describe quantities? Which describe rates?

a. 4 points for each credit hour for which a grade of A is earned
b. 17% protein
c. 3 coins
d. 9 gallons
e. \$55 per share
f. 60 pounds
g. \$5000 invested
h. \$4.99 per pound for antibiotic-free hamburger
i. $4\frac{1}{4}$% interest
j. \$4.99 in change

SOLUTION Parts c, d, f, g, and j describe quantities. The rest are rates. ◀

---

*In *Where the Sidewalk Ends*

A key idea in this section is that quantity and rate take different forms, so problems that seem different may be the same mathematically.

Building Quantity-Rate Tables

When we place the quantity and rate from a problem setting into a table, called a **quantity-rate table**, we find that the product, quantity times rate, has meaning.

EXAMPLE 2 Completing a quantity-rate table Use the information in the poem "SMART" to fill in Table 6. The first two rows have been completed.

TABLE 6

Item	Quantity	Rate (per bill or coin)	Quantity · Rate
Dollar bill	1	$1.00	$1.00
Quarters	2	$0.25	$0.50
Dimes			
Nickels			
Pennies			

SOLUTION There are 3 dimes at $0.10 per dime, for a total of $0.30.

There are 4 nickels at $0.05 per nickel, for a total of $0.20.

There are 5 pennies at $0.01 per penny, for a total of $0.05. ◀

THINK ABOUT IT 1: Does the sum of the numbers in any column of Table 6 have meaning?

Table 6 illustrates the pattern of multiplying the quantity and the rate to obtain a total item value.

> The product of the quantity, Q, of an item and the item's rate, R, is given in the last column, $Q \cdot R$, of the quantity-rate table.

▶ The sum of the quantity column has no meaning in the "SMART" table, nor does the sum of the quantity · rate column.

In Example 3, both the sum of the quantity column and the sum of the $Q \cdot R$ column have meaning.

EXAMPLE 3 Building a quantity-rate table: lunch for four Make a table for this meal at Jamie's Hamburgers, identifying quantity and rate for each item ordered: 4 hamburgers at $4.99 each, 2 orders of fries at $2.79 each, 2 coffees at $0.99 each, and 2 milks at $1.89 each. Calculate the item total and the total cost of the meal.

SOLUTION The completed table is shown in Table 7. The sum of the $Q \cdot R$ column, $31.30, is the cost of the meal.

TABLE 7

Item	Quantity, Q	Rate, R	Quantity · Rate, $Q \cdot R$
Hamburger	4	$4.99 each	$19.96
Fries	2	2.79 each	5.58
Coffee	2	0.99 each	1.98
Milk	2	1.89 each	3.78
Total	10 items		$31.30

◀

THINK ABOUT IT 2: What is the meaning of the sum of the quantity column?

▶ Average Rate

In Example 3, dividing $31.30 by 10 gives the average cost, $3.13, of the 10 items. This average cost is meaningful but not particularly useful. The average cost may be written as the last entry in the rate column. The rate column is never added.

FINDING AVERAGE RATE

The **average rate** may be included as the last entry in the rate column. To find the average rate, divide the sum of the $Q \cdot R$ column by the sum of the quantity column.

▶ In Example 4, we divide total rate by total quantity to get an average that is both meaningful and useful. To organize our thinking, we return to the four-step problem-solving process.

▶ **EXAMPLE 4** **Building a quantity-rate table and finding average rate: investment earnings** Silvia invests $2000 at $4\frac{1}{4}\%$ interest and $2500 at $2\frac{1}{2}\%$ interest. Build and complete a quantity-rate table. Estimate her total earnings and the average interest rate earned as part of your four-step process for finding the total earnings and average rate. What is the meaning of the average rate?

SOLUTION *Understand:* To estimate total earnings, we look for easy mental multiplications. Noting that $4\frac{1}{4}\%$ is close to 4% and $2\frac{1}{2}\%$ is close to 2%, we calculate

$$(\$2000 \cdot 4\%) + (\$2500 \cdot 2\%) = \$80 + \$50 = \$130$$

The amounts invested at the two rates are similar, so the average interest rate will be near 3%; but because more money is invested at $2\frac{1}{2}\%$, the rate will be slightly closer to $2\frac{1}{2}\%$ than to $4\frac{1}{4}\%$.

Plan: Set up a quantity-rate table, as shown in Table 8.

TABLE 8 Investment

Quantity (money invested)	Rate (interest rate)	$Q \cdot R$ (interest earned)
$2000	0.0425	$85.00
$2500	0.025	$62.50
Total: $4500		$147.50

Carry out the plan: The total earnings are \$147.50.

Check and extend: The quantity column sum gives the total money invested, \$4500. The quotient of the total interest earned and the total money invested (\$147.50 ÷ \$4500) gives the average interest rate earned on the total invested, approximately 3.3%. ◀

Writing and Solving Equations with the Help of Quantity-Rate Tables

What makes the quantity-rate approach important is that it is not limited to purchases and investments. Quantities and rates appear in grade point averages, medicine, business, science, engineering, transportation, and other fields.

Avoid the temptation to learn this material on an application-by-application basis. Try to see the product, quantity · rate, in each problem.

STOCKS In Example 4, the money invested was the quantity because we were multiplying by an interest rate to find the interest earned. In Example 5, the money invested is in the last, or $Q \cdot R$, column because it is the product of number of shares (quantity) and cost per share (rate).

▶ EXAMPLE 5 Finding one quantity Inez has \$29,000 to invest in the stock market. Abercrombie and Fitch stock costs \$55 per share. Hershey stock costs \$60 per share. Estimate how many shares of Hershey she can buy if she has already decided to buy 200 shares of Abercrombie and Fitch. Use a quantity-rate table to build an equation, and solve for the number of shares. What is the average rate and what does it mean?

SOLUTION ***Understand:*** Inez will spend a little over \$10,000 on Abercrombie and Fitch stock. The prices of the two stocks are close, so she should be able to purchase more than 200 shares of Hershey stock. The number of shares is a quantity, and the price per share is a rate. Let x be the number of shares of Hershey stock.

Plan: We set up a quantity-rate table, as shown in Table 9.

TABLE 9 Investment

Stock	Quantity (shares)	Rate (price per share)	$Q \cdot R$
Abercrombie and Fitch	200	\$55	200(55)
Hershey	x	\$60	$x(60)$
Total	$200 + x$	Average purchase price per share	\$29,000

Carry out the plan: We find the equation by adding the $Q \cdot R$ products in the last column and setting the resulting expression equal to the total investment, \$29,000.

$$200(55) + x(60) = 29{,}000$$

$$11{,}000 + 60x = 29{,}000 \qquad \text{Subtract 11,000 from both sides.}$$

$$60x = 18{,}000 \qquad \text{Divide by 60.}$$

$$x = 300$$

She buys 300 shares of Hershey stock, for a total of 500 shares. The average rate is $\frac{\$29,000}{500} = \$58$ average cost per share.

Check: 200 shares at \$55 plus 300 shares at \$60 gives \$29,000. ◀

THINK ABOUT IT 3: Which word tells us that *price per share* is the rate for Table 9 in Example 5?

AVERAGE RATE Verify these with your calculator:

In Example 3,

$$\$19.96 + \$5.58 + \$1.98 + \$3.78 = (10 \text{ items})(\$3.13 \text{ average price per item})$$

In Example 4,

$$\$85 + \$62.50 \approx (\$4500)(3.3\% \text{ average return})$$

In Example 5,

$$(200 \text{ shares})(\$55 \text{ per share}) + (300 \text{ shares})(\$60 \text{ per share}) = (500 \text{ shares})(\$58 \text{ average price per share})$$

In each example, the sum of the $Q \cdot R$ column equals the total quantity times the average rate.

BUILDING EQUATIONS USING AVERAGE RATE

> When the average rate is meaningful, the sum of the $Q \cdot R$ column equals the product across the total row.

WATER TEMPERATURE We use this idea in Example 6, where we blend water of two different temperatures. Although the example can be modeled quite well in the classroom with small quantities of water and inexpensive thermometers (see Exercise 56), the results are not accurate enough for scientific purposes. The laws of heat transfer in the field of thermodynamics are far too complex to be studied here.

The units on the temperature rate in Example 6 are degrees, so when we multiply by gallons we get degree•gallons, a very unusual unit.

▶ **EXAMPLE 6** Finding a missing quantity, given an average rate: a hot bath One cold winter evening, you plan to take a leisurely bath, but you get called away after you turn the water on. When you get back, you discover that only the cold water was turned on. The tub is one-fourth full (9 gallons). The cold water temperature in winter is about 35°. The hot water heater is set at 140°. You would rather not waste water by draining the tub and starting over. Use a quantity-rate table to find how much hot water must be added to correct the temperature to the desired 103°.

SOLUTION **TABLE 10** Bath Water

Item	Quantity	Rate (temperature)	$Q \cdot R$
Cold water	9 gal	35°	9(35)
Hot water	x gal	140°	$140x$
Total	$9 + x$	103°	(Two expressions)

The two expressions for the lower right corner of Table 10 are *the sum down*, $9(35) + 140x$, and *the product across*, $(9 + x)(103°)$. We set them equal and solve for x:

$$9(35) + 140x = (9 + x)(103) \quad \text{Apply the distributive property.}$$

$$315 + 140x = 927 + 103x \quad \text{Subtract 315 and } 103x \text{ from both sides.}$$

$$37x = 612 \quad \text{Divide by 37.}$$

$$x \approx 16.5 \text{ gal}$$

You need to add approximately 16.5 gallons of hot (140°) water. Will the added water raise the water level too high (either before or after you step into the tub)? ◀

GRADE POINT AVERAGE In Example 7, look carefully at how the two expressions for the lower right corner are used to build the equation.

▶ **EXAMPLE 7** Finding a missing quantity, given an average rate: GPAs A student with a 1.875 GPA has 24 credit hours. Find how many credit hours of Bs, at 3 points per credit hour, he would need to raise his GPA to a 2.5. (*Hint:* Let x be the number of credit hours of Bs.)

SOLUTION ***Plan:*** The desired GPA, 2.50, is placed in the last row of the rate column. The sum of the last column (the total points from current and future grades) gives one expression for the lower right corner. The product of the total quantity of credit hours and the desired GPA gives a second expression for the lower right corner.

TABLE 11 Grade Point Average

Item	Quantity (credit hours)	Rate (GPA)	$Q \cdot R$ (grade points)
Current GPA status	24	1.875 points	24(1.875)
Future B grades	x	3.00 points	$3x$
Total	$24 + x$	2.50 average	(Two expressions: the sum down and the product across)

Carry out the plan: The two expressions for the lower right corner of Table 11 are *the sum down*, $24(1.875) + 3x$, and *the product across*, $(24 + x)(2.50)$. We set them equal and solve for x:

$$24(1.875) + 3x = (24 + x)(2.50)$$

$$45 + 3x = 60 + 2.50x$$

$$0.5x = 15$$

$$x = 30 \text{ hr}$$

Check: $24(1.875) + 3(30) \stackrel{?}{=} (24 + 30)(2.50)$ ✓ ◀

▶ **EXAMPLE 8** Finding a missing quantity, given an average rate: more GPAs Suppose the student in Example 7 wants to raise his GPA to a 3.00. How many credit hours of Bs does he need to raise his GPA to a 3.00? Use Table 12 to write the two expressions that describe the lower right corner.

TABLE 12 Grade Point Average

Item	Quantity (credit hours)	Rate (GPA)	$Q \cdot R$
Current GPA	24	1.875 points per credit	24(1.875)
Desired grades	x	3.00 points per credit	$3x$
Total	$24 + x$	3.00 average	(Write two expressions)

SOLUTION The two expressions for the lower right corner are *the sum down*, $24(1.875) + 3x$, and *the product across*, $(24 + x)(3.00)$. We set them equal and solve for x:

$$24(1.875) + 3x = (24 + x)(3.00)$$

$$45 + 3x = 72 + 3.00x$$

$$0 = 27 \qquad \text{No real-number solution}$$

There is no real number x that makes the equation true, so the solution set is empty, { }. ◀

THINK ABOUT IT 4: What does "No real-number solution" mean in terms of the problem setting?

ANSWER BOX

Warm-up: 1. $x = 300$ **2.** $x \approx 16.5$ **3.** $x = 30$ **4.** No real-number solution **5.** $x = 3200$ **Think about it 1:** No **Think about it 2:** The sum of the quantity column is the total number of items ordered. The total is the number of items ordered. **Think about it 3:** *Per* suggests a rate or value. **Think about it 4:** "No real-number solution" means that it is impossible for the student to raise his GPA to a 3.00 by earning only Bs. Some As will be needed.

▶ 5.5 Exercises

No algebraic expressions or equations are needed in Exercises 1 to 28. In Exercises 1 to 10, which phrases describe quantities? Which describe rates?

1. 25 cents per quarter
2. $85 per share
3. 10 liters
4. 12 coins
5. 3 points per credit hour of Bs
6. 6% interest rate
7. A molarity rate of 12
8. $2.88 per pound
9. 40% boric acid solution
10. $2000

For Exercises 11 to 28, do the following:

(a) Set up a quantity-rate table.
(b) Write the meaning, if any, of the sum of the quantity column.
(c) Find the average rate for the table by dividing the sum of the $Q \cdot R$ column by the sum of the quantity column.
(d) Write the meaning of the sum of the $Q \cdot R$ column.

11. Demi's snack shop has 5 kilograms of peanuts at $8.80 per kilogram and 2 kilograms of cashews at $24.20 per kilogram.

Blue numbers are core exercises.

12. Reuel's café has 100 pounds of coffee at \$9.00 per pound and 100 pounds of coffee at \$10.80 per pound.

13. Abraham buys 3 pounds of grapes at \$0.98 per pound, 5 pounds of potatoes at \$0.49 per pound, and 2 pounds of broccoli at \$0.89 per pound. What is the total cost of his purchase?

14. Ludvina buys 15 international stamps at \$0.84, 50 first-class postage stamps at \$0.39, and 10 postcards at \$0.24. What is the total cost of her purchase?

15. Serena has 15 dimes and 20 quarters.

16. Andrzej has 12 half-dollars and 30 nickels.

17. Erin has \$1500 invested at 9% interest and \$1500 at 6% interest.

18. Kim has \$2500 invested at 8% interest and \$3000 at 4% interest.

19. Amel, a veterinarian, has 100 pounds of dog food containing 12% protein and 50 pounds of dog food containing 15% protein.

20. La Deane, a veterinarian, has 30 kilograms of cat food containing 8% fat and 40 kilograms of cat food containing 14% fat.

21. JuLeah, a horse trainer, has 150 pounds of alfalfa at 10% protein and 25 pounds of straw at 0% protein.

22. Ward, a horse trainer, has 1 gallon of coat conditioner in a 5% solution to be added to 20 gallons of water (0% solution).

23. Lenny has 5 hours of Ds (1 point per credit hour), 4 hours of Cs (2 points per credit hour), and 3 hours of Bs (3 points per credit hour).

24. Bruce has 4 hours at a 2.00 GPA from fall term, 4 hours at a 1.75 GPA from winter term, and 12 hours at a 1.83 GPA from spring term.

25. Loki drives 3 hours at 80 kilometers per hour and 2 hours at 30 kilometers per hour.

26. Kana drives 4 hours at 50 miles per hour and 3 hours at 18 miles per hour.

27. Li, a chemist, has 150 milliliters of sulfuric acid with a molarity of 18 and 100 milliliters of sulfuric acid with a molarity of 3. (*Molarity* is a chemical term; the larger the molarity rate, the more concentrated the acid.)

28. Ingrid, a chemist, has 200 milliliters of nitric acid with a molarity of 16 and 500 milliliters of nitric acid with a molarity of 6.

Blue numbers are core exercises.

Maria has \$15,000 to invest for one year. Investment A pays 5% interest for the year, and Investment B pays 8% interest. In Exercises 29 and 30, set up a quantity-rate table, find how much she earns, and predict and find the average rate of return.

29. **a.** Maria invests all at 8% and nothing at 5%.

b. Maria invests half the money at 8% and half at 5%.

30. **a.** Maria invests all the money at 5% and nothing at 8%.

b. Maria invests $\frac{1}{3}$ of the money at 5% and $\frac{2}{3}$ of the money at 8%.

In Exercises 31 to 48, solve by setting up a quantity-rate table and building an equation from the table.*

31. Georgia purchases 200 shares of Boeing stock at \$30 per share. How many shares of Walgreen stock can she purchase at \$46 per share if she has a total of \$19,800 to invest?

32. Mikhail has \$21,250 to invest. How many shares of Ford stock at \$8.50 per share can he buy if he also buys 300 shares of General Electric at \$34 per share?

33. If 200 shares of Bayer AG at \$40 each have already been purchased, how many shares of La Z Boy stock can be purchased at \$25 per share on a total budget of \$7200?

34. A hundred shares of Gap stock have been purchased at \$20 per share. How many shares of Carnival at \$50 per share are needed to raise the average price of all shares to \$50 per share? Explain your result.

35. How many gallons of cold water (summer temperature 60°) would need to be added to a bathtub containing 12 gallons of hot water at 140° to lower the temperature to 103°?

36. How many gallons of cold water (winter temperature 35°) would need to be added to a bathtub containing 12 gallons of hot water at 140° to lower the temperature to 103°?

37. How many gallons of hot water at 140° would need to be added to a bathtub containing 15 gallons of cold water at 35° to raise the temperature to 103°?

38. How many gallons of hot water at 140° would need to be added to a bathtub containing 15 gallons of cold water at 60° to raise the temperature to 103°?

39. Predict the grade point average if equal numbers of credit hours of Bs (3 points per credit hour) are added to Cs (2 points per credit hour). Is this result true for any equal quantities?

40. Predict the grade point average if equal numbers of credit hours of As (at 4 points per credit hour) are added to Bs (at 3 points per credit hour). Is this result true for any equal quantities?

*These are not investment recommendations. The author owns none of these stocks.

41. How many credit hours of Bs (at 3 points per credit hour) does Lenny need to raise his GPA to a 2.00? He has 20 credit hours at a 1.85 GPA.

42. How many credit hours of Bs (at 3 points per credit hour) does Alex need to raise 40 credit hours at a 1.85 GPA to a 2.25 GPA?

43. How many credit hours of Cs (at 2 points per credit hour) will raise 24 credit hours at a 1.875 GPA to a 2.00 GPA?

44. How many credit hours of As (at 4 points per credit hour) will raise 24 credit hours at a 1.875 GPA to a 2.5 GPA?

45. One week, Neva works 30 hours on one job at $5.80 per hour. Her second job pays $7.20 per hour. If she needs to earn an average of $6.36 per hour, estimate and then find how many hours she must work at the second job.

46. Florence works 36 hours one week at $7.50 per hour. She is offered overtime at $11 per hour. Estimate and then find how many hours she must work overtime to average $8 per hour.

47. Estimate and then find how many hours Neva (Exercise 45) needs to work each week at the second job to earn an average of $7 per hour. Is this a reasonable expectation? Explain.

48. Estimate and then find how many hours Florence (Exercise 46) needs to work overtime to average $9.10 per hour.

▶ Writing

49. Explain why total money invested appears in different places in the quantity-rate tables in Examples 4 and 5.

50. Two students both have a 3.00 GPA. This semester, each earns a 4.00 GPA on 12 credit hours. Which student will see more improvement in overall GPA: the second-year full-time student or the fourth-year full-time student near graduation? State any assumptions you make.

51. Explain how to estimate the quantities of two items, given the rate of each and the average rate.

52. Explain how to estimate the average wage, given two different wages and the number of hours worked at each.

53. Explain how to estimate whether a quantity will have a large effect on an average rate.

54. **Connecting with Section 5.1: Percent Change** In the poem "SMART," the money decreased from $1.00 to $0.50 to $0.30 to $0.20 to $0.05. Estimate which change reflected the largest percent decrease and which reflected the smallest percent decrease. What is the percent decrease with each change?

55. This exercise proves that when we have equal quantities of two items, the average rate for the table is the average of the rates for the two items.

  a. Explain how you know that the table below shows equal quantities of the two items.

  b. Copy and complete the table.

  c. Set up an equation containing A.

  d. Solve for A and explain the result.

Item	Quantity	Rate	$Q \cdot R$
First item	x	m	
Second item	x	n	
Total		A	Two expressions set equal.

▶ Projects

56. **Water-Temperature Experiment** Model the bathtub water temperature problem in Example 6. Take two plastic containers marked in either fluid ounces or tenths of a liter. Place a known quantity of hot tap water in one container. Place a known quantity of cold water without ice (say, from a refrigerated drinking fountain) in the other container. Leave enough room in one container to pour in the water from the other container. Record the temperature in each container. Make a table for the water (quantity) and temperatures (rate). With the table, predict the average temperature that will result from mixing the water. "Check" the table results by putting the water, together with both thermometers, into one container. Read the temperature when the two thermometers agree.

Blue numbers are core exercises.

57. Spreadsheet Create a computer spreadsheet similar to the table opposite to model the quantity-rate tables. Use it to do the relevant exercises in this section. To run the spreadsheet, enter numbers in the positions A2, A3, B2, and B3. If A2, A3, B2, or B3 information is missing in the problem, solve the problem by guessing and then use the spreadsheet to check.

Row	Column A	Column B	Column C
1	Quantity	Rate	$Q \cdot R$
2	A2	B2	=A2 * B2
3	A3	B3	=A3 * B3
4	=A2 + A3	=(C2 + C3)/A4	=C2 + C3
5	Calculated Average Rate =		=B4
6	Product Across =		=A4 * B4
7	$Q \cdot R$ Sum Down =		=C2 + C3

▶ 5 Chapter Summary

Vocabulary

For definitions and page references, see the Glossary/Index.

average rate
bimodal
centroid
continued ratio
corresponding
cross multiplication
equivalent ratios
inner-outer product
mean
median
midpoint of a line segment
midpoint on a coordinate graph
mode
percent
percent change
percent grade
proportion
quantity
quantity-rate table
rate
ratio
similar figures
similar triangles
vertex (vertices)
weight
weighted average

Concepts

5.1 Ratios, Percents, and Rates

Two ratios are equivalent if they simplify to the same number or if they divide to the same decimal value.

Ratios containing variables or units are simplified in the same way as fractions—by eliminating common variables or units where

$$\frac{a}{a} = 1 \quad \text{or} \quad \frac{\text{inches}}{\text{inches}} = 1$$

Applying unit analysis to rates may require changing both the units in the numerator and the units in the denominator.

To find percent change, calculate the difference between the new number and the original number and then divide by the original number.

5.2 Proportions and Proportional Reasoning

If $\frac{a}{b} = \frac{c}{d}$, then $ad = bc$.

In percents, if a is $n\%$ of base b, then

$$\frac{n}{100} = \frac{a}{b}$$

When a proportion contains units, the units must match either side to side or within numerator and denominator pairs.

5.3 Proportions in Similar Figures and Similar Triangles

Apply proportions to find the lengths of the sides in similar figures. When you know the total length or quantity of something that is separated into two parts, use one variable to describe the two parts. If the whole is S and one part is x, the other part is $S - x$.

The sum of the angle measures in a triangle is 180°.

Similar triangles have corresponding angles that are equal, and similar triangles have corresponding sides that are proportional.

5.4 Averages

To find the mean of a set of numbers, add the numbers and divide by the number of numbers in the set.

To find the median, select the middle number when the numbers are arranged in numerical order (from smallest to largest or largest to smallest). If there is no single middle number, the median is the mean of the two middle numbers.

To find the mode, select the number appearing the highest number of times.

The ordered pair giving the midpoint of a line segment is the average of the x-coordinates and of the y-coordinates of the endpoints.

The ordered pair giving the centroid of a square, rectangle, or triangle is the average of the x-coordinates and of the y-coordinates of the vertices.

5.5 Writing Equations with Quantity-Rate Tables

The average rate is found by dividing the sum of the $Q \cdot R$ column by the sum of the quantity column. Average rate is placed at the bottom of the rate column.

Finding two expressions for the lower right corner of the quantity-rate table is the key to building an equation. The sum of the $Q \cdot R$ column must equal the product across the total row (the sum of the quantities multiplied by the average rate).

When equal quantities of two items are used, the average rate is the average of the separate rates.

▶ 5 Review Exercises

For Exercises 1 and 2, simplify the ratios.

1. $16x^2$ to $4x^4$

2. $\dfrac{48(x - 3)}{16}$

Simplify the ratios in Exercises 3 and 4.

3. 5 feet : 18 inches

4. 50 centimeters to 2 meters

5. What is the percent decrease when a price drops from \$49.99 to \$42.49?

6. What is the percent increase when a wage rises from \$5.40 per hour to \$6.30 per hour?

7. How many miles per hour is 100 meters in 9 seconds?

12 inches = 1 foot
39.37 inches = 1 meter
5280 feet = 1 mile
3600 seconds = 1 hour

8. On a trip to Canada, you pay \$0.85 Canadian per liter for gasoline. How much is this in U.S. dollars per gallon? Estimate first.

1 Canadian dollar $\approx$ 0.87 U.S. dollar
1.0567 quarts = 1 liter
4 quarts = 1 gallon

In Exercises 9 and 10, describe in a complete sentence an appropriate situation for each rate.

9. hours per revolution

10. dollars per day

11. Sales of beverages are in a 7 : 7 : 1 ratio for Pepsi products, Coca-Cola products, and other brands, respectively. How many of each brand were sold if a total of 120,000 cases were sold?

12. What percent of 56 is 84?

13. 75% of what number is 108?

14. 108% of 75 is what number?

In Exercises 15 and 16, use the fact that nutrition experts recommend that no more than 30% of a person's daily calorie intake come from fat. The remaining calories should come from carbohydrates and protein.

15. In a 1500-calorie diet, how many calories should come from fat? At 9 calories per gram of fat, how many grams of fat would this be?

16. In a 2000-calorie diet, how many calories should come from carbohydrates and protein together? At 4 calories per gram of protein and carbohydrate, how many grams of protein and carbohydrate would this be?

17. An access ramp needs to rise 30 inches. If the ramp is to have a slope ratio of 1 : 8, what horizontal distance is needed to build the ramp? Give the answer in feet.

18. Some stairs need to have a slope of 6.5 to 11. What horizontal distance is needed for an 8.5-foot vertical distance between floors?

19. Kei's research team catches and tags 250 bats. The next week, they catch 300 bats from the same cave, and only 15 carry a tag. Estimate the total bat population from this cave.

20. Orange juice is a mixture of water and frozen orange juice concentrate. The ratio of water to concentrate is 3 to 1. How much water is needed to mix with 1.5 gallons of concentrate?

For Exercises 21 to 25, solve the equations.

21. $\frac{2}{3} = \frac{x}{17}$

22. $\frac{2}{6} = \frac{3}{x}$

23. $\frac{2}{x} = 2{,}000{,}000$

24. $\frac{x+1}{2} = \frac{4x-1}{6}$

25. $\frac{2x+5}{7} = \frac{2x-1}{5}$

For Exercises 26 and 27, solve the formulas.

26. $\frac{a}{b} = \frac{c}{d}$ for d

27. $\frac{V_1}{C_2} = \frac{V_2}{C_1}$ for V_2

Solve the equations in Exercises 28 and 29. Explain why they are not proportions.

28. $\frac{5x}{3} = \frac{7}{3} + x$

29. $\frac{x}{2} + \frac{1}{4} = 3 + \frac{x}{7}$

30. Find a value or expression for x.

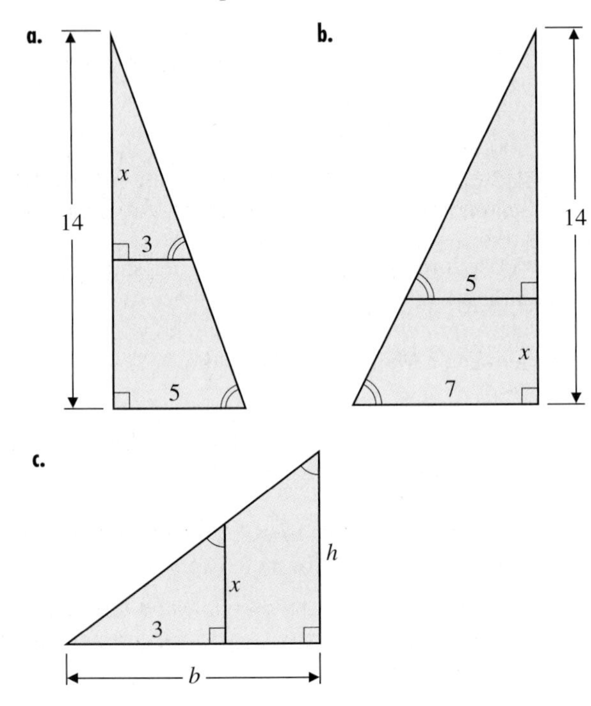

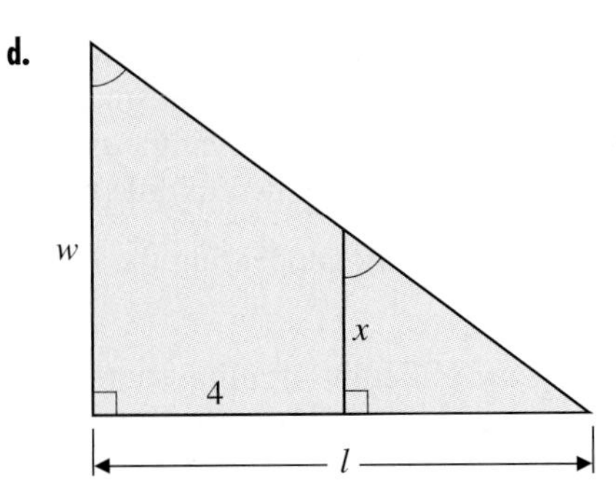

For Exercises 31 and 32, find the coordinates of *A* and *B* on the graph. Use slope, proportions, or counting, as necessary.

31.

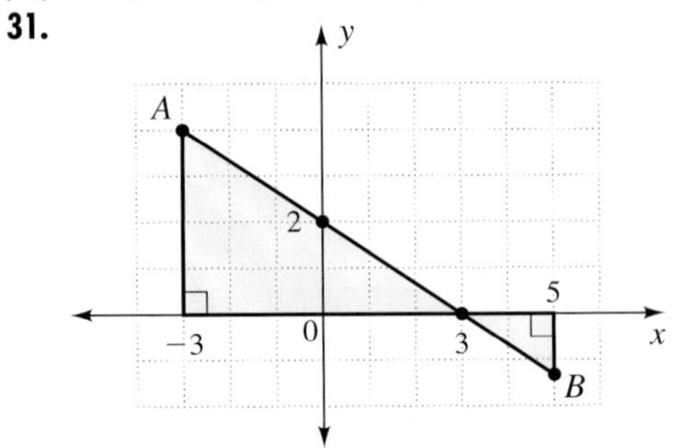

32.

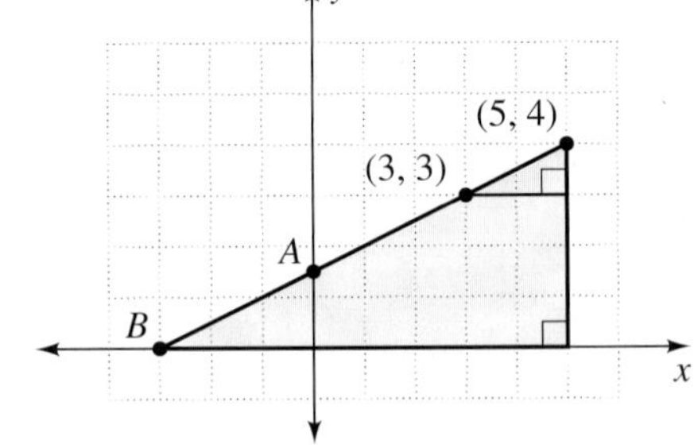

Find the mean, median, and mode of each set of measurements in Exercises 33 to 36. Round to the nearest hundredth.

33. 4.2 grams, 4.3 grams, 4.3 grams

34. 31.7 cm, 31.8 cm, 31.6 cm, 31.8 cm

35. 6.9 miles, 6.9 miles, 6.8 miles, 7.0 miles, 6.8 miles, 6.8 miles

36. 45.0 mL, 44.0 mL, 45.5 mL, 45.0 mL, 44.5 mL

37. Refer to the figure.

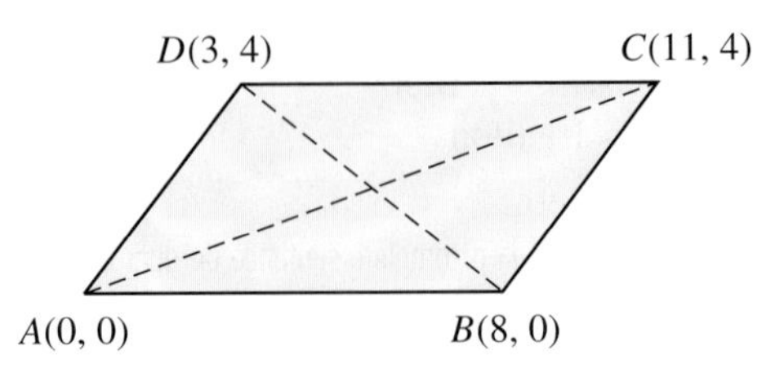

a. Find the midpoint of each line: AD, BC, AB, CD, AC, and BD.

b. Find the slope of each line: AD, AB, CD, and BC.

c. What is the name of the shape $ABCD$?

38. Refer to the figure.

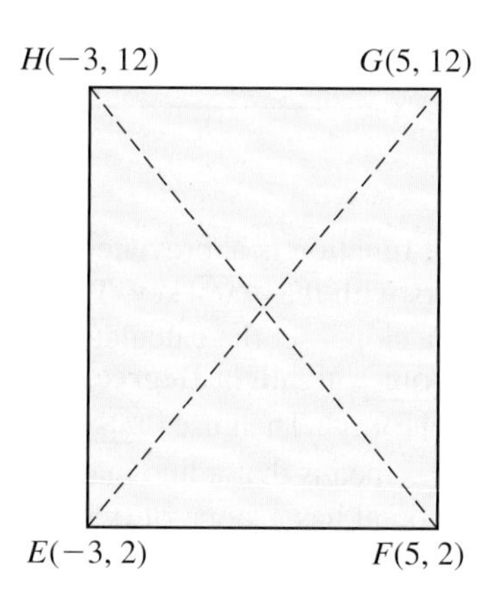

a. Find the midpoint of each line: EG, EF, HE, and HF.

b. Find the slope of each line: HF and EG.

c. What concept explains why the slope of HE is undefined? (*Hint:* Calculate the slope.)

d. What is the centroid of figure $EFGH$?

For Exercises 39 and 40, find the centroid.

39.

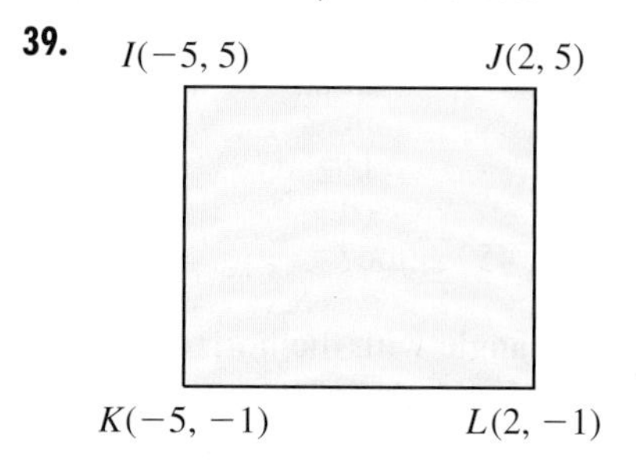

40.

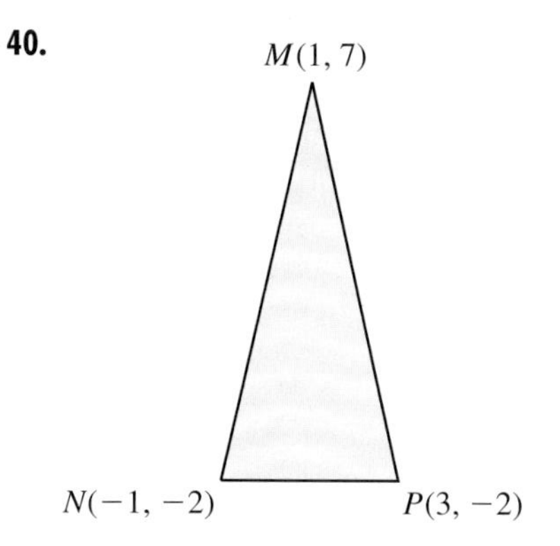

41. Staples sells large paper clips in groups of 5 boxes marked 100 clips each. The actual count in 5 boxes is 96, 101, 98, 102, 102.

a. Find the mean.

b. Discuss the content from the point of view of the buyer.

c. Discuss the content from the point of view of Staples.

d. Would you accept the count as representative of all boxes?

42. Langche's math instructor weights the term project and the final exam each as 2 and the two midterms each as 1. He gets 100 on the project, 85 on the final exam, and 95 and 75 on the midterms. What is his course average?

43. The table shows several examples of quantities and rates. Fill in the missing units.

Quantity	Rate	$Q \cdot R$
Number of coins	Cents per coin	
Time in hours		Distance in miles
Dollars invested	Percent interest	
Credit hours	Points per credit hour	
Number of cans	Cost per can	
Number of pounds	Cost per pound	
	Wages per hour	Wages
Liters of acid	Moles per liter	

Make a quantity-rate table for each problem situation in Exercises 44 to 48. Find appropriate totals and averages. Indicate units and whether any of the totals are meaningful. Suggest questions that could be asked.

44. Debi invests \$10,000 in a tax-free bond at 5% and \$5000 in savings earning 3.5%.

45. A plane flies 4 hours at 125 mph and 7.5 hours at 200 mph.

46. An aquarium is to be filled with 8 liters of hot water at 90°C and 50 liters of cold water at 5°C.

47. Max buys 6 gallons of 88-octane fuel and 10 gallons of 92-octane fuel.

48. Arlan earns an A on a 12-credit course and a B on a 4-credit course. Assume an A is 4 points per credit and a B is 3 points per credit.

49. Because of heat loss in uninsulated pipes, hot water at the bathtub faucet is 130°. How many gallons of hot water need to be added to 9 gallons of cold water (35°) to raise the average temperature to 103°?

50. Rafael wants to keep his loans at an average of 9% interest. He has $3000 in credit card debt at 14%. What is the most he can borrow for a car at 7%?

51. Shawna has borrowed a total of $25,000 through school loans and her credit card. Her school loans are at 8% interest, and her credit card debt is at 18% interest. Her total interest paid is $2600 per year. Find how much money Shawna has borrowed from each source. Then find her average interest rate.

▶ Chapter Project

52. Sleep-driving On March 8, 2006, Stephanie Saul of the *New York Times* raised concern about driving under the influence of the perscription sedative Ambien. Wisconsin reported 187 arrested drivers with Ambien in their bloodstream during 1999 to 2004. The state of Washington reported 56 similar driving arrests in 2004 and 78 in 2005.

a. Suppose you are a forensic toxicology supervisor or public safety official in your state. What action might you take when reading the information? Use the numbers to make a case. Gather information about your state or other nearby states to support your recommendation.

b. Suppose you are the public relations official for Sanofi-Aventis, the maker of Ambien. There were 26.5 million prescriptions of Ambien in 2005. How might you argue against special action?

Possibly relevant facts include populations of Wisconsin in 2000, of Washington in 2005, and of the United States in 2005 and the number of U.S. drivers in 2005. Reports are made to American Academy of Forensic Sciences.

53. Slope and Trigonometry In trigonometry, we relate angle measures and ratios of the sides of a right triangle. One of the ratios, a to b, is the same as the slope of a line, $\frac{\Delta y}{\Delta x}$. The relationship is described by the tangent equation

$$\tan A = \frac{a}{b}$$

This equation says: For the right triangle ABC, the tangent of angle A (measured in degrees) is the ratio of the length of side a to the length of side b.

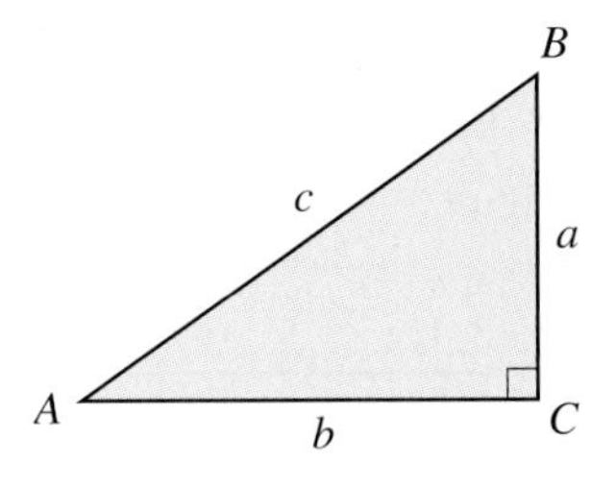

The tangent function is abbreviated *tan* and is built into calculators with the [TAN] key. To change an angle measure into a slope, set the calculator to degree measure, press [MODE], highlight **Degree**, and press [ENTER] [2nd] [QUIT]. Then use [TAN] A. The equation $\tan 26.6° \approx 0.5$ means that a line making a 26.6° angle with the horizontal has a slope of 0.5, or $\frac{1}{2}$.

What is the slope (tangent) of each angle in parts a to f? Round to the nearest hundredth.

a. 22° **b.** 45° **c.** 76°

d. 10° **e.** 20° **f.** 30°

To change slopes back into angle measures, we use the shifted tangent key, [2nd] [TAN]. To change a slope of $\frac{1}{2}$ to its angle with the horizontal axis, we enter [2nd] [TAN] (1/2). The answer is 26.6°.

What is the angle for each slope in parts g to l? Round to the nearest tenth.

g. $\frac{1}{10}$ **h.** $\frac{3}{10}$ **i.** $\frac{7}{10}$

j. 1 **k.** 1.5 **l.** 2

m. What percent grade is a 45° angle?

n. What are the slope and angle with the horizontal for a mountain highway with a 6% grade?

o. What are the slope and angle for the 62.5% grade on the sides of Wizard Island in Crater Lake, Oregon?

p. What are the slope and angle for the 72.7% grade on the Lookout Mountain Incline Railway in Chattanooga, Tennessee?

▶ 5 Chapter Test

For Exercises 1 and 2, give three ratios that are equivalent to these ratios.

1. $\frac{6}{14}$ **2.** $\frac{7.5}{100}$

Simplify the ratios in Exercises 3 to 5.

3. $\frac{7ab^3}{28a^2b^2}$ **4.** $\frac{36(a+b)}{9}$ **5.** $\frac{10 \text{ ft}}{24 \text{ in.}}$

6. 125% of what number is 105?

7. Arrange these facts into a unit analysis to find how many dollars per year are spent on 2 packs per day.

20 packs per carton
$50 per carton
365 days per year

8. A $3\frac{1}{2}$-inch by 5-inch photo is to be enlarged into a poster so that its longer side is 2 feet. How long will the other side be?

In Exercises 9 and 10, solve the equations for x.

9. $\frac{2}{5} = \frac{13}{x}$ **10.** $\frac{x+1}{4} = \frac{x-3}{3}$

11. Rosa wants a 90% average in math. She has 85% and 91% on her first two of three equal-value tests. What does she need on the third test to obtain the 90%?

12. What is the length of the line labeled x in the triangle?

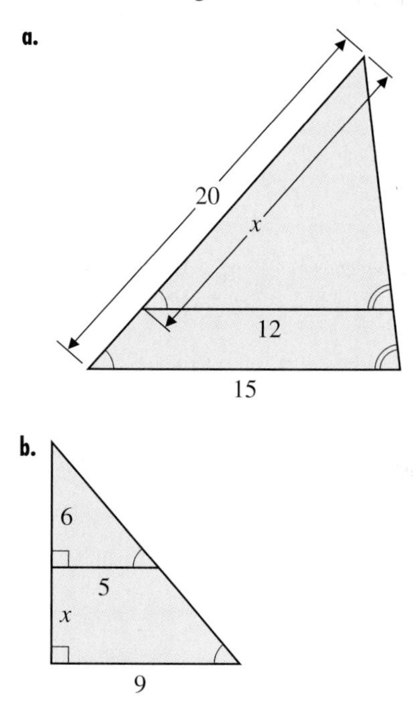

What are the median and the mean of each set of data in Exercises 13 and 14?

13. Five ballpoint pens: $1.29, $1.29, $1.29, $1.29, $2.79

14. Five ballpoint pens: $1.69, $1.29, $1.19, $1.69, $2.09

15. Make some observations about the medians and means in Exercises 13 and 14 and what might have caused these results. Under what circumstances would each be a good description of the average cost of the pens?

For Exercises 16 and 17, refer to the figure.

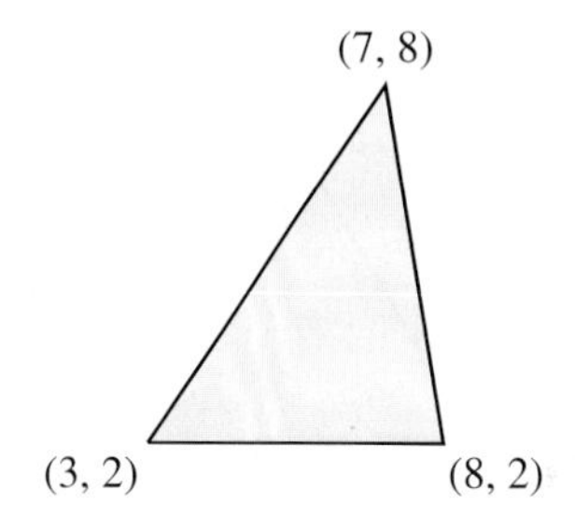

16. What is the midpoint of each side of the triangle?

17. What is the centroid of the triangle?

18. Eugene borrows $17,000 for a car at 5.8% interest and has a $2000 credit card balance at 14.9%. Make a quantity-rate table. Explain whether the totals are meaningful and, if so, why. Is there a meaningful average value?

19. Pavel wants to raise his grade point average (GPA). He currently has 60 credits, with a GPA of 3.40. How many credits of As, at 4.00 points per credit, does he need to raise his GPA to a 3.50? What is the percent change from 3.40 to 3.50?

20. Solve for x:

$$\frac{8x}{3} - 1 = \frac{13}{3} + x$$

Explain why the equation is not a proportion.

21. Without changing the width of the paper, a newspaper increases the number of columns in the classified ads from 9 to 10 per page. Because ads are sold by the vertical inch, income increases with an increase in the number of columns. Is the newspaper's revenue increased by $\frac{1}{9}$ or $\frac{1}{10}$? Which increase is larger, $\frac{1}{9}$ or $\frac{1}{10}$?

▶ Cumulative Review of Chapters 1 to 5

1. Simplify the ratio of 12 days to 3 weeks.

2. Simplify:

a. $\dfrac{3 - 8.5}{6 + 1.5}$ **b.** $4 + 4 \cdot 5 - 6 \div 3$

c. Place parentheses in part b to obtain 6 as an answer.

3. Solve:

a. $15 + 2x = 8$ **b.** $\dfrac{2}{3}x = 0.5$

c. Explain two different ways to solve part b.

4.

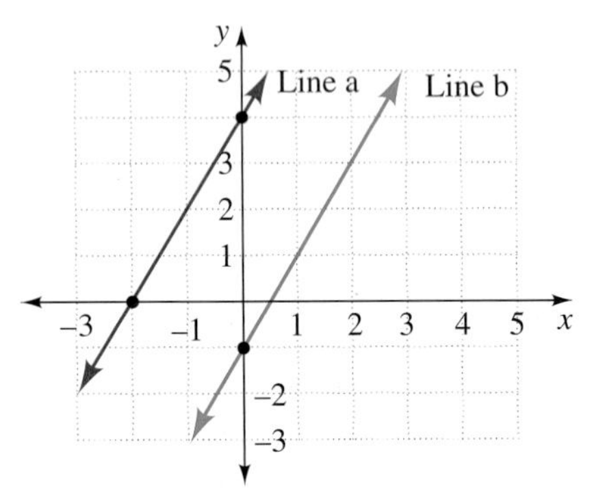

What is the equation for line a? line b?

5. Make a table for $y = f(x)$ with domain $\{-2, -1, 0, 1, 2\}$:

a. $f(x) = -3x$ **b.** $f(x) = -(3x)$

c. $f(x) = \dfrac{1}{x}$

6. A square is formed by $A(1, 2)$, $B(2, -1)$, $C(-1, -2)$, and $D(-2, 1)$.

a. Name the four sides.

b. Name two parallel segments and give their slopes.

c. Name two segments that are perpendicular and explain why in terms of slopes.

d. What are the slopes of the diagonals, AC and BD? Comment.

7. Will each the following patients live or die under your care?

a. A patient is to receive 0.02 gram of a medication from 20-mg tablets. Fact: 1 gram = 1000 mg. What should be given to the patient?

b. A vial contains 10 milliliters (mL). The label on the bottle indicates that the solution is 500 mg per 5 mL. A patient is to receive 200 mg. How many milliliters should the patient receive?

c. How many doses are in the vial in part b?

8. You start to put on your swimsuit and discover that it is inside out. What two steps are needed to set it right? How is this like solving an equation?

9. Give one answer and explain why there is more than one answer to each:

a. Write an equation that is equivalent to $3x + 2y = 6$.

b. Write an equation of a line that is parallel to $3x + 2y = 6$.

c. Write an equation that is perpendicular to $3x + 2y = 6$.

CHAPTER 6

Systems of Equations and Inequalities

Figure 1 shows a color chart with the three primary colors—red, yellow, and blue—each shading half the hexagon. What colors appear in the regions where the colors overlap? We will look at overlapping regions in mathematics as we study graphs of systems of inequalities in Section 6.6.

This chapter is about sets (or systems) of equations and inequalities. There are five ways to solve a system of equations: with graphs (Section 6.1), by guess and check (Section 6.2), with tables (Section 6.2), by substitution (Sections 6.3 and 6.5), and by elimination (Section 6.4). There are three different possible outcomes for a system of linear equations: no solution, one solution, and an infinite number of solutions.

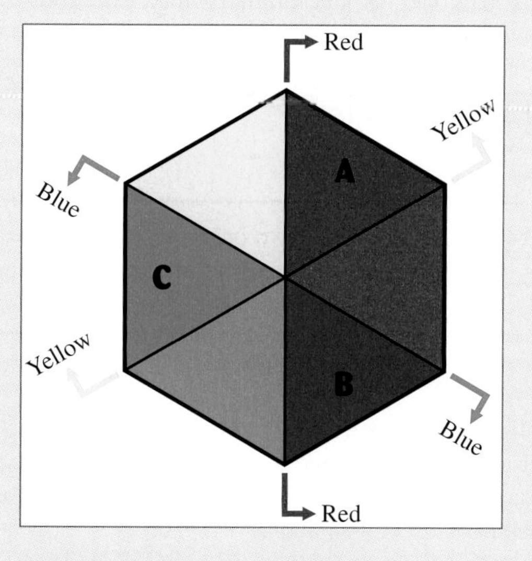

FIGURE 1 Primary colors blend to make secondary colors

▶ 6.1 Solving Systems of Equations with Graphs

Objectives

- Graph a system of equations.
- Solve a system of equations by graphing.
- Give the geometric meaning for a system of equations having no solution, one solution, or infinitely many solutions.

WARM-UP

For each of these equations, name the slope ratio and y-intercept point and sketch a graph.

1. $y = 3x + 2$

2. $y = -x - 2$

3. $y = 3x - 2$

4. $y = -2x + 4$

5. $2y = 6 - 4x$

6. $y = \frac{1}{2}x - 2$

IN CHAPTERS 3, 4, and 5 we wrote linear equations to compare prices or costs. By writing two equations as a system, we can find two quantities or rates. We will find solutions in this section by graphing. We will start with systems of two linear equations and solve systems with one, no, and infinitely many solutions.

▶ Systems of Linear Equations

DEFINITION OF SYSTEM OF EQUATIONS

A **system of equations** is a set of two or more equations that are to be solved for the values of the variables that make all the equations true, if such values exist.

Taken together, the equations $x + y = 4$ and $x - y = 6$ form a system of two linear equations. Because $5 + (-1) = 4$ and $5 - (-1) = 6$, the ordered pair $(5, -1)$ makes both $x + y = 4$ and $x - y = 6$ true. The ordered pair $(5, -1)$ is the point of intersection of the graphs of $x + y = 4$ and $x - y = 6$.

DEFINITION OF SOLUTION TO A SYSTEM

A **solution to a system** of two linear equations is the ordered pair that makes both equations true.

Because every point on the graph of an equation makes the equation true, the point of intersection makes both equations true. A system of two or more linear equations may have one, no, or infinitely many solutions, or intersections. The exploration in Example 1 shows how these three outcomes might occur.

SOLUTION **a.** Plan A: $y = 0.06x + 300$
Plan B: $y = 0.10x + 150$

b. The equations are graphed in Figure 4.

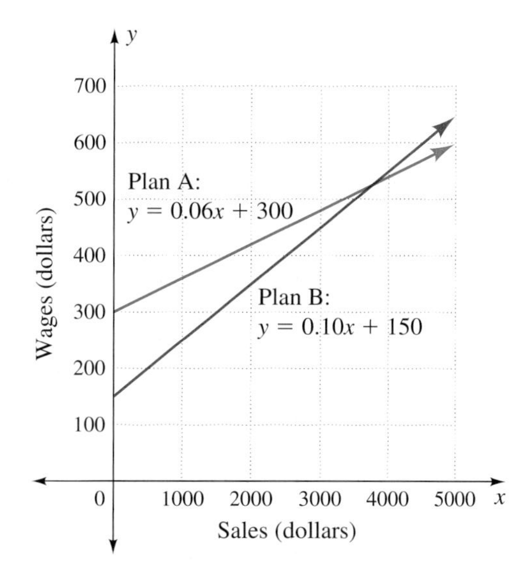

FIGURE 4 Electronics Works wage options

c. The lines intersect where the plans give the same total income: about \$500 to \$550 on sales of between \$3500 and \$4000. This occurs where

$$0.06x + 300 = 0.10x + 150 \quad \text{Subtract 150 and } 0.06x \text{ from both sides.}$$

$$150 = 0.04x \quad \text{Divide both sides by 0.04.}$$

$$3750 = x$$

d. Plan A yields more money with low sales. If business is slow or a salesperson is new, Plan A might be better. A good salesperson would do well with Plan B. ◀

On a graphing calculator, the Intersection option gives the solution as an ordered pair.

GRAPHING CALCULATOR TECHNIQUE: USING INTERSECTION

Enter the equations from Example 3: $Y_1 = 0.06X + 300$ and $Y_2 = 0.10X + 150$. Graph with window settings in the same scale as shown in Figure 4. To find the intersection of the two graphs, select the Intersection option, 2nd [CALC] **5 : intersect**. Press ENTER twice to choose the two graphs. For a guess, move the cursor close to the point of intersection and press ENTER. The intersection, (3750, 525), is shown in Figure 5.

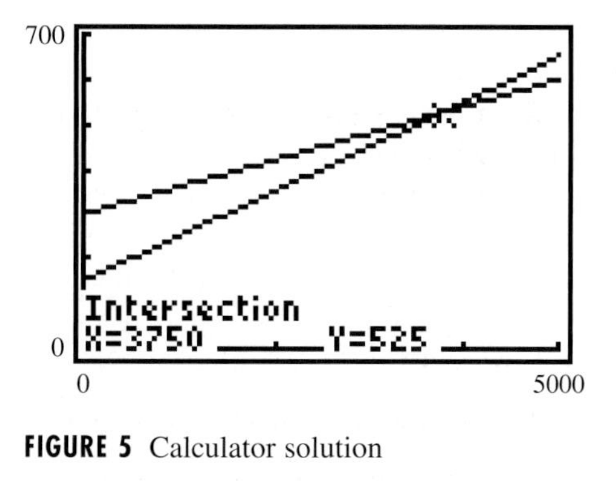

FIGURE 5 Calculator solution

THINK ABOUT IT:

a. Explain the meaning of the slopes of the lines in Example 3.

b. Explain the meaning of the y-intercept of each line.

Systems with No Solution

We now consider examples in which there are no solutions to the system. In Example 4, we graph equations to solve the system.

EXAMPLE 4 Solving a system of equations by graphing Graph the equations $2x + y = 4$ and $2y = 6 - 4x$.

SOLUTION First we solve the equations for $y = mx + b$, and then we graph them.

$$2x + y = 4 \qquad \text{Subtract } 2x \text{ on both sides of the first equation.}$$

$$y = -2x + 4$$

$$2y = 6 - 4x \qquad \text{Divide by 2 on both sides of the second equation.}$$

$$y = 3 - 2x \qquad \text{Rearrange the right side.}$$

$$y = -2x + 3$$

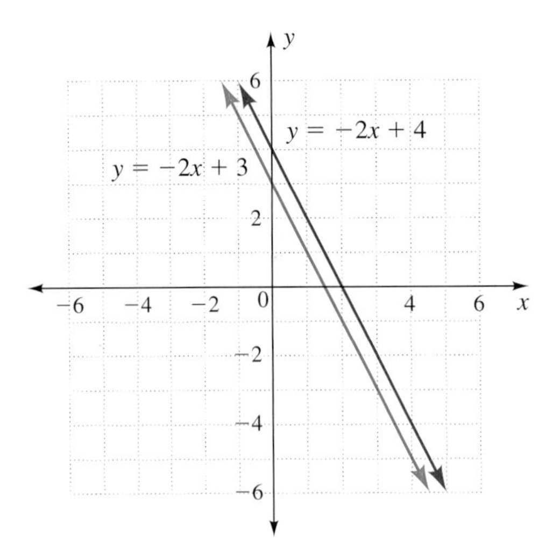

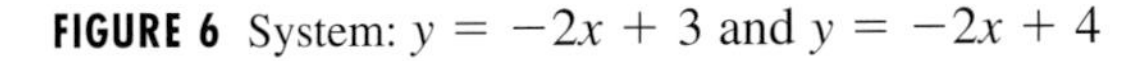

FIGURE 6 System: $y = -2x + 3$ and $y = -2x + 4$

The lines have the same slope but different y-intercepts, as shown in Figure 6. They are parallel lines and have no point of intersection. ◀

Example 4 agrees with our work on parallel lines in Section 4.5. When a system of equations contains parallel lines, there is no point of intersection and no ordered pair makes both equations true. In this case, we say *there is no solution to the system of equations.*

SYSTEMS OF EQUATIONS WITH NO SOLUTION

- A system containing parallel lines has no solution.
- When there is no solution, the solution set is empty, $\{\ \}$ or $\varnothing$.

That parallel lines have no solution is perhaps easier to understand in application settings. In Example 5, we extend the wage option application from Example 3.

EXAMPLE 5 Interpreting solutions for parallel lines: more wage options Suppose Wage Plan A is given by $y = 0.06x + 300$ and Wage Plan C is given by $y = 0.06x + 150$.

a. Graph the wage plans.
b. Find the point of intersection.
c. Compare the two plans.

SOLUTION **a.** The equations are graphed in Figure 7.

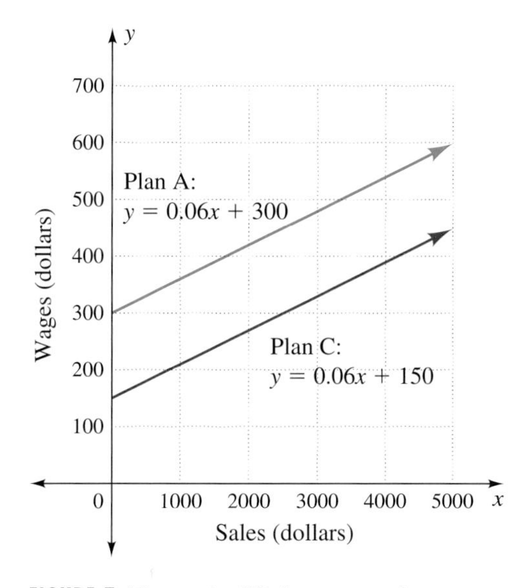

FIGURE 7 Electronics Works wage options

b. The two plans give parallel lines with no point of intersection; hence, there is no real-number solution to the system. The solution set is empty: $\{\ \}$.

c. The \$150 salary plus 6% of sales would never be better than the \$300 salary plus 6% of sales. ◀

Attempting to use the calculator Intersection option to find the point of intersection for parallel lines, such as those in Examples 4 and 5, results in an error message: NO SIGN CHNG. This message indicates that there is no point of intersection.

Systems with an Infinite Number of Solutions

EXAMPLE 6 Solving a system of equations by graphing Graph the system of equations below and describe the solution set.

$$y = \tfrac{1}{2}x - 2$$
$$2y + 4 = x$$

SOLUTION The equation $y = \frac{1}{2}x - 2$ has a slope of $\frac{1}{2}$ and a y-intercept point at $(0, -2)$. The graph of $y = \frac{1}{2}x - 2$ is shown in Figure 8. To graph $2y + 4 = x$, we solve for y.

$$2y + 4 = x \quad \text{Subtract 4 on both sides.}$$
$$2y = x - 4 \quad \text{Divide by 2 on both sides.}$$
$$y = \tfrac{1}{2}x - 2$$

The equation $2y + 4 = x$ is the same as the first equation. The graphs will coincide. Every ordered pair that makes $y = \frac{1}{2}x - 2$ true will also make $2y + 4 = x$ true.

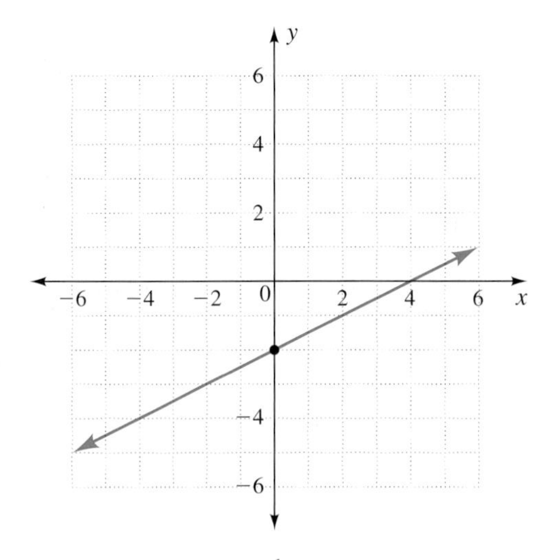

FIGURE 8 System: $y = \frac{1}{2}x - 2$ and $x = 2y + 4$ ◀

▶ *Two lines that have the same equation* are **coincident lines**. Because coincident lines have all points in common, we say that as a system *they have an infinite number of solutions*.

SYSTEMS OF EQUATIONS WITH COINCIDENT LINES

> A system whose graphs are coincident lines has an infinite number of solutions.

▶ **EXAMPLE 7** Interpreting solutions for coincident lines Suppose management of the Electronics Works in Example 3 offers employees a wage package described by $2y = 0.12x + 600$, instead of $y = 300 + 0.06x$. How should the employees react?

SOLUTION We solve the new offer for y:

$$2y = 0.12x + 600 \quad \text{Divide by 2.}$$

$$y = 0.06x + 300$$

The equations are the same. The graphs will coincide, and there will be an infinite number of points in common. The new wage offer is identical to the old offer. The employees should be insulted by the meaningless new offer. ◀

Attempting to use the calculator Intersection option to find the point of intersection for coincident lines, such as those in Examples 6 and 7, results in an intersection wherever the cursor is left after the guess prompt. For Example 7, enter $Y = 0.06X + 300$ in both Y_1 and Y_2 and set the viewing window to give the same scale as is shown in Figure 7. Press [2nd] [CALC] and choose **5 : intersect**. Then press [ENTER] three times. The intersection will be given as (2500, 450) (see Figure 9). Choose Intersection again. At the guess prompt, press the right cursor key 10 times and then press [ENTER]. The result will be as shown in Figure 10. The calculator is indicating that every point on the first line is a point of intersection with the second line.

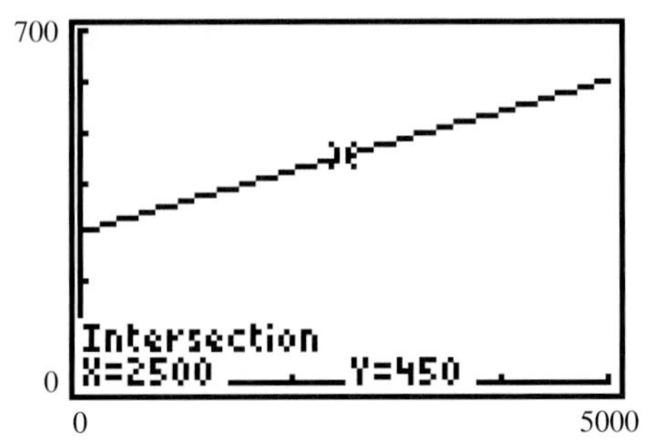

FIGURE 9 Intersection on coincident lines

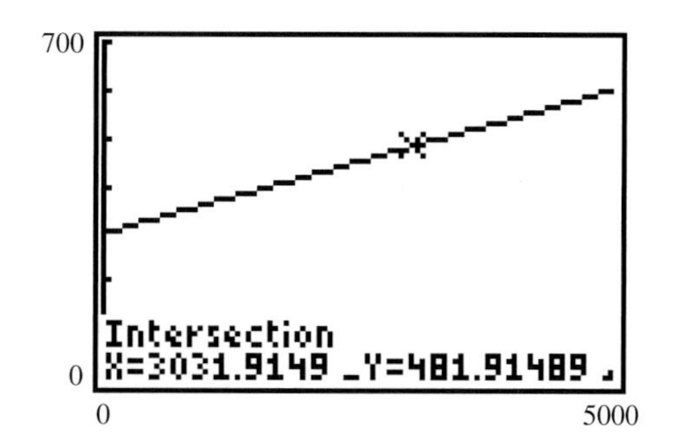

FIGURE 10 Intersection on coincident lines

Table 1 summarizes solution sets, graphs, and properties of equations in systems of two linear equations.

TABLE 1 Systems of Two Linear Equations

Systems of Two Equations	One Solution (x, y)	No Solution { } or $\varnothing$	Infinite Number of Solutions
Graph of System	Lines intersect exactly once	Lines are parallel	Lines coincide
Intersection	One intersection	No intersection	All points in common
Equation: $y = mx + b$ Slope, m; y-intercept, b	Unequal slopes	Equal slopes Unequal intercepts	Equal slopes Equal intercepts

ANSWER BOX

Chapter Opener: When the colors overlap, region A is orange, region B is violet (or purple), and region C is green. **Warm-up: 1.** $m = \frac{3}{1}$, (0, 2) **2.** $m = -\frac{1}{1}$, (0, −2) **3.** $m = \frac{3}{1}$, (0, −2) **4.** $m = -\frac{2}{1}$, (0, 4) **5.** $m = -\frac{2}{1}$, (0, 3) **6.** $m = \frac{1}{2}$, (0, −2). For graphs, see Figures 2, 6, and 8. **Example 1: a.** line 2 **b.** line 1 **c.** line 3 **d.** (−1, −1) **e.** $-1 = 3(-1) + 2$ and $-1 = -(-1) - 2$ both are true, so (−1, −1) makes both equations true and is the solution set. **f.** There is no point of intersection; lines 2 and 3 are parallel. **g.** The ordered pairs (−1, −1) and (0, 2) lie on line 2. This suggests that $y - 3x - 2 = 0$ and $y = 3x + 2$ are coincident lines having the same equation. Solving $y - 3x - 2 = 0$ for y gives $y = 3x + 2$. **Think about it: a.** The slope is the percent of sales received as a commission. **b.** The y-intercept is the wage received if no sales are made.

▶ 6.1 Exercises

Find the slope and y-intercept for each equation in the systems listed in Exercises 1 to 16. Sketch graphs for each system, and find the point of intersection. To show that the point of intersection is a solution to the system, substitute it into each equation of the system.

1. $y = -2x$
$y = 1 - x$

2. $y = -x$
$y = x + 2$

3. $y = x$
$y = 2x + 3$

4. $y = 2x - 2$
$y = 1 - x$

5. $y = 3$
$x = -1$

6. $x = -3$
$y = -1$

7. $x + y = 5$
$y = -x + 5$

8. $x - y = 4$
$y = x - 8$

9. $x - y = 3$
$y = x - 6$

10. $y = x + 2$
$x - y = -2$

11. $y = x - 4$
$3x + y = 4$

12. $4x - y = -10$
$y + 2x = 4$

13. $4x - 2y = 2$
$4x + 3 = y$

14. $3y - 9x = -6$
$y + 2x = 13$

15. $2x + 3y = 12$
$x + y = 5$

16. $x + 2y = 10$
$x + y = 4$

Which of the pairs of equations in Exercises 17 to 22 form parallel lines? It is not necessary to graph them.

17. $2x + 2y = 100$
$y = 20 - x$

18. $y = x + 55$
$y = x + 25$

19. $y = 55x$
$y = 25x$

20. $y = 10 - x$
$y = 5 + x$

21. $y = 4x$
$y = \frac{1}{4}(12 + 16x)$

22. $y = 300 - 60x$
$y = 60 - 300x$

In Exercises 23 to 28, which of the pairs of equations form coincident lines?

23. $y + 60x = 300$
$y = 60 - 300x$

24. $y + x = 20$
$2x = 40 - 2y$

25. $y = x + 0.15x$
$y = 1.15x$

26. $y = 150 + 0.10x$
$y = 150x + 0.10$

27. $2x + y = 10$
$2y + x = 10$

28. $y = x + 0.08x$
$y = 1.08x$

In Exercises 29 to 34, what system of equations is represented by each graph? Write the equations in $y = mx + b$ form. Find the point of intersection, and show that it is a solution to the system.

29.

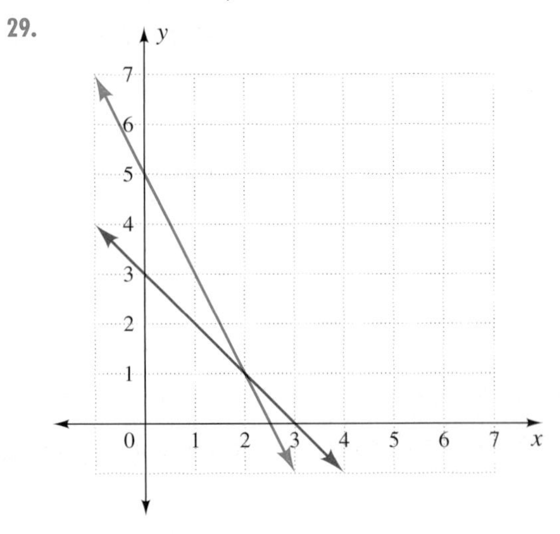

30.

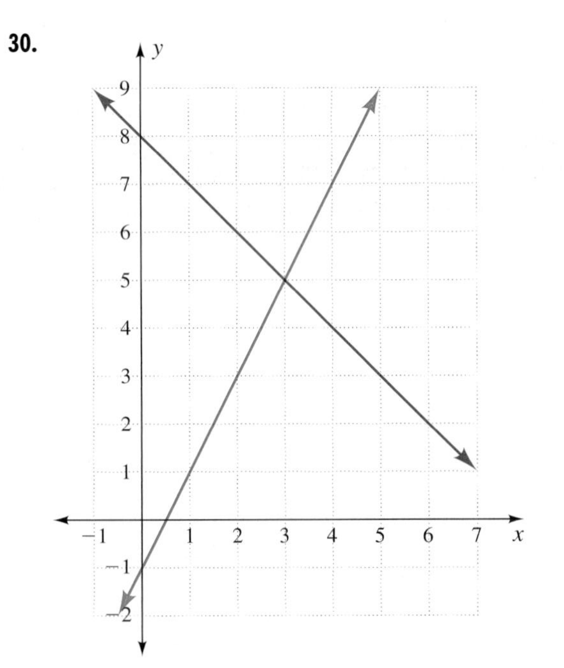

Blue numbers are core exercises.

31.

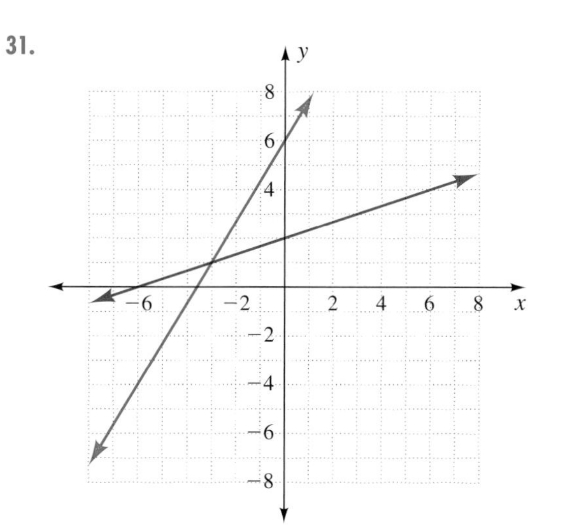

32.

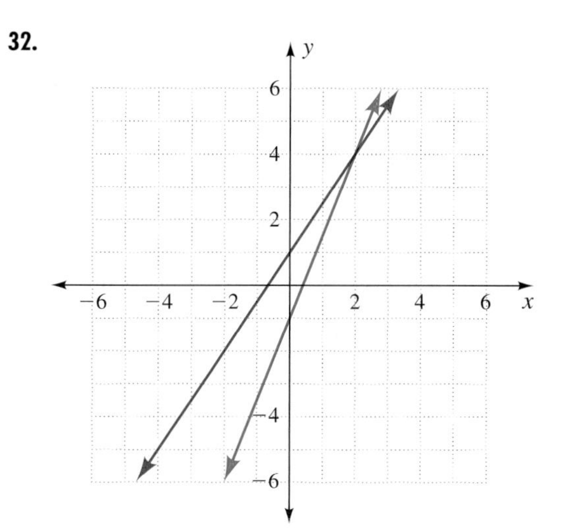

33.

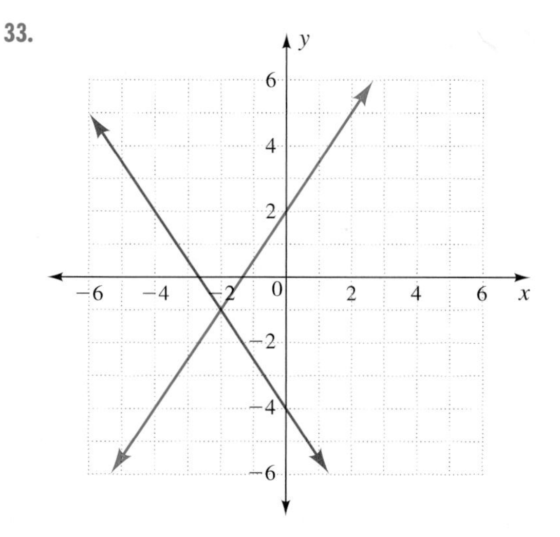

34.

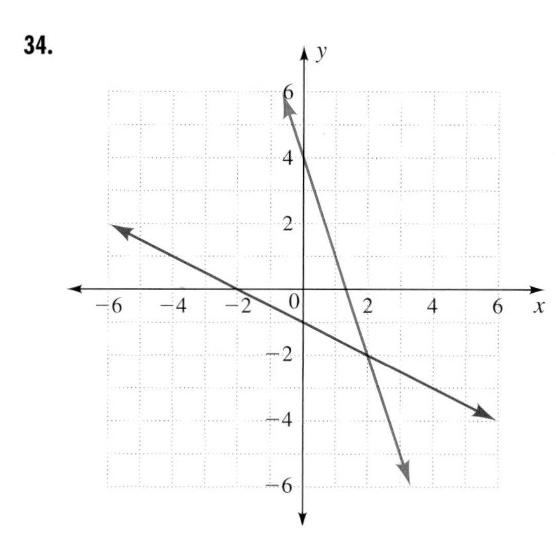

35. The Electronics Works wants to increase income, y, for people with large sales, x.

a. Which plan rewards big sales: $y = 0.04x + 350$ or $y = 0.12x + 50$?

b. Graph the two plans in part a.

c. What is the point of intersection of the graphs of the equations in part a?

d. Under what conditions would an employee prefer $y = 0.04x + 350$?

e. What are the new equations if the employer doubles the rate per dollar of sales?

The figure for Exercises 36 to 40 shows two lines: revenue and cost. Revenue is from registration fees at a student career conference in 1995. Costs are from printing the promotional flyer and ordering lunches for x participants. In business, the point of intersection of a cost graph and a revenue graph is the break-even point, B.E. Profit is equal to revenue minus cost.

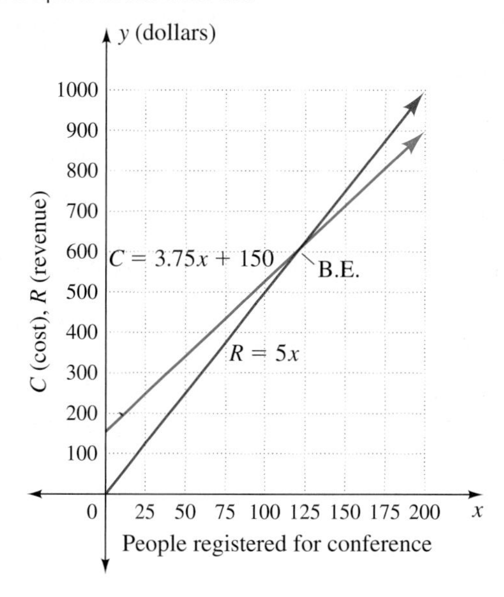

Blue numbers are core exercises.

36. a. What is the registration fee?

b. What is the printing cost?

c. What is the cost of each lunch?

d. Estimate the ordered pair for the break-even point.

37. Suppose 100 students attend the conference.

a. What is the total cost for the people putting on the conference?

b. What is the total revenue to the conference?

c. What is the profit?

d. If the total registration is below the break-even point, which graph is on top?

38. Suppose 150 students attend the conference.

a. What is the total cost for the people putting on the conference?

b. What is the total revenue to the conference?

c. What is the profit?

d. If the total registration is above the break-even point, which graph is on top?

39. When this same conference was held again in 2003, a local sponsor gave \$200 and the charge for registration was \$10. The planning group paid \$250 for printing and \$8.50 for each lunch.

a. What is the equation for total cost?

b. What is the equation for total revenue?

c. Draw the graph for the new data.

d. What is the new break-even point?

40. Repeat Exercise 39 with an \$8.50 fee for registration instead of a \$10 fee.

▶ Projects

41. Travel Graphs Marti leaves home at 7 a.m. on her bicycle and travels 20 miles per hour. Jan leaves from the same house two hours later by car and travels 45 miles per hour along Marti's route. Let the input, t, be the number of hours after 7 a.m. Let the output, d, be in miles, with $d = rt$.

a. What equation describes Marti's distance from home?

b. What equation describes Jan's distance from home?

c. Draw a graph for each person that shows the distance in miles each travels between 7 a.m. and noon.

d. What is the intersection of the two graphs?

e. What does the intersection mean?

f. Who has the steeper graph, Marti or Jan? Why?

g. Why does Jan's graph start at (2, 0) and not at the origin?

Blue numbers are core exercises.

▶ 6.2 Setting Up Systems of Equations

Objectives

- Use guess and check to set up a system of equations.
- Use a quantity-rate table to set up a system of equations.

WARM-UP

Sort these units into quantities and rates:

5 pounds per bag	100 bags	60 nickels
\$0.25 per quarter	30 hours	\$7.50 per hour
20 quarters	\$0.05 per nickel	10 pounds per bag
20 hours	\$9.75 per hour	400 bags

IN THIS SECTION, we examine some strategies for setting up the system of equations necessary to solve a problem. We use tables to organize a guess-and-check process for building equations. We return to the quantity and rate concept of Section 5.5.

Problem-Solving Steps in Setting Up Equations

To review the process for setting up equations from word problems, we return to the four problem-solving steps in Section 1.1:

1. Move toward *understanding* by reading carefully.
2. *Plan* by considering what might be reasonable inputs and preparing tables, charts, or pictures to organize the information. Guess and check and estimate the output are good strategies.
3. *Carry out the plan* by working through the problem with the chosen input.
4. *Check* by comparing the result with the conditions or requirements of the original problem.

Building Systems of Equations with Guess-and-Check Tables

In the first example, we will use guess and check to help us understand the problem and then write the system of equations. Guess-and-check tables have *a column for each unknown and a row for each guess.*

EXAMPLE 1 Using guess and check to understand the problem and write a system of equations: job sharing Cal and Joe share a job that takes a total of 243 hours each month. Cal works 17 hours more than Joe. Because no rates are given, use a guess-and-check table to write a system of equations to find out how many hours each works.

SOLUTION The unknowns are the numbers of hours Cal and Joe work, so we have a column for Cal's hours, another for Joe's hours, and a third for total hours. For a first guess of Cal's hours, we pick 160, a number that is more than half of 243. Because Cal works 17 more hours than Joe, our guess for Joe's hours is 143. Table 2 shows the total hours to be 303, which is too high.

TABLE 2

Cal's Hours	Joe's Hours	Total Hours
160	$160 - 17 = 143$	$160 + 143 = 303$
150	$150 - 17 = 133$	$150 + 133 = 283$
x	$x - 17 = y$	$x + y = 243$

Our next guess is slightly lower: 150. Subtracting 17, we get 133 for Joe's hours. We add the hours to get 283 in total. In the third row, we let $x =$ Cal's hours and $y =$ Joe's hours and write $x - 17 = y$ for Joe's hours and $x + y = 243$ for total hours. We then write the two-equation system:

$y = x - 17$ The difference in their work hours

$x + y = 243$ The total hours worked ◀

THINK ABOUT IT 1: Describe a pattern in Table 2. Use guess and check to solve for the number of hours.

Building Systems of Equations with Quantity-Rate Tables

EQUATIONS WITH TWO UNKNOWN QUANTITIES For our exploration, we use a puzzle problem, because it is easier to think about than more realistic applications.

EXAMPLE 2 Setting up a quantity-rate table with guess and check

Understand the problem; read it carefully. Roberta likes to keep nickels and quarters in the jar by her front door to use for tolls and parking meters. Her husband notices that there are no coins in the jar, so he gets \$30 in change. Their 4-year-old daughter proudly announces that there are now 152 coins in the jar. How many of each coin is in the jar?

Make a plan:

What information is not known?

What information is assumed?

What facts (the constants) do not change?

Can we estimate an answer?

Can we use a quantity-rate table?

Carry out the plan: First check your understanding by explaining why 80 quarters might be a reasonable guess. Then complete Table 3 for 80 quarters. Make another guess for the number of quarters, and complete the table again.

TABLE 3 Guessing Quantities.

Item	Quantity	Rate	$Q \cdot R$
Nickels		0.05	
Quarters		0.25	
Total	152		\$30

SOLUTION There are two unknowns: the number of nickels and the number of quarters. The problem assumes that we know that nickels are worth \$0.05 per coin and quarters are worth \$0.25 per coin. The total number of coins, their rates, and the total value are constants.

If we divide the total value, \$30, by the number of coins, 152, we get about \$0.20. This is closer to \$0.25 than to \$0.05, so we might guess that there are more quarters than nickels. If there are 152 coins and more are quarters than nickels, 80 quarters is a reasonable starting guess.

Because we have both the number (quantity) of coins and the value (rate) of each coin to consider, a quantity-rate table is appropriate for the problem. A table is in the Answer Box. ◀

THINK ABOUT IT 2: How can the completed quantity-rate tables for Example 2 be used to predict the numbers of coins? Suggest another guess and explain your reasoning.

In Example 3, we use two variables to set up equations.

EXAMPLE 3 Finding equations in two variables from a quantity-rate table Set up a system of two equations for the problem in Example 2 by letting $x =$ the number of nickels in the quantity-rate table and $y =$ the number of quarters. Place a dollar sign in the column headings where appropriate.

SOLUTION We write x, y, and 152 (the total number of coins) in the quantity column in Table 4. Next we write the values (\$ per coin) of the coins in the rate column. The total value of the money is \$30, and so we put 30 in the lower right corner, as the total of the $Q \cdot R$ column.

TABLE 4 Writing Equations from a Table

Item	Quantity	Rate ($/coin)	$Q \cdot R$ ($)
Nickels	x	0.05	$0.05x$
Quarters	y	0.25	$0.25y$
Total	152		30

As in Section 5.5, we multiply across the table (Quantity · Rate, or $Q \cdot R$) to obtain the total value of the nickels, $0.05x$, and the total value of the quarters, $0.25y$. From the columns, we write a system of two equations in two variables. The quantity column gives $x + y = 152$. The $Q \cdot R$ column gives $0.05x + 0.25y = 30$. ◀

THINK ABOUT IT 3: What is the advantage of placing the dollar sign in column headings?

As you work through the remaining examples, do extra guess-and-check steps. Only experience can show you the importance of guess and check in reading problems, setting up tables, observing patterns, and then writing equations.

In Example 3, the unknowns were the numbers of coins. In Example 4, the unknowns are the numbers of hours.

▶ **EXAMPLE 4** **Finding equations from a quantity-rate table: wages** Jaynine works two different jobs for one employer. Last week she worked 50 hours altogether and earned a total of $435.75. When she works in the shipping department, she earns $7.50 per hour. When she is filling orders, she earns $9.75 per hour. She wants to know how many hours she was paid for at each job.

a. List the constants and variables.

b. Find the average wage, and use it to come up with a guess for a quantity-rate table.

c. Make another table with two variables, and write a system of two equations that can be solved to find the number of hours for which she was paid at each job.

SOLUTION

a. The constants are the total hours of work, 50; the total earnings, $435.75; and the hourly wages, $7.50 and $9.75, which reflect the value of each hour worked. The variables are the number of hours worked in shipping and the number of hours spent filling orders.

b. We find the average wage by dividing earnings by hours: $435.75 ÷ 50 = $8.715 per hour. This is slightly closer to $9.75 than to $7.50, so we estimate that the number of hours spent filling orders will be slightly higher than the number spent shipping. In Table 5, we try 30 hours for filling orders and 20 hours for shipping.

TABLE 5 Guessing Quantities

Item	Quantity (hr)	Rate ($/hr)	$Q \cdot R$ ($)
Shipping	20	7.50	150
Filling orders	30	9.75	292.50
Total	50		442.50

c. If we let x = the number of hours worked in shipping and y = the number of hours spent filling orders, then the table will look like Table 6.

TABLE 6 Writing Equations from a Table

Item	Quantity (hr)	Rate ($/hr)	$Q \cdot R$ ($)
Shipping	x	7.50	$7.50x$
Filling orders	y	9.75	$9.75y$
Total	50		435.75

The system of equations comes from the columns:

$$x + y = 50 \quad \text{Sum of the quantity column}$$

$$7.50x + 9.75y = 435.75 \quad \text{Sum of the } Q \cdot R \text{ column}$$ ◀

THINK ABOUT IT 4: Continue guessing to find the numbers of hours spent shipping and filling orders in Example 4.

Example 5 has a business decision setting. This time, the known "rate" is the weight per bag instead of cents per coin or dollars per hour.

▶ **EXAMPLE 5** Finding equations from a quantity-rate table: onion bags A produce distributor supplies fresh fruit and vegetables to grocery stores. The distributor receives a shipment of 4050 pounds of onions and needs to bag them for retail sale. Past history shows that stores sell 5-pound bags and 10-pound bags in a 4-to-1 ratio.

Set up a quantity-rate table for a guess of one hundred 10-pound bags. Set up a second quantity-rate table with two variables. Write a system of equations to describe the problem setting.

SOLUTION The variables are the numbers of the two sizes of bags. The weights of the bags and the total weight of the onions are the constants.

Setting up the quantity-rate table, as shown in Table 7, helps us read the problem carefully. We guess a number of 10-pound bags and, according to the 4-to-1 ratio, multiply by 4 to get the number of 5-pound bags. The total number of pounds we come up with is about 1000 less than the required number of pounds, so we must raise the number of 10-pound bags if we continue to guess.

TABLE 7 Guessing Number of Bags

Item	Quantity (bags)	Rate (lb/bag)	$Q \cdot R$ (lb)
5-pound bags	400	5	2000
10-pound bags	100	10	1000
Total	500		3000

If we let x = the number of 5-pound bags and y = the number of 10-pound bags, we have Table 8 relating the facts.

TABLE 8 Writing Equations from a Table

Item	Quantity (bags)	Rate (lb/bag)	$Q \cdot R$ (lb)
5-pound bags	x	5	$5x$
10-pound bags	y	10	$10y$
Total	Not given		4050

We get one equation from the total weight:

$$5x + 10y = 4050$$

The second equation comes from the store sales:

$$\frac{x}{y} = \frac{4}{1}$$ ◀

THINK ABOUT IT 5: First explain why $x = 4y$ describes the number of 5-pound bags. Then use guess and check to solve Example 5.

EQUATIONS WITH UNKNOWN RATES In the next problem, the rates are the unknowns. We will still be able to write an equation from the sum of the $Q \cdot R$ column, but—because adding rates has no meaning—we will need a separate table for each fact.

▶ **EXAMPLE 6** Finding equations from a quantity-rate table: ticket prices Build a quantity-rate table from each fact, and write an equation describing the total $Q \cdot R$.

a. 2 adult movie tickets and 4 student tickets cost $31.

b. 3 adult tickets and 2 student tickets cost $28.50.

SOLUTION Let a = the cost of an adult ticket and s = the cost of a student ticket.

a. Table 9 is the quantity-rate table. The equation is $2a + 4s = 31$.

TABLE 9 Writing One $Q \cdot R$ Equation

Item	Quantity	Rate: Cost per Ticket	$Q \cdot R$ ($)
Adult tickets	2	a	$2a$
Student tickets	4	s	$4s$
Total			31

b. Table 10 is the quantity-rate table. The equation is $3a + 2s = 28.50$.

TABLE 10 Writing a Second $Q \cdot R$ Equation

Item	Quantity	Rate: Cost per Ticket	$Q \cdot R$ ($)
Adult tickets	3	a	$3a$
Student tickets	2	s	$2s$
Total			28.50

◀

THINK ABOUT IT 6: Why can we not add a and s in Example 6 to obtain a second equation from each table? Use guess and check to solve the system formed by parts a and b.

In Example 6, a and s were rates (cost per ticket), and their coefficients were the quantities of the two types of tickets. The two equations can be summarized with the quantity-rate equation:

QUANTITY-RATE EQUATION

$$\text{Quantity} \cdot \text{rate} + \text{quantity} \cdot \text{rate} = \text{total } Q \cdot R$$

EQUATIONS WITH UNKNOWN RATES: WIND SPEED AND CURRENT SPEED Have you ever walked up an escalator and been surprised at how fast you were traveling at the top? You have added your rate of walking to the rate of the escalator. A similar addition of rates happens with tailwinds and airplanes, as well as river currents and boats going downstream.

A strong wind called the *jet stream* flows eastward across the United States, as shown in Figure 11. The jet stream is at an altitude of about 12 kilometers (40,000 feet). Airplanes traveling eastbound take advantage of the jet stream in that their speed is increased by the wind speed. Those flying westbound must allow extra time and fuel for the flight because their speed is decreased by the wind speed.

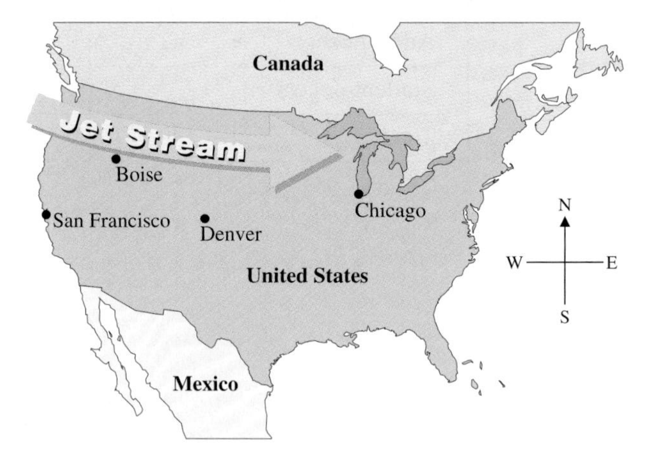

FIGURE 11 Jet stream : Wind at 40,000 feet

Example 7 shows how wind can affect the speed of an airplane.

▶ **EXAMPLE 7** Finding rates given wind speed In winter, the jet stream blows at about 130 km/hr across the southern part of the country. In summer, the jet stream blows at about 65 km/hr across the northern part of the country. What is the speed of each airplane?

a. An eastbound airplane (San Francisco to Denver) flying 500 km/hr in winter
b. A westbound airplane (Chicago to Boise) flying 500 km/hr in winter
c. An eastbound airplane flying 500 km/hr in summer
d. A westbound airplane flying 500 km/hr in summer
e. An eastbound airplane flying 500 km/hr in still air

SOLUTION We assume that the airplane is flying parallel to the direction of the jet stream.

a. Net speed is $(500 + 130)$ km/hr $= 630$ km/hr.
b. Net speed is $(500 - 130)$ km/hr $= 370$ km/hr.
c. Net speed is $(500 + 65)$ km/hr $= 565$ km/hr.
d. Net speed is $(500 - 65)$ km/hr $= 435$ km/hr.
e. Net speed is 500 km/hr. In still air, there is no wind effect on the airplane. ◀

FINDING RATES GIVEN WIND SPEED

To find the speed of an airplane flying directly into the wind (a head wind), we reduce the airspeed, r, by the wind, w:

$$\text{Net speed} = r - w$$

To find the speed of an airplane flying with the wind (a tail wind), we increase the airspeed, r, by the wind, w:

$$\text{Net speed} = r + w$$

The general travel formula is

$$\text{Time} \cdot \text{rate} = \text{distance}$$

We can write travel information in a quantity-rate table, letting quantity = time, rate = net speed, and $Q \cdot R$ = distance. When we seek information about wind (or current), we place the net speed ($r - w$ or $r + w$) in the rate column.

▶ **EXAMPLE 8** Finding equations from a quantity-rate table: travel An airplane flying westbound from Chicago to Boise, approximately 1400 miles (against the wind), takes 4 hours. Another airplane flying the same airspeed travels from Boise to Chicago (with the wind) in 3.5 hours. Write a system of two equations for finding the airspeed (r) and the wind speed (w).

SOLUTION The distance traveled is 1400 miles. The time spent traveling against the wind is 4 hours; the time spent traveling with the wind is 3.5 hours. Net speed against the wind is $r - w$ and with the wind is $r + w$. We enter these facts into the quantity-value table shown in Table 11.

TABLE 11 Wind and Airspeed

Quantity: Time (hr)	Rate: Net Speed Relative to Ground (mph)	Time · Net Speed: Distance (mi)
4.0	$r - w$	1400
3.5	$r + w$	1400

By multiplying across Table 11, we obtain equations containing r and w:

$$4.0(r - w) = 1400$$

$$3.5(r + w) = 1400$$ ◀

THINK ABOUT IT 7: Use guess and check to solve the system of equations in Example 8.

The current in a river affects boats in the same way as wind affects airplanes. Going upstream—*against* the current—subtracts from the speed at which a boat travels in still water. Going downstream—*with* the current—adds to the speed at which a boat travels in still water. (See Figure 12.)

If the head wind exceeds the airspeed, the airplane will move backwards. Similarly, if the current exceeds the speed in still water, the boat heading upstream will move backwards. This may sound like fantasy, but just imagine swimming 3 miles per hour against a flood traveling 20 miles per hour.

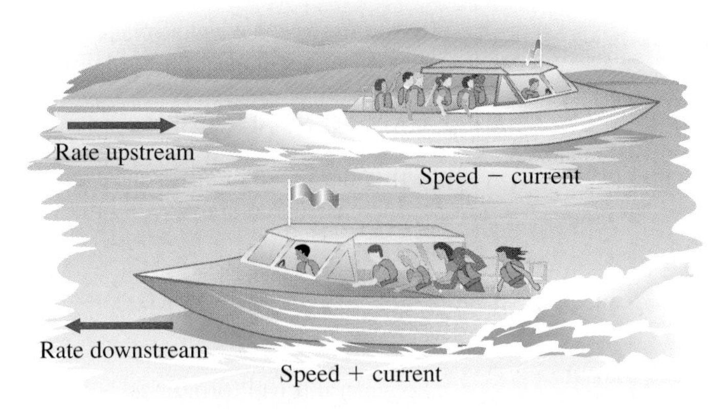

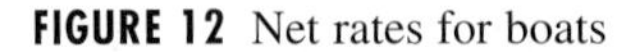

FIGURE 12 Net rates for boats

THINK ABOUT IT 8: Describe a situation where the speed of a person and the rate of an escalator are subtracted.

We assumed here that the direction of travel is parallel to the wind and river currents. In Section 8.1, we will assume that the direction of travel is perpendicular to the wind and river currents. Other angles related to travel and wind (or currents) are addressed in trigonometry and navigation courses.

ANSWER BOX

Warm-up: Quantities are 100 bags, 60 nickels, 30 hours, 20 quarters, 20 hours, and 400 bags. Rates are 5 pounds per bag, \$0.25 per quarter, \$7.50 per hour, \$0.05 per nickel, 10 pounds per bag; and \$9.75 per hour. **Think about it 1:** For each 10-hour drop in Cal's hours, total hours drop by 20; 130 hours and 113 hours.

Example 2:

Item	Quantity	Rate	$Q \cdot R$
Nickels	72	\$0.05	\$3.60
Quarters	80	\$0.25	\$20
Total	152		\$23.60

The total value is too low, so we need more quarters and fewer nickels. Here is a table for 100 quarters.

Item	Quantity	Rate	$Q \cdot R$
Nickels	52	\$0.05	\$2.60
Quarters	100	\$0.25	\$25
Total	152		\$27.60

This time the total value is \$27.60 instead of \$30, so we need more quarters. **Think about it 2:** Changing 20 coins from nickels to quarters increased the quarter $Q \cdot R$ by \$5 and decreased the nickel $Q \cdot R$ by \$1. Try 120 quarters. **Think about it 3:** No \$ in equations **Think about it 4:** 23 hours shipping and 27 hours filling orders **Think about it 5:** Multiply by y on both sides of $\frac{x}{y} = \frac{4}{1}$. 135 ten-pound bags and 540 five-pound bags **Think about it 6:** a and s are rates; $a =$ \$7.50 per ticket, $s =$ \$4 per ticket **Think about it 7:** $r = 375$ miles per hour, $w = 25$ miles per hour **Think about it 8:** Walking up the down escalator or down the up escalator. Not a recommended action!

▶ 6.2 Exercises

In Exercises 1 to 4, set up a guess-and-check table. Make three guesses, and then label the columns in the table with variables. Write a system of equations that will solve the problem. It is not necessary to solve the problem.

1. Renee cuts a 10-meter hose into two lengths. One piece is 5 meters longer than the other. What is the length of each piece?

2. Jacques cuts a 12-decimeter submarine sandwich into two pieces. One piece is 6 decimeters longer than the other. What is the length of each piece?

3. Ed works on two projects during one month. He works 28 hours longer on one project than on the other. He works a total of 176 hours during the month. How many hours does he work on each project?

4. Alexandra works for two clients during the month. She works a total of 170 hours during the month. She works 62 hours more for one client than for the other. How many hours does she work for each client?

Set up quantity-rate tables for Exercises 5 to 14. Complete one table with a guess as one input, and then complete a second table with variables as inputs. Write a system of equations to describe the problem situation. It is not necessary to solve the problem.

5. Celesta has \$20.20 in quarters and nickels. She has 104 coins altogether. How many of each does she have?

6. Larry has 133 coins in nickels and quarters. He has \$16.25 altogether. How many nickels and how many quarters does he have?

7. Casey has 140 coins. He has \$31.10 altogether. If he has only dimes and quarters, how many of each does he have?

8. Nancy has \$21.90. She has only dimes and quarters. She has 120 coins altogether. How many of each does she have?

9. Lindsay has 159 coins in nickels and dimes. She has \$12.10 altogether. How many nickels and how many dimes does she have?

10. Martina has 164 coins in nickels and dimes. She has \$11.60 altogether. How many nickels and how many dimes does she have?

11. Se Ri earns \$406 from working 43 total hours at two jobs. She earns \$10.75 per hour at the first job and \$8.50 per hour at the second job. How many hours does she work at each job?

12. Paolo earns \$569 from working 54 total hours at two jobs. He earns \$11.25 per hour at the first job and \$9.50 per hour at the second job. How many hours does he work at each job?

13. The I. R. Rabbit Company sells carrots in a 9 to 1 ratio to grocery stores and restaurants. It has 5250 pounds of carrots. How many pounds should be packaged for grocery stores and how many for restaurants?

14. Restaurants buy 10-pound and 25-pound bags of carrots. They want 10-pound bags and 25-pound bags in a 5 to 1 ratio. If the company estimates that restaurants will buy 1050 pounds of carrots, how many bags of each size should be packaged?

In Exercises 15 to 20, write a system of equations to describe the problem situation. Use guess and check to solve the problem.

15. On a weekend, two adult movie tickets and three student tickets cost \$38.50. Three adult and two student tickets cost \$41.50. What is the price of each ticket?

16. On a weekday, three children's movie tickets and two adult tickets cost \$29.25. Five children's tickets and one adult ticket cost \$31.25. What is the price of each ticket?

17. Tickets for five kids and two adults to a Saturday morning magic show cost \$30. Tickets for six kids and one adult cost \$32.50. What is the price of each ticket?

18. Three adult and five student tickets for the alternative music hall cost \$67.50. Five adult and eight student tickets cost \$110. What is the price of each ticket?

19. Two of the top-price tickets and five of the lowest-price tickets for a Tim McGraw concert cost \$800. Three of the top-price and three of the lowest-price tickets cost \$1020. What is the price of each ticket?

20. Two of the top-price tickets and eight of the lowest-price tickets to see Norah Jones in concert cost \$2245. Three of the top-price and four of the lowest-price tickets cost \$2887.50. What is the price of each ticket?

21. Find the rate of travel for a fishing boat with a speed of 8 miles per hour (mph) in each of these settings:

    a. Upstream against a 3-mph current

    b. Upstream against a 6-mph current

    c. Upstream against a 15-mph flood

    d. Downstream with a 3-mph current

    e. Downstream with a 5-mph current

Blue numbers are core exercises.

22. Find the rate of travel for the following ships, with the given speeds, should they have to travel against a 16-knot current, as is found in the Nakwakto Rapids, Slings-by Channel, British Columbia, Canada. (*Note:* A knot is a nautical mile (6080 ft) per hour.)

a. Human-powered craft, 6 knots

b. Passenger liner, 18 knots

c. Catamaran car ferry, 35 knots

d. Russian alpha-class submarine, 45 knots

In Exercises 23 to 26, write a system of equations to describe the problem situation. Use guess and check to solve the problem.

23. An airplane flies eastbound from Boise to Chicago, covering the 1400 miles with the wind in 5 hours. A return flight against the same wind takes 5.6 hours. Find the airspeed of the airplane and the speed of the wind.

24. An airplane flies eastbound from San Francisco to Denver (950 miles) with the wind in 2.5 hours. On a return flight against the same wind, the travel time is 3.8 hours. Find the airspeed of the airplane and the speed of the wind.

25. A boat travels 12 miles upstream to a fishing spot. The trip takes 1.2 hours. The return trip, downstream, takes 0.8 hour. Find the speed of the boat and the speed of the current.

26. A boat travels 10 miles downstream in an hour. While its passengers are fishing, the tide comes in, and tidewater reaches 10 miles upstream. The trip home on still water takes $1\frac{1}{4}$ hours. Find the speed of the boat and the speed of the current.

▶ Writing

27. How would you explain to another student the difference between quantity and rate?

28. How would you explain to another student why we can add (or subtract) rates related to an airplane, a boat, or an escalator?

29. Explain how the $Q \cdot R$ equation is based on a $Q \cdot R$ table.

▶ Projects

30. Using Manipulatives to Solve Problems *Part I:* One day Mr. McFadden decides to count his farm animals. He counts strangely and reports 10 heads and 28 feet. He has cows and ducks. (*Note:* It is permissible to have zero animals of one type.) How many of each does Mr. McFadden have? Use 10 coins, rubber bands, or buttons for heads. Use 28 toothpicks, paper clips, safety pins, or cotton swabs for feet. Model the animals.

a. Is it possible to have 10 heads and 24 feet?

b. Is it possible to have 10 heads and 30 feet?

c. What is the largest number of feet possible with the 10 heads?

d. What is the smallest number of feet possible with the 10 heads?

e. What patterns do you observe if you organize your data into a table? Use the following table, adding more rows or columns as needed.

Ducks	Cows	Total Heads	Total Feet

f. Write a system of equations to solve Mr. McFadden's problem.

Part II: Mr. McFadden's neighbor is Mr. Schaaf. Mr. Schaaf counts his animals the same way. He reports 12 heads and 28 feet. How many cows and ducks does Mr. Schaaf have?

g. What is the largest number of heads possible with 28 feet?

h. What is the smallest number of heads possible with 28 feet?

i. Organize your results in a copy of the table in part e. Look for patterns. Add more rows or columns as needed.

j. Write a system of equations to solve Mr. Schaaf's problem.

k. How might heads be considered quantities?

l. How might feet be considered rates?

Blue numbers are core exercises.

▶ 6.3 Solving Systems of Equations by Substitution

Objectives

- Solve an equation for one variable in terms of a second variable.
- Solve a system of equations by substitution.
- Solve a system of equations when both equations are in $y = mx + b$ form.
- Use geometry facts in setting up a system of equations.

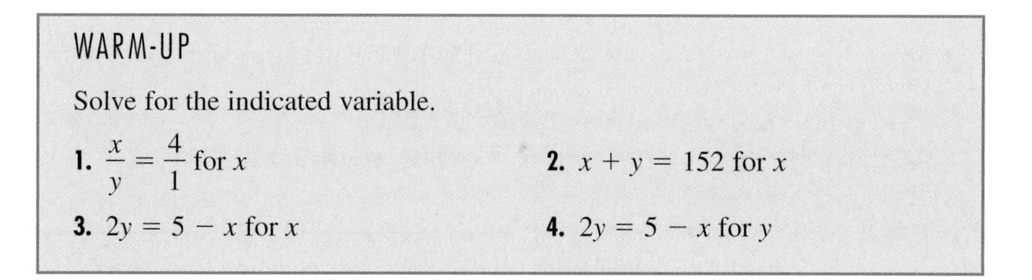

WARM-UP

Solve for the indicated variable.

1. $\frac{x}{y} = \frac{4}{1}$ for x

2. $x + y = 152$ for x

3. $2y = 5 - x$ for x

4. $2y = 5 - x$ for y

THERE ARE MANY WAYS of solving systems of equations, and some work better than others in particular situations. We have used graphs, guess and check, and tables to solve systems. In this section, we solve systems of equations using a process called substitution. We return to several examples from Sections 6.1 and 6.2 and work some new examples involving geometry facts.

▶ Solving Systems of Equations by Substitution

SUBSTITUTION METHOD

> A process for solving a system of equations by replacing variables with equivalent expressions or numbers in order to eliminate one variable.

Mathematical substitution resembles the substitution of players in a sports event or the substitution of ingredients in a recipe. In sports and cooking, as well as in mathematics, there is a "taking out" and a "putting in" process. However, substitutions differ in that mathematics uses an equal replacement, whereas sports and cooking use a replacement that only approximates the player or ingredient removed.

▶ **EXAMPLE 1** Solving a system of equations by substitution We return to the onion bag problem from Section 6.2 (Example 5, page 336), where we had the system

$$x = 4y$$

$$5x + 10y = 4050$$

In a graph of these two equations, the point of intersection locates the ordered pair that makes both equations true. To solve by the substitution method:

a. Replace x in the second equation with $4y$, and solve the new equation for y.
b. Substitute the y value from part a into $x = 4y$ to find x.
c. State the solution as an ordered pair, (x, y).
d. Check that the ordered pair makes both equations true.

SOLUTION **a.**

$5x + 10y = 4050$ Substitute $4y$ for x.

$5(4y) + 10y = 4050$ Simplify.

$20y + 10y = 4050$ Add like terms.

$30y = 4050$ Divide both sides by 30.

$y = 135$

b. $x = 4y$

$x = 4(135)$

$x = 540$

c. (540, 135); the distributor fills 540 five-pound bags and 135 ten-pound bags.

d. $540 \stackrel{?}{=} 4(135)$ ✓

$5(540) + 10(135) \stackrel{?}{=} 4050$ ✓ ◀

STEPS IN SOLVING A SYSTEM OF EQUATIONS BY SUBSTITUTION

1. Solve one equation for a variable equal to an expression.
2. In the other equation, substitute the expression from step 1 for the variable, placing the expression in parentheses. Solve the second equation for the remaining variable.
3. Substitute the value from step 2 into the first equation.
4. State the solution as an ordered pair, (x, y), or as a system, $x = \quad, y = \quad$.
5. Check the solution in both equations.

▶ In Example 2, we solve the equations for the coins in the jar problem from Section 6.2 (Example 3, page 334).

▶ **EXAMPLE 2** Solving by substitution Using substitution, solve the system of equations for Roberta's coins:

$x + y = 152$ Number of coins

$0.05x + 0.25y = 30$ Value of coins

SOLUTION ***Step 1:*** We solve one equation for a variable equal to an expression.

$x + y = 152$ Subtract y from both sides.

$x = 152 - y$

Step 2: In the other equation, we substitute the expression from step 1 for the variable, placing the expression in parentheses. We then solve for the remaining variable.

$0.05x + 0.25y = 30$ Substitute $x = 152 - y$.

$0.05(152 - y) + 0.25y = 30$ Apply the distributive property.

$0.05(152) - 0.05y + 0.25y = 30$ Simplify.

$7.60 + 0.20y = 30$ Subtract 7.60 from both sides.

$0.20y = 22.40$ Divide both sides by 0.20.

$y = 112$

Step 3: We substitute the value from step 2 into the first equation.

$x = 152 - y$ Let $y = 112$

$x = 152 - 112$

$x = 40$

Step 4: Stating the solution as an ordered pair and as a system, we have (40, 112) and $x = 40$, $y = 112$, respectively. Roberta has 40 nickels and 112 quarters.

Step 5: We check the solution in both equations.

$x + y = 152$

$40 + 112 \stackrel{?}{=} 152$ ✓

$0.05x + 0.25y = 30$

$0.05(40) + 0.25(112) \stackrel{?}{=} 30$ ✓ ◀

CHOOSING SUBSTITUTION 1 When one variable in the system has a coefficient of 1 or -1, the first step in substitution is easy to find: We start by isolating the variable with coefficient 1 or -1.

▶ **EXAMPLE 3** Solving by substitution Name the equation that can be easily solved for one variable. Then solve the system by substitution.

$3x - 4y = -25$ (1)

$2y = 5 - x$ (2)

SOLUTION The variable x in $2y = 5 - x$ has -1 as a coefficient, so we choose to isolate the variable in equation (2).

$2y = 5 - x$ Add x to both sides.

$x + 2y - 5$ Subtract $2y$ from both sides.

$x = 5 - 2y$

Next we substitute $5 - 2y$ for x in equation (1).

$3x - 4y = -25$ Substitute $x = 5 - 2y$.

$3(5 - 2y) - 4y = -25$ Apply the distributive property.

$15 - 6y - 4y = -25$ Add like terms.

$15 - 10y = -25$ Subtract 15 on both sides.

$-10y = -25 - 15$ Add like terms.

$-10y = -40$ Divide both sides by -10.

$y = 4$

Then we substitute $y = 4$ into equation (2).

Student Note:
Substituting into an original equation prevents repeating errors.

$2y = 5 - x$ Substitute $y = 4$.

$2(4) = 5 - x$ Simplify.

$8 = 5 - x$ Subtract 5 on both sides.

$3 = -x$ Multiply by -1 on both sides.

$-3 = x$

Finally, we check that $x = -3$ and $y = 4$ make both equations true.

$3(-3) - 4(4) \stackrel{?}{=} -25$ ✓

$2(4) \stackrel{?}{=} 5 - (-3)$ ✓ ◀

THINK ABOUT IT 1: What one step is needed to solve $2y = 5 - x$ for y? Why might substituting the resulting expression for y into $3x - 4y = -25$ be more difficult than substituting $5 - 2y$ for x?

CHOOSING SUBSTITUTION 2 Equations in $y = mx + b$ form are the most natural to solve by substitution. In Example 4, we replace y in one equation with $mx + b$ from the other equation.

▶ **EXAMPLE 4** Solving a system of equations in $y = mx + b$ form Using substitution, solve this system of equations from the wage options problem in Section 6.1 (Example 3, page 324):

$y = 0.06x + 300$

$y = 0.10x + 150$

SOLUTION To solve this system of equations by substitution, we replace y in the second equation with $0.06x + 300$ from the first equation and then solve for x.

$y = 0.10x + 150$	Let $y = 0.06x + 300$.
$0.06x + 300 = 0.10x + 150$	Subtract $0.06x$ on both sides.
$300 = 0.04x + 150$	Subtract 150 on both sides.
$150 = 0.04x$	Divide both sides by 0.04.
$3750 = x$	

Then we solve for the second variable.

$y = 0.06x + 300$	Replace x with 3750.
$y = 0.06(3750) + 300$	
$y = 525$	

The solution is $x = 3750$, $y = 525$. The wages, \$525, are equal for sales of \$3750.

Finally, we check the solution in both equations:

$y = 0.06x + 300$

$525 \stackrel{?}{=} 0.06(3750) + 300$ ✓

$y = 0.10x + 150$

$525 \stackrel{?}{=} 0.10(3750) + 150$ ✓ ◀

THINK ABOUT IT 2: What is it about the problem situation that allows us to set $0.06x + 300$ equal to $0.10x + 150$?

▶ Algebraic Results for Special Systems

In Section 6.1, we saw that a system of equations that graphs as parallel lines has no solution and a system that graphs as coincident lines has an infinite number of solutions. In the next two examples, we look at the algebraic results of solving these special systems. Example 5 is from Example 5 of Section 6.1 (page 327) on more wage options.

▶ **EXAMPLE 5** Finding the special algebraic results of solving a system that graphs as parallel lines Solve the system of equations

$y = 0.06x + 300$

$y = 0.06x + 150$

SOLUTION We substitute $0.06x + 300$ for y in the second equation.

$y = 0.06x + 150$	Let $y = 0.06x + 300$.
$0.06x + 300 = 0.06x + 150$	Subtract 150 from both sides.
$0.06x + 150 = 0.06x$	Subtract $0.06x$ from both sides.
$150 = 0$	The variables drop out, leaving a false statement.

Because the statement $150 = 0$ is false, we say that the system of equations has no solution. The wages will never be the same. ◀

In Example 6, we return to the coincident lines system of Example 6 in Section 6.1 (page 327).

▶ **EXAMPLE 6** Finding the special algebraic results of solving a system that graphs as coincident lines

Solve the following system of equations, which forms coincident lines:

$$y = \tfrac{1}{2}x - 2$$

$$2y + 4 = x$$

SOLUTION We substitute $2y + 4$ for x in the first equation.

$y = \frac{1}{2}x - 2$	Let $x = 2y - 4$.
$y = \frac{1}{2}(2y + 4) - 2$	Simplify using the distributive property.
$y = y + 2 - 2$	Add 2 and -2.
$y = y$	Subtract y from both sides.
$0 = 0$	The variables drop out, leaving a true statement.

The statement $0 = 0$ is always true. Thus, *any solution to the first equation is also a solution to the second equation. There are an infinite number of solutions.* ◀

THINK ABOUT IT 3: In Example 6, we could have substituted $y = \frac{1}{2}x - 2$ into $2y + 4 = x$ as a starting step. Do so. Do you come to the same conclusion as in Example 6?

Table 12 summarizes the geometric and algebraic results for various types of systems of two linear equations.

TABLE 12 Geometric and Algebraic Results

Geometric Results	Intersecting lines	Parallel lines	Coincident lines
Algebraic Results	The equations can be solved for x and y: (x, y).	The variables drop out, and the remaining statement is false: $0 = 4$.	The variables drop out, and the remaining statement is true: $0 = 0$.
Solution	One ordered pair is the solution.	There is no real-number solution to the system.	An infinite number of ordered pairs satisfy the system.

SUPPLEMENTARY ANGLES IN GEOMETRY A traditional protractor, like the one in Figure 13, shows 180° for a half circle. The numbers on a protractor permit measuring angles from either the left side or the right side.

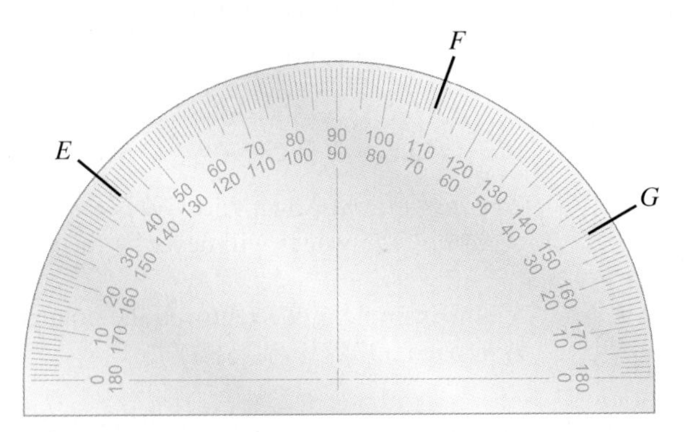

FIGURE 13 Protractor; angles in degrees

▶ **EXAMPLE 7** Finding patterns on a protractor What pattern can you find in the pairs of numbers labeled with the letters E, F, and G on the protractor in Figure 13?

SOLUTION At position E, 40 matches with 140. At position F, 110 matches with 70. At position G, 150 matches with 30. Each pair of numbers adds to 180. ◀

When we look at the two angles being described by each pair of numbers on a protractor, we find that *the angles share a side and their other sides form a straight line*. These angles are often called a **linear pair**. Figure 14a shows a linear pair.

The drawings of a parallelogram in Figure 14b and parallel lines in Figure 14c show angles that may be rearranged to form a linear pair. *Two angles whose measures add to 180°*, forming a linear pair, are **supplementary angles**. The justification for the fact that the angles in Figure 14 are supplementary is left to a geometry course.

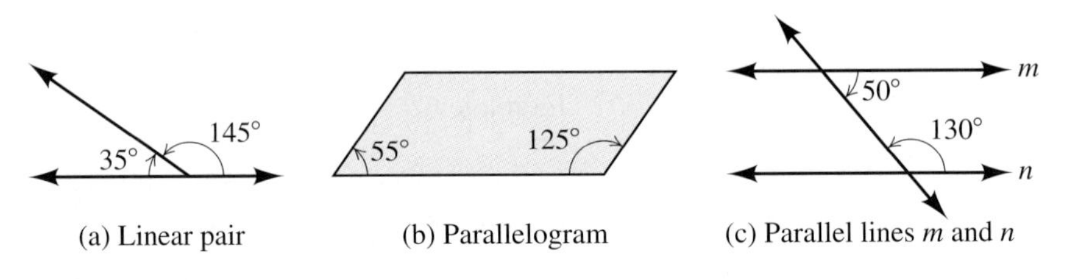

FIGURE 14 Three sources of supplementary angles

▶ **EXAMPLE 8** Solving by substitution: supplementary angles Angles M and N in Figure 15 are supplementary. The measure of angle N is 24° more than five times the measure of angle M. What is the measure of each angle? Write a system of equations and solve by substitution.

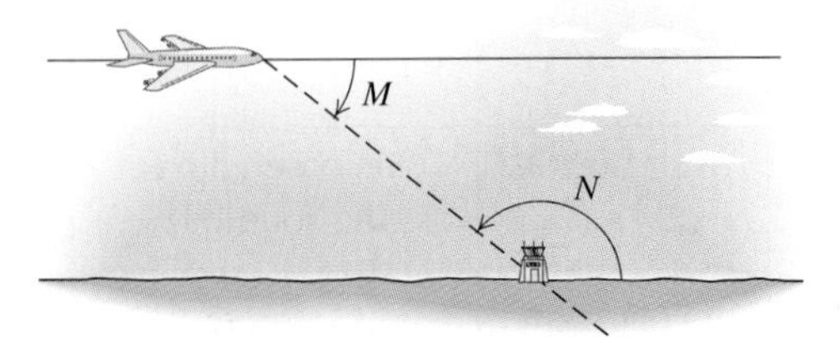

FIGURE 15 Supplementary angles formed by parallel lines

SOLUTION The angles are supplementary, so they add to 180°:

$$M + N = 180$$

We translate the other sentence into a second equation,

$$N = 24 + 5M$$

We use the second equation to substitute for N in the first equation:

$$M + (24 + 5M) = 180 \quad \text{Let } N = 24 + 5M\text{. Add like terms.}$$
$$6M + 24 = 180 \quad \text{Subtract 24.}$$
$$6M = 156 \quad \text{Divide by 6.}$$
$$M = 26$$

We substitute $M = 26$ into the second equation:

$$N = 24 + 5(26)$$
$$N = 154$$

The angles are $M = 26°$ and $N = 154°$.

Check:

$$26 + 154 \stackrel{?}{=} 180 \ \checkmark$$
$$154 \stackrel{?}{=} 24 + 5(26) \ \checkmark$$

◀

COMPLEMENTARY ANGLES IN GEOMETRY The measures of **complementary angles** *add to 90°*. Any two angles forming a right angle are complementary; see Figure 16a. In a right triangle, because one angle is 90°, the two smaller angles (called *acute angles*) add to 90° and are therefore complementary, as shown in Figure 16b.

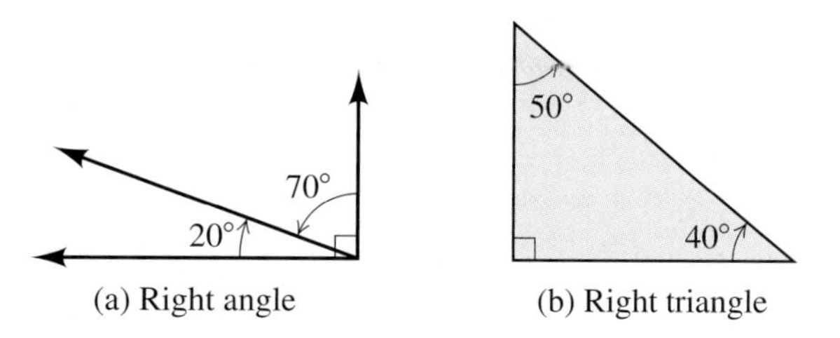

(a) Right angle (b) Right triangle

FIGURE 16 Two sources of complementary angles

▶ **EXAMPLE 9** Solving by substitution: complementary angles In Figure 17, the axes are perpendicular, and the measure of angle D is twice the measure of angle C. What is the measure of each angle? Write a system of equations and solve by substitution.

SOLUTION Because the axes are perpendicular, $C + D = 90°$. From the second phrase, $D = 2C$. We substitute $D = 2C$ into the first equation.

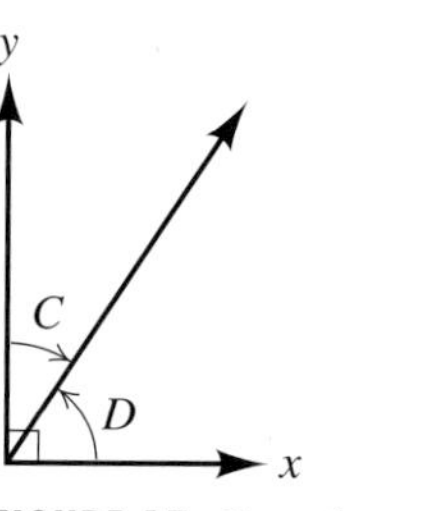

FIGURE 17 Complementary angles

$$C + D = 90 \quad \text{Let } D = 2C.$$
$$C + 2C = 90 \quad \text{Add like terms.}$$
$$3C = 90 \quad \text{Divide by 3.}$$
$$C = 30°$$

We substitute $C = 30$ into $D = 2C$.

$$D = 60°$$

Check:

$$C + D = 90$$

$$30 + 60 \stackrel{?}{=} 90 \quad ✓$$

$$D = 2C$$

$$60 \stackrel{?}{=} 2(30) \quad ✓$$

◀

▶ Summary of Strategies

Here is a guide to some of the strategies we have used so far in this chapter.

STRATEGIES FOR SOLVING SYSTEMS OF EQUATIONS

Use guess and check to help you work through an unfamiliar word problem—reading, understanding, and writing equations. You may even want to guess the correct answer before writing the equations.

Solve by graphing if the two equations are easily written in $y = mx + b$ form.

Solve by substitution if one equation has a variable with a coefficient of 1 and the other equation is somewhat complicated or if the equations are in $y = mx + b$ form.

ANSWER BOX

Warm-up: 1. $x = 4y$ **2.** $x = 152 - y$ **3.** $x = 5 - 2y$ **4.** $y = \frac{5}{2} - \frac{1}{2}x$ **Think about it 1:** We divide $2y = 5 - x$ by 2. The result, $y = \frac{5}{2} - \frac{1}{2}x$, contains fractions, and some students do not like to work with fractions in equation solutions. **Think about it 2:** We want the income levels, y, to be equal. **Think about it 3:** Upon substituting, we obtain $2(\frac{1}{2}x - 2) + 4 = x$. Applying the distributive property, we have $x - 4 + 4 = x$. Simplifying and subtracting x on both sides gives $0 = 0$. This always-true statement implies that the system is true for all x.

▶ 6.3 Exercises

Exercises 1 to 22 provide practice with skills needed to solve equations by substitution. Solve each of the equations in Exercises 1 to 10 for the indicated variable.

1. $L = 2W$ for W **2.** $L = 3W$ for W

3. $a + b = c$ for b **4.** $x + 3 = y$ for x

5. $C = 2\pi r$ for r **6.** $D = 2r$ for r

7. $x - y = 5$ for y **8.** $a - b = c$ for b

9. $C = \pi d$ for d **10.** $P = 4x$ for x

In Exercises 11 to 14, solve for x.

11. $2x + 3(2) = 12$ **12.** $9x - 2(6) = -3$

13. $-5x - 6(-3) = 3$ **14.** $7 = -2x - 13(-1)$

In Exercises 15 to 22, name the variable with a coefficient of 1 or −1. Solve for that variable in terms of the other variable.

15. $3x + y = 4$ **16.** $x + 3y = 26$

17. $x - 4y = 5$ **18.** $2y - x = 7$

19. $5y - x = 9$ **20.** $x - 3y = 3$

21. $3x - y = -2$ **22.** $2x - y = 13$

Use substitution to solve the systems of equations in Exercises 23 to 52.

23. $y = x - 8$
$3x + y = 4$

24. $x = 26 - 3y$
$x - 4y = 5$

25. $5x + 5 = y$
$y - 3x = 9$

26. $4x - 2y = 20$
$x = 2 - y$

Blue numbers are core exercises.

27. $4x + 5y = 11$
$x = 9 + 5y$

28. $-x + 2y = 7$
$x = 3 + 3y$

29. $2x - y = 1$
$2y = 3x + 3$

30. $3y + x = -1$
$2x + 6 = -5y$

31. $2x + 3y = 0$
$3x + y = 7$

32. $y = 3x - 2$
$y = -2x + 13$

33. $y = \frac{4}{3}x$
$y = -\frac{8}{3}x + 8$

34. $y = -\frac{8}{3}x + 8$
$y = -\frac{2}{3}x + 4$

35. $y = 3x + 4$
$3x - y = 8$

36. $y = 4x - 3$
$4x - y = 3$

37. $y = 2x - 3$
$y - 2x = 5$

38. $y = -3x + 4$
$y + 3x = 4$

39. $x + y = 7$
$x = 7 - y$

40. $x - y = 5$
$x = y + 3$

41. $y = 3x + 2$
$y = -x - 2$

42. $y = 3x - 2$
$y = 3x + 2$

43. $y = -x - 2$
$y = 3x - 2$

44. $y = 3x + 2$
$y - 3x - 2 = 0$

45. $2x + 3y = 6$
$2y - x = -10$

46. $2x + y = 4$
$2y = 6 - 4x$

47. $y = \frac{1}{2}x - 2$
$2y + 4 = x$

48. $y = 0.10x + 40$
$y = 0.20x + 20$

49. $x + y = 50$
$7.50x + 9.75y = 435.75$

50. $y = 55 + 6(x - 10)$
$y = 85 + 4.75(x - 10)$

51. $y = 45 + 5(x - 10)$
$y = 30 + 3x$

52. $x + y = 60$
$0.10x + 0.25y = 9.75$

For Exercises 53 to 64, define variables, build equations, and then solve the resulting system.

53. Stephan cuts a 20-yard ribbon. One piece is 3 yards longer than the other. What is the length of each piece?

54. Delores cuts a 16-inch salami. One piece is 4 inches longer than the other. What is the length of each piece?

55. The perimeter of the front of a 15-ounce Cheerios® box is 40 inches. The height is 4 inches more than the width. What are the width and height of the front of the box?

56. The perimeter of the front of a 3-ounce Jello® box is 32 centimeters. The height is 2 centimeters less than the width. What are the width and height of the front of the box?

57. The perimeter of the front of a videotape box is 58 centimeters. The height of the front is 1 less than twice the width. What are the height and width of the box front?

58. The perimeter of the front of a 7-ounce Jiffy Muffin® Mix box is 44 centimeters. The height of the front is 2 less than twice the width. What are the height and width of the box front?

In Exercises 59 to 68, use a quantity-rate table as needed to set up a system of two equations. Then solve the system.

59. Yoko has \$13.80 in quarters and nickels. She has 72 coins altogether. How many of each does she have?

60. Bart has 120 coins in nickels and quarters. He has \$14.80 altogether. How many of each coin does he have?

61. Chen Chen has 84 coins. She has \$13.35 altogether. If she has only dimes and quarters, how many of each does she have?

62. Janice has \$30.30. She has only dimes and quarters. She has 150 coins altogether. How many of each does she have?

63. An organic potato supply weighing 6300 pounds is to be placed in bags. Stores need the bags in a ratio of 1 to 4, 5-pound bags to 10-pound bags, respectively. How many of each bag are needed?

64. A shipment of 74,000 pounds of apples is to be packed in 5-pound bags or 30-pound boxes. The sales ratio of bags to boxes is 1 to 6. How many of each packaging unit are needed?

65. Janelle has two kinds of chocolates with which to fill 1-pound boxes. The chocolate truffles sell for \$36 per pound, and the chocolate creams sell for \$20 per pound. She wants to make 60 of the 1-pound boxes to sell for \$24 per pound. How many pounds of each chocolate should she use?

66. The Healthy Options Company is mixing 8000 pounds of cat food from two ingredients. The first ingredient has 7% protein, and the second has 12% protein. The mixture needs to have 10% protein. How many pounds of each ingredient are needed?

67. Peanuts are 27% protein (by weight), and cashews are 16% protein (by weight). How many grams of each need to be blended to make 270 grams of a mixture containing 66 grams of protein?

Blue numbers are core exercises.

68. Shrimp is 24.7% protein (by weight), and cooked brown rice is 2.6% protein (by weight). How many grams of shrimp and brown rice need to be blended to make 1145 grams of a mixture containing 67 grams of protein?

69. Write one equation from the figure and the other from the sentence. Solve by substitution.

a. The measure of angle A is 24° more than that of angle B.

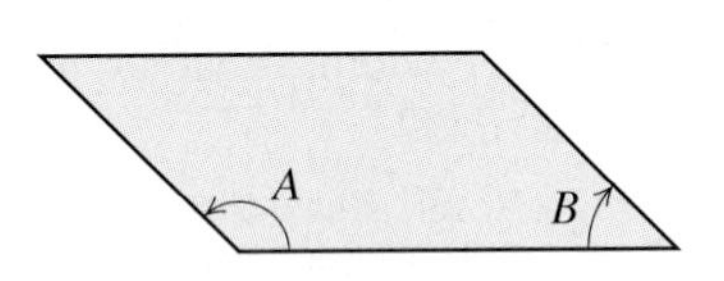

Parallelogram

b. The measure of angle D is 26° more than that of angle C.

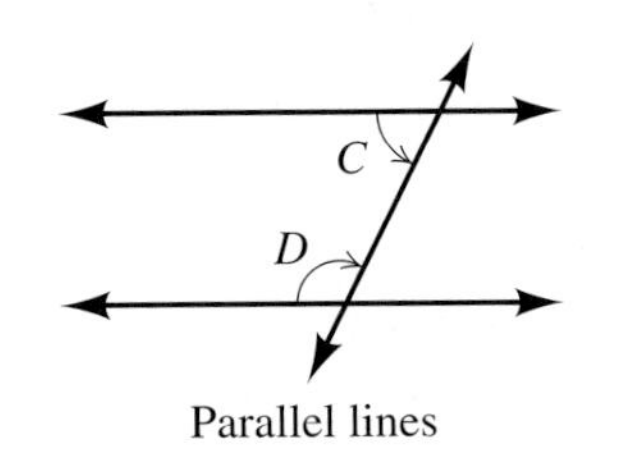

Parallel lines

c. The measure of angle F is 2° more than that of angle E.

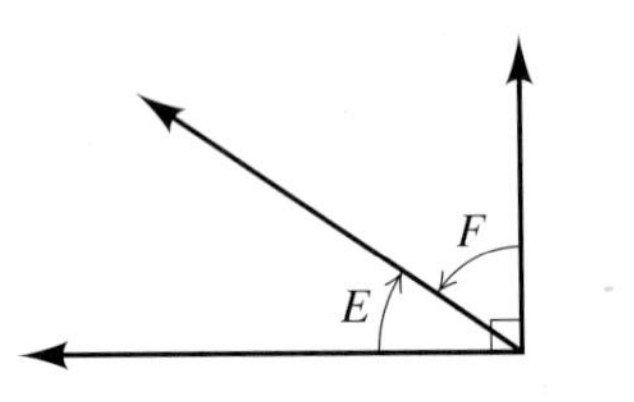

d. The measure of angle H is 3° more than twice that of angle G.

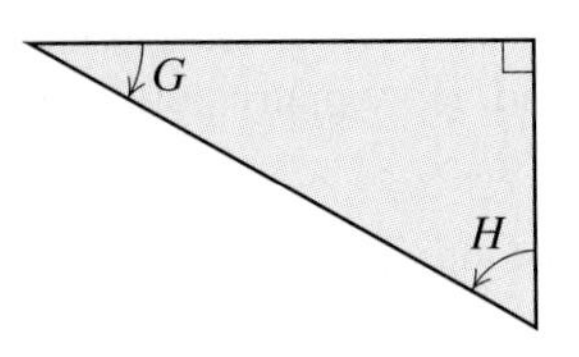

e. The measure of angle J is 45° less than twice that of angle I.

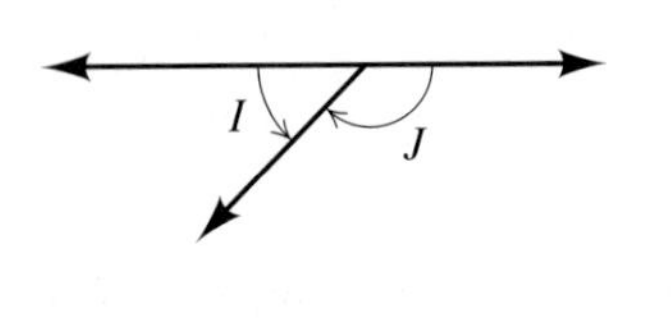

70. Write one equation from the figure and the other from the sentence. Solve by substitution.

a. The measure of angle B is 21° less than twice that of angle A.

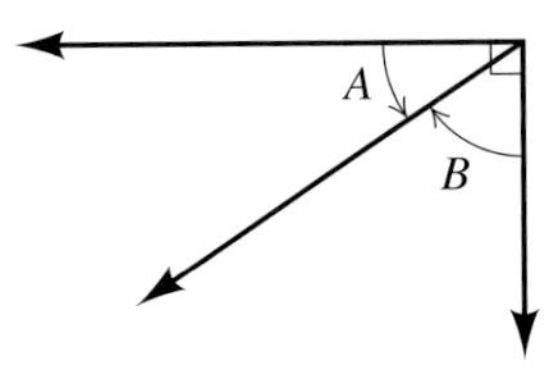

b. The measure of angle D is 40° less than three times that of angle C.

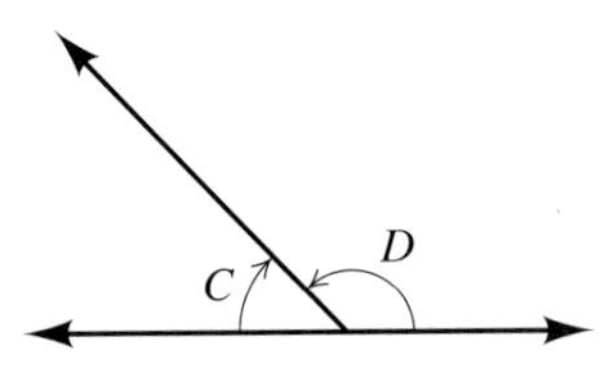

c. The measure of angle E is 18° less than three times that of angle F.

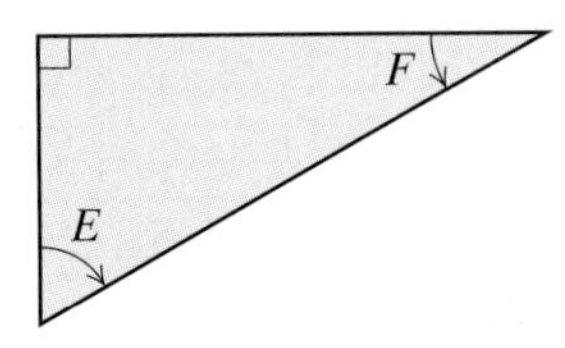

d. The measure of angle G is 3° less than twice that of angle H.

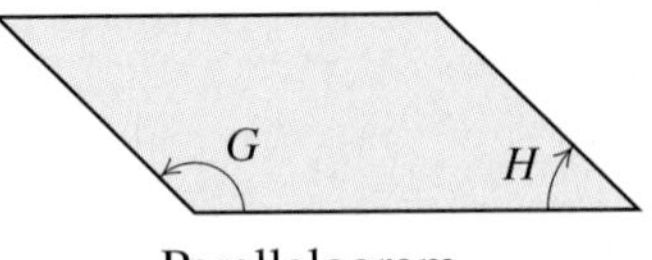

Parallelogram

e. The measure of angle J is 21° more than three times that of angle I.

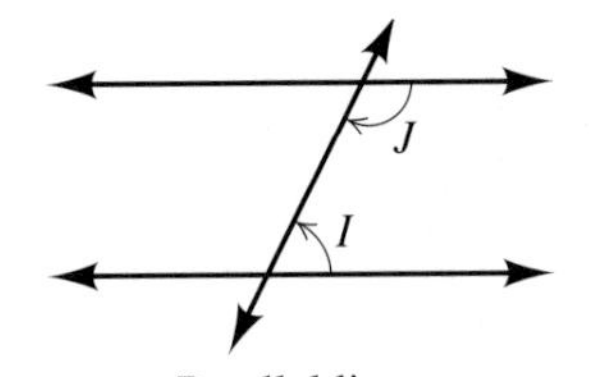

Parallel lines

▶ Writing

71. Explain how to solve a system of two equations by substitution.

72. Explain how to choose a variable to use for the substitution.

73. Explain how, after looking at a system of equations, you would choose substitution, guess and check, or graphing as the solution process.

Blue numbers are core exercises.

74. The Metro subway fast pass permits adding value at any time. The value remaining on the fast pass is a function of the number of trips taken. Your regular trip costs \$4. Your office partner has a regular trip of \$2. Suppose your fast pass starts with \$40 and your office partner's starts with \$30.

a. Write an equation for the value of each fast pass as a function of the number of trips taken.

b. Describe a problem setting with one solution.

c. Describe a problem setting with no solution.

d. Describe a problem setting with an infinite number of solutions.

▶ Projects

75. Matching Solutions with Results *Part I:* Match each numerical, geometric, and algebraic result with the appropriate one of the following phrases:

(a) The system has no solution.

(b) The system has one solution, (x, y).

(c) The system has an infinite number of solutions.

Numerical results:

$0 = 0$

$0 = 4$

$x = 3, y = 5$

Geometric results:

The lines intersect at a point.

The lines are coincident.

The lines are parallel.

Algebraic results:

The algebra always yields a false statement.

The algebra always yields a true statement.

The algebra yields a unique x and unique y value.

Part II: Write a system of two equations in two unknowns that fits each of the descriptions in parts a to c. Make a graph for each of the three systems to prove that your equations fit the requirements.

76. Formulas, Functions, and Substitution In the formula $A = \pi r^2$, area is a function of radius. In $C = 2\pi r$, circumference is a function of radius. If we solve $C = 2\pi r$ for r and substitute the result into $A = \pi r^2$, we get $A = \pi\left(\frac{C}{2\pi}\right)^2$ or $A = \frac{C^2}{4\pi}$, which gives area as a function of circumference.

a. Write diameter as a function of circumference, using $d = 2r$ and $C = 2\pi r$.

b. Write the area of a square as a function of perimeter, using $A = x^2$ and $P = 4x$.

c. Write area as a function of height, using $A = \frac{bh}{2}$ and $b = h$.

d. Write area as a function of diameter, using $A = \pi r^2$ and $d = 2r$.

▶ 6 Mid-Chapter Test

In Exercises 1 to 3, solve for the indicated variable.

1. $2x - y = 5000$ for y

2. $V = \frac{\pi r^2 h}{3}$ for h

3. $\frac{3}{2}x + \frac{2}{3}y = \frac{1}{4}$ for y

Solve with substitution the systems in Exercises 4 and 5 for the x and y that make both equations true.

4. $x + y = 5000$

$3x - 2y = -2500$

5. $x + y = 5000$

$3x - 2y = 2500$

6. a. The point (3, 4) is the solution to which two equations in the figure at right?

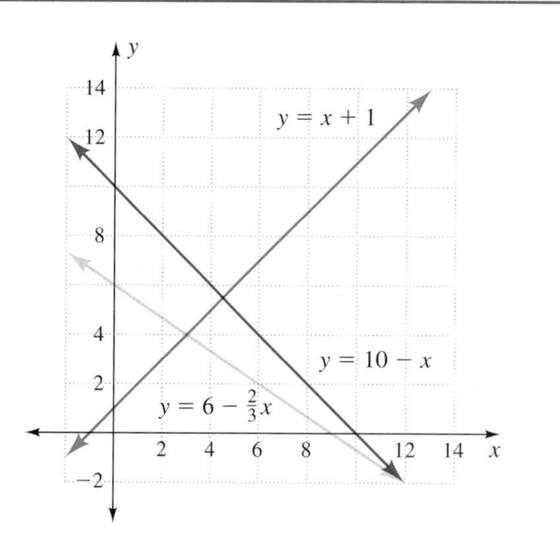

b. Estimate the point of intersection of $y = x + 1$ and $y = 10 - x$.

c. Use substitution to find the coordinates of the point of intersection of $y = 10 - x$ and $y = x + 1$.

d. Use substitution to find the coordinates of the point of intersection of $y = 10 - x$ and $y = 6 - \frac{2}{3}x$.

7. a. Graph the system

$$2x + y = 6$$
$$x - 2 = y$$

b. Estimate the point of intersection from the graph.

c. Solve the system by substitution.

8. Solve by substitution:

$$2x + y = 4$$
$$y = 4 - 2x$$

Comment on the result. What does the result imply about the graphs of the lines?

Use guess and check to start a solution to Exercises 9 and 10. Define your variables, and write a system of equations. Solve the system if your guess and check did not reach a solution.

9. A green turtle lays 12 times more eggs than an ostrich. If the sum of the eggs laid is 195, how many did each lay?

10. A produce distributor receives 10,000 pounds of potatoes and needs to bag them for retail sale. Local grocery stores sell 5-pound bags and 10-pound bags in a 3 to 1 ratio. How many bags of each weight should be prepared?

11. To cut costs, the new manager at Healthy Options pet food orders 6000 pounds of cat food to have 8% protein. One of two ingredients contains 7% protein. The other has 12% protein. How many pounds of each are needed?

12. Why does the intersection of two graphs solve the system of equations describing the graphs?

13. Solving a system of two equations gives $4 = 5$. What can you conclude about the graphs of the equations?

14. In a comparison of total cost as a function of number of gallons purchased, will there be one, no, or an infinite number of solutions in each of the following cases?

a. Two gas stations have the same price per gallon.

b. One station charges a fee for use of the debit card but has the same price per gallon as another station that does not charge for using the debit card.

c. The stations have different prices per gallon. The station with a cheaper price per gallon charges a fee for use of the debit card.

d. The stations have different prices per gallon.

e. Two customers buy the same type of fuel at the same pump. One pays outside, and the other buys something in the minimart.

15. Explain how it is possible for nonparallel lines to have no point of intersection in an application setting.

▶ 6.4 Solving Systems of Equations by Elimination

Objectives

- Solve a system of equations by eliminating one variable with addition and substituting to find the second variable.
- Change a system of equations to make terms with opposite coefficients.
- Solve systems based on geometry facts.
- Solve systems related to quantity and rate.
- Solve systems involving wind and current.

WARM-UP

The following exercises review the distributive property.

1. $3(2x + 4y) - 2(3x + 2y)$
2. $3(5x + y) - 5(3x - 2y)$
3. $1(2x + 4y) + 4(3x - y)$
4. $4(x - 3y) + 3(2x + 4y)$
5. $c(ax + by) - a(cx + dy)$
6. $d(ax + by) + b(cx - dy)$

THIS SECTION INTRODUCES a fifth method of solving systems of equations: elimination. We will apply elimination to equations with like and unlike numerical coefficients. Applications will include complementary and supplementary angles in geometry, quantity and rate problems, and travel problems.

Principles Underlying Solving Systems by Elimination

Thus far, we have solved equations by graphing, guess and check, tables, and substitution. We now consider a fifth method: elimination.

In the following explorations, we look at graphical reasons why elimination works.

EXAMPLE 1 Exploring equations and point of intersection*

a. Which line in Figure 18 is the graph of each equation?

$$y = 2x + 3$$
$$y = -2x - 1$$
$$y = 1$$
$$x = -1$$

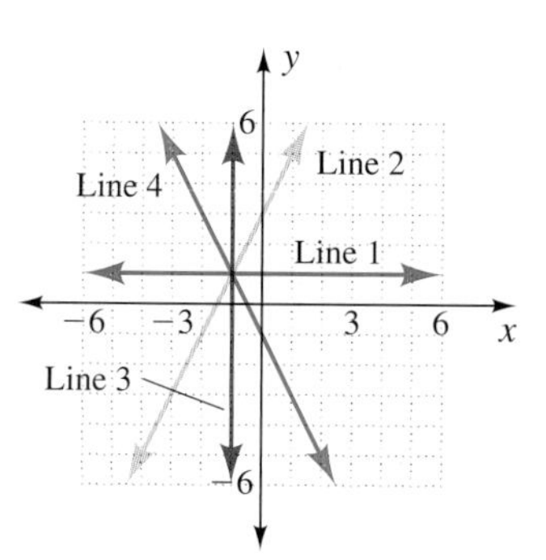

FIGURE 18 Equations with common point of intersection

b. What is the point of intersection of the four graphs?

c. Add the terms on the left sides of the following equations from part a, and then add the terms on the right sides:

$$y = \quad 2x + 3$$
$$y = -2x - 1$$

Solve the resulting equation.

d. Subtract the terms on the left sides of the following equations from part a, and then subtract the terms on the right sides:

$$y = \quad 2x + 3$$
$$y = -2x - 1$$

Solve the resulting equation.

e. What do you observe in parts c and d?

SOLUTION See the Answer Box. ◀

In Example 1, the equations in parts c and d were arranged so that like terms were lined up vertically. When we added the equations in part c, the terms $2x$ and $-2x$ added to zero and we were able to solve for the value of y. When we subtracted the equations in part d, the terms y and y subtracted to zero and we were able to solve for the value of x. The x and y make an ordered pair naming the point of intersection.

In Example 2, we will both add and subtract the equations.

EXAMPLE 2 Exploring elimination

a. Which line in Figure 19 is the graph of each equation?

$$x + y = 7$$
$$x - y = -1$$
$$x = 3$$
$$y = 4$$

b. What is the point of intersection of the two graphs?

c. Substitute the values of x and y from the point of intersection into each equation.

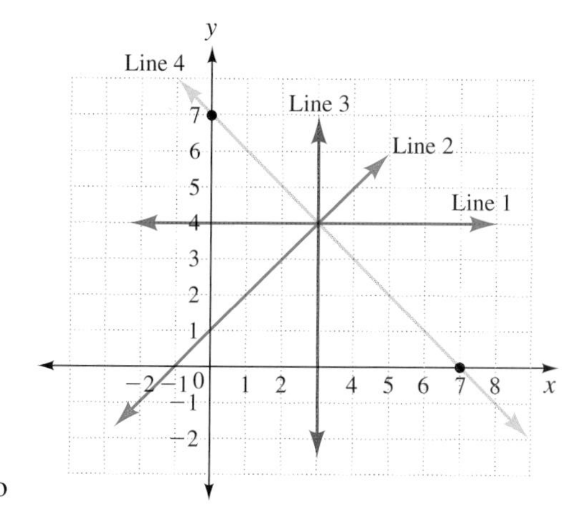

FIGURE 19 Equations with common point of intersection

*The structure of this and the next exploration was suggested by my colleague Jill McKenney, Lane Community College, Eugene, Oregon.

d. Add the left sides and right sides of the following equations:

$$x + y = 7$$
$$x - y = -1$$

What equation results?

e. Subtract the left sides and right sides of the following equations:

$$x + y = 7$$
$$x - y = -1$$

What equation results?

f. Compare the intersections of the graphs for parts d and e with those for part a. What do you observe?

SOLUTION See the Answer Box. ◀

The key idea in Examples 1 and 2 is that *adding or subtracting equations in a system preserves the solution to the system of equations.*

ELIMINATION METHOD

> **Elimination** is a process in which one variable is removed from a system of equations by adding (or subtracting) the respective sides of two equations.

It may be helpful to call the process *elimination of one variable* or *elimination of like terms.*

The addition property of equations is the principle underlying the elimination method.

ADDITION PROPERTY OF EQUATIONS

> Adding the same number to both sides of an equation produces an equivalent equation. In algebraic notation,
>
> If $a = b$, then $a + c = b + c$.

The explorations suggest this extension of the addition property of equations:

EXTENSION TO ADDITION PROPERTY OF EQUATIONS

> Adding equal values to both sides of an equation produces an equivalent equation. In algebraic notation,
>
> If $a = b$ and $c = d$, then $a + c = b + d$.

The extended addition property says that if we add equal values to both sides of an equation, the solution set to the equation (or system of equations) remains the same.

▶ Elimination with Like Coefficients

▶ **EXAMPLE 3** Solving by elimination (and substitution) Use the addition property of equations to eliminate a variable from the system of equations

$$2x - 3y = 16 \quad (1)$$
$$x + 3y = -1 \quad (2)$$

SOLUTION We write one equation below the other, with like terms lined up, and then add.

Student Note:
The equations are numbered for easier reference.

$$\begin{aligned} 2x - 3y &= 16 && (1) \\ x + 3y &= -1 && \text{Add equation (2) to eliminate } y. \\ \hline 3x \quad &= 15 && \text{Divide both sides by 3.} \\ x &= 5 \end{aligned}$$

We then substitute $x = 5$ into the first equation and solve for y:

$$\begin{aligned} 2x - 3y &= 16 && \text{Let } x = 5 \text{ in (1).} \\ 2(5) - 3y &= 16 \\ 10 - 3y &= 16 && \text{Subtract 10 on both sides.} \\ -3y &= 6 && \text{Divide both sides by } -3. \\ y &= -2 \end{aligned}$$

The solution is $x = 5$, $y = -2$ or, as a point of intersection of the graphs, $(5, -2)$.

Check:

$$\begin{aligned} 2x - 3y &= 16 \\ 2(5) - 3(-2) &\stackrel{?}{=} 16 \ \checkmark \\ x + 3y &= -1 \\ 5 + 3(-2) &\stackrel{?}{=} -1 \ \checkmark \end{aligned}$$

◀

▶ We now look more closely at features in the equations that make the elimination process possible.

In Section 1.3, we defined the **numerical coefficient** as *the sign and number multiplying the variable(s)*. The reason addition eliminated $3y$ and $-3y$ in Example 3 was that the numerical coefficients were opposites and so the terms added to zero. *To identify like terms and terms with opposite numerical coefficients, we always line up the like terms in the equations*. Changing to $ax + by = c$ may be convenient.

In Example 4, we solve a system of equations by elimination (and substitution). Example 4 returns to the subject of Cal and Joe's job sharing (Section 6.2, page 333).

▶ **EXAMPLE 4** Solving by elimination (and substitution) Cal and Joe's job-sharing equations are $x + y = 243$ and $y = x - 17$. Arrange the equations to line up like terms, and solve by elimination.

SOLUTION When the second equation is rearranged, the terms containing x have opposite signs, and we add the equations.

$$\begin{aligned} x + \ y &= 243 \\ -x + \ y &= -17 && \text{Add the equations to eliminate } x. \\ \hline 2y &= 226 && \text{Divide by 2 on both sides.} \\ y &= 113 \end{aligned}$$

We substitute $y = 113$ into the first equation and solve for x:

$$\begin{aligned} x + y &= 243 \\ x + 113 &= 243 && \text{Subtract 113 on both sides.} \\ x &= 130 \end{aligned}$$

The solution is $x = 130$, $y = 113$.

Check:

$$x + y = 243$$
$$130 + 113 \stackrel{?}{=} 243 \quad \checkmark$$
$$y = x - 17$$
$$113 \stackrel{?}{=} 130 - 17 \quad \checkmark$$

◀

THINK ABOUT IT 1: Subtract the equations in Example 4: $x + y = 243$ and $-x + y = -17$. What do you observe?

▶ Elimination with Unlike Coefficients

In the examples above, terms had like or opposite coefficients. The system

$$5x - y = 4$$
$$4x + 2y = -1$$

does not have like coefficients. In order to use elimination, we must change the equations to obtain like coefficients.

EQUATIONS IN WHICH ONE VARIABLE HAS A 1 OR −1 COEFFICIENT When a variable in one equation has a coefficient of 1, we may multiply that equation by the opposite of the coefficient of the like term in the other equation and then add the equations. When a variable in one equation has a coefficient of −1, we may multiply that equation by the coefficient of the like term in the other equation and then add the equations.

▶ EXAMPLE 5 Solving a system with a 1 or −1 coefficient Use multiplication on one equation in this system to obtain terms with opposite coefficients:

$$5x - y = 4$$
$$4x + 2y = -1$$

SOLUTION The system can be written with a -1 coefficient on y in the first equation:

$5x - 1y = 4$	Multiply by 2.	$10x - 2y = 8$	
$4x + 2y = -1$		$4x + 2y = -1$	Add the equations.
		$14x = 7$	Divide by 14.
		$x = \frac{1}{2}$ or $x = 0.5$	

We substitute $x = 0.5$ into the first equation and solve for y:

$5x - y = 4$	
$5(0.5) - y = 4$	Let $x = 0.5$.
$2.5 - y = 4$	Subtract 2.5 on both sides.
$-y = 1.5$	Multiply both sides by -1.
$y = -1.5$	

The solution to the system is $x = 0.5$, $y = -1.5$.

Check:

$$5x - y = 4$$
$$5(0.5) - (-1.5) \stackrel{?}{=} 4 \quad \checkmark$$
$$4x + 2y = -1$$
$$4(0.5) + 2(-1.5) \stackrel{?}{=} -1 \quad \checkmark$$

◀

EQUATIONS IN WHICH ALL VARIABLES HAVE UNLIKE COEFFICIENTS In most systems of equations, the variables have unlike coefficients. In this case, we use a process similar to that of finding a common denominator in order to add or subtract fractions.

CHANGING UNLIKE COEFFICIENTS TO LIKE COEFFICIENTS

> When all variables have unlike coefficients, multiply both equations by numbers that make opposite coefficients on one variable.

The variables in Example 6 have unlike coefficients.

▶ **EXAMPLE 6** Solving a system with unlike coefficients Use multiplication on both equations to obtain opposite coefficients on like terms:

$$4x - 2y = 11$$
$$5x + 3y = 0$$

SOLUTION The y terms have opposite signs. We multiply by numbers that make opposite coefficients.

$$4x - 2y = 11 \quad \text{Multiply by 3.} \quad 12x - 6y = 33$$
$$5x + 3y = 0 \quad \text{Multiply by 2.} \quad \underline{10x + 6y = 0} \quad \text{Add the equations.}$$
$$22x = 33 \quad \text{Divide by 22.}$$
$$x = 1.5$$

We substitute $x = 1.5$ into the first equation and solve for y:

$$4x - 2y = 11$$
$$4(1.5) - 2y = 11 \quad \text{Simplify.}$$
$$6 - 2y = 11 \quad \text{Subtract 6 on both sides.}$$
$$-2y = 5 \quad \text{Divide by } -2 \text{ on both sides.}$$
$$y = -2.5$$

The solution is $x = 1.5$, $y = -2.5$.

Check:

$$4x - 2y = 11$$
$$4(1.5) - 2(-2.5) \stackrel{?}{=} 11 \quad \checkmark$$
$$5x + 3y = 0$$
$$5(1.5) + 3(-2.5) \stackrel{?}{=} 0 \quad \checkmark$$

◀

THINK ABOUT IT 2: By what numbers could we multiply to give the x terms in Example 6 opposite coefficients? Do the multiplications and solve for y.

STEPS IN SOLVING EQUATIONS BY ELIMINATION (AND SUBSTITUTION)

1. Arrange the equations so that like terms line up. (You might want to write each equation in standard form, $ax + by = c$.)
2. As needed, multiply one or both equations by numbers that make opposite coefficients on one variable.
3. Add the sides of the equations to eliminate one variable, and solve for the first variable.
4. Use substitution to find the second variable.
5. Check the solution in both equations.

Applications

Elimination is the foundation for techniques programmed into calculators and computers for solving large systems in applications. In business, economics, mathematics, and sociology, techniques using matrices, linear programming, and the simplex method build upon the elimination method.

SPECIAL ANGLES IN GEOMETRY, CONTINUED In Example 7, we return to supplementary and complementary angles (Section 6.3).

EXAMPLE 7 Solving by elimination (and substitution): special angles The measures of two angles add to 200°. The supplement of the first less the complement of the second is 20°. Write a system of equations, and solve the measures of the two angles.

SOLUTION Figure 20 shows the supplements and complements. Let A be the measure of the first angle in degrees and B be the measure of the second angle. The equations are numbered for easier reference.

$$A + B = 200 \quad (1)$$
$$(180 - A) - (90 - B) = 20 \quad (2)$$

Simplifying the second equation and lining up like variables, we have

$$A + B = 200 \quad (1)$$
$$-A + B = -70 \quad (2)$$

Adding the equations gives

$$2B = 130$$
$$B = 65$$

Substituting $B = 65$ into equation (1) then gives

$$A + 65 = 200$$
$$A = 135$$

Check:

$$135 + 65 \stackrel{?}{=} 200 \checkmark$$
$$(180 - 135) - (90 - 65) \stackrel{?}{=} 20 \checkmark$$

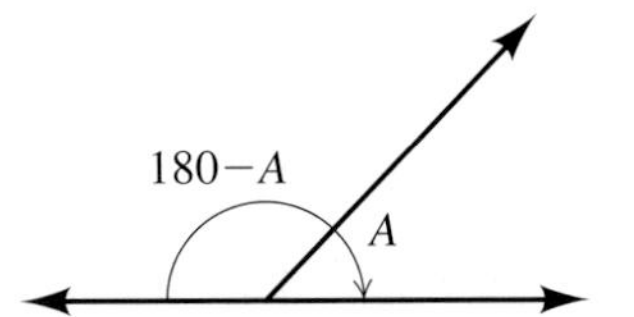

Supplementary angles

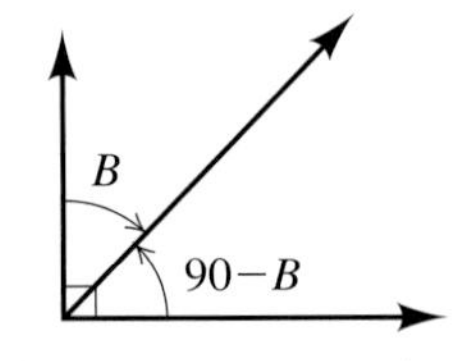

Complementary angles

FIGURE 20 Special angle pairs

TICKET PRICES In Example 8, we solve the ticket price equations from Section 6.2 (Example 6, page 337).

▶ **EXAMPLE 8** Solving by elimination (and substitution): ticket prices Solve the system of equations

$$2a + 4s = 31$$
$$3a + 2s = 28.50$$

SOLUTION If we multiply the second equation by -2, the terms containing s will have opposite coefficients.

$$2a + 4s = 31 \qquad\qquad 2a + 4s = 31$$
$$3a + 2s = 28.50 \quad \text{Multiply by } -2. \quad \underline{-6a + -4s = -57}$$
$$-4a = -26$$
$$a = 6.50$$

Substituting $a = 6.50$ into the first equation gives

$$2(6.50) + 4s = 31$$
$$s = 4.50$$

The adult tickets cost \$6.50, and the student tickets cost \$4.50.

Check:

$$2(6.50) + 4(4.50) \stackrel{?}{=} 31 \quad ✓$$
$$3(6.50) + 2(4.50) \stackrel{?}{=} 28.50 \quad ✓$$

◀

In Example 8, a and s were the rates of the tickets, and their coefficients were the quantities of the two types of tickets. The two equations in that example can be summarized with the quantity-rate equation:

$$\text{Quantity} \cdot \text{rate} + \text{quantity} \cdot \text{rate} = \text{total } Q \cdot R$$

TRAVEL PROBLEMS The formula $d = rt$ relates distance, rate, and time. A $d = rt$ problem is a special type of quantity and rate problem. The time of travel is the quantity, and the speed of travel is the rate. The distance traveled is the product across the table: time · rate.

▶ **EXAMPLE 9** Solving by elimination (and substitution): time, rate, and distance Build a quantity-rate table for the following setting:

Maxine drives 60 miles per hour (mph) on the toll road and 30 mph in town. If her total travel time for a 210-mile journey was 4 hours, how long did she drive at each speed?

Solve the resulting system of equations by elimination.

SOLUTION We start by setting up the quantity-rate table in Table 13.

TABLE 13 Time · Rate = Distance

	Quantity: Time (hr)	Rate (mph)	$Q \cdot R$: Distance (mi)
On the toll road	x	60	$60x$
In town	y	30	$30y$
	4		210

The system of equations is

$x + y = 4$ From the quantity column

$60x + 30y = 210$ From the $Q \cdot R$ column

To get opposite coefficients on y, we multiply the first equation by -30. We then add the equations.

$$\begin{aligned} -30x - 30y &= -120 \\ 60x + 30y &= 210 \\ \hline 30x \quad &= 90 \\ x &= 3 \end{aligned}$$

We substitute 3 for x in the first equation and solve for y:

$$3 + y = 4$$
$$y = 1$$

Maxine drove 3 hours on the toll road and 1 hour in town.

Check:

$$3 + 1 \stackrel{?}{=} 4 \quad ✓$$

$$60(3) + 30(1) \stackrel{?}{=} 210 \quad ✓$$

◀

THINK ABOUT IT 3: How many miles did Maxine drive on the toll road and in town? What was her average rate for the trip?

In Example 10, we return to the airspeed and wind problem in Section 6.2 (Example 8, page 339).

▶ **EXAMPLE 10** Solving with elimination (and substitution): rates of travel Solve the following system of equations for airspeed r, in mph, and wind speed w, in mph:

$$3.5(r + w) = 1400$$
$$4.0(r - w) = 1400$$

SOLUTION Rather than use the distributive property, we divide both sides of the equations by the leading number and obtain

$$r + w = 400$$
$$r - w = 350$$

Adding these equations gives

$$2r = 750$$
$$r = 375$$

We then substitute the rate into the first equation:

$$375 + w = 400$$
$$w = 25$$

Check:

$$3.5(375 + 25) = 1400$$
$$4.0(375 - 25) = 1400$$

The airspeed is $r = 375$ mph, and the wind speed is $w = 25$ mph. ◀

ANSWER BOX

Warm-up: **1.** $8y$ **2.** $13y$ **3.** $14x$ **4.** $10x$ **5.** $bcy - ady$ **6.** $adx + bcx$
Example 1: **a.** line 2, line 4, line 1, line 3, respectively **b.** $(-1, 1)$
c. $2y = 2$, so $y = 1$. **d.** $0 = 4x + 4$, so $x = -1$. **e.** When we added (or subtracted) the equations, we eliminated one variable and were able to solve for the other variable. The values of the variables gave the point of intersection. The original two equations, as well as $x = -1$ and $y = 1$, all intersect at $(-1, 1)$. **Example 2:** **a.** line 4, line 2, line 3, line 1, respectively
b. $(3, 4)$ **c.** Substituting $x = 3$ and $y = 4$ makes each equation true.
d. $x = 3$ **e.** $y = 4$ **f.** The graphs of the equations in parts d and e pass through the same point of intersection as the original two lines.
Think about it 1: When we subtract the equations, we get the x value directly: $2x = 260$, so $x = 130$. This suggests an optional solution process for some elimination problems. **Think about it 2:** We could multiply the first equation by 5 and the second equation by -4. The results would be the same. **Think about it 3:** She drove $3 \cdot 60 = 180$ mi on the toll road and $1 \cdot 30 = 30$ mi in town. Her average rate was $\frac{210}{4} = 52.5$ mph.

▶ 6.4 Exercises

1. a. Solve by graphing:

$$x + 2y = 1$$
$$x - y = 4$$

b. Add the equations and graph the resulting equation.

c. Subtract the equations and graph the resulting equation.

d. Comment on the results in parts b and c.

2. a. Solve by graphing:

$$y = 3x + 1$$
$$y = x + 3$$

b. Add the equations and graph the resulting equation.

c. Subtract the equations and graph the resulting equation.

d. Comment on the results in parts b and c.

Solve the systems of equations in Exercises 3 to 28 by the elimination method.

3. $x + y = -2$
$x - y = 8$

4. $x + y = -15$
$x - y = 3$

5. $m + n = 3$
$-m + n = -11$

6. $p - q = 9$
$p + q = -5$

7. $2x + y = -1$
$x + 2y = 4$

8. $3x + y = -6$
$x + 2y = -7$

9. $2a + b = -5$
$a + 3b = 35$

10. $3c + d = 28$
$c + 3d = -12$

11. $2x + 3y = 3$
$3x - 4y = -21$

12. $3m - 2n = 22$
$2m + 3n = -7$

13. $5p - 2q = -6$
$2p + 3q = 9$

14. $4x - 3y = -8$
$3x + 5y = -6$

15. $x + y = 6$
$x + y = 10$

16. $x - y = 5$
$x - y = 7$

17. $x - y = 7$
$2x - 2y = 14$

18. $x + y = 3$
$3x + 3y = 9$

19. $x + y = 5$
$y - x = -13$

20. $x + y = 3$
$y - x = 7$

21. $2x + 3y = 0$
$3x + 2y = 5$

22. $2x - 3y = -2$
$3x + 2y = -16$

23. $7 = 5m + b$
$3 = 3m + b$

24. $7 = -m + b$
$-1 = 3m + b$

25. $0.2x + 0.6y = 2.2$
$0.4x - 0.2y = 1.6$

26. $0.8x + 0.3y = 5.3$
$0.4x - 0.2y = 0.2$

27. $0.5x + 0.2y = 1.8$
$0.2x - 0.3y = -0.8$

28. $0.4x - 0.3y = 3.0$
$0.5x + 0.2y = 2.6$

Use the alternative methods described in Exercises 29 and 30 to solve the system in Example 6. The equations are $4x - 2y = 11$ and $5x + 3y = 0$.

29. Multiply the first equation by 5 and the second equation by -4.

Blue numbers are core exercises.

30. Multiply the first equation by -5 and the second equation by 4.

31. Solve the system from Example 4 by substitution:

$$x + y = 243$$
$$y = x - 17$$

32. Solve the system from Example 9 by substitution:

$$x + y = 4$$
$$60x + 30y = 210$$

In Exercises 33 to 36, define variables for the measures of the angles. Write equations based on the given facts as well as the complementary angles and supplementary angles. Solve the resulting system of equations by elimination.

33. **a.** Two angles are complementary. The measure of one angle is 50° larger than that of the other.

b. Two angles are supplementary. The measure of one angle is 50° larger than that of the other.

34. **a.** Two angles are supplementary. The measure of one angle is 60° less than that of the other.

b. Two angles are complementary. The measure of one angle is 60° less than that of the other.

35. Write one equation from the figure and the other from the sentence. Solve by elimination.

a. The measure of angle A is 22° less than that of angle B.

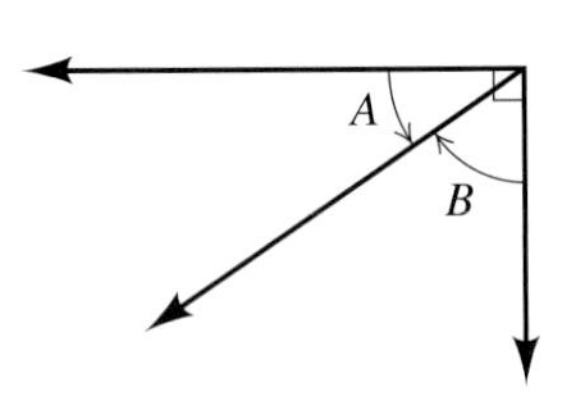

b. The measure of angle D is 5° more than four times that of angle C.

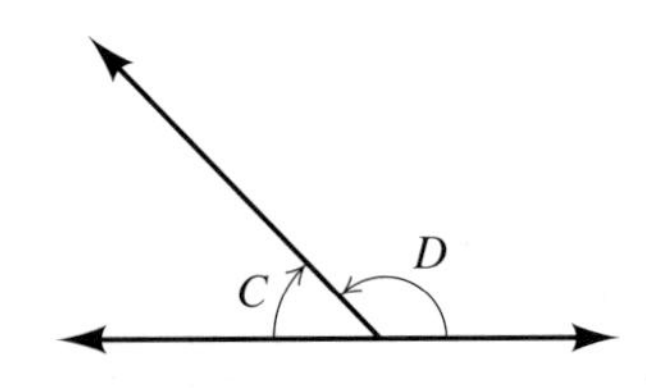

c. The measure of angle E is 10° more than four times that of angle F.

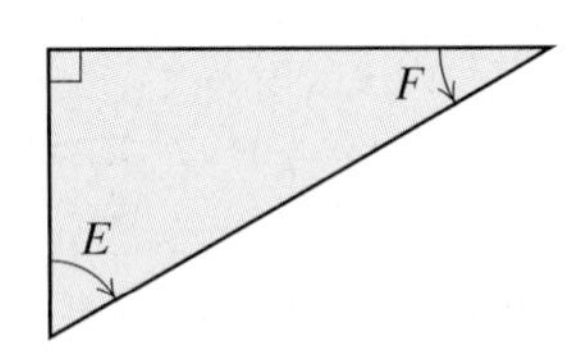

d. The measure of angle G is 15° less than twice that of angle H.

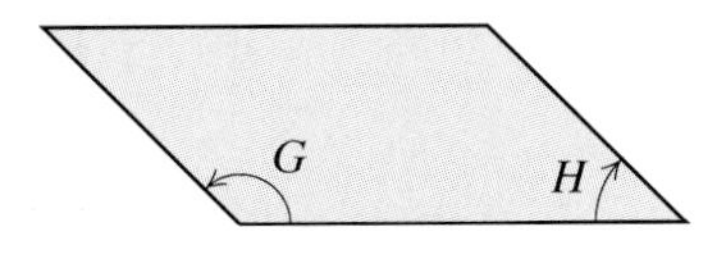

Parallelogram

e. The measure of angle J is 20° more than three times that of angle I.

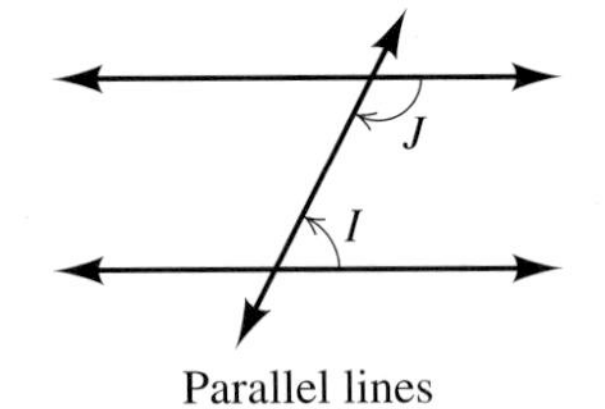

Parallel lines

36. Write one equation from the figure and the other from the sentence. Solve by elimination.

a. The measure of angle A is 30° more than that of angle B.

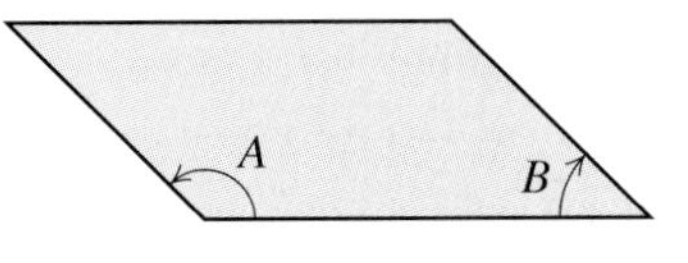

Parallelogram

b. The measure of angle D is 15° more than twice that of angle C.

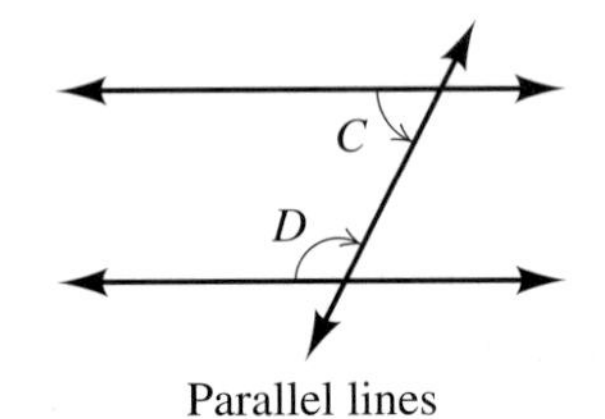

Parallel lines

c. The measure of angle E is 10° less than that of angle F.

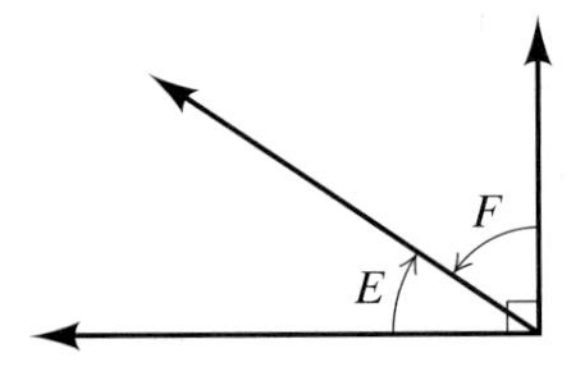

d. The measure of angle H is 5° less than four times that of angle G.

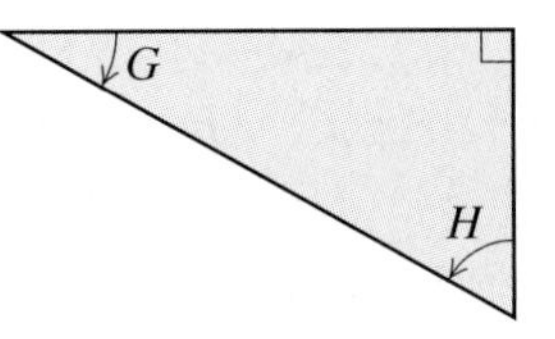

Blue numbers are core exercises.

e. The measure of angle J is 60° less than twice that of angle I.

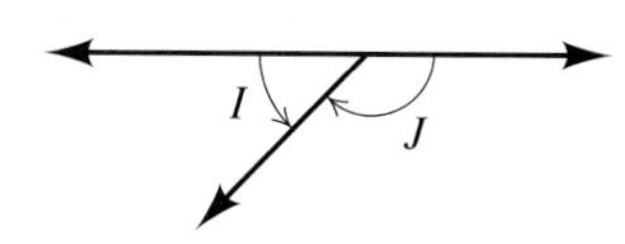

In Exercises 37 to 40, identify two appropriate variables, and use the variables to write equations. Use either the elimination method or the substitution method to solve the equations.

37. The sum of two numbers is 25. Their difference is 8. Find the numbers.

38. The sum of two numbers is 35. Their difference is 10. Find the numbers.

39. The sum of two numbers is 20. Twice the second less twice the first is 21. Find the numbers.

40. The sum of two numbers is 12. Twice the larger less four times the smaller is 2. Find the numbers.

For Exercises 41 to 60, define two variables. Then write equations. Quantity-rate equations may be helpful. Use either the elimination method or the substitution method to solve the equations.

41. A set of 6 adult tickets and 3 student tickets to an afternoon movie costs $58.50. Another set, consisting of 5 adult tickets and 4 student tickets, costs $54. What is the cost of one of each type of ticket?

42. A group of 3 adult tickets and 8 student tickets to an evening movie costs $67.50. For 4 adult tickets and 5 student tickets at the same time, the cost is $64.50. What is the cost of one of each type of ticket?

43. Jordan buys 3 identical shirts and 2 identical ties for $109.95. Gabe buys 4 of the same shirts and a tie for $119.95. What is the cost of one of each?

44. At the bookstore, 2 notebooks and 5 pens cost $14.91. At the same time, 6 notebooks and 3 pens cost $32.85. What is the cost of one of each?

45. Two CDs and three game disks cost $137.95. Four CDs and one game disk cost $75.95. What does each cost?

46. Three DVDs and four cassettes cost $60.36. Two DVDs and six cassettes cost $61.24. What does each cost?

47. A snack of 4 sugar cookies and 2 ginger snaps has 296 calories. Another of 3 sugar cookies and 10 ginger snaps has 329 calories. How many calories are in one of each kind of cookie?

48. A candy selection of 5 large gumdrops and 8 caramels has 511 calories. Another selection of 3 large gumdrops and 10 caramels has 525 calories. How many calories are in a piece of each kind of candy?

49. A fruit plate of 15 large cherries and 22 red grapes has 126 calories. Another assortment with 20 large cherries and 11 red grapes has 113 calories. How many calories are in one of each kind of fruit?

50. A bag of dried fruit, 10 dates and 3 figs, has 415 calories. Another bag of 6 dates and 5 figs has 425 calories. How many calories are in one of each fruit?

51. An appetizer of 8 green olives and 5 ripe olives has 285 calories. Another of 4 green olives and 10 ripe olives has 330 calories. How many calories are in one of each type of olive?

52. A vegetable snack of 6 small carrots and 4 stalks of celery has a total of 132 calories. Another snack of 3 small carrots and 10 stalks of celery has 90 calories. How many calories are in a small carrot? In a stalk of celery?

53. Ned drives a total of 7 hours and travels 405 miles. He drives 40 mph on gravel roads and 65 mph on paved roads. How long does he drive on each surface?

54. Nerine combines walking (at 6 km/hr) and running (at 18 km/hr) in her fitness program. Last week she exercised a total of 11 hours and traveled 126 kilometers. How many hours did she walk and how many hours did she run?

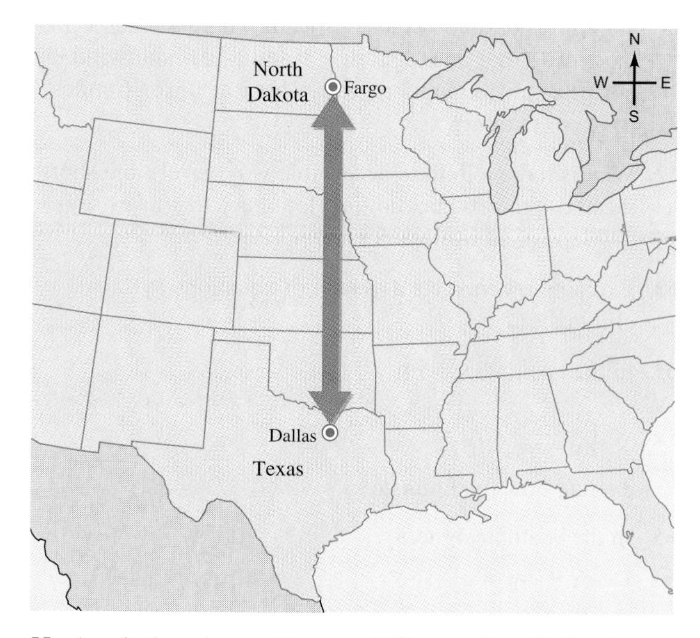

55. An airplane leaves Fargo and flies south to Dallas (see the figure). Another airplane flies the opposite route. The airplanes have the same airspeed, r, but there is a wind, w, from the north. The Fargo-to-Dallas flight takes 4 hours for the 1100-mile trip. The Dallas-to-Fargo flight takes 5 hours to cover the same distance. What is the airspeed of the two airplanes, and what is the wind speed?

56. An airplane makes a round trip between Dallas and Fargo, a one-way distance of 1100 miles. The flight with the wind takes 2 hours, and the flight against the wind takes 4 hours. What is the speed of the wind? What is the airspeed of the airplane?

Blue numbers are core exercises.

57. An airplane flies between San Francisco and Denver. Going east with the wind, the airplane travels 950 miles in 2 hours. Returning west (along a slightly different route) against the wind, the airplane travels 975 miles in 3 hours. What airspeed and what constant wind speed would give these results?

58. A flight from Lincoln, Nebraska, to Dallas takes 2.5 hours with the wind. The return flight takes 3.4 hours against the same wind. The distance between the cities is 612 miles. Find the airspeed of the airplane and the speed of the wind.

59. A fishing boat goes 20 miles upstream, against a current, in 5 hours. The same boat goes 20 miles downstream, with the current, in 2 hours. What is the speed of the boat (in still water)? What is the speed of the current?

60. A jet boat goes 24 miles upstream, against a current, in 3 hours. It travels downstream the same distance in 2 hours. What is the speed of the boat (in still water)? What is the speed of the current?

▶ Writing

61. Why is it possible to make forward progress when walking 3 miles per hour against a 40-mile-per-hour wind but not when swimming 3 miles per hour against a 6-mile-per-hour current?

62. Why is there a limit to acceptable wind levels in a short race (100 meters) but not in a long race that takes several laps of an oval track to complete?

63. Explain how to solve a system of equations by elimination.

64. In the example system

$$ax + by = c$$
$$dx + ey = f$$

explain how to eliminate x.

65. In the example system

$$ax + by = c$$
$$dx + ey = f$$

explain how to eliminate y.

66. Explain how, after looking at a system of equations, you would choose substitution, elimination, or graphing as the solution process.

▶ Projects

67. Changing Repeating Decimals to Fractions We change repeating decimals into fractions with a process similar to the elimination method. Instead of using multiplication and subtraction to eliminate a variable, we use them to eliminate decimal portions of a number.

Example a: To change $0.3333\overline{3}$ to a fraction, we let f represent the fraction equivalent to the given repeating decimal:

$$f = 0.3333\overline{3}$$

Because only one digit is being repeated, we multiply both sides of the equation by 10. This shifts the decimal point one place to the right. Then we subtract the original equation.

$$\begin{aligned} 10f &= 3.3333\overline{3} \\ -f &= -0.3333\overline{3} \\ \hline 9f &= 3 \\ f &= \tfrac{3}{9}, \text{ or } \tfrac{1}{3} \end{aligned}$$

If the decimal repeated two digits, we would multiply both sides by 100 and repeat the process in Example a.

Example b: To change $0.454545\overline{45}$ to a fraction, we let $f = 0.454545\overline{45}$. Because two digits are repeated, we multiply both sides by 100 to move the decimal point two places to the right. Then we subtract the original equation.

$$\begin{aligned} 100f &= 45.4545\overline{45} \\ -f &= -0.4545\overline{45} \\ \hline 99f &= 45 \\ f &= \frac{45}{99} = \frac{9 \cdot 5}{9 \cdot 11} = \frac{5}{11} \end{aligned}$$

Use the process shown to change these repeating decimals to fractions. Simplify the fractions to lowest terms.

a. $0.444\overline{4}$ **b.** $0.777\overline{7}$

c. $0.1515\overline{15}$ **d.** $0.1616\overline{16}$

e. $0.243243\overline{243}$ **f.** $0.270270\overline{270}$

g. $0.567\overline{567}$

68. Mr. Hall's Farm*

a. Mr. Hall has a farm where he raises cows and ducks. There are 20 total heads among the animals and 50 total feet. How many cows and how many ducks are there?

b. Return to Exercise 30 in Section 6.2 (page 342). Set up and solve systems of equations for Mr. McFadden's farm and Mr. Schaaf's farm.

*This problem is dedicated to the memory of three great problem solvers who were my mentors.

c. Just for fun, now that you know how to do the problems mathematically, ask a child as young as kindergarten age to model cows and ducks with the equipment suggested in Exercise 30 in Section 6.2. Give the child exactly 10 heads (coins) and 28 feet (paper clips) with which to model cows and ducks, and tell him or her that no pieces should be left over. With the child, explore the other two problems you just solved in the text. Make up and explore your own problem.

d. Report on your observations of the problem-solving skills of young children.

▶ 6.5 Solving Systems of Equations in Three Variables

Objectives

- Use guess and check to describe a problem setting.
- Use substitution to solve a system of three equations.
- Identify whether a system has one, no, or an infinite number of solutions.
- Use guess and check to set up a system of equations in three variables.

WARM-UP

In Exercises 1 to 3, solve each equation.

1. $(5 - 2x) + (2 - 2x) = 1$

2. $(c + 12) + c + (2c) = 64$

3. $A + A + (A - 15) = 180$

In Exercises 4 to 6, evaluate the following three equations with the set of numbers given:

$$x + z = 3, \quad x - y = 4, \quad y + z = -1$$

4. $(x, y, z) = (8, 4, -5)$, where the notation means that $x = 8$, $y = 4$, and $z = -5$.

5. $(x, y, z) = (10, 6, -7)$

6. $(x, y, z) = (-1, -5, 4)$

IN THIS SECTION, we extend our work with guess and check and substitution to solving systems of equations in three variables.

▶ Systems of Equations in Three Variables

In Example 1, we use guess and check to set up a system of equations to solve for three unknowns.

▶ EXAMPLE 1 Writing a system of three equations in three variables: schedule planning Herman needs to include work, class, and study time in his schedule. He spends 12 more hours per week at work than he does in class. He studies 2 hours for each hour he is in class. He spends a total of 64 hours each week on the three activities. Set up a guess-and-check table to find the number of hours spent at each activity. In the last row of the table, show variables. Then write a system of equations to describe the problem.

SOLUTION The guess-and-check table appears in Table 14. The table contains a heading for each unknown: work time, class time, and study time.

TABLE 14 Scheduling Time

Work Time (hr)	Class Time (hr)	Study Time (hr)	Total (hr)
20	$20 - 12 = 8$	$8 \times 2 = 16$	$20 + 8 + 16 = 44$
24	$24 - 12 = 12$	$12 \times 2 = 24$	$24 + 12 + 24 = 60$
w	c	s	64

The system is

$$c = w - 12$$

$$s = 2c$$

$$w + c + s = 64$$

◀

THINK ABOUT IT 1: Continue to guess and check until you find a solution.

▶ Solving Systems of Equations in Three Variables by Substitution

SYSTEMS WITH EXACTLY ONE SOLUTION Calculator technology now makes it easy to solve complicated systems of three or more linear equations in three or more unknowns. Thus, we will consider here only selected systems that can be solved reasonably quickly by hand.

▶ **EXAMPLE 2** Solving a system of equations in three variables by substitution Solve

$$2x + y = 5 \quad (1)$$

$$2x + z = 2 \quad (2)$$

$$y + z = 1 \quad (3)$$

SOLUTION Equations (1) and (2) can be solved in terms of the variable x. The expressions for the variables y and z can then be substituted into equation (3).

$$2x + y = 5 \qquad \text{Solve (1) for } y.$$

$$y = 5 - 2x$$

$$2x + z = 2 \qquad \text{Solve (2) for } z.$$

$$z = 2 - 2x$$

$$y + z = 1 \qquad \text{Substitute } y = 5 - 2x \text{ and } z = 2 - 2x \text{ into (3).}$$

$$(5 - 2x) + (2 - 2x) = 1 \qquad \text{Simplify.}$$

$$7 - 4x = 1 \qquad \text{Subtract 7 from both sides.}$$

$$-4x = -6 \qquad \text{Divide both sides by } -4.$$

$$x = 1.5$$

Substitute $x = 1.5$ into equations (1) and (2).

$$2(1.5) + y = 5 \qquad \text{Solve (1) for } y.$$

$$y = 2$$

$$2(1.5) + z = 2 \qquad \text{Solve (2) for } z.$$
$$z = -1$$

The solution to the system is $x = 1.5$, $y = 2$, $z = -1$, or the ordered triple $(1.5, 2, -1)$. Check by substituting in all equations. ◀

The solution to a system of two equations in two variables is an ordered pair, (x, y). The solution to Example 2 is an ordered triple, (x, y, z). This suggests that graphing would require a third dimension, a z-axis, as well as an x-axis and a y-axis. For this reason, we will not solve systems of three equations graphically.

SYSTEMS WITH NO OR AN INFINITE NUMBER OF SOLUTIONS As with systems of equations in two unknowns, not all systems of equations in three unknowns have one solution.

▶ **EXAMPLE 3** Solving a system with no solution Solve

$$x + y = -2 \qquad (1)$$
$$z - x = 4 \qquad (2)$$
$$y + z = 3 \qquad (3)$$

SOLUTION Equations (1) and (2) can be solved in terms of the variable x. The expressions for the variables y and z can then be substituted into equation (3).

$$x + y = -2 \qquad \text{Solve (1) for } y.$$
$$y = -2 - x$$

$$z - x = 4 \qquad \text{Solve (2) for } z.$$
$$z = 4 + x$$

$$y + z = 3 \qquad \text{Substitute } y = -2 - x \text{ and } z = 4 + x \text{ into (3).}$$
$$(-2 - x) + (4 + x) = 3 \qquad \text{Simplify.}$$
$$2 = 3 \qquad \text{False}$$

When the result is false, the system has no solution. ◀

▶ **EXAMPLE 4** Solving a system with an infinite number of solutions Solve

$$x + z = 3 \qquad (1)$$
$$x - y = 4 \qquad (2)$$
$$y + z = -1 \qquad (3)$$

SOLUTION Equations (1) and (2) can be solved in terms of the variable x. The expressions for the variables y and z can then be substituted into equation (3).

$$x + z = 3 \qquad \text{Solve (1) for } z.$$
$$z = 3 - x$$

$$x - y = 4 \qquad \text{Solve (2) for } y.$$
$$-y = 4 - x$$
$$y = -4 + x$$

$$y + z = -1 \qquad \text{Substitute } z = 3 - x \text{ and } y = -4 + x \text{ into (3).}$$
$$(-4 + x) + (3 - x) = -1 \qquad \text{Simplify.}$$
$$-1 = -1 \qquad \text{True}$$

When the result is always true, the system has an infinite number of solutions. ◀

The three ordered triples in Warm-up Exercises 4 to 6 make the three equations in Example 4 true. They are three of an infinite number of triples that do so.

THINK ABOUT IT 2: Find the value of x when $y = 2$ and $z = -3$ in the system of equations in Example 4. Is the resulting ordered triple a solution?

SOLVING A SYSTEM OF THREE EQUATIONS IN THREE VARIABLES

1. Look for a common variable in all three equations; or, if possible, solve two of the equations for one variable *in terms of* the common variable.
2. Use the equations in step 1 to substitute for all but the common variable in the third equation.
3. Solve for the value of the common variable.
4. Substitute the value of the common variable into the first two equations to find the values of the other variables.

Applications

SCHEDULE PLANNING In Example 5, we return to the system of equations for Herman's work, study, and class time, which we set up in Example 1. We solve the system by substitution.

▶ **EXAMPLE 5** Solving a system of three equations: schedule planning Solve this system of equations by substitution:

$$c = w - 12 \quad (1)$$
$$s = 2c \quad (2)$$
$$w + c + s = 64 \quad (3)$$

SOLUTION Each equation contains the variable c, so we solve the first two equations for the variables w and s in terms of c.

$$w = c + 12 \quad (1)$$
$$s = 2c \quad (2)$$

We then substitute for w and s in the third equation, to build an equation containing only c. We place the substitutions in parentheses so that they show up clearly.

$$w + c + s = 64 \quad \text{Substitute for } w \text{ and } s \text{ in (3).}$$
$$(c + 12) + c + (2c) = 64 \quad \text{Simplify.}$$
$$4c + 12 = 64 \quad \text{Subtract 12 on both sides.}$$
$$4c = 52 \quad \text{Divide by 4 on both sides.}$$
$$c = 13$$

Next we substitute $c = 13$ into the other two equations and solve for w and s.

$$w = c + 12 \quad (1)$$
$$w = 13 + 12$$
$$w = 25$$

$$s = 2c \qquad (2)$$
$$s = 2(13)$$
$$s = 26$$

Finally, we check in all three equations.

$$c = w - 12 \qquad (1)$$
$$13 \stackrel{?}{=} 25 - 12 \quad ✓$$
$$s = 2c \qquad (2)$$
$$26 \stackrel{?}{=} 2(13) \quad ✓$$
$$w + c + s = 64 \qquad (3)$$
$$25 + 13 + 26 \stackrel{?}{=} 64 \quad ✓$$

◀

ANGLE MEASURES The fact that *the interior angle measures of a triangle sum to 180°* is an important geometric result. This fact is well enough known that it usually is assumed rather than stated in algebra problems. In Example 6, we solve a puzzle problem based on the sum of the angle measures.

▶ **EXAMPLE 6** Using substitution to solve problems involving the sum of angle measures in a triangle

If one angle of an isosceles triangle is 15° smaller than another angle, what are the angle measures? Use guess and check to build equations, and then solve the equations by substitution. (Geometry note: An **isosceles triangle** has *two equal sides with equal angles opposite these sides.*) Figure 21 shows an isosceles triangle with $AC = BC$ and angle A = angle B.

SOLUTION We build the guess-and-check table shown in Table 15, with a column for each angle measure and a column for the angle sum. As indicated earlier, the sum of the measures of all three angles of the triangle must be 180°. We use 60° as our starting guess because that is the angle measure when all the angles are equal (180° divided by 3). We assume that angle C is 15° smaller than angles A and B and fill in the first row of the table.

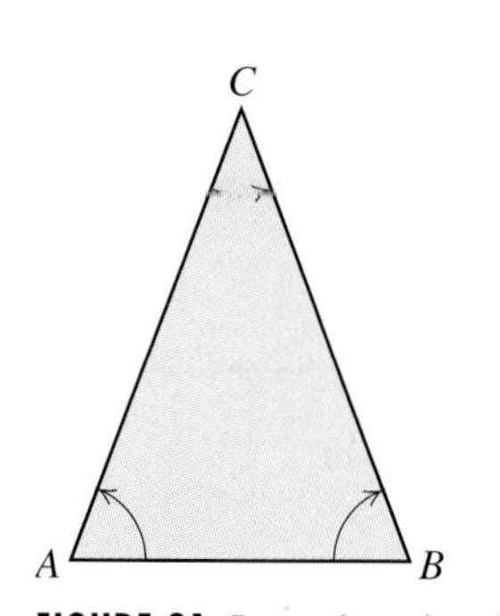

FIGURE 21 Isosceles triangle

TABLE 15 Guess-and-Check Table: Sum of Angles in a Triangle

Angle A	Angle B	Angle C	Total Degrees
60°	60°	45°	165°
A	B	C	180°

The table helps us to see how the angles are related and to write equations.

Two angles are equal: $A = B$. (1)

One angle is 15° smaller than another: $C = A - 15$. (2)

The three angles add to 180°: $A + B + C = 180$. (3)

Scanning the three equations, we see that each equation contains A, so we solve the system *in terms of A*.

$$A + B + C = 180 \qquad \text{Substitute values from (1) and (2) into (3).}$$
$$A + A + (A - 15) = 180 \qquad \text{Simplify.}$$
$$3A - 15 = 180 \qquad \text{Add 15 to both sides.}$$
$$3A = 195 \qquad \text{Divide by 3.}$$
$$A = 65$$

We substitute $A = 65$ into the first equation to solve for B:

$$B = A \qquad (1)$$

$$B = 65$$

We substitute $A = 65$ into the second equation to solve for C:

$$C = A - 15 \qquad (2)$$

$$C = 65 - 15 = 50$$

We check by replacing the variables with their values in all three equations.

$$A = B \qquad (1)$$

$$65 \stackrel{?}{=} 65 \checkmark$$

$$C = A - 15 \qquad (2)$$

$$50 \stackrel{?}{=} 65 - 15 \checkmark$$

$$A + B + C = 180 \qquad (3)$$

$$65 + 65 + 50 \stackrel{?}{=} 180 \checkmark$$

◀

ANSWER BOX

Warm-up: 1. $x = 1.5$ **2.** $c = 13$ **3.** $A = 65$ **4.** $10 + (-7) = 3$, $8 - 4 = 4$, $4 + (-5) = -1$, all true **5.** $10 + -7 = 3$, $10 - 6 = 4$, $6 + (-7) = -1$, all true **6.** $-1 + 4 = 3$, $-1 - (-5) = 4$, $-5 + 4 = -1$, all true **Think about it 1:** $c = 13$, $s = 26$, $w = 25$ **Think about it 2:** Substitute $y = 2$ or $z = -3$ into equation (1) or (2) and solve for x; $x = 6$; $(6, 2, -3)$ is a solution.

▶ 6.5 Exercises

Solve the systems of equations in Exercises 1 to 10. Write the answers as ordered triples, (x, y, z).

1. $x + z = 2$; $x - y = 1$; $y - z = -2$

2. $x + y = 3$; $z - x = 2$; $y - z = -9$

3. $x + y = 2$; $z - x = 7$; $z - y = 2$

4. $x + z = 2$; $x - y = 5$; $z - y = 2$

5. $x + z = 3$; $y + x = 4$; $y - z = 1$

6. $x + y = 5$; $y - z = 5$; $x + z = 5$

7. $x - y = 3$; $y + z = 4$; $x + z = 5$

8. $y + z = -2$; $z + x = 6$; $y - x = -8$

9. $x + y = 5$; $y + z = 5$; $z + x = 5$

10. $x - y = 3$; $y - z = 3$; $x + z = 3$

11. To find ordered triples for a system that has an infinite number of solutions, pick any two numbers making x and y true; then take the x (or y) value and solve one of the other equations for z. Substitute into the other equations to check for true statements.

a. Find another ordered triple for Example 4.

b. Find two ordered triples for Exercise 5.

c. Find two ordered triples for Exercise 8.

12. There are three different ways to solve the systems in Exercises 1 to 10. Find a second way to solve two of the exercises.

Blue numbers are core exercises.

13. The number of equations we write to help solve a problem is related to the number of variables we use. Look for a pattern in the examples and exercises, and complete these statements:

a. When we use one variable, we need ____ equation(s).

b. When we use two variables, we need ____ equation(s).

c. When we use three variables, we need ____ equation(s).

For Exercises 14 to 17, set up a guess-and-check table and solve using guess and check. Write a system of equations and solve using substitution.

14. Marielena spent \$620 on a watch, locket, and chain. She paid \$20 more for the locket than for the chain. She paid twice as much for the watch as for the locket. How much did she pay for each?

15. Katreen earns \$1375 one summer. She earns twice as much mowing lawns as shopping for the elderly. She earns \$500 more scraping old paint on houses than mowing lawns. How much does she earn at each job?

16. Andre buys notebooks, paperback books, and hardbound books for his classes. He buys 3 more paperback books than hardbound books. He buys 1 fewer notebooks than paperback books. He buys 20 items altogether. How many of each does he buy?

17. The I. R. Rabbit Company is packaging carrots. It has 2400 pounds of carrots to be packaged into bags for grocery stores. The bags are 1 pound, 2 pounds, and 5 pounds. The company sells twice as many 1-pound bags as 2-pound bags and ten times as many 1-pound bags as 5-pound bags. How many bags of each size should the company fill?

18. In Example 6, we assumed that angle C was the smallest angle. Assume that the two equal angles are smaller than angle C. Write the new system of equations, and solve it by substitution.

Solve the systems in Exercises 19 and 20.

19.
$$A + B + C = 180$$
$$A = 2B$$
$$B = 3C$$

20.
$$A + B + C = 180$$
$$A = 3B$$
$$B = 2C$$

In Exercises 21 to 26, write and solve a system of equations.

21. The measure of angle A of a triangle is twice that of angle B. The measure of angle C is 20° more than that of angle A. What is the measure of each angle?

22. The two equal angles of an isosceles triangle are each twice the size of the third angle. What is the measure of each angle?

23. The two equal angles of an isosceles triangle are each half the size of the third angle. What is the measure of each angle?

24. The two equal angles of an isosceles triangle are each four times the size of the third angle. What is the measure of each angle?

25. The length of a large gift box is 7 inches more than the width. The width is four times the height. The sum of the length, width, and height is 34 inches. Find the length, width, and height.

26. A gift box for a shirt is 5.5 inches longer than it is wide. The sum of the length, width, and height is 26.5 inches. The height of the box is 1 less than a fifth of the length. Find the length, width, and height.

▶ Projects

27. U.S. Currency Who is pictured on each denomination of U.S. currency? Currency is printed in denominations of \$1, \$2, \$5, \$10, \$20, \$50, \$100, \$500, \$1000, \$5000, and \$10,000. Select an appropriate variable for each person pictured on the bills, as named in the following clues. Where two people's names start with the same letter, hints for variables are suggested. Write an equation for each relevant clue. Solve by substitution.

a. 10 Clevelands (C_l) equal 1 Chase (C_h).

b. Cleveland, Franklin, and Chase total \$11,100.

c. 10 Hamiltons make a Franklin.

d. 3 Washingtons plus a Jefferson (J_e) equal a Lincoln.

e. 4 Lincolns make a Jackson (J_a).

f. Hamilton, Franklin, and Chase were never president.

g. 2 Washingtons make a Jefferson.

h. 10 Grants make a McKinley (M_c).

i. 2 Grants make a Franklin.

j. Cleveland was born on March 18.

k. 2 Hamiltons make a Jackson.

l. Madison's picture is on the \$5000 bill.

m. Washington is on the \$1 bill.

Blue numbers are core exercises.

▶ 6.6 Solving Systems of Linear Inequalities by Graphing

Objectives

- Find out if an ordered pair is in the solution set to a system of inequalities.
- Solve a system of linear inequalities by graphing.

> **WARM-UP**
>
> Graph these inequalities in two variables. (If you need help, review Section 4.6.)
>
> **1.** $x + y \geq 5$ **2.** $x - y < -2$ **3.** $x \geq -1$ **4.** $y < 1$

HAVE YOU EVER BLENDED paints or pigments? The colors that result from the blend are similar to solution sets to inequalities. In this section, we combine the concept of systems of equations, from this chapter, with the process of graphing inequalities in two variables, from Section 4.6.

▶ Solutions to Systems of Inequalities

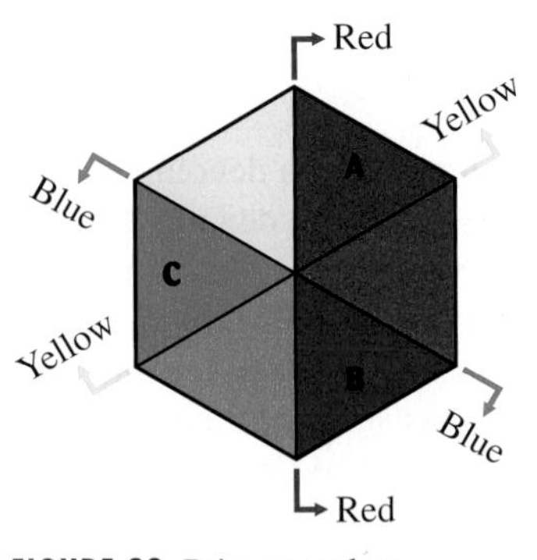

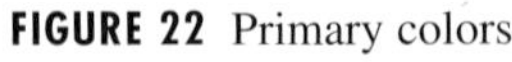

FIGURE 22 Primary colors

In Figure 22, which repeats the figure on the chapter opener, each of the three lines through the six-sided figure is a boundary line for one of the primary colors—red, yellow, or blue. Each color produces a shaded half-plane in the figure, just as the solution set to a linear inequality forms a half-plane.

The overlapping of two primary colors creates another color; for example, the overlapping of yellow and red creates orange. One could say that orange is the solution set to the colors that contain both yellow and red. Orange is in the half-plane containing yellow and the half-plane containing red.

A **system of inequalities** is *a set of two or more inequalities*. The **solution set to a system of inequalities** is *the set of ordered pairs that make all inequalities in the system true*. To test a single point, we substitute its ordered pair into each inequality in the system.

▶ EXAMPLE 1 Identifying solutions to systems of inequalities Find out which ordered pairs are solutions to the system of inequalities

$$x + y \geq 5$$
$$x - y < -2$$

a. $(0, 0)$ **b.** $(5, 0)$ **c.** $(0, 5)$

d. $(-2, 8)$ **e.** $(1, 3)$

SOLUTION We substitute the ordered pairs into the inequalities. A solution must make both inequalities true.

a. $0 + 0 \geq 5$ is false; $0 - 0 < -2$ is false; not a solution

b. $5 + 0 \geq 5$ is true; $5 - 0 < -2$ is false; not a solution

c. $0 + 5 \geq 5$ is true; $0 - 5 < -2$ is true; a solution

d. $-2 + 8 \geq 5$ is true; $-2 - 8 < -2$ is true; a solution

e. $1 + 3 \geq 5$ is false; $1 - 3 < -2$ is false; not a solution ◀

We find solution sets to systems of inequalities by graphing equations, shading half-planes, and finding the overlapping regions.

Finding Solution Sets to Systems of Inequalities

> To find the solution set to a system of inequalities:
>
> **1. a.** Graph the first boundary line.
> **b.** Use a test point to find the half-plane that is the solution set.
> **c.** Shade that half-plane.
>
> **2. a.** Graph the second boundary line.
> **b.** Use a test point to find the half-plane that is the solution set.
> **c.** Shade that half-plane.
>
> **3.** The region that lies in both half-planes is the solution set to the system.
>
> (The process can be extended to find solution sets to systems of three or more inequalities.)

Two-Variable Boundary Lines

In Example 2, we graph two inequalities that include boundary lines.

EXAMPLE 2 Finding the solution set to a system of inequalities Graph the inequalities and find the portions of their graphs that overlap:

$$x - y \leq 6 \quad (1)$$
$$x + y \leq 3 \quad (2)$$

SOLUTION We solve each inequality for y, check a test point, and state the boundary line.

$x - y \leq 6$	(1)	Solve for y.	$x + y \leq 3$	(2)
$x \leq 6 + y$			$y \leq -x + 3$	
$x - 6 \leq y$				
$y \geq x - 6$				
$0 \geq 0 - 6$	True	Test point (0, 0)	$0 \leq -0 + 3$	True
$y = x - 6$		Boundary line	$y = -x + 3$	
$m = 1, b = -1$			$m = -1, b = 3$	

Inequality (1), $y \geq x - 6$, is graphed in Figure 23. Inequality (2), $y \leq -x + 3$, is graphed in Figure 24. The inequalities are graphed together in Figure 25. The overlapping region is the solution set.

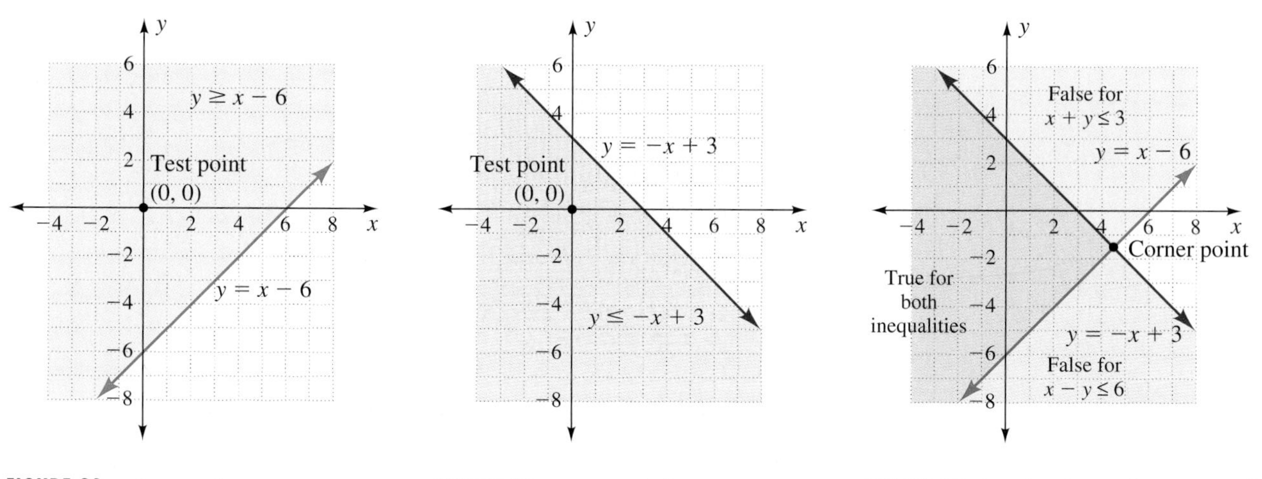

FIGURE 23 $y \geq x - 6$

FIGURE 24 $y \leq -x + 3$

FIGURE 25 Overlapping region ◀

▶ In linear programming and other applications in business and economics, the coordinates of the corner point are important.

DEFINITION OF CORNER POINT

> The **corner point** marks the point of intersection of the boundary lines of two inequalities.

To locate the corner point, find the ordered pair for the point of intersection of the two boundary lines.

▶ **EXAMPLE 3** Finding a corner point Solve the system consisting of the boundary lines in Figure 25 to find the corner point. Is the corner point in the solution set?

SOLUTION We solve the two equations to find the point of intersection of the two boundary lines.

$$x - y = 6$$

$$\underline{x + y = 3} \qquad \text{Add the equations.}$$

$$2x \qquad = 9 \qquad \text{Divide by 2 on both sides.}$$

$$x = 4.5$$

We substitute $x = 4.5$ into the first equation and solve for y.

$$x - y = 6$$

$$4.5 - y = 6 \qquad \text{Subtract 4.5 on both sides.}$$

$$-y = 1.5 \qquad \text{Multiply by } -1 \text{ on both sides.}$$

$$y = -1.5$$

The corner point is $(4.5, -1.5)$. Both boundary lines are in the solution set, so the corner point is also in the solution set. ◀

Student Note:
See page 246 for shading options.

Figure 26 is a calculator version of the graph in Figure 25. We shade above $Y_1 = X - 6$ and below $Y_2 = -X + 3$. We can use Intersection on the graphs of $Y_1 = X - 6$ and $Y_2 = -X + 3$ to find the corner point $(4.5, -1.5)$, as shown in Figure 26.

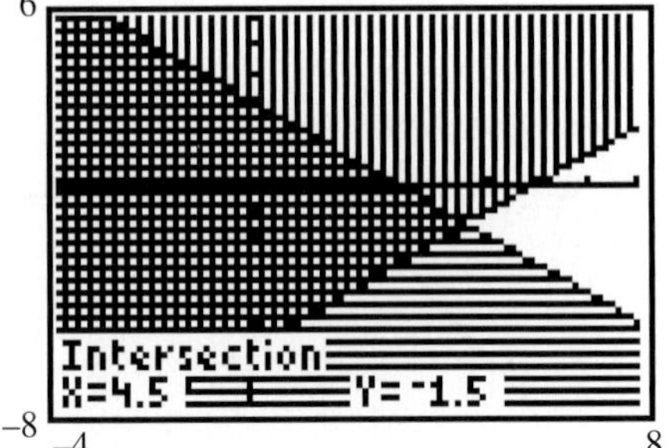

FIGURE 26 Examples 2 and 3 viewscreen

In Example 4, both boundary lines are dashed lines.

▶ **EXAMPLE 4** Finding the solution set to a system of inequalities Graph the inequalities and find the portions of their graphs that overlap:

$$y > x + 3$$
$$y < -x + 2$$

SOLUTION We solve each inequality for y (if needed), check a test point, and state the boundary line.

$y > x + 3$	(1)	Solve for y.	$y < -x + 2$	(2)
$0 > 0 + 3$	False	Test point (0, 0)	$0 < -0 + 2$	True
$y = x + 3$		Boundary line	$y = -x + 2$	
$m = 1, b = 3$			$m = -1, b = 2$	

The inequalities are graphed in Figure 27. The boundary lines are dashed because the inequalities are $>$ and $<$ (with no equal sign). The overlapping region is the solution set. The corner point is $(-0.5, 2.5)$ and is not in the solution set because the boundary lines are not in the solution set.

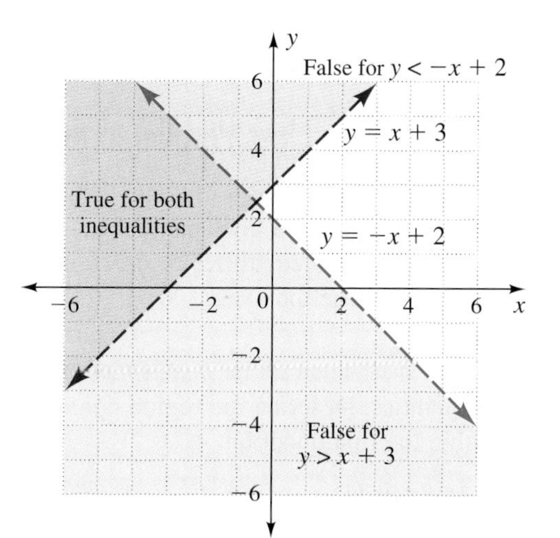

FIGURE 27 Solution set for $y > x + 3$ and $y < -x + 2$ ◀

In Example 5, one of the boundary lines is solid and the other is dashed.

▶ **EXAMPLE 5** Finding the solution set to a system of inequalities Graph the inequalities, find the portions of their graphs that overlap, and state the corner point:

$$x + y \leq 5$$
$$2x + y > 5$$

SOLUTION We solve each inequality for y (if needed), check a test point, and state the boundary line.

$x + y \leq 5$	(1)	Solve for y.	$2x + y > 5$	(2)
$y \leq -x + 5$			$y > -2x + 5$	
$0 \leq -0 + 5$	True	Test point (0, 0)	$0 > -0 + 5$	False
$y = -x + 5$		Boundary line	$y = -2x + 5$	
$m = -1, b = 5$			$m = -2, b = 5$	

The inequalities are graphed in Figure 28. The boundary line for (1) is solid and for (2) is dashed. The overlapping region is the solution set. The corner point is (0, 5) and is not in the solution set because one boundary line is not in the solution set.

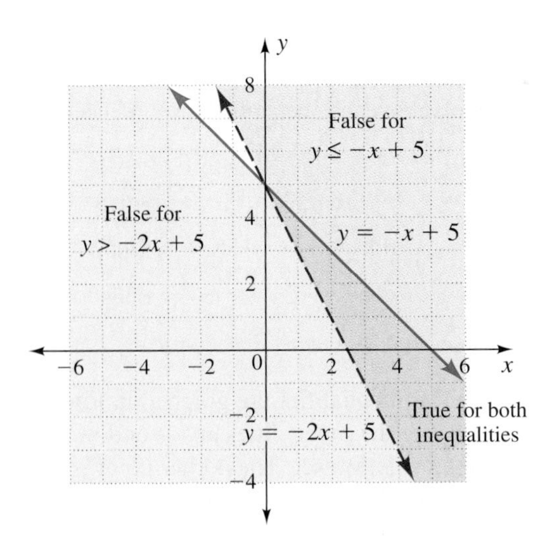

FIGURE 28 Solution set for $x + y \le 5$ and $2x + y > 5$ ◀

▶ Boundary Lines Parallel to the Axes

Some inequalities have only one variable. Recall that lines parallel to the x-axis have equations $y = c$ and lines parallel to the y-axis have equations $x = c$. In Example 6, we graph four inequalities to create a region. Which inequalities have boundaries parallel to or on the axes?

▶ EXAMPLE 6 Finding the solution set for a system of inequalities Graph the following system of inequalities. Describe the region containing the solution set. Label the corner point within the first quadrant.

$$x \ge 0$$
$$y \ge 0$$
$$y \le 6$$
$$2y + 5x \le 16$$

SOLUTION The inequality $x \ge 0$ describes quadrants 1 and 4 and the y-axis. The inequality $y \ge 0$ describes quadrants 1 and 2 and the x-axis. The inequality $y \le 6$ describes all points on or below the line $y = 6$.

To graph $2y + 5x \le 16$, we solve for y, check a test point, and state the boundary line.

$$2y + 5x \le 16 \qquad \text{Solve for } y.$$
$$2y \le -5x + 16$$
$$y \le -\frac{5x}{2} + 8$$
$$0 \le -0 + 8 \quad \text{True} \qquad \text{Test point } (0, 0)$$
$$y = -\frac{5x}{2} + 8 \qquad \text{Boundary line}$$
$$m = -\frac{5}{2}, b = 8$$

The inequalities are graphed in Figure 29. All boundary lines are solid. The region containing the solution set is in the first quadrant, bounded above by $y = 6$ and on the right by the boundary line for $2y + 5x \leq 16$. The corner point is (0.8, 6).

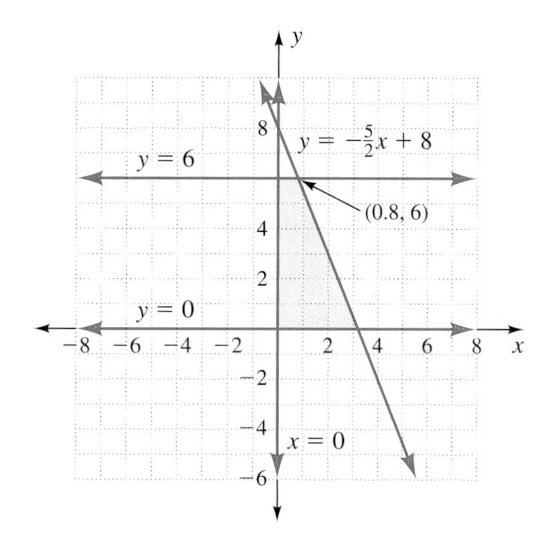

FIGURE 29 Solution set for Example 6 ◀

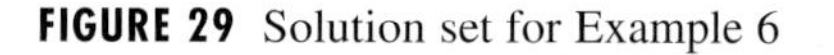

GRAPHING CALCULATOR TECHNIQUE: SHADING INEQUALITIES AND GRAPHING SYSTEMS

Student Note:
See page 246 for shading options.

To graph the system of inequalities in Example 6, set the window for the first quadrant: $0 \leq X \leq 10$ and $0 \leq Y \leq 10$. With this setting, the conditions $x \geq 0$ and $y \geq 0$ will be satisfied by the viewing window itself.

Enter $Y_1 = 6$, and set the shading on Y_1 to below the line (see Figure 30). Enter $Y_2 = -5X/2 + 8$, and set the shading on Y_2 to below the line. Graph (see Figure 31).

Optional: Use Intersection to find the corner point, (0.8, 6).

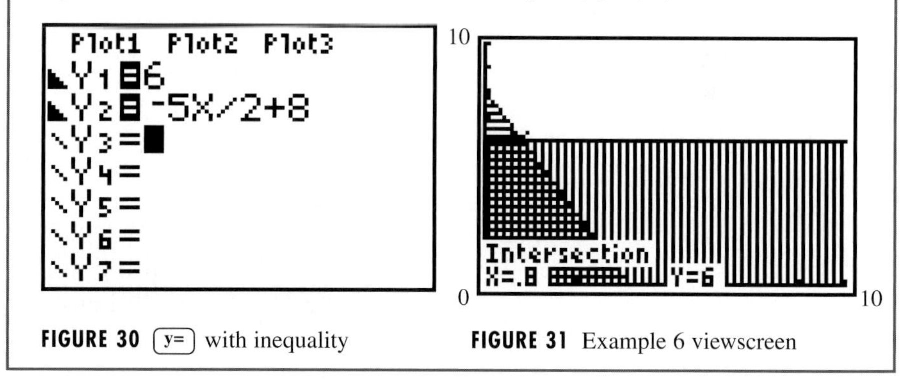

FIGURE 30 [y=] with inequality **FIGURE 31** Example 6 viewscreen

Applications

In Example 7, we return to the setting in Example 6 of Section 4.6 (page 243).

EXAMPLE 7 **Applying a system of inequalities: Drive-thru Mart cash** Drive-thru Mart keeps the currency in its overnight cash drawer limited to \$100, with a maximum of 10 five-dollar bills. Suppose the cash drawer has x one-dollar bills, y five-dollar bills, and no larger bills. Write a system of inequalities to describe the possible numbers of each type of currency. Graph and solve the system. Label the corner point within the first quadrant.

SOLUTION It is a good idea to set up the boundary equations before the graph to make it easier to estimate the scales needed on the axes.

Because x is the number of one-dollar bills, $x \geq 0$, and because y is the number of five-dollar bills, $y \geq 0$. These conditions suggest a first-quadrant graph. The 10 five-dollar bills imply that $y \leq 10$. We will shade below $y = 10$.

The \$100 limit is a quantity and rate statement: $1x + 5y \leq 100$. Solving the boundary equation, $1x + 5y = 100$, for y gives $y = -\frac{1}{5}x + 20$. We use the slope and y-intercept to graph this solid line: $m = -\frac{1}{5}$ and $b = 20$. Because (0, 0) makes the inequality true, we shade below the boundary line.

If the cash register has only ones, then x can be as large as 100. If the cash register has only fives, then y can be as large as 20. However there is another limit, $y \leq 10$. We set the axes with $0 \leq x \leq 140$ and $0 \leq y \leq 30$. The graph is in Figure 32. The corner point is (50, 10).

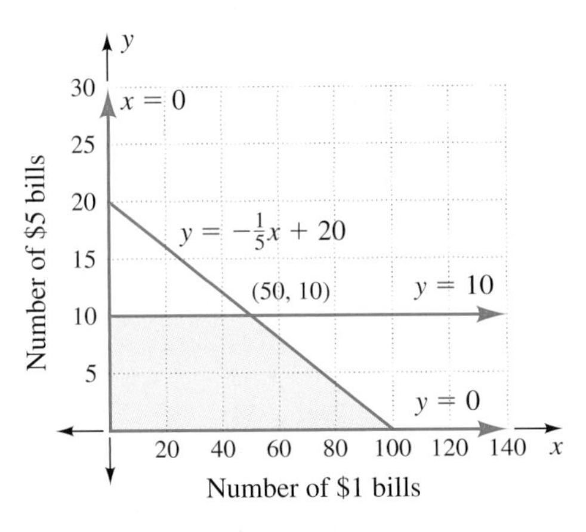

FIGURE 32 Solution set for Example 7

GRAPHING CALCULATOR SOLUTION

We solve the boundary equation, $1x + 5y = 100$, for y: $y = -\dfrac{x}{5} + 20$. We enter the equation, and because (0, 0) makes the inequality true, we shade below the line. We enter the equation $Y_2 = 10$ and set the shading below the line. A window with X on the interval [0, 140], scale of 10, and Y on the interval [0, 30], scale of 2, gives a first-quadrant graph and a complete view of the solution region, as shown in Figure 33.

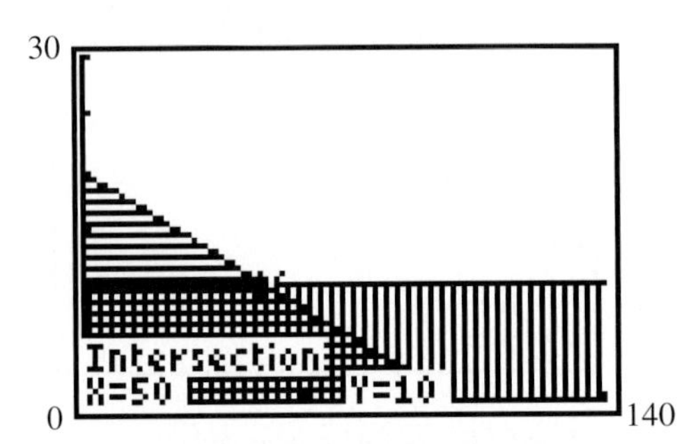

FIGURE 33 Example 7 viewscreen. ◀

THINK ABOUT IT: What is the meaning of the intersection in Figure 33?

In Figures 32 and 33, the solution set should be dots because we can have only whole numbers of one-dollar and five-dollar bills. The large number of bills makes showing the dots difficult, however, so we use shading.

In Example 8, we return to the setting in Example 7 of Section 4.6 (page 244). Because of the small quantity of whole-number answers, the solution set in Example 8 will be shown with dots.

▶ EXAMPLE 8 **Applying a system of inequalities: calories in snacks** You are allowed 160 calories in snacks. Caramel candies have 40 calories each, and ginger snaps have 20 calories each. You prefer no more than 5 ginger snaps each day. Let x be the number of caramels and y be the number of ginger snaps. Write a system of inequalities to describe the possible numbers of each type of snack. Graph and solve the system. Label the corner point within the first quadrant.

SOLUTION Because the number of snacks must be zero or positive, this will be a first-quadrant graph with $x \geq 0$ and $y \geq 0$. The ginger snaps are limited, so $y \leq 5$.

The calorie information is a quantity and rate statement: $40x + 20y \leq 160$. Solving the boundary equation, $40x + 20y = 160$, for y gives $y = -2x + 8$. We use the slope and y-intercept to graph: $m = -2$ and $b = 8$.

To set the numbers on the axes in Figure 34, we look at the limits on the calories. If no caramels are eaten, then the limit is $\frac{160}{20} = 8$ ginger snaps. If no ginger snaps are eaten, then the limit is $\frac{160}{40} = 4$ caramels. We set $0 \leq x \leq 9$ and $0 \leq y \leq 9$. The corner point is not a whole number. Either (1, 5) or (2, 4) is close.

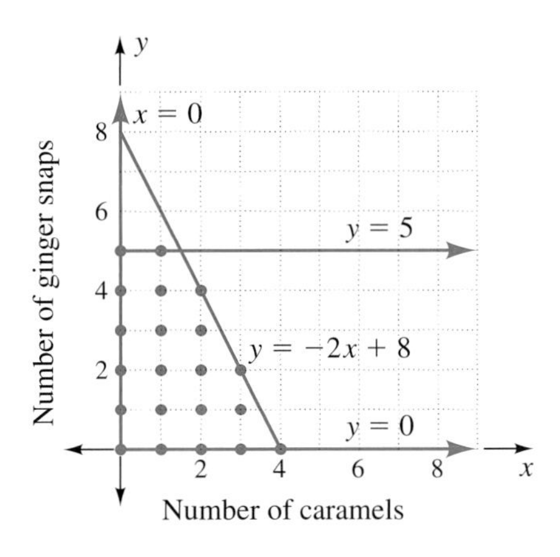

FIGURE 34 Solution set as points ◀

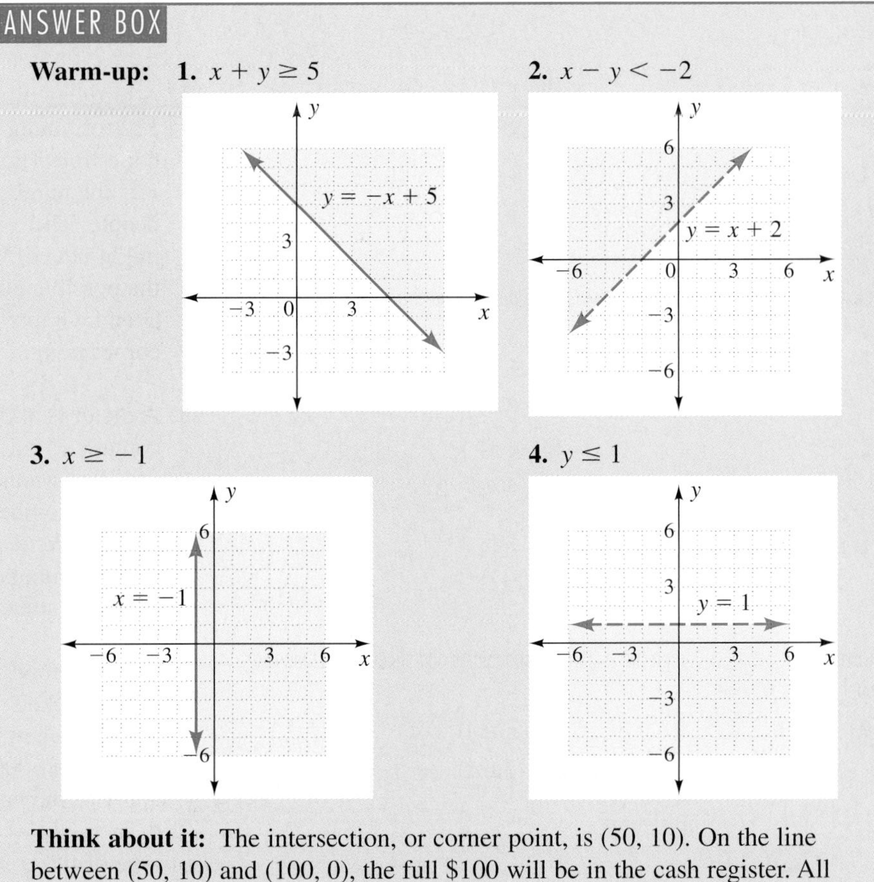

Think about it: The intersection, or corner point, is (50, 10). On the line between (50, 10) and (100, 0), the full \$100 will be in the cash register. All other shaded points will satisfy the requirements but will add to less than \$100.

▶ 6.6 Exercises

In Exercises 1 to 4, which ordered pairs satisfy the system of inequalities given?

1. $y \le x + 2$
$y \ge x - 2$

a. $(0, 0)$ **b.** $(1, 1)$ **c.** $(1, -2)$ **d.** $(-3, 3)$

2. $x + y < 3$
$x + y > -3$

a. $(0, 0)$ **b.** $(2, 2)$ **c.** $(-1, 4)$ **d.** $(-2, -1)$

3. $x + y > 4$
$x - y > 4$

a. $(0, 0)$ **b.** $(2, 2)$ **c.** $(5, 0)$ **d.** $(0, 5)$

4. $y \le -x - 2$
$y \ge x$

a. $(0, 0)$ **b.** $(-2, -2)$ **c.** $(-2, 2)$ **d.** $(0, -2)$

In Exercises 5 to 12, graph the system of inequalities and show the solution set. Label the corner point.

5. $y \le 2x + 2$
$y \ge -x + 4$

6. $y \ge 3x + 3$
$y \le -2x - 1$

7. $y \le 2x + 2$
$y \ge 3x - 3$

8. $y \le 2x + 5$
$y > -2x + 1$

9. $y \ge 2x - 3$
$y > -x + 3$

10. $y \ge 2x + 4$
$y > -2x - 3$

11. $y > 3x - 3$
$y < 2x - 2$

12. $y < -x + 1$
$y < -2x - 2$

In Exercises 13 to 18, graph the system of inequalities and show the solution set. Label the corner point.

13. $x \ge 2$
$y \ge 3$

14. $x \le -2$
$y \ge -2$

15. $x > -2$
$y \le 3$

16. $x \le 1$
$y < -2$

17. $x > -2$
$y < -1$

18. $x < 1$
$y > 2$

In Exercises 19 and 20, graph the system of inequalities and shade the solution set. Label the corner point.

19. $x \ge 0$
$y \ge 0$
$y \le 4$
$y \le -2x + 5$

20. $x \ge 0$
$y \ge 0$
$x \le 3$
$y \le -x + 4$

21. $x \ge 0$
$y \ge 0$
$x + y \ge 3$
$x + y \le 5$

22. $x \le 0$
$y \le 0$
$x - y \ge 2$
$x - y \le 4$

23. $x \le 0$
$y \ge 0$
$x \ge -3$
$y \le -x + 2$

24. $x \ge 0$
$y \le 0$
$y \ge -x - 1$
$x \le 3$

25. Johanna has time for at most 12 workouts each week. She can jog, at most, 4 times each week. Let x = the number of times she jogs and y = the number of times she swims. Write a system of inequalities to describe the possible numbers of times she carries out each activity. Graph and show the solution set. Label the corner point.

26. Jaime has 40 hours per week for doing math and reading. He can spend, at most, 20 hours reading. Let x = the number of hours spent doing math and y = the number of hours spent reading. Write a system of inequalities to describe the possible numbers of hours spent at each activity. Graph and show the solution set. Label the corner point.

27. A birthday brunch caterer charges \$20 for adults and \$12 for young people between 6 and 15. Children under 6 are free. The total budget for a party is \$360. Let x = the number of adults and y = the number of young people. There are a maximum of 12 young people who might attend. Write a system of inequalities to describe the possible numbers of people age 6 and over who can attend. Graph and show the solution set. Label the corner point.

28. A dieter is allowed 140 calories for a snack. Apricots contain 20 calories each, and tangerines contain 35 calories. Sue wants no more than 2 tangerines each day. Let x = the number of apricots and y = the number of tangerines. Write a system of inequalities to describe the possible numbers of each item she can eat. Graph and show the solution set. Label the corner point.

29. For each football game, the athletic department can sell 45,000 tickets. There are regular admission tickets and student tickets. The number of student tickets can be no more than 5000. Let x = the number of regular tickets and y = the number of student tickets. Write a system of inequalities to describe the possible numbers of each type of ticket sold. Graph and show the solution set. Label the corner point.

Blue numbers are core exercises.

30. Annette, the chief financial officer for a Fortune 500 company, has a \$200,000 bonus to invest. She wants to invest no more than \$40,000 in the stock market and the rest in municipal (tax-free) bonds. The bonds cost \$1000 each. Assume the shares of stocks cost \$100 each. Let $x =$ the number of shares and $y =$ the number of bonds. Write a system of inequalities to describe the possible numbers of shares and bonds. Graph and show the solution set. Label the corner point.

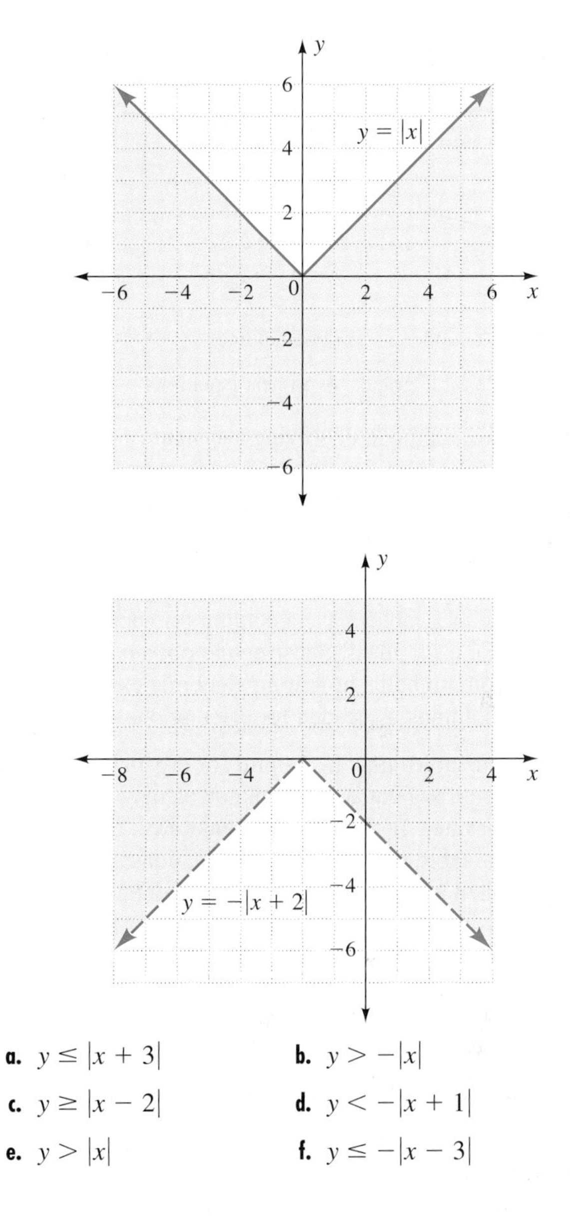

▶ Projects

These projects introduce new material and extend various ideas about inequalities.

31. Absolute Value Graphs For the rules listed in parts a to f, build a table and graph with the integers -3 to 3 as inputs. Do operations inside the absolute value signs first.

a. $y = |x|$

b. $y = -|x|$

c. $y = |x + 1|$

d. $y = |x - 2|$

e. $y = |x| - 2$

f. $y = |x| + 1$

32. Absolute Value Inequalities The absolute value graph has a V-shape. Make a short input-output table to find where the V is located for each inequality in parts a to f. Graph the ordered pairs from the table and connect to form the V. Show the solution set to the inequality by shading above or below the graph, depending on the truth of the inequality for a test point. As examples, see the graphs of $y \le |x|$ and $y > -|x + 2|$ opposite.

a. $y \le |x + 3|$

b. $y > -|x|$

c. $y \ge |x - 2|$

d. $y < -|x + 1|$

e. $y > |x|$

f. $y \le -|x - 3|$

▶ 6 Chapter Summary

Vocabulary

For definitions and page references, see the Glossary/Index.

coincident lines
complementary angles
corner point
elimination method
isosceles triangle
linear pair
numerical coefficient
quantity-rate equation
solution set to a system of inequalities
solution to a system
substitution method
supplementary angles
system of equations
system of inequalities

Concepts

6.1 Solving Systems of Equations with Graphs

A system of linear equations can have one, no, or an infinite number of solutions. (See the summary charts on page 329 and page 347.)

One solution means that the graphs of the two linear equations intersect in a single point; see Figure 35a. A unique ordered pair describes the point of intersection.

No solution means that the graphs of the two linear equations are parallel; see Figure 35b. No ordered pair makes both equations true. An algebraic solution gives a false statement.

An infinite number of solutions means that the graphs of the two linear equations are coincident; see Figure 35c. The lines have every point in common. An algebraic solution gives a true statement.

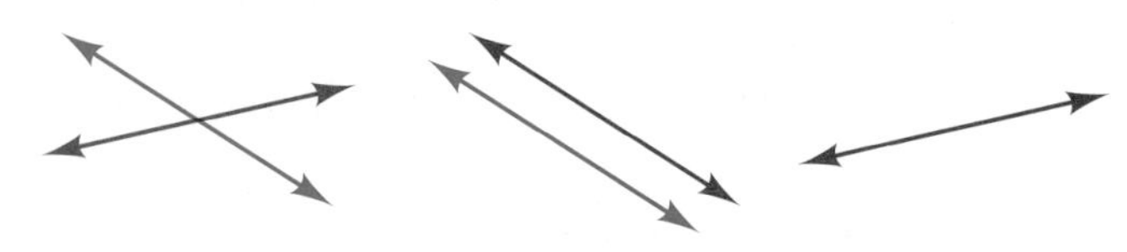

FIGURE 35 Possible graphs, two linear equations

There are five methods of solving a system of two linear equations: graphing (visual), guess and check (numerical), tables (numerical), substitution (algebraic notation), and elimination (algebraic notation).

6.2 Setting Up Systems of Equations

Guess-and-check tables have a column for each unknown and a row for each guess. Showing the operations within the unknown column helps in writing algebraic expressions. Add other columns as needed for calculations.

Quantity-rate tables may be helpful for organizing information or for building equations. Unknowns may be either quantities or rates. If quantities are unknown, then a quantity-rate table will contain two equations. If rates are unknown, then a quantity-rate table will contain only one equation. As needed, use a second table or write an equation directly with the quantity-rate equation.

$$\text{Quantity} \cdot \text{rate} + \text{quantity} \cdot \text{rate} = \text{total } Q \cdot R$$

Distance, rate, and time are related by the formula $d = r \cdot t$.

The net speed against a wind (or current) is $r - w$ (or $r - c$).

The net speed with a wind (or current) is $r + w$ (or $r + c$).

6.3 Solving Systems of Equations by Substitution

Solving by substitution is convenient when one variable has a coefficient of 1 or -1. For the steps in substitution, see page 344.

The measures of supplementary angles add to 180°. The measures of complementary angles add to 90°.

6.4 Solving Systems of Equations by Elimination

The elimination method is based on an extension to the addition property of equations: Adding equal expressions to both sides of an equation produces an equivalent equation. If $a = b$ and $c = d$, then $a + c = b + d$. For the steps in elimination, see page 360.

6.5 Solving Systems of Equations in Three Variables

Solving a system containing three variables requires three equations. The number of unknowns in a problem situation determines the number of equations needed to solve for the unknowns.

Guess and check assists in building equations and, frequently, finding a solution.

A system of three equations can have one, no, or an infinite number of solutions. If algebra leads to a false statement ($5 = 0$), the system has no solution. If algebra leads to an always-true statement ($5 = 5$), the system has an infinite number of solutions.

The measures of the interior angles of a triangle add to 180°.

6.6 Solving Systems of Linear Inequalities by Graphing

A system on the coordinate plane may contain any number of inequalities. Inequalities describe half-planes. Boundary lines may be any line, including the axes or lines parallel to the axes. Many application settings assume first-quadrant graphs with $x \geq 0$ and $y \geq 0$. For the steps in solving systems of inequalities, see page 375.

▶ 6 Review Exercises

In Exercises 1 to 6, solve for the indicated variable.

1. $y - 2x = 5000$ for y

2. $P = 2b + 2h$ for h

3. $C = 2\pi r$ for r

4. $x - 3y = 3$ for y

5. $\dfrac{a}{b} = \dfrac{5}{8}$ for b

6. $\dfrac{x}{y} = \dfrac{4}{5}$ for x

In Exercises 7 to 10, solve the system of equations by graphing.

7. $y = x - 2$, $y = \dfrac{-2x}{3} + 3$

8. $y = -x$, $y = x - 3$

9. $x = 2$, $y = 4$

10. $y = -\dfrac{1}{2}x + 5$, $y = \dfrac{3x}{2} - 3$

In Exercises 11 to 14, solve the system of equations by substitution.

11. $y = x + 2$, $y = 3x + 3$

12. $x + y = 5$, $3x + 5y = 27$

13. $x + y = 5000$, $3x - 2y = 12{,}500$

14. $x + y = 5000$, $3x - 2y = -7500$

In Exercises 15 to 20, solve the system of equations by elimination.

15. $3 = -2m + b$, $2 = 3m + b$

16. $2x - 3y = 7$, $-2x + 3y = 4$

17. $2x - 3y = 7$, $3y + 4x = -1$

18. $x - 2y = 11$, $3x + 4y = -7$

19. $5x + 3y = -18$, $3x - 2y = -7$

20. $2x - 4 = 6y$, $3y - x = -2$

In Exercises 21 to 24, solve the system of equations by the method of your choice.

21. $x + y = 15$
$y = -6x$

22. $x + y = 12$
$y + 4x = -3$

23. $3y - 4x = 2$
$8x = 6y + 4$

24. $3y - 4x = 2$
$8x = 6y - 4$

25. What does an algebraic result of $0 = 0$ tell us about the graph of a system of two linear equations?

26. Write an equation of a line parallel to $x + y = 5$. Solve the system formed by $x + y = 5$ and your new equation, and tell at what point in the algebra you know that there are no solutions to the system.

For Exercises 27 to 38, define variables and write a system of equations. Use guess and check as needed. Quantity-rate tables or equations may be helpful in many exercises. Solve the system.

27. The record centipede has 356 fewer legs than the record millipede. Together the two have 1064 legs. How many legs does each have?

28. A Boeing 747 holds 279 more people than a Boeing 707. If the two planes together carry 721 people, how many does each plane carry? (The total number of people includes crew.)

29. An English muffin and two fried eggs contain 330 calories. Three English muffins and one egg contain 515 calories. How many calories are in each item?

30. Suppose 5 grams of carbohydrate and 2 grams of fat contain 38 calories. Furthermore, 2 grams of carbohydrate and 6 grams of fat contain 62 calories. How many calories are in 1 gram of carbohydrate? in 1 gram of fat?

31. Mr. McFadden has a farm where he raises cows and ducks. There are 20 total heads among the animals and 64 total feet. How many cows and how many ducks are there? State your assumptions.

32. Mr. Schaaf builds 3-legged stools and 4-legged tables. He has finished 19 pieces of furniture altogether and counts 64 legs. How many of each type does he have?

33. A trout travels upstream 8 miles in 0.8 hour. Another trout travels downstream 16 miles in the same time. Assume that both trout travel at the same speed in still water. What is their speed in still water? What is the speed of the current?

34. An airplane makes a round trip between Atlanta and New Orleans. The trip is 450 miles each way. The flight takes 1.5 hours with the wind and 2.25 hours against the wind. What is the airspeed of the plane, and what is the speed of the wind?

35. Wayne owes \$800 on a car loan and has \$2200 charged on a credit card. Wayne's total annual interest would be \$452. Valorie owes \$2000 on a car loan at the same rate and has \$1000 charged on a credit card at the same rate. Valorie's total annual interest would be \$320. Find the rate of interest for the car loan and the rate of interest for the credit card.

36. Janet has borrowed \$3500. Some of it is an 8% auto loan, and the rest is at 21% on her credit card. She pays \$345 per year interest on the loans. How much has she borrowed from each source?

37. Angles A and B are complementary. The measure of angle A is 10° more than twice that of angle B. Find the angle measures.

38. Two angles are supplementary. Twelve degrees more than the measure of the larger angle is twice the measure of the smaller angle. Find the angle measures.

In Exercises 39 to 44, solve by substitution.

39. $A + B + C = 180$
$2A = B$
$6A = C$

40. $A + B + C = 180$
$B = C$
$\frac{4}{7}A = B$

41. $x + y = 6$
$y + z = 8$
$x + z = 7$

42. $y - x = 6$
$y + z = 5$
$x + z = 0$

43. $x + y = 6$
$y + z = 8$
$z - x = 2$

44. $y - x = 6$
$y + z = 5$
$x - z = -7$

For Exercises 45 to 49, start with guess and check. Use your guesses to define variables and write a system of three equations to describe each problem situation.

45. The perimeter of an isosceles triangle is 32 inches. The two equal sides are each 2.5 inches longer than the third side. What is the length of each side?

46. The measure of one angle of a triangle is 16° less than that of a second angle. The measure of a third angle equals the sum of those of the first two angles. The total measure of all three angles is 180°. What is the measure of each angle?

47. One serving of 7 medium shrimp, fried in vegetable shortening, contains 198 calories. Protein and carbohydrates contain 4 calories per gram; fat contains 9 calories per gram. The total weight of protein, fat, and carbohydrates in the shrimp is 37 grams. There are 5 more grams of protein than of carbohydrates. Find the number of grams of each (protein, fat, and carbohydrates) in this serving of shrimp.

48. The record number of children born to one mother is 69. There were 27 sets of births, including quadruplets, triplets, and twins. There were no single births. The woman bore four times as many sets of twins as sets of quadruplets. There were nine fewer sets of triplets than sets of twins. How many sets of twins, triplets, and quadruplets were delivered?

49. A cup of peanuts roasted in oil contains 903 calories. The total weight in protein, fat, and carbohydrates is 137 grams. Fat contains 9 calories per gram; protein and carbohydrates each contain 4 calories per gram. There are 44 more grams of fat than of carbohydrates. How many grams of each (protein, fat, and carbohydrates) are there in the peanuts?

50. Why does the intersection of two graphs solve the system of equations describing the graphs?

51. For each of the situations described in parts a to e, explain which of the following methods would be most appropriate for solving the system of equations:

Graphing
Table and graph on a graphing calculator
Guess and check
Substitution
Elimination

a. One variable has a 1 or -1 coefficient.

b. We are trying to understand a problem.

c. We have no algebraic method for a solution.

d. The coefficients are integers or can be made into integers with multiplication.

e. The equations are easily placed into $y = mx + b$ form.

In Exercises 52 to 57, graph the system of inequalities. Shade the solution set. Label the corner point.

52. $y > x + 3$
$y \leq -x + 2$

53. $x + y < 4$
$y \geq x$

54. $x > 3$
$y \geq -2$

55. $x \leq 2$
$y < 1$

56. $x \geq 0$
$y \geq 0$
$x \leq 5$
$y \leq -\frac{1}{2}x + 6$

57. $x \geq 0$
$y \geq 0$
$y \leq 4$
$y \leq -x + 6$

58. A wedding caterer charges \$36 for adults and \$20 for young people between 6 and 15. Children under 6 are free. The total food budget for a wedding is \$7200. Let $x =$ the number of adults and $y =$ the number of young people. At most, 100 young people will attend. Write a system of inequalities to describe the possible numbers of people over age 6 who can attend. Graph and show the solution set. Label the corner point.

59. A dieter is allowed 715 calories for a snack. Popcorn (salted, popped in vegetable oil) contains 55 calories per cup, and large pretzels contain 65 calories. Jean wants no more than 4 pretzels each day. Let $x =$ the number of cups of popcorn and $y =$ the number of pretzels. Write a system of inequalities to describe the possible numbers of each item she can eat. Graph and show the solution set. Label the corner point.

60. For each basketball game, the athletic department can sell 15,000 tickets. There are regular admission tickets and student tickets. The number of student tickets can be no more than 3000. Let $x =$ the number of regular tickets and $y =$ the number of student tickets. Write a system of inequalities to describe the possible numbers of each type of ticket sold. Graph and show the solution set. Label the corner point.

61. In comparing total cost as a function of number of miles driven, will there be one, no, or an infinite number of solutions in these cases?

a. Two rental truck companies have the same cost per mile but different basic charges.

b. Two rental truck companies have different costs per mile and the same basic charge.

c. Two rental truck companies have the same cost per mile and the same basic charge.

d. Two rental truck companies have different costs per mile and different basic charges. Give two answers and explain with a sketch of a graph.

62. Chapter Project: Racing Times

a. In a 500-mile Indy-car street race, Danica Patrick averaged 150 miles per hour for the entire race. Her main competitor, Dan Wheldon, averaged 160 miles per hour for the first 250 miles but, because of a slight brush with the safety wall, averaged 140 miles per hour for the second 250 miles of the race. Guess which driver won the race.

b. Draw a graph for Danica with $d = 150t$. Place time on the horizontal axis and distance on the vertical axis. Suppose $t_1 =$ time for Dan to drive the first 250 miles at 160 miles per hour. Calculate t_1. Draw a graph for Dan with $d = 160t$ for $t < t_1$ and $d = 140(t - t_1) + 250$ for $t > t_1$.

c. Assume each driver completed the 500 miles. How long did it take Danica to finish the race? How long did it take Dan to finish the race?

d. What is the diffrence in elapsed time, if any, between the drivers? Use unit analysis to change the time to seconds.

e. Who won? At what time and distance after the race started did the winner take the lead?

▶ 6 Chapter Test

In Exercises 1 and 2, solve for the indicated variable.

1. $3x - y = 400$ for y

2. $x - 2y = 3$ for y

In Exercises 3 to 8, solve the system of equations using any method.

3. $a + 2b = 6$
$3a - b = -17$

4. $2 = 5m + b$
$-1 = m + b$

5. $x + y = 5$
$y - 5 = -x$

6. $y = 2x - 1$
$y - 2x = 3$

7. $3x - 4y = 3$
$5x - 2y = 47$

8. $3x + 2y = 5700$
$x - y = 900$

9. What does "infinitely many solutions" tell us about the graph of a system of two linear equations?

10. What does an algebraic result of $0 = 1$ tell us about the graph of a system of two linear equations?

Use the figure for Exercises 11 to 13.

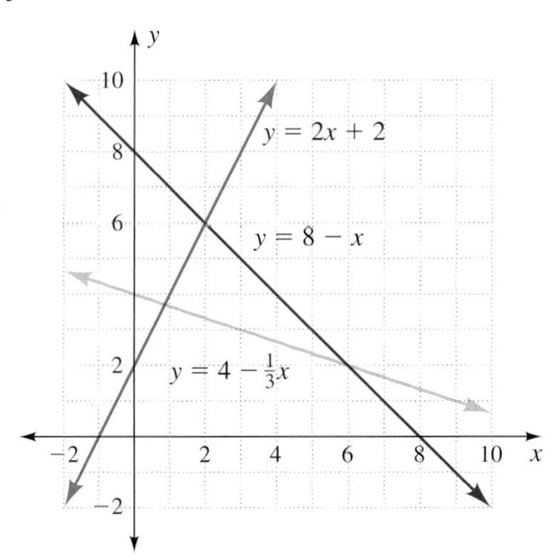

11. The point (6, 2) in the figure is the point of intersection of which two lines? What will be the result when you substitute the coordinates of the intersection into the two equations?

12. Shade the region in the figure that satisfies this system of inequalities. Label the corner point.

$y > 4 - \frac{1}{3}x$
$y \leq 8 - x$

13. Is (7, 1) in the solution set for the system containing $y < 2x + 2$ and $y \leq 8 - x$? Explain your answer.

In Exercises 14 to 17, identify variables and write equations to describe the problem situation. Solve the problem either by guess and check or by using your equations.

14. A typical caterpillar has ten more legs than a butterfly. Eight butterflies and six caterpillars have a gross (144) of legs. How many does each have?

15. A snack of 16 peanuts and 5 cashews contains 135 calories. Another snack of 20 peanuts and 25 cashews contains 405 calories. How many calories are in each peanut? in each cashew?

16. A dolphin travels 135 miles in 3 hours with an ocean current. Another dolphin travels 100 miles in 4 hours against the same current. Assume that the dolphins travel at the same speed in still water. What is the speed of the dolphins in still water, and what is the speed of the current?

17. Two angles are complementary. The measure of the larger angle less 12° equals 24° more than three times that of the smaller angle. Find the angle measures.

18. There are 216 coach seats on an airplane. The airline sells regular economy tickets and discount economy tickets. The airline limits the number of discount tickets on any flight to 50. Let $x =$ the number of regular tickets and $y =$ the number of discount tickets. Write a system of inequalities to describe the possible numbers of each type of ticket. Graph and show the solution set. Label the corner point.

19. Bridget's blend of coffee combines Colombian coffee beans at \$8.35 per pound with Sumatran beans at \$9.35 per pound. She wishes to blend 300 pounds to sell at \$9.15 per pound. How many pounds of each does she need?

20. Describe how you decide whether to use elimination or substitution to solve a system of equations.

21. Solve by substitution:

$y - z = 3$
$x + y = 2$
$x - z = -5$

22. Describe how one solution, no solutions, and an infinite number of solutions can arise in comparing locations for a birthday party.

▶ Cumulative Review of Chapters 1 to 6

Simplify the expressions in Exercises 1 to 4.

1. $14 - 6(-3 - 5) - 7$ **2.** $4^3 - 3(-2)$

3. $(x^4)^2$ **4.** $\dfrac{x^7}{x^4}$

5. Find the area of a triangle with base 3 inches and height 2 feet.

6. A continuous-flow pool measuring 8 feet by 15 feet permits swimming "in place" against a current with a rate up to 1 mile in 20 minutes. What is the rate in miles per hour?

7. Write in symbols and solve: A number less 6 is 8 more than twice the number.

8. Solve for x: $6x - 5 = 4x + 2$

9. Solve for y: $5x - 3y = 24$

10. Solve for h: $A = \dfrac{h(a + b)}{2}$

11. If $f(x) = 5 - 4x$, what are $f(3)$ and $f(-2)$?

12. What are the slope and y-intercept of the equation in Exercise 9?

13. What is a line parallel to $x = 2y - 3$ that passes through the origin?

14. How do the slopes of perpendicular lines compare?

15. Set up a proportion and solve:

a. 53% of a number is 140.45

b. What percent of 15 is 6?

16. A 48-inch child grows to 51 inches by her next birthday.

a. What is the percent change?

b. The new height is what percent of the old height?

17. Alan blends two types of coffee beans to sell at a price of \$8.75 per pound. Colombian coffee beans sell for \$8.35 per pound. Sumatran beans sell for \$9.35 per pound. To make 300 pounds of blend, how many pounds of each are needed?

Maria has \$15,000 to invest for one year. Investment A pays 5% interest for the year, and Investment B pays 8% interest. In Exercises 18 and 19, estimate the investments in A and B. Then set up a quantity-rate table and an equation to find how much is invested in each.

18. a. If her total earnings are \$1060, how much is invested in each?

b. If her total earnings are \$825, how much is invested in each?

19. a. If her total earnings are \$1005, how much is invested in each?

b. If her total earnings are \$1140, how much is invested in each?

20. Local gasoline stations have posted the per-gallon prices summarized in the table.

2006 Prices	Shell	Chevron	Arco*
Regular Octane	3.169	3.159	3.139
Medium Octane	3.269	3.259	3.239
High Octane	3.389	3.359	3.339

*Cash or debit card only with a \$0.45 debit card charge

a. You need gasoline. Quickly, what is your first choice?

b. If you knew about the \$0.45 charge in advance, would you go to the Arco station?

c. Write an equation for Shell Regular, for Chevron Regular, and for Arco Regular (with a debit card).

d. Write an inequality to compare Shell and Arco. For how many gallons is Arco a lower cost?

e. Write an inequality to compare Chevron with Arco. For how many gallons is Arco a lower cost?

f. Will the results for parts d and e change when buying Medium Octane?

g. Will the results for parts d and e change when buying High Octane?

h. For \$10 in Regular gas, where do you go?

21. A classic study of sparrow migration was conducted by Margaret Morse Nice in Columbus, Ohio, between 1930 and 1935. The following data are from page 54 (Table IV) and page 55 (Table V) *Studies in the Life History of the Song Sparrow*, Vol. I, (New York: Dover, 1964). Make observations about the data and support your statements with mathematics.

Bird	1931	1932	Sex of Bird
#2	Mar 20	Mar 1	male
#10	Apr 3	Mar 26	male
#23	Apr 3	Mar 21	male
#24	Mar 23	Feb 26	male
#47	Mar 10	Feb 26	male
#62	Mar 23	Mar 19	male
#64	Mar 30	Feb 26	male
#11	Apr 3	Mar 25	female
#14	Apr 3	Mar 29	female
#41	Mar 24	Mar 20	female
#46	Apr 3	Mar 25	female
#52	Apr 1	Mar 1	female
#58	Mar 24	Mar 3	female
#60	Apr 3	Mar 28	female

CHAPTER 7

Polynomial Expressions and Integer Exponents

In 1974, the coded message in Figure 1 was sent by the transmitter at Arecibo, Puerto Rico, toward the star cluster M13, located 27,000 light-years from earth. The coded message is a set of 1679 ones and zeros. The number 1679 is factorable into only two primes, 23 and 73. When the ones and zeros are arranged as black and white squares in a rectangular grid of dimensions 23 by 73, the pictorial message appears.

In this chapter, we turn our attention to computation with algebraic notation. In Sections 7.1 to 7.4, factoring (such as that needed by distant life forms) is reviewed and extended to expressions called polynomials. Use of the basic operations—addition, subtraction, multiplication, and division—with polynomials is the main focus of the early sections. In Section 7.5, we investigate how many miles the coded message will travel in 27,000 years. Among the skills covered are working with exponents and working with large and small numbers written in scientific notation.

FIGURE 1 Arecibo coded message

▶ 7.1 Operations on Polynomials

Objectives

- Identify monomials, binomials, and trinomials.
- Arrange terms in descending order of exponents.
- Add and subtract polynomials.
- Multiply polynomials.
- Factor the greatest common monomial factor from an expression.
- Distinguish between terms and factors.

Student Note:
According to the positive integer definition of exponents, $x^3 = x \cdot x \cdot x$, so $x^3 = x^2 \cdot x$.

WARM-UP

These exercises review ideas in Section 2.3. In Exercises 1 to 3, use the distributive property to multiply.

1. $7(x + 5)$

2. $-6a(a - b - 2c)$

3. Complete the table by multiplying each expression on the top by the expression on the left.

Multiply	$2a$	$-2b$	$+c$
$5a$			

In Exercises 4 and 5, divide the expressions.

4. $\dfrac{6a + 3b}{3}$ **5.** $\dfrac{8x^3 + 4x^2 - 6x}{2x}$

Add or subtract like terms or like units in Exercises 6 to 8.

6. 5 kg + 11 kg − 2 kg

7. $\frac{1}{2}x^2 + 3x^2 + 0.5x^2$

8. 5 mL + 12 mL − 3 mL

9. Which numbers in Exercises 6, 7, and 8 are not integers?

10. Give an example of a negative integer.

THIS SECTION RETURNS to the algebra tile model and ideas introduced in Section's 2.3 and 2.4—applying the distributive property, adding like terms, dividing, factoring, and the definition of integer exponents. We use like terms and factoring with special types of algebraic expressions called polynomials.

▶ Polynomials

▶ EXAMPLE 1 **Exploring with tiles: width, length, perimeter, and area** The shapes on the next page represent algebra tiles. Arrange each set into a rectangle, and sketch the rectangle in the first column of Table 1. Then complete the rest of the table for each rectangle. Let a = the side of the large square, let b = the side of the small square, and let a and b be the large and small sides of the rectangle, respectively.

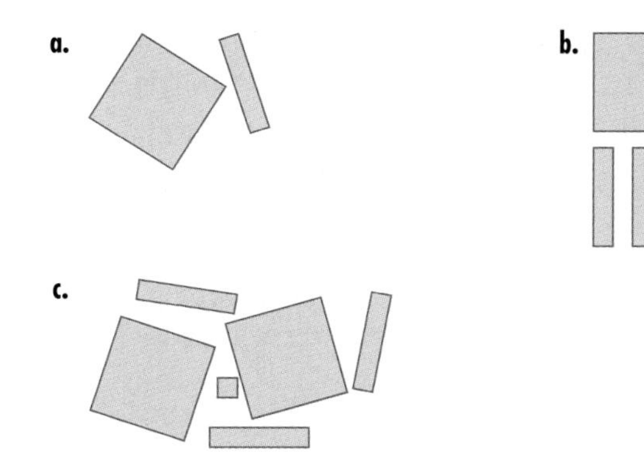

TABLE 1

Sketch of Rectangle	Width (height)	Length (base)	Perimeter	Area
a.				$a^2 + ab$
b.		$a + b$		
c.			$6a + 4b$	

SOLUTION See the Answer Box. ◀

Expressions such as $a^2 + ab$, $a^2 + 2ab + b^2$, and $2a^2 + 3ab + b^2$ (the areas in Example 1) are called polynomials. Polynomials may be used to describe many things: consecutive numbers, the path of a rocket, the height of the rocket given the time after launch, the surface area of a cylinder, the resistance to bending by a structural beam, and the amount of deflection of a beam when loaded.

DEFINITION OF POLYNOMIAL

> A **polynomial** is an expression with one or more terms being added or subtracted. The exponents on the variables in each term must be non-negative (zero or positive) integers.

SPECIAL NAMES The prefix *poly* means "many." *Polynomial* means "many terms." A *polynomial expression with one, two, or three terms* is called, respectively, a **monomial**, **binomial**, or **trinomial**.

Number of terms	Example	Name
one	$2a^2b$	monomial
two	$x^2 + xy$	binomial
three	$x^2 + 2x + 1$	trinomial

No special names are given to expressions containing four, five, or more terms—all are polynomials. The expression $1x^5 + 2x^4 + 3x^3 + 4x^2 + 5x + 6$ is a six-term polynomial.

▶ **EXAMPLE 2** Identifying monomials, binomials, and trinomials How many terms are in each expression and what is the polynomial name for each?

a. $4 + 3x + x^2$ **b.** $3a + a^2$ **c.** x^2

d. $2b + 3b^2$ **e.** $3 - 2x - x^2$ **f.** $-\frac{1}{2}gt^2$

SOLUTION

a. 3 terms; trinomial **b.** 2 terms; binomial **c.** 1 term; monomial

d. 2 terms; binomial **e.** 3 terms; trinomial **f.** 1 term; monomial ◀

THINK ABOUT IT 1: Write a three-term polynomial with variable x. Write a four-term polynomial with variable y.

ORDER OF TERMS The expressions in Example 2 are written in **ascending order**; that is, the terms are listed with *the exponents from lowest to highest*. When we list polynomials with *the highest exponents first*, we have **descending order**. The expression $1x^5 + 2x^4 + 3x^3 + 4x^2 + 5x + 6$ is in descending order of exponents on x. Descending order makes it easy to compare polynomials (for example, to compare your answers to the exercises with those in the back of the book).

▶ **EXAMPLE 3** Arranging terms Write the terms in descending order of exponents.

a. $4 + 3x + x^2$ **b.** $3a + a^2$ **c.** $2b + 3b^2$

d. $3 - 2x - x^2$ **e.** $3x^2 + x^4 - x^3 + 5 - x$

SOLUTION **a.** $x^2 + 3x + 4$ **b.** $a^2 + 3a$ **c.** $3b^2 + 2b$

d. $-x^2 - 2x + 3$ **e.** $x^4 - x^3 + 3x^2 - x + 5$ ◀

▶ If there are two variables, list the variables in each term alphabetically.

▶ **EXAMPLE 4** Arranging terms Write the terms in descending order of exponents for the indicated variable.

a. $x^2y^2 + x^3y + xy^3$ for x

b. $ab^2 - a^3b^2 + a^2b^2$ for a

SOLUTION **a.** $x^3y + x^2y^2 + xy^3$

b. $-a^3b^2 + a^2b^2 + ab^2$ ◀

ARRANGING TERMS IN POLYNOMIALS

Unless told to do otherwise, arrange the variables in each term in alphabetical order and then arrange the terms so that the exponents on the first variable are in descending order:

- $x^5 + x^4y + x^2y^2 - xy$
- $x^2 + xy + xz + y^2 + yz + z^2$

▶ Adding and Subtracting Polynomials

To add or subtract polynomials, combine like terms. When all additions and subtractions of like terms within a polynomial have been completed and the terms are arranged in descending order of exponents, the polynomial is said to be *simplified*.

▶ **EXAMPLE 5** Adding and subtracting polynomials Simplify by adding or subtracting like terms.

a. $3a^2 + 4b^2 - 2a^2 - 8b^2 + 4a^2$

b. $(x^2 + 2x - 1) - 2(x^2 - 2x + 1)$

c. $(-8a^2 + 2ab + 5b^2) - (-9a^2 - 5ab + 6b^2)$

d. $(x^3 - 6x^2 + 9x) + (-3x^2 + 18x - 27)$

SOLUTION **a.** $3a^2 + 4b^2 - 2a^2 - 8b^2 + 4a^2$ Arrange the terms in alphabetical order.

$= 3a^2 - 2a^2 + 4a^2 + 4b^2 - 8b^2$ Add like terms.

$= 5a^2 - 4b^2$

b. $(x^2 + 2x - 1) - 2(x^2 - 2x + 1)$ — Change the subtraction to addition of the opposite.

$= x^2 + 2x - 1 + (-2)(x^2 - 2x + 1)$ — Apply the distributive property.

$= x^2 + 2x - 1 - 2x^2 + 4x - 2$ — Add like terms.

$= -x^2 + 6x - 3$

c. $(-8a^2 + 2ab + 5b^2) - (-9a^2 - 5ab + 6b^2)$ — Change the subtraction to addition of the opposite.

$= (-8a^2 + 2ab + 5b^2) + (-1)(-9a^2 - 5ab + 6b^2)$
Apply the distributive property.

$= -8a^2 + 2ab + 5b^2 + 9a^2 + 5ab - 6b^2$ — Add like terms.

$= a^2 + 7ab - b^2$

d. $(x^3 - 6x^2 + 9x) + (-3x^2 + 18x - 27)$ — The distributive property is not needed.

$= x^3 - 6x^2 + 9x - 3x^2 + 18x - 27$ — Add like terms.

$= x^3 - 9x^2 + 27x - 27$ ◀

THINK ABOUT IT 2: Give the polynomial name of each answer in Example 5.

Multiplying Polynomials

▶ **EXAMPLE 6** Exploring with tiles Lengths and widths for six rectangles are shown. Sketch in the tiles needed to complete the rectangles, and write the total area for each part. In parts a, b, and c, the side of the large square tile is a, and the side of the small square tile is b; the rectangle is a by b. Another way to name the tiles is shown in parts d, e, and f: The side of the large square tile is x, and the side of the small square tile is 1; the rectangle is x by 1.

a. $a(a + b)$ **b.** $2a(a + 2b)$ **c.** $b(a + b)$

d. $2x(x + 1)$ **e.** $3(x + 1)$ **f.** $x(x + 3)$

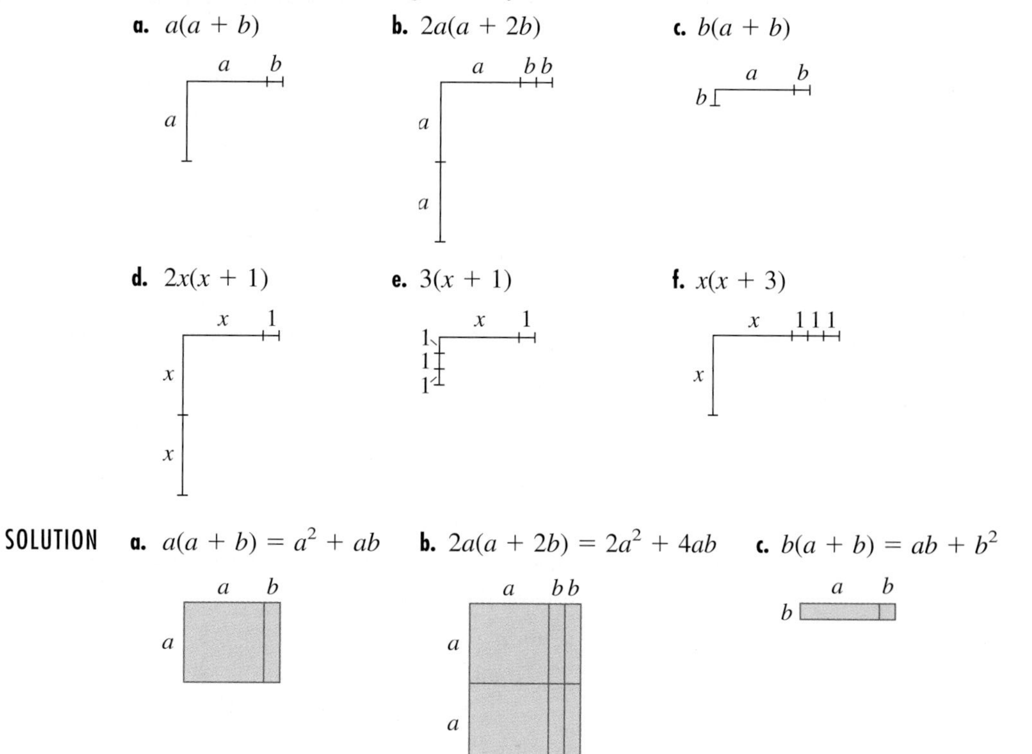

SOLUTION **a.** $a(a + b) = a^2 + ab$ **b.** $2a(a + 2b) = 2a^2 + 4ab$ **c.** $b(a + b) = ab + b^2$

d. $2x(x + 1) = 2x^2 + 2x$ **e.** $3(x + 1) = 3x + 3$ **f.** $x(x + 3) = x^2 + 3x$

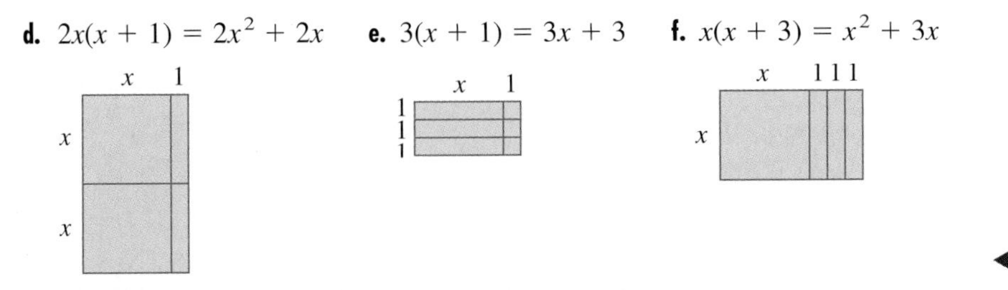

◀

In Example 6, we found that the total area of the tiles was the product of the length and width of the rectangle. We have seen this multiplication before in the distributive property: $a(b + c) = ab + ac$.

MULTIPLYING A MONOMIAL AND A BINOMIAL The distributive property is the basis for all multiplication of polynomials. In Example 7, we multiply a monomial and a binomial using the distributive property and observe how the product relates to the area of the algebra tiles.

▶ **EXAMPLE 7** Multiplying a monomial and a binomial Apply the distributive property to multiply these expressions. Explain how the algebra tile figures show the same product.

a. $a(a + b)$ **b.** $3b(a + b)$

c. $a(a + 3b)$ **d.** $b(a + 3b)$

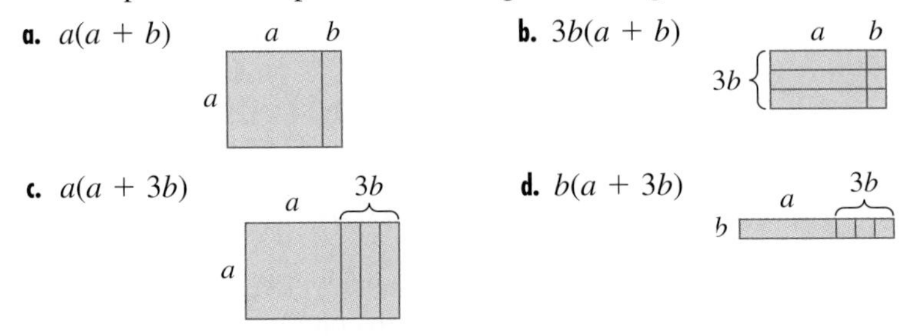

SOLUTION **a.** $a(a + b) = a \cdot a + a \cdot b = a^2 + ab$
The rectangle formed by the tiles has a length of $a + b$. The width of the rectangle is a. The area of the tiles is a large square plus a rectangle: $a^2 + ab$. The area agrees with the product from the distributive property.

b. $3b(a + b) = 3b \cdot a + 3b \cdot b = 3ab + 3b^2$
The rectangle formed by the tiles has a length of $a + b$. The width of the rectangle is $3b$. The area of the tiles is three rectangles plus three small squares: $3ab + 3b^2$. This area agrees with the product from the distributive property.

c. $a(a + 3b) = a \cdot a + a \cdot 3b = a^2 + 3ab$
We have one large square and three rectangles: $a^2 + 3ab$. This area agrees with the product.

d. $b(a + 3b) = b \cdot a + b \cdot 3b = ab + 3b^2$
We have one rectangle and three small squares: $ab + 3b^2$. This area agrees with the product. ◀

MULTIPLYING A MONOMIAL AND A POLYNOMIAL Example 8 shows that the distributive property may be applied to three or more terms. Part d gives a multiplication table form of the distributive property.

▶ **EXAMPLE 8** Multiplying a monomial and a trinomial or other polynomial Use the distributive property on these expressions.

a. $6a(a^2 + 2a + 1)$ **b.** $x(x^4 + 2x^3 + 3x + 1)$

c. $-4ab(a^2 - 2ab + b^2)$ **d.**

Multiply	x	$+2x^2$	$-4x^3$
$3x^2$			

SOLUTION **a.** $6a(a^2 + 2a + 1) = 6a^3 + 12a^2 + 6a$

b. $x(x^4 + 2x^3 + 3x + 1) = x^5 + 2x^4 + 3x^2 + x$

c. $-4ab(a^2 - 2ab + b^2) = -4a^3b + 8a^2b^2 - 4ab^3$

d.

Multiply	x	$+2x^2$	$-4x^3$
$3x^2$	$3x^3$	$+6x^4$	$-12x^5$

◀

Factoring

When we write the distributive property in reverse, we are said to be factoring the sum of two terms, $ab + ac$, into a product $a(b + c)$.

FACTORING WITH THE DISTRIBUTIVE PROPERTY

For real numbers a, b, c,

$$ab + ac = a(b + c)$$

The factor a is called the **common monomial factor** because it is *a factor that appears in each term* of the polynomial (left side). The **greatest common factor (gcf)** is *the largest possible common factor of all terms.*

▶ **EXAMPLE 9** Finding the greatest common factor, or gcf What is the greatest common factor for each expression?

a. $3x + 3$ **b.** $ab + b^2$ **c.** $a^2b - a^2b^2 + a^2b^3$ **d.** $x^2 - 1$

SOLUTION **a.** gcf $= 3$ **b.** gcf $= b$ **c.** gcf $= a^2b$

d. gcf $= 1$; there is no other common factor. ◀

▶ To factor $3ab + 3b^2$, we reverse the distributive property, which we can do in three ways.

First, we can circle the greatest common factor in each term and use what remains to find the other factor:

$$3ab + 3b^2 = (3)a(b) + (3b)b = 3b(a + b)$$

Second, for many expressions, we can use tiles. As shown in Figure 2, we can build a rectangle with the greatest common factor, $3b$, as the width. Again, we have

$$3ab + 3b^2 = 3b(a + b)$$

FIGURE 2 The gcf, $3b$, appears as a length or width.

Third, we can use a table. We place the gcf on the left, and the second factor appears across the top of the table, as in the tile model. In the table, the operation sign goes with the following term.

Factor		
	$3ab$	$+3b^2$

	Factor	a	$+b$
gcf →	$3b$	$3ab$	$+3b^2$

EXAMPLE 10 Factoring, given the gcf

Student Note:
Given the gcf, our factoring is a division. Division is usually what we consider to be opposite to multiplication.

a. Factor $a^2 + 3ab + a$, given that gcf $= a$.

b. Factor $x^2y - 4y^3$, given that gcf $= y$.

c. Using the following table, factor $4x^2y^2 - 4x^3y^3 + 2x^4y^4$, given that gcf $= 2x^2y^2$.

Factor			
$2x^2y^2$	$4x^2y^2$	$-4x^3y^3$	$+2x^4y^4$

SOLUTION

a. $a^2 + 3ab + a = a(a + 3b + 1)$

b. $x^2y - 4y^3 = y(x^2 - 4y^2)$

c.

Factor	2	$-2xy$	$+x^2y^2$
$2x^2y^2$	$4x^2y^2$	$-4x^3y^3$	$+2x^4y^4$

$$4x^2y^2 - 4x^3y^3 + 2x^4y^4 = 2x^2y^2(2 - 2xy + x^2y^2)$$

To check, multiply using the distributive property. ◀

When the terms are all negative, the gcf is negative. In Example 11, we find both the gcf and the other factor.

EXAMPLE 11 Finding the greatest common factor and factoring the expression Identify the greatest common factor, or gcf, and then apply the distributive property to factor each sum into a product.

Student Note:
We are undoing a multiplication. If we can find a gcf, then we "divide" to find the other factor. Other factoring strategies will be covered in Sections 7.3 abnd 7.4.

a. $4x + 6$ **b.** $-4x - 8$ **c.** $x^2 - x$

d. $2ab - 8a^2b + 6ab^2$ **e.** $-10x^2 - 10x - 5$ **f.** $x^2y^2 + 2xy^3 - y^3$

In parts g and h, place the gcf on the left and the second factor will appear across the top of the table.

g.

Factor			
	$8a^3$	$+12a^2$	$-16a$

h.

Factor		
	x^2	$+3x$

SOLUTION

a. gcf $= 2$; $4x + 6 = 2(2x + 3)$

b. gcf $= -4$; $-4x - 8 = -4(x + 2)$

c. gcf $= x$; $x^2 - x = x(x - 1)$

d. gcf $= 2ab$; $2ab - 8a^2b + 6ab^2 = 2ab(1 - 4a + 3b)$

e. gcf $= -5$; $-10x^2 - 10x - 5 = -5(2x^2 + 2x + 1)$

f. gcf $= y^2$; $x^2y^2 + 2xy^3 - y^3 = y^2(x^2 + 2xy - y)$

g. gcf $= 4a$

Factor	$2a^2$	$+3a$	-4
$4a$	$8a^3$	$+12a^2$	$-16a$

h. gcf $= x$

Factor	x	$+3$
x	x^2	$+3x$

◀

Factors versus Terms

As mentioned in Chapter 3, confusing the words *terms* and *factors* is a common error. Factors are the numbers, variables, or expressions being multiplied in a product. Terms are the expressions being added or subtracted. The product $(a + b)(a + b)(a - b)$ has three factors, and each factor is a two-term expression.

In Example 12, we identify terms, and in Example 13, we identify factors.

EXAMPLE 12 Identifying terms How many terms are there in these expressions?

a. $x^2 + 4x + 2$ **b.** $-b + d$ **c.** $3 - 2x$

d. $1 + 3y + 3y^2 + y^3$ **e.** $x^2 - y^2$

SOLUTION See the Answer Box. ◀

EXAMPLE 13 Identifying factors How many factors are there in each of these products?

a. $wxyz$ **b.** $(x + y)(x - y)$ **c.** $x(x + 1)(x + 2)$

d. $n(n - 1)$ **e.** $(x - 1)(x^2 + x + 1)$ **f.** πr^2

SOLUTION See the Answer Box. ◀

ANSWER BOX

Warm-up: **1.** $7x + 35$ **2.** $-6a^2 + 6ab + 12ac$ **3.** $10a^2 - 10ab + 5ac$ **4.** $2a + b$ **5.** $4x^2 + 2x - 3$ **6.** 14 kg **7.** $4x^2$ **8.** 14 mL **9.** $\frac{1}{2}$ and 0.5 **10.** any number in $\{\ldots, -3, -2, -1\}$

Example 1:

Sketch of Rectangle	Width (height)	Length (base)	Perimeter	Area
a. $a + b$; a	a	$a + b$	$4a + 2b$	$a^2 + ab$
b. $a + b$; a, b	$a + b$	$a + b$	$4a + 4b$	$a^2 + 2ab + b^2$
c. $2a + b$; a, b	$a + b$	$2a + b$	$6a + 4b$	$2a^2 + 3ab + b^2$

Think about it 1: For example, $x^2 + x + 1$, $y^3 + y^2 + y + 1$
Think about it 2: **a.** binomial **b.** trinomial **c.** trinomial **d.** polynomial
Example 12: **a.** 3 terms **b.** 2 terms **c.** 2 terms **d.** 4 terms **e.** 2 terms
Example 13: **a.** 4 factors **b.** 2 factors **c.** 3 factors **d.** 2 factors **e.** 2 factors **f.** 3 factors

▶ 7.1 Exercises

Exercises 1 and 2 review integer addition and subtraction.

1. a. $-9 + 4$ **b.** $-8 + 3$ **c.** $-5 - 8$

d. $-7 - 3$ **e.** $-9 - (-5)$ **f.** $3 - (-4)$

2. a. $-4 + 6$ **b.** $-5 + 1$ **c.** $-4 - 3$

d. $-3 - 9$ **e.** $9 - (-8)$ **f.** $-5 - (-1)$

Exercises 3 and 4 review integer multiplication.

3. a. $8(-7)$ **b.** $-6(7)$ **c.** $-4(-9)$

4. a. $-9(6)$ **b.** $-7(-9)$ **c.** $-6(-8)$

In Exercises 5 and 6, arrange the terms in descending order of exponents on *x*.

5. a. $5 - 3x^2 + 5x - x^3$ **b.** $4x + 3x^3 - 3 + 2x^2$

6. a. $x^2 - x^4 + 1 - x$ **b.** $-3 + x^5 - 4x^2 + x^3$

In Exercises 7 to 10, use the distributive property as needed and add or subtract like terms. Arrange the terms alphabetically, with exponents on the first variable in descending order. Identify each answer as a monomial, a binomial, a trinomial, or a polynomial with more than three terms.

7. a. $5a + 3b - 2c - 4a - 6c + 9b$

b. $6m + 2n - 6p - 3m - 3n + 6p$

c. $8y + 5y + 5y - 5y + 8y$

d. $(x^2 + 3x) - (4x + 12)$

e. $(x^2 - 2x + 3) - (2x^2 - 4x + 6)$

8. a. $a - 8c - d + a + 4d - 6c$

b. $5m - 4p + 8n - 9p + 8m - 7n$

c. $9y + 7y + 7y + 8y - 2y$

d. $(2x^2 + 4x) - (3x - 4)$

e. $(y^2 + 3y - 2) - (3y^2 + 9y - 6)$

9. a. $x^2 + 2x + 3x + 6$

b. $3x^2 + 6x + x + 2$

c. $5 - 4(x - 3)$

d. $6x^2 + 2x + 3x + 1$

e. $a^2 - ab + ab - b^2$

10. a. $x^2 + x + 2x + 2$

b. $2x^2 + 3x + 4x + 6$

c. $3 + 4(x - 5)$

d. $2x^2 - 2x + 3x - 3$

e. $a^2 + b^2 + ab + ab$

Blue numbers are core exercises.

Exercises 11 to 14 show rectangles made from algebra tiles. Find the length, width, perimeter, and area of each rectangle.

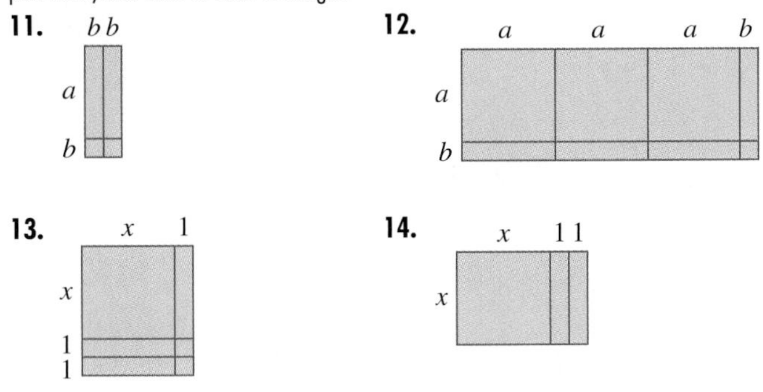

In Exercises 15 and 16, multiply the expressions. The exercises can be done with algebra tiles.

15. a. $2a(a + 2b)$ **b.** $2b(2a + b)$

c. $x(x + 3)$ **d.** $x(2x + 1)$

16. a. $a(a + 3b)$ **b.** $b(3a + b)$

c. $x(3x + 1)$ **d.** $2x(x + 2)$

In Exercises 17 to 20, multiply, using the definition of exponents as needed.

17.

Multiply	x^2	$-3xy$	$-4y^2$
$4x^2y^4$			

18.

Multiply	x^2	$-5xy$	$+y^2$
$3x^3y^5$			

19.

Multiply	x^2	$-2xy$	$-y^2$
$-5x^3y^4$			

20.

Multiply	x^2	$-4xy$	$-y^2$
$-2x^2y^3$			

In Exercises 21 and 22, multiply and simplify as needed.

21. a. $2x(x^2 + 3x)$ **b.** $x^2(x - 1)$

c. $x^2(x^2 + 2x + 1)$ **d.** $ab(b^2 - 1)$

e. $b^2(a - b)$ **f.** $a^2(1 + b - b^2)$

22. a. $3x(x - 3x^2)$ **b.** $x(x^2 + 4)$

c. $x^2(1 - x - x^2)$ **d.** $b^2(a - a^2)$

e. $ab(b - a)$ **f.** $a^2(1 - b + b^2)$

Simplify the expressions in Exercises 23 to 28.

23. **a.** $5x - 3x(1 - x)$ **b.** $4a + 2a(3 - a)$

24. **a.** $8b + 2b(b + 1)$ **b.** $7a - 2a(4 - a)$

25. **a.** $4b - 2b(5 - b)$ **b.** $6x - x(x + 1)$

26. **a.** $3x - x(2 - x)$ **b.** $5b - 2b(7 - b)$

27. **a.** $2y(-x + 2y) + x(x - 2y)$

b. $x^2 - 4y^2 - 2xy + 2xy$

c. $x^3 + 2x^2 + x + x^2 + 2x + 1$

d. $x(4 + 4x + x^2) - 2(4 + 4x + x^2)$

e. $b^3 - ab^2 + a^2b + a^3 - a^2b + ab^2$

28. **a.** $3y(3y + x) - x(x + 3y)$

b. $-9y^2 + 3xy + x^2 - 3xy$

c. $-x^2 + 2x - 1 + x^3 - 2x^2 + x$

d. $2(4 - 2x + x^2) + x(4 - 2x + x^2)$

e. $a^3 - a^2b + ab^2 - b^3 - ab^2 + a^2b$

In Exercises 29 to 32, the greatest common factor for the expression inside the table is on the left. Write the other factor on the top of the table.

29.

Factor			
y	y^3	$+2xy^2$	$+x^2y$

30.

Factor			
x^2y^2	x^2y^3	$-x^3y^2$	$+x^2y^2$

31.

Factor			
$2ab$	$2a^2b$	$-4ab^2$	$+6ab^3$

32.

Factor			
$4ab^2$	$4ab^3$	$-8a^2b^3$	$-16a^2b^2$

In Exercises 33 to 36, write the gcf on the left and then write the other factors above.

33.

Factor		
	$2a$	$-4ab$

34.

Factor		
	x^2y	$-xy^2$

35.

Factor			
	ab^2	$-a^2b^2$	$-a^2b^3$

36.

Factor			
	$6xy$	$-3y^2$	$+9y^3$

For the expressions in Exercises 37 to 46, name the greatest common factor and then factor the expression.

37. $x^3 + 4x^2 + 4x$

38. $x^2y + xy^2 + y^3$

39. $a^2b + ab^2 + b^3$

40. $a^3 - a^2b + ab^2$

41. $6x^2 + 2x$

42. $12y^2 + 6y$

43. $15y^2 - 3y$

44. $8x^2 - 4x$

45. $15x^2y + 10xy^2$

46. $9x^2y^2 + 18xy^2 - 24y^2$

Factor the expressions in Exercises 47 to 54 in two ways. First, use a positive sign on the common factor. Second, use a negative sign on the common factor.

47. $-4x - 12$

48. $-12y - 3$

49. $-2xy + 4y^2$

50. $-6x^2 + 18xy$

51. $-12x^2 - 8x - 8$

52. $-x^2y - xy^2 - y^3$

53. $-y^2 + 4y^3 - 8y^4$

54. $-10x^2 + 15xy - 20y^2$

How many terms are in each expression in Exercises 55 to 60?

55. $x^2 + 5xy$

56. $2xyz$

57. $\frac{1}{2}gt^2$

58. mc^2

59. $4x^2 + 8x + 8$

60. $2\pi rh + 2\pi r^2$

How many factors are in each expression in Exercises 61 to 66?

61. $4(x + 2)(x + 2)$

62. $x(x + 5y)$

63. $2\pi r(h + r)$

64. mc^2

65. $2xyz$

66. $\frac{1}{2}gt^2$

Blue numbers are core exercises.

▶ Writing

67. Why are the expressions in Exercises 55 and 62 equal?

68. Why are the expressions in Exercises 60 and 63 equal?

69. Complete the sentence: The distributive property, $a(b + c) = ab + ac$, changes the product of two ________ into the ________ of two ________.

70. Complete the sentence: Factoring $ab + ac = a(b + c)$ changes the sum of ________ terms into the ________ of two ________.

71. Describe how to tell if two terms are "like" terms.

72. There is a saying "You cannot add apples and oranges." How does this saying apply to adding like terms?

73. Describe how to find the greatest common factor.

74. Explain the difference between terms and factors.

▶ Number Review

In Exercises 75 and 76, use guess and check on a calculator to factor the numbers into prime factors.

75. a. 111 **b.** 91

76. a. 1001 **b.** 129

What is the greatest common factor of the numerator and denominator in each of the expressions in Exercises 77 and 78? Simplify by eliminating common factors.

77. a. $\frac{36}{99}$ **b.** $\frac{66}{990}$

c. $\frac{185}{999}$ **d.** $\frac{mn}{mp}$

e. $\frac{4np}{24mn}$

78. a. $\frac{407}{999}$ **b.** $\frac{108}{999}$

c. $\frac{39}{1001}$ **d.** $\frac{xz}{xy}$

e. $\frac{3xy}{9yz}$

▶ Connecting with Chapter 2

79. Find the perimeter of each shape.

a.

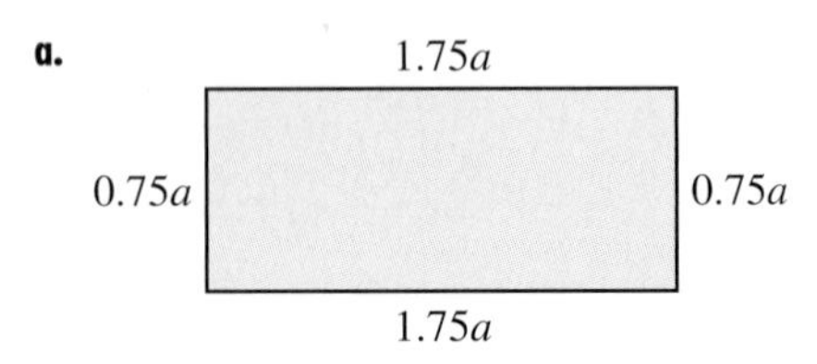

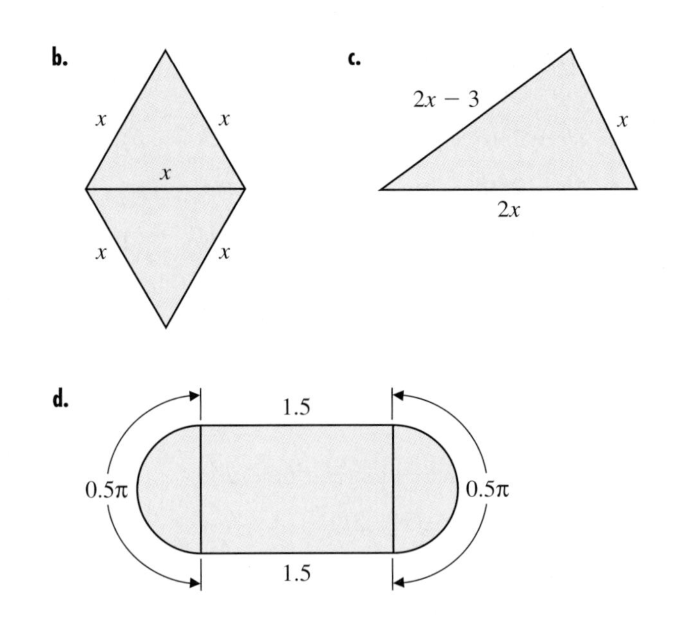

80. Find the perimeter of each shape.

a.

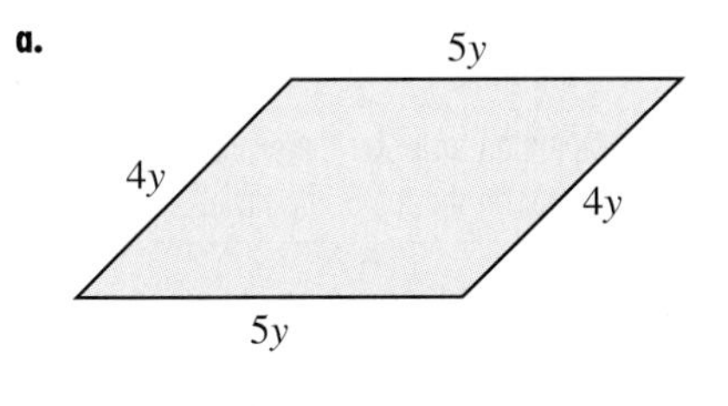

b.

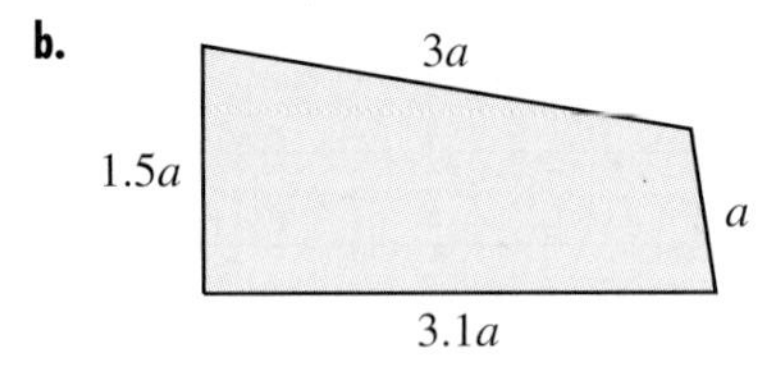

c.

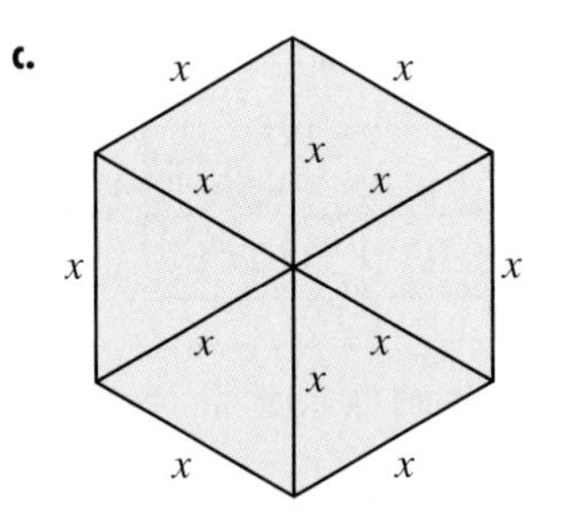

d.

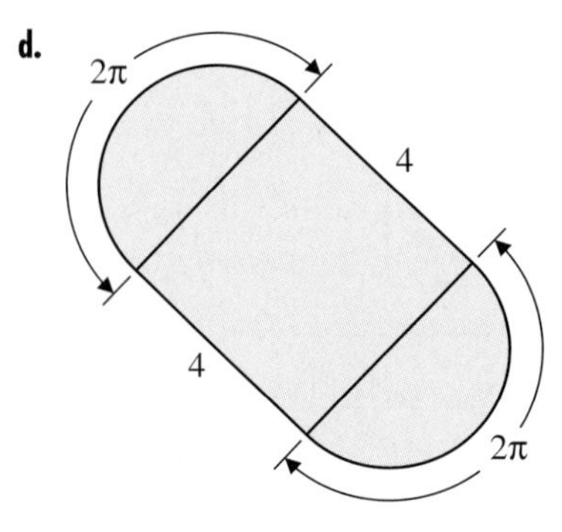

In Exercises 81 and 82, use the given perimeter, P, to find the missing side in each figure. Then find the area of each figure.

81. **a.** **b.** **c.**

82. **a.** **b.** **c.**

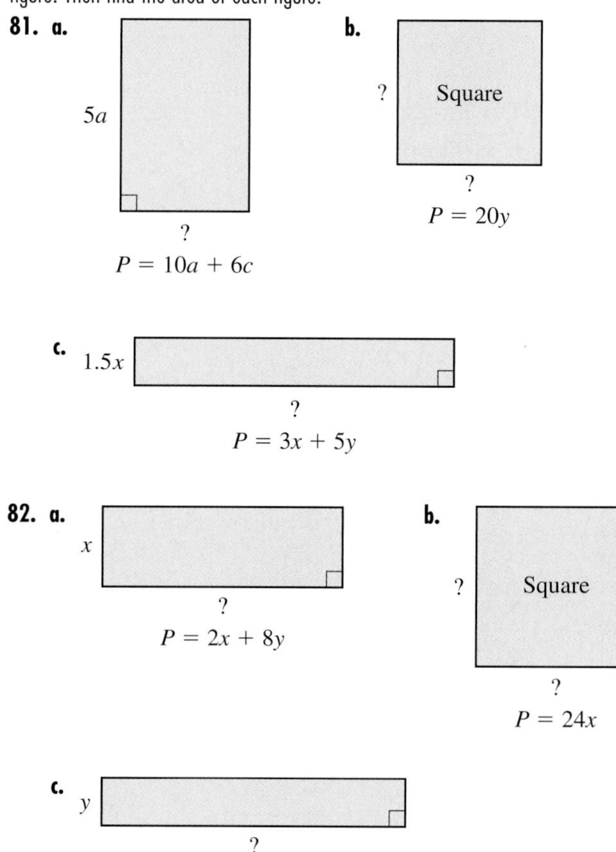

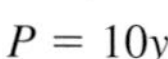

▶ Problem Solving

83. Find the perimeter and area of the squares in parts a and b.

a. **b.**

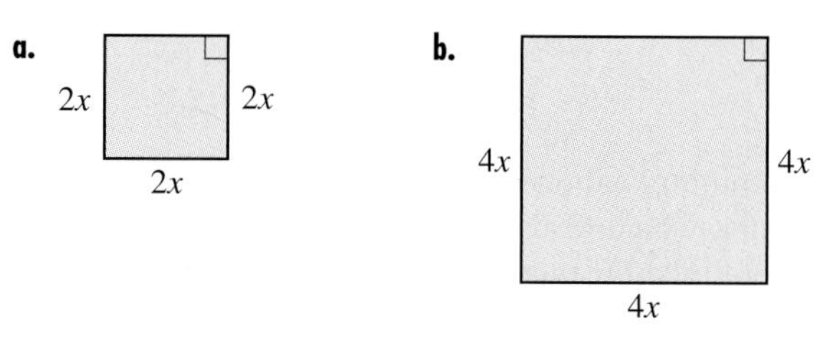

c. Using a fraction, compare the perimeter of the square in part a to that of the square in part b.

d. Using a fraction, compare the area of the square in part a to that of the square in part b.

84. Find the surface area and volume of the cubes in parts a and b.

a. **b.**

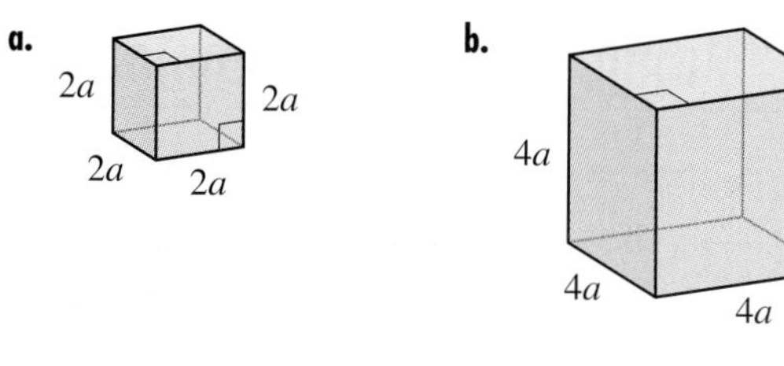

c. Using a fraction, compare the surface area of the cube in part a to that of the cube in part b.

d. Using a fraction, compare the volume of the cube in part a to that of the cube in part b.

▶ Projects

85. Coded Message Scientists assume that intelligent extraterrestrials will factor 1679 into $23 \cdot 73$ and find the message in Figure 1 on the chapter-opening page.

a. Using zeros and ones to transmit data is common on Earth. Dot matrix printers, an early form of printer for personal computers, relied on a rectangular array of dots to form each letter or character. Suppose a dot matrix printer head has a rectangular array, 8 squares wide by 9 squares tall. Arrange the following message into an 8 by 9 array, and then decode the message by shading each 1 and leaving each zero unshaded.

001111000100001001000010001111000100
001001000010001111000000000000000
00000

b. Suppose the following 15-digit coded message, simi lar to the one on the opening page of this chapter, was sent. What are the factors of 15? Into what rectangular shapes could the data be arranged? Only one of the possible rectangles will make sense.

101011110110101

c. Repeat part b for the following 50-digit message.

1100010100010001010111000101110100010001
1101010001

d. Why was the coded message on the chapter opener sent in symbols instead of words?

e. Research the meaning of the message shown in Figure 1 of the chapter opener. The Arecibo message was written by Frank Drake (and others) of Cornell University in 1974.

▶ 7.2 Multiplication of Binomials and Special Products

Objectives

- Multiply binomials with the tile model and the table method.
- Identify patterns in the products of binomial factors.
- Multiply binomials mentally.
- Identify binomials that multiply to perfect square trinomials and differences of squares.

WARM-UP

Complete the table.

	m	n	Sum $m + n$	Product $m \cdot n$
1.	3	4		
2.			8	12
3.			8	15
4.	1	15		
5.	4	6		
6.			11	24
7.	−4	6		
8.	2	−12		
9.			5	−24
10.			−11	24

IN THIS SECTION, we focus on the multiplication of two binomials called binomial factors. We start with the tile model, which leads to multiplication in a table. We then look at examples that can be multiplied mentally and at special products called perfect square trinomials and differences of squares.

▶ Binomial Multiplication

▶ **EXAMPLE 1** Exploring with tiles Let the large square algebra tile have side x, the rectangle have sides x and 1, and the small square tile have side 1. Arrange the sets of tiles into rectangles that show these products.

a. $(x + 2)(x + 3)$

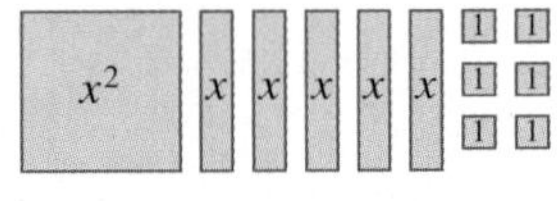

b. $(2x + 1)(x + 2)$

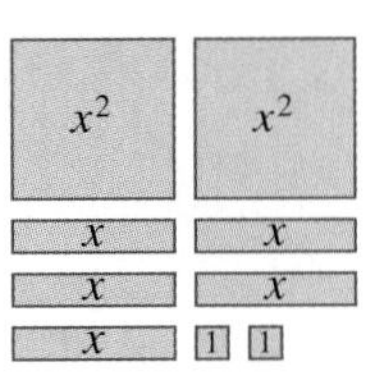

SOLUTION One possible rectangle for part a appears in Example 2. The rectangle for part b appears in Example 3. ◀

USING TILES TO BUILD TABLES We can arrange tiles into a rectangular shape that matches that of a multiplication table. To make this special arrangement, we start with the large squares in the upper left corner. We put some of the rectangles in the lower left corner and some in the upper right corner. Then we group all of the small squares in the lower right corner.

▶ EXAMPLE 2 **Showing multiplication with a table** Multiply $(x + 2)(x + 3)$.

SOLUTION

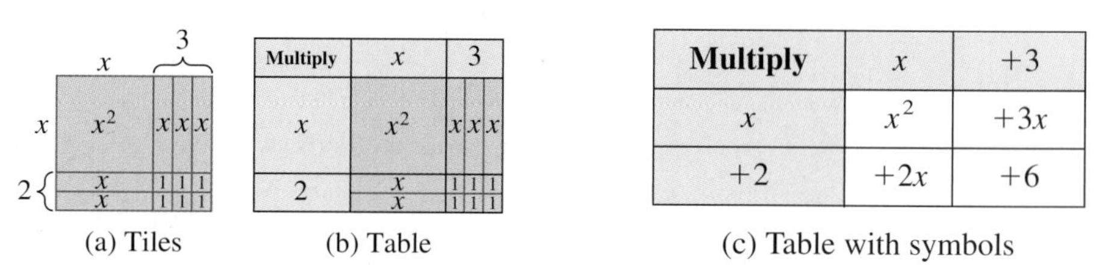

Multiply	x	$+3$
x	x^2	$+3x$
$+2$	$+2x$	$+6$

(a) Tiles (b) Table (c) Table with symbols

We count the tiles in figure (a) and find that the area is the sum of 1 large square, 5 rectangles, and 6 small squares, or $x^2 + 5x + 6$. Thus,

$$(x + 2)(x + 3) = x^2 + 5x + 6$$

Figure (b) shows the tiles in a table. In (c), the multiplication table on the right, we use the distribution property twice, to get the product of $x(x + 3)$ in the upper row and $2(x + 3)$ in the lower row. We add like terms $x^2 + 3x + 2x + 6$ to get

$$(x + 2)(x + 3) = x^2 + 5x + 6$$ ◀

Example 3 shows the multiplication in part b of Example 1 with tiles and a table.

▶ EXAMPLE 3 **Showing multiplication with a table** Show the product $(2x + 1)(x + 2)$ with tiles and a table.

SOLUTION

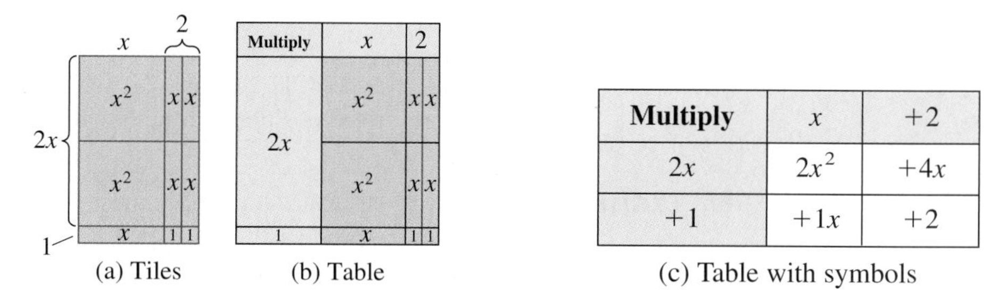

Multiply	x	$+2$
$2x$	$2x^2$	$+4x$
$+1$	$+1x$	$+2$

(a) Tiles (b) Table (c) Table with symbols

From the tiles in figure (a), the product is the sum of 2 large squares, 5 rectangles, and 2 small squares, or $2x^2 + 5x + 2$. Thus,

$$(2x + 1)(x + 2) = 2x^2 + 5x + 2$$

Figure (b) shows the tiles in a table. In (c), the multiplication table on the right, we use the distributive property twice, to get the product of $2x(x + 2)$ in the upper row and $1(x + 2)$ in the lower row. We add like terms in $2x^2 + 4x + 1x + 2$ to get $2x^2 + 5x + 2$.

Both the tiles and the table give the same product. ◀

USING TABLES AND THE TRADITIONAL METHOD One advantage of the table method is that we can use terms with either positive or negative coefficients. Example 4 shows a table solution along with a traditional multiplication.

▶ **EXAMPLE 4** Multiplying binomials containing negative coefficients Multiply $(x - 3)(2x - 1)$ with a table and with traditional multiplication.

SOLUTION **Table method:**

Multiply	$2x$	-1
x	$2x^2$	$-x$
-3	$-6x$	$+3$

Traditional method:

$$\begin{array}{rl} 2x - 1 & \\ \underline{x - 3} & \\ -6x + 3 & \text{Multiply } -3(2x - 1). \\ \underline{2x^2 - 1x \quad\quad} & \text{Multiply } x(2x - 1). \\ 2x^2 - 7x + 3 & \text{Add like terms.} \end{array}$$

Using both methods, we have $(x - 3)(2x - 1) = 2x^2 - 7x + 3$. ◀

▶ In the next example, we repeat both methods—table and traditional.

▶ **EXAMPLE 5** Multiplying binomials containing negative coefficients Multiply $(3x + 1)(2x - 3)$ with a table and with traditional multiplication.

SOLUTION **Table method:**

Multiply	$2x$	-3
$3x$	$6x^2$	$-9x$
$+1$	$+2x$	-3

Traditional method:

$$\begin{array}{rl} 2x - 3 & \\ \underline{3x + 1} & \\ 2x - 3 & \text{Multiply } 1(2x - 3). \\ \underline{6x^2 - 9x \quad\quad} & \text{Multiply } 3x(2x - 3). \\ 6x^2 - 7x - 3 & \text{Add like terms.} \end{array}$$

Using both methods, we have $(3x + 1)(2x - 3) = 6x^2 - 7x - 3$. ◀

Many students like the table method because it helps them remember that multiplying is like finding area. You may have a favorite way to multiply already. If your method works every time, don't switch. Just use the table method in this and the next section to help you understand the meanings of the operations and identify some useful patterns.

FINDING PATTERNS IN THE POSITIONS OF LIKE TERMS In Example 6, we multiply binomials and observe the positions of the like terms.

▶ **EXAMPLE 6** Multiplying binomials Multiply these binomials with a table. Where are the like terms in the products?

a. $(5x - 3)(2x + 3)$ **b.** $(ax + b)(cx + d)$

SOLUTION **a.**

Multiply	$2x$	$+3$
$5x$	$10x^2$	$+15x$
-3	$-6x$	-9

$(5x - 3)(2x + 3)$

$= 10x^2 + 9x - 9$

b.

Multiply	cx	$+d$
ax	acx^2	$+adx$
$+b$	$+bcx$	$+bd$

$(ax + b)(cx + d)$

$= acx^2 + adx + bcx + bd$

$= acx^2 + (ad + bc)x + bd$

In part a, the like terms are on the diagonal in the table from lower left to upper right. The binomial in part b has letters instead of numbers. It shows that, no matter what the coefficients on x, the like terms will be on the diagonal from lower left to upper right. ◀

PATTERNS IN LIKE TERMS

> Within a table for multiplication of binomials, the like terms will be on the diagonal from lower left to upper right.

FINDING PATTERNS IN THE PRODUCTS OF DIAGONALS The next pattern gives us a way to check whether the multiplication is correct. The pattern also gives us information that we will need in the next section, on factoring trinomials.

In Example 7, we return to the tables from Example 6.

▶ **EXAMPLE 7** Comparing diagonal products Multiply the two diagonal terms in each table. What pattern do you observe?

Student Note:
Look at diagonal products in all prior examples. Are they equal? The red and blue in the tables highlight diagonals.

a.

Multiply	$2x$	$+3$
$5x$	$10x^2$	$+15x$
-3	$-6x$	-9

b.

Multiply	cx	$+d$
ax	acx^2	$+adx$
$+b$	$+bcx$	$+bd$

SOLUTION **a.** $(10x^2)(-9) = -90x^2$

$(-6x)(+15x) = -90x^2$

The products of the diagonals are equal.

b. $(acx^2)(bd) = abcdx^2$

$(bcx)(adx) = abcdx^2$

Because a, b, c, and d represent any numbers, this pair of products shows that the products of the diagonals are equal for all tables. ◀

PATTERNS IN DIAGONAL PRODUCTS

> In a table showing multiplication of binomials, the products of the diagonal terms are equal.

Diagonal products will be a key to factoring in Sections 7.3 and 7.4. The patterns we observed in binomial products apply to trinomial and other products as well. See Exercise 83.

Multiply		
	First	Outside
	Inside	Last

MENTAL MULTIPLICATION You are encouraged to multiply easy products mentally. In earlier examples, each time we multiplied two binomials, the product had four terms and two of the terms could be added. To mentally multiply a product such as $(x + 5)(x - 7)$, follow three steps:

1. Find the product of the first terms: $x \cdot x$.
2. Find the products that form like terms and add them: $-7x + 5x$.
3. Find the product of the last terms: $5(-7)$.

$$(x + 5)(x - 7) = \underbrace{x^2}_{1} + \underbrace{-7x + 5x}_{2} \underbrace{- 35}_{3} = x^2 - 2x - 35$$

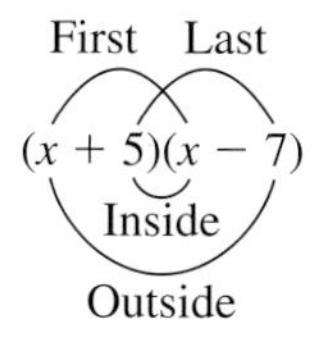

FIGURE 3 Mental shortcut

This method is often called FOIL because we multiply the first terms, then outside terms, inside terms, and last terms. Some people use the visual memory aid shown in Figure 3.

Remember: FOIL is limited to multiplying binomials; it does not work for monomials, trinomials, or other polynomials. FOIL is a memory device and a shortcut, not a general procedure.

Special Products

After some practice, you should be able to recognize and mentally carry out the multiplications leading to the two special types of products explored in Example 8.

EXAMPLE 8 Exploring special products Multiply these binomials without a table. Describe patterns in the problems or answers.

a. $(x - 1)(x - 1)$ **b.** $(x - 1)(x + 1)$ **c.** $(a + 3)(a - 3)$
d. $(a + 3)(a + 3)$ **e.** $(3x + y)(3x - y)$ **f.** $(3x + y)(3x + y)$

SOLUTION **a.** $(x - 1)(x - 1) = x^2 - x - x + 1$
$= x^2 - 2x + 1$

b. $(x - 1)(x + 1) = x^2 + x - x - 1$
$= x^2 - 1$

c. $(a + 3)(a - 3) = a^2 - 3a + 3a - 9$
$= a^2 - 9$

d. $(a + 3)(a + 3) = a^2 + 3a + 3a + 9$
$= a^2 + 6a + 9$

e. $(3x + y)(3x - y) = 9x^2 - 3xy + 3xy - y^2$
$= 9x^2 - y^2$

f. $(3x + y)(3x + y) = 9x^2 + 3xy + 3xy + y^2$
$= 9x^2 + 6xy + y^2$

Parts a, d, and f are the squares of binomials, $(x - 1)^2$, $(a + 3)^2$, and $(3x + y)^2$. The answers always start and end with a perfect square. The middle term is twice the product of the terms in one binomial factor.

Parts b, c, and e are binomials of the form $(a + b)(a - b)$. The answers are binomials containing the differences of squared terms. The middle terms add to zero and drop out. ◀

We now elaborate on the patterns in Example 8.

PERFECT SQUARE TRINOMIALS The products resulting from squaring binomials are called perfect square trinomials.

DEFINITION OF PERFECT SQUARE TRINOMIAL

> The expression $a^2 + 2ab + b^2$ is a **perfect square trinomial** because its terms are the square of a, twice the product of a and b, and the square of b.

A perfect square trinomial is created when we multiply any binomial times itself—that is, *square* a binomial:

$$(a + b)(a + b) = (a + b)^2 = a^2 + 2ab + b^2$$

In Example 9, we multiply $(a + b)^2$ with both the tile and the table method.

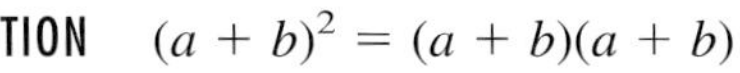

▶ **EXAMPLE 9** Showing a perfect square trinomial Multiply $(a + b)^2$ with tiles and a table.

SOLUTION $(a + b)^2 = (a + b)(a + b)$

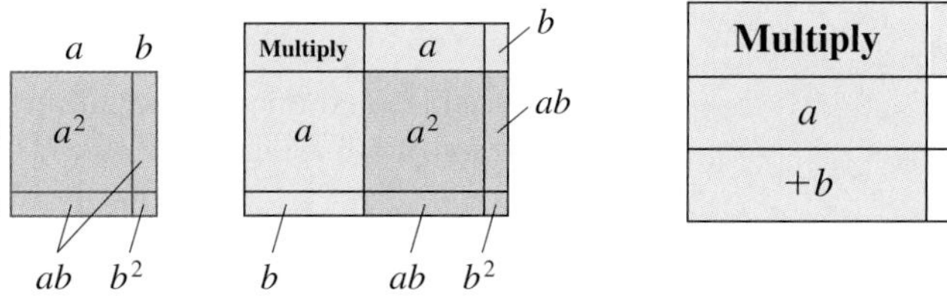

Multiply	a	$+b$
a	a^2	$+ab$
$+b$	$+ab$	$+b^2$

With both the tiles and table,

$$(a + b)(a + b) = a^2 + ab + ab + b^2 = a^2 + 2ab + b^2$$

The tiles form a square with $a + b$ on each side. ◀

The phrase *perfect square trinomial* describes the answer or product. The phrase *square of a binomial* describes the original multiplication, $(a + b)^2$ or $(x + 2)^2$. Both phrases describe square tile designs, as shown in Example 9.

DIFFERENCES OF SQUARES Expressions such as $x^2 - 1$ and $a^2 - 9$ in Example 8 are called differences of squares. The phrase *difference of squares* describes the answer.

DEFINITION OF DIFFERENCE OF SQUARES

> The expression $a^2 - b^2$ is a **difference of squares** because its two terms are the square of a and the square of b.

The original binomials contain the same variables but have opposite operations between the variables:

$$(a + b)(a - b) = a^2 - b^2$$

In Example 10, we look at why multiplying these binomials gives a difference of squares.

▶ **EXAMPLE 10** Finding differences of squares Multiply $(a + b)(a - b)$ with a table. What happens to the like terms?

SOLUTION

Multiply	a	$-b$
a	a^2	$-ab$
$+b$	$+ab$	$-b^2$

Adding terms in the table gives $a^2 - ab + ab - b^2 = a^2 - b^2$. The like terms $-ab$ and $+ab$ add to zero. The remaining terms a^2 and b^2 are squares separated by a subtraction—hence the name *difference of squares*. ◀

In Example 11, we practice identifying special products.

▶ **EXAMPLE 11** Identifying expressions that multiply to special products Predict whether the product from each of these expressions will be a perfect square trinomial, a difference of squares, or neither.

a. $(y - 6)(y + 6)$ **b.** $(y - 6)(y - 6)$ **c.** $(2x - 3y)(2x - 3y)$

d. $(2x + 3y)(2x - 3y)$ **e.** $(a + b)(-a - b)$

SOLUTION **a.** The variables are alike, but the operations between the terms are different; the answer will be a difference of squares.

$$(y - 6)(y + 6) = y^2 + 6y - 6y - 36 = y^2 - 36$$

b. The variables and operations within the binomials are alike. The expression is the square of a binomial; the answer will be a perfect square trinomial.

$$(y - 6)(y - 6) = y^2 - 6y - 6y + 36 = y^2 - 12y + 36$$

c. The variables and operations within the binomials are alike. The expression is the square of a binomial; the answer will be a perfect square trinomial.

$$(2x - 3y)(2x - 3y) = 4x^2 - 6xy - 6xy + 9y^2 = 4x^2 - 12xy + 9y^2$$

d. The variables are alike, but the operations are different; the answer will be a difference of squares.

$$(2x + 3y)(2x - 3y) = 4x^2 - 6xy + 6xy - 9y^2 = 4x^2 - 9y^2$$

e. The variables are alike, but both signs in the second factor are different.

$$(a + b)(-a - b) = -a^2 - ab - ab - b^2 = -a^2 - 2ab - b^2$$

The square terms in the answer are negative, so the expression is not a perfect square trinomial. The like terms do not add to zero, so the expression is not a difference of squares. ◀

Recognizing perfect square trinomials and differences of squares is useful in mental multiplication, factoring, and applications.

ANSWER BOX

Warm-up:

	m	n	Sum $m + n$	Product $m \cdot n$
1.	3	4	7	12
2.	2	6	8	12
3.	3	5	8	15
4.	1	15	16	15
5.	4	6	10	24
6.	3	8	11	24
7.	−4	6	2	−24
8.	2	−12	−10	−24
9.	−3	8	5	−24
10.	−3	−8	−11	24

▶ 7.2 Exercises

In Exercises 1 to 6, what binomial factors are shown in the figures? What is the product of each pair of factors?

1. **2.** **3.** **4.**

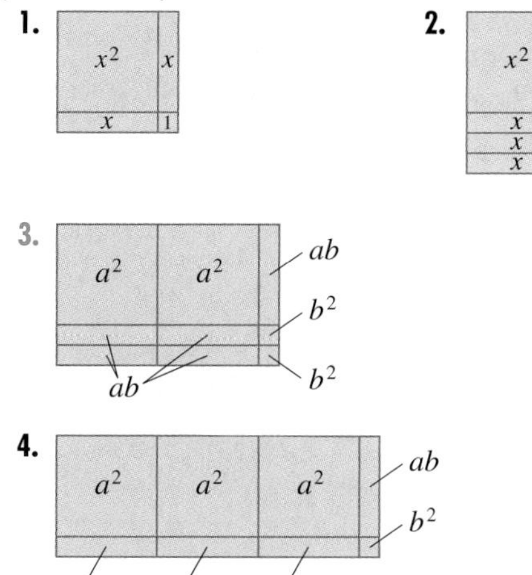

5. **6.**

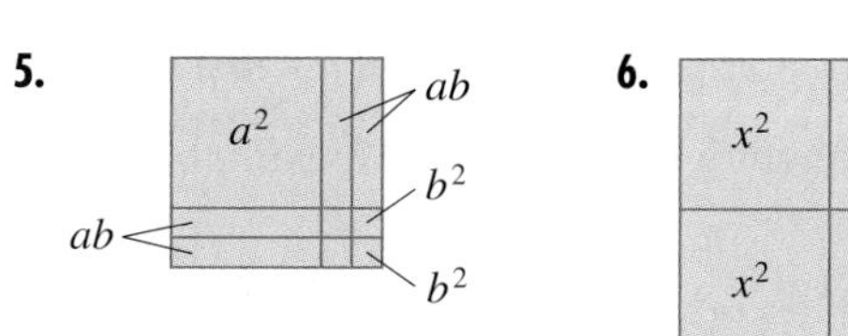

In Exercises 7 to 14, complete the table and write the product.

7.

Multiply	$2x$	$+5$
x		
-4		

$(x - 4)(2x + 5) =$

8.

Multiply	x	-1
$3x$		
$+1$		

$(3x + 1)(x - 1) =$

9.

Multiply	x	-2
$2x$		
$+3$		

$(2x + 3)(x - 2) =$

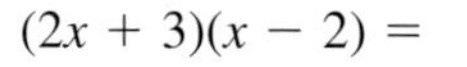

10.

Multiply	$2x$	-3
x		
$+3$		

$(x + 3)(2x - 3) =$

Blue numbers are core exercises.

11.

Multiply	x	-2
$2x$		
-1		

$(2x - 1)(x - 2) =$

12.

Multiply	$4x$	-1
x		
-5		

$(x - 5)(4x - 1) =$

13.

Multiply	x	-4
$5x$		
-1		

$(5x - 1)(x - 4) =$

14.

Multiply	x	-3
$3x$		
-1		

$(3x - 1)(x - 3) =$

Mentally multiply the expressions in Exercises 15 to 26.

15. **a.** $(x - 2)(x - 2)$ **b.** $(x + 2)(x - 2)$

16. **a.** $(x - 1)(x - 1)$ **b.** $(x - 1)(x + 1)$

17. **a.** $(a + 5)(a + 5)$ **b.** $(b + 5)(b - 5)$

18. **a.** $(b - 4)(b + 4)$ **b.** $(a + 4)(a + 4)$

19. **a.** $(a + b)(a - b)$ **b.** $(a - b)(a - b)$

20. **a.** $(a - b)(a + b)$ **b.** $(a + b)(a + b)$

21. **a.** $(x + 1)(x + 7)$ **b.** $(x + 1)(x - 7)$

22. **a.** $(x - 1)(x + 7)$ **b.** $(x - 1)(x - 7)$

23. **a.** $(b + 7)(b + 7)$ **b.** $(a + 7)(a - 7)$

24. **a.** $(a - 7)(a + 7)$ **b.** $(b - 7)(b - 7)$

25. **a.** $(x + y)(x + y)$ **b.** $(x - y)(x - y)$

26. **a.** $(x + y)(x - y)$ **b.** $(x - y)(x + y)$

27. Identify ten of the expressions in Exercises 15 to 26 that multiply to a perfect square trinomial.

28. Identify ten of the expressions in Exercises 15 to 26 that multiply to a difference of squares.

In Exercises 29 to 42, predict whether the expressions will multiply to a perfect square trinomial (pst), a difference of squares (ds), or neither. Then find the products.

29. $(2x + 3)(2x + 3)$
30. $(2x - 3)(2x - 3)$
31. $(2x - 3)(2x + 3)$
32. $(2x + 3)(2x - 3)$
33. $(2x - 3)(3 - 2x)$
34. $(3x - 2)(2 - 3x)$
35. $(3x - 2)(3x - 2)$
36. $(3x + 2)(3x + 2)$
37. $(3x + 2)(3x - 2)$
38. $(3x - 2)(3x + 2)$
39. $(x + 5)^2$
40. $(x + 7)^2$
41. $(a - 6)^2$
42. $(a - 8)^2$

Finish Exercises 43 to 48 with a mental multiplication.

	Binomial Factors	Trinomial Product
43.	$(x - 1)(x + 12)$	$= x^2 + \_\_\_ x - 12$
44.	$(x - 2)(x + 6)$	$= x^2 + \_\_\_ x - 12$
45.	$(x - 3)(x + 4)$	$= x^2 + \_\_\_ x - 12$
46.	$(x + 3)(x - 4)$	$= x^2 - \_\_\_ x - 12$
47.	$(x + 2)(x - 6)$	$= x^2 - \_\_\_ x - 12$
48.	$(x + 1)(x - 12)$	$= x^2 - \_\_\_ x - 12$

49. All of the products in Exercises 43 to 48 end with -12. What are the six sets of binomial factors that multiply to $x^2 \pm \_\_\_ x + 12$? Assume the factors contain only whole numbers.

50. Look again at Exercises 43 to 48. Are there any other binomial factors that multiply to $x^2 \pm \_\_\_ x - 12$? Assume the factors contain only whole numbers.

51. What are the six sets of binomial factors that multiply to $x^2 \pm \_\_\_ x + 20$?

52. What are the six sets of binomial factors that multiply to $x^2 \pm \_\_\_ x - 20$?

53. What are the possible numbers n in $x^2 \pm nx + 24$?

54. What are the possible numbers n in $x^2 \pm nx - 24$?

Multiply the factors in Exercises 55 to 66. Look for patterns.

55. a. $(x + 1)(x + 8)$ **b.** $(x + 1)(x - 8)$
56. a. $(x + 8)(x - 1)$ **b.** $(x - 8)(x - 1)$
57. a. $(x + 2)(x + 4)$ **b.** $(x + 2)(x - 4)$
58. a. $(x + 4)(x - 2)$ **b.** $(x - 4)(x - 2)$
59. a. $(2x - 3)(3x - 2)$ **b.** $(2x + 3)(3x - 2)$
60. a. $(2x - 3)(3x + 2)$ **b.** $(2x + 3)(3x + 2)$
61. a. $(6x + 1)(x + 6)$ **b.** $(6x - 1)(x + 6)$
62. a. $(6x - 1)(x - 6)$ **b.** $(6x + 1)(x - 6)$
63. a. $(2x + 5)(x + 1)$ **b.** $(2x + 5)(x - 1)$
64. a. $(2x - 5)(x + 1)$ **b.** $(2x - 5)(x - 1)$
65. a. $(2x - 1)(x + 5)$ **b.** $(2x + 1)(x + 5)$
66. a. $(2x - 1)(x - 5)$ **b.** $(2x + 1)(x - 5)$

For $5(x - 4)^2$, the order of operations requires that we multiply $(x - 4)(x - 4)$ before multiplying by the numerical coefficient 5. Use the order of operations in Exercises 67 to 74.

67. $2(x + 3)^2$
68. $3(x - 2)^2$
69. $3(x - 5)^2$
70. $2(x + 6)^2$
71. $4(5 - x)^2$
72. $4(3 - x)^2$
73. $5(3 - 2x)^2$
74. $5(2 - 3x)^2$

▶ Writing

75. Explain how to multiply $(x + 3)(x + 2)$ with tiles.

76. Explain how to multiply $(x + 3)(x + 2)$ with a table.

77. Explain how to multiply $(x + 3)(x + 2)$ with the traditional method.

78. Explain how to multiply $(x + 3)(x + 2)$ with FOIL.

79. Explain how we obtain the coefficient -2 on the middle term of the product in $(x + 3)(x - 5) = x^2 - 2x - 15$.

80. Explain how we obtain the coefficient -8 on the middle term of the product in $(x - 4)(x - 4) = x^2 - 8x + 16$.

▶ Error Analysis

81. Explain what is wrong here:

$$(x - a)^2 = x^2 - a^2$$

82. Explain what is wrong here:

$$3(a - b)^2 = 9a^2 - 18ab + 9b^2$$

▶ Projects

83. Exploring Binomial and Trinomial Multiplication Multiply these expressions. Use a table, such as the one shown in part a.

a. $(x - 2)(x^2 + 2x + 4)$

Multiply	x^2	$+2x$	$+4$
x			
-2			

b. $(x - 1)(x^2 - 2x + 1)$

c. $(a - b)(a^2 + ab + b^2)$

Blue numbers are core exercises.

d. $(x + y)(x^2 - xy + y^2)$

e. $(x + y)(x^2 + 2xy + y^2)$

f. $(a + b)(a^2 + 2ab + b^2)$

g. Where are like terms located in the table?

84. **Algebra Tiles on the Internet** Mathematics educators have written websites to give practice with algebra tiles, including tiles for integers (those natural numbers and their opposites, the negative integers). Enjoy! **http://argyll.epsb.ca/jreed/math9/strand2/factor1.htm**

▶ 7.3 Factoring Trinomials

Objectives

- Factor trinomials with algebra tiles.
- Factor trinomials with a table.
- Factor trinomials with guess and check.

WARM-UP

The whole-number factor pairs of 28 are $1 \cdot 28$, $2 \cdot 14$, and $4 \cdot 7$. Find all the whole-number factor pairs for these numbers:

1. 45

2. 60

3. 36

4. 24

5. 56

AN ARABIC BOOK written around the year 825 is the source of the word *algebra*. The author, Muḥammad ibn Mūsa al-Khwārizmī, lived in Baghdad. Al-Khwārizmī's book begins with verbal and visual study of $x^2 + 10x - 39$. The tile model in Section 7.2 is similar to his visual approach. In this section, we reverse multiplication in order to factor trinomials, such as $x^2 + 10x - 39$. Tiles, tables, and guess and check will all play a part.

▶ The Tile Model of Factoring

When we arrange tiles into a rectangle or square, we can find the length and width from the area. We call this process *factoring*. In addition to being a basic algebraic process, factoring changes everyday formulas into more usable forms; see the project in Exercise 57.

We have factored expressions such as $3ab + 3b^2$ into $3b(a + b)$, the product of the gcf (a monomial) and a binomial. We now factor trinomials into two binomials. We begin with the tile model to review how length, width, and area are related to multiplication and factoring.

▶ **EXAMPLE 1** Exploration: finding width and length, given area Arrange these sets of tiles into rectangles. Describe the width and length of each rectangle. Write the binomial factors that give each set of tiles as a product.

a. $x^2 + 3x + 2$ **b.** $2x^2 + 7x + 3$

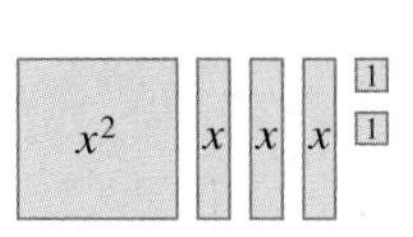

SOLUTION **a.** For the tile area $x^2 + 3x + 2$ (Figure 4a), the width and length are $x + 1$ and $x + 2$, respectively. Using binomial factors, we have

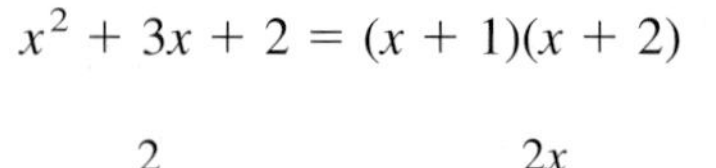

$$x^2 + 3x + 2 = (x + 1)(x + 2)$$

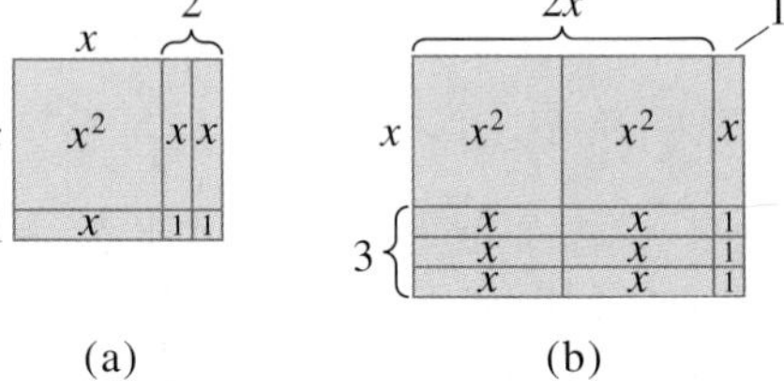

FIGURE 4 Tiles for $x^2 + 3x + 2$ and $2x^2 + 7x + 3$

b. For the tile area $2x^2 + 7x + 3$ (Figure 4b), the width and length are $x + 3$ and $2x + 1$, respectively. Using binomial factors, we have

$$2x^2 + 7x + 3 = (x + 3)(2x + 1)$$ ◀

THINK ABOUT IT 1: Is there another way to arrange the rectangles in Figure 4? Will the factors change?

▶ The Table Model of Factoring

Table 2 shows the factoring $x^2 + 3x + 2 = (x + 1)(x + 2)$. The table has the same appearance as the one for multiplying $(x + 1)(x + 2)$. The terms in the white part of the table are arranged exactly the same way as the tiles in Figure 5 (a repeat of Figure 4a).

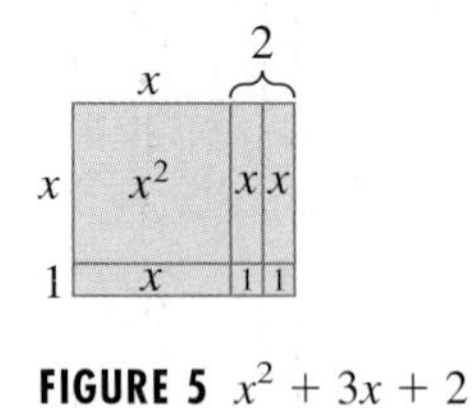

FIGURE 5 $x^2 + 3x + 2$

TABLE 2 Factoring $x^2 + 3x + 2$

Factor	x	$+2$
x	x^2	$+2x$
$+1$	$+1x$	$+2$

As with the multiplication table,

- The product of the first terms, x and x, is in the upper left corner.
- The product of the last terms, 1 and 2, is in the lower right corner.
- The like terms on the diagonal from lower left to upper right add to $3x$. The terms on the diagonals multiply to the same value:

$$(x^2)(2) = 2x^2 \quad \text{and} \quad (2x)(1x) = 2x^2$$

Example 2 shows the factoring steps in five tables. The following examples show the steps in four and then three tables. In the Exercises, it is sufficient to sketch the table once.

▶ **EXAMPLE 2** Factoring by table Use a table to factor $2x^2 + 7x + 3$.

SOLUTION ***Step 1:*** We write the first term, $2x^2$, in the upper left corner. We write the last term, 3, in the lower right corner. We multiply the first and last terms to get the diagonal product, $6x^2$.

Factor		
	$2x^2$	
		$+3$

Diagonal product, $2x^2 \cdot 3 = 6x^2$

Step 2: From the diagonal product, $6x^2$, we list all the factors of the coefficient, 6.

$$6 \cdot 1 = 6$$
$$3 \cdot 2 = 6$$

Using this list, we find the factors that add to the middle term, $7x$. The factors are $6x \cdot 1x$. We write the factors in the other diagonal, from lower left to upper right.

Factor		
	$2x^2$	$+1x$
	$+6x$	$+3$

Sum of like terms, $6x + 1x = 7x$

Step 3: To the left of $2x^2$, we write the greatest common factor of the first row:

Factor		
x	$2x^2$	$+1x$
gcf $= x$	$+6x$	$+3$

Step 4: We use the gcf to find the next two factors. If the gcf is x, the factor in the top left must be $2x$ and the factor in the top right must be $+1$.

Factor	$2x$	$+1$
x	$2x^2$	$+1x$
	$+6x$	$+3$

If the factor in the top left is $2x$, the factor in the bottom left must be $+3$.

Factor	$2x$	$+1$
x	$2x^2$	$+1x$
$+3$	$+6x$	$+3$

We multiply $(+3)(+1) = +3$ as a check.

Step 5: Check. The table shows the binomial factors $(x + 3)$ and $(2x + 1)$. We multiply the binomials to check the answer:

$$(x + 3)(2x + 1) = 2x^2 + 7x + 3$$ ◀

THINK ABOUT IT 2: If we switch the positions of $6x$ and $1x$ in the table, does that change the factors? Sketch the table and tile design for this set of factors.

THINK ABOUT IT 3: Use the five steps in Example 2 to factor the al-Khwārizmī trinomial $x^2 + 10x - 39$ in a table. As needed, use the Summary box.

Following is a summary of factoring by table, using the general trinomial $ax^2 + bx + c$. You will see this trinomial again in Chapter 8 in the quadratic equation and the quadratic formula.

SUMMARY: FACTORING BY TABLE

To find the binomial factors of $ax^2 + bx + c$ by table:

1. **a.** Write the first term of the trinomial, ax^2, in the upper left corner of the table.
   b. Write the last term of the trinomial, c, in the lower right corner.
   c. Multiply the first and last terms to get the diagonal product, acx^2.
2. **a.** From the diagonal product, list the factors of the coefficient, ac.
   b. Find the factors that add to the middle term of the trinomial, bx.
   c. Write the factors adding to bx in the diagonal from lower left to upper right (although order does not matter).
3. Write the greatest common factor of the first row to the left of ax^2.
4. **a.** Use the gcf to find the remaining factors at the top and left of the table.
   b. Check that the top right and bottom left factors multiply to the last term, c.
5. Check the factoring. Multiply the binomial factors from the table.

Factor			
Greatest common factor (gcf) of this row	First term of trinomial, ax^2	——	Sum of like terms equals middle term of trinomial, bx
	——	Last term of trinomial, c	Diagonal product, acx^2

We again factor by table in Example 3.

▶ **EXAMPLE 3** **Factoring by table** Use a table to factor $3x^2 + 13x + 12$. Sketch a tile model that shows the factors.

SOLUTION ***Step 1:*** We write the first term, $3x^2$, in the upper left corner. We write the last term, 12, in the lower right corner. We multiply the first and last terms to get the diagonal product, $36x^2$.

Factor		
	$3x^2$	
		$+12$

Diagonal product, $3x^2 \cdot 12 = 36x^2$

Step 2: From the diagonal product, $36x^2$, we list all the factors of the coefficient, 36.

$1 \cdot 36 = 36$

$2 \cdot 18 = 36$

$3 \cdot 12 = 36$

$(4 \cdot 9) = 36$ $\quad 4x + 9x = 13x$

$6 \cdot 6 = 36$

Using this list, we find the factors that add to the middle term, $13x$. The factors are $4x \cdot 9x$. We write the factors in the diagonal from lower left to upper right.

Factor		
	$3x^2$	$+9x$
	$+4x$	$+12$

Sum of like terms, $4x + 9x = 13x$

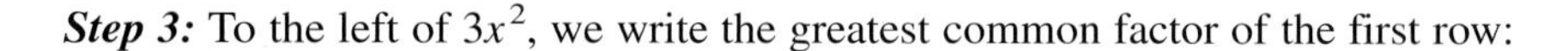

Step 3: To the left of $3x^2$, we write the greatest common factor of the first row:

Factor		
$3x$	$3x^2$	$+9x$
gcf = $3x$	$+4x$	$+12$

Step 4: We use the gcf to find the next two factors. If the gcf is $3x$, the factor in the top left must be x and the factor in the top right must be $+3$. If the factor in the top left is x, the factor in the bottom left must be $+4$. We multiply $(+4)(+3) = +12$ to check.

Factor	x	$+3$
$3x$	$3x^2$	$+9x$
$+4$	$+4x$	$+12$

Step 5: Check. The table shows the binomial factors $(3x + 4)$ and $(x + 3)$. We multiply the binomials to check the answer:

$$(3x + 4)(x + 3) = 3x^2 + 13x + 12$$

Figure 6 shows the tile model of $(3x + 4)(x + 3)$. ◀

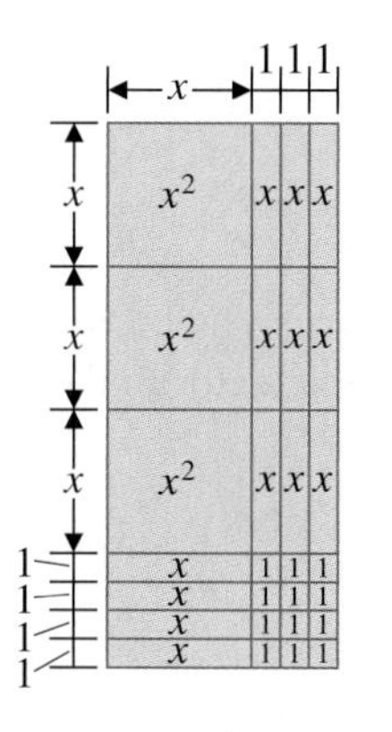

FIGURE 6 $3x^2 + 13x + 12$

▶ In Example 4, we factor a trinomial containing negative terms.

▶ **EXAMPLE 4** Factoring trinomials with negative terms Use a table to factor $6x^2 - 11x - 10$.

SOLUTION ***Step 1:*** We write the first term, $6x^2$, in the upper left corner. We write the last term, -10, in the lower right corner. We multiply the first and last terms to get the diagonal product, $-60x^2$.

Factor		
	$6x^2$	
		-10

Diagonal product, $6x^2 \cdot (-10) = -60x^2$

Student Note:
You may find that only one list of factors of 60 is sufficient to find a pair of factors that sum to -11. Once you find $+11$, you can swap signs on -4 and 15, to obtain -11.

Step 2: From the diagonal product, $-60x^2$, we list all the factors of the coefficient, -60. One of the two numbers will be negative.

$-1 \cdot 60 = -60$ $\quad 1 \cdot (-60) = -60$

$-2 \cdot 30 = -60$ $\quad 2 \cdot (-30) = -60$

$-3 \cdot 20 = -60$ $\quad 3 \cdot (-20) = -60$

$-4 \cdot 15 = -60$ $\quad 4 \cdot (-15) = -60$ $\quad 4x + (-15x) = -11x$

$-5 \cdot 12 = -60$ $\quad 5 \cdot (-12) = -60$

$-6 \cdot 10 = -60$ $\quad 6 \cdot (-10) = -60$

Using this list, we find the factors that add to the middle term, $-11x$. The factors are $4x \cdot (-15x)$. We write the factors in the diagonal from lower left to upper right.

Factor		
	$6x^2$	$-15x$
	$+4x$	-10

Sum of like terms, $4x + (-15x) = -11x$

Step 3: To the left of $6x^2$, we write the greatest common factor of the first row:

Factor		
$3x$	$6x^2$	$-15x$
gcf = $3x$	$+4x$	-10

Step 4: We use the gcf to find the next two factors. If the gcf is $3x$, the factor in the top left must be $2x$ and the factor in the top right must be -5. If the factor in the top left is $2x$, the factor in the bottom left must be $+2$.

Factor	$2x$	-5
$3x$	$6x^2$	$-15x$
$+2$	$+4x$	-10

We use the last term, -10, to check the bottom left and top right factors, $2(-5)$.

Step 5: Check. The binomial factors are $(3x + 2)$ and $(2x - 5)$. We multiply the binomials to check the answer:

$$(3x + 2)(2x - 5) = 6x^2 - 11x - 10$$ ◀

▶ In the next example, along with the table method, we use the traditional method of factoring trinomials (called the factoring-by-grouping or *ac*-product method).

▶ **EXAMPLE 5** Factoring by table and the traditional method Use a table to factor $6x^2 - 11x + 3$. Compare the steps with those in the traditional method.

TABLE SOLUTION ***Step 1:*** We write the first and last terms, $6x^2$ and $+3$, in the upper left and lower right corners. We multiply the terms to get the diagonal product, $18x^2$.

Factor		
	$6x^2$	
		$+3$

Diagonal product, $6x^2 \cdot 3 = 18x^2$

Step 2: We list the factors of the coefficient of the diagonal product:

Student Note:
You may find that only one list of factors of 18 is sufficient.

$1 \cdot 18 = 18 \qquad -1 \cdot (-18) = 18$

$2 \cdot 9 = 18 \qquad -2 \cdot (-9) = 18 \qquad -2x + (-9x) = -11x$

$3 \cdot 6 = 18 \qquad -3 \cdot (-6) = 18$

We write the factors that add to $-11x$ in the diagonal from lower left to upper right.

Student Note:
The factors $-2x$ and $-9x$ may be placed in the diagonal in either order.

Factor		
	$6x^2$	$-2x$
	$-9x$	$+3$

Sum of like terms, $-11x$

Step 3: To the left of $6x^2$, we write the gcf, $2x$, of the first row.

Step 4: We use the gcf to find the remaining factors at the top and left. We use the last term, $+3$, to check the bottom left and top right factors.

Factor	$3x$	-1
$2x$	$6x^2$	$-2x$
-3	$-9x$	$+3$

Step 5: Check. We multiply the binomial factors to check the answer:

$$(3x - 1)(2x - 3) = 6x^2 - 11x + 3$$

TRADITIONAL SOLUTION The following traditional solution is called the *ac-product* or *factoring-by-grouping* method. The steps are numbered to show how they match those in the table method.

Student Note:
The comments to the right tell what was done to get to that step.

Step 1: $6x^2 - 11x + 3$

$a = 6 \qquad$ Find a and c in $ax^2 + bx + c$.

$c = 3$

$6 \cdot 3 = 18 \qquad$ Multiply a and c.

Step 2:	$1 \cdot 18, 2 \cdot 9, 3 \cdot 6$	List all the factors of ac.
	2 and 9 add to 11	Find the factors that add to bx.
	$2x + 9x = 11x$	
	$6x^2 - (2x + 9x) + 3$	Write $ax^2 + bx + c$ as a four-term expression by replacing bx with the two factors $2x$ and $9x$.
	$6x^2 - 2x - 9x + 3$	Apply the distributive property to $-(2x + 9x)$.
Step 3:	$2x(3x - 1) - 9x + 3$	Find the gcf and factor the first two terms.
Step 4:	$2x(3x - 1) - 3(3x - 1)$	Find the gcf and factor the last two terms.
	$(3x - 1)(2x - 3)$	Factor the $3x - 1$ from the two expressions.
Step 5:	$(3x - 1)(2x - 3) = 6x^2 - 11x + 3$	Multiply the binomial factors to check the factoring. ◀

The advantage of the table method is that we do the factoring in a table that is identical to the multiplication table. The traditional factoring method does not *look* like the traditional multiplication method. However, if you look closely, you will see that the traditional method is the same as the table method. The differences lie in how and where we write the information.

▶ Factoring with Guess and Check

Even with the availability of algebraic calculators and alternative traditional procedures (see the quadratic formula in Chapter 8), it is important to be able to factor some expressions mentally.

FACTORING BY GUESS AND CHECK

When $a = 1$ is the leading coefficient on the trinomial $ax^2 + bx + c$, the way to guess the factors is to look at the numbers, b and c. The numbers in the blanks must multiply to c and add to b:

$$x^2 + bx + c = (x \pm \_\_\_)(x \pm \_\_\_)$$

The $\pm$ sign indicates *addition or subtraction* because the operations depend on the signs of b and c.

In Example 6, we practice factoring mentally. In all examples, we consider factors with whole numbers only.

▶ **EXAMPLE 6** Factoring mentally Find the missing numbers in these statements. The addition and subtraction signs have been placed in the factors.

a. $x^2 + 8x + 12 = (x + \_\_\_)(x + \_\_\_)$

b. $x^2 + 8x + 15 = (x + \_\_\_)(x + \_\_\_)$

c. $x^2 + 5x - 24 = (x - \_\_\_)(x + \_\_\_)$

d. $x^2 - 11x + 24 = (x - \_\_\_)(x - \_\_\_)$

SOLUTION **a.** The numbers that multiply to 12 are $1 \cdot 12$, $2 \cdot 6$, and $3 \cdot 4$. Only 2 and 6 add to 8.

$$x^2 + 8x + 12 = (x + 2)(x + 6)$$

b. The numbers that multiply to 15 are $1 \cdot 15$ and $3 \cdot 5$. Only 3 and 5 add to 8.

$$x^2 + 8x + 15 = (x + 3)(x + 5)$$

c. The numbers that multiply to 24 are $1 \cdot 24$, $2 \cdot 12$, $3 \cdot 8$, and $4 \cdot 6$. Because the product is -24, one number must be negative. The sum is $+5$, so the numbers are -3 and 8.

$$x^2 + 5x - 24 = (x - 3)(x + 8)$$

d. Using the same factors of 24 as in part c, we get a sum of -11 with -3 and -8.

$$x^2 - 11x + 24 = (x - 3)(x - 8)$$ ◀

ANSWER BOX

Warm-up: 1. $1 \cdot 45, 3 \cdot 15, 5 \cdot 9$ **2.** $1 \cdot 60, 2 \cdot 30, 3 \cdot 20, 4 \cdot 15, 5 \cdot 12, 6 \cdot 10$ **3.** $1 \cdot 36, 2 \cdot 18, 3 \cdot 12, 4 \cdot 9, 6 \cdot 6$ **4.** $1 \cdot 24, 2 \cdot 12, 3 \cdot 8, 4 \cdot 6$ **5.** $1 \cdot 56, 2 \cdot 28, 4 \cdot 14, 7 \cdot 8$ **Think about it 1:** Yes, the $x + 2$ is on the side and $x + 1$ on the top; the $2x + 1$ is on the side and $x + 3$ on the top. See **Think about it 2** below. No, the factors are the same expressions.
Think about it 2: The factor $(2x + 1)$ would be on the left, and the factor $(x + 3)$ would be on the top of the table.

Factor	x	$+3$
$2x$	$2x^2$	$+6x$
$+1$	$+1x$	$+3$

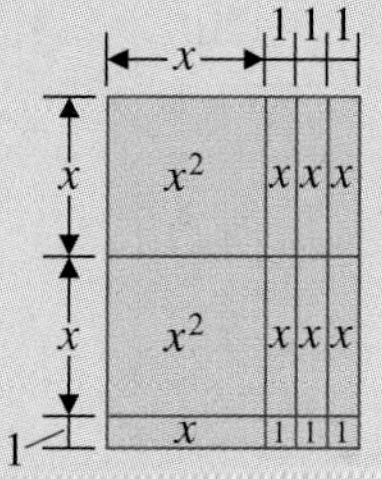

Think about it 3:

Factor		
	x^2	$+13x$
	$-3x$	-39

Factor	x	$+13$
x	x^2	$+13x$
-3	$-3x$	-39

$(x + 13)(x - 3) = x^2 + 10x - 39$

▶ 7.3 Exercises

In Exercises 1 to 4, arrange the tiles into a rectangle and state the area, which is the product of width and length.

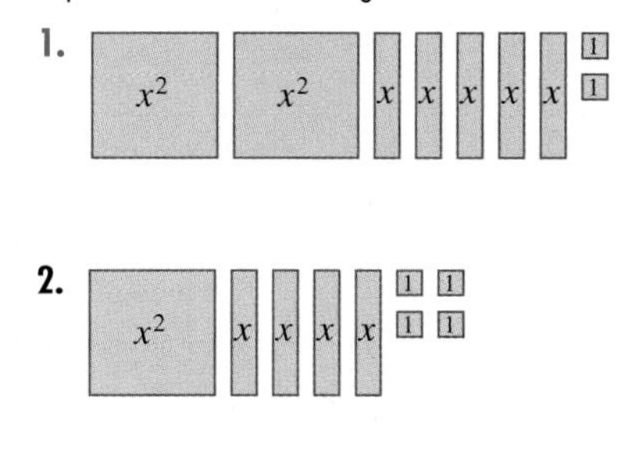

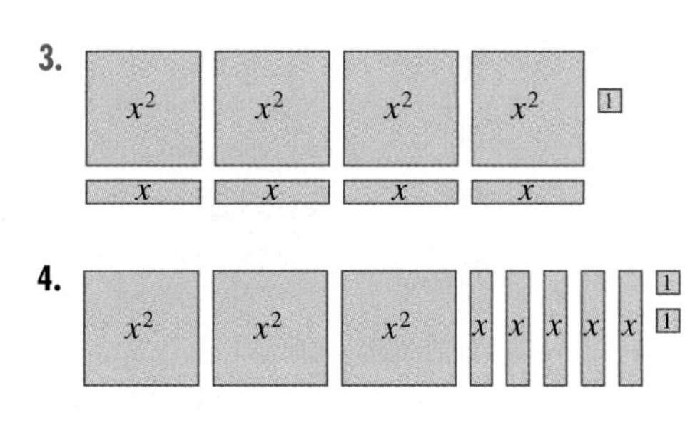

Blue numbers are core exercises.

In Exercises 5 to 10, complete the table and state the problem and factors described by the table.

5.

Factor		
	x^2	$+5x$
	$+4x$	$+20$

6.

Factor		
	x^2	$+5x$
	$-4x$	-20

7.

Factor		
	x^2	$+2x$
	$-10x$	-20

8.

Factor		
	x^2	$-2x$
	$+10x$	-20

9.

Factor		
	$6x^2$	$+x$
	$-18x$	-3

10.

Factor		
	$6x^2$	$-2x$
	$+9x$	-3

11. Describe why we get the coefficient -2 on the middle term of the product in $(x + 3)(x - 5) = x^2 - 2x - 15$.

12. Describe why we get the coefficient $+1$ on the middle term of the product in $(x - 4)(x + 5) = x^2 + x - 20$.

Mentally factor the trinomials in Exercises 13 to 24.

13. $x^2 + 8x + 12$
14. $x^2 - 8x + 12$
15. $x^2 - 13x + 12$
16. $x^2 - 7x + 12$
17. $x^2 + 4x - 12$
18. $x^2 + 13x + 12$
19. $a^2 + 7a + 12$
20. $a^2 + 11a - 12$
21. $x^2 + x - 12$
22. $x^2 - 4x - 12$
23. $x^2 - 11x - 12$
24. $x^2 - x - 12$

In Exercises 25 to 56, factor with any of the methods in this section.

25. $x^2 + 6x + 9$
26. $x^2 + 10x + 25$
27. $x^2 + 11x + 30$
28. $x^2 + 17x + 30$
29. $x^2 + 13x - 30$
30. $x^2 + x - 30$
31. $x^2 - 6x - 16$
32. $x^2 + 6x - 16$
33. $x^2 + 15x - 16$
34. $x^2 - 15x - 16$
35. $x^2 + 0x - 25$
36. $x^2 + 0x - 36$
37. $2x^2 + 11x + 12$
38. $3x^2 + 13x + 12$
39. $2x^2 - 3x - 9$
40. $2n^2 + 3n - 9$
41. $2n^2 + n - 3$
42. $2x^2 - 5x - 3$
43. $3x^2 + 5x - 2$
44. $3a^2 + a - 2$
45. $3a^2 - 11a - 4$
46. $3x^2 - 4x - 4$
47. $9x^2 + 0x - 49$
48. $9x^2 + 0x - 25$
49. $16x^2 + 0x - 9$
50. $16x^2 + 0x - 81$
51. $6x^2 + x - 2$
52. $6x^2 - 7x - 5$
53. $6x^2 + 5x - 6$
54. $6x^2 - 13x + 5$
55. $2n^2 + 9n - 5$
56. $2n^2 + 11n - 6$

▶ Projects

57. CD Earnings* Eli and Shana are calculating the value of a certificate of deposit (CD) at the end of years 1 to 4. The rate of interest, r, is 4%. The starting amount of the CD, P, is \$500. Eli's formulas are shown in the first table. Shana's formulas are shown in the second table.

a. Find the amount of money after each year using the two formulas.

b. What do you observe about the two formulas? Can you prove your observation with multiplication?

c. Which table (set of formulas) would you rather use?

d. Use Shana's table to create a formula for t years.

Eli's Formulas

Years, t	Rule	Amount (\$)
1	$P + Pr$	
2	$P + 2Pr + Pr^2$	
3	$P + 3Pr + 3Pr^2 + Pr^3$	
4	$P + 4Pr + 6Pr^2 + 4Pr^3 + Pr^4$	

Shana's Formulas

Years, t	Rule	Amount (\$)
1	$P(1 + r)$	
2	$P(1 + r)^2$	
3	$P(1 + r)^3$	
4	$P(1 + r)^4$	

Blue numbers are core exercises.

*This project was suggested by Charlotte Hutt, Rogue Community College.

58. Hardy-Weinberg Research Research the Hardy-Weinberg principle in genetics or biology. Related information may also be found under the *Punnett-square method*. Explain how these relate to trinomials and multiplication of binomials as presented in this section.

▶ 7 Mid-Chapter Test

1. Simplify each of the following expressions. Arrange the terms alphabetically, with exponents on the first letter in descending order. Indicate whether the result is a monomial, binomial, trinomial, or other polynomial.

a. $7b - 8c + 3a + 4b - 5c - 6a$

b. $x^2 - 4x + x^2 - 6$

c. $4xy^2 + x^3y^2 - 3x^2y + x^3y^2 - 2xy^2$

d. $(5a - 3b - 2c) - (3a + 4b - 6c)$

e. $x(x^2 + 5x + 25) - 5(x^2 + 5x + 25)$

f. $9 - 3x(x + 3)$

g. $8 - 4(4 - x)$

2. Name the greatest common factor for each expression, and then factor the expression.

a. $6x^2 - 2x + 8$ **b.** $2abc - 3ac + 4ab$

3. Explain the difference between terms and factors. Use expressions like ab and $a + b$ in your explanation.

4. Find the length, width, perimeter, and area of each rectangle.

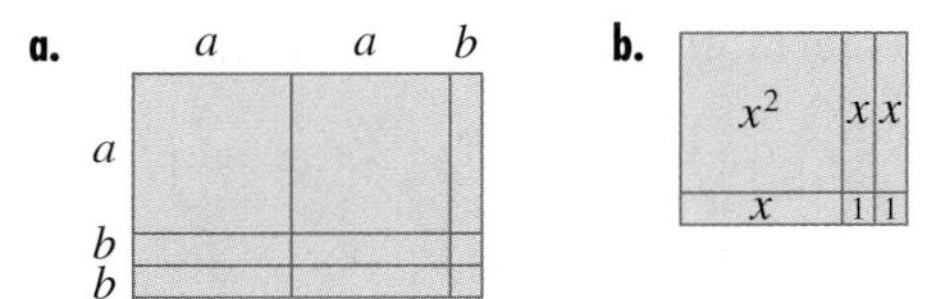

5. Arrange these tiles into a rectangle. What multiplication and answer are shown?

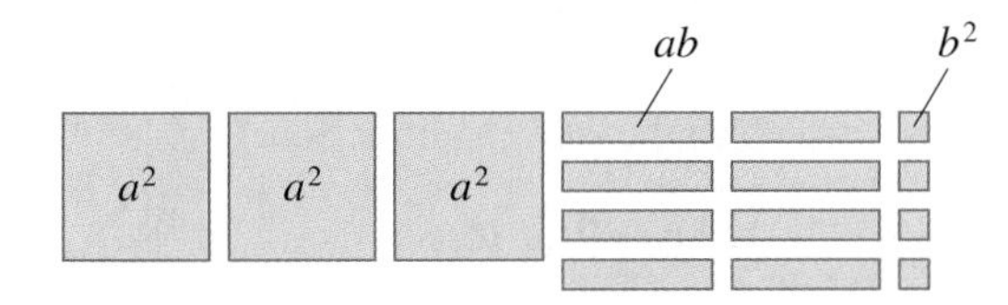

6. Complete the table. Write the binomial factors and trinomial product.

Multiply	$2x$	-5
$3x$		
$+2$		

7. What is the diagonal product in the table in Exercise 6?

8. Complete the table. Write the factors and the resulting product.

Multiply	$3x^2$	$-2x$	$+1$
x			
-2			

9. What are the like terms in the table in Exercise 8?

In Exercises 10 to 14, find the products.

10. $(x + 3)(x + 5)$

11. $(x - 4)(3x + 5)$

12. $(2x + 7)(3x - 1)$

13. $(x - 3)(x + 3)$

14. $(2x - 5)^2$

15. Identify each expression as a perfect square trinomial (pst), a difference of squares (ds), or neither.

a. $x^2 - 2x - 1$ **b.** $x^2 + 9$

c. $x^2 - 49$ **d.** $x^2 + 6x + 9$

16. What are all the possible numbers that could complete this statement?

$$(x \pm \_\_\_\_)(x \pm \_\_\_\_) = x^2 \pm \_\_\_\_\, x \pm 10$$

17. Tell what trinomial and its factors are shown in each figure in Exercise 4.

18. What trinomial is shown in each table? Complete the factoring set up in the table.

a.

Factor		
	$6x^2$	$-8x$
	$+15x$	-20

b.

Factor		
	$6x^2$	$-24x$
	$+5x$	-20

19. Factor.

a. $x^2 + 12x + 35$

b. $x^2 - 5x - 14$

c. $6x^2 - 17x - 10$

20. Looking ahead Explain why $2x^2 + 11x - 15$ does not factor and show that $2x^2 - 13x + 15$ does.

7.4 Factoring Special Products and Greatest Common Factors

Objectives

- Factor special products.
- Factor out the greatest common factor.
- Identify trinomials that cannot be factored.

WARM-UP

Predict whether each product will be a perfect square trinomial (pst) or a difference of squares (ds) and then multiply.

1. $(2x + 1)(2x - 1)$ **2.** $(5 - x)(5 - x)$ **3.** $(2 - 3x)(2 - 3x)$

4. $(3x + 5)(3x + 5)$ **5.** $(6 - x)(6 + x)$ **6.** $(3 - 2x)(3 + 2x)$

THIS SECTION CONTINUES the focus on factoring, as we look at the special products from Section 7.2 and the step of factoring out the greatest common factor. We also identify trinomials that do not factor.

Factoring Special Products

We now factor special products. In Section 7.2, we multiplied identical binomial factors and got perfect square trinomials:

$$(a + b)(a + b) = a^2 + 2ab + b^2 \quad \text{Perfect square trinomial}$$

When we use tiles, perfect square trinomials form squares, as shown in Figure 7.

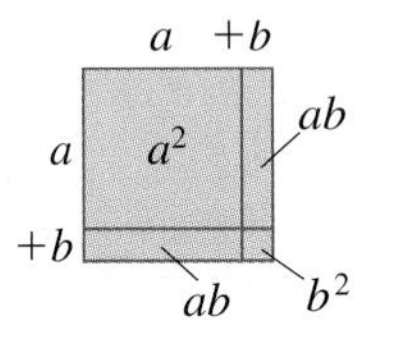

FIGURE 7 Perfect square trinomial

A difference of squares has binomial factors with the same numbers and variables, but one factor contains addition and the other subtraction:

$$(a + b)(a - b) = a^2 - b^2 \quad \text{Difference of squares}$$

Table 3 shows how the like terms add to zero in the product $(a + b)(a - b)$.

TABLE 3 Difference of squares

Multiply	a	$-b$
a	a^2	$-ab$
$+b$	$+ab$	$-b^2$

$ab + (-ab)$ sums to zero.

EXAMPLE 1 Identifying and factoring special products Identify and factor each expression that is a perfect square trinomial (pst) or a difference of squares (ds).

a. $x^2 - 2x + 1$ **b.** $y^2 - 4y + 4$ **c.** $x^2 - 5x + 25$

d. $a^2 - 49$ **e.** $9x^2 + 25$ **f.** $x^2 + 4x + 8$

g. $y^2 + 2yz + z^2$ **h.** $4x^2 - 25$

SOLUTION **a.** $x^2 - 2x + 1 = (x - 1)(x - 1) = (x - 1)^2$; pst

a^2 $2ab$ $b^2, b = -1$

b. $y^2 - 4y + 4 = (y - 2)(y - 2) = (y - 2)^2$; pst

a^2 $2ab$ $b^2, b = -2$

c. $x^2 - 5x + 25$; neither. The middle term would need to be $-10x$ for a pst.

a^2 $b^2, b = -5$

d. $a^2 - 49 = (a - 7)(a + 7)$; ds

a^2 $b^2, b = 7$

e. $9x^2 + 25$; neither. The operation would need to be subtraction for a ds.

$a^2, a = 3x$ $b^2, b = 5$

f. $x^2 + 4x + 8$; neither. The last term would need to be 4 for a pst.

g. $y^2 + 2yz + z^2 = (y + z)(y + z) = (y + z)^2$; pst

h. $4x^2 - 25 = (2x - 5)(2x + 5)$; ds ◀

We now summarize factoring by adding one step to each type of factoring.

Factoring Out the Greatest Common Factor

Before proceeding with traditional factoring processes, it is best to first factor out any greatest common factor from the binomial or trinomial.

EXAMPLE 2 Factoring out the greatest common factor Factor $2x^2 + 4x + 2$ by guess and check and with tiles.

GUESS-AND-CHECK SOLUTION

$2x^2 + 4x + 2 = 2(x^2 + 2x + 1)$ Remove the greatest common factor.

$= 2(x + \_\_\_)(x + \_\_\_)$ The numbers in the blanks must multiply to 1 and add to 2.

$= 2(x + 1)(x + 1)$ The numbers are 1 and 1.

$= 2(x + 1)^2$

You should recognize that $x^2 + 2x + 1$ is a perfect square trinomial and is equal to $(x + 1)^2$.

TILES SOLUTION The tile method provides a more flexible way to find the greatest common factor. The rectangle shown in Figure 8a represents

$$\begin{aligned} 2x^2 + 4x + 2 &= (x + 1)(2x + 2) \\ &= (x + 1) \cdot 2(x + 1) \\ &= 2(x + 1)^2 \end{aligned}$$

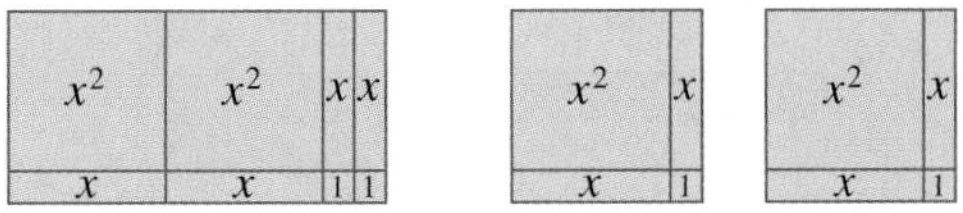

(a) x^2 x^2 x x / x x 1 1

(b) x^2 x / x 1 — x^2 x / x 1

FIGURE 8 Separating a rectangle into two squares

The tiles can be split into two identical sets to form two squares, as shown in Figure 8b. This shows that

$$2x^2 + 4x + 2 = 2(x + 1)^2$$ ◀

▶ In Example 3, the guess-and-check process requires more than finding numbers that add to b and multiply to c, because the coefficient a is not equal to 1, as it was in earlier examples.

▶ EXAMPLE 3 Factoring out the greatest common factor Factor $4x^2 - 10x - 6$ by guess and check and with a table.

GUESS-AND-CHECK SOLUTION $4x^2 - 10x - 6 = 2(2x^2 - 5x - 3)$ Remove the greatest common factor.

The 2 on the x^2 term means that the guess-and-check process used earlier must be modified. The only way to get $2x^2$ is with $2x$ and $1x$. Thus, we have

$$2(2x \pm \_\_\_\_)(x \pm \_\_\_\_)$$

We must have 1 and 3 as last terms to have the product -3, so the expression must be of the form

$$2(2x \pm 3)(x \pm 1) \quad \text{or} \quad 2(2x \pm 1)(x \pm 3)$$

We multiply each possible option until a pair of factors works.

$$2(2x - 3)(x + 1) = 2(2x^2 - 1x - 3) \qquad -1x \neq -5x$$
$$2(2x + 3)(x - 1) = 2(2x^2 + 1x - 3) \qquad +1x \neq -5x$$
$$2(2x - 1)(x + 3) = 2(2x^2 + 5x - 3) \qquad +5x \neq -5x$$
$$2(2x + 1)(x - 3) = 2(2x^2 - 5x - 3) \quad ✓$$

Thus, $4x^2 - 10x - 6 = 2(2x + 1)(x - 3)$.

TABLE SOLUTION ***Step 1:*** From $4x^2 - 10x - 6$ we write $4x^2$ and -6 in the diagonal. We multiply $4x^2$ and -6 to obtain the diagonal product, $-24x^2$.

Factor		
	$4x^2$	
		-6

Step 2: We list all the factors of the coefficient, -24:

$-1 \cdot 24 \qquad 1 \cdot (-24)$

$-2 \cdot 12 \qquad 2 \cdot (-12)$

$-3 \cdot 8 \qquad 3 \cdot (-8)$

$-4 \cdot 6 \qquad 4 \cdot (-6)$

The middle term is $-10x$. After finding the factors that add to $-10x$, we place $+2x$ and $-12x$ in the other diagonal.

Factor		
	$4x^2$	$-12x$
	$+2x$	-6

Step 3: The greatest common factor of $4x^2$ and $-12x$ is $4x$, so we write $4x$ to the left of $4x^2$.

Step 4: Starting with $4x$, we find the remaining table factors: $x - 3$ on top and $+2$ on the lower left. The factors $+2$ and -3 multiply to -6 in the lower right.

Factor	x	-3
$4x$	$4x^2$	$-12x$
$+2$	$+2x$	-6

Student Note:
With the table method, we can remove the greatest common factor after factoring into binomials.

Step 5: The factors $4x + 2$ and $x - 3$ multiply to $4x^2 - 10x - 6$. We can now remove the common factor from $4x + 2$ and complete the factoring:

$$4x^2 - 10x - 6 = (4x + 2)(x - 3)$$
$$= 2(2x + 1)(x - 3)$$ ◀

THINK ABOUT IT 1: How does removing the gcf from $4x^2 - 10x - 6$ change factoring with a table?

In Examples 4 and 5, we use the greatest common factor with special products to factor a binomial.

▶ **EXAMPLE 4** Factoring out the greatest common factor Factor $3x^2 - 3$ by observing patterns.

SOLUTION $3x^2 - 3 = 3(x^2 - 1)$ When we remove the greatest common factor, the parentheses contain a difference of squares.

$$= 3(x - 1)(x + 1)$$ ◀

▶ **EXAMPLE 5** Factoring out the greatest common factor Factor $4a^3 + 24a^2 + 36a$.

SOLUTION $4a^3 + 24a^2 + 36a = 4a(a^2 + 6a + 9)$ The remaining factor is a pst.

a^2 $2ab$ $b^2, b = 3$

$$= 4a(a + 3)(a + 3)$$
$$= 4a(a + 3)^2$$ ◀

Trinomials That Cannot Be Factored

In the next two examples, we try to factor trinomials that do not have whole-number factors. We need to be able to recognize these cases.

▶ **EXAMPLE 6** Identifying a trinomial that does not factor Using both tiles and a table, show that $x^2 + x + 1$ does not factor.

SOLUTION With tiles, we see in Figure 9 that we have one large square, a rectangle, and one small square. There is no way to build a rectangle with these tiles.

x^2 x 1

FIGURE 9 No rectangle can be formed.

With a table, we go through the following steps:

Step 1: We write x^2 and 1 in the diagonal. We multiply x^2 and 1 to obtain the diagonal product, $1x^2$.

Factor		
	x^2	
		$+1$

Diagonal product, $+1x^2$

Step 2: From the diagonal product, $+1x^2$, we list the factors of the coefficient, 1.

$1 \cdot 1$ Sum = 2

$-1 \cdot (-1)$ Sum = -2

We cannot get a sum of $1x$ from whole-number factors of the diagonal product. The trinomial $x^2 + x + 1$ does not factor. ◀

THINK ABOUT IT 2: Some students prefer to use guess and check to try to factor $x^2 + x + 1$. Multiply these guesses to see if any give the product $x^2 + x + 1$.

a. $(x - 1)(x + 1)$ **b.** $(x - 1)(x - 1)$ **c.** $(x - 1)(-x - 1)$

d. $(x + 1)(x + 1)$ **e.** Try any other factors.

Do any of the binomial factors make special products?

▶ **EXAMPLE 7** Identifying a trinomial that does not factor Using a table, show that $4x^2 - 3x - 3$ does not factor.

SOLUTION With a table, we need two steps.

Step 1: We enter $4x^2$ and -3 in the diagonal. We multiply $(4x^2)(-3)$ to obtain the diagonal product, $-12x^2$.

Factor		
	$4x^2$	
		-3

Diagonal product, $-12x^2$

Step 2: We list the factors of the coefficient, -12:

$-12 \cdot 1$	Sum = -11	$12 \cdot (-1)$	Sum = 11
$-6 \cdot 2$	Sum = -4	$6 \cdot (-2)$	Sum = 4
$-4 \cdot 3$	Sum = -1	$4 \cdot (-3)$	Sum = 1

None of the factors add to -3. The trinomial $4x^2 - 3x - 3$ does not factor. ◀

ANSWER BOX

Warm-up: 1. ds; $4x^2 - 1$ **2.** pst; $25 - 10x + x^2$ **3.** pst; $4 - 12x + 9x^2$ **4.** pst; $9x^2 + 30x + 25$ **5.** ds; $36 - x^2$ **6.** ds; $9 - 4x^2$ **Think about it 1:** A simpler table results; the diagonal product for $2x^2 - 5x - 3$ is $2x^2(-3) = -6x^2$ instead of the $-24x^2$ in the example. There are only four factors for -6: $-6 \cdot 1$, $-3 \cdot 2$, $6 \cdot (-1)$, and $3 \cdot (-2)$. **Think about it 2: a.** This is a difference of squares, $x^2 - 1$. **b.** This is a square of a binomial and gives a perfect square trinomial, $x^2 - 2x + 1$. **c.** $-x^2 + 1$ or $1 - x^2$, a difference of squares. **d.** This is a square of a binomial and gives a perfect square trinomial, $x^2 + 2x + 1$.

▶ 7.4 Exercises

Look for special products and greatest common factors in Exercises 1 to 12. Factor mentally. Two of the expressions cannot be factored.

1. $x^2 - 4$
2. $x^2 + 36$
3. $4x^2 - 16$
4. $4x^2 - 36$
5. $x^2 + 12x + 36$
6. $x^2 - 12x + 36$
7. $4x^2 + 8x + 4$
8. $9x^2 + 36x + 36$
9. $x^2 + 4$
10. $x^2 - 8x + 16$
11. $x^2 - 6x + 9$
12. $5x^2 - 20$

In Exercises 13 to 42, factor using any of the methods in this section. Some of the expressions cannot be factored.

13. $3x^2 + 12x + 9$
14. $2x^2 + 6x + 4$
15. $3x^2 - 27$
16. $2x^2 - 18$
17. $3a^2 + 5a - 4$
18. $3n^2 - 9n - 4$
19. $25x^2 - 36$
20. $100x^2 - 49$
21. $3x^2 + 10x - 5$
22. $3x^2 - 5x - 5$
23. $5x^2 - 10x + 5$
24. $3x^2 - 18x + 27$
25. $18x^2 - 50$
26. $5x^2 - 80$
27. $3x^2 - 30x + 75$
28. $12x^2 + 4x - 60$
29. $3x^2 + 6x + 12$
30. $10x^2 - 20x + 40$
31. $8x^2 - 5x - 1$
32. $6x^2 - 4x + 1$
33. $6x^2 - 4x - 3$
34. $8x^2 - 5x - 2$
35. $2x^3 + x^2 - 3x$
36. $3x^3 - 7x^2 - 6x$
37. $9a^3 + 3a^2 - 20a$
38. $6a^3 - 11a^2 - 10a$
39. $6x^3 - 2x^2 - 8x$
40. $9x^3 - 3x^2 - 6x$
41. $12a^3 - 3a$
42. $2a^3 - 18a$
43. Set up a table and show at what point we know the trinomial $6x^2 - 11x - 6$ cannot be factored.
44. Sketch tiles and show that $4x^2 + 3x + 1$ cannot be factored.

▶ Problem Solving

45. **Factoring Patterns** Give several numbers that may be used in place of the b in the expression in order to make a trinomial that factors.

a. $x^2 + bx + 12$

b. $x^2 + bx - 15$

c. $x^2 + bx - 20$

d. $x^2 + bx + 18$

Blue numbers are core exercises.

▶ Projects

46. **Cubic Polynomials** Polynomials with x^3 or a^3 as the highest term are called cubic expressions. Multiply the expressions in parts a to f to obtain cubic expressions, copying the original factors and then writing the result of the multiplication.

a. $(x + 1)(x^2 + 2x + 1)$

b. $(x - 2)(x^2 + 2x + 4)$

c. $(x + 3)(x^2 + 6x + 9)$

d. $(a + b)(a^2 + 2ab + b^2)$

e. $(a - 3)(a^2 + 3a + 9)$

f. $(a - b)(a^2 + ab + b^2)$

Complete the tables in parts g to j to factor the expressions in the tables. Write the factors, as well as the original expression, in a statement below each table.

g.

Factor			
	x^3	$-2x^2$	$+x$
	$-x^2$	$+2x$	-1

____________ = (   )(   )

h.

Factor			
	a^3	$-2a^2b$	$+ab^2$
	$-a^2b$	$+2ab^2$	$-b^3$

____________ = (   )(   )

i.

Factor			
	a^3	$-a^2b$	$+ab^2$
	a^2b	$-ab^2$	$+b^3$

____________ = (   )(   )

j.

Factor			
	x^3	$-2x^2$	$+4x$
	$+2x^2$	$-4x$	$+8$

____________ = (   )(   )

k. When we multiply $(x + 1)(x^2 + 2x + 1)$, our answer is like the answer to $(a - b)(a^2 - 2ab + b^2)$. What other three problems, a to j, have the same pattern? Use the pattern to factor $n^3 + 3n^2 + 3n + 1$ and $y^3 - 3y^2 + 3y - 1$.

l. When we multiply $(x - 2)(x^2 + 2x + 4)$, our answer is like the answer to $(a + b)(a^2 - ab + b^2)$. What other three problems, a to j, have the same pattern? Use the pattern to factor $x^3 - 1$, $x^3 + 64$, and $x^3 - 27$.

47. Polynomials and Volume The volume of a cube with side x is x^3. The volume of a cube with side $(x + y)$ is $(x + y)^3$. The parts making up the volume may also be shown geometrically with stacks of boxes.

a. The figure below illustrates $(x + y)^3$. It is split into a front stack of boxes and a back stack of boxes, shown separately. Write out the volume for $(x + y)^3$.

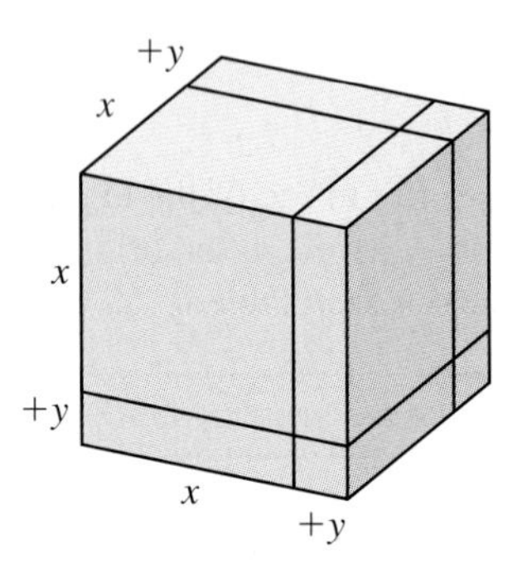

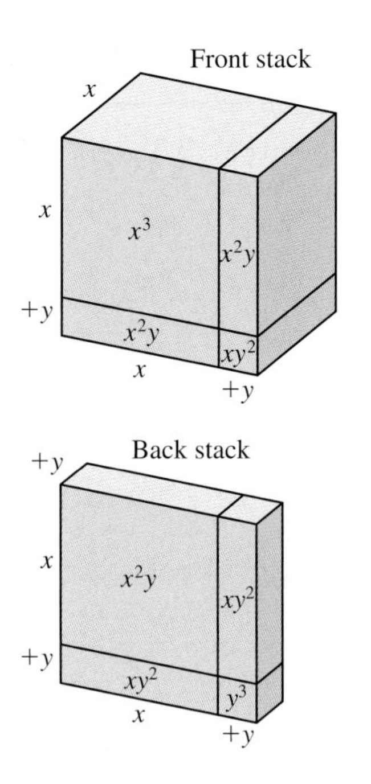

b. Sketch two stacks of boxes (front and back) that show $(x + 2)^3$. Multiply out $(x + 2)^3$ to find another expression for the volume.

c. Sketch three stacks of boxes that show $(x + y + z)^3$. Multiply out $(x + y + z)^3$ to find another expression for the volume.

▶ 7.5 Exponents

Objectives

- Simplify expressions containing zero and negative exponents.
- Simplify multiplication, division, and powers of exponential expressions.

WARM-UP

Complete the table. Look for patterns.

TABLE 4

Exponent Expression	Value	Exponent Expression	Value	Exponent Expression	Value
10^3	1000	5^3	125	4^3	64
10^2	100	5^2	25	4^2	16
10^1		5^1		4^1	
$10^{\square}$		$5^{\square}$		$4^{\square}$	
$10^{\square}$		$5^{\square}$		$4^{\square}$	
$10^{\square}$		$5^{\square}$		$4^{\square}$	
$10^{\square}$		$5^{\square}$		$4^{\square}$	

THE POLYNOMIAL EXPRESSIONS of Sections 7.1 to 7.4 are defined for only non-negative integer exponents. In this section, we return to exponents to explore properties and to extend the definition of exponents from positive integer exponents to zero and negative exponents.

Definition of Exponents I

Section 2.4 introduced the following definition of exponents.

POSITIVE INTEGER EXPONENTS

> In the power expression x^n, the positive integer exponent n indicates the number of times the base, x, is used as a factor. Thus, $x^n = x \cdot x \cdot x \cdot x \cdot \dots \cdot x$ has n factors of x.

Properties of Exponents

Because the polynomials we have used so far have contained only positive integers, we have been able to simplify all expressions using the factor definition of exponents and the simplification property of fractions:

$$x^2 \cdot x^3 = x \cdot x \cdot x \cdot x \cdot x \qquad \frac{a^2b}{ac} = \frac{a}{a} \cdot \frac{ab}{c}$$

$$= x^5 \qquad = 1 \cdot \frac{ab}{c} = \frac{ab}{c}$$

To work with other exponents, we need additional properties. Examples 1 to 4 suggest five properties of exponents. The explorations in Examples 1 and 2 are based on the numbers in Figure 10. (To simplify counting, every fifth number in each set in Figure 10 is underlined.)

Squares (Red): These are the squares of the first 30 integers, 1^2 to 30^2.
1, 4, 9, 16, 25, 36, 49, 64, 81, 100, 121, 144, 169, 196, 225, 256, 289, 324, 361, 400, 441, 484, 529, 576, 625, 676, 729, 784, 841, 900, . . .

Cubes (Green): These are the cubes of the first 25 integers, 1^3 to 25^3.
1, 8, 27, 64, 125, 216, 343, 512, 729, 1000, 1331, 1728, 2197, 2744, 3375, 4096, 4913, 5832, 6859, 8000, 9261, 10648, 12167, 13864, 15625, . . .

FIGURE 10 Squares and cubes

▶ **EXAMPLE 1** Reading the lists Don't make these hard.

a. Find the 10th red number, and compare it to 10^2. From the red list, complete the blanks: $(\square)^2 = 324$ and $(\square)^2 = 676$. Count, don't calculate! What changes in this list, the base or the exponent?

b. Find the 5th green number, and compare it to 5^3. From the green list, $(\square)^3 = 1000$ and $(\square)^3 = 2744$. Again, count, don't calculate! What changes in this list, the base or the exponent?

SOLUTION

a. 324 is the 18th red number, $324 = 18^2$; 676 is 26^2. The base changes.

b. 1000 is the 10th number, 10^3, and 2744 is the 14th number, 14^3. In the squares and cubes, the bases change: the exponents are 2 (square or red list) and 3 (cube or green list). ◀

When we look carefully at the lists, we can observe important properties about exponents and, possibly, why they work.

EXAMPLE 2 Exploring squares and cubes Look at each number by its position on the lists, and guess whether the answer is somewhere on the same list.

a. Write each pair as a square (or cube) of a single number: $25 \cdot 36$ (squares), $8 \cdot 216$ (cubes)

b. Write a rule for the products of numbers with either 2 or 3 as an exponent.

c. Divide these pairs of numbers: $\frac{784}{49}$ (squares), $\frac{1728}{27}$ (cubes).

d. Write a rule for the quotients of numbers with either 2 or 3 as an exponent.

SOLUTION **a.** $25 \cdot 36 = 5^2 \cdot 6^2 = (5 \cdot 6)^2 = 30^2$

$8 \cdot 216 = 2^3 \cdot 6^3 = (2 \cdot 6)^3 = 12^3$

b. $m^2n^2 = (m \cdot n)^2$ The product of the squares of two numbers is the square of the product.

$m^3n^3 = (m \cdot n)^3$ The product of the cubes of two numbers is the cube of the product.

c. $\frac{784}{49} = \frac{28^2}{7^2} = \left(\frac{28}{7}\right)^2 = 4^2$

$\frac{1728}{27} = \frac{12^3}{3^3} = \left(\frac{12}{3}\right)^3 = 4^3$

d. $\frac{m^2}{n^2} = \left(\frac{m}{n}\right)^2$ The quotient of the squares of two numbers is the square of the quotient.

$\frac{m^3}{n^3} = \left(\frac{m}{n}\right)^3$ The quotient of the cubes of two numbers is the cube of the quotient. ◀

The properties suggested by Example 2 are summarized for all exponents in the following box:

PROPERTIES OF EXPONENTS I

Student Note:

$(x \cdot y)^n$

Unlike bases

An exponent outside the parentheses applies to all parts of a product or quotient inside the parentheses:

$(x \cdot y)^m = x^m \cdot y^m$ Power of a product property

$\left(\frac{x}{y}\right)^m = \frac{x^m}{y^m}$ for $y \neq 0$ Power of a quotient property

Examples 3 and 4 explore with the powers of numbers shown in Figure 11.

Powers of Two (Orange): These are the first 20 powers of two, 2^1 to 2^{20}.
2, 4, 8, 16, 32, 64, 128, 256, 512, 1024, 2048, 4096, 8192, 16,384, 32,768, 65,536, 131,072, 262,144, 524,288, 1,048,576, . . .

Powers of Three (Black): These are the first 15 powers of three, 3^1 to 3^{15}.
3, 9, 27, 81, 243, 729, 2187, 6561, 19,683, 59,049, 177,147, 531,441, 1,594,323, 4,782,969, 14,348,907, . . .

FIGURE 11 Powers of two and three

▶ **EXAMPLE 3** Exploring powers of numbers with like bases Use Figure 11 to find the numbers.

a. Find two 3-digit powers of two (orange) whose product is a power of two.
b. Find two powers of three (black) whose product is a power of three.
c. Find two powers of two (orange) whose quotient is a power of two.
d. Find two powers of three (black) whose quotient is a power of three.
e. Find a pattern in the exponents for the products in parts a and b.
f. Find a pattern in the exponents for the quotients in parts c and d.

SOLUTION Every pair of numbers has a product or quotient within the same color.

a. For example, $128 \times 256 = 32{,}768$, $2^7 \times 2^8 = 2^{15}$

b. For example, $81 \times 729 = 59{,}049$, $3^4 \times 3^6 = 3^{10}$

c. For example, $65{,}536 \div 2048 = 32$, $2^{16} \div 2^{11} = 2^5$

d. For example, $1{,}594{,}323 \div 6561 = 243$, $3^{13} \div 3^8 = 3^5$

e. The sum of the exponents in the factors equals the exponent in the product.

f. The difference between the exponents in the numerator and the denominator is the exponent in the quotient. ◀

▶ **EXAMPLE 4** Exploring powers of numbers with like bases Use Figure 11 to find the numbers.

a. Find the cube of a power of two (orange). Is it a power of two (orange)?
b. Find the square of a power of three (black). Is it a power of three (black)?
c. Find a pattern in the exponents for the powers in parts a and b.

SOLUTION Every black or orange number works.

a. For example, $64^3 = 262{,}144$, $(2^6)^3 = 2^{18}$

b. For example, $81^2 = 6{,}561$, $(3^4)^2 = 3^8$

c. The exponent in the power is the product of the exponents in the original expression. ◀

Example 3 suggests the multiplication of like bases property and the division of like bases property. Example 4 suggests the power of a power property. These properties are summarized in the following box.

PROPERTIES OF EXPONENTS II

To multiply numbers with like bases, add the exponents:

$$x^m \cdot x^n = x^{m+n} \quad \text{Multiplication of like bases property}$$

To divide numbers with like bases, subtract the exponents:

$$\frac{x^m}{x^n} = x^{m-n} \text{ for } x \neq 0 \quad \text{Division of like bases property}$$

To apply an exponent to a power expression, multiply the exponents:

$$(x^m)^n = x^{m \cdot n} \quad \text{Power of a power property}$$

Student Note:

$x^m \cdot x^n$
↑ ↑
Like bases

▶ Definition of Exponents II

Example 5 suggests the meaning of exponents other than positive integers.

▶ **EXAMPLE 5** Exploring exponents Complete Table 4. The exponent on each number is 1 less than it was in the prior row. Look for a pattern between rows in the value column.

TABLE 4

Exponent Expression	Value	Exponent Expression	Value	Exponent Expression	Value
10^3	1000	5^3	125	4^3	64
10^2	100	5^2	25	4^2	16
10^1		5^1		4^1	
$10^{\square}$		$5^{\square}$		$4^{\square}$	
$10^{\square}$		$5^{\square}$		$4^{\square}$	
$10^{\square}$		$5^{\square}$		$4^{\square}$	
$10^{\square}$		$5^{\square}$		$4^{\square}$	

SOLUTION *Hint:* By what number do we divide 1000 to get 100? Does the pattern continue? By what number do we divide 125 to get 25? Does the pattern continue? By what number do we divide 64 to get 16? Does the pattern continue? See the Answer Box. ◀

Example 5 suggests the following definitions:

ZERO EXPONENTS

For zero exponents,

$$x^0 = 1$$

The expression 0^0 is not defined.

NEGATIVE INTEGER EXPONENTS

For negative-one exponents,

$$x^{-1} = \frac{1}{x}$$

the reciprocal of x. The expression 0^{-1} is not defined.

For negative-two exponents,

$$x^{-2} = \frac{1}{x^2}$$

the square of the reciprocal of x. The expression 0^{-2} is not defined.

For any negative real-number exponents,

$$x^{-n} = \frac{1}{x^n}$$

the nth power of the reciprocal of x. The expression 0^{-n} is not defined.

THINK ABOUT IT 1: Name the reciprocals of 4, $\frac{1}{2}$, $\frac{3}{4}$, 0.25, and 1.25.

EXAMPLE 6 Applying definitions of exponents Complete parts a to l with the definitions of exponents. Check your answers four parts at a time with a calculator, using [^] to indicate an exponent and [(−)] for negatives.

a. 5^0 **b.** $(-2)^0$
c. 1.5^0 **d.** 0^0
e. 4^{-1} **f.** 3^{-1}
g. $(-2)^{-1}$ **h.** $\left(\frac{2}{3}\right)^{-1}$
i. 0.25^{-1} **j.** 0.25^{-2}
k. $\left(\frac{3}{4}\right)^{-2}$ **l.** 0^{-2}

SOLUTION **a.** to **d.** The solutions to parts a to d appear on the calculator screen in Figure 12. For part d, as soon as [ENTER] is pressed after 0^0, the error message in Figure 13 appears. Selecting 2 returns the working screen and places the cursor after 0^0. Zero is not a permissible input when you are raising to the zero power.

Student Note:
If zero is not a permissible input for $f(x) = x^0$ or $f(x) = x^{-2}$, then it is not in the *domain* of either function. Hence the calculator shows a DOMAIN ERROR.

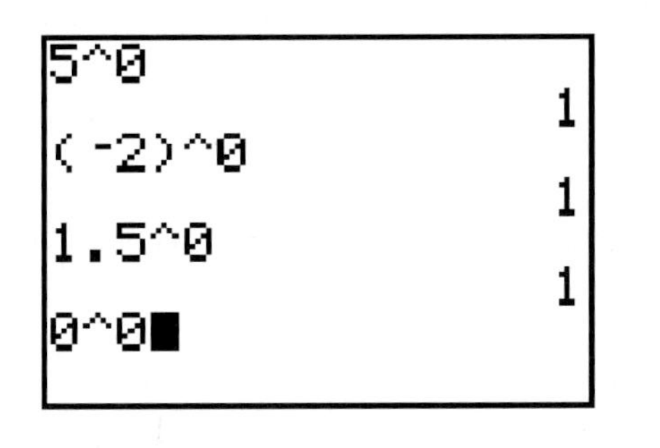

FIGURE 12 Zero as exponent

FIGURE 13 Results from 0^0 and 0^{-2}

e. to **h.** The solutions to parts e to h appear in Figure 14. Pressing [ENTER] returns $\frac{3}{2}$ for part h. The reciprocal of $\frac{2}{3}$ is $\frac{3}{2}$.

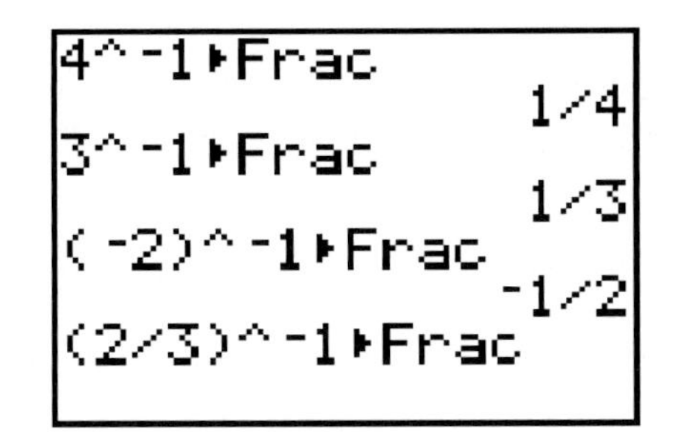

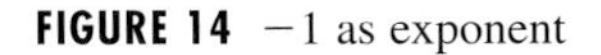
FIGURE 14 −1 as exponent

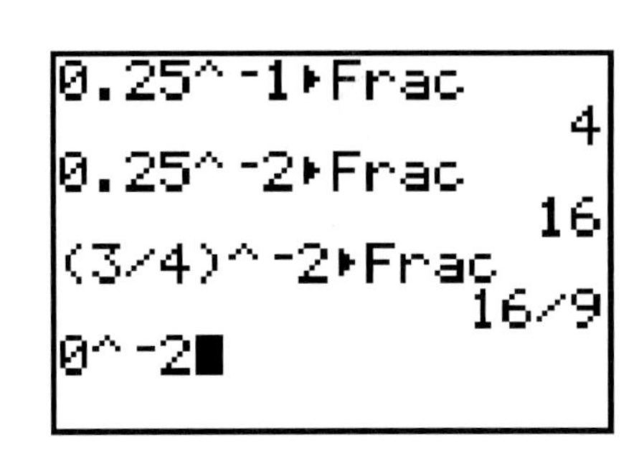

FIGURE 15 −1 and −2 as exponents

i. to **l.** The solutions to parts i to l appear in Figure 15. Since $0.25 = \frac{1}{4}$, its reciprocal in part i is $\frac{4}{1}$ and in part j, its square is $\frac{16}{1}$. In part k, the reciprocal of $\frac{3}{4}$ is $\frac{4}{3}$ and its square is $\frac{16}{9}$. For part l, pressing [ENTER] will return domain error message identical to the one in Figure 13. ◀

THINK ABOUT IT 2: Explain why parentheses are needed in some parts of Example 2 and not in others.

Applying Properties of Exponents

In Example 7, we look at how the properties of exponents show the meaning of zero and negative exponents.

▶ **EXAMPLE 7** **Finding the value of zero and negative exponents** First simplify each expression to a number by using properties of fractions. Then simplify to an exponent expression by using the appropriate property of exponents.

a. $\dfrac{5^3}{5^3}$ **b.** $\dfrac{x^n}{x^n}$, where $x \neq 0$ **c.** $\dfrac{4^3}{4^5}$ **d.** $\dfrac{x^5}{x^6}$, $x \neq 0$

SOLUTION **a.** $\dfrac{5^3}{5^3} = \dfrac{5 \cdot 5 \cdot 5}{5 \cdot 5 \cdot 5} = 1$ by the simplification property of fractions.

$\dfrac{5^3}{5^3} = 5^{3-3} = 5^0$ by the division property.

The value of $5^0 = 1$.

b. $\dfrac{x^n}{x^n} = 1$, where $x \neq 0$, by the simplification property of fractions.

$\dfrac{x^n}{x^n} = x^{n-n} = x^0$, where $x \neq 0$, by the division property.

The value of $x^0 = 1$, where $x \neq 0$.

c. $\dfrac{4^3}{4^5} = \dfrac{4 \cdot 4 \cdot 4}{4 \cdot 4 \cdot 4 \cdot 4 \cdot 4} = \dfrac{1}{4^2}$ by the simplification property of fractions.

$\dfrac{4^3}{4^5} = 4^{3-5} = 4^{-2}$ by the division property.

Thus, $4^{-2} = \dfrac{1}{4^2}$.

d. $\dfrac{x^5}{x^6} = \dfrac{x \cdot x \cdot x \cdot x \cdot x}{x \cdot x \cdot x \cdot x \cdot x \cdot x} = \dfrac{1}{x}$ by the simplification property of fractions.

$\dfrac{x^5}{x^6} = x^{5-6} = x^{-1}$ by the division property.

Thus, $x^{-1} = \dfrac{1}{x}$, $x \neq 0$. ◀

When we simplify exponent expressions, we leave answers with positive exponents. In Example 8, we perform a variety of operations with negative exponents.

▶ **EXAMPLE 8** **Simplifying exponent expressions** Use the appropriate definition or property to simplify these expressions. Assume all expressions are defined.

a. $x^{-4} \cdot x^7$ **b.** $\dfrac{x}{x^{-2}}$ **c.** $\dfrac{b^3}{b^{-3}}$ **d.** $\dfrac{xy^{-3}}{x^2y}$ **e.** $\dfrac{a^3b^{-1}}{a^{-2}b^3}$

f. $(x^{-5})^2$ **g.** $\left(\dfrac{4x^2}{y^3}\right)^2$ **h.** $\left(\dfrac{b}{a}\right)^{-3}$ **i.** $\left(\dfrac{2x^3}{3}\right)^{-2}$

SOLUTION **a.** $x^{-4} \cdot x^7 = x^{-4+7} = x^3$

b. $\dfrac{x}{x^{-2}} = x^{1-(-2)} = x^3$

c. $\dfrac{b^3}{b^{-3}} = b^{3-(-3)} = b^6$

d. $\dfrac{xy^{-3}}{x^2y} = x^{1-2}y^{-3-1} = x^{-1}y^{-4} = \dfrac{1}{xy^4}$

e. $\dfrac{a^3b^{-1}}{a^{-2}b^3} = a^{3-(-2)}b^{-1-3} = a^5b^{-4} = \dfrac{a^5}{b^4}$

f. $(x^{-5})^2 = x^{-5 \cdot 2} = x^{-10} = \dfrac{1}{x^{10}}$

g. $\left(\dfrac{4x^2}{y^3}\right)^2 = \dfrac{4^2x^{2 \cdot 2}}{y^{3 \cdot 2}} = \dfrac{16x^4}{y^6}$

h. $\left(\dfrac{b}{a}\right)^{-3} = \left(\dfrac{a}{b}\right)^3 = \dfrac{a^3}{b^3}$

i. $\left(\dfrac{2x^3}{3}\right)^{-2} = \left(\dfrac{3}{2x^3}\right)^2 = \dfrac{3^2}{2^2x^{3 \cdot 2}} = \dfrac{9}{4x^6}$ ◀

ANSWER BOX

Warm-up: 1. and **Example 5:**

Exponent Expression	Value	Exponent Expression	Value	Exponent Expression	Value
10^3	1000	5^3	125	4^3	64
10^2	100	5^2	25	4^2	16
10^1	10	5^1	5	4^1	4
10^0	1	5^0	1	4^0	1
10^{-1}	$\frac{1}{10}$	5^{-1}	$\frac{1}{5}$	4^{-1}	$\frac{1}{4}$
10^{-2}	$\frac{1}{100}$	5^{-2}	$\frac{1}{25}$	4^{-2}	$\frac{1}{16}$
10^{-3}	$\frac{1}{1000}$	5^{-3}	$\frac{1}{125}$	4^{-3}	$\frac{1}{64}$

Think about it 1: $\frac{1}{4}$; 2; $\frac{4}{3}$; $0.25 = \frac{1}{4}$, so $\frac{4}{1}$; $1.25 = \frac{5}{4}$, so $\frac{4}{5}$
Think about it 2: To group the negative sign with the 2 in parts b and g; to apply the exponent to both the numerator and the denominator in parts h and k.

▶ 7.5 Exercises

Complete the input-output tables in Exercises 1 to 4. Look for a pattern down the output column. Change decimals to fractions.

1.

Input: x	Output: 3^x
2	
1	
0	
−1	
−2	
−3	

2.

Input: x	Output: 2^x
2	
1	
0	
−1	
−2	
−3	

3.

Input: x	Output: 4^x
2	
1	
0	
−1	
−2	
−3	

4.

Input: x	Output: 5^x
2	
1	
0	
−1	
−2	
−3	

Blue numbers are core exercises.

Simplify the expressions in Exercises 5 to 10 without a calculator.

5. **a.** 2^{-2} **b.** 2^{-1} **c.** 2^0

6. **a.** $\left(\frac{1}{2}\right)^{-1}$ **b.** $\left(\frac{1}{2}\right)^0$ **c.** $\left(\frac{1}{2}\right)^{-2}$

7. **a.** $\left(\frac{1}{4}\right)^0$ **b.** $\left(\frac{1}{4}\right)^{-2}$ **c.** $\left(\frac{1}{4}\right)^{-1}$

8. **a.** $\left(\frac{3}{4}\right)^0$ **b.** $\left(\frac{3}{4}\right)^{-1}$ **c.** $\left(\frac{3}{4}\right)^{-2}$

9. **a.** $(0.5)^{-1}$ **b.** $(0.5)^0$ **c.** $(0.5)^{-2}$

10. **a.** $(0.25)^{-2}$ **b.** $(0.25)^{-1}$ **c.** $(0.25)^0$

11. Which exponent gives the reciprocal of a number?

12. Which exponent gives the square of the reciprocal of a number?

13. Which exponent gives 1 unless the base is zero?

14. Name a number with a reciprocal smaller than the original number. (*Hint:* Look at the reciprocal problems in Exercises 5 to 10.)

15. Name a number with a reciprocal larger than the original number.

16. Name a number with a reciprocal equal to the original number.

Simplify Exercises 17 to 24 using the appropriate property and then check with a calculator. Leave answers without denominators.

17. **a.** $5^3 \cdot 5^{-7}$ **b.** $6^5 \cdot 6^{-2}$ **c.** $10^3 \cdot 10^{-6}$ **d.** $10^2 \cdot 10^{-7}$

18. **a.** $10^5 \cdot 10^{-4}$ **b.** $10^{-4} \cdot 10^6$ **c.** $2^6 \cdot 2^{-3}$ **d.** $2^{-4} \cdot 2^2$

19. **a.** $10^{-15} \cdot 10^{-15}$ **b.** $10^{-28} \cdot 10^{19}$ **c.** $2^{24} \cdot 2^{-16}$ **d.** $2^{13} \cdot 2^{-5}$

20. **a.** $5^{-14} \cdot 5^6$ **b.** $6^8 \cdot 6^{-15}$ **c.** $10^{-22} \cdot 10^{14}$ **d.** $10^{12} \cdot 10^{-21}$

21. **a.** $\dfrac{3^5}{3^{-2}}$ **b.** $\dfrac{10^{-5}}{10^{-2}}$ **c.** $\dfrac{10^{-4}}{10^{12}}$ **d.** $\dfrac{1}{10^2}$

22. **a.** $\dfrac{10^{15}}{10^{-8}}$ **b.** $\dfrac{2^{-3}}{2^{-4}}$ **c.** $\dfrac{10^{-6}}{10^7}$ **d.** $\dfrac{1}{5^2}$

23. **a.** $(2^3)^4$ **b.** $(2^3)^{-4}$ **c.** $(10^5)^2$ **d.** $(10^{-6})^3$

24. **a.** $(2^{-3})^4$ **b.** $(2^{-3})^{-4}$ **c.** $(10^{-5})^{-1}$ **d.** $(10^4)^{-5}$

Blue numbers are core exercises.

Simplify Exercises 25 to 30 using the appropriate property. Leave answers without denominators. Assume all expressions are defined.

25. **a.** m^3m^5 **b.** n^4n^4 **c.** a^6a^2 **d.** a^7a^1 **e.** $a^5 \cdot a^{-12}$ **f.** $x^{-5} \cdot x^{13}$ **g.** $n^{-6} \cdot n^2$

26. **a.** m^3m^3 **b.** m^1m^5 **c.** b^2b^4 **d.** n^2n^3 **e.** $x^{-2} \cdot x^{-8}$ **f.** $n^{-9} \cdot n^{15}$ **g.** $a^{12} \cdot a^{-7}$

27. **a.** $\dfrac{1}{x^2}$ **b.** $\dfrac{1}{a^{-1}}$ **c.** $\dfrac{b}{b^{-2}}$ **d.** $\dfrac{a^3}{a^{-6}}$ **e.** $\dfrac{a^{-6}}{a^2}$ **f.** $\dfrac{x^4}{x^{-2}}$

28. **a.** $\dfrac{x}{x^2}$ **b.** $\dfrac{a}{a^{-1}}$ **c.** $\dfrac{1}{b^{-3}}$ **d.** $\dfrac{x^{-2}}{x^{-3}}$ **e.** $\dfrac{a^6}{a^{-3}}$ **f.** $\dfrac{a^{-5}}{a^3}$

29. **a.** $(x^2)^3$ **b.** $(xy)^2$ **c.** $(x^2y^3)^2$ **d.** $(x^2)^{-4}$ **e.** $(x^{-4})^{-3}$ **f.** $(b^{-2})^3$

30. **a.** $(a^3)^2$ **b.** $(a^2b)^2$ **c.** $(a^3b^2)^3$ **d.** $(b^{-4})^{-1}$ **e.** $(x^{-2})^5$ **f.** $(x^6)^{-3}$

In Exercises 31 to 38, simplify the expressions. Remove all negative and zero exponents. Assume all expressions are defined.

31. **a.** $\dfrac{x^5}{x^2}$ **b.** $\dfrac{a^8}{a^5}$ **c.** $\left(\dfrac{x}{y}\right)^2$ **d.** $\left(\dfrac{2x}{y}\right)^3$ **e.** $\dfrac{-6b}{9b}$ **f.** $\dfrac{x^5y^2}{xy^2}$ **g.** $\dfrac{-2a^7b^3}{-8ab^2}$

32. **a.** $\dfrac{x^6}{x^3}$ **b.** $\dfrac{b^8}{b^4}$ **c.** $\left(\dfrac{x}{3y}\right)^2$ **d.** $\left(\dfrac{2a}{3b}\right)^3$ **e.** $\dfrac{-4ab}{6a}$ **f.** $\dfrac{a^3b^2}{ab}$ **g.** $\dfrac{-4x^4y^3}{-10x^2y}$

33. **a.** x^{-1} **b.** $\left(\dfrac{x}{y}\right)^{-1}$ **c.** $\left(\dfrac{y}{x}\right)^{-1}$ **d.** $\left(\dfrac{a}{b}\right)^0$ **e.** $\left(\dfrac{a}{c}\right)^0$ **f.** $\left(\dfrac{a}{bc}\right)^{-1}$

34. **a.** y^{-1} **b.** $\left(\dfrac{b}{c}\right)^{-1}$ **c.** $\left(\dfrac{c}{b}\right)^{-1}$ **d.** $\left(\dfrac{c}{b}\right)^0$ **e.** $\left(\dfrac{x}{y}\right)^0$ **f.** $\left(\dfrac{c}{ab}\right)^{-1}$

35. **a.** y^{-3} **b.** $\left(\dfrac{y}{x}\right)^{-2}$ **c.** $\dfrac{1}{b^{-3}}$ **d.** $\left(\dfrac{a}{b}\right)^{-3}$ **e.** $\left(\dfrac{4a^2}{c}\right)^{-2}$ **f.** $\left(\dfrac{a}{b^2}\right)^{-3}$

36. **a.** x^{-3} **b.** $\left(\dfrac{c}{b}\right)^{-2}$ **c.** $\dfrac{1}{x^{-2}}$

d. $\left(\frac{c}{b}\right)^{-3}$ **e.** $\left(\frac{2x^2}{y}\right)^{-2}$ **f.** $\left(\frac{2c}{a^2b}\right)^{-3}$

37. a. $\frac{xy^{-2}}{x^2y^3}$ **b.** $\frac{x^{-1}y}{x^{-1}y^{-2}}$ **c.** $\frac{a^{-2}b^2}{a^3b^{-1}}$

d. $\frac{x^{-2}y}{x^3y^2}$ **e.** $\frac{x^2y^{-1}}{x^{-2}y^{-3}}$ **f.** $\frac{a^3b^{-1}}{a^{-1}b^2}$

38. a. $\frac{a^{-3}b^2}{ab^4}$ **b.** $\frac{x^2y^{-2}}{x^{-3}y^{-2}}$ **c.** $\frac{x^3y^{-2}}{x^{-1}y^3}$

d. $\frac{a^3b^{-2}}{a^{-1}b}$ **e.** $\frac{a^{-2}b}{a^{-1}b^{-1}}$ **f.** $\frac{x^{-1}y^{-2}}{x^{-1}y}$

39. Evaluate the expressions in Exercise 37 for $x = 3$, $y = -2$, $a = 3$, $b = -2$. Check by substituting into the answers for the exercise.

40. Repeat Exercise 39 using the expressions in Exercise 38.

41. Extending multiplication.

a. $(1 + x^{-1})(1 + x^{-2})$ **b.** $(x^{-1} + y)(x^{-1} - y)$

42. Extending factoring.

a. $x^{-2}y^2 - x^{-2}$ **b.** $x^{-1}y^2 - 2x^{-1}y + x^{-1}$

43. Explain how the multiplication property of like bases lets us simplify $a^5 \cdot a^8$ but not a^2b^2.

44. Explain how the division property of like bases lets us simplify a^8/a^5 but not a^2/b^2.

45. Explain how the power of a power property lets us simplify $(a^4)^3$.

▶ Problem Solving

46. Exponent Puzzle The variables A, B, C, and D in the equation $A^BC^D = ABCD$ represent numbers from the set $\{0, 1, 2, 3, 4, 5, 6, 7, 8, 9\}$. The expression $ABCD$ in the equation represents a number with four digits, not a product of the numbers A and B and C and D. Use a calculator to guess the numbers. (For example: $7^3 \cdot 8^3$ gives the six-digit answer 175,616, not 7383.) Digits may be used twice.

Blue numbers are core exercises.

▶ Projects

47. Interest on CDs The interest on certificates of deposit (CDs) at many banks and credit unions is calculated with annual compounding. If you want A dollars n years in the future, you need to invest P dollars now at interest rate r (expressed as a decimal).

a. Use the formula $P = A(1 + r)^{-n}$ in the table to calculate how much, P, you need to save now to have \$50,000 in the future. Round your answers to the nearest dollar.

Interest Rate	$P = 50000(1 + r)^{-10}$ (n = 10 years)	$P = 50000(1 + r)^{-20}$ (n = 20 years)
$r = 0.03$		
$r = 0.04$		
$r = 0.05$		
$r = 0.06$		
$r = 0.07$		
$r = 0.08$		

b. Use the completed table to predict the interest rate needed to reach \$50,000 in 10 years if you start with an investment of \$25,000.

c. Use the completed table to predict the interest rate needed to reach \$50,000 in 20 years if you start with an investment of \$25,000.

▶ 7.6 Scientific Notation

Objectives

- Change numbers between decimal notation and scientific notation.
- Perform operations in scientific notation, both by estimating and by using a calculator.

WARM-UP

1. Apply the properties of exponents from Section 7.5 to these expressions.

a. $10^3(10^2)$ **b.** $10^{-2}10^5$ **c.** $10^8 \div 10^{-3}$ **d.** $10^{-4} \div 10^{-8}$

2. Normally, the order of operations directs us to do the brackets before the parentheses.

a. Why can we arrange $[(1.2)(10^3)] \cdot [(3.0)(10^2)]$ into $1.2(3.0)(10^3)(10^2)$?

b. Why can we arrange $\dfrac{1.5(10^{-4})}{5.0(10^2)}$ into $\dfrac{1.5}{5.0}\left(\dfrac{10^{-4}}{10^2}\right)$?

IN THIS SECTION, we introduce scientific notation, an important tool for describing the vastness of space in astronomy, the tiny bacteria and viruses in medicine, and the subatomic particles in chemistry and physics. We practice operations with exponents (from Section 7.5) while doing operations with scientific notation. A summary of the earliest uses of exponents and scientific notation is at
http://members.aol.com/jeff570/operation.html

Scientific Notation

One of the most common uses of negative exponents is in scientific notation. Scientific notation is a short way of writing large and small numbers, such as the distance to a star or the size of a virus. Calculators automatically change into scientific notation when the number is too large or too small to fit the answer display, as in Example 1.

EXAMPLE 1 Exploring large and small numbers Use a calculator to evaluate these numbers.

a. 3600^4 **b.** 0.00063^5

SOLUTION Figure 16 shows the calculator answers written in scientific notation.

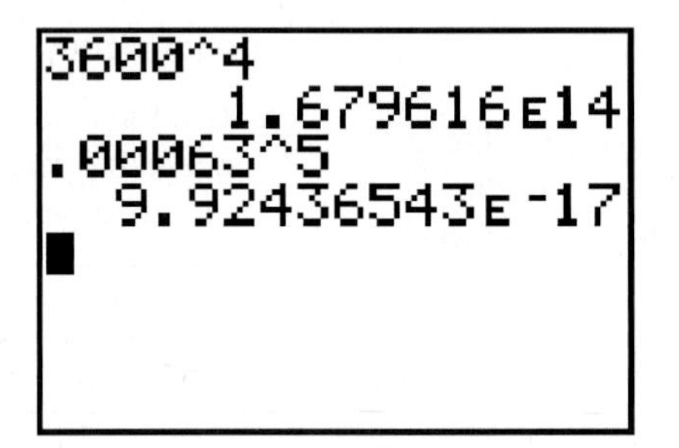

FIGURE 16 Computations yielding scientific notation ◀

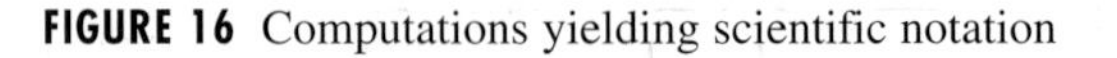

DEFINITION OF SCIENTIFIC NOTATION

Numbers written in scientific notation are the product of a decimal between 1 and 10 with a power of ten.

__ . _____ × 10^—

Integer exponent

Decimal number between 1 and 10

The first part is a decimal with one nonzero number to the left of the decimal point.

Scientific notation is based on the powers of ten. We use the $\times$ sign for multiplication in scientific notation; the $\times$ is not a variable. Rounding the first part to two nonzero digits, we write the answers in Example 1 as

$$1.7 \times 10^{14} \qquad \text{and} \qquad 9.9 \times 10^{-17}$$

Calculator displays show the decimal part as well as the exponent but omit the symbol for multiplication and the base 10.

Table 5 gives examples of scientific notation in a variety of settings.

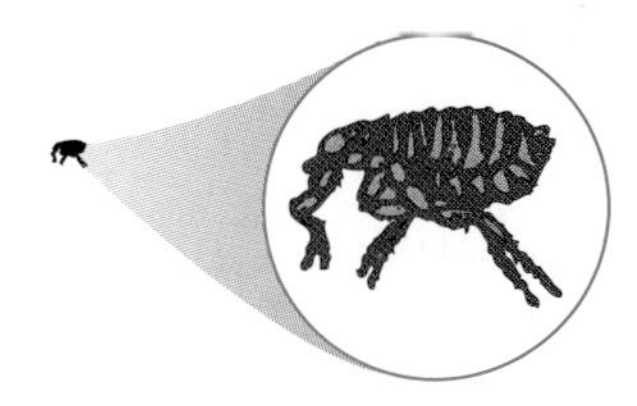

TABLE 5 Applications of Scientific Notation

Power of 10	Value	Scientific Notation	Application
10^9	1,000,000,000 (billion)	4.6×10^9 years ago	Estimated formation of Earth
10^8	100,000,000	9.83×10^8 feet per second	Speed of light in a vacuum
10^7	10,000,000	2.46×10^7 years	Half-life of ^{236}U, radioactive uranium
10^6	1,000,000	6.38×10^6 meters	Radius of Earth
10^5	100,000	5.256×10^5 minutes	Minutes in a 365-day year
10^4	10,000	8.64×10^4 seconds	Seconds in one day
10^3	1,000	5.73×10^3 years	Half-life of ^{14}C, radioactive carbon
10^2	100	1×10^2 meters	Length of 100-meter dash
10^1	10	1.049×10^1 seconds	Women's 100-meter record, 1988
10^0	1	8.04×10^0 days	Half-life of ^{131}I, radioactive iodine
10^{-1}	$1/10 = 0.1$	1.2×10^{-1} meter	DVD diameter
10^{-2}	$1/100 = 0.01$	2.54×10^{-2} meter	1 inch
10^{-3}	$1/1000 = 0.001$	3×10^{-3} meter	Size of a flea (3 mm)
10^{-7}	0.000 000 1	1×10^{-7} meter	Length of HIV virus
10^{-10}		1×10^{-10} meter	1 angstrom, measure of distance between atoms
10^{-27}		1.66×10^{-27} kilogram	Unit of atomic mass

DECIMAL NOTATION TO SCIENTIFIC NOTATION We need to be able to change between decimal notation and scientific notation in order to interpret answers from a calculator and enter large and small numbers into a calculator. Here are some things to keep in mind.

Changing Decimal Notation to Scientific Notation

To change to scientific notation:

1. Place the decimal point after the first nonzero digit.
2. Count the number of places the decimal point has moved, and use that number as the number exponent on ten.
3. **a.** If the original number is greater than 10, the exponent on ten is positive.
   b. If the original number is between 10 and 1, the exponent on ten is zero.
   c. If the original number is between 0 and 1, the exponent on ten is negative.

In Example 2, we apply the rules for changing to scientific notation.

▶ **EXAMPLE 2** *Changing numbers to scientific notation* Predict the exponent on ten needed to change each number to scientific notation. Then change the number.

a. The minimum distance from the Sun to Mercury is 28,600,000 miles.
b. The maximum distance from Earth to Pluto is 4,644,000,000 miles.
c. The mass of an electron is

0.000 000 000 000 000 000 000 000 000 000 910 9 kilogram

d. The mass of a proton is

0.000 000 000 000 000 000 000 000 001 672 6 kilogram

SOLUTION

a. The first nonzero digit is 2. The decimal point will move 7 places. The distance is greater than 10. The exponent on ten will be positive.

2.86×10^7 miles

b. The first nonzero digit is 4. The decimal point will move 9 places. The distance is greater than 10. The exponent on ten will be positive.

4.644×10^9 miles

c. The first nonzero digit is 9. The decimal point will move 31 places. The mass is between 0 and 1. The exponent on ten will be negative.

9.109×10^{-31} kilogram

d. The first nonzero digit is 1. The decimal point will move 27 places. The mass is between 0 and 1. The exponent on ten will be negative.

1.6726×10^{-27} kilogram ◀

SCIENTIFIC NOTATION TO DECIMAL NOTATION Here are some things to keep in mind.

Changing Scientific Notation to Decimal Notation

To change from scientific notation to decimal notation:

1. Look at the exponent on ten to find the number of decimal places and the direction in which to move the decimal point.
2. **a.** If the exponent is negative, move the decimal point left to make a number between 0 and 1.
   b. If the exponent is zero, do not change the decimal point.
   c. If the exponent is positive, move the decimal point right to make a number greater than 10.

In Example 3, we change scientific notation to decimal notation.

▶ EXAMPLE 3 Changing scientific notation to decimal notation Predict the direction in which the decimal point will move and then change each number to decimal notation.

a. The minimum distance from the Sun to Earth is 9.14×10^7 miles.

b. In chemistry, Avogadro's number measures the number of molecules in a mole: 6.022×10^{23}.

c. The estimated human population of the world in 1990 was 5.333×10^9.

d. The human population on Earth in 1650 is estimated at 5.5×10^8.

e. The mass of a house spider is about 1×10^{-4} kilogram.

SOLUTION *Remember:* Positive exponents on ten go with numbers greater than 10, and negative exponents on ten go with numbers between 0 and 1.

a. 91,400,000 miles

b. 602,200,000,000,000,000,000,000 molecules per mole

c. 5,333,000,000 people

d. 550,000,000 people

e. 0.0001 kilogram ◀

▶ Operations with Scientific Notation

It is important to know how to multiply and divide numbers in scientific notation both by estimating and by using a calculator. Both are important skills, as estimation is often easier with scientific notation and the data for many application problems are written in scientific notation.

In estimating, we perform operations on the two parts of the scientific notation separately.

ESTIMATING USING SCIENTIFIC NOTATION

1. a. To estimate the product of two numbers in scientific notation, multiply the decimals and then add the exponents on ten.

b. To estimate the quotient of two numbers in scientific notation, divide the decimals and then subtract the exponents on ten.

2. If the product or the quotient of the decimals is greater than 10 or between 0 and 1, adjust the decimal point and the exponent on ten.

▶ EXAMPLE 4 Performing operations with scientific notation mentally Use estimation techniques to do these operations mentally.

a. $(1.2 \times 10^3) \cdot (3.0 \times 10^2)$

b. $(1.5 \times 10^{-2}) \cdot (4.0 \times 10^5)$

c. $(9.0 \times 10^5) \cdot (8.0 \times 10^{-3})$

d. $(1.5 \times 10^{-4}) \div (5.0 \times 10^2)$

e. $(2.4 \times 10^8) \div (3.0 \times 10^{-3})$

f. $\dfrac{1.6 \times 10^{-4}}{(2.0 \times 10^{12})(5.0 \times 10^{-8})}$

SOLUTION

a. 1.2 times 3.0 is 3.6, and 10^3 times 10^2 is 10^5. The answer is 3.6×10^5, or 360,000.

b. 1.5 times 4.0 is 6.0, and 10^{-2} times 10^5 is 10^3. The answer is 6.0×10^3, or 6000.

c. 9 times 8 is 72, and 10^5 times 10^{-3} is 10^2. The answer is

$$72 \times 10^2 = 7.2 \times 10^1 \times 10^2 = 7.2 \times 10^3$$

d. 1.5 divided by 5.0 is 0.3, and 10^{-4} divided by 10^2 is 10^{-6}. The answer is

$$0.3 \times 10^{-6} = 3.0 \times 10^{-1} \times 10^{-6} = 3.0 \times 10^{-7}$$

e. 2.4 divided by 3.0 is 0.8, and 10^8 divided by 10^{-3} is 10^{11}. The answer is

$$0.8 \times 10^{11} = 8.0 \times 10^{-1} \times 10^{11} = 8.0 \times 10^{10}$$

f. 1.6 divided by the product of 2 and 5 is 0.16, and 10^{-4} divided by the product of 10^{12} and 10^{-8} is $10^{-4-(12+(-8))} = 10^{-8}$. The answer is

$$0.16 \times 10^{-8} = 1.6 \times 10^{-1} \times 10^{-8} = 1.6 \times 10^{-9}$$ ◀

Calculators can do operations with scientific notation. We use [2nd] [EE] to enter a number in scientific notation. The letters EE represent "*e*nter the *e*xponent on ten." Negative exponents need to have the negative sign, [(−)], entered before the exponent. Change answers to scientific notation mode with [MODE] ▶ [ENTER] [2nd] [QUIT].

Student Note: [EE] replaces $\times 10$ [^]. Only E shows in the viewscreen.

▶ **EXAMPLE 5** Using a calculator for scientific notation operations Repeat the operations from Example 4 with a calculator.

SOLUTION Parts a to c are shown in Figure 17. Parts d to f are shown in Figure 18.

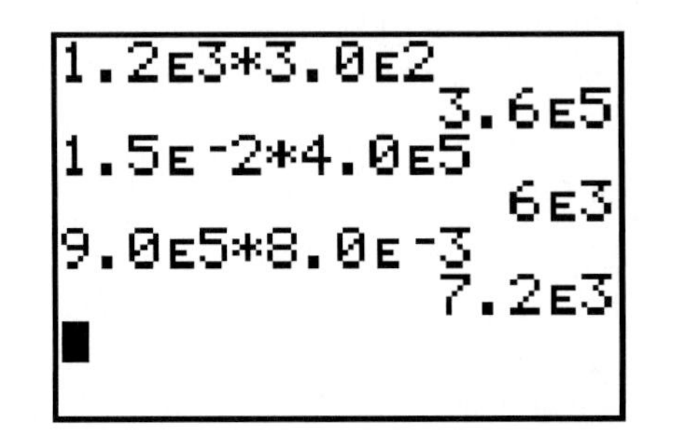

FIGURE 17 Multiplication, Example 4

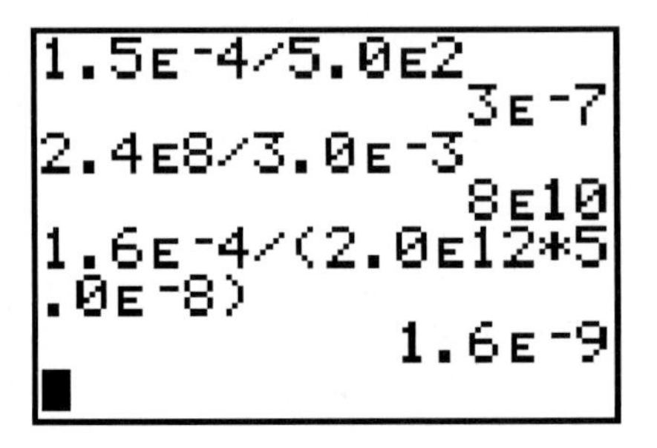

FIGURE 18 Division, Example 4 ◀

THINK ABOUT IT: When a number is placed into a calculator in scientific notation using the [EE] key, the entire number is considered to be one number, as in parts d and e of Example 5. Repeat these two parts using multiplication by 10 with the [^] key and no parentheses around the denominator. Explain the results.

ANSWER BOX

Warm-up: 1. a. 10^5 **b.** 10^3 **c.** 10^{11} **d.** 10^4 **2a.** The commutative property of multiplication **2b.** The commutative property of multiplication, or by regrouping parentheses, the associative property of multiplication
Think about it: Incorrect answers are obtained. For d, we get 0.003; for e, 80,000. Without any parentheses around the denominator, the last multiplication is performed as though it were in the numerator.

▶ 7.6 Exercises

In Exercises 1 and 2, predict how many decimal places to the left or right the decimal point will move in each case. Do the operations without a calculator.

1. a. 34.6×10

b. 1.6(0.000 001)

c. 1.6(100,000)

d. $219.1 \div 10$

e. $219.1 \div 1000$

f. $219.1 \div 0.0001$

2. **a.** 34.6×100

b. $2.78(0.0001)$

c. $2.78(1000)$

d. $57.8 \div 0.01$

e. $57.8 \div 0.0001$

f. $57.8 \div 100$

3. Complete the table.

x	10^x as fraction	10^x as decimal
0		
-1		
-2		
-3		
-4		
-5		

In Exercises 4 to 6, use a calculator and record answers in scientific notation. Round the decimal part to two decimal places.

4. **a.** 0.0012^2 **b.** $500{,}000^2$

5. **a.** 5280^3 **b.** 0.00008^3

6. **a.** 0.0054^2 **b.** $123{,}456^2$

In Exercises 7 to 12, change each number to scientific notation. (*Hint:* 1 million = 1,000,000.) Round the decimal part to two decimal places.

7. The diameter of the Sun is 1,391,400 kilometers.

8. The maximum distance from the Sun to Earth is 94.6 million miles.

9. The minimum distance from the Sun to Pluto is 2756.4 million miles.

10. The mass of a bacteria is 0.000 000 000 000 1 kilogram.

11. The mass of a chicken is 1800 grams.

12. The mass of an average polar bear is 322 kilograms.

In Exercises 13 to 20, change from scientific notation to decimal notation.

13. The mass of the Sun is 1.99×10^{30} kilograms.

14. The mass of Earth is 5.98×10^{24} kilograms.

15. An electron charge is -1.602×10^{-19} coulomb.

16. The gravitational constant is 6.672×10^{-11} N·m^2/kg^2.

17. Dinosaurs first appeared on Earth about 2.0×10^8 years ago.

18. Dinosaurs were extinct by 6.5×10^7 years ago.

19. The mass of a neutron is 1.6750×10^{-27} kilogram.

20. The projected human population of Earth for 2025 is 8.17×10^9.

In Exercises 21 to 28, estimate the products and quotients without a calculator.

21. **a.** $(2 \times 10^{15})(3 \times 10^{12})$

b. $(4 \times 10^{13})(3 \times 10^{18})$

22. **a.** $(6 \times 10^{14})(2 \times 10^{-19})$

b. $(8 \times 10^{-12})(5 \times 10^{17})$

23. **a.** $(5 \times 10^{-16})(6 \times 10^{-11})$

b. $(8 \times 10^{-12})(6 \times 10^{-13})$

24. **a.** $(8 \times 10^{14}) \div (2 \times 10^{11})$

b. $(2.0 \times 10^{12}) \div (5 \times 10^{17})$

25. **a.** $(2.8 \times 10^{-14}) \div (7 \times 10^{-18})$

b. $(9 \times 10^{-25}) \div (4.5 \times 10^{-12})$

26. **a.** $(6 \times 10^{13}) \div (3 \times 10^{-18})$

b. $(1.5 \times 10^{-12}) \div (3 \times 10^{17})$

27. $\dfrac{4.8 \times 10^{-7}}{(1.6 \times 10^{8})(3.0 \times 10^{-18})}$

28. $\dfrac{5.6 \times 10^{15}}{(1.4 \times 10^{-8})(4.0 \times 10^{4})}$

Change the "calculator outputs" in Exercises 29 and 30 to regular scientific notation and to decimals.

29. **a.** 2.34E−2 **b.** 3.14E3 **c.** 6.28E7

30. **a.** 9.01E−3 **b.** 4.56E8 **c.** 2.41E12

▶ Error Analysis

The numbers in Exercises 31 to 34 are not in standard scientific notation. Correct them.

31. **a.** 34×10^3 **b.** 560×10^{-2}

32. **a.** 450×10^{-1} **b.** 6700×10^4

33. **a.** 0.432×10^4 **b.** 0.567×10^{-5}

34. **a.** 0.025×10^{-3} **b.** 0.0042×10^4

▶ Comparing Scientific Notations

Place an inequality sign between the numbers in Exercises 35 to 42 to make a true statement.

35. **a.** $(2 \times 10^5) \square (3 \times 10^4)$

b. $(4 \times 10^3) \square (3 \times 10^4)$

36. **a.** $(2.5 \times 10^5) \square (3.5 \times 10^5)$

b. $(4.6 \times 10^5) \square (3.6 \times 10^3)$

Blue numbers are core exercises.

37. a. $(2 \times 10^{-5}) \square (3 \times 10^{-4})$

b. $(4 \times 10^{-3}) \square (3 \times 10^{-4})$

38. a. $(2 \times 10^{-15}) \square (2 \times 10^{-14})$

b. $(3 \times 10^{-13}) \square (3 \times 10^{-14})$

39. a. $(3.2 \times 10^{-14}) \square (2.8 \times 10^{-14})$

b. $(4.3 \times 10^{-13}) \square (5.3 \times 10^{-13})$

40. a. $(1.2 \times 10^{-19}) \square (2.4 \times 10^{-19})$

b. $(1.8 \times 10^{-18}) \square (1.2 \times 10^{-18})$

41. a. $(3.2 \times 10^{5}) \square (-3.2 \times 10^{5})$

b. $(-1.2 \times 10^{-2}) \square (-1.2 \times 10^{2})$

42. a. $(-1.5 \times 10^{2}) \square (1.5 \times 10^{-1})$

b. $(-2.5 \times 10^{3}) \square (-2.5 \times 10^{2})$

43. Which has a smaller mass, an electron at 9.101×10^{-31} kilogram or a proton at 1.6726×10^{-27} kilogram?

44. Which is closer to the Sun, Neptune at 1.860×10^{9} miles or Saturn at 9.375×10^{8} miles?

Scientific notation commonly results when we work with units of measure. In Exercises 45 to 52, solve with a unit analysis. Assume 365 days per year. Round the decimal part to two decimal places.

45. The speed of light is 186,000 miles per second. How many miles does light travel in one year? This distance is called a light-year.

speed of light = 186,000 mi per sec
60 sec = 1 min
60 min = 1 hr
24 hr = 1 day
365 days = 1 yr (rounded)

46. If a giant sequoia (redwood) is 360 feet tall and is estimated to be 6000 years old, what is its average growth rate in miles per hour?

speed of growth = 360 ft per 6000 yr
1 mi = 5280 ft
24 hr = 1 day
365 days = 1 yr (rounded)

47. How many miles does light travel in the 27,000 light-years to the M13 star cluster mentioned in the chapter opener?

48. How fast in feet per second is a silver birch tree growing if it is 30 feet tall after 15 years of growth?

49. The Ohio class submarine displaces 18,700 tons of water. How many quarts of water is this? (*Hint:* 2000 lb = 1 ton, 1 gal ≈ 8.3 lb of water. List other facts as needed.)

Blue numbers are core exercises.

50. Suppose a 20-foot record-height tomato plant took 6 months to grow. What was the tomato's growth rate in inches per minute? Assume 30 days per month.

51. If an average boy grows from 20 inches to 5 feet 9 inches in 18 years, what is the average growth rate in miles per hour?

52. If an average girl grows from 19 inches to 5 feet 4 inches in 17 years, what is the average growth rate in miles per hour?

▶ Writing

53. Explain how to change a number from scientific notation to decimal notation. Make up your own example.

54. Explain how to change a number from decimal notation to scientific notation. Make up your own example.

55. Explain how $(x + 3)(x + 4)$ is a different multiplication from $(2 \times 10^{3})(3 \times 10^{4})$.

▶ Project

56. Black Holes and Scientific Notation* No light escapes from a black hole. Thus, the formula relating the mass and radius of a black hole and the speed of light is the same as the formula for a rocket ship leaving Earth. The radius, r_B, of a black hole depends on its mass, m, according to the formula

$$r_B = \frac{2Gm}{c^2}$$

G is the gravitational constant $6.672 \times 10^{-11}\ \text{N}\cdot\text{m}^2/\text{kg}^2$, where $\text{N} = \text{kg}\cdot\text{m}/\text{sec}^2$. The mass is in kilograms. The letter c is the speed of light, 2.998×10^{8} m/sec.

a. Substitute just the units into the formula to see what units describe the radius.

For parts b to e, find the radius to which each of these objects in our solar system would have to be compressed in order to be a black hole.

b. Sun: mass = 1.99×10^{30} kg

c. Jupiter: mass = 1.90×10^{27} kg

d. Earth: mass = 5.98×10^{24} kg

e. Moon: mass = 7.35×10^{22} kg

f. The diameters for the four objects in parts b through e are, respectively, 1,391,400 kilometers, 142,800 kilometers, 12,756 kilometers, and 3476 kilometers. To find a number describing the number of times each object is compressed, find the actual radius and divide it by the black hole radius.

*Astronomy information from William K. Hartman, *The Cosmic Voyage Through Time and Space* (Belmont, CA: Wadsworth Publishing Company, 1992).

▶ 7 Chapter Summary

Vocabulary

For definitions and page references, see the Glossary/Index.

ascending order
binomial
common factor
descending order
difference of squares
division of like bases property
greatest common factor
monomial
multiplication of like bases property
perfect square trinomial
polynomial
power of a power property
power of a product property
power of a quotient property
scientific notation
trinomial

Concepts

7.1 Operations on Polynomials

Multiplying two expressions is like finding the area of a rectangle given its base and height. Arrange answers in descending order of exponents.

Factoring an expression is like finding the base and height of a rectangle given its area.

The first step in any factoring is to factor the greatest common factor.

7.2 Multiplication of Binomials and Special Products

In multiplication of binomials by table, there are two patterns: The terms on the diagonal from lower left to upper right add to the middle term of the trinomial answer, and the terms on the two diagonals multiply to the same product.

A perfect square trinomial describes $(a + b)^2$. The product $(a + b)(a + b)$ has three terms: the square of a, twice the product of a and b, and the square of b:

$$(a + b)(a + b) = a^2 + 2ab + b^2$$

A difference of squares describes $a^2 - b^2$. The factors of $a^2 - b^2$ are $(a - b)(a + b)$.

7.3 Factoring Trinomials

To factor $ax^2 + bx + c$ by table:

1. **a.** Write the first term of the trinomial, ax^2, in the upper left corner of the table.

   b. Write the last term of the trinomial, c, in the lower right corner.

   c. Multiply the first and last terms to get the diagonal product, acx^2.

2. **a.** From the diagonal product, list the factors of the coefficient, ac.

   b. Find the factors that add to the middle term of the trinomial, bx.

   c. Write the factors adding to bx in the diagonal from lower left to upper right.

3. Write the greatest common factor of the first row to the left of ax^2.

4. **a.** Use the gcf to find the remaining factors at the top and left of the table.

   b. Check that the top right and bottom left factors multiply to the last term, c.

5. Multiply the binomial factors from the table to check the factoring.

Factor			
Greatest common factor (gcf) of this row	First term of trinomial, ax^2	____	Sum of like terms equals middle term of trinomial, bx
	____	Last term of trinomial, c	Diagonal product, acx^2

7.4 Factoring Special Products and Greatest Common Factors

Use the patterns $a^2 + 2ab + b^2$ and $a^2 - b^2$ to factor into special products.

The first step in any factoring is to factor the greatest common factor. If you forget to factor the gcf first, look for a gcf in the factors from the table as your last step.

A trinomial $ax^2 + bx + c$ factors if a pair of factors of ac add to b.

7.5 Exponents

For zero exponents, $x^0 = 1$. The expression 0^0 is not defined.

For negative-one exponents, $x^{-1} = \frac{1}{x}$, the reciprocal of x.

For negative-two exponents, $x^{-2} = \frac{1}{x^2}$, the square of the reciprocal of x.

In general, $x^{-n} = \frac{1}{x^n}$ or $\left(\frac{a}{b}\right)^{-n} = \left(\frac{b}{a}\right)^n$, the nth power of the reciprocal.

7.6 Scientific Notation

To change decimal notation to scientific notation:

1. Place the decimal point after the first nonzero digit.
2. Count the number of places the decimal point has moved, and use that number as the number exponent on ten.
3. **a.** If the original number is greater than 10, the exponent on ten is positive.
   b. If the original number is between 10 and 1, the exponent on ten is zero.
   c. If the original number is between 0 and 1, the exponent on ten is negative.

To change from scientific notation to decimal notation:

1. Look at the exponent on ten to find the number of decimal places and the direction in which to move the decimal point.
2. **a.** If the exponent is negative, move the decimal point left to make a number between 0 and 1.
   b. If the exponent is zero, do not change the decimal point.
   c. If the exponent is positive, move the decimal point right to make a number greater than 10.

▶ 7 Review Exercises

1. Simplify each of the following expressions. Arrange the terms alphabetically, with exponents on the first letter in descending order. Indicate whether the result is a monomial, binomial, trinomial, or other polynomial.
   a. $3a^2 - 5ab + 4a^2 - 3b^2 + 2ab - 7b^2$
   b. $3 + 4x^2 + 5x - 2(x - 1)$
   c. $x(x^2 + 4x + 16) - 4(x^2 + 4x + 16)$
   d. $11x - 4x(1 - x)$
   e. $9 - 4(x - 3)$

2. Name the greatest common factor for each expression, and then factor the expression.
   a. $6x^2 + 9x + 6$
   b. $14x^2 + 7xh + 49h^2$
   c. $32xy - 24xy^2 + 16x^2y^2$
   d. $25abc - 15ac + 35bc$

3. Find the perimeter and area of each shape.

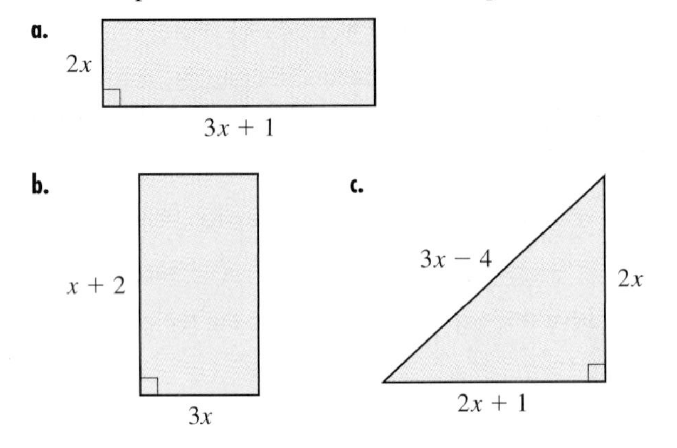

4. Find the missing side of each shape.

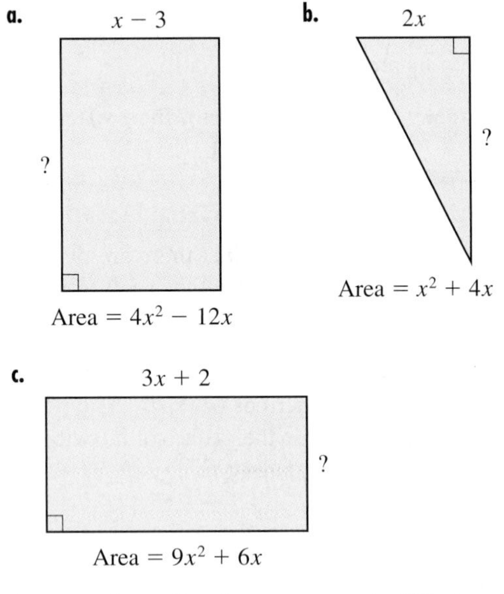

5. Arrange these tiles into a rectangle. What multiplication is shown?

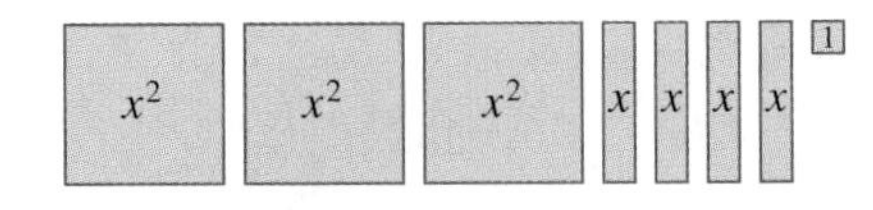

6. Complete these tables. Write the multiplication problem each table describes.

a.

Multiply	$2x$	$+5$
$3x$		

b.

Multiply		
	$3a^2$	-9

In Exercises 7 to 16, find the products.

7. $(x + 4)(x + 3)$

8. $(x + 4)(x - 3)$

9. $(2x - 5)(2x - 5)$

10. $(2x - 5)(2x + 5)$

11. $(3x + 5)(3x - 2)$

12. $(2x - 3)(3x - 2)$

13. $(2x - 3)(3x + 2)$

14. $(a + b)(a - b)$

15. $(a - b)(a - b)$

16. $(a - b)(b - a)$

17. In which of Exercises 7 to 16 are the answers perfect square trinomials?

18. In which of Exercises 7 to 16 are the answers differences of squares?

19. Complete these tables. Write the multiplication problem each table describes.

a.

Factor		
	x^2	$+7x$
	$-2x$	-14

b.

Factor		
	$6x^2$	$+10x$
	$+9x$	$+15$

20. Find the missing operations and numbers in these factorizations.

a. $x^2 + 2x - 24 = (x \_\_\_)(x \_\_\_)$

b. $x^2 - 10x + 24 = (x \_\_\_)(x \_\_\_)$

c. $x^2 - 10x - 24 = (x \_\_\_)(x \_\_\_)$

d. $x^2 - 14x + 24 = (x \_\_\_)(x \_\_\_)$

21. Complete these tables. Then write the original trinomial and its factors.

a. Factor $x^2 - 9x + 14$.

Factor		
	x^2	
		$+14$

b. Factor $2x^2 + 11x - 21$.

Factor		
	$2x^2$	
		-21

c. Factor $12x^2 - x - 1$.

Factor		
	$12x^2$	
		-1

d. Factor $20x^2 + 24x - 9$.

Factor		

Factor each expression in Exercises 22 to 51. Identify it as a difference of squares (ds), a perfect square trinomial (pst), or neither.

22. $x^2 - 16$

23. $x^2 - 3x + 2$

24. $x^2 - 6x + 9$

25. $9x^2 - 16$

26. $4x^2 + 3x - 1$

27. $2x^2 + x - 6$

28. $7x^2 - 5x - 2$

29. $9x^2 + 3x - 2$

30. $4x^2 - 4x + 1$

31. $x^2 + 6x + 8$

32. $x^2 - 8x + 16$

33. $x^2 - 11x + 10$

34. $y^2 + 12y + 36$

35. $25 - 9x^2$

36. $x^2 + 8x + 5$

37. $x^2 - 4x + 6$

38. $4x^2 - 1$

39. $y^2 + 8y + 12$

40. $2x^2 + 10x - 12$

41. $2x^2 - 3x - 35$

42. $4x^2 - 4x - 35$

43. $4x^2 - 8x + 4$

44. $4x^2 - 4$

45. $x^3 + 4x^2 + 4x$

46. $x^3 + 6x^2 + 9x$

47. $3x^2 - 27$

48. $5x^2 - 20x + 15$

49. $x^3 - 7x^2 + 10x$

50. $x^3 + 3x^2 + 5x$

51. $2x^2 + 8x + 14$

Write the expressions in Exercises 52 to 65 without zero or negative exponents. Assume no undefined expressions.

52. a. 5^0 **b.** 5^{-1} **c.** 5^{-2}

d. $\left(\frac{1}{3}\right)^{-2}$ **e.** $\left(\frac{1}{3}\right)^{-1}$ **f.** $\left(\frac{1}{3}\right)^{0}$

53. a. 3^{-1} **b.** 3^0 **c.** 3^{-2}

d. $\left(\frac{2}{3}\right)^{0}$ **e.** $\left(\frac{2}{3}\right)^{-1}$ **f.** $\left(\frac{2}{3}\right)^{-2}$

54. a. b^4b^{-3} **b.** $b^{-4}b^{-7}$ **c.** $x^{-5}x^5$

55. a. x^7x^{-2} **b.** x^3x^{-3} **c.** $b^{-5}b^{-5}$

56. a. $\dfrac{n^4}{n^5}$ **b.** $\dfrac{n^{-4}}{n^5}$ **c.** $\dfrac{1}{n^{-9}}$

57. a. $\dfrac{n^4}{n^{-5}}$ **b.** $\dfrac{n^{-4}}{n^{-5}}$ **c.** n^{-9}

58. a. $(x^{-2})^3$ **b.** $(b^{-1})^{-2}$ **c.** $(x^{-4})^0$

59. a. $(b^2)^{-4}$ **b.** $(x^{-2})^{-3}$ **c.** $(b^0)^{-2}$

60. a. $\left(\frac{a}{b}\right)^{-2}$ **b.** $\left(\frac{a}{b^2}\right)^0$ **c.** $\left(\frac{a}{b^2}\right)^{-1}$

61. a. $\left(\frac{a^2}{b^2}\right)^{-1}$ **b.** $\left(\frac{2a}{b^2c}\right)^{-3}$ **c.** $\left(\frac{3b}{a^2}\right)^0$

62. a. $\frac{x^3y^{-2}}{x^2y^{-1}}$ **b.** $\frac{a^4b^{-5}}{a^{-2}b^3}$

c. $\frac{a^0}{a^{-1}}$ **d.** $\frac{x^{-4}}{x^0}$

63. a. $\frac{x^{-3}y^{-3}}{x^5y^{-2}}$ **b.** $\frac{a^3b^{-6}}{a^6b^{-4}}$

c. $\frac{x^3x^0}{x^0}$ **d.** $\frac{a^{-2}a^2}{a^0}$

64. Evaluate the expressions in Exercise 62 for $x = 2$, $y = -3$, $a = 2$, $b = -3$. Check by substituting into the answers.

65. Repeat Exercise 64 for the expressions in Exercise 63.

66. The property $\frac{x^a}{x^b} = x^{a-b}$ can be written "When dividing expressions with like bases, keep the base and subtract the exponents." Write in words the following properties.

a. $x^a \cdot x^b = x^{a+b}$

b. $(x^a)^b = x^{ab}$

c. $(xy)^a = x^ay^a$

d. $\left(\frac{x}{y}\right)^a = \frac{x^a}{y^a}$

67. Change each number in the national debt* and population columns of the table to scientific notation. Calculate the last column (divide national debt by population). Round to three non-zero digits

Year	United States Treasury: National Debt	Census Bureau Population	Debt per Person
1900	1,200,000,000	76,200,000	
1920	24,200,000,000	106,000,000	
1950	256,100,000,000	151,300,000	
1990	3,233,300,000,000	248,700,000	
2000	5,674,200,000,000	281,400,000	
2005	8,170,000,000,000	297,800,000	

* http://www.publicdebt.treas.gov/opd/opdpenny.htm

68. Complete the table by changing each number to decimal notation.

Chemical	Symbol for Isotope	Half-life	Half-life as Decimal
Potassium	^{40}K	1.4×10^9 years	
Calcium	^{41}Ca	1.2×10^5 years	
Radon	^{219}Rn	1.243×10^{-7} year	
Polonium	^{212}Po	3.0×10^{-7} second	

69. Use unit analysis to change the radon half-life to seconds.

70. Use unit analysis to change the polonium half-life to years.

In Exercises 71 to 74, find the product or quotient mentally and then check with a calculator.

71. a. $(1.5 \times 10^{-4})(3.0 \times 10^9)$

b. $(2.5 \times 10^{-9})(4.0 \times 10^3)$

72. a. $(6.0 \times 10^8)(7.0 \times 10^{-2})$

b. $(8.0 \times 10^7)(4.0 \times 10^{-3})$

73. a. $\frac{6.4 \times 10^{-12}}{1.6 \times 10^{-3}}$

b. $\frac{7.5 \times 10^{-13}}{2.5 \times 10^{-6}}$

74. a. $\frac{7.2 \times 10^9}{8.0 \times 10^{-8}}$

b. $\frac{5.4 \times 10^9}{9.0 \times 10^{-6}}$

In Exercises 75 and 76, change the number to standard scientific notation.

75. a. 38.5×10^{-3} **b.** 0.48×10^{-2}

76. a. 234×10^5 **b.** 0.0436×10^8

77. The moon travels about 1.5×10^6 miles as it orbits Earth. It completes one orbit in about 27.3 days. What is its speed to the nearest ten miles per hour?

78. In a fishbowl, a snail crawls 2 inches in 15 minutes. Using scientific notation, write its speed in miles per hour.

▶ Chapter Projects

79. Number Patterns Use a calculator to evaluate the expressions in parts a to d.

a. $17^2 - 14^2 - (16^2 - 13^2)$

b. $16^2 - 13^2 - (15^2 - 12^2)$

c. $15^2 - 12^2 - (14^2 - 11^2)$

d. $14^2 - 11^2 - (13^2 - 10^2)$

e. What do you notice about the values of the expressions?

f. Write a symbolic expression for the number pattern in part a. Let $x = 17$, $x - 3 = 14$, $x - 1 = 16$, and $x - 4 = 13$. Does the same expression describe parts b, c, and d?

g. By simplifying your expression, prove that the number pattern works for all inputs x. (*Hint:* Multiply the squared terms and combine like terms.)

h. Write a description of the following pattern in symbols, and simplify to prove that the pattern always works.

$$1 \cdot 3 - 2 \cdot 2 = -1$$
$$2 \cdot 4 - 3 \cdot 3 = -1$$
$$3 \cdot 5 - 4 \cdot 4 = -1$$
$$4 \cdot 6 - 5 \cdot 5 = -1$$
$$5 \cdot 7 - 6 \cdot 6 = -1$$
$$\vdots$$
$$49 \cdot 51 - 50 \cdot 50 = -1$$
$$\vdots$$

(*Hint:* Let x be the first number.)

80. Binomial Powers Multiply the binomials, using a table as needed.

a. $(a + b)^2 = (a + b)(a + b)$

b. $(a + b)^3 = (a + b)(a + b)^2$

c. $(a + b)^4 = (a + b)(a + b)^3$

d. $(a + b)^5 = (a + b)(a + b)^4$

To check, multiply your answer in part d by $(a + b)$ to obtain

$$(a + b)^6 = a^6 + 6a^5b + 15a^4b^2 + 20a^3b^3 + 15a^2b^4 + 6ab^5 + 1b^6$$

Multiply the binomials, using a table as needed.

e. $(x + 1)^2 = (x + 1)(x + 1)$

f. $(x + 1)^3 = (x + 1)(x + 1)^2$

g. $(x + 1)^4 = (x + 1)(x + 1)^3$

h. $(x + 1)^5 = (x + 1)(x + 1)^4$

To check, multiply your answer in part h by $(x + 1)$ to obtain

$$(x + 1)^6 = x^6 + 6x^5 + 15x^4 + 20x^3 + 15x^2 + 6x + 1$$

i. Predict $(x + y)^4$, $(x + 2)^4$, and $(x - 1)^4$.

▶ 7 Chapter Test

1. Choose from *factors* or *terms*: The expression $3x^2 + 4x + 1$ has three ________.

2. The $2ab$ that divides evenly into the expression $4a^2b - 8ab + 10ab^2$ is called the ________.

3. Choose from *dividing*, *factoring*, or *multiplying*: When we find the base and height of a rectangle with area $x^2 + 3x + 2$, we are ________.

4. Which exponent gives the reciprocal of a number?

5. What is the value of the exponent in $4^x = 1$?

6. What is the value of the exponent in $4^x = \frac{1}{16}$?

7. Write 3.482×10^{-4} in decimal notation.

8. Write 45,000,000,000 in scientific notation.

In Exercises 9 to 11, do the operation and indicate whether the result is a monomial, binomial, trinomial, or other polynomial.

9. $(a + 3b - 3c - d) + (3a - 5b + 8c - d)$

10. $x(x^2 + 3x + 9) - 3(x^2 + 3x + 9)$

11. $-4x(x - 5)$

12. Factor $6x^2y + 14xy - 18y^2$.

13.

Factor		
	$6x^2$	$+8x$
	$-15x$	-20

14. Find the perimeter and area of the rectangle in the figure.

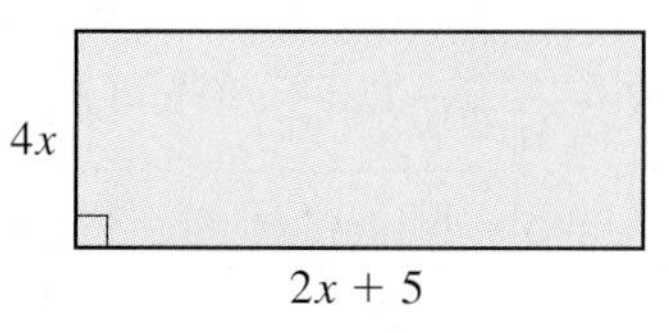

15. Find the width of the rectangle in the figure.

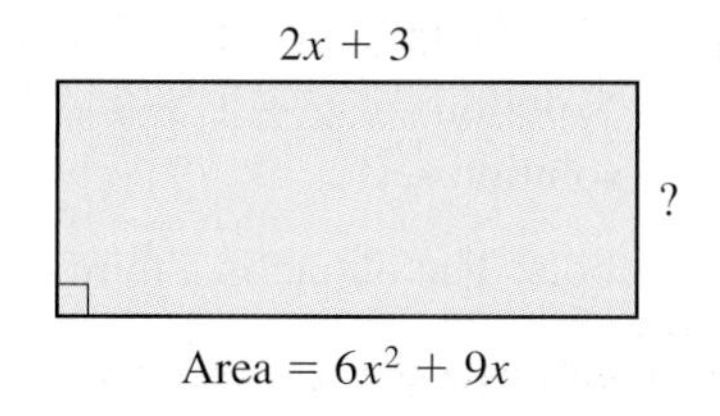

Multiply the expressions in Exercises 16 to 19.

16. $(x - 4)(x + 7)$

17. $(x - 7)(x - 7)$

18. $(2x - 7)(2x + 7)$

19. $2(x - 4)(x - 4)$

Factor the expressions in Exercises 20 to 25.

20. $x^2 - 9x + 20$

21. $2x^2 - 3x - 2$

22. $2x^2 - 8$

23. $x^2 - 8x + 16$

24. $9x^2 + 6x + 1$

25. $3x^2 + 6x + 15$

26. Name one of Exercises 16 to 25 that has a difference of squares as a problem or answer.

Simplify each expression in Exercises 27 to 29. Assume none of the variables equals zero.

27. a. $b^{-2}\, b^3$ **b.** $(x^3)^{-2}$

28. a. $\dfrac{b^3}{b^{-2}}$ **b.** $\dfrac{ab^0}{a^2b^{-1}}$

29. a. $\left(\dfrac{a}{2b}\right)^0$ **b.** $\left(\dfrac{9x^2}{25y^2}\right)^{-2}$

30. Evaluate the expressions in the Exercises 27 to 29 for $a = 3, b = -2, x = -1, y = -2$

31. Multiply $(2.5 \times 10^{-15})(4.0 \times 10^2)$. Write the answer in decimal notation.

32. Divide $\dfrac{1.25 \times 10^{-8}}{2.5 \times 10^{-4}}$. Write the answer in scientific notation.

Choose one of Exercises 33 to 35 to answer.

33. What are the possible sets of factors (containing whole numbers) and trinomials that could complete this statement?

$$(x \pm \_\_\_)(x \pm \_\_\_) = x^2 \pm \_\_\_ x \pm 21$$

Explain how you know that your listing is complete.

34. Explain how the table multiplication model shows that $(a + b)^2$ cannot equal $a^2 + b^2$.

35. Why does entering 10 [EE] 3 on a calculator give 10,000, whereas entering 10 [^] 3 gives 1000?

36. Explain with tiles, table, or other reasoning why $x^2 + 2x + 5$ does not factor.

▶ Cumulative Review of Chapters 1 to 7

1. For each credit card phone call, one AT&T plan charges \$0.25 plus \$0.25 per minute. Make an input-output table and a graph. Use inputs of 0 to 30 minutes, in steps of 5.

Simplify the expressions in Exercises 2 to 6.

2. $-5 - (-12) - 7$

3. $5 - 2(4 - x)$

4. $3^3 - 2 \cdot 5 + 4(7 - 2)$

5. $x^3 \cdot x^8$

6. $\dfrac{a^9}{a^4}$

7. Evaluate $R = \dfrac{ab}{a + b}$ if $a = 30$ and $b = 20$.

8. Write in symbols: A number plus 3 equals six less than twice the number.

9. Solve for x: $3x - 13 = 13 - 2x$.

10. Solve for x: $9 - 4(x - 3) = -2 - 3x$.

11. Solve for the solution set: $x + 1 \geq 9 - 3x$. Sketch a line graph of the solution set.

12. Write in symbols: The output is a third of the input.

13. Solve for y: $3x + 5y = 15$.

14. Solve for y: $5x - 3y = 8$.

15. Solve for a: $\dfrac{a + b + c}{3} = m$.

16. If $f(x) = 3x - 5$, what are $f(2)$ and $f(5)$?

17. On a rectangular coordinate grid, first sketch a line (anywhere) with slope $\frac{5}{3}$ and then sketch a line with slope $-\frac{2}{3}$ that passes through (3, 5).

18. What is the equation of a line with slope 4 and y-intercept -5?

19. What is the equation of a line parallel to $3x + 5y = 15$ (from Exercise 13) through the origin?

20. What is the simplified ratio of 15 minutes to 3 hours?

21. Quarts of motor oil cost \$0.92 in 1986 and \$1.19 in 1995. What is the percent change in price between the two years?

22. Set up a proportion or equation and solve:

a. 15% of what number is 5.4?

b. 42 is what percent of 48?

c. What is 40% of 65?

23. Solve for x: $\dfrac{x - 1}{2} = \dfrac{x + 3}{3}$.

24. Suppose you earned \$6.50 per hour from age 15 to age 65. At this rate, how many dollars would you earn? State the facts needed to set up a unit analysis. State all your assumptions.

25. Match the lines with their slopes:

a. $m = -1$ **b.** slope is undefined

c. $m = -5$ **d.** $m = 1$

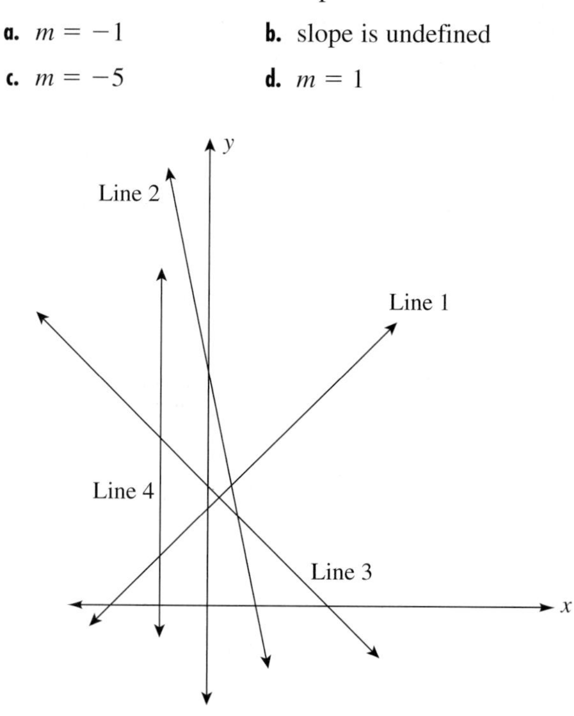

26. Solve the system of equations by substitution and by elimination:

$2x + 3y = 2$

$y - 2x = -18$

27. Beatriz and Ricardo's Shop mixes cashews and peanuts. Cashews sell for \$24 a kilogram, and peanuts sell for \$10 a kilogram. Find the number of kilograms of each needed to make a 50-kilogram mixture to sell for the prices listed below. As needed, use quantity-rate tables. Round to the nearest tenth.

a. \$12 per kilogram

b. \$16 per kilogram

c. Explain how you would estimate the number of kilograms of each needed to sell the mix for \$20 based on your results in parts a and b.

28. Multiply, then add like terms:

$x(x^2 - 2x + 1) - (x^2 - 2x + 1)$.

29. Factor $8xy - 4xyz + 12yz$.

30. Multiply $(2x + 1)(2x + 3)$.

31. Factor $x^2 - 6x + 9$. What special product is this?

32. Factor $4x^2 - 25$. What special product is this?

33. Factor $6x^2 - 2x - 8$.

34. In 1993, Georgia farmers produced 1,360,000,000 pounds of peanuts. If there are 16 ounces per pound and 160 calories per ounce for dry-roasted unsalted peanuts, how many calories were in this peanut crop? How many calories per person did the crop represent if the population of the world at that time was 5.6 billion?

35. a. What pairs of exponents make $x^{\square}x^{\circ} = x^{12}$?

b. What pairs of exponents make $x^{\square}x^{\circ} = x^{10}$?

36. Explain each use of parentheses shown in this Cumulative Review.

CHAPTER 8

Squares and Square Roots: Expressions and Equations

The squares in parts (a) and (b) of Figure 1 both have sides of length $a + b$. What is the area of the square in (a)? Write the area of each of the five parts of the square in (b). What is the sum of the areas?

Because both figures are squares with sides $a + b$, their areas are equal. Set the expressions for their areas equal and simplify. What do you observe? We will return to these questions in Example 2 of Section 8.1.

The chapter opens with an introduction to the Pythagorean theorem (Section 8.1), moves to formal work with radicals and radical equations (Sections 8.2 and 8.3), and then goes on to quadratic expressions and equations (Sections 8.4 to 8.6). It closes with an application of square roots in statistics (Section 8.7).

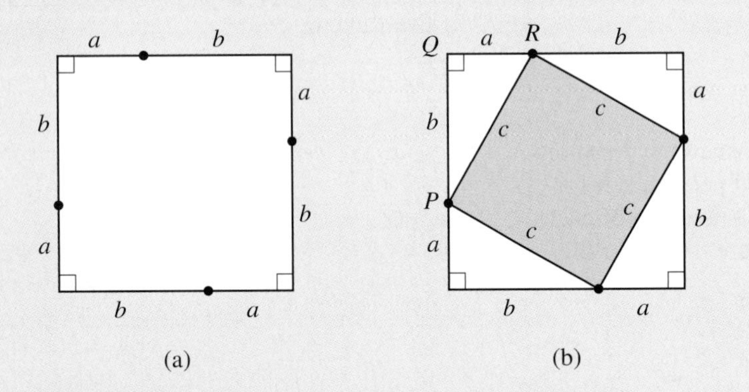

FIGURE 1 Areas of the same square

▶ 8.1 Pythagorean Theorem

Objectives

- Use the Pythagorean theorem to write equations and to solve for missing sides of right triangles.
- Use the converse of the Pythagorean theorem to find out whether triangles are right triangles.
- Apply the Pythagorean theorem in a variety of settings.

WARM-UP

1. List the values of these perfect squares:

1^2	6^2	11^2	16^2
2^2	7^2	12^2	20^2
3^2	8^2	13^2	25^2
4^2	9^2	14^2	30^2
5^2	10^2	15^2	40^2

2. Multiply out these binomials.

a. $(x + 3)(x + 3)$ **b.** $(x - y)(x - y)$

c. $(a + b)(a + b)$ **d.** $(x - 1)^2$

3. Your friend says $(x + y)^2 = x^2 + y^2$. Explain the error.

THE PYTHAGOREAN THEOREM, introduced in this section, will be applied to work with both square root functions (Sections 8.2 and 8.3) and quadratic functions (Sections 8.4, 8.5, and 8.6). The section provides practice in numerical work with squares and square roots. Familiarity with the numbers here will help you understand the concepts in the rest of the chapter.

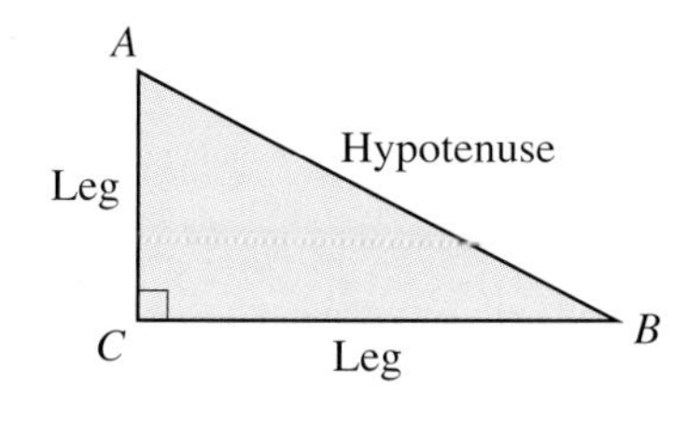

FIGURE 2 Right triangle

▶ Introduction to the Pythagorean Theorem

RIGHT TRIANGLES The **Pythagorean theorem** relates *the lengths of the perpendicular sides* (legs) *of a right triangle to the length of the longest side* (hypotenuse). The legs and hypotenuse are shown on the right triangle in Figure 2.

▶ EXAMPLE 1 Exploring the right triangle Place the measures of the sides of the triangles in Table 1, and then complete the remaining columns. What patterns do you observe?

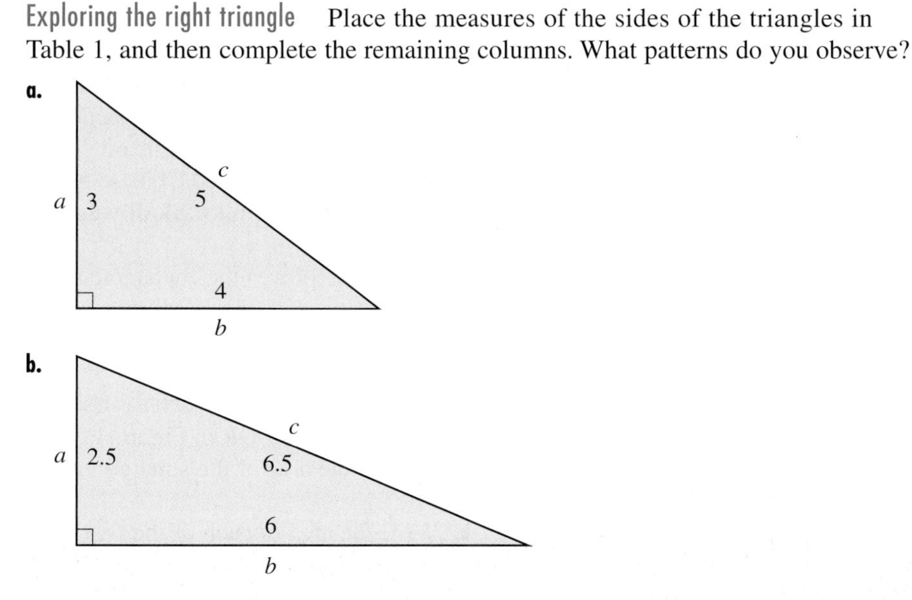

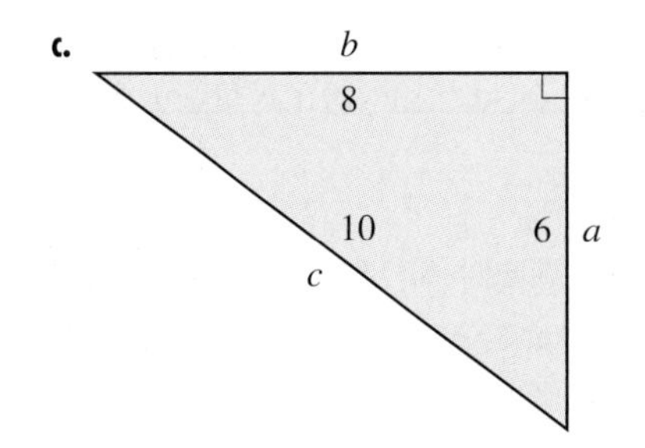

TABLE 1

Triangle	Leg a	Leg b	Hypotenuse c	a^2	b^2	$a^2 + b^2$	c^2
a							
b							
c							

SOLUTION The Pythagorean theorem appears in the last two columns of Table 1, where $a^2 + b^2 = c^2$. See the Answer Box for the completed table. ◀

THE PYTHAGOREAN THEOREM

If a triangle is a right triangle, then the sum of the squares of the lengths of the two shorter sides (legs) is equal to the square of the length of the longest side (hypotenuse):

$$a^2 + b^2 = c^2$$

PROOF OF THE PYTHAGOREAN THEOREM A **proof** is a *logical argument that demonstrates the truth of a statement.* It is based on previously known facts or agreed upon statements. The name of the Greek mathematician Pythagoras is given to the theorem because he or one of his followers was the first to record a proof of this relationship for right triangles. Proofs of the Pythagorean theorem have come from many cultures. A well-documented collection of 64 proofs is at **http://www.cut-the-knot.org/pythagoras/index.shtml** by Alexander Bobomolny. Number 28 in this collection is the illustrated Chinese proof by Liu Hui (third century). For details, see **http://www.staff.hum.ku.dk/dbwagner/Pythagoras/Pythagoras.html** by Donald B. Wagner.

Example 2 proves the Pythagorean theorem by showing that it is true for any a, b, and c that satisfy the conditions of the theorem: a and b are legs of a right triangle, and c is the hypotenuse.

▶ **EXAMPLE 2** Proving the Pythagorean theorem Use the following steps to prove the Pythagorean theorem, where $\triangle PQR$ in Figure 3(b) is a right triangle.

a. What is the area of the square in part (a) of Figure 3, which has $a + b$ on each side?

b. Write the area of each of the five parts of the square in part (b) of Figure 3. Write the sum of the areas.

c. Because both figures are squares with sides $a + b$, their areas are equal. Set the expressions for their areas equal, and show that $a^2 + b^2 = c^2$.

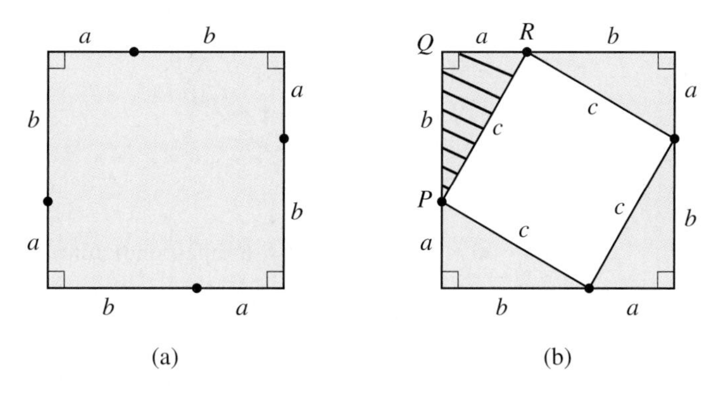

FIGURE 3 Areas of the same square

SOLUTION **a.** The square in (a), with side $a + b$, has area $(a + b)^2$.

b. The square in (b) has four right triangles and an inner square of side c. The area of each right triangle is $\frac{1}{2}ab$. The area of the inner square is c^2. Thus, the total area is $4\left(\frac{1}{2}ab\right) + c^2$.

c. We set the areas equal:

$$(a + b)^2 = 4\left(\tfrac{1}{2}ab\right) + c^2 \quad \text{Multiply to remove the parentheses.}$$

$$a^2 + 2ab + b^2 = 2ab + c^2 \quad \text{Subtract } 2ab \text{ from both sides.}$$

$$a^2 + b^2 = c^2$$

Therefore, for right triangle PQR, $a^2 + b^2 = c^2$. ◀

SOLVING FOR MISSING SIDES In Example 3, we solve for missing sides using the Pythagorean theorem. In the drawing in part c, there is a right angle on the circle and the hypotenuse is coincident with the diameter of the circle. The drawing has applications in navigation. In the drawing in part d, n is the length of the diagonal of the rectangle.

▶ **EXAMPLE 3** Finding missing sides What is the missing side, n, in each of the following drawings?

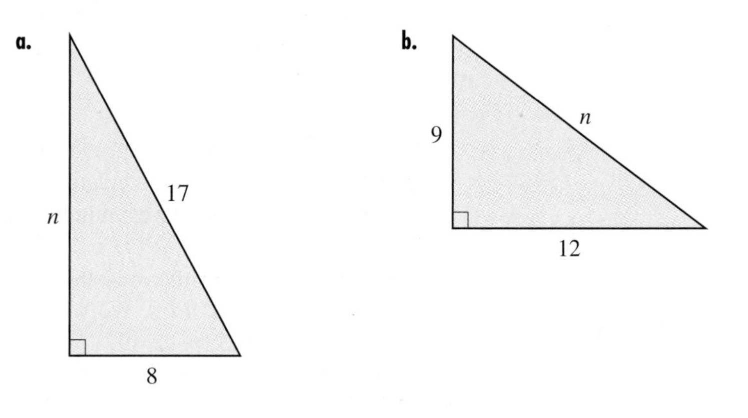

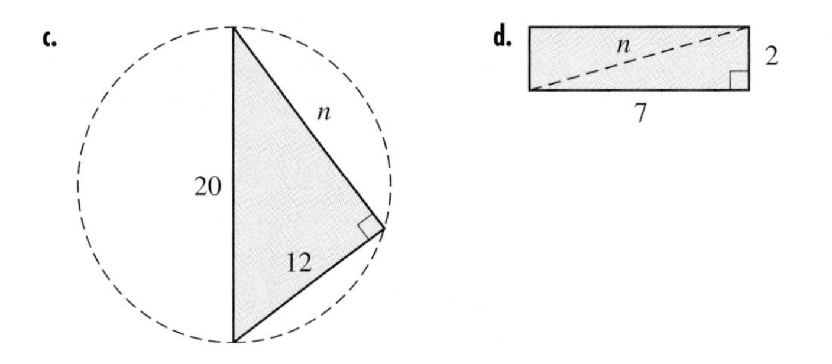

SOLUTION **a.** The sides n and 8 are perpendicular and are the legs, a and b.

$$n^2 + 8^2 = 17^2$$
$$n^2 + 64 = 289$$
$$n^2 = 225$$
$$n = 15$$

Although $n = -15$ also makes $n^2 = 225$ true, length must be positive, so we disregard $n = -15$.

b. The sides 9 and 12 are perpendicular and are the legs.

$$9^2 + 12^2 = n^2$$
$$81 + 144 = n^2$$
$$225 = n^2$$
$$15 = n$$

c. The sides n and 12 are perpendicular.

$$n^2 + 12^2 = 20^2$$
$$n^2 + 144 = 400$$
$$n^2 = 256$$
$$n = 16$$

d. The sides 2 and 7 are perpendicular.

$$2^2 + 7^2 = n^2$$
$$4 + 49 = n^2$$
$$53 = n^2$$
$$\sqrt{53} = n$$

When a square root does not give an exact value, we generally leave it in square-root form or use the calculator to give a decimal approximation. In this case, $n \approx 7.28$ to the nearest hundredth. ◀

When we solve for the missing side, we are using a new process: *taking the square root of both sides*. We will discuss this operation more formally in Example 2 of Section 8.5 (page 497).

▶ Converse of the Pythagorean Theorem

Changing the position of the if *and* then *statements* in the Pythagorean theorem gives us the **converse of the Pythagorean theorem**. The converse is also true.

Converse of the Pythagorean Theorem

> If the sum of the squares of the lengths of the two shorter sides (legs) is equal to the square of the length of the longest side (hypotenuse), then the triangle is a right triangle.

In Example 4, we apply the converse of the Pythagorean theorem to find out if a triangle is a right triangle.

▶ **EXAMPLE 4** Finding right triangles Each triangle below is drawn to look like a right triangle. Which *is* a right triangle?

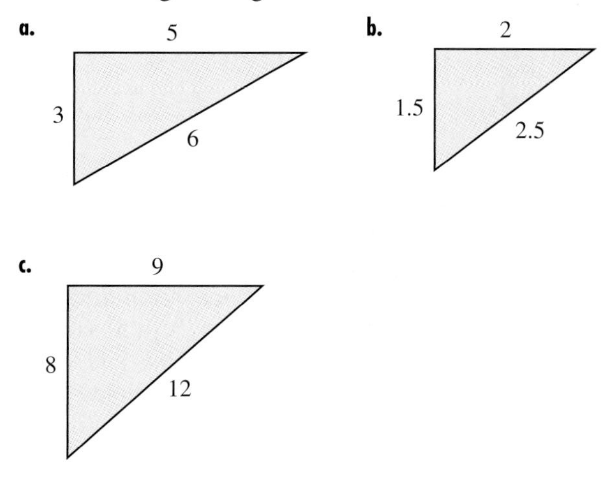

SOLUTION

a. $3^2 + 5^2 = 9 + 25$, which does not equal 36, the square of the third side. This triangle is not a right triangle.

b. $1.5^2 + 2^2 = 2.25 + 4$, which equals the square of the third side, 6.25. This is a right triangle, with the right angle where the shorter sides meet.

c. $8^2 + 9^2 = 64 + 81$, which does not equal 144, the square of the third side. This triangle is not a right triangle. ◀

▶ **EXAMPLE 5** Identifying right triangles Show whether or not these sets of three numbers could be used as sides of right triangles. The numbers may not be in order from smallest to largest.

a. $\{4, 5, 6\}$

b. $\{1, 1, \sqrt{2}\}$

c. $\{5, \sqrt{11}, 6\}$

d. $\{8, \sqrt{17}, 9\}$

e. $\{\frac{1}{3}, \frac{1}{4}, \frac{1}{5}\}$

SOLUTION The two smaller sides of a right triangle can be placed into a^2 and b^2 in any order. The longest side (the hypotenuse) must be substituted into c^2. If $a^2 + b^2 = c^2$ is satisfied, the set of numbers is a Pythagorean triple.

Student Note:
The square of $\sqrt{2}$ is $\sqrt{2} \cdot \sqrt{2} = 2$. More about this in Section 8.2.

a. $a^2 + b^2 = c^2$

$4^2 + 5^2 \stackrel{?}{=} 6^2$

$16 + 25 \neq 36$

The numbers do not satisfy $a^2 + b^2 = c^2$.

HISTORICAL NOTE

To form a tool for making a right angle in construction and survey work, early Egyptians tied eleven equally spaced knots in a rope. They then tied the rope in a loop with a twelfth knot. The rope formed a {3, 4, 5} right triangle when staked to the ground.

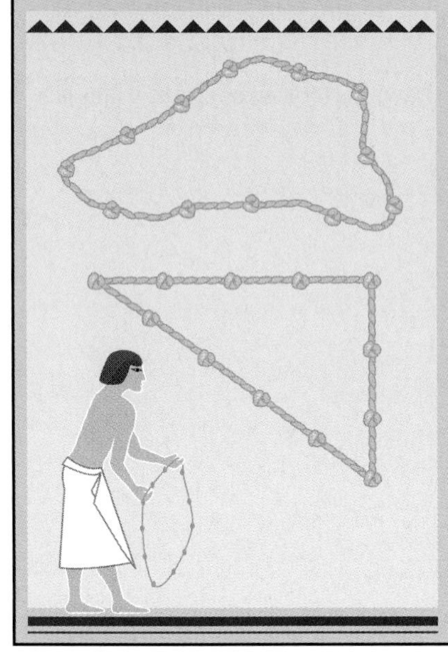

Student Note:
A Pythagorean triple is a continued ratio of the sides of a right triangle. Thus, the set {3, 4, 5} is the continued ratio 3 : 4 : 5.

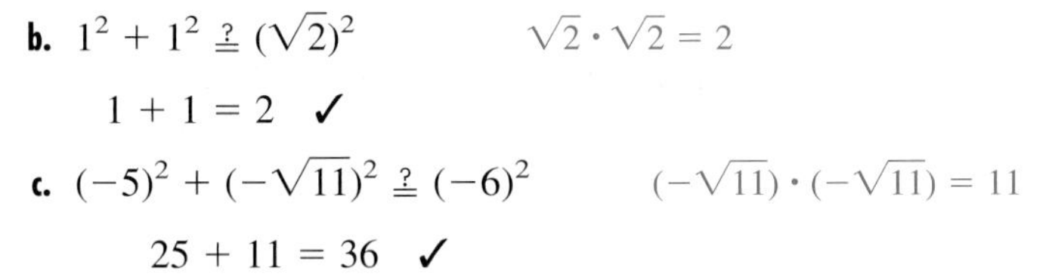

b. $1^2 + 1^2 \stackrel{?}{=} (\sqrt{2})^2$ $\qquad \sqrt{2} \cdot \sqrt{2} = 2$

$1 + 1 = 2$ ✓

c. $(-5)^2 + (-\sqrt{11})^2 \stackrel{?}{=} (-6)^2$ $\qquad (-\sqrt{11}) \cdot (-\sqrt{11}) = 11$

$25 + 11 = 36$ ✓

The numbers satisfy $a^2 + b^2 = c^2$ but are not positive and cannot form a triangle.

d. $8^2 + (\sqrt{17})^2 \stackrel{?}{=} 9^2$ $\qquad \sqrt{17} \cdot \sqrt{17} = 17$

$64 + 17 = 81$ ✓

e. The side measuring $\frac{1}{3}$ is the longest, so it replaces c.

$$\left(\tfrac{1}{4}\right)^2 + \left(\tfrac{1}{5}\right)^2 \stackrel{?}{=} \left(\tfrac{1}{3}\right)^2$$

$$\frac{1}{16} + \frac{1}{25} \stackrel{?}{=} \frac{1}{9}$$

$$\frac{41}{400} \neq \frac{1}{9}$$

The numbers do not satisfy $a^2 + b^2 = c^2$.

Parts b and d describe right triangles. ◀

Pythagorean triples are *sets of three numbers that make the Pythagorean theorem true*. The set of numbers {3, 4, 5} satisfies the Pythagorean theorem: $3^2 + 4^2 = 5^2$.

Before calculators, students and instructors saved considerable time by learning selected sets of Pythagorean triples. Engineering textbooks still make frequent use of triples, as do nationally standardized exams for entry into college or graduate school. The triple {3, 4, 5} is the basis for a rope surveying instrument used by the early Egyptians.

▶ Applications

LADDER SAFETY In Section 5.2, we used the 4-to-1 safe-ladder ratio (height of the ladder on the wall to distance of the base of the ladder from the wall). We now use the Pythagorean theorem to find the length of the ladder needed for a particular job (Example 6) and safe positions for the ladder given its length (Example 7).

▶ **EXAMPLE 6** Finding ladder length A ladder needs to reach 21 feet up a wall (see Figure 4). For a safe ratio of 4 to 1, the base of the ladder must be 5.25 feet from the wall. To the nearest foot, what length ladder is needed?

FIGURE 4 Ladder ratio: $4/1 = 21/5.25$

SOLUTION The ground and the wall form the legs of a right triangle, and the ladder is the hypotenuse. The length of the ladder is c.

$$c^2 = (21 \text{ ft})^2 + (5.25 \text{ ft})^2$$
$$c^2 \approx 468.56 \text{ ft}^2$$
$$c \approx 21.6 \text{ ft}$$

Thus, a 22-foot ladder will be needed. ◀

In Example 7, we are given the length of the ladder. Both the base and the height are unknown, but we know the ratio of base to height. Look carefully at how the example is solved in one variable and then try the "Think about it" with two variables.

▶ **EXAMPLE 7** Finding ladder positions given ladder length What is the safe ladder position for a 16-foot ladder (see Figure 5)?

FIGURE 5 Ladder ratio: $4/1 = 4x/1x$

SOLUTION The base-height ratio must be 1 to 4. Thus, the legs of the right triangle are x and $4x$. By the Pythagorean theorem,

$$a^2 + b^2 = c^2$$
$$x^2 + (4x)^2 = (16 \text{ ft})^2$$
$$17x^2 = 256 \text{ ft}^2$$
$$x^2 = \frac{256}{17}$$
$$x^2 = 15.059$$
$$x \approx 3.88 \text{ ft from the base of the wall}$$
$$4x \approx 15.52 \text{ ft up the wall}$$
◀

THINK ABOUT IT: Set up Example 7 in two variables. First, write a proportion for the 4 to 1 ratio, with l = length and w = width. Second, solve your proportion for l. Third, write the Pythagorean equation, using l and w for the wall height and base and 16 for the ladder length. Last, substitute for l in the Pythagorean equation.

FLIGHT AND WIND In the examples in Section 6.2, the wind and the direction of airplane flight were parallel. We now consider the result when the wind and the direction of flight are perpendicular. The figure in Example 8 shows a northbound plane and a cross wind from the west. The effect of the wind is to blow the plane off course to the right (or northeast).

The data in the example are given in miles per hour. To simplify, we will use a flying time of 1 hour. Here is how we change units to miles:

$$D = rt$$

$$D = \frac{x \text{ mi}}{\text{hr}} \cdot 1 \text{ hr} = x \text{ mi}$$

EXAMPLE 8 Finding distance traveled with a cross wind If the wind is blowing at 68 miles per hour from the west and the plane is traveling 100 miles per hour due north, calculate the distance flown after 1 hour (as shown in Figure 6).

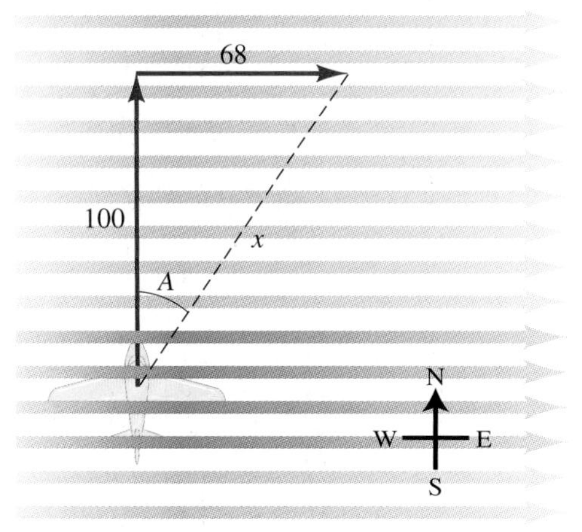

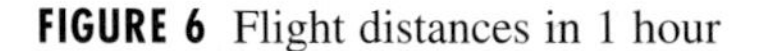

FIGURE 6 Flight distances in 1 hour

SOLUTION In 1 hour, the plane will travel 100 miles north and 68 miles east. Because these directions are at right angles, the actual flight path will be the hypotenuse.

$$100^2 + 68^2 = x^2$$

$$14{,}624 = x^2$$

$$x = \sqrt{14{,}624} \approx 121 \text{ mi}$$

◀

ANSWER BOX

Warm-up: 1. 1, 4, 9, 16, 25, 36, 49, 64, 81, 100, 121, 144, 169, 196, 225, 256, 400, 625, 900, 1600. A pattern can help you learn the squares: The last digits in 13^2 are reversed in 14^2. **2. a.** $x^2 + 6x + 9$ **b.** $x^2 - 2xy + y^2$ **c.** $a^2 + 2ab + b^2$ **d.** $x^2 - 2x + 1$ **3.** There is no middle term given; $(x + y)^2 = x^2 + 2xy + y^2$. Using a table to multiply $(x + y)(x + y)$ will show him the terms he missed. **Example 1:**

Triangle	Leg a	Leg b	Hypotenuse c	a^2	b^2	$a^2 + b^2$	c^2
a	3	4	5	9	16	25	25
b	2.5	6	6.5	6.25	36	42.25	42.25
c	6	8	10	36	64	100	100

Think about it: $\frac{l}{w} = \frac{4}{1}$, $l = 4w$, $l^2 + w^2 = 16^2$, $(4w)^2 + w^2 = 16^2$

▶ 8.1 Exercises

In Exercises 1 to 6, find the length of the side marked with an x. Round to the nearest hundredth.

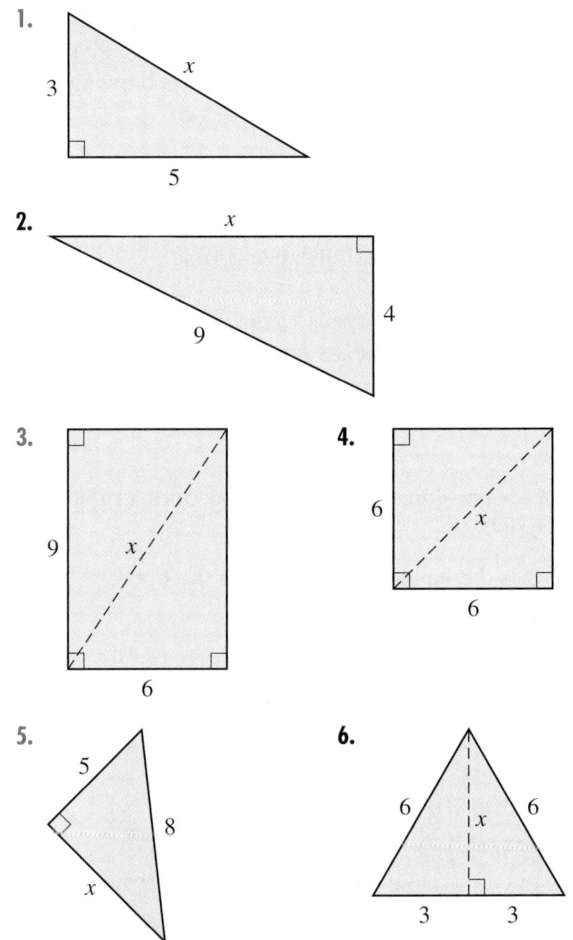

Which of the sets of three numbers in Exercises 7 to 12 could represent the sides of a right triangle?

7. $\{7, 8, 9\}$ **8.** $\{12, 15, 18\}$

9. $\{11, 60, 61\}$ **10.** $\{20, 21, 29\}$

11. $\{7, 24, 25\}$ **12.** $\{9, 40, 41\}$

In Exercises 13 to 16 the numbers might not be listed in a, b, c order. Which sets satisfy $a^2 + b^2 = c^2$ and yet cannot satisfy the Pythagorean theorem? Explain why.

13. a. $\{2.1, 2.9, 2\}$ **b.** $\{2, \sqrt{3}, 1\}$

c. $\{1, 2, 3\}$ **d.** $\{\frac{3}{7}, \frac{4}{7}, \frac{5}{7}\}$

14. a. $\{-1, 0, 1\}$ **b.** $\{-5, -4, -3\}$

c. $\{7, \sqrt{15}, 8\}$ **d.** $\{\frac{5}{8}, 1\frac{1}{2}, 1\frac{5}{8}\}$

15. a. $\{3, \sqrt{7}, 4\}$

b. $\{-5, -12, -13\}$

c. $\{6, \sqrt{13}, 7\}$

d. $\{\frac{3}{4}, \frac{4}{5}, \frac{5}{6}\}$

16. a. $\{10, \sqrt{21}, 11\}$

b. $\{8, 15, 17\}$

c. $\{6, 6\sqrt{3}, 12\}$

d. $\{\frac{3}{5}, \frac{4}{5}, 1\}$

Use proportions to find the missing sides of the similar triangles shown in Exercises 17 and 18. Show, with the converse of the Pythagorean theorem, that they are all right triangles.

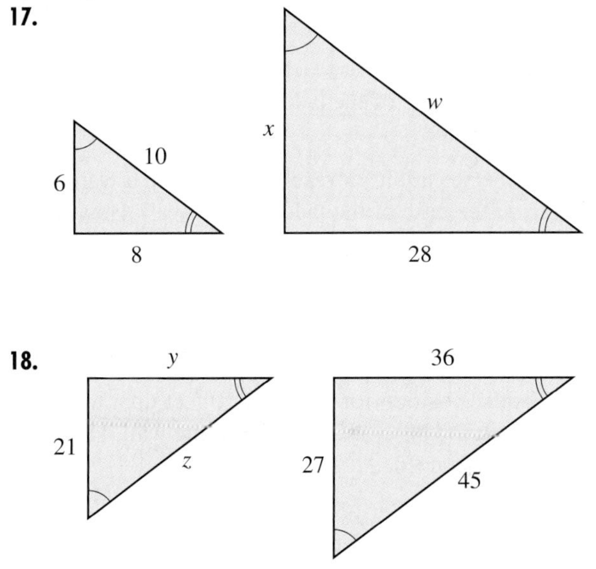

In Exercises 19 and 20, complete the tables, which show enlargements or reductions of $\{3, 4, 5\}$ right triangles. Each row represents the sides of a different triangle. All triangles are similar to those pictured in Exercise 17 and 18.

19.

Leg	Leg	Hypotenuse
3	4	5
6		
		30
	12	
1		

Blue numbers are core exercises.

20.

Leg	Leg	Hypotenuse
3	4	5
12		
		40
	20	
		1

Use the variable expressions in Exercises 21 to 26 in the Pythagorean theorem and solve for x. Round answers to the nearest tenth. The sides are listed smallest to largest.

21. $\{x, x, 8\}$

22. $\{x, x, 12\}$

23. $\{x, 5x, 52\}$

24. $\{x, 3x, 20\}$

25. $\{x, 2x, 15\}$

26. $\{x, 7x, 50\}$

In Exercises 27 to 32, assume a safe-ladder ratio, height to base, of 4 to 1. Round to the nearest tenth.

27. A safe ladder position for reaching 12 feet up a wall is 3 feet from the base of the ladder to the wall. How long a ladder is needed?

28. A safe ladder position for reaching 8 feet up a wall is 2 feet from the base of the ladder to the wall. How long a ladder is needed?

29. A safe ladder position for reaching 9 feet up a wall is 2.25 feet from the base of the ladder to the wall. How long a ladder is needed?

30. A safe ladder position for reaching 10 feet up a wall is 2.5 feet from the base of the ladder to the wall. How long a ladder is needed?

31. What is the safe ladder position for reaching 14 feet up a wall? How long a ladder is needed?

32. What is the safe ladder position for reaching 19 feet up a wall? How long a ladder is needed?

Find the safe ladder positions for the ladder lengths in Exercises 33 to 36. Use a safe-ladder ratio, height to base, of 4 to 1. Give your answers first with decimal portions of a foot and then in feet and inches to the nearest inch.

33. 12-foot ladder

34. 22-foot extension ladder

35. 18-foot extension ladder

36. 25-foot extension ladder

In Exercises 37 to 40, draw triangles and show the direction actually flown in a cross wind.

37. An airplane flies south for 1 hour at 250 miles per hour from Montreal, Canada. A 30-mile-per-hour wind is blowing from the east. How far does the airplane travel?

Blue numbers are core exercises.

38. An airplane flies east for 1 hour from Missoula, Montana, at 210 miles per hour. A 35-mile-per-hour wind from the north blows the plane off course. How far does the airplane travel?

39. An airplane flies for 3 hours on a heading due north from San Francisco. The airplane is flying at 200 miles per hour while a 32-mile-per-hour wind is blowing from the west. What is the total distance flown?

40. An airplane flies for 3 hours on a heading due north from Atlanta. The airplane travels 225 miles per hour. A wind is blowing 20 miles per hour from the west during the trip. What is the total distance flown?

In Exercises 41 and 42, use the Pythagorean theorem to find all or part of unknown lengths. Look around the figure for numbers that suggest other lengths.

41. a. What are the width and the length of one side of the roof shown in the figure? What is the total area of the roof?

b. What is the slope of the roof? How does a right triangle help?

c. What is the height of the house in the figure?

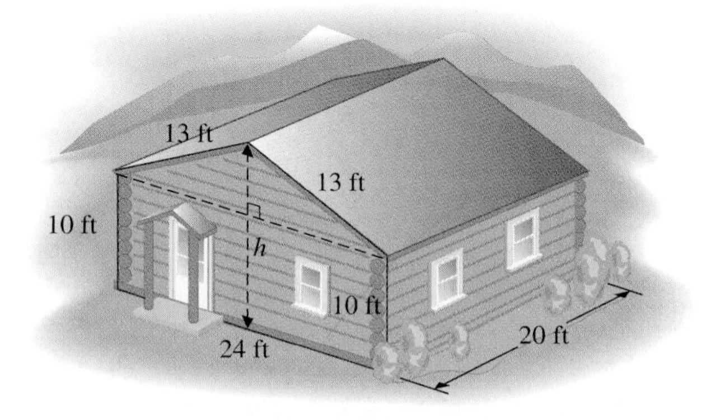

42. What are the lengths of the crossed pieces, AB and CD, needed to make the kite in the figure?

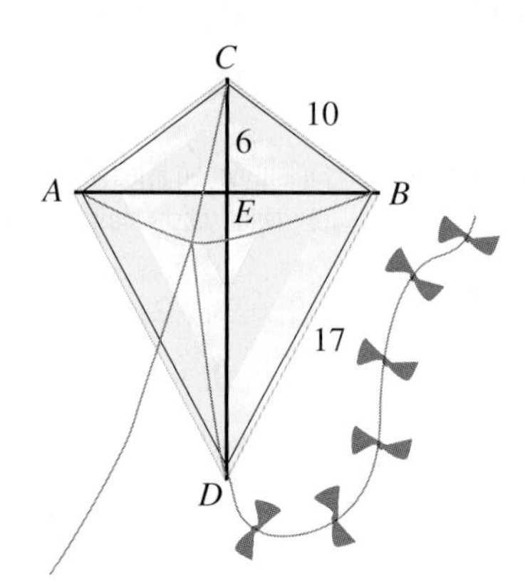

A rectangular house has a 28-foot by 40-foot floor plan (see the figure). For Exercises 43 to 46, assume that the roof ends at the edge of the wall. Assume that both sides of the roof have the same area and slope.

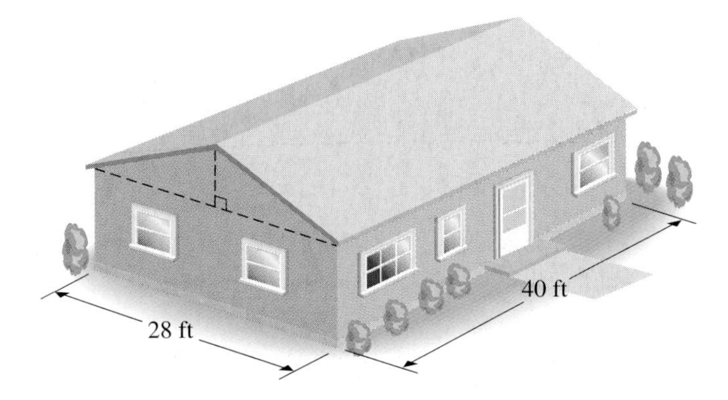

43. How many square feet of roofing material are needed for a slope of 3 to 14?

44. How many square feet of roofing material are needed for a slope of 5 to 14?

45. How many square feet of roofing material are needed for a slope of 9 to 14?

46. How many square feet of roofing material are needed for a slope of 10 to 14?

▶ Writing

47. In Example 7, the equation was $x^2 + (4x)^2 = 16^2$. Suppose we made a mistake and wrote $x^2 + 4x^2 = 16^2$. Solve the second equation, and explain how the resulting base (x) and wall height ($4x$) cannot match with a 16-foot ladder. Include a sketch with your explanation.

48. When we multiply any two square numbers, we get another square number. Is this true, sometimes true, or false? Explain. If false, give an example showing it to be false.

49. After completing work with the 3-4-5 right triangles in Exercises 17 to 20, a classmate suggests that "there must be other sets of similar right triangles." You ask for examples. The first set suggested is 5-12-13, 7-24-25, and 9-40-41. What is the pattern they follow and do they form similar triangles? Find examples to prove or disprove your classmate's statement.

50. It is told that Pythagorus was deeply troubled by not finding a natural number as a hypotenuse for a triangle with equal natural numbers as legs (say, n and n). Explain the problem. What kind of numbers did he need?

51. For the same house, which has a greater area, a steeper roof or a flatter roof? Explain in terms of rise and hypotenuse.

52. Problem Solving with Square Numbers

a. The author's grandmother was x years old in the year x^2. Grandmother was still alive in 1945. Find the age x and the year x^2. When was she born?

b. The author's great, great, great grandmother was x years old in the year x^2. She was alive during the American Civil War. Find the age x and the year x^2. When was she born?

c. Would it be possible for someone you know to be alive in 2010 and be x years old in the year x^2? If so, give the person's age, the year, and the person's year of birth.

▶ Projects

53. Garfield Proof James A. Garfield, 20th president of the United States, is credited with a proof of the Pythagorean theorem based on the areas of a trapezoid (see the figure). The area of a trapezoid is given by

$$A = \frac{1}{2}(\text{height})(\text{top} + \text{bottom})$$

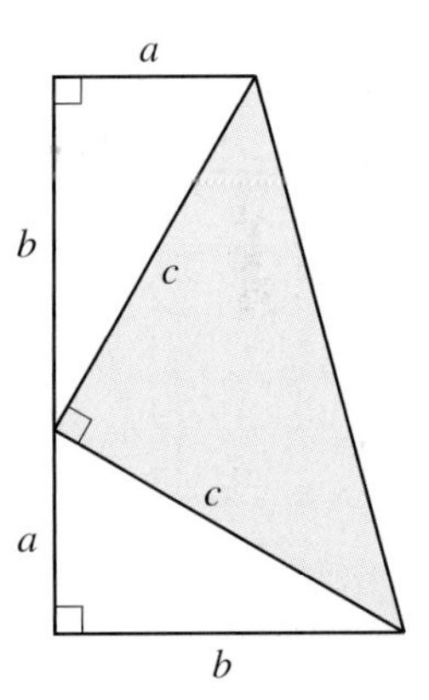

a. Find the area of the trapezoid using the formula.

b. Find the area of the trapezoid by adding the areas of the three triangles.

c. Because the areas in parts a and b both describe the trapezoid, we can set the areas equal. Show that the resulting equation simplifies to $a^2 + b^2 = c^2$.

Blue numbers are core exercises.

8.2 Square Root Expressions and Properties and the Distance Formula

Objectives

- Name the radicand in an expression.
- Give the principal square root of a number.
- Simplify square root expressions using the properties for products and quotients.
- Apply the distance formula.

> **WARM-UP**
>
> **1.** Find these square roots without a calculator.
>
> **a.** $\sqrt{64}$ **b.** $\sqrt{121}$ **c.** $\sqrt{49}$ **d.** $\sqrt{256}$
>
> **e.** $\sqrt{400}$ **f.** $\sqrt{225}$ **g.** $\sqrt{144}$ **h.** $\sqrt{625}$
>
> **2.** Practice mental estimation by giving the two consecutive integers above and below each square root—for example, $2 < \sqrt{6} < 3$.
>
> **a.** $\sqrt{45}$ **b.** $\sqrt{75}$ **c.** $\sqrt{200}$ **d.** $\sqrt{20}$
>
> **e.** $\sqrt{110}$ **f.** $\sqrt{12}$ **g.** $\sqrt{905}$ **h.** $\sqrt{395}$

THIS SECTION STARTS by defining radicals, radicands, and the principal square root. After a brief review of irrational numbers, the square root function and graph are introduced. We then practice multiplication and division with square root expressions. Finally, square roots help us use the distance formula to find the length of line segments.

Radicals and Radicands

The more general name for a square root (and higher index root) is **radical**. The *radical sign* is the symbol we use for square root.

$$\sqrt[n]{\text{radicand}}$$

index → n; radical sign → $\sqrt{\ }$

The *index* is the small n on the radical sign. For square roots, the index is 2. We assume $n = 2$ and don't write it. See the project in Exercise 76 of Section 8.3 for examples of radicals with indices of 3, 4, and 5.

The **radicand** is *the number or expression under the radical sign*. Radical vocabulary is useful because it is easier to say "radicand" than "the expression under the square root sign."

EXAMPLE 1 Identifying radicands What is the radicand in each of these expressions?

a. $5\sqrt{2}$ **b.** $b\sqrt{bc}$ **c.** $3\sqrt{x+2}$ **d.** $3\sqrt{x}+2$ **e.** $(\sqrt{x+2})^2$ **f.** $\sqrt{(x+2)^2}$

SOLUTION **a.** 2 **b.** bc **c.** $x + 2$

d. x. The 2 is added after the square root is taken, so the 2 is not part of the radicand.

e. $x + 2$. The square applies to the entire expression, so the radicand is just the $x + 2$ under the radical sign.

f. $(x + 2)^2$. The square applies to the $x + 2$ and not the radical sign. ◀

Square Roots

The **principal square root** of a number is *the positive number that, multiplied by itself, produces the given number.* In algebraic notation, $x = \sqrt{n}$ if $x^2 = n$ for $x \geq 0$.

> The principal square root of a real number is always zero or positive:
>
> $\sqrt{x} \geq 0$

EXAMPLE 2 Naming principal square roots What are the principal square roots?

a. 25 **b.** 256 **c.** 2.25 **d.** 10 **e.** 15

SOLUTION **a.** 5 **b.** 16 **c.** 1.5 **d.** $\sqrt{10} \approx 3.162$ **e.** $\sqrt{15} \approx 3.873$ ◀

Many square roots are irrational numbers: $\sqrt{2}$, $\sqrt{3}$, $\sqrt{5}$, $\sqrt{6}$, and so forth. An **irrational number** is *a real number that cannot be written as the quotient of two nonzero integers.* In Example 2, the numbers $\sqrt{10}$ and $\sqrt{15}$ are irrational numbers. Unless otherwise noted, round irrational numbers to three decimal places.

SQUARE ROOT GRAPHS The square root equation is $y = \sqrt{x}$. We graph the equation in Example 3 and look at how important features of the square root graph are related to inputs and outputs.

EXAMPLE 3 Plotting a square root graph

a. Use a calculator as needed to evaluate $\sqrt{-4}$, $\sqrt{-1}$, $\sqrt{0}$, $\sqrt{1}$, $\sqrt{2}$, $\sqrt{3}$, $\sqrt{4}$, and $\sqrt{5}$, and write the results as ordered pairs of $y = \sqrt{x}$.
b. Graph the ordered pairs, and graph $y = \sqrt{x}$ with a calculator.
c. Does the graph pass the vertical-line test for a function?

SOLUTION **a.** The calculator gives an error message (NONREAL ANS) for $\sqrt{-4}$ and $\sqrt{-1}$. Table 2 shows the ordered pairs for the remaining expressions.

TABLE 2 Evaluating $y = \sqrt{x}$

Input x	Output $y = \sqrt{x}$	Ordered Pairs (x, y)
0	0	(0, 0)
1	1	(1, 1)
2	≈1.414	(2, 1.414)
3	≈1.732	(3, 1.732)
4	2	(4, 2)
5	≈2.236	(5, 2.236)

b. In graphing by hand, we can only estimate the positions of irrational numbers to about a tenth, so the other decimal places are excessive. The graph of $y = \sqrt{x}$ is shown in Figure 7.

In the graphing calculator solution in Figure 8, we set $Y_1 = \sqrt{X}$, with the window for X at [−4, 10], scale 1, and the window for Y at [−4, 4].

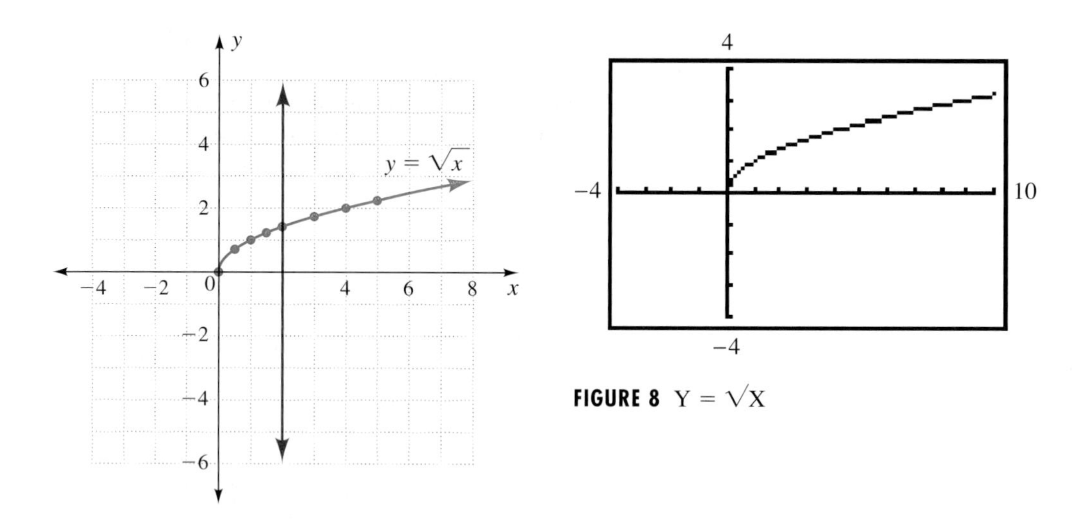

FIGURE 7 Principal square root graph

FIGURE 8 $Y = \sqrt{X}$

c. In Figure 7, for each input on the x-axis, there is only one output for the equation $y = \sqrt{x}$. Because its graph passes the vertical-line test, the equation $y = \sqrt{x}$ describes a function. ◀

THINK ABOUT IT 1: Are irrational numbers also real numbers?

The square root graphs shown in Example 3 have two important features:

1. There is no graph to the left of the y-axis (in the second or third quadrant), where x is negative. Thus, *there are no outputs for inputs less than zero. We cannot take the square root of a negative number and obtain a real-number output.**

Square Root of a Negative Number

The square root of a negative number is *undefined in the real numbers.*

2. There is no graph below the x-axis (in the third or fourth quadrant), where y is negative. Thus, *there are no negative outputs to a square root; $\sqrt{x}$ is always positive*. That is why the principal square root is defined as the positive root of a number.

SQUARE ROOT FUNCTION The square root function describes the principal square root.

Square Root Function

The square root function is $f(x) = \sqrt{x}$. The input set (or domain) is $x \geq 0$. The output set (or range) is $y \geq 0$ or $f(x) \geq 0$.

The graph of the square root function is shown in Figures 7 and 8.

*Calculator Note:
In response to $\sqrt{-1}$ and $\sqrt{-4}$, a calculator in $a + bi$ (complex) mode will give i and $2i$. These numbers are not real numbers. They contain the imaginary unit $i = \sqrt{-1}$. Other calculators will give ordered pairs: (0, 1) and (0, 2). In this case, the ordered pair (a, b) describes the complex number $a + bi$. The term a in $a + bi$ is the real-number part and is zero for these square roots. The term bi is the imaginary part. Thus, (0, 1) indicates $0 + 1i$, and (0, 2) is $0 + 2i$. The ordered pair (0, 1) cannot be graphed on a Cartesian coordinate graph, which has real numbers on both axes. Graphs for imaginary numbers are called Argand diagrams.

NEGATIVE SIGNS WITH SQUARE ROOTS *Simplify* can mean to evaluate a square root expression. To simplify, we need to interpret negative signs correctly. Note that a negative sign outside the square root sign means the opposite of a square root (which is defined), whereas a negative sign inside the square root sign means the square root of a negative (which is not defined in the real numbers). We use the *plus or minus sign* ($\pm$) outside the square root sign to indicate that the answer is both the positive root and its opposite.

EXAMPLE 4 Working with negatives in and on square root expressions Simplify these expressions.

a. $\sqrt{25}$ **b.** $-\sqrt{49}$ **c.** $\pm\sqrt{121}$

d. $\sqrt{-16}$ **e.** $\pm\sqrt{1.44}$ **f.** $-\sqrt{25}$

SOLUTION

a. $\sqrt{25} = 5$. We give the positive, or principal, root.

b. $-\sqrt{49} = -7$. The negative sign outside the square root sign means the opposite of the principal square root.

c. $\pm\sqrt{121} = \pm 11$. Both positive and negative numbers are given.

d. $\sqrt{-16}$. There is no real-number solution to the square root of a negative number.

e. $\pm\sqrt{1.44} = \pm 1.2$. Although a decimal, 1.44 has an exact square root. Both positive and negative numbers are given.

f. $-\sqrt{25} = -5$. The negative sign outside the square root sign means the opposite of the principal square root. ◀

EXPONENTS AND SQUARE ROOTS Example 5 shows a surprising result with exponents. Recall that one property of exponents is that $a^x a^y = a^{x+y}$.

EXAMPLE 5 Exploring the meaning of one-half as exponent Use a calculator exponent key, [^], to find the values in parts a to d. Find the values in parts e to h without a calculator.

a. $25^{0.5}$ **b.** $16^{1/2}$ **c.** $225^{0.5}$ **d.** $1.44^{1/2}$

e. $36^{1/2} \cdot 36^{1/2}$ **f.** $\sqrt{36} \cdot \sqrt{36}$ **g.** $25^{1/2} \cdot 25^{1/2}$ **h.** $\sqrt{25} \cdot \sqrt{25}$

SOLUTION The decimal 0.5 is equal to the fraction $\frac{1}{2}$. The meaning of the exponent is the same in both decimal and fraction notation: the square root.

a. 5 **b.** 4 **c.** 15 **d.** 1.2

In parts e and g, we can add the exponents. In parts b and d, we take the square roots first and then multiply.

e. $36^{1/2} \cdot 36^{1/2} = 36^{(1/2+1/2)} = 36^1 = 36$ **f.** $\sqrt{36} \cdot \sqrt{36} = 6 \cdot 6 = 36$

g. $25^{1/2} \cdot 25^{1/2} = 25^{(1/2+1/2)} = 25^1 = 25$ **h.** $\sqrt{25} \cdot \sqrt{25} = 5 \cdot 5 = 25$ ◀

ONE-HALF AS EXPONENT

When one-half is used as an exponent, it means to take the principal square root of the base. Thus, if a is zero or a positive real number,

$$a^{1/2} = a^{0.5} = \sqrt{a}$$

Multiplication and Division with Square Roots

PRODUCT AND QUOTIENT PROPERTIES In Example 6, we explore the order in which we do multiplication, division, and square roots.

▶ EXAMPLE 6 Exploring the order of operations Do these without a calculator.

a. $\sqrt{4} \cdot \sqrt{9}$ **b.** $\sqrt{4 \cdot 9}$ **c.** $\sqrt{9 \cdot 25}$ **d.** $\sqrt{9} \cdot \sqrt{25}$

e. $\dfrac{\sqrt{16}}{\sqrt{4}}$ **f.** $\sqrt{\dfrac{16}{4}}$ **g.** $\dfrac{\sqrt{144}}{\sqrt{16}}$ **h.** $\sqrt{\dfrac{144}{16}}$

SOLUTION In parts a and d, we take the square roots first. In parts b and c, we multiply first. In parts e and g, we do the square roots first. In parts f and h, we do the division first. See the Answer Box. ◀

Example 6 suggests that the results for positive numbers are the same whether we (1) take the square roots and then multiply or (2) multiply and then take the square root. Similarly, the results are the same whether we (1) take the square roots first and then divide or (2) divide and then take the square root.

PRODUCT PROPERTY FOR SQUARE ROOTS

The square root of a product equals the product of the square roots. In algebraic notation,

$$\sqrt{a \cdot b} = \sqrt{a} \cdot \sqrt{b}$$

if a and b are positive numbers or zero.

QUOTIENT PROPERTY FOR SQUARE ROOTS

The square root of a quotient equals the quotient of the square roots. In algebraic notation,

$$\sqrt{\frac{a}{b}} = \frac{\sqrt{a}}{\sqrt{b}}$$

if a is zero or positive and b is positive.

THINK ABOUT IT 2: What word describes the set containing zero and the positive numbers?

▶ We prove the product property of square roots by going back to the product property of exponents and applying the fact that $\frac{1}{2}$ as an exponent means the square root.

▶ EXAMPLE 7 Proving the product property of square roots Use the product property of exponents to show that

$$\sqrt{a \cdot b} = \sqrt{a} \cdot \sqrt{b}$$

SOLUTION We assume a and b are zero or positive.

$$\sqrt{a \cdot b} = (a \cdot b)^{1/2} \qquad \text{Definition of } \tfrac{1}{2} \text{ as exponent.}$$

$$(a \cdot b)^{1/2} = a^{1/2} \cdot b^{1/2} \qquad \text{Product property for exponents}$$

$$a^{1/2} \cdot b^{1/2} = \sqrt{a} \cdot \sqrt{b} \qquad \text{Definition of } \tfrac{1}{2} \text{ as exponent}$$

Thus, $\sqrt{a \cdot b} = \sqrt{a} \cdot \sqrt{b}$. ◀

We prove the quotient property of square roots by using the product property of exponents and the fact that $\frac{1}{2}$ as an exponent means the square root. This proof is left as an exercise.

SIMPLIFYING NUMERICAL EXPRESSIONS In Example 8, we practice simplifying radical expressions without a calculator. These simplifications are useful when we want exact—not rounded—answers. That we can *square a radical* and *take a square root of some but not all factors* are key ideas.

EXAMPLE 8 Using the properties of square roots Simplify without a calculator.

a. $\sqrt{8}\sqrt{8}$ **b.** $(\sqrt{2})^2$ **c.** $(2\sqrt{3})^2$ **d.** $\sqrt{3 \cdot 9}$

e. $\sqrt{\dfrac{36}{81}}$ **f.** $\sqrt{25 \cdot 3}$ **g.** $\sqrt{\dfrac{50}{18}}$ **h.** $\sqrt{50}$ **i.** $\sqrt{12}$

SOLUTION **a.** $\sqrt{8}\sqrt{8} = \sqrt{64} = 8$ by the product property.

b. To square a number, we can write it twice and multiply. Thus,

$$(\sqrt{2})^2 = \sqrt{2} \cdot \sqrt{2} = \sqrt{2 \cdot 2} = \sqrt{4} = 2$$

c. We can write $(2\sqrt{3})$ twice:

$$(2\sqrt{3})^2 = 2\sqrt{3} \cdot 2\sqrt{3} = 4\sqrt{9} = 4 \cdot 3 = 12$$

d. $\sqrt{3 \cdot 9} = \sqrt{3} \cdot \sqrt{9} = \sqrt{3} \cdot 3 = 3\sqrt{3}$

e. $\sqrt{\dfrac{36}{81}} = \dfrac{6}{9} = \dfrac{2}{3}$

f. $\sqrt{25 \cdot 3} = \sqrt{25} \cdot \sqrt{3} = 5\sqrt{3}$

g. $\sqrt{\dfrac{50}{18}} = \sqrt{\dfrac{25 \cdot 2}{9 \cdot 2}} = \sqrt{\dfrac{25}{9}} = \dfrac{5}{3}$

h. We obtain an exact square root expression by factoring 50 into $25 \cdot 2$ and taking the square root of one factor:

$$\sqrt{50} = \sqrt{25 \cdot 2} = \sqrt{25} \cdot \sqrt{2} = 5\sqrt{2}$$

i. We factor 12 to obtain $4 \cdot 3$ and take the square root of the perfect square factor:

$$\sqrt{12} = \sqrt{4 \cdot 3} = \sqrt{4} \cdot \sqrt{3} = 2\sqrt{3}$$

Obtaining $2\sqrt{3}$ confirms our work in part c. ◀

From the fact that we took the square root of certain factors in Example 8, we can conclude the following:

A simplified square root expression contains no perfect square factors.

SIMPLIFYING WITH VARIABLES IN SQUARE ROOT EXPRESSIONS Because the square root of a negative number is not defined in the real numbers, we must state assumptions about the inputs when working with square roots and variables. In this section, we make the assumption that the variables have zero or positive value.

When the variables are positive,

$$\sqrt{x^2} = x, \qquad \sqrt{x^4} = x^2, \qquad \sqrt{x^6} = x^3$$

In Example 9, we work with roots containing variables.

EXAMPLE 9 Working with variables in square root expressions Simplify these expressions. Assume that all variables represent a positive number.

a. $\sqrt{x}\sqrt{x}$ **b.** $\sqrt{y^5}$ **c.** $-\sqrt{4x^2}$ **d.** $\sqrt{\dfrac{64x^2}{16}}$

e. $\sqrt{\dfrac{4x^3}{36x}}$ **f.** $\sqrt{25x^4y^3}$ **g.** $\sqrt{x}\sqrt{x^3}$

SOLUTION If x and y are zero or positive, we have the following solutions.

a. $\sqrt{x}\sqrt{x} = \sqrt{x^2} = x$

b. $\sqrt{y^5} = \sqrt{y^4 \cdot y} = y^2\sqrt{y}$

c. $-\sqrt{4x^2} = -2x$ (The negative sign here means "opposite," not the square root of a negative.)

d. $\sqrt{\dfrac{64x^2}{16}} = \dfrac{8x}{4} = 2x$ or $\sqrt{\dfrac{64x^2}{16}} = \sqrt{4x^2} = 2x$

e. $\sqrt{\dfrac{4x^3}{36x}} = \sqrt{\dfrac{4 \cdot x^2 \cdot x}{4 \cdot 9 \cdot x}} = \sqrt{\dfrac{x^2}{9}} = \dfrac{x}{3}$

f. $\sqrt{25x^4y^3} = 5x^2y\sqrt{y}$

g. $\sqrt{x}\sqrt{x^3} = \sqrt{x^4} = x^2$ ◀

▶ Application: Distance Formula

When we did coordinate graphing, we found the slopes and equations of lines, the midpoints of lines, and the intersections of lines with the axes and other lines. We now find the length of lines with the distance formula. As in calculating slope, the choice of which point is (x_1, y_1) and which is (x_2, y_2) is arbitrary.

The **distance formula** permits us to *find the length*, or distance, *between two coordinate points* (x_1, y_1) and (x_2, y_2), as shown in Figure 9. The distance formula is derived from the Pythagorean theorem. As in finding slope, we find the change in x, Δx, to be $(x_2 - x_1)$ and the change in y, Δy, to be $(y_2 - y_1)$. The angle at (x_2, y_1) (see Figure 9) is a right angle, so the Pythagorean theorem holds:

$$d^2 = (\Delta x)^2 + (\Delta y)^2$$

$$d^2 = (x_2 - x_1)^2 + (y_2 - y_1)^2$$

$$d = \sqrt{(x_2 - x_1)^2 + (y_2 - y_1)^2}$$

DISTANCE FORMULA

The distance between (x_1, y_1) and (x_2, y_2) is

$$d = \sqrt{(x_2 - x_1)^2 + (y_2 - y_1)^2}$$

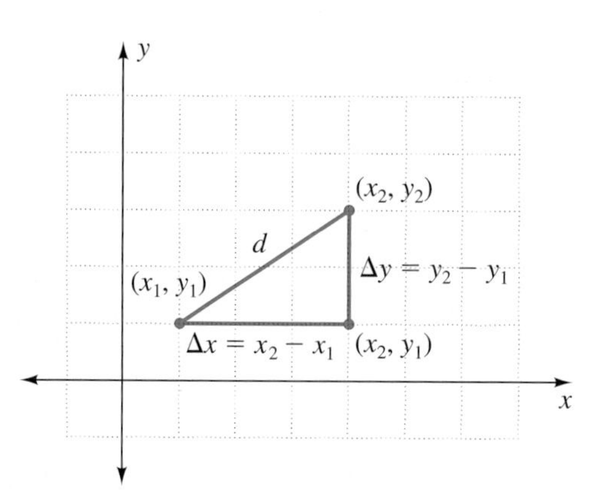

FIGURE 9 d is the distance between (x_1, y_1) and (x_2, y_2)

▶ **EXAMPLE 10** Finding distance Find the three distances between these coordinates: (1, 1), (3, 0), and (5, 4). Determine whether the three points form a right triangle. Make a sketch of the points and lines on coordinate axes. Identify the location of the right angle, if one exists.

SOLUTION Figure 10 shows the triangle. Between (1, 1) and (3, 0),

$$d = \sqrt{(3 - 1)^2 + (0 - 1)^2} = \sqrt{2^2 + 1^2} = \sqrt{5}$$

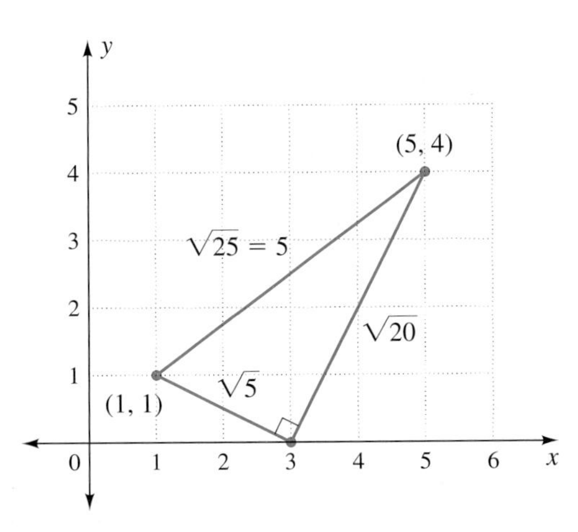

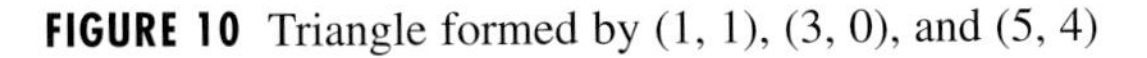

FIGURE 10 Triangle formed by (1, 1), (3, 0), and (5, 4)

Between (3, 0) and (5, 4),

$$d = \sqrt{(5 - 3)^2 + (4 - 0)^2} = \sqrt{2^2 + 4^2} = \sqrt{20}$$

Between (1, 1) and (5, 4),

$$d = \sqrt{(5 - 1)^2 + (4 - 1)^2} = \sqrt{4^2 + 3^2} = \sqrt{25} = 5$$

By the Pythagorean theorem,

$$(\sqrt{5})^2 + (\sqrt{20})^2 = 25 = 5^2$$

The right angle is at (3, 0). ◀

ANSWER BOX

Warm-up: **1. a.** 8 **b.** 11 **c.** 7 **d.** 16 **e.** 20 **f.** 15 **g.** 12 **h.** 25 **2. a.** $6 < \sqrt{45} < 7$ **b.** $8 < \sqrt{75} < 9$ **c.** $14 < \sqrt{200} < 15$ **d.** $4 < \sqrt{20} < 5$ **e.** $10 < \sqrt{110} < 11$ **f.** $3 < \sqrt{12} < 4$ **g.** $30 < \sqrt{905} < 31$ **h.** $19 < \sqrt{395} < 20$ **Think about it 1:** Yes; together, the irrational and rational numbers make up the set of real numbers. **Example 6:** **a.** $\sqrt{4} \cdot \sqrt{9} = 2 \cdot 3 = 6$ **b.** $\sqrt{4 \cdot 9} = \sqrt{36} = 6$ **c.** $\sqrt{9 \cdot 25} = \sqrt{225} = 15$ **d.** $\sqrt{9} \cdot \sqrt{25} = 3 \cdot 5 = 15$ **e.** $\frac{\sqrt{16}}{\sqrt{4}} = \frac{4}{2} = 2$ **f.** $\sqrt{\frac{16}{4}} = \sqrt{4} = 2$ **g.** $\frac{\sqrt{144}}{\sqrt{16}} = \frac{12}{4} = 3$ **h.** $\sqrt{\frac{144}{16}} = \sqrt{9} = 3$ **Think about it 2:** non-negative

▶ 8.2 Exercises

What are the radicands in the expressions in Exercises 1 and 2?

1. a. $\sqrt{-3}$ **b.** $a\sqrt{ab}$ **c.** $\sqrt{4x^2}$
d. $5\sqrt{2}$ **e.** $\sqrt{x} + 2$ **f.** $\sqrt{x+2}$

2. a. $\sqrt{-5}$ **b.** $\sqrt{9y^2}$ **c.** $x\sqrt{xy}$
d. $3\sqrt{5}$ **e.** $\sqrt{x-3}$ **f.** $\sqrt{x} - 3$

In Exercises 3 to 8, find as many square roots as you can mentally, and then finish with a calculator. Round numbers to three decimal places. Identify all irrational numbers with the letter I.

3. a. $\sqrt{81}$ **b.** $\sqrt{15}$ **c.** $\sqrt{25}$ **d.** $\sqrt{2.25}$

4. a. $\sqrt{169}$ **b.** $\sqrt{12.25}$ **c.** $\sqrt{36}$ **d.** $\sqrt{6}$

5. a. $\sqrt{400}$ **b.** $\sqrt{35}$ **c.** $\sqrt{16}$ **d.** $\sqrt{0.01}$

6. a. $\sqrt{14}$ **b.** $\sqrt{0.25}$ **c.** $\sqrt{34}$ **d.** $\sqrt{64}$

7. a. $\sqrt{24}$ **b.** $\sqrt{144}$ **c.** $\sqrt{6.25}$ **d.** $\sqrt{0.1}$

8. a. $\sqrt{28}$ **b.** $\sqrt{121}$ **c.** $\sqrt{49}$ **d.** $\sqrt{0.4}$

Between which two consecutive integers is each square root in Exercises 9 and 10? Do not use a calculator.

9. a. $\sqrt{80}$ **b.** $\sqrt{54}$ **c.** $\sqrt{210}$ **d.** $\sqrt{18}$

10. a. $\sqrt{15}$ **b.** $\sqrt{115}$ **c.** $\sqrt{895}$ **d.** $\sqrt{405}$

Simplify the expressions in Exercises 11 to 14.

11. a. $\sqrt{-36}$ **b.** $-\sqrt{81}$ **c.** $\pm\sqrt{144}$

12. a. $\pm\sqrt{256}$ **b.** $\sqrt{-49}$ **c.** $-\sqrt{64}$

13. a. $\sqrt{49}$ **b.** $-\sqrt{225}$ **c.** $\pm\sqrt{400}$

14. a. $\sqrt{121}$ **b.** $\pm\sqrt{64}$ **c.** $-\sqrt{169}$

Simplify the expressions in Exercises 15 to 30 without a calculator. Assume the variables are positives.

15. a. $\sqrt{5} \cdot \sqrt{3}$ **b.** $(3\sqrt{5})^2$ **c.** $\sqrt{16 \cdot 3}$

16. a. $(5\sqrt{3})^2$ **b.** $\sqrt{11}\sqrt{11}$ **c.** $\sqrt{32}$

17. a. $(3\sqrt{7})^2$ **b.** $\sqrt{13}\sqrt{13}$ **c.** $\sqrt{36 \cdot 2}$

18. a. $\sqrt{3} \cdot \sqrt{7}$ **b.** $(4\sqrt{2})^2$ **c.** $\sqrt{75}$

19. a. $\sqrt{3} \cdot \sqrt{27}$ **b.** $(2\sqrt{5})^2$ **c.** $\sqrt{18}$

20. a. $\sqrt{8} \cdot \sqrt{18}$ **b.** $(5\sqrt{2})^2$ **c.** $\sqrt{98}$

21. a. $\sqrt{a}\sqrt{a}$ **b.** $\sqrt{b^2}$ **c.** $\sqrt{121a^4}$

22. a. $\sqrt{2x} \cdot \sqrt{18x^3}$ **b.** $\sqrt{3y}\sqrt{3y}$ **c.** $\sqrt{256x^2}$

23. a. $\sqrt{32a} \cdot \sqrt{2a}$ **b.** $\sqrt{2x}\sqrt{2x}$ **c.** $\sqrt{16b^2}$

24. a. $\sqrt{b}\sqrt{b}$ **b.** $\sqrt{c^2}$ **c.** $\sqrt{400x^2}$

25. a. $\sqrt{\dfrac{x^2}{9}}$ **b.** $\sqrt{\dfrac{4}{25}}$ **c.** $\sqrt{x^3}$

26. a. $\sqrt{\dfrac{y^2}{4}}$ **b.** $\sqrt{\dfrac{64}{9}}$ **c.** $\sqrt{y^4}$

27. a. $\sqrt{\dfrac{28x}{7x^3}}$ **b.** $\sqrt{\dfrac{3x^2}{27}}$ **c.** $\sqrt{y^5}$

28. a. $\sqrt{\dfrac{6y^3}{24y}}$ **b.** $\sqrt{\dfrac{3x^2}{12}}$ **c.** $\sqrt{y^7}$

29. a. $\sqrt{\dfrac{8x^2y}{2y}}$ **b.** $\sqrt{\dfrac{2x^4}{50y^2}}$
c. $-\sqrt{\dfrac{3xy^4}{27x}}$ **d.** $\dfrac{\sqrt{8xy^5}}{\sqrt{2x^3y}}$

30. a. $\sqrt{\dfrac{18x^2y}{2y}}$ **b.** $-\sqrt{\dfrac{32x^4y}{2x^2y}}$
c. $\sqrt{\dfrac{20xy^2}{5x}}$ **d.** $\dfrac{\sqrt{2xy}}{\sqrt{50x^5y^3}}$

For each pair of points in Exercises 31 to 38,
(a) Find the distance between the two points. Simplify the square root.
(b) Find the slope of the line segment connecting the two points.
(c) Find the equation of the line passing through the points.

31. (2, 3) and (4, 9)

32. (3, 2) and (9, 4)

33. (2, 2) and (5, −1)

34. (5, 3) and (6, −2)

35. (−3, 3) and (4, 2)

36. (−3, 3) and (2, 4)

37. (−3, −1) and (3, −3)

38. (−2, −1) and (2, −4)

In Exercises 39 to 44, the vertices of a triangle are given. Use the distance formula to find what kind of triangle each is. A sketch of the graph may be helpful. Note: An **isosceles triangle** has *two equal sides.* An **equilateral triangle** has *three equal sides.* A right triangle satisfies the Pythagorean theorem.

39. (3, 4), (0, 1), (6, 1) **40.** (3, 1), (0, 4), (−3, 1)

41. (6, 5), (4, 2), (8, 2) **42.** (3, 6), (6, 4), (3, 2)

43. (4, 8), (1, 6), (5, 0) **44.** (5, 8), (0, 7), (2, −3)

Blue numbers are core exercises.

▶ Writing

45. True or false: The square root of x is smaller than x. Explain your reasoning and give an example.

46. True or false: Only whole numbers can have exact square roots. Explain your reasoning and give an example.

47. Using properties of exponents, prove the quotient property of square roots,

$$\sqrt{\frac{a}{b}} = \frac{\sqrt{a}}{\sqrt{b}}$$

where a is zero or positive and b is positive.

48. How do we know whether a question is asking for both the square roots or just the principal square root?

49. Evaluate and simplify $\sqrt{x^n}$ for $n = 1$ to 10. Assume $x \geq 0$. State at least two patterns.

50. Looking ahead:

a. For $\sqrt{x^4} = x^2, x \geq 0$, discuss whether the condition $x \geq 0$ is needed.

b. For $\sqrt{x^6} = x^3, x \geq 0$, discuss whether the condition $x \geq 0$ is needed.

▶ Connecting to Chapter 7

Evaluate the expressions in Exercises 51 to 58 without a calculator. *Hint:* Change decimals to fractions.

51. a. 25^0 **b.** 25^{-1} **c.** $25^{1/2}$ **d.** $25^{0.5}$

52. a. 4^{-1} **b.** 4^0 **c.** $4^{0.5}$ **d.** $4^{1/2}$

53. a. $9^{1/2}$ **b.** 9^0 **c.** $9^{0.5}$ **d.** 9^{-1}

54. a. $36^{1/2}$ **b.** $36^{0.5}$ **c.** 36^0 **d.** 36^{-1}

55. a. $(\frac{1}{4})^{-1}$ **b.** $(\frac{1}{4})^0$ **c.** $(\frac{1}{4})^{1/2}$ **d.** $(\frac{1}{4})^{0.5}$

56. a. $(\frac{1}{9})^{1/2}$ **b.** $(\frac{1}{9})^{0.5}$ **c.** $(\frac{1}{9})^0$ **d.** $(\frac{1}{9})^{-1}$

57. a. $(0.25)^{-1}$ **b.** $(0.01)^{0.5}$ **c.** $(6.25)^{0.5}$
d. $(0.25)^{0.5}$ **e.** $(0.02)^{-1}$ **f.** $(0.05)^{-1}$

58. a. $(2.25)^{0.5}$ **b.** $(0.36)^{0.5}$ **c.** $(0.01)^{-1}$
d. $(0.1)^{-1}$ **e.** $(0.125)^{-1}$ **f.** $(0.0001)^{0.5}$

▶ Projects

59. Rounding Numbers When we do several steps on a calculator, rounding at a middle step can cause us to overlook patterns or arrive at unacceptable answers. Simplify these calculator expressions and describe the differences in the answers.

a. Exact: $\sqrt{(5^2 + (\sqrt{11})^2)}$
Rounded: $\sqrt{(5^2 + 3.3^2)}$

b. Exact: $\sqrt{(6^2 + (\sqrt{13})^2)}$
Rounded: $\sqrt{(6^2 + 3.6^2)}$

c. Exact: $\sqrt{(7^2 + (\sqrt{15})^2)}$
Rounded: $\sqrt{(7^2 + 3.9^2)}$

d. What general number pattern is shown in parts a to c? *Hint:* $20^2 + (\sqrt{\_\_\_\_})^2 = 21^2$ and $21^2 + (\sqrt{\_\_\_\_})^2 = 22^2$.

The error from rounding gets worse as exponents get higher in compounded interest problems. Describe the differences in the answers to parts e and f.

e. Exact: $1000(1 + 0.08/12)^{12 \cdot 10}$
Rounded: $1000(1.007)^{12 \cdot 10}$

f. Exact: $1000(1 + 0.05/12)^{12 \cdot 30}$
Rounded: $1000(1.004)^{12 \cdot 30}$

You can avoid working with rounded numbers by entering the entire expression into the calculator at once.

60. Square Root Patterns

a. Find a pattern in the values of these square roots. Describe your findings in words. (*Hint:* Your word description should relate to the number of zeros or decimal places in the radicand.)

$\sqrt{4}$	$\sqrt{9}$	$\sqrt{25}$
$\sqrt{40}$	$\sqrt{90}$	$\sqrt{250}$
$\sqrt{400}$	$\sqrt{900}$	$\sqrt{2500}$
$\sqrt{4000}$	$\sqrt{0.09}$	$\sqrt{0.25}$
$\sqrt{40,000}$	$\sqrt{0.0009}$	$\sqrt{0.025}$
$\sqrt{4,000,000}$	$\sqrt{0.000\ 000\ 9}$	$\sqrt{0.000\ 25}$

b. Test your pattern statement on the following square roots. Check with a calculator.

$\sqrt{0.000\ 000\ 4}$	$\sqrt{9,000,000}$	$\sqrt{250,000,000}$
$\sqrt{0.000\ 000\ 000\ 04}$	$\sqrt{0.000\ 09}$	$\sqrt{0.000\ 000\ 25}$

c. Use your pattern on the following square roots. Suppose you know that $\sqrt{8} = 2.8284$ and $\sqrt{80} = 8.9443$. Do not use a calculator.

$\sqrt{800,000}$	$\sqrt{0.8}$	$\sqrt{8000}$
$\sqrt{8,000,000}$	$\sqrt{0.0008}$	$\sqrt{0.000\ 000\ 8}$

Blue numbers are core exercises.

61. Calculator Investigations with Squaring To answer the questions, use the x^2 key to guess and check. In part d, write both the number and its square.

a. Write $\sqrt{10}$ to 6 decimal places.

b. What is the largest three-digit number that gives a five-digit number when squared?

c. What is the largest four-digit number that gives a seven-digit number when squared?

d. What is the largest five-digit number that gives a nine-digit number when squared?

e. Compare the digits above with those in $\sqrt{10}$. Is this a coincidence? Explain.

▶ 8.3 Solving Square Root Equations and Simplifying Expressions

Objectives

- Graph square root equations.
- Find sets of inputs for square root functions.
- Solve square root equations by graphing.
- Solve equations and formulas by squaring both sides.
- Use absolute value in simplifying square root expressions.

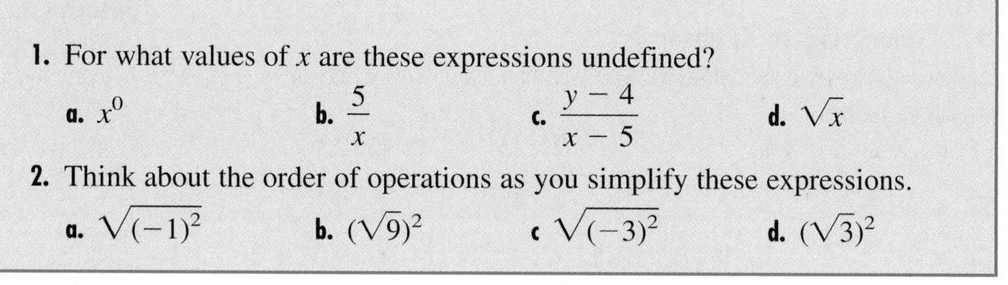

WARM-UP

1. For what values of x are these expressions undefined?

a. x^0 **b.** $\dfrac{5}{x}$ **c.** $\dfrac{y-4}{x-5}$ **d.** $\sqrt{x}$

2. Think about the order of operations as you simplify these expressions.

a. $\sqrt{(-1)^2}$ **b.** $(\sqrt{9})^2$ **c** $\sqrt{(-3)^2}$ **d.** $(\sqrt{3})^2$

IN THIS SECTION, we solve square root equations by graphing and by squaring both sides of the equation. We look at two applications in which square roots appear in formulas. We close by considering assumptions about square roots of powers.

▶ Graphing Square Root Equations

In Figure 11, the principal square root graph has no points to the left of the origin because negative numbers have no real-number square roots. When graphing square roots containing expressions, though, we must look carefully for negative numbers. We graph two square root equations in Examples 1 and 2.

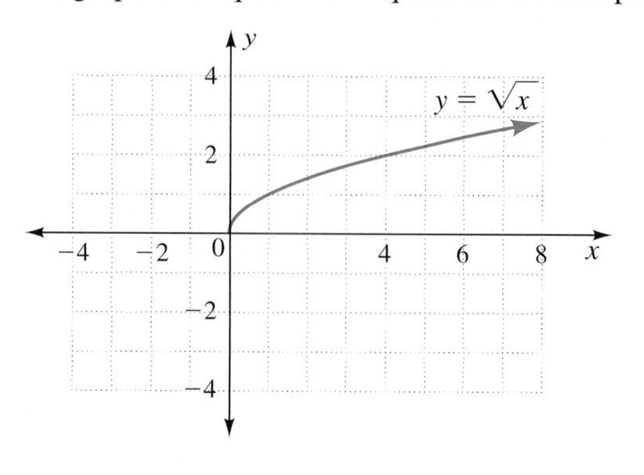

FIGURE 11 $y = \sqrt{x}$

▶ EXAMPLE 1 Graphing square root equations Make a table and graph for $y = \sqrt{x+2}$. Use integer inputs on the interval $[-4, 7]$.

SOLUTION The table is Table 3, and the graph appears in Figure 12. For $y = \sqrt{x+2}$, there are defined values to the left of the y-axis. The expression $x + 2$ causes the inputs between -2 and 0 to be acceptable inputs.

TABLE 3 $y = \sqrt{x+2}$ for $-4 \leq x \leq 7$

x	$y = \sqrt{x+2}$	x	$y = \sqrt{x+2}$
-4	error	2	2
-3	error	3	2.236
-2	0	4	2.449
-1	1	5	2.646
0	1.414	6	2.828
1	1.732	7	3

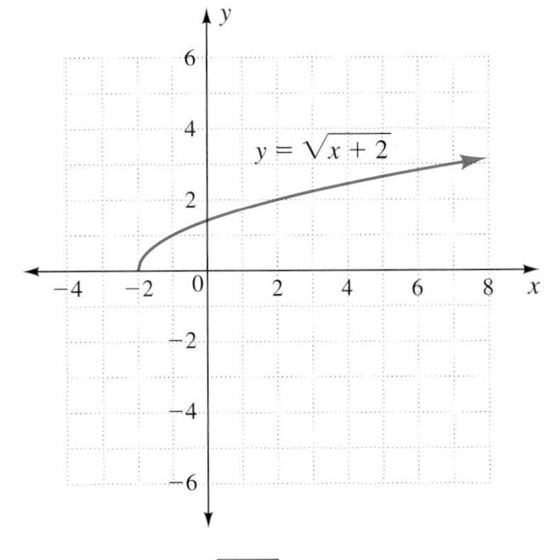

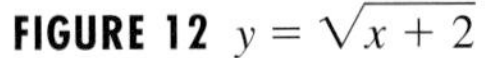
FIGURE 12 $y = \sqrt{x+2}$

GRAPHING CALCULATOR SOLUTION Enter $Y_1 = \sqrt{(X + 2)}$. Set the table feature, with **TblStart** $= -4$ and **ΔTbl** $= 1$. Choose [TABLE] and compare the result to Table 3. Set the window for X at $[-4, 8]$, scale 1, and for Y at $[-6, 6]$, scale 1. Graph. The graph should look like the one in Figure 12. ◀

▶ EXAMPLE 2 Graphing square root equations Make a table and graph for $y = \sqrt{x-3}$.

SOLUTION The table is Table 4, and the graph appears in Figure 13. For $y = \sqrt{x-3}$, there are no defined values to the left of $x = 3$. The expression $x - 3$ causes all the inputs from -4 to 2 to give square roots of negative numbers.

TABLE 4 $y = \sqrt{x-3}$ for $-4 \leq x \leq 7$

x	$y = \sqrt{x-3}$	x	$y = \sqrt{x-3}$
-4	error	2	error
-3	error	3	0
-2	error	4	1
-1	error	5	1.414
0	error	6	1.732
1	error	7	2

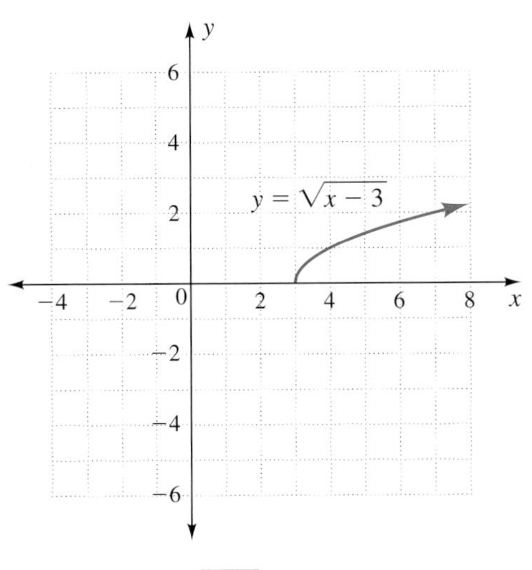

FIGURE 13 $y = \sqrt{x-3}$

GRAPHING CALCULATOR SOLUTION Enter $Y_1 = \sqrt{(X - 3)}$. Set the table feature, with **TblStart** = −4 and **ΔTbl** = 1. Choose [TABLE] and compare the result to Table 4. Set the window for X at [−4, 8], scale 1, and for Y at [−6, 6], scale 1. Graph. The graph should look like the one in Figure 13. ◀

▶ Solving Square Root Equations

SETTING CONDITIONS BEFORE SOLVING When solving square root equations, we keep the radicand positive by limiting the input set (domain) with an inequality.

> To keep the radicand positive, limit the input set by solving this inequality for the variable:
>
> radicand ≥ 0

▶ **EXAMPLE 3** Finding the input set (domain) For what inputs, x, is $\sqrt{x + 2}$ defined?

SOLUTION $\sqrt{x + 2}$ is defined if $x + 2$ is zero or positive.

$$x + 2 \geq 0 \quad \text{Find when the radicand is zero or positive.}$$

$$x + 2 - 2 \geq 0 - 2 \quad \text{Subtract 2 from both sides.}$$

$$x \geq -2$$

Thus, $\sqrt{x + 2}$ is defined when $x \geq -2$. ◀

The result in Example 3 agrees with the graph in Example 1. We will find conditions as we solve square root equations, beginning with Example 5.

SOLVING BY GRAPHING We can solve a square root equation by graphing.

▶ **EXAMPLE 4** Solving square root equations Solve $\sqrt{2 - x} = 3$ by graphing.

SOLUTION The graphs of $y_1 = \sqrt{2 - x}$ and $y_2 = 3$ are shown in Figure 14. The point of intersection is $(-7, 3)$, so $\sqrt{2 - x} = 3$ at $x = -7$.

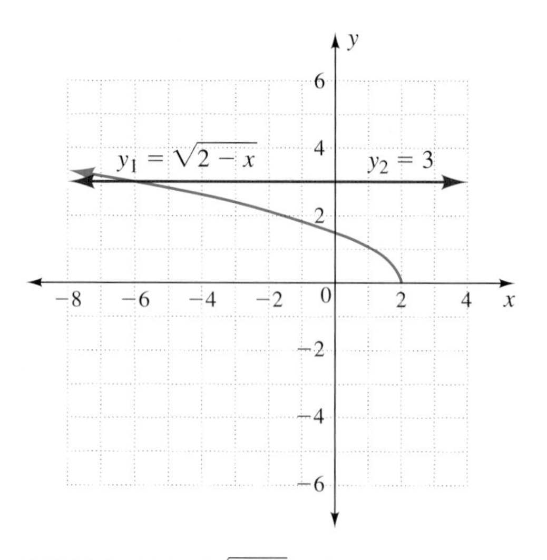

FIGURE 14 Solve $\sqrt{2 - x} = 3$

Check: $\sqrt{2 - 1} \stackrel{?}{=} 1$ ✓ ◀

THINK ABOUT IT 1: How does the condition $x \le 2$ show on the graph in Figure 14?

SOLVING BY SQUARING BOTH SIDES We now add another technique to our tools for solving equations with algebraic notation: squaring both sides.

> To solve an equation of the form $\sqrt{x} = a$, we square both sides.

Note that $(\sqrt{x})^2 = \sqrt{x}\sqrt{x} = x$ if $x \ge 0$. Again, the x represents any number or expression and, as a radicand, must be positive. In Example 5, the radicand is $2 - x$, and it must be zero or positive: $2 - x \ge 0$.

▶ **EXAMPLE 5** Solving by squaring both sides

a. For what x is $\sqrt{2 - x}$ defined?

b. Solve $\sqrt{2 - x} = 3$.

SOLUTION

a. To make the radicand zero or positive, $2 - x \ge 0$, we must have $2 \ge x$, or $x \le 2$.

b.

$\sqrt{2 - x} = 3$ Square both sides.

$(\sqrt{2 - x})^2 = 3^2$ Simplify.

$2 - x = 9$ Add x to both sides.

$2 = x + 9$ Subtract 1 on both sides.

$-7 = x$

Check: $x = -7$ satisfies the condition $x \le 2$.

$\sqrt{2 - (-7)} \stackrel{?}{=} 3$ ✓ ◀

THINK ABOUT IT 2: In Example 5, we squared both sides. Explain why multiplying the left side of $\sqrt{2 - x} = 3$ by $\sqrt{2 - x}$ and the right side by 3 is the same as multiplying each side by 3.

We repeat the squaring of both sides of an equation in Example 6.

▶ **EXAMPLE 6** Solving a square root equation

a. Find inputs for which the equation $y = \sqrt{x - 5}$ is defined.

b. Solve $\sqrt{x - 5} = 3$ by squaring both sides.

c. Solve with a graph.

SOLUTION

a. The expression $x - 5$ must be zero or positive: $x - 5 \ge 0$. By adding 5 to both sides, we have $x \ge 5$.

b. We solve the equation by squaring both sides.

$\sqrt{x - 5} = 3$ Square both sides.

$(\sqrt{x - 5})^2 = 3^2$

$x - 5 = 9$ $(\sqrt{x - 5})^2 = x - 5$ because $x - 5$ is positive.

$x = 14$

Check: $\sqrt{14 - 5} \stackrel{?}{=} 3$ ✓

c. The intersection of $y_1 = \sqrt{x - 5}$ with $y_2 = 3$ is at (14, 3) in Figure 15. Thus, $x = 14$ is the solution to $\sqrt{x - 5} = 3$. There are no points on the square root graph to the left of 5 because the expression $\sqrt{x - 5}$ is undefined for $x < 5$.

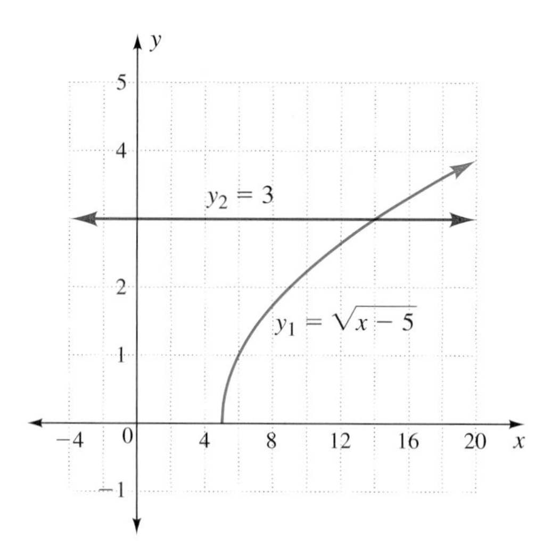

FIGURE 15 Solve $\sqrt{x - 5} = 3$ ◀

▶ Applications

DISTANCE TO THE HORIZON In Examples 7 and 8, we use the square root formula for distance to the horizon.

▶ **EXAMPLE 7** Using square root formulas: distance to the horizon at the seashore On Earth, the distance to the horizon is given by

$$d \approx \sqrt{\frac{3h}{2}}$$

where d is in miles to the horizon and h is the height in feet above sea level.

a. Find the distance seen from a height of 5 feet.
b. Find the distance seen from a height of 10 feet (Figure 16).
c. How high would we have to be to see 12 miles?

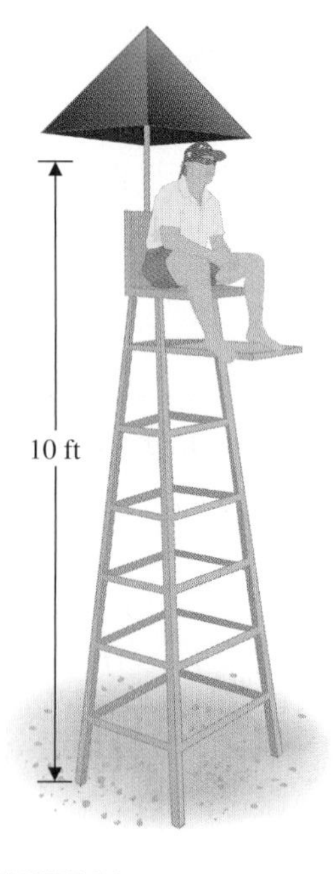

FIGURE 16

SOLUTION **a.** $d \approx \sqrt{\frac{3h}{2}} \approx \sqrt{\frac{3 \cdot 5}{2}} \approx 2.7$ mi

b. $d \approx \sqrt{\frac{3 \cdot 10}{2}} \approx 3.9$ mi

c. $12 \approx \sqrt{\frac{3h}{2}}$ Square both sides.

$144 \approx \frac{3h}{2}$ Multiply both sides by 2.

$288 \approx 3h$ Divide both sides by 3.

$h \approx 96$ ft ◀

The unit change from h feet to d miles is not shown in the formula. The units are included in the coefficient of $\frac{3}{2}$ on h. See the project in Exercise 77 for the steps in deriving this formula and the formula for the distance seen on the moon.

Squaring both sides can be used to solve formulas as well as equations.

EXAMPLE 8 Solving formulas Solve the formula $d = \sqrt{\frac{3h}{2}}$ for h.

SOLUTION

$$d = \sqrt{\frac{3h}{2}} \quad \text{Square both sides.}$$

$$d^2 = \left(\sqrt{\frac{3h}{2}}\right)^2 \quad \text{Simplify.}$$

$$d^2 = \frac{3h}{2} \quad \text{Multiply both sides by 2.}$$

$$2d^2 = 3h \quad \text{Divide both sides by 3.}$$

$$\frac{2d^2}{3} = h$$

◀

FALLING TIME Examples 9 and 10 involve the square root formula for the time required for a falling object to reach the ground.

EXAMPLE 9 Using square root formulas: Washington Monument As a child, I dropped a paper-wrapped sugar cube from an open window of the Washington Monument in our nation's capital. Fortunately, it was a cold winter day in the 1950s, and no one was at the base of the 555-foot tower.

The windows are at 500 feet. The time required for a dropped object to hit the ground is given by

$$t = \sqrt{\frac{2h}{g}}$$

where h is the vertical distance traveled and g is the acceleration due to gravity, assumed to be 32.2 feet per second squared. The downward speed, or velocity v, such an object will be traveling is given by $v = gt$.

a. How long did it take for the cube to hit the ground?
b. What speed in feet per second was the cube traveling when it hit the ground?
c. Use unit analysis to change the speed to miles per hour.

SOLUTION **a.** Because the distance from the window to the ground is 500 feet, we have

$$t = \sqrt{\frac{2h}{g}} = \sqrt{\frac{2 \cdot 500 \text{ ft}}{32.2 \frac{\text{ft}}{\text{sec}^2}}} \approx 5.6 \text{ sec}$$

b. Speed is $v = gt$.

$$v = gt = \frac{32.2 \text{ ft}}{\text{sec}^2} \cdot 5.6 \text{ sec} \approx 180 \text{ ft per sec}$$

c. Use unit analysis to change feet per second to miles per hour, we have

$$\frac{180 \text{ ft}}{\text{sec}} \cdot \frac{1 \text{ mi}}{5280 \text{ ft}} \cdot \frac{60 \text{ sec}}{1 \text{ min}} \cdot \frac{60 \text{ min}}{1 \text{ hr}} \approx 123 \text{ mi per hr}$$

P.S. My parents were furious! The Washington Monument windows are now sealed.

◀

EXAMPLE 10 Solving formulas Solve the formula $t = \sqrt{\frac{2h}{g}}$ for h.

SOLUTION

$$t = \sqrt{\frac{2h}{g}} \qquad \text{Square both sides.}$$

$$t^2 = \left(\sqrt{\frac{2h}{g}}\right)^2 \qquad \text{Simplify.}$$

$$t^2 = \frac{2h}{g} \qquad \text{Multiply both sides by } g.$$

$$gt^2 = 2h \qquad \text{Divide both sides by 2.}$$

$$\frac{gt^2}{2} = h$$

◀

▶ Square Roots of Powers of x

WHEN X IS ANY REAL NUMBER IN $y = \sqrt{x^2}$ The function $f(x) = \sqrt{x^2}$ is an example of a square root function that is defined for all real numbers (yes, even negative numbers, $x < 0$). In Example 11, we evaluate and graph $f(x) = \sqrt{x^2}$.

▶ **EXAMPLE 11** Exploring a square root function

a. For $f(x) = \sqrt{x^2}$, find $f(-3)$, $f(-1)$, $f(0)$, $f(1)$, and $f(3)$.

b. List the facts from part a as ordered pairs and graph the ordered pairs. Connect them from left to right.

c. What function describes the graph?

SOLUTION **a.** $f(-3) = \sqrt{(-3)^2} = \sqrt{9} = 3$

$f(-1) = \sqrt{(-1)^2} = \sqrt{1} = 1$

$f(0) = \sqrt{0^2} = \sqrt{0} = 0$

$f(1) = \sqrt{1^2} = \sqrt{1} = 1$

$f(3) = \sqrt{3^2} = \sqrt{9} = 3$

b. The ordered pairs are $(-3, 3)$, $(-1, 1)$, $(0, 0)$, $(1, 1)$, and $(3, 3)$. The graph is in Figure 17.

c. The graph is the absolute value function, $y = |x|$.

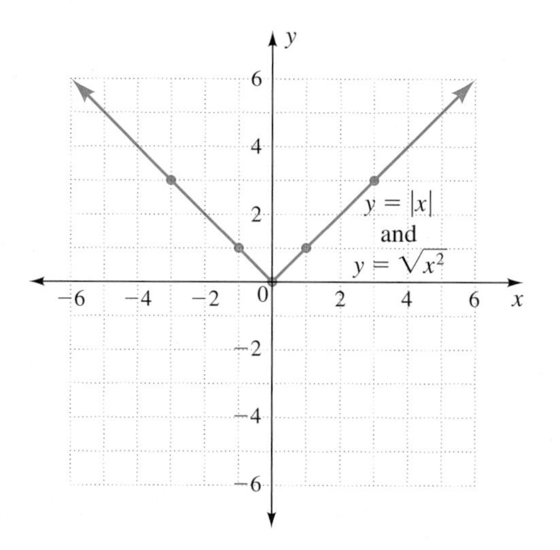

FIGURE 17 $\sqrt{x^2} = |x|$

GRAPHING CALCULATOR SOLUTION Let $Y_1 = \sqrt{(X^2)}$ for X in $[-5, 5]$ and Y in $[-5, 5]$. Compare the table and the graph with that of the absolute value, $Y_2 = \text{abs}(X)$. The absolute value function is under 2nd [CATALOG]. The two graphs should be identical. ◀

> The function $f(x) = \sqrt{x^2}$ is the same as the absolute value function, $f(x) = |x|$.

TABLE 5 Square Roots When $x \geq 0$

$\sqrt{x^2} = x$	$\sqrt{x^3} = x\sqrt{x}$
$\sqrt{x^4} = x^2$	$\sqrt{x^5} = x^2\sqrt{x}$
$\sqrt{x^6} = x^3$	

WHEN x IS POSITIVE OR ZERO IN $y = \sqrt{x^n}$ Because the square roots of negative numbers are undefined in the set of real numbers, we avoid having a negative number in the radicand (the expression under the square root sign). In Section 8.2, we avoided negative numbers by assuming that all variables represented positive numbers or zero. We wrote $x \geq 0$. See Table 5.

WHEN x IS ANY REAL NUMBER IN $y = \sqrt{x^n}$ If the exponent of the radicand is even (as in x^2 in Example 11), a negative input will be transformed into a positive number before the square root is taken. In this case, the input can be negative. Note that Table 6 provides for negative inputs in the case of even powers.

The output from the square root function must be a positive (or zero). There are two ways to ensure a positive output: by having an even exponent on the output, $\sqrt{x^4} = x^2$, or by having the absolute value symbol around the output, $\sqrt{x^2} = |x|$. Note the even exponents or absolute value on the outputs in Table 6.

TABLE 6 Square Roots When x Is Any Real Number

$\sqrt{x^2} = \|x\|$	Because x could be negative
$\sqrt{x^3} = x\sqrt{x}, x \geq 0$	Undefined if $x < 0$
$\sqrt{x^4} = x^2$	Because an even power is positive
$\sqrt{x^5} = x^2\sqrt{x}, x \geq 0$	Undefined if $x < 0$
$\sqrt{x^6} = \|x^3\|$	Because x could be negative

THINK ABOUT IT 3: What is $\sqrt{x^7}$ and $\sqrt{x^8}$?

In Example 12, we use Table 6 and properties from Section 8.2 to simplify expressions.

▶ **EXAMPLE 12** Simplifying radical expressions Let the variables be any real number. Simplify these expressions, and state any conditions that must be added.

a. $\sqrt{4x^2}$ **b.** $\sqrt{x^6y^4}$ **c.** $\sqrt{a \cdot b^2}$ **d.** $\sqrt{c^5}$

SOLUTION **a.** $\sqrt{4x^2} = \sqrt{4}\sqrt{x^2} = 2|x|$

b. $\sqrt{x^6y^4} = |x^3|y^2$; y^2 is positive and does not need absolute value.

c. $\sqrt{a \cdot b^2} = |b|\sqrt{a}, a \geq 0$

d. $\sqrt{c^5} = \sqrt{c^4 \cdot c} = c^2\sqrt{c}, c \geq 0$ ◀

ANSWER BOX

Warm-up: **1. a.** 0 **b.** 0 **c.** 5 **d.** $x < 0$ **2. a.** $\sqrt{(-1)^2} = \sqrt{1} = 1$ **b.** 9 **c.** $\sqrt{(-3)^2} = \sqrt{9} = 3$ **d.** 3 **Think about it 1:** The graph is only to the left of $x = 2$. **Think about it 2:** Because $\sqrt{2 - x} = 3$, we are replacing 3 with $\sqrt{2 - x}$. **Think about it 3:** $x^3\sqrt{x}, x \geq 0$; x^4

▶ 8.3 Exercises

Evaluate the functions in Exercises 1 to 4 for $f(-4)$, $f(-1)$, $f(0)$, $f(4)$, and $f(6)$.

1. $f(x) = \sqrt{4 - x}$
2. $f(x) = \sqrt{2 - x}$
3. $f(x) = \sqrt{3 - x}$
4. $f(x) = \sqrt{1 - x}$
5. Is $y = \sqrt{x + 3}$ defined for $x = -2$? Explain.
6. Is $y = \sqrt{x - 2}$ defined for $x = 1$? Explain.
7. Is $y = \sqrt{-2x}$ ever defined? Explain.
8. Is $y = \sqrt{|x|}$ ever defined? Explain.

In Exercises 9 to 14, make a table and graph for inputs in $[-4, 4]$, counting by 2s.

9. $y = \sqrt{x + 4}$
10. $y = \sqrt{x - 1}$
11. $y = \sqrt{x - 2}$
12. $y = \sqrt{x + 3}$
13. $y = \sqrt{2x}$
14. $y = \sqrt{\frac{1}{2}x}$

In Exercises 15 to 18, for what inputs is each expression defined?

15. a. $\sqrt{x - 1}$ b. $\sqrt{x + 3}$
16. a. $\sqrt{x - 3}$ b. $\sqrt{x + 1}$
17. a. $\sqrt{4 - x}$ b. $\sqrt{3 - x}$
18. a. $\sqrt{2 - x}$ b. $\sqrt{1 - x}$

For Exercises 19 and 20, use the graphs to solve the equations.

19. $\sqrt{x - 2} = 2$

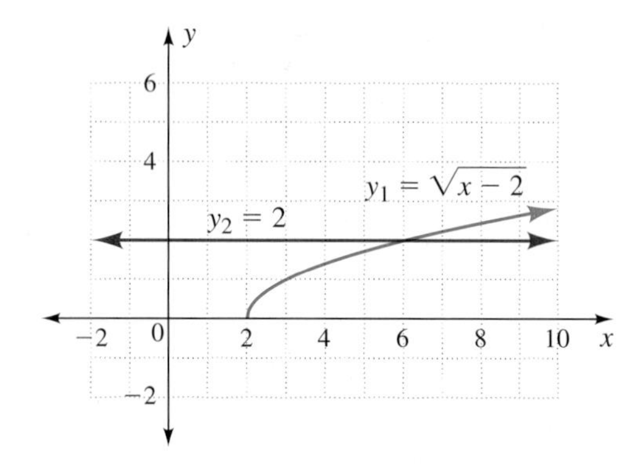

20. $\sqrt{x + 2} = 2$

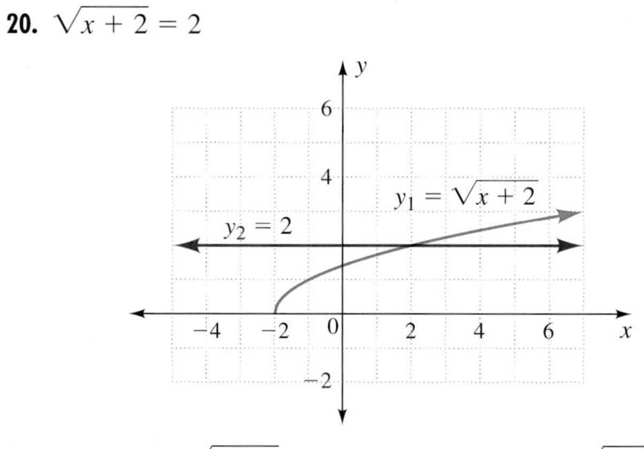

21. Graph $y = \sqrt{x + 3}$ and $y = 1$. For what x is $\sqrt{x + 3}$ defined? Solve $\sqrt{x + 3} = 1$ from your graph and with algebraic notation.
22. Graph $y = \sqrt{x - 3}$ and $y = 1$. For what x is $\sqrt{x - 3}$ defined? Solve $\sqrt{x - 3} = 1$ from your graph and with algebraic notation.
23. Graph $y = \sqrt{3 - x}$ and $y = 2$. For what x is $\sqrt{3 - x}$ defined? Solve $\sqrt{3 - x} = 2$ from your graph and with algebraic notation.
24. Graph $y = \sqrt{4 - x}$ and $y = 3$. For what x is $\sqrt{4 - x}$ defined? Solve $\sqrt{4 - x} = 3$ from your graph and with algebraic notation.

Solve the equations in Exercises 25 to 34. For what inputs, x, are the radical expressions defined?

25. $\sqrt{x - 2} = 3$
26. $\sqrt{x - 3} = 2$
27. $\sqrt{2x + 2} = 4$
28. $\sqrt{3x + 1} = 5$
29. $\sqrt{3x - 3} = 6$
30. $\sqrt{2x - 1} = 7$
31. $\sqrt{5 - x} = 3$
32. $\sqrt{6 - 2x} = 4$
33. $\sqrt{x - 2} = -1$
34. $\sqrt{x - 3} = -1$
35. How can a graph explain the result in Exercise 33?
36. How does the range (set of outputs) of a square root function explain the result in Exercise 34?

Assume a clear day and flat terrain in the nearby region. How far could you see, in miles, from the top of each of the buildings whose heights are given in Exercises 37 to 40? See text examples for formulas.

37. Sears Tower, Chicago, 1454 feet
38. Bank of China, Hong Kong, 1209 feet
39. Texas Commerce Tower, Houston, 1002 feet
40. Transamerica Pyramid, San Francisco, 853 feet

Blue numbers are core exercises.

On the moon, the distance seen in miles from a height of h feet is given by

$$d \approx \sqrt{\frac{3h}{8}}$$

For Exercises 41 to 44, calculate how far can be seen from the given heights. Why might the distance be different from that on Earth?

41. 24 feet

42. 5 feet

43. 96 feet

44. 10 feet

45. **a.** How high do we need to be on Earth to see 30 miles?

b. How high do we need to be on the moon to see 30 miles?

c. How many times as high do we need to be in part b to see the same distance as in part a?

d. Extension: Explain why. A picture may be helpful.

46. **a.** How high do we need to be on Earth to see 5 miles?

b. How high do we need to be on the moon to see 5 miles?

c. How many times as high do we need to be in part b to see the same distance as in part a?

d. Extension: Explain why. A picture may be helpful.

How long would it take an object to fall from the top of each of the places whose heights are given in Exercises 47 to 50? To the nearest whole number, how fast would the object be traveling in feet per second when it hit the ground? See text examples for formulas.

47. C. N. Tower, Toronto, 1821 feet

48. Torre Commercial America, Monterrey, Mexico, 427 feet

49. C&C Plaza, Atlanta, 1063 feet

50. Eiffel Tower, Paris, 984 feet

51. How tall a building is required for an object to take 12 seconds to fall to the ground?

52. How tall a building is required for an object to take 10 seconds to fall to the ground?

53. How long would it take an object to fall from one mile up? (Assume no air resistance, wind, or other complicating factors.)

54. If a piece of ice falls off an airplane at 35,000 feet, how long until it will hit the ground? (Assume the ice does not melt.)

In Exercises 55 to 62, solve the formula for the indicated variable. Exercises 55 to 57 are from the Electric Formulas Wheel. What is the third equivalent formula for each exercise?

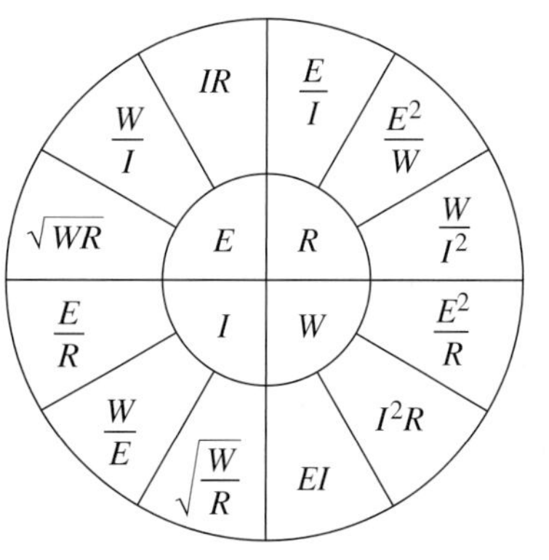

Electric Formulas Wheel

E = Electromagnetic force (volts)
R = Resistance (ohms)
I = Intensity (amps)
W = Watts (or power)

55. Electricity: $E = \sqrt{W \cdot R}$ for R

56. Electricity: $I = \sqrt{\frac{W}{R}}$ for W

57. Electricity: $I = \sqrt{\frac{W}{R}}$ for R

58. Falling objects: $t = \sqrt{\frac{2d}{g}}$ for g

59. Distance seen on the moon: $d = \sqrt{\frac{3h}{8}}$ for h

60. Orbital velocity: $V_o = \sqrt{\frac{GM}{R}}$ for M

61. Orbital velocity: $V_o = \sqrt{\frac{GM}{R}}$ for R

62. Escape velocity: $V_e = \sqrt{\frac{2GM}{R}}$ for M

Simplify each expression in Exercises 63 to 70. The variables may represent any real number, so absolute values or conditions may be needed.

63. **a.** $\sqrt{ab^2}$ **b.** $\sqrt{a^2b}$ **c.** $\sqrt{a^2b^2}$

64. **a.** $\sqrt{a^2b^4}$ **b.** $\sqrt{a^4b^2}$ **c.** $\sqrt{a^6b^4}$

65. **a.** $\sqrt{49x^2}$ **b.** $\sqrt{121y^2}$ **c.** $\sqrt{x^6}$

66. **a.** $\sqrt{x^2y}$ **b.** $\sqrt{y^2x}$ **c.** $\sqrt{x^8}$

67. **a.** $\sqrt{p^5q^2}$ **b.** $\sqrt{p^{10}}$ **c.** $\sqrt{r^6 \cdot s^8}$

68. **a.** $\sqrt{p^9}$ **b.** $\sqrt{p^1q^7}$ **c.** $\sqrt{a \cdot b^6}$

69. **a.** $\sqrt{a \cdot b^4}$ **b.** $\sqrt{b \cdot c^2}$ **c.** $\sqrt{c^7}$

70. **a.** $\sqrt{c \cdot d^2}$ **b.** $\sqrt{k \cdot m^6}$ **c.** $\sqrt{b^5}$

Blue numbers are core exercises.

▶ Writing

71. Explain why $\sqrt{x^2y} = |x|\sqrt{y}$ ($y \geq 0$) needs absolute value.

72. For x, y any real number, explain why $\sqrt{x^8y} = x^4\sqrt{y}$ ($y \geq 0$) does not need absolute value. Is ($y \geq 0$) an assumption or a condition?

73. Solve $\sqrt{\frac{3h_e}{2}} = \sqrt{\frac{3h_m}{8}}$ for the ratio $\frac{h_e}{h_m}$, the ratio of the height needed on Earth to the height needed on the moon to see the same distance. Explain the results.

74. How do we find the sets of inputs for which a radicand is defined?

75. How do we know when to use absolute value on an expression after taking the square root?

▶ Projects

76. **Radicals with Index Other than 2** If $2^3 = 8$, then 2 is a cube root of 8. If $3^4 = 81$, then the fourth root of 81 is 3. In radical notation, these facts are written $\sqrt[3]{8} = 2$ and $\sqrt[4]{81} = 3$, where 3 and 4 are the *indices* (plural of *index*). What are the values of these expressions? Do the problems mentally by thinking "What number raised to the [index] power gives [the radicand]?" For example, part a would be "What number raised to the 4th power gives 625?"

a. $\sqrt[4]{625}$ **b.** $\sqrt[4]{16}$ **c.** $\sqrt[3]{64}$

d. $\sqrt[5]{32}$ **e.** $\sqrt[3]{27}$ **f.** $\sqrt[3]{\frac{1}{8}}$

g. $\sqrt[5]{243}$ **h.** $\sqrt[4]{256}$ **i.** $\sqrt[3]{\frac{1}{27}}$

j. $\sqrt[3]{125}$ **k.** $\sqrt[3]{0.001}$ **l.** $\sqrt[3]{0.125}$

77. **Distance to the Horizon**

a. We are given the formula $d = \sqrt{3h/2}$ to estimate the distance d seen, in miles, from a height h, in feet, on Earth. The Earth's radius at the equator is 3963 miles. There are 5280 feet in 1 mile. Follow the steps below to show where the formula came from and why the unit change from feet to miles doesn't appear in the formula.

In the figure, the circle is the Earth at the equator. Let r be the radius in miles, $h/5280$ be the height above the surface in miles (so that h is in feet), and d be the distance from the given height to the horizon. (*Geometry fact:* There is a right angle between the line to the horizon and the radius at that point.)

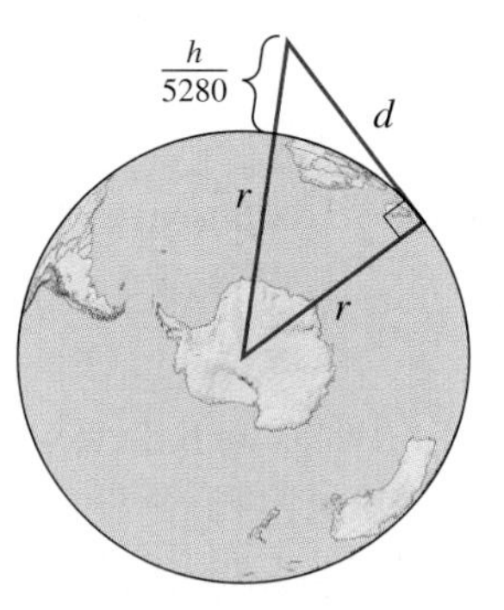

Step 1: Use the Pythagorean theorem to write the formula relating r, $h/5280$, and d. (*Hint:* The hypotenuse is $r + h/5280$ in miles.)

Step 2: Simplify the formula. (Don't forget that binomials like $(a + b)^2$ have three terms.) Note that r^2 drops from each side. Because the $h^2/5280^2$ term is virtually zero for heights of up to 1000 feet, cross off that term. (Scientists and engineers throw out squared terms all the time!) Take the square root of both sides.

Step 3: Substitute the radius of the Earth for r. Divide the resulting coefficient on the h term. Round to two decimal places. The result should be approximately $\frac{3}{2}$. Mission accomplished!

b. To check the formula $d = \sqrt{3h/8}$ for the moon, repeat step 3 for the radius of the moon. The radius of the moon at the equator is 1090 miles.

Blue numbers are core exercises.

▶ 8 Mid-Chapter Test

1. Which of these sets of numbers represent the sides of a right triangle?

a. {4, 6, 8} **b.** {10, 15, 20} **c.** {15, 20, 25}

2. A plane flies due north at 120 miles per hour (mph). There is a 50-mph cross wind from the west. How many miles will the plane travel in 1 hour?

3. The sides of a rectangle are 6 inches and 15 inches. What is the length of the diagonal of the rectangle?

4. Simplify these exponent and radical expressions.

a. $\sqrt{2} \cdot \sqrt{18}$ **b.** $(3\sqrt{2})^2$ **c.** $\sqrt{72}$

d. $(2\sqrt{3})^2$ **e.** $\sqrt{48}$ **f.** $-\sqrt{4}$

g. $\sqrt{-16}$ **h.** $\pm\sqrt{\frac{25}{16}}$ **i.** $\frac{\sqrt{2}}{\sqrt{32}}$

5. Simplify. Assume that $x \geq 0$ and $y > 0$.

a. $\sqrt{3x}\sqrt{27x^3}$ **b.** $\sqrt{x^2y^7}$

c. $\dfrac{\sqrt{x^3y^3}}{\sqrt{x^2y^5}}$ **d.** $\sqrt{\dfrac{x^2y}{xy^5}}$

6. Simplify. Assume that x and y are any real numbers (except for zero in the denominator).

a. $\sqrt{x^3y^7}$ **b.** $\sqrt{\dfrac{3x^2}{12y^4}}$

c. $\sqrt{\dfrac{196x}{x^3}}$ **d.** $\sqrt{x^2 + y^2}$

7. Find the distance between $(2, -3)$ and $(-4, 6)$.

8. Between what two consecutive integers is $\sqrt{135}$?

9. Solve $I = \sqrt{\dfrac{W}{R}}$ for R.

10. Solve by graphing and with algebraic notation: $\sqrt{x + 7} = 4$. For what inputs is the equation defined?

11. Solve $\sqrt{2x} + 3 = 5$. Note any assumptions or conditions.

12. How does the graph of the square root function, $y = \sqrt{x}$, show that the square root of a negative number is undefined in the real numbers?

13. How does the graph of the square root function, $y = \sqrt{x}$, show that the principal square root of a number is positive?

14. One reference shows the circumference of an ellipse as $C \approx 2\pi\sqrt{\frac{1}{2}(R^2 + r^2)}$. Another shows $C \approx \pi\sqrt{2(R^2 + r^2)}$.

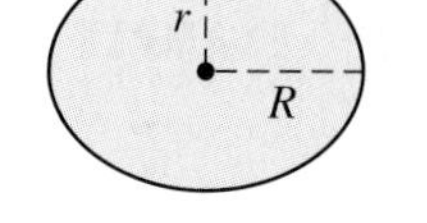

a. For the ellipse shown, with $R = 4$ and $r = 2$, find the circumference with each formula.

b. What can be concluded about the two formulas? Explain.

15. Find the distance between $(-2, 3)$ and $(6, -3)$. At what point does the line between the points cross the x-axis?

▶ 8.4 Graphing and Solving Quadratic Equations

Objectives

- Make a table and graph for a quadratic function.
- Find the x-intercepts, y-intercept, vertex, and axis of symmetry of the graph of a quadratic function.
- Identify a, b, and c from a quadratic equation in standard form, and build a quadratic equation given a, b, and c.
- Solve a quadratic equation from a table and a graph.

WARM-UP

Make an input-output table for each equation, using the integers on the interval $[-2, 4]$ as inputs.

1. $y = x^2 - 3x - 4$ **2.** $y = 4 - x^2$

THIS SECTION INTRODUCES quadratic functions, special features of their graphs, and solving quadratic equations from tables and graphs.

▶ Quadratic Functions

Linear functions are polynomials in which the highest power on the input variable is 1: $y = mx^1 + b$. When the highest power (called the *degree*) on a polynomial equation is 2, we have a quadratic function.

DEFINITION OF QUADRATIC FUNCTION

A **quadratic function** may be written $f(x) = ax^2 + bx + c$, where a, b, and c are real numbers and a is not zero.

The equations in the Warm-up are quadratic equations because the highest power on x is 2, in the x^2 term. The set of inputs, or domain, for a quadratic function is all real numbers. The set of outputs, or range, for a quadratic function depends on the location of the vertex (more on that later this section).

Graphing Quadratic Functions

EXAMPLE 1 Exploring patterns in tables and graphs

a. Plot the points from Tables 7 and 8. Use $y = f(x)$.

TABLE 7

Input x	Output $f(x) = x^2 - 3x - 4$
-2	6
-1	0
0	-4
1	-6
2	-6
3	-4
4	0

TABLE 8

Input x	Output $f(x) = 4 - x^2$
-2	0
-1	3
0	4
1	3
2	0
3	-5
4	-12

b. What is the name of the point on the graph where $x = 0$?
c. What is the name of the point(s) on the graph where $y = 0$?
d. Are any output numbers within a table the same?
e. What is the lowest or highest point on each graph?
f. What is the Δy pattern in each table? What is the same about the patterns in the two tables?
g. Use the patterns in the tables to predict the output for each function when $x = -3$.

SOLUTION See Examples 2 and 3 for answers to parts a to e. See the Answer Box for parts f and g. ◀

VERTEX, AXIS OF SYMMETRY, AND INTERCEPTS The *graph of a quadratic function* is a curve called a **parabola**. The *highest or lowest point on the parabola* is called the **vertex**. The path of the bouncing ball in Figure 18 (repeated from Figure 1 in Chapter 2) is a series of parabolas. The vertex of each parabola is the highest point in the bounce.

FIGURE 18 Height of bouncing ball

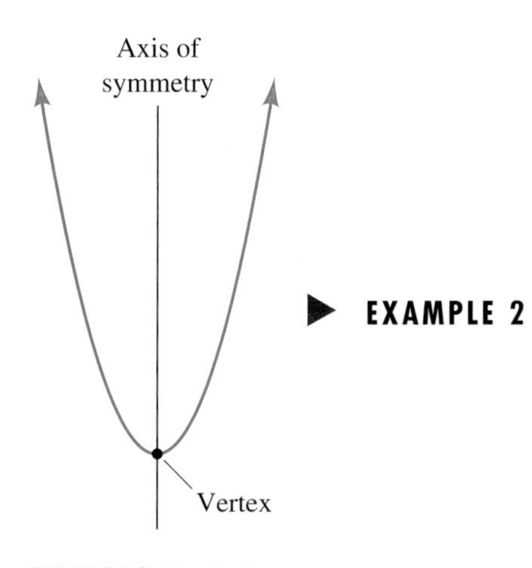

FIGURE 19 Parabola

An **axis**, or *line*, **of symmetry** is a *line across which the graph can be folded so that points on one side of the graph match up with points on the other side of the graph*. The parabolic graphs of quadratic functions have an axis of symmetry with equation $x = n$; see Figure 19. The axis of symmetry is halfway between any pair of inputs having equal outputs (such as the x-intercepts having output $y = 0$). The axis of symmetry passes through the vertex.

▶ **EXAMPLE 2** Reading a graph of a quadratic function The graph of $f(x) = x^2 - 3x - 4$ is shown in Figure 20.

FIGURE 20 $y = x^2 - 3x - 4$

a. What is the y-intercept point?
b. What are the x-intercept points?
c. Give two or three sets of ordered pairs that have the same output. Will this pattern continue? What is the axis of symmetry?
d. What is the vertex?

SOLUTION

a. The y-intercept is -4. The y-intercept point is $(0, -4)$.
b. From Table 7 at $y = 0$, the x-intercepts are $x = -1$ and $x = 4$. The x-intercept points are $(-1, 0)$ and $(4, 0)$.
c. Ordered pairs with the same output include $(0, -4)$ and $(3, -4)$, $(1, -6)$ and $(2, -6)$, as well as the x-intercept points. The axis of symmetry is halfway between the x-intercepts:

$$\frac{-1 + 4}{2} = \frac{3}{2} = 1\frac{1}{2}$$

or $x = 1.5$.
d. The vertex lies on the axis of symmetry, so $x = 1.5$. To find y, we have

$$y = x^2 - 3x - 4 \qquad \text{Substitute } x = 1.5.$$
$$y = (1.5)^2 - 3(1.5) - 4 \qquad \text{Simplify.}$$
$$y = 2.25 - 4.5 - 4$$
$$y = -6.25$$

The vertex is $(1.5, -6.25)$. 

Student Note:
The next time you ride a roller coaster look for parabolas that lift you out of your seat or press you down into your seat!

THINK ABOUT IT 1: When we substitute $x = 0$ or $x = 1.5$ into a function $f(x)$, what notation do we write to describe the substitution?

▶ In many cases, the intercepts, vertex, and axis of symmetry are all you need to sketch the graph of a quadratic function. If needed, one or two other points can be found by substituting x-values into the function.

▶ EXAMPLE 3 Graphing quadratic functions

a. What is the y-intercept point for the graph of $y = 4 - x^2$?
b. What are the x-intercept points?
c. What is the axis of symmetry?
d. What is the vertex?
e. Graph $y = 4 - x^2$.

SOLUTION **a.** Let $x = 0$ in $y = 4 - x^2$. The y-intercept is $y = 4$. The y-intercept point is $(0, 4)$.

b. Let $y = 0$ in $y = 4 - x^2$.

$$0 = 4 - x^2$$
$$x^2 = 4$$

The x-intercepts are $x = -2$ and $x = 2$. The x-intercept points are $(-2, 0)$ and $(2, 0)$.

c. The axis of symmetry is halfway between the x-intercepts:

$$\frac{-2 + 2}{2} = 0$$

so $x = 0$, the y-axis.

d. The vertex lies on the axis of symmetry (y-axis), so $x = 0$.

$$y = 4 - x^2 \qquad \text{Let } x = 0.$$
$$y = 4$$

The vertex is $(0, 4)$.

e. The graph is in Figure 21.

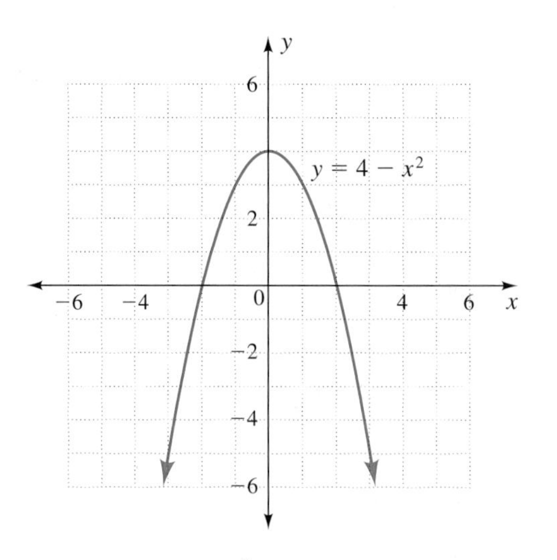

FIGURE 21 $y = 4 - x^2$ ◀

THINK ABOUT IT 2: What part of the function in Example 3 gives the y-intercept? What part of the function causes the graph to open downward?

Before we summarize graphing, we practice identifying the parts a, b, and c in the quadratic equation.

Interpreting *a*, *b*, and *c* in a Quadratic Equation

With linear functions, we write $y = mx + b$ to identify the slope m and the y-intercept b. Similarly, with quadratic functions, we obtain certain information from the letters a, b, and c. In some applications, we are given an equation and need to identify a, b, and c.

> To identify a, b, and c, arrange a quadratic equation into standard form. For a one-variable equation, standard form is
>
> $$ax^2 + bx + c = 0$$
>
> For a two-variable equation, standard form is
>
> $$y = ax^2 + bx + c$$

EXAMPLE 4 Identifying *a*, *b*, and *c* in the quadratic equation What are the values of a, b, and c in each equation? In parts e and f, the input variable is t.

a. $3x^2 + 5x + 2 = 0$ **b.** $x^2 = 16$ **c.** $y = x^2 + 3x - 4$
d. $4 - y = x^2$ **e.** $h = -16t^2 + 48t$ **f.** $h = -\frac{1}{2}gt^2 + v_0t + h_0$

SOLUTION **a.** $a = 3, b = 5, c = 2$

b. Rearranging to $x^2 - 16 = 0$ gives $a = 1, b = 0, c = -16$.

c. $a = 1, b = 3, c = -4$

d. Rearranging to $y = -x^2 + 4$ gives $a = -1, b = 0, c = 4$.

e. $a = -16, b = 48, c = 0$

f. $a = -\frac{1}{2}g, b = v_0, c = h_0$ ◀

In other applications, we are given a, b, and c and must place them into a quadratic equation. We practice this process in Example 5.

EXAMPLE 5 Finding quadratic equations given *a*, *b*, and *c* Write two-variable equations given the following information.

a. $a = 1, b = -2, c = 1$
b. $a = 1, b = 0, c = -9$
c. $a = 4, b = 0, c = 0$
d. $a = -1, b = 0, c = 0$

SOLUTION **a.** $y = x^2 - 2x + 1$

b. $y = x^2 - 9$

c. $y = 4x^2$

d. $y = -x^2$ ◀

SUMMARY EXAMPLE OF GRAPHING In Example 6, we compare two graphs to observe what causes a graph to open upward or open downward.

EXAMPLE 6 Graphing quadratic functions

a. What is the y-intercept point for $y = x^2$ and for $y = -x^2$?
b. What are the x-intercept points?
c. For each function, give two or three sets of ordered pairs that have the same output. Will this pattern continue? What is the axis of symmetry?
d. What is the vertex?
e. What part of the function $f(x) = ax^2 + bx + c$ causes the parabola to open upward or open downward?
f. Graph $y = x^2$ and $y = -x^2$.

SOLUTION **a.** At $x = 0$, $y = 0$. The y-intercept point is the origin, (0, 0) for both graphs.

b. The x-intercept point (where $x^2 = 0$ and $-x^2 = 0$) is also the origin.

c. Any pair of opposite inputs gives matching outputs; for example, -2 and 2 both give 4. The ordered pairs $(-2, 4)$ and $(2, 4)$ are in $y = x^2$. The ordered pairs $(-2, -4)$ and $(2, -4)$ are in $y = -x^2$. The axis of symmetry is the y-axis, $x = 0$.

d. The vertex for both graphs is the origin, (0, 0).

e. The $a = -1$ coefficient of x^2 in $y = -x^2$ gives the negative values, causing the graph to open downward instead of upward, as $y = x^2$ does.

f. The graphs are shown in Figure 22.

Student Note:
We can say that the graph for $y = -x^2$ is a mirror reflection of the graph for $y = x^2$ across the x-axis.

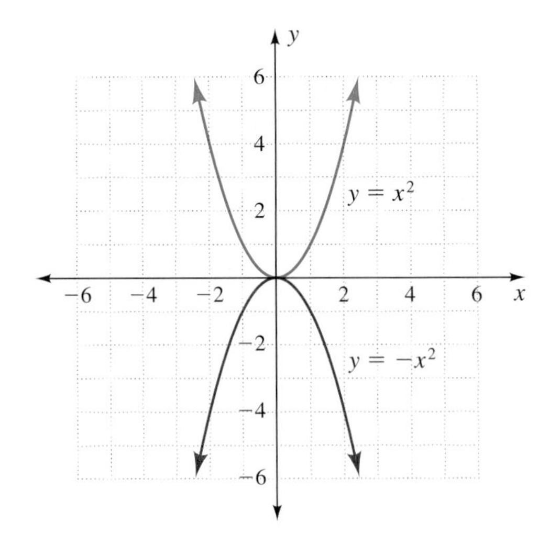

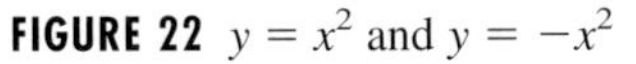

FIGURE 22 $y = x^2$ and $y = -x^2$ ◀

The special parts of the graph of the quadratic function, $f(x) = ax^2 + bx + c$, are summarized below.

SPECIAL FEATURES OF THE QUADRATIC FUNCTION GRAPH

1. The y-intercept is at $x = 0$. In function notation, the y-intercept is $f(0) = c$.
2. The x-intercepts are where $y = 0$. In function notation, the x-intercepts are where $f(x) = 0$.
3. The graph has an axis (or line) of symmetry, $x = n$. To find the axis of symmetry, average the inputs for two ordered pairs with the same output. Symmetry or the change in y can be used to extend the table once one side of the parabola is known.
4. The lowest (or highest) point on the parabolic graph is the vertex. The x-coordinate for the vertex is the same as x in the axis of symmetry.
5. The coefficient of x^2, a, gives the direction of the graph. A positive value for the letter a indicates that the graph opens upward. A negative value for the letter a indicates that the graph opens downward.

RANGE The vertex of the parabola helps us to find the set of outputs, or range, for the quadratic function.

FINDING THE RANGE FOR A QUADRATIC FUNCTION

> If the vertex is (x, n), then the range is
>
> $y \geq n$ for a parabola that opens up
>
> $y \leq n$ for a parabola that opens down

▶ **EXAMPLE 7** Finding the range Use the vertex to find the range for these two quadratic functions.

a. $f(x) = x^2 - 3x - 4$ **b.** $f(x) = 4 - x^2$

SOLUTION **a.** From Example 2, the vertex for $f(x) = x^2 - 3x - 4$ is $(1.5, -6.25)$. The parabola opens upward, so the range of the function includes all y values from a low of -6.25: $y \geq -6.25$.

b. From Example 3, the vertex for $f(x) = 4 - x^2$ is $(0, 4)$. The parabola opens downward, so the range includes all y values from a high of 4: $y \leq 4$. ▶

Solving Quadratic Equations with Tables and Graphs

There are at least five different ways to solve a quadratic equation. In this section, we solve equations from tables and graphs. In the next two sections, we use algebraic notation to solve quadratic equations in three other ways.

Like other equations, quadratic equations can be solved with tables by looking for inputs that give the required output.

To solve with graphs, we look for inputs that correspond to the intersection of the graphs of the left and right sides of the equation.

▶ **EXAMPLE 8** Solving quadratic equations by table and graph Use a table and graph to solve these equations.

a. $x^2 - 3x - 4 = 0$ **b.** $x^2 - 3x - 4 = -6$ **c.** $x^2 - 3x - 4 = 6$

SOLUTION The table is Table 9, and the graph is shown in Figure 23.

TABLE 9 Solve from a table

Input x	Output $y = x^2 - 3x - 4$
-2	6
-1	0
0	-4
1	-6
2	-6
3	-4
4	0

FIGURE 23 Solve from a graph

a. In Table 9, $y = 0$ when $x = -1$ and $x = 4$. On the graph in Figure 23, $y_2 = 0$ is the x-axis, and the intersections with the parabola are at $x = -1$ and $x = 4$.

Check:

$(-1)^2 - 3(-1) - 4 \stackrel{?}{=} 0$ ✓

$(4)^2 - 3(4) - 4 \stackrel{?}{=} 0$ ✓

b. In Table 9, $y = -6$ when $x = 1$ and $x = 2$. On the graph in Figure 23, $y_3 = -6$ intersects the parabola at $x = 1$ and $x = 2$.

Check:

$(1)^2 - 3(1) - 4 \stackrel{?}{=} -6$ ✓

$(2)^2 - 3(2) - 4 \stackrel{?}{=} -6$ ✓

c. Table 9 shows only one place, $x = -2$, where the output is 6. However, by symmetry, there will be another 6 at $x = 5$. On the graph in Figure 23, $y_4 = 6$ intersects the parabola at $x = -2$ and $x = 5$.

Check:

$(-2)^2 - 3(-2) - 4 \stackrel{?}{=} 6$ ✓

$(5)^2 - 3(5) - 4 \stackrel{?}{=} 6$ ✓

GRAPHING CALCULATOR SOLUTION

Enter $Y_1 = X^2 - 3X - 4$ and $Y_2 = 0$. To solve by table, set up the table with **TblStart** $= -2$ and **ΔTbl** $= 1$. View the table, and compare it with Table 9.

To solve by graph, set the window for X at $[-6, 6]$, scale 1, and for Y at $[-7, 7]$, scale 1. Graph. Compare the result with Figure 23. Use Intersection to solve the equation. Change Y_2 to -6 and then 6 to find the other solutions. ◀

▶ Application: Height of Falling Items

The height h, at any given time t, of an object dropped from a starting height h_0 is described by the formula

$$h = h_0 - \tfrac{1}{2}gt^2$$

The letter g represents the acceleration due to gravity and may be approximated by 32.2 feet per second squared. (The square is on the second, not on the feet.)

▶ **EXAMPLE 9** Solving quadratic equations Write equations to find at what times a sugar cube dropped from a height of 500 feet will be at 400 feet, 300 feet, 200 feet, 100 feet, and zero feet. Solve the equations by table and graph.

SOLUTION The table is Table 10, and the graph is shown in Figure 24.

Student Note:
Because the horizontal axis shows time, the graph does not show the path of the sugar cube.

TABLE 10

Time (sec)	Height (ft)
0	500
1	483.9
2	435.6
3	355.1
4	242.4
5	97.5
6	−79.6

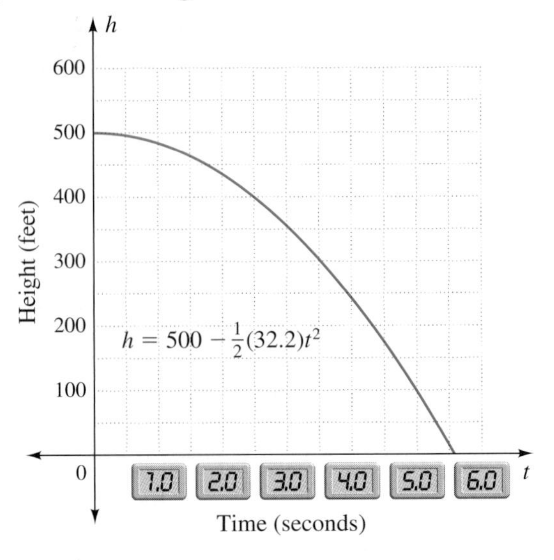

FIGURE 24 Height as a function of time

From $h = h_0 - \frac{1}{2}gt^2$, we get the following equations to be solved. Estimated solutions following each equation are from the table and graph.

Student Note:
The height $h = 0$ is ground level.

	Corresponding Input Value:
$400 = 500 - \frac{1}{2}(32.2)t^2$	$t \approx 2\frac{1}{2}$ sec
$300 = 500 - \frac{1}{2}(32.2)t^2$	$t \approx 3\frac{1}{2}$ sec
$200 = 500 - \frac{1}{2}(32.2)t^2$	$t \approx 4\frac{1}{2}$ sec
$100 = 500 - \frac{1}{2}(32.2)t^2$	$t \approx 5$ sec
$0 = 500 - \frac{1}{2}(32.2)t^2$	$t \approx 5\frac{1}{2}$ sec

For related information, see Examples 9 and 10 of Section 8.3 (page 479). ◀

THINK ABOUT IT 3: Does the sugar cube drop the same amount each second? What part of the equation would explain any change in speed?

ANSWER BOX

Warm-up: See Tables 7 and 8. **Example 1: f.** The Δy pattern is consecutive even numbers for $f(x) = x^2 - 3x - 4$ and consecutive odd numbers for $f(x) = 4 - x^2$. Both patterns show consecutive numbers. **g.** The consecutive even number pattern for Δy suggests that $f(x) = 14$ at $x = -3$ for $f(x) = x^2 - 3x - 4$. The matching of outputs in the table suggests that $f(x) = -5$ at $x = -3$ for $f(x) = 4 - x^2$. **Think about it 1:** $f(0), f(1.5)$ **Think about it 2:** $f(0)$ in $f(x) = 4 - x^2$ or 4; The subtraction of x^2 creates large negative values as x takes on numbers away from the origin. This causes the parabola to open downward. **Think about it 3:** The sugar cube drops about 16 feet in the first second, 48 feet in the second, 81 feet in the third, 113 feet in the fourth, and 144 feet in the fifth. The cube is speeding up. The t^2 is causing this acceleration.

▶ 8.4 Exercises

Find the x- and y-intercept points, axis of symmetry, and vertex for each of the parabolas shown in Exercises 1 to 4.

1. $y = x^2 + x - 6$

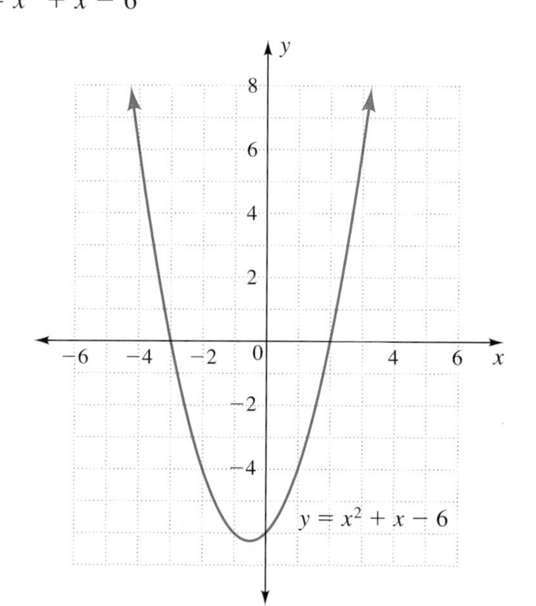

2. $y = x^2 + 2x - 3$

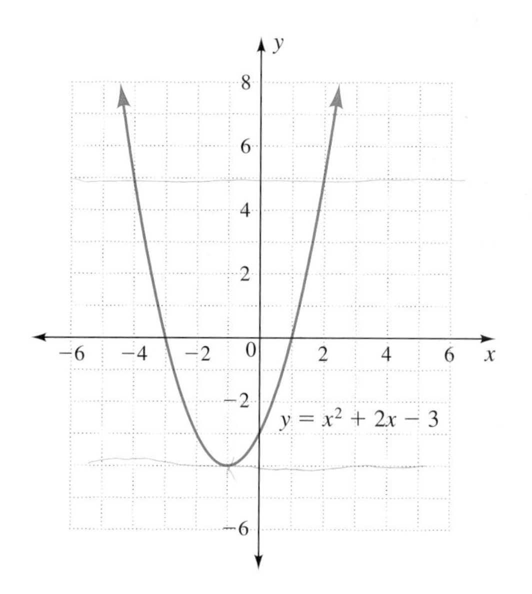

Blue numbers are core exercises.

3. $y = 5x - x^2$

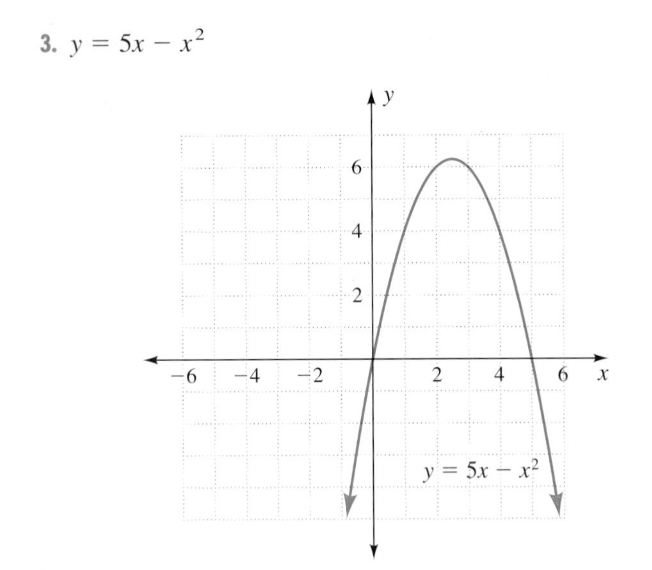

4. $y = 2 - x - x^2$

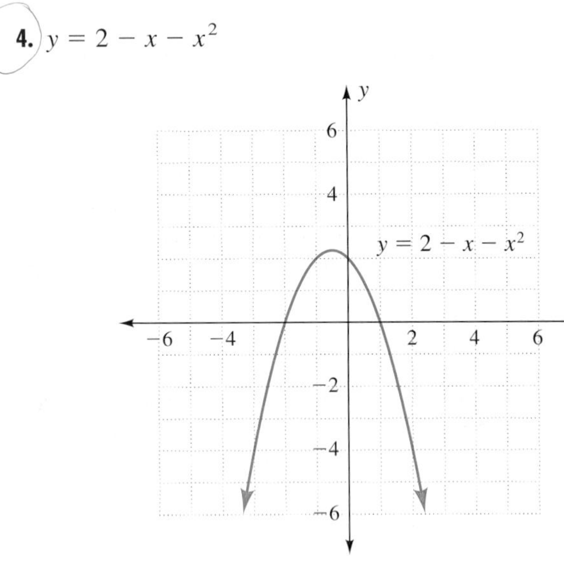

Make a table and graph for the functions in Exercises 5 to 12. Find the *x*- and *y*-intercept points, axis of symmetry, and vertex for each.

5. $y = x^2 - 6x + 7$ **6.** $y = x^2 - 3x - 10$

7. $y = x^2 + x - 12$ **8.** $y = x^2 - 2x - 8$

9. $y = x - x^2$ **10.** $y = 2x - x^2$

11. $y = 4x - 2x^2$ **12.** $y = 3x - x^2$

For Exercises 13 to 16, find the range of the function graphed in the given exercise.

13. Exercise 1 **14.** Exercise 2

15. Exercise 3 **16.** Exercise 4

In Exercises 17 to 32, name *a*, *b*, and *c*.

17. $y = 2x^2 + 3x + 1$ **18.** $y = 3x^2 + 2x - 1$

19. $r^2 - 4r + 4 = 0$ **20.** $r^2 - 6r + 9 = 0$

21. $y = x^2 - 4$ **22.** $y = x^2 - 25$

23. $4t^2 - 8t = 0$ **24.** $3t^2 - 27t = 0$

25. $4 = x - x^2$ **26.** $x = 6 - 2x^2$

27. $x^2 = x - 1$ **28.** $x^2 = 3 - x$

29. Height after time t: $h = -0.5gt^2 + vt + s$

30. Angle swept in time t for circular motion: $A = 0.5\alpha t^2 + wt$

31. Area of circle: $A = \pi r^2$

32. Surface area of cylinder of height 1: $SA = 2\pi r^2 + 2\pi r$

In Exercises 33 to 38, write a two-variable quadratic equation with these coefficients.

33. $a = 4, b = 4, c = 1$ **34.** $a = 2, b = -1, c = -6$

35. $a = 9, b = 0, c = -16$ **36.** $a = 25, b = 0, c = -1$

37. $a = 3, b = 6, c = 0$ **38.** $a = 5, b = 15, c = 0$

Solve the equations in Exercises 39 to 42 from the graphs in Exercises 1 to 4.

39. a. $x^2 + x - 6 = 6$ **b.** $x^2 + x - 6 = -4$

c. $x^2 + x - 6 = -8$ **d.** $x^2 + x - 6 = 0$

40. a. $x^2 + 2x - 3 = 5$ **b.** $x^2 + 2x - 3 = -3$

c. $x^2 + 2x - 3 = -4$ **d.** $x^2 + 2x - 3 = 0$

41. a. $5x - x^2 = 6$ **b.** $5x - x^2 = 8$

c. $5x - x^2 = 4$ **d.** $5x - x^2 = 0$

42. a. $2 - x - x^2 = 2$ **b.** $2 - x - x^2 = 4$

c. $2 - x - x^2 = -4$ **d.** $2 - x - x^2 = 0$

In Exercises 43 and 44, assume that the formula $h = h_0 - \frac{1}{2}gt^2$ holds true. Let g equal 32.2 feet per second squared.

43. Suppose a drop of water starts at the top of Ribbon Falls in Yosemite National Park and falls 1612 feet. How long will it take the drop to reach halfway? To reach the bottom?

44. Suppose a drop of water starts at the top of the American side of Niagara Falls and falls 182 feet. How long will it take the drop to reach halfway? To reach the bottom?

▶ Projects

45. Largest Product

a. List ten pairs of numbers that add to 12. Multiply the numbers in each pair to get a product, y. Describe any patterns.

b. Let x be the first number in each pair. Let y be the product of each pair. Graph the ordered pairs (x, y).

c. Let x be the first number in each pair. Write an expression for the second number in terms of x. Write an equation describing the products, y.

d. What is the largest product?

Blue numbers are core exercises.

46. Fixed Perimeter and Area Cut a piece of string 30 inches long. Form it into a rectangle with width 1 inch.

a. Measure the length of the rectangle, and record it in the table below. Calculate the area of the rectangle.

Width	Length	Area
1 in.		
2 in.		
3 in.		
4 in.		
5 in.		
6 in.		

b. Form another rectangle with width 2 inches. Measure the length of the new rectangle, and record it. Calculate the area.

c. Repeat this procedure to complete the table. Add more rows until you reach a largest possible area.

d. On one set of axes, graph the length and width pairs, using length as x and width as y. What is the equation of the graph?

e. On another set of axes, graph the length and area pairs, using length as x and area as y. What is the equation of the graph?

f. Where on the length and area graph is the point representing the largest area?

47. A Number Pattern

a. What do bowling, pocket billiards, and stacks of paint cans at a paint store have in common? (*Hint:* How many bowling pins are in a full rack? How many balls are in a rack of pocket billiards? How many cans are pictured in the illustration?)

b. Complete the table, where the input represents the number of rows in the stack of cans and the output represents the total number of cans in the stack for that many rows.

Row of Cans x	Total Number of Cans y
1	
2	
3	
4	
5	
6	
7	

c. Find a rule for the table. (*Hint:* It is a quadratic function and involves consecutive numbers.) To find out if your rule is correct, see if it gives 300 cans for 24 rows. Use your rule to find the number of cans in 15 rows.

48. Lines and Points

a. Complete this table for the number of distinct straight lines defined by each of the following numbers of points. (Two points define one straight line.)

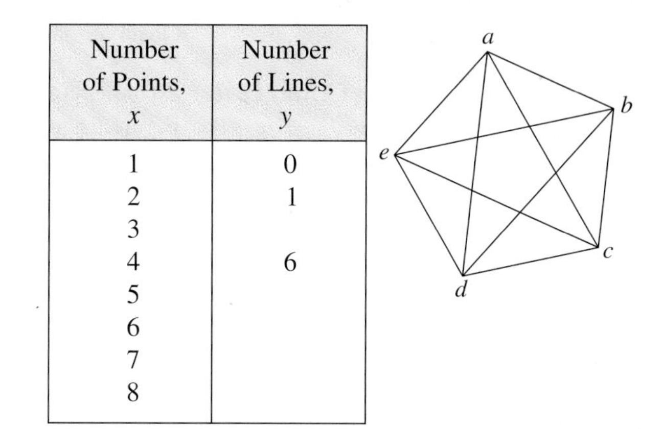

Number of Points, x	Number of Lines, y
1	0
2	1
3	
4	6
5	
6	
7	
8	

b. What is the pattern in the outputs?

c. What is the input-output rule (in words, if not a function)? (*Hint:* The rule is quadratic.)

▶ 8.5 Solving Quadratic Equations by Taking the Square Root or by Factoring

Objectives

- Solve quadratic equations by taking the square root of both sides.
- Solve quadratic equations by factoring and applying the zero product rule.

WARM-UP

1. Simplify these absolute value expressions.

a. $|-2|$ **b.** $|2|$ **c.** $|-3|$

2. Use guess and check to solve these absolute value equations.

a. $|x| = 2$ **b.** $|x| = 3$

3. Factor these expressions.

a. $x^2 - 4$ **b.** $x^2 - 3x - 4$

c. $t^2 - 3t + 2$ **d.** $25x^2 - 16$

e. $3x^2 - 75$ **f.** $x^2 - 3x$

g. $-16t^2 + 48t$ **h.** $x^2 - x - 2$

THIS SECTION INTRODUCES two methods for solving quadratic equations with algebraic notation: taking the square root of both sides and factoring. We use the zero product rule and again apply the techniques to applications.

▶ Solving Quadratic Equations by Taking the Square Root

Taking the square root is a way to solve quadratic equations for which $b = 0$ in $ax^2 + bx + c = 0$. These quadratic equations are in the form $ax^2 + c = 0$. In the first example, we look at the graphical results from solving equations with $b = 0$. Observe that some solutions include both positive and negative numbers.

▶ **EXAMPLE 1** Exploring graphical solutions After completing Table 11, use the graph of $y = x^2$ in Figure 25 to solve the equations.

TABLE 11

Input x	Output $y = x^2$
−3	
−2	
−1	
0	
1	
2	
3	

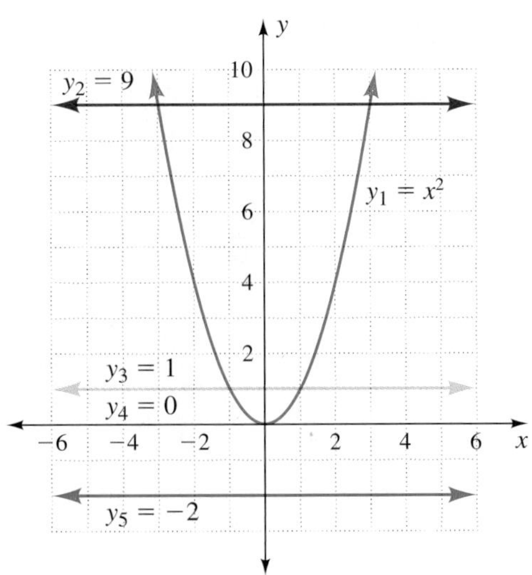

FIGURE 25 Solve $x^2 = n$

a. $x^2 = 9$ **b.** $x^2 = 1$ **c.** $x^2 = 0$ **d.** $x^2 = -2$

SOLUTION See the Answer Box. ◀

▶ As noted in Sections 8.2 and 8.3, so long as we observe the restriction that the input to a square root is zero or positive and the output from taking the square root is zero or positive, we can take the square root of both sides of an equation.

> To solve an equation, we can take the square root of both sides.

Recall from page 481 that $\sqrt{x^2} = |x|$ for any real number x.

▶ **EXAMPLE 2** Solving by taking the square root of both sides Solve $3x^2 - 75 = 0$.

SOLUTION

$3x^2 - 75 = 0$	Add 75 to both sides.
$3x^2 = 75$	Divide both sides by 3.
$x^2 = 25$	Take the square root of both sides.
$\sqrt{x^2} = \sqrt{25}$	Simplify.
$\lvert x\rvert = 5$	$\sqrt{x^2} = \lvert x\rvert$, the formal step
$x = 5$ or $x = -5$	There are two solutions.

Check:

$3(5)^2 - 75 \stackrel{?}{=} 0$ ✓

$3(-5)^2 - 75 \stackrel{?}{=} 0$ ✓ ◀

In earlier sections of this chapter, we ignored the negative solution because we were working with triangles or other objects, which have only positive dimensions. In this section, we want both solutions. In Example 2, the absolute value gives a formally correct answer. If it is too formal, just remember this:

> The equation $x^2 = n$ has two solutions:
>
> $x = +\sqrt{n}$ and $x = -\sqrt{n}$

▶ **EXAMPLE 3** Solving by taking the square root of both sides Solve $25x^2 - 16 = 0$.

SOLUTION We solve first for x^2:

$25x^2 - 16 = 0$	Add 16 to both sides.
$25x^2 = 16$	Divide both sides by 25.
$x^2 = \frac{16}{25}$	Take the square root of both sides.
$\sqrt{x^2} = \sqrt{\frac{16}{25}}$	Simplify.
$\lvert x\rvert = \frac{4}{5}$	$\sqrt{x^2} = \lvert x\rvert$
$x = \frac{4}{5}$ or $x = -\frac{4}{5}$	

Check:

$25(\frac{4}{5})^2 - 16 \stackrel{?}{=} 0$ ✓

$25(-\frac{4}{5})^2 - 16 \stackrel{?}{=} 0$ ✓ ◀

SOLVING BY TAKING A SQUARE ROOT

To solve by taking a square root:

1. Look for $b = 0$ in $ax^2 + bx + c = 0$.
2. Solve for x^2.
3. Take the square root of both sides. Simplify, and write the solutions.
4. Check.

Solving Quadratic Equations by Factoring

In our second solution method, we start by looking at products that multiply to zero.

EXAMPLE 4 Exploring zero products Describe the role of zero in these equations and expressions.

a. $A \cdot 5 = 0$

b. $-5 \cdot B = 0$

c. $(x - 4)(x + 1)$ if $x = 4$

d. $(x - 4)(x + 1)$ if $x = -1$

SOLUTION

a. $A = 0$, because zero multiplied by 5 gives a zero product.

b. $B = 0$, because -5 multiplied by zero gives a zero product.

c. $(4 - 4)(4 + 1) = 0 \cdot 5 = 0$
The product $(x - 4)(x + 1)$ is zero if $x = 4$.

d. $(-1 - 4)(-1 + 1) = (-5) \cdot 0 = 0$
The product $(x - 4)(x + 1)$ is zero if $x = -1$.

Parts a and b show that if two numbers multiply to zero, then one of the numbers is zero. Parts c and d show that if one of two expressions in a product is zero, then the product is zero. ◀

Example 4 suggests the **zero product rule**.

ZERO PRODUCT RULE

If the product of two expressions is zero, then either one or the other expression is zero; that is,

$$\text{if } A \cdot B = 0, \text{ then either } A = 0 \text{ or } B = 0.$$

A and B usually represent monomial or binomial factors such as x, $(x + 2)$, $(x - 3)$, or $(2x - 5)$.

▶ In Example 5, look for numbers in the factors that make the entire factor have a zero value.

EXAMPLE 5 Finding numbers that make a factor equal to zero For what values of x are these expressions true?

a. $(x + 2)(x - 2) = 0$

b. $(x - 4)(x + 7) = 0$

c. $(2x + 1)(3x - 2) = 0$

SOLUTION

a. Either $x + 2 = 0$ or $x - 2 = 0$
$x = -2$ or $x = 2$

b. Either $x - 4 = 0$ or $x + 7 = 0$
$x = 4$ or $x = -7$

c. Either $2x + 1 = 0$ or $3x - 2 = 0$
$2x = -1$ $3x = 2$
$x = -\frac{1}{2}$ or $x = \frac{2}{3}$ ◀

▶ **EXAMPLE 6** Solving by factoring Solve $x^2 + 3x - 28 = 0$.

SOLUTION We will factor the left side and then set each of the factors equal to zero.

$$x^2 + 3x - 28 = 0$$

$$(x + 7)(x - 4) = 0$$

Either $x + 7 = 0$ or $x - 4 = 0$

$x = -7$ or $x = 4$

Check:

$(-7)^2 + 3(-7) - 28 \stackrel{?}{=} 0$ ✓

$4^2 + 3(4) - 28 \stackrel{?}{=} 0$ ✓ ◀

SOLVING BY FACTORING

To solve by factoring:

1. Write the equation in standard form, $ax^2 + bx + c = 0$.
2. Factor the expression on the left.
3. Use the zero product rule to write two factor equations.
4. Solve each factor equation.
5. Check.

▶ **EXAMPLE 7** Solving by factoring Solve $x^2 = 3x$ by factoring.

SOLUTION

$x^2 = 3x$ Change to standard form.

$x^2 - 3x = 0$ Factor the left side.

$x(x - 3) = 0$ Use the zero product rule to set each factor equal to zero.

Either $x = 0$ or $x - 3 = 0$ Solve the factor equations.

$x = 0$ or $x = 3$

Check:

$(0)^2 \stackrel{?}{=} 3 \cdot 0$ ✓

$3^2 \stackrel{?}{=} 3 \cdot 3$ ✓ ◀

▶ In Examples 8 and 9, equations from Examples 2 and 3 are solved by factoring. These examples show that many equations can be solved in several ways.

▶ **EXAMPLE 8** Solving by factoring Solve $3x^2 - 75 = 0$ by factoring.

SOLUTION We factor the left side by first removing the greatest common factor.

$3(x^2 - 25) = 0$

$3(x - 5)(x + 5) = 0$ Because $3 \neq 0$, set the other two factors equal to zero.

Either $x - 5 = 0$ or $x + 5 = 0$ Solve the factor equations.

$x = 5$ or $x = -5$

Check:

$3(5)^2 - 75 \stackrel{?}{=} 0$ ✓

$3(-5)^2 - 75 \stackrel{?}{=} 0$ ✓ ◀

▶ Example 9 shows that extra steps are needed to solve for x when the equation contains factors of the form $ax + b$.

▶ EXAMPLE 9 Solving by factoring Solve $25x^2 = 16$ by factoring.

SOLUTION

$25x^2 = 16$	First change to standard form.
$25x^2 - 16 = 0$	To factor, observe that $25x^2 - 16$ is the difference of squares.
$(5x - 4)(5x + 4) = 0$	Set the factors equal to zero.
Either $5x - 4 = 0$ or $5x + 4 = 0$	Solve the factor equations.
$5x = 4$ $\quad$ $5x = -4$	
$x = \frac{4}{5}$ or $x = -\frac{4}{5}$	

Check:

$25(\frac{4}{5})^2 \stackrel{?}{=} 16$ ✓

$25(-\frac{4}{5})^2 \stackrel{?}{=} 16$ ✓ ◀

▶ Applications

HEIGHT AND TIME The falling sugar cube in Example 9 of Section 8.4 (page 492) gave height in terms of time for an object already a given distance above the ground. In the following example, the formula gives height in terms of time for an object traveling up from the ground and back down again.

▶ EXAMPLE 10 Finding the time required to reach given heights: spraying water Suppose a connector in a high-pressure water system has broken and a fine spray of water is shooting straight up into the air, according to the formula

$$h = -16t^2 + 48t$$

where h is height and t is time. In how many seconds after leaving the break will a droplet of water reach the 32-foot level? Solve from the graph in Figure 26 and by factoring.

Student Note:
The parabola is not the path of the water, because the horizontal axis is in units of time.

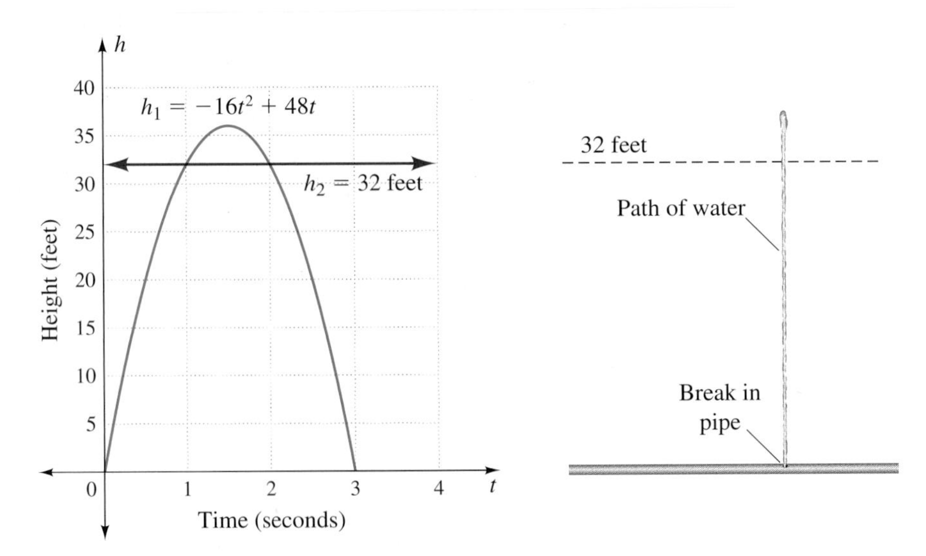

FIGURE 26 Height as a function of time

SOLUTION According to the graph, a water droplet reaches 32 feet at $t = 1$ sec on the way up and again at $t = 2$ sec on the way down.

The equation to solve by factoring is $32 = -16t^2 + 48t$.

$32 = -16t^2 + 48t$ — Change to equal zero.

$16t^2 - 48t + 32 = 0$ — Factor out the greatest common factor.

$16(t^2 - 3t + 2) = 0$ — Factor the remaining trinomial.

$16(t - 2)(t - 1) = 0$ — Use the zero product rule.

Either $(t - 2) = 0$ or $(t - 1) = 0$ — Solve the factor equations.

$t = 2$ or $t = 1$

Check:

$32 \stackrel{?}{=} -16(2)^2 + 48(2)$

$32 \stackrel{?}{=} -16(1)^2 + 48(1)$ ✓ ◀

The graph in Figure 26 does not show the path of the water. For the graph to show such a path, the x-axis would have to be labeled in feet instead of seconds. The path of the water in this example goes straight up and straight back down, as shown in the drawing to the right of the graph.

USING QUADRATIC EQUATIONS IN SOLVING RADICAL EQUATIONS In Example 11, we return to solving equations by squaring both sides. The example shows why it is important to check solutions to an equation.

▶ EXAMPLE 11 Solving an equation containing square roots

a. Find the inputs for which $y = \sqrt{3 - x}$ is defined.

b. Solve $\sqrt{3 - x} = x - 1$ using symbols.

c. Solve $\sqrt{3 - x} = x - 1$ by graphing.

SOLUTION **a.** The equation is defined for positive radicands: $3 - x \geq 0$. The solution to this inequality is $3 \geq x$, or $x \leq 3$.

b. Solve $\sqrt{3 - x} = x - 1$ by squaring both sides.

$\sqrt{3 - x} = x - 1$ — Square both sides.

$(\sqrt{3 - x})^2 = (x - 1)^2$ — Simplify.

$3 - x = x^2 - 2x + 1$ — Change to equal zero.

$0 = x^2 - x - 2$ — Factor.

$0 = (x + 1)(x - 2)$ — Use the zero product rule.

Either $(x + 1) = 0$ or $(x - 2) = 0$ — Solve the factor equations.

$x = -1$ or $x = 2$

Check:

$\sqrt{3 - (-1)} \stackrel{?}{=} (-1) - 1$ False

$\sqrt{3 - (2)} \stackrel{?}{=} 2 - 1$ ✓

The input $x = 2$ gives a true statement. The input $x = -1$ gives $2 = -2$, which is false. The solution $x = -1$ must be discarded.

c. Figure 27 shows the intersection of $y_1 = \sqrt{3 - x}$ and $y_2 = x - 1$ at $x = 2$, but it shows no intersection at $x = -1$. This confirms the single solution found symbolically.

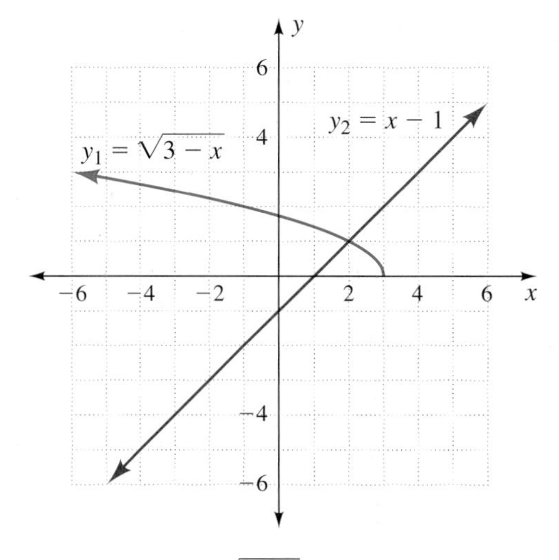

FIGURE 27 Solve $\sqrt{3 - x} = x - 1$

A solution that does not satisfy the original equation is an **extraneous root**. In Example 11, $x = -1$ made a false statement in $\sqrt{3 - x} = x - 1$.

ANSWER BOX

Warm-up: **1. a.** 2 **b.** 2 **c.** 3 **2. a.** $x = 2, x = -2$ **b.** $x = 3, x = -3$ **3. a.** $(x - 2)(x + 2)$ **b.** $(x - 4)(x + 1)$ **c.** $(t - 1)(t - 2)$ **d.** $(5x - 4)(5x + 4)$ **e.** $3(x - 5)(x + 5)$ **f.** $x(x - 3)$ **g.** $-16t(t - 3)$ **h.** $(x - 2)(x + 1)$ **Example 1:** The missing table entries are 9, 4, 1, 0, 1, 4, and 9. **a.** $x = \pm 3$ **b.** $x = \pm 1$ **c.** $x = 0$ **d.** no real-number solutions

▶ 8.5 Exercises

Solve the equations in Exercises 1 to 18. The variable x is any real number (except zero in a denominator).

1. $x^2 = 5$

2. $x^2 = 10$

3. $2x^2 = 14$

4. $3x^2 = 33$

5. $x^2 = \frac{4}{25}$

6. $x^2 = \frac{36}{49}$

7. $100x^2 = 4$

8. $100x^2 = 25$

9. $49x^2 - 225 = 0$

10. $144x^2 = 169$

11. $36x^2 = 121$

12. $121x^2 - 64 = 0$

13. $75x^2 - 27 = 0$

14. $75x^2 - 12 = 0$

15. $\frac{3}{x} = \frac{x}{27}$

16. $\frac{2}{x} = \frac{x}{32}$

17. $\frac{x}{4} = \frac{9}{x}$

18. $\frac{x}{6} = \frac{24}{x}$

For what values of x are the equations in Exercises 19 to 24 true?

19. $(x - 4)(x + 4) = 0$

20. $(x + 3)(x - 5) = 0$

21. $(2x - 1)(3x + 2) = 0$

22. $(3x + 4)(2x - 1) = 0$

23. $(2x - 5)(x + 2) = 0$

24. $(2x + 1)(3x - 5) = 0$

Solve the equations in Exercises 25 to 52 by factoring.

25. $x^2 - 4 = 0$

26. $x^2 - 3x - 4 = 0$

27. $x^2 + x - 6 = 0$

28. $x^2 + 2x - 3 = 0$

29. $x^2 - 2x - 15 = 0$

30. $x^2 - 8x + 15 = 0$

31. $x^2 + 3x = 0$

32. $x^2 - 4x = 0$

Blue numbers are core exercises.

33. $2x^2 = -x$

34. $3x^2 = 2x$

35. $x^2 - 4x = 12$

36. $x^2 + x = 12$

37. $2x^2 = x + 3$

38. $2x^2 = 5x + 3$

39. $x^2 = x + 12$

40. $x^2 = 15 - 2x$

41. $x^2 + 6 = 7x$

42. $x^2 = 4x - 3$

43. $2x^2 + 3x = 5$

44. $2x^2 + 3 = 7x$

45. $4x^2 - 25 = 0$

46. $9x^2 - 4 = 0$

47. $3x^2 - 12 = 0$

48. $5x^2 - 80 = 0$

49. $\dfrac{x + 2}{2} = \dfrac{5}{x - 1}$

50. $\dfrac{x - 3}{5} = \dfrac{1}{x + 1}$

51. $\dfrac{1}{x - 3} = \dfrac{x - 2}{2}$

52. $\dfrac{3}{x - 4} = \dfrac{x + 1}{2}$

53. In Example 10, the water is at ground level when the height, h, is zero. Use factoring to find the inputs, t, when $-16t^2 + 48t = 0$. Do the results agree with the graph?

54. Estimate the maximum height, h, reached by the water in Example 10, where $h = -16t^2 + 48t$. Estimate the input that gives maximum height. How could the input be used to find the maximum height?

55. An air-powered rocket is launched, and at time t it is at height $h = -16t^2 + 64t$. Use the graph in the figure to find the times when the rocket is at 48 feet. Use factoring to find the times. Do the results agree?

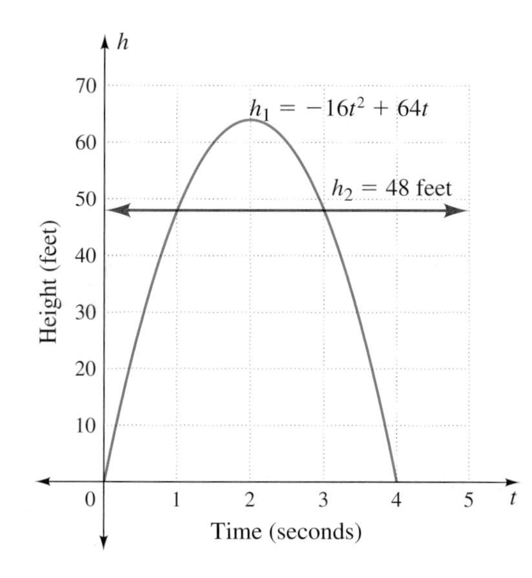

56. Estimate the maximum height reached by the rocket in Exercise 55 (see the figure). Estimate the input that gives maximum height. How could the input be used to find the maximum height?

57. Use factoring to find when the rocket in Exercise 55 is at ground level, $y = 0$. Does your result agree with the graph?

58. Use the graph to estimate how far the rocket in Exercise 55 travels in the first second and how far it travels in the second second. If rate, or speed, is distance divided by time, what is the average speed of the rocket during the first second? What is the average speed of the rocket in the second second?

Exercises 59 to 62 return to equations containing square roots (Section 8.3). Use factoring as needed to solve the quadratic equations. Show your check for each solution.

59. $\sqrt{4 - x} = x - 2$

60. $\sqrt{3 - x} = x - 1$

61. $\sqrt{x + 5} = x + 3$

62. $\sqrt{3 - x} = x - 3$

63. Refer to the figure below.

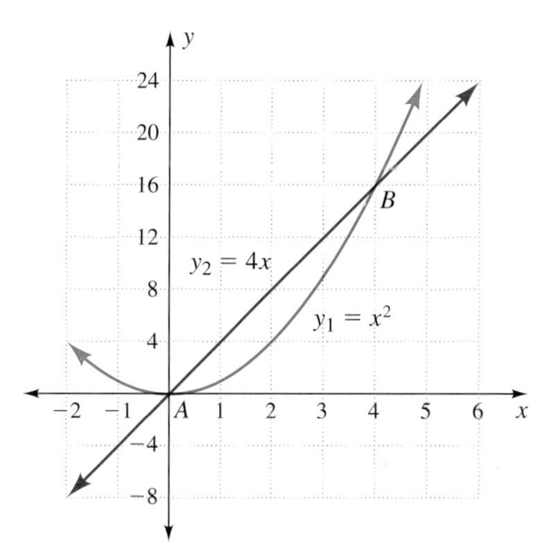

a. What are the ordered pairs at A and B?

b. The intersections at A and B solve what equation?

c. Solve your equation with algebra. Do you obtain the same results as with the graph?

Blue numbers are core exercises.

64. Refer to the figure below.

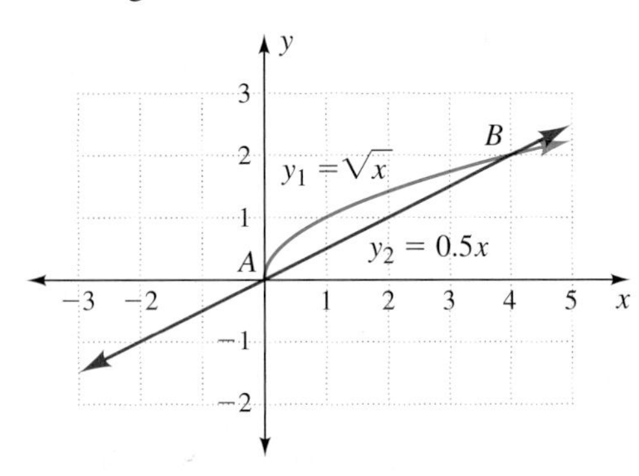

a. What are the ordered pairs at A and B?

b. The intersections at A and B solve what equation?

c. Solve you equation with algebra. Do you obtain the same results as with the graph?

▶ Writing

65. Explain how to solve $ax^2 + b = 0$ by the square root method.

66. Explain for what values of a and b the equation $ax^2 + b = 0$ can be solved by factoring.

67. Explain why $\sqrt{(x + 2)^2} = |x + 2|$.

68. Explain the difference between factoring a trinomial expression and solving a quadratic equation by factoring.

▶ Projects

69. Consecutive Integers and Right Triangles The following steps prove that there is only one set of consecutive positive integers that satisfies the Pythagorean theorem.

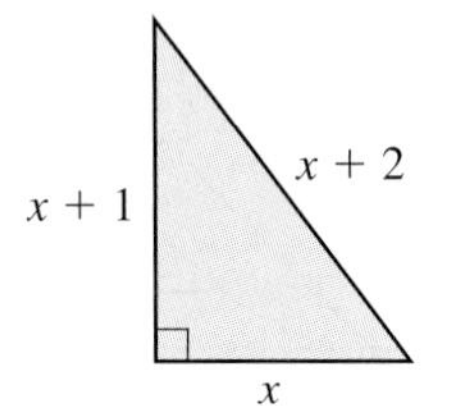

a. As shown in the figure, let x, $x + 1$, and $x + 2$ be the three consecutive integers representing the lengths of the sides of a right triangle. Write the Pythagorean theorem using consecutive integer notation, and simplify.

b. Find the solutions to the resulting equation from a table.

c. Find the solutions to the resulting equation by factoring.

d. Substitute the solutions into the expressions in part a, and state your conclusions.

70. Quadratic Formula Preview Not all quadratic equations factor easily. Some do not factor at all. When a quadratic equation is written in the form $ax^2 + bx + c = 0$, it may be solved for x with the quadratic formula. The quadratic formula is shown as two equations:

$$x = \frac{-b + \sqrt{b^2 - 4ac}}{2a} \quad \text{and}$$

$$x = \frac{-b - \sqrt{b^2 - 4ac}}{2a}$$

a. Use the quadratic formula on these equations to check that you are using it correctly. They are equations from earlier exercises.

$x^2 - 4x - 12 = 0$ Answer: $x = -2, x = 6$

$2x^2 - x - 3 = 0$ Answer: $x = 1.5, x = -1$

$4x^2 - 25 = 0$ Answer: $x = 2.5, x = -2.5$

b. Use the quadratic formula to solve these equations.

$2x^2 + 5x - 3 = 0$ $\quad$ $3x^2 + 5x - 4 = 0$

$3x^2 - 2x - 4 = 0$ $\quad$ $3x^2 - 2x - 3 = 0$

Blue numbers are core exercises.

▶ 8.6 Solving Quadratic Equations with the Quadratic Formula

Objectives

- Simplify expressions based on the quadratic formula.
- Estimate the value of expressions containing radicals.
- Solve quadratic equations with the quadratic formula.
- Give reasons in a proof of the quadratic formula.

WARM-UP

Multiply these expressions.

1. $(2x - y)(2x - y)$ **2.** $(a + b)(a + b)$

3. $(ax + b)(ax + b)$ **4.** $(2ax + b)(2ax + b)$

Factor these expressions.

5. $x^2 + 2x + 1$ **6.** $x^2 - 2x + 1$ **7.** $x^2 - 1$

8. $x^2 - x - 1$ **9.** $2x^2 + 2x + 1$ **10.** $x^2 - x - 2$

These radicals are between what two consecutive integers?

11. $\sqrt{5}$ **12.** $\sqrt{28}$ **13.** $\sqrt{84}$

THIS SECTION INTRODUCES the final method of solving quadratic equations: the quadratic formula. We review operations needed to simplify expressions obtained from the quadratic formula and solve equations with the quadratic formula.

The Quadratic Formula

Two expressions in the Warm-up could not be factored: $x^2 - x - 1$ and $2x^2 + 2x + 1$. We explore $x^2 - x - 1$ further in Example 1 and $2x^2 + 2x + 1$ in Example 5.

EXAMPLE 1 Exploring solutions to quadratic equations Graph $y = x^2 - x - 1$. Solve the equation $x^2 - x - 1 = 0$ from your graph. Estimate solutions, if they exist.

SOLUTION The graph is shown in Figure 28. The solutions are $x \approx -0.5$ and $x \approx 1.5$.

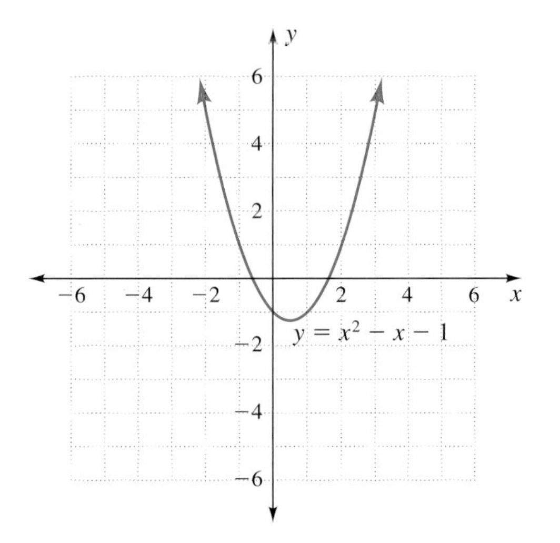

FIGURE 28 Solve $x^2 - x - 1 = 0$ ◀

Our prior methods for solving quadratic equations don't work well—if at all—with the equation in Example 1. The expression does not factor. Graphing gives approximate results. The graph crosses the x-axis, but not at an integer. The expression has $b \neq 0$, so we cannot use the square root method.

To solve an equation like the one in Example 1, mathematicians developed *a more general procedure:* the **quadratic formula**. We may use the quadratic formula to solve for the variable x in any equation of the form $ax^2 + bx + c = 0$.

The quadratic formula may be shown as two equations: For $ax^2 + bx + c = 0$,

$$x = \frac{-b + \sqrt{b^2 - 4ac}}{2a} \quad \text{or} \quad x = \frac{-b - \sqrt{b^2 - 4ac}}{2a}$$

The plus or minus sign ($\pm$) allows us to write two equations with one statement.

QUADRATIC FORMULA

For $ax^2 + bx + c = 0$,

$$x = \frac{-b \pm \sqrt{b^2 - 4ac}}{2a}$$

Another useful form of the quadratic formula is

$$x = \frac{-b}{2a} \pm \frac{\sqrt{b^2 - 4ac}}{2a}$$

In this form, the fraction is separated into two parts over the common denominator, $2a$. The separation prevents many simplification errors. This form also contains graphical information, which is explored in the project in Exercise 72.

SOLVING WITH THE QUADRATIC FORMULA

To solve with the quadratic formula:

1. Place the equation in standard form, $ax^2 + bx + c = 0$.
2. Substitute a, b, and c into the quadratic formula.
3. Simplify the resulting expression, following the order of operations.
4. State and check the solutions.

In Example 2, we solve the equation from Example 1.

▶ **EXAMPLE 2** Applying the quadratic formula Solve $x^2 - x - 1 = 0$ with the quadratic formula.

SOLUTION We use the formula, with $a = 1$, $b = -1$, and $c = -1$:

$$x = \frac{-b \pm \sqrt{b^2 - 4ac}}{2a} \qquad \text{Substitute into the formula.}$$

$$x = \frac{-(-1) \pm \sqrt{(-1)^2 - 4(1)(-1)}}{2(1)} \qquad \text{Simplify.}$$

$$x = \frac{1 \pm \sqrt{1 + 4}}{2} \qquad \text{Write the two solutions.}$$

$$x = \frac{1 + \sqrt{5}}{2} \quad \text{or} \quad x = \frac{1 - \sqrt{5}}{2}$$

These solutions are irrational numbers. ◀

To check, we first estimate the values and see that the solutions agree with our estimates from the graph. Then we substitute on a calculator.

ESTIMATION AND CALCULATOR USE In Example 3, we will first estimate and then use a calculator to substitute the solutions into the equation.

▶ **EXAMPLE 3** Estimating and calculating irrational numbers Using an integer estimate for each radical, estimate the value of the expression. Compare the results with the graph of $y = x^2 - x - 1$. Then substitute the original expressions into the equation, using a calculator.

a. $x = \dfrac{1 + \sqrt{5}}{2}$ **b.** $x = \dfrac{1 - \sqrt{5}}{2}$

SOLUTION Since $\sqrt{5}$ is closest to 2, we replace the $\sqrt{5}$ with 2 in each part and simplify:

a. $x = \dfrac{1 + \sqrt{5}}{2}$

$x \approx \dfrac{1 + 2}{2} \approx 1.5$

b. $x = \dfrac{1 - \sqrt{5}}{2}$

$x \approx \dfrac{1 - 2}{2} \approx -0.5$

The answer to part a is about 1.5, in agreement with the right x-intercept in Figure 28, shown enlarged in Figure 29. The answer to part b is about -0.5, in agreement with the left x-intercept.

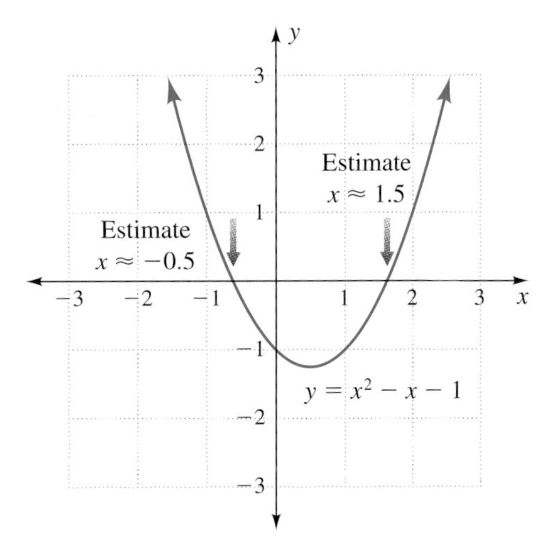

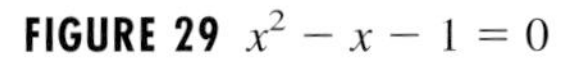
FIGURE 29 $x^2 - x - 1 = 0$

On a calculator, we place the expression from part a into $x^2 - x - 1$, as shown in Figure 30. We put a set of parentheses around the entire expression before squaring. The result is zero, as expected for a solution to $x^2 - x - 1 = 0$.

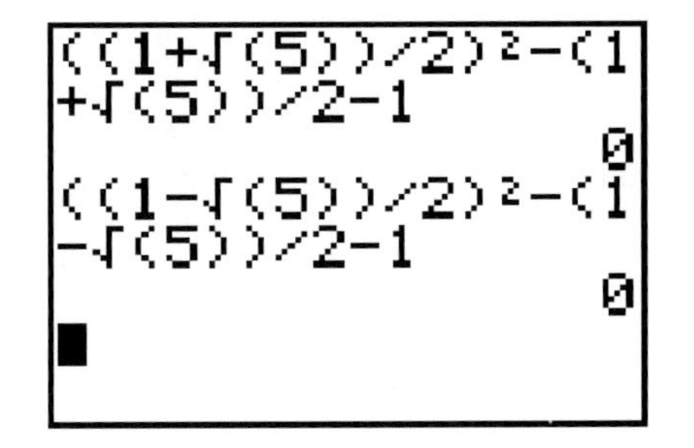

FIGURE 30 Check solutions

Using [2nd] [ENTER], we can replay the check for part a in order to check part b. Changing from the addition in part a to the subtraction in part b requires only two operation-sign changes. See Figure 30. Again, the result is zero. ◀

THINK ABOUT IT 1: Use a calculator to evaluate the expressions in parts a and b of Example 3. Make sure your results agree with the estimates in Example 3.

Types of Solutions Found with the Quadratic Formula

In Examples 4 and 5, we return to solving equations with the quadratic formula. We examine the solutions to the equations and the graphs of the related functions.

EXAMPLE 4 Solving equations by the quadratic formula Use the quadratic formula to solve these equations. We will check with graphs in Example 5.

a. $2x^2 + 7x + 3 = 0$ **b.** $4x^2 = 5x + 2$ **c.** $2x^2 + 2x = -1$

SOLUTION **a.** In $2x^2 + 7x + 3 = 0$, $a = 2$, $b = 7$, and $c = 3$.

$$x = \frac{-b \pm \sqrt{b^2 - 4ac}}{2a} = \frac{-7 \pm \sqrt{7^2 - 4(2)(3)}}{2(2)}$$

$$= \frac{-7 \pm \sqrt{49 - 24}}{4} = \frac{-7 \pm \sqrt{25}}{4}$$

The solutions are rational:

$$x = \frac{-7 + 5}{4} = \frac{-2}{4} = \frac{-1}{2} \quad \text{or} \quad x = \frac{-7 - 5}{4} = \frac{-12}{4} = -3$$

b. The equation $4x^2 = 5x + 2$ must be rearranged to equal zero. In $4x^2 - 5x - 2 = 0$, $a = 4$, $b = -5$, and $c = -2$.

$$x = \frac{-b \pm \sqrt{b^2 - 4ac}}{2a} = \frac{-(-5) \pm \sqrt{(-5)^2 - 4(4)(-2)}}{2(4)}$$

$$= \frac{-(-5) \pm \sqrt{25 + 32}}{8} = \frac{5 \pm \sqrt{57}}{8}$$

The solutions are irrational:

$$x = \frac{5 + \sqrt{57}}{8} \approx 1.569 \quad \text{or} \quad x = \frac{5 - \sqrt{57}}{8} \approx -0.319$$

c. The equation $2x^2 + 2x = -1$ must be rearranged to equal zero. In $2x^2 + 2x + 1 = 0$, $a = 2$, $b = 2$, and $c = 1$.

$$x = \frac{-b \pm \sqrt{b^2 - 4ac}}{2a} = \frac{-2 \pm \sqrt{2^2 - 4(2)(1)}}{2(2)}$$

$$= \frac{-2 \pm \sqrt{4 - 8}}{4} = \frac{-2 \pm \sqrt{-4}}{4}$$

There are no real-number solutions because the square root of a negative number, such as $\sqrt{-4}$, is not defined in the real numbers. ◀

In later courses, you will learn about other types of numbers that can be used to solve problems such as the one in part c.

▶ We now examine the graphs related to the equations in Example 4.

EXAMPLE 5 Identifying solutions on graphs Graph these functions and identify the solutions to $y = 0$ on the graphs.

a. $y = 2x^2 + 7x + 3$

b. $y = 4x^2 - 5x - 2$

c. $y = 2x^2 + 2x + 1$

SOLUTION The solutions to $y = 0$ are the x-intercepts.

a. The graph in Figure 31 confirms that the solutions to $2x^2 + 7x + 3 = 0$ are at $x = -\frac{1}{2}$ and $x = -3$.

b. The graph in Figure 32 confirms that one solution to $4x^2 - 5x - 2 = 0$ is near $x = 1\frac{1}{2}$ and the other is between $x = -\frac{1}{2}$ and $x = 0$.

c. The graph in Figure 33 does not cross the x-axis, so it has no x-intercepts. This confirms that the equation $2x^2 + 2x + 1 = 0$ has no real-number solutions.

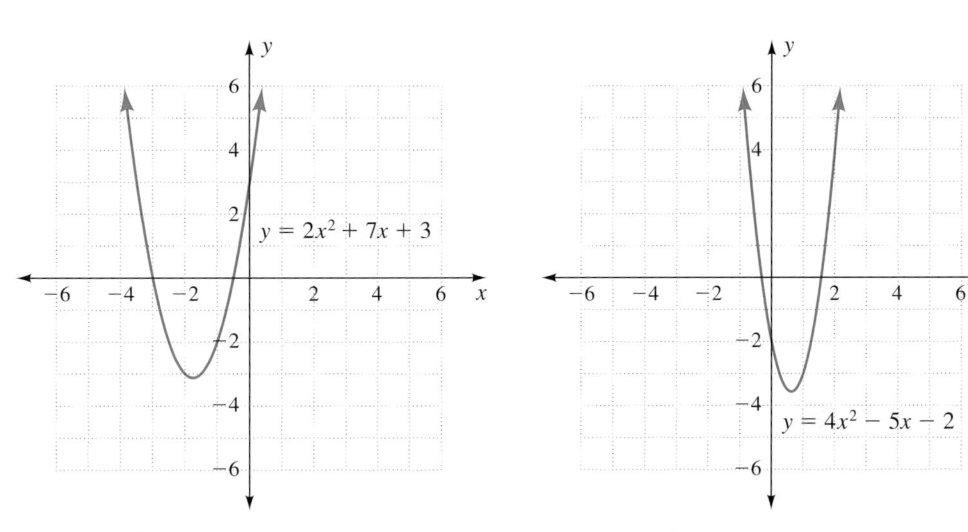

FIGURE 31 $2x^2 + 7x + 3 = 0$

FIGURE 32 $4x^2 - 5x - 2 = 0$

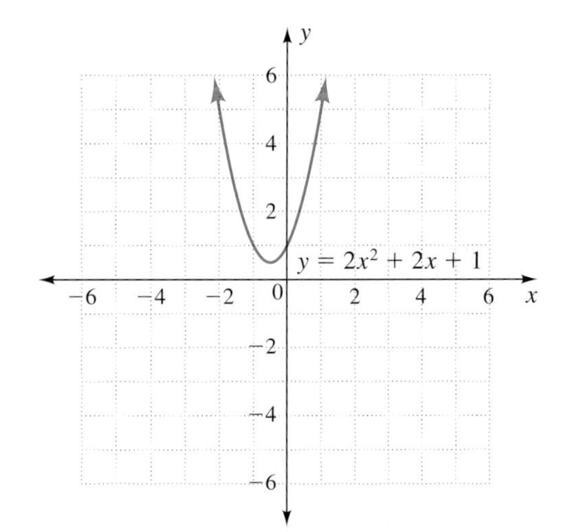

FIGURE 33 $2x^2 + 2x + 1 = 0$ ◀

THINK ABOUT IT 2: What part of the quadratic formula shows whether there are no, one, or two solutions to a quadratic equation?

▶ Proof of the Quadratic Formula

In the proof of the Pythagorean theorem, we squared a binomial, $(a + b)$, to obtain the perfect square trinomial, $a^2 + 2ab + b^2$. In proving the quadratic formula, we will build a perfect square trinomial and factor it back to the square of the binomial: $4a^2x^2 + 4abx + b^2 = (2ax + b)(2ax + b) = (2ax + b)^2$.

▶ **EXAMPLE 6** Proving the quadratic formula The steps below show how we obtain the quadratic formula from a quadratic equation in the form $ax^2 + bx + c = 0$. Explain what was done to obtain each step.

$$ax^2 + bx + c = 0$$

a. $4a^2x^2 + 4abx + 4ac = 0$

b. $4a^2x^2 + 4abx = -4ac$

c. $4a^2x^2 + 4abx + b^2 = b^2 - 4ac$

d. $(2ax + b)^2 = b^2 - 4ac$

e. Either $2ax + b = \sqrt{b^2 - 4ac}$ or $2ax + b = -\sqrt{b^2 - 4ac}$

f. Either $2ax = -b + \sqrt{b^2 - 4ac}$ or $2ax = -b - \sqrt{b^2 - 4ac}$

g. Either $x = \dfrac{-b + \sqrt{b^2 - 4ac}}{2a}$ or $x = \dfrac{-b - \sqrt{b^2 - 4ac}}{2a}$

SOLUTION **a.** Multiply both sides of $ax^2 + bx + c = 0$ by $4a$.

b. Subtract $4ac$ from both sides of the equation.

c. Add b^2 to both sides. This creates a perfect square trinomial on the left.

d. Factor the left side.

e. Find the square root of both sides. Because $2ax + b$ may be either positive or negative, we obtain two equations.

f. Subtract b from both sides of each equation.

g. Divide by $2a$ in both equations. ◀

Our purpose is not to justify why each step was done; rather, it is simply to present a proof.

▶ Deciding on a Solution Method

Regardless of which method you choose, you should either graph the equation or estimate answers so that you know your final results are sensible. Make a sketch showing the y-intercept and two or three other easy-to-find points, and use the sign on the coefficient of x^2 to decide whether the graph opens upward ($a > 0$) or downward ($a < 0$).

- Use a *graphing calculator graph* if the numbers in the equation are hard to work with or you need several digits in your answer. The graph makes it easy to find out if no, one, or two answers are sensible.
- Use the *square root* if the equation is $ax^2 + c = 0$ (that is, if $b = 0$).
- Use *factoring* if it takes 30 seconds or less to factor the expression.
- Use the *quadratic formula* in all other cases.

ANSWER BOX

Warm-up: 1. $4x^2 - 4xy + y^2$ **2.** $a^2 + 2ab + b^2$ **3.** $a^2x^2 + 2abx + b^2$ **4.** $4a^2x^2 + 4abx + b^2$ **5.** $(x + 1)(x + 1)$ **6.** $(x - 1)(x - 1)$ **7.** $(x - 1)(x + 1)$ **8.** does not factor **9.** does not factor **10.** $(x + 1)(x - 2)$ **11.** $2 < \sqrt{5} < 3$ **12.** $5 < \sqrt{28} < 6$ **13.** $9 < \sqrt{84} < 10$ **Think about it 1: a.** $(1 + \sqrt{(5)}) \div 2 \approx 1.618$ **b.** $(1 - \sqrt{(5)}) \div 2 \approx -0.618$ **Think about it 2:** The expression under the radical: $b^2 - 4ac$. See the project in Exercise 71.

▶ 8.6 Exercises

In Exercises 1 to 8, identify *a*, *b*, and *c* from the quadratic equations.

1. $9x^2 + 6x + 1 = 0$
2. $3x^2 - 7x + 2 = 0$
3. $3x^2 - 9x = 0$
4. $x^2 + 4 = 0$
5. $x^2 = 4x - 3$
6. $x^2 + 6 = 5x$
7. $x^2 = -9$
8. $2x^2 = 6x$

Simplify the expressions in Exercises 9 to 16 without a calculator.

9. $\frac{-(-4) - \sqrt{(-4)^2 - 4(1)(-12)}}{2 \cdot 1}$
10. $\frac{-(-4) + \sqrt{(-4)^2 - 4(1)(-12)}}{2 \cdot 1}$
11. $\frac{-(-5) + \sqrt{(-5)^2 - 4(6)(1)}}{2(6)}$
12. $\frac{-(-5) - \sqrt{(-5)^2 - 4(6)(1)}}{2(6)}$
13. $\frac{-1 - \sqrt{(1)^2 - 4(3)(-4)}}{2(3)}$
14. $\frac{-1 + \sqrt{(1)^2 - 4(3)(-4)}}{2(3)}$
15. $\frac{-6 + \sqrt{(6)^2 - 4(16)(-1)}}{2(16)}$
16. $\frac{-6 - \sqrt{(6)^2 - 4(16)(-1)}}{2(16)}$

Replace the radical with the nearest whole number and estimate the value of each expression in Exercises 17 to 24. Then use a calculator to approximate the value to three decimal places.

17. a. $\frac{2 - \sqrt{2}}{2}$ b. $\frac{3 + \sqrt{6}}{3}$ c. $\frac{3\sqrt{6}}{3}$
18. a. $\frac{3 - \sqrt{6}}{3}$ b. $\frac{2 + \sqrt{2}}{2}$ c. $\frac{2\sqrt{2}}{2}$
19. a. $\frac{3 - \sqrt{5}}{3}$ b. $\frac{2\sqrt{10}}{2}$ c. $\frac{2 + \sqrt{10}}{2}$
20. a. $\frac{3\sqrt{5}}{3}$ b. $\frac{2 - \sqrt{10}}{2}$ c. $\frac{3 + \sqrt{5}}{3}$
21. a. $\frac{-5 - \sqrt{13}}{2}$ b. $\frac{-9 + \sqrt{57}}{4}$
22. a. $\frac{-9 - \sqrt{57}}{4}$ b. $\frac{-5 + \sqrt{13}}{2}$
23. a. $\frac{-2 - \sqrt{28}}{6}$ b. $\frac{-2 + \sqrt{84}}{10}$
24. a. $\frac{7 + \sqrt{73}}{12}$ b. $\frac{7 - \sqrt{29}}{10}$

Solve the equations in Exercises 25 to 44 with the quadratic formula.

25. $4x^2 + 3x - 1 = 0$
26. $2x^2 + x - 6 = 0$
27. $7x^2 = 5x + 2$
28. $9x^2 = 2 + 3x$
29. $x = 2 - 10x^2$
30. $x = 2 - 6x^2$
31. $2x^2 - x - 1 = 0$
32. $2x^2 - 3x - 2 = 0$
33. $3x^2 + x - 2 = 0$
34. $2x^2 - 2x + 1 = 0$
35. $x^2 - 2x + 2 = 0$
36. $3x^2 + 5x - 2 = 0$
37. $3x^2 - 2x = 6$
38. $2x^2 - 3x = 6$
39. $5x^2 + 4x + 1 = 0$
40. $7x^2 = 3 - 4x$
41. $5x^2 = 4 - 2x$
42. $5x^2 + 2x + 1 = 0$
43. $6x^2 - 3x - 2 = 0$
44. $2x^2 - 3x - 6 = 0$

The equations in Exercises 45 to 52 are in the form $h = at^2 + bt + c$. The input is the time *t*, in seconds. The output is the height *h* of an object tossed straight up from a height *c*, in feet, with an initial velocity *b*, in feet per second. Solve the equations for *t*, using the quadratic formula. Round to the nearest hundredth. Which answers are acceptable in the setting?

45. $-16t^2 + 30t + 150 = 0$
46. $-16t^2 + 20t + 50 = 0$
47. $-16t^2 + 15t + 150 = 0$
48. $-16t^2 + 25t + 100 = 0$
49. $-16t^2 + 30t + 150 = 200$
50. $-16t^2 + 15t + 150 = -50$
51. $-16t^2 + 20t + 50 = -40$
52. $-16t^2 + 25t + 100 = 150$

Choose a reasonable method for solving each of the equations in Exercises 53 to 64. Round to three decimal places.

53. $4x^2 = 10$
54. $9x^2 = 14$
55. $4x^2 - 9 = 0$
56. $9x^2 - 25 = 0$
57. $x^2 + 5 = -2x$
58. $x^2 + 4x = -5$
59. $x^2 + 9 = 6x$
60. $x^2 + 16 = 8x$
61. $4x^2 = 4x + 7$
62. $3x^2 + 4x = 7$
63. $4x^2 = 12x - 9$
64. $9x^2 + 4 = 12x$

Blue numbers are core exercises.

▶ Writing

65. Explain how to tell the number of solutions to $ax^2 + bx + c = 0$ from a graph of $y = ax^2 + bx + c$.

66. Explain how to estimate the square root of a number under 250.

67. Show the keystrokes for entering the quadratic formula for $2x^2 + 3x + 4$ into a calculator. Tell what calculator you use.

68. **a.** Compare the graphs for $y = x^2 - 4$ and $y = 4 - x^2$. Explain why the graphs are the same or different.

b. Solve $x^2 - 4 = 0$, and solve $4 - x^2 = 0$. Explain why the solutions are the same or different.

69. Write an equation that describes "The output is the product of two consecutive numbers." Solve your equation for $y = 0$. In what interval is the output negative?

70. Write an equation that describes "The output is the sum of the squares of two consecutive numbers." Solve your equation for $y = 0$. What does this say about the squares of real numbers?

▶ Projects

71. **The Discriminant** We can predict whether there will be no real solution, one real solution, or two real solutions to a quadratic equation from $b^2 - 4ac$, the discriminant.

a. Which part of $x = \dfrac{-b \pm \sqrt{b^2 - 4ac}}{2a}$ causes there to be two answers?

b. How many real numbers do we obtain from $\pm\sqrt{\text{(negative)}}$?

c. From $\pm\sqrt{\text{(positive)}}$?

d. From $\pm\sqrt{\text{(zero)}}$?

e. How many real-number solutions will there be to a quadratic equation if $b^2 - 4ac = 0$?

f. If $b^2 - 4ac > 0$?

g. If $b^2 - 4ac < 0$?

h. If c is negative and a is positive?

i. If a and c are both positive and $4ac > b^2$?

Use the discriminant to find how many different real-number solutions there are to these equations:

j. $4x^2 + 4x + 1 = 0$

k. $5x^2 - 8x + 2 = 0$

l. $3x^2 - 9x = 0$

m. $x^2 + 4 = 0$

72. **Quadratic Formula and Graphs** The highest or lowest point on the graph of a quadratic equation $y = ax^2 + bx + c$ is called the *vertex*. The quadratic formula in the form

$$x = \frac{-b}{2a} \pm \frac{\sqrt{b^2 - 4ac}}{2a}$$

shows why the vertex is symmetrically placed between the x-intercepts. The distance between the axis of symmetry and each x-intercept is

$$\frac{\sqrt{b^2 - 4ac}}{2a}, \quad a > 0$$

as shown in the figure. The axis of symmetry, which passes through the vertex V, has equation

$$x = \frac{-b}{2a}$$

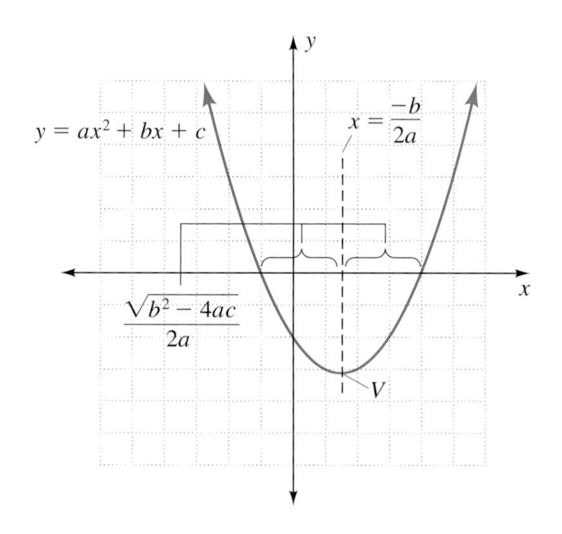

Using either the x-intercepts or the axis of symmetry, find the coordinates of the vertex for each quadratic equation:

a. $y = x^2 + 6x - 16$

b. $y = 2x^2 + 3x - 4$

c. $y = -16t^2 + 30t + 150$

d. $y = -16t^2 + 20t + 50$

e. If the graphs for parts c and d represent the height of a ball at time t, what is the meaning of the vertex?

f. Find the coordinates of the vertex for each part in Example 5. Check that each vertex agrees with the graphs.

▶ 8.7 Range, Box and Whisker Plots, and Standard Deviation

Objectives

- Find the range from a set of data.
- Calculate the median and quartiles of a set of data, and summarize with a box and whisker plot.
- Calculate the standard deviation for a set of data.

WARM-UP

Exercises 1 to 3 each contain a list of monthly averages of airborne lead (the chemical Pb in nanograms per cubic meter) as measured in Manila (Philippines, 2002), Sydney (Australia, 2002) and Prague* (Czechoslovakia, 2004). Calculate the mean and median of each set. What do these averages mean?

1. 50, 60, 50**, 40, 70, 70, 90, 100, 90, 20, 10, 10

2. 3, 5, 10, 22, 25, 19, 31, 22, 19, 6, 3, 5

3. 25, 20, 29, 23, 25, 26, 22, 16, 23, 20, 29, 34

Calculate the distance between these points.

4. $(3, 7)$ and $(-2, -5)$

5. $(-3, 5)$ and $(5, -10)$

SECTION 5.4 INTRODUCED the mean, median, and mode. These calculations are called *measures of central tendency* because they indicate the center of a set of data. In this section, we introduce a related application of square root—standard deviation—as we extend the work in Section 5.4 with range and box and whisker plots.

▶ Measures of Variation or Dispersion

Suppose you were choosing among Sydney, Manila and Prague for travel and were concerned about airborne lead. Would you make a decision based solely on the mean or median? Is Sydney always better than the other choices? The variation in the monthly data indicates there are months in Manila and Prague that are comparable to Sydney. **Measures of variation or dispersion** are *measures that describe how close to the middle or how scattered a set of data is*. The range, box and whisker plot, and standard deviation are three such measures.

▶ Range

Student Note:
This is a different use of range than in functions.

Finding the range is a first step in determining how sets of numbers differ. The range is calculated by subtracting the lowest number from the highest number.

RANGE

The **range** is the difference between the largest and the smallest number in a data set:

$$\text{Range} = \text{maximum number} - \text{minimum number}$$

*Three other locations in Prague had consistently lower numbers.
**Data for March, month 3, for Manila were not available and were filled in with the average of adjacent months.

▶ **EXAMPLE 1** Finding range for lead The measures are in nanograms per cubic meter. Find the range of each set.

a. Manila: 50, 60, 50, 40, 70, 70, 90, 100, 90, 20, 10, 10
b. Sydney: 3, 5, 10, 22, 25, 19, 31, 22, 19, 6, 3, 5
c. Prague: 25, 20, 29, 23, 25, 26, 22, 16, 23, 20, 29, 34

SOLUTION **a.** The range is $100 - 10 = 90$ ng/m^3.
b. The range is $31 - 3 = 28$ ng/m^3.
c. The range is $34 - 16 = 18$ ng/m^3. ◀

The range indicates the spread of the data, which is considerably greater in Manila than in Sydney or Prague.

Pollution (Australian), box plots illustrate the Project in Exercise 31:
http://old-www.ansto.gov.au/nugeo/iba/news/news_letters/Newsletter33.pdf
Data source: Pollution, Czechoslovakia:
http://www.chmi.cz/uoco/isko/tab_roc/2004_enh/eng/index.html
Pollution (Philippines): E-mail correspondence with the Manila Observatory.

THINK ABOUT IT 1: What city has the lowest lead each month? State your assumptions.

Box and Whisker Plots

Box and whisker plots are a relatively new means of comparing data. The plots were invented by John Wilder Tukey (1915–2000) and introduced in his textbook *Exploratory Data Analysis* (Reading, MA: Addison Wesley, 1977). Less than 30 years later, box and whisker plots (or modified box plots) may be found worldwide in summarizing data from the testing of optical fingerprint devices, the treatment of hearing loss, the measurement of water quality in Hawaii and electric fields in office buildings (Sweden), and the comparisons of air pollution in Australia and the Philippines. (Do an Internet search on the underlined words with "box and whisker" to find these and other addresses.)

Optical fingerprint devices:
http://kitkat.wvu.edu:8080/files/3980/Rosiek_Travis_thesis.pdf
Reports on treatment of hearing loss:
http://www.ncbi.nlm.nih.gov/entrez/query.fcgi?cmd=Retrieve&db=PubMed&list_uids=9627261&dopt=Abstract
Water quality:
http://www.hawaii.edu/ssri/hcri/files/laws_noaa_final_report_01-02.pdf
Electric fields:
http://emconsite.tt.ltu.se/admin/artiklar/Greece2002.pdf

The **box and whisker plot** is *a 5-number visual summary of data.* The plot is drawn to scale (like a number line). It includes (in order) the minimum value, the middle (Q_1) of the numbers below the median, the median, the middle (Q_3) of the numbers above the median, and the maximum value. Directions are given in the box.

DRAWING A BOX AND WHISKER PLOT

Build a box and whisker plot with the Manila data for 2002: {50, 60, 50, 40, 70, 70, 90, 100, 90, 20, 10, 10}. (Data courtesy of Urban Air Quality group, Manila Observatory, Philippines.)

1. Find the median—the middle number when data are arranged from smallest to largest order. If there is no single middle number, average the two

middle numbers. In the set {10, 10, 20, 40, 50, 50, 60, 70, 70, 90, 90, 100} the median is 55, the average of 50 and 60. Locate the median on a horizontal line with a scale appropriate to the data set (see Figure 34).

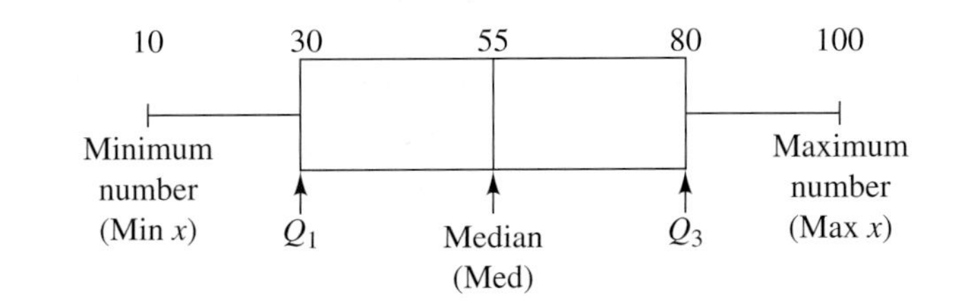

FIGURE 34 Manila 2002 lead

2. Find the **quartiles,** Q_1 and Q_3, as follows: Q_1 is *the middle of the numbers below the median.* Q_3 is *the middle of the numbers above the median.* As with the median, if there is no single middle number, average the two middle numbers. Q_1 is 30, the average of 20 and 40, the middle two numbers in {10, 10, 20, 40, 50, 50}. Q_3 is 80, the average of 70 and 90, the middle two numbers in {60, 70, 70, 90, 90, 100}. Draw a box (rectangle) from Q_1 to Q_3, on the horizontal line with length to scale (see Figure 34).
3. Draw a line from the left end of the box to the minimum number, 10. Draw another line from the right end of the box to the maximum number, 100. These lines form the whiskers (see Figure 34).

▶ **EXAMPLE 2** Building a box and whisker plot Calculate the median, Q_1, and Q_3 for Sydney and Prague (Example 1). Draw a box and whisker plot for each.

a. Sydney: 3, 5, 10, 22, 25, 19, 31, 22, 19, 6, 3, 5 (Data from Summary Sheets in **http://old-www.ansto.gov.au/nugeo/iba**)

b. Prague: 25, 20, 29, 23, 25, 26, 22, 16, 23, 20, 29, 34 (Data from **www.chmi.cz/uoco/isko/tab_roc/2004_enh/eng/index.html**)

SOLUTION In both sets, the median, Q_1, and Q_3 are between two numbers. Those two numbers must be averaged.

a. In the set {3, 3, 5, 5, 6, 10, 19, 19, 22, 22, 25, 31} the median is 14.5 (see Figure 35) and is the average of the middle two numbers, 10 and 19. Q_1 is 5 and is the average of 5 and 5, the middle two numbers in {3, 3, 5, 5, 6, 10}. Q_3 is 22 and is the average of 22 and 22, the middle two numbers in {19, 19, 22, 22, 25, 31}.

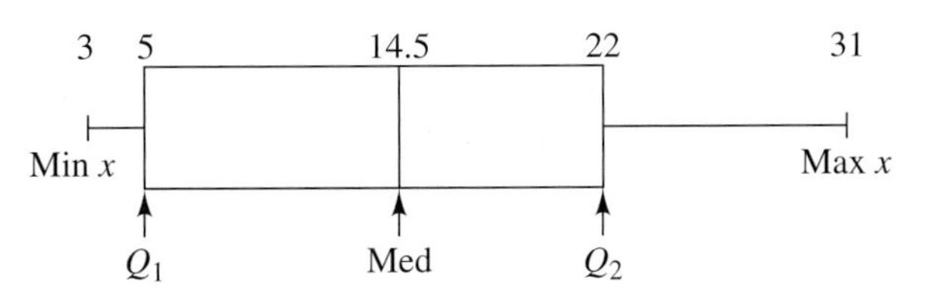

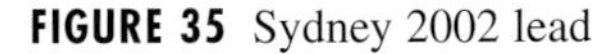

FIGURE 35 Sydney 2002 lead

b. In the set {16, 20, 20, 22, 23, 23, 25, 25, 26, 29, 29, 34}, the median is 24 (see Figure 36) and is the average of the middle two numbers, 23 and 25. Q_1 is 21, the average of 20 and 22, the middle two numbers in {16, 20, 20, 22, 23, 23}. Q_3 is 27.5, the average of 26 and 29, the middle two numbers in {25, 25, 26, 29, 29, 34}

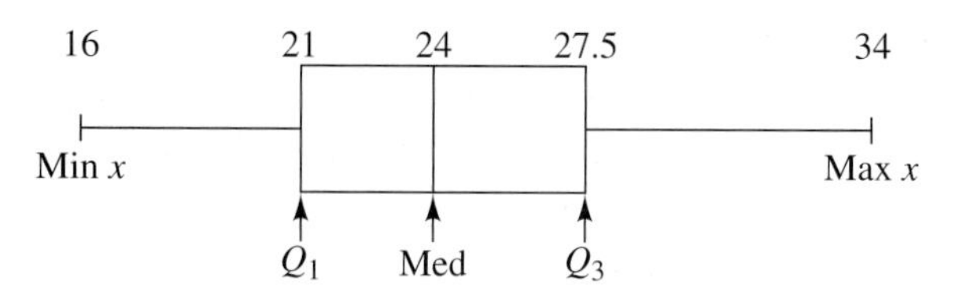

FIGURE 36 Prague 200 Lead

Standard Deviation

Student Note:
The letter $\bar{x}$, x with a bar over it, denotes the **sample mean** and is read "x-bar" or "the mean."

The box and whisker plot shows the spread of the data visually. A disadvantage of the box and whisker plot is that it does not give us a numerical description. One way to find a numerical measure of the spread of the data is to consider the differences between each data point, x_i, and the mean, $\bar{x}$. *The difference between each number and the mean* is called the **deviation**.

In developing the formula for what British mathematician Karl Pearson was to call "standard deviation" in 1894, statisticians may have thought about how we find the distance between two points. The distance between (x_1, y_1) and (x_2, y_2) is given by

$$d = \sqrt{(x_2 - x_1)^2 + (y_2 - y_1)^2}$$

The distance formula finds the deviation between the x-coordinates and between the y-coordinates. To assure that the distances are positive, we square the deviations and then add. To compensate for squaring the deviations, we then take the square root of the sum to obtain the distance. Observe the similarity between the distance formula and the formula for the **sample standard deviation**, *a measure of the variation between the mean of a set and each number in the set.*

SAMPLE STANDARD DEVIATION

For a set of n numbers $x_1, x_2, x_3, \ldots, x_n$ with mean $\bar{x}$, drawn randomly from a population, the **sample standard deviation** is

$$s_x = \sqrt{\frac{(x_1 - \bar{x})^2 + (x_2 - \bar{x})^2 + (x_3 - \bar{x})^2 + \cdots + (x_n - \bar{x})^2}{n - 1}}$$

The division by $n - 1$ in the sample standard deviation formula creates an "average" deviation.

▶ **EXAMPLE 3** Finding the standard deviation Find the standard deviation, as a whole number, for the set of lead measures from Manila, {50, 60, 50, 40, 70, 70, 90, 100, 90, 20, 10, 10}. The mean is 55 ng/m^3.

SOLUTION The standard deviation, s_x:

$$s_x = \sqrt{\frac{(50-55)^2+(60-55)^2+(50-55)^2+(40-55)^2+(70-55)^2+(70-55)^2+(90-55)^2+(100-55)^2+(90-55)^2+(20-55)^2+(10-55)^2+(10-55)^2}{12-1}}$$

$$s_x = \sqrt{\frac{(-5)^2+5^2+(-5)^2+(-15)^2+15^2+15^2+35^2+45^2+35^2+(-35)^2+(-45)^2+(-45)^2}{11}}$$

$$s_x = \sqrt{\frac{10500}{11}} \approx 30.896$$

The standard deviation for the Manila data is 31 ng/m^3. ◀

THINK ABOUT IT 2: In each part of the solution to Example 3, how might the standard deviation expressions be simplified?

▶ EXAMPLE 4 Finding the mean and standard deviation For the city lead measures as listed, find the mean (round to a whole number) and the sample standard deviation (round to a whole number) in nanograms per cubic meter.

a. Sydney: 3, 5, 10, 22, 25, 19, 31, 22, 19, 6, 3, 5

b. Prague: 25, 20, 29, 23, 25, 26, 22, 16, 23, 20, 29, 34

SOLUTION **a.** The mean for Sydney is 14 ng\m^3. The standard deviation, s_x, is

$$\sqrt{\frac{(3-14)^2+(5-14)^2+(10-14)^2+(22-14)^2+(25-14)^2+(19-14)^2+(31-14)^2+(22-14)^2+(19-14)^2+(6-14)^2+(3-14)^2+(5-14)^2}{11}}$$

$$\sqrt{\frac{(-11)^2+(-9)^2+(-4)^2+(8)^2+(11)^2+(5)^2+(17)^2+(8)^2+(5)^2+(-8)^2+(-11)^2+(-9)^2}{11}}$$

$$\sqrt{\frac{1072}{11}} \approx 9.87$$

The standard deviation for Sydney is 10 ng/m^3.

b. The mean for Prague is 24.

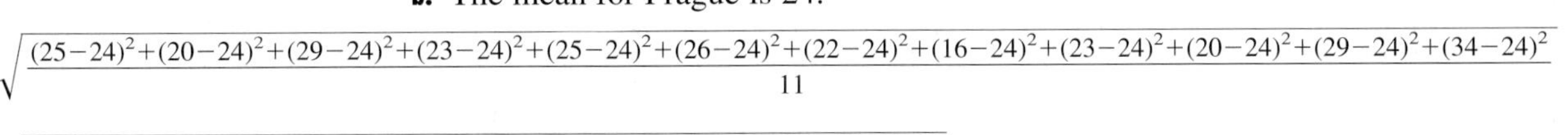

$$\sqrt{\frac{(25-24)^2+(20-24)^2+(29-24)^2+(23-24)^2+(25-24)^2+(26-24)^2+(22-24)^2+(16-24)^2+(23-24)^2+(20-24)^2+(29-24)^2+(34-24)^2}{11}}$$

$$\sqrt{\frac{(1)^2+(-4)^2+(5)^2+(-1)^2+(1)^2+(2)^2+(-2)^2+(-8)^2+(-1)^2+(-4)^2+(5)^2+(10)^2}{11}}$$

$$\sqrt{\frac{258}{11}} \approx 4.84$$

The standard deviation for Prague is 5 ng/m^3. ◀

The standard deviation is largest for widely-scattered data (31 for Manila, Example 3) and smallest for less widely-scattered data (10 for Sydney and 5 for Prague, Example 4).

▶ Calculator Techniques

The mean, standard deviation, and box plots are available on the graphing calculator. The data are entered in the edit option in the STAT key menu. The calculation of the mean and so on is done under the calc(ulate) option in the STAT key menu.

Graphing Calculator Technique: One-Variable Statistics and Box and Whisker Plots

Finding One-Variable Statistics:

1. Enter the data into the first list: Choose STAT **1 : Edit** and enter the data under L1. (See Figures 37a and b.)
2. To calculate the one-variable statistics on L1, choose STAT CALC **1 : 1–Var Stats** 2nd [L1] ENTER. The mean, $\bar{x}$, will appear first in the list. The sample standard deviation, s_x, is fourth in the list. Move the cursor down the screen for the information below the arrow: minimum X, quartiles (Q_1 and Q_3), median, and maximum X. (See Figures 38a and b.)

Creating a Box and Whisker Plot for L1:
Always turn off the graphs under Y = with the cursor on = and press ENTER.

1. To set up the box plot, choose 2nd [STATPLOT]1 and then press ENTER twice to turn on **Plot1**. Move the cursor to the fifth choice under **Type:** and press ENTER. Press 2nd [L1] as needed after **Xlist**. Leave 1 after **Freq**. (See Figure 39.)

(continued)

(continued)

2. Go to [WINDOW]. Enter the smallest number in L1 (or in all the lists) as **Xmin** and the largest number in L1 (or in all the lists) as **Xmax**. Set **Xscl** = $\frac{(\textbf{Xmax} - \textbf{Xmin})}{10}$. To show the x-axis and its scale, set **Ymin** = 0 (**Ymax** = any positive number).
3. Press [GRAPH] to view the plot.
4. Explore with [TRACE]. Move the cursor left and right.

For additional plots, set up **Plot2** with L2 and **Plot3** with L3. Graph. Trace up and down as well as left and right. (See Figure 40.)

When finished, always turn on the graphs under [Y =] with the cursor on = and press [ENTER]. Turn off the plots with the cursor on **Plot1** and press [ENTER]. Repeat for **Plot2** and **Plot3**.

L1	L2	L3
50	3	25
60	5	20
50	10	29
40	22	23
70	25	25
70	19	26
90	31	[illegible]

L3(7) =22

FIGURE 37a Upper portion, lists Manila, Sydney, Prague

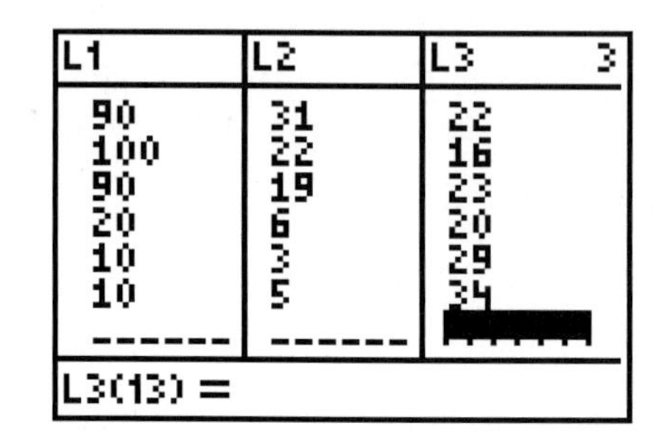

FIGURE 37b Lower portion, lists

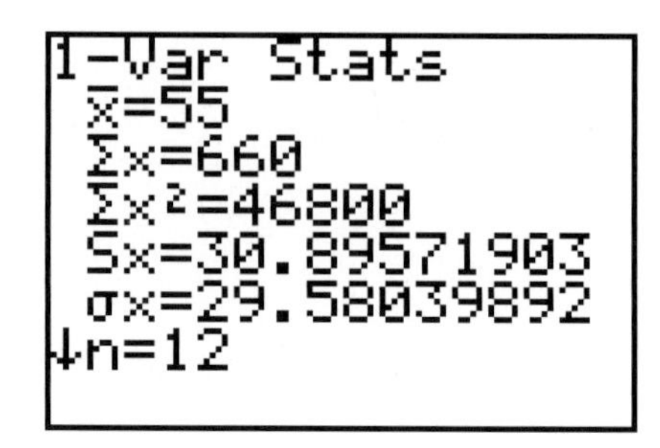

FIGURE 38a Upper portion, L1, Manila

Figures 37a and b show the data (Manila, Sydney, Prague) entered in lists L1, L2, and L3. Figure 38a shows the upper half of the screen containing the one-variable statistics for L1, and Figure 38b shows the lower half of the same screen. Figure 39 shows the Plot 1 screen with box plot highlighted. Figure 40 shows the box and whisker plots for all three lists. Trace has been activated, and the cursor is on the median (Med = 55) for the plot P1 (data in L1).

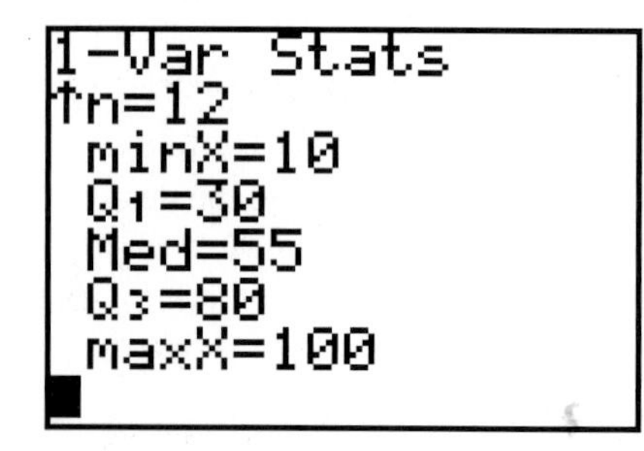

FIGURE 38b Lower portion, L1, Manila

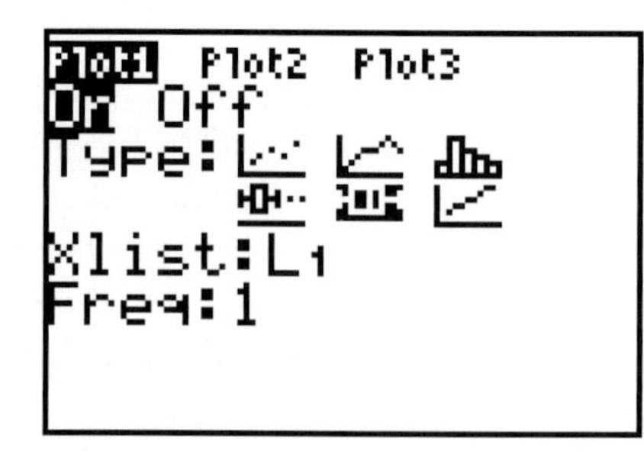

FIGURE 39 Plot type highlight

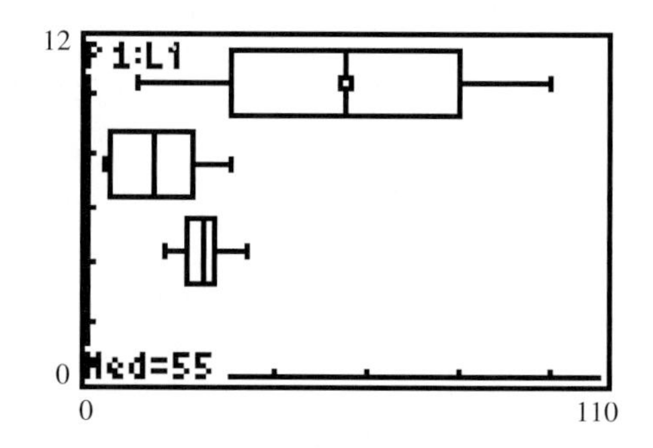

FIGURE 40 Box plots Manila, Sydney, Prague Trace on Manila, Med = 55

ANSWER BOX

Warm-up the average annual airborne lead 1: 55, 55 **2.** 14.2, 14.5 **3.** 24.3, 24 **4.** 13 **5.** 17 **Think about it 1:** Sydney, except for May, July, and August, when Prague is tied or lower. Assume monthly readings are consistent relative to other cities. If visiting Manila, go in November or December. **Think about it 2:** Rearrange to add like terms, such as $(50 - 55)^2 + (50 - 55)^2 = 2(50 - 55)^2$. In the next, $(-5)^2 = 5^2$, so the numerator equals $3(5^2) + 3(15^2) + 3(35^2) + 3(45^2)$, or $3(5^2 + 15^2 + 35^2 + 45^2)$.

▶ 8.7 Exercises

Calculate the mean, the median, and the range of each set of incomes in Exercises 1 to 4.

1. \$8000, \$10,000, \$12,000, \$13,000, \$13,000, \$100,000

2. \$4000, \$8000, \$9000, \$9000, \$100,000

3. \$20,000, \$27,500, \$27,500, \$27,500, \$27,500

4. \$23,000, \$25,000, \$26,000, \$27,000, \$29,000

In Exercises 5 to 8, find Q_1, and Q_3 for each set and draw a box and whisker plot.

5. Data from Exercise 1

6. Data from Exercise 2

7. Data from Exercise 3

8. Data from Exercise 4

In Exercises 9 to 12, find the sample standard deviation, s_x. Round to the nearest hundred. Which would you expect to have the smallest s_x?

9. Data from Exercise 1

10. Data from Exercise 2.

11. Data from Exercise 3.

12. Data from Exercise 4.

Exercises 13 to 16 contain elemental carbon (airborne soot) readings in ng/m^3. For each:

a. Find the mean, median, and range.

b. Find Q_1 and Q_3 and draw a box and whisker plot.

13. Sydney, 2002: 1118, 983, 1318, 2255, 2735, 2898, 2932, 2278, 1790, 1655, 1211, 1160

14. Sydney, 2005: 738, 814, 892, 1251, 2466, 2512, 2087, 2088, 1301, 1024, 889, 833

15. Manila, 2002: 24180, 20810, 20530, 15470, 13460, 15650, 15990, 16370, 14880, 13490, 11810, 10780

16. Wollongong (Australia), 2005: 560, 584, 569, 1047, 729, 935, 553, 1031, 744, 746, 764, 920

17. Code requires that the risers of staircases in homes have no more than $\frac{3}{8}$ inch variation in a flight of stairs. Suppose a flight of stairs has a riser-to-tread ratio of 7.5 in. to 10.5 in.

a. What is the largest riser if the smallest is 7.5 inches?

b. What is the smallest riser if the largest is 7.5 inches?

c. Will the following sets of steps pass inspection? Explain in terms of the concepts of this section. $7\frac{1}{2}$, $7\frac{1}{2}$, $7\frac{1}{4}$, $7\frac{3}{4}$, $7\frac{1}{2}$

d. List a set of five steps with an average 7.5-inch rise and a $\frac{1}{4}$-inch variation.

e. List a set o five steps with an average 7.5-inch rise and a $\frac{3}{8}$-inch variation.

18. The average sale price of a home in one particular month was \$146,843. This average was calculated from sales of 325 homes. The median home price was \$129,400.

a. Why might the average be larger than the median?

b. What was the total value of homes sold?

c. Are we able to calculate the standard deviation from the given information?

d. How many homes sold for less than \$129,400?

In Exercises 19 to 22, find the mean, the standard deviation, and the five box plot numbers (Min, Q_1, Med, Q_3, Max) for the given sets of average monthly temperatures in these pairs of cities. Compare the cities.

19. a. Atlanta, GA: 43, 47, 54, 62, 70, 77, 80, 79, 73, 63, 53, 45

b. Seattle, WA: 41, 43, 46, 50, 56, 61, 65, 66, 61, 53, 45, 41

Blue numbers are core exercises.

20. a. Chicago, IL: 22, 27, 37, 48, 59, 68, 73, 72, 64, 52, 39, 27

b. Hartford, CT: 26, 29, 38, 49, 60, 69, 74, 72, 63, 52, 42, 31

21. a. Honolulu, HI: 73, 73, 74, 76, 77, 80, 81, 82, 82, 80, 78, 75

b. Miami, FL: 68, 69, 72, 76, 80, 82, 84, 84, 82, 79, 74, 70

22. a. Portland, OR: 40, 43, 47, 51, 57, 63, 68, 69, 64, 54, 46, 40

b. Philadelphia, PA: 32, 35, 43, 53, 64, 72, 78, 76, 69, 57, 47, 37

Connecting with Chapter 7: Negative Exponents Exercises 23 to 26 are related to airborne lead measured in nanograms (10^{-9}) or micrograms (10^{-6}) per cubic meter.

23. The U.S. Environmental Protection Agency (EPA) quarterly standard for lead is 1.5 micrograms per cubic meter. Change this to nanograms per cubic meter. The Czech limit is 500 nanograms per cubic meter. Change this to micrograms per cubic meter. Which is the more strict limit?

24. The United States, Australia, and many other countries introduced unleaded gasoline and cut in half the lead in leaded gasoline. In the early 1990s in Sydney, lead readings were 1.7 micrograms per cubic meter in a 24-hour period. Today a high reading is 0.3 micrograms/m^3. What is the percent change?

25. Change these quarterly averages for lead readings in Illinois to nanograms/m^3. Which exceed the EPA limit?

a. Chicago (Cermak site), 2004; 0.06 micrograms/m^3

b. Granite City (15th & Madison), 2003: 0.34 micrograms/m^3

c. Chemetco 5N* (rural Madison county), 1999: 1.36 micrograms per m^3

d. Chemetco 5N*, 2001: 2.26 micrograms per m^3

26. Change these Missouri maximum 3-month averages to micrograms/m^3. Which exceed the EPA limit?

a. Glover (Big Creek site 5), 2001: 1200 ng/m^3

b. Dunklin High school, Herculaneum, 2001: 4500 ng/m^3

c. Doe Run's Broadway St. site 7, 2000: 6860 ng/m^3

d. Doe Run's site 7, 2005: 1930 ng/m^3

▶ Writing

27. What is it about shoe sizes of men and women that makes the topic a popular classroom investigation worldwide? Who needs to know about the distribution of sizes? (Check classroom investigations for yourself: Search "box and whisker" along with shoe size.)

28. Two box plots have the same numbers for Q_1 and Q_3, but one plot has longer whiskers on each side. Compare the data.

29. One piece of data was missing in the lead readings from Manila (see the Warm-up). Suggest two other ways to resolve missing-data problems, and give advantages or disadvantages of each.

▶ Projects

30. Penny Plot

a. Plot the data given below, with date on the horizontal axis. (The first number is the date; the second is the weight of the penny in grams.)

1983D, 2.501; 1994D, 2.510; 1969S, 3.161; 1982D, 2.518; 1972D, 3.107; 1964, 3.070; 1974, 3.130; 1967, 3.135; 1994D, 2.497; 1968D, 3.085; 1960D, 3.111; 1966, 3.100; 1977D, 3.084; 1963, 3.078; 1981D, 3.051; 1985D, 2.515; 1984D, 2.515; 1988D, 2.548; 1984D, 2.628; 1989D, 2.440; 1962, 3.037; 1973D, 3.134; 1979, 3.055; 1970, 3.140; 1991D, 2.538; 1978D, 3.100; 1980D, 3.119

b. What do you observe from your graph? Find a way to use the mean and standard deviation to justify your observation.

31. Alternative Box Plot Because box plots are relatively new, most applications on the Web explain how they defined their box plot. One variation on the box plots has whiskers stretching 1.5 times the distance ($Q_3 - Q_1$) away from the box, or the maximum and minimum values, whichever is closer to the box. *Data outside the whiskers* are called outliers. The Australians used this box plot for their air pollution comparisons (see Figure 41). This box plot is also available on the graphing calculator. On the plot menu, look for the box and whiskers option with dots to the right of the whisker, the fourth choice after **Type:** in Figure 42.

a. If $Q_3 - Q_1 = 40$, what is the maximum length of a whisker?

Blue numbers are core exercises.

*Readings were discontinued at this site in 4th quarter, 2001.

b. Draw both types of box plots for the data in Exercise 1. Name the outliers.

c. Find an example in another field on the Internet.

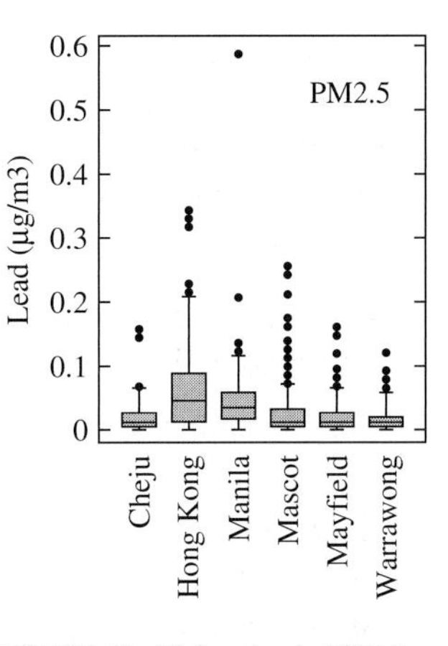

FIGURE 41 $PM_{2.5}$ lead, 2001*

Plot1 Plot2 Plot3
On Off
Type:
Xlist:L2
Freq:1
Mark:

FIGURE 42 Box plot with outlier

▶ 8 Chapter Summary

Vocabulary

For definitions and page references, see the Glossary/Index.

axis of symmetry
box and whisker plot
converse of the Pythagorean theorem
deviation
distance formula
equilateral triangle
extraneous roots
irrational number
isosceles triangle
measures of variation or dispersion
one-half as exponent
parabola
principal square root
proof
Pythagorean theorem
Pythagorean triples
quadratic formula
quadratic function
quartiles
radicals
radicand
range (of a data set)
sample standard deviation
vertex
zero product rule

Concepts

8.1 Pythagorean Theorem

Pythagorean theorem: If a triangle is a right triangle, then the sum of the squares of the lengths of the two shorter sides (legs) is equal to the square of the length of the longest side (hypotenuse). If the legs are of lengths a and b and the hypotenuse is of length c, then $a^2 + b^2 = c^2$.

Converse of the Pythagorean theorem: If the squares of the lengths of the sides of a triangle satisfy $a^2 + b^2 = c^2$, the triangle is a right triangle.

Two common Pythagorean triples are {3, 4, 5} and {5, 12, 13}.

8.2 Square Root Expressions and Properties and the Distance Formula

The principal square root of a number is the positive root.

The square root of a negative number is undefined in the set of real numbers.

A negative root or both positive and negative roots of a number are given only when specified with $-$ or $\pm$, respectively.

No portion of the graph of the square root function lies to the left of the vertical axis because the square root of a negative number is not a real number. No portion of the graph is below the x-axis because the output of the square root function is the principal, or positive, square root.

When one-half is used as an exponent, take the principal square root of the base:

$$a^{1/2} = \sqrt{a}, \quad a \geq 0$$

The square root property for products is given by

$$\sqrt{a \cdot b} = \sqrt{a} \cdot \sqrt{b}$$

The square root property for quotients is given by

$$\sqrt{\frac{a}{b}} = \frac{\sqrt{a}}{\sqrt{b}}$$

A simplified square root expression contains no perfect square factors inside the radical sign.

*Box and whisker plots for particulate matter appear in most Fine Particulate Aerosol Sampling Newsletters. See http://old-www.ansto.gov.au/nugeo/iba/news/news_letters/Newsletter33.pdf for a study that compares samples in 4 Australian cities. Permission was granted to use Figure 41 from Newsletter 26.

Distance formula: The distance between (x_1, y_1) and (x_2, y_2) is given by

$$d = \sqrt{(x_2 - x_1)^2 + (y_2 - y_1)^2}$$

8.3 Solving Square Root Equations and Simplifying Expressions

If x is zero or a positive number $(x \geq 0)$,

$$\sqrt{x^2} = x$$
$$\sqrt{x^4} = x^2$$
$$\sqrt{x^6} = x^3$$
$$\sqrt{x^8} = x^4$$

If x is any real number, the function $f(x) = \sqrt{x^2}$ is the same as the absolute value function, $f(x) = |x|$.

If x is any real number,

$$\sqrt{x^2} = |x|$$
$$\sqrt{x^4} = x^2$$
$$\sqrt{x^6} = |x^3|$$
$$\sqrt{x^8} = x^4$$

Square roots of odd powers of x always have positive inputs, $x \geq 0$.

To find the domain for a radical expression, limit the inputs to those that make the radicand zero or positive.

We can solve radical equations by table, graph, or squaring both sides.

8.4 Graphing and Solving Quadratic Equations

We can solve quadratic equations by table or graph (Section 8.4), taking the square root or factoring (Section 8.5), or the quadratic formula (Section 8.6).

The intercepts, vertex, and axis of symmetry are usually all you need to sketch the graph of a quadratic function, $y = ax^2 + bx + c$. If a, the coefficient on x^2, is positive, the graph opens upward; if a is negative, the graph opens downward. The vertical axis intercept is $f(0)$. The horizontal axis intercept (s) is solved from $f(x) = 0$.

8.5 Solving Quadratic Equations by Taking the Square Root or by Factoring

So long as we observe the restriction that the radicand is zero or positive and the output is zero or positive, we can take the square root of both sides of an equation.

The zero product rule permits us to set the factors on the left side of $ax^2 + bx + c = 0$ equal to zero. It may not always be possible to factor the left side.

8.6 Solving Quadratic Equations with the Quadratic Formula

The quadratic formula, $x = \dfrac{-b \pm \sqrt{b^2 - 4ac}}{2a}$, solves

$$ax^2 + bx + c = 0$$

To solve with the quadratic formula,

1. Place the equation in standard form, $ax^2 + bx + c = 0$.
2. Substitute a, b, and c into the quadratic formula.
3. Simplify the resulting expression, following the order of operations.
4. State and check the solutions.

See page 510 for hints on deciding on a solution method.

8.7 Range, Box and Whisker Plots, and Standard Deviation

The mean, median, and mode are measures of central tendency (Section 5.4). The mean is denoted by $\bar{x}$. The topics of this section are measures of variation or dispersion.

The range is the difference between the largest and smallest numbers in a data set.

To draw a box and whisker plot, find the median (after arranging the data in order from smallest to largest). Locate the median on a horizontal line with a scale appropriate to the data set. The quartiles are the midpoints of the first half and of the second half of the data. Locate the quartiles, Q_1 and Q_3, on the line. The box extends from Q_1 to Q_3. The whiskers extend out of the box to the minimum and maximum data points.

For a set of n numbers $x_1, x_2, x_3, \ldots, x_n$ with mean $\bar{x}$, drawn randomly from a population, the sample standard deviation is

$$s_x = \sqrt{\frac{(x_1 - \bar{x})^2 + (x_2 - \bar{x})^2 + (x_3 - \bar{x})^2 + \cdots + (x_n - \bar{x})^2}{n - 1}}$$

The sample standard deviation, s_x, uses division by $n - 1$.

▶ 8 Review Exercises

1. Which of these sets of numbers could represent the sides of a right triangle?

   a. $\{2, 3, 4\}$ b. $\{5, 17, 18\}$ c. $\{8, 15, 17\}$

2. Find the length of the side marked with an x.

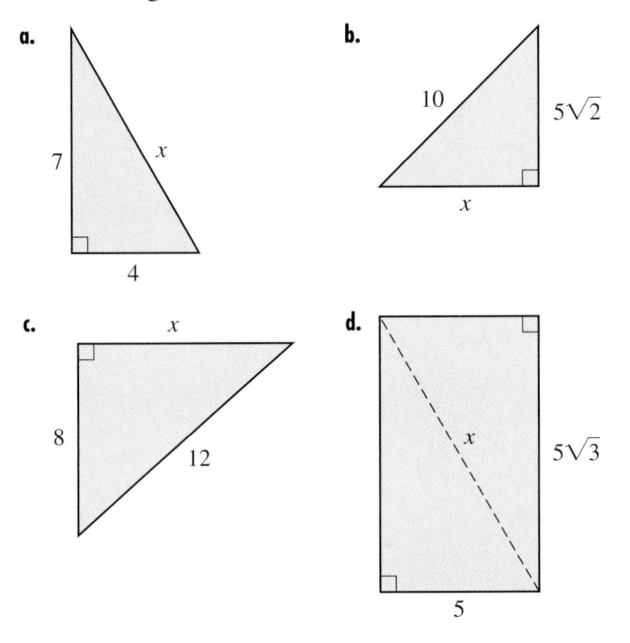

3. An extension ladder is to reach 24 feet up a wall. The safe ladder position for the base is 6 feet from the wall. How long a ladder is needed?

4. Simplify these exponent and radical expressions.

   a. $\sqrt{6} \cdot \sqrt{24}$ b. $(5\sqrt{2})^2$

   c. $\sqrt{72} \cdot \sqrt{2}$ d. $(2\sqrt{5})^2$

   e. $\sqrt{3} \cdot \sqrt{12}$

5. Which of the three expressions in each set has the same value as the radical given first?

   a. $\sqrt{60}$ $\{6\sqrt{10}, 10\sqrt{6}, 2\sqrt{15}\}$

   b. $\sqrt{63}$ $\{7\sqrt{3}, 9\sqrt{7}, 3\sqrt{7}\}$

   c. $\sqrt{54}$ $\{6\sqrt{3}, 3\sqrt{6}, 9\sqrt{6}\}$

Use the definitions of exponents to simplify each expression in Exercises 6 to 10. Try them without a calculator first.

6. a. 49^{-1} b. $49^{1/2}$ c. $49^{0.5}$ d. 49^0

7. a. $144^{1/2}$ b. 144^{-1} c. 144^0 d. $144^{0.5}$

8. a. $\left(\frac{1}{25}\right)^{-1}$ b. $\left(\frac{1}{25}\right)^{1/2}$ c. $\left(\frac{1}{25}\right)^0$ d. $\left(\frac{1}{25}\right)^{0.5}$

9. a. $(0.36)^{1/2}$ b. $(0.36)^{-1}$ c. $(0.36)^{0.5}$ d. $(0.36)^0$

10. a. $(0.25)^{-1}$ b. $(0.25)^0$ c. $(0.25)^{1/2}$ d. $(0.25)^{0.5}$

11. Use the distance formula to find the lengths of the sides and the diagonals of the four-sided shapes below. (These figures are not drawn to scale.) If the diagonals are equal, then the shape is a rectangle (or square). Which are rectangles?

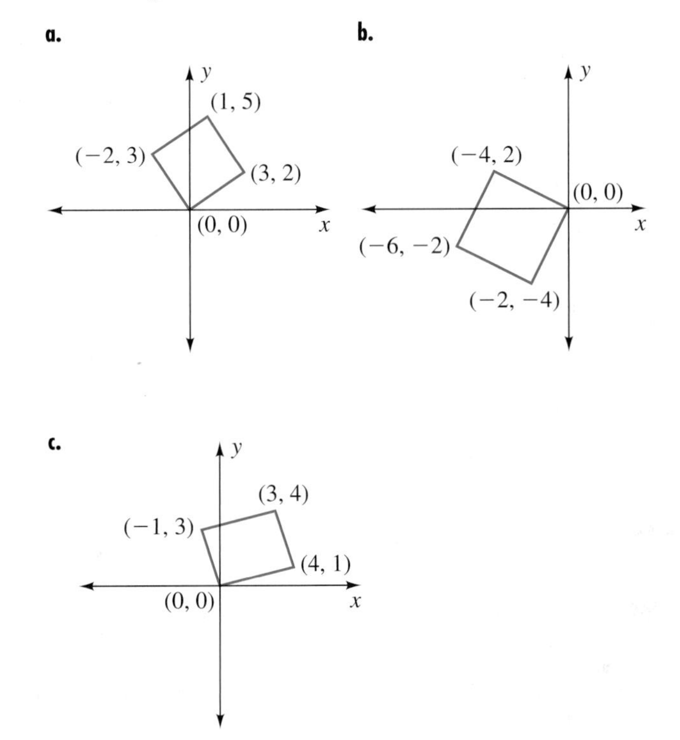

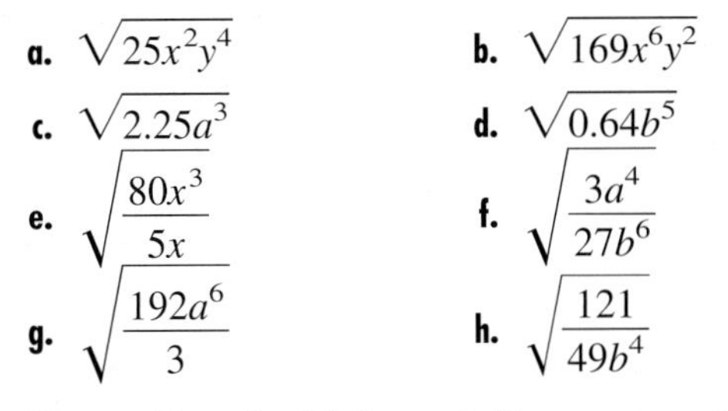

12. The sides of a square or rectangle are perpendicular. Find the slopes of two consecutive sides of each shape in Exercise 11. Verify that the two rectangles do have two consecutive perpendicular sides and the other shape does not.

13. Simplify these expressions. Assume that the variables represent only positive numbers.

   a. $\sqrt{25x^2y^4}$ b. $\sqrt{169x^6y^2}$

   c. $\sqrt{2.25a^3}$ d. $\sqrt{0.64b^5}$

   e. $\sqrt{\dfrac{80x^3}{5x}}$ f. $\sqrt{\dfrac{3a^4}{27b^6}}$

   g. $\sqrt{\dfrac{192a^6}{3}}$ h. $\sqrt{\dfrac{121}{49b^4}}$

14. Repeat Exercise 13 for variables representing any real number.

15. The f-stops on a camera lens are the numbers 1.4, 2, 2.8, 4, 5.6, 8, 11, and 16. Complete the table, and compare the results with the f-stops.

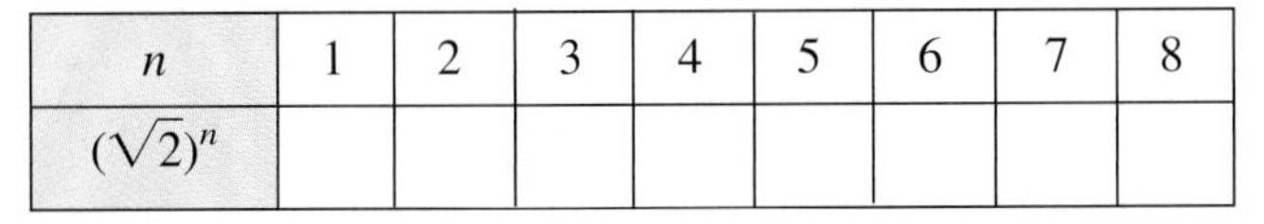

n	1	2	3	4	5	6	7	8
$(\sqrt{2})^n$								

16. Instead of building by "tradition," British engineer William Froude (1810–1879) made models of boat hull designs and towed them in a canal in order to observe and quantify their performance. One discovery was that the maximum speed, in knots, of displacement hulls is a function of the waterline length of the boat: $v = 1.34\sqrt{\text{length}}$.

a. What is the maximum speed for a 30-foot boat (at waterline)? A 60-foot boat?

b. When the length doubles, does the maximum speed double?

c. What length boat would have half the maximum speed of a 60-foot boat?

d. An international knot is 1852 meters per hour. A mile is 1609 meters. How many miles per hour are the answers in part a?

Note: Simplifying the idea that speed of a displacement hull is a function of length: at maximum speed there is one wave at the bow (front) and another at the stern (back) with none in between. The longer the distance between these waves, the greater the speed. Naval engineers watching the movie *Titanic* could tell by wave position that the *Titanic* they were seeing was a model.

17. Solve these equations. For what inputs is each radical expression defined?

a. $\sqrt{7x - 3} = 5$

b. $\sqrt{2 - x} = x - 2$

c. $\sqrt{4x} - 3 = 7$

d. $\sqrt{2x - 1} = x - 2$

18. Solve these formulas for the indicated variable.

a. $E = \sqrt{W \cdot R}$ for W

b. $V_o = \sqrt{\dfrac{GM}{R}}$ for R

c. $n = \dfrac{1}{2l}\sqrt{\dfrac{T}{m}}$ for T

d. $n = \dfrac{1}{2rl}\sqrt{\dfrac{T}{\pi d}}$ for T

19. On the moon, the distance seen in miles from a height of h feet is given by

$$d \approx \sqrt{\frac{3h}{8}}$$

a. How far can be seen from a height of 20 feet?

b. How high would an astronaut need to climb to see 4 miles?

In Exercises 20 and 21, solve for x with algebraic notation. Describe how to find the solution on the graph.

20. $\sqrt{x + 1} = 1$

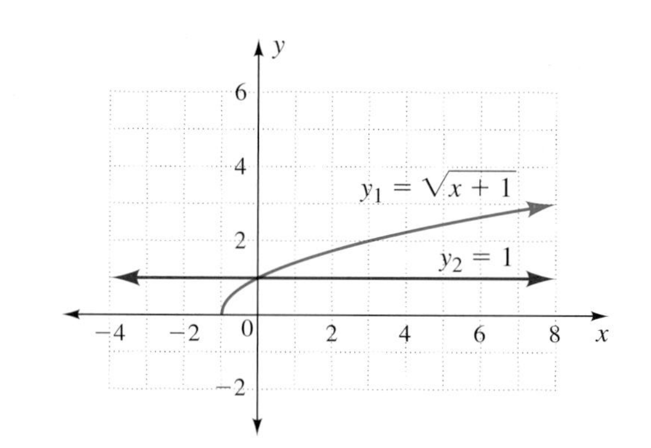

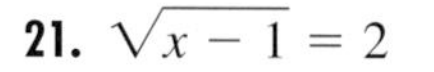

21. $\sqrt{x - 1} = 2$

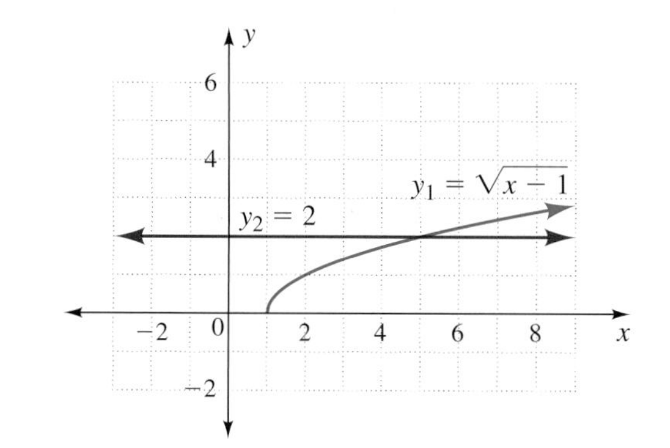

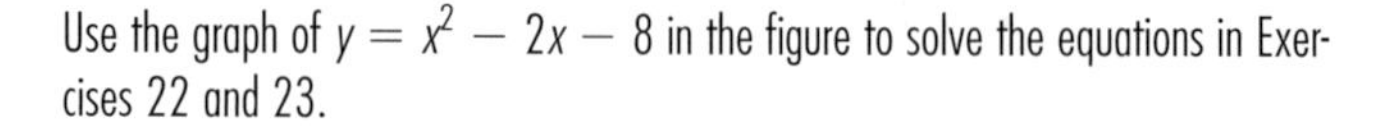

Use the graph of $y = x^2 - 2x - 8$ in the figure to solve the equations in Exercises 22 and 23.

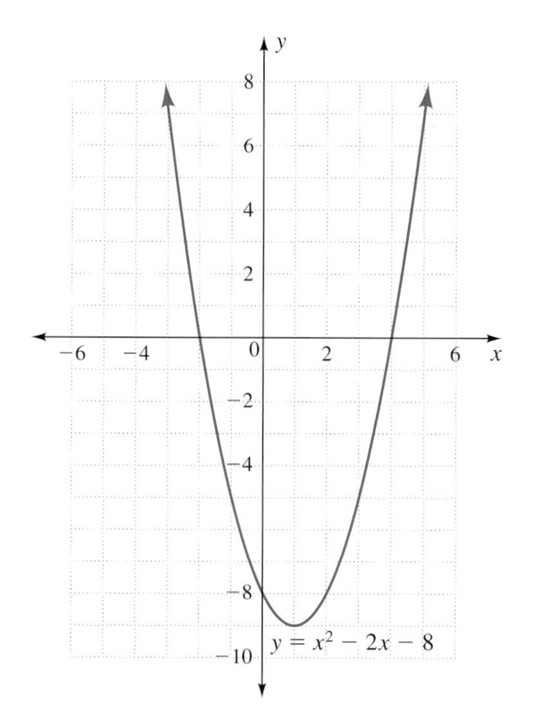

22. $x^2 - 2x - 8 = 8$

23. $x^2 - 2x - 8 = 0$

24. What are the vertex, axis of symmetry, y-intercept point, and x-intercept points of the graph for Exercises 22 and 23?

In Exercises 25 to 38, solve for x. Round irrational roots to the nearest thousandth. Use two methods on each.

25. $x^2 = \frac{16}{144}$

26. $2x^2 - 18 = 0$

27. $x^2 - 5x + 4 = 0$

28. $x^2 - 4x - 5 = 0$

29. $2x^2 + 5x - 6 = 0$

30. $2x^2 + 5x + 1 = 0$

31. $x^2 + 3x - 18 = 0$

32. $x^2 - 4x - 21 = 0$

33. $3x^2 = 4x + 7$

34. $4x - 3 + 4x^2 = 0$

35. $2x^2 + 3x + 5 = 0$

36. $8 - 4x + x^2 = 0$

37. $x^2 - 6x + 9 = 0$

38. $x^2 + 12x = -36$

39. Solve these formulas for the indicated variable. Assume variables and outputs take on positive values only.

a. $A = \pi r^2$ for r

b. $p = \frac{1}{2}dv^2$ for v

c. $S = 4\pi r^2$ for r

d. $h = \dfrac{v^2}{2g}$ for v

40. In traffic accident investigations, tire skid tests are used to find the coefficient of friction between tires and the road surface near an accident scene. An investigator measures the following tire skid marks, in feet, for a skid at 30 miles per hour:

Test 1: Left front, 50; right front, 49; left rear, 47; right rear, 48

Test 2: Left front, 47; right front, 50; left rear, 48; right rear, 51

a. Find the mean, range, and sample standard deviation (s_x) of the skid marks for each of the two tests. Round the standard deviation to two decimal places.

b. Find the coefficient of friction for each test with $f = \frac{S^2}{30D}$, where f is the coefficient of friction, S is the speed of the car making the tests, and D is the mean skid mark distance for the four tires.

41. Following are the winning times (in seconds) for the 500-meter event in speed skating at the past eight Olympic Games.

Men: 39.17, 38.03, 38.19, 36.45, 37.14, 36.33, 35.59, 34.42, 34.82

Women: 42.76, 41.78, 41.02, 39.10, 40.33, 39.25, 38.21, 37.30, 38.23

a. Find the range of each set of data.

b. Make a box and whisker plot for each set.

c. Find the mean and sample standard deviation (s_x) for each set.

42. Discuss how the average age of the set of automobiles observed in an athletic club parking lot might compare with that of the set of automobiles found in the lot of a major chain grocery store. Discuss the standard deviation in ages expected for the two locations.

▶ Chapter Projects

43. Dot Paper Areas Copy, trace, or draw several grids of dots like that shown below. The 25 dots are 1 unit apart in the horizontal and vertical directions.

. . . . .
. . . . .
. . . . .
. . . . .
. . . . .

a. By connecting dots, make squares on the grid. Find 8 squares of different sizes. Your squares must fit inside the 25-dot grids. Label each square with the length of its side (this is where the Pythagorean theorem comes in) and its area.

b. Think of the grid as being the first quadrant in a coordinate graph. Label each side of each square with its slope. Explain how the slopes of the sides of a square are related.

44. From Solutions to Equations The solutions to five different quadratic equations are given in parts a to e. Find an equation $y = x^2 + bx + c$ for each set of solutions.

a. $x = -3$ and $x = -2$

b. $x = -1$ and $x = 2$

c. $x = 4$ and $x = -3$

d. $x = 5$ and $x = 2$

e. $x = \frac{1}{2}$ and $x = 5$

▶ 8 Chapter Test

1. Which of these sets of numbers could represent the sides of a right triangle?

a. $\{3, 4, 5\}$ **b.** $\{1, 2, 3\}$

c. $\{1, \sqrt{3}, 2\}$ **d.** $\{\sqrt{2}, \sqrt{2}, 2\}$

2. Which of the three expressions in each set has the same value as the radical given first?

a. $\sqrt{45}$ $\{9\sqrt{5}, 5\sqrt{9}, 3\sqrt{5}\}$

b. $\sqrt{44}$ $\{4\sqrt{11}, 2\sqrt{11}, 11\sqrt{4}\}$

3. Romeo plans to use an extension ladder to reach Juliet's balcony, 20 feet above the ground. He will set the base of the ladder 5 feet away from the wall below the edge of the balcony. To preserve this safe ladder position, how long a ladder does he need?

4. Simplify these expressions. Assume that the variables represent positive numbers.

a. $5^{\frac{1}{2}} \cdot 20^{\frac{1}{2}}$ **b.** $\sqrt{270}$ **c.** $(3\sqrt{6})^2$

d. $(3\sqrt{a})^2$ **e.** $\sqrt{36x^2y}$ **f.** $\sqrt{0.81x^4y^3}$

g. $\sqrt{\dfrac{147}{3}}$ **h.** $\dfrac{\sqrt{18}}{\sqrt{32}}$ **i.** $\sqrt{\dfrac{a^5b^2}{a}}$

5. Simplify. Assume the variables represent any real number.

a. $\sqrt{a^5}$ **b.** $\sqrt{x^6}$

c. $(\sqrt{a})^2$ **d.** $\sqrt{y^7}$

6. Find the missing sides of the similar right triangles below. Use the marks in the angles to determine which sides are proportional. Round decimals to the nearest tenth.

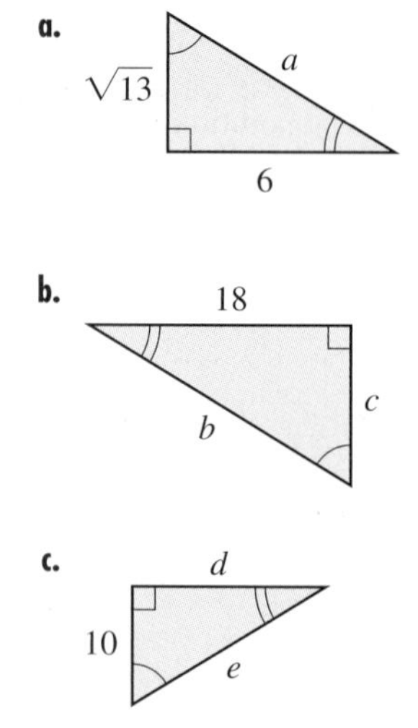

7. Solve with the graph: $x^2 - 3 = 1$.

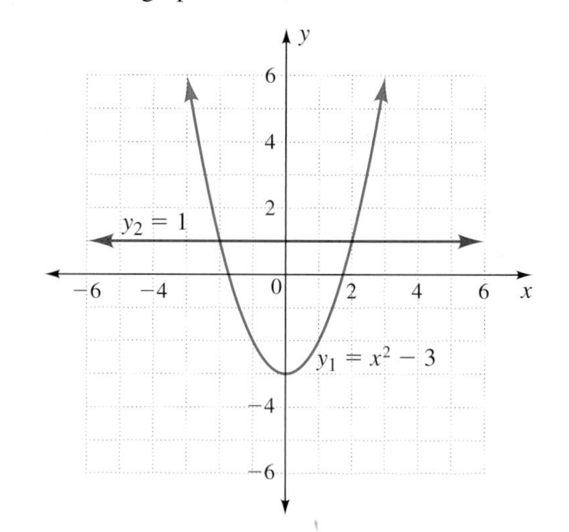

8. Solve for all inputs, x, that make these equations true. Use an inequality or an interval to indicate where inputs are defined.

a. $x^2 = \frac{36}{121}$ **b.** $\sqrt{3 - x} = x + 3$

c. $x^2 + x - 2 = 0$ **d.** $\sqrt{5x - 6} = 12$

e. $2x^2 = 8 - 15x$ **f.** $8x^2 - 5x = 4$

g. $(3x + 8)(3x - 8) = 0$ **h.** $4x^2 + 8 = 0$

i. $x^2 - 6x + 9 = 0$ **j.** $4x^2 - 25 = 0$

k. $4x^2 - 10x - 24 = 0$ **l.** $3x^2 + 34x - 24 = 0$

9. Solve these formulas for the indicated variable. Assume all variables and outputs represent positive numbers.

a. $V_e = \sqrt{\dfrac{2GM}{R}}$ for R **b.** $E = \frac{1}{2}mv^2$ for v

c. $E = \dfrac{kH^2}{8\pi}$ for H

10. Hydroplaning occurs when a tire slides on the surface of the water on a pavement instead of gripping the pavement's surface. The *Advanced Pilot's Flight Manual* gives the relationship between the minimum hydroplaning speed s, in miles per hour, and tire pressure t, in pounds per square inch, as

$$s = 8.6\sqrt{t}$$

The implication of this formula may not be obvious. Perhaps these thoughts will help: The softer the tire, the greater the surface area and the greater the tendency to slide along the surface, or hydroplane. A harder tire tends to cut through the water to the paved surface.

a. If the tire pressure is 36 pounds per square inch, what is the speed at which the tire will hydroplane?

b. If the tire pressure is 100 pounds per square inch, what is the speed at which the tire will hydroplane?

c. If a plane lands at 120 miles per hour on wet pavement, what is the tire pressure at which hydroplaning will occur?

11. Each of the following shapes has an area of 50 square feet. What is the length of x in each case? Round to the nearest thousandth.

a. Square, side x:

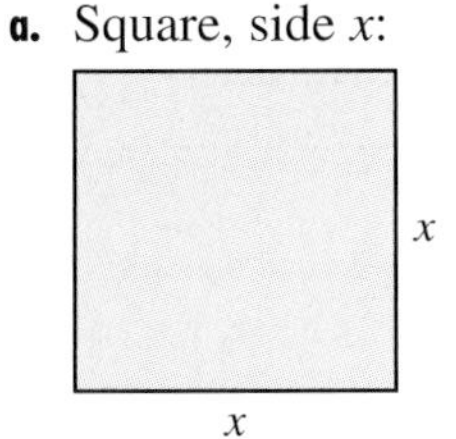

b. Circle, diameter x:

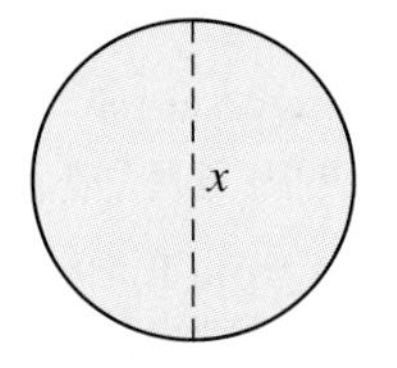

c. Equilateral triangle with area $A = \dfrac{x^2\sqrt{3}}{4}$:

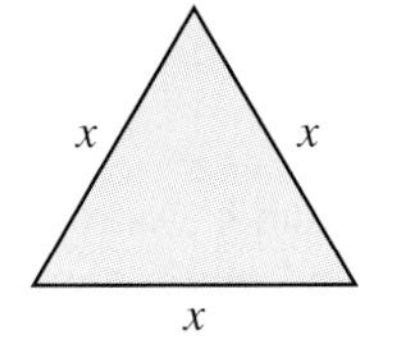

12. The coordinates (3, 5), (6, 8), and (3, 11) form a triangle.

a. Sketch them on coordinate axes.

b. What is the equation of a line passing through each pair of points?

c. What is the length of each side of the triangle?

d. Use slope to determine whether any two sides are perpendicular.

e. What kind of a triangle is this triangle? Explain why.

13. Explain when the distance between two points on a graph can be found from the slope of the line segment connecting them. Give an example of when it is possible and when it is not.

14. For the two equations $y_1 = \sqrt{(2 - x)^2}$ and $y_2 = (\sqrt{2 - x})^2$:

a. Are they equivalent equations? Simplify each.

b. Describe the difference in their graphs.

15. a. Do these sets of numbers satisfy the Pythagorean theorem?

$\{6, \sqrt{13}, 7\}, \{8, \sqrt{17}, 9\}, \{10, \sqrt{21}, 11\}$

b. Does this set satisfy the Pythagorean theorem?

$\{a, \sqrt{a + b}, b\}$

c. Does this set satisfy the Pythagorean theorem?

$\{n, \sqrt{n + (n + 1)}, n + 1\}$

d. Why is part c possible to answer, whereas part b is not?

16. The table below gives food energy and sodium content for a variety of dry cereals.

Dry Cereal	Food Energy (calories)	Sodium Content (milligrams)
Cap'n Crunch®	120	145
Froot Loops®	110	145
Super Golden Crisp®	105	25
Sugar Frosted Flakes®	110	230
Sugar Smacks®	105	75
Trix®	110	181

a. Find the mean and sample standard deviation (s_x) for food energy.

b. Find the median and make a box and whisker plot for sodium content.

▶ Cumulative Review of Chapters 1 to 8

Describe each of the situations in Exercises 1 and 2 with an expression showing the appropriate operation. Then answer the question.

1. A surgical procedure takes $1\frac{1}{4}$ hours. How many procedures can be scheduled in a 12-hour day?

2. An attorney spends $3\frac{5}{6}$ hours on a client's business. What is the total bill to the client at $180 per hour?

3. Simplify 3($15) + 2(−$20) − ($10) − (−$25).

4. Simplify $-3(4) - 4(-5) + (-4)^2$.

5. Simplify $\dfrac{3x^3y^2}{27xy^3}$. State any conditions.

6. a. What are the perimeter and area of a square with sides of 90 feet (a baseball infield, or diamond)?

b. What are the perimeter and area of a square with sides of 12 meters (a gymnastics mat)?

c. Use unit analysis to write a ratio comparing the perimeter in part a to the perimeter in part b.

d. Use unit analysis to write a ratio comparing the area in part a with the area in part b.

Solve the equations in Exercises 7 to 10.

7. $2 - 3x = 26$

8. $3(x - 5) = x + 9$

9. $3(x + 4) = 2(1 - x)$

10. $\frac{2}{3}x - 4 = 26$

11. Solve $2x - 1 \leq 2 - x$ with a graph and with symbols.

12. Solve $A = \frac{1}{2}bh$ for h.

13. If $f(x) = 3x - 2x^2$, what are $f(-1)$, $f(0)$, $f(1)$, and $f(2)$?

14. Sketch a line with slope $\frac{2}{3}$ that passes through $(2, -1)$.

15. Write a linear equation with slope $= \frac{4}{3}$ and y-intercept -2.

16. a. Make a table and graph for $y = \frac{1}{3}x - 2$. Circle the point that shows the solution for each of these equations. Solve the equation.

b. $\frac{1}{3}x - 2 = 0$

c. $\frac{1}{3}x - 2 = -1$

d. $\frac{1}{3}x - 2 = -4$

17. Build an input-output table, graph, and equation for the value remaining on a \$20 prepaid transit card where each trip costs \$2.25. Let the input be the number of trips from 0 to 10, counting by 2.

18. Payment for e-mail began when Yahoo and AOL offered businesses delivery that bypassed spam filters for $\frac{1}{4}$ cent to 1 cent per e-mail. Write an equation to find the number of e-mails that could be sent at each price for a \$10,000 cost.

Write proportions for Exercises 19 to 22. Make an estimate of the answer before solving.

19. What is 125% of 52?

20. 35% of what number is 21?

21. 33 is what percent of 88?

22. Brawny® Paper Towels were \$0.89 a roll in 1996. In 2003, they were \$1.69 a roll. What is the percent change in price?

23. A sneeze may reach 100 miles per hour. How far in feet could sneezed bacteria travel in 2 seconds? Solve with a unit analysis.

24. Solve for x: $\frac{3x + 2}{7} = \frac{5(x - 2)}{9}$.

25. Multiply.

a. $(x + 3)(2x - 1)$

b. $x(x + 1)(3x - 1)$

26. Factor.

a. $x^2 - 4x - 21$

b. $x^2 - 4x$

c. $4x^2 - 10x$

d. $4x^2 - 11x - 3$

e. $4x^2 - 4x + 1$

f. $4x^2 - 9$

27. a. The ratio of the distance the sun is from Earth to the diameter of the sun is 1.496×10^8 kilometers to 1.3914×10^6 kilometers. Divide the ratio.

b. The ratio of the distance the moon is from Earth to the diameter of the moon is 384,000 kilometers to 3480 kilometers. Divide the ratio.

c. Compare the quotients in parts a and b. Astronomers suggest that this is why the sun and moon appear to be of similar size in the sky.

Solve the systems of equations in Exercises 28 and 29.

28. $x + y = -13$
$x - y = 23$

29. $7x - 8y = -25$
$3x + 4y = -7$

30. Two angles are supplementary. The measure of one angle is 5° greater than four times that of the other. Write a system of equations to find the angle measures, and then solve the system.

31. The length of one leg of a right triangle is three more than the length of the other leg. The hypotenuse is 15. What are the lengths of the legs?

32. Solve $4x^2 - 20 = 0$.

33. a. Sketch $y = 2x^2 - 5x - 3$.

b. Solve $2x^2 - 5x - 3 = 0$ from the graph.

c. Solve $2x^2 - 5x - 3 = 0$ by factoring.

d. Solve $2x^2 - 5x - 3 = 0$ with the quadratic formula.

34. For what values is each equation defined? Solve the equation.

a. $\sqrt{x + 24} = 3$

b. $\sqrt{2x + 3} = -x$

35. The data below are for 6″ Subway sandwiches.

Sandwiches	Fat (g)	Cholesterol (mg)	Calories
Subway club®	5	26	312
Turkey breast & ham	5	24	295
Veggie delite™	3	0	237
Turkey breast	4	19	289
Ham	5	28	302
Roast beef	5	20	303
Roasted chicken breast	6	48	348

SUBWAY® regular 6″ subs include bread, veggies, and meat. Addition of condiments or cheese alters nutrition content.

a. What are the mean, median, and mode for the grams of fat?

b. Draw a box and whisker plot for the milligrams of cholesterol.

c. What are the range, mean, and sample standard deviation for the calories?

36. A coupon for Gaterade gives you a price of 4 for \$3 on the first 12 bottles. Additional bottles cost \$1.19. Write an equation showing how many bottles can be purchased for \$20. How much change is left?

CHAPTER 9

Rational Expressions and Equations

You are behind schedule. You have 18 minutes to travel 10 miles. The posted speed is 55 miles per hour. Traffic is flowing just under the posted speed. Are you going to make up time by exceeding the posted speed limit? This chapter will provide examples to show situations where little time can be saved at high speeds (as compared to traveling at the speed limit).

In this chapter, we work with operations on expressions and equations containing fraction notation. We look at zero in the denominator in Section 9.1, simplifying expressions in Section 9.2, multiplication and division in Section 9.3, adding and subtracting in Section 9.4, and solving equations in Section 9.5.

FIGURE 1 To speed or not to speed

▶ 9.1 Rational Functions: Graphs and Applications

Objectives

- Identify rational expressions.
- Identify inputs that create zero denominators.
- Explore graphical behavior near inputs that create undefined expressions.
- Use graphs to investigate applications of rational expressions.

> **WARM-UP**
>
> Suppose you have 10 miles to travel. How long will it take? The distance, rate, and time formula is $D = rt$. Make an input-output table showing the rate of travel as input and the time required to go 10 miles as output. Use inputs from 0 to 10 mph, counting by 2. Does it look hopeful for traveling 10 miles in 18 minutes?

THIS SECTION INTRODUCES rational functions. We look at places where rational functions are undefined, graphs of rational functions, and a selection of applications.

▶ Rational Functions

Recall that **rational numbers** are *the set of numbers that may be written as the ratio of two integers a/b, with b not equal to zero*. The **rational function** $f(x)$ is *a function that may be written as the ratio of two polynomials* $\frac{p(x)}{q(x)}$, *where the denominator is not zero.*

Examples of rational functions are

$$f(x) = \frac{x}{x+1}, \qquad f(x) = \frac{x^2 + 2x + 1}{x - 3}, \qquad \text{and} \qquad f(a) = \frac{a+1}{a-1}$$

Some formal definitions of rational functions exclude fractions that have only whole numbers in the denominator, such as $\frac{x+3}{2}$. We will include such expressions because the algebraic techniques are the same. Expect lots of other fraction work as well.

▶ Division by Zero

▶ EXAMPLE 1 Evaluating a rational function By hand or with a calculator, evaluate $f(x) = \frac{x}{x-4}$ and $f(x) = \frac{x+3}{(x+2)(x-3)}$ for $-2 \le x \le 4$. Is there anything unusual about the answers?

SOLUTION Extra parentheses are required to enter the functions in Y= (see Figure 2). Error messages then appear in several places in the table (see Figure 3). These errors indicate values of $y = f(x)$ that are undefined because of a zero denominator.

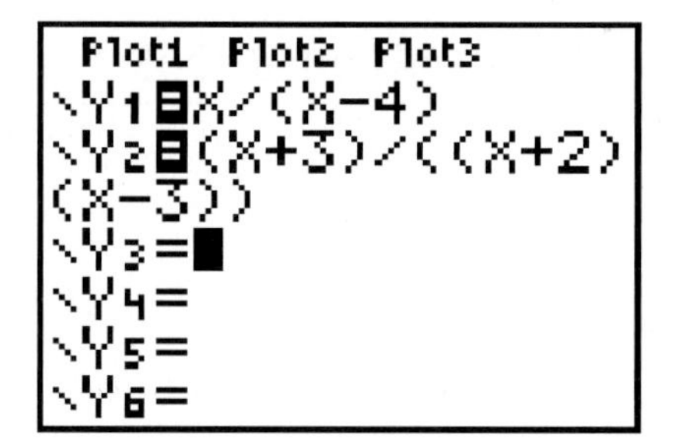

FIGURE 2 Rational functions

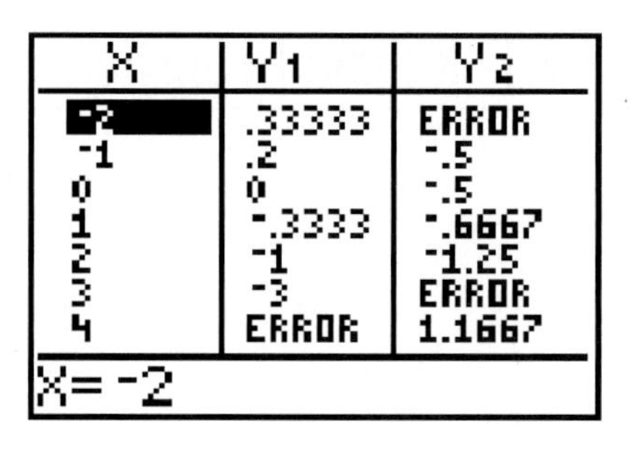

FIGURE 3 Table with ERROR ◀

In the table in Figure 3, ERROR appears when the input creates a zero in the denominator. In rational functions, we must always identify numbers that create a zero denominator, for which the function is undefined, and exclude them from the set of inputs (the domain).

▶ **EXAMPLE 2** Finding the domain (set of inputs) What inputs must be excluded for each rational function to be defined? State the domain.

a. $f(x) = \dfrac{1}{x}$ **b.** $f(x) = \dfrac{x}{x - 4}$ **c.** $f(x) = \dfrac{x + 3}{x^2 - x - 6}$

SOLUTION Inputs that give zero denominators lead to undefined expressions.

a. When $x = 0$,

$$f(x) = \frac{1}{x} = \frac{1}{0}$$

The domain, or set of inputs, is the set of all real numbers x, $x \neq 0$.

b. When $x = 4$,

$$f(x) = \frac{x}{x - 4} = \frac{4}{4 - 4} = \frac{4}{0}$$

The domain is the set of all real numbers x, $x \neq 4$.

c. To see where $x^2 - x - 6 = 0$, we need to solve the quadratic equation. We factor the denominator:

$$f(x) = x^2 - x - 6 = (x + 2)(x - 3)$$

When $x = -2$,

$$f(x) = \frac{x + 3}{(x + 2)(x - 3)} = \frac{-2 + 3}{(-2 + 2)(-2 - 3)} = \frac{1}{(0)(-5)} = \frac{1}{0}$$

or when $x = 3$,

$$f(x) = \frac{x + 3}{(x + 2)(x - 3)} = \frac{3 + 3}{(3 + 2)(3 - 3)} = \frac{6}{(5)(0)} = \frac{6}{0}$$

The domain is the set of all real numbers x, $x \neq -2$, $x \neq 3$. ◀

The notation for the set of all real numbers is $\mathbb{R}$. The domains in Example 2 are $\mathbb{R}$, $x \neq 0$ for part a, $\mathbb{R}$, $x \neq 4$ for part b, and $\mathbb{R}$, $x \neq -2$, $x \neq 3$ for part c.

When we work with graphing and solving equations in this chapter, we will state numbers that must be excluded from the set of inputs. To save time and space, however, we will not state excluded numbers when working with expressions unless specifically requested to do so.

Graphs of Rational Expressions

If an expression is undefined for a certain input, there is no point on the graph of the expression for that input. Furthermore, the graph near such a point has unusual features, as shown in Example 3.

EXAMPLE 3 Exploring the graph as the denominator value approaches zero What happens to the graph of $y = \frac{2}{x}$ as x gets close to zero?

SOLUTION Tables 1 and 2 show outputs for $y = \frac{2}{x}$ as x gets close to zero. Set a calculator window with ZOOM **4 : ZDecimal** and TRACE to find the table values from the graph. The data in each table form a curve when graphed (see Figure 4). If we trace the third-quadrant curve from left to right, we find that as the inputs, x, approach zero, the curve turns downward. Although the curve gets close to the y-axis, it never touches the y-axis. We describe this behavior by saying *y approaches negative infinity as x approaches zero.*

TABLE 1 $y = 2/x$ as x approaches zero from the left

Input x	Output $y = 2/x$
-4	$2/-4 = -0.5$
-2	$2/-2 = -1.0$
-1	$2/-1 = -2.0$
-0.5	$2/-0.5 = -4.0$
-0.2	$2/-0.2 = -10.0$
-0.1	$2/-0.1 = -20.0$

TABLE 2 $y = 2/x$ as x approaches zero from the right

Input x	Output $y = 2/x$
4	$2/4 = 0.5$
2	$2/2 = 1.0$
1	$2/1 = 2.0$
0.5	$2/0.5 = 4.0$
0.2	$2/0.2 = 10.0$
0.1	$2/0.1 = 20.0$

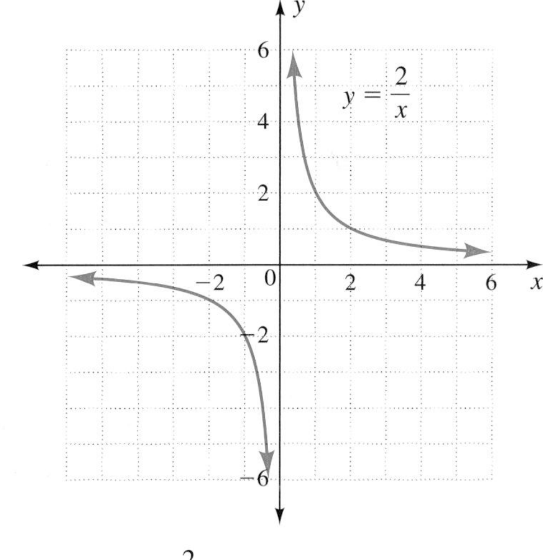

FIGURE 4 $y = \frac{2}{x}$

If we trace the first-quadrant graph from right to left, we see that as the inputs, x, approach zero, the curve rises. As the curve rises, it gets closer to the y-axis, but, like the third-quadrant curve, it never touches the y-axis. We describe this behavior by saying *y approaches positive infinity as x approaches zero.*

Because the equation $y = \frac{2}{x}$ has no output at $x = 0$, we say *the equation is defined for the set of real numbers x,* $x \neq 0$, *or* $\mathbb{R}$, $x \neq 0$. ◀

THINK ABOUT IT 1: In Example 3 for what inputs is $y < -2$? $y > 2$?

> The graph of a rational expression (simplified to lowest terms) approaches infinity in a vertical direction whenever the denominator approaches zero.

Together, the two curves in Figure 4 form a *rectangular hyperbola.* The name comes from the fact that the curve approaches but does not intersect the rectangular coordinate axes. Look for these curves and changes in their position in the remainder of this chapter.

EXAMPLE 4 Exploring the graph as the denominator value approaches zero Make a table and graph for $y = \frac{-2}{x + 4}$. What inputs must be excluded? What do you observe about the behavior of the graph near that input?

SOLUTION The expression has a zero denominator for $x = -4$; thus, $x \neq -4$. The graph in Figure 5 becomes nearly vertical as we approach $x = -4$ from either the left (Table 3) or the right (Table 4).

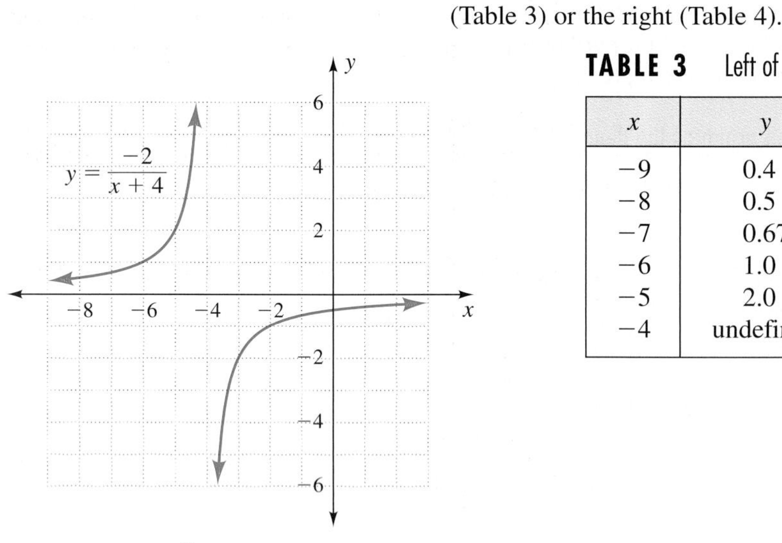

FIGURE 5 $y = \frac{-2}{x + 4}$

TABLE 3 Left of $x = -4$

x	y
−9	0.4
−8	0.5
−7	0.67
−6	1.0
−5	2.0
−4	undefined

TABLE 4 Right of $x = -4$

x	y
−4	undefined
−3	−2
−2	−1
−1	−0.67
0	−0.5
1	−0.4

◀

Before we examine applications with rational expressions, two facts should be noted about graphing calculators and zero denominators. First, as we trace along the graph of a rational expression, the output will be blank each time the input creates a zero denominator. It often takes considerable adjustment of the viewing window (Figure 6) to see the blank (Figure 7). (ZOOM **4 : ZDecimal** may help.)

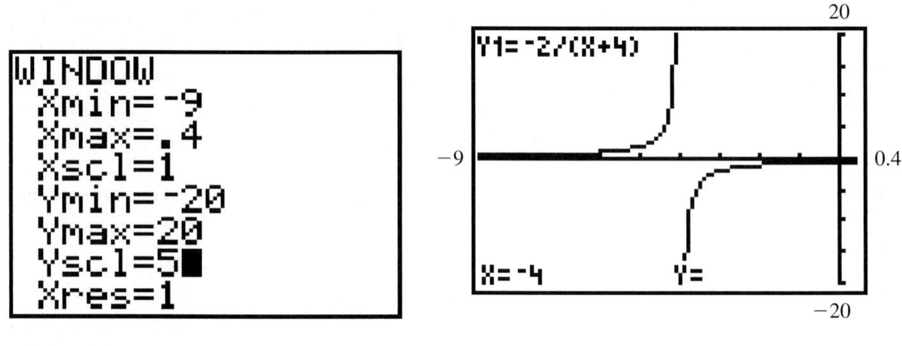

FIGURE 6 Window for trace in tenths

FIGURE 7 Correct graph

Second, in certain viewing-window settings (Figure 8), the calculator draws an almost vertical line on the graph (Figure 9) at an undefined point. This line is an error made by the calculator. The calculator evaluates the functions to the left and right of the undefined point and connects the ordered pairs. The line will disappear if the calculator is set on **Dot** mode rather than **Connected** mode, under MODE.

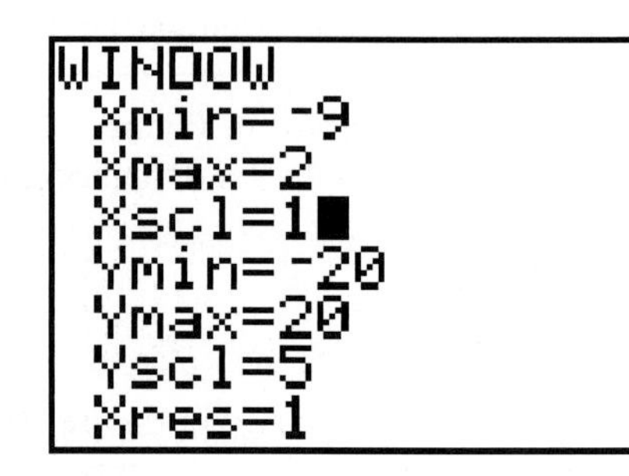

FIGURE 8 Window causing connection across $x = -4$

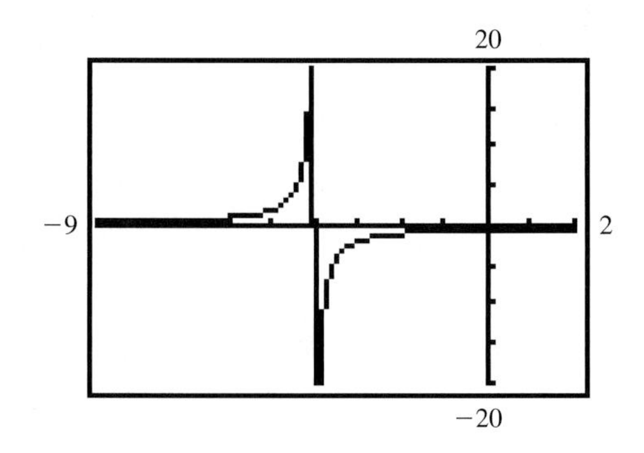

FIGURE 9 False connection across $x = -4$

Thus, the length and width are 16 centimeters and 3 centimeters, respectively. The mathematical results do not specify that length be the larger number, as is customary in everyday use. ◀

RESOURCE MANAGEMENT In the next example, we examine the application of rational expressions in resource management. Many resources, such as minerals, are limited in quantity. Strategic planners estimate the total quantity available and attempt to plan for future shortages.

▶ **EXAMPLE 7** Finding the number of years' supply of a resource: resource management Silver is used to produce electrical and electronic products and photographic film, as well as flatware and jewelry. The estimated world reserves of silver were 570,000 metric tons in 2004.

a. If the rate of use of silver reserves is x metric tons per day, write an equation for how many years the reserves will last.

b. If the 2004 rate of use of silver was 40 metric tons per day, how many years will the 2004 reserves last?

c. Graph the equation, letting x be use in metric tons per day and y be supply in years.

SOLUTION **a.** The reserves will last y years, as shown by the equation

$$y = \frac{570{,}000}{365x}, \quad x \neq 0$$

b. If the world silver reserves are used up at the rate of 40 metric tons per day, the 2004 reserves will last y years:

$$y = \frac{570{,}000 \text{ metric tons}}{\dfrac{365 \text{ days}}{1 \text{ year}} \cdot \dfrac{40 \text{ metric tons}}{1 \text{ day}}}$$

$$y = 39 \text{ yr}$$

c. The graph of $y = 570{,}000/365x$ is shown in Figure 13.

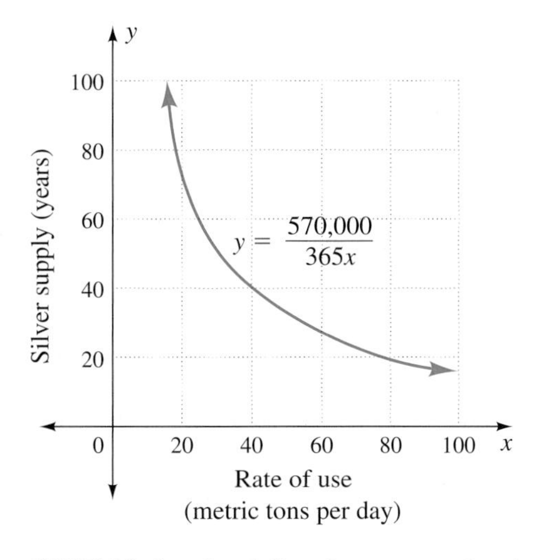

FIGURE 13 Supply of silver in years as a function of daily use ◀

In Sections 9.2 and 9.3, we will practice working with expressions containing units, such as those in part b of Example 7. For now, we observe that *metric tons* and *days* both cancel and only *years* remain.

Reproduce the graph in Figure 13 on a calculator with **Xmin** = 0, **Xmax** = 125, **Ymin** = 0, and **Ymax** = 125. Place parentheses around $365x$. Trace to answer the following "Think about it."

THINK ABOUT IT 3: If we use 16 metric tons per day, how many years will the silver reserves last? If we use 100 metric tons per day, how many years will the silver reserves last? Finish this sentence: As the rate of use increases, the number of years' supply ____.

ANSWER BOX

Warm-up: The ordered pairs for the table are (0, (undefined)), (2, 5), (4, 2.5), $(6, 1\frac{2}{3})$, $(8, 1\frac{1}{4})$, (10, 1). **Think about it 1:** x is between -1 and 0; x between 0 and 1. **Think about it 2:** Between (1, 10) and (2, 5), the "slope," $\frac{\Delta t}{\Delta r}$, is $-5/1$. Between (2, 5) and (5, 2), $\frac{\Delta t}{\Delta r} = \frac{-3}{3}$, or -1. Between (5, 2) and (10, 1), $\frac{\Delta t}{\Delta r} = \frac{-1}{5}$. The "slope" changes from $\frac{-5}{1}$ to $\frac{-1}{5}$. The slope of a linear equation is constant. **Think about it 3:** From the calculator graph we can estimate an output of 74 years for an input near 16 metric tons per day. For 100 metric tons per day, the number of years drops to about 12. As the rate of use increases, the number of years' supply decreases.

▶ 9.1 Exercises

1. **a.** Complete the table for $y = \frac{1}{x+1}$.

-4	-3	-2	-1.5	-1.25	-1	-0.75	-0.5	0	1	2	3

b. Is there an input that makes $y = 0$?

c. Does the graph of the equation cross the x-axis?

d. What is "wrong" with the output at $x = -1$?

2. **a.** Complete the table for $y = \frac{3}{x-1}$.

-3	-2	-1	0	0.25	0.5	1	1.5	1.75	2	3	4

b. Is there an input that makes $y = 0$?

c. Does the graph of the equation cross the x-axis?

d. What is "wrong" with the output at $x = 1$?

3. For the equation $y = \frac{20}{x}$,

a. List three numbers that make y larger than 20.

b. List three numbers that make y smaller than -20.

4. For the equation $y = \frac{5}{x-1}$,

a. List three numbers that make y larger than 5.

b. List three numbers that make y smaller than -5.

5. For the equation $y = \frac{2}{x+2}$,

a. What interval of numbers makes y larger than 2?

b. What interval of numbers makes y smaller than -2?

6. For the equation $y = \frac{-8}{x}$,

a. What interval of numbers makes y smaller than -8?

b. What interval of numbers makes y larger than 8?

7. When entering $y = \frac{x}{x-5}$ into a calculator, what must we do to the denominator?

8. If we write $y = x/x - 5$, what is the value of y at $x = 1$? at $x = 2$? at any x, x not equal to 0? Explain.

What inputs are defined for each expression in Exercises 9 to 16?

9. $\frac{2}{x+1}$

10. $\frac{3}{x-1}$

11. $\frac{x}{2x-1}$

12. $\frac{x}{2x+3}$

Blue numbers are core exercises.

13. $\dfrac{x}{3x + 1}$

14. $\dfrac{x}{3x - 2}$

15. $\dfrac{5}{x^2 + 4x + 3}$

16. $\dfrac{3x}{x^2 - 6x + 5}$

Using a calculator to build a table and a graph or using a graphing calculator, describe the behavior of the graph of each equation in Exercises 17 to 20 as x approaches the indicated number.

17. $y = \dfrac{x}{x - 5}$, as x approaches 5

18. $y = \dfrac{-x}{x + 3}$, as x approaches -3

19. $y = \dfrac{x + 3}{(x + 2)(x - 3)}$, as x approaches -2

20. $y = \dfrac{x + 3}{(x + 2)(x - 3)}$, as x approaches 3

21. In Example 5, $r = 0$ makes the equation $t = \dfrac{10}{r}$ undefined. How do we interpret $r = 0$ in the problem setting?

22. In Example 5, Figure 10 shows a first-quadrant graph only. Give five (r, t) coordinates from other quadrants that make the equation $t = \dfrac{10}{r}$ true. Why are these points most likely not relevant to the problem situation?

23. In the equation $t = \dfrac{10}{r}$, is there an input r that makes $t = 0$?

24. As time, t, changes from 0.1 to 0.01 hour, what happens to the rate, r, in the equation $t = \dfrac{10}{r}$?

25. Do you gain enough time by driving fast to risk getting a speeding ticket? Find how long in hours it takes to travel 20 miles at these rates. Next, change your answers to minutes. (First solve $D = rt$ for t.)

a. 30 mph　**b.** 40 mph

c. 50 mph　**d.** 60 mph

e. 70 mph　**f.** 80 mph

26. Suppose you have 6 miles to travel on a congested freeway. How long in hours would it take you to travel that distance at each of these speeds? Change to minutes.

a. 3 mph (walk)　**b.** 6 mph (jogging)

c. 10 mph　**d.** 20 mph (bicycle)

e. 60 mph　**f.** 0 mph

27. At what rate, in miles per hour, would you need to drive to travel 3 miles in these times. (First solve $D = rt$ for r.)

a. $\frac{1}{4}$ hour (female long distance runner)

b. $\frac{1}{5}$ hour　**c.** 7 minutes (bicycle)

d. 30 minutes

e. 13.25 minutes (male long-distance runner)

28. Suppose you have 60 miles to travel. At what rate would you need to drive to travel the distance in these times? (*Hint:* First solve $D = rt$ for r.)

a. 2 hours　**b.** 1 hour

c. $\frac{5}{6}$ hour　**d.** $\frac{3}{4}$ hour

e. 40 minutes　**f.** 90 minutes

29. a. List ten possible widths and lengths of rectangles with area of 30 square inches. Use the table headings shown. Calculate the perimeter for each rectangle.

Width	Length	Area	Perimeter
		30	
		30	

b. Graph the length (x-axis) and the perimeter (y-axis).

c. Where is the location of the point on the graph describing the smallest perimeter?

d. Find the equation (perimeter, y, in terms of length, x) describing the graph.

30. Refer to the table completed in Exercise 29.

a. Graph the length and width pairs on coordinate axes, placing width on the x-axis and length on the y-axis. What shape is formed?

b. What length and width give a perimeter smaller than 22 for an area of 30?

c. Find the equation (length y in terms of width x) describing the graph.

31. Estimated world crude oil reserves in 2004 were about 1000 billion barrels. If the average world production of crude oil is x barrels per day, what equation describes how many days the 2004 oil reserves will last? How many years?

32. A potential landfill site contains 1,000,000 cubic yards of space. If the average fill per day is x cubic yards, what equation describes how many days the landfill site may be used? How many years?

Blue numbers are core exercises.

33. World production of crude oil uses up oil reserves. If crude oil production in 2004 was 76 million barrels per day, how many years will the oil reserves last? (See Exercise 31.)

34. If a city with a population of 100,000 produces 600 cubic yards of garbage each day, how many years will the landfill in Exercise 32 last?

35. A two-year college has financial aid available for a total of 90 credits. The input, x, is the number of credits taken each term.

a. What is the meaning of the output, y, for the equation $y = \dfrac{90}{x}$?

b. Describe the problem situation if $x = 12$ credits per term.

36. A student saves a total of \$2800 over the summer to spend during the school year. The input, x, is dollars spent per month.

a. What is the meaning of the output, y, for the equation $y = \dfrac{2800}{x}$?

b. Describe the problem situation if $x = \$400$.

In Exercises 37 and 38, use $D = rt$ with unit analysis.

37. The movie *Rabbit-proof Fence* depicts the taking of a 14-year-old aboriginal girl, Molly, from her mother's home in the 1930s because her father was white. Molly escaped from the boarding school where she was placed and, with her younger sister, walked 1200 miles across the Australian outback to her home. The walk took 9 weeks. What was the girls' average speed? Assume that they walked 16 hours each day, 7 days each week. (At the end of the movie, the now-elderly Molly is shown walking with her younger sister and the audience is told that, after recapture, Molly escaped from the same school a second time and walked home again.)

38. Traveling the 2000-mile Oregon Trail was a 6-month trip for immigrants of the 1800s. Assume that the immigrants traveled 10 hours each day, 7 days a week. What was their average speed?

Blue numbers are core exercises.

▶ Writing

39. Explain why these questions are the same concept: "Is there an input that makes $y = 0$?" and "Does the graph of the equation cross the x-axis?"

40. Show that $y = \dfrac{5}{x}$ is equivalent to $xy = 5$. Explain: If $xy = 5$, neither x nor y can equal zero.

▶ Project

41. Reciprocals of x and x^2 The equations $y_1 = \dfrac{1}{x}$ and $y_2 = \dfrac{1}{x^2}$ are graphed in the figure below.

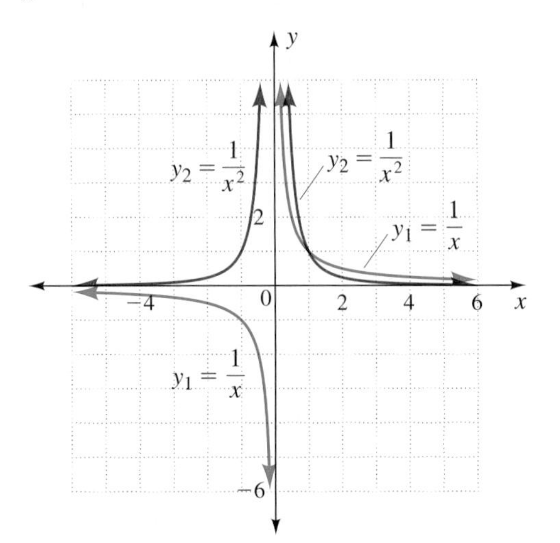

a. Draw the line of symmetry for each graph and write its equation.

b. Why does $y = \dfrac{1}{x^2}$ not have points in the third quadrant?

c. The coordinates $\left(2, \frac{1}{2}\right)$ and $\left(\frac{1}{2}, 2\right)$ both lie on $y = \dfrac{1}{x}$. Is there such a reversal for all points on the graph? Why may the coordinate be reversed?

d. For inputs $x > 1$, which graph is on top: $y = \dfrac{1}{x}$ or $y = \dfrac{1}{x^2}$? Why?

e. For x between 0 and 1, y_1 appears to be left of y_2. How world you explain that $y_2 > y_1$ in this interval?

▶ 9.2 Simplifying Rational Expressions

Objectives

- Simplify rational expressions.
- Simplify rational expressions containing units of measure.
- Find when a rational expression simplifies to 1 or to −1.

WARM-UP

Factor these binomials and trinomials by guess and check or the table method.

1. $x^2 - 4$ **2.** $x^2 + 4x - 5$ **3.** $2x^2 + 7x + 6$

4. $x^2 + 2x$ **5.** $x^2 - 5x + 6$ **6.** $2x^2 + 6x$

7. $x^2 + 3x + 2$ **8.** $x^2 + 6x + 9$ **9.** $4x^2 + 6x$

10. $x^2 - 9$ **11.** $6 - x - x^2$ **12.** $4 - 3x - x^2$

IN THIS SECTION, we simplify rational expressions containing variables and units of measure. We investigate the simplified result when the numerator is the opposite of the denominator.

Factoring Before Simplifying

As mentioned in Section 2.3, when we *simplify a fraction to an equivalent fraction,* we are using the **simplification property of fractions**.

SIMPLIFICATION PROPERTY OF FRACTIONS

For all real numbers, a not zero and c not zero,

$$\frac{ab}{ac} = \frac{a}{a} \cdot \frac{b}{c} = 1 \cdot \frac{b}{c} = \frac{b}{c}$$

When we simplify a fraction, we factor the numerator and denominator and remove the common factors, such as $\frac{a}{a}$. If there are no common factors, the fraction cannot be simplified and is said to be in lowest terms.

▶ **EXAMPLE 1** Simplifying rational expressions Assume there are no zero denominators and simplify the following:

a. $\frac{15}{35}$ **b.** $\frac{28}{18}$ **c.** $\frac{2a}{a^2}$ **d.** $\frac{6xy}{2y^2}$

SOLUTION **a.** $\frac{15}{35} = \frac{3 \cdot 5}{5 \cdot 7} = \frac{3}{7}$ **b.** $\frac{28}{18} = \frac{2 \cdot 2 \cdot 7}{2 \cdot 3 \cdot 3} = \frac{14}{9}$

c. $\frac{2a}{a^2} = \frac{2 \cdot a}{a \cdot a} = \frac{2}{a}$ **d.** $\frac{6xy}{2y^2} = \frac{2 \cdot 3 \cdot x \cdot y}{2 \cdot y \cdot y} = \frac{3x}{y}$ ◀

The expressions in Example 2 need to be factored before we can simplify to lowest terms. The factors may be a monomial and binomial or two binomials.

▶ **EXAMPLE 2** Simplifying rational expressions Assume there are no zero denominators and simplify these rational expressions.

a. $\frac{2x - 4}{3x - 6}$ **b.** $\frac{x^2 - 4x}{x^2 + 2x}$ **c.** $\frac{x - 3}{(x + 3)(x - 3)}$

d. $\frac{x + x^2}{1 + x}$ **e.** $\frac{x^2 - 4}{x^2 + 3x + 2}$ **f.** $\frac{x^2 + 4x - 5}{x^2 - 2x + 1}$

SOLUTION

a. $\dfrac{2x-4}{3x-6} = \dfrac{2(x-2)}{3(x-2)} = \dfrac{2}{3}$

b. $\dfrac{x^2-4x}{x^2+2x} = \dfrac{x(x-4)}{x(x+2)} = \dfrac{x-4}{x+2}$

c. $\dfrac{x-3}{(x+3)(x-3)} = \dfrac{1(x-3)}{(x+3)(x-3)} = \dfrac{1}{(x+3)}$

d. $\dfrac{x+x^2}{1+x} = \dfrac{x(1+x)}{(1+x)} = \dfrac{x}{1} = x$

e. $\dfrac{x^2-4}{x^2+3x+2} = \dfrac{(x+2)(x-2)}{(x+1)(x+2)} = \dfrac{(x-2)}{(x+1)}$

f. $\dfrac{x^2+4x-5}{x^2-2x+1} = \dfrac{(x+5)(x-1)}{(x-1)(x-1)} = \dfrac{(x+5)}{(x-1)}$ ◀

Equivalent Fractions

To add or subtract fractions (Section 9.4), we need fractions with common denominators. Once we know the common denominator, we change each fraction to an equivalent fraction. This change uses the **equivalent fraction property**, which is the *simplification property in reverse*. We multiply the numerator and denominator of the fraction by a common factor.

EQUIVALENT FRACTION PROPERTY

For all real numbers, a not zero and c not zero,

$$\frac{b}{c} = \frac{b}{c} \cdot \frac{a}{a} = \frac{ab}{ac}$$

▶ **EXAMPLE 3** Finding equivalent expressions Change each of the following to an equivalent expression with the indicated numerator or denominator. Assume there are no zero denominators.

a. $\dfrac{4}{5} = \dfrac{12}{}$ b. $\dfrac{a}{2} = \dfrac{}{2b}$ c. $\dfrac{3}{x} = \dfrac{3x}{}$

d. $\dfrac{2}{x+2} = \dfrac{}{2x(x+2)}$ e. $\dfrac{x}{x+2} = \dfrac{}{x^2+3x+2}$ f. $\dfrac{}{10x} = \dfrac{1}{2}$

SOLUTION

a. $\dfrac{4}{5} = \dfrac{4 \cdot 3}{5 \cdot 3} = \dfrac{12}{15}$ b. $\dfrac{a}{2} = \dfrac{a \cdot b}{2 \cdot b} = \dfrac{ab}{2b}$ c. $\dfrac{3}{x} = \dfrac{3 \cdot x}{x \cdot x} = \dfrac{3x}{x^2}$

d. $\dfrac{2}{x+2} = \dfrac{2x \cdot 2}{2x(x+2)} = \dfrac{4x}{2x(x+2)}$

e. $\dfrac{x}{x+2} = \dfrac{x(x+1)}{(x+2)(x+1)} = \dfrac{x^2+x}{x^2+3x+2}$ f. $\dfrac{1 \cdot 5x}{2 \cdot 5x} = \dfrac{5x}{10x}$ ◀

We are not always told whether to simplify a fraction or to change it to an equivalent fraction in unsimplified form.

Special Expressions

EXPRESSIONS CONTAINING UNITS As noted earlier, many expressions involve units of measurement. The simplification property indicates that fractions containing units, such as

$$\frac{\text{meters}}{\text{meters}}, \frac{\text{inches}}{\text{inches}}, \frac{\text{gallons}}{\text{gallons}}, \frac{\text{miles}}{\text{miles}}, \text{ and } \frac{\text{hours}}{\text{hours}}$$

all simplify to 1.

EXAMPLE 4 Simplifying fractions containing units Simplify the following:

a. $\dfrac{48 \text{ cm}^3}{16 \text{ cm}}$ **b.** $\dfrac{5000 \text{ foot-pounds}}{10 \text{ feet}}$ **c.** $\dfrac{24 \text{ degree-days}}{6 \text{ days}}$

SOLUTION

a. $\dfrac{48 \text{ cm}^3}{16 \text{ cm}} = \dfrac{16 \cdot 3 \text{ cm} \cdot \text{cm} \cdot \text{cm}}{16 \text{ cm}} = \dfrac{3 \text{ cm}^2}{1}$

b. $\dfrac{5000 \text{ foot-pounds}}{10 \text{ feet}} = \dfrac{500 \cdot 10 \text{ foot-pounds}}{10 \text{ feet}} = \dfrac{500 \text{ pounds}}{1}$

c. $\dfrac{24 \text{ degree-days}}{6 \text{ days}} = \dfrac{6 \cdot 4 \text{ degree-days}}{6 \text{ days}} = \dfrac{4 \text{ degrees}}{1}$ ◀

EXPRESSIONS CONTAINING OPPOSITES The concept of opposites is central to the next two examples, so we restate the definition here.

DEFINITION OF OPPOSITE EXPRESSIONS

Opposite expressions add to zero.

We use opposites as we look at some special forms of simplifying. The examples provide a shortcut and a caution.

EXAMPLE 5 Working with opposites In parts a to c, fill in the blank.

a. The opposite of 7 is ____. **b.** The opposite of x is ____.

c. The opposite of $7 + x$ is ____.

In parts d to k, add the two expressions in parentheses and note whether they are opposites.

d. $(7 + x) + (-7 - x)$ **e.** $(7 + x) + (7 - x)$ **f.** $(7 - x) + (7 + x)$

g. $(7 - x) + (x - 7)$ **h.** $(a - b) + (a + b)$ **i.** $(a - b) + (b - a)$

j. $(a - b) + (-a + b)$ **k.** $(a - b) + (a - b)$

Simplify the expressions in parts l and m.

l. $\dfrac{3 - 4}{4 - 3}$ **m.** $\dfrac{5 - (-2)}{-2 - 5}$

n. Why do parts l and m have the same answer?

SOLUTION

a. -7 **b.** $-x$ **c.** $-7 - x$ **d.** 0; $7 + x$ and $-7 - x$ are opposites

e. 14; not opposites **f.** 14; not opposites

g. 0; $7 - x$ and $x - 7$ are opposites

h. $2a$; not opposites **i.** 0; $a - b$ and $b - a$ are opposites

j. 0; $a - b$ and $-a + b$ are opposites

k. $2a - 2b$; not opposites **l.** -1 **m.** -1

n. The fractions have opposites in the numerator and denominator. Both simplify to -1. ◀

Here are some things to remember when simplifying fractions (see Figure 14):

- When the numerator and denominator of a fraction are the same, the fraction simplifies to 1.
- When the numerator and denominator of a fraction are opposites, the fraction simplifies to -1.

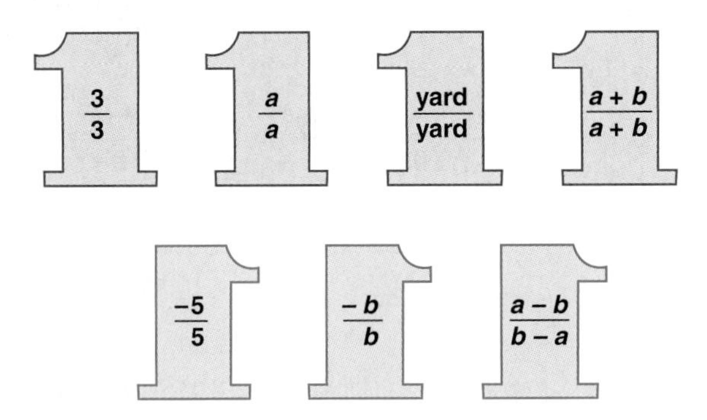

FIGURE 14 One and negative one as opposites

Example 6 shows why rational expressions containing opposite numerators and denominators simplify to -1.

▶ **EXAMPLE 6** Simplifying fractions containing opposites Show that $\dfrac{a-b}{b-a} = -1$, $a \neq b$.

SOLUTION At least three methods are possible.

Method 1: Multiply the numerator and denominator of the fraction by -1:

$$\frac{a-b}{b-a} = \frac{(-1)(a-b)}{(-1)(b-a)} = \frac{(-1)(a-b)}{(a-b)} = -1$$

Method 2: Multiply either the numerator or the denominator by $(-1)(-1)$, which equals 1 and will not change the fraction:

$$\frac{a-b}{b-a} = \frac{(-1)(-1)(a-b)}{(b-a)} = \frac{(-1)(b-a)}{(b-a)} = -1$$

Method 3: Factor -1 from either the numerator or the denominator:

$$\frac{a-b}{b-a} = \frac{(-1)(b-a)}{(b-a)} = -1$$

With the first two methods, we multiplied out only one of the -1 factors. This changed one of the expressions to its opposite and thus permitted simplification. ◀

It does not matter which method is used to simplify fractions containing opposites to -1. Choose a method that makes sense and consistently gives you the correct result.

ANSWER BOX

Warm-up: **1.** $(x-2)(x+2)$ **2.** $(x+5)(x-1)$ **3.** $(2x+3)(x+2)$ **4.** $x(x+2)$ **5.** $(x-2)(x-3)$ **6.** $2x(x+3)$ **7.** $(x+1)(x+2)$ **8.** $(x+3)(x+3)$ **9.** $2x(2x+3)$ **10.** $(x+3)(x-3)$ **11.** $(3+x)(2-x)$ **12.** $(4+x)(1-x)$

▶ 9.2 Exercises

Simplify the expressions in Exercises 1 to 20. Assume there are no zero denominators.

1. $\dfrac{2ab}{6a^2b}$

2. $\dfrac{3ac}{15ac^2}$

3. $\dfrac{12cd^2}{8c^2d}$

4. $\dfrac{15b^2c^3}{10b^3c}$

5. $\dfrac{(x-2)(x+3)}{x+3}$

6. $\dfrac{(x-2)(x+3)}{(x+3)(x+2)}$

7. $\dfrac{2-x}{(x+2)(x-2)}$

8. $\dfrac{1-x}{(x-1)(x+2)}$

9. $\dfrac{3ab+3ac}{5b^2+5bc}$

10. $\dfrac{2ac+4bc}{4ad+8bd}$

11. $\dfrac{4x^2+8x}{2x^2-4x}$

12. $\dfrac{3x^2-6x}{6x^2+12x}$

13. $\dfrac{x^2-4}{x^2+5x+6}$

14. $\dfrac{x^2-1}{x^2+4x+3}$

15. $\dfrac{x^2+x-6}{x^2-2x}$

16. $\dfrac{x^2+x-2}{2x+4}$

17. $\dfrac{x-3}{6-2x}$

18. $\dfrac{x-4}{12-3x}$

19. $\dfrac{x^2-5x+6}{x^2-9}$

20. $\dfrac{x^2-3x-4}{x^2-16}$

Find the missing number or expression in each pair of equivalent fractions in Exercises 21 to 30. Assume there are no zero denominators.

21. $\dfrac{6}{9}, \dfrac{}{45}$

22. $\dfrac{15}{10}, \dfrac{3}{}$

23. $\dfrac{24x}{3x^2}, \dfrac{8}{}$

24. $\dfrac{24x}{3x^2}, \dfrac{72x^3}{}$

25. $\dfrac{}{10a^2}, \dfrac{b}{2a}$

26. $\dfrac{6cd}{9cd^2}, \dfrac{2}{}$

27. $\dfrac{x+2}{}, \dfrac{2x+4}{2x-6}$

28. $\dfrac{x+3}{x+5}, \dfrac{x^2+3x}{}$

29. $\dfrac{a+b}{}, \dfrac{a^2+2ab+b^2}{a^2-b^2}$

30. $\dfrac{}{a^2+4a+4}, \dfrac{a-2}{a+2}$

In Exercises 31 to 36, simplify.

31. $\dfrac{108\text{ m}^2}{6\text{ m}}$

32. $\dfrac{125\text{ in}^3}{5\text{ in.}}$

33. $\dfrac{144\text{ in}^2}{1728\text{ in}^3}$

34. $\dfrac{27\text{ yd}^3}{9\text{ yd}^2}$

35. $\dfrac{2060\text{ degree-gallons}}{103\text{ degrees}}$

36. $\dfrac{1200\text{ foot-pounds}}{200\text{ pounds}}$

Simplify the fractions in Exercises 37 to 40. What do you observe? Why?

37. $\dfrac{7-4}{4-7}$

38. $\dfrac{15-9}{9-15}$

39. $\dfrac{-3-4}{4-(-3)}$

40. $\dfrac{6-(-2)}{-2-6}$

What is the opposite of each expression in Exercises 41 to 46?

41. $a-b$

42. $a+b$

43. $-a+b$

44. $b-a$

45. $b+a$

46. $-a-b$

In Exercises 47 to 50, use both of the following statements to test whether the two expressions are opposites:
(a) Opposites add to zero.
(b) If we multiply an expression by -1, we get its opposite.

47. $n-m$ and $n+m$

48. $n-m$ and $m-n$

49. $x-2$ and $2-x$

50. $x-2$ and $x+2$

In Exercises 51 to 58, what numerator or denominator is needed in the equation to make a true statement? State any inputs that must be excluded.

51. $\dfrac{x+2}{} = 1$

52. $\dfrac{x+2}{} = -1$

53. $\dfrac{x-2}{} = -1$

54. $\dfrac{x-2}{} = 1$

55. $\dfrac{}{a-b} = -1$

56. $\dfrac{b-a}{} = 1$

57. $\dfrac{3-x}{} = 1$

58. $\dfrac{3-x}{} = -1$

In Exercises 59 to 62, simplify. Assume there are no zero denominators.

59. a. $\dfrac{2-3x}{3+2x}$ b. $\dfrac{2-3x}{3x-2}$

60. a. $\dfrac{5x-3}{3+5x}$ b. $\dfrac{5x-3}{3-5x}$

61. a. $\dfrac{2x-4}{2-x}$ b. $\dfrac{2x-1}{3-6x}$

62. a. $\dfrac{5-15x}{3x-1}$ b. $\dfrac{x-3}{6-2x}$

Blue numbers are core exercises.

In Exercises 63 and 64, indicate whether the statement is true or false. If true, explain why; if false, give an example that shows why it is false.

63. A rational expression must be factorable in order to be simplified to lowest terms.

64. A rational expression can have the same variables in the numerator and denominator and still not be simplifiable.

65. If $\frac{a}{a} = 1$, list three fractions containing positive or negative a's that would simplify to -1.

66. What happens when we divide opposites?

67. Describe the role of factoring in simplifying fractions.

68. What may be concluded about $-\frac{a}{b}$, $\frac{-a}{b}$, and $\frac{a}{-b}$?

▶ Projects

69. Wheat Harvest Two farmers are helping a neighbor who fell off his barn roof. Suppose each large rectangle in the following figure represents the entire wheat harvest. If Terry harvests the wheat in 18 days, then each day he harvests $\frac{1}{18}$ of the crop. The shading in the first rectangle represents the amount Terry harvests in one day. The shading in the second rectangle is the amount Lee harvests in one day: $\frac{1}{12}$. Together they harvest the amount shown in the third rectangle. The second day of harvest is shown by the second row of rectangles.

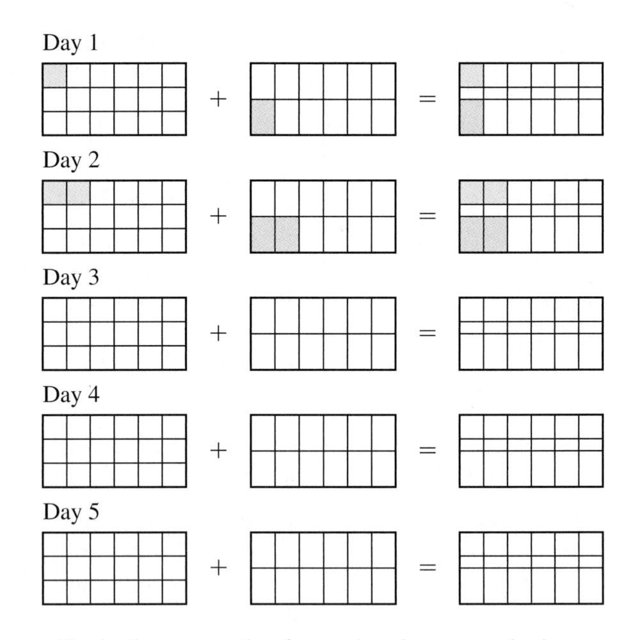

a. Shade the rectangles for each subsequent day's accumulative harvest. Add more rectangles as needed.

b. How will the rectangle look when the harvest is complete?

c. Estimate the total number of days needed to complete the harvest.

d. What is a better way to show the fraction parts $\frac{1}{12}$ and $\frac{1}{18}$ so that we can improve our estimate of the total number of days needed to complete the harvest?

▶ 9.3 Multiplication and Division of Rational Expressions

Objectives

- Factor, simplify, and multiply rational expressions.
- Change division problems to multiplication problems and complete the multiplication.
- Simplify complex rational expressions by changing to division.
- Apply multiplication and division principles to expressions containing units of measure.

WARM-UP

Translate each phrase into symbols, and perform the indicated operation.

1. The product of 3 and $\frac{1}{2}$

2. The quotient of 3 and $\frac{1}{2}$

3. The quotient of $\frac{1}{2}$ and 3

4. The quotient of $\frac{1}{2}$ and $\frac{1}{3}$

5. The product of $\frac{1}{3}$ and $\frac{3}{4}$

6. The quotient of $\frac{3}{4}$ and $\frac{3}{5}$

Factor these numbers or expressions.

7. 35

8. $cx + c$

9. $x^2 - 5x - 6$

10. $x^2 + 3x + 2$

11. $x^2 - 3x - 4$

12. $x^2 - 16$

IN THIS SECTION, we apply principles of multiplication and division to rational expressions, complex rational expressions, and expressions containing units of measure.

Multiplication of Rational Expressions

In the first three examples, observe the role of $\frac{a}{a} = 1$ in simplifying the multiplication and division of rational expressions. In each case, *the solution is found most easily by simplifying expressions before doing the final multiplication.*

Example 1 reviews multiplying and simplifying fractions.

EXAMPLE 1 Multiplying fractions Multiply $\frac{8}{35} \cdot \frac{25}{32}$.

SOLUTION

$$\frac{8}{35} \cdot \frac{25}{32} = \frac{8 \cdot 25}{35 \cdot 32} = \frac{2 \cdot 2 \cdot 2 \cdot 5 \cdot 5}{5 \cdot 7 \cdot 2 \cdot 2 \cdot 2 \cdot 2 \cdot 2} = \frac{5}{7 \cdot 2 \cdot 2} = \frac{5}{28}$$

In this solution, we use the multiplication property to change the two fractions to a single fraction with one numerator and one denominator. We factor, simplify, and then multiply the remaining factors in the numerator and denominator. ◀

MULTIPLICATION OF RATIONAL EXPRESSIONS

To multiply proper or improper fractions, multiply the numerators and divide by the product of the denominators:

$$\frac{a}{b} \cdot \frac{c}{d} = \frac{a \cdot c}{b \cdot d}, \quad b \neq 0, d \neq 0$$

EXAMPLE 2 Multiplying rational expressions Multiply $\frac{ax^2}{cx} \cdot \frac{cx + c}{a^2}$. Assume there are no zero denominators.

SOLUTION

$$\frac{ax^2}{cx} \cdot \frac{cx + c}{a^2} = \frac{ax^2(cx + c)}{cx \cdot a^2} = \frac{a \cdot x \cdot x \cdot c(x + 1)}{c \cdot x \cdot a \cdot a} = \frac{x(x + 1)}{a}$$

In this solution, we again write the numerators and denominators as products and factor them completely. Simplification with $\frac{n}{n} = 1$ eliminates any further need for multiplication. ◀

EXAMPLE 3 Multiplying rational expressions Multiply $\frac{x + 2}{x + 6} \cdot \frac{x^2 - 5x - 6}{x^2 + 3x + 2}$. Assume there are no zero denominators.

SOLUTION

$$\frac{x + 2}{x + 6} \cdot \frac{x^2 - 5x - 6}{x^2 + 3x + 2} = \frac{(x + 2)(x^2 - 5x - 6)}{(x + 6)(x^2 + 3x + 2)}$$

$$= \frac{(x + 2)(x + 1)(x - 6)}{(x + 6)(x + 1)(x + 2)} = \frac{x - 6}{x + 6}$$

We apply the multiplication property of fractions by writing the two fractions as one fraction. We then factor the numerator and denominator expressions and simplify the fraction. ◀

Caution: If we first multiply the fractions in Example 3, the resulting expression is not easily simplified. It requires factoring techniques beyond the level of this or the next mathematics course.

$$\frac{x + 2}{x + 6} \cdot \frac{x^2 - 5x - 6}{x^2 + 3x + 2} = \frac{(x + 2)(x^2 - 5x - 6)}{(x + 6)(x^2 + 3x + 2)} = \frac{x^3 - 3x^2 - 16x - 12}{x^3 + 9x^2 + 20x + 12} = ?$$

If your homework solutions contain similar expressions, go back to the original problem, factor, and simplify before multiplying.

▶ In all multiplication problems, keep in mind that we are eliminating factors, not terms. Any units are treated the same way as factors. We used this idea earlier in unit analysis.

▶ **EXAMPLE 4** Multiplying and simplifying units: water flow A shower head permits a flow of 5 gallons per minute. How many gallons of water are used in a $3\frac{1}{2}$-minute shower?

SOLUTION

$$\frac{5 \text{ gal}}{1 \text{ min}} \cdot 3.5 \text{ min} = \frac{5(3.5)}{1} \frac{\text{gal} \cdot \text{min}}{\text{min}} = 17.5 \text{ gal}$$ ◀

▶ Here are some things to remember when multiplying rational expressions:

- Factor, simplify to lowest terms, and then multiply, as needed.
- Simplify only like factors using $\frac{a}{a} = 1$. The x's in $\frac{x+5}{x}$ are terms, not factors.
- Any units are treated the same way as factors.
- No common denominator is needed for multiplication.

▶ Division of Rational Expressions

Division of rational expressions is based on the same property as division of fractions.

DIVISION OF RATIONAL EXPRESSIONS

To divide rational expressions, multiply the first fraction by the reciprocal of the second:

$$\frac{a}{b} \div \frac{c}{d} = \frac{a}{b} \cdot \frac{d}{c}, \quad b \neq 0, c \neq 0, d \neq 0$$

▶ **EXAMPLE 5** Dividing expressions in fraction notation Assume there are no zero denominators.

a. Divide $\frac{1}{18}$ and $\frac{1}{25}$.

b. Divide $\frac{ax}{b}$ and $\frac{cx}{d}$.

SOLUTION To divide, we change the division to multiplication by the reciprocal of the second fraction.

a. $\frac{1}{18} \div \frac{1}{25} = \frac{1}{18} \cdot \frac{25}{1} = \frac{25}{18}$

b. $\frac{ax}{b} \div \frac{cx}{d} = \frac{ax}{b} \cdot \frac{d}{cx} = \frac{adx}{bcx} = \frac{ad}{bc}$ ◀

▶ **EXAMPLE 6** Dividing rational expressions Divide $\frac{x^2 - 3x - 4}{x - 3} \div \frac{x^2 - 16}{x^2 - 9}$. Assume there are no zero denominators.

SOLUTION

$$\frac{x^2 - 3x - 4}{x - 3} \div \frac{x^2 - 16}{x^2 - 9} = \frac{x^2 - 3x - 4}{x - 3} \cdot \frac{x^2 - 9}{x^2 - 16}$$

$$= \frac{(x-4)(x+1)(x-3)(x+3)}{(x-3)(x-4)(x+4)}$$

$$= \frac{(x+1)(x+3)}{x+4}$$

After changing the problem to a multiplication problem, we factor the expression and simplify. No further simplification is possible because there are no common factors in the numerator and denominator. ◀

▶ **EXAMPLE 7** Dividing rational expressions containing opposites Divide $\frac{x^2 - 4}{x^3} \div \frac{2 - x}{x}$. Assume there are no zero denominators.

SOLUTION

$$\frac{x^2 - 4}{x^3} \div \frac{2 - x}{x} = \frac{(x + 2)(x - 2) \cdot x}{x \cdot x^2 \cdot (2 - x)}$$

$$= \frac{-1(x + 2)}{x^2}$$

After changing the problem to a multiplication, we factor and simplify. The factors $(x - 2)$ and $(2 - x)$ are opposites and simplify to -1. ◀

▶ **EXAMPLE 8** Dividing expressions containing units Divide 450 miles by 60 miles per hour.

SOLUTION

$$450 \text{ mi} \div \frac{60 \text{ mi}}{1 \text{ hr}} = 450 \text{ mi} \cdot \frac{1 \text{ hr}}{60 \text{ mi}} = 7.5 \text{ hr}$$ ◀

▶ Here are some things to remember when dividing rational expressions:

- Change division to multiplication by the reciprocal.
- Factor, simplify to lowest terms, and then multiply.

▶ Complex Rational Expressions (Method 1, Change Division to Multiplication by the Reciprocal Number)

The technique of changing division to multiplication by a reciprocal may be applied to more complicated forms of rational expressions—the complex rational expressions. *Fractions that contain fractions in either the numerator or the denominator* are called **complex fractions**. *Rational expressions that contain fractions in either the numerator or the denominator* are called **complex rational expressions**. When the numerator, the denominator, or both are single fractions, recall that the fraction bar means division and change the complex fraction to a division problem.

▶ **EXAMPLE 9** Dividing complex fractions Simplify $\frac{\frac{8}{15}}{\frac{4}{5}}$.

SOLUTION The fraction bar means division, so we change the notation to two fractions separated by the ÷ sign:

$$\frac{\frac{8}{15}}{\frac{4}{5}} = \frac{8}{15} \div \frac{4}{5} = \frac{8}{15} \cdot \frac{5}{4} = \frac{2 \cdot 4 \cdot 5}{3 \cdot 5 \cdot 4} = \frac{2}{3}$$ ◀

▶ **EXAMPLE 10** Simplifying complex rational expressions Assume there are no zero denominators and simplify the complex rational expression $\frac{\frac{a}{b}}{\frac{c}{d}}$.

SOLUTION

$$\frac{\frac{a}{b}}{\frac{c}{d}} = \frac{a}{b} \div \frac{c}{d} = \frac{a}{b} \cdot \frac{d}{c} = \frac{ad}{bc}$$

We simplify the complex rational expression by writing it as a division problem, with the longer fraction bar replaced by a division sign. The division is then changed to multiplication by a reciprocal. ◀

▶ We may include units in simplifying complex fractions.

▶ **EXAMPLE 11** Applying complex fractions Answer the question by writing a complex fraction and simplifying.

a. How many half-dollars are in \$5.00?

b. How many fourths are in $\frac{5}{8}$?

c. If a 19-passenger jet flying at 459 nautical miles (nm) per hour consumes 397 gallons of fuel per hour, how many nautical miles does it get per gallon?

SOLUTION In parts a and b, *how many* implies division.

a. $$\frac{5.00 \text{ dollars}}{\frac{1}{2} \text{ dollar}} = 5 \div \frac{1}{2} = 5 \cdot \frac{2}{1} = 10$$

b. $$\frac{\frac{5}{8}}{\frac{1}{4}} = \frac{5}{8} \div \frac{1}{4} = \frac{5}{8} \cdot \frac{4}{1} = \frac{20}{8} = \frac{5}{2} = 2\tfrac{1}{2}$$

In part c, we are looking for nautical miles per gallon, so the expression containing nautical miles should be placed in the numerator.

c. $$\frac{\frac{459 \text{ nm}}{\text{hr}}}{\frac{397 \text{ gal}}{\text{hr}}} = \frac{459 \text{ nm}}{\text{hr}} \div \frac{397 \text{ gal}}{\text{hr}} = \frac{459 \text{ nm}}{\text{hr}} \cdot \frac{\text{hr}}{397 \text{ gal}} \approx 1.16 \frac{\text{nm}}{\text{gal}}$$

Unit analysis can also be used to solve this problem. We start with 459 nm per hour and multiply by an expression containing hours in the numerator in order to cancel the hours:

$$\frac{459 \text{ nm}}{\text{hr}} \cdot \frac{\text{hr}}{397 \text{ gal}} \approx 1.16 \frac{\text{nm}}{\text{gal}}$$ ◀

ANSWER BOX

Warm-up: **1.** 1.5 **2.** 6 **3.** $\frac{1}{6}$ **4.** $\frac{3}{2}$ **5.** $\frac{1}{4}$ **6.** $\frac{5}{4}$ **7.** $5 \cdot 7$ **8.** $c(x + 1)$ **9.** $(x - 6)(x + 1)$ **10.** $(x + 2)(x + 1)$ **11.** $(x - 4)(x + 1)$ **12.** $(x - 4)(x + 4)$

▶ 9.3 Exercises

Multiply and divide each pair of fractions in Exercises 1 to 4. Assume there are no zero denominators.

1. a. $\frac{1}{3}$ and $\frac{1}{4}$ **b.** $\frac{1}{2}$ and $\frac{1}{5}$

2. a. $\frac{1}{4}$ and $\frac{1}{12}$ **b.** $\frac{1}{8}$ and $\frac{1}{12}$

3. a. $\frac{3}{4}$ and $\frac{1}{6}$ **b.** $\frac{2}{3}$ and $\frac{1}{6}$

4. a. $\frac{a}{b}$ and $\frac{c}{d}$ **b.** $\frac{w}{x}$ and $\frac{y}{z}$

5. Calculate these fractional expressions. What may be observed about each pair of problems? What do they tell us about multiplication and division?

a. $100 \div \frac{4}{1}$ and $100 \cdot \frac{1}{4}$

b. $100 \div \frac{1}{5}$ and $100 \cdot \frac{5}{1}$

6. For $\frac{5}{8} \div \frac{1}{4}$, the answer is $2\frac{1}{2}$. The figure below shows $\frac{5}{8}$ of one rectangle shaded and $\frac{1}{4}$ of an identical rectangle shaded. Trace these rectangles and show why there are $2\frac{1}{2}$ fourths in $\frac{5}{8}$.

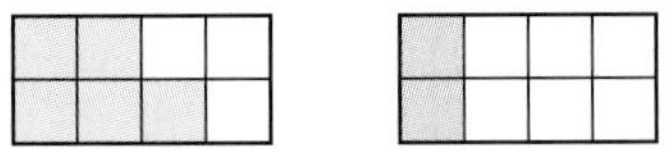

In Exercises 7 and 8, multiply or divide, as indicated. Assume there are no zero denominators.

7. a. $\frac{1}{x} \cdot \frac{x^2}{1}$ **b.** $\frac{1}{a} \div \frac{a^2b^2}{1}$

c. $\frac{a}{b} \cdot \frac{b^2}{a^2}$ **d.** $\frac{a}{b} \div \frac{a^2}{b^3}$

Blue numbers are core exercises.

8. **a.** $\frac{1}{x} \cdot \frac{x^3}{1}$ **b.** $\frac{1}{b} \div \frac{a^2b^2}{1}$

c. $\frac{b}{a} \div \frac{a^2}{b^2}$ **d.** $\frac{a^2}{b^3} \div \frac{a}{b}$

In Exercises 9 to 14, multiply or divide, as indicated. Assume there are no zero denominators.

9. **a.** $\frac{x^2 + 2x + 1}{x + 1} \cdot \frac{x}{x^2 + x}$

b. $\frac{x^2 - 4}{x + 2} \cdot \frac{1}{x^2 - x}$

c. $\frac{x + 2}{x^2 - 4x + 4} \div \frac{x^2 + 2x}{x - 2}$

d. $\frac{x^2 - 5x}{x^2 + 5x} \div \frac{x^2 - 10x + 25}{x}$

e. $\frac{x^2 - 6x + 9}{x^2 + 3x} \div \frac{x^2 - 9}{x}$

10. **a.** $\frac{x^2 - 7x + 12}{x^2 - 4} \cdot \frac{x^2 + 2x}{x - 3}$

b. $\frac{x^2 - 2x}{x} \cdot \frac{x^2}{x^2 - 3x + 2}$

c. $\frac{x - 3}{x^2 - 4x + 3} \div \frac{x^2 + x}{x - 1}$

d. $\frac{x^2 - 6x + 9}{x^2 + 3x} \cdot \frac{x + 3}{x - 3}$

e. $\frac{x^2 + 3x}{x} \cdot \frac{x^2 - x - 6}{x^2 - 9}$

11. **a.** $\frac{x^2 + x}{x - 1} \cdot \frac{x^2 - 1}{x + 1}$

b. $\frac{x - 3}{x^2 + 6x + 9} \cdot \frac{x + 3}{x^2 - 9}$

c. $\frac{4 - 8x}{x + 1} \div \frac{1 - 2x}{x^2 - 1}$

d. $\frac{x - x^2}{x + 1} \cdot \frac{x - 1}{1 - x}$

e. $\frac{x^2 - x}{x^2 - 3x + 2} \div \frac{1 - x^2}{x^2 - 2x + 1}$

12. **a.** $\frac{3x + 3}{1 - x} \cdot \frac{1 - x^2}{x + 1}$

b. $\frac{2 - x}{x + 2} \div \frac{4 - 2x}{x^2 + 4x + 4}$

c. $\frac{x^2 - x}{x^2} \div \frac{x - 1}{x}$

d. $\frac{x + 3}{3x} \cdot \frac{9x^2}{x^2 - 9}$

e. $\frac{x^2 + 4x + 4}{x^2 - 4} \div \frac{x^2 + 2x}{2 - x}$

Blue numbers are core exercises.

13. **a.** $\frac{x}{1 - x} \cdot \frac{x - 1}{x^2}$

b. $\frac{c^2 - d^2}{d} \cdot \frac{cd}{d - c}$

c. $\frac{x^2}{4 - x} \div \frac{4x}{x - 4}$

d. $\frac{a^2 - 2ab + b^2}{a + b} \div \frac{b - a}{a + b}$

14. **a.** $\frac{a}{a - b} \cdot \frac{b - a}{b}$

b. $\frac{x - 3}{3x} \cdot \frac{9}{9 - x^2}$

c. $\frac{3 - x}{x^3} \div \frac{x - 3}{3x}$

d. $\frac{c + d}{d - c} \div \frac{c + d}{c^2 - 2cd + d^2}$

Exercises 15 and 16 offer more challenging factoring.

15. **a.** $\frac{12x^2 - 5x - 3}{3x + 3} \cdot \frac{2x^2 - x - 3}{4x^2 + 9x - 9}$

b. $\frac{12x^2 + 35x - 3}{8x^2 + 26x + 21} \div \frac{2 - 24x}{4x^2 + 15x + 14}$

16. **a.** $\frac{18x^2 - 17x - 15}{24x^2 - 7x - 6} \cdot \frac{9 + 12x - 32x^2}{9 - 18x + 8x^2}$

b. $\frac{18x^2 + 19x - 12}{36 - 18x} \div \frac{9x^2 + 14x - 8}{18x^2 - 35x - 2}$

In Exercises 17 to 28, use properties of fractions to simplify the expressions. The word *per* means division and may be replaced by a fraction bar.

17. $\dfrac{\text{miles}}{\dfrac{\text{miles}}{\text{hour}}}$ **18.** $\dfrac{\text{kilometers}}{\dfrac{\text{kilometers}}{\text{minute}}}$

19. $\frac{\text{93,000,000 miles}}{\text{186,000 miles per second}}$

20. $\frac{\text{5280 feet}}{\text{1130 feet per second}}$

21. $\frac{\text{300 miles per hour}}{\text{100 gallons per hour}}$

22. $\frac{\text{60 miles per hour}}{\text{25 miles per gallon}}$

23. $\dfrac{\text{12 cookies per dozen}}{\text{\$2.98 per dozen}}$

24. $\dfrac{\text{12 stitches per inch}}{\dfrac{\text{1 yd}}{\text{36 inches}}}$

25. $\dfrac{\text{85 words per minute}}{\text{300 words per page}}$

26. $\dfrac{\text{880 cycles per second}}{\text{344 meters per second}}$

27. $\dfrac{\text{40 moles}}{\text{12 moles per liter}}$

28. $\dfrac{\text{186 days}}{\text{5 days per week}}$

29. Choose one of the expressions from Exercises 17 to 28, and give a situation in which it would make sense.

30. Make up a division problem containing units of measure, and explain what the answer means.

31. The current, I, in an electrical circuit is found by dividing the voltage, V, by the resistance, R. Suppose the voltage and resistance vary with time as in the equations

$$V = \frac{t^2 - 4}{2t^2 - 3t - 2} \quad \text{and} \quad R = \frac{t + 2}{t^2}$$

Assume there are no zero denominators. Find a formula in terms of t that gives the current, I.

32. In economics, total quantity sold is the product of the price and the demand. Find the total quantity, Q.

$$\text{Price} = 3x + 6 \quad \text{and} \quad \text{Demand} = \frac{800}{x^2 + 2x}$$

Assume there are no zero denominators.

▶ Writing

33. A student familiar with simplifying fractions such as $\frac{12}{15}$ is puzzled by the canceling of threes in $\frac{3}{5} \cdot \frac{4}{3}$. Explain why the product can be simplified when the threes are in different fractions.

34. Explain why

$$\frac{x(x + 1)(x - 3)}{x(x - 3)} = x + 1$$

is correct and

$$\frac{x^2 + x + 1}{x^2} = x + 1$$

is not correct.

35. Name the properties that explain why we can eliminate factors in different fractions in a product.

▶ Project

36. Exiting a Theater, I A movie theater has two exits. One door, by itself, can empty the theater in 5 minutes. The second door, by itself, can empty the theater in 8 minutes. Suppose we wish to answer this question: If both doors are available, how many minutes will it take to empty the theater?

Each large rectangle in the figure below represents a full theater. During each minute, the first door permits $\frac{1}{5}$ of the theater to empty. During each minute, the second door permits $\frac{1}{8}$ of the theater to empty.

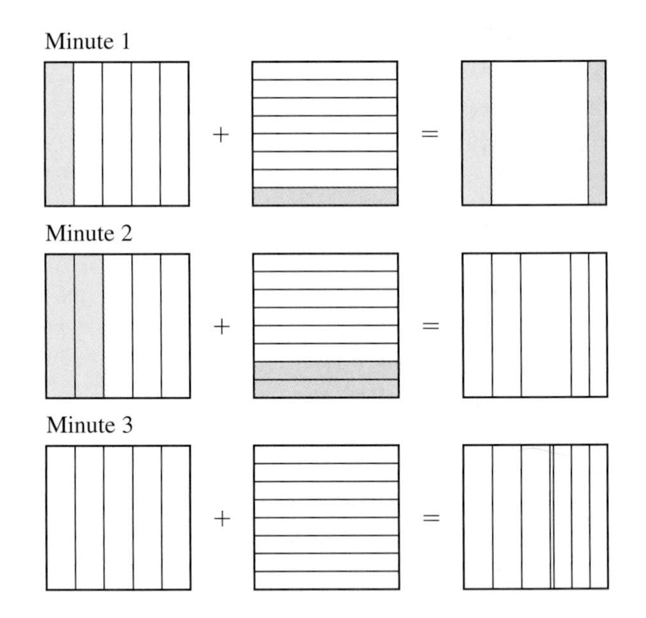

a. What does the shading in the first rectangle for Minute 1 represent?

b. What does the shading in the second rectangle for Minute 1 represent?

c. What does the shading in the third rectangle for Minute 1 represent?

d. Shade the rectangles for each subsequent minute's departures. Add more rectangles as needed.

e. How will the last rectangle look when the theater is empty?

f. Estimate the total number of minutes needed to empty the theater.

g. What is a better way to draw the $\frac{1}{5}$ and $\frac{1}{8}$ fractions to improve the addition?

Blue numbers are core exercises.

▶ 9 Mid-Chapter Test

1. For what inputs, x, will the expression $\dfrac{1}{(x+2)(x-1)}$ have a zero denominator? Show the parentheses needed to enter the expression into a calculator.

For Exercises 2 and 3, assume that the budget for a credit union's annual meeting is \$4800. The budget covers food, chair set-up charges, and a souvenir gift for each member attending.

2. Make a table and graph for the possible spending per person for zero to 1000 members. Use number of members attending as input and spending per member as output.

3. What equation describes the relationship between the number of members attending the meeting and the spending per member?

4. Describe the behavior of the graph in the figure for the situations specified.

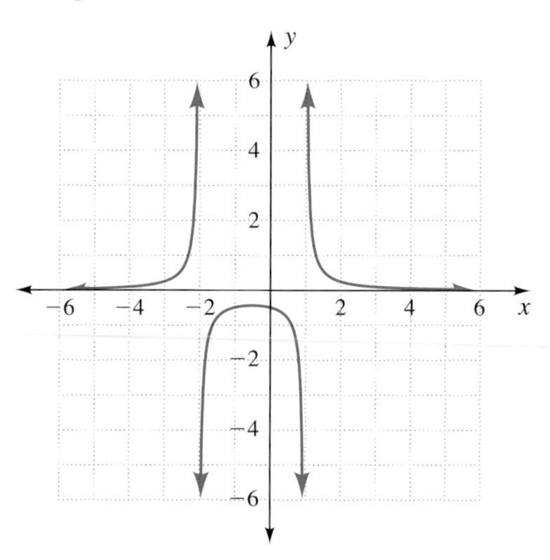

a. As x approaches -2 from the left

b. As x approaches 1 from the right

c. As x approaches 1 from the left

5. Evaluate the expression in Exercise 1.

−2.2	−2.1	−2	0.7	0.8	0.9

6. Round to the nearest tenth. In Exercise 1, for what positive interval of x is $y > 1$?

In Exercises 7 and 8, simplify. If the expression does not simplify, explain why. Assume there are no zero denominators.

7. a. $\dfrac{24ac}{28a^2}$ **b.** $\dfrac{a+2}{a-2}$

8. a. $\dfrac{x+2}{x^2-4}$ **b.** $\dfrac{x^2-2x+1}{x^2-3x+2}$

In Exercises 9 and 10, find the missing numerator or denominator. Assume there are no zero denominators.

9. a. $\dfrac{3x}{5y} = \dfrac{}{10xy}$

b. $\dfrac{2a}{3b} = \dfrac{8a^2b}{}$

10. a. $\dfrac{3}{x+5} = \dfrac{}{5(x+5)}$

b. $\dfrac{2}{x-3} = \dfrac{}{(x+2)(x-3)}$

Perform the indicated operations in Exercises 11 to 15. Simplify the answers. Assume there are no zero denominators.

11. $\dfrac{3x}{y^2} \div \dfrac{x^2}{y}$ **12.** $\dfrac{2x}{y} \div \dfrac{x^2}{y}$

13. $\dfrac{x^2-5x-6}{x+2} \cdot \dfrac{x^2-4}{2-x}$

14. $\dfrac{x^2-3x}{x^2-16} \div \dfrac{x-3}{x+4}$

15. $\dfrac{\frac{2}{3x}}{\frac{x^2}{6}}$

In Exercises 16 to 18, simplify these expressions containing units of measure.

16. $\dfrac{\text{63 days}}{\text{7 days per week}}$

17. $\dfrac{\text{16 stitches per second}}{\text{8 stitches per inch}}$

18. From dosage computation: $\dfrac{\frac{10\text{ mL}}{100\text{ mL}} \cdot 400\text{ mL}}{\frac{50\text{ mL}}{100\text{ mL}}}$

▶ 9.4 Finding the Common Denominator and Addition and Subtraction of Rational Expressions

Objectives

- Determine the common denominator for two or more rational expressions.
- Add and subtract rational expressions with like denominators.
- Convert rational expressions to like denominators, and complete the addition or subtraction.
- Simplify complex fractions, using multiplication by a common denominator.

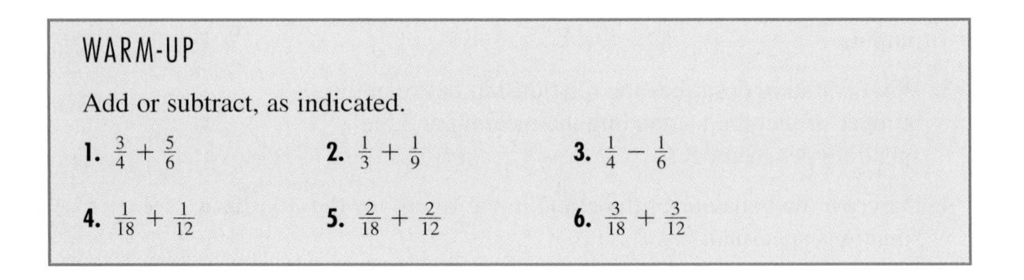

WARM-UP

Add or subtract, as indicated.

1. $\frac{3}{4} + \frac{5}{6}$ **2.** $\frac{1}{3} + \frac{1}{9}$ **3.** $\frac{1}{4} - \frac{1}{6}$

4. $\frac{1}{18} + \frac{1}{12}$ **5.** $\frac{2}{18} + \frac{2}{12}$ **6.** $\frac{3}{18} + \frac{3}{12}$

IN THIS SECTION, we find the common denominator and use it to add and subtract rational expressions and to simplify complex fractions.

▶ Least Common Denominator

Factoring plays an important role in finding the **least common denominator (LCD),** *the smallest number into which both denominators divide evenly.*

▶ **EXAMPLE 1** Finding the least common denominator and adding fractions Add $\frac{3}{4} + \frac{5}{6}$.

SOLUTION To find the least common denominator, we list the factors of each denominator:

$$4 = 2 \cdot 2$$
$$6 = 2 \cdot 3$$

The least common denominator needs to be divisible by both denominators and needs two factors of 2 and one factor of 3. (See Figure 15.)

The product of these factors, $2 \cdot 2 \cdot 3$, gives the least common denominator: 12. To add, we change each fraction to an equivalent fraction with the common denominator.

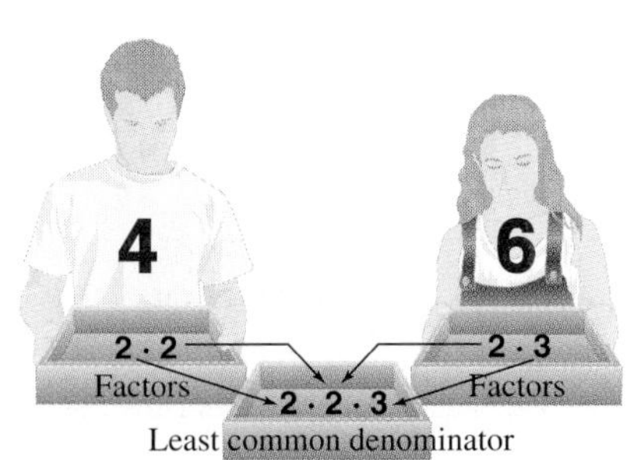

FIGURE 15 Finding the LCD

$$\frac{3}{4} = \frac{3 \cdot 3}{4 \cdot 3} = \frac{9}{12}$$
$$+\frac{5}{6} = \frac{5 \cdot 2}{6 \cdot 2} = \frac{10}{12}$$
$$\frac{19}{12}$$

◀

▶ Although any common denominator can be used to add or subtract fractions, there are two advantages to using the least common denominator. First, the fractions are simpler. Second, the answer is less likely to need simplifying. In the next example, we extend the common denominator to rational expressions.

LEAST COMMON DENOMINATOR (LCD)

To find the least common denominator:

1. List the prime factors of each denominator.
2. Compare the factored denominators.
3. a. If the denominators have no common factors, the LCD is the product of the denominators.
   b. If the denominators have common factors, list each factor the highest number of times it appears in any one denominator.
4. Write the LCD as the product of the listed factors.

▶ **EXAMPLE 2** Finding the least common denominator Assume there are no zero denominators and find the LCD for these pairs of rational expressions.

a. $\dfrac{3}{ab^2}, \dfrac{5}{abc}$ **b.** $\dfrac{2}{x+1}, \dfrac{x}{(x+1)^2}$

SOLUTION **a.** To find the least common denominator, we list the factors of each denominator:

$$ab^2 = a \cdot b \cdot b$$

$$abc = a \cdot b \cdot c$$

The least common denominator needs to be divisible by both denominators and needs two factors of b as well as one each of a and c.

$$\text{LCD} = a \cdot b \cdot b \cdot c = ab^2c$$

b. To find the least common denominator, we list the factors of each denominator:

$$x + 1 = (x + 1)$$

$$(x + 1)^2 = (x + 1)(x + 1)$$

The least common denominator needs to be divisible by both denominators and needs two factors of $(x + 1)$.

$$\text{LCD} = (x + 1)(x + 1)$$ ◀

▶ Addition and Subtraction of Rational Expressions

ADDING AND SUBTRACTING RATIONAL NUMBERS

To add or subtract rational numbers:

1. Find a common denominator, if necessary.
2. Change each rational number to the common denominator by multiplying numerator and denominator by any factor missing from the common denominator.
3. Add (or subtract) the numerators and place over the common denominator.
4. Simplify the numerator.
5. Eliminate any *common factors* (not common terms) in the numerator and denominator to simplify the expression to lowest terms.

LIKE DENOMINATORS When rational expressions have the same denominator, they can be added or subtracted as written.

▶ **EXAMPLE 3** Adding or subtracting rational expressions Add or subtract these rational expressions containing like denominators. Assume there are no zero denominators.

a. $\frac{8}{y} - \frac{1}{y}$ **b.** $\frac{2x}{x+2} + \frac{5}{x+2}$ **c.** $\frac{3}{x-2} - \frac{x+1}{x-2}$

SOLUTION **a.** $\frac{8}{y} - \frac{1}{y} = \frac{8-1}{y} = \frac{7}{y}$ **b.** $\frac{2x}{x+2} + \frac{5}{x+2} = \frac{2x+5}{x+2}$

In parts a and b, the numerators and denominators contain no common factors, and therefore the expressions are in lowest terms.

Student Note:
In part c the $x + 1$ is subtracted and becomes $-x - 1$.

c. $\frac{3}{x-2} - \frac{x+1}{x-2} = \frac{3-(x+1)}{x-2}$

$$= \frac{2-x}{x-2} = -1$$

The factors $(2 - x)$ and $(x - 2)$ are opposites, and the fraction simplifies to -1. ◀

UNLIKE DENOMINATORS When rational expressions have unlike denominators, first we must find a common denominator.

▶ **EXAMPLE 4** Adding rational expressions Add $\frac{3}{ab^2} + \frac{5}{abc}$. Assume there are no zero denominators.

SOLUTION From Example 2, the least common denominator is ab^2c.

$$\frac{3}{ab^2} + \frac{5}{abc} = \frac{3 \cdot c}{ab^2 \cdot c} + \frac{5 \cdot b}{abc \cdot b}$$ Set up the common denominator.

$$= \frac{3c}{ab^2c} + \frac{5b}{ab^2c}$$ Add the numerators.

$$= \frac{3c + 5b}{ab^2c}$$ ◀

▶ **EXAMPLE 5** Subtracting rational expressions Subtract $\frac{2}{x+1} - \frac{x}{(x+1)^2}$. Assume there are no zero denominators.

SOLUTION From Example 2, the least common denominator is $(x + 1)^2$.

$$\frac{2}{x+1} - \frac{x}{(x+1)^2} = \frac{2(x+1)}{(x+1)(x+1)} - \frac{x}{(x+1)^2}$$ Set up the common denominator and combine numerators.

$$= \frac{2(x+1) - x}{(x+1)^2}$$ Apply the distributive property.

$$= \frac{2x + 2 - x}{(x+1)^2}$$ Add like terms.

$$= \frac{x+2}{(x+1)^2}$$ ◀

▶ Subtraction problems must be worked carefully because there may be a sign change when numerators are subtracted. The next example illustrates both finding the common denominator and changing signs with subtraction.

▶ **EXAMPLE 6** Subtracting rational expressions Subtract $\frac{x}{x^2 + 3x + 2} - \frac{3}{2x + 2}$. Assume there are no zero denominators.

SOLUTION To find the least common denominator, we factor the denominators:

$$x^2 + 3x + 2 = (x + 1)(x + 2)$$
$$2x + 2 = 2(x + 1)$$

The least common denominator will be the product of the three factors, $(x + 1)$, $(x + 2)$, and 2.

$$\text{LCD} = 2(x + 1)(x + 2)$$

$$\frac{x}{x^2 + 3x + 2} - \frac{3}{2x + 2} = \frac{x}{(x + 1)(x + 2)} - \frac{3}{2(x + 1)}$$ Factor the denominators and set up the LCD.

$$= \frac{2 \cdot x}{2(x + 1)(x + 2)} - \frac{3(x + 2)}{2(x + 1)(x + 2)}$$ Combine numerators.

$$= \frac{2x - 3(x + 2)}{2(x + 1)(x + 2)}$$ Apply the distributive property.

$$= \frac{2x - 3x - 6}{2(x + 1)(x + 2)}$$ Add like terms.

$$= \frac{-x - 6}{2(x + 1)(x + 2)}$$ Factor the numerator.

$$= \frac{-1(x + 6)}{2(x + 1)(x + 2)}$$

Note the sign change from the distributive property. Each time we subtract rational expressions, we must watch for such sign changes. ◀

Applications

FORMULAS An important application for students is checking to see whether their answers match those in the back of the book. In Example 7, we work with two solutions to the problem "Solve $A = h(a + b)/2$ for b."

▶ **EXAMPLE 7** Working with formulas containing fractions Use common denominators and subtraction of fractions to show that $b = \frac{2A}{h} - a$ is equivalent to $b = \frac{2A - ha}{h}$, $h \neq 0$.

SOLUTION

$$b = \frac{2A}{h} - a$$ Change a to a fraction.

$$b = \frac{2A}{h} - \frac{a}{1}$$ Build an equivalent fraction with a common denominator.

$$b = \frac{2A}{h} - \frac{h}{h} \cdot \frac{a}{1}$$ Combine numerators.

$$b = \frac{2A - ha}{h}$$ ◀

THINK ABOUT IT: Solve the formula $A = h(a + b)/2$ for b. Do you obtain one of the formulas in Example 7 or yet another formula?

THEATER EXITS Example 8 introduces a formula for emptying a theater.

▶ **EXAMPLE 8** Working with formulas containing fractions: theater exits In architecture, the rate of flow of traffic through doors is important. A movie theater has two exit doors of slightly different sizes. The first can empty the theater in t_1 minutes, and the second can empty the theater in t_2 minutes. The formula to find the number of minutes, t, required to empty the theater if both doors are functioning is

$$\frac{1}{t_1} + \frac{1}{t_2} = \frac{1}{t}$$

Add the two fractions on the left side of the equation. Assume there are no zero denominators.

SOLUTION

$$\frac{1}{t_1} + \frac{1}{t_2} = \frac{1}{t} \qquad \text{The common denominator is } t_1t_2.$$

$$\frac{1 \cdot t_2}{t_1 \cdot t_2} + \frac{1 \cdot t_1}{t_2 \cdot t_1} = \frac{1}{t}$$

$$\frac{t_2 + t_1}{t_1t_2} = \frac{1}{t}$$ ◀

If we solve for t by multiplying by t_1t_2 and dividing by $(t_2 + t_1)$, we obtain

$$t = \frac{t_1t_2}{t_2 + t_1}$$

Thus, the time required to empty the theater is not a simple sum of the individual door times.

Simplifying Complex Rational Expressions (Method 2: Multiply by LCD)

We may use the least common denominator of two fractions within a fraction to simplify a complex fraction.

▶ **EXAMPLE 9** Simplifying complex fractions Simplify $\dfrac{\frac{3}{4}}{\frac{5}{8}}$.

a. Find the least common denominator of the fractions in the numerator and the denominator.

b. Simplify the complex fraction by multiplying the numerator and denominator by the least common denominator.

SOLUTION **a.** The least common denominator of 4 and 8 is 8.

b. $$\frac{\frac{3}{4}}{\frac{5}{8}} = \frac{\frac{3}{4} \cdot 8}{\frac{5}{8} \cdot 8} = \frac{\frac{3}{1} \cdot 2}{\frac{5}{1} \cdot 1} = \frac{6}{5}$$ ◀

▶ **EXAMPLE 10** Simplifying complex rational expressions Find the slope of the line connecting $\left(\frac{a}{2}, \frac{b}{2}\right)$ with (a, b).

SOLUTION A sketch of the coordinates is shown in Figure 16. We substitute the coordinates directly into the slope formula:

$$\text{Slope} = m = \frac{y_2 - y_1}{x_2 - x_1} = \frac{b - \frac{b}{2}}{a - \frac{a}{2}} = \frac{\frac{b}{2}}{\frac{a}{2}}$$

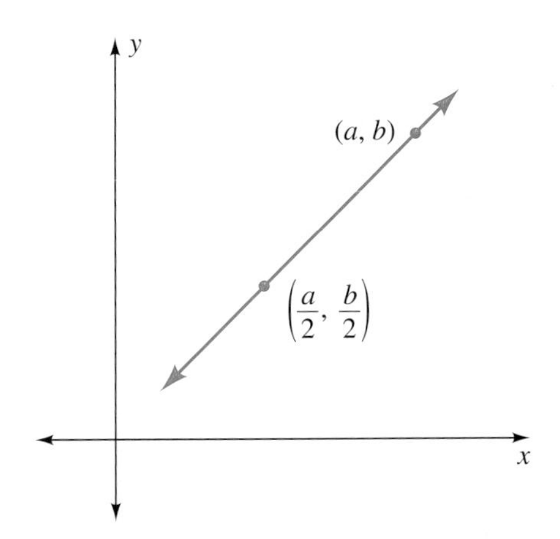

FIGURE 16 Finding slope

We have two ways to simplify the resulting complex expression. We can change the fraction bar into division, as in Section 9.3, or we can multiply the numerator and denominator by the common denominator of the two fractions.

Method 1:

$$\frac{\frac{b}{2}}{\frac{a}{2}} = \frac{b}{2} \div \frac{a}{2} = \frac{b}{2} \cdot \frac{2}{a} = \frac{b}{a}$$

Method 2: The LCD is 2.

$$\frac{\frac{b}{2}}{\frac{a}{2}} = \frac{\frac{b}{2} \cdot \frac{2}{1}}{\frac{a}{2} \cdot \frac{2}{1}} = \frac{b}{a}$$

Thus, the slope of the line connecting $\left(\frac{a}{2}, \frac{b}{2}\right)$ with (a, b) is $\frac{b}{a}$. ◀

ANSWER BOX

Warm-up: 1. $\frac{19}{12}$ **2.** $\frac{4}{9}$ **3.** $\frac{1}{12}$ **4.** $\frac{5}{36}$ **5.** $\frac{5}{18}$ **6.** $\frac{5}{12}$ **Think about it:** One solution method is to multiply both sides by 2, divide both sides by h, and then subtract a from both sides. These steps give the starting equation in Example 7.

▶ 9.4 Exercises

Add or subtract the fractions or rational expressions in Exercises 1 and 2. Assume there are no zero denominators.

1. a. $\frac{11}{7} - \frac{4}{7}$ **b.** $\frac{2}{3} + \frac{x}{3}$ **c.** $\frac{3}{2x} - \frac{5}{2x}$

d. $\frac{4}{x^2 + 1} - \frac{x^2}{x^2 + 1}$ **e.** $\frac{2}{x - 1} - \frac{x + 1}{x - 1}$

2. a. $\frac{13}{6} - \frac{5}{6}$ **b.** $\frac{2}{5} - \frac{x}{5}$ **c.** $\frac{5}{3x} - \frac{8}{3x}$

d. $\frac{4}{x^2 - 4} - \frac{x^2}{x^2 - 4}$ **e.** $\frac{2}{x + 1} + \frac{x - 1}{x + 1}$

What is the least common denominator for each set of rational expressions in Exercises 3 and 4? Assume there are no zero denominators.

3. a. $\frac{5}{12} + \frac{7}{20}$ **b.** $\frac{2}{x} + \frac{5}{2x}$

c. $\frac{8}{y} - \frac{1}{y^2}$ **d.** $\frac{3}{b} + \frac{2}{a} + \frac{5}{b} - \frac{3}{a}$

Blue numbers are core exercises.

e. $\dfrac{4}{x-3} - \dfrac{2}{x^2-9}$

f. $\dfrac{4}{x^2+5x+6} - \dfrac{2}{x^2-9}$

4. a. $\dfrac{1}{2} + \dfrac{3}{8}$ b. $\dfrac{5}{8} + \dfrac{7}{18}$ c. $\dfrac{2}{3} - \dfrac{3}{a}$

d. $\dfrac{b}{a} - \dfrac{c}{a^2}$ e. $\dfrac{2}{x^2+2x} + \dfrac{5}{x^2-4}$

f. $\dfrac{4}{x^2-2x+1} - \dfrac{3}{x^2-1}$

5. Add or subtract, as indicated, the expressions in Exercise 3.

6. Add or subtract, as indicated, the expressions in Exercise 4.

In Exercises 7 to 20, add or subtract the rational expressions, as indicated. Assume there are no zero denominators.

7. $\dfrac{3}{2b} + \dfrac{3}{4a}$ **8.** $\dfrac{5}{3a} + \dfrac{7}{6b}$

9. $\dfrac{1}{x-1} + \dfrac{1}{x}$ **10.** $\dfrac{1}{x+1} + \dfrac{1}{x-1}$

11. $\dfrac{8}{x+1} - \dfrac{3}{x}$ **12.** $\dfrac{5}{x+1} - \dfrac{8}{x}$

13. $\dfrac{1}{a} - \dfrac{1}{a^2}$ **14.** $\dfrac{2}{b^2} - \dfrac{3}{b}$

15. $\dfrac{2}{ab} - \dfrac{3}{2b}$ **16.** $\dfrac{3}{ac} - \dfrac{5}{2a}$

17. a. $\dfrac{2}{x+3} + \dfrac{3}{(x+3)^2}$ b. $\dfrac{2}{x^2-1} - \dfrac{1}{x-1}$

18. a. $\dfrac{4}{x-2} + \dfrac{2}{(x-2)^2}$

b. $\dfrac{-5}{x^2-x-6} - \dfrac{1}{x+2}$

19. a. $\dfrac{3}{x^2-x} + \dfrac{x}{x^2-3x+2}$

b. $\dfrac{1}{x^2-3x} - \dfrac{2}{x^2-9}$

20. a. $\dfrac{x}{x^2-6x+9} + \dfrac{5}{x^2-3x}$

b. $\dfrac{4}{x^2-4x} - \dfrac{7}{x^2-x-12}$

Add the fractions on the right side of each question in Exercises 21 to 28 to obtain a single fraction for the application formula. Assume there are no zero denominators.

21. Temperature change: $\Delta T = \dfrac{T_0}{T} - 1$

22. Resistors in parallel in an electrical circuit:

$$\frac{1}{R} = \frac{1}{R_1} + \frac{1}{R_2}$$

23. Condensers in series in an electrical circuit:

$$\frac{1}{C} = \frac{1}{C_1} + \frac{1}{C_2}$$

24. Days to complete a wheat harvest with two machines:
$\dfrac{1}{D} = \dfrac{1}{D_1} + \dfrac{1}{D_2}$

25. Total time for a round trip: $t = \dfrac{D}{r_1} + \dfrac{D}{r_2}$

26. Traffic accident analysis, preliminary to calculating vehicle speed: $R = \dfrac{C^2}{8M} + \dfrac{M}{2}$

27. Radius of curvature of a surface: $F = \dfrac{L^2}{6d} + \dfrac{d}{2}$

28. Approximating an exponential function:

$$e^x \approx 1 + x + \frac{x^2}{2} + \frac{x^3}{6} + \frac{x^4}{24}$$

For Exercises 29 to 34, find the slope of the line connecting the points. Simplify to eliminate fractions from the numerator and the denominator.

29.

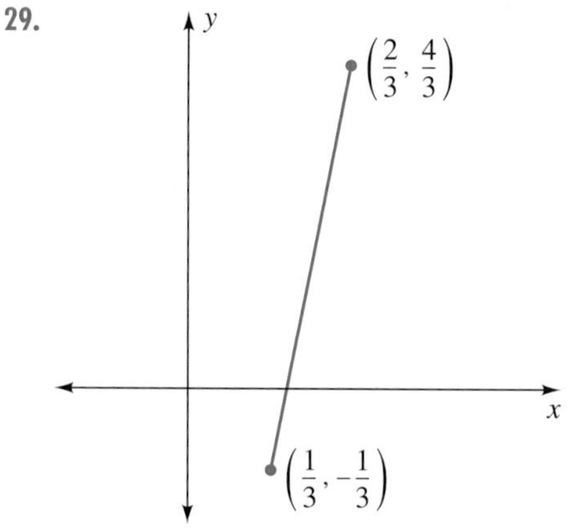

30.

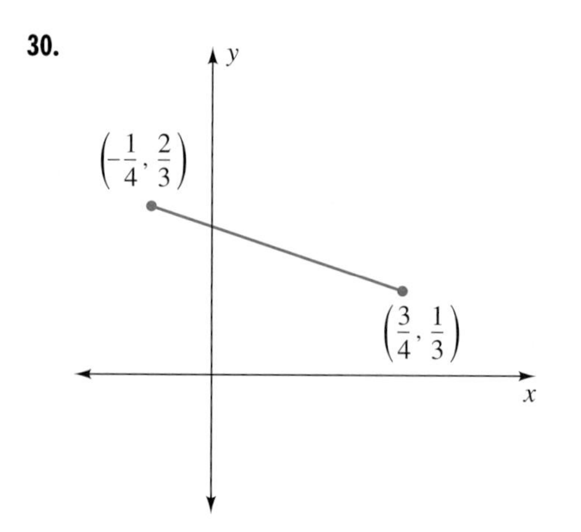

Blue numbers are core exercises.

31. $\left(\frac{1}{5}, \frac{2}{5}\right), \left(\frac{-4}{5}, \frac{1}{5}\right)$

32. $\left(\frac{2}{3}, \frac{1}{4}\right), \left(\frac{-2}{3}, \frac{3}{4}\right)$

33. $\left(-\frac{b}{2}, a\right), \left(b, \frac{a}{2}\right)$

34. $\left(-\frac{b}{2}, -a\right), \left(b, \frac{a}{2}\right)$

In Exercises 35 and 36, find the slopes of the line segments connecting the indicated points in each figure. What do you observe about the slopes?

35. Line segment (a, b) to $(c, 0)$

Line segment $\left(\frac{a}{2}, \frac{b}{2}\right)$ to $\left(\frac{c}{2}, 0\right)$

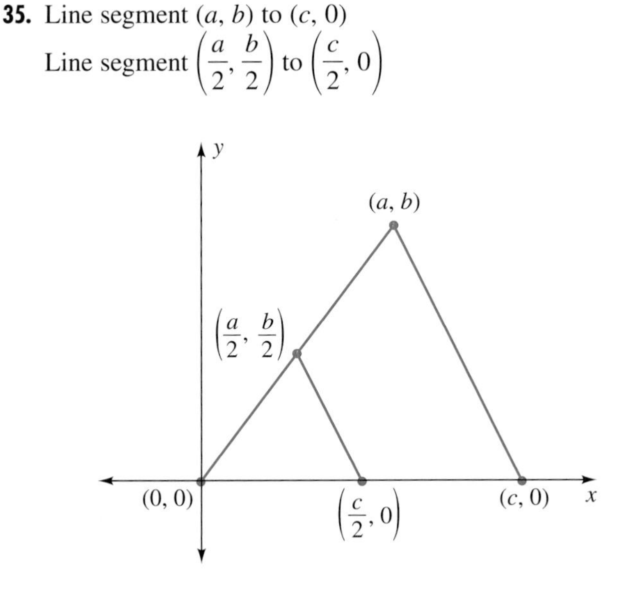

36. Line segment $(0, 0)$ to $(c, 0)$

Line segment $\left(\frac{a}{2}, \frac{b}{2}\right)$ to $\left(\frac{a+c}{2}, \frac{b}{2}\right)$

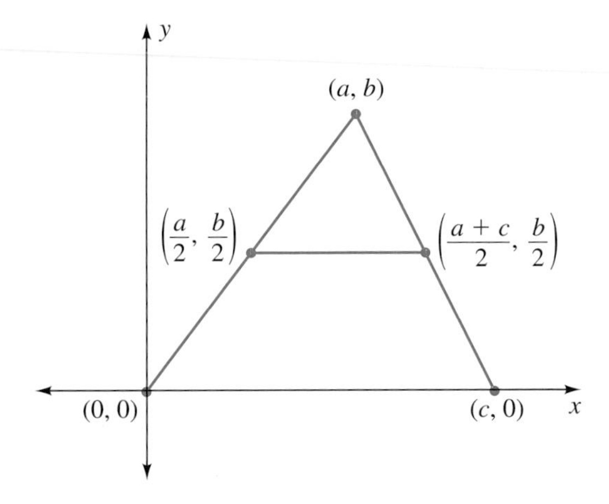

Simplify the complex fractions in Exercises 37 to 44 to eliminate fractions from the numerators and denominators. Assume there are no zero denominators.

37. $\dfrac{5}{\frac{1}{5} + 1}$

38. $\dfrac{3}{1 + \frac{1}{3}}$

39. $h = \dfrac{A}{\frac{1}{2}b}$

40. $h = \dfrac{V}{\frac{1}{3}\pi r^2}$

41. $\dfrac{x + \dfrac{x}{2}}{2 - \dfrac{x}{3}}$

42. $\dfrac{\dfrac{x}{3} + x}{3 - \dfrac{x}{2}}$

43. Refrigeration cycle: $\dfrac{1}{\dfrac{Q_H}{Q_L} - 1}$

44. Heat transfer: $\dfrac{1}{1 - \dfrac{Q_L}{Q_H}}$

▶ Graphing Calculator Exploration

45. a. Graph the expression in Exercise 41 before and after simplifying.

b. Compare the graphs before and after simplifying.

c. Where is the graph nearly vertical? Why?

46. a. Graph the expression in Exercise 42 before and after simplifying.

b. Compare the graphs before and after simplifying.

c. Where is the graph nearly vertical? Why?

▶ Problem Solving: Fraction Operations

Identify the missing operation symbol (+, −, •, or ÷) in Exercises 47 to 50. Assume there are no zero denominators.

47. a. $\frac{3}{4} \square \frac{2}{5} = \frac{6}{20} = \frac{3}{10}$

b. $\frac{3}{4} \square \frac{2}{5} = \frac{15}{8}$

c. $\frac{3}{4} \square \frac{2}{5} = \frac{23}{20}$

48. a. $\frac{2}{5} \square \frac{3}{4} = \frac{8}{15}$

b. $\frac{3}{4} \square \frac{2}{5} = \frac{7}{20}$

c. $\frac{5}{4} \square \frac{2}{3} = \frac{23}{12}$

49. a. $\frac{1}{a} \square \frac{1}{b} = \frac{a+b}{ab}$

b. $\frac{1}{a} \square \frac{1}{b} = \frac{1}{ab}$

50. a. $\frac{1}{a} \square \frac{1}{b} = \frac{b-a}{ab}$

b. $\frac{1}{b} \square \frac{1}{a} = \frac{a}{b}$

▶ Projects

51. Exiting a Theater, II Return to the movie theater project in Exercise 36 of Section 9.3. The movie theater has two exits. One door, by itself, can empty the theater in 5 minutes. The second door, by itself, can empty the theater in 8 minutes. Suppose we wish to answer this question: If both doors are available, how many minutes will it take to empty the theater?

Blue numbers are core exercises.

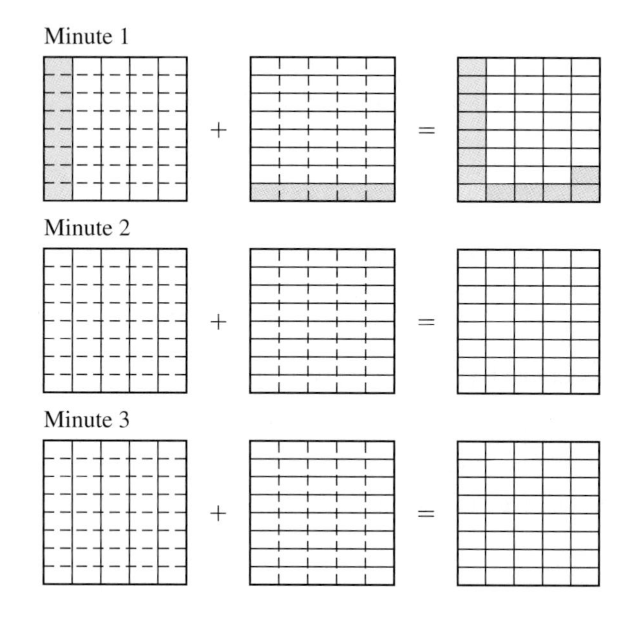

a. How are the rectangles in the figure the same as those in the figure in Exercise 35 of Section 9.3?

b. How are the rectangles different?

c. How many small pieces are there in each large rectangle? Why is this number important to adding the pieces together?

d. What does the shading in the first rectangle in the first row represent?

e. What does the shading in the second rectangle in the first row represent?

f. What does the shading in the third rectangle in the first row represent?

g. Shade the rectangles for each subsequent minute's departures. Add more rectangles as needed.

h. How will the rectangle look when the theater is empty?

i. Estimate the total number of minutes needed to empty the theater.

j. Why might the rectangles in this figure be a better way to represent the fractions $\frac{1}{5}$ and $\frac{1}{8}$ than those in the figure in Exercise 36 of Section 9.3?

k. Make up a problem of your own involving three doors.

▶ 9.5 Solving Rational Equations

Objectives

- Find the least common denominator of rational expressions in an equation.
- Multiply by the least common denominator to eliminate the denominators in an equation.
- Solve equations containing rational expressions.
- Solve application problems related to $\frac{1}{a} + \frac{1}{b} = \frac{1}{c}$.

WARM-UP

Multiply these expressions using the distributive property.

1. $36d\left(\frac{1}{12} + \frac{1}{18}\right)$

2. $2x\left(\frac{2}{x} + \frac{3x}{2}\right)$

3. $x(x + 1)\left(\frac{2}{x + 1} + \frac{3}{x}\right)$

Find the least common denominator for each pair of fractions.

4. $\frac{1}{12}, \frac{1}{18}$

5. $\frac{1}{9}, \frac{1}{15}$

6. $\frac{1}{8}, \frac{1}{18}$

IN THIS SECTION, we apply the distributive property to rational expressions. We solve equations containing rational expressions. We also examine one application of rational equations that has particular appeal to the mathematician: a common formula to describe patterns that occur in a variety of applications.

The Harvest Problem

In Examples 1 and 2, we solve the wheat harvest problem in the project, Exercise 69, Section 9.2.

EXAMPLE 1 Exploring work done together: wheat harvest Farmers Terry and Lee can separately harvest a property in 18 days and 12 days, respectively. How might we calculate the number of days needed to harvest the crop if they work together?

Consider the following proposed methods. Comment on the results.

Student Note:
Harvesters have headers (cutting surfaces) of different widths and run at different speeds.

Proposed method 1: Suppose we add the 18 days and 12 days.

Proposed method 2: Suppose we average the number of days.

Proposed method 3: Suppose we subtract 12 days from the 18 days.

Proposed method 4: Suppose we find how much the farmers harvest each day.

SOLUTION ***Proposed method 1:*** Adding the 18 days and 12 days gives 30 days, which is not reasonable. If the farmers work together, the task should take fewer than the 12 days Lee requires alone.

Proposed method 2: If we average the number of days, we have

$$\text{Average} = \frac{18 \text{ days} + 12 \text{ days}}{2} = 15 \text{ days}$$

Again, this result is not reasonable, as the two farmers should not take longer than Lee working by herself.

Proposed method 3: If we subtract 12 days from the 18 days, we get 6 days. Although subtraction might give a reasonable answer in this setting, consider subtraction in the case where both farmers harvest in 18 days. Subtraction would give zero days working together, whereas a reasonable answer might be 9 days.

Proposed method 4: Finding how much the farmers harvest each day is another approach. On the first day, Terry harvests $\frac{1}{18}$ of the total while Lee harvests $\frac{1}{12}$ (see Figure 17). Together they harvest $\frac{1}{12} + \frac{1}{18}$. Each day, they harvest these fractions of the total job. The following summarizes the fraction of the total crop harvested by the end of each day, for 8 days:

Day 1: $\frac{1}{18} + \frac{1}{12} = \frac{2}{36} + \frac{3}{36} = \frac{5}{36}$

Day 2: $2\left(\frac{1}{18} + \frac{1}{12}\right) = 2\left(\frac{5}{36}\right) = \frac{10}{36}$

Day 3: $3\left(\frac{1}{18} + \frac{1}{12}\right) = 3\left(\frac{5}{36}\right) = \frac{15}{36}$

Day 4: $4\left(\frac{5}{36}\right) = \frac{20}{36}$

Day 5: $5\left(\frac{5}{36}\right) = \frac{25}{36}$

Day 6: $6\left(\frac{5}{36}\right) = \frac{30}{36}$

Day 7: $7\left(\frac{5}{36}\right) = \frac{35}{36}$

Day 8: $8\left(\frac{5}{36}\right) = \frac{40}{36}$

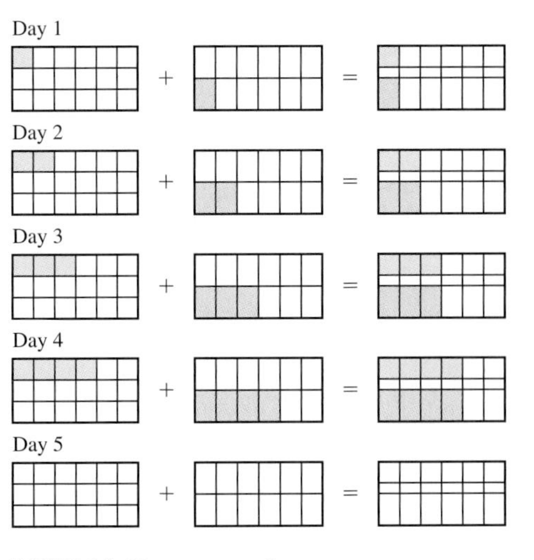

FIGURE 17 Harvest together

The harvest is complete when the fraction reaches $\frac{36}{36} = 1$. The harvest is nearly complete at the end of day 7. It is completely finished during day 8, when the fraction exceeds 1. The exact time to finish is d days, where

$$d\left(\frac{1}{18} + \frac{1}{12}\right) = 1$$

Traditionally, this equation is divided on both sides by d and written

$$\frac{1}{18} + \frac{1}{12} = \frac{1}{d}, \quad d \neq 0$$

◀

Our first step in solving the harvest equation will be to eliminate the denominators. (Yes, we just divided by d and put it in the denominator. Now we are going to take it out again. Keep in mind that the traditional form of the equation has the variable in the denominator.)

▶ In the Warm-up, we applied the distributive property to fractions. When we multiplied by the least common denominator, we eliminated the denominators. We use this approach to eliminate denominators in equations.

ELIMINATING DENOMINATORS

> To eliminate denominators from an equation, multiply both sides of the equation by the least common denominator.

▶ **EXAMPLE 2** Solving a rational equation Solve $\frac{1}{18} + \frac{1}{12} = \frac{1}{d}$ for d, $d \neq 0$.

SOLUTION The least common denominator for all fractions in the equation is $36d$.

$$\frac{1}{18} + \frac{1}{12} = \frac{1}{d}$$

$$36d\left(\frac{1}{18} + \frac{1}{12}\right) = 36d\left(\frac{1}{d}\right)$$

$$\frac{36d \cdot 1}{18} + \frac{36d \cdot 1}{12} = \frac{36d \cdot 1}{d}$$

$$2d + 3d = 36$$

$$5d = 36$$

$$d = \frac{36}{5} = 7.2 \text{ days}$$

Calculator check: $\frac{1}{18} + \frac{1}{12} \approx 0.0556 + 0.0833 \approx 0.1389 \approx \frac{1}{7.2}$ ✓

Working together, the two farmers harvest the crop in 7.2 days. ◀

▶ Solving Rational Equations

Examples 3, 4, and 5 provide more illustrations of solving equations containing rational expressions. All three use factoring to solve a quadratic equation.

▶ **EXAMPLE 3** Solving a rational equation List any inputs that must be excluded, and then solve $\frac{2}{x} + \frac{3x}{2} = 4$ for x.

SOLUTION The expressions in the equation are undefined for a zero denominator; thus, $x \neq 0$.

$$\frac{2}{x} + \frac{3x}{2} = 4 \qquad \text{Multiply by the LCD.}$$

$$2x\left(\frac{2}{x} + \frac{3x}{2}\right) = 2x \cdot 4 \qquad \text{Distribute the LCD.}$$

$$\frac{2x \cdot 2}{x} + \frac{2x \cdot 3x}{2} = 2x \cdot 4 \qquad \text{Simplify the fractions.}$$

$$4 + 3x^2 = 8x \qquad \text{Solve for the zero form.}$$

$$3x^2 - 8x + 4 = 0 \qquad \text{Factor.}$$

$$(3x - 2)(x - 2) = 0 \qquad \text{Apply the zero product rule and solve.}$$

$$\text{Either} \quad 3x - 2 = 0 \quad \text{or} \quad x - 2 = 0$$

$$x = \tfrac{2}{3} \quad \text{or} \quad x = 2$$

We check by substituting each x into the original equation. The check is left as an exercise. ◀

▶ **EXAMPLE 4** Solving a rational equation List any inputs that must be excluded, and then solve $\frac{2}{x + 1} + \frac{3}{x} = -2$ for x.

SOLUTION The expressions in the equation are undefined for a zero denominator; thus, $x \neq -1$ and $x \neq 0$.

$$\frac{2}{x + 1} + \frac{3}{x} = -2 \qquad \text{Multiply by the LCD.}$$

$$x(x + 1)\left(\frac{2}{x + 1} + \frac{3}{x}\right) = -2 \cdot x(x + 1) \qquad \text{Distribute the LCD.}$$

$$\frac{x(x + 1) \cdot 2}{x + 1} + \frac{x(x + 1) \cdot 3}{x} = -2x(x + 1) \qquad \text{Simplify the fractions.}$$

$$2x + 3(x + 1) = -2x^2 - 2x \qquad \text{Solve for the zero form.}$$

$$2x + 3x + 3 + 2x^2 + 2x = 0 \qquad \text{Combine like terms.}$$

$$2x^2 + 7x + 3 = 0 \qquad \text{Factor.}$$

$$(2x + 1)(x + 3) = 0 \qquad \text{Apply the zero product rule and solve.}$$

$$\text{Either} \quad x = -\tfrac{1}{2} \quad \text{or} \quad x = -3$$

The check by substitution is left as an exercise. ◀

Checking our work not only improves our accuracy but also prevents us from applying solutions that are not acceptable. In Example 5, one answer, an *extraneous root*, gives an undefined expression when we substitute it into the equation.

▶ **EXAMPLE 5** Solving a rational equation Solve $x + \frac{12}{x - 4} = \frac{3x}{x - 4}$, where $x \neq 4$.

SOLUTION

$$x + \frac{12}{x-4} = \frac{3x}{x-4} \quad \text{Multiply both sides by } x - 4.$$

$$x(x-4) + 12 = 3x \quad \text{Solve for the zero form.}$$

$$x^2 - 4x - 3x + 12 = 0 \quad \text{Combine like terms.}$$

$$x^2 - 7x + 12 = 0 \quad \text{Factor.}$$

$$(x-4)(x-3) = 0 \quad \text{Apply the zero product rule.}$$

$$\text{Either } x - 4 = 0 \quad \text{or} \quad x - 3 = 0$$

$$x = 4 \quad \text{or} \quad x = 3$$

Check: In checking our solutions, we find that $x = 3$ satisfies the equation:

$$3 + \frac{12}{3-4} \stackrel{?}{=} \frac{3(3)}{3-4} \quad ✓$$

However, $x = 4$ gives a zero denominator:

$$4 + \frac{12}{4-4} \stackrel{?}{=} \frac{3(4)}{4-4}$$

As noted in the original problem, $x = 4$ has been excluded from the set of possible inputs. The solution $x = 4$ is an extraneous root. ◀

▶ Sum of Rates Applications

We now return to the setting in Example 1. When a task is completed in a days, the rate is $\frac{1}{a}$. When a task is completed in b days, the rate is $\frac{1}{b}$. A sum of rates is used in Example 6 for the wheat harvest and in Example 7 for exiting a theater. The same thinking holds for ventilation fans moving air and pipes filling swimming pools.

▶ **EXAMPLE 6** Building and solving a rational equation: wheat harvest When it is time to harvest Terry's wheat, Lee moves her equipment to the south end of the county. Terry normally harvests his wheat in 15 days. Lee's equipment could do the work in 9 days. If they work together, how long will it take to harvest Terry's wheat? Estimate an answer, and then write an equation and solve it.

SOLUTION We have the same situation as before but with different numbers of days of work. The harvest should take fewer than 9 days (Lee's time) and more than $4\frac{1}{2}$ days (if they both worked at Lee's rate).

The equation is

$$\frac{1}{9} + \frac{1}{15} = \frac{1}{d}, \quad d \neq 0$$

The least common denominator of the fractions within the equation is $45d$.

$$\frac{1}{9} + \frac{1}{15} = \frac{1}{d}$$

$$45d\left(\frac{1}{9} + \frac{1}{15}\right) = 45d\left(\frac{1}{d}\right)$$

$$\frac{45d \cdot 1}{9} + \frac{45d \cdot 1}{15} = \frac{45d \cdot 1}{d}$$

$$5d + 3d = 45$$

$$8d = 45$$

$$d = \frac{45}{8} = 5.625 \text{ days}$$

Student Note:
Describe what is done at each step in solving for d.

Calculator check: $\frac{1}{9} + \frac{1}{15} \approx 0.1111 + 0.0667 \approx 0.1778 \approx \frac{1}{5.625}$ ✓

Working together, the farmers can harvest the wheat in 5.625 days. ◀

The individual times to harvest the wheat may be described as d_1 and d_2. Working individually, the farmers could harvest the whole crop at a rate of $\frac{1}{d_1}$ and $\frac{1}{d_2}$, respectively, each day. Because both are working to finish 1 job, we add the rates for the two farmers to obtain the rate for working together, $\frac{1}{d}$:

$$\frac{1 \text{ wheat crop}}{d_1 \text{ days}} + \frac{1 \text{ wheat crop}}{d_2 \text{ days}} = \frac{1 \text{ wheat crop}}{d \text{ days together}}$$

This relationship can be described with the *sum of rates* formula:

Sum of Rates

$$\frac{1}{d_1} + \frac{1}{d_2} = \frac{1}{d}$$

▶ **EXAMPLE 7** **Building an equation: exiting a theater** Suppose a movie theater has two exits. One is a double-wide door that, by itself, can empty the theater in 5 minutes. The other is a single door that, by itself, can empty the theater in 8 minutes. Estimate the time required to empty the theater when both doors are open, and then build an equation.

SOLUTION The time will be less than 5 minutes (double-wide door) and greater than $2\frac{1}{2}$ minutes (if both were double-wide doors).

We change exit times to rates for emptying the theater. During each minute, the doors permit $\frac{1}{5}$ and $\frac{1}{8}$ of the theater, respectively, to empty. The following expressions represent the portion of the theater that has been emptied by the end of each minute:

First minute:

$$\frac{1}{5} + \frac{1}{8} = \frac{8}{40} + \frac{5}{40} = \frac{13}{40}$$

Second minute:

$$2\left(\frac{1}{5} + \frac{1}{8}\right) = 2\left(\frac{13}{40}\right) = \frac{26}{40}$$

Third minute:

$$3\left(\frac{13}{40}\right) = \frac{39}{40}$$

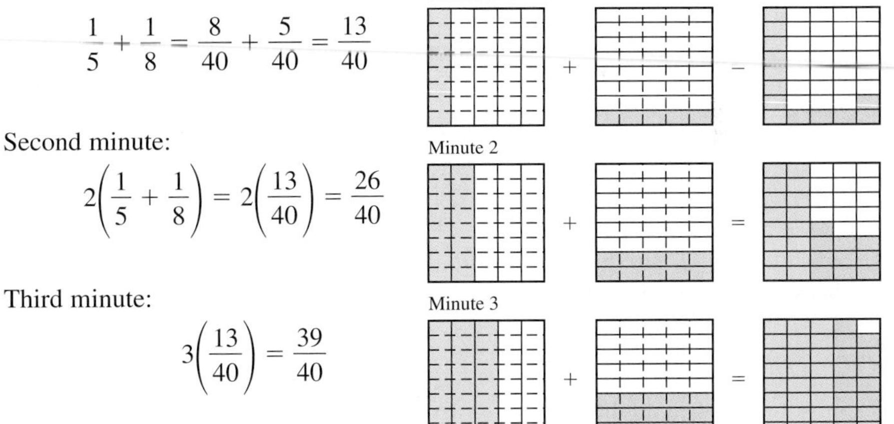

FIGURE 18 Theater exit rates

The theater will be entirely empty during the fourth minute, when the fraction $\frac{40}{40} = 1$ is reached. To find the exact time, x, needed to empty the theater, we solve the equation

$$x\left(\frac{1}{5} + \frac{1}{8}\right) = 1$$

Again, the traditional form of this equation is

$$\frac{1}{5} + \frac{1}{8} = \frac{1}{x}, \quad x \neq 0$$

◀

▶ **EXAMPLE 8** Solving a rational equation Solve $\frac{1}{5} + \frac{1}{8} = \frac{1}{x}$ for x, where $x \neq 0$.

SOLUTION The least common denominator for the equation is $40x$.

$$\frac{1}{5} + \frac{1}{8} = \frac{1}{x}$$

$$40x\left(\frac{1}{5} + \frac{1}{8}\right) = 40x\left(\frac{1}{x}\right)$$

$$8x + 5x = 40$$

$$13x = 40$$

$$x = \frac{40}{13} \approx 3.1 \text{ min}$$

Calculator check: $\frac{1}{5} + \frac{1}{8} = 0.200 + 0.125 = 0.325 \approx \frac{1}{3.1}$ ✓

The two-exit system permits the theater to be cleared rapidly. ◀

Student Note:
Congratulations; this is the last section of the textbook! You are to be commended for finishing the course.

ANSWER BOX

Warm-up: **1.** $5d$ **2.** $4 + 3x^2$ **3.** $5x + 3$ **4.** 36 **5.** 45 **6.** 72

▶ 9.5 Exercises

Multiply the expressions in Exercises 1 to 6, and simplify the results. Assume there are no zero denominators.

1. $12x\left(\frac{1}{12} + \frac{2}{3x}\right)$

2. $8x\left(\frac{1}{4x} + \frac{1}{2}\right)$

3. $4x^2\left(\frac{1}{2x} + \frac{3}{x^2}\right)$

4. $6x^2\left(\frac{2}{3x^2} + \frac{1}{6}\right)$

5. $x(x - 1)\left(\frac{1}{x - 1} + \frac{1}{x}\right)$

6. $x(x + 2)\left(\frac{1}{x} + \frac{3}{x + 2}\right)$

7. Substitute $x = 2$ into this equation from Example 3, and check that it is a solution:

$$\frac{2}{x} + \frac{3x}{2} = 4$$

8. Substitute $x = \frac{2}{3}$ into the equation in Exercise 7, and check that it is a solution. Use a calculator as needed.

9. Substitute $x = -\frac{1}{2}$ into this equation from Example 4, and check that it is a solution:

$$\frac{2}{x + 1} + \frac{3}{x} = -2$$

Use a calculator as needed.

10. Substitute $x = -3$ into the equation in Exercise 9, and check that it is a solution.

Find the least common denominator for each equation in Exercises 11 to 26.

11. $\frac{x}{4} + \frac{x}{6} = 28$

12. $\frac{x}{14} + \frac{x}{8} = 11$

13. $\frac{3}{4} + \frac{1}{5} = \frac{1}{x}$

14. $\frac{2}{3} + \frac{2}{5} = \frac{1}{x}$

15. $\frac{1}{8} + \frac{1}{x} = \frac{1}{2}$

16. $\frac{1}{10} + \frac{1}{x} = \frac{1}{6}$

17. $\frac{1}{x} = \frac{1}{3x} + \frac{1}{3}$

18. $\frac{4}{x} + \frac{2}{x} = \frac{3}{x}$

19. $\frac{3}{x} - \frac{2}{x} = \frac{4}{x}$

20. $\frac{1}{x} = \frac{1}{2} - \frac{1}{2x}$

21. $\frac{1}{x - 1} = \frac{2}{x + 3}$

22. $\frac{2}{x - 1} = \frac{1}{x - 4}$

23. $\frac{2}{x^2} - \frac{3}{x} + 1 = 0$

24. $1 + \frac{5}{x} - \frac{14}{x^2} = 0$

25. $\frac{1}{x - 3} - 3 = \frac{4 - x}{x - 3}$

26. $\frac{1}{x - 4} = \frac{5 - x}{x - 4} + 4$

For Exercises 27 to 42, solve the given exercise for x, using multiplication by the LCD. Indicate any inputs that must be excluded.

27. Exercise 11

28. Exercise 12

Blue numbers are core exercises.

29. Exercise 13 **30.** Exercise 14

31. Exercise 15 **32.** Exercise 16

33. Exercise 17 **34.** Exercise 18

35. Exercise 19 **36.** Exercise 20

37. Exercise 21 **38.** Exercise 22

39. Exercise 23 **40.** Exercise 24

41. Exercise 25 **42.** Exercise 26

In Exercises 43 to 54, indicate any inputs that must be excluded and then solve the equation. Factoring or the quadratic equation may be helpful.

43. $5x + \frac{13}{2} = \frac{3}{2x}$ **44.** $\frac{1}{2} + 3x = \frac{1}{x}$

45. $\frac{1}{x} + 3 = \frac{5}{x(x+1)}$ **46.** $4x + 7 = \frac{1}{x+1}$

47. $x = 6 - \frac{6}{x-1}$ **48.** $x + 1 = \frac{x+1}{x}$

49. $\frac{x-1}{x-2} = \frac{1}{x-2}$ **50.** $\frac{3-x}{x-1} = \frac{2}{x-1}$

51. $\frac{10}{x-5} = x + \frac{2x}{x-5}$ **52.** $x + \frac{x}{x-3} = \frac{24}{x-3}$

53. $2x + \frac{2x}{x+1} = \frac{5}{x+1}$ **54.** $x + \frac{4x}{x-6} = \frac{24}{x-6}$

In Exercises 55 to 58, solve for the indicated variable. Assume $d \neq 0$, $x \neq 0$.

55. Solve $\frac{1}{15} + \frac{1}{18} = \frac{1}{d}$ for d.

56. Solve $\frac{1}{12} + \frac{1}{15} = \frac{1}{d}$ for d.

57. Solve $\frac{1}{8} + \frac{1}{20} = \frac{1}{x}$ for x.

58. Solve $\frac{1}{5} + \frac{1}{6} = \frac{1}{x}$ for x.

Set up and solve equations for Exercises 59 to 66. Round to the nearest tenth.

59. One farmer harvests his barley in 14 days. A second farmer harvests the same crop in 12 days. How many days will it take them working together?

60. One farmer bales her hay in 6 days. A second farmer does it in 8 days. How many days will it take them working together?

61. One ventilation fan changes the air in a house in 4 hours. A second fan vents the same volume of air in 5 hours. How long will it take to vent the house if both fans are working? If building code requires a complete change of air every 3 hours, will the two fans be sufficient?

62. A $\frac{5}{8}$-inch garden hose fills a child's pool in 1 hour. A $\frac{1}{2}$-inch garden hose fills a child's pool in 1.5 hours. How long will it take to fill the pool if both hoses are used at once?

63. One ventilation fan changes the air in a house in 5 hours. A second fan is to be installed. In order to satisfy code, both fans working together must change the air in 3 hours. Describe the fan needed to satisfy code.

64. A $1\frac{1}{2}$-inch pipe fills a swimming pool directly from a farm well in 5 days. A $\frac{5}{8}$-inch hose feeding water from the cistern is added, and the pool fills in 4 days. How long would the hose take to fill the pool by itself?

65. One door can clear a theater in 9 minutes. A second door can clear the theater in 6 minutes. How fast can the theater be emptied if both doors are available?

66. A large theater has three exits. Operating individually, each door can clear the theater in 6 minutes. How fast can the theater be emptied if all three doors are available?

In Exercises 67 to 70, solve for the indicated letter. Assume there are no zero denominators.

67. $\frac{1}{a} + \frac{1}{b} = \frac{1}{c}$ for b

68. $\frac{1}{a} + \frac{1}{b} = \frac{1}{c}$ for a

69. $\frac{1}{a} + \frac{1}{b} = \frac{1}{c}$ for c

70. $\frac{1}{R} = \frac{1}{R_1} + \frac{1}{R_2}$ for R_1

The reciprocal key, x^{-1} or $1/x$, gives a way to obtain decimals for fractions with 1 in the numerator. In Exercises 71 to 74, use the reciprocal key to solve the equation in the given exercise; list your keystrokes. Round to three decimal places.

71. Exercise 55 **72.** Exercise 56

73. Exercise 57 **74.** Exercise 58

75. Compare the roles of multiplying by the least common denominator in simplifying complex fractions and in solving equations.

76. What was done wrong in the following simplification?

$$\overset{4}{12x}\left(\frac{2}{\underset{1}{3x}} + \frac{1}{4}\right) = \frac{4 \cdot 2}{1} + \frac{4}{4} = 8 + 1 = 9$$

▶ Projects

77. Exiting a Theater, III A movie theater has three exit doors of slightly different sizes. The first can empty the theater in t_1 minutes; the second, in t_2 minutes; and the third, in t_3 minutes. The formula to determine the

Blue numbers are core exercises.

number of minutes, t, required to empty the theater if all three doors are functioning is

$$\frac{1}{t_1} + \frac{1}{t_2} + \frac{1}{t_3} = \frac{1}{t}$$

a. Add the three fractions on the left side of the equation.

b. Solve the equation in part a for t. Assume none of the times are zero.

c. Make up reasonable exit times for a theater with three doors, as described, and find the total exiting time.

78. Graphing Calculator

a. Enter $y_1 = \frac{2}{x} + \frac{1}{x - 2}$ and $y_2 = \frac{2(x - 2) + x}{x(x - 2)}$.

b. Use the table function to compare the outputs. Try TblSet with **TblMin** $= -3$, **ΔTbl** $= 1$.

c. Why are there no outputs at $x = 0$ and $x = 2$?

d. Graph the equations.

e. What may be concluded about the two equations? Why?

▶ 9 Chapter Summary

Vocabulary

For definitions and page references, see the Glossary/Index.

complex fractions
complex rational expressions
equivalent fraction property
least common denominator (LCD)
opposites
rational functions
rational numbers
simplification property of fractions
sum of rates

Concepts

9.1 Rational Functions: Graphs and Applications

A zero denominator means division by zero, an undefined operation.

The symbol $\mathbb{R}$ is the notation for the set of all real numbers.

The graph of a rational expression (simplified to lowest terms) approaches infinity in a vertical direction whenever the denominator approaches zero.

9.2 Simplifying Rational Expressions

A rational expression must be factored in order to simplify to lowest terms.

Simplify expressions containing units of measure, using facts such as feet/feet = 1.

When the numerator and denominator of a fraction are the same, the fraction equals 1.

When the numerator and denominator of a fraction are opposites, the fraction equals -1.

9.3 Multiplication and Division of Rational Expressions

Advice from Angie Cowles, student at Lane Community College: "Common factors will be more apparent if you leave the final expressions in factored form."

To multiply rational expressions, simplify first. Then multiply the numerators. Then multiply the denominators.

To divide rational expressions, multiply the first expression by the reciprocal of the second expression.

To simplify complex fractions or complex expressions, multiply the numerator by the reciprocal of the denominator.

9.4 Finding the Common Denominator and Addition and Subtraction of Rational Expressions

To find the least common denominator, form a product by including each factor the highest number of times it appears in any denominator.

To add or subtract rational expressions, rewrite the expressions with a least common denominator (if needed). Combine the numerators over the common denominator and then simplify by eliminating common factors.

To simplify a complex fraction or rational expression, find the LCD for the numerator and denominator fractions and multiply the LCD times both the numerator and the denominator.

9.5 Solving Rational Equations

To solve an equation containing rational expressions, multiply each side by the least common denominator.

Check answers to eliminate extraneous roots.

▶ 9 Review Exercises

1. For what input, x, is the expression $\frac{2 - x}{x + 3}$ undefined?

2. Find ten coordinate points that satisfy $y = \frac{6}{x + 3}$, and sketch a graph. Is there an input that gives $y = 0$? Why or why not?

3. World reserves of natural gas were estimated at 6800 trillion cubic feet in 2004. Write an equation that describes the number of years the gas reserves will last if x cubic feet are used per day.

In Exercises 4 and 5, describe the output behavior of the graph in the figure in the given situations.

4.

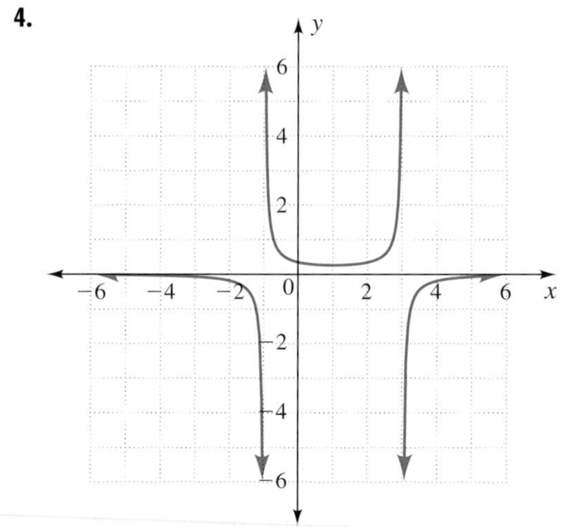

a. As x approaches -1 from the left

b. As x approaches 3 from the right

c. As x approaches -1 from the right

5.

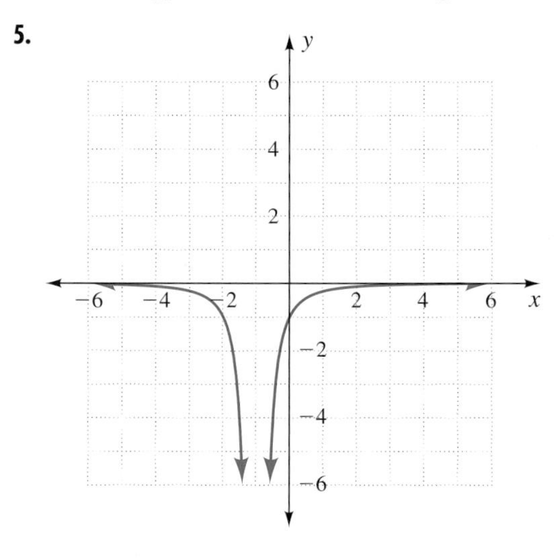

a. As x approaches -1 from the right

b. As x approaches -1 from the left

6. Simplify each expression. Which simplify to fractions that are equivalent to $\frac{6}{8}$? Why?

a. $\frac{6 \cdot 2}{8 \cdot 2}$ **b.** $\frac{6 \div 2}{8 \div 2}$ **c.** $\frac{6 - 2}{8 - 2}$ **d.** $\frac{6 + 2}{8 + 2}$

Simplify the rational expressions in Exercises 7 to 12, if possible. If not, explain why. Assume there are no zero denominators.

7. $\frac{2xy}{x^2}$ **8.** $\frac{x + y}{xy}$ **9.** $\frac{a^2 - b^2}{a + b}$

10. $\frac{ab}{a^2 + b}$ **11.** $\frac{ab}{a^2 + b^2}$ **12.** $\frac{ab}{a^2 + ab}$

Simplify the expressions in Exercises 13 and 14. Indicate any inputs that must be excluded.

13. a. $\frac{1 - a}{a - 1}$ **b.** $\frac{6x^2 - 7x - 10}{6x^2 - 2x - 20}$

14. a. $\frac{x^2 + 4x + 4}{x^2 + 3x + 2}$ **b.** $\frac{15x^2 - 34x + 15}{15x^2 - 24x + 9}$

In Exercises 15 to 20, what numerator or denominator is needed to make a true statement? Assume there are no zero denominators.

15. $\frac{x - 3}{} = -1$ **16.** $\frac{}{2 - x} = -1$

17. $\frac{}{a - b} = 1$ **18.** $\frac{}{a + b} = 1$

19. $\frac{}{4 - x} = -1$ **20.** $\frac{b - 5}{} = -1$

21. Is the pair of fractions $\frac{16}{9}$ and $\frac{\sqrt{16}}{\sqrt{9}}$ equal or not equal? Explain why.

22. Explain why $\frac{x - 6}{x + 6}$ will not simplify.

23. a. Calculate the fractional expressions $300 \div \frac{3}{1}$ and $300 \cdot \frac{1}{3}$.

b. What may be observed about the pair of problems?

c. What do they tell us about multiplication and division?

Identify the word clues that indicate the necessary operations (addition, subtraction, multiplication, or division), and then answer the questions in Exercises 24 to 29.

24. John ate $\frac{1}{3}$ of the pie; Sue ate $\frac{1}{4}$ of the pie. Altogether they ate what portion of the pie?

25. How many servings are there in 4 large pizzas if each person eats one-third of a pizza?

26. Sred stitches $\frac{2}{3}$ of the hem. Evelyn rips out $\frac{1}{2}$ of Sred's work. What fraction remains to be finished?

27. Sally ran $\frac{3}{4}$ mile. Jim ran half as far. How far did Jim run?

28. A box of corn flakes contains 18 ounces. A serving is $1\frac{1}{10}$ ounces. How many servings are in the box?

29. Half an animal shelter's funding comes from cat and dog licensing fees. A third comes from property taxes. What fraction remains to be raised from private donations?

In Exercises 30 to 36, multiply or divide, as indicated. Factor and simplify. Assume there are no zero denominators.

30. $\dfrac{x}{x-1} \cdot \dfrac{x^2-1}{x^2}$

31. $\dfrac{1-x}{x+1} \div \dfrac{x^2-1}{x^2}$

32. $\dfrac{x^2-9}{3-x} \div \dfrac{x^2}{(x+3)}$

33. $\dfrac{n-2}{n(n-1)} \cdot \dfrac{(n+1)n(n-1)}{n-2}$

34. $\dfrac{2x+6}{x-2} \cdot \dfrac{x^2-4}{x^2+3x}$

35. $\dfrac{9-3x}{x+3} \cdot \dfrac{x}{x^2-6x+9}$

36. $\dfrac{6x^2+17x-28}{6x^2+28x-10} \div \dfrac{x^2+11x+28}{5-14x-3x^2}$

Add or subtract the expressions in Exercises 37 to 41, as indicated. Assume there are no zero denominators.

37. $\dfrac{x}{x+2} - \dfrac{2}{x^2-4}$

38. $\dfrac{x}{x+1} - \dfrac{2}{(x+1)^2}$

39. $\dfrac{a}{a+b} - \dfrac{b}{a-b}$

40. $\dfrac{4}{5} - \dfrac{3}{x} + \dfrac{2}{x^2}$

41. $1 - \dfrac{x^2}{2} + \dfrac{x^4}{24} - \dfrac{x^6}{720}$

42. Explain the role of factoring in the addition or subtraction of rational expressions.

Use the order of operations to simplify the expressions in Exercises 43 and 44. Assume $a \neq 0$ and $c \neq 0$.

43. $\dfrac{1}{a} + \dfrac{2a}{3} \cdot \dfrac{6}{a} \div \dfrac{1}{3} - \dfrac{a}{3}$

44. $\dfrac{3}{c} - \dfrac{4}{3c} \div \dfrac{2}{3} + \dfrac{1}{4} \cdot \dfrac{8}{c}$

In Exercises 45 and 46, simplify the expression or equation containing units of measurement.

45. $\dfrac{\text{4 buttons per card}}{\text{12 buttons per shirt}}$

46. $d = -\dfrac{1}{2}\left(\dfrac{9.81\text{ m}}{\text{sec}^2}\right)(5\text{ sec})^2 + \left(\dfrac{8\text{ m}}{\text{sec}}\right)(5\text{ sec}) + 50\text{ m}$

In Exercises 47 to 49, simplify the expressions from dosage computation. The abbreviation *gr* is for grain and precedes the number.

47. $\text{gr}\,\dfrac{1}{2} \cdot \dfrac{1\text{ tab}}{\text{gr}\,\frac{1}{6}}$

48. $\dfrac{\dfrac{1\text{ mL}}{25\text{ mL}} \cdot 400\text{ mL}}{\dfrac{1\text{ mL}}{4\text{ mL}}}$

49. $\text{gr}\,\dfrac{1}{150} \cdot \dfrac{1\text{ mL}}{\text{gr}\,\frac{1}{750}}$

50. Find the slope of the line connecting the points $\left(\dfrac{a}{2}, b\right)$ and $\left(\dfrac{b}{2}, \dfrac{a}{3}\right)$.

51. Simplify the expression on the right side of the equation so that it contains no fractions in the numerator or denominator:

$$t^2 = \frac{d}{\frac{1}{2}g}, \quad g \neq 0$$

Multiply the expressions in Exercises 52 and 53, and simplify the result. Assume there are no zero denominators.

52. $x^2\left(\dfrac{1}{x} + \dfrac{2}{x^2}\right)$

53. $(x+2)(x-2)\left(\dfrac{2}{x-2} + \dfrac{1}{x+2}\right)$

Solve the equations in Exercises 54 to 58. What inputs must be excluded?

54. $\dfrac{1}{5} = \dfrac{1}{9} + \dfrac{1}{x}$

55. $\dfrac{1}{5} + \dfrac{1}{x} = \dfrac{8}{5x}$

56. $\dfrac{3}{x} + \dfrac{1}{x-1} = \dfrac{17}{4x}$

57. $\dfrac{1}{2x} = \dfrac{2}{x} - \dfrac{x}{24}$

58. $\dfrac{3x}{x-6} + x = \dfrac{18}{x-6}$

Add the expressions on the right side of each equation in Exercises 59 and 60.

59. Traffic flow through parallel doors: $\dfrac{1}{m} = \dfrac{1}{m_1} + \dfrac{1}{m_2}$

60. Focal distance for a lens, in optics: $\dfrac{1}{F} = \dfrac{1}{f_1} + \dfrac{1}{f_2}$

61. The hot water line fills the clothes washer in 3 minutes. The cold water line fills the washer in 2 minutes. How long will it take to fill the washer on the "warm" setting, using both water lines?

62. One clerk is able to issue all advance mail-order tickets for an event in 30 days. An additional clerk is hired. The two clerks together get the job finished in 20 days. In how many days could the second clerk have done the job alone?

63. A can of paint covers 300 square feet. Mark four points on the graph in the figure that show possible lengths and widths of a rectangular floor to be painted. Connect your points. Where on the graph will points be located that represent floor sizes requiring less than a full can of paint?

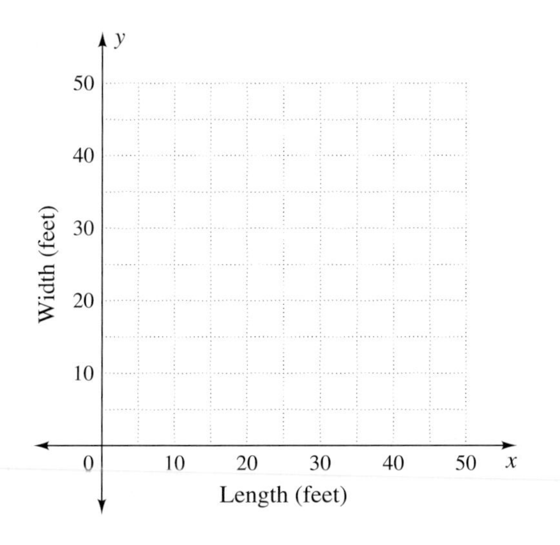

▶ Chapter Projects

64. Fraction Pattern, I

a. Simplify:

$$\frac{1+2+3}{4+5+6} \quad \frac{7+8+9}{10+11+12} \quad \frac{13+14+15}{16+17+18}$$

b. Predict the values of these fractions:

$$\frac{50+51+52}{53+54+55} \quad \frac{100+101+102}{103+104+105}$$

c. Show that your pattern always works by building a rational expression (a fraction) with x as the first number, $x + 1$ as the second number, and so on. Simplify the expression.

65. Fraction Pattern, II

a. Add these two sets of fractions.

$$\tfrac{1}{6} + \tfrac{1}{9} + \tfrac{1}{18} \qquad \tfrac{1}{10} + \tfrac{1}{15} + \tfrac{1}{30}$$

b. Look for a shortcut for finding the sum, and use it to add the next two sets of fractions.

$$\tfrac{1}{22} + \tfrac{1}{33} + \tfrac{1}{66} \qquad \tfrac{1}{18} + \tfrac{1}{27} + \tfrac{1}{54}$$

c. Write a formula for the pattern, and show why the pattern always works.

▶ 9 Chapter Test

1. For what input is the expression $\frac{4-x}{x-4}$ undefined?

2. Explain why the expression $\frac{4-x}{x-4}$ may or may not be simplified?

3. Evaluate $y = \frac{4}{x-2}$ for x as integers between -5 and 5. Sketch a graph.

4. Simplify each expression. Which simplify to fractions equivalent to $\frac{6}{9}$? Why?

a. $\frac{6 \div 3}{9 \div 3}$ **b.** $\frac{6-3}{9-3}$

c. $\frac{6+3}{9+3}$ **d.** $\frac{6 \cdot 3}{9 \cdot 3}$

Simplify the expressions in Exercises 5 to 8, if possible. If not, explain why. Assume there are no zero denominators.

5. $\frac{b^2}{3ab}$ **6.** $\frac{a-b}{ab}$

7. $\frac{a^2-b^2}{a-b}$ **8.** $\frac{xy}{xy+y}$

Multiply or divide the expressions in Exercises 9 to 12, as indicated. Factor and simplify. Assume there are no zero denominators.

9. $\frac{1-x}{x+1} \cdot \frac{x^2-1}{x^2}$

10. $\frac{x-1}{1-x^2} \div \frac{x^2-2x+1}{1+x}$

11. $\frac{n+1}{n} \div \frac{n(n-1)}{n^2}$

12. $\frac{x^2-2x}{x^2-4} \cdot \frac{x-2}{x}$

13. Multiply this expression, and simplify:

$$(x + 1)(x - 1)\left(\frac{1}{x + 1} + \frac{2}{x - 1}\right)$$

Simplify each expression or equation in Exercises 14 to 16.

14. 4 yd per shirt · \$8.98 per yd

15. $\dfrac{\dfrac{\$2.50}{1 \text{ gal}}}{\dfrac{25 \text{ mi}}{1 \text{ gal}}}$

16. $3 \text{ mg} \cdot \dfrac{\text{gr } 1}{60 \text{ mg}} \cdot \dfrac{1 \text{ tab}}{\text{gr } \frac{1}{120}}$

17. One door can empty a meeting room in 16 minutes. A different width door is to be added. Both doors working together must be able to empty the room in 5 minutes. How fast must the new door empty the room?

18. Write the expression on the left side of this optics equation as a single fraction:

$$\frac{1}{p} + \frac{1}{q} = \frac{1}{f}$$

19. Simplify the right side:

$$h = \frac{A}{\frac{1}{2}b}$$

Add or subtract the expressions in Exercises 20 to 23. Assume there are no zero denominators.

20. $\dfrac{2}{9} + \dfrac{7}{15}$

21. $\dfrac{2}{ab^2} - \dfrac{a}{2b}$

22. $\dfrac{x}{x + 2} + \dfrac{2}{x^2 - 4}$

23. $\dfrac{2}{3} + \dfrac{3}{x} - \dfrac{4}{x^2}$

24. Simplify $\dfrac{a}{2} + \dfrac{2}{3} \cdot \dfrac{a}{2} - \dfrac{3a}{2} \div \dfrac{9}{2} + \dfrac{1}{4}$.

Solve the equations in Exercises 25 to 29. State any inputs that must be excluded.

25. $\dfrac{1}{3} + \dfrac{1}{6} = \dfrac{1}{x}$

26. $\dfrac{1}{x} = \dfrac{1}{2x} + \dfrac{1}{2}$

27. $\dfrac{3}{x} = \dfrac{2}{x - 3}$

28. $\dfrac{x - 2}{4} + \dfrac{1}{x} = \dfrac{19}{4x}$

29. $x + \dfrac{10}{x - 2} = \dfrac{5x}{x - 2}$

30. Explain the role of factoring in multiplying and dividing rational expressions.

Final Exam Review

1. Simplify these expressions without a calculator.

a. $-7 + (-5)$ **b.** $-3 - (-8)$ **c.** $-6 + 11$

d. $(-5)(12)$ **e.** $(-4)(-16)$ **f.** $|4 + (-8)|$

g. $\left|\frac{-45}{9}\right|$ **h.** $\frac{3}{2} \cdot \frac{4}{15}$ **i.** $\frac{-2}{3} \div \frac{5}{6}$

2. Evaluate these expressions if $n = -3$.

a. $-4n$ **b.** $-n$ **c.** $-n^2$

d. $(2n)^2$ **e.** $(-n)^2$ **f.** $|n - 4|$

g. $2n - 3$ **h.** $3 - 2n$ **i.** $\frac{1}{n}$

3. Simplify these expressions. Leave answers without negative or zero exponents.

a. $\frac{bcd}{bdf}$ **b.** $\frac{-3rs}{12sx}$ **c.** $\frac{a^3}{a^2}$

d. $\frac{x - 2}{2 - x}$ **e.** $\frac{(4m^2n)^3}{6n^2}$ **f.** $\frac{ab}{c} \div \frac{ac}{b}$

g. n^3n^4 **h.** $(mn^2)^3$ **i.** $m^{-1}m^{-2}$

j. $(3x^3)^0$ **k.** $\frac{6x^{-3}}{2x^2y^{-2}}$ **l.** $(3x^{-3}y^2)^3$

4. Simplify these expressions.

a. $3 - 4(5 - 6) + 7(8)$

b. $|3 - 4| + |\sqrt{4} - 3|$

c. $\sqrt{9} + \sqrt{16} - \sqrt{49}$

d. $3x + 4y - 5x + 2y$

e. $3x^2 + 2x - 1 - (x^2 - 3x + 2)$

f. $3(x - 1) + 4(x + 2) - 5(x - 3)$

5. Solve for the indicated variable.

a. $2x + 3y = 12$ for y

b. $2x - 3y = 7$ for x

c. $ax + by = c$ for y

d. $\frac{x}{x + 1} = \frac{3}{8}$ for x

e. $\frac{P_1V_1}{T_1} = \frac{P_2V_2}{T_2}$ for P_2

f. $3(x + 2) = 7(x - 10)$ for x

g. $4^2 + x^2 = 8^2$ for x

h. $d^2 = [4 - (-3)]^2 + (5 - 2)^2$ for d

i. $6 = \sqrt{3x - 3}$ for x

6. Multiply these polynomials.

a. $3(x^2 - 6x + 4)$ **b.** $-4x(x^2 - 2x - 3)$

c. $(x - 4)(x + 4)$ **d.** $(x - 4)^2$

e. $(x - 4)(x^2 - 8x + 16)$ **f.** $(2x - 3)(3x - 2)$

7. Factor these expressions, using the table method as needed.

a. $12xy + 3xy^2 + 6x^2y^2$ **b.** $x^2 + x + 1$

c. $x^2 + x - 20$ **d.** $4x^2 - 9$

e. $6x^2 + x - 2$ **f.** $6x^2 - 11x - 2$

g. $x^2 - 10x + 25$ **h.** $6x^2 + 3x - 9$

8. Multiply out each of these binomial squares.

a. $(n - 3)^2$

b. $(n - 1)^2$

c. $(n - 4)^2$

d. Simplify $(n - 1)^2 - (n - 4)^2$.

e. Show that $n^2 - (n - 3)^2 - [(n - 1)^2 - (n - 4)^2] = 6$.

9. Graph $x + y = 0$ and $x - y = 2$. Solve the system of equations from the graph.

Solve the systems in Exercises 10 to 15.

10. $x + y = 6$
$y = -\frac{2}{3}x$

11. $3x + 4y = 10$
$x = 6 - 2y$

12. $x + y = 8$
$x - 3y = 4$

13. $3x + 4y = -20$
$2x - 3y = -2$

14. $x + y = 8$
$x + y = 10$

15. $x + y = 8$
$x = 8 - y$

In Exercises 16 to 19, find the slope of the line through the points and the distance between the points.

16. $(2, -3)$ and $(3, 4)$

17. $(2, -3)$ and $(2, 5)$

18. $(2, -3)$ and $(-5, -3)$

19. $(2, -3)$ and $(-5, -2)$

20. Use the slopes found in Exercises 16 to 19 to identify pairs of perpendicular lines.

Solve the equations in Exercises 21 to 32. Indicate any inputs that must be excluded.

21. $\frac{4}{x} = \frac{6}{x}$ **22.** $\frac{x}{4} = \frac{x}{6}$

23. $2x + \frac{3x}{x - 4} = \frac{12}{x - 4}$

24. $\frac{x - 1}{5} = \frac{3x + 3}{18}$

25. $\frac{x}{1} = \frac{14}{x + 5}$

26. $\frac{2x}{5} - \frac{x}{10} = -4.5$

27. $\frac{2x}{5} - \frac{x}{10} = \frac{3x}{10}$

28. $\frac{3}{x + 4} + \frac{6}{x} = 1$

29. $\sqrt{x + 13} = 5$

30. $\sqrt{x + 13} = x - 7$

31. $\sqrt{x + 13} = 0$

32. $\sqrt{x + 13} = -7$

33. The graph of $y = \sqrt{4 - x}$ is shown. Use it to answer the questions.

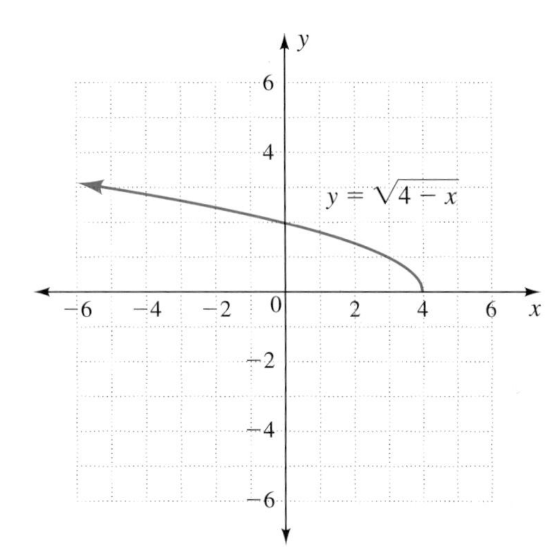

a. What is the x-intercept of the graph?

b. What is the y-intercept of the graph?

c. If $\sqrt{4 - x} = 1$, then $x =$ ____.

d. If $x = 13$, then $\sqrt{4 - x} =$ ________.

e. Why does the graph show no negative values for y?

34. Graph $y = \frac{-24}{x}$.

a. What is the y-intercept?

b. What is the x-intercept?

c. Does the graph of $y = x + 4$ intersect your graph?

d. How does solving $x + 4 = \frac{-24}{x}$, $x \neq 0$, with algebraic notation show whether or not the graphs in part c intersect?

35. The following graph shows a dieting experience.

a. Describe what happened.

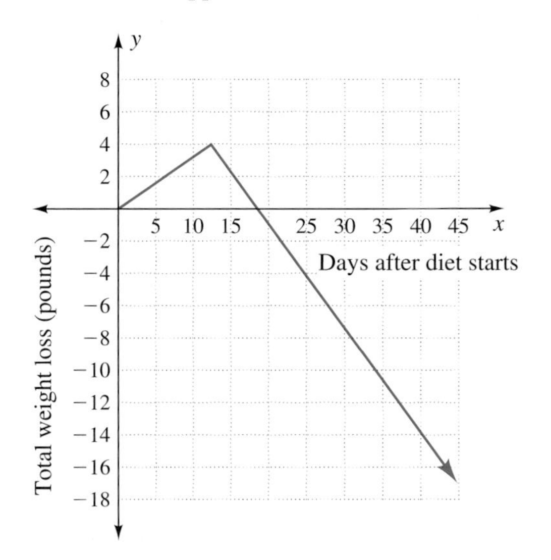

b. In how many days will the dieter have lost 4 pounds?

c. In how many days will the dieter gain 4 pounds over the starting weight?

d. Describe the situation after 40 days.

36. Solve these equations.

a. $x^2 + 3x - 4 = 0$

b. $x^2 - 4x + 3 = 0$

c. $2x^2 + 8x + 8 = 0$

d. $x(2x - 3) = 0$

37. Simplify these expressions using properties of exponents and radicals.

a. $3^0 + 4^2$

b. $5^2 + 5^{-1}$

c. $\sqrt{25} + 16^{1/2}$

d. $\sqrt{225} + 25^{1/2}$

e. $\sqrt{64x^2}$, assuming x is either positive or negative.

f. $\sqrt{64x^2}$, assuming x is only positive.

g. 0^0

h. $\sqrt{x}$, where x is a negative number

38. Draw an example that shows why calculating the slope of a vertical line results in division by zero.

39. Factor and simplify $\frac{4x^2 - 1}{2x^2 + 5x + 2}$.

40. For what inputs, x, will $\frac{-1}{(x - 3)(x + 1)}$ be undefined?

41. Simplify these expressions containing units of measure.

a. $\frac{\$6.00 \text{ per hour}}{3 \text{ rooms cleaned per hour}}$

b. $d = -\frac{1}{2}\left(\frac{32.2 \text{ ft}}{\text{sec}^2}\right)(3 \text{ sec})^2 + \left(\frac{22 \text{ ft}}{\text{sec}}\right)(3 \text{ sec}) + 100 \text{ ft}$

42. Simplify these expressions. Assume there are no zero denominators.

a. $\frac{2a + 2b}{a^2 - b^2}$ **b.** $\frac{4 - x}{6x - 24}$

43. Perform the indicated operations. Leave the answers in reduced form. Assume there are no zero denominators.

a. $\frac{2 - x}{xy^2} \cdot \frac{x^2}{x - 2}$

b. $\frac{x^2 - 3x - 4}{x^2 - 2x} \cdot \frac{x^2 - 4}{4 - x}$

c. $\frac{x - 4}{x + 2} + \frac{x^2 - 3x - 4}{x^2 - 4}$

d. $\frac{x^2 + 3x + 2}{x^2 + 6x + 9} \div \frac{x + 2}{x + 3}$

e. $\frac{1}{3x} - \frac{2}{5x}$ **f.** $8x\left(\frac{1}{2x} - \frac{3}{x}\right)$

g. $3(x - 2)\left(\frac{4}{x - 2} + \frac{x}{3}\right)$

44. a. Make an input-output table for $y = 3x - 4$ for inputs in the interval $[-1, 3]$.

b. Graph the equation.

c. Where do we find the slope in a linear equation? Explain how to find the slope from a table. Show the slope on the graph, with rise and run.

d. Where is the y-intercept in the equation? How do we find the y-intercept from the table? Indicate the y-intercept on the graph.

e. If we set $y = 0$ and solve $0 = 3x - 4$ for x, what point on the graph have we found?

f. Solve the equation $3x - 4 = -6.4$. Show the steps.

g. Describe how we might estimate the solution to $3x - 4 = -3$ from the graph.

45. A credit card payment schedule is shown in the table.

Charge Balance, x	Payment Due, y
(\$0, \$30]	Full amount
(\$30, \$100]	\$30 + 20% of amount in excess of \$30
(\$100, $+\infty$)	\$50 + 50% of amount in excess of \$100

a. How much is paid for charge balances of \$25, \$35, \$95, \$100, and \$105?

b. Write an equation that describes the payment due in terms of the charge balance for each of the three categories. The equations should use y in terms of x.

c. Explain the difference between the brackets, [ ], and the parentheses, ( ), in the charge balance column.

d. Does it matter which interval notation—brackets or parentheses—is used with the infinity symbol, ∞?

e. Write each charge balance interval using an inequality expression.

46. Describe each set of numbers. Choose from the following:

real numbers, rational numbers, irrational numbers, integers, whole numbers, natural numbers

a. $\{1, 2, 3, 4, \ldots\}$ **b.** $\{0, 1, 2, 3, 4, \ldots\}$

c. $\{-3, -2, -1, 0, 1, 2, 3, \ldots\}$

d. $\{\text{pi}, \sqrt{2}, \sqrt{3}\}$

47. Use the given facts and unit analysis, as needed, to answer the following questions.

The speed of light is 186,000 miles per second.
The minimum distance from Pluto to the sun is 2756.4×10^6 miles.
The maximum distance from Pluto to the sun is 4551.4×10^6 miles.
There are 60 seconds in a minute.
There are 60 minutes in an hour.
The speed limit on a freeway is 65 miles per hour.
There are 5280 feet per mile.
There are 1609 meters in a mile.
There are 1000 meters in a kilometer.

a. What is the minimum distance from Pluto to the sun, written in the standard form of scientific notation?

b. How long does it take light to travel the maximum distance from the sun to Pluto?

c. Mary Meagher set a world record in the butterfly in 1981. She swam 200 meters in 125.96 seconds. What was her average speed in miles per hour?

d. The orbit velocity of the space shuttle is 28,300 km/hr. To the nearest hundred, how many miles per hour is this?

48. Set up equations for each problem. If the problem describes a system of equations, solve using substitution or elimination or a graphing calculator.

a. Three strings of holiday lights and two packages of giftwrap cost \$18.90. Two strings of holiday lights and five packages of giftwrap cost \$19.86. What is the cost of each item?

b. Four cups of milk and a cup of cottage cheese contain 685 calories. Three cups of milk and two cups of cottage cheese contain 770 calories. How many calories are in each?

c. The three angles of a triangle have measures that add to 180°. The measure of the largest angle is four times that of the smallest. The measure of the middle angle is 9° less than that of the largest. Find the measure of each angle.

d. The equations $y = \frac{3}{5}x$ and $y = -\frac{3}{5}x + 6$ form the diagonals of a rectangle. What is the point of intersection of the two lines? If two sides of the rectangle lie on the coordinate axes, what is the area of the rectangle? What is the length of the diagonal of the rectangle?

e. A hamburger birthday party at McDonald's costs $75 for a party of 10 and $7.50 for each additional child. What equation describes the cost for x children?

f. An ice cream birthday party at McDonald's costs $55 for a party of 10 and $5.50 for each additional child. What equation describes the cost for x children?

g. Will a graph of the equations for parts e and f be parallel for $0 < x \leq 10$? For $x > 10$? Explain.

h. A jumbo package contains 20 ounces of potato chips. If the input is the number of ounces eaten by each person, what equation describes the number of people served by the package?

49. The volume of a cube, or box, with all edges equal to x is found with the formula $V = x^3$. The surface area of the box is the amount of area covering all six faces. The surface area of a cube is $6x^2$.

a. Fill in this table.

Length of Edge of Box, x	Volume of Cube, x^3	Surface Area, $6x^2$
10		
20		
40		
n		
$2n$		

b. If we double the length of an edge of the box, what happens to the volume?

c. If we double the length of an edge of the box, what happens to the surface area?

d. Use the Pythagorean theorem to determine the length of the diagonal on the face of a cubical box if the length of the edge is 20.

50. Call Trace vs Caller ID On phone lines without caller ID, a *69 last-call trace costs $0.75 for each of the first 8 traces each month and is free after that. Graph the total cost of each trace for 0 to 10 call traces. Caller ID costs $6.50 per month. Discuss the advantages and disadvantages of each option.

Appendix Graphing Calculator Basics for the TI-83/84

Introduction

Your graphing calculator may have ten rows of keys in five columns, giving up to 50 keys (written here as boxes, such as ON and, with the two shift keys, 2ND and ALPHA, up to 100 options (printed on the calculator and written here in brackets, such as [OFF]). Before your eyes glaze over, prepare your own calculator guide. Preparing the guide will take a few minutes and focus your attention on important sets of keys. The guide will permit you to select what you need and to ignore other keys and options until later.

Student Note:
For key locations, include phrases such as left column (or first column); right column (fifth column); bottom row, top row; first row, tenth row. The suggested numbering corresponds to the first quadrant in graphing with the ON key at or near the origin in the lower-left corner.

A Dozen Sets of Keys for Introductory Algebra

On lined paper, write the heading and, below it, the name of each key or option listed within the numbered set. Write one key or option per line. Find the key or option, and describe its location in your own words. Answer the questions as you locate the keys.

Student Note:
Calculator numbers are in a different order than on the cell phone.

1. **Find keys to turn the calculator on and off:** ON, 2ND, [OFF].
2. **Find numbers**: zero 0; the numbers 1, 2, 3, 4, 5, 6, 7, 8, and 9 as a block; the number pi, [π], (as in the formula for the area of a circle). Are the calculator numbers 1 to 9 in the same order as on a cell phone?
3. **Find basic operations:** +, −, ×, ÷; ENTER.
4. **Find cursor control and edit keys:** up [△], right [▷], down [▽], left [◁], delete character DEL, insert character [INS].
5. **Find helpful shortcuts:** Return prior [ENTRY]; return prior answer [ANS]; CLEAR line or screen; [QUIT] and return to the computation or home screen.
6. **Find keys to access options printed on the calculator:** 2ND and ALPHA What color is the 2ND key? What color is the ALPHA key? Name the two options printed above the ENTER key. Guess which option is obtained with 2ND? Why? Which option is obtained with ALPHA?
7. **Find more operations with numbers:** Parentheses (and), opposite or negative (−), reciprocal X^{-1}, exponent ^, square X^2, and square root [$\sqrt{\ }$]
8. **Find keys leading to lists of options or menus:** [CATALOG] for "absolute value" and MATH for "change to fraction notation."
9. **Find keys for making a table with a rule:** The variable X, T, θ, n key is needed to enter an equation or expression under Y=. We then set up the table with [TBLSET] and show it with [TABLE]. Our work requires only x. The other variables—t, θ, and n—appear when the calculator is in special "modes."
10. **Find keys for graphing with a rule:** The variable X, T, θ, n key is needed to enter an equation or expression under Y=. We then set axes in the WINDOW, draw the GRAPH, TRACE the graph, and adjust the window with the ZOOM menu. Where else did we use the variable and Y=?
11. **Find keys for making lists and calculating with sets of numbers:** Entering lists STAT **EDIT**; choosing lists [L_1], [L_2], ..., [L_6]; and comma ,.
12. **Find keys for making graphs with sets of numbers:** Set dimensions WINDOW, and choose type of graph, [STAT PLOT]. Where else did we use WINDOW?

Student Note:
2ND is blue and ALPHA is green. Option [ENTRY] is blue; use 2ND. Option [SOLVE] is green; use ALPHA. 2ND is yellow on the TI-83, as is option [ENTRY].

Student Note:
Variable and Y= let us enter the rule in making tables and in graphing.

Student Note:
A WINDOW is needed for graphing with lines and with numbers.

Now, the following instructions return to the keys you have found and give you practice using the keys and options.

1. **Turn the calculator on and off.** Press ON. If your calculator is new, you see the blinking cursor in the upper-left corner of the home screen. If your calculator is used, press 2ND [QUIT], then CLEAR twice to return to and clear the home screen. Press 2ND [OFF] to shut off the calculator.
2. **Numbers:** From here on, boxes will not be used around numbers.
3. **Basic Operations:** Boxes also will be omitted around $+$, $-$, $\times$, and $\div$. Turn the calculator on, and then press 3×4 ENTER. Describe the multiplication sign as it appears on the screen. Add to your notes. Press $15 \div 5$ ENTER. Describe the division sign as it appears on the screen. Turn the calculator off.

Student Note:
An asterisk for multiplication; a slash for division.

4. **Control cursor and edit entries:** Turn calculator on. Type in $31 - 2 \times 4$. Edit your entry. Move left to the 1 with the cursor. Press DEL to delete the 1. Write over the $-$ with $+$. Move right with the cursor to the 4. Press 2ND [INS] and insert a 1 before the 4 to get $3 + 2 \times 14$. Press ENTER to obtain 31. Repeat until this feels natural.
5. **Helpful shortcuts:**
   a. **Obtain the prior** [ENTRY] **with** 2ND [ENTRY]**:** Type in $31 - 2 \times 4$ ENTER to obtain 23. Press 2ND [ENTRY]. Edit as in part 4 and press ENTER.
   b. **Obtain the prior answer with** 2nd [ANS]**:** Type in 4×4 ENTER 2ND [ANS]. How does the prior entry compare with the prior answer?
   c. **Clear a line with** CLEAR before pressing ENTER; **Clear the screen with** CLEAR after pressing ENTER.

Student Note:
2ND [ENTRY] gives the expression, and 2ND [ANS] gives only the answer. Just for fun: 2 ENTER $\times$ 2 ENTER ENTER ENTER. This is repeated multiplication by 2.

6. **HELP! If you get "lost," go to the home screen with** 2nd [QUIT]**.**
7. **More operations with numbers:** These keys are frequently written without the box.
   a. **Parentheses** (**and**)**.** Type in $3(4 + 5)$ ENTER and $(3 \times 4) + (3 \times 5)$ ENTER.
   b. **Opposite or negative** (–)**. Press $-(-3)$** ENTER **to get 3.** Type in 9 (–) 6 ENTER to get an error message. Press 2 to go to the error and replace the opposite sign with a subtraction sign: $9 - 6$. Press ENTER to do the subtraction. Describe the difference in screen appearance between the opposite, (–), sign and the subtraction sign in your notes.
   c. **Reciprocal** x^{-1}**:** Press 4 x^{-1} ENTER to get .25. The calculator omits the 0 in 0.25. You should always write the zero.
   d. **Exponent** ^**:** Press 4 ^ 2 ENTER for 16.
   e. **Square** x^2**:** Press 4 x^2, which appears as 4^2, then ENTER for 16
   f. **Square root** [√]**:** You must enter the sign first; 2nd [√] 16) ENTER to get 4. When the calculator writes an opening parenthesis, you should always write a closing parenthesis. For more practice, copy the first two parts of Figure A.1.

Student Note:
The opposite sign is 3 pixels in length, and the subtraction is 5 pixels in length and slightly lower.

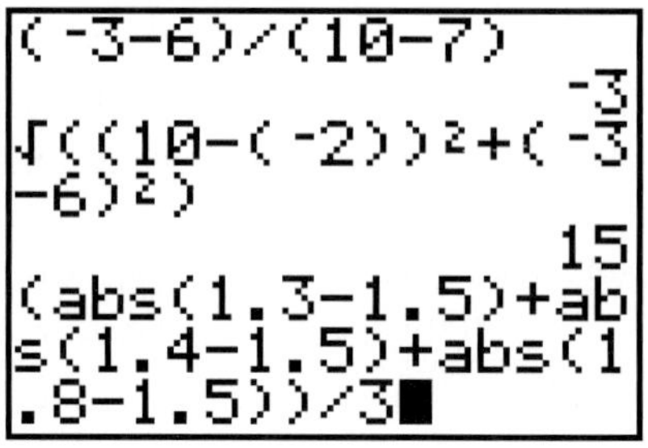

```
(-3-6)/(10-7)
                -3
√((10-(-2))²+(-3
-6)²)
                15
(abs(1.3-1.5)+ab
s(1.4-1.5)+abs(1
.8-1.5))/3█
```

Use 2ND [√] for square root.
Use 2ND [CATALOG] ENTER for absolute value.

FIGURE A.1

8. Keys leading to menus or lists of options:

a. [CATALOG] for "absolute value": [CATALOG] is above the number zero. Press (2ND [CATALOG] ENTER. Because abs (for absolute value) is the first entry in the list, abs will appear on your screen after the parenthesis. Continue by copying the third expression in Figure A.1. When complete, press ENTER and compare your answer with 0.2, the correct answer. (As mentioned, although the calculator drops the zero before the decimal point, you should always write the zero.) Use 2ND [ENTRY] with insert or delete as needed to edit until you get .2 on the screen.

b. MATH for "change to fraction notation": Enter the first line of Figure A.2 and press ENTER for the answer. Change the answer to fraction notation in two steps: **Obtain the prior answer** with 2ND [ANS], and then press MATH 1 to choose "into fraction notation" from the MATH menu. The fraction $\frac{3}{5}$ appears in Figure A.2.

Try the expressions in Figure A.3 to practice reciprocal, X^{-1}, and "change to fraction notation." The answer to the last product is 1.

Obtain prior answer with 2ND [ANS]. Obtain fraction notation with MATH **1:Frac** ENTER.

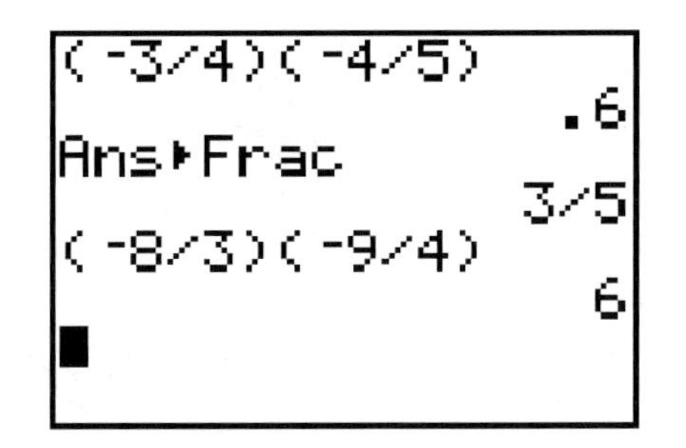

FIGURE A.2

8⁻¹▸Frac
1/8
Ans*8
1
(-6)⁻¹▸Frac
-1/6
Ans*-6

FIGURE A.3

9. Making a table with an equation or expression: Use the variable key, X, T, θ, n, for X or *x*.

Step 1: Press Y= to obtain Figure A.4. Let $Y_1 = 3X - 2$.

Step 2: Press 2ND [TBLSET]. Let **TblStart** = 1 and △ **Tbl** = 1 ÷ 3; see Figure A.5.

Step 3: Press 2ND [TABLE] to see the table in Figure A.6.

Plot1 Plot2 Plot3
\Y1=
\Y2=
\Y3=
\Y4=
\Y5=
\Y6=
\Y7=

FIGURE A.4

FIGURE A.5

X	Y1
1	1
1.3333	2
1.6667	3
2	4
2.3333	5
2.6667	6
3	7

X=1.3333333333

FIGURE A.6

10. Graphing with an equation or expression: Use the variable key, X, T, θ, n, for X or *x*.

Step 1: Press Y=. To enter the equations graphed in Figure A.7, let $Y_1 = 3X - 2$ ENTER and $Y_2 = 9$.

Step 2: Press WINDOW to set axes: **Xmin** = −10 ENTER **Xmax** = 10 ENTER **Xscl** = 1 ENTER **Ymin** = −10 ENTER **Ymax** = 10 ENTER **Yscl** = 1 ENTER. Let **Xres** = 1 for all graphs.

Step 3: Press [GRAPH] to do graphs. Press [TRACE] to view Figure A.7, the graphs with cursor on $Y_1 = 3X - 2$ at the point $(0, -2)$. Press up or down, to place the cursor on $Y_2 = 9$.

Step 4: If there is a point you wish to examine more closely, such as where $Y_1 = 3X - 2$ intersects $Y_2 = 9$, move the cursor to the right along Y_1 until Y is approximately 9. Press [ZOOM] to see the menu in Figure A.8. Press 2 to select **Zoom In**, then [ENTER] [TRACE] to see that $x \approx 3.67$ when $y \approx 9$. Press [WINDOW] to see how the screen has moved away from the intervals $[-10, 10]$ and $[-10, 10]$ for x and y. See instruction 13 below for other hints on [ZOOM].

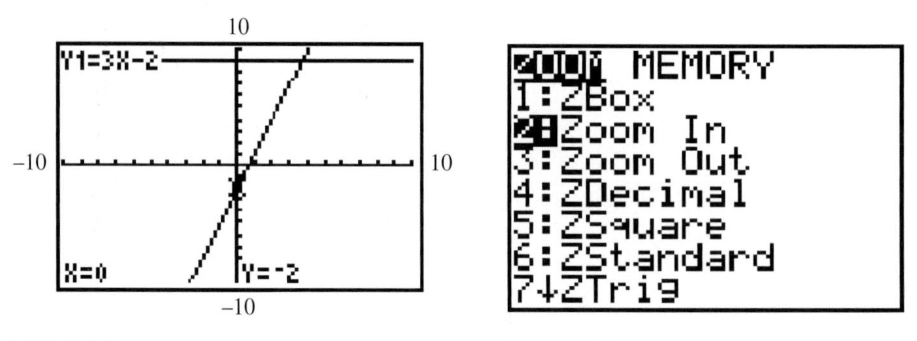

FIGURE A.7

FIGURE A.8

11. Making lists and calculating with sets of numbers.

Step 1: Clear lists as needed: [STAT] **4: ClrList** [2ND] [L1], [2ND] [L2] [ENTER].

Step 2: Press [STAT] **1: Edit** to obtain the list screen, Figure A.9. Fill in L1: -4 [ENTER] 3 [ENTER]. Use right cursor to move to L2. Fill in L2: 2 [ENTER] -1 [ENTER].

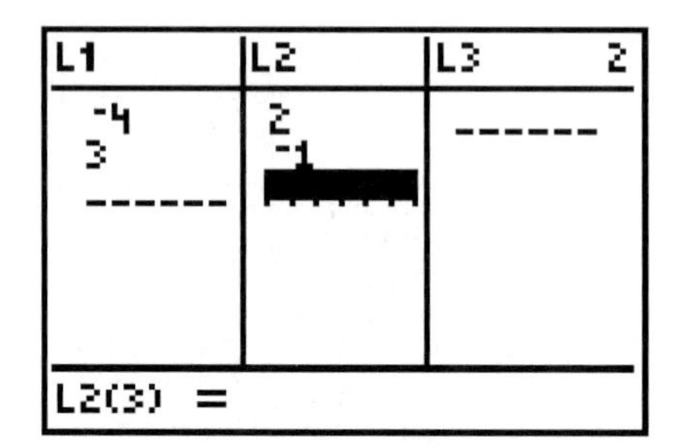

FIGURE A.9

Step 3: Enter other numbers or up to six lists, as needed. Press [2ND] [QUIT] and then choose calculation options:

a. If the lists are ordered pairs, $(-4, 2)$ and $(3, -1)$, the data can be fit with an appropriate equation (say, linear regression, page 231, Section 4.5).

b. If the lists are grouped data in tables, we can find means or medians (page 296 in Section 5.4).

c. If the lists are sets of numbers, we can compare the lists (see one-variable statistics, page 517 in Section 8.7).

12. Making graphs with sets of numbers.

Step 1: Clear or shut off equations: In [Y=], place cursor on = and press [ENTER].

Clear prior lists: [STAT] **4: ClrList** [2ND] [L1] [,] [2ND] [L2] [ENTER].

Step 2: Press (STAT) **1: Edit** to obtain the list screen and enter data. Place x in L1 and y in L2, as shown in Figure A.10. (Figure A.10 also shows another set of ordered pairs with x in L1 and y in L3.)

Step 3: Press (WINDOW) and set axes, as shown in Figure A.11.

Set **Xmin** and **Xmax** to include the smallest and largest x.

Set **Xscl** to 10% or 20% of (**Xmax** − **Xmin**), rounded to an easy counting number.

Repeat for **Ymin**, **Ymax**, and **Yscl**.

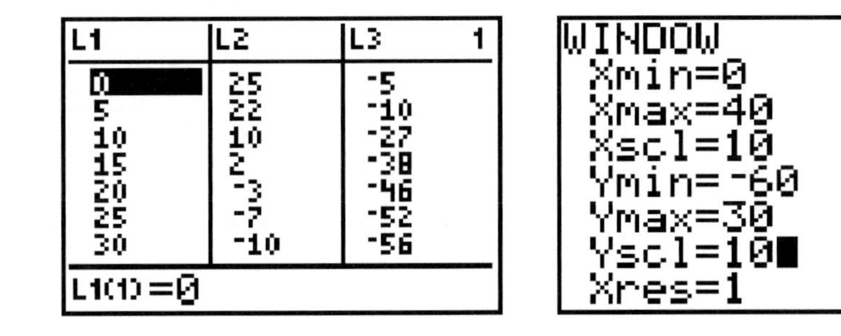

FIGURE A.10

FIGURE A.11

Step 4: Choose the statistical plot: (2ND) [STAT PLOT]. Press 1 and (ENTER) to turn on the first plot (see Figure A.12).

Choose the type of graph (scatter plot is shown).

Choose the source of the data, L1 for x and L2 for y.

Choose the type of mark for the graph (square is shown).

(To graph a second set of ordered pairs, move the cursor to Plot2 at the top of the Plot1 page and press (ENTER). Turn on Plot2 with (ENTER), then choose the scatter plot and the source of data, x in L1 and y in L3.)

Step 5: Press (GRAPH) to plot the data (see Figure A.13).

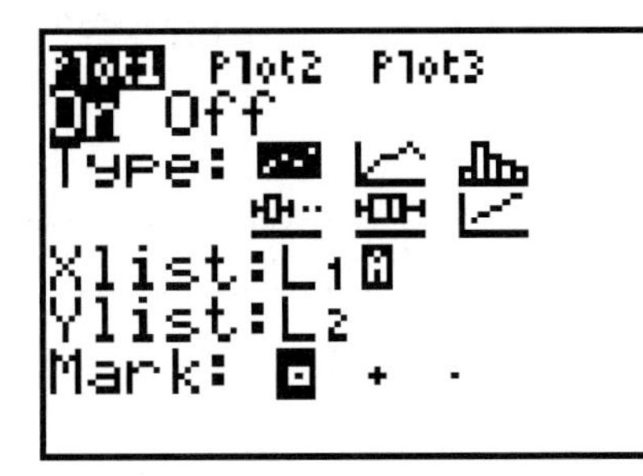

FIGURE A.12

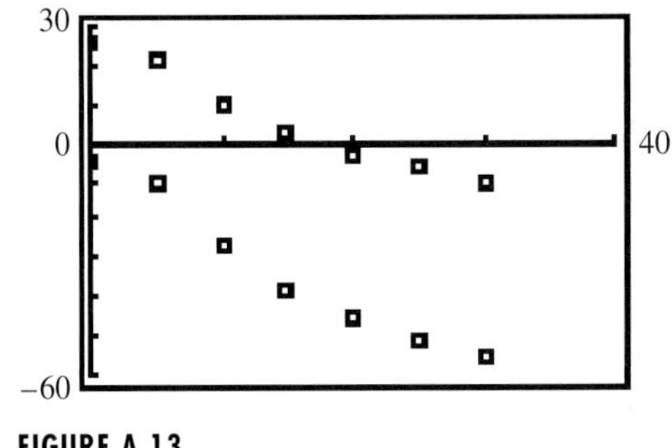

FIGURE A.13

13. (ZOOM) **Menu Notes.**

a. To draw a rectangle around a region and make it larger, choose **1: ZBox**. Move the cursor to upper-left corner of your region, and press (ENTER). Move the cursor down and to the right to create a rectangle around your region. Press (ENTER) and (TRACE).

b. To eliminate distortion, choose **5: ZSquare**.

c. To obtain the interval [−10, 10] on both x and y axes, choose **6: ZStandard**.

d. Author's advice: A more general way (than **4: ZDecimal**) to obtain x as a decimal in (TRACE) is to set the difference **Xmax − Xmin, divisible by 94**. For example, the interval [**Xmin, Xmax**] = [−9.4, 9.4], [−4.7, 4.7], [0, 94],

[−47, 47], or (0, 940]. There are 94 pixels across the width of the screen. Because the height of the window is about two-thirds the width of the window, setting (**Ymax − Ymin**) = (2⁄3)(**Xmax − Xmin**) will minimize distortion. Matching settings to the intervals [**Xmax, Xmin**] above are intervals [**Ymin, Ymax**] = [−6.3, 6.3], [−3.1, 3.1], [0, 63], [−31, 31], or [0, 630], respectively. See page 534, in Section 9.1, for the divisibility by 94 in the interval [−9, 0.4] and a corresponding graph. See page 292 in Section 5.3 for a related project.

14. Other calculator skills (see also the Glossary/Index).

a. Checking a solution, 2ND [TEST], Section 3.2, page 141.

b. Solving an inequality with a graph, Section 3.5, page 167.

c. Shading inequalities: on Y=, left of = sign, Section 4.6, page 242, and Section 6.6, page 379.

d. Domain error, 0^0, Section 7.5, page 433, and Section 9.1, page 532.

e. Scientific notation 1,000,000 = 1 2ND [EE] 6, Section 7.6, page 442.

f. Finding a point of intersection with 2ND [CALC]**5: intersect**, Section 6.1, page 325.

g. Finding one-variable statistics (mean and sample standard deviation) and box and whisker plots, Section 8.7, page 517.

Answers to Selected Odd-Numbered Exercises and Tests

As you compare your answers to those listed here, keep these hints in mind:

- Don't give up too quickly.
- Have confidence that you worked the exercise correctly.
- Check that you looked up the right answer.
- Check that you copied the exercise correctly.
- See if you can use algebraic notation or simplification to change your answer to match the text's answer.

If the answers still don't match, try working the exercise again:

- Make sure you thoroughly understand the exercise. Read the exercise aloud. Shut the book and say it in your own words.
- On a separate piece of paper, copy the exercise from the text.
- Work the exercise without looking at your first attempt.
- Let the problem rest for an hour or two or overnight. Sometimes the solutions to problems become clear when you step away from them.
- Compare your work with that of another student. (Do this only after you have done the problem twice.)
- Review the text material and related examples.
- Go on to another problem. You can continue doing homework without having completed each and every exercise.
- At the next opportunity, ask your instructor to review your work. Note that this does not mean that you should ask him or her to show you how to do the exercise.

Only after trying the above steps should you assume either that your work is wrong or that the four to six human beings who worked every exercise for this book made an error. The latter is possible, and the author and publisher would appreciate corrections.

EXERCISES 1.1

1. You will be successful; your income will increase with education; you will be able to pay back the loan.
3. Your teacher knows your last name and in which class you are enrolled.
5. 41 **7.** The animals may be tied to the panels; 41 **9.** 60
15. draw a picture, try a simpler problem, look for pattern, use table, make estimate

EXERCISES 1.2

1.

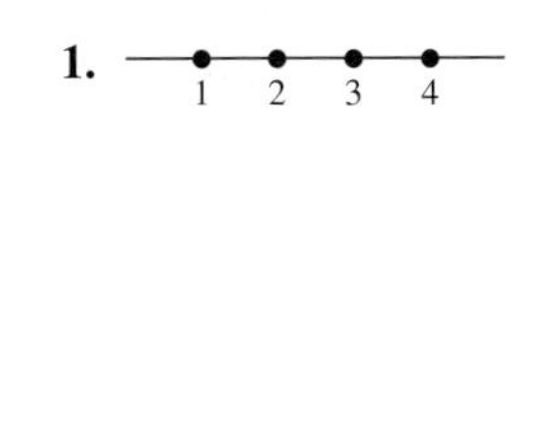

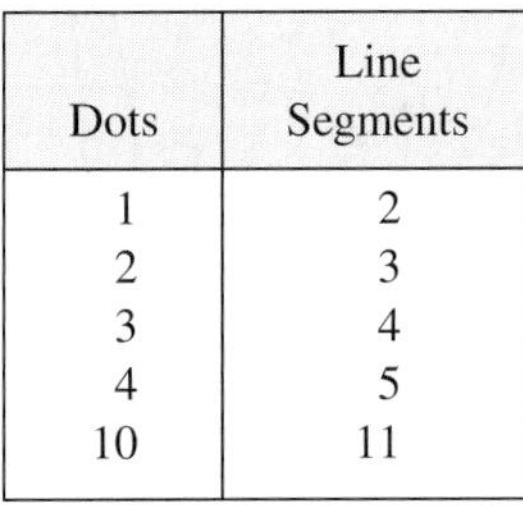

Dots	Line Segments
1	2
2	3
3	4
4	5
10	11

3.

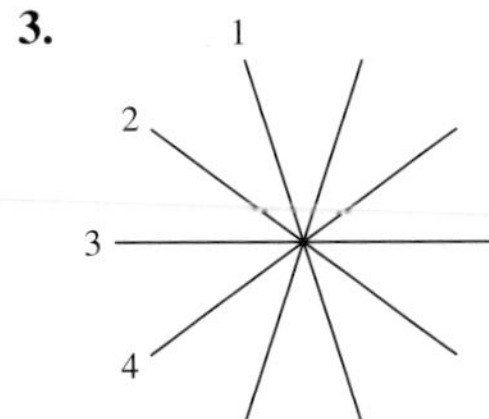

Lines	Regions
1	2
2	4
3	6
4	8
10	20

5.

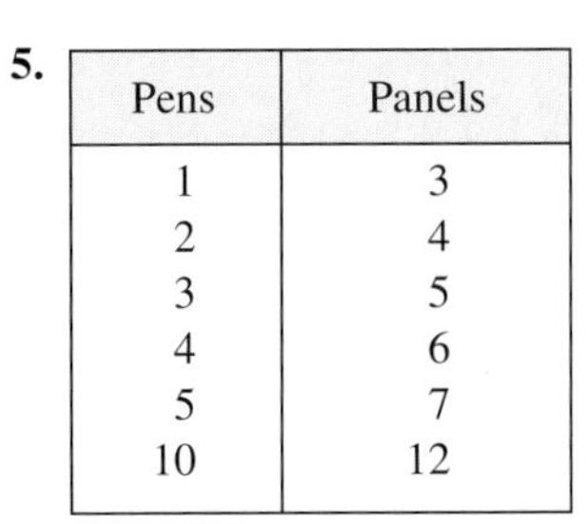

Pens	Panels
1	3
2	4
3	5
4	6
5	7
10	12

Panels inside figure are not counted in perimeter.

7.

Credit Hours	Tuition (dollars)
14	938
15	1005
16	1072
17	1139
18	1206
19	1500
20	1500

9. a. \$5.95 **b.** \$7.95 **c.** \$11.95 **d.** \$3.95

17. a. natural numbers **b.** negative real numbers **c.** rational numbers **d.** integers

19. positive

21. a. $\frac{3}{10}$ **b.** $\frac{1}{20}$ **c.** $\frac{1}{8}$ **d.** $\frac{1}{50}$

23. a. 0.4375 **b.** 0.8 **c.** 0.72 **d.** 0.45

25. a. $1\frac{7}{9}$ **b.** $3\frac{3}{8}$ **c.** $2\frac{6}{7}$ **d.** $2\frac{2}{3}$

27. a. $\frac{9}{5}$ **b.** $\frac{7}{3}$ **c.** $\frac{19}{4}$ **d.** $\frac{7}{2}$

29. a. $\frac{5}{7}$ **b.** in lowest terms **c.** $\frac{2}{3}$ **d.** $\frac{3}{4}$

31.

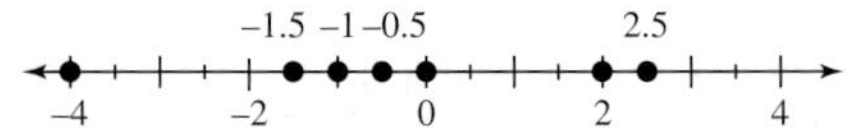

33.

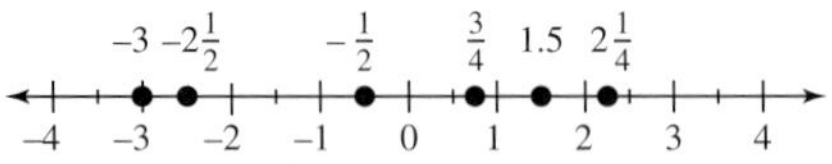

35.

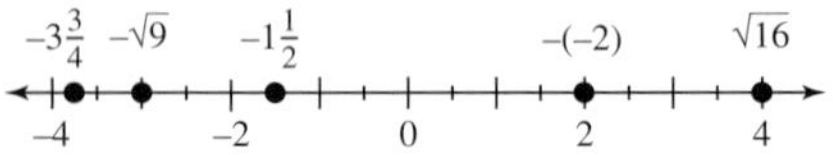

37.

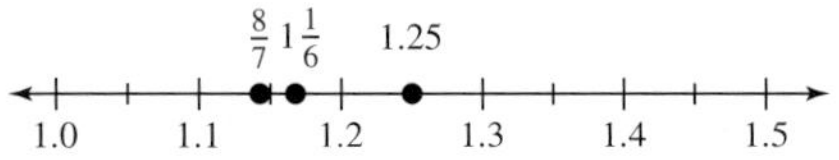

39. $\frac{5}{6}, \frac{1}{6}, \frac{1}{6}, \frac{3}{2}$

41. $\frac{14}{15}, \frac{4}{15}, \frac{1}{5}, \frac{9}{5}$

43. $\frac{13}{6}, \frac{1}{2}, \frac{10}{9}, \frac{8}{5}$

45. $\frac{57}{10}, \frac{7}{10}, 8, \frac{32}{25}$

47. a. $4\frac{4}{9}$ or 4 shows, 15 minutes
b. $6\frac{6}{7}$ or 6 shows, 15 minutes

48 to 53. no answers are provided to writing, problem-solving, or project exercises.

EXERCISES 1.3

1. $\frac{3}{8}$, 0.375 **3.** $\frac{5}{6}$, 0.833 **5.** $37\frac{1}{2}$, 37.5 **7.** $\frac{4}{15}$, 0.267
9. $1\frac{5}{12}$, 1.417 **11.** $2\frac{2}{15}$, 2.133 **13.** $\frac{7}{15}$, 0.467 **15.** $3\frac{3}{4}$, 3.75
17. 4 **19.** $\frac{7}{8}$, 0.875

21. The quotient of a and b is $\frac{a}{b}$.

23. The letters a, b, and c are often used as constants for agreements, definitions, and properties. The letters x, y, and z and the first letter or sound of key words are often used as variables in word problems.

25. a. 2 and π, constants and numerical coefficients; r, variable
b. 1.5, constant and numerical coefficient; x, variable
c. −4, constant and numerical coefficient; 3, constant; n, variable
d. 1, constant and numerical coefficient; −9, constant; x, variable

31. a, c, f

33.

x	y
1	3
2	6
3	9

35.

x	y
1	2.5
2	5.0
3	7.5

37.

Input	Output
1	55
2	110
3	165
t	$55t$

39. a. $3n$ **b.** $8 \div n$ or $\frac{8}{n}$ **c.** $n - 4$ **d.** $n \div 5$ or $\frac{n}{5}$ **e.** $\$15n$

41. a. $3 + 2n$ **b.** $4 - 3n$ **c.** $7n + 4$ **d.** $n \cdot n$ or n^2 **e.** $\$0.79n$

43. The output is five times the input. 500, $y = 5x$

45. The output is two less than the input. 98, $y = x - z$

47 to 53. No answers are provided for writing, problem-solving or project exercises.

MID-CHAPTER 1 TEST

1. a. 4.5 **b.** 30; 150 **c.** The output is 1.5 times the input.
d. $y = 1.5x$

2.

Sides	Triangles
3	1
4	2
5	3
6	4
7	5
20	18

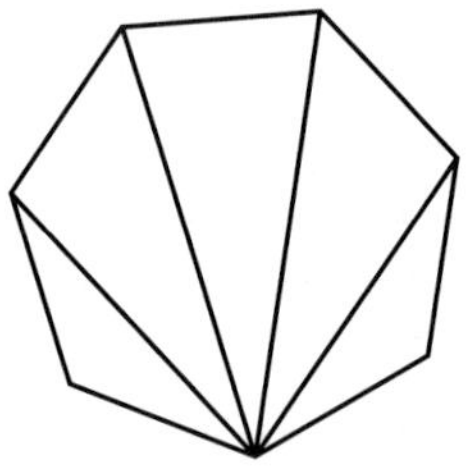

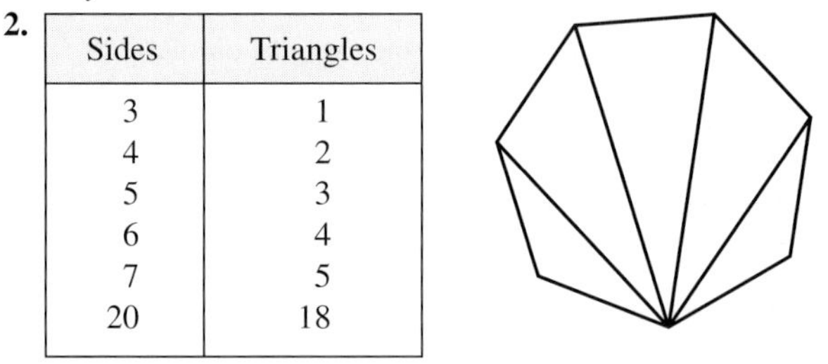

d. $t = s - 2$ **e.** t = no. of triangles; s = no. of sides
f. natural (or whole) numbers greater than or equal to 3

3. a. true **b.** true
c. false; zero is added to the set of natural numbers to make the set of whole numbers.
d. false; 3 divided by 4 is not an integer.

4. a. $5 - n$ **b.** $3.5n = 105$ **c.** $n + 7$ or $7 + n$ **d.** $\frac{6}{n} = 30$
An equation contains a verb (such as *is*) indicating where the equals sign goes. A sentence ends with a period and suggests an equation. An expression is described by a phrase and has no period at the end.

5. a. variable **b.** numerical coefficient **c.** expression
d. constant

6. The reciprocal of $\frac{a}{b}$ is $\frac{b}{a}$ because their product is 1 when a and b are integers and neither $a = 0$ nor $b = 0$.

7. $\frac{12}{5}$, 2.4

8. $\frac{13}{20}$

9. $\frac{32}{21}, \frac{4}{21}, \frac{4}{7}, \frac{9}{7}$

10. a. $6\frac{1}{12}$ **b.** $1\frac{7}{8}$ **c.** $1\frac{1}{2}$ **d.** 2

EXERCISES 1.4

1. output is three times the input plus one; $A = 3n + 1$

Input	Output
4	13
50	151
100	301

3. output is four times the input plus three; $A = 4n + 3$

Input	Output
4	19
50	203
100	403

5.

n	p
1	10
2	12
3	14
4	16
50	108
100	208

The perimeter (output) is eight more than twice the input. $p = 2n + 8$

7.

n	p
1	12
2	14
3	16
4	18
50	110
100	210

The perimeter (output) is ten more than twice the input. $p = 2n + 10$

9.

Input	Output
1	$325
2	$400
3	$475
4	$550
5	$625

$y = 250 + 75n$

11.

Input	Output
1	6
2	10
3	14
4	18
5	22

$y = 2 + 4n$

13. **a.** The output is ten less the quotient of the input and five.
b. The area of a circle is pi times one fourth of the diameter times itself.
c. The circumference of a circle is pi times twice the radius.
d. The surface area of a box is twice the product of the length and width plus twice the product of the height and length plus twice the product of the height and width.

15. **a.** $S = 6n^2$ **b.** $n + (-n) = 0$
c. $n \cdot \frac{1}{n} = 1$ **d.** $d = 2r$
The parentheses in part b group the sign with the letter.

17. **a.** The area, A, of a triangle is the base, b, times half the height, h.
b. The area, A, of a triangle is half the product of the base, b, and the height, h.
c. The area, A, of a triangle is half the base, b, times the height, h.
d. The area, A, of a triangle is the product of the base, b, and height, h. divided by 2.

19. **a.** $0.15, \frac{3}{20}$ **b.** $0.005, \frac{1}{200}$ **c.** $0.48, \frac{12}{25}$ **d.** 2.5, or $\frac{5}{2}$ $2\frac{1}{2}$

21. $\frac{100}{100} = 1$, 1.00

23. $\frac{9}{20}$, 0.45

25. **a.** 90%, **b.** $66\frac{2}{3}\%$ **c.** 50% **d.** 490% **e.** 625% **f.** 900%

27. **a.** 150% **b.** 75% **c.** 36% **d.** 560% **e.** 225% **f.** 1500%

29. **a.** $0.35n$ **b.** $0.10x$ **c.** $0.875n$ **d.** $0.375x$ **e.** $0.005n$ **f.** $1.08x$

31. **a.** $0.015m$ **b.** $0.985m$ **c.** $0.01m$ **d.** $0.99m$

33. **a.** $y = \$5$; $y = 0.03n$; $y = \$75$
b. $5 **c.** $9 **d.** $30 **e.** $75
f. Cash advance equals fee.
g. $5,0001, $74.9997; amounts for which $5 and 3% of n are equal and $75 and 3% of n are equal.

35. 22, $y = 5x + 2$, 52, 127; the number of panels is two more than five times the number of pens

37. 8, 14, $3n + 2$, 32

39.

Input	0	1	2	3	4	5	6	7	8
Output	5	2	5	6	5	10	5	14	5

41.

$a + b$	$a - b$	$a \cdot b$	$a \div b$
20	10	75	3
$1\frac{3}{20}$	$\frac{7}{20}$	$\frac{3}{10}$	$1\frac{7}{8}$
0.42	0.30	0.0216	6
6.3	4.9	3.92	8
3.75	0.75	3.375	1.5

EXERCISES 1.5

1. $A(-2, 2)$; $B(-5, 0)$; $C(-3, -4)$; $D(0, -2)$; $E(4, -4)$; $F(2, -6)$; $G(4, 0)$; $H(4, 6)$; $I(0, 5)$

3. **a.** 2 **b.** 4 **c.** 3 **d.** 3

5. **a.** vertical **b.** vertical **c.** horizontal

EXERCISES 1.5

7.

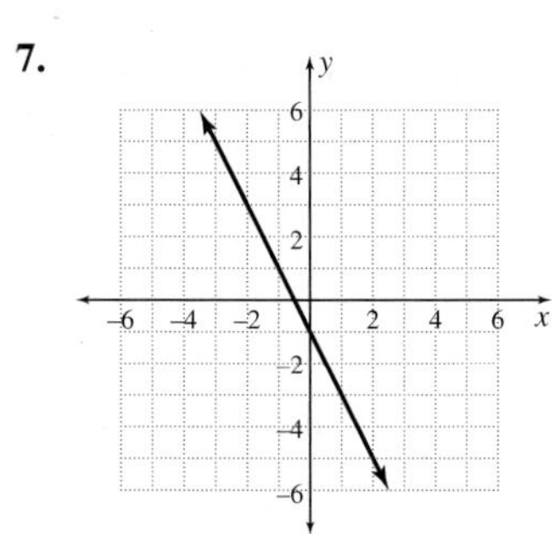

9.

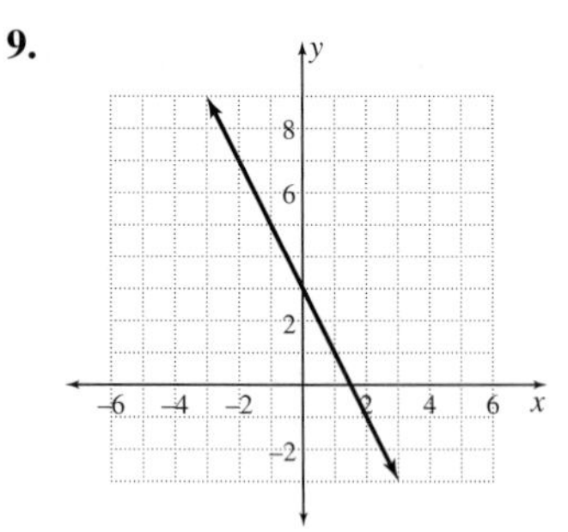

11–17.

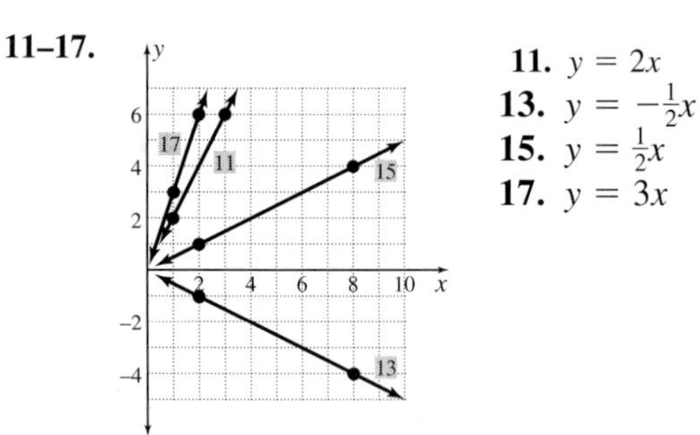

11. $y = 2x$
13. $y = -\frac{1}{2}x$
15. $y = \frac{1}{2}x$
17. $y = 3x$

19.

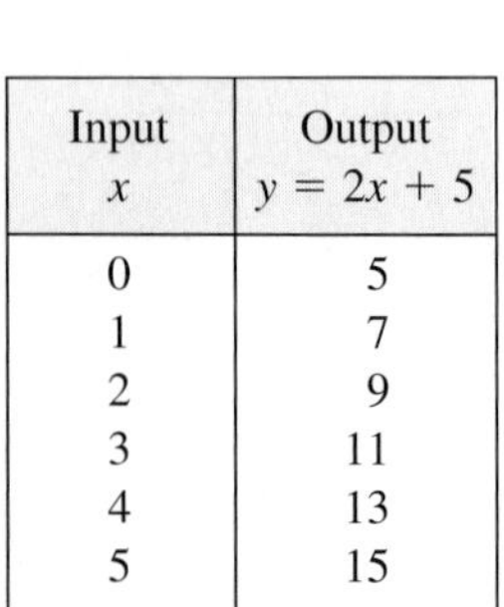

Input x	Output $y = 2x + 5$
0	5
1	7
2	9
3	11
4	13
5	15

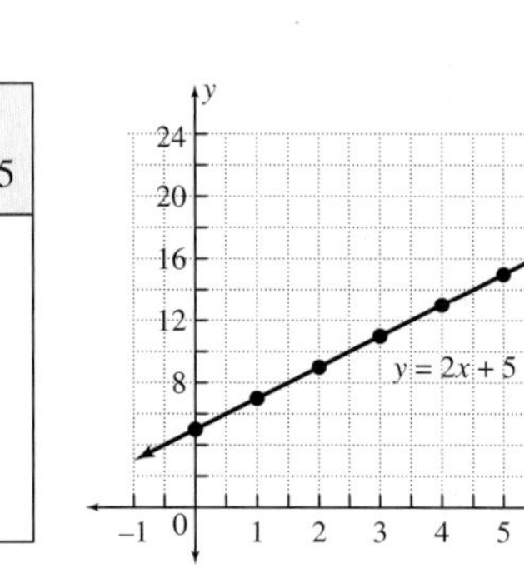

21.

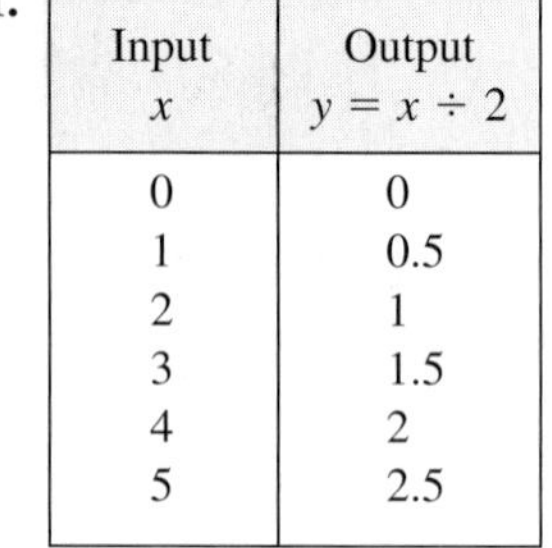

Input x	Output $y = x \div 2$
0	0
1	0.5
2	1
3	1.5
4	2
5	2.5

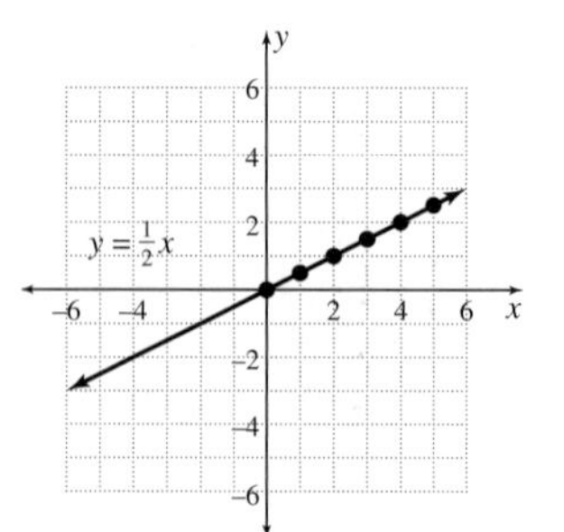

23.

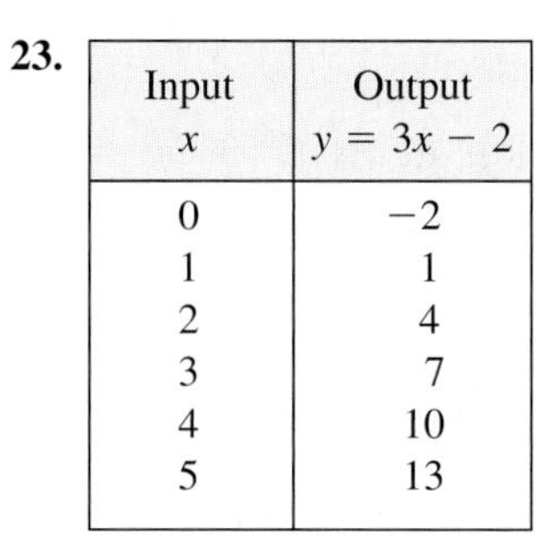

Input x	Output $y = 3x - 2$
0	−2
1	1
2	4
3	7
4	10
5	13

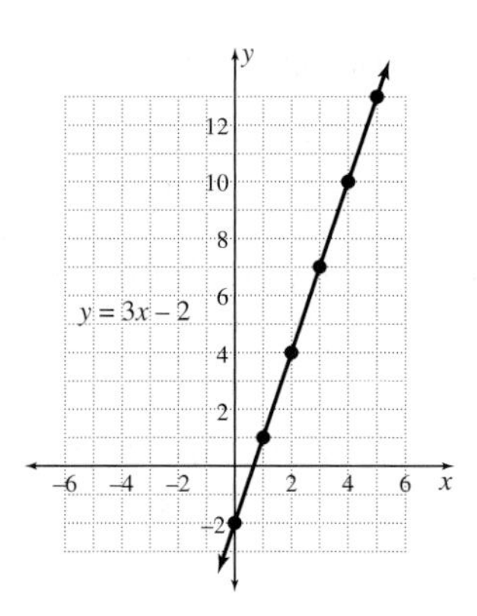

25.

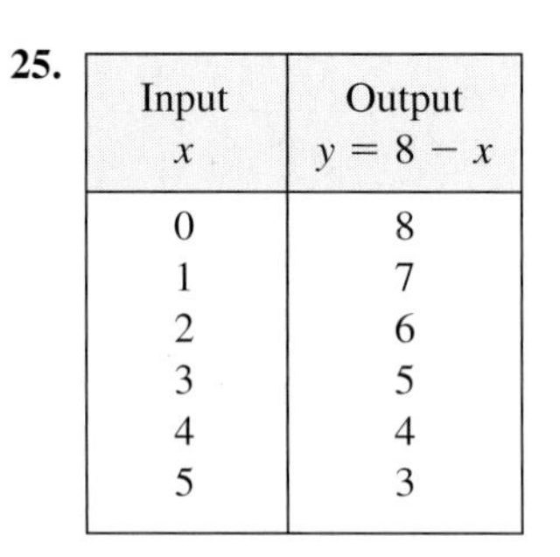

Input x	Output $y = 8 - x$
0	8
1	7
2	6
3	5
4	4
5	3

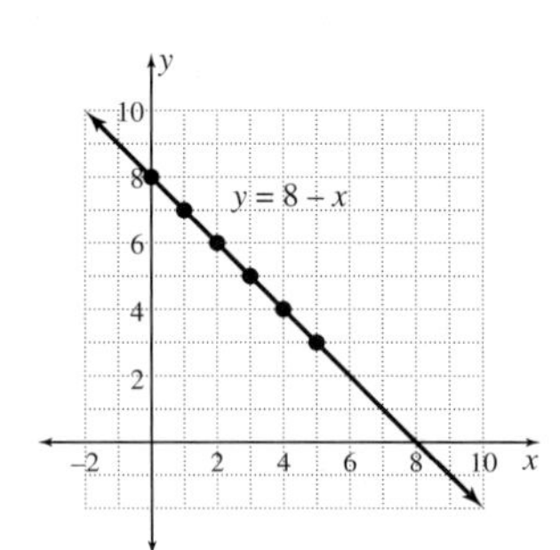

27. a

Weight (pounds)	Cost (dollars)
0	0
1	6.50
2	13.00
3	19.50
4	26.00

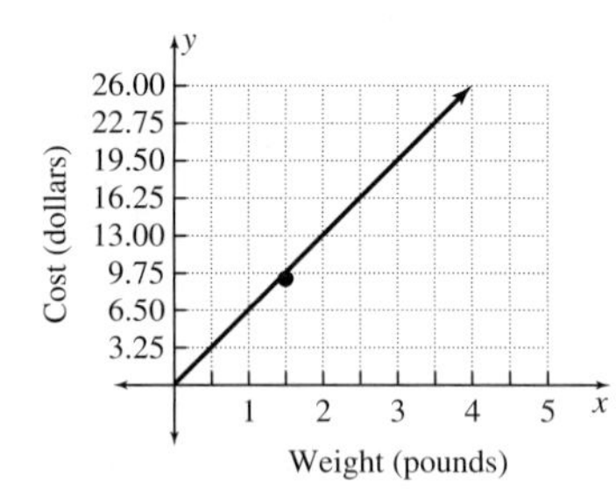

b. ≈\$16.25; ≈\$11.38; ≈\$21.13 **c.** packaged

29. a. and b.

Minutes	Card Value (dollars)
0	6.00
40	4.80
80	5.60
120	2.40
160	1.20
200	0

c. \$6 is the value of the card before any calls are made.
d. After 200 minutes of calls, the card value is zero.

31. a. (1, 15.5), (2, 15.75), (3, 15.25), (4, 15.25), (5, 15), (6, 15.25), (7, 16), (8, 15.5), (9, 15.75), (10, 16), (11, 15.75), (12, 16), (13, 16.5), (14, 16.75), (15, 16.5)

CHAPTER 1 REVIEW EXERCISES

1. Answers will vary, some possibilities are given:
a. You have time to reach the bus stop.
b. Write the 8 first and subtract 5.
c. Write the 12 first and divide by 8.
d. Your bicycle is in good working condition.

3. a. D **b.** N **c.** N **d.** D **e.** N **f.** D **g.** D

5. a. real, rational **b.** real, rational, integer, whole number **c.** real, rational **d.** real, rational **e.** real, rational **f.** real, rational, integer

7. 3

9. a. Four less than three times the input.
b. Three more than the input times itself.
c. The input divided by three.

11.

Input	Output
0	0
1	0.5
2	1
3	1.5
4	2
5	2.5
6	3

$y = \frac{1}{2}x$

13.

Input	Output
0	0
1	4
2	4
3	4
4	8
5	4
6	12

$y = 2x$ if x is even
$y = 4$ if x is odd

15.

Kilowatt Hours	Cost (dollars)
10	7.50
20	9.50
50	15.50
h	$5.50 + 0.20h$

17.

Pens	Panels
4	13
20	61
50	151

19. Outputs: 5, 9, 13; outputs: 21, 201, 401; The area, A, is one more than four times the design position, x; $y = 4x + 1$.

21. a. $\frac{11}{24}$ **b.** $\frac{1}{5}$ **c.** $\frac{94}{15}$ or $6\frac{4}{15}$ **d.** $\frac{11}{8}$ or $1\frac{3}{8}$ **e.** $8\frac{1}{8}$ **f.** $\frac{3}{8}$

23. $0 \cdot n = a$; false: a cannot be both zero and not zero.

25. a. $\frac{1}{4}$, 25% **b.** $\frac{7}{20}$, 35% **c.** $\frac{7}{25}$, 28% **d.** $\frac{3}{8}$, 37.5% **e.** $\frac{1}{250}$, 0.4% **f.** $\frac{6}{5}$, 120%

27. a. 36 **b.** 72 **c.** 180 **d.** 90 **e.** 10.8 **f.** 1.8

29. For x between \$0 and \$149.99, $y = 0.05x$; for x between \$150 and \$499.99, $y = 0.06x$; for $x = \$500$ and over, $y = 0.08x$, or \$50, whichever is greater.

31. a. Quadrant 2, **b.** vertical or y-axis, **c.** origin, **d.** (4, 0), **e.** (3, −2), **f.** (0, −3), **g.** (−3, −2)

33.

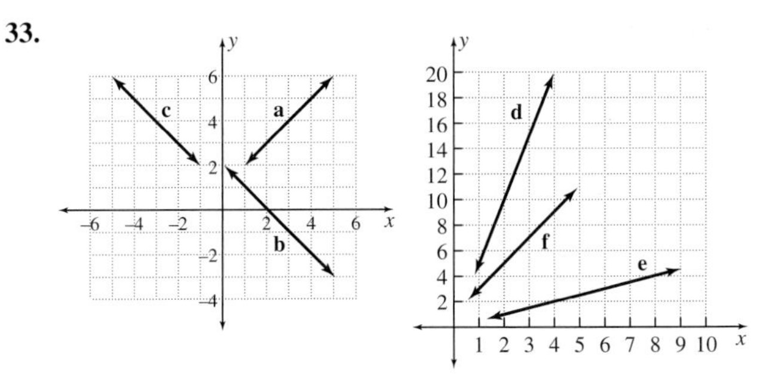

35. a. $y = x + 1$ **b.** $y = -x + 2$ **c.** $y = -x + 1$ **d.** $y = 5x$ **e.** $y = \frac{x}{2}$ **f.** $y = 2x + 1$

37. a. 1, 4, 9, and 16 lb **b.** 1, 3, 5, and 7 lb **c.** The weight loss started slowly and increased as days went by.

39. a. 3, 6, 9 and 12 lb **b.** 3, 3, 3, and 3 lb **c.** The weight loss is 3 lb every 10 days.

41.

Input (pounds)	Output: Cost (dollars)
0	0
1	24
2	48
3	72
4	96
5	120
6	144
7	168
8	192

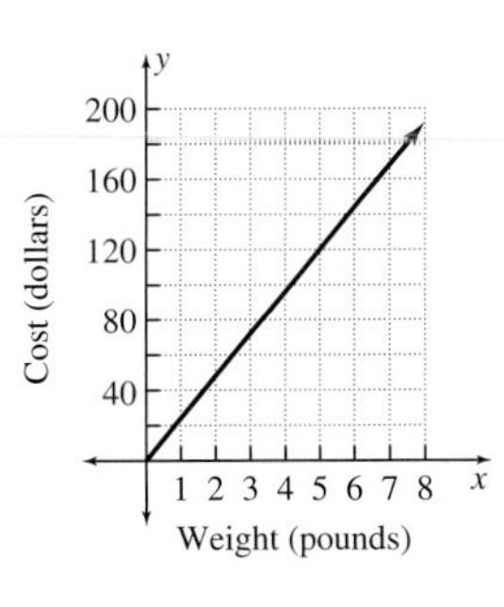

43. a. 3.3 cents
b.

Input (min)	Output: Value (dollars)
0	9.99
50	8.325
100	6.66
150	4.995
200	3.33
250	1.665
300	0

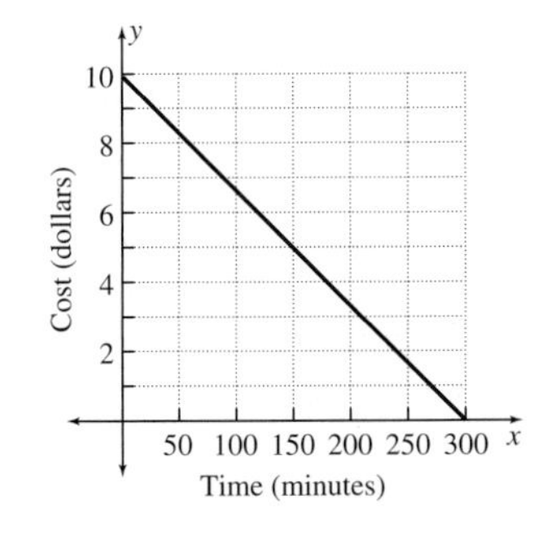

45. Inputs can be any positive number.

47.

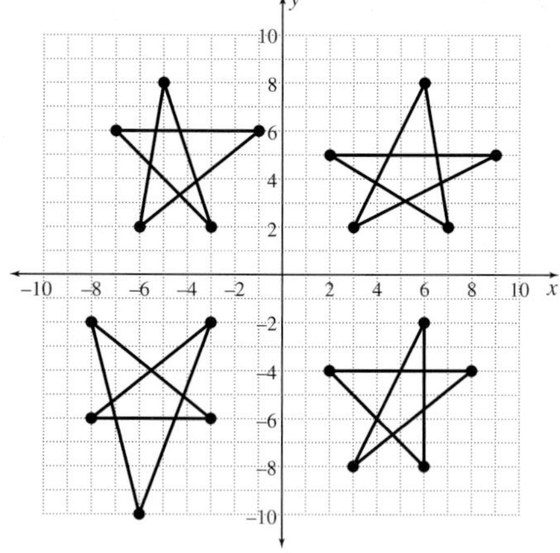

d. At 39 minutes the two methods are equal. Above 39 minutes, EMBARQ is the better deal.

10.

Input	Output
0	0
1	2
2	1
3	6
4	2
5	10
6	3
7	14
8	4

11. $\frac{19}{24}$; $\frac{11}{24}$ **12.** $1\frac{31}{54}$; $\frac{40}{51}$ **13.** 0.64 and 64% **14.** $\frac{21}{5}$ and 420%

15. 3.15

16.

Input	Output
5	20
100	400

The output is the product of four and the input. $y = 4x$

17. b and d

18. $A(-5, 4)$, $B(-2, 5)$

19. to group information, to show multiplication, to write an ordered pair.

20. The input variable, x, goes on the horizontal axis, and each is respectively earlier in the alphabet than the output variable, y, on the vertical axis.

CHAPTER 1 TEST

1. Possible answers: Material on the test will be similar to the chapter exercises and chapter review; the test will cover the entire chapter with about the same number of questions from each section.
2. Possible answers: The test has a time limit; students must work individually on the test; the test counts about the same amount toward a grade as other chapter tests.
3. difference **4.** set **5.** integers
6. origin, vertical axis or y-axis, horizontal axis or x-axis, Quadrant 3
7. $(-4, 4)$, $(2, -3)$, $(0, -5)$, $(-2, -7)$
8. 

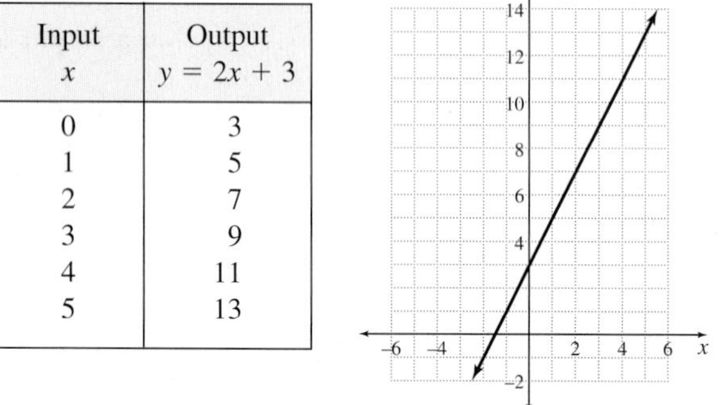

Input x	Output $y = 2x + 3$
0	3
1	5
2	7
3	9
4	11
5	13

9. a. time in minutes on x and cost in dollars on y; $y = 0.04x + 0.39$

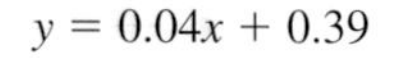

c.

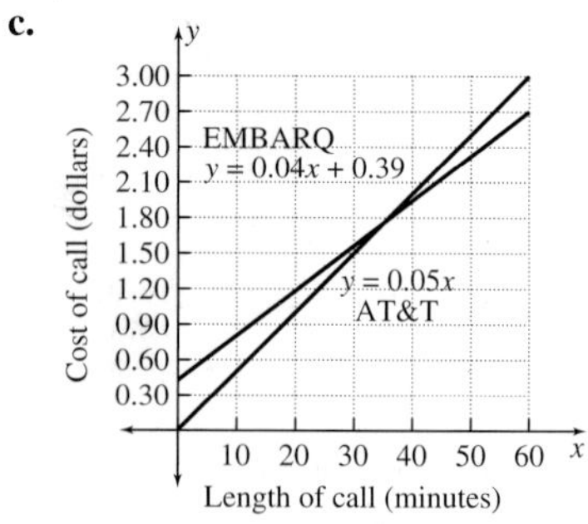

EXERCISES 2.1

1. $+4 + (-3) = 1$ or $-3 + 4 = 1$

3. $+5 + (-5) = 0$ or $-5 + 5 = 0$

5. $+5 + (-7) = -2$ or $-7 + 5 = -2$

7. a. -5 **b.** -3 **c.** 0

9. a. 0 **b.** -11 **c.** -4

11. a. -1 **b.** 2 **c.** -20

13. a. -25 **b.** 7 **c.** 16

15. a. -5 **b.** $\frac{1}{2}$ **c.** -0.4 **d.** $-x$ **e.** $2x$

17. a. 4 **b.** 6 **c.** 5 **d.** 2

19. a. -7 **b.** -8 **c.** 3 **d.** 7

21. a. 4 **b.** 5 **c.** 5 **d.** -7

23. a. 4; -4 **b.** -4; -4

25. the opposite of the absolute value of x

27. The variables represent any integer. If $x = 5$ and $y = -2$, then $x - y = x + (-y)$ means $5 - (-2) = 5 + (+2)$. If $x = 5$ and $y = 2$, then it means $5 - 2 = 5 + (-2)$.

29. a. $1 - (-2) = 3$ **b.** $2 - (+3) = -1$

31. a. $0 - (+2) = -2$ **b.** $1 - (-3) = 4$

33. a. -3 **b.** 3 **c.** -19 **35. a.** 5 **b.** -19 **c.** -13

37. a. 0 **b.** 4 **c.** 5 **39. a.** −4 **b.** 1 **c.** −13
41. a. −5 **b.** −3 **c.** 7 **43. a.** −12 **b.** −12 **c.** 2
45. a. 7000 m **b.** 6052 m **c.** 2244 m
d. −8600 m − (−10,930 m) = 2330 m
47. true
55.

Day	Google Closing Price	Change	Percent Change
1	440		
2	400	−40	−9.1%
3	425	+25	+6.25%
4	450	+25	+5.88%
5	435	−15	−3.33%
6	435	0	0%

Stock fell sharply but seems to have recovered.

EXERCISES 2.2

1. −2(+150) = −$300 **3.** +2(+400) = +$800
5. −8(−40) = +$320 **7.** +3(−90) = −$270

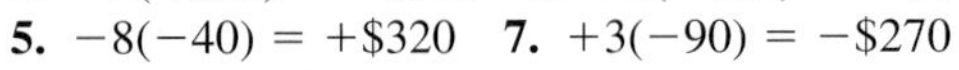

9.

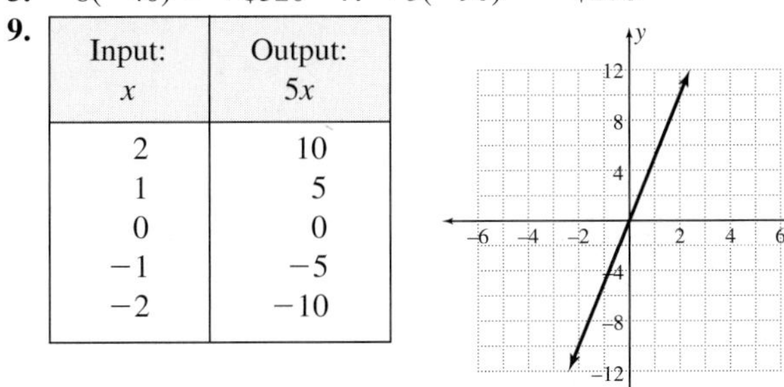

Input: x	Output: $5x$
2	10
1	5
0	0
−1	−5
−2	−10

11.

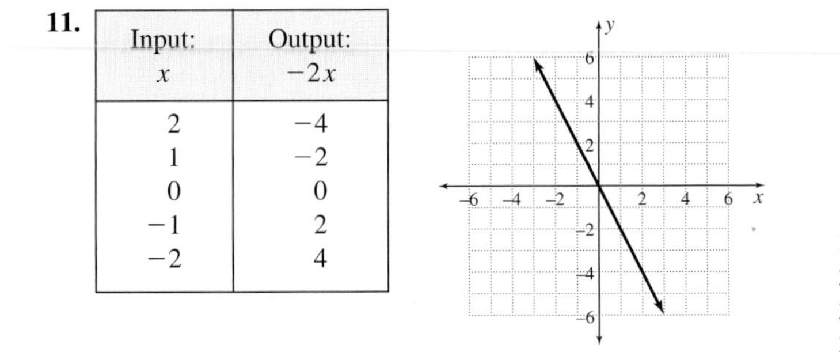

Input: x	Output: $-2x$
2	−4
1	−2
0	0
−1	2
−2	4

13. a. 42 **b.** −48 **c.** −45 **d.** 35
15. a. −48 **b.** −64 **c.** 60 **d.** 81
17. a. −1 **b.** 1 **c.** −1 **d.** 1
19. a. −6 **b.** −1 **c.** 3
21. a. $\frac{10}{7}$ **b.** $-\frac{3}{2}$ **c.** 0
23. a. $\frac{1}{4}$ **b.** $-\frac{1}{2}$ **c.** 2 **d.** $-\frac{4}{3}$ **e.** 2
25. a. $\frac{3}{10}$ **b.** $\frac{2}{13}$ **c.** $\frac{1}{x}$ **d.** $\frac{b}{a}$ **e.** $-\frac{1}{x}$
27. a. −3 **b.** −5 **c.** −7
29. a. −7 **b.** 7 **c.** −7
31. a. 5 **b.** −7 **c.** −3
33. a. 36 **b.** −16 **c.** $-\frac{128}{3}$
35. a. −10 **b.** −32 **c.** $\frac{7}{8}$

37.

a	b	$-\left(\frac{b}{a}\right)$	$\frac{-b}{a}$	$\frac{b}{-a}$
5	35	−7	−7	−7
−27	3	$\frac{1}{9}$	$\frac{1}{9}$	$\frac{1}{9}$

The expression $-\left(\frac{b}{a}\right)$ is equal to either $\frac{-b}{a}$ or $\frac{b}{-a}$. The position of the negative sign makes no change in the value of the expression.

39. $\frac{-b}{a}, -\frac{b}{a}$

41.

Input	Output
−3	3
−2	2
−1	1
0	0
1	1
2	2
3	3

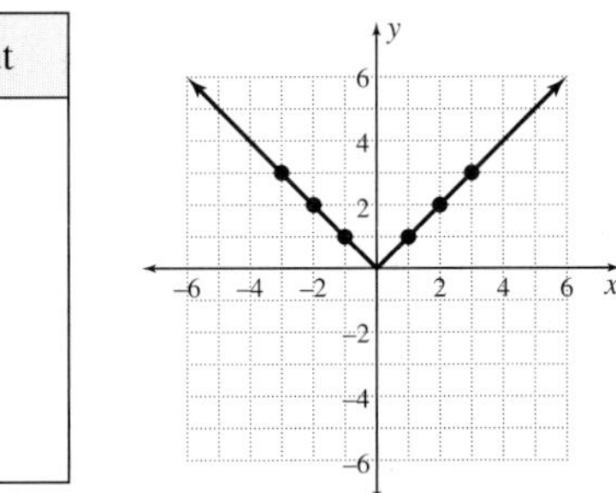

The rule is for absolute value.

45. $\frac{1}{\frac{2}{3}} = 1 \div \frac{2}{3} = 1 \cdot \frac{3}{2} = \frac{3}{2}$

EXERCISES 2.3

1. a. associative property of multiplication
b. associative property of addition
c. associative property of addition
d. commutative properties **e.** associative property of addition
f. associative property of multiplication
g. distributive property of multiplication over addition
3. a. 13 **b.** $9\frac{1}{3}$ **5. a.** 8.98 **b.** 7.30 **7. a.** −2 **b.** −4
9. a. −7 **b.** −3 **11.** −14 **13.** −1 **15. a.** −1 **b.** 2
17. a. 9 **b.** 1 **19. a.** 200 **b.** 8 **21. a.** 48 **b.** 25 **c.** −24
23. a. −210 **b.** −108 **c.** 0
25. 4(5.00 − 0.03) = 20.00 − 0.12 = $19.88
27. 3(11.00 − 0.02) = 33.00 − 0.06 = $32.94
29. $4\left(2 + \frac{3}{4}\right) = 8 + 3 = 11$ **31.** $6\left(3 + \frac{5}{6}\right) = 18 + 5 = 23$
33. 60 = 60; commutative property of multiplication
35. 6 = 6; commutative property of addition
37. 12 = 12; associative property of addition
39. a. $6x + 12$ **b.** $-3x + 9$ **c.** $-6x - 24$
41. a. $-3x - 3y + 15$ **b.** $-x + y + z$
43. a. $-x + 3$ **b.** $4y + y^2$ **c.** $-2 + y$
45. a. 3 · 3 · 5 **b.** prime **c.** 2 · 2 · 2 · 3 · 3 **d.** 3 · 37
47. a. $\frac{1}{3}$ **b.** $-\frac{2}{3}$ **c.** $\frac{2}{5yz}$ **49. a.** $\frac{5x}{7}$ **b.** $\frac{3b}{d}$ **c.** $\frac{-y}{4z}$
51. a. $\frac{a}{c}$ **b.** $\frac{b}{c}$ **c.** $\frac{a}{b}$ **53. a.** $\frac{3}{4}x + 1$ **b.** $x + 2$ **c.** $x + y$
d. $a - c$ **55.** $2a^2 + 4ab + 6b^2$ **57.** $3x^2 + 3x + 6$
59. $9x$ **61. a.** −4 **b.** 1 **c.** −1
63. a. $9a^2$ **b.** no like terms **c.** $-1a^2$ or $-a^2$
65. a. $3x$ **b.** $-11y^2$ **c.** $5a + 15$ **d.** $x + 5$ **e.** $-x + 4$
f. $4x - 6y$
67. a. $\frac{1}{4}x + \frac{3}{4}y$ **b.** $2a + 0.25b$ **c.** $-12x + 3y$ **d.** $4c$

69. a. 6 **b.** 0 **c.** 8 **d.** 2 **e.** no
f. no None of the expressions are equal.
71. b and c need to be added or subtracted.
73. $23 \neq 35$; associative properties don't apply.
75. $60 \neq 180$; distributive property does not apply.
77. a. 3 **b.** 2 **c.** 4 **d.** 3 **e.** 3
87. a. $0.015m$, $0.985m$ **b.** $1.000m - 0.015m = m(1 - 0.015)$ or $0.985m$ **c.** $0.01m$; $1.00m - 0.01m = 0.99m$

MID-CHAPTER 2 TEST

1. a. -3 **b.** 2 **c.** 2 **2. a.** 7 **b.** -7 **c.** 3.5
3. a. -6 **b.** 15 **c.** 4 **4. a.** -9 **b.** 14 **c.** -8
5. a. $2x + 6y$ **b.** $-x$ **6. a.** $4z$ **b.** $\dfrac{3}{yz}$ **c.** $-\dfrac{1}{2x}$
7. $46.11c - 42.39d$ **8.** $24x - 18$ **9.** $\frac{3}{4}$ **10.** 780
11. a. $2x + y$ **b.** $a + c$

12.

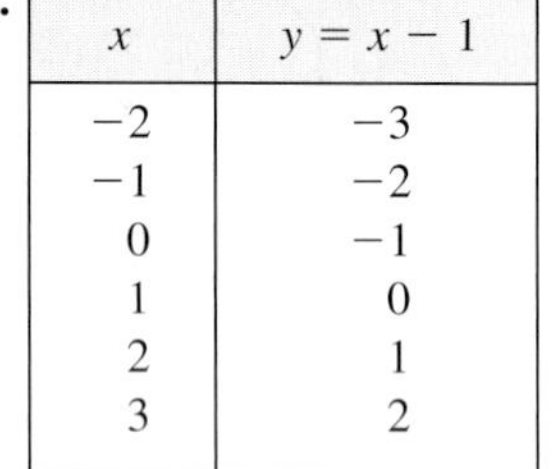

x	$y = x - 1$
-2	-3
-1	-2
0	-1
1	0
2	1
3	2

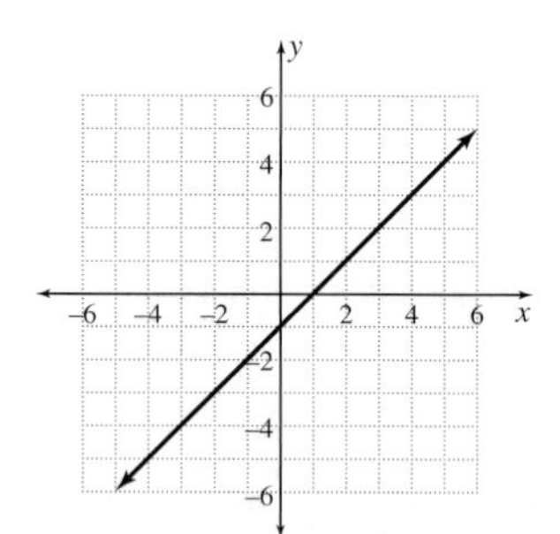

13.

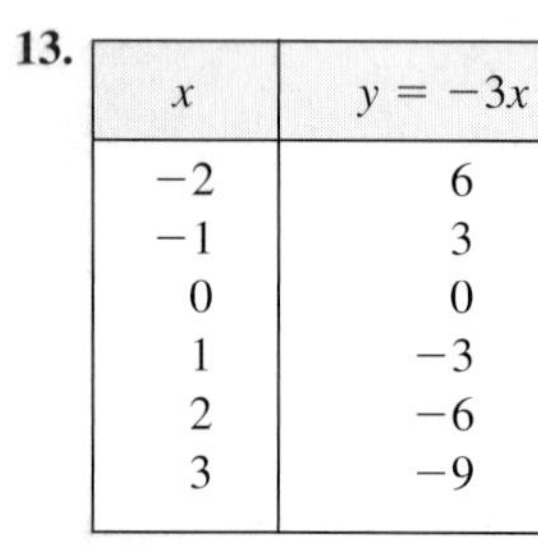

x	$y = -3x$
-2	6
-1	3
0	0
1	-3
2	-6
3	-9

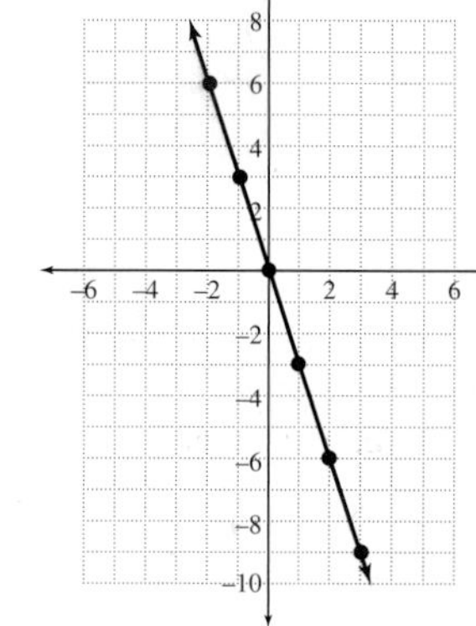

14.

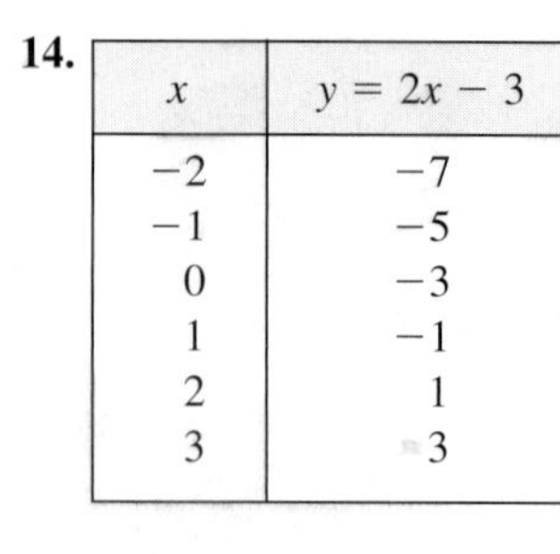

x	$y = 2x - 3$
-2	-7
-1	-5
0	-3
1	-1
2	1
3	3

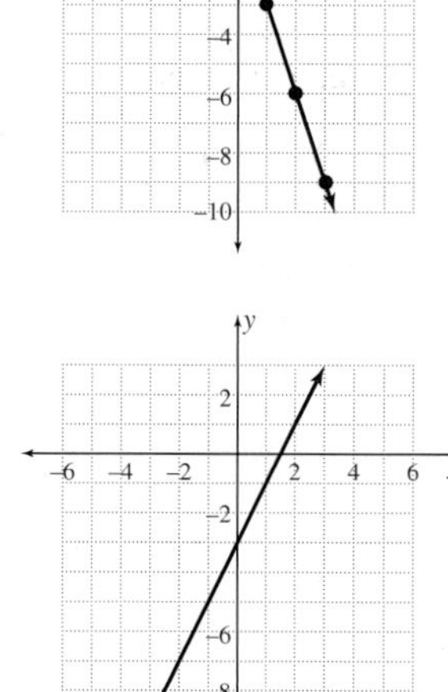

15.

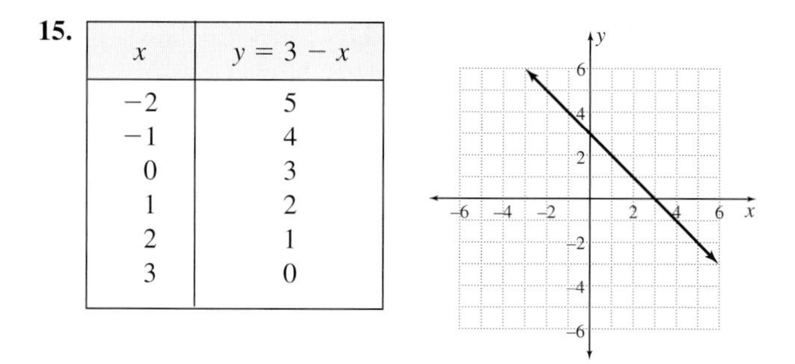

x	$y = 3 - x$
-2	5
-1	4
0	3
1	2
2	1
3	0

16. 9250 m **17.** 6280 m
18. Mauna Kea, 30,110 ft $-$ 29,028 ft $=$ 1082 ft
19. Subtraction is not commutative.
20. Change subtraction to addition of the opposite number.

EXERCISES 2.4

1. a. x; $3 \cdot x \cdot x$ **b.** x; $-3 \cdot x \cdot x$ **c.** $-3x$; $(-3x)(-3x)$
d. x; $a \cdot x \cdot x$ **e.** x; $-1 \cdot x \cdot x$ **f.** $-x$; $(-x)(-x)$
3. a. 243 **b.** 64 **c.** 4 **d.** -27
5. a. $\frac{1}{27}$ **b.** $\frac{64}{125}$ **c.** $-\frac{8}{27}$ **d.** $-\frac{1}{9}$
7. a. 32 **b.** -4 **c.** 12 **d.** -36
9. a. x^3 **b.** x^5 **c.** a^9 **d.** b^3
11. a. $\dfrac{1}{x}$ **b.** a **c.** $\dfrac{1}{y}$ **d.** $\dfrac{a^2}{b}$
13. a. $16x^2$ **b.** $27x^3$ **c.** $0.04x^2$ **d.** $-8x^3$
15. a. a^2b^2 **b.** x^3y^3 **c.** $4a^2c^2$
17. a. $-32a^2$ **b.** $-27x^2$ **c.** $16x^3$
19. a. $\dfrac{x^3}{64}$ **b.** $\dfrac{4a^2}{25}$ **c.** $\dfrac{9x^2}{25}$
21. a. $\dfrac{16n^2}{9}$ **b.** $\dfrac{-27x^3}{64}$ **c.** $\dfrac{-125n^3}{27}$
23. a. 64 in.^{2} **b.** 0.09 sec^2 **c.** 216 m^3
25. a. $\dfrac{1}{2\text{ yd}^2}$ **b.** $\frac{1}{6}$ m^2 **c.** $\dfrac{5}{32\text{ in.}}$
27. a. -128 **b.** -243 **c.** 432
29. a. $-\frac{1}{8}$ **b.** $\frac{4}{25}$ **c.** $\frac{36}{25}$ **31. a.** $\frac{16}{9}$ **b.** $\frac{27}{8}$ **c.** $\frac{125}{27}$
33. 35 **35.** 100 **37.** 169 **39.** 1 **41.** 15 **43.** $-3x + 18$
45. $3x - 12$ **47.** 36 **49.** 4 **51.** 6
53. a. $-\frac{3}{5}$ **b.** -5 **c.** $\frac{3}{7}$ **55. a.** 2 **b.** $\frac{3}{2}$ **c.** 4
57. 15 **59.** 10 **61.** 2.83 **63.** 75.36 **65.** 33.49
67. 18 **69.** 3 cm **73.** subtracted 4 from 7 first **75.** $x^3 \cdot x^3 = x^6$
77. $x^3 \cdot x^3 = x^6$; decimal error: $(0.2)^2 = 0.04$
79. $x^3 \cdot x^3 = x^6$

EXERCISES 2.5

1. $\frac{1}{2}$ lb **3.** 3600 in. **5.** ft^2 **7.** $\dfrac{1}{\text{inch}}$ **9.** 3281 ft
11. 54,545 g **13.** 86,400 sec **15.** 11.6 days
17. 2,270,592,000 sec, assuming 365 days/yr **19.** 1728 in.^{3}
21. 1296 in.^{2} **23.** 8.3 ft^2 **25.** 65,000 ft
27. a. $P = 9.9$ cm; $A = 4.8$ cm^2 **b.** $P = 12.6$ cm; $A = 6.8$ cm^2
c. $P = 12.6$ ft; $A = 12.6$ ft^2 **d.** $P = 9.6$ cm; $A = 4.0$ cm^2
29. a. $P = 60$ yd; $A = 120$ yd^2 **b.** $P = 6.8$ cm; $A = 2.9$ cm^2
c. $P = 85$ m; $A = 270.8$ m^2 **d.** $P = 28.3$ in.; $A = 63.6$ in.^{2}
31. a. $S = 280$ ft^2; $V = 300$ ft^3 **b.** $S = 256$ in.^{2}; $V = 240$ in.^{3}
33. a. $S = 527.5$ cm^2; $V = 904.3$ cm^3
b. $S = 131.9$ cm^2; $V = 113.0$ cm^3
35. a. $S = 803.8$ cm^2; $V = 2143.6$ cm^3
b. $S = 201.0$ cm^2; $V = 267.9$ cm^3 **37.** 125.6 ft^3; 939.6 gal

3. a. conditional, 2 **b.** conditional, 1 **c.** conditional, 3 **d.** identity **e.** identity **f.** identity
5. $3x - 4 = 17$ **7.** $26 - 4x = 2$
9. $y = 2x + 5$ **11.** $y = \frac{1}{2}x - 5$
13. Five less than the product of three and a number is 16.
15. Two thirds of a number is 24.
17. Six less twice a number is 10.
19. Twice the difference between a number and three is ten.
21. Half the sum of a number and two is six.
23. $y = 0.15x$, x is cost of meal in dollars, y is tip in dollars
25. $y = 0.0145x$, x is wages in dollars, y is payment in dollars
27. $y = 2.49x$, x is weight in pounds, y is cost in dollars
29. $y = 3x$, x is speed in mph or kph, y is distance in miles or km
31. $y = 75x + 32$, x is no. of credit hours, y is total cost in dollars
33. $y = -4.25x + 26$, x is no. of beverages, y is value in dollars
35. $3(x + 5) = 21$
37. $5(9 - x) = -35$ **39.** $\frac{1}{2}(8 + x) = y$
41. $y = 0.80(200 - x), x \le 200$; $y = 0, x > 200$
43. $y = 0.15x, 0 < x \le 2$; $y = 0.15(2) + 0.05(x - 2), x > 2$
45. $y = 5, x = 1$; $y = 5 + 2(x - 1), x > 1$
47. $y = 125$ for 1 to 10 children, $y = 125 + 12.50(x - 10)$ for more than 10 children

EXERCISES 3.2

1. a. correct **b.** correct **c.** Add 3.
3. a. Divide by 55. **b.** Multiply by 2. **c.** correct
5. Divide by 2. **7.** Subtract 5. **9.** Add 4. **11.** Multiply by 2.
13. Add 5; $x = 13$ **15.** Subtract 12; $x = -3$
17. Add 6; $x = -4$ **19.** Divide by 2; $x = 13$
21. Divide by 8; $x = \frac{3}{8}$ **23.** Multiply by 4; $x = 64$
25. Multiply by -12; $x = -48$ **27.** Multiply by 2, $x = -8$
29. Multiply by $-\frac{4}{3}$; $x = -16$ **31.** Multiply by $\frac{4}{3}$; $x = 24$
33. Subtract $\frac{2}{3}$; $x = 15\frac{1}{3}$
35. Subtract $\frac{3}{4}$ and divide by -1; $x = -17\frac{1}{4}$ **37.** f, $x = \frac{1}{2}$
39. c, $x = 14$ **41.** b, $x = 20$ **43.** $x = 5$ **45.** $x = 15$
47. $x = 12$ **49.** $x = 7$ **51.** $x = 1$ **53.** $x = -2$ **55.** $x = -10$
57. $x = -2$ **59.** $x = -3$ **61.** $x = -1.5$ **63.** $x = 2.5$
65. $x = 300$ **67.** $x = 1050$
69. $x = 12$ children
71. $x = 8$ children, \$2.40
73. $x = 11$ children, \$2.25
83. 7 is added twice to one side, instead of once to both sides.
85. $t = 2$ **87.** $C = 100$ **89.** $t = 3\frac{7}{11}$ **91.** $r = 66\frac{2}{3}$
93. $p \approx \$1.67$ **95.** $x \approx 846.5$ kwh **97.** $h = 10$

EXERCISES 3.3

1. a. $x = 2$ **b.** $x = 4$ **c.** $x = 7$ **d.** $x = 12$
3. a. $x = 1$ **b.** $x = 2$ **c.** $x = -1$ **d.** $x = 2\frac{1}{2}$

5.

x	$y = 2x + 1$
3	7
3.5	8
4	9
4.5	10
5	11
5.5	12

7.

x	$y = 5x - 4$
3	11
3.2	12
3.4	13
3.6	14
3.8	15
4	16

9. a. $x = -4$ **b.** $x = -6$ **c.** $x = -10$
11. a. $x = 1$ **b.** $x = -1$ **c.** $x = 2$

13. a. $\{-2, 2\}$ **b.** $\{-1, 1\}$ **c.** $\{0\}$ **d.** $\{\ \}$
15. a.

x	$y = 8 - 3(x + 2)$
−3	11
−2	8
−1	5
0	2
1	−1
2	−4
3	−7

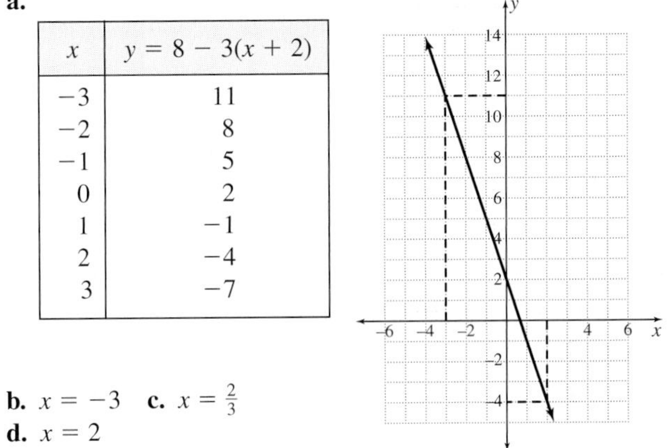

b. $x = -3$ **c.** $x = \frac{2}{3}$
d. $x = 2$

17. $x = 4$ **19.** $x = -1.6$ **21.** $x = 4.5$ **23.** $x = -4\frac{1}{3}$
25. $x = -4$ **27.** $x = 2$ **29.** $x = -5$ **31.** $x = -7$
33. $x = -4$ **35.** $2(x + 5) = 14$; $x = 2$
37. $-2(x - 4) = 6$; $x = 1$ **39.** $\frac{1}{2}(x - 5) = 24$; $x = 53$
41. $150 = 29.95 + 0.89x + 0.40(x - 50)$ for $x > 50$ miles; 108 miles
43. $11 = 5 + 1(x - 1)$; $x = 7$ copies
45. $150 = 80 + 10(x - 8)$ for $x > 8$ children; $x = 15$ children. For $x > 8$ children, the \$80 is same as \$10 per child, so divide the total cost by \$10.
47. $175 = 125 + 12.50(x - 10)$ for $x > 10$ children; $x = 14$ children. For $x > 10$ children, \$125 is \$12.50 per child, so divide the total cost by \$12.50.
55. true **57.** Don't multiply the right side by 2.

MID-CHAPTER 3 TEST

1. a. conditional **b.** identity **c.** identity **d.** conditional
2. a. equivalent **b.** not equivalent **c.** not equivalent **d.** not equivalent **3.** yes **4.** no **5.** yes **6.** $2x + 5 = 10$
7. $\$545 = \$85x + \$35$ **8.** $\$7.20 = -\$0.04x + \$20$
9. Five is four less than three times a number.
10. Twice the sum of 2 and a number is -3.
11. $x = 7$ **12.** $x = 36$ **13.** $x = -5$ **14.** $x = 14$
15. $x = \frac{1}{6}$ **16.** $x = -2.5$ **17.** $x = 4$ **18.** $x = 5$
19. a. $x = 2$ **b.** $x = 8$ **c.** $x = 5$ **d.** $x = 0$
20. a. $x = 2$ **b.** $x = -4$ **c.** $x = -1$
21. a. $\{-4, 3\}$ **b.** $\{-3, 2\}$ **c.** $\{\ \}$
22. It says that $10 = 5$. There should be no equal sign between 10 and x; a comma would be correct.
23. $A = \frac{h}{2}(a + b)$ or $A = \frac{1}{2}h(a + b)$
24. $32 = 0.80(200 - x)$ for $x < 200$; 160
25. $24 = 5 + 2.19(x - 4)$ for $x > 4$ half-gallons; 12 half-gallon cartons; \$1.48

EXERCISES 3.4

1.

x	$5x - 8$
−1	−13
0	−8
1	−3
2	2
3	7
4	12

x	$2(x + 2)$
−1	2
0	4
1	6
2	8
3	10
4	12

(4, 12); $x = 4$

3.

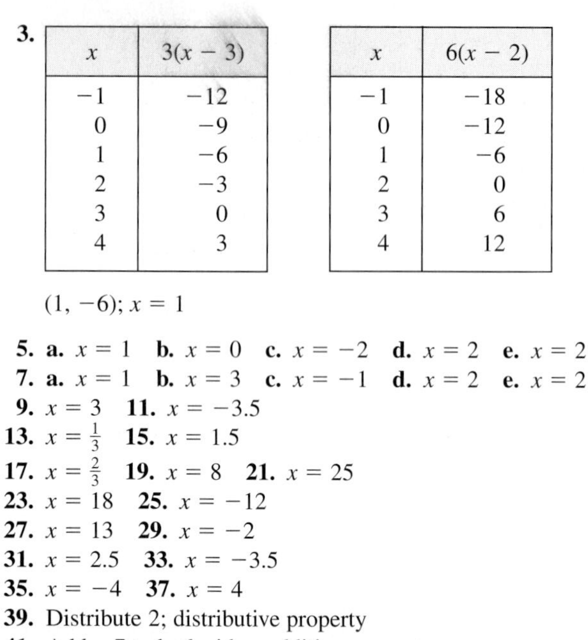

x	$3(x-3)$
−1	−12
0	−9
1	−6
2	−3
3	0
4	3

x	$6(x-2)$
−1	−18
0	−12
1	−6
2	0
3	6
4	12

$(1, -6)$; $x = 1$

5. **a.** $x = 1$ **b.** $x = 0$ **c.** $x = -2$ **d.** $x = 2$ **e.** $x = 2$
7. **a.** $x = 1$ **b.** $x = 3$ **c.** $x = -1$ **d.** $x = 2$ **e.** $x = 2$
9. $x = 3$ **11.** $x = -3.5$
13. $x = \frac{1}{3}$ **15.** $x = 1.5$
17. $x = \frac{2}{3}$ **19.** $x = 8$ **21.** $x = 25$
23. $x = 18$ **25.** $x = -12$
27. $x = 13$ **29.** $x = -2$
31. $x = 2.5$ **33.** $x = -3.5$
35. $x = -4$ **37.** $x = 4$
39. Distribute 2; distributive property
41. Add −7 to both sides; addition property
43. Add x to both sides; addition property
45. **a.** $y = 49.95 + 0.59x$; $y = 39.95 + 0.99x$
b. $49.95 + 0.59x = 39.95 + 0.99x$, at $x = 25$ miles the costs are the same
c. (25, 64.70) The rental cost at 25 miles is \$64.70.
d. U-Haul has a greater cost per mile.
e. Below 25 miles
47. **a.** $39.95 + 0.59x = 59.95 + 0.57x$, $x = 1000$ miles
b. It is impossible to drive a truck 1000 miles in 6 hours.
c. U-Haul is the better rental for the 6 hours.
55. Subtraction is not inverse for $4(x - 3)$; division is.

EXERCISES 3.5

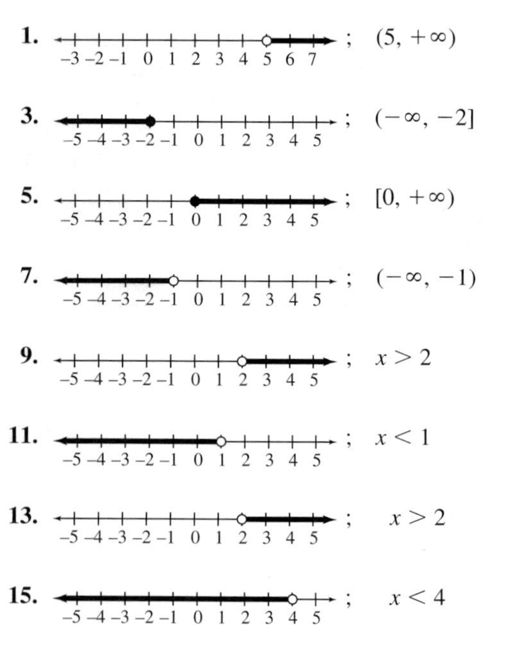

1. ; $(5, +\infty)$
3. ; $(-\infty, -2]$
5. ; $[0, +\infty)$
7. ; $(-\infty, -1)$
9. ; $x > 2$
11. ; $x < 1$
13. ; $x > 2$
15. ; $x < 4$

17.

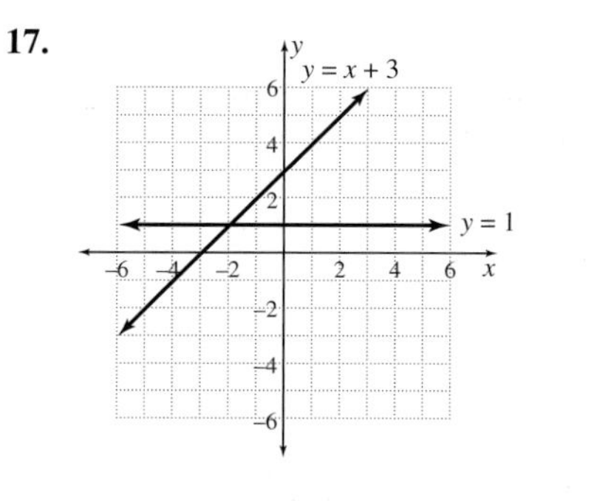

$x > -2$

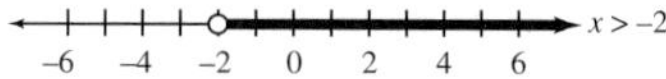

19.

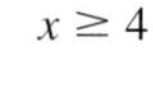

$x \geq 4$

21.

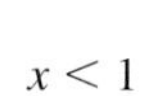

$x < 1$

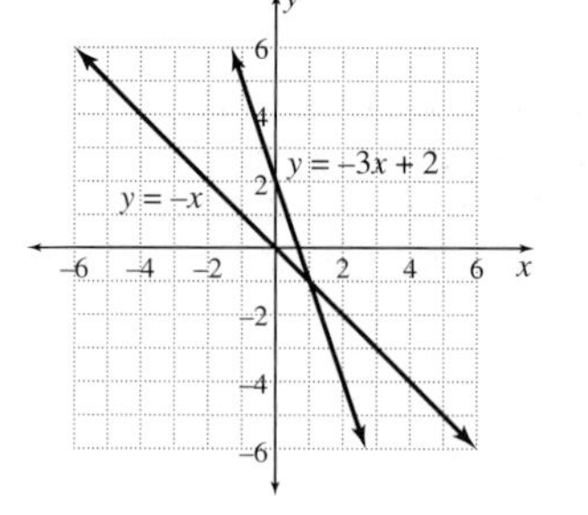

23.

$x < 1$

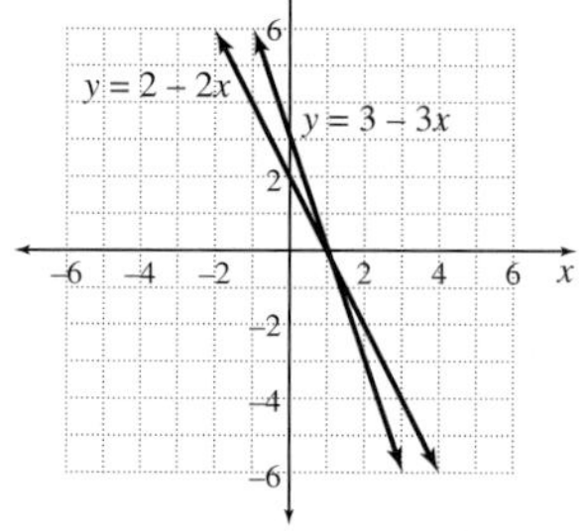

25. $x > 2, (2, +\infty)$ **27.** $x < 1, (-\infty, 1)$ **29.** $x > 2, (2, +\infty)$
31. $x < 4, (-\infty, 4)$ **33.** $x < 1, (-\infty, 1)$ **35.** $x \leq 2, (-\infty, 2]$
37. $x \geq 4, [4, +\infty)$ **39.** $x < -9, (-\infty, -9)$
41. $x > 1.25$; $(1.25, +\infty)$ **43.** $x < -1.5, (-\infty, -1.5)$
45. $x < 5, (-\infty, 5)$ **47.** $x \leq 2, (-\infty, 2]$ **49.** $x < 1, (-\infty, 1)$
51. $x > 2, (2, +\infty)$ **53.** $x < -8, (-\infty, -8)$
55. $x < -2, (-\infty, -2)$

57. a. (16, 184) **b.** $10.25x + 20 \le 184$
c. −2 0 2 4 6 8 10 12 14 16 18 $0 < x \le 16$; (0, 16]

59. $15x + 25 \le 250$, $0 < x \le 15$ **61.** $38.50x + 350 \le 3430$, $0 < x \le 80$ **63.** $38.50x + 350 < 37.50x + 0.2(37.50)x$, $x > 53$ **65.** no; $x = 177$ is not possible

67. $x \ge 263$ **69.** $x \ge 182$

CHAPTER 3 REVIEW EXERCISES

1. equivalent equations
3. two-variable equation, conditional equation
5. conditional equation **7.** $4 = 5x - 11$
9. $\frac{x}{6} = 12$ **11.** $2x - 5 = x$ **13.** $2(7 - x) = -4$
15. Three less than four times a number is 29.
17. The product of two and the difference between a number and one, subtracted from five, is seven more than the number.
19. yes **21.** yes **23.** $x = -7$ **25.** $x = 9$ **27.** $x = 24$
29. $x = 6$ **31.** $x = -3$ **33.** $x = -5$ **35.** $x = 7$ **37.** $x = 1$
39. $x = 1.2$ **41. a.** $x = 2$ **b.** $x = 5$ **c.** $x = 3\frac{1}{3}$ **d.** $x = 6$
43. a. {0, 4} **b.** {1, 3} **c.** {2} **d.** { }
45. a. $x = 2$ **b.** $x = -3$ **c.** $x = -1$
47. x = speed, y = distance; $y = 4x$
49. x = cost of meal, y = tax; $y = 0.075x$
51. x = no. of credits, y = total cost in dollars; $y = 300x + 150$
53. x = no. of visits, y = remaining value in dollars; $y = -10x + 520$
55. $440 = 20(x + 3)$; $x = 19$ seats
57. Let x = number of kWh, $10.39 + 0.176707(241.7) + 0.183146(x - 241.7) = 64.51$; $x = 304$ kWh
59. $n + 0.001n + 0.0515n + 0.024(55.88) = 62.56$ or $(1 + 0.001 + 0.0515)n + 0.024(55.88) = 62.56$; \$58.16
61. a −3 −2 −1 0 1 2 3 4 5 6 7
63. $x > 7$
65. $x \le 2$

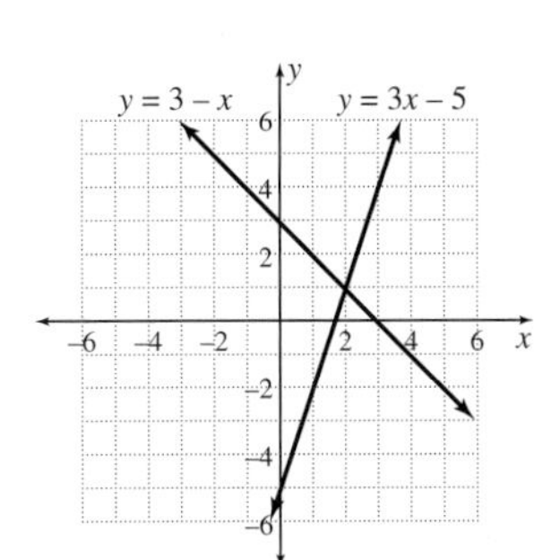

67. $5 < x$, $(5, +\infty)$ **69.** $x < -16$, $(-\infty, -16)$
71. $x < -2\frac{2}{3}$, $(-\infty, -2\frac{2}{3})$ **73.** $x \le -18$, $(-\infty, -18]$
75. $x > -1$, $(-1, +\infty)$ **77.** $-3 > x$, $(-\infty, -3)$
79. $x \le 1$, $(-\infty, 1]$ **81.** $4 > x$, $(-\infty, 4)$
83. $x \ge 0.2$, $[0.2, +\infty)$
85. $$\frac{82 + 72 + 20 + 0 + 20 + 20 + 18 + 12 + x}{100 + 100 + 20 + 20 + 20 + 20 + 20 + 70 + 150} \ge 0.80;$$ $x \ge 172$; no
87. x = no. of people; $3500 \ge 25x + 550$; $x \le 118$
89. x = linear feet; $24x + 500 < 16x + 640$; $x < 17.5$ ft

CHAPTER 3 TEST

1. $\frac{1}{2}x + 6 = 15$; $x = 18$ **2.** $3x - 7 = -31$; $x = -8$
3. $y = \frac{1}{3}x$ **4.** $y = 2x - 2$
5. $3(x + (-3)) = -12$; $x = -1$
6. $x = -11$ **7.** $x = -1$ **8.** $x = 75$ **9.** $x = -\frac{1}{2}$
10. $x = 1$ **11.** $x = -16$ **12.** $x = 1$ **13.** $x = 4\frac{1}{2}$
14. $x = 3$ **15.** $x = 2$ **16.** (1, 3) in table and graph
17. Zero is halfway between $y = -1$ and $y = 1$, so answer is halfway between $x = 2$ and $x = 3$.
18. Find where the graph crosses the x-axis.
19. a. (1, −2) **b.** $-2 = 2(1) - 4$; $-2 = -(1) - 1$ **c.** $x = 1$ **d.** The solution is $x = 1$. **e.** $x > 1$
20. −5 −4 −3 −2 −1 0 1 2 3 4 5 $x < -4$, $(-\infty, -4)$
21. −5 −4 −3 −2 −1 0 1 2 3 4 5 $-4 > x$, $(-\infty, -4)$
22. −8 −7 −6 −5 −4 −3 −2 −1 0 1 2 ; $x \ge -5$, $[-5, +\infty)$
23. a.

x (therms)	$y = 1.36828x + 6$ (dollars)
0	6.00
40	60.74
80	115.47
120	170.20
160	224.93
200	279.66

Costs rounded up because we are paying.

b.

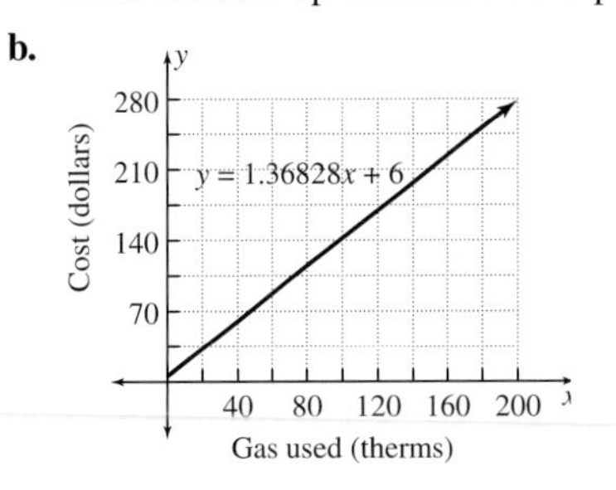

c. $y = 1.36828x + 6.00$ **d.** 70 therms
24. $29.95 + 0.59x > 21.95 + 0.99x$; $x < 20$ miles; \$41.75
25. Solving gives another identity, $3 = 3$ or $0 = 0$. The two graphs are identical, giving one line on top of another line.
26. For x = budget in dollars, $0.03x = 1.1$ million; x is approximately \$36 million.
27. Expressions showing adding like terms, using $\frac{a}{a} = 1$ property, using commutative property, etc.
28. Adding (or subtracting, multiplying, or dividing) with the same number on both sides of an equation
29. Using any step in answer to Exercise 27 within an equation or multiplying, say, $3(x + c) = d$ to $3x + 3c = d$
30. No

CUMULATIVE REVIEW OF CHAPTERS 1 TO 3

1. a.

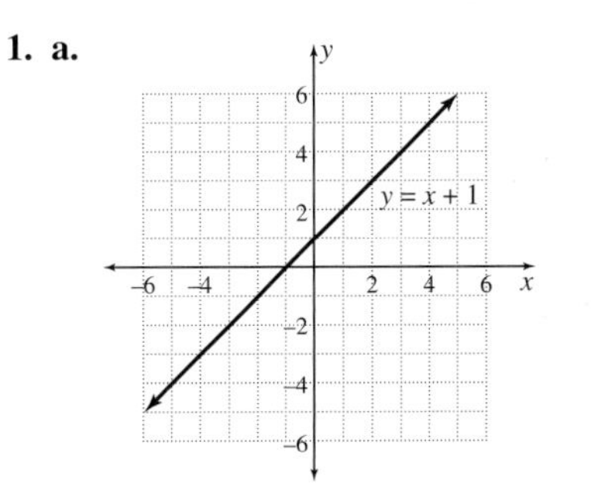

b.

c.

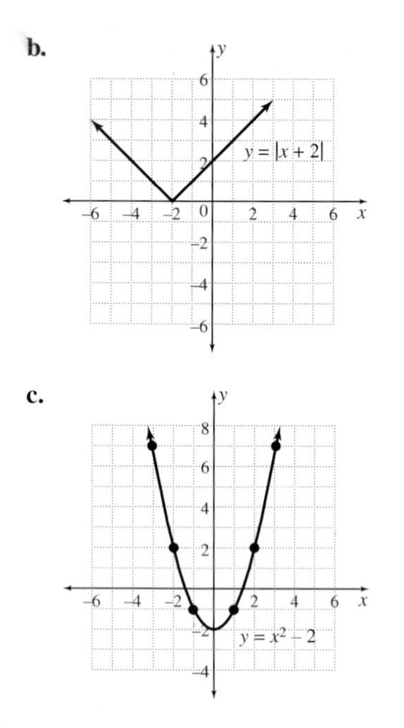

3. a. numbers greater than 5
b. opposite of the absolute value of x
c. the difference between 3 and the product of 2 with the sum of a and 8

5. $-9, -1, 20, 1.25$

7. $0.75, 1.75, -0.625, -2.5$

9. a. $4x^3$ **b.** $(4x)^3$ **c.** The base in part a is x and the base in part b is $4x$.

11. a. $x = -7$ **b.** $x = -2$

13. $15 + 12x = 100$; 7 dozen in Miami; $12 + 10x = 100$; 8 dozen in Chicago

15. 1.7% times \$176 is \$2.99. 1.7% times \$177 is \$3.01. For all balances up to \$176, 1.7% of the balance is below \$3.00; \$300; 20.4%

EXERCISES 4.1

1. interest in terms of principal, rate, and time
3. rate in terms of distance and time
5. grade percent in terms of tests, homework, final exam, and total points possible

7. $A = lw$ **9.** $A = \pi r^2$ **11.** $A = \frac{1}{2}bh$ **13.** $P = 2l + 2w$

15. $C = fm$ **17.** $n = \dfrac{p}{5}$ **19.** $A = H + P$ **21.** $r = \dfrac{C}{2\pi}$

23. $h = \dfrac{A}{b}$ **25.** $t = \dfrac{I}{pr}$ **27.** $d = \dfrac{C}{\pi}$ **29.** $C = R - P$

31. $n = \dfrac{PV}{RT}$ **33.** $V_1 = \dfrac{C_2V_2}{C_1}$ **35.** $c = P - a - b$

37. $h = \dfrac{2A}{a + b}$ **39.** $r^2 = \dfrac{3V}{\pi h}$ **41.** $b = -2ax$

43. $b = y - mx$ **45.** $h = \dfrac{2d^2}{3}$ **47.** $g = \dfrac{2d}{t^2}$

49. a. $r = \dfrac{A - P}{Pt}$ **b.** $r = 0.0525$ or $5\frac{1}{4}\%$

51. a. $F = \frac{9}{5}C + 32$ **b.** $F = 98.6°$

53. a. 48 **b.** $A = 200 - \dfrac{H}{0.8}$ **c.** 170

55. a. $b = y - mx$ **b.** $b = -2$ **c.** $b = 10$ **d.** $b = 2.5$ **e.** $b = 5.5$

57. $y = -\dfrac{6}{x}$ **59.** $y = 2x - 3$ **61.** $y = \dfrac{2x - 4}{3}$ or $y = \dfrac{2x}{3} - \dfrac{4}{3}$

63. $y = -\frac{2}{3}x + 3$

EXERCISES 4.2

1. no **3.** no **5. a.** yes **b.** no **7. a.** yes **b.** no

9. function; inputs {Eden, Tuckman, McClintock}; output {Barbara}

11. not a function

13. a. function; inputs $\{-6, -5, 5, 6\}$; outputs $\{5, 6\}$
b. not a function

15. a. function; inputs $\{2, 3, 4\}$; outputs $\{\frac{1}{2}, \frac{1}{3}, \frac{1}{4}\}$
b. not a function

17. a

19. input {units of time}; output {cost of call}

21. input {distance from equator}; output {hours of sunlight}

23. A = amount earned (\$), x = no. of hr worked; $A(x) = 8x$

25. C = circumference, d = diameter, $C = f(d)$ or $C(d) = \pi d$

27.

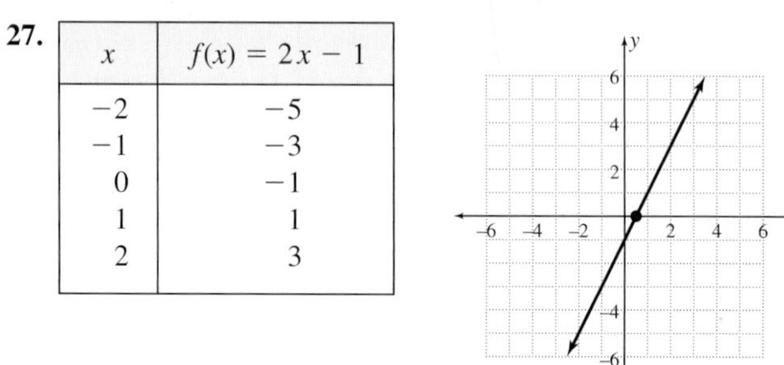

x	$f(x) = 2x - 1$
-2	-5
-1	-3
0	-1
1	1
2	3

29.

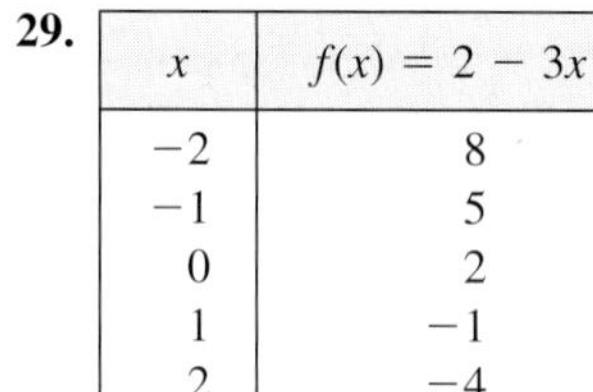

x	$f(x) = 2 - 3x$
-2	8
-1	5
0	2
1	-1
2	-4

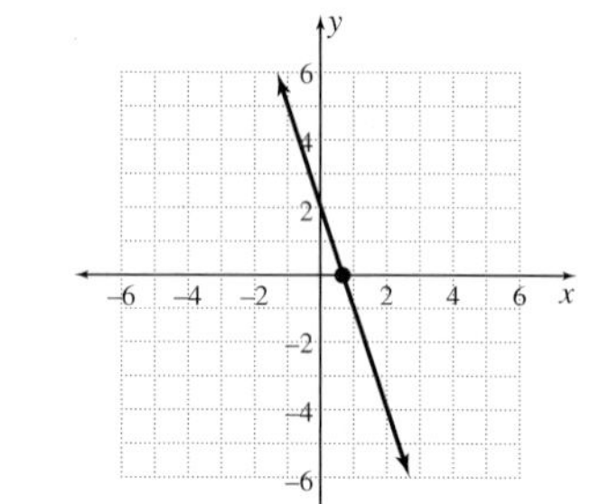

31.

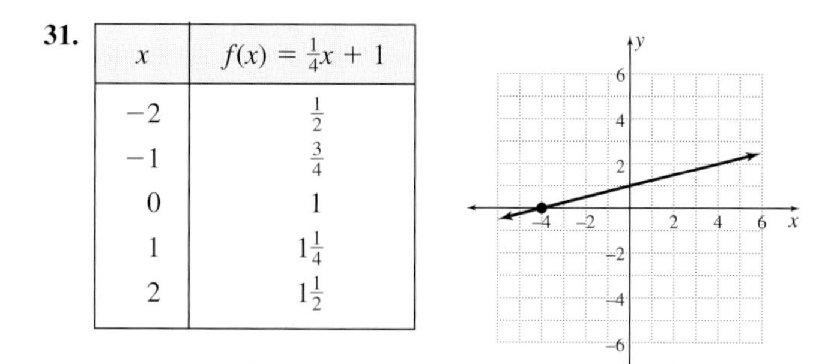

x	$f(x) = \frac{1}{4}x + 1$
-2	$\frac{1}{2}$
-1	$\frac{3}{4}$
0	1
1	$1\frac{1}{4}$
2	$1\frac{1}{2}$

33. $h(4) = 2, h(-4) = 2$ **35.** $H(4) = -12, H(-4) = -20$

37. $g(-2) = 5, g(1) = 2$ **39.** $G(-2) = -2, G(1) = 1$

41. function **43.** not a function **45.** function of x

47. function of x **49.** function of x **51.** not a function of x

53. not a function of x

59. Not a function. The cost of admission does not depend on the position in line but, rather, on the size of the group. Each person in line will pay \$15, \$12.50, or \$10.

EXERCISES 4.3

1. a. positive, 3 **b.** positive, $\frac{2}{3}$ **c.** zero, 0
3. a. negative, -2 **b.** positive, $\frac{3}{2}$ **c.** undefined
5. a. negative, -50 **b.** positive, 40 **c.** negative, -20
7. $\frac{1}{4}$ **9.** $-\frac{7}{2}$ **11.** undefined
13. 0 **15.** $-\frac{2}{3}$ **17.** 0 **19.** $-\frac{4}{5}$
21. undefined **23.** negative, linear, slope $= -3$, no units
25. positive, linear, slope $= 7$, no units
27. positive, linear, slope $= 9$, earnings per hour
29. positive, linear, slope $= 0.50$, cost per kg
31. positive, linear, slope $= 0.32$, cost per pound
33. positive, nonlinear, feet per second
35. negative, linear, slope $= -0.25$, dollar value per copy
37. positive, nonlinear, miles per hour **39.** $\dfrac{b-d}{a-c}$

41. $\dfrac{-b}{a}$ **43.** undefined **45.** $\frac{1}{2}$ **47.** $\frac{1}{2}$

49. a.

Gallons, g	Cost, c
0	0
1	\$1.55
2	\$3.10

b. slope $= 1.55$, dollars per gallon **c.** $c = \$1.55g$

51. a.

Hours, h	Earnings, e
0	0
1	\$6.25
2	\$12.50

b. slope $= 6.25$, dollars per hour **c.** $e = \$6.25h$

53. a.

Hours, h	Distance, d (in km)
0	0
1	80
2	160

b. slope $= 80$, kilometers per hour **c.** $d = 80h$

55. a.

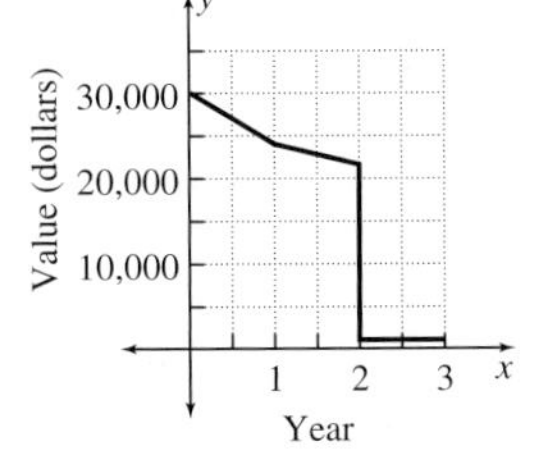

b. $-6{,}000$ **c.** $-2{,}400$

d. Extreme loss of value happened suddenly **e.** 0
f. One possible answer: After driving it for two years, she had a wreck in the third year.
57. $-\frac{1}{3}, -1, -\frac{3}{1}$ **59.** $\frac{2}{5}, \frac{1}{2}, \frac{6}{5}$ **61.** $-\frac{1}{2}, -\frac{2}{3}, -\frac{3}{2}, -2$
63. The correct formula is the reciprocal of the student's answer.
71. a. 0 **b.** 0.03 or $\frac{3}{100}$ **c.** 0

MID-CHAPTER 4 TEST

1. temperature in Celsius in terms of temperature in Fahrenheit
2. volume in terms of radius **3.** $b = 11$ **4.** $b = 0$
5. $K = C + 273$ **6.** $h = \dfrac{8d^2}{3}$ **7.** $d = \dfrac{l-a}{n-1}$
8. a. yes **b.** $\{2, 3, 4, 5, 6, 7\}$ **c.** $\{3, 4\}$

9. a.

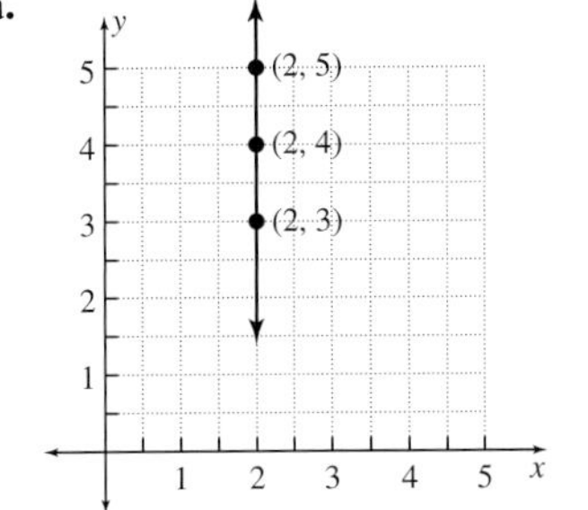

b. one **c.** One input has 3 outputs. **d.** does not pass
10. a. 18 **b.** 5 **c.** 4 **d.** 34

11.

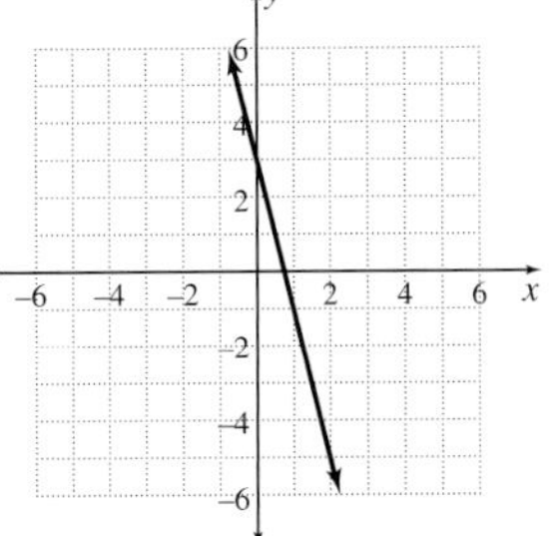

12. $-\frac{2}{7}$ **13.** 0
14. -2, no. of participants eliminated each episode
15. AB, -1; BC, 1
16. a. $\frac{5}{4}$ **b.** $\frac{-2}{1}$ **c.** $-\frac{1}{2}$ **d.** $\frac{3}{4}$

EXERCISES 4.4

1. (5, 0), (0, 3) **3.** (8, 0), (0, -24) **5.** (6, 0), (0, 3)
7. (3, 0), (0, -4)
9. a. $y = \frac{1}{2}x$, $y = 2x$, $y = 4x$
b. $y = -\frac{1}{3}x$, $y = -x$, $y = -3x$
c. $y = x$, $y = 1.5x$, $y = 2.5x$, $y = 3x$
d. $y = -0.25x$, $y = -0.5x$, $y = -x$, $y = -4x$
11. $m = 2, b = -\frac{1}{2}$ **13.** $m = -4, b = 15$ **15.** $m = -\frac{3}{4}, b = 0$
17. $m = 2, b = -4$ **19.** $m = -\frac{2}{3}, b = 4$ **21.** $m = \frac{2}{5}, b = 2$
23. $m = \frac{1}{4}, b = -1$ **25.** $m = -0.30, b = 12$
27. $m = 55, b = 0$ **29.** $m = 2\pi, b = 8$
31. $m = 2.98, b = 0.50$ **33.** $m = -0.29, b = 50$
35. $m = -0.8, b = 160$ **37.** $m = 0.15, b = 50$

39. a.

Hours, x	Cost, c
0	\$3
1	\$4
2	\$5

b. slope $= 1$, dollars per hour
c. \$3, the fee
d. $c = \$1x + \$3$

41. a.

Meal Cost, x	Total Cost, c
0	0
1	\$1.15
2	\$2.30

b. slope $= 1.15$, total cost (dollars) per meal cost (dollar)
c. \$0, no meal cost means no total cost
d. $c = 1.15x$

43. and 45.

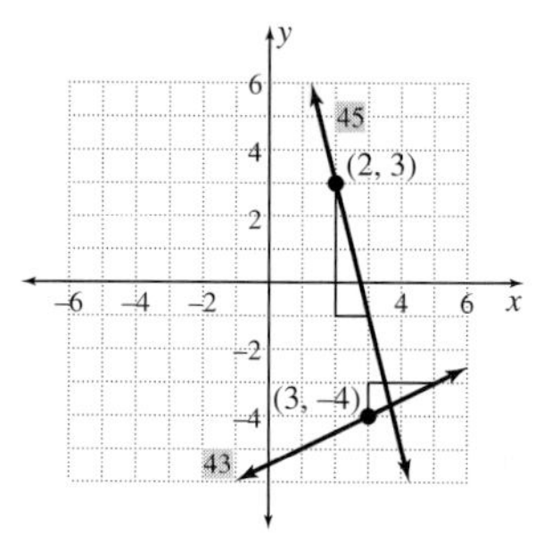

47., 49., and 51.

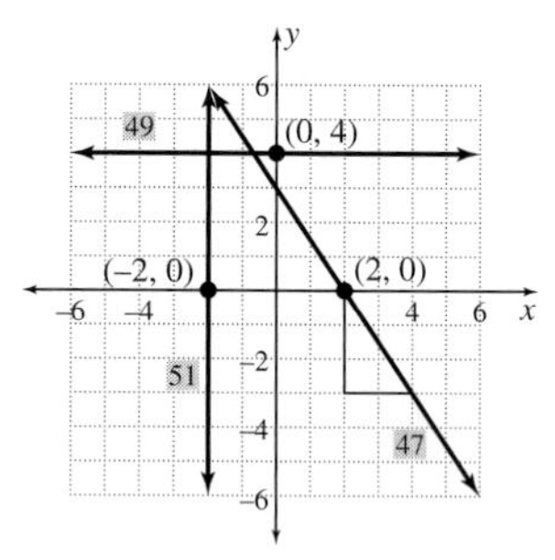

53. $y = 4$ **55.** $x = 4$ **57.** $x = 3$ **59.** $y = 0$

61. $x = 3$, $(3, 0)$; $f(0) = -12$, $(0, -12)$
63. $x = 4$, $(4, 0)$; $f(0) = 12$, $(0, 12)$
65. $x = 3$, $(3, 0)$; $f(0) = -15$, $(0, -15)$
67. $f(0) = 48$, the fees included in tuition; $f(x) = 0$, $x = -\frac{48}{115}$, no meaning
69. $f(0) = \$19.50$, original value of phone card; $f(x) = 0$, $x = 390$, total minutes the card will buy

EXERCISES 4.5

1. $y = \frac{1}{2}x + 3$ **3.** $y = \frac{2}{3}x - 2$ **5.** $y = 5x + \frac{1}{4}$
7. $y = -\frac{3}{2}x + 1$ **9.** $y = 4x - 3$ **11.** $y = -2x + 2$
13. $y = -\frac{2}{3}x - 1$ **15.** $y = 2x - 20$ **17.** $y = -8x - 34$
19. $y = -\frac{1}{2}x + \frac{9}{2}$ **21.** $y = 4x - 13$ **23.** $y = -x + 2$
25. $y = \frac{1}{2}x + 3$ **27.** $y = \frac{4}{5}x + 11$
29. $y = \frac{5}{3}x + 6$ **31.** $y = -2x + 6$
33. $y = -x + 1$ **35.** $y = 2x + 2$
37. $y = -1$ **39.** $C = \frac{1}{30}n + 1$
41. $C = -4t + 160$
43. $y = 0.05x$
45. $y = -x + 17$
47. a. freezing $(0, 32)$, boiling $(100, 212)$, slope $= \frac{9}{5}$, °F/°C
b. $b = 32$ **c.** $F = \frac{9}{5}C + 32$
49. *GH* **51.** *NO*
53. $y = \frac{1}{3}x + 1$; $y = \frac{1}{3}x - \frac{4}{3}$; $x = -1$; $x = 1$
55. b and d are parallel, a and c are perpendicular
57. a and c are parallel, b and d are perpendicular
59.

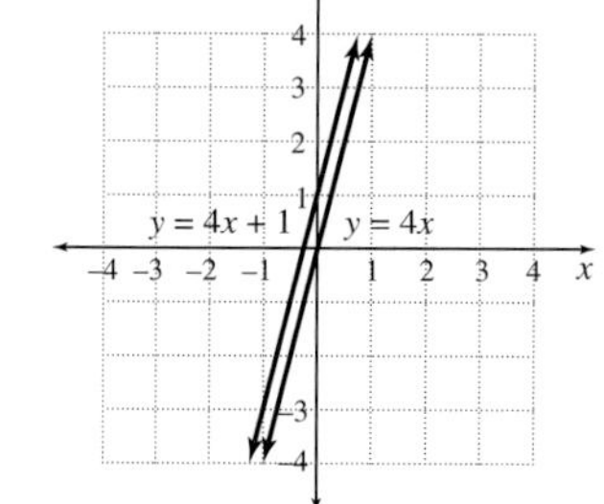

61.

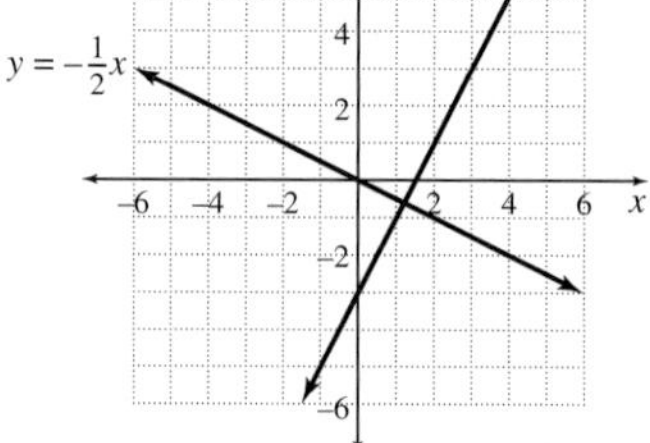
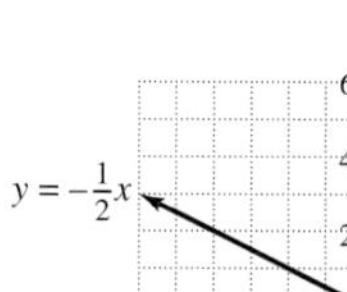

63.

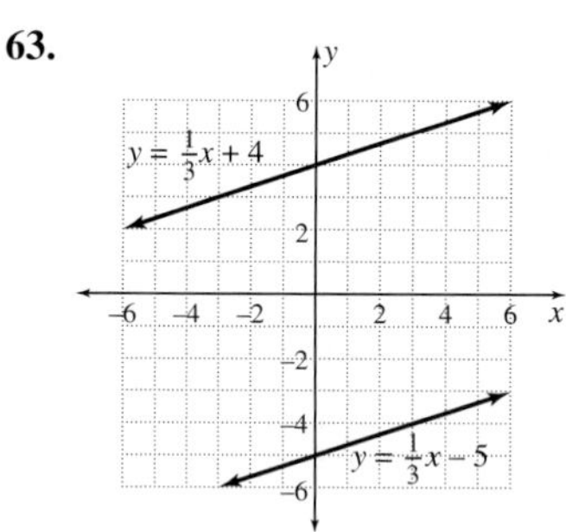

65.

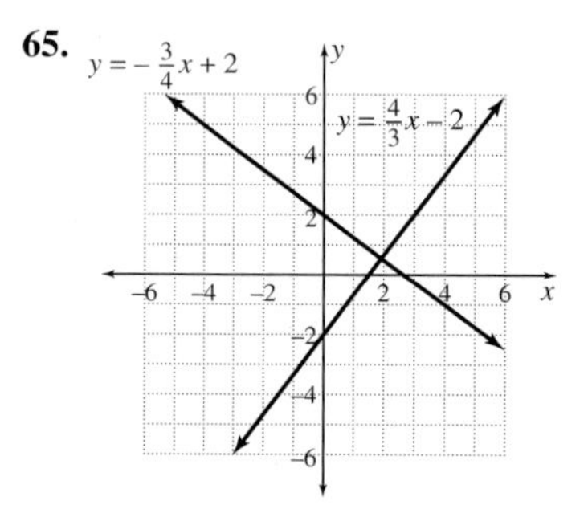

67. a. \$2 rental per hour **b.** \$50 prepaid amount
c. $y = -2x + 50$
69. a. \$126 rental per hour **b.** \$28 insurance fee
c. $y = 126x + 28$
71. per hour, percent of **73.** parallel line, $C = 0.79n + 45$
75. steeper line, $C = 3.20g$
77. parallel line, $C = 0.01x + 5$

EXERCISES 4.6

1. Multiply both sides by -1; reverse the inequality sign.
3. Multiply both sides by -2; reverse the inequality sign.
5. Subtract $2x$ from both sides; divide both sides by -3; reverse the inequality sign.

7. (4, 0), (1, 4) **9.** (−2, 3), (6, −1) **11.** (1, −1), (3, 0)
13. (0, −2)
15.

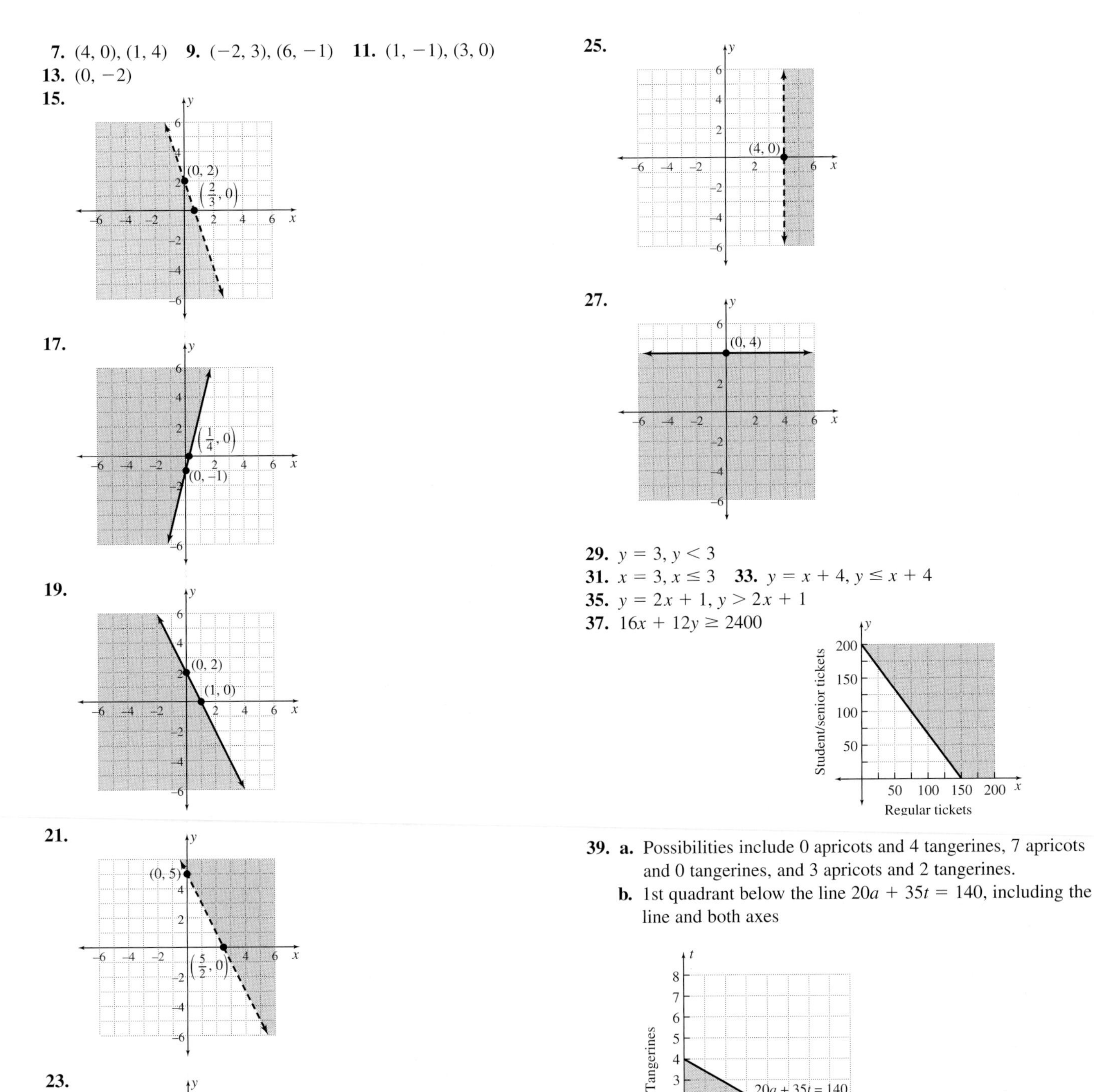

23.

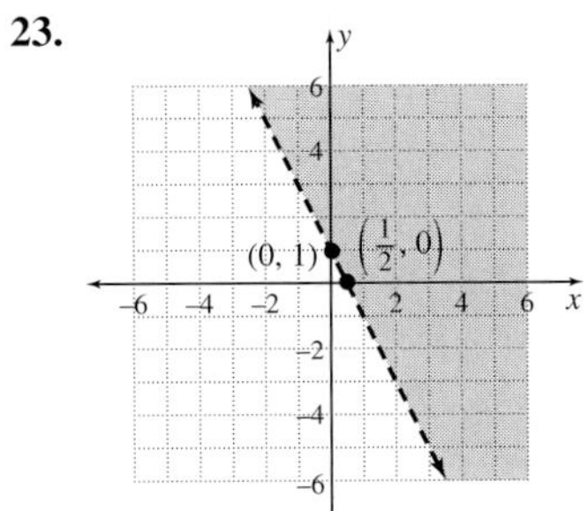

25.

27.

29. $y = 3, y < 3$
31. $x = 3, x \le 3$ **33.** $y = x + 4, y \le x + 4$
35. $y = 2x + 1, y > 2x + 1$
37. $16x + 12y \ge 2400$

39. a. Possibilities include 0 apricots and 4 tangerines, 7 apricots and 0 tangerines, and 3 apricots and 2 tangerines.
b. 1st quadrant below the line $20a + 35t = 140$, including the line and both axes

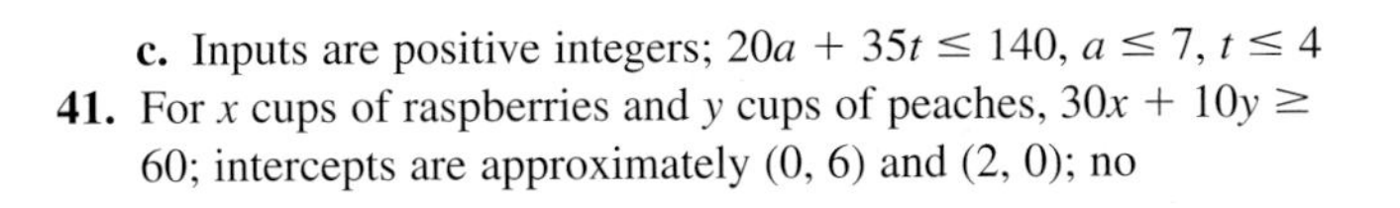

c. Inputs are positive integers; $20a + 35t \le 140$, $a \le 7$, $t \le 4$
41. For x cups of raspberries and y cups of peaches, $30x + 10y \ge 60$; intercepts are approximately (0, 6) and (2, 0); no

43. a. (10, 4); (20, 0) **b.** (15, 2); (0, 8) **c.** $0.10x$ **d.** $0.25y$
e. $0.10x + 0.25y \le 2$
f. $0 \le x \le 20$, $0 \le y \le 8$, x and y are integers

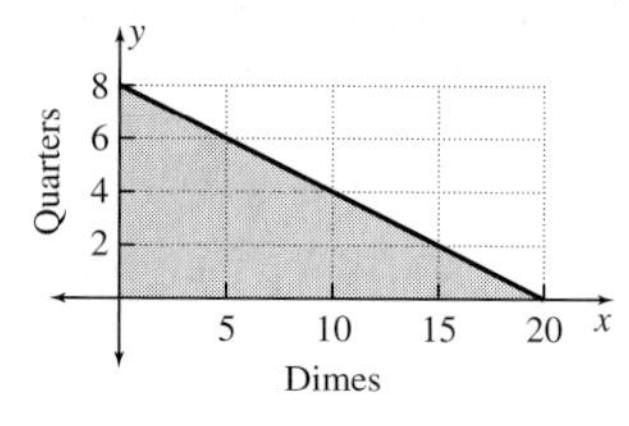

CHAPTER 4 REVIEW EXERCISES

1. $b = 10$ **3.** $F = 98.6$ **5.** area in terms of base and height
7. $h = \dfrac{W}{p}$ **9.** $T = \dfrac{AH}{I}$ **11.** $T = \dfrac{PV}{nR}$
13. $x = \dfrac{c - by}{a}$ **15.** $k = \dfrac{C}{5} + 93$ **17.** $m = \dfrac{y_2 - y_1}{x_2 - x_1}$
19. $V_2 = \dfrac{P_1T_2V_1}{P_2T_1}$ **21. a.** $2P = l + w$ **b.** $dr = t$ **23.** $b = 9$
25. $f(0) = 7, f(3) = 1, f(-5) = 17, f(a) = 7 - 2a$
27. $f(0) = 0, f(3) = 12, f(-5) = 20, f(a) = a^2 + a$
29. a. {2, 4, 6, 8} **b.** {25, 50, 75, 100} **c.** value of bits in cents
31. a. {1, 2, 3, 4, 5, 6} **b.** {2, 6, 10, 20, 40, 60}
c. "pennies," usually abbreviated d
33. function **35.** function **37.** not a function
39.

	Function	Sketch	Domain	Range
a.	$f(x) = x^2$	$y = x^2$	All real numbers $\mathbb{R}$	$y \ge 0$
b.	$f(x) = x$	$y = x$	All real numbers $\mathbb{R}$	All real numbers $\mathbb{R}$
c.	$f(x) = 2$	$y = 2$	All real numbers $\mathbb{R}$	$y = 2$

41. a. $\frac{3}{1}$ **b.** $-\frac{7}{10}$ **c.** $\frac{1}{5}$ **d.** $\frac{2}{15}$
43. and 45.

47. Line goes down from left to right.

49. Locate (x, y) and count a spaces up (or down) and b spaces to the right.
51. b **53.** a **55.** y-intercept **57.** $x = \frac{5}{2}$ **59.** $x = -\frac{4}{3}$
61. $x = 12$ **63.** $y = -\frac{3}{5}x + 3, m = -\frac{3}{5}, b = 3$
65. $y = \frac{5}{2}x - 5, m = \frac{5}{2}, b = -5$
67. $y = \frac{3}{4}x + 2, m = \frac{3}{4}, b = 2$ **69.** $y = -3, m = 0, b = -3$
71. undefined slope, no y-intercept **73.** $m = 1.5, b = 0.5$
75. $x = 4$ **77.** $x = -2$ **79.** $y = -2$ **81.** $y = 0$

83. $y = 2x + 4$ **85.** $y = 3x$ **87.** $y = \frac{1}{2}x + 2, 2y = x + 4$
89. $y = -3x + \frac{1}{4}, 4y = -12x + 1$
91. $m = -\frac{1}{2}, b = 1, y = -\frac{1}{2}x + 1$
93. linear, $m = -\frac{3}{2}$; $y = -\frac{3}{2}x + \frac{13}{2}$ **95.** nonlinear
97. a. (14, 0.95), (5, 0.50); x = minutes, y = cost in dollars
b. $m = 0.05$ **c.** cost per minute **d.** $y = 0.05x + 0.25$
99. a. (12, 12), (17, 32); x = hours, y = feet **b.** $m = 4$
c. feet per hour **d.** $y = 4x - 36$ **101.** d, $y + b = mx$
103. $y = 3x$

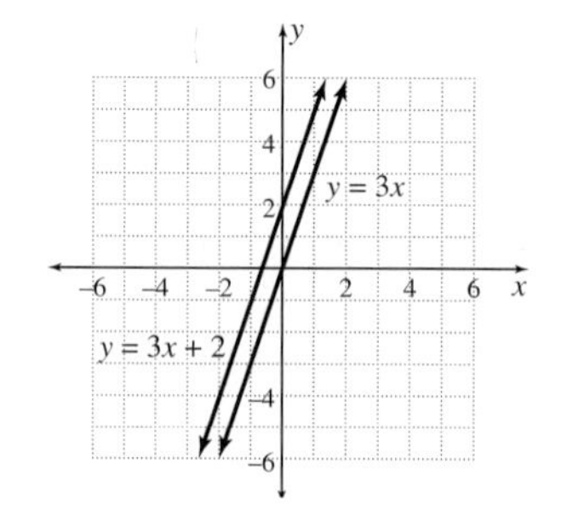

105. $y = \frac{3}{2}x - 2$

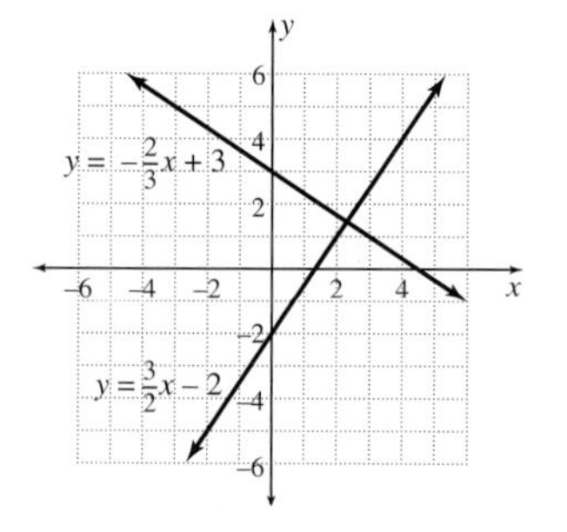

107. $C = 15w - 50$ **109.** $C = 20w - 150$
111. a. x, miles **b.** y, cost in dollars **c.** yes **d.** \$31.12
e. 41 mi **f.** \$19.99; basic cost, fee, or insurance
g. 0.89; cost/mi **h.** $y = 0.89x + 19.99$, y in dollars
i. -22.6; no meaning
113. c and d

115.

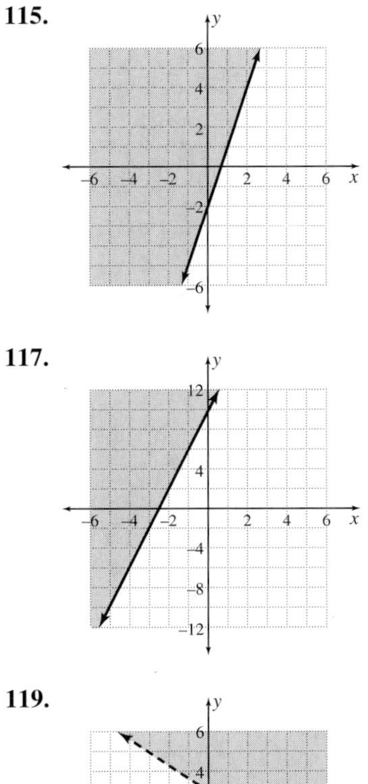

117.

119.

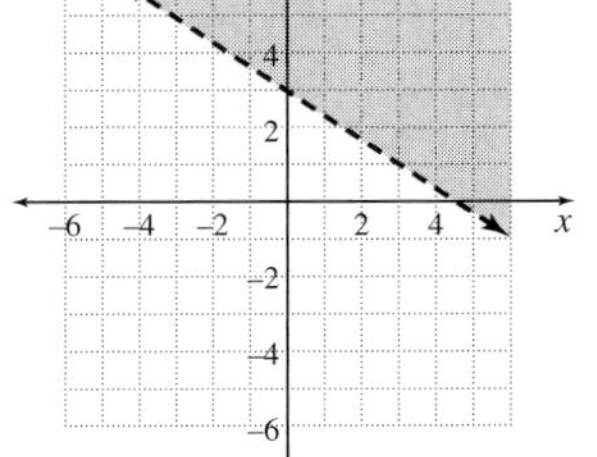

121. **a.** $y \geq 0$ **b.** $x \geq 0$ **c.** $x < 0$ **d.** $y < 0$
e. $x > 0$ **f.** $y \leq 0$

CHAPTER 4 TEST

1. $b = 9$ **2.** Possible answer: $G = \dfrac{T_1 + T_2 + H + F}{P}$

3. a. $d = \dfrac{C}{\pi}$ **b.** $h = \dfrac{2A}{b}$ **c.** $b = y - mx$ **d.** $V_2 = \dfrac{P_1V_1}{P_2}$

4. a. function **b.** not a function
c. function **d.** not a function

5. a. graph 1, $m = 0$; graph 2, $m = 10$; graph 3, $m = 50$; graph 4, undefined
b. graph 2

6. a. $-\frac{2}{3}$ **b.** 0 **c.** -7 **d.** undefined

7. a. $m = -2, b = 5$ **b.** $m = \frac{1}{3}, b = -\frac{1}{2}$

8. **9.**

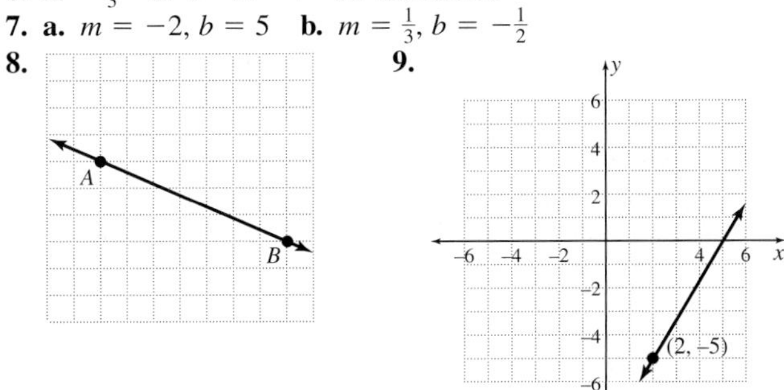

10. a. $f(x) = 0$ **b.** $f(0)$ **c.** $f(a)$ **11.** $y = 5x - 1$

12. $y = \frac{1}{3}x + 2$ or $3y = x + 6$

13. a. neither **b.** parallel **c.** perpendicular **d.** neither

14. Substitute slope and one ordered pair into $b = y - mx$. Solve for b. Write $y = mx + b$ using slope and b.

15. slope $= a$, $b = k$, $y = ax + k$ **16.** $y = k$

17.

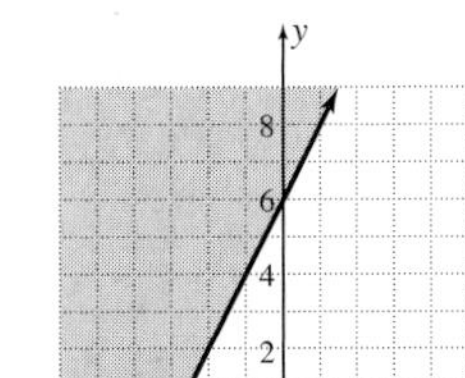

18.

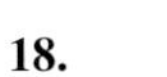

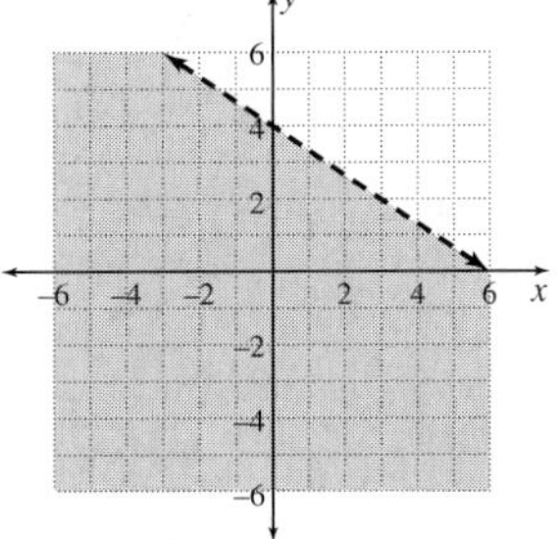

19. a. x, miles **b.** y, cost in dollars **c.** yes **d.** \$42.93
e. 45 miles **f.** 29.95, base price in dollars for rental
g. 0.59, cost per mile **h.** $y = 29.95 + 0.59x$
i. -50.8, no meaning

20. $f(x)$ is function notation; the $(x - 3)$ is multiplied by 2; the function is defined on the interval $(-2, 5]$.

CUMULATIVE REVIEW OF CHAPTERS 1 TO 4

1. $\frac{16}{21}, -\frac{2}{21}, \frac{1}{7}, \frac{7}{9}$ **3.** $-1.6, 11.2, -30.72, -0.75$

5. $4x^2 + 6x, 4x^2 - 6x, 24x^3, \frac{2}{3}x$ **7.** $\approx$694.4 days

9. $h = 14$ in. **11.** $x = -7$ **13.** $x = -3$ **15. a.** $-\frac{3}{2}$ **b.** 9

17.

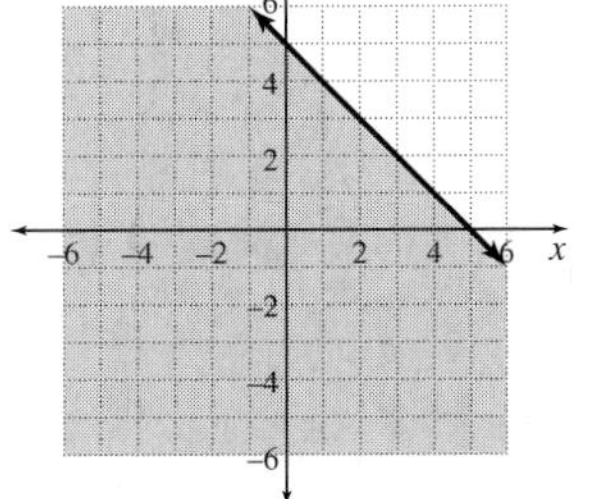

19. vertical-axis intercept **21.** $f(-2) = 4$

23. $m = -1, b = -4$; $y = -x - 4$

25.

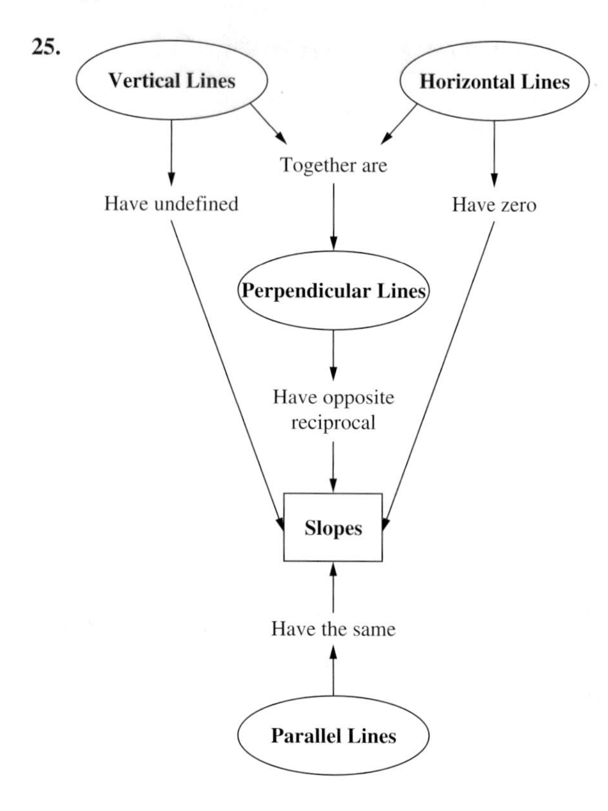

EXERCISES 5.1

1. a. $\frac{5}{3}$; 5 : 3 **b.** 3 : 2, 3 to 2 **c.** $\frac{4}{9}$, 4 to 9
3. a. 5, $\frac{3}{7}$ **b.** 16, $\frac{3}{1}$ **c.** 240, $\frac{22}{15}$ **5. a.** $\frac{1}{5}$ **b.** $\frac{3}{4}$ **c.** $\frac{9}{14}$ **d.** $\frac{29}{39}$
7. a. $\frac{3}{1}$ **b.** $\frac{2}{1}$ **c.** $-\frac{4}{1}$
9. a. $\frac{2.98}{1}$ or $\frac{149}{50}$ **b.** $\frac{-1.50}{1}$ or $\frac{-3}{2}$
11. a. 6 to 1 **b.** 3 to 1 **c.** 2 to 5 **d.** 4 to 1 **e.** 1 to 100 **f.** 4 to 25 **g.** 4 to 3
13. 3 to 1 **15.** 1 to 16 **17.** 15 cc to 1 hr
19. a. $3 : x$ **b.** 1 to $3x$ **c.** $1 : 3x^3$
21. a. $4x + 8 : 1$ **b.** $5x - 5 : 1$ **c.** $3x + 9 : 1$
25. b, c, and d **27. a.** 44 ft/sec **b.** $80\frac{2}{3}$ ft/sec
29. a. 60 mph **b.** 45 mph **31.** 352 ft; more than a city block
33. ≈56.8 mi **35.** 3 mi **37.** 20.8 microdrops/min **39.** 15 mL
41. $\frac{5}{8}$ **43.** $8\frac{1}{3}$% **45.** $12\frac{1}{2}$% **47.** ≈0.6%
49. 66.6%; 166.6%; −10%; 90%
51. 119.8%; 219.8%; 8.2%; 108.2% **53. a.** $18.75 **b.** $14.06
55. a. 18 in. **b.** $9
57. original value of which percent is taken is different number
59. a. $540 **b.** $375
61. 30%, 11%, 20%, 7%, 8%, 4%, 5%, 2%, 13%
63. a. 70% **b.** 64% **c.** 60% **d.** 65% **e.** 86%
65. 4 : 3 : 1; 50%, 37.5%, 12.5%; $500, $375, $125
67. a. 9 carbon : 8 hydrogen : 4 oxygen
b. 1 sodium : 1 hydrogen : 1 carbon : 3 oxygen
c. 1 calcium : 1 carbon : 3 oxygen
d. 12 carbon : 22 hydrogen : 11 oxygen
e. 8 carbon : 10 hydrogen : 4 nitrogen : 2 oxygen
69. $\frac{5}{12}$, $\frac{6.75}{10.75}$, $\frac{7.5}{10.5}$; divide and compare decimals

EXERCISES 5.2

1. 3 to 1, unsafe, slip **3.** 4.5 to 1, unsafe, tip **5.** 4 to 1, safe
7. proportion **9.** proportion **11.** false statement
15. $x = 11\frac{1}{4}$ **17.** $x = 3\frac{1}{3}$ **19.** $x = 2\frac{1}{7}$ **21.** $x = 9\frac{1}{3}$ **23.** 80
25. 66.7% **27.** 25.2 **29.** 140% **31.** 4.5 **33.** 25 **35.** 130%
37. 70% **39.** 1.65 m **41.** 24.15 km **43.** 27,878,400 ft^2
45. 26 ft **47.** 36 ft **49.** 12.57 ft **51.** $\frac{1}{6864}$ **53.** 130,000 units
55. 37.4 pound **57.** 1875 ft **59.** 950 ft **61.** ≈488 birds
63. a. 40,500 bats **b.** 607,500,000 mosquitoes
71. $x = 6$ **73.** $x = -\frac{8}{9}$ **75.** $x = -\frac{5}{13}$ **77.** $x = 9$
79. $x = 2$ **81.** $x = 5$ **83.** $x = 2$ **85.** $x = 26$ **87.** $x = -3$
89. $x = 8$

EXERCISES 5.3

1. 5 cm **3.** height = $1\frac{1}{4}$ in. **5.** height ≈ $\frac{3}{16}$ in.
7. $\frac{16}{25}$; similar triangles; sides AB and ND, AT and NE, BT and DE
9. $\frac{29}{34} \neq \frac{10}{29}$; not similar
11. $\frac{6}{12} = \frac{16}{32}$; similar figures; sides FG and HK, GO and KW, and others
13. $\frac{22}{16} = \frac{11}{8}$; similar figures; radii RT and OE, diameters EN and VD
15. $n = 31.5$ **17.** $n = 48$ **19.** 4.7 ft **21.** 20 ft **23.** 40.4 ft
25. $x = 6\frac{2}{3}$ **27.** $x = 9$ **29.** $x = 3$ **31.** $A(6, 0)$, $B(3, 2)$
33. $A(0, 2)$, $B(0, 4.8)$ **35.** 8 ft **37.** $AB = 10 - x$
39. $BC = x - 10$

MID-CHAPTER 5 TEST

1. $4x$ to $5yz$ **2.** $\dfrac{3x - 6}{1}$ **3.** 1 : 4 **4.** 4 to 1 **5.** 70 **6.** 144
7. 25% **8.** ≈12.5% **11.** $x = \frac{1}{100{,}000}$ **12.** $x = 26\frac{2}{3}$
13. $x = 15$ **14.** $x = 17$ **15.** $x = 2$
16. not in $\frac{a}{b} = \frac{c}{d}$ form; $x = 5.5$ **17.** $b = \frac{ad}{c}$ **18.** $x = 11.25$ **19.** $A(5, 0)$, $B(7, 5.6)$ **20.** 20%, 30%, 50%; 300, 450, 750
21. 260 min or 4 hrs 20 min **22.** 48 ft **23.** 400 **24.** 8 min
25. a. $295,000,000 **b.** $9.35/sec

EXERCISES 5.4

1. a. 85; 82.5; 80 **b.** 85; 82.5; 80
c. 52; 72.5; 0
d. 72.5; 85; 85
3. a. 477; 475; 475
b. 537; 525; 525 Apartment mean prices are lower in the smaller town by about $60.
5. add the elements; divide by the number of elements
7. no; for example, mean of 6 and 8 = 7 = mean of 5 and 9.
9. $A = \frac{h(a + b)}{2}$; area is height times average of lengths of parallel sides.
11. is not **13.** Both describe the "middle" of something.
15. Find most-often-occurring number.
17. can increase mean; no effect on mode or median
19. when we want to reduce the effect of large or small data
21. Some data are comparatively very large.
25. a. 18; 17; 17 **b.** It became more uniform.
c. 1991: 17 and 17.03; 2002: 17 and 16.84; 2006: 17 and 17.01
d. fewer younger drivers on the road in some states
27. GRCC, 31; LCC, 29; BCC, 30 **29.** 0.85 **31.** 0.70
33. ≈0.86 **35.** not possible **37. a.** (2.5, 3) **b.** (1, 0)
39. a. $\left(\frac{a}{2}, \frac{b}{2}\right)$ **b.** $\left(\frac{a}{2}, \frac{a}{2}\right)$ **c.** $\left(0, \frac{b}{2}\right)$
41. (0, 3), (4.5, 3), (4.5, 0); (3, 2)
43. (0, 3), (4, 3), (4, 0); $\left(2\frac{2}{3}, 2\right)$

EXERCISES 5.5

1. rate **3.** quantity **5.** rate **7.** rate **9.** rate

11. a.

	Quantity	Rate	$Q \cdot R$
Peanuts Cashews	5 kg 2 kg	\$8.80/kg \$24.20/kg	\$44.00 \$48.40
	7 kg	\$13.20/kg	\$92.40

b. Sum of quantity column is total kg of peanuts and cashews.
c. \$13.20/kg
d. Sum of $Q \cdot R$ column is total worth of peanuts and cashews.

13. a.

	Quantity	Rate	$Q \cdot R$
Grapes Potatoes Broccoli	3 lb 5 lb 2 lb	\$0.98 \$0.49 \$0.89	\$2.94 \$2.45 \$1.78
	10 lb	\$0.717	\$7.17

b. Sum of quantity column is total lb of vegetables and fruit purchased.
c. average price \$0.72
d. Sum of $Q \cdot R$ column is total purchase amount.

15. a.

	Quantity	Rate	$Q \cdot R$
Dimes Quarters	15 20	\$0.10 \$0.25	\$1.50 \$5.00
	35	\$0.1857	\$6.50

b. Sum of quantity column is total coins.
c. average coin rate \$0.19
d. Sum of $Q \cdot R$ column is total amount of money.

17. a.

Quantity	Rate	$Q \cdot R$
\$1500 \$1500	0.09 0.06	\$135 \$ 90
\$3000	0.075	\$225

b. Sum of quantity column is total invested.
c. average rate 7.5%
d. Sum of $Q \cdot R$ column is total return on investment.

19. a.

Quantity	Rate	$Q \cdot R$
100 lb 50 lb	0.12 0.15	12.0 7.5
150 lb	0.13	19.5

b. Sum of quantity column is total lb of dog food.
c. average protein content 13%
d. Sum of $Q \cdot R$ column is total lb of protein.

21. a.

Quantity	Rate	$Q \cdot R$
150 lb 25 lb	0.10 0	15 0
175 lb	≈0.086	15

b. Sum of quantity column is total weight.
c. average protein content 8.6%
d. Sum of $Q \cdot R$ column is total lb of protein.

23. a.

	Quantity	Rate	$Q \cdot R$
Ds Cs Bs	5 hr 4 hr 3 hr	1 2 3	5 8 9
	12 hr	1.83	22

b. Sum of quantity column is total time. **c.** average ≈1.83
d. Sum of $Q \cdot R$ column is total points.

25. a.

Quantity	Rate	$Q \cdot R$
3 hr 2 hr	80 kph 30 kph	240 km 60 km
5 hr	60 kph	300 km

b. Sum of quantity column is total time. **c.** average 60 kph
d. Sum of $Q \cdot R$ column is total distance.

27. a.

Quantity	Rate	$Q \cdot R$
150 mL 100 mL	18 3	2700 300
250 mL	12	3000

b. Sum of quantity column is total mL.
c. average molarity rate 12
d. Sum of $Q \cdot R$ column, divided by 1000, is total moles of sulfuric acid. (The division is needed because molarity is in terms of liters, not milliliters.)

29. a. earns \$1200, average rate is 8%
b. earns \$975, average rate is 6.5%

31.

	Quantity	Rate	$Q \cdot R$
Boeing Walgreen	200 x	\$30 \$46	\$6000 \$46x
	$200 + x$		\$19,800

$6000 + 46x = 19{,}800$
$x = 300$ shares

33. negative; no purchase possible; over budget already
35. ≈10.3 gal **37.** ≈27.6 gal **39.** 2.5; yes **41.** 3 hr
43. no real-number solution

45.

Quantity	Rate	$Q \cdot R$
30 x	\$5.80 \$7.20	\$174.00 \$7.20x
$30 + x$	\$6.36	2 expressions

$(30 + x)6.36 = 174 + 7.20x$
$x = 20$ hours

47. $(30 + x)7 = 174 + 7.20x$
$x = 180$; not reasonable; only 168 hr in week

49. In Example 4, money invested is multiplied by interest rate to get earnings; in Example 5, number of shares is multiplied by price per share to get total money invested.

CHAPTER 5 REVIEW EXERCISES

1. 4 to x^2 **3.** 10 : 3 **5.** ≈15% **7.** ≈24.85 mph
11. 56,000, 56,000, 8,000 **13.** 144 **15.** 450 cal; 50 g
17. 20 ft **19.** 5000 **21.** $x = 11\frac{1}{3}$ **23.** $x = \frac{1}{1{,}000{,}000}$
25. $x = 8$ **27.** $V_2 = C_1V_1/C_2$ **29.** $x = 7.7$, not in $\frac{a}{b} = \frac{c}{d}$ form
31. $A(-3, 4)$, $B(5, -1\frac{1}{3})$ **33.** 4.27; 4.3; 4.3 **35.** 6.87; 6.85; 6.8
37. a. (1.5, 2), (9.5, 2), (4, 0), (7, 4), (5.5, 2), (5.5, 2)
b. $\frac{4}{3}$, 0, 0, $\frac{4}{3}$ **c.** parallelogram
39. (−1.5, 2) **41.** 99.8
43. cents; miles per hour; interest in dollars; points; cost; cost; hours worked; moles

45.

Quantity	Rate	$Q \cdot R$
4 hr 7.5 hr	125 mph 200 mph	500 mi 1500 mi
11.5 hr	≈174 mph	2000 mi

47.

Quantity	Rate	$Q \cdot R$
6 gal 10 gal	88 octane 92 octane	528 gal·octane 920 gal·octane
16 gal	90.5 octane	1448 gal·octane

49. $22\frac{2}{3}$ gal **51.** school: $19,000; card: $6,000; 10.4%

CHAPTER 5 TEST

3. $\frac{b}{4a}$ **4.** $4a + 4b$ **5.** $\frac{5}{1}$ **6.** 84 **7.** $1277.50 per year
8. 1.4 ft **9.** $x = 32.5$ **10.** $x = 15$ **11.** 94%
12. a. $x = 16$ **b.** $x = 4.8$ **13.** $1.29; $1.59 **14.** $1.69; $1.59
16. clockwise from (3, 2): (5, 5), (7.5, 5), (5.5, 2) **17.** (6, 4)

18.

Quantity	Rate	$Q \cdot R$
$17,000 $ 2,000	0.058 0.149	$986 $298
$19,000 (total debt)	≈0.068 (average interest rate on total debt)	$1284 (total interest on debt)

19. 12 credits; 2.9% **20.** $x = 3.2$ **21.** $\frac{1}{9}$; $\frac{1}{9}$

CUMULATIVE REVIEW OF CHAPTERS 1 TO 5

1. $\frac{4}{7}$ **3. a.** $x = -3.5$ **b.** $x = 0.75$ **c.** multiply both sides by $\frac{3}{2}$ or multiply both sides by 3 then divide both sides by 2

5. a.

x	$f(x)$
−2	6
−1	3
0	0
1	−3
2	−6

b. same as **a.**

c.

x	$f(x)$
−2	$-\frac{1}{2}$
−1	−1
0	undefined
1	1
2	$\frac{1}{2}$

7. a. $0.02 \text{ g}\left(\frac{1000 \text{ mg}}{1\text{g}}\right)\left(\frac{1 \text{ tab}}{20 \text{ mg}}\right) = 1$ tablet

b. $200 \text{ mg}\left(\frac{5 \text{ mL}}{500 \text{ mg}}\right) = 2$ mL

c. $\left(\frac{10 \text{ mL}}{\text{vial}}\right)\left(\frac{1 \text{ dose}}{2 \text{ mL}}\right) = 5$ doses per vial

9. a. $2y = -3x + 6$ or $y = -\frac{3x}{2} + 3$

b. $3x + 2y$ = any number except 6; needs $-\frac{3}{2}$ slope

c. $y = \frac{2x}{3}$ + any number; needs $\frac{2}{3}$ slope

EXERCISES 6.1

1.

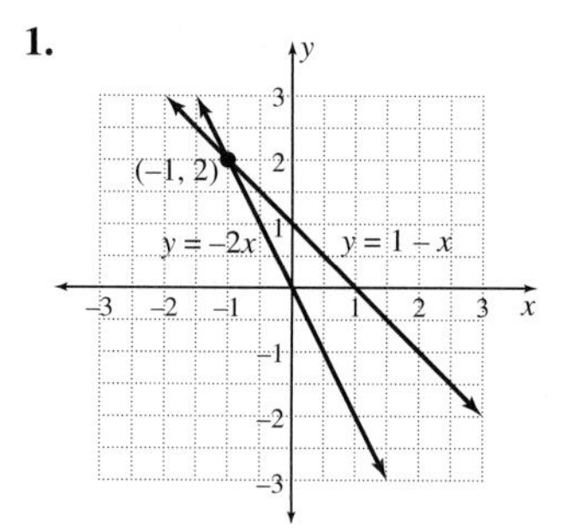

3.

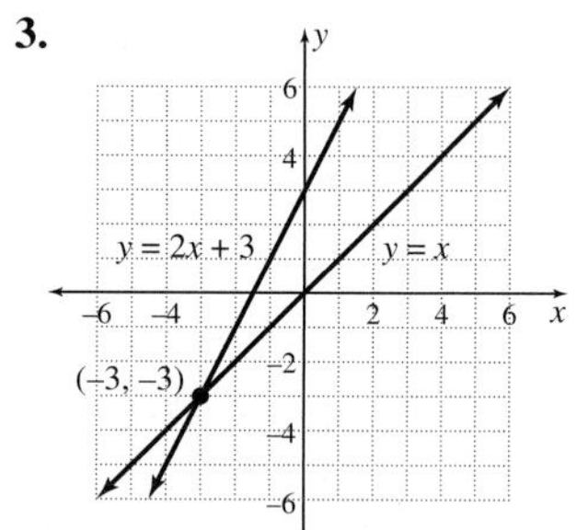

5.

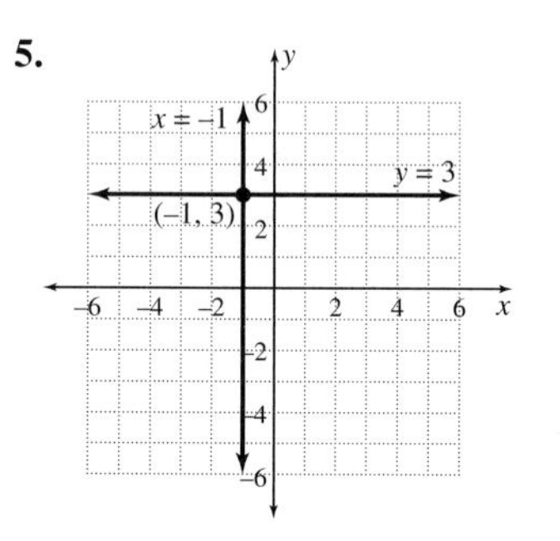

7.

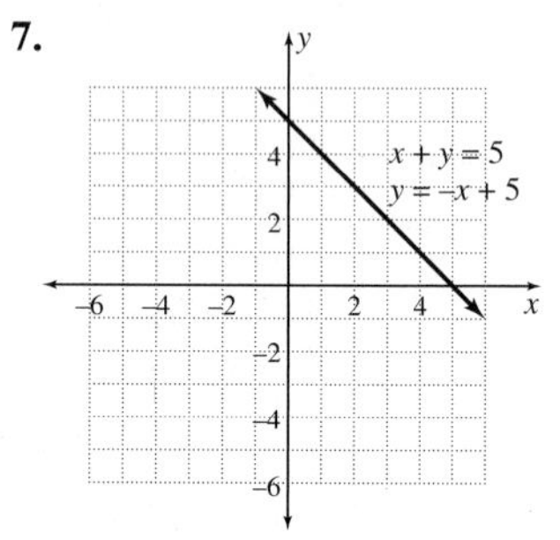

coincident lines; infinite number of solutions

9.

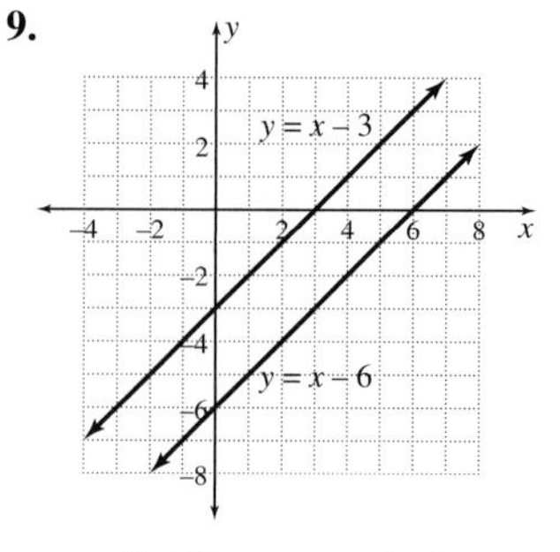

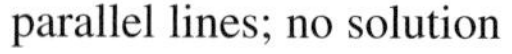
parallel lines; no solution

11.

y = x − 4
(2, −2)
y = −3x + 4

13.

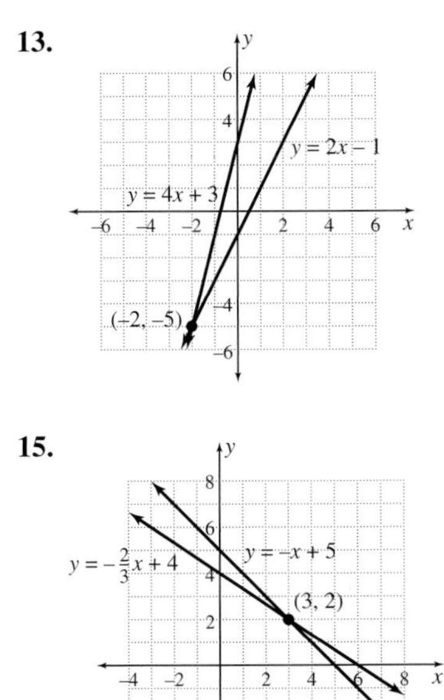

15.

17. parallel **19.** not parallel **21.** parallel **23.** not coincident
25. coincident **27.** not coincident
29. $y = -2x + 5$, $y = -x + 3$; intersection (2, 1)
31. $y = \frac{5x}{3} + 6$, $y = \frac{x}{3} + 2$, $(-3, 1)$
33. $y = \frac{3x}{2} + 2$, $y = \frac{-3x}{2} - 4$, $(-2, -1)$
35. **a.** $y = 0.12x + 50$ rewards large sales
b.

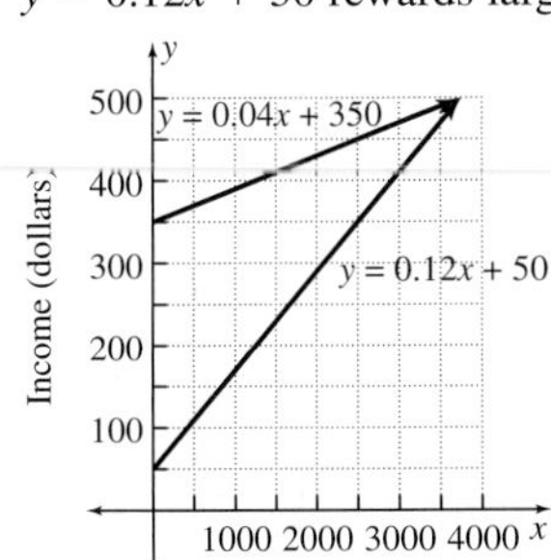

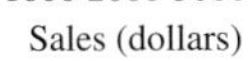

c. (3750, 500) **d.** sales less than \$3750
e. $y = 0.08x + 350$, $y = 0.24x + 50$
37. **a.** \$525 **b.** \$500 **c.** −\$25 **d.** cost
39. **a.** $C = 8.50x + 250$ **b.** $R = 10x + 200$
Other forms of the equations are possible.
c.

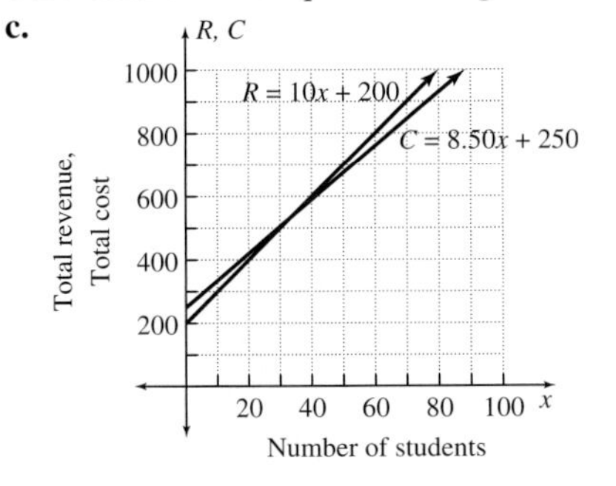

d. ≈(34, 540) (rounded to nearest whole person)

EXERCISES 6.2

For Exercises 1 through 13, guesses will vary.
1. x = long segment, y = short segment; $x + y = 10$, $x = y + 5$
3. x = first project hours, y = second project hours; $x + y = 176$, $y = x + 28$

5.

Item	Quantity	Rate (\$)	$Q \cdot R$ (\$)
Nickels	x	0.05	$0.05x$
Quarters	y	0.25	$0.25y$
Total	104		20.20

$x + y = 104$, $0.05x + 0.25y = 20.20$

7.

Item	Quantity	Rate (\$)	$Q \cdot R$ (\$)
Dimes	x	0.10	$0.10x$
Quarters	y	0.25	$0.25y$
Total	140		31.10

$x + y = 140$, $0.10x + 0.25y = 31.10$

9.

Item	Quantity	Rate (\$)	$Q \cdot R$ (\$)
Nickels	x	0.05	$0.05x$
Dimes	y	0.10	$0.10y$
Total	159		12.10

$x + y = 159$, $0.05x + 0.10y = 12.10$

11.

Job	Hours	Wage (\$)	Earnings (\$)
A	x	10.75	$10.75x$
B	y	8.50	$8.50y$
Total	43		406

$x + y = 43$, $10.75x + 8.50y = 406$

13. Not a quantity-rate setting; use guess and check as needed. $x + y = 5250$, $x = 9y$
15. \$9.50, \$6.50 **17.** kid, \$5; adult, \$2.50
19. \$300, \$40
21. 5 mph; 2 mph; −7 mph; 11 mph; 13 mph
23. 265 mph, 15 mph **25.** 12.5 mph, 2.5 mph

EXERCISES 6.3

1. $W = \frac{L}{2}$ **3.** $b = c - a$ **5.** $r = \frac{C}{2\pi}$ **7.** $y = x - 5$
9. $d = \frac{C}{\pi}$ **11.** $x = 3$ **13.** $x = 3$ **15.** $y = 4 - 3x$
17. $x = 4y + 5$ **19.** $x = 5y - 9$ **21.** $y = 3x + 2$
23. $x = 3, y = -5$ **25.** $x = 2, y = 15$ **27.** $x = 4, y = -1$
29. $x = 5, y = 9$ **31.** $x = 3, y = -2$ **33.** $x = 2, y = \frac{8}{3}$
35. no solution **37.** no solution
39. infinite number of solutions **41.** $x = -1, y = -1$
43. $x = 0, y = -2$ **45.** $x = 6, y = -2$
47. infinite number of solutions **49.** $x = 23, y = 27$
51. $x = 17.5, y = 82.5$

For Exercises 53 to 71, the answers are listed in the order in which items appear in the exercises.

53. 11.5 yd, 8.5 yd **55.** 8 in., 12 in. **57.** 19 cm, 10 cm
59. 51, 21 **61.** 51, 33
63. x = no. of 5-1b bags, y = no. of 10-lb bags; $y = 4x$, $5x + 10y = 6300$; $x = 140$ bags (5-lb), $y = 560$ bags (10-lb)
65. x is pounds of truffles, y is pounds of creams; $x + y = 60$, $36x + 20y = 60(24)$; $x = 15$ pounds, $y = 45$ pounds
67. ≈ 207 g, ≈ 63 g
69. a. $A = 102°$, $B = 78°$ **b.** $C = 77°$, $D = 103°$ **c.** $E = 44°$, $F = 46°$ **d.** $G = 29°$, $H = 61°$ **e.** $I = 75°$, $J = 105°$

MID-CHAPTER 6 TEST

1. $y = 2x - 5000$ **2.** $h = \dfrac{3V}{\pi r^2}$ **3.** $y = \frac{3}{8} - \frac{9}{4}x$
4. $x = 1500$, $y = 3500$ **5.** $x = 2500$, $y = 2500$
6. a. $y = 6 - \frac{2}{3}x$, $y = x + 1$ **b.** (4.5, 5.5) **c.** (4.5, 5.5) **d.** (12, −2)
7. a.

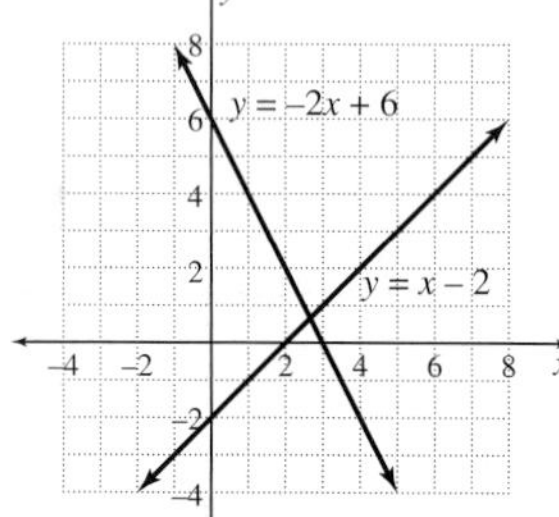

b. (2.5, 1) **c.** $\approx(2.7, 0.7)$ or $\left(\frac{8}{3}, \frac{2}{3}\right)$
8. A true statement is obtained. There are an infinite number of solutions. The lines are coincident.
9. 180, 15 **10.** 1200, 400 **11.** 4800 lb of 7%, 1200 lb of 12%
12. The ordered pair for the point of intersection is where the same input gives the same output in each equation.
13. Parallel
14. a. Infinite number of solutions
b. No solution
c. One solution
d. One solution, at $x = 0$
e. No solution
15. The point of intersection may be in a quadrant that is irrelevant to the application. For example, with number of people on the x-axis, an intersection of the lines in the second quadrant suggests a negative number of people.

EXERCISES 6.4

1. a.

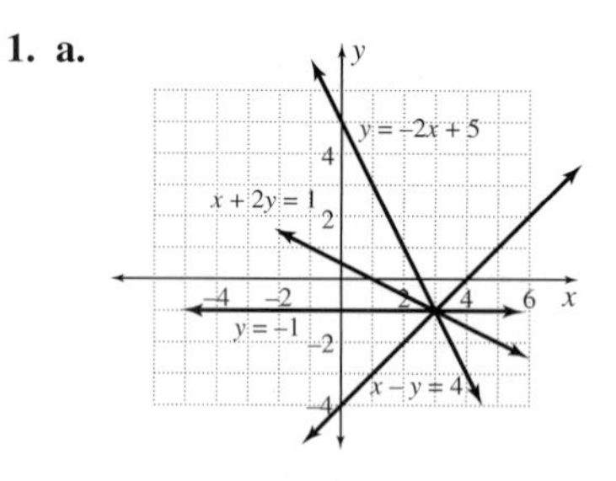

b. $2x + y = 5$ **c.** $y = -1$
d. All graphs intersect at the same point, (3, −1).
3. $x = 3$, $y = -5$ **5.** $m = 7$, $n = -4$ **7.** $x = -2$, $y = 3$
9. $a = -10$, $b = 15$ **11.** $x = -3$, $y = 3$ **13.** $p = 0$, $q = 3$
15. no solution **17.** infinite number of solutions
19. $x = 9$, $y = -4$ **21.** $x = 3$, $y = -2$
23. $b = -3$, $m = 2$ **25.** $x = 5$, $y = 2$ **27.** $x = 2$, $y = 4$
29. $x = 1.5$, $y = -2.5$ **31.** $x = 130$, $y = 113$
33. a. 70°, 20° **b.** 115°, 65°
35. a. $A = 34°$, $B = 56°$ **b.** $C = 35°$, $D = 145°$ **c.** $E = 74°$, $F = 16°$ **d.** $G = 115°$, $H = 65°$ **e.** $I = 40°$, $J = 140°$

For Exercises 37 to 59, answers are listed in the order in which the items are listed in the exercise.

37. 16.5, 8.5 **39.** 4.75, 15.25
41. \$8.00, \$3.50 **43.** \$25.99, \$15.99 **45.** \$8.99, \$39.99
47. ≈ 67.7 cal, ≈ 12.6 cal **49.** 4 cal, 3 cal **51.** 20 cal, 25 cal
53. 2 hr, 5 hr **55.** 247.5 mph, 27.5 mph **57.** 400 mph, 75 mph
59. 7 mph, 3 mph

EXERCISES 6.5

1. (0.5, −0.5, 1.5) **3.** $\left(-1\frac{1}{2}, 3\frac{1}{2}, 5\frac{1}{2}\right)$
5. infinite number of solutions **7.** no solution
9. (2.5, 2.5, 2.5) **13. a.** 1 **b.** 2 **c.** 3
15. \$350, \$175, \$850; $m + p + s = 1375$, $p = m + 500$, $m = 2s$
17. 960, 480, 96; $x + 2y + 5z = 2400$, $x = 2y$, $x = 10z$
19. $A = 108$, $B = 54$, $C = 18$ **21.** $A = 64°$, $B = 32°$, $C = 84°$
23. $A = 45°$, $B = 45°$, $C = 90°$
25. length = 19 in., width = 12 in., height = 3 in.

EXERCISES 6.6

1. a, b **3.** c
5.

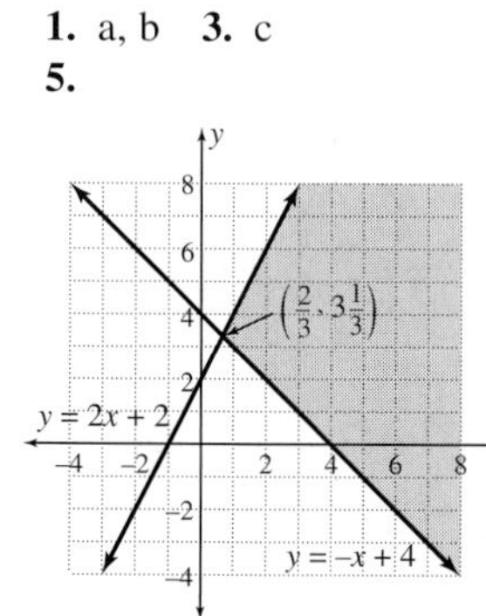

7.

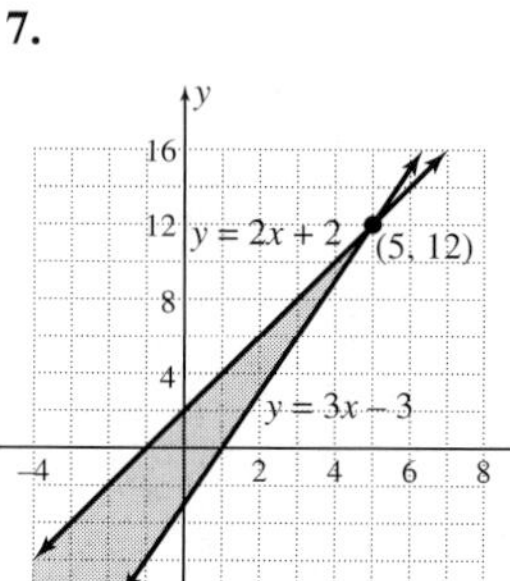
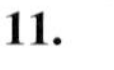

9.

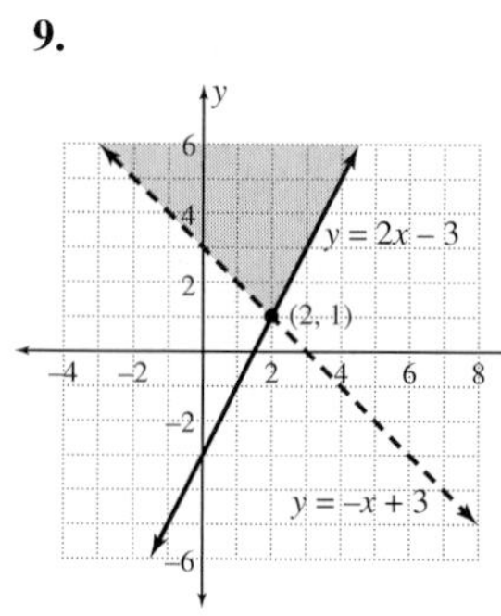

11.

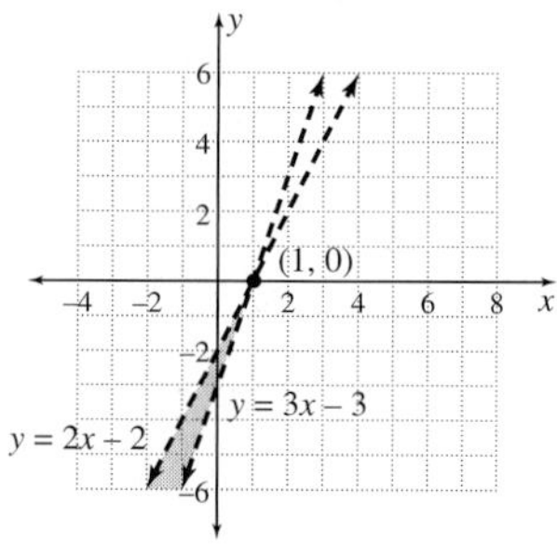

13.

15.

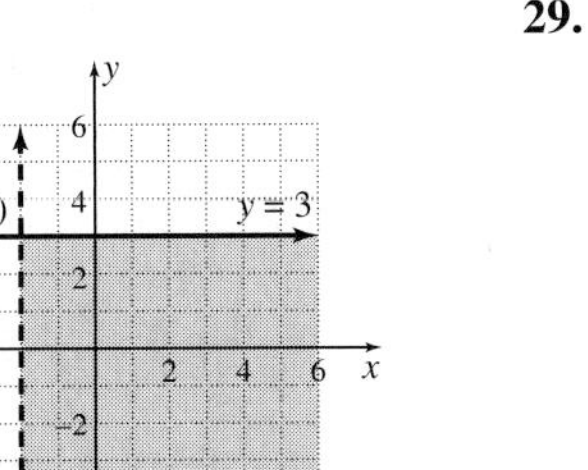

17.

19.

21.

23.

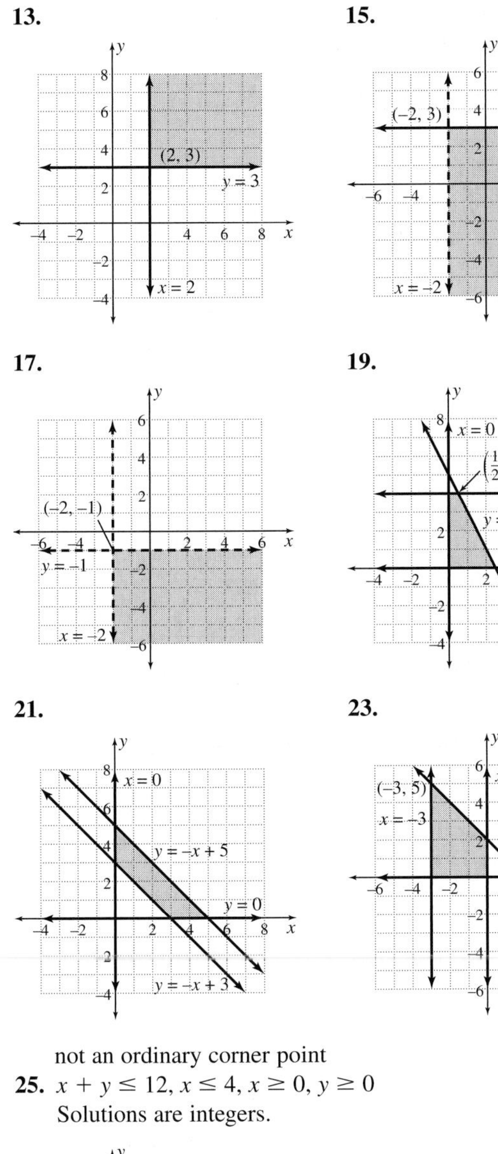

not an ordinary corner point

25. $x + y \leq 12, x \leq 4, x \geq 0, y \geq 0$
Solutions are integers.

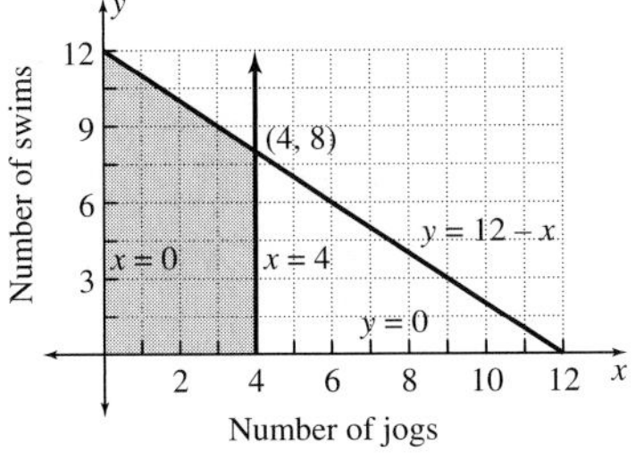

27. $20x + 12y \leq 360, y \leq 12, x \geq 0, y \geq 0$
Solutions are integers.

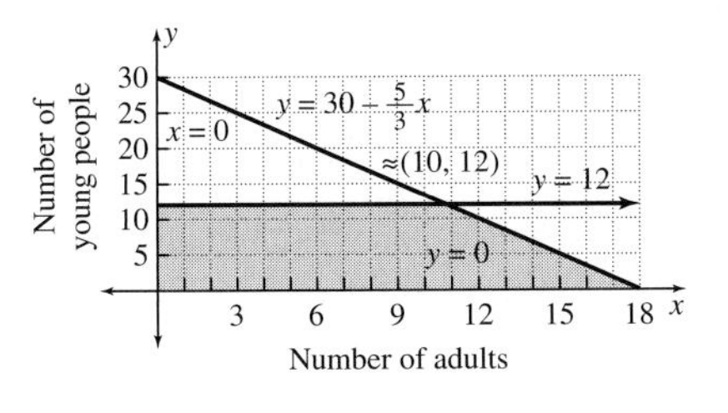

29. $x + y \leq 45{,}000, y \leq 5000, x \geq 0, y \geq 0$
Solutions are integers.

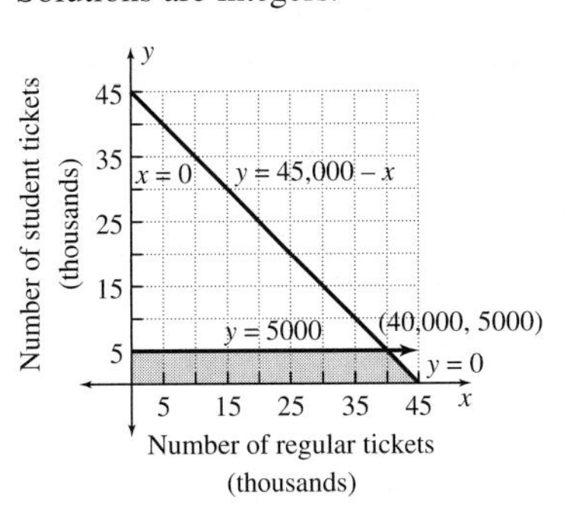

CHAPTER 6 REVIEW EXERCISES

1. $y = 2x + 5000$ 3. $r = \dfrac{C}{2\pi}$ 5. $b = \dfrac{8a}{5}$

7.

9.

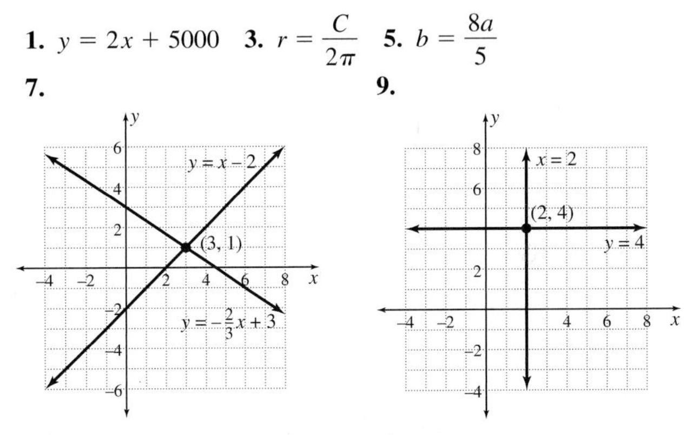

11. $x = -\frac{1}{2}, y = \frac{3}{2}$ 13. $x = 4500, y = 500$
15. $m = -\frac{1}{5}, b = \frac{13}{5}$ 17. $x = 1, y = -\frac{5}{3}$
19. $x = -3, y = -1$ 21. $x = -3, y = 18$ 23. no solution
25. The lines are coincident. 27. centipede, 354; millipede, 710
29. muffin, 140 cal; egg, 95 cal 31. 12 cows, 8 ducks
33. 15 mph, 5 mph 35. 7%, 18% 37. $63\frac{1}{3}, 26\frac{2}{3}$
39. 20, 40, 120 41. 2.5, 3.5, 4.5
43. infinite number of solutions 45. 11.5 in., 11.5 in., 9 in.
47. 16 g, 10 g, 11 g 49. 39 g, 71 g, 27 g

53.

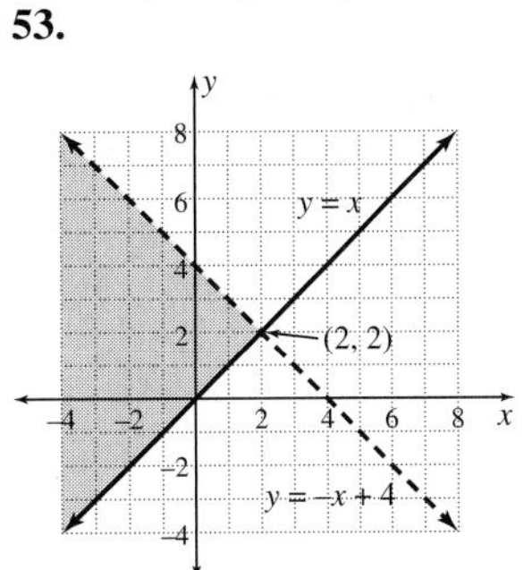

55.

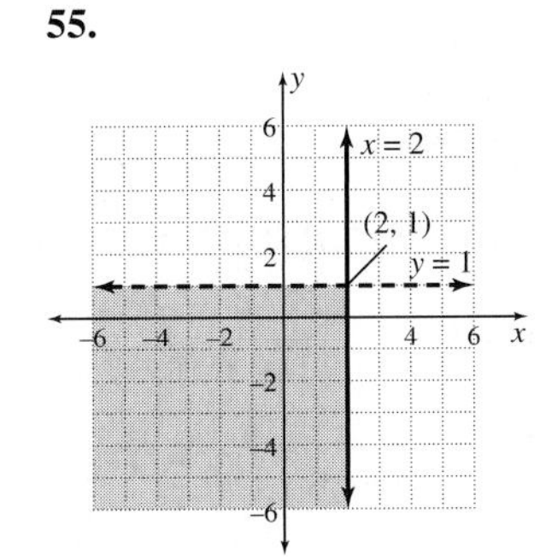

57.

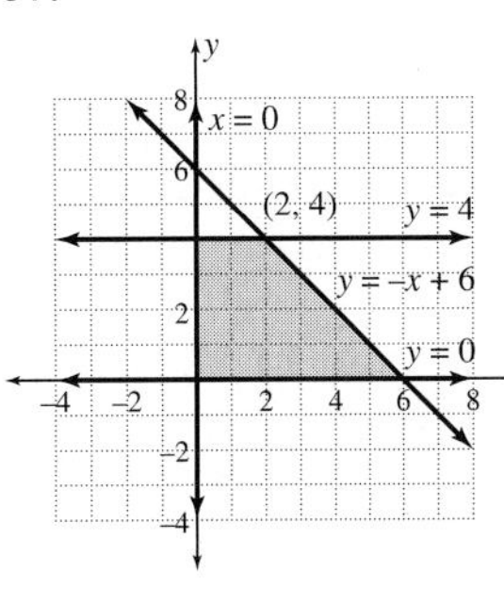

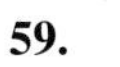

59.

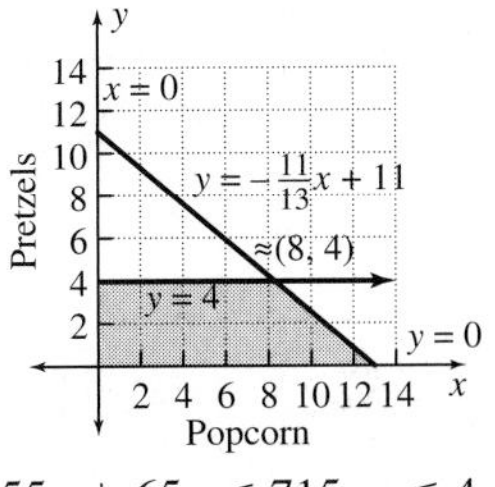

$55x + 65y \leq 715, y \leq 4, x \geq 0, y \geq 0$

61. a. no solution
b. one solution (on the y-axis)
c. infinite number of solutions
d. At most one solution; if one company has a higher basic charge and a lower cost per mile, the graphs will intersect in the first quadrant. If one company has a higher basic charge and a higher cost per mile, the graphs will intersect in the second quadrant (negative miles), resulting in no reasonable solution. Graph sketch will vary.

CHAPTER 6 TEST

1. $y = 3x - 400$ **2.** $y = \frac{1}{2}x - \frac{3}{2}$ **3.** $a = -4, b = 5$
4. $m = 0.75, b = -1.75$ **5.** infinite number of solutions
6. no solution **7.** $x = 13, y = 9$ **8.** $x = 1500, y = 600$
9. The graphs are coincident lines. **10.** The lines are parallel.
11. $y = 4 - \frac{1}{3}x$ and $y = 8 - x$; the ordered pair makes both equations true.
12. 

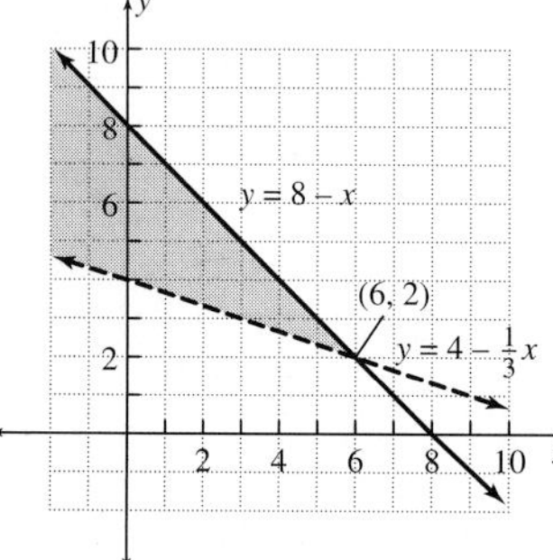

13. (7, 1) is a solution; the ordered pair makes both inequalities true.
14. 16, 6 **15.** 4.5 cal, 12.6 cal
16. dolphins, 35 mph; current, 10 mph **17.** 76.5°, 13.5°
18.

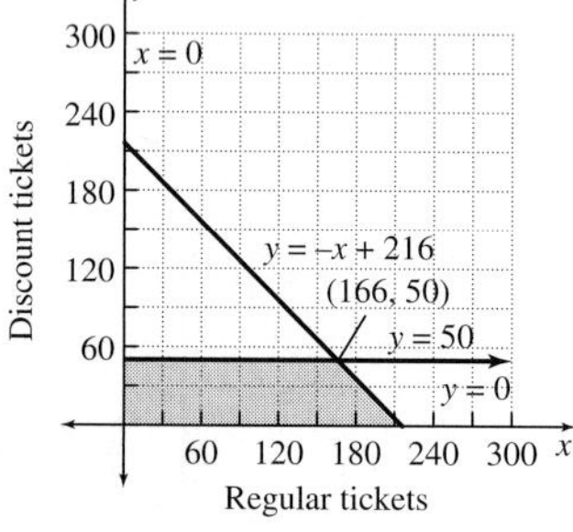

$x + y \leq 216, y \leq 50, x \geq 0, y \geq 0$
19. 60 lb, 240 lb
20. Use elimination if two equations have opposite coefficients on one variable. Use substitution if a variable has 1 as a coefficient.
21. $(-3, 5, 2)$
22. One solution: there is a number of persons for which the price at two locations is the same.
No solutions: there are two locations where the cost per person is the same, but one has a higher fixed cost (such as a room cost).
Infinite number of solutions: there are two locations where the costs are identical for all numbers of persons.

CUMULATIVE REVIEW OF CHAPTERS 1 TO 6

1. 55 **3.** x^8 **5.** $A = 36$ sq. in. **7.** $x - 6 = 2x + 8, x = -14$
9. $y = \frac{5}{3}x - 8$ **11.** $-7, 13$ **13.** $y = \dfrac{x}{2}$
15. a. 265 **b.** 40% **17.** 180 lb, 120 lb
19. a. $A = \$6500, B = \$8500$ **b.** $A = \$2000, B = \$13{,}000$
21. Males arrive earlier than females. Change the dates to days from January 1 and average the dates. 1932 is a leap year. On average, male birds arrived before females (1931, 83.9 days into year compared to 89.9 days; 1932, 68.3 days compared to 78.7 days), and both sexes of birds arrived earlier in 1932 than in 1931.

EXERCISES 7.1

1. a. -5 **b.** -5 **c.** -13 **d.** -10 **e.** -4 **f.** 7
3. a. -56 **b.** -42 **c.** 36
5. a. $-x^3 - 3x^2 + 5x + 5$ **b.** $3x^3 + 2x^2 + 4x - 3$
7. a. $a + 12b - 8c$; trinomial **b.** $3m - n$; binomial
c. $21y$; monomial **d.** $x^2 - x - 12$; trinomial
e. $-x^2 + 2x - 3$; trinomial
9. a. $x^2 + 5x + 6$; trinomial **b.** $3x^2 + 7x + 2$; trinomial
c. $-4x + 17$; binomial **d.** $6x^2 + 5x + 1$; trinomial
e. $a^2 - b^2$; binomial
11. length $= 2b$, width $= a + b$, $P = 2a + 6b$, $A = 2ab + 2b^2$
13. length $= x + 1$, width $= x + 2$, $P = 4x + 6$, $A = x^2 + 3x + 2$
15. a. $2a^2 + 4ab$ **b.** $4ab + 2b^2$ **c.** $x^2 + 3x$ **d.** $2x^2 + x$
17. $4x^4y^4 - 12x^3y^5 - 16x^2y^6$ **19.** $-5x^5y^4 + 10x^4y^5 + 5x^3y^6$
21. a. $2x^3 + 6x^2$ **b.** $x^3 - x^2$ **c.** $x^4 + 2x^3 + x^2$
d. $ab^3 - ab$ **e.** $ab^2 - b^3$ **f.** $a^2 + a^2b - a^2b^2$
23. a. $3x^2 + 2x$ **b.** $-2a^2 + 10a$
25. a. $2b^2 - 6b$ **b.** $-x^2 + 5x$
27. a. $x^2 - 4xy + 4y^2$ **b.** $x^2 - 4y^2$ **c.** $x^3 + 3x^2 + 3x + 1$
d. $x^3 + 2x^2 - 4x - 8$ **e.** $a^3 + b^3$
29. $y^2 + 2xy + x^2$ **31.** $a - 2b + 3b^2$ **33.** $2a$; $1 - 2b$
35. ab^2; $1 - a - ab$ **37.** gcf $= x$; $x(x^2 + 4x + 4)$
39. gcf $= b$; $b(a^2 + ab + b^2)$ **41.** gcf $= 2x$; $2x(3x + 1)$
43. gcf $= 3y$; $3y(5y - 1)$ **45.** gcf $= 5xy$; $5xy(3x + 2y)$
47. $4(-x - 3), -4(x + 3)$ **49.** $2y(-x + 2y), -2y(x - 2y)$
51. $4(-3x^2 - 2x - 2), -4(3x^2 + 2x + 2)$
53. $y^2(-1 + 4y - 8y^2), -y^2(1 - 4y + 8y^2)$ **55.** 2 **57.** 1
59. 3 **61.** 3 **63.** 4 **65.** 4 **75. a.** $3 \cdot 37$ **b.** $7 \cdot 13$
77. a. gcf $= 9$; $\frac{4}{11}$ **b.** gcf $= 66$; $\frac{1}{15}$ **c.** gcf $= 37$; $\frac{5}{27}$
d. gcf $= m$; $\dfrac{n}{p}$ **e.** gcf $= 4n$; $\dfrac{p}{6m}$
79. a. $5a$ **b.** $4x$ **c.** $5x - 3$ **d.** $3 + \pi$
81. a. side $= 3c$, $A = 15ac$
b. side $= 5y$, $A = 25y^2$
c. side $= 2.5y$, $A = 3.75xy$

EXERCISES 7.2

1. $(x + 1)(x + 1) = x^2 + 2x + 1$
3. $(a + 2b)(2a + b) = 2a^2 + 5ab + 2b^2$
5. $(a + 2b)(a + 2b) = a^2 + 4ab + 4b^2$ **7.** $2x^2 - 3x - 20$
9. $2x^2 - x - 6$ **11.** $2x^2 - 5x + 2$ **13.** $5x^2 - 21x + 4$
15. a. $x^2 - 4x + 4$ **b.** $x^2 - 4$
17. a. $a^2 + 10a + 25$ **b.** $b^2 - 25$
19. a. $a^2 - b^2$ **b.** $a^2 - 2ab + b^2$
21. a. $x^2 + 8x + 7$ **b.** $x^2 - 6x - 7$

23. **a.** $b^2 + 14b + 49$ **b.** $a^2 - 49$
25. **a.** $x^2 + 2xy + y^2$ **b.** $x^2 - 2xy + y^2$
27. 15a, 16a, 17a, 18b, 19b, 20b, 23a, 24b, 25a, 25b
29. pst; $4x^2 + 12x + 9$ **31.** ds; $4x^2 - 9$
33. neither; $-4x^2 + 12x - 9$ **35.** pst; $9x^2 - 12x + 4$
37. ds; $9x^2 - 4$ **39.** pst; $x^2 + 10x + 25$ **41.** pst; $a^2 - 12a + 36$
43. 11 **45.** 1 **47.** 4
49. $(x + 1)(x + 12)$; $(x - 1)(x - 12)$; $(x + 2)(x + 6)$; $(x - 2)(x - 6)$; $(x + 3)(x + 4)$; $(x - 3)(x - 4)$
51. $(x + 1)(x + 20)$; $(x - 1)(x - 20)$; $(x + 2)(x + 10)$; $(x - 2)(x - 10)$; $(x + 4)(x + 5)$; $(x - 4)(x - 5)$
53. 25, 14, 11, 10 **55.** **a.** $x^2 + 9x + 8$ **b.** $x^2 - 7x - 8$
57. **a.** $x^2 + 6x + 8$ **b.** $x^2 - 2x - 8$
59. **a.** $6x^2 - 13x + 6$ **b.** $6x^2 + 5x - 6$
61. **a.** $6x^2 + 37x + 6$ **b.** $6x^2 + 35x - 6$
63. **a.** $2x^2 + 7x + 5$ **b.** $2x^2 + 3x - 5$
65. **a.** $2x^2 + 9x - 5$ **b.** $2x^2 + 11x + 5$
67. $2x^2 + 12x + 18$ **69.** $3x^2 - 30x + 75$
71. $4x^2 - 40x + 100$ **73.** $20x^2 - 60x + 45$
81. The exponent was applied incorrectly: $(x - a)^2 = (x - a)(x - a) = x^2 - 2ax + a^2$

EXERCISES 7.3

1.

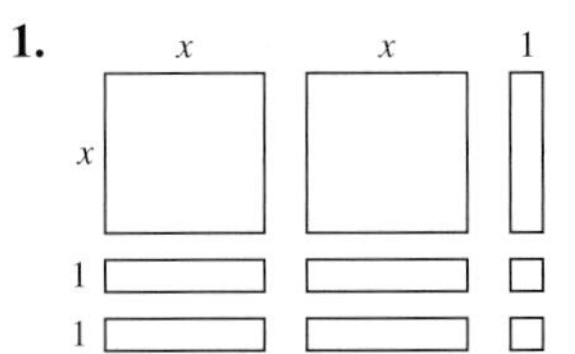

$(2x + 1)(x + 2) = 2x^2 + 5x + 2$

3.

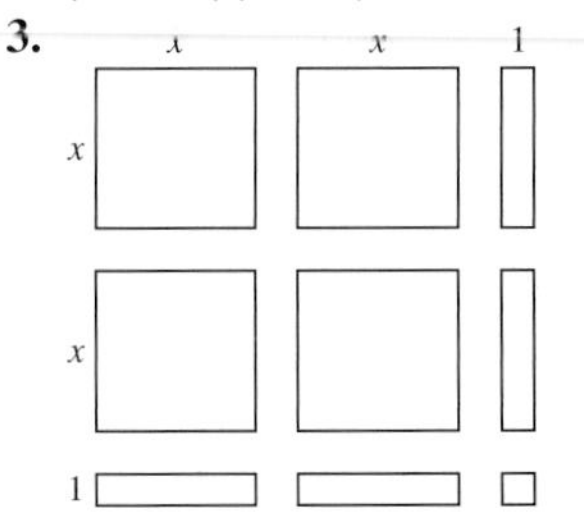

$(2x + 1)(2x + 1) = 4x^2 + 4x + 1$

5. $(x + 5)(x + 4) = x^2 + 9x + 20$
7. $(x + 2)(x - 10) = x^2 - 8x - 20$
9. $(6x + 1)(x - 3) = 6x^2 - 17x - 3$
11. The product gives $x^2 + 3x - 5x - 15$, and $3x - 5x = -2x$.
13. $(x + 2)(x + 6)$ **15.** $(x - 12)(x - 1)$ **17.** $(x + 6)(x - 2)$
19. $(a + 3)(a + 4)$ **21.** $(x + 4)(x - 3)$ **23.** $(x - 12)(x + 1)$
25. $(x + 3)^2$ **27.** $(x + 6)(x + 5)$ **29.** $(x + 15)(x - 2)$
31. $(x + 2)(x - 8)$ **33.** $(x + 16)(x - 1)$ **35.** $(x + 5)(x - 5)$
37. $(2x + 3)(x + 4)$ **39.** $(2x + 3)(x - 3)$ **41.** $(2n + 3)(n - 1)$
43. $(3x - 1)(x + 2)$ **45.** $(3a + 1)(a - 4)$ **47.** $(3x + 7)(3x - 7)$
49. $(4x + 3)(4x - 3)$ **51.** $(3x + 2)(2x - 1)$
53. $(3x - 2)(2x + 3)$ **55.** $(2n - 1)(n + 5)$

MID-CHAPTER 7 TEST

1. **a.** $-3a + 11b - 13c$; trinomial **b.** $2x^2 - 4x - 6$; trinomial **c.** $2x^3y^2 - 3x^2y + 2xy^2$; trinomial **d.** $2a - 7b + 4c$; trinomial **e.** $x^3 - 125$; binomial **f.** $-3x^2 - 9x + 9$; trinomial **g.** $4x - 8$; binomial
2. **a.** gcf = 2; $2(3x^2 - x + 4)$ **b.** gcf = a; $a(2bc - 3c + 4b)$
3. Terms are added (or subtracted), as in $a + b$; factors are multiplied, as in $a \cdot b$.
4. **a.** length = $2a + b$, width = $a + 2b$, $P = 6a + 6b$, $A = 2a^2 + 5ab + 2b^2$ **b.** length = $x + 2$, width = $x + 1$, $P = 4x + 6$, $A = x^2 + 3x + 2$
5. $(a + 2b)(3a + 2b) = 3a^2 + 8ab + 4b^2$
6. $(3x + 2)(2x - 5) = 6x^2 - 11x - 10$
7. $-60x^2$ **8.** $(x - 2)(3x^2 - 2x + 1) = 3x^3 - 8x^2 + 5x - 2$
9. $-6x^2$ and $-2x^2$, $4x$ and x **10.** $x^2 + 8x + 15$
11. $3x^2 - 7x - 20$ **12.** $6x^2 + 19x - 7$ **13.** $x^2 - 9$
14. $4x^2 - 20x + 25$ **15.** **a.** neither **b.** neither **c.** ds **d.** pst
16. 1 and 10, 2 and 5; ± 11, ± 9, ± 7, ± 3
17. $2a^2 + 5ab + 2b^2 = (a + 2b)(2a + b)$; $x^2 + 3x + 2 = (x + 1)(x + 2)$
18. **a.** $6x^2 + 7x - 20 = (2x + 5)(3x - 4)$ **b.** $6x^2 - 19x - 20 = (6x + 5)(x - 4)$
19. **a.** $(x + 5)(x + 7)$ **b.** $(x - 7)(x + 2)$ **c.** $(2x + 1)(3x - 10)$
20. $2x^2 - 13x + 15 = (2x - 3)(x - 5)$; numbers multiplying to -30 are ± 1 and ∓ 30, ± 2 and ∓ 15, ± 3 and ∓ 10, and ± 5 and ∓ 6. Their sums are ± 29, ± 13, ± 7, and ± 1. None add to 11.

EXERCISES 7.4

1. $(x + 2)(x - 2)$; ds **3.** $4(x + 2)(x - 2)$; ds **5.** $(x + 6)^2$; pst
7. $4(x + 1)^2$; pst **9.** cannot be factored **11.** $(x - 3)^2$; pst
13. $3(x + 3)(x + 1)$ **15.** $3(x + 3)(x - 3)$
17. cannot be factored **19.** $(5x - 6)(5x + 6)$
21. cannot be factored **23.** $5(x - 1)^2$ **25.** $2(3x + 5)(3x - 5)$
27. $3(x - 5)^2$ **29.** $3(x^2 + 2x + 4)$ **31.** cannot be factored
33. cannot be factored **35.** $x(2x + 3)(x - 1)$
37. $a(3a - 4)(3a + 5)$ **39.** $2x(x + 1)(3x - 4)$
41. $3a(2a - 1)(2a + 1)$

EXERCISES 7.5

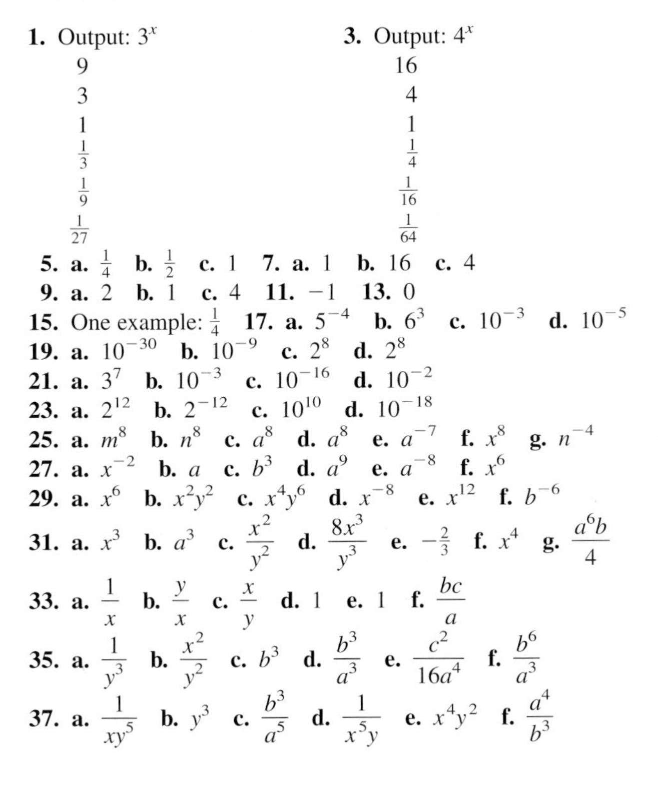

1. Output: 3^x	**3.** Output: 4^x
9	16
3	4
1	1
$\frac{1}{3}$	$\frac{1}{4}$
$\frac{1}{9}$	$\frac{1}{16}$
$\frac{1}{27}$	$\frac{1}{64}$

5. **a.** $\frac{1}{4}$ **b.** $\frac{1}{2}$ **c.** 1 **7.** **a.** 1 **b.** 16 **c.** 4
9. **a.** 2 **b.** 1 **c.** 4 **11.** -1 **13.** 0
15. One example: $\frac{1}{4}$ **17.** **a.** 5^{-4} **b.** 6^3 **c.** 10^{-3} **d.** 10^{-5}
19. **a.** 10^{-30} **b.** 10^{-9} **c.** 2^8 **d.** 2^8
21. **a.** 3^7 **b.** 10^{-3} **c.** 10^{-16} **d.** 10^{-2}
23. **a.** 2^{12} **b.** 2^{-12} **c.** 10^{10} **d.** 10^{-18}
25. **a.** m^8 **b.** n^8 **c.** a^8 **d.** a^8 **e.** a^{-7} **f.** x^8 **g.** n^{-4}
27. **a.** x^{-2} **b.** a **c.** b^3 **d.** a^9 **e.** a^{-8} **f.** x^6
29. **a.** x^6 **b.** x^2y^2 **c.** x^4y^6 **d.** x^{-8} **e.** x^{12} **f.** b^{-6}
31. **a.** x^3 **b.** a^3 **c.** $\dfrac{x^2}{y^2}$ **d.** $\dfrac{8x^3}{y^3}$ **e.** $-\frac{2}{3}$ **f.** x^4 **g.** $\dfrac{a^6b}{4}$
33. **a.** $\dfrac{1}{x}$ **b.** $\dfrac{y}{x}$ **c.** $\dfrac{x}{y}$ **d.** 1 **e.** 1 **f.** $\dfrac{bc}{a}$
35. **a.** $\dfrac{1}{y^3}$ **b.** $\dfrac{x^2}{y^2}$ **c.** b^3 **d.** $\dfrac{b^3}{a^3}$ **e.** $\dfrac{c^2}{16a^4}$ **f.** $\dfrac{b^6}{a^3}$
37. **a.** $\dfrac{1}{xy^5}$ **b.** y^3 **c.** $\dfrac{b^3}{a^5}$ **d.** $\dfrac{1}{x^5y}$ **e.** x^4y^2 **f.** $\dfrac{a^4}{b^3}$

39. **a.** $-\frac{1}{96}$ **b.** -8 **c.** $-\frac{8}{243}$ **d.** $-\frac{1}{486}$ **e.** 324 **f.** $-\frac{81}{8}$

41. **a.** $1 + x^{-1} + x^{-2} + x^{-3}$ or $1 + \dfrac{1}{x} + \dfrac{1}{x^2} + \dfrac{1}{x^3}$
b. $x^{-2} - y^2$ or $\dfrac{1}{x^2} - y^2$

EXERCISES 7.6

1. **a.** 1 right; 346 **b.** 6 left; 0.000 001 6 **c.** 5 right; 160,000
d. 1 left; 21.91 **e.** 3 left; 0.2191 **f.** 4 right; 2,191,000

3.

x	10^x as fraction	10^x as decimal
0	1	1
-1	$\frac{1}{10}$	0.1
-2	$\frac{1}{100}$	0.01
-3	$\frac{1}{1000}$	0.001
-4	$\frac{1}{10,000}$	0.0001
-5	$\frac{1}{100,000}$	0.00001

5. **a.** 1.47×10^{11} **b.** 5.12×10^{-13} **7.** 1.39×10^6 km
9. 2.76×10^9 miles **11.** 1.8×10^3 g
13. 1,990,000,000,000,000,000,000,000,000,000 kg
15. $-0.000\ 000\ 000\ 000\ 000\ 000\ 160\ 2$ coulomb
17. 200,000,000 years
19. 0.000 000 000 000 000 000 000 000 001 675 kg
21. **a.** 6×10^{27} **b.** 1.2×10^{32}
23. **a.** 3×10^{-26} **b.** 4.8×10^{-24}
25. **a.** 4×10^3 **b.** 2×10^{-13} **27.** 1×10^3
29. **a.** 0.0234 **b.** 3140 **c.** 62,800,000
31. **a.** 3.4×10^4 **b.** 5.6×10^0
33. **a.** 4.32×10^3 **b.** 5.67×10^{-6} **35.** **a.** $>$ **b.** $<$
37. **a.** $<$ **b.** $>$ **39.** **a.** $>$ **b.** $<$ **41.** **a.** $>$ **b.** $>$
43. electron **45.** 5.87×10^{12} mi **47.** 1.58×10^{17} mi
49. 1.80×10^7 qt **51.** 4.90×10^{-9} mph

CHAPTER 7 REVIEW EXERCISES

1. **a.** $7a^2 - 3ab - 10b^2$; trinomial **b.** $4x^2 + 3x + 5$; trinomial
c. $x^3 - 64$; binomial **d.** $4x^2 + 7x$; binomial
e. $-4x + 21$; binomial
3. **a.** $P = 10x + 2, A = 6x^2 + 2x$
b. $P = 8x + 4, A = 3x^2 + 6x$ **c.** $P = 7x - 3, A = 2x^2 + x$
5. $(3x + 1)(x + 1) = 3x^2 + 4x + 1$ **7.** $x^2 + 7x + 12$
9. $4x^2 - 20x + 25$ **11.** $9x^2 + 9x - 10$ **13.** $6x^2 - 5x - 6$
15. $a^2 - 2ab + b^2$ **17.** 9, 15
19. **a.** $(x + 7)(x - 2) = x^2 + 5x - 14$
b. $(3x + 5)(2x + 3) = 6x^2 + 19x + 15$
21. **a.** $(x - 2)(x - 7) = x^2 - 9x + 14$
b. $(x + 7)(2x - 3) = 2x^2 + 11x - 21$
c. $(3x - 1)(4x + 1) = 12x^2 - x - 1$
d. $(2x + 3)(10x - 3) = 20x^2 + 24x - 9$
23. $(x - 2)(x - 1)$; neither **25.** $(3x + 4)(3x - 4)$; ds
27. $(2x - 3)(x + 2)$; neither **29.** $(3x + 2)(3x - 1)$; neither
31. $(x + 2)(x + 4)$; neither **33.** $(x - 1)(x - 10)$; neither
35. $(5 + 3x)(5 - 3x)$; ds **37.** does not factor; neither
39. $(y + 2)(y + 6)$; neither **41.** $(2x + 7)(x - 5)$; neither
43. $4(x - 1)^2$; pst **45.** $x(x + 2)^2$; contains pst
47. $3(x + 3)(x - 3)$; contains ds **49.** $x(x - 2)(x - 5)$; neither
51. $2(x^2 + 4x + 7)$; neither
53. **a.** $\frac{1}{3}$ **b.** 1 **c.** $\frac{1}{9}$ **d.** 1 **e.** $\frac{3}{2}$ **f.** $\frac{9}{4}$
55. **a.** x^5 **b.** 1 **c.** $\dfrac{1}{b^{10}}$ **57.** **a.** n^9 **b.** n **c.** $\dfrac{1}{n^9}$
59. **a.** $\dfrac{1}{b^8}$ **b.** x^6 **c.** 1 **61.** **a.** $\dfrac{b^2}{a^2}$ **b.** $\dfrac{b^6c^3}{8a^3}$ **c.** 1
63. **a.** $\dfrac{1}{x^8y}$ **b.** $\dfrac{1}{a^3b^2}$ **c.** x^3 **d.** 1
65. **a.** $-\frac{1}{768}$ **b.** $\frac{1}{72}$ **c.** 8 **d.** 1
67. 15.7; 228; 1690; 13,000; 20,200; 27,400 **69.** $\approx$3.9 sec
71. **a.** 4.5×10^5 **b.** 1.0×10^{-5}
73. **a.** 4.0×10^{-9} **b.** 3.0×10^{-7}
75. **a.** 3.85×10^{-2} **b.** 4.8×10^{-3} **77.** 2290 mph

CHAPTER 7 TEST

1. terms **2.** greatest common factor **3.** factoring
4. -1 **5.** $x = 0$ **6.** $x = -2$ **7.** 0.000 348 2
8. 4.5×10^{10} **9.** $4a - 2b + 5c - 2d$; four-term polynomial
10. $x^3 - 27$; binomial **11.** $-4x^2 + 20x$; binomial
12. $2y(3x^2 + 7x - 9y)$ **13.** $(3x + 4)(2x - 5) = 6x^2 - 7x - 20$
14. $P = 12x + 10, A = 8x^2 + 20x$ **15.** $3x$ **16.** $x^2 + 3x - 28$
17. $x^2 - 14x + 49$ **18.** $4x^2 - 49$ **19.** $2x^2 - 16x + 32$
20. $(x - 4)(x - 5)$ **21.** $(2x + 1)(x - 2)$ **22.** $2(x + 2)(x - 2)$
23. $(x - 4)^2$ **24.** $(3x + 1)^2$ **25.** $3(x^2 + 2x + 5)$ **26.** 18 or 22
27. **a.** b **b.** $\dfrac{1}{x^6}$ **28.** **a.** b^5 **b.** $\dfrac{b}{a}$
29. **a.** 1 **b.** $\dfrac{625y^4}{81x^4}$ **30.** -2, 1; -32, $-\frac{2}{3}$; 1, $\frac{10,000}{81}$
31. 0.000 000 000 001 **32.** 5.0×10^{-5}
33. factors: $(x \pm 1)(x \pm 21)$; $(x \pm 3)(x \pm 7)$; trinomials:
$x^2 \pm 22x + 21$; $x^2 \pm 10x + 21$; $x^2 \pm 20x - 21$;
$x^2 \pm 4x - 21$; 1 and 21, 3 and 7 are the only factors of 21
34. There are two places where a and b multiply each other:

	a	$+b$
a	a^2	$+ab$ (circled)
$+b$	$+ab$ (circled)	b^2

35. 10 [EE] 3 is $10 \times 10^3 = 10^4 = 10,000$.
10 [^] 3 is $10^3 = 1000$.

CUMULATIVE REVIEW OF CHAPTERS 1 TO 7

1.

Input x	Output $y = 0.25x + 0.25$
0	0.25
5	1.50
10	2.75
15	4.00
20	5.25
25	6.50
30	7.75

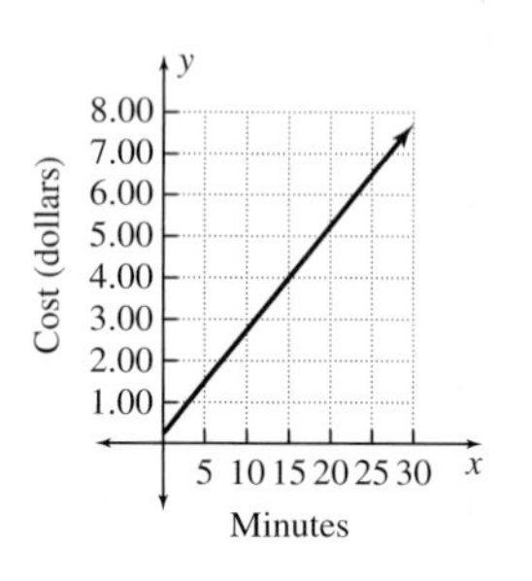

3. $2x - 3$ **5.** x^{11} **7.** $R = 12$ **9.** $x = 5.2$
11. $x \geq 2$

13. $y = -\frac{3}{5}x + 3$ **15.** $a = 3m - b - c$
17.

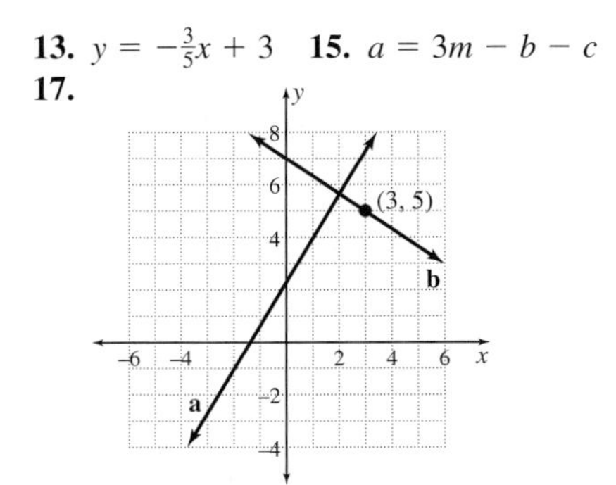

19. $y = -\frac{3}{5}x$ **21.** $\approx 29.3\%$ **23.** $x = 9$
25. **a.** line 3 **b.** line 4 **c.** line 2 **d.** line 1
27. **a.** 7.1 kg, 42.9 kg **b.** 21.4 kg, 28.6 kg
c. The higher the price of the mix, the greater the amount of cashews. $20 is close to $24, so most of the mixture is cashews.
29. $4y(2x - xz + 3z)$
31. $(x - 3)^2$; pst
33. $2(3x - 4)(x + 1)$
35. **a.** 1, 11; 2, 10; 3, 9; 4, 8; 5, 7; 6, 6
b. 1, 9; 2, 8; 3, 7; 4, 6; 5, 5

EXERCISES 8.1

1. $x \approx 5.83$ **3.** $x \approx 10.82$ **5.** $x \approx 6.24$ **7.** no **9.** yes
11. yes **13.** **a.** yes **b.** yes **c.** no **d.** yes
15. **a.** yes **b.** yes; not a triangle **c.** yes **d.** no
17. $x = 21$; $w = 35$
19.

Leg	Leg	Hypotenuse
3	4	5
6	8	10
18	24	30
9	12	15
1	$\frac{4}{3}$	$\frac{5}{3}$

21. $x \approx 5.7$ **23.** $x \approx 10.2$ **25.** $x \approx 6.7$ **27.** 12.4 ft **29.** 9.3 ft
31. 3.5 ft; 14.4 ft
33. $\approx$11.64 ft, $\approx$11 ft 8 in.; $\approx$2.91 ft, $\approx$2 ft 11 in.
35. $\approx$17.46 ft, $\approx$17 ft 6 in.; $\approx$4.37 ft, $\approx$4 ft 4 in.
37. $\approx$251.8 mi **39.** $\approx$607.6 mi
41. **a.** 13 ft by 20 ft; 520 ft^2
b. 5 to 12; heights above 10 ft, 12 ft, and 13 ft make right triangle.
c. 15 ft
43. $\approx$1146 ft^2 **45.** $\approx$1332 ft^2 **47.** $x \approx 7.16$ ft; $4x \approx 28.6$ ft

EXERCISES 8.2

1. **a.** -3 **b.** ab **c.** $4x^2$ **d.** 2 **e.** x **f.** $x + 2$
3. **a.** 9 **b.** 3.873, I **c.** 5 **d.** 1.5
5. **a.** 20 **b.** 5.916, I **c.** 4 **d.** 0.1
7. **a.** 4.899, I **b.** 12 **c.** 2.5 **d.** 0.316, I
9. **a.** 8, 9 **b.** 7, 8 **c.** 14, 15 **d.** 4, 5
11. **a.** no real number **b.** -9 **c.** ± 12
13. **a.** 7 **b.** -15 **c.** ± 20
15. **a.** $\sqrt{15}$ **b.** 45 **c.** $4\sqrt{3}$ **17.** **a.** 63 **b.** 13 **c.** $6\sqrt{2}$
19. **a.** 9 **b.** 20 **c.** $3\sqrt{2}$ **21.** **a.** a **b.** b **c.** $11a^2$
23. **a.** $8a$ **b.** $2x$ **c.** $4b$ **25.** **a.** $\frac{x}{3}$ **b.** $\frac{2}{5}$ **c.** $x\sqrt{x}$
27. **a.** $\frac{2}{x}$ **b.** $\frac{x}{3}$ **c.** $y^2\sqrt{y}$
29. **a.** $2x$ **b.** $\frac{x^2}{5y}$ **c.** $\frac{-y^2}{3}$ **d.** $\frac{2y^2}{x}$
31. **a.** $2\sqrt{10}$ **b.** 3 **c.** $y = 3x - 3$
33. **a.** $3\sqrt{2}$ **b.** -1 **c.** $y = -x + 4$
35. **a.** $5\sqrt{2}$ **b.** $-\frac{1}{7}$ **c.** $y = -\frac{1}{7}x + \frac{18}{7}$
37. **a.** $2\sqrt{10}$ **b.** $-\frac{1}{3}$ **c.** $y = -\frac{1}{3}x - 2$
39. isosceles right **41.** isosceles **43.** right
51. **a.** 1 **b.** $\frac{1}{25}$ **c.** 5 **d.** 5 **53.** **a.** 3 **b.** 1 **c.** 3 **d.** $\frac{1}{9}$
55. **a.** 4 **b.** 1 **c.** $\frac{1}{2}$ **d.** $\frac{1}{2}$
57. **a.** 4 **b.** 0.1 **c.** 2.5 **d.** 0.5 **e.** 50 **f.** 20

EXERCISES 8.3

1. $f(-4) = 2\sqrt{2}, f(-1) = \sqrt{5}, f(0) = 2, f(4) = 0, f(6)$ is not a real number
3. $f(-4) = \sqrt{7}, f(-1) = 2, f(0) = \sqrt{3}, f(4)$ is not a real number, $f(6)$ is not a real number
5. yes; $y = 1$ **7.** yes; if $x \leq 0$
9.

x	$y = \sqrt{x + 4}$
-4	0
-2	1.414
0	2
2	2.449
4	2.828

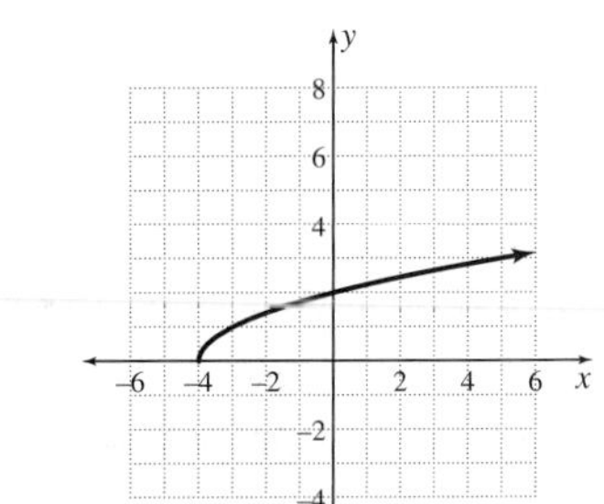

11.

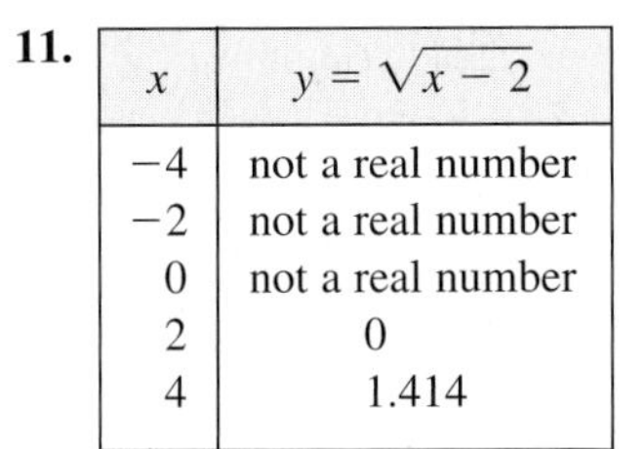

x	$y = \sqrt{x - 2}$
-4	not a real number
-2	not a real number
0	not a real number
2	0
4	1.414

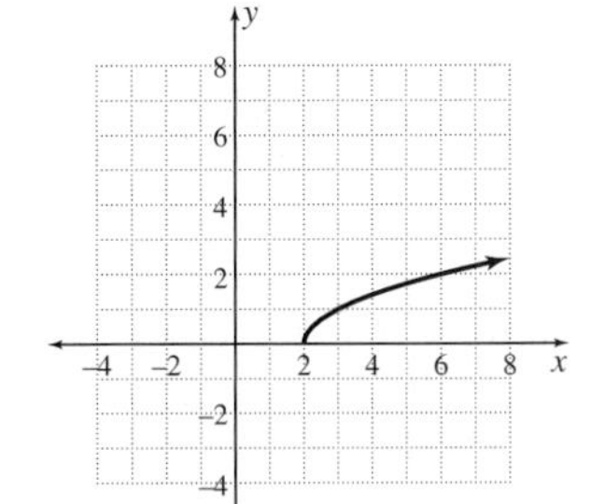

13.

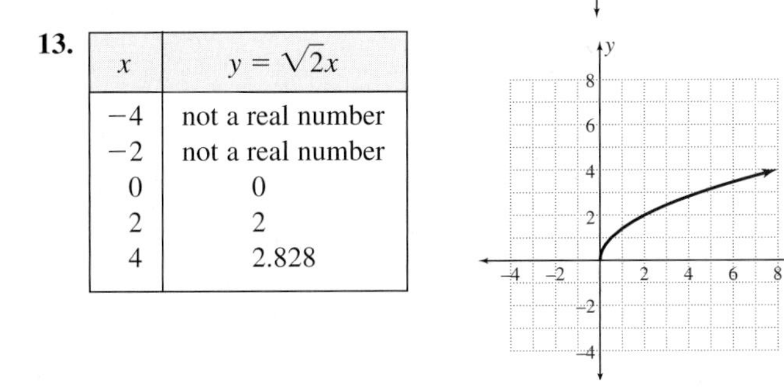

x	$y = \sqrt{2x}$
-4	not a real number
-2	not a real number
0	0
2	2
4	2.828

15. **a.** $x \geq 1$ **b.** $x \geq -3$ **17.** **a.** $x \leq 4$ **b.** $x \leq 3$ **19.** $x = 6$

21. **23.**

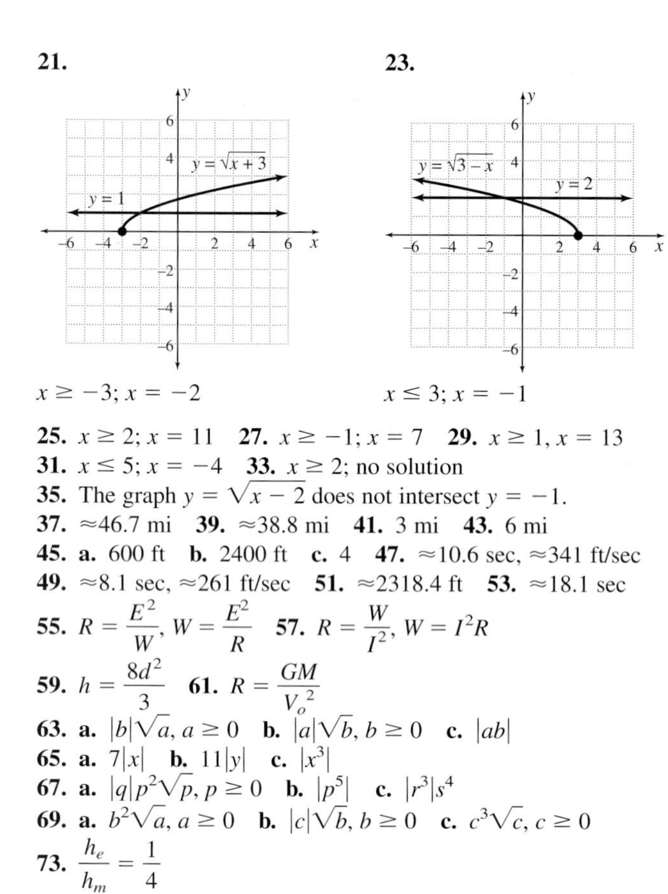

$x \geq -3;\ x = -2$ $\quad x \leq 3;\ x = -1$

25. $x \geq 2;\ x = 11$ **27.** $x \geq -1;\ x = 7$ **29.** $x \geq 1,\ x = 13$
31. $x \leq 5;\ x = -4$ **33.** $x \geq 2$; no solution
35. The graph $y = \sqrt{x - 2}$ does not intersect $y = -1$.
37. $\approx$46.7 mi **39.** $\approx$38.8 mi **41.** 3 mi **43.** 6 mi
45. a. 600 ft **b.** 2400 ft **c.** 4 **47.** $\approx$10.6 sec, $\approx$341 ft/sec
49. $\approx$8.1 sec, $\approx$261 ft/sec **51.** $\approx$2318.4 ft **53.** $\approx$18.1 sec
55. $R = \dfrac{E^2}{W},\ W = \dfrac{E^2}{R}$ **57.** $R = \dfrac{W}{I^2},\ W = I^2R$
59. $h = \dfrac{8d^2}{3}$ **61.** $R = \dfrac{GM}{V_o^{\,2}}$
63. a. $|b|\sqrt{a},\ a \geq 0$ **b.** $|a|\sqrt{b},\ b \geq 0$ **c.** $|ab|$
65. a. $7|x|$ **b.** $11|y|$ **c.** $|x^3|$
67. a. $|q|p^2\sqrt{p},\ p \geq 0$ **b.** $|p^5|$ **c.** $|r^3|s^4$
69. a. $b^2\sqrt{a},\ a \geq 0$ **b.** $|c|\sqrt{b},\ b \geq 0$ **c.** $c^3\sqrt{c},\ c \geq 0$
73. $\dfrac{h_e}{h_m} = \dfrac{1}{4}$

MID-CHAPTER 8 TEST

1. a. no **b.** no **c.** yes **2.** 130 mi **3.** $\sqrt{261}$ in.
4. a. 6 **b.** 18 **c.** $6\sqrt{2}$ **d.** 12 **e.** $4\sqrt{3}$ **f.** -2
g. not a real number **h.** $\pm\frac{5}{4}$ **i.** $\frac{1}{4}$
5. a. $9x^2$ **b.** $xy^3\sqrt{x}$ **c.** $\dfrac{\sqrt{x}}{y}$ **d.** $\dfrac{\sqrt{x}}{y^2}$
6. a. $|xy^3|\sqrt{xy}$ **b.** $\dfrac{|x|}{2y^2}$ **c.** $\dfrac{14}{|x|}$ **d.** already simplified
7. $3\sqrt{13} \approx 10.8$ **8.** 11, 12 **9.** $R = \dfrac{W}{I^2}$
10. $x = 9,\ x \geq -7$

11. Assume $x \geq 0$, $x = 2$.
12. no graph in second or third quadrant
13. no graph in third or fourth quadrant
14. a. $C \approx 2\pi\sqrt{10}$
b. equal; The letters and π are all the same, so compare the numbers: $2\sqrt{\frac{1}{2}} = \sqrt{\frac{4}{2}} = \sqrt{2}$.
15. 10, (2, 0)

EXERCISES 8.4

1. $(-3, 0), (2, 0); (0, -6)$; axis of symmetry $x = -\frac{1}{2}$, vertex $(-0.5, -6.25)$
3. $(0, 0), (5, 0); (0, 0)$; axis of symmetry $x = 2.5$; vertex $(2.5, 6.25)$
5.

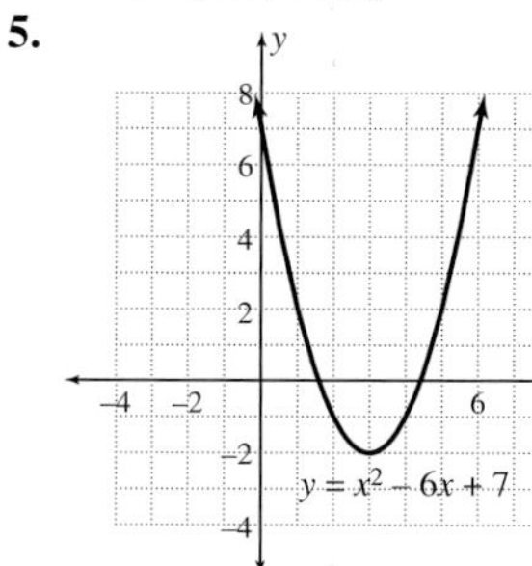

x	$y = x^2 - 6x + 7$
−4	47
−2	23
0	7
2	−1
4	−1

$(4.4, 0), (1.6, 0); (0, 7); x = 3; (3, -2)$

7.

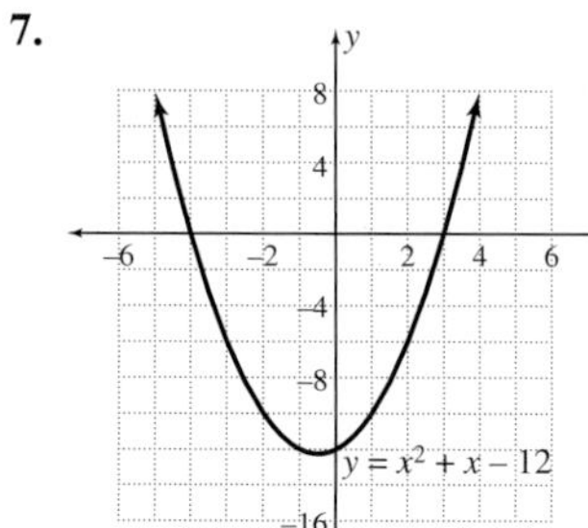

x	$y = x^2 + x - 12$
−3	−6
−1	−12
0	−12
1	−10
3	0

$(-4, 0), (3, 0); (0, -12); x = -0.5; (-0.5, -12.25)$

9.

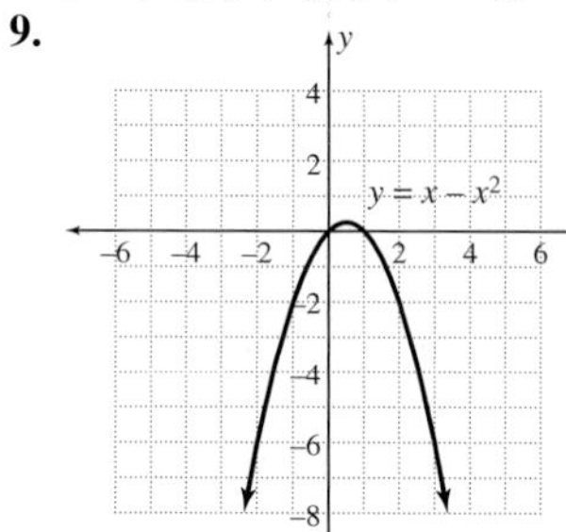

x	$y = x - x^2$
−3	−12
−1	−2
0	0
1	0
3	−6

$(0, 0), (1, 0); (0, 0); x = 0.5; (0.5, 0.25)$

11.

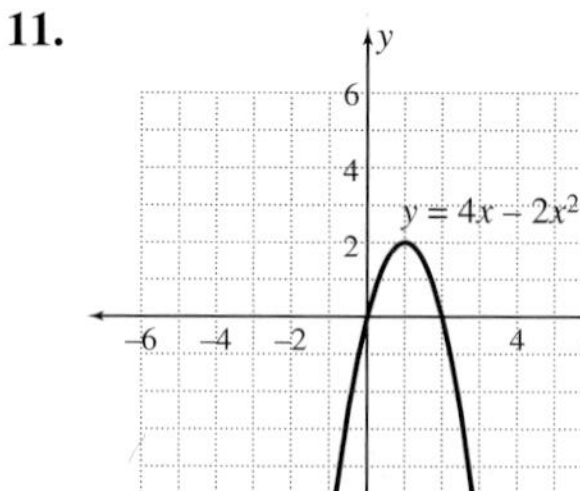

x	$y = 4x - 2x^2$
−2	−16
1	−6
0	0
1	2
2	0

$(0, 0), (2, 0); (0, 0); x = 1; (1, 2)$
13. $y \geq -6.25$ **15.** $y \leq 6.25$ **17.** $a = 2, b = 3, c = 1$
19. $a = 1, b = -4, c = 4$ **21.** $a = 1, b = 0, c = -4$
23. $a = 4, b = -8, c = 0$ **25.** $a = 1, b = -1, c = 4$
27. $a = 1, b = -1, c = 1$ **29.** $a = -0.5g, b = v, c = s$
31. $a = \pi, b = 0, c = 0$ **33.** $y = 4x^2 + 4x + 1$
35. $y = 9x^2 - 16$ **37.** $y = 3x^2 + 6x$
39. a. $x = -4, x = 3$ **b.** $x = -2, x = 1$
c. no real-number solution **d.** $x = -3, x = 2$

41. a. $x = 2, x = 3$ **b.** no real-number solution
c. $x = 1, x = 4$ **d.** $x = 0, x = 5$
43. ≈ 7.1 sec; ≈ 10 sec

EXERCISES 8.5

1. $x = \pm\sqrt{5}$ **3.** $x = \pm\sqrt{7}$ **5.** $x = \pm\frac{2}{5}$ **7.** $x = \pm\frac{1}{5}$
9. $x = \pm\frac{15}{7}$ **11.** $x = \pm\frac{11}{6}$ **13.** $x = \pm\frac{3}{5}$ **15.** $x = \pm 9$
17. $x = \pm 6$ **19.** $x = 4$ or $x = -4$ **21.** $x = \frac{1}{2}$ or $x = -\frac{2}{3}$
23. $x = \frac{5}{2}$ or $x = -2$ **25.** $x = 2$ or $x = -2$
27. $x = 2$ or $x = -3$ **29.** $x = 5$ or $x = -3$
31. $x = 0$ or $x = -3$ **33.** $x = 0$ or $x = -\frac{1}{2}$
35. $x = -2$ or $x = 6$ **37.** $x = \frac{3}{2}$ or $x = -1$
39. $x = 4$ or $x = -3$ **41.** $x = 1$ or $x = 6$
43. $x = -\frac{5}{2}$ or $x = 1$ **45.** $x = \pm\frac{5}{2}$ **47.** $x = \pm 2$
49. $x = -4$ or $x = 3$ **51.** $x = 4$ or $x = 1$
53. $t = 0$ sec or $t = 3$ sec; yes **55.** $t = 1$ sec or $t = 3$ sec; yes
57. $t = 0$ sec or $t = 4$ sec; yes **59.** $x = 3$; $x = 0$, extraneous
61. $x = -1$; $x = -4$, extraneous
63. a. $A(0, 0)$, $B(4, 16)$ **b.** $4x = x^2$ **c.** $x = 0$ or $x = 4$; yes

EXERCISES 8.6

1. $a = 9, b = 6, c = 1$ **3.** $a = 3, b = -9, c = 0$
5. $a = 1, b = -4, c = 3$ **7.** $a = 1, b = 0, c = 9$
9. -2 **11.** $\frac{1}{2}$ **13.** $-\frac{4}{3}$ **15.** $\frac{1}{8}$
17. a. $\frac{1}{2}$; 0.293 **b.** $\frac{5}{3}$; 1.816 **c.** 2; 2.449
19. a. $\frac{1}{3}$; 0.255 **b.** 3; 3.162 **c.** $\frac{5}{2}$; 2.581
21. a. $-\frac{9}{2}$; -4.303 **b.** $-\frac{1}{4}$; -0.363
23. a. $-\frac{7}{6}$; -1.215 **b.** $\frac{7}{10}$; 0.717 **25.** $x = -1, x = \frac{1}{4}$
27. $x = -\frac{2}{7}, x = 1$ **29.** $x = -\frac{1}{2}, x = \frac{2}{5}$ **31.** $x = 1, x = -\frac{1}{2}$
33. $x = -1, x = \frac{2}{3}$ **35.** no real-number solution
37. $x \approx 1.786, x \approx -1.120$ **39.** no real-number solution
41. $x \approx -1.117, x \approx 0.717$ **43.** $x \approx -0.379, x \approx 0.879$
45. $t \approx 4.14$; discard $t \approx -2.26$ (negative time is not acceptable)
47. $t \approx 3.57$; discard $t \approx -2.63$ (negative time is not acceptable)
49. no real-number solution
51. $t \approx 3.08$; discard $t \approx -1.83$ (negative time is not acceptable)
53. $x = \pm\frac{\sqrt{10}}{2} \approx \pm 1.581$ **55.** $x = \pm\frac{3}{2}$
57. no real-number solution **59.** $x = 3$
61. $x \approx -0.914, x \approx 1.914$ **63.** $x = 1.5$

EXERCISES 8.7

1. \$26,000; \$12,500; \$92,000 **3.** \$26,000; \$27,500; \$7,500
5. \$10,000; \$13,000

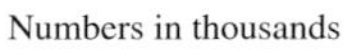

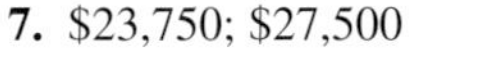

7. \$23,750; \$27,500

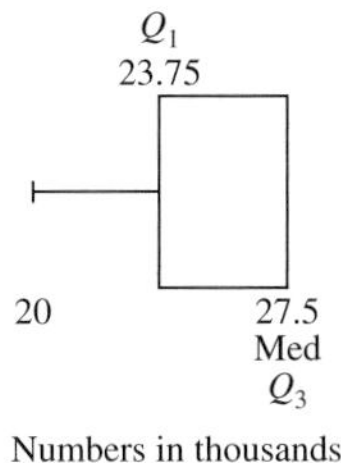

9. \$36,300 **11.** \$3,400
13. a. 1861; 1722.5; 1949 **b.** 1185.5; 2506.5

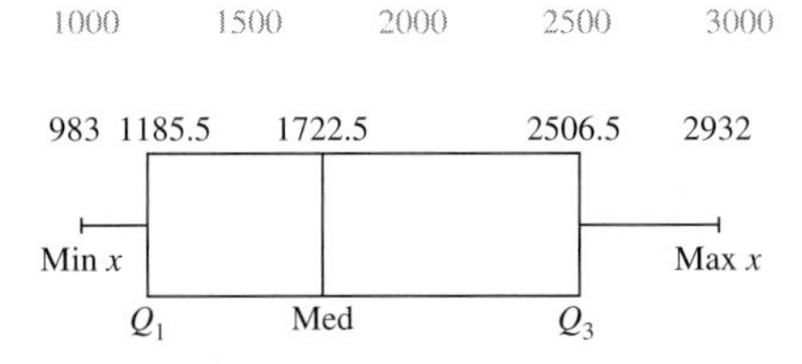

15. a. 16118; 15560; 13400 **b.** 13475; 18450

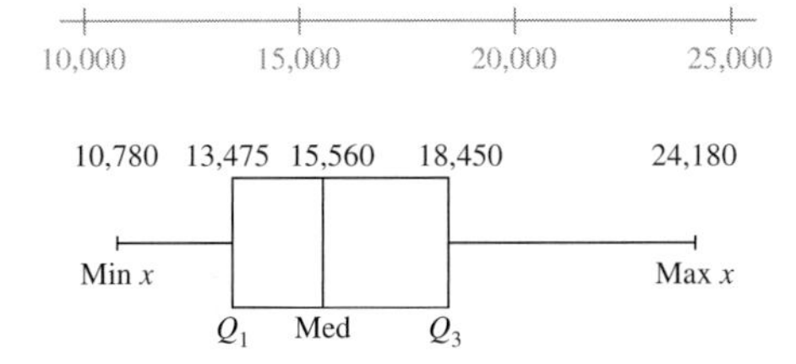

17. a. $7.5 + \frac{3}{8}$ or $7\frac{7}{8}$ in. **b.** $7.5 - \frac{3}{8}$ or $7\frac{1}{8}$ in.
c. The $\frac{3}{8}$-in. variation is a range. The steps will not pass; the range between the smallest and largest is $\frac{1}{2}$ in; $\frac{1}{2}$ in. is greater than $\frac{3}{8}$ in.
19. a. 62.2; 13.6; 43, 50, 62.5, 75, 80
b. 52.3; 9.3; 41, 44, 51.5, 61, 66 Cities have similar lows, but Atlanta has greater mean and high temperatures.
21. a. 77.6; 3.4; 73, 74.5, 77.5, 80.5, 82
b. 76.7; 5.9; 68, 71, 77.5, 82, 84 Cities have nearly the same average temperature, but Miami has greater range and lower low temperatures.
23. 1500 ng/m^3; 0.5 micrograms/m^3; the Czech limit
25. a. 60 ng/m^3
b. 340 ng/m^3
c. 1360 ng/m^3
d. 2260 ng/m^3, exceeds limit

CHAPTER 8 REVIEW EXERCISES

1. a. no **b.** no **c.** yes **3.** ≈ 24.7 ft
5. a. $2\sqrt{15}$ **b.** $3\sqrt{7}$ **c.** $3\sqrt{6}$
7. a. 12 **b.** $\frac{1}{144}$ **c.** 1 **d.** 12 **9. a.** 0.6 **b.** $\frac{25}{9}$ **c.** 0.6 **d.** 1
11. a. sides all $\sqrt{13}$; diagonals $\sqrt{26}$; square
b. sides all $2\sqrt{5}$; diagonals $2\sqrt{10}$; square
c. sides $\sqrt{10}$ and $\sqrt{17}$; diagonals $3\sqrt{3}$ and 5; not a rectangle
13. a. $5xy^2$ **b.** $13x^3y$ **c.** $1.5a\sqrt{a}$ **d.** $0.8b^2\sqrt{b}$ **e.** $4x$
f. $\frac{a^2}{3b^3}$ **g.** $8a^3$ **h.** $\frac{11}{7b^2}$
15. ≈ 1.41, 2, ≈ 2.83, 4, ≈ 5.66, 8, ≈ 11.3, 16
17. a. $x = 4, x \geq \frac{3}{7}$ **b.** $x = 2$, discard $x = 1, x \leq 2$
c. $x = 13, x \geq \frac{3}{4}$ **d.** $x = 5$, discard $x = 1, x \geq \frac{1}{2}$
19. a. ≈ 2.7 min **b.** $42\frac{2}{3}$ ft **21.** $x = 5$ **23.** $x = -2, x = 4$

25. $x = \pm\frac{1}{3}$ **27.** $x = 1, x = 4$ **29.** $x \approx -3.386, x \approx 0.886$
31. $x = -6, x = 3$ **33.** $x = -1, x = 2\frac{1}{3}$
35. no real-number solution **37.** $x = 3$

39. a. $r = \sqrt{\dfrac{A}{\pi}}$ **b.** $v = \sqrt{\dfrac{2p}{d}}$ **c.** $r = \dfrac{1}{2}\sqrt{\dfrac{S}{\pi}}$
d. $v = \sqrt{2gh}$

41. a. 4.75, 5.46
b.

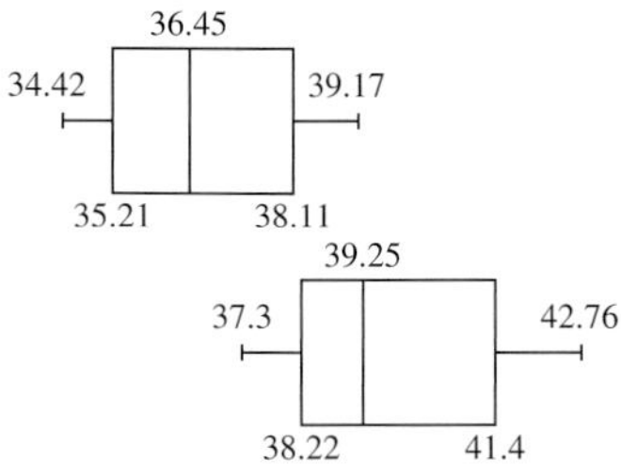

Men: 34.42, 35.205, 36.45, 38.11, 39.17
Women: 37.3, 38.22, 39.25, 41.4, 42.76
c. 36.68, 1.60; 39.78, 1.82

CHAPTER 8 TEST

1. a. yes **b.** no **c.** yes **d.** yes
2. a. $3\sqrt{5}$ **b.** $2\sqrt{11}$ **3.** ≈ 20.6 ft
4. a. 10 **b.** $3\sqrt{30}$ **c.** 54 **d.** $9a$ **e.** $6x\sqrt{y}$ **f.** $0.9x^2y\sqrt{y}$
g. 7 **h.** $\frac{3}{4}$ **i.** a^2b
5. a. $a^2\sqrt{a}, a \geq 0$ **b.** $|x^3|$, **c.** $a, a \geq 0$ **d.** $y^3\sqrt{y}, y \geq 0$
6. a. $a = 7$ **b.** $b = 21, c \approx 10.8$ or $3\sqrt{13}$
c. $d \approx 16.6, e \approx 19.4$
7. $x = -2, x = 2$
8. a. $x = \pm\frac{6}{11}$ **b.** $x = -1$, discard $x = -6, x \leq 3$
c. $x = -2, x = 1$ **d.** $x = 30, x \geq \frac{6}{5}$ **e.** $x = \frac{1}{2}, x = -8$
f. $x \approx -0.46, x \approx 1.09$ **g.** $x = \pm\frac{8}{3}$
h. no real-number solution **i.** $x = 3$ **j.** $x = \pm\frac{5}{2}$
k. $x = -\frac{3}{2}$ or $x = 4$ **l.** $x = \frac{2}{3}$ or $x = -12$

9. a. $R = \dfrac{2GM}{V_e^2}$ **b.** $v = \sqrt{\dfrac{2E}{m}}$ **c.** $H = 2\sqrt{\dfrac{2\pi E}{k}}$
10. a. 51.6 mph **b.** 86 mph **c.** ≈ 194.7 psi
11. a. $x = 5\sqrt{2}$ or 7.071 ft **b.** $x = 10\sqrt{\dfrac{2}{\pi}}$ or 7.979 ft
c. $x = 10.746$ ft
12. a.

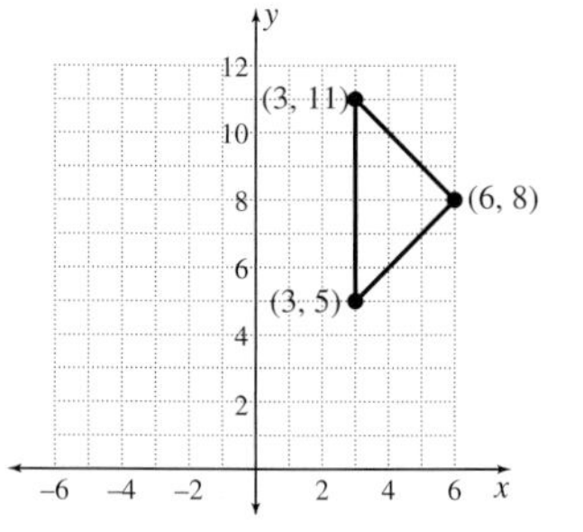

b. $y = x + 2, y = -x + 14, x = 3$ **c.** 6, 4.24, 4.24
d. yes; slopes are $1, -1$, and undefined.
e. isosceles right triangle; 2 equal sides and 1 right angle

13. The rise and run are perpendicular and are the legs of a right triangle. When we chose the rise and run connecting the points, the rise^2 + run^2 = distance^2. When we can reduce the slope, say from $\frac{8}{6}$ to $\frac{4}{3}$, we create a similar right triangle half the size of the original: $8^2 + 6^2 = 10^2$ compared with $4^2 + 3^2 = 5^2$.

14. a. No. Because $(2 - x)$ is squared, $(2 - x)$ could be any real number, so after taking the root, we must place $2 - x$ within absolute values symbols: $y_1 = |2 - x|$ for x as any real number. Because $(2 - x)$ is under the radical in y_2, it must be positive (or zero). Thus $y_2 = 2 - x, x \leq 2$.
b. $y = \sqrt{(2 - x)^2}$ is an absolute value graph with the point of V at (2, 0). $y_2 = (\sqrt{2 - x})^2$ is only the left side of the V starting at (2, 0) and passing through (0, 2).

15. a. Yes for all **b.** $a^2 + a + b = b^2$, cannot tell
c. Yes, $n^2 + (\sqrt{n + (n + 1)})^2 \stackrel{?}{=} (n + 1)^2$ simplifies on each side to the same expression: $n^2 + 2n + 1$. It makes an identity and, being true, satisfies the Pythagorean theorem.
d. We were able to see the algebraic relationship, numbers n and $n + 1$ being one apart in part c that was not visible in part b.

16. a. 110, ≈ 5.5 **b.** 145

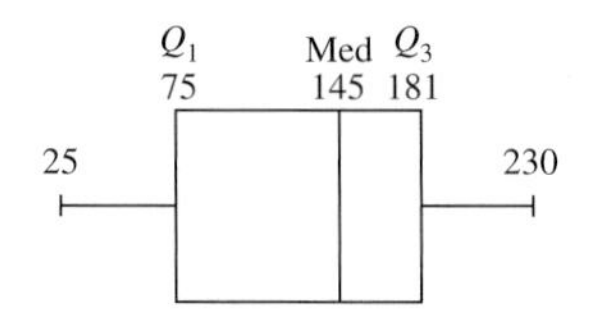

CUMULATIVE REVIEW OF CHAPTERS 1 TO 8

1. 9 **3.** \$20 **5.** $\dfrac{x^2}{9y}, x \neq 0, \; y \neq 0$ **7.** $x = -8$ **9.** $x = -2$
11. $x \leq 1$ **13.** $f(-1) = -5; f(0) = 0; f(1) = 1; f(2) = -2$
15. $y = \frac{4}{3}x - 2$
17.

Trips	Value (dollars)
0	20.00
2	15.50
4	11.00
6	6.50
8	2.00
10	−2.50

$y = 20 - 2.25x$

19. 65 **21.** 37.5% **23.** $293\frac{1}{3}$ ft
25. a. $2x^2 + 5x - 3$ **b.** $3x^3 + 2x^2 - x$
27. a. $\approx 1.075 \times 10^2$ **b.** $\approx 1.103 \times 10^2$
29. $x = -3, y = \frac{1}{2}$
31. 9, 12
33. a.

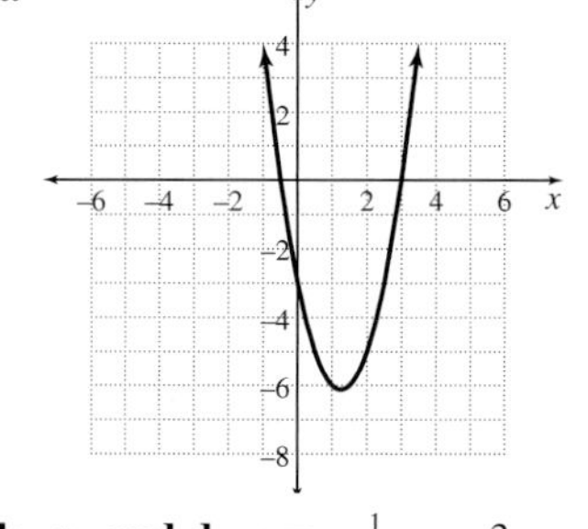

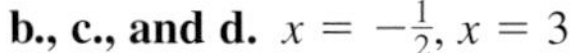

b., c., and d. $x = -\frac{1}{2}, x = 3$

35. a. 4.7, 5, 5

b. median = 24; $Q_1 = 19$; $Q_3 = 28$

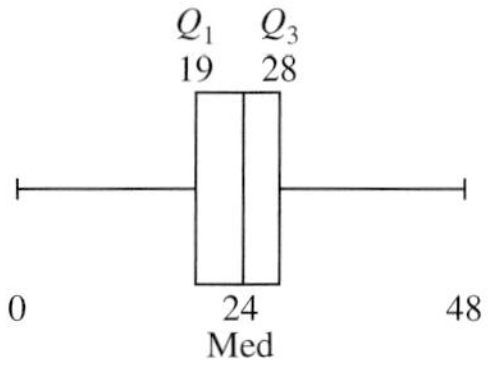

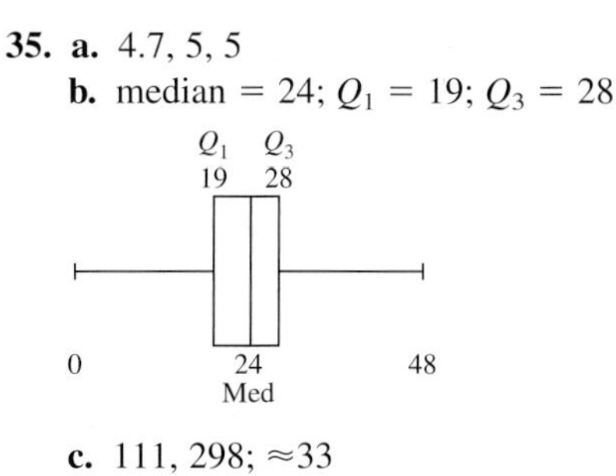

c. 111, 298; ≈ 33

EXERCISES 9.1

1. a. $-\frac{1}{3}, -\frac{1}{2}, -1, -2, -4$, undefined, $4, 2, 1, \frac{1}{2}, \frac{1}{3}, \frac{1}{4}$

b. No **c.** No

d. We have a zero denominator, and division by zero is undefined.

3. a. Any number between 0 and 1

b. Any number between -1 and 0

5. a. $(-2, -1)$ **b.** $(-3, -2)$

7. Place parentheses around the denominator.

9. $\mathbb{R}, x \neq -1$ **11.** $\mathbb{R}, x \neq \frac{1}{2}$ **13.** $\mathbb{R}, x \neq -\frac{1}{3}$

15. $\mathbb{R}, x \neq -1, x \neq -3$

17. y approaches $-\infty$ from the left and $+\infty$ from the right.

19. y approaches $+\infty$ from the left and $-\infty$ from the right.

21. You are not moving. **23.** no

25. a. $\frac{2}{3}$ hr, 40 min **b.** $\frac{1}{2}$ hr, 30 min **c.** $\frac{2}{5}$ hr, 24 min

d. $\frac{1}{3}$ hr, 20 min **e.** $\frac{2}{7}$ hr, 17 min **f.** $\frac{1}{4}$ hr, 15 min

The time saved between 30 mph and 40 mph is 10 minutes. The time saved between 60mph and 80 mph is only 5 minutes. Once you exceed 60 mph, the time saved drops slowly.

27. a. 12 mph **b.** 15 mph **c.** 25.7 mph **d.** 6 mph **e.** 13.6 mph.

29. a.

Width	Length	Area	Perimeter
1	30	30	62
2	15	30	34
3	10	30	26
4	7.5	30	23
5	6	30	22
6	5	30	22
7.5	4	30	23
10	3	30	26
15	2	30	34
30	1	30	62

b.

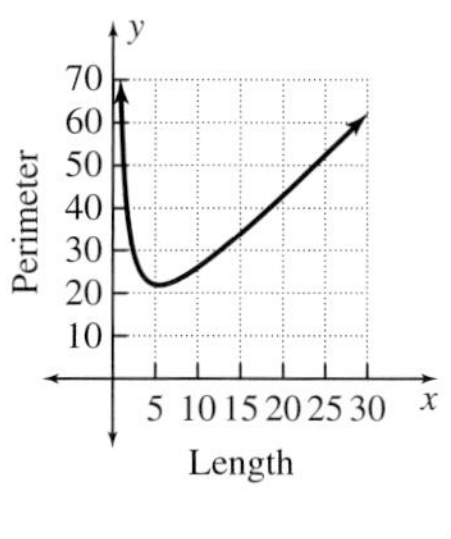

c. $\approx(5.5, 21.9)$

d. $y = 2x + 2\left(\frac{30}{x}\right)$

31. For $x \neq 0$, $y = \frac{1000 \cdot 10^9}{x}$ days; $y = \frac{1000 \cdot 10^9}{365x}$ years

33. ≈ 36 yr

35. a. number of terms the aid will last

b. Available aid will last $7\frac{1}{2}$ terms.

37. 1.2 mph

EXERCISES 9.2

1. $\frac{1}{3a}$ **3.** $\frac{3d}{2c}$ **5.** $x - 2$ **7.** $\frac{-1}{x + 2}$ **9.** $\frac{3a}{5b}$

11. $\frac{2(x + 2)}{x - 2}$ **13.** $\frac{x - 2}{x + 3}$ **15.** $\frac{x + 3}{x}$ **17.** $-\frac{1}{2}$

19. $\frac{x - 2}{x + 3}$ **21.** 30 **23.** x **25.** $5ab$ **27.** $x - 3$

29. $a - b$ **31.** 18 m **33.** $\frac{1}{12\text{ in.}}$ **35.** 20 gal

37. -1; opposite numerator and denominator simplify to -1.

39. -1 **41.** $-a + b$ **43.** $a - b$ **45.** $-a - b$

47. not opposites **49.** opposites **51.** $x + 2; x \neq -2$

53. $-x + 2; x \neq 2$ **55.** $b - a; a \neq b$ **57.** $3 - x; x \neq 3$

59. a. no **b.** -1

61. a. -2 **b.** $-\frac{1}{3}$

63. True; only factors may be simplified.

65. $\frac{a}{-a}, \frac{-a}{a}, -\frac{a}{a}$

67. Fractions simplify only if the numerator and denominator contain common factors.

EXERCISES 9.3

1. a. $\frac{1}{12}, \frac{4}{3}$ **b.** $\frac{1}{10}, \frac{5}{2}$ **3. a.** $\frac{1}{8}, \frac{9}{2}$ **b.** $\frac{1}{9}, 4$

5. a. 25 **b.** 500

The expressions within each part are equal; division by a number is the same as multiplication by its reciprocal.

7. a. x **b.** $\frac{1}{a^3b^2}$ **c.** $\frac{b}{a}$ **d.** $\frac{b^2}{a}$

9. a. 1 **b.** $\frac{x - 2}{x(x - 1)}$ **c.** $\frac{1}{x(x - 2)}$ **d.** $\frac{x}{(x + 5)(x - 5)}$

e. $\frac{x - 3}{(x + 3)^2}$

11. a. $x(x + 1)$ **b.** $\frac{1}{(x + 3)^2}$ **c.** $4(x - 1)$ **d.** $\frac{x(x - 1)}{x + 1}$

e. $-\frac{x(x - 1)}{(x - 2)(x + 1)}$

13. a. $-\frac{1}{x}$ **b.** $-c(c + d)$ **c.** $\frac{-x}{4}$ **d.** $-(a - b)$ or $b - a$

15. a. $\frac{(3x + 1)(2x - 3)}{3(x + 3)}$ **b.** $-\frac{(x + 3)(x + 2)}{2(2x + 3)}$

17. hour **19.** 500 sec **21.** 3 mpg **23.** ≈ 4 cookies/dollar

25. ≈ 0.28 page/min **27.** $3\frac{1}{3}$ L **31.** $I = \frac{t^2}{2t + 1}$

MID-CHAPTER 9 TEST

1. $x = -2, x = 1$; $1/((x + 2)(x - 1))$

2.

Input	Output	Budget
0	undefined	4800
10	480	4800
20	240	4800
100	48	4800
1000	4.80	4800

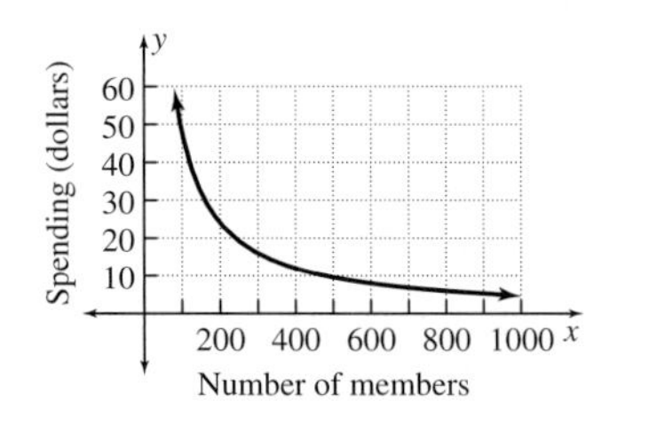

3. $y = \dfrac{4800}{x}, x \neq 0$ **4. a.** $+\infty$ **b.** $+\infty$ **c.** $-\infty$

5. 1.56, 3.23, undefined, −1.23, −1.79, −3.45 **6.** $1 < x < 1.3$

7. a. $\dfrac{6c}{7a}$ **b.** no common factors **8. a.** $\dfrac{1}{x - 2}$ **b.** $\dfrac{x - 1}{x - 2}$

9. a. $6x^2$ **b.** $12ab^2$ **10. a.** 15 **b.** $2(x + 2)$

11. $\dfrac{3}{xy}$ **12.** $\dfrac{2}{x}$ **13.** $-(x - 6)(x + 1)$ **14.** $\dfrac{x}{x - 4}$

15. $\dfrac{4}{x^3}$ **16.** 9 weeks **17.** 2 in. per sec **18.** 80 mL

EXERCISES 9.4

1. a. 1 **b.** $\dfrac{2 + x}{3}$ **c.** $-\dfrac{1}{x}$ **d.** $\dfrac{4 - x^2}{x^2 + 1}$ **e.** −1

3. a. 60 **b.** $2x$ **c.** y^2 **d.** ab **e.** $(x + 3)(x - 3)$
f. $(x + 3)(x + 2)(x - 3)$

5. a. $\frac{23}{30}$ **b.** $\dfrac{9}{2x}$ **c.** $\dfrac{8y - 1}{y^2}$ **d.** $\dfrac{8a - b}{ab}$

e. $\dfrac{4x + 10}{(x + 3)(x - 3)}$ **f.** $\dfrac{2x - 16}{(x + 3)(x + 2)(x - 3)}$

7. $\dfrac{6a + 3b}{4ab}$ **9.** $\dfrac{2x - 1}{x(x - 1)}$ **11.** $\dfrac{5x - 3}{x(x + 1)}$ **13.** $\dfrac{a - 1}{a^2}$

15. $\dfrac{4 - 3a}{2ab}$ **17. a.** $\dfrac{2x + 9}{(x + 3)(x + 3)}$ **b.** $\dfrac{-1}{x + 1}$

19. a. $\dfrac{x^2 + 3x - 6}{x(x - 1)(x - 2)}$ **b.** $\dfrac{-1}{x(x + 3)}$ **21.** $\dfrac{T_0 - T}{T}$

23. $\dfrac{C_2 + C_1}{C_1C_2}$ **25.** $\dfrac{D(r_2 + r_1)}{r_1r_2}$ **27.** $\dfrac{L^2 + 3d^2}{6d}$

29. 5 **31.** $\frac{1}{5}$ **33.** $-\dfrac{a}{3b}$ **35.** $\dfrac{b}{a - c}$; slopes are equal

37. $\frac{25}{6}$ **39.** $\dfrac{2A}{b}$ **41.** $\dfrac{9x}{2(6 - x)}$ **43.** $\dfrac{Q_L}{Q_H - Q_L}$

45. a. and b. Both graphs are the same:

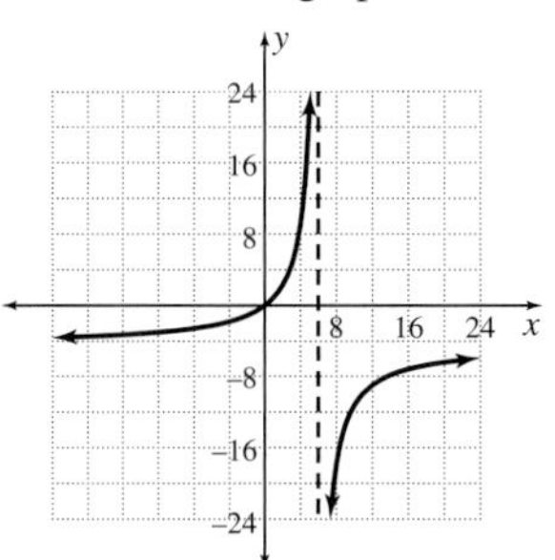

c. near $x = 6$; undefined when $x = 6$

47. a. · (mult.) **b.** ÷ **c.** + **49. a.** + **b.** · (mult.)

EXERCISES 9.5

1. $x + 8$ **3.** $2x + 12$ **5.** $2x - 1$ **11.** 12 **13.** $20x$
15. $8x$ **17.** $3x$ **19.** x **21.** $(x - 1)(x + 3)$ **23.** x^2
25. $x - 3$ **27.** $x = 67.2$ **29.** $x = \frac{20}{19}; x \neq 0$
31. $x = \frac{8}{3}; x \neq 0$ **33.** $x = 2; x \neq 0$ **35.** no solution; $x \neq 0$
37. $x = 5; x \neq 1, x \neq -3$ **39.** $x = 1$ or $x = 2; x \neq 0$
41. no solution; $x \neq 3$ **43.** $x \neq 0; x = \frac{1}{5}, x = -\frac{3}{2}$
45. $x \neq 0, x \neq -1; x = \frac{2}{3}, x = -2$ **47.** $x \neq 1; x = 3, x = 4$
49. $x \neq 2$, no solution; discard $x = 2$
51. $x \neq 5; x = -2$; discard $x = 5$

53. $x \neq -1; x = \dfrac{-2 \pm \sqrt{14}}{2}$ **55.** $d = \frac{90}{11}$ **57.** $x = \frac{40}{7}$

59. 6.5 days **61.** 2.2 hr; yes **63.** one that vents in $7\frac{1}{2}$ hr

65. 3.6 min **67.** $b = \dfrac{ac}{a - c}$ **69.** $c = \dfrac{ab}{a + b}$ **71.** $d = 8.182$

73. $x = 5.714$

CHAPTER 9 REVIEW EXERCISES

1. $x = -3$ **3.** $y = \dfrac{6800 \text{ trillion}}{365x}$ **5. a.** $-\infty$ **b.** $-\infty$

7. $\dfrac{2y}{x}$ **9.** $a - b$ **11.** no common factors

13. a. $-1; a \neq 1$ **b.** $\dfrac{6x + 5}{2(3x + 5)}; x \neq -\dfrac{5}{3}, x \neq 2$

15. $3 - x$ **17.** $a - b$ **19.** $x - 4$
21. $\frac{16}{9} \neq \frac{4}{3}$; do not simplify to the same fraction
23. a. 100, 100 **b.** Both equal 100.
c. Division by a is the same as multiplication by its reciprocal, $\dfrac{1}{a}$.

25. each, one-third; 12 **27.** half as far; $\frac{3}{8}$ mi **29.** remains; $\frac{1}{6}$

31. $\dfrac{-x^2}{(x + 1)^2}$ **33.** $n + 1$ **35.** $\dfrac{-3x}{(x + 3)(x - 3)}$

37. $\dfrac{x^2 - 2x - 2}{(x + 2)(x - 2)}$ **39.** $\dfrac{a^2 - 2ab - b^2}{(a + b)(a - b)}$

41. $\dfrac{720 - 360x^2 + 30x^4 - x^6}{720}$ **43.** $\dfrac{3 + 36a - a^2}{3a}$

45. $\frac{1}{3}$ shirt per card **47.** 3 tab **49.** 5 mL **51.** $\dfrac{2d}{g}$

53. $3x + 2$ **55.** $x = 3; x \neq 0$ **57.** $x = -6$ or $x = 6; x \neq 0$

59. $\dfrac{m_2 + m_1}{m_1m_2}$ **61.** 1.2 min

63.

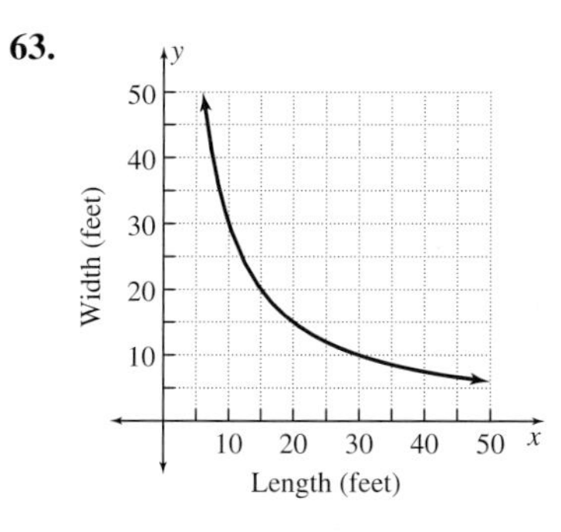

CHAPTER 9 TEST

1. $x = 4$
2. Because $4 - x$ is the opposite of $x - 4$, the expression simplifies to -1, as long as $x \neq 4$
3.

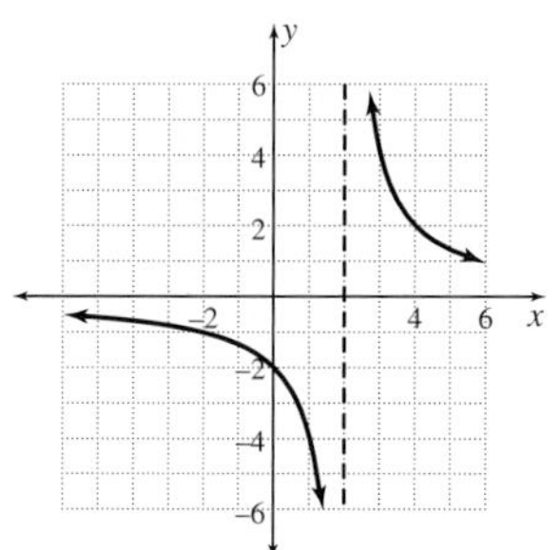

4. a. $\frac{2}{3}$; equivalent b. $\frac{1}{2}$ c. $\frac{3}{4}$ d. $\frac{2}{3}$; equivalent
Only a and d satisfy the simplification property of fractions.
5. $\frac{b}{3a}$ 6. no common factors 7. $a + b$ 8. $\frac{x}{x+1}$
9. $\frac{(1-x)(x-1)}{x^2}$ or $\frac{-(x-1)^2}{x^2}$
10. $\frac{1}{(1-x)(x-1)}$ or $\frac{-1}{(x-1)^2}$ 11. $\frac{n+1}{n-1}$ 12. $\frac{x-2}{x+2}$
13. $3x + 1$ 14. \$35.92/shirt 15. \$0.10/mi 16. 6 tab
17. ≈ 7.3 min 18. $\frac{q+p}{pq}$ 19. $\frac{2A}{b}$ 20. $\frac{31}{45}$
21. $\frac{4 - a^2b}{2ab^2}$ 22. $\frac{x^2 - 2x + 2}{(x+2)(x-2)}$
23. $\frac{2x^2 + 9x - 12}{3x^2}$ 24. $\frac{2a+1}{4}$ 25. $x = 2; x \neq 0$
26. $x = 1; x \neq 0$ 27. $x = 9; x \neq 0, x \neq 3$
28. $x = 5$ or $x = -3; x \neq 0$ 29. $x = 5; x \neq 2$; discard $x = 2$
30. Factoring permits us to simplify before multiplying.

FINAL EXAM REVIEW

1. a. -12 b. 5 c. 5 d. -60 e. 64 f. 4
g. 5 h. $\frac{2}{5}$ i. $-\frac{4}{5}$
3. a. $\frac{c}{f}$ b. $\frac{-r}{4x}$ c. a d. -1 e. $\frac{32m^6n}{3}$ f. $\frac{b^2}{c^2}$
g. n^7 h. m^3n^6 i. $\frac{1}{m^3}$ j. 1 k. $\frac{3y^2}{x^5}$ l. $\frac{27y^6}{x^9}$
5. a. $y = 4 - \frac{2}{3}x$ b. $x = \frac{7}{2} + \frac{3}{2}y$ c. $y = \frac{c}{b} - \frac{a}{b}x$
d. $x = \frac{3}{5}$ e. $P_2 = \frac{P_1V_1T_2}{T_1V_2}$ f. $x = 19$ g. $x \approx \pm 6.9$
h. $d \approx \pm 7.6$ i. $x = 13$
7. a. $3xy(4 + y + 2xy)$ b. does not factor c. $(x + 5)(x - 4)$
d. $(2x - 3)(2x + 3)$ e. $(2x - 1)(3x + 2)$
f. $(x - 2)(6x + 1)$ g. $(x - 5)^2$ h. $3(x - 1)(2x + 3)$
9.

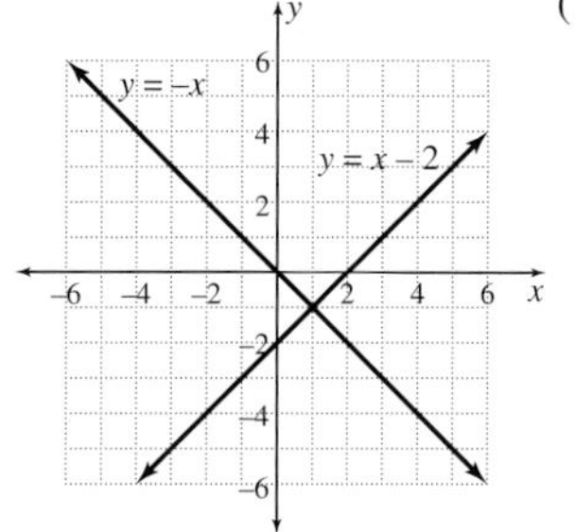

11. $x = -2, y = 4$ 13. $x = -4, y = -2$
15. infinite number of solutions
17. slope is undefined; distance = 8
19. slope = $-\frac{1}{7}$; distance = $5\sqrt{2}$ 21. no solution; $x \neq 0$
23. $x = -\frac{3}{2}; x \neq 4$; discard $x = 4$ 25. $x = 2$ or $x = -7; x \neq -5$
27. true for all real numbers 29. $x = 12; x < -13$
31. $x = -13; x < -13$
33. a. (4, 0) b. (0, 2) c. 3 d. undefined
e. because $\sqrt{4 - x}$ is always positive or zero
35. a. Person went off the diet after about 13 days and gained weight.
b. 12 to 13 days
c. 25 days
d. The dieter will have gained 14 pounds.
37. a. 17 b. 25.2 c. 9 d. 20 e. $8|x|$ f. $8x$
g. undefined h. undefined in the real numbers
39. $\frac{2x-1}{x+2}$ 41. a. \$2/room cleaned b. 21.1 ft
43. a. $-\frac{x}{y^2}$ b. $-\frac{(x+1)(x+2)}{x}$ c. $\frac{(2x-1)(x-4)}{(x-2)(x+2)}$
d. $\frac{x+1}{x+3}$ e. $\frac{-1}{15x}$ f. -20 g. $x^2 - 2x + 12$
45. a. \$25, \$31, \$43, \$44, \$52.50
b. $y = x$; $y = 30 + 0.20(x - 30)$; $y = 50 + 0.50(x - 100)$
c. [ ] includes endpoints; ( ) excludes endpoints
d. yes; use parentheses
e. $0 < x \leq 30$; $30 < x \leq 100$; $x > 100$
47. a. 2.76×10^9 mi b. ≈ 6.8 hr c. ≈ 3.55 mph
d. 17,600 mph
49.

x^3	$6x^2$
1000	600
8000	2400
64,000	9600
n^3	$6n^2$
$8n^3$	$24n^2$

b. 8 times as large c. 4 times as large d. ≈ 28.28 in.

Glossary/Index

Absolute value The nonnegative distance a number is from zero. 57
addition using, 57–58
and square root, 481
Absolute value symbol, 90
Acute angles, 349
Adding like terms Adding the numerical coefficients of terms with identical variable factors and exponents. 79
Addition
associative property of, 74
commutative property of, 75
of integers, 56–58
of like terms, 79
of polynomials, 392–393
of rational expressions, 555–556
using absolute value, 57–58
Addition property of equations The property that says that adding the same number to both sides of an equation produces an equivalent equation. 136, 356
extension to, 356
Addition property of inequalities If $a < b$, then $a + c < b + c$. 170
Additive inverse The number that, when added to n, gives zero; the opposite. 55
Agreement A common way to do something. 19
Algebra, 18
Algebraic fractions Expressions in fractional notation. 73. *See also* Rational expressions.
Algebraic notation, 21–23
solving equations with, 159–160
solving inequalities with, 169–171, 173
Algebra tiles Objects used to represent algebraic symbols. 79, 390–391, 411–412
factoring with, 395, 402–403, 411–413, 422
negative terms and, 411
perfect square trinomial and, 407, 422
Area The measure of the surface enclosed within a flat object. 18, 97, 99–100
Ascending order For a polynomial, arrangement of terms so that exponents run from lowest to highest. 392
Associative properties The properties that permit numbers to be grouped in any way that is convenient for adding or multiplying. 74–75
Assumption Something not stated but taken as a fact. 2
Average rate In a quantity-rate table, the number derived by dividing the sum of the $Q \cdot R$ column by the sum of the quantity column. 307, 309
Averages, 294–296
in geometry, 299–301
weighted, 297–299
Axes Number lines that divide a coordinate plane into four quadrants. 39
Axis of symmetry A line across which a graph can be folded so that points on one side of the graph match up with points on the other side of the graph. 487

Base The number x in the power expression x^n. 85
negative sign and, 85
of parallelogram, 98
and properties of exponents, 431
of triangle, 98
Bimodal Having two modes. 296
Binomial A polynomial expression with two terms. 391
multiplication of, 402–405
Boundary line The line between two half-planes. 241
of inequalities, 375–378
parallel to axes, 378–379
Box and whisker plot A 5-number visual summary of data. 514
Braces, 90
Brackets, 90

Calculator
angles from slope on, 2ND TAN, 318
division and parentheses on, 95
grouping symbols on, 91
negative exponents on, 433
scientific notation on, 438, 441–442

Calculator (*cont.*)
- scientific notation mode on, 442
- tangent on, (TAN) A, 318
- zero exponents on, 433

Calculator, graphing
- absolute value on, (2ND) [CATALOG], 91
- box and whisker plots on, 517–518
- checking a solution on, 141, 507
- clearing list 1 on, (STAT) **4 : ClrList** (2ND) [L1], 231
- decimal windows near zero on, 533
- displaying division by zero with tables on, 531–532
- displaying graphs when denominator is zero on, 534
- equals sign on, (2ND) [TEST] **1 : =**, 141
- evaluating denominators on, 266
- evaluating an expression with replay on, 102
- evaluating with [TABLE] on, 103
- finding corner point on, 376
- finding mean on, 296–297
- finding median on, 296–297
- finding one-variable statistics on, 517–518
- fractions on, (MATH) **1 : ▶ Frac** (ENTER), 71
- graphing two-variable inequalities on, 245
- intersection and coincident lines on, 328
- intersection and parallel lines on, 327
- linear regression on, (STAT) **CALC 4 : LinReg**, 231
- negative on, (–), 62
- obtaining linear regression variables on, (VAR) **5 : Statistics EQ**, 232
- obtaining list screen on, (STAT) **1 : Edit**, 231
- opposite on, (–), 62
- pi on, (2ND) [π], 99
- prior answer on, (2ND) [ANS], 71
- reciprocal on, (x^{-1}), 71
- replay option on, (2ND) (ENTER), 102
- setting a viewing window on, 148
- shading options on, 246
- slope on, 206
- solving an equation from a graph on, 148, 159
- solving an equation from a table on, 146, 159
- solving an inequality on, 173–174
- solving systems of inequalities on, 379, 380
- turning off axes on, 174
- using intersection on, 325
- vertical line graphs on, 223
- viewing right angle on perpendicular lines on, 234
- *y*-intercept on, 218

Capacity The amount (especially of liquids) a container holds. 100

Cartesian coordinates A pair of numbers that indicate the position of a point on a flat surface by the point's distance from two lines. 38, 39

Center of gravity, 300

Center of mass, 300

Centroid The center of a flat or solid geometric shape; the average of the coordinates of the vertices of a rectangle, square, or triangle. 300–301

Circle, formulas for, 98

Circle graphs, 269

Circumference The perimeter of a circle. 97

Coefficients
- like, 356–358
- positive, 171
- unlike, changing to like, 359

Coincident lines Two lines that have the same equation. 328, 347

Common factor A factor that appears in each term of an expression or numerator and denominator of a fraction. 78, 395

Common monomial factor A factor that appears in each term of a polynomial. 395

Commutative properties The properties that let us change the order in which addition or multiplication is done. 75–76

Complementary angles Two angles whose measures add to 90º; any two angles that form a right angle. 349–360

Complex fractions Fractions that contain fractions in either the numerator or the denominator. 549
- simplifying, 558–559

Complex rational expressions Rational expressions that contain fractions in either the numerator or the denominator. 549, 558

Composite number A number with factors other than 1 and itself. 81

Compound inequality Two inequalities in one statement. 108, 176

Condition A requirement or restriction stated within a problem setting. 2

Conditional equation An equation that is true for only certain values of the variable(s) in the equation. 129

Conditional inequality, 176

Consecutive integers Integers that follow one after another without interruption. 134

Constant A number, letter, or symbol whose value is fixed. 22

Continued equality Three or more equivalent expressions separated by two or more equals signs. 68

Continued ratio A ratio of three or more quantities. 258–259

Converse of the Pythagorean theorem If the sum of the squares of the lengths of the two shorter sides (legs) in a triangle is equal to the square of the length of the longest side (hypotenuse), then the triangle is a right triangle. 456–457

Coordinate graph A method of identifying position by two numbers. 38

Coordinate plane A flat surface containing the coordinate axes. 38, 39
Coordinates Ordered pairs of numbers. *See* Cartesian coordinates.
Corner point The point of intersection of the boundary lines of two inequalities. 376
Corresponding In the same relative position. 282
Cross multiplication Setting the diagonal products in a proportion equal. 272
and solving equations, 276–277
Cross multiplication property of proportions The property that says that the proportion $a/b = c/d$ implies $ad = bc$, where $b \neq 0$ and $d \neq 0$. 271
Cubed Having an exponent of 3. 86
Cylinder, formulas for, 101

Decimal notation, 439–441
Degree, 485
Delta (Δ) Symbol indicating a change in the variable that follows. 146, 209
Denominator The bottom number in fraction notation. 11
eliminating, 564
like and unlike, 555–556
Dependent variable The output variable in an application setting, frequently y. 130
Descartes, René, 27
Descending order For a polynomial, arrangement of terms so that exponents run from highest to lowest. 392
Deviation The difference between each number in a set and the mean of the set. 516
Diagonal products, 405
Difference The answer to a subtraction problem. 18
Difference of squares An expression of the form $a^2 - b^2$, which is the result of multiplying the binomials $(a - b)(a + b)$. 407
Distance formula A formula that enables one to find the length, or distance, between two coordinate points, (x_1, y_1) and (x_2, y_2):

$$d = \sqrt{(x_2 - x_1)^2 + (y_2 - y_1)^2}.\ 470$$

Distribution of multiplication over subtraction, 77
Distributive property of multiplication over addition The property that lets us change certain products into sums: $a(b + c) = ab + ac$. 76–77
factoring with, 395
solving equations with, 149–150
Division
as fraction, 13
of integers, 69–70
of rational expressions, 548–549
by zero, 531–532
Division of like bases property, 431
Domain The set of inputs to a function. 195
for rational functions, 532
for square root function, 476
Double slash, 43

Electrical charge model, 56–57
Elevation, 62
Elimination method A process for solving a system of equations in which one variable is removed from the system by adding (or subtracting) the respective sides of the two equations. 356
of denominators, 564
steps in solving equations by, 360
Empty set A set with nothing in it, denoted by { } or ∅. 149
Equals sign A symbol indicating that two or more numbers or expressions have the same value. 11, 129
Equation A statement of equality between two quantities. 19, 127
Equation in one variable A statement of equality containing only a single variable. 129
Equation in two variables, 128
Equilateral triangle A triangle with three equal sides. 472
Equivalent equations Two or more equations that have the same solution set. 136
Equivalent expressions Expressions that have the same value for all replacements of the variables. 61
Equivalent fraction property The reverse of the simplification property of fractions; for all real numbers, $a \neq 0$ and $c \neq 0$:

$$\frac{b}{c} = \frac{b}{c} \cdot \frac{a}{a} = \frac{ab}{ac}.\ 542$$

Equivalent inequalities Two or more inequalities that have the same solution set. 169
Equivalent ratios Ratios that simplify to the same number. 259
Estimated scale for an axis, 42
Estimation
of populations, 275–276
simplifying radical expressions and, 506–507
Evaluate To substitute numbers in place of the variables in a formula or expression and simplify. 86
Even numbers The integers divisible by two. 35, 134
Exponent The number to which a base is raised: n in b^n. 85, 429
negative, 432
one-half as, 467
operations with, 86–87
positive integer, 85

properties of, 430, 431
and square roots, 467
zero, 432
Expression Any combination of signs, numbers, constants, and variables with operations such as addition, subtraction, multiplication, or division. 22
Extraneous root A solution found algebraically that does not satisfy the original equation. 502, 565

Factoring, 395–396
ac-product method of, 417–418
with distributive property, 395
and greatest common factor, 423–425
by grouping, 417
by guess and check, 418–419
before simplifying, 541–542
solving quadratic equations by, 498–500
of special products, 422–423
by table, 412–415, 417
with tiles, 411–412
Factors Two or more signed numbers, variables, or expressions being multiplied. 12, 76, 397
Feynman, Richard, 37
FOIL method, 405–406
Formula A rule or principle written in mathematical language. 97
for geometric figures, 98, 101
solving, 187, 188–189, 479–480
for sum of rates, 567
using, 99
Four-step cycle for problem solving, 4–5, 333
Fractions. *See also* Rational expressions.
operations with, 13–14
signs on, 70
simplification property of, 77, 541
Function A relationship or association where for each input x there is exactly one output y. 194
evaluating, 197–198
Function notation Notation for writing functions in symbols. 196

Garfield, James A., 463
Grade The slope of a road as a percent. 263
Graph (noun) The points that have been plotted on a coordinate plane or the line or curve drawn through those points; (verb) to locate points on the coordinate plane as described by ordered pairs. 40
finding slope from, 205–207
four steps for creating, 40–42, 44
of rational expressions, 533–534
reading, 43
solving equations from, 144, 147–149, 157–159
solving inequalities from, 110–112, 167–169
Greatest common factor (gcf) The largest possible factor that appears in both the numerator and the denominator of a fraction; the largest possible common factor of all terms in an expression or polynomial. 78, 395
factoring out, 423–425
Grouping symbols Symbols used so as to apply the order of operations. 90
Guess and check
building systems of equations by, 333
factoring by, 418

Half-plane The region on one side of a line in a coordinate plane. 241
Horizontal axis The number line on a coordinate plane that goes left to right. 39
Horizontal-axis intercept point The point where a graph crosses the horizontal axis, defined by the ordered pair $(a, 0)$. 217, 218
Horizontal fraction bar, 90
Horizontal line
equation of, 223
slope of, 209–210
Hypotenuse, 453

Identity A type of equation formed when the expression on the left side of the equals sign is equal to the expression on the right side for all values of the variable(s). 129
In terms of Phrase that indicates the named variable is by itself on one side of the equals sign and terms containing the other variables, numbers, and operations are on the other side. 187, 200
Independent variable The input variable in an application setting, frequently x. 130
Index (pl. *indices*), 464
Indirect measurement, 285–288
Inequality A statement that one quantity is greater than or less than another quantity. 107
addition property of, 170
boundary lines of, 375–378
compound, 108
multiplication property of, 169, 170
solving with algebraic notation, 169–170, 173
solving from a graph, 110–112, 167–169
Infinite Without bound. 11, 110
Infinity sign The symbol ∞, indicating that numbers get large without bound. 110
Inner-outer product method, 279
Input-output rule A rule that tells what to do to the input to get the output. 20
Input-output table A tabular form for describing or summarizing numerical relationships. 4, 20

Integers The numbers in the set $\{\ldots, -3, -2, -1, 0, 1, 2, 3, \ldots\}$; the set of natural numbers, their opposites, and zero. 9, 55
addition of, 56–58
consecutive, 134
division of, 69–70
multiplication of, 66–68
subtraction of, 59–62
Intercepts
horizontal-axis, 223
vertical-axis, 223
Interval A set containing all the numbers between its endpoints as well as one endpoint, both endpoints, or neither endpoint. 109
Inverse, 55
Inverse operations, 136–140
Irrational numbers Real numbers that cannot be written as the quotient of two nonzero integers. 11, 465
Isosceles triangle A triangle with two equal sides and equal angles opposite these sides. 371, 472

Jet stream, 338

Least common denominator (LCD) The smallest number into which two or more denominators divide evenly. 554, 555
Legs, of right triangle, 453
L'Enfant, Pierre Charles, 37
Light-year, 389
Like terms Terms with identical variable factors and exponents. 79
patterns in, 404
Line graph, 110–112
Line of symmetry, 487
Linear equation, 218
slope and, 218–219, 228
y-intercept and, 220–221, 228
Linear equation in one variable An equation that can be written in the form $ax + b = 0$, where x is the variable and a and b are real-number constants. 128–129
Linear equation in two variables An equation that can be written in the form $y = ax + b$, where x and y are variables and a and b are real-number constants; an equation describing a set of input-output pairs, or ordered pairs (x, y), whose graph makes a straight line. 128
Linear function, 224
equation of, 224
Linear inequality in one variable An inequality that can be written in the form $ax + b < c$ (using $<, \leq, >$, or $\geq$), where a, b, and c are real numbers and a is not zero. 167, 173
Linear inequality in two variables An inequality that can be written in the form $ax + b < c$ (using $<, \leq, >$, or $\geq$), where a, b, and c are real numbers and a and b are not both zero. 239
graphing, 241–242
Linear pair Two angles that share a side and whose other sides form a straight line. 348
Linear regression A statistical function for finding a linear equation from data or ordered pairs. 231
Lowest terms The condition of a fraction when the numerator and denominator have no common factors. 12, 77

Maps, 37–38
Mean The sum of a set of numbers divided by the number of items in the set. 294, 516
Measures of central tendency, 513
Measures of variation or dispersion Measures that describe how close to the middle or how scattered a set of data is. 513
Median The middle number in a set of numbers arranged in numerical order. 294, 514–515
Midpoint of a line segment The center; the point halfway between the endpoints of the line segment. 299
Midpoint on a coordinate graph The mean in both the x and the y direction. 299
Mode The number that occurs most often in a set of numbers. 294
Molarity, 312
Monomial A polynomial expression with one term. 391
Multiplication, 69
associative property of, 74
commutative property of, 75
of integers, 66–68
notation for, 21
of polynomials, 393–395
of rational expressions, 547
Multiplication of like bases property, 431
Multiplication property of equations The property that says that multiplying both sides of an equation by the same nonzero number produces an equivalent equation. 136
Multiplication (by a negative number) property of inequalities The property that says that the direction of the inequality sign must be changed when both sides of an inequality are multiplied (or divided) by the same negative number. If $c < 0$ and $a < b$, then $ac > bc$. 169
Multiplication (by a positive number) property of inequalities The property that says that the direction of the inequality sign is not changed when both sides of an inequality are multiplied (or divided) by the same positive number. If $c > 0$ and $a < b$, then $ac < bc$. 170

Multiplicative inverse The number that, when multiplied by n, gives 1; the reciprocal. 69

Natural numbers The numbers in the set {1, 2, 3, 4, . . .}. 10
Negative integer exponents, 432
Negative numbers The numbers less than zero. 10, 55
Negative sign, and base, 85
Negative slope, 205, 207
Net charge, 56
Nonlinear equation An equation whose graph does not form a straight line. 128
Nonnegative Positive or zero. 57
Notation, 11. *See also* Algebraic notation.
Numerator The top number in fraction notation. 11
Numerical coefficient The sign and number multiplying a variable or variables. 22, 357

Odd numbers The integers not divisible by two. 35
One-half as exponent Alternative notation for the principal square root. 467
Operations, words for, 18
Opposite of an opposite, 61
Opposites Numbers on different sides of zero and the same distance from zero on a number line; two expressions that add to zero. 9, 55–56, 543
Order of operations An agreed-upon order in which mathematical operations are performed. 88–89, 91
Ordered pair The pair of numbers that identify a position on a coordinate plane. 39
  finding linear equations from, 228–232
  finding slope from, 208
Origin The point where the axes on a coordinate plane cross. 39

Parabola The name given to the graph of a quadratic equation. 486
Parallel lines Two lines in the coordinate plane with equal slope but different y-intercepts. 232, 326–327, 346–347
Parallelogram, 348
  formulas for, 98
Parentheses, 90, 130–131, 151
Pearson, Karl, 516
Percent Per hundred or division by 100; the ratio $n\%$ means n per hundred or division of n by 100. 30, 263, 265, 273–274
Percent change An expression of change found by subtracting the original number from the new number and dividing the difference by the original number. 264
Percentage The part of arithmetic dealing with percents. 30
Perfect square trinomial An expression of the form $a^2 + 2ab + b^2$ whose terms are the square of a, twice the product of a and b, and the square of b; the result of squaring a binomial. 406
Perimeter The distance around the outside of a flat object. 18, 97, 98–99
Perpendicular lines Two lines that cross at a right angle. 97, 233–234
Pi The number found by dividing the circumference of any circle by its diameter. 99
Polya, George, 2
Polynomial An expression with only nonnegative integer exponents on the variables and with one or more terms being added or subtracted. 391
  adding and subtracting, 392–393
  arranging terms in, 392
  multiplying, 393–395
Positive integer exponent, 85, 429
Positive numbers The numbers greater than zero. 55
Power A base and an exponent together: b^n. 85
Power of a power property, 431
Power of a product property, 430
Power of a quotient property, 430
Prime number A number greater than one with no integer factors except 1 and itself. 12, 81
Principal square root The positive number that, multiplied by itself, produces a given number. 465
Problem solving
  four steps in, 2–5, 333
  strategy for, 3, 135–136
Product The answer to a multiplication problem. 18
Product property for square roots, 468
Proof A logical argument that demonstrates the truth of a statement. 454
Property A statement that is always true for a given set of numbers. 19
Proportion An equation formed by two equal ratios. 271
  matching units in, 272
  solving application problems with, 274–276
  solving percent problems with, 273–274
Protractor, 348
  and slope, 269
Pythagoras, 454
Pythagorean theorem If a triangle is a right triangle, then the sum of the squares of the lengths of the two shorter sides (legs) is equal to the square of the length of the longest side (hypotenuse). 453, 454–455
Pythagorean triples Sets of three numbers that make the Pythagorean theorem true. 458

Quadrants The four sections of a coordinate plane. 39
Quadratic equation
  interpreting a, b, and c in, 489
  solving by factoring, 498–500
  solving with tables and graphs, 491–492

solving by taking square root, 496–498
special features of graph of, 481, 490
Quadratic formula For $ax^2 + bx + c = 0$,

$$x = \frac{-b \pm \sqrt{b^2 - 4ac}}{2a}.$$ 505, 506

proof of, 509–510
solving equations with, 506–509
Quadratic function A function that may be written as $f(x) = ax^2 + bx + c$, where a, b, and c are real numbers and a is not zero. 486
graphing, 486–488
range of, 491
Quantity A number, word, or phrase that answers the question "How many?" or "How much?" 305
Quantity-rate equation, 338
Quantity-rate table A table that displays the quantities and rates from a particular problem setting, along with their products. 306–307
writing and solving equations using, 308–311, 333–337
Quartiles Q_1 is the middle number of the numbers below the median; Q_3 is the middle number of the numbers above the median. The median is Q_2. 515
Quotient The answer to a division problem. 18
Quotient property for square roots, 468

Radical (radical sign) The square root symbol; more generally, the name for square roots and higher-degree roots. 90, 464
Radicand The number or expression under the radical sign. 464
keeping positive, 476
Range The set of outputs from a function; also, the difference between the largest and the smallest number in a data set. 195, 491, 513
Rate A special ratio for comparing quantities with different units; a number that expresses monetary worth, percent, or a specific result. 261, 305
Ratio A comparison of two (or more) like or unlike quantities. 258
simplifying, 259–261
Rational expressions
addition and subtraction of, 555–557
complex, 549
division of, 548–549
graphs of, 533–534
multiplication of, 547–548
Rational equations, solving, 564–566
Rational function A function that may be written as the ratio of two polynomials $\frac{p(x)}{q(x)}$, where the denominator is not zero. 531, 535–537
Rational numbers The set of numbers that may be written as a quotient, or ratio, of two integers $\frac{a}{b}$, $b \neq 0$. 10, 11, 531
Real numbers The set containing both rational and irrational numbers. 11
using properties of, 73–77
Reciprocals Two numbers whose product is 1; also the multiplicative inverse. 14, 69
Rectangle, formulas for, 98
Rectangular prism (box), formulas for, 101
Rise The vertical change in a line. 207, 210
Run The horizontal change in a line. 207, 210

Sample standard deviation A measure of the variation between the mean of a set and each number in the set. 516
Scale The value assigned to the spacing between numbers on the coordinate axes. 42
Scientific notation Describes numbers written as the product of a decimal between 1 and 10 and a power of ten. 438
converting from and to decimal notation, 440–441
estimating with, 441
operations with, 441–442
Set A collection of objects or numbers. 9
Similar figures Geometric figures with corresponding sides that are proportional and corresponding angles that are equal. 282–283
Similar triangles, 283, 288
Simplification property of fractions If the numerator and denominator of a fraction contain the same factor (a common factor), those factors can be eliminated:

$$\frac{ab}{ac} = \frac{a}{a} \cdot \frac{b}{c} = \frac{b}{c}, a \neq 0, c \neq 0.$$ 77, 541

Simplify To use the simplification property of fractions to eliminate common factors as well as to do the given operations; to combine expressions with like bases and use the properties of exponents to do the indicated operations; to use the order of operations to calculate the value of an expression; to apply the properties of exponents; to add like terms; to change fractions to lowest terms; or to use the real number properties. 78, 89, 259–261, 467, 558–559
Slope The steepness of a line, expressed as the vertical change relative to the horizontal change. 205, 212
finding, 205–210
formula for, 208
and graphing lines, 221–224
and linear equations, 218–221
Solution A value of the variable that makes the equation true. 135
Solution set The set of all solutions to an equation. 135

Solution set of a one-variable inequality The set of values of the input variable that make the inequality a true statement. 167
Solution set to a system of inequalities The set of ordered pairs that make all inequalities in the system true. 374, 375
Solution to a system An ordered pair that makes both equations in a system of two linear equations true. 322
Solving an equation The process of finding the values of an input variable for a given output. 127, 135
using algebraic notations, 150, 159–161
using distributive property, 149–150
using graphs, 144, 146–149, 157–159
using tables, 144–146, 157–159
Solving a one-variable inequality, 167–171
Solving rational equations, 564–566
Special products, 406–408
factoring, 422–423
Sphere, formulas for, 101
Square, formulas for, 98
Square numbers, 86
Square root(s)
and absolute value, 481
exponents and, 467
multiplication and division with, 467–468
of a negative number, 466
negative signs with, 467
of powers of x, 480–481
simplifying expressions with, 468–469
solving quadratic equations with, 496–498
Square root equations
graphing, 474–476
solving, 476–478
Square root function, 466
Square root symbol The symbol $\sqrt{\ }$, which asks for the positive number that, when multiplied by itself, gives the number inside the symbol. 9
Squared Having an exponent of 2. 86
Squaring both sides, 477–478
Standard deviation, 516
Standard form of quadratic equation, 489
Statistics, 275
Subscripts Small numbers or letters placed to the right of variables to distinguish a particular item from a group of similar items. 147
Substitution method A process for solving a system of equations by replacing variables with equivalent expressions or numbers in order to eliminate one variable. 135, 343
solving systems of equations by, 356–358, 360–362, 368–370
Subtraction
by adding the opposite, 60–61
of integers, 59–61
of polynomials, 392–393
of rational expressions, 555–557
Sum The answer to an addition problem. 18
Sum of rates, 566–568
Supplementary angles Two angles whose measures add to 180º. 348, 360
Surface area The area needed to cover a three-dimensional object. 100, 101
System of equations A set of two or more equations that are to be solved for the values of the variables that make all the equations true, if such values exist. 322
algebraic and geometric results for, 346–347
building with guess and check, 333
building with quantity-rate tables, 333–337, 339
with an infinite number of solutions, 327–328, 329, 369–370
with no solution, 326–327, 329, 369
with one solution, 323–324, 329
solving by elimination, 355–358, 360
solving by substitution, 343–346, 360
strategies for solving, 350
in three variables, 367–371
System of inequalities A set of two or more inequalities. 374
finding solution sets to, 375

Tables
for binomial multiplication, 402–405
finding averages from, 295–296
finding slope from, 208–209
quantiy-rate, 306–311, 333–337
solving equations from, 144–146, 157–159
Term A signed number, variable, or expression being added or subtracted. 76, 397
Test point, 111
Trapezoid, formulas for, 98
Travel formula, 339, 361–362, 535–536
Triangles
equilateral, 472
formulas for, 98
isosceles, 371, 472
missing side of, 455–456
similar, 283–288
Trinomial A polynomial expression with three terms. 391
without whole-number factors, 425–426

Undefined Having no mathematical meaning. 9
Undefined slope The slope of a vertical line. 210
Unit analysis A method for changing from one unit of measure to another or from one rate to another. 95–97
Units of measure, 87, 542
Unlike coefficients, changing to like coefficients, 358–359

Variable A letter or symbol that can represent any number from some set. 19
Vertex A corner of a triangle (pl. *vertices*); also, the highest or lowest point on a parabola. 283, 486
Vertical axis The number line on a coordinate plane that runs up and down. 39
Vertical-axis intercept point The point where a graph crosses the vertical axis, defined by the ordered pair $(0, b)$. 217, 218
Vertical line
  equation of, 223
  slope of, 209–210
Vertical-line test A graph shows a function if every vertical line intersects the graph no more than once. 199
Volume The space taken up by a three-dimensional object. 100, 428

Web sales model, 66
Weight A value assigning importance to a number. 297
Weighted average An average found by multiplying each number in a set by its weight, adding the products, and dividing by the total weight. 297–299
Whole numbers The set of natural numbers along with the number zero: $\{0, 1, 2, 3, \ldots\}$. 10
Wind speed, 339
Working backwards, 135–136

***x*-axis** The horizontal axis on a coordinate plane. 39
***x*-intercept** The number a in $(a, 0)$. 217
***x*-intercept point** The ordered pair $(a, 0)$, which defines the horizontal-axis intercept point. 217

***y*-axis** The vertical axis on a coordinate plane. 39
***y*-intercept** The number b in $(0, b)$. 217
  finding from linear equation, 220–221
***y*-intercept formula** If m is the slope and (x, y) is any point on the line, the vertical-axis intercept is $b = y - mx$. 229
***y*-intercept point** The ordered pair $(0, b)$, which defines the vertical-axis intercept point. 217

Zero, division by, 531–532
Zero product rule If the product of two expressions is zero, then either one or the other expression is zero. If $A \cdot B = 0$, then either $A = 0$ or $B = 0$. 498
Zero slope The slope of a horizontal line. 210

(continued from front endpaper)
Bicycling, 267
Bouncing ball, 54
Bowling, 132, 156, 191
Cross-country skiing, 204–205
Figure skating, 301
Paintball, 182, 203
Racquetball, 238
Running, 365
Soccer, 99
Speed skating, 525
Street racing, 386
Swimming, 366
Tennis, 94
Track and field, 303

Travel & Transportation

Accident skid marks, 525
Airplanes, length of flight of, 293
  loading of, 385
  rental of, 238
  seating on, 180
  speed of, 339, 342, 365–366, 385
Boat speed, 341–342, 366, 524
Car rental, 165, 176, 183–184, 238
Distance traveled/time required, 180, 186, 238, 267, 315, 332, 340, 362, 460, 462, 539, 540
Gasoline costs, 238, 315, 354, 388
Hydroplaning, 526
Mileage between towns, 123
Motor oil costs, 450
Prepaid mass transit ticket, 132, 198, 238, 353, 361, 365, 528
Railway grade, 267, 318
Subway rides, 23
SUV prices, 302
Taxi rides, 225

Symbols

Symbol	Meaning
$a + b$	addition of a and b
$\frac{a}{b}$	division of a by b
$a \cdot b$, $a(b)$, ab, $(a)(b)$	multiplication of a and b
$a - b$	subtraction of a and b
-3	negative 3
$-b$	opposite of b
$\pm$	plus or minus (add or subtract)
$+3$	positive three
$\lvert\ \rvert$	absolute value
{ }	braces
[ ]	brackets
( )	parentheses
○	circle on a graph: the point is excluded from the graph
•	dot or filled-in circle on a graph: the point is included in the graph
//	double slash on a graph: the spacing between the origin and the first number on the axis is different from the spacing between the other numbers
$-\infty$	negative infinity
$+\infty$	positive infinity

Symbol	Meaning
. . .	repeats or continues, as in a pattern of numbers
$\mathbb{R}$	set of real numbers
$\varnothing$, { }	the empty set
$a \overset{?}{=} b$	is a equal to b?
$\approx$	is approximately equal to
$=$	is equal to
$>$	is greater than
$\geq$	is greater than or equal to
$<$	is less than
$\leq$	is less than or equal to
$\neq$	is unequal to
b^n	base b with exponent n
^{2}	square (exponent 2)
^{3}	cube (exponent 3)
$\sqrt{\ }$	principal square root, radical sign
$^\circ$	degree (temperature or angle)
Δ	delta: change
%	percent
$\perp$	perpendicular
π	pi, approximately 3.14
∟	square corner at perpendicular lines
x_1	variable x with subscript 1

Formulas

(See pages 98 and 101 for perimeter, area, volume, and surface area formulas.)

Distance formula (when traveling): Distance = rate • time, $d = rt$

Distance formula (length of a line segment on a graph): $d = \sqrt{(x_2 - x_1)^2 + (y_2 - y_1)^2}$

Interest formula: Interest = principal • rate • time, $i = prt$

Linear equation: $ax + by = c$ (standard), $y = mx + b$ (slope-intercept)

Pythagorean theorem, where c is always the hypotenuse: $a^2 + b^2 = c^2$

Quadratic equation: $y = ax^2 + bx + c$

Quadratic formula: If $ax^2 + bx + c = 0$, then $x = \frac{-b + \sqrt{b^2 - 4ac}}{2a}$ or $x = \frac{-b - \sqrt{b^2 - 4ac}}{2a}$

Slope formula: $m = \frac{y_2 - y_1}{x_2 - x_1}$

y-intercept: $b = y - mx$